W9-AXU-541

Thomas L. Floyd

ELECTRONICS FUNDAMENTALS: CIRCUITS, DEVICES AND APPLICATIONS

Merrill Publishing Company
A Bell & Howell Information Company
Columbus Toronto London Melbourne

To the electronics instructor, in the hope that this book will make your job a little easier

Published by
Merrill Publishing Company
A Bell & Howell Information Company
Columbus, Ohio 43216

This book was set in Century Schoolbook.

Administrative Editor: Tim McEwen
Developmental Editor: Don Thompson
Production Coordinator: Constantina Geldis
Cover Designer: Cathy Watterson
Text Designer: Cynthia Brunk
Cover Photo: Merrill Publishing/Larry Hamill

Credits: Part 1 opening photo by Merrill Publishing/Bruce Johnson; Part 2 opening and Chapter 1, 8, 9, 11, and 18 opening photos courtesy of Tektronix, Inc.; Part 3 opening and Chapter 5 opening photos courtesy of Motorola Semiconductor Products, Inc. Chapter 2 opening photo courtesy of University of Akron; Chapter 3 and 16 opening photos courtesy of Texas Instruments, Inc.; Chapter 4 and 6 opening photos courtesy of Hawaii Visitors Bureau; Chapter 7 opening photo courtesy of ADT Security Systems; Chapter 10, 15, 19, and 20 opening photos by Jo Hall; Chapter 12 opening photo courtesy of The Coliseum, Richfield, Ohio; Chapter 13 opening photo courtesy of Tandy Corp.; Chapter 14 opening photo courtesy of the Bureau of Sport Fisheries and Wildlife, Washington, D.C.; Chapter 17 opening photo courtesy of Hewlett-Packard; Chapter 21 opening photo courtesy of Grumman Corp.; Chapter 22 opening photo courtesy of E. I. duPont de Nemours and Co.; and four-color insert photos by Merrill Publishing/Bruce Johnson.

Library of Congress Catalog Card Number: 86-61567
International Standard Book Number: 0-675-20714-2
Printed in the United States of America
3 4 5 6 7 8 9—91 90 89 88

MERRILL'S INTERNATIONAL SERIES IN ELECTRICAL AND ELECTRONICS TECHNOLOGY

BATESON	*Introduction to Control System Technology, Second Edition,* 8255-2
BEACH/JUSTICE	*DC/AC Circuit Essentials,* 20193-4
BERLIN	*Experiments in Electronic Devices,* 20234-5
	The Illustrated Electronics Dictionary, 20451-8
BOGART	*Electronic Devices and Circuits,* 20317-1
BOGART/BROWN	*Experiments in Electronic Devices and Circuits,* 20488-7
BOYLESTAD	*Introductory Circuit Analysis, Fifth Edition,* 20631-6
BOYLESTAD/KOUSOUROU	*Experiments in Circuit Analysis, Fifth Edition,* 20743-6
BREY	*Advanced Microprocessors (16-Bit),* 20443-7
	Microprocessor/Hardware Interfacing and Applications, 20158-6
BUCHLA	*Experiments in Electric Circuits Fundamentals,* 20736-3
	Digital Experiments: Emphasizing Systems and Design, 20562-X
COX	*Digital Experiments: Emphasizing Troubleshooting,* 20518-2
FLOYD	*Electric Circuits Fundamentals,* 20756-8
	Electronics Fundamentals: Circuits, Devices, and Applications, 20714-2
	Digital Fundamentals, Third Edition, 20517-4
	Electronic Devices, 20157-8
	Essentials of Electronic Devices, 20062-8
	Principles of Electric Circuits, Second Edition, 20402-X
	Electric Circuits, Electron Flow Version, 20037-7
GAONKAR	*Microprocessor Architecture, Programming, and Applications with the 8085/8080A,* 20159-4
LAMIT/LLOYD	*Drafting for Electronics,* 20200-0
LAMIT/WAHLER/HIGGINS	*Workbook in Drafting for Electronics,* 20417-8
MARUGGI	*Technical Graphics: Electronics Worktext,* 20311-2
NASHELSKY/BOYLESTAD	*BASIC Applied to Circuit Analysis,* 20161-6
ROSENBLATT/FRIEDMAN	*Direct and Alternating Current Machinery, Second Edition,* 20160-8
SCHWARTZ	*Survey of Electronics, Third Edition,* 20162-4
SEIDMAN/WAINTRAUB	*Electronics: Devices, Discrete and Integrated Circuits,* 8494-6
STANLEY, B. H.	*Experiments in Electric Circuits, Second Edition,* 20403-8
STANLEY, W. D.	*Operational Amplifiers with Linear Integrated Circuits,* 20090-3
TOCCI	*Fundamentals of Electronic Devices, Third Edition,* 9887-4
	Electronic Devices, Third Edition, Conventional Flow Version, 20063-6
	Fundamentals of Pulse and Digital Circuits, Third Edition, 20033-4
	Introduction to Electric Circuit Analysis, Second Edition, 20002-4
WARD	*Applied Digital Electronics,* 9925-0
YOUNG	*Electronic Communication Techniques,* 20202-7

Preface

Electronics Fundamentals: Circuits, Devices and Applications covers the essential topics in dc and ac circuits and electronic devices with emphasis on applications and troubleshooting. The coverage provides a thorough foundation on which to develop skills in electronics and related fields.

This book is designed particularly for those students in technician-level programs in technical institutes, community colleges, and vocational/technical schools. Emphasis is on the understanding and application of basic concepts. Mathematics is held to a minimum and is used only in a supportive role where it is essential for the development of a complete and meaningful coverage of a topic or for providing a means of analyzing basic circuit operation.

Some of the specific features are:

- Chapter objectives
- Application assignments
- Application notes
- Calculator sequences
- Section reviews with answers at the end of each chapter
- Two-color format throughout; use of color in the art program is *functional,* emphasizing concepts
- Full-color section keyed to figures, problems, section reviews, and examples
- Many photographs
- Over 1,200 illustrations
- Over 250 numbered examples
- Emphasis on applications and troubleshooting
- End-of-chapter summaries
- Self-tests at the end of each chapter with solutions at the end of the book
- Two end-of-chapter problem sets (one easier, the other more difficult); answers to odd-numbered problems at the end of the book
- Glossary at the end of the book
- Availability of a coordinated lab manual and a set of transparencies

Chapter Organization

At the beginning of each chapter, the introduction provides an overview of the material to be covered, and the *objectives* list itemizes the specific things that the student will learn.

The *Application Assignment* at the beginning of each chapter places the students in a practical on-the-job situation as technicians and requires them to complete the assignment based on knowledge gained during the study of the chapter. These Application Assignments are intended to be not only educational, but also motivational. They put the students in a situation to which they can relate but which they are generally unable to handle before studying the chapter. After completing the material in the chapter, however, they can successfully carry out the assignment and thus derive a sense of accomplishment. The *Application Note* at the end of the chapter provides a suggested approach or solution to the assignment.

Calculator Sequences are provided for selected examples to show the student how to use a typical calculator (TI55-II; sequences may vary for other calculators) to arrive at the solution. Generally, these appear where an unfamiliar mathematical procedure is first introduced.

The *Section Reviews* consist of questions and problems which focus on key concepts presented in that section in order to provide students with frequent feedback on their comprehension of the material. Answers are given at the end of the chapter.

The *Summary* at the end of the chapter provides a concise listing of facts, definitions, new symbols, and a formula list.

The *Self-test* at the end of the chapter allows the students to check their mastery of the general concepts covered in the chapter. The self-test consists of essay-type questions, definitions, and multiple-choice questions as well as basic problems requiring some calculation. Solutions to all self-tests appear at the end of the book.

In each of the first 16 chapters, there are two problem sets. Problem Set A is a sectionalized series of relatively straightforward exercises, and Problem Set B contains exercises that are more demanding and thought provoking. Answers to the odd-numbered problems appear at the end of the book.

In the first six chapters, a number of figures, problems, section reviews, and examples relate to a unique full-color section positioned between pages 128 and 129. In these exercises, the students are required to identify simple component relationships in actual breadboard assemblies or on printed circuit boards, to determine resistance values directly from the color bands, and to read instruments connected to operating circuits in order to analyze or troubleshoot the circuit.

Overall Organization of the Text

The text is divided into three parts: DC Circuits, consisting of Chapters 1 through 8; AC Circuits, consisting of Chapters 9 through 16; and Electronic Devices consisting of Chapters 17 through 22.

For those wishing to introduce capacitance and inductance in the dc portion of the course, all of Chapter 10 except Sections 10–6 and 10–7 and all of Chapter 11 except Sections 11–6 and 11–7 can be added to the dc coverage.

Then, Sections 10–6, 10–7, 11–6, and 11–7 can be covered as part of the ac course.

The organization and content of this book are suited particularly to those programs which cover dc, ac, and devices during the first year in a two-term or three-term sequence.

Acknowledgments

This book is the result of the efforts of many people. In particular, I want to express my appreciation to Don Thompson, Connie Geldis, Tim McEwen, Cindy Brunk, Bruce Johnson, Jim Hubbard, Terry Tietz, and Cathy Watterson at Merrill Publishing Company for their work in making this book a reality. As always, the people at Merrill are progressive, creative, and dedicated to quality. My thanks also go to the following instructors who reviewed the manuscript and provided many valuable suggestions: Roman R. Braun, Terrence D. Nelson, and Kenneth J. Dreistadt, Lincoln Technical Institute; Robert A. Ciuffetti, Sylvania Technical Institute; John Colyer, ITT Technical Institute—Austin; Kenneth Edwards, International Brotherhood of Electrical Workers; William Greer, Albuquerque Technical Vocational Institute; Jill Harlamert, DeVry Institute of Technology—Columbus; Steve Kalina, DeVry, Inc.; Arnold Kroeger, Hillsborough Community College; Floyd Martin, Santa Ana College; W. A. McIntyre, DeVry Institute of Technology—Chicago; and Tim Staley, DeVry Institute of Technology—Dallas. In addition, I thank Morris McCarthy, DeVry—Atlanta for suggesting the four-color insert and Dean Gay Farmer, DeVry—Columbus, for his help in implementing this suggestion. Art Vildavs, Joyce Mielke, and Sam McCord, all students at DeVry Institute of Technology—Columbus, helped to build and to set up the circuits for the full-color photographs. I am grateful to the following industrial organizations that contributed photographs and other technical material for use in this book: Tektronix, Texas Instruments, Hewlett-Packard, Bell Laboratories, B&K Precision/Dynascan Corporation, Grumman Corporation, Burroughs Corporation, Motorola Semiconductor Products, E. I. duPont de Nemours and Company, Triplett, Bussman, Eaton, Grayhill, Bourns Trimpot, Dale Electronics, Radio Shack, Ford, Murata Erie, Delevan, and Sprague Electric. Finally, my wife, Sheila, deserves a great deal of credit for her help and support during the development of this book.

Thomas L. Floyd

Contents

PART ONE DC CIRCUITS

PULSE RESPONSE OF *RC* AND *RL* CIRCUITS 576

16

PART THREE DEVICES

INTRODUCTION TO SEMICONDUCTOR DEVICES 616

17

DIODES AND APPLICATIONS 644

18

TRANSISTORS AND THYRISTORS 694

19

APPENDICES

This special 16-page full-color insert illustrates actual color-coded resistors and real-world circuits that are used in examples, section reviews, and end-of-chapter problems, thereby taking the student "into the lab" and placing him or her "on the job."

PART ONE DC CIRCUITS

Introduction

1–1 History of Electricity and Electronics
1–2 Careers in Electronics
1–3 Applications of Electricity and Electronics
1–4 Circuit Components and Measuring Instruments
1–5 Electrical Units
1–6 Scientific Notation
1–7 Metric Prefixes

This chapter presents a brief history of the fields of electricity and electronics and discusses some of the many areas of application. Also, to aid you throughout the book, the basics of scientific notation and metric prefixes are reviewed, and the quantities and units commonly used in electronics are introduced.

In this chapter you will learn:

- A brief history of electricity and electronics.
- Some of the important areas in which electronics technology is applied.
- How to recognize some important electrical components and measuring instruments.
- The electrical quantities and their units.
- How to use scientific notation (powers of ten).
- The metric prefixes and how to use them.

APPLICATION ASSIGNMENT

At the beginning of each chapter starting with Chapter 2, you will find an Application Assignment that relates to that chapter. These assignments present a variety of practical job situations that a technician might encounter in industry.

As you study each chapter, think about how to approach the Application Assignment. When you have completed each chapter, you should have a sufficient knowledge of the topics covered to enable you to carry out the assignment. An Application Note at the end of the chapter suggests an approach or offers a solution to the assignment.

HISTORY OF ELECTRICITY AND ELECTRONICS

1–1

One of the first important discoveries about static electricity is attributed to William Gilbert (1540–1603). Gilbert was an English physician who, in a book published in 1600, described how amber differs from magnetic loadstones in its attraction of certain materials. He found that when amber was rubbed with a cloth, it attracted only lightweight objects, whereas loadstones attracted only iron. Gilbert also discovered that other substances, such as sulfur, glass, and resin, behave as amber does. He used the Latin word *elektron* for amber and originated the word *electrica* for the other substances that acted similarly to amber. The word *electricity* was used for the first time by Sir Thomas Browne (1605–82), an English physician.

Another Englishman, Stephen Gray (1696–1736), discovered that some substances conduct electricity and some do not. Following Gray's lead, a Frenchman named Charles du Fay experimented with the conduction of electricity. These experiments led him to believe that there were two kinds of electricity. He called one type *vitreous electricity* and the other type *resinous electricity*. He found that objects charged with vitreous electricity repelled each other and those charged with resinous electricity attracted each other. It is known today that two types of electrical *charge* do exist. They are called *positive* and *negative.*

Benjamin Franklin (1706–90) conducted studies in electricity in the mid-1700s. He theorized that electricity consisted of a single *fluid,* and he was the first to use the terms *positive* and *negative.* In his famous kite experiment, Franklin showed that lightning is electricity.

Charles Augustin de Coulomb (1736–1806), a French physicist, in 1785 proposed the laws that govern the attraction and repulsion between electrically charged bodies. Today, the unit of electrical charge is called the *coulomb.*

Luigi Galvani (1737–98) experimented with current electricity in 1786. Galvani was a professor of anatomy at the University of Bologna in Italy. Electrical current was once known as *galvanism* in his honor.

In 1800, Alessandro Volta (1745–1827), an Italian professor of physics, discovered that the chemical action between moisture and two different metals produced electricity. Volta constructed the first battery, using copper and zinc plates separated by paper that had been moistened with a salt solution. This battery, called the *voltaic pile,* was the first source of steady electric current. Today, the unit of electrical potential energy is called the *volt* in honor of Volta.

A Danish scientist, Hans Christian Oersted (1777–1851), is credited with the discovery of electromagnetism, in 1820. He found that electrical current flowing through a wire caused the needle of a compass to move. This finding showed that a magnetic field exists around a current-carrying conductor and that the field is produced by the current.

The modern unit of electrical current is the *ampere* (also called *amp*) in honor of the French physicist André Ampère (1775–1836). In 1820, Ampère measured the magnetic effect of an electrical current. He found that two wires carrying current can attract and repel each other, just as magnets can. By 1822, Ampère had developed the fundamental laws that are basic to the study of electricity.

One of the most well known and widely used laws in electrical circuits today is *Ohm's law.* It was formulated by Georg Simon Ohm (1789–1854), a

German teacher, in 1826. Ohm's law gives us the relationship among the three important electrical quantities of resistance, voltage, and current.

Although it was Oersted who discovered electromagnetism, it was Michael Faraday (1791–1867) who carried the study further. Faraday was an English physicist who believed that if electricity could produce magnetic effects, then magnetism could produce electricity. In 1831 he found that a moving magnet caused an electric current in a coil of wire placed within the field of the magnet. This effect, known today as *electromagnetic induction,* is the basic principle of electric generators and transformers.

Joseph Henry (1797–1878), an American physicist, independently discovered the same principle in 1831, and it is in his honor that the unit of inductance is called the *henry*. The unit of capacitance, the *farad,* is named in honor of Michael Faraday.

In the 1860s, James Clerk Maxwell (1831–79), a Scottish physicist, produced a set of mathematical equations that expressed the laws governing electricity and magnetism. These formulas are known as *Maxwell's equations*. Maxwell also predicted that electromagnetic waves (radio waves) that travel at the speed of light in space could be produced.

It was left to Heinrich Rudolph Hertz (1857–94), a German physicist, to actually produce these waves that Maxwell predicted. Hertz performed this work in the late 1880s. Today, the unit of frequency is called the *hertz*.

The Beginning of Electronics

The early experiments in electronics involved electric currents flowing in glass tubes. One of the first to conduct such experiments was a German named Heinrich Geissler (1814–79). Geissler found that when he removed most of the air from a glass tube, the tube glowed when an electrical potential was placed across it.

Around 1878, Sir William Crookes (1832–1919), a British scientist, experimented with tubes similar to those of Geissler. In his experiments, Crooke found that the current flowing in the tubes seemed to consist of particles.

Thomas Edison (1847–1931), experimenting with the carbon-filament light bulb that he had invented, made another important finding. He inserted a small metal plate in the bulb. When the plate was positively charged, a current flowed from the filament to the plate. This device was the first *thermionic diode.* Edison patented it but never used it.

The electron was discovered in the 1890s. The French physicist Jean Baptiste Perrin (1870–1942) demonstrated that the current in a vacuum tube consists of negatively charged particles. Some of the properties of these particles were measured by Sir Joseph Thomson (1856–1940), a British physicist, in experiments he performed between 1895 and 1897. These negatively charged particles later became known as *electrons*. The charge on the electron was accurately measured by an American physicist, Robert A. Millikan (1868–1953), in 1909. As a result of these discoveries, electrons could be controlled, and the electronic age was ushered in.

Putting the Electron to Work

A vacuum tube that allowed electrical current to flow in only one direction was constructed in 1904 by John A. Fleming, a British scientist. The tube was used to detect electromagnetic waves. Called the *Fleming valve,* it was the forerunner of the more recent vacuum diode tubes.

Major progress in electronics, however, awaited the development of a device that could boost, or *amplify,* a weak electromagnetic wave or radio signal. This device was the *audion,* patented in 1907 by Lee de Forest, an American. It was a triode vacuum tube capable of amplifying small electrical signals.

Two other Americans, Harold Arnold and Irving Langmuir, made great improvements in the triode tube between 1912 and 1914. About the same time, de Forest and Edwin Armstrong, an electrical engineer, used the triode tube in an *oscillator circuit.* In 1914, the triode was incorporated in the telephone system and made the transcontinental telephone network possible.

The tetrode tube was invented in 1916 by Walter Schottky, a German. The tetrode, along with the pentode (invented in 1926 by Tellegen, a Dutch engineer), provided great improvements over the triode. The first television picture tube, called the *kinescope,* was developed in the 1920s by Vladimir Zworykin, an American researcher.

During World War II, several types of microwave tubes were developed that made possible modern microwave radar and other communications systems. In 1939, the *magnetron* was invented in Britain by Henry Boot and John Randall. In the same year, the *klystron* microwave tube was developed by two Americans, Russell Varian and his brother Sigurd Varian. The *traveling-wave* tube was invented in 1943 by Rudolf Komphner, an Austrian-American.

The Computer

The computer probably has had more impact on modern technology than any other single type of electronic system. The first electronic digital computer was completed in 1946 at the University of Pennsylvania. It was called the Electronic Numerical Integrator and Computer (ENIAC). One of the most significant developments in computers was the *stored program* concept, developed in the 1940s by John von Neumann, an American mathematician.

Solid State Electronics

The crystal detectors used in the early radios were the forerunners of modern solid state devices. However, the era of solid state electronics began with the invention of the *transistor* in 1947 at Bell Labs. The inventors were Walter Brattain, John Bardeen, and William Shockley. Figure 1–1 shows these three men, along with the notebook entry describing the historic discovery.

In the early 1960s, the integrated circuit was developed. It incorporated many transistors and other components on a single small *chip* of semiconductor material. Integrated circuit technology continues to be developed and improved, allowing more complex circuits to be built on smaller chips. The introduction of the microprocessor in the early 1970s created another electronics revolution: the entire processing portion of a computer placed on a single, small, silicon chip. Continued development brought about complete computers on a single chip by the late 1970s.

(a)

(b)

FIGURE 1–1

(a) Nobel Prize winners Drs. John Bardeen, William Shockley, and Walter Brattain, shown left to right, with apparatus used in their first investigations that led to the invention of the transistor. The trio received the 1956 Nobel Physics award for their invention of the transistor, which was announced by Bell Laboratories in 1948. (b) The laboratory notebook entry of scientist Walter H. Brattain recorded the events of December 23, 1947, when the transistor effect was discovered at Bell Telephone Laboratories. The notebook entry describes the event and adds, "This circuit was actually spoken over and by switching the device in and out a distinct gain in speech level could be heard and seen on the scope presentation with no noticeable change in quality." ((a) and (b) courtesy of Bell Laboratories)

SECTION REVIEW 1–1

1. Who developed the first battery?
2. The unit of what electrical quantity is named after André Ampère?
3. What contribution did Georg Simon Ohm make to the study of electricity?
4. In what year was the transistor invented?
5. What major development followed the invention of the transistor?

CAREERS IN ELECTRONICS

1–2

The field of electronics is very diverse, and career opportunities are available in many areas. Because electronics is currently found in so many different applications and new technology is being developed at a fast rate, its future is limitless. There is hardly an area of our lives that is not enhanced to some degree by electronics technology. Those who acquire a sound, basic knowledge of electrical and electronic principles and are willing to continue learning will always be in demand and will be able to command a very good salary.

The importance of obtaining a thorough understanding of the basic principles contained in this text cannot be overemphasized. Most employers prefer to hire people who have both a thorough grounding in the basics and the ability and eagerness to grasp new concepts and techniques. If you have a good training in the basics, an employer will train you in the specifics of the job to which you are assigned.

There are many types of job classifications for which a person with training in electronics technology may qualify. A few of the most common job functions are discussed briefly in the following paragraphs.

Service Shop Technician

Technical personnel in this category are involved in the repair or adjustment of both commercial and consumer electronic equipment that is returned to the dealer or manufacturer for service. Specific areas include TVs, VCRs, stereo equipment, CB radios, computer equipment, and so on. This area also offers opportunities for self-employment.

Manufacturing Technician

Manufacturing personnel are involved either in the testing of electronic products at the assembly-line level or in the maintenance and troubleshooting of electronic and electromechanical systems used in the testing and manufacturing of products. Virtually every type of manufacturing plant, regardless of its product, uses automated equipment that is electronically controlled.

Laboratory Technician

These technicians are involved in breadboarding, prototyping, and testing new or modified electronic systems in research and development laboratories. They generally work closely with engineers during the development phase of a product.

Field Service Technician

Field service personnel service and repair electronic equipment—for example, computer systems, radar installations, automatic banking equipment, and security systems—at the user's location.

Engineering Assistant/Associate Engineer

Personnel in this category work closely with engineers in the implementation of a concept and in the basic design and development of electronic systems.

Engineering assistants are frequently involved in a project from its initial design through the early manufacturing stages.

Technical Writer

Technical writers compile technical information and then use the information to write and produce manuals and audiovisual materials. A broad knowledge of a particular system and the ability to clearly explain its principles and operation are essential.

Technical Sales

Technically trained people are in demand as sales representatives for high-technology products. The ability both to understand technical concepts and to communicate the technical aspects of a product to a potential customer is very valuable. In this area, as in technical writing, competency in expressing yourself orally and in writing is essential. Actually, being able to communicate well is very important in any technical job category because you must be able to record data clearly and explain procedures, conclusions, and actions taken so that others can readily understand what you are doing.

Figure 1–2 shows technical personnel at work in typical settings to help give you a more complete picture of this exciting career field.

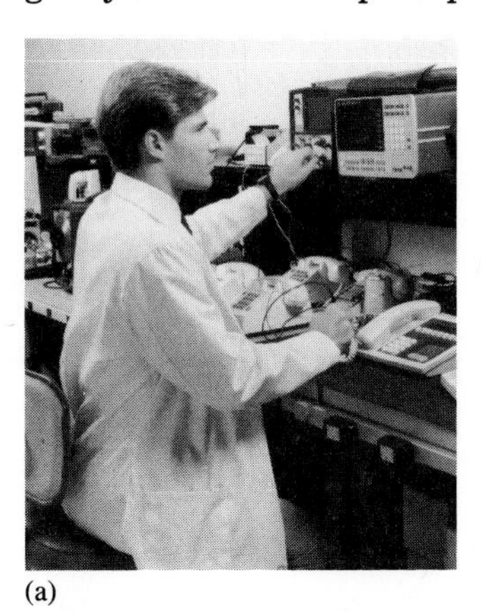

(a)

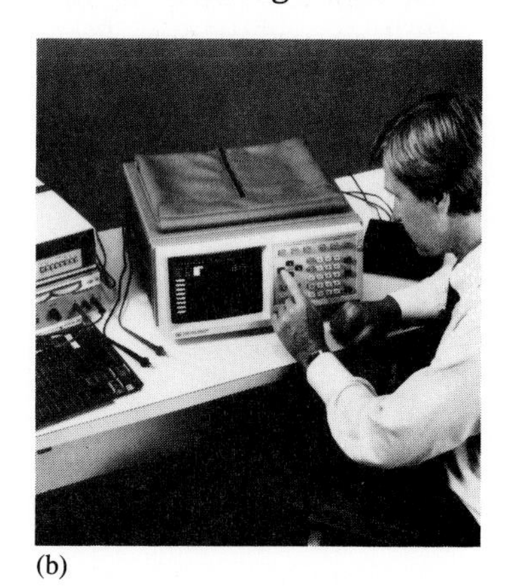

(b)

FIGURE 1–2
Technicians on the job ((a) courtesy of Burroughs Corp. (b) courtesy of Hewlett-Packard Co.).

SECTION REVIEW 1–2

1. For which of the following reasons have you selected or are you considering selecting electronics as a career?
   - **(a)** An interesting and exciting field.
   - **(b)** Good opportunities for advancement.
   - **(c)** Good income.
   - **(d)** Clean and comfortable working environment.
   - **(e)** Very promising future.
   - **(f)** Like to work with electronic equipment.
   - **(g)** Like to understand how things work.
2. Name several ways in which electronics affects your everyday life.

APPLICATIONS OF ELECTRICITY AND ELECTRONICS

1–3

Electricity and electronics are diverse technological fields with very broad applications. There is hardly an area of our lives that is not dependent to some extent on electricity and electronics. Some of the applications are discussed here in a general way to give you an idea of the scope of the fields.

Computers

One of the most important electronic systems is the digital computer; its applications are broad and diverse. For example, computers have applications in business for record keeping, accounting, payrolls, inventory control, market analysis, and statistics, to name but a few.

Scientific fields utilize the computer to process huge amounts of data and to perform complex and lengthy calculations. In industry, the computer is used for controlling and monitoring intricate manufacturing processes. Communications, navigation, medical, military, and home uses are a few of the other areas in which the computer is used extensively.

The computer's success is based on its ability to perform mathematical operations extremely fast and to process and store large amounts of information.

Computers vary in complexity and capability, ranging from very large systems with vast capabilities down to a computer on a chip with much more limited performance. Figure 1–3 shows some typical computers of varying sizes.

(a)

(b)

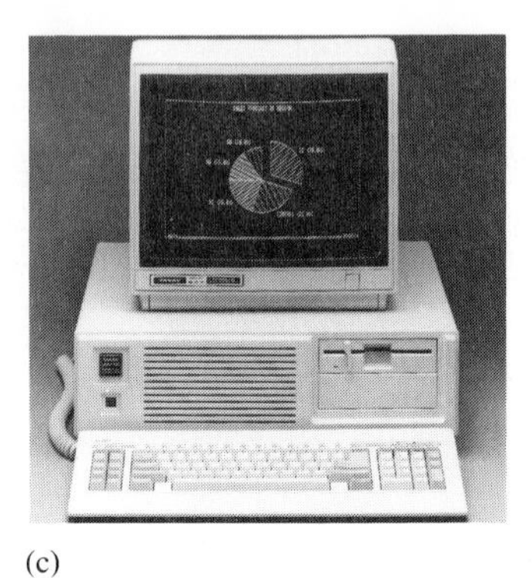

(c)

FIGURE 1–3
Typical computer systems. (a) Large computer system (courtesy of Burroughs Corp.). (b) PDP-11V23 computer (courtesy of Digital Equipment Corp.). (c) Personal computer (courtesy of Tandy Corp.).

Communications

Electronic communications encompasses a wide range of specialized fields. Included are space and satellite communications, commercial radio and television, citizens' band and amateur radio, data communications, navigation systems, radar, telephone systems, military applications, and specialized radio applications such as police, aircraft, and so on. Computers are used to a great extent in many communications systems. Figure 1–4 shows a telephone switching system as an example of electronic communications.

FIGURE 1–4
A telephone switching system (courtesy of GTE Automatic Electric Inc.).

Automation

Electronic systems are employed extensively in the control of manufacturing processes. Computers and specialized electronic systems are used in industry for various purposes, for example, control of ingredient mixes, operation of machine tools, product inspection, and control and distribution of power. Figure 1–5 shows an example of automation in a manufacturing facility using robots.

FIGURE 1–5
Robots on an automobile assembly line (courtesy of Ford Motor Co.).

Medicine

Electronic devices and systems are finding ever-increasing applications in the medical field. The familiar *electrocardiograph* (ECG), used for the diagnosis of heart and other circulatory ailments, is a widely used medical electronic instrument. A closely related instrument is the *electromyograph,* which uses a cathode ray tube display rather than an ink trace.

The *diagnostic sounder* uses ultrasonic sound waves for various diagnostic procedures in neurology, for heart chamber measurement, and for detection of certain types of tumors. The *electroencephalograph* (EEG) is similar to the electrocardiograph. It records the electrical activity of the brain rather than heart activity. Another electronic instrument used in medical procedures is the *coagulograph.* This instrument is used in analysis of blood clots.

Electronic instrumentation is also used extensively in intensive-care facilities. Heart rate, pulse, body temperature, respiration, and blood pressure can be monitored on a continuous basis. Monitoring equipment is also used a great deal in operating rooms. Some typical medical electronic equipment is pictured in Figure 1–6.

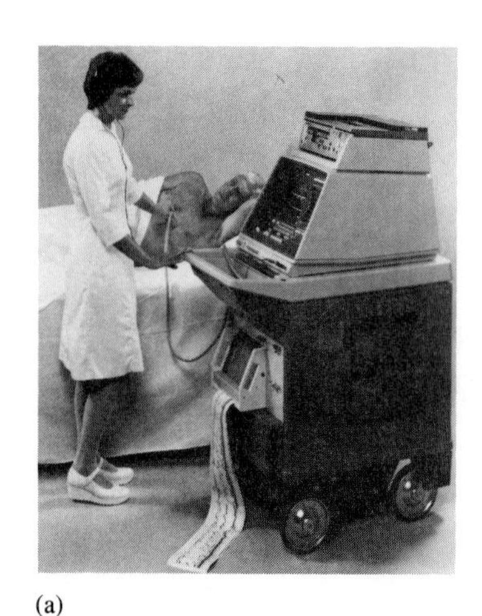

(a)

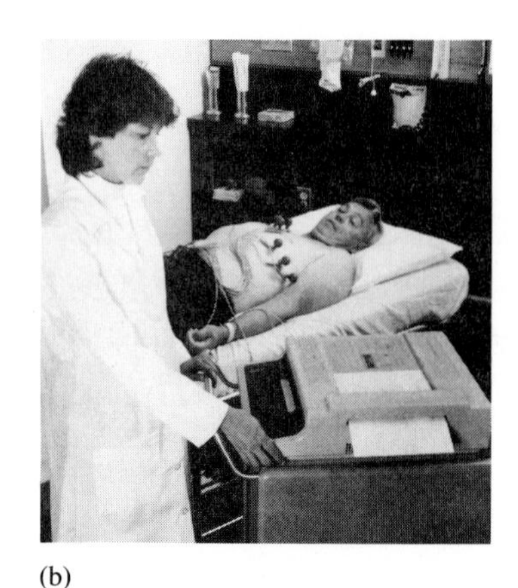

(b)

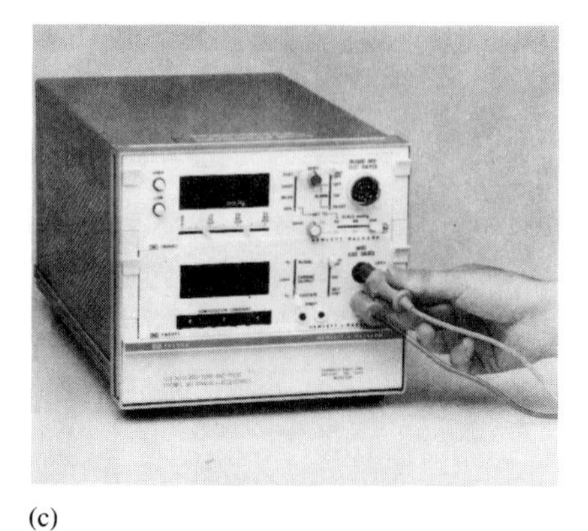

(c)

FIGURE 1–6
Typical medical instrumentation. (a) Doppler enhancement provides quantified measurements of blood-flow direction and velocity through the heart. (b) Page-Writer cardiographs can provide analysis reports, measurements, and/or interpretation at the bedside in under 90 seconds. (c) Cardiac output module computes cardiac output and measures continuous pulmonary artery blood pressure (courtesy of Hewlett-Packard Co.).

Consumer Products

Electronic products used directly by the consumers for information, entertainment, recreation, or work around the home are an important segment of the total electronics market. For example, the electronic calculator and digital watch are popular examples of consumer electronics. The small personal computer is used widely by hobbyists and is also becoming a common household appliance.

Electronic systems are used in automobiles to control and monitor engine functions, control braking, provide entertainment, and display useful information to the driver.

Most appliances such as microwave ovens, washers, and dryers are available with electronic controls. Home entertainment, of course, is largely electronic. Examples are television, radio, stereo, and recorders. Also, many new games for adults and children incorporate electronic devices.

SECTION REVIEW 1–3

1. Name some of the areas in which electronics is used.
2. The computerization of manufacturing processes is an example of ______.

CIRCUIT COMPONENTS AND MEASURING INSTRUMENTS

1–4

In this text you will study many types of electrical components and measuring instruments. This coverage of dc/ac circuit fundamentals provides the foundation for understanding electronic devices and circuits. Several types of electrical and electronic components and instruments that you will be studying in detail in this and in other courses are introduced briefly in this section.

Resistors

These components *resist,* or limit, the flow of electrical current in a circuit. Several common types of resistors are shown in Figure 1–7.

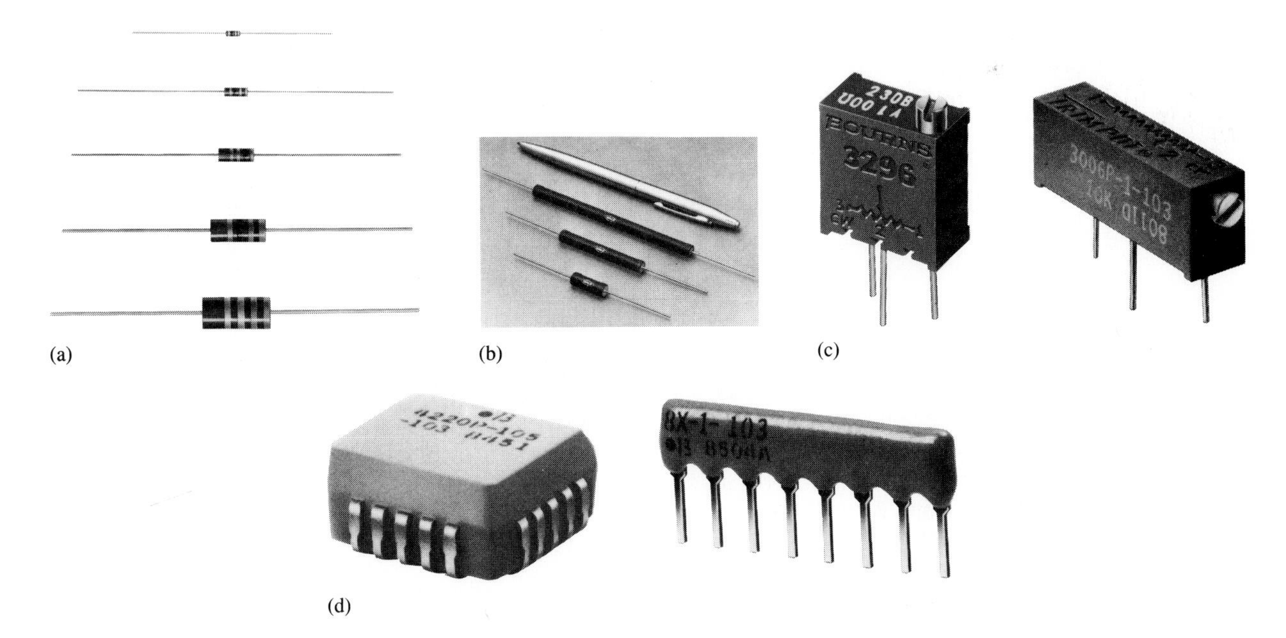

FIGURE 1–7
Typical fixed and variable resistors. (a) Carbon-composition resistors with standard power ratings of ⅛W, ¼W, ½W, 1W, and 2W (courtesy of Allen-Bradley Co.). (b) Wirewound resistors (courtesy of Dale Electronics, Inc.). (c) Variable resistors or potentiometers (courtesy of Bourns, Inc.). (d) Resistor networks (courtesy of Bourns, Inc.).

Capacitors

These components *store* electrical charge; they are found in a variety of applications. Figure 1–8 shows several typical capacitors.

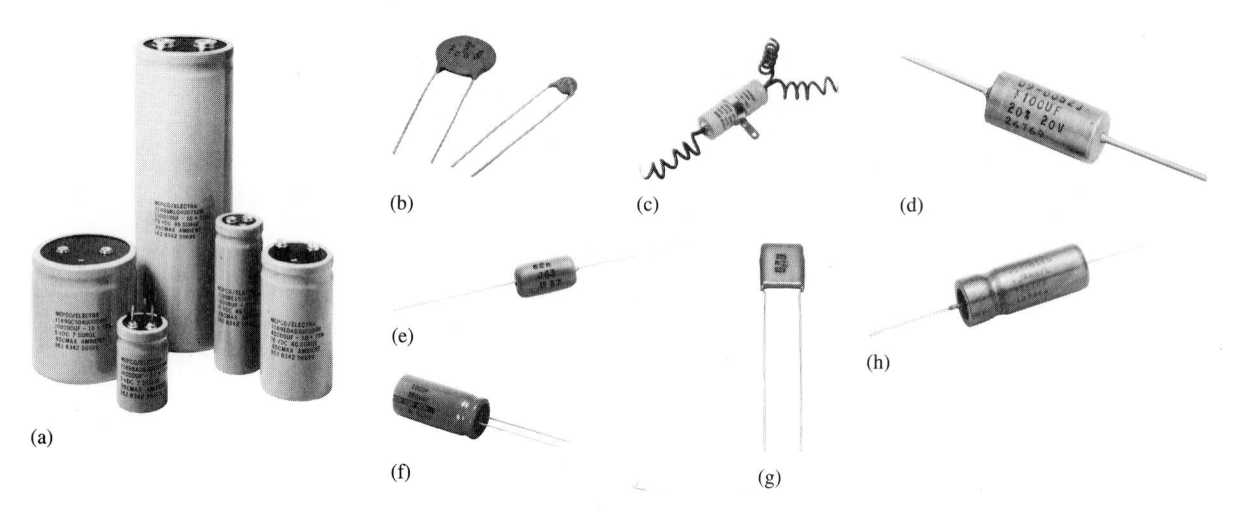

FIGURE 1–8
Typical capacitors. ((a,d,e,h) courtesy of Mepco/Centralab. (b) courtesy of Murata Erie, North America. (c,f,g) courtesy of Sprague Electric Co.).

Inductors

These components, also known as *coils,* are used to store energy in an electromagnetic field; they serve many useful functions in an electrical circuit. Figure 1–9 shows several typical inductors.

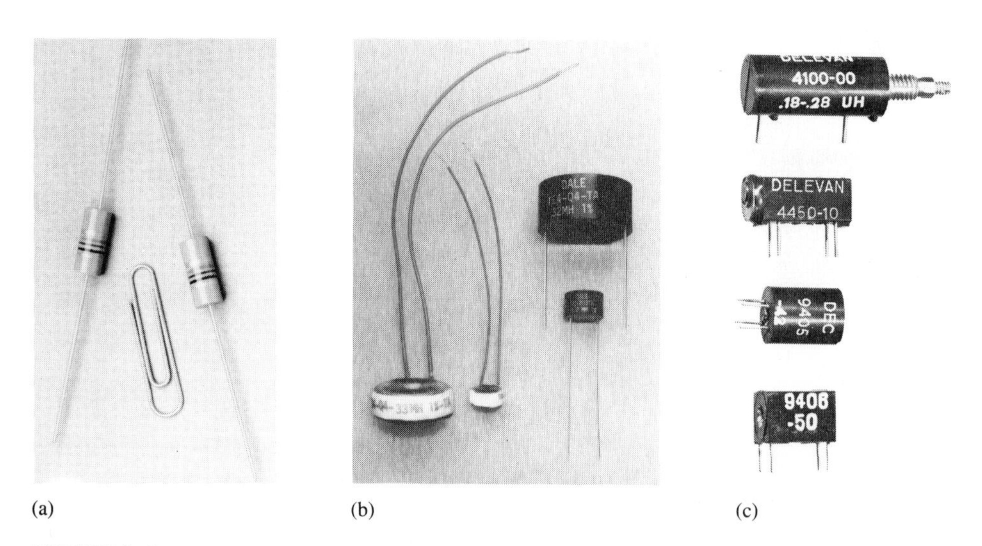

FIGURE 1–9
Typical inductors. ((a) and (b) courtesy of Dale Electronics, Inc. (c) courtesy of Delevan.).

Transformers

These components are sometimes used to couple ac voltages from one point in a circuit to another, or to increase or decrease the ac voltage. Several types of transformers are shown in Figure 1–10.

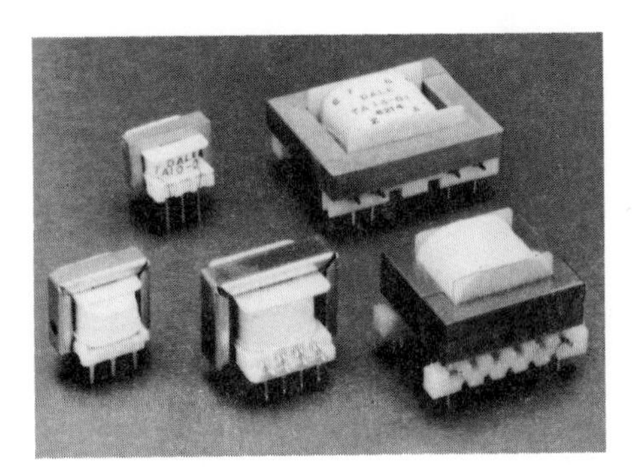

FIGURE 1–10
Typical transformers (courtesy of Dale Electronics, Inc.).

Semiconductor Devices

Several varieties of diodes, transistors, and integrated circuits are shown in Figure 1–11.

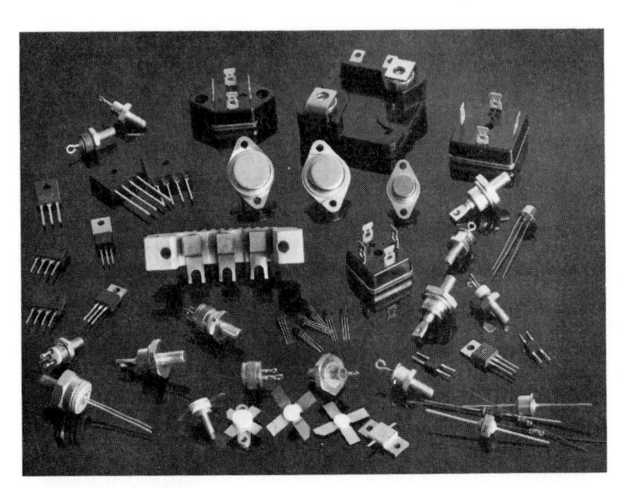

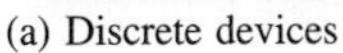

(a) Discrete devices

(b) Integrated circuits

FIGURE 1–11
A grouping of typical semiconductor devices (courtesy of Motorola Semiconductor Products, Inc.).

Electronic Instruments

Figure 1–12 shows a variety of instruments that are discussed throughout the text. Typical instruments include the power supply, for providing voltage and current; the voltmeter, for measuring voltage; the ammeter, for measuring current; the ohmmeter, for measuring resistance; the wattmeter, for measuring power; and the oscilloscope for observing and measuring ac voltages.

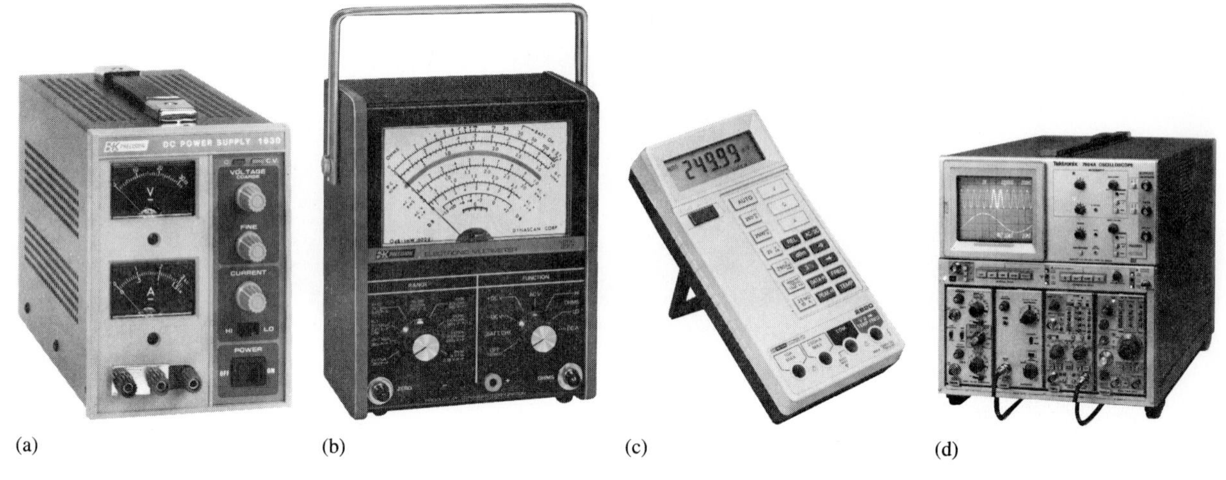

(a) (b) (c) (d)

FIGURE 1–12
Typical instruments. (a) dc power supply (courtesy of B&K Precision/Dynascan Corp.). (b) Analog multimeter (courtesy of B&K Precision/Dynascan Corp.). (c) Digital multimeter (courtesy of B&K Precision/Dynascan Corp.). (d) Oscilloscope (courtesy of Tektronix, Inc.).

SECTION REVIEW 1–4

1. Name three types of common electrical components.
2. What instrument is used for measuring electrical current?

ELECTRICAL UNITS

1–5

In electronics work, you must deal with measurable quantities. For example, you must be able to express how many volts are measured at a certain test point, how much current is flowing through a wire, or how much power a certain amplifier produces.

In this section you will learn the units and symbols for many of the electrical quantities that are used throughout the book. Definitions of the quantities are presented as they are needed in later chapters.

Symbols are used in electronics to represent both quantities and their units. One symbol is used to represent the name of the quantity, and another is used to represent the unit of measurement of that quantity. For example, P stands for *power,* and W stands for *watts,* which is the unit of power. Table 1–1 lists the most important quantities, along with their SI units and symbols. The term *SI* is the French abbreviation for *International System* (*Système International* in French).

TABLE 1–1
Electrical quantities and units with SI symbols.

Quantity	Symbol	Unit	Symbol
Capacitance	C	farad	F
Charge	Q	coulomb	C
Conductance	G	siemen	S
Current	I	ampere	A
Energy	W	joule	J
Frequency	f	hertz	Hz
Impedance	Z	ohm	Ω
Inductance	L	henry	H
Power	P	watt	W
Reactance	X	ohm	Ω
Resistance	R	ohm	Ω
Time	t	second	s
Voltage	V	volt	V

SECTION REVIEW 1–5

1. What does *SI* stand for?
2. Without referring to Table 1–1, list as many electrical quantities as possible, including their symbols, units, and unit symbols.

SCIENTIFIC NOTATION

1–6

In electronics work, you will encounter both very small and very large numbers. For example, it is common to have electrical current values of only a few thousandths or even a few millionths of an ampere. On the other hand, you will find resistance values of several thousand or several million ohms. This range of values is typical of many other electrical quantities also.

Powers of Ten

Scientific notation uses *powers of ten,* a method that makes it much easier to express large and small numbers and to do calculations involving such numbers.

Table 1–2 lists some powers of ten, both positive and negative. The power of ten is expressed as an exponent of the base 10 in each case. The exponent indicates the number of decimal places to the right or left of the decimal point in the expanded number. If the power is *positive,* the decimal point is moved to the *right.* For example,

$$10^4 = 1 \times 10^4 = 1.0000. = 10,000.$$

If the power is *negative,* the decimal point is moved to the *left.* For example,

$$10^{-4} = 1 \times 10^{-4} = .0001. = 0.0001$$

TABLE 1–2
Some positive and negative powers of ten.

$1,000,000 = 10^6$	$0.000001 = 10^{-6}$
$100,000 = 10^5$	$0.00001 = 10^{-5}$
$10,000 = 10^4$	$0.0001 = 10^{-4}$
$1,000 = 10^3$	$0.001 = 10^{-3}$
$100 = 10^2$	$0.01 = 10^{-2}$
$10 = 10^1$	$0.1 = 10^{-1}$
$1 = 10^0$	

EXAMPLE 1–1

Express each number as a positive power of ten.
(a) 200 **(b)** 5000 **(c)** 85,000 **(d)** 3,000,000

Solution
In each case there are many possibilities for expressing the number in powers of ten. We do not show all possibilities in this example but include the most common powers of ten used in electrical work.
(a) $200 = 0.0002 \times 10^6 = 0.2 \times 10^3$
(b) $5000 = 0.005 \times 10^6 = 5 \times 10^3$
(c) $85,000 = 0.085 \times 10^6 = 8.5 \times 10^4 = 85 \times 10^3$
(d) $3,000,000 = 3 \times 10^6 = 3000 \times 10^3$

EXAMPLE 1–2

Express each number as a negative power of ten.
(a) 0.2 **(b)** 0.005 **(c)** 0.00063 **(d)** 0.000015

Solution
Again, all the possible ways to express each number as a power of ten are not given. The most commonly used powers are included, however.
(a) $0.2 = 2 \times 10^{-1} = 200 \times 10^{-3} = 200,000 \times 10^{-6}$
(b) $0.005 = 5 \times 10^{-3} = 5000 \times 10^{-6}$
(c) $0.00063 = 0.63 \times 10^{-3} = 6.3 \times 10^{-4} = 630 \times 10^{-6}$
(d) $0.000015 = 0.015 \times 10^{-3} = 1.5 \times 10^{-5} = 15 \times 10^{-6}$

EXAMPLE 1–3

Express each of the following powers of ten as a regular decimal number:
(a) 10^5 **(b)** 2×10^3 **(c)** 3.2×10^{-2} **(d)** 250×10^{-6}

Solution
(a) $10^5 = 1 \times 10^5 = 100,000$ **(b)** $2 \times 10^3 = 2000$
(c) $3.2 \times 10^{-2} = 0.032$ **(d)** $250 \times 10^{-6} = 0.000250$

Calculating with Powers of Ten

The advantage of scientific notation is in addition, subtraction, multiplication, and division of very small or very large numbers.

Rules for Addition

The rules for adding numbers in powers of ten are as follows:

1. Convert the numbers to be added to the *same* power of ten.
2. Add the numbers directly to get the sum.
3. Bring down the common power of ten, which is the power of ten of the sum.

EXAMPLE 1–4 Add 2×10^6 and 5×10^7.

Solution

1. Convert both numbers to the same power of ten:

$$(2 \times 10^6) + (50 \times 10^6)$$

2. Add $2 + 50 = 52$.
3. Bring down the common power of ten (10^6), and the sum is 52×10^6.

Rules for Subtraction

The rules for subtracting numbers in powers of ten are as follows:

1. Convert the numbers to be subtracted to the *same* power of ten.
2. Subtract the numbers directly to get the difference.
3. Bring down the common power of ten, which is the power of ten of the difference.

EXAMPLE 1–5 Subtract 25×10^{-12} from 75×10^{-11}.

Solution

1. Convert each number to the same power of ten:

$$(75 \times 10^{-11}) - (2.5 \times 10^{-11})$$

2. Subtract $75 - 2.5 = 72.5$.
3. Bring down the common power of ten (10^{-11}), and the difference is 72.5×10^{-11}.

Rules for Multiplication

The rules for multiplying numbers in powers of ten are as follows:

1. Multiply the numbers directly.
2. Add the powers of ten algebraically (the powers do not have to be the same).

EXAMPLE 1–6 Multiply 5×10^{12} and 3×10^{-6}.

Solution

Multiply the numbers, and algebraically add the powers:

$$(5 \times 10^{12})(3 \times 10^{-6}) = 15 \times 10^{12+(-6)} = 15 \times 10^{6}$$

Rules for Division

The rules for dividing numbers in powers of ten are as follows:

1. Divide the numbers directly.
2. Subtract the power of ten in the denominator from the power of ten in the numerator.

EXAMPLE 1–7 Divide 50×10^8 by 25×10^3.

Solution
The division problem is written with a numerator and denominator as

$$\frac{50 \times 10^8}{25 \times 10^3}$$

Dividing the numbers and subtracting 3 from 8, we get

$$\frac{50 \times 10^8}{25 \times 10^3} = 2 \times 10^{8-3} = 2 \times 10^5$$

SECTION REVIEW 1–6

1. Scientific notation uses powers of ten (T or F).
2. Express 100 as a power of ten.
3. Do the following operations:
   (a) $(1 \times 10^5) + (2 \times 10^5)$ **(b)** $(3 \times 10^6)(2 \times 10^4)$
   (c) $(8 \times 10^3) \div (4 \times 10^2)$

METRIC PREFIXES

1–7

In electrical and electronics work, certain powers of ten are used more often than others. The most frequently used powers of ten are 10^9, 10^6, 10^3, 10^{-3}, 10^{-6}, 10^{-9}, and 10^{-12}.

It is common practice to use *metric prefixes* to represent these quantities. Table 1–3 lists the metric prefix for each of the commonly used powers of ten.

TABLE 1–3
Metric prefixes and their symbols.

Power of Ten	Value	Metric Prefix	Metric Symbol	Power of Ten	Value	Metric Prefix	Metric Symbol
10^9	one billion	giga	G	10^{-6}	one-millionth	micro	μ
10^6	one million	mega	M	10^{-9}	one-billionth	nano	n
10^3	one thousand	kilo	k	10^{-12}	one-trillionth	pico	p
10^{-3}	one-thousandth	milli	m				

Use of Metric Prefixes

Now we use examples to illustrate use of metric prefixes. The number 2000 can be expressed in scientific notation as 2×10^3. Suppose we wish to represent 2000 watts (W) with a metric prefix. Since $2000 = 2 \times 10^3$, the metric prefix *kilo* (k) is used for 10^3. So we can express 2000 W as 2 kW (2 kilowatts).

As another example, 0.015 ampere (A) can be expressed as 15×10^{-3} A. The metric prefix *milli* (m) is used for 10^{-3}. So 0.015 becomes 15 mA (15 milliamperes).

EXAMPLE 1–8

Express each quantity using a metric prefix.
(a) 50,000 V **(b)** 25,000,000 Ω **(c)** 0.000036 A

Solution
(a) $50{,}000\ \text{V} = 50 \times 10^3\ \text{V} = 50\ \text{kV}$
(b) $25{,}000{,}000\ \Omega = 25 \times 10^6\ \Omega = 25\ \text{M}\Omega$
(c) $0.000036\ \text{A} = 36 \times 10^{-6}\ \text{A} = 36\ \mu\text{A}$

Entering Numbers with Metric Prefixes on the Calculator

To enter a number expressed in scientific notation on the calculator, use the EE key (the EXP key on some calculators). The following example shows how to enter numbers with metric prefixes on a typical scientific calculator (TI 55–II).

EXAMPLE 1–9

(a) Enter 3.3 kΩ ($3.3 \times 10^3\ \Omega$) on the calculator.
(b) Enter 450 μA (450×10^{-6} A) on the calculator.

Solution
(a) **Step 1:** Enter [3] [.] [3]. The display shows 3.3.
Step 2: Press [EE]. The display shows 3.3 00.
Step 3: Enter [3]. The display shows 3.3 03.
(b) **Step 1:** Enter [4] [5] [0]. The display shows 450.
Step 2: Press [EE]. The display shows 450 00.
Step 3: Press [+/−]. Enter [6]. The display shows 450 − 06.

SECTION REVIEW 1–7

1. List the metric prefix for each of the following powers of ten: 10^6, 10^3, 10^{-3}, 10^{-6}, 10^{-9}, and 10^{-12}.
2. Use an appropriate metric prefix to express 0.000001 ampere.

SUMMARY

Facts

- Electronics is a varied field with many opportunities.
- Areas of application include computers, communications, automation, medicine, military, and consumer products.
- Resistors limit the flow of electrical current.
- Capacitors store electrical charge.

- Inductors store energy electromagnetically.
- Inductors are also known as *coils.*
- Transformers magnetically couple ac voltages.
- Semiconductor devices include diodes, transistors, and integrated circuits.
- Power supplies provide current and voltage.
- Voltmeters measure voltage.
- Ammeters measure current.
- Ohmmeters measure resistance.

Units

- Coulomb (C)—the unit of charge *(Q).*
- Farad (F)—the unit of capacitance *(C).*
- Henry (H)—the unit of inductance *(L).*
- Hertz (Hz)—the unit of frequency *(f).*
- Siemen (S)—the unit of conductance *(G).*
- Watt (W)—the unit of power *(P).*

SELF-TEST

1. List the units of the following electrical quantities: current, voltage, resistance, power, and energy.
2. List the symbol for each unit in Question 1.
3. List the symbol for each quantity in Question 1.
4. Express the following using metric prefixes:
   (a) ten milliamperes **(b)** five kilovolts
   (c) fifteen microwatts **(d)** twenty Megohms
5. Express each of the following metric quantities as a decimal number with the appropriate unit:
   (a) 8 μA **(b)** 25 MW **(c)** 100 mV

PROBLEMS

Section 1–6

1–1 Express each of the following numbers as a power of ten:
(a) 3000 **(b)** 75,000 **(c)** 2,000,000

1–2 Express each number as a power of ten.
(a) 1/500 **(b)** 1/2000 **(c)** 1/5,000,000

1–3 Express each of the following numbers in three ways, using 10^3, 10^4, and 10^5:
(a) 8400 **(b)** 99,000 **(c)** 0.2×10^6

1–4 Express each of the following numbers in three ways, using 10^{-3}, 10^{-4}, and 10^{-5}:
(a) 0.0002 **(b)** 0.6 **(c)** 7.8×10^{-2}

1–5 Express each power of ten in regular decimal form.
(a) 2.5×10^{-6} **(b)** 50×10^2 **(c)** 3.9×10^{-1}

1–6 Express each power of ten in regular decimal form.
(a) 45×10^{-6} **(b)** 8×10^{-9} **(c)** 40×10^{-12}

1–7 Add the following numbers:
(a) $(92 \times 10^6) + (3.4 \times 10^7)$ **(b)** $(5 \times 10^3) + (85 \times 10^{-2})$
(c) $(560 \times 10^{-8}) + (460 \times 10^{-9})$

1–8 Perform the following subtractions:
(a) $(3.2 \times 10^{12}) - (1.1 \times 10^{12})$ **(b)** $(26 \times 10^{8}) - (1.3 \times 10^{9})$
(c) $(150 \times 10^{-12}) - (8 \times 10^{-11})$

1–9 Perform the following multiplications:
(a) $(5 \times 10^{3})(4 \times 10^{5})$ **(b)** $(12 \times 10^{12})(3 \times 10^{2})$
(c) $(2.2 \times 10^{-9})(7 \times 10^{-6})$

1–10 Divide the following:
(a) $(10 \times 10^{3}) \div (2.5 \times 10^{2})$ **(b)** $(250 \times 10^{-6}) \div (50 \times 10^{-8})$
(c) $(4.2 \times 10^{8}) \div (2 \times 10^{-5})$

Section 1–7

1–11 Express each of the following as a quantity having a metric prefix:
(a) 31×10^{-3} A **(b)** 5.5×10^{3} V **(c)** 200×10^{-12} F

1–12 Express the following using metric prefixes:
(a) 3×10^{-6} F **(b)** 3.3×10^{6} Ω **(c)** 350×10^{-9} A

ANSWERS TO SECTION REVIEWS

Section 1–1

1. Volta. 2. Current.
3. He established the relationship among current, voltage, and resistance as expressed in Ohm's law.
4. 1947. 5. Integrated circuits.

Section 1–3

1. Computers, communications, automation, medicine, and consumer products.
2. Automation.

Section 1–4

1. Resistors, capacitors, inductors, and transformers.
2. Ammeter.

Section 1–5

1. It is the abbreviation for Système International.
2. Refer to Table 1–1 after you have compiled your list.

Section 1–6

1. T. 2. 10^{2} 3. **(a)** 3×10^{5}; **(b)** 6×10^{10}; **(c)** 2×10^{1}.

Section 1–7

1. Mega (M), kilo (k), milli (m), micro (μ), nano (n), and pico (p).
2. 1 μA (one microampere).

The Basic Electrical Quantities

Three basic electrical quantities are presented in this chapter: voltage, current, and resistance. No matter what type of electrical or electronic equipment you may work with, these quantities will always be of primary importance.

To help you understand voltage and current, the basic structure of the atom is discussed and the concept of charge is introduced. The basic electric circuit is studied, along with techniques for measuring voltage, current, and resistance.

Specifically, you will learn:

- The concept of the atom.
- That the electron is the basic particle of electrical charge.
- What voltage, current, and resistance are.
- How voltage causes current to flow.
- How resistance restricts the flow of current.
- Various types of voltage sources.
- Various types of fixed and variable resistors.
- How to determine resistance value by color code.
- What a basic electric circuit consists of.
- The concepts of closed and open circuits.
- Various types of switches and how they are used.
- Various types of fuses and circuit breakers.
- How to measure current, voltage, and resistance.
- How some of the components introduced in this chapter are used in a specific application.

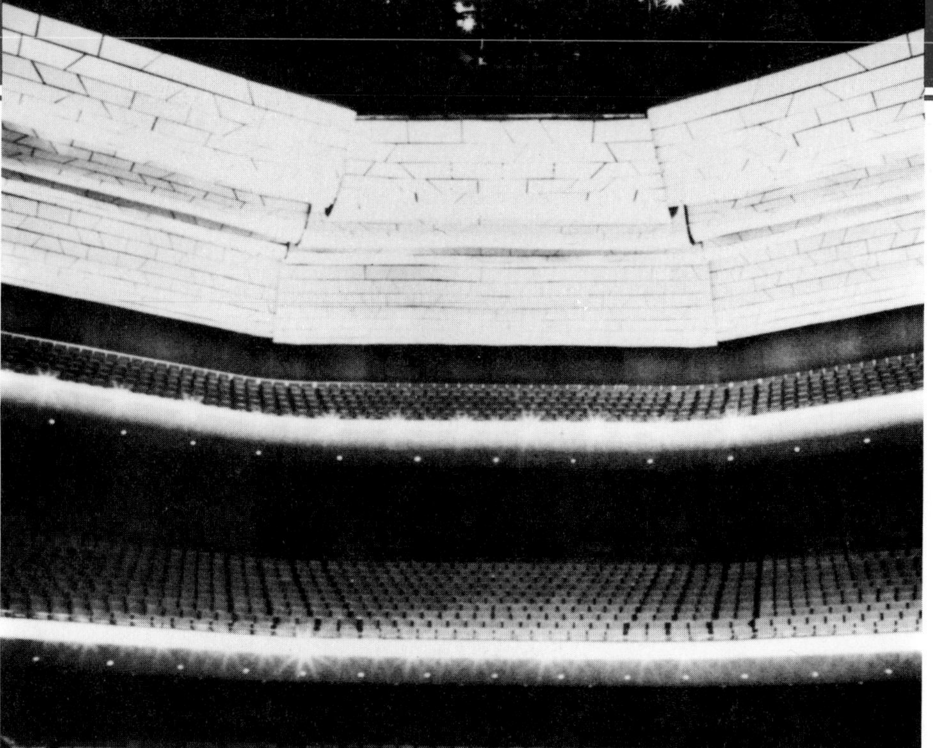

2

APPLICATION ASSIGNMENT

As head technician for special effects, you are asked to devise a system of six lamps for use in the stage lighting for a play. There are three requirements for the system:

1. Each of the lamps can be turned on or off one at a time in any sequence.
2. All the lamps can be turned on or off simultaneously at specified times.
3. A single control can brighten or dim all of the lamps that are on.

After studying this chapter, you should be able to solve this problem. The Application Note at the end of the chapter gives one solution.

ATOMS

2–1

An atom is the smallest particle of an element that still retains the characteristics of that element. Different elements have different types of atoms. In fact, every element has a unique atomic structure.

Atoms have a planetary type of structure, consisting of a central nucleus surrounded by orbiting electrons. The nucleus consists of positively charged particles called *protons* and uncharged particles called *neutrons*. The electrons are the basic particles of negative charge.

Each type of atom has a certain number of electrons and protons that distinguishes the atom from all other atoms of other elements. For example, the simplest atom is that of hydrogen. It has one proton and one electron, as pictured in Figure 2–1(a). The helium atom, shown in Figure 2–1(b), has two protons and two neutrons in the nucleus, which is orbited by two electrons.

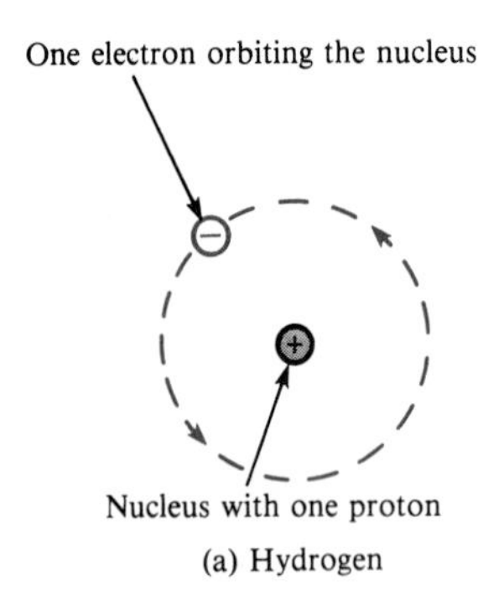

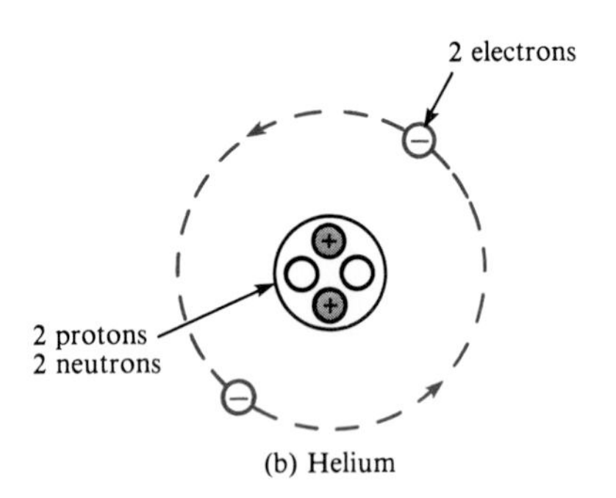

FIGURE 2–1
Hydrogen and helium atoms.

Atomic Weight and Number

All elements are arranged in the periodic table of the elements in order according to their atomic number, which is the number of protons in the nucleus. The elements can also be arranged by their atomic weight, which is approximately the number of protons and neutrons in the nucleus. For example, hydrogen has an atomic number of one and an atomic weight of one. The atomic number of helium is two, and its atomic weight is four.

In their normal, or neutral, state, all atoms of a given element have the same number of electrons as protons. So the positive charges cancel the negative charges, and the atom has a net charge of zero.

The Copper Atom

Since copper is the most commonly used metal in electrical applications, let us examine its atomic structure. The copper atom has 29 electrons in orbit around the nucleus. They do not all occupy the same orbit, however. They move in orbits at varying distances from the nucleus. The orbits in which the electrons revolve are called *shells*. The number of electrons in each shell follows a predictable pattern.

The first shell of any atom can have up to two electrons, the second shell up to eight electrons, the third shell up to 18 electrons, and the fourth shell up to 32 electrons. A copper atom is shown in Figure 2–2. Notice that the fourth or outermost shell has only one electron, called the *valence* electron.

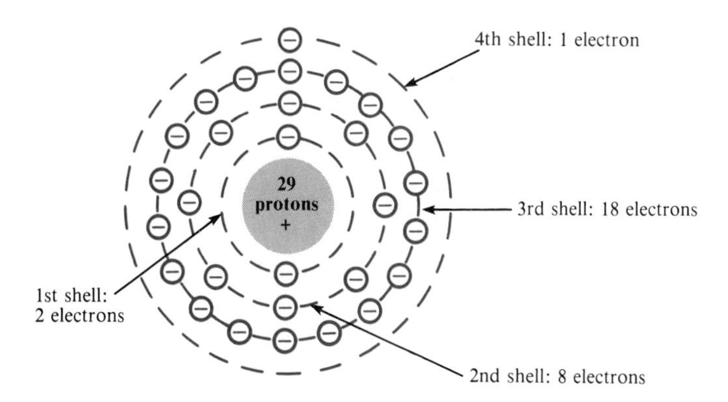

FIGURE 2–2
The copper atom.

Free Electrons

When the electron in the outer shell of the copper atom gains sufficient energy from the surrounding media, it can break away from the parent atom and become what is called a *free electron.* The free electrons in the copper material are capable of moving from one atom to another in the material. In other words, they drift randomly from atom to atom within the copper. As you will see, the free electrons make electrical *current* possible.

Three categories of materials are used in electronics: *conductors, semiconductors,* and *insulators.*

Conductors

Conductors are materials that allow current to flow easily. They have a large number of free electrons in their structure. Most metals are good conductors. Silver is the best conductor, and copper is next. Copper is the most widely used conductive material because it is less expensive than silver.

Semiconductors

These materials are classed below the conductors in their ability to carry current because they have fewer free electrons in their structure than do conductors. However, because of their unique characteristics, certain semiconductor materials are the basis for modern electronic devices such as the diode, transistor, and integrated circuit. Silicon and germanium are common semiconductor materials.

Insulators

Insulating materials are poor conductors of electric current. In fact, they are used to *prevent* current where it is not wanted. Compared to conductive materials, insulators have very few free electrons.

SECTION REVIEW 2–1

1. What is the basic particle of negative charge?
2. Define *atom.*
3. What does a typical atom consist of?
4. Do all elements have the same types of atoms?
5. What is a free electron?

ELECTRICAL CHARGE

2–2

As you saw in the previous discussion of the structure of an atom, there are two types of charge: *positive* and *negative.* The electron is the smallest particle that exhibits negative electrical charge. When an excess of electrons exists in a material, there is a net negative electrical charge. When an excess of protons exists, there is a net positive electrical charge. The charge of an electron and that of a proton are equal in magnitude. Electrical charge is symbolized by Q.

Static electricity is the presence of a net positive or negative charge in a material. Everyone has experienced the effects of static electricity from time to time, for example, when attempting to touch a metal surface or another person or when the clothes in a dryer cling together.

Materials with charges of opposite polarity are attracted to each other, and materials with charges of the same polarity are repelled, as indicated in Figure 2–3. There is a force acting between charges, as evidenced by the attraction or repulsion. This force, called an *electric field,* consists of invisible lines of force as shown in Figure 2–4.

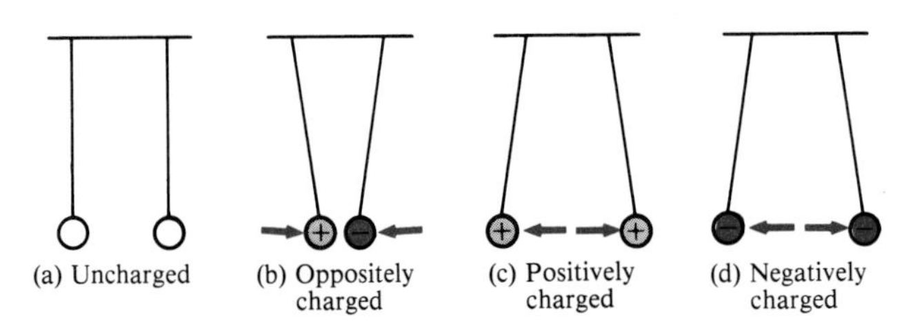

FIGURE 2–3
Attraction and repulsion of electrical charges.

FIGURE 2–4
Electric field between oppositely charged surfaces.

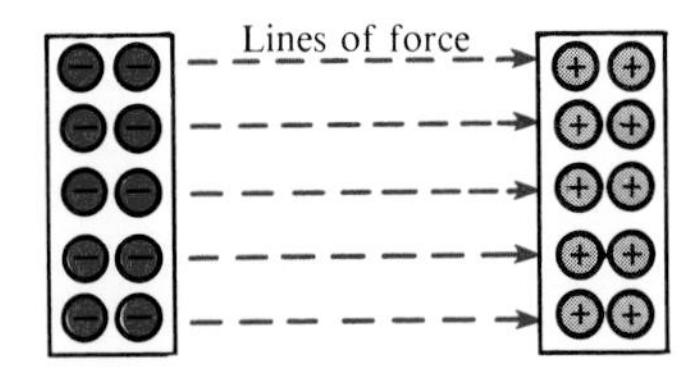

The Coulomb: The Unit of Charge

Electrical charge is measured in *coulombs,* abbreviated C. *One coulomb is the total charge possessed by* 6.25×10^{18} *electrons.* A single electron has a charge of 1.6×10^{-19} C. This unit is named for Charles Coulomb (1736–1806), a French scientist.

How Positive and Negative Charges Are Created

Consider a neutral atom—that is, one that has the same number of electrons and protons and thus has no net charge. If a valence electron is pulled away from the atom by the application of energy, the atom is left with a net positive charge (more protons than electrons) and becomes a *positive ion.* If an atom acquires an extra electron in its outer shell, it has a net negative charge and becomes a *negative ion.*

The amount of energy required to free a valence electron is related to the number of electrons in the outer shell. An atom can have up to eight valence electrons. The more complete the outer shell, the more stable the atom and thus the more energy is required to release an electron. Figure 2–5 illustrates the creation of a positive and a negative ion when sodium chloride dissolves and the sodium atom gives up its single valence electron to the chloride atom.

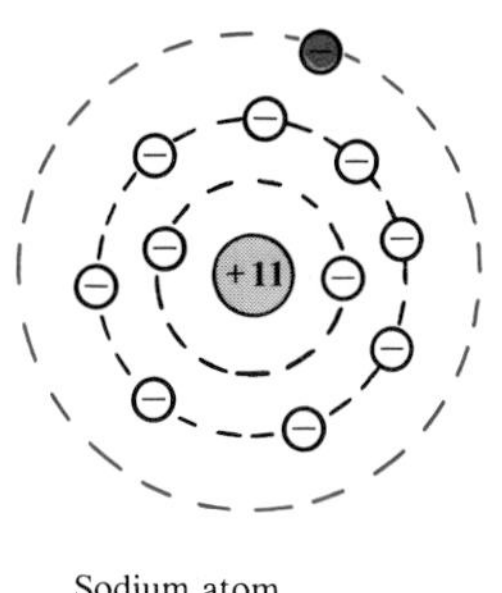

Sodium atom
(11 protons, 11 electrons)

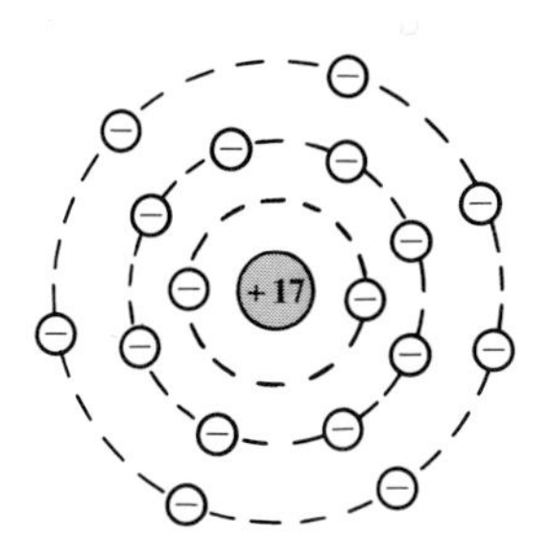

Chlorine atom
(17 protons, 17 electrons)

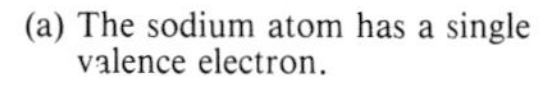

(a) The sodium atom has a single valence electron.

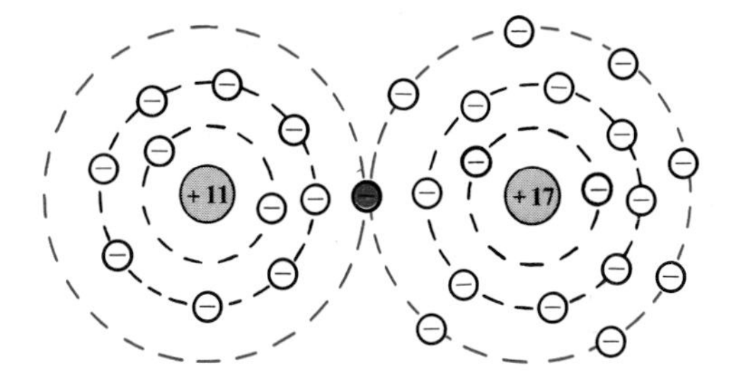

(b) The atoms combine by sharing the valence electron to form sodium chloride (table salt).

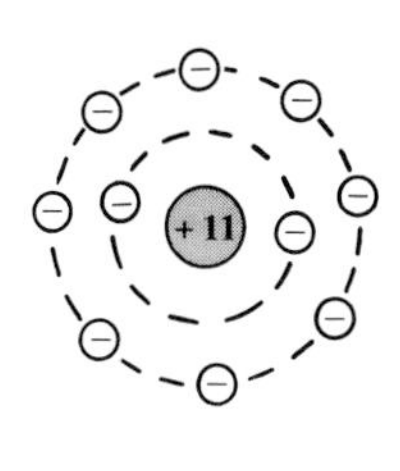

Positive sodium ion
(11 protons, 10 electrons)

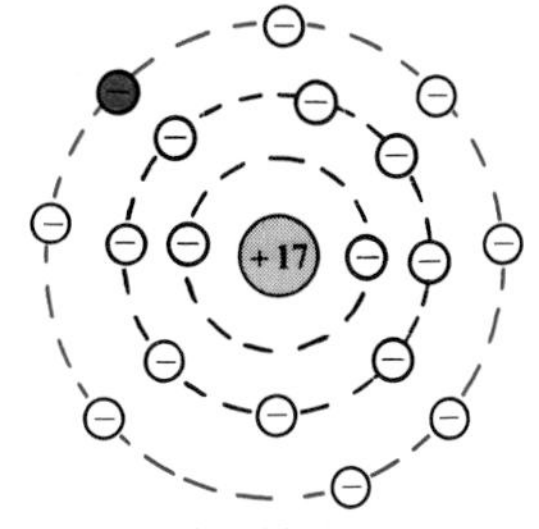

Negative chlorine ion
(17 protons, 18 electrons)

(c) When dissolved, the sodium atom gives up the valence electron to become a positive ion, and the chlorine atom retains the extra valence electron to become a negative ion.

FIGURE 2–5
Example of the formation of positive and negative ions.

EXAMPLE 2–1 How many coulombs do 93.75×10^{16} electrons represent?

Solution

$$Q = \frac{\text{number of electrons}}{\text{number of electrons in one coulomb}} = \frac{93.75 \times 10^{16} \text{ electrons}}{6.25 \times 10^{18} \text{ electrons/C}} = 0.15 \text{ C}$$

SECTION REVIEW 2–2

1. What is the symbol for charge?
2. What is the unit of charge, and what is its symbol?
3. How much charge, in coulombs, is there in 10×10^{12} electrons?

VOLTAGE

2–3

As you have seen, there is a force of attraction between a positive and a negative charge. A certain amount of energy (work) is required to overcome the force and move the charges a given distance apart. All opposite charges possess a certain potential energy because of the separation between them. The difference in potential energy of the charges is the *potential difference.* For example, consider a water tank that is supported several feet above the ground. A given amount of energy (work) is required to pump water up to fill the tank. Once the water is stored in the tank, it has a certain potential energy which, if released, can be used to perform work. For example, the water can be allowed to fall down a chute to turn a water wheel.

Potential difference in electrical terms is more commonly called *voltage* *(V)* and is expressed as energy *(W)* per unit charge *(Q)*:

$$V = \frac{W}{Q} \tag{2–1}$$

where W is expressed in joules (J) and Q is in coulombs (C).

The unit of voltage is the *volt,* symbolized by V. *One volt is the potential difference (voltage) between two points when one joule of energy is used to move one coulomb of charge from one point to the other.*

EXAMPLE 2–2 If 50 joules of energy are available for every 10 coulombs of charge, what is the voltage?

Solution

$$V = \frac{W}{Q} = \frac{50 \text{ J}}{10 \text{ C}} = 5 \text{ V}$$

Sources of Voltage

The Battery

A voltage source is a source of potential energy that is also called *electromotive force* (emf). The battery is one type of voltage source that converts chemical energy into electrical energy. A voltage exists between the electrodes (terminals) of a battery, as shown by a voltaic cell in Figure 2–6. One electrode is positive and the other negative as a result of the separation of charges caused by the chemical action when two different conducting materials are dissolved in the electrolyte.

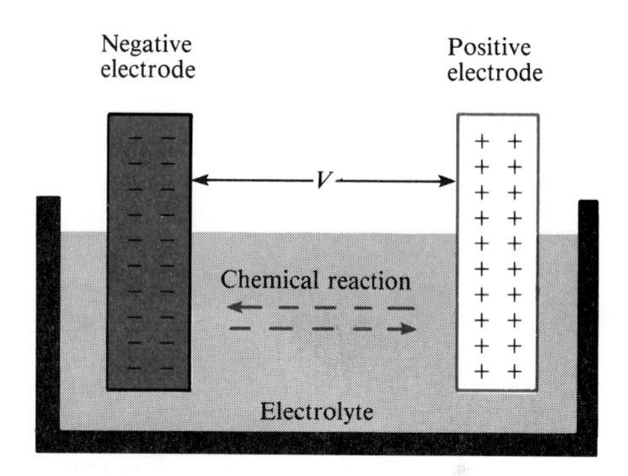

FIGURE 2–6
A voltaic cell converts chemical energy into electrical energy.

Batteries are generally classified as *primary cells,* which cannot be recharged, and *secondary cells,* which can be recharged by reversal of the chemical action. The amount of voltage provided by a battery varies. For example, a flashlight battery is 1.5 V and an automobile battery is 12 V. Some typical batteries are shown in Figure 2–7.

FIGURE 2–7
Typical batteries. (courtesy of Gould, Inc.)

The Electronic Power Supply

These voltage sources convert the ac voltage from the wall outlet to a constant (dc) voltage which is available across two terminals, as indicated in Figure 2–8(a). Typical commercial power supplies are shown in part (b).

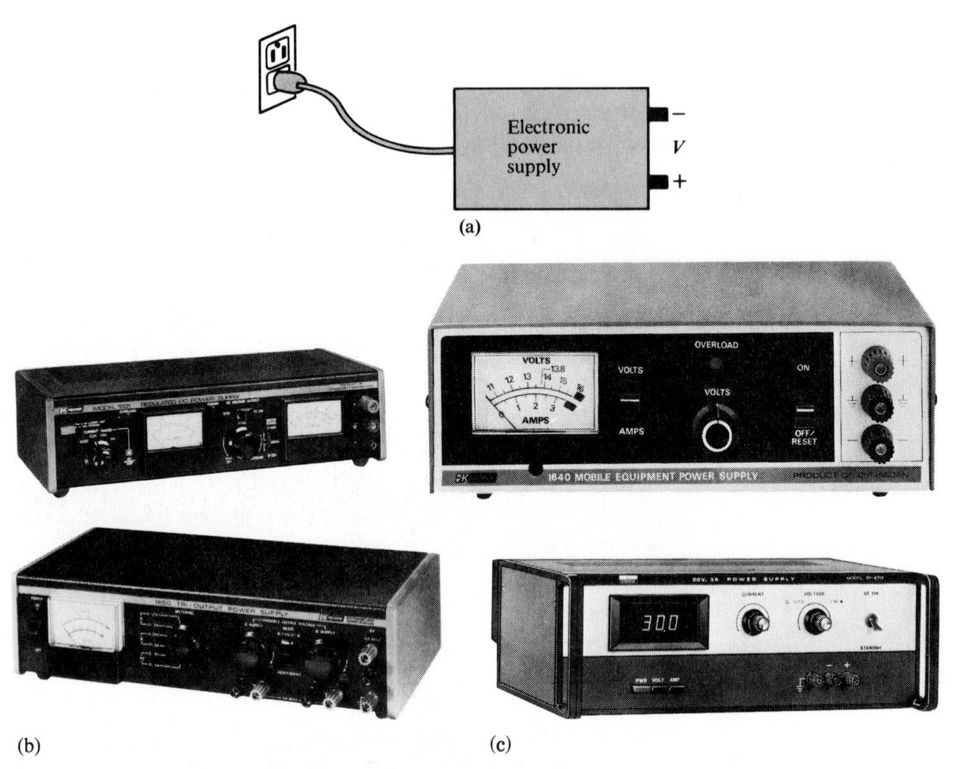

FIGURE 2–8
Electronic power supplies. ((b) courtesy of B&K Precision/Dynascan Corp. (c) courtesy of B&K Precision/Dynascan Corp. (top); courtesy of Health Schlumberger Instruments (bottom)).

The Solar Cell

The operation of solar cells is based on the principle of *photovoltaic action,* which is the process whereby light energy is converted directly into electrical energy. A basic solar cell consists of two layers of different semiconductive materials joined together to form a junction. When one layer is exposed to light, many electrons acquire enough energy to break away from their parent atoms and cross the junction. This process forms negative ions on one side of the junction and positive ions on the other, and thus a potential difference (voltage) is developed. Construction of a basic solar cell is shown in Figure 2–9.

The Generator

Generators convert mechanical energy into electrical energy using a principle called *electromagnetic induction* (to be studied later). A conductor is rotated through a magnetic field, and a voltage is produced across the conductor. A typical generator is pictured in Figure 2–10.

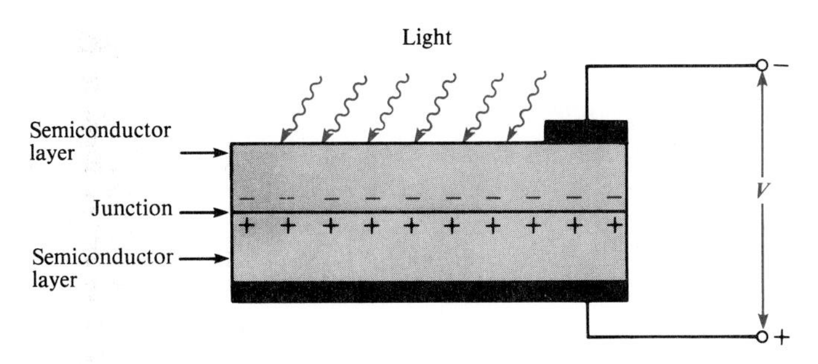

FIGURE 2–9
Construction of basic solar cell.

FIGURE 2–10
Cutaway view of dc generator (courtesy of Pacific Scientific Motor and Control Division).

SECTION REVIEW 2–3

1. Define *voltage.*
2. How much is the voltage when there are 24 joules of energy for 10 coulombs of charge?
3. List four sources of voltage.

CURRENT

2–4

As you have seen, there are free electrons available in all conductive and semiconductive materials. These electrons drift randomly in all directions, from atom to atom, within the structure of the material, as indicated in Figure 2–11.

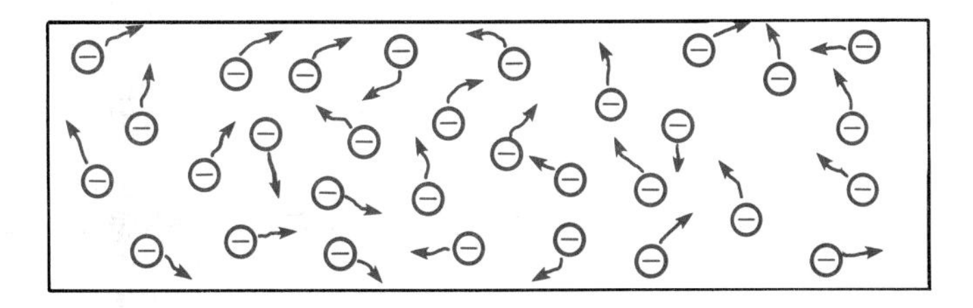

FIGURE 2–11
Random motion of free electrons in a material.

Now, if a voltage is placed across the conductive or semiconductive material, one end becomes positive and the other negative, as indicated in Figure 2–12. The repulsive force between the negative voltage at the left end causes the free electrons (negative charges) to move toward the right. The attractive force between the positive voltage at the right end pulls the free electrons to the right. The result is a net movement of the free electrons from the negative end of the material to the positive end, as shown in Figure 2–12.

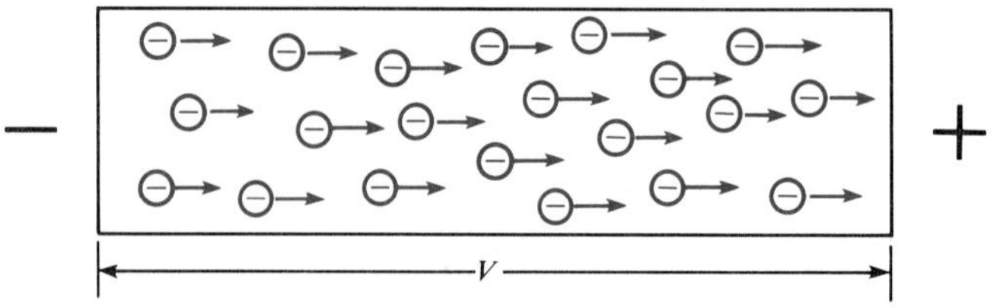

FIGURE 2–12
Electrons flow from negative to positive when a voltage is applied across a conductive or semiconductive material.

The movement of these free electrons from the negative end of the material to the positive end is the *electrical current,* symbolized by *I. Electrical current is defined as the rate of flow of electrons in a conductive or semiconductive material.* It is measured by the number of electrons (amount of charge, *Q*) that flow past a point in a unit of time:

$$I = \frac{Q}{t} \tag{2–2}$$

where *I* is current, *Q* is the charge of the electrons, and *t* is the time.

The Ampere: The Unit of Current

Current is measured in a unit called the *ampere* or *amp* for short, symbolized by A. It is named after André Ampère (1775–1836), a French physicist whose work contributed to the understanding of electrical current and its effects.

One ampere (1 A) is the amount of current flowing when a number of electrons having one coulomb (1 C) of charge move past a given point in one second (1 s). (See Figure 2–13.) One coulomb is the charge carried by 6.25×10^{18} electrons.

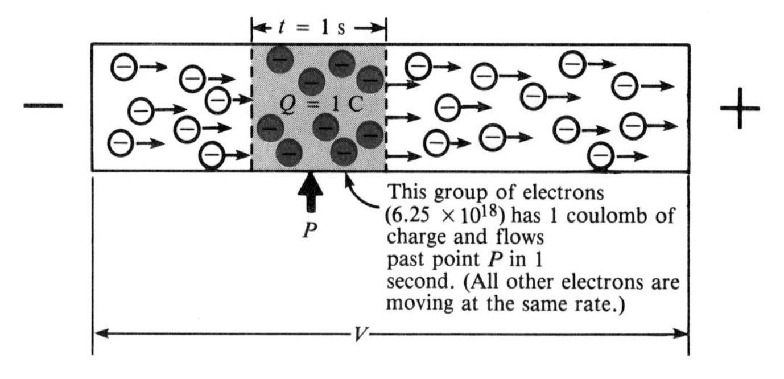

FIGURE 2–13
Illustration of one ampere of current in a material (1 C/s).

EXAMPLE 2–3

Ten coulombs of charge flow past a given point in a wire in 2 seconds. How many amperes of current are flowing?

Solution

$$I = \frac{Q}{t} = \frac{10\ \text{C}}{2\ \text{s}} = 5\ \text{A}$$

SECTION REVIEW 2–4

1. Define *current* and state its unit.
2. How many electrons make up one coulomb of charge?
3. What is the current in amperes when 20 coulombs flow past a point in a wire in 4 s?

RESISTANCE

2–5

When current flows in a material, the free electrons move through the material and occasionally collide with atoms. These collisions cause the electrons to lose some of their energy, and thus their movement is restricted. The more collisions, the more the flow of electrons is restricted. This restriction varies and is determined by the type of material. The property of a material that restricts the flow of electrons is called *resistance,* designated R. *Resistance is the opposition to current.* The schematic symbol for resistance is shown in Figure 2–14.

FIGURE 2–14 Resistance/resistor symbol.

When current flows through any material that has resistance, heat is produced by the collisions of electrons and atoms. Therefore, wire, which typically has a very small resistance, becomes warm when there is current through it. A discussion of how the size of wire affects its resistance is given in Appendix A.

The Ohm: The Unit of Resistance

Resistance, R, is expressed in the unit of *ohms,* named after Georg Simon Ohm (1789–1854) and symbolized by the Greek letter omega (Ω). *There is one ohm (1 Ω) of resistance when one ampere (1 A) of current flows in a material with one volt (1 V) applied.*

Resistors

Components that are specifically designed to have a certain amount of resistance are called *resistors.* The principal applications of resistors are to limit the flow of current and, in certain cases, to generate heat. Although there are a variety of different types of resistors that come in many shapes and sizes, they can all be placed in one of two main categories: *fixed* or *variable.*

Fixed Resistors

This kind of resistor is available with a large selection of ohmic values that are set during manufacturing and cannot be changed easily. Fixed resistors are constructed using various methods and materials. Several common types are shown in Figure 2–15.

One common fixed resistor is the *carbon-composition* type, which is made with a mixture of finely ground carbon, insulating filler, and a resin binder. The

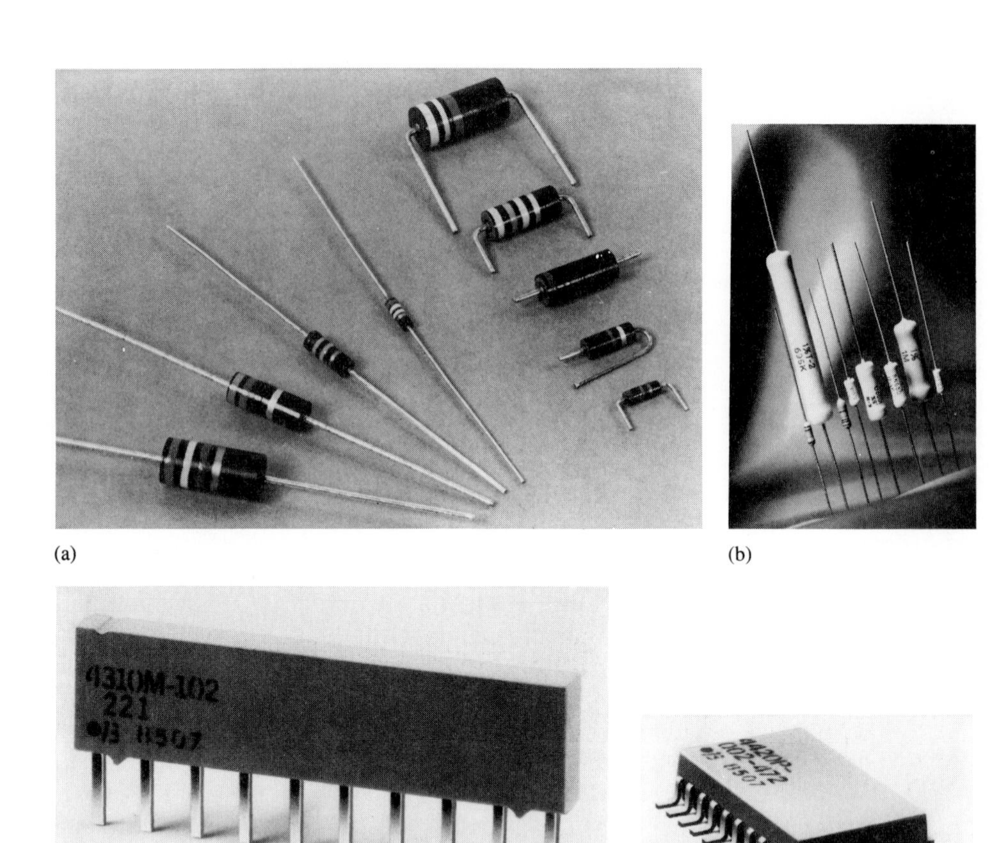

FIGURE 2–15
Typical fixed resistors ((a) and (b) courtesy of Stackpole Carbon Co. (c) courtesy of Bourns, Inc.).

ratio of carbon to insulating filler sets the resistance value. The mixture is formed into rods, and lead connections are made. The entire resistor is then encapsulated in an insulated coating for protection. Figure 2–16 shows the construction of a typical carbon-composition resistor.

Other types of fixed resistors include carbon film, metal oxide, metal film, metal glaze, and wire-wound. In film resistors, a resistive material is deposited evenly onto a high-grade ceramic rod. The resistive film may be carbon (carbon film), nickel chromium (metal film), a mixture of metals and glass (metal glaze), or metal and insulating oxide (metal oxide). In these types of resistors, the desired resistance value is obtained by removing part of the resistive material in a helical pattern along the rod using a *spiraling* technique. Very close tolerance can be achieved with this method.

Wire-wound resistors are constructed with resistive wire wound around an insulating rod and then sealed. Normally, wire-wound resistors are used because of their relatively high power ratings.

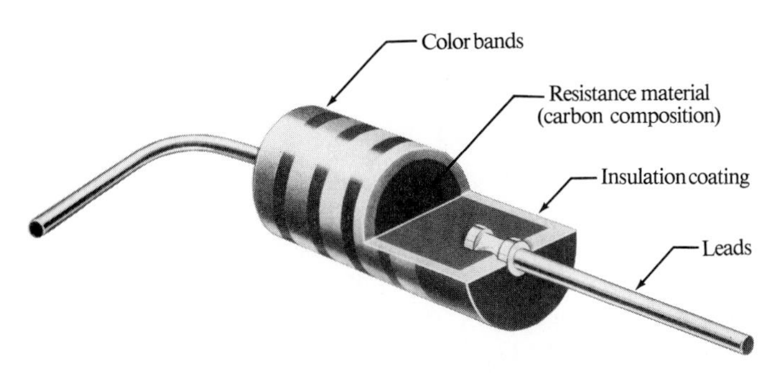

FIGURE 2–16
Cutaway view of carbon-composition resistor (courtesy of Allen-Bradley Co.).

Some typical fixed resistors are shown in the construction views of Figure 2–17.

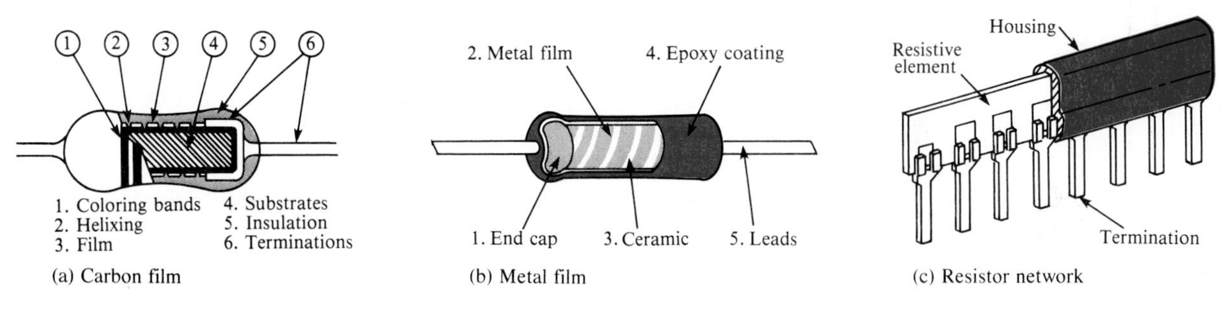

FIGURE 2–17
Construction views of typical fixed resistors [(a) and (b) courtesy of Stackpole Carbon Co. (c) courtesy of Bourns, Inc.]

Resistor Color Codes

Fixed resistors with value tolerances of 5%, 10%, or 20% are color coded with four bands to indicate the resistance value and the tolerance. This color-code band system is shown in Figure 2–18, and the color code is listed in Table 2–1.

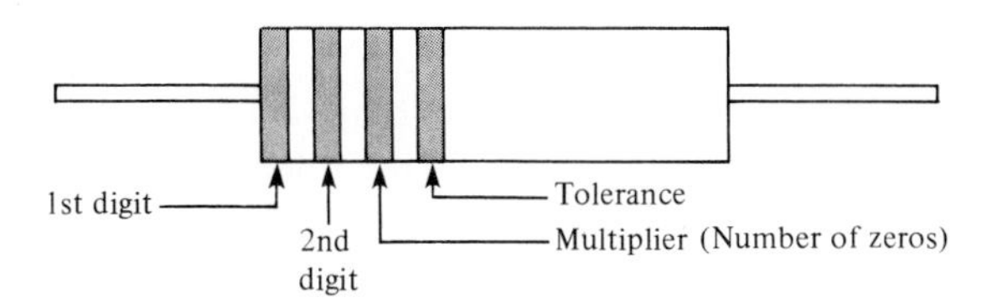

FIGURE 2–18
Color-code bands on a resistor.

The color code is read as follows:

1. Beginning at the banded end, the first band is the first digit of the resistance value.
2. The second band is the second digit.
3. The third band is the number of zeros, or the *multiplier*.
4. The fourth band indicates the tolerance.

TABLE 2–1
Resistor color code.

	Digit	Color
Resistance value, first three bands	0	Black
	1	Brown
	2	Red
	3	Orange
	4	Yellow
	5	Green
	6	Blue
	7	Violet
	8	Gray
	9	White
Tolerance, fourth band	5%	Gold
	10%	Silver
	20%	No band

For example, a 5% tolerance means that the *actual* resistance value is within ±5% of the color-coded value. Thus, a 100-Ω resistor with a tolerance of ±5% can have acceptable values as low as 95 Ω and as high as 105 Ω.

For resistance values less than 10 Ω, the third band is either gold or silver. Gold represents a multiplier of 0.1, and silver represents 0.01. For example, a color code of red, violet, gold, and silver represents 2.7 Ω with a tolerance of ±10%.

EXAMPLE 2–4

Find the resistance values in ohms and the percent tolerance for each of the color coded resistors shown in Figure 2–19 (a full-color photo appears between pages 128 and 129).

Solution

(a) First band is red = 2, second band is violet = 7, third band is orange = 3 zeros, fourth band is silver = 10% tolerance.

$$R = 27{,}000\ \Omega\ \pm 10\%$$

(b) First band is brown = 1, second band is black = 0, third band is brown = 1 zero, fourth band is silver = 10% tolerance.

$$R = 100\ \Omega\ \pm 10\%$$

(c) First band is green = 5, second band is blue = 6, third band is green = 5 zeros, fourth band is gold = 5% tolerance.

$$R = 5{,}600{,}000\ \Omega\ \pm 5\%$$

Certain precision resistors with tolerances of 1% or 2% are color coded with five bands. Beginning at the banded end, the first band is the first digit of the resistance value, the second band is the second digit, the third band is the third digit, the fourth band is the multiplier, and the fifth band indicates the tolerance. Table 2–1 applies, except that gold indicates 1% and silver indicates 2%.

Numerical labels are also commonly used on certain types of resistors where the resistance value and tolerance are stamped on the body of the resistor. For example, a common system uses R to designate the decimal point and letters to indicate tolerance as follows:

$$F = \pm 1\%, \quad G = \pm 2\%, \quad J = \pm 5\%, \quad K = \pm 10\%, \quad M = \pm 20\%$$

For values above 100 Ω, three digits are used to indicate resistance value, followed by a fourth digit that specifies the number of zeros. For values less than 100 Ω, R indicates the decimal point.

Some examples are as follows: 6R8M is a 6.8-Ω ±20% resistor; 3301F is a 3300-Ω ±1% resistor; and 2202J is a 22000-Ω ±5% resistor.

Resistor Reliability Band

The fifth band on some color-coded resistors indicates the resistor's reliability in percent of failures per 1000 hours of use. The fifth-band reliability color code is listed in Table 2–2. For example, a brown fifth band means that if a group of like resistors are operated under standard conditions for 1000 hours, 1% of the resistors in that group will fail.

TABLE 2–2
Fifth-band reliability color code.

Color	Failures (%) during 1000 hours of operation
Brown	1.0%
Red	0.1%
Orange	0.01%
Yellow	0.001%

Variable Resistors

Variable resistors are designed so that their resistance values can be changed easily with a manual or an automatic adjustment.

Two basic types of manually adjustable resistors are the *potentiometer* and the *rheostat*. Schematic symbols for these types are shown in Figure 2–20. The potentiometer is a *three-terminal device,* as indicated in Part (a). Terminals 1

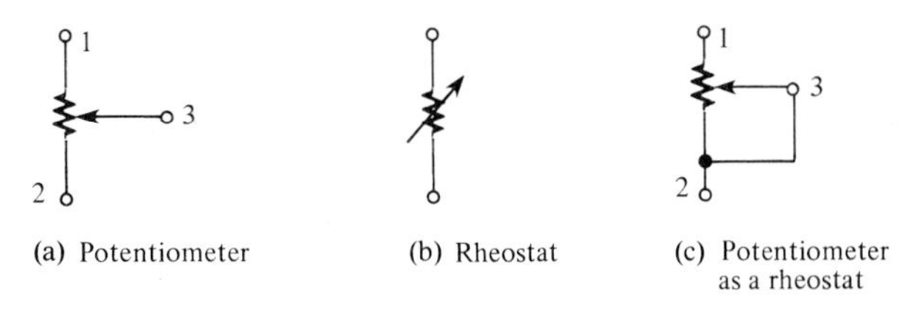

FIGURE 2–20
Potentiometer and rheostat symbols.

and 2 have a fixed resistance between them, which is the total resistance. Terminal 3 is connected to a moving contact (wiper). We can vary the resistance between 3 and 1 or between 3 and 2 by moving the contact up or down.

Figure 2–20(b) shows the rheostat as a *two-terminal* variable resistor. Part (c) shows how we can use a potentiometer as a rheostat by connecting terminal 3 to either terminal 1 or terminal 2. Some typical potentiometers are pictured in Figure 2–21.

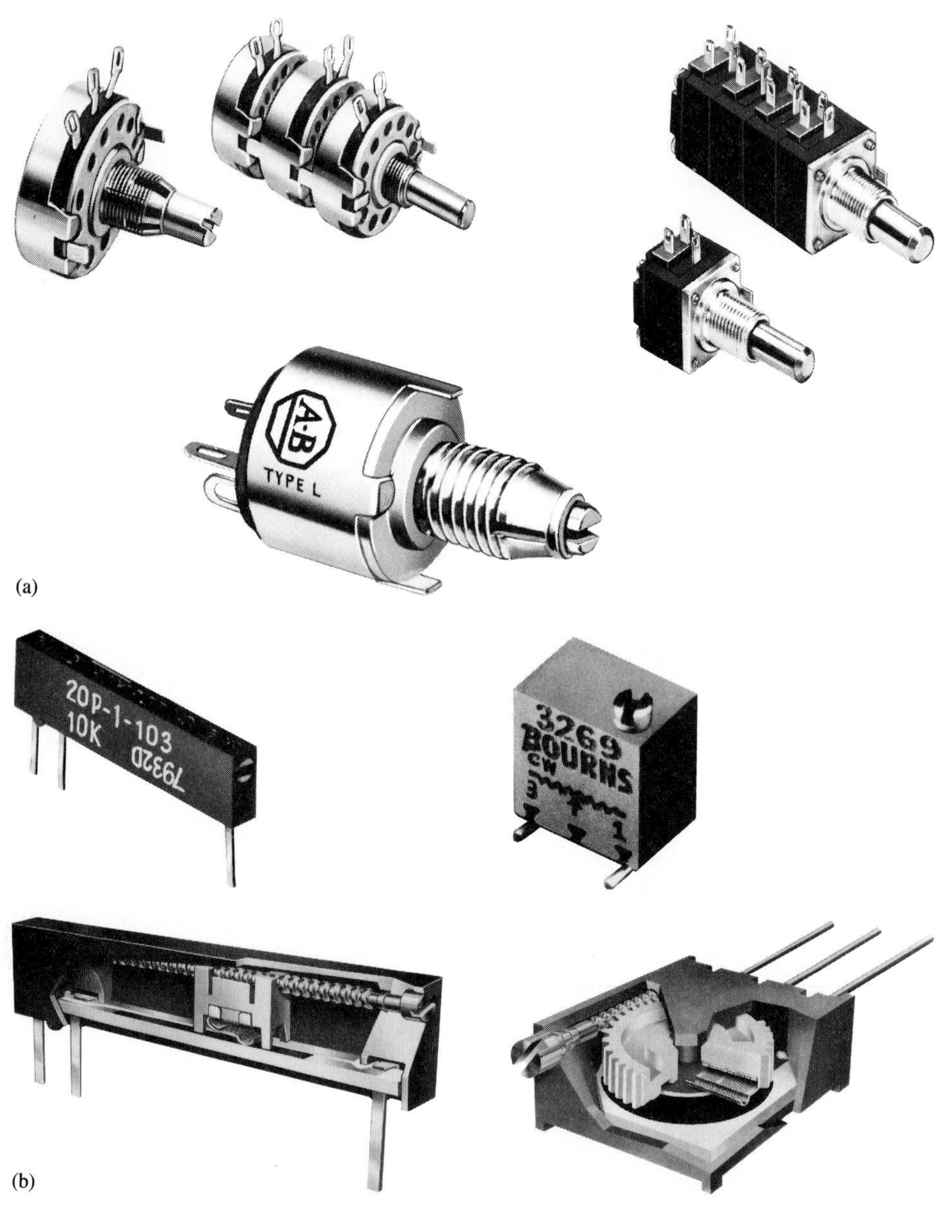

FIGURE 2–21
(a) Typical potentiometers (courtesy of Allen-Bradley Co.). (b) Trimmer potentiometers with construction views (courtesy of Bourns Trimpot).

Potentiometers and rheostats can be classified as *linear* or *tapered,* as shown in Figure 2–22, where a potentiometer with a total resistance of 100 Ω is used as an example. As shown in Part (a), in a linear potentiometer, the resistance between either terminal and the moving contact varies linearly with the position of the moving contact. For example, one-half of a turn results in one-half the total resistance. Three-quarters of a turn results in three-quarters of the total resistance between the moving contact and one terminal, or one-quarter of the total resistance between the other terminal and the moving contact.

In the tapered potentiometer, the resistance varies nonlinearly with the position of the moving contact, so that one-half of a turn does not necessarily result in one-half the total resistance. This concept is illustrated in Figure 2–22(b), where the nonlinear values are arbitrary.

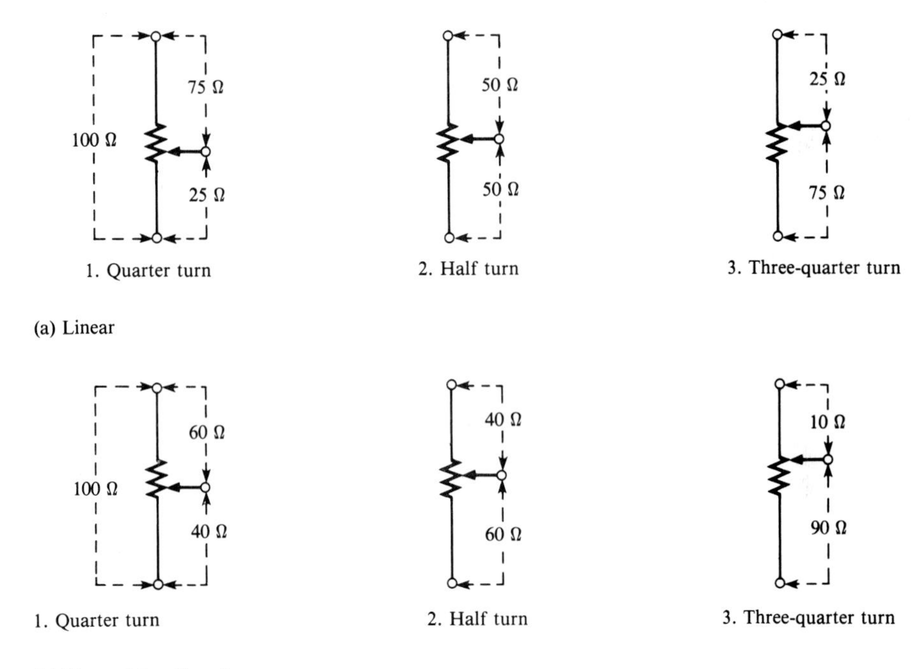

FIGURE 2–22
Examples of (a) linear and (b) tapered potentiometers.

SECTION REVIEW 2–5

1. Define *resistance* and name its unit.
2. What are the two main categories of resistors? Briefly explain the difference between them.
3. In the resistor color code, what does each band represent?
4. Determine the resistance and tolerance for each of the following color codes:
   (a) yellow, violet, red, gold.
   (b) blue, red, orange, silver.
   (c) brown, gray, black, gold.
5. What is the basic difference between a rheostat and a potentiometer?

THE ELECTRIC CIRCUIT

Basically, an electric circuit consists of a *voltage source,* a *load,* and a *path for current* between the source and the load. Figure 2–23 shows an example of a simple electric circuit: a battery connected to a lamp with two wires. The battery is the voltage source, the lamp is the load on the battery because it draws current from the battery, and the two wires provide the current path from the negative terminal of the battery to the lamp and back to the positive terminal of the battery, as shown in Part (b). Current flows through the filament of the lamp (which has a resistance), causing it to emit visible light. Current flows through the battery by chemical action. In many practical cases, one terminal of the battery is connected to a *ground* point. For example, in automobiles, the negative battery terminal is connected to the metal chassis of the car. The chassis is the ground for the automobile electrical system. The concept of circuit ground is covered in detail in a later chapter.

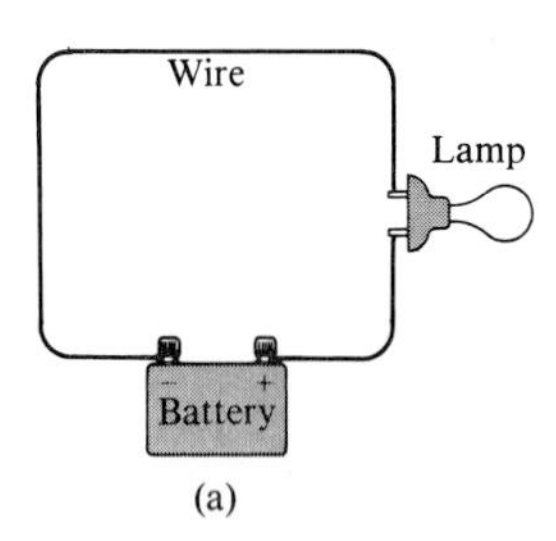

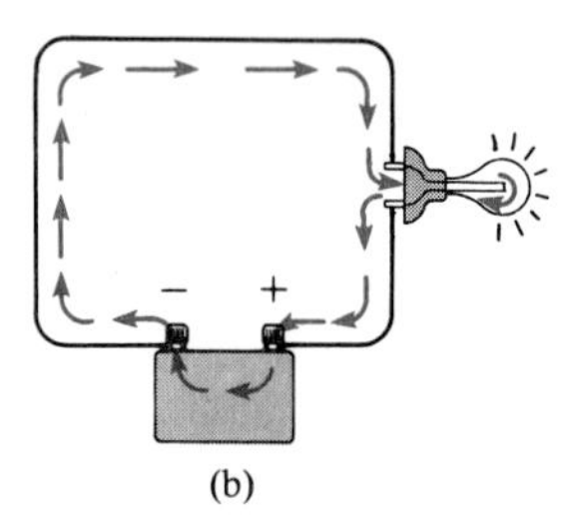

FIGURE 2–23
A simple electric circuit.

The Electrical Schematic

An electric circuit can be represented by a *schematic diagram* using standard symbols for each element, as shown in Figure 2–24 for the simple circuit in Figure 2–23. The purpose of a schematic diagram is to show in an organized manner how the various components in a given circuit are interconnected so that the operation of the circuit can be determined.

Closed and Open Circuits

The example circuit in Figure 2–23 illustrated a *closed circuit*—that is, a circuit in which the current has a complete path through which to flow. When the current path is broken so that current cannot flow, the circuit is called an *open circuit.*

Switches

Switches are commonly used for controlling the opening or closing of circuits by either mechanical or electronic means. For example, a switch is used to turn a lamp on or off as illustrated in Figure 2–25. Each circuit pictorial is shown with its associated schematic diagram. The type of switch indicated is a *single-pole–single-throw* (SPST) toggle switch.

Figure 2–26 shows a somewhat more complicated circuit using *single-pole–double-throw* (SPDT) type of switch to control the current to two different lamps. When one lamp is on, the other is off, and vice versa, as illustrated by the two schematic diagrams which represent each of the switch positions.

The term *pole* refers to the movable arm in a switch, and the term *throw* indicates the number of contacts that are affected (either opened or closed) by a single switch action (a single movement of a pole).

In addition to the SPST and the SPDT switches already introduced, several other types are of importance:

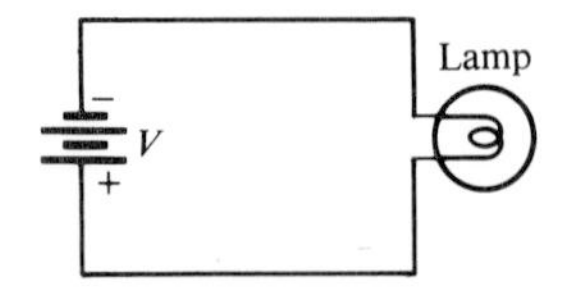

FIGURE 2–24
Schematic diagram for the circuit in Figure 2–23.

- *Double-pole–single-throw (DPST).* The DPST switch permits simultaneous opening or closing of two sets of contacts. The symbol is shown in Figure 2–27(a). The dashed line indicates that the contact arms are mechanically linked so that both move with a single switch action.

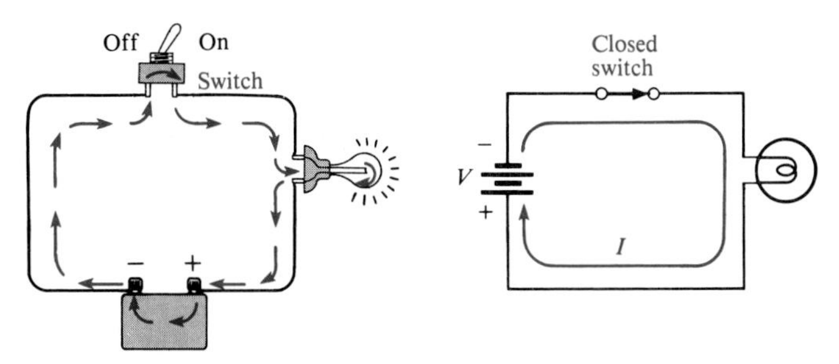

(a) Current flows in a *closed* circuit (switch is ON or in the *closed* position).

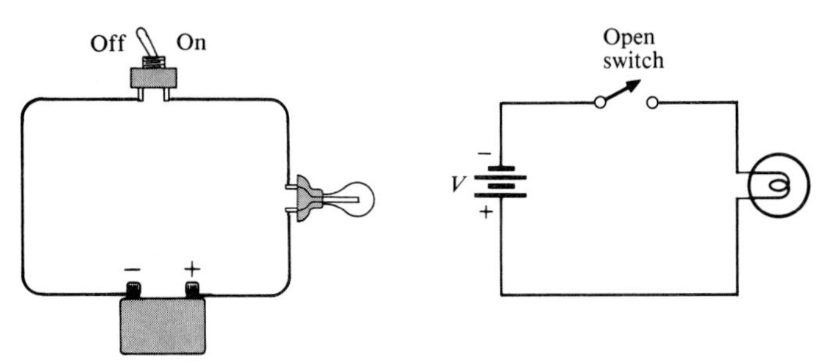

(b) No current flows in an *open* circuit (switch is OFF or in the *open* position).

FIGURE 2–25
Basic closed and open circuits using an SPST switch for control.

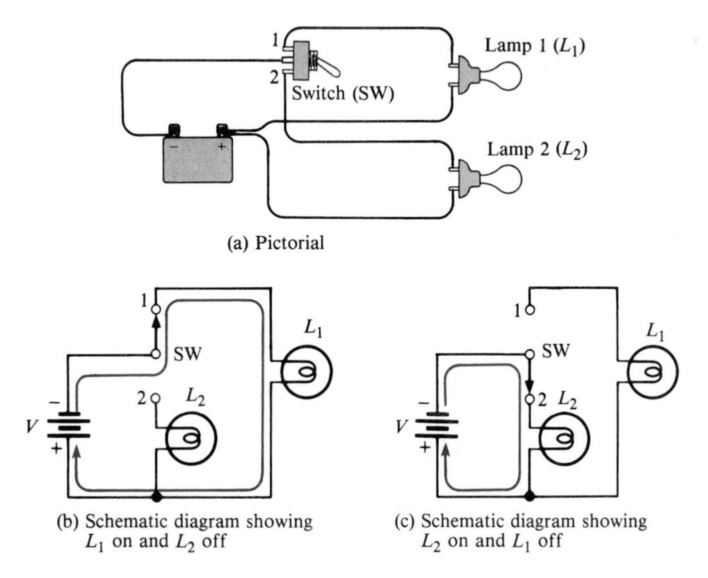

(a) Pictorial

(b) Schematic diagram showing L_1 on and L_2 off

(c) Schematic diagram showing L_2 on and L_1 off

FIGURE 2–26
An example of an SPDT switch controlling two lamps.

- *Double-pole–double-throw (DPDT).* The DPDT switch provides connection from one set of contacts to either of two other sets. The schematic symbol is shown in Figure 2–27(b).

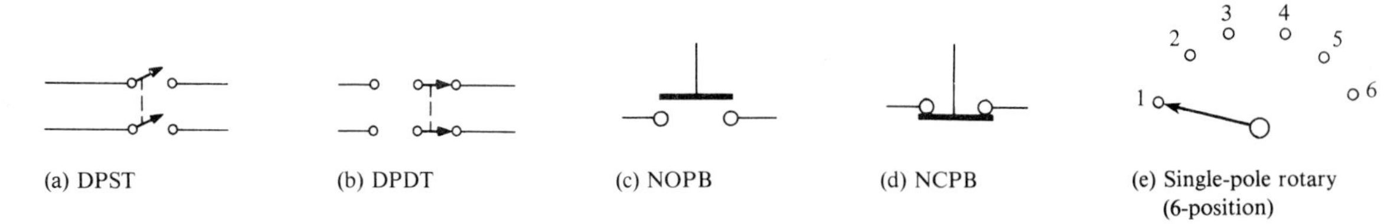

FIGURE 2–27
Switch symbols.

- *Push-button (PB).* In the normally open push-button switch (NOPB), shown in Figure 2–27(c), connection is made between two contacts when the button is depressed, and connection is broken when the button is released. In the normally closed push-button switch (NCPB), shown in Figure 2–27(d), connection between the two contacts is broken when the button is depressed.
- *Rotary.* In a rotary switch, a knob is turned to make connection between one contact and any one of several others. A symbol for a simple six-position rotary switch is shown in Figure 2–27(e).

Several varieties of switches are shown in Figure 2–28.

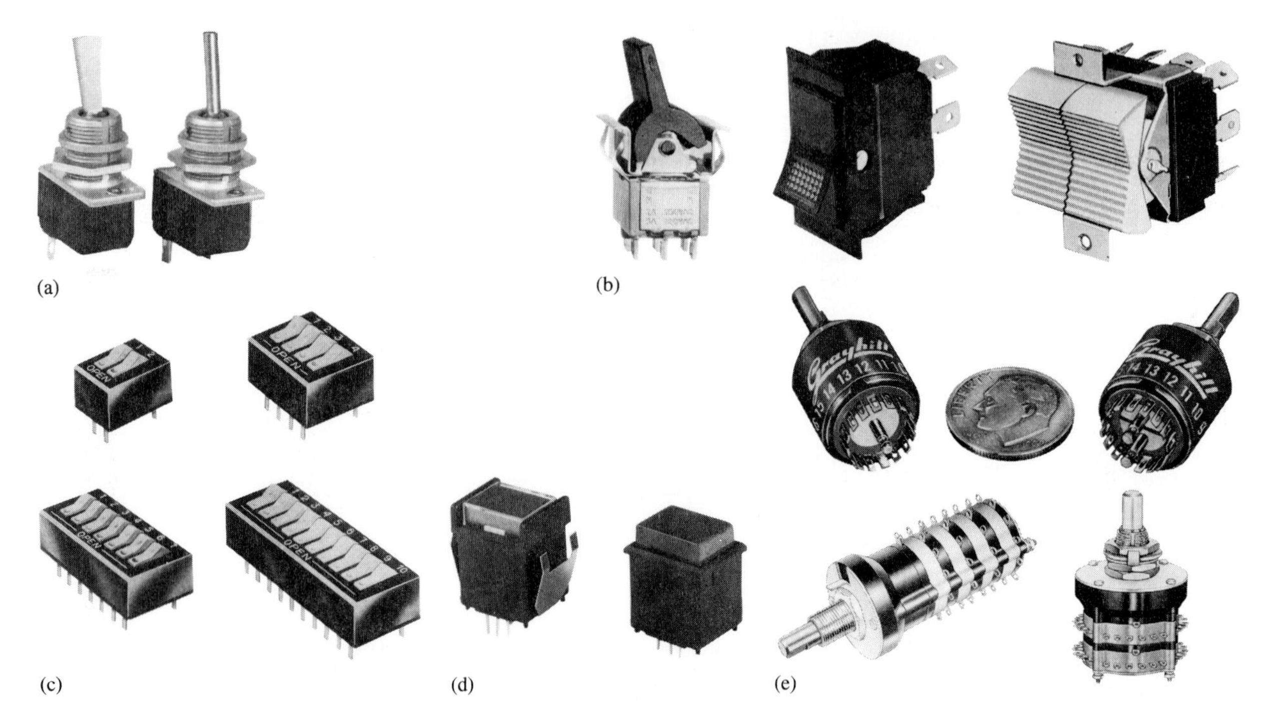

FIGURE 2–28
Switches. (a) Typical toggle-lever switches (courtesy of Eaton Corporation). (b) Rocker switches (courtesy of Eaton Corporation). (c) Rocker DIP (dual-in-line package) switches (courtesy of Amp, Inc. and Grayhill, Inc.). (d) Push-button switches (courtesy of Eaton Corporation). (e) Rotary-position switches (courtesy of Grayhill, Inc.).

Protective Devices

Fuses and *circuit breakers* are used to deliberately create an open circuit when the current exceeds a specified number of amperes due to a malfunction or other abnormal condition in a circuit. For example, a 20-A fuse or circuit breaker will open a circuit when the current exceeds 20 A.

The basic difference between a fuse and a circuit breaker is that when a fuse is "blown," it must be replaced, but when a circuit breaker opens, it can be reset and reused repeatedly. The purpose of both of these devices is to protect against damage to a circuit due to excess current or to prevent a hazardous condition created by the overheating of wires and other components when the current is too great. Several typical fuses and circuit breakers, along with their schematic symbols, are shown in Figure 2–29.

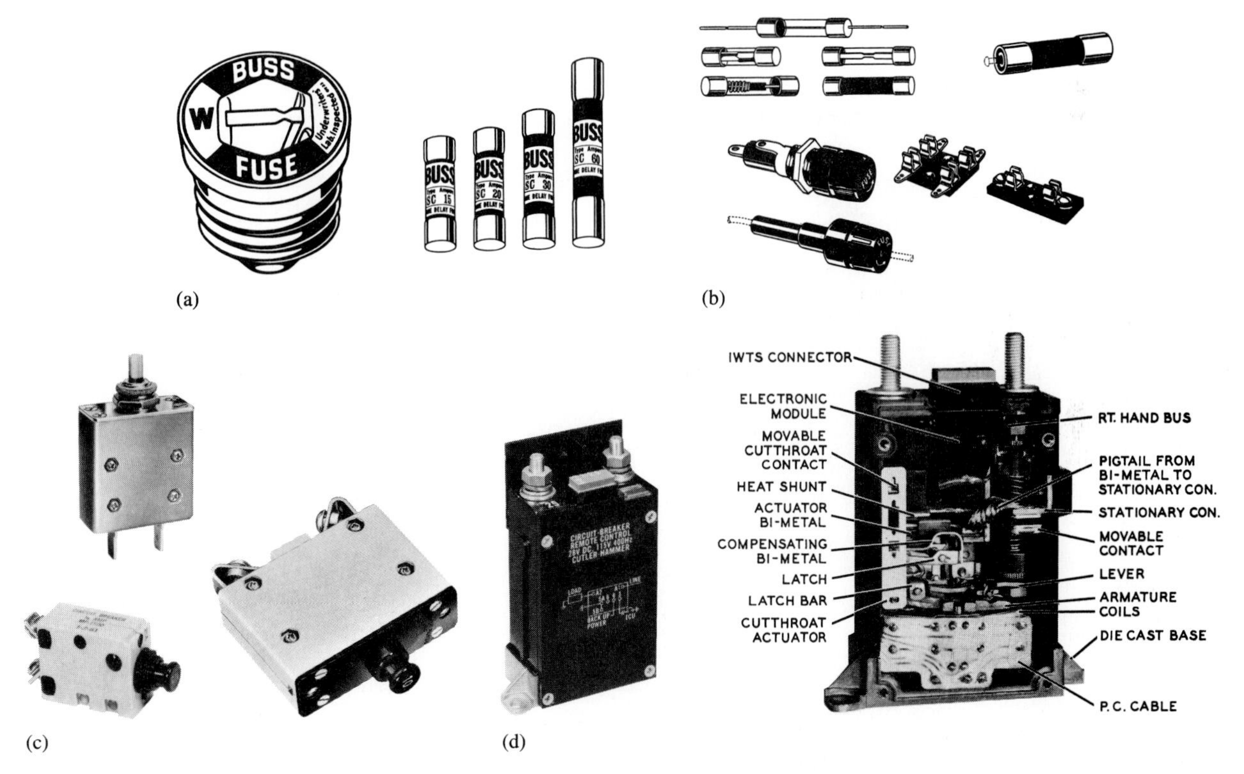

FIGURE 2–29
Fuses and circuit breakers. (a) Power fuses (courtesy of Bussman Manufacturing Corp.). (b) Fuses and fuse holders (courtesy of Bussman Manufacturing Corp.). (c) Circuit breakers (courtesy of Bussman Manufacturing Corp.). (d) Remote control circuit breaker with cutaway showing construction (courtesy of Eaton Corp.).

SECTION REVIEW 2–6

1. What are the basic elements of an electric circuit?
2. Define *open circuit.*
3. Define *closed circuit.*

BASIC CIRCUIT MEASUREMENTS

2–7

Voltage, current, and resistance measurements are commonly required in electronics work. Special types of instruments are used to measure these basic electrical quantities.

The instrument used to measure voltage is a *voltmeter,* the instrument used to measure current is an *ammeter,* and the instrument used to measure resistance is an *ohmmeter.* Commonly, all three instruments are combined into a single instrument known as a *multimeter,* or VOM (volt-ohm-milliammeter), in which you can choose what specific quantity to measure by selecting the switch setting.

Typical multimeters are shown in Figure 2–30. Part (a) shows an analog meter, that is, with a needle pointer, and Part (b) shows a digital multimeter (DMM), which provides a digital readout of the measured quantity.

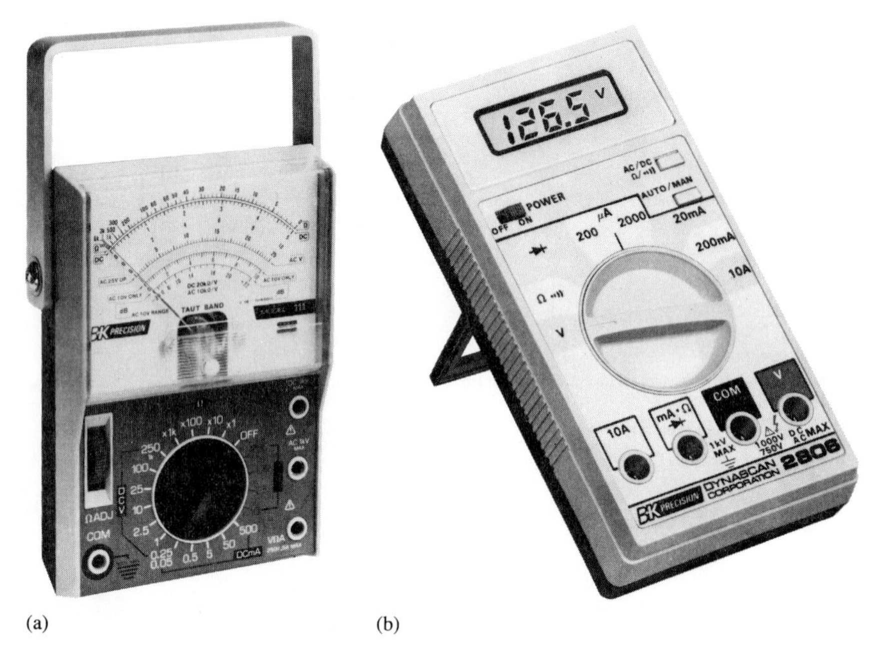

(a) (b)

FIGURE 2–30

Typical multimeters. (a) Analog and (b) digital (courtesy of B&K Precision/Dynascan Corp.).

Meter Symbols

Throughout this book, certain symbols will be used to represent the different meters, as shown in Figure 2–31. You may see any of three types of symbols for voltmeters, ammeters, and ohmmeters, depending on which symbol most effectively conveys the information required. Generally, the pictorial analog symbol is used when relative measurements or changes in quantities are to be depicted by the position or movement of the pointer. The pictorial digital symbol is used when fixed values are to be indicated in a circuit. The general schematic symbol is used to indicate placement of meters in a circuit when no values or value changes need to be shown.

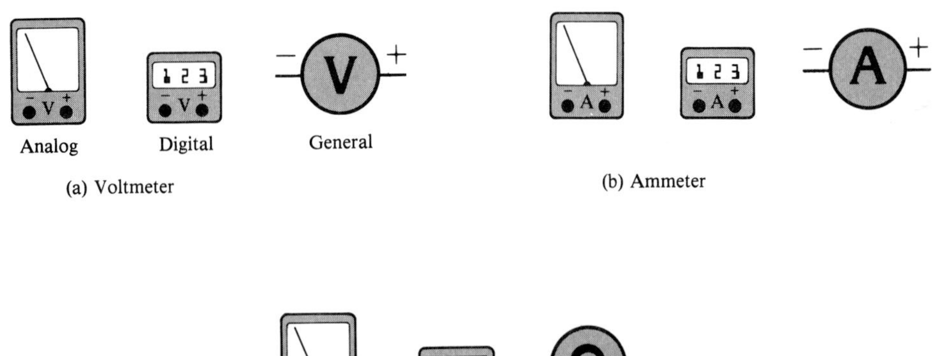

(c) Ohmmeter

FIGURE 2–31
Meter symbols.

How to Measure Current with an Ammeter

Figure 2–32 illustrates how to measure current with an ammeter. Part (a) shows the simple circuit in which the current through the resistor is to be measured. Connect the ammeter *in the current path* by first opening the circuit, as shown in Part (b). Then insert the meter as shown in Part (c). As you will learn later, such a connection is a *series* connection. The polarity of the meter must be such that the current flows in at the negative terminal and out at the positive.

How to Measure Voltage with a Voltmeter

To measure voltage, connect the voltmeter *across the component* for which the voltage is to be found. As you will learn later, such a connection is a *parallel* connection. The negative terminal of the meter must be connected to the negative side of the circuit, and the positive terminal of the meter to the positive side of the circuit. Figure 2–33 shows a voltmeter connected to measure the voltage across the resistor.

How to Measure Resistance with an Ohmmeter

To measure resistance, connect the ohmmeter across the resistor. *The resistor must first be removed or disconnected from the circuit.* This procedure is shown in Figure 2–34.

SECTION REVIEW 2–7

1. Name the meters for measurement of (a) current, (b) voltage, and (c) resistance.
2. Place two ammeters in the circuit of Figure 2–26 to measure the current through either lamp (be sure to observe the polarities). How can the same measurements be accomplished with only one ammeter?
3. Show how to place a voltmeter to measure the voltage across lamp 2 in Figure 2–26.

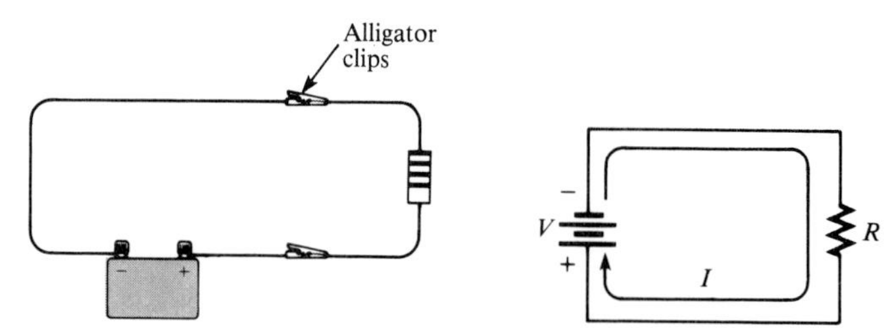

(a) Circuit in which the current is to be measured

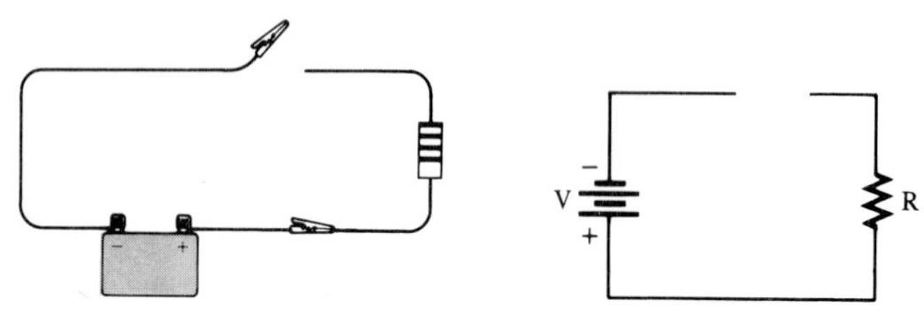

(b) Open the circuit either between the resistor and the negative terminal or between the resistor and the positive terminal of source.

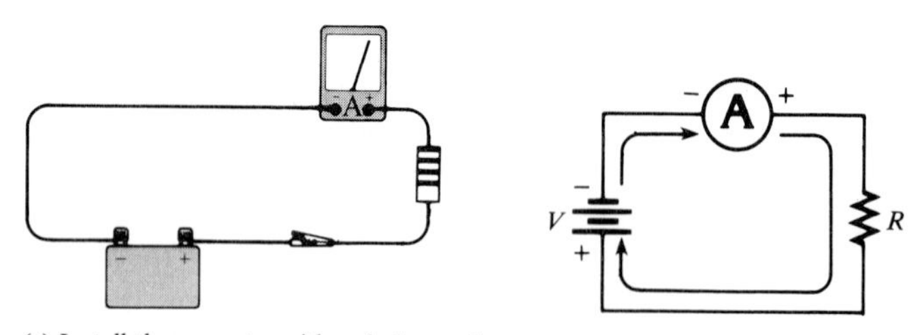

(c) Install the ammeter with polarity as shown (negative to negative–positive to positive).

FIGURE 2–32
Example of an ammeter connection.

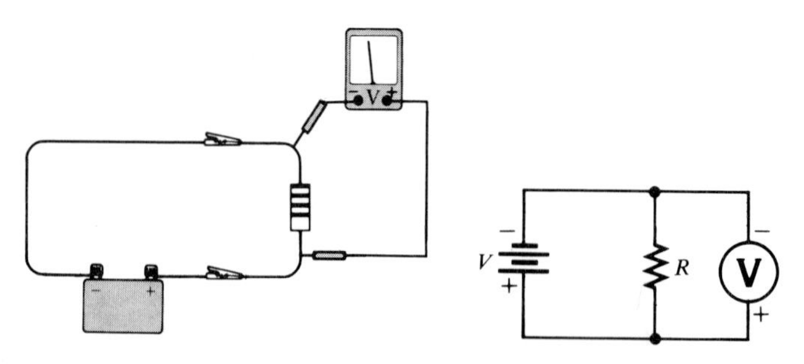

FIGURE 2–33
Example of a voltmeter connection.

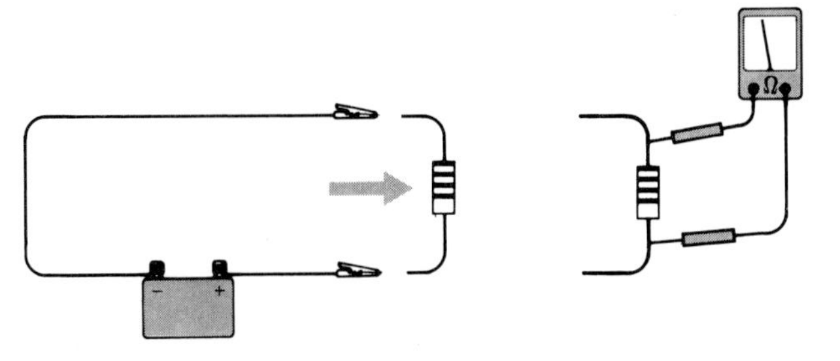

(a) Disconnect the resistor from the circuit to avoid damage to the meter and/or incorrect measurement.

(b) Measure the resistance. (Polarity is not important.)

FIGURE 2–34
Example of using an ohmmeter.

APPLICATION NOTE

The Application Assignment at the beginning of this chapter required a stage lighting system consisting of six lamps that (1) can be turned on or off individually, (2) can all be turned on or off simultaneously, and (3) can all be simultaneously brightened or dimmed by a single control. Figure 2–35 is the schematic diagram for one implementation of this system. Each lamp is individually controlled by an SPST switch. An SPST (SW7) is located in the main line and can turn all the lamps on or off simultaneously, provided that all the individual switches are closed. A *rheostat* is connected in the main line to provide a variable resistance for controlling the current to all the lamps. By increasing or decreasing the resistance, you can increase or decrease the current. More current through a lamp brightens it, and less current dims it.

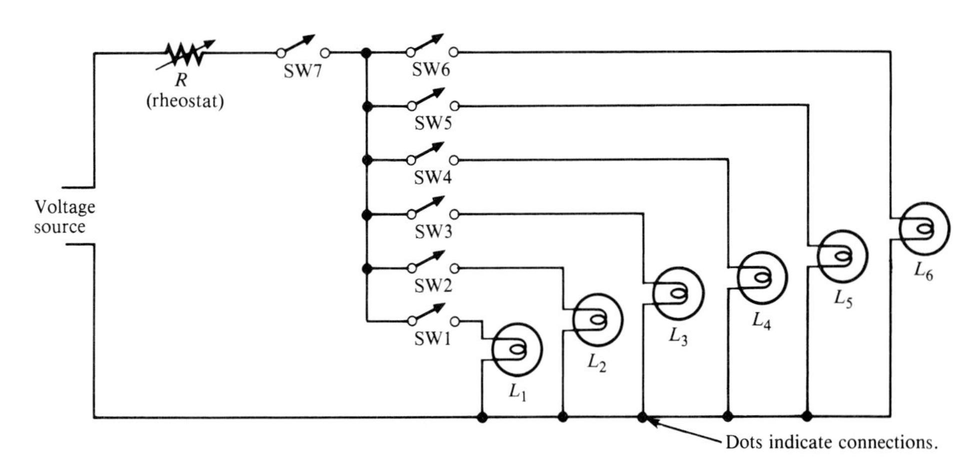

FIGURE 2–35
Lighting system with switch and dimmer controls.

SUMMARY

Facts

- An atom is the smallest particle of an element that retains the characteristics of that element.
- The electron is the basic particle of negative electrical charge.
- The proton is the basic particle of positive charge.
- An ion is an atom that has gained or lost an electron and is no longer neutral.
- When electrons in the outer orbit of an atom (valence electrons) break away, they become free electrons.
- Free electrons make current possible.
- Like charges repel each other, and opposite charges attract each other.
- Voltage must be applied to a circuit before current can flow.
- Resistance limits the current.
- Basically, an electric circuit consists of a source, a load, and a current path.
- An open circuit is one in which the current path is broken.
- A closed circuit is one which has a complete current path.
- An ammeter is connected in line with the current path.
- A voltmeter is connected across the current path.
- An ohmmeter is connected across a resistor after removal from circuit.

Units

- Ampere (A)—the unit of current (I).
- Joule (J)—the unit of energy (W).
- Ohm (Ω)—the unit of resistance (R).
- Second (s)—a unit of time (t).
- Volt (V)—the unit of voltage or potential difference (V).

Definitions

- *One coulomb*—the charge of 6.25×10^{18} electrons.
- *Voltage*—the amount of energy available to move a charge from one point to another.
- *One volt*—the potential difference (voltage) between two points when one joule of energy is used to move one coulomb from one point to the other.
- *Current*—the rate of flow of electrons.
- *One ampere*—the amount of current flowing when one coulomb of charge passes a given point in one second.
- *Resistance*—the opposition to current.
- *One ohm*—the resistance when one ampere of current flows in a material with one volt applied across the material.

Formulas

$$V = \frac{W}{Q} \quad \text{Voltage equals energy divided by charge.} \tag{2–1}$$

$$I = \frac{Q}{t} \quad \text{Current equals charge divided by time.} \tag{2–2}$$

Symbols

Some basic electric and electronic symbols are shown in Figure 2–36.

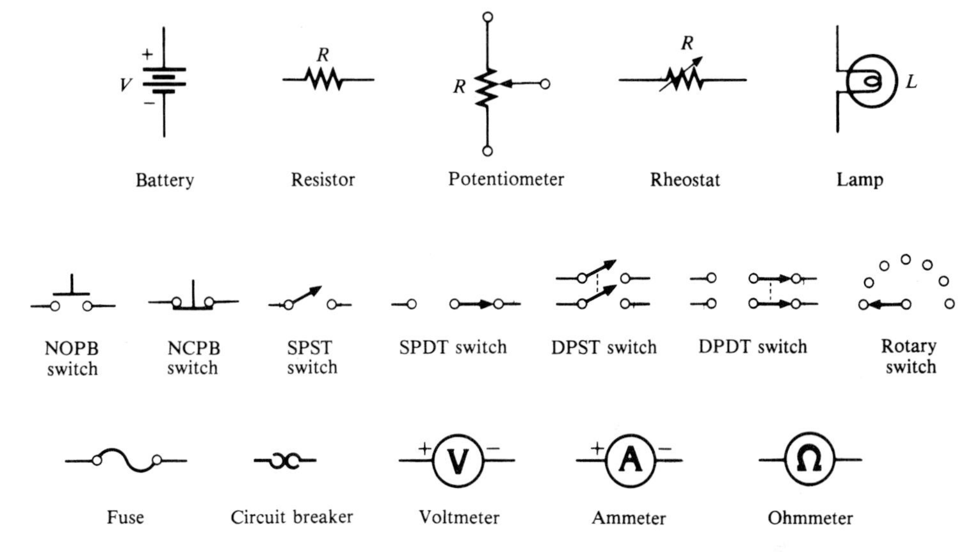

FIGURE 2–36

SELF-TEST

1. How many electrons are in a neutral atom with an atomic number of three?
2. In atomic theory, what are *shells?*
3. Discuss how conductors, semiconductors, and insulators basically differ.
4. What is static electricity?
5. When a positively charged material is placed near a negatively charged material, what will happen? Why?
6. What is the charge on a single electron?
7. Describe how a positive ion is created.
8. What is the more common term for *potential difference?* What is its unit?
9. What is the unit of energy?
10. Batteries, power supplies, solar cells, and generators are types of ________.
11. Can voltage exist between two points when there is no current?
12. Explain how current flows in a material when voltage is applied across the material.
13. Define *one ampere* of current.
14. Name two applications of resistors.
15. Name two types of variable resistor.
16. How do linear and tapered potentiometers differ?
17. A lamp and switch are connected to a voltage source to form a simple circuit. The switch is in a position such that the lamp is off. Is the circuit open or closed?
18. Is a voltage measurement or a current measurement more difficult to make? Why?

PROBLEM SET A

Section 2–2

2–1 How many coulombs of charge do 50×10^{31} electrons possess?

2–2 How many electrons does it take to make 80 μC (microcoulombs) of charge?

Section 2–3

2–3 Determine the voltage in each of the following cases:
(a) 10 J/C **(b)** 5 J/2 C **(c)** 100 J/25 C

2–4 Five hundred joules of energy are used to move 100 C of charge through a resistor. What is the voltage across the resistor?

2–5 What is the voltage of a battery that uses 800 J of energy to move 40 C of charge through a resistor?

2–6 How much energy does a 12-V battery use to move 2.5 C through a circuit?

Section 2–4

2–7 Determine the current in each of the following cases:
(a) 75 C in 1 s **(b)** 10 C in 0.5 s **(c)** 5 C in 2 s

2–8 Six-tenths coulomb passes a point in 3 seconds. What is the current in amperes?

2–9 How long does it take 10 C to flow past a point if the current is 5 A?

2–10 How many coulombs pass a point in 0.1 s when the current is 1.5 A?

Section 2–5

2–11 Figure 2–37(a) shows color-coded resistors (a full-color photo appears between pages 128 and 129). Determine the resistance value and the tolerance of each.

2–12 Find the minimum and the maximum resistance within the tolerance limits for each resistor in Figure 2–37(a).

2–13 **(a)** If you need a 270-Ω resistor, what color bands should you look for?
(b) From the selection of resistors in Figure 2–37(b) (a full-color photo appears between pages 128 and 129), choose the following values: 330 Ω, 2.2 kΩ, 56 kΩ, 100 kΩ, and 39 kΩ.

2–14 The adjustable contact of a linear potentiometer is set at the mechanical center of its adjustment. If the total resistance is 1000 Ω, what is the resistance between each end terminal and the adjustable contact?

Section 2–6

2–15 Trace the current path in Figure 2–26(a) with the switch in position 2.

2–16 With the switch in either position, redraw the circuit in Figure 2–26(b) with a fuse connected to protect the circuit against excessive current.

Section 2–7

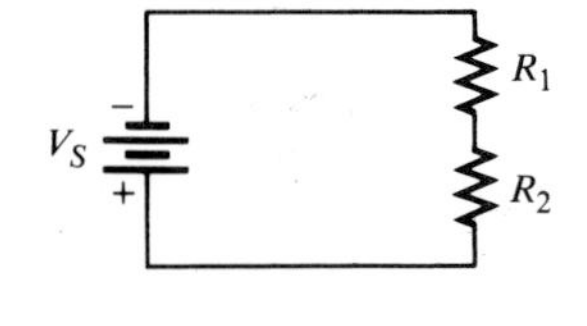

FIGURE 2–38

2–17 Show the placement of an ammeter and a voltmeter to measure the current and the source voltage in Figure 2–38.

2–18 Show how you would measure the resistance of R_2 in Figure 2–38.

2–19 In Figure 2–39, how much voltage does each meter indicate when the switch is in position 1? In position 2?

2–20 In Figure 2–39, indicate how to connect an ammeter to measure the current flowing from the voltage source regardless of the switch position.

FIGURE 2–39

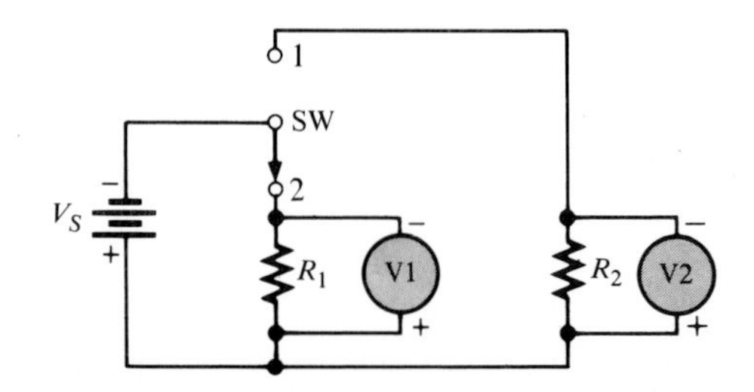

PROBLEM SET B

2–21 A resistor with a current of 2 A through it converts 1000 J of electrical energy to heat energy in 15 s. What is the voltage across the resistor?

2–22 If 574×10^{15} electrons flow through a wire in 250 ms, what is the current in amperes?

2–23 A 120-V source is to be connected to a 1500-Ω resistive load by two lengths of wire as shown in Figure 2–40. The voltage source is to be located 50 ft from the load. Using the wire table in Appendix A, determine the gage number of the *smallest* wire that can be used if the *total* resistance of the two lengths of wire is not to exceed 6 Ω.

2–24 Determine the resistance and tolerance of each resistor labeled as follows:
(a) 4R7J **(b)** 5602M **(c)** 1501F

2–25 There is only one circuit in Figure 2–41 in which it is possible to have all lamps on at the same time. Determine which circuit it is.

2–26 Through which resistor in Figure 2–42 does current always flow, regardless of the position of the switches?

2–27 In Figure 2–42, show the proper placement of ammeters to measure the current flowing through each resistor and the current flowing out of the battery.

2–28 Show the proper placement of voltmeters to measure the voltage across each resistor in Figure 2–42.

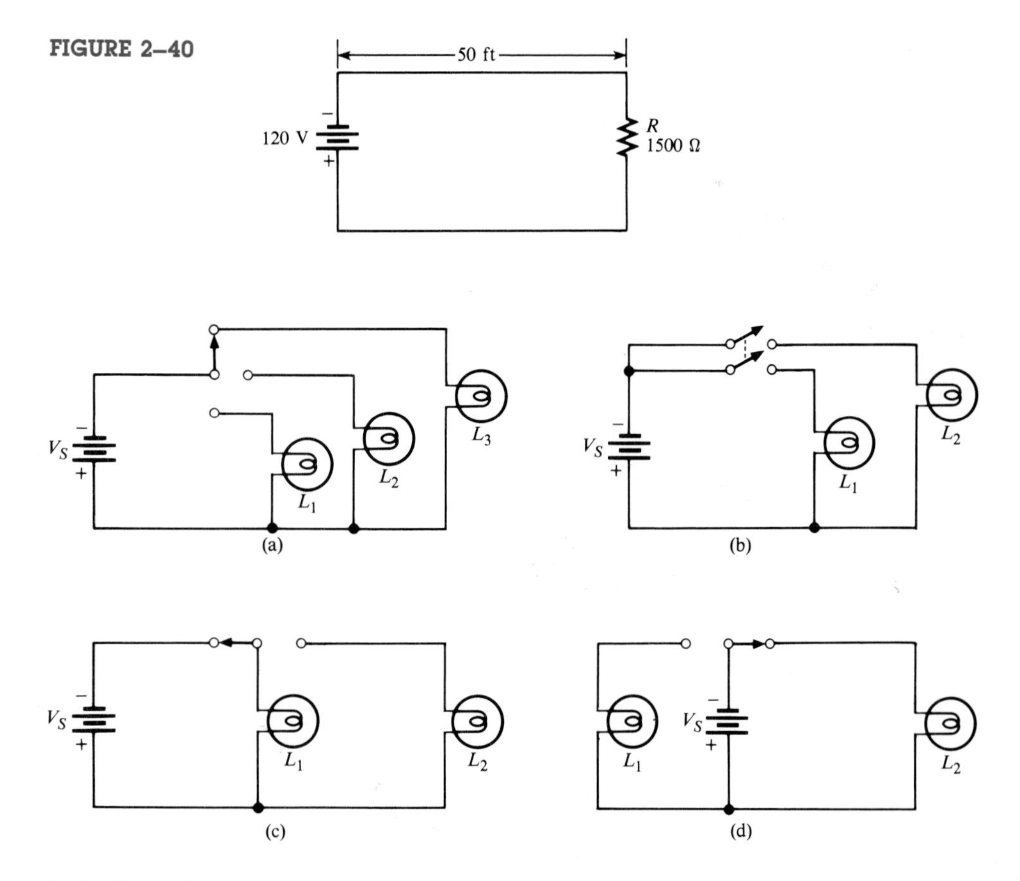

FIGURE 2–40

FIGURE 2–41

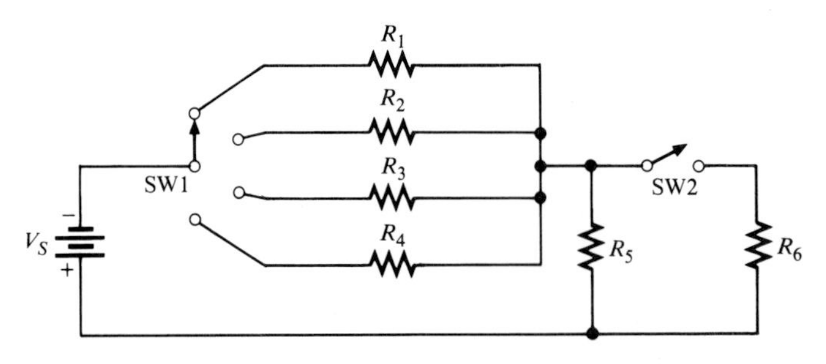

FIGURE 2–42

2–29 Devise a switch arrangement whereby two voltage sources (V_{S1} and V_{S2}) can be connected simultaneously to either of two resistors (R_1 and R_2) as follows:

V_{S1} connected to R_1 and V_{S2} connected to R_2 or
V_{S1} connected to R_2 and V_{S2} connected to R_1

2–30 The different sections of a stereo system are represented by the blocks in Figure 2–43. Show how a single switch can be used to connect the phonograph, the CD (compact disk) player, the tape deck, the AM tuner, or the FM tuner to the amplifier by a single knob control. Only one section can be connected to the amplifier at any time.

FIGURE 2–43

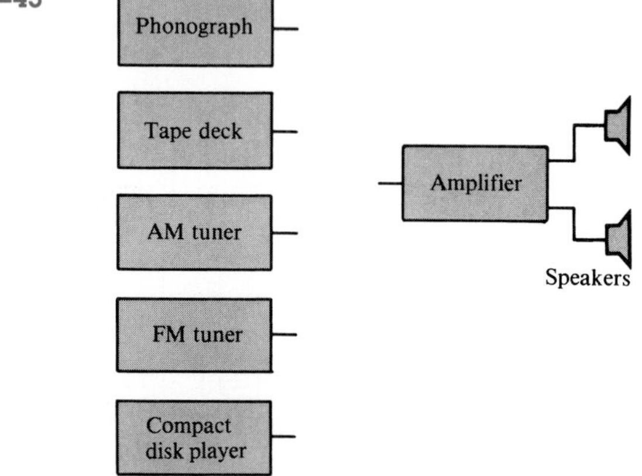

ANSWERS TO SECTION REVIEWS

Section 2–1

1. Electron.
2. The smallest particle of an element that retains the unique characteristics of the element.
3. A positively charged nucleus surrounded by orbiting electrons.
4. No.
5. An outer-shell electron that has drifted away from the parent atom.

Section 2–2

1. Q. **2.** Coulomb, C. **3.** 1.6×10^{-6} C.

Section 2–3

1. Energy per unit charge. 2. 2.4 V.
3. Battery, power supply, solar cell, and generator.

Section 2–4

1. Rate of flow of electrons, ampere (A). 2. 6.25×10^{18}
3. 5 A.

Section 2–5

1. Opposition to current, Ω.
2. Fixed and variable. The value of a fixed resistor cannot be changed, but that of a variable resistor can.
3. *First band:* first digit of resistance value. *Second band:* second digit of resistance value. *Third band:* number of zeros. *Fourth band:* tolerance.
4. **(a)** 4700 Ω ±5%; **(b)** 62,000 Ω ±10%; **(c)** 18 Ω ±5%.
5. A rheostat has two terminals; a potentiometer has three terminals.

Section 2–6

1. Source, load, and current path between source and load.
2. A circuit that has no path for current.
3. A circuit that has a complete path for current.

Section 2–7

1. **(a)** Ammeter; **(b)** voltmeter; **(c)** ohmmeter.
2. See Figure 2–44.
3. See Figure 2–45.

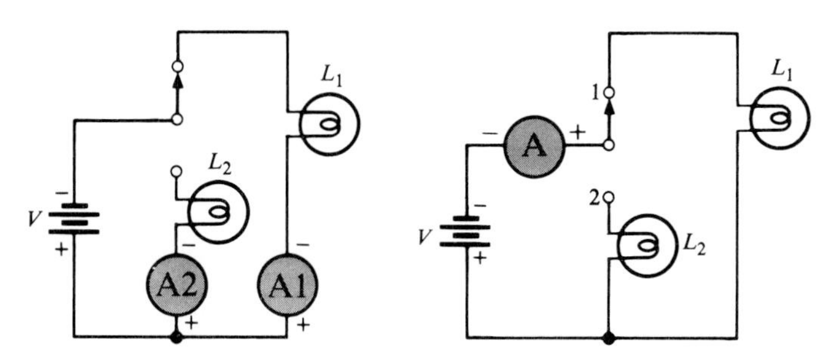

FIGURE 2–44

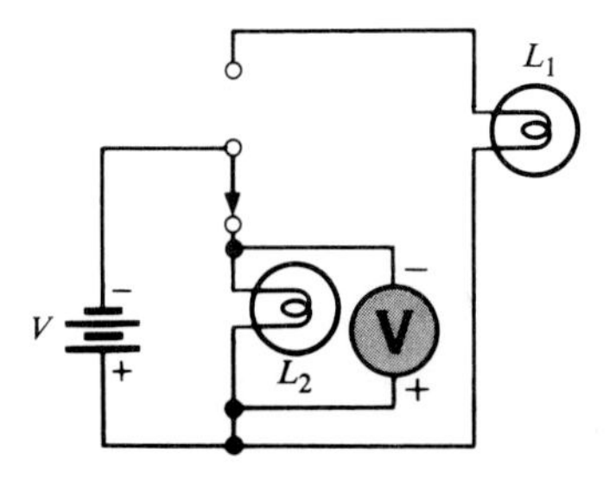

FIGURE 2–45

Ohm's Law and Power

Georg Simon Ohm (1789–1854) experimentally found that voltage, current, and resistance are all related in a specific way. This basic relationship, known as *Ohm's law,* is one of the most fundamental and important laws in the fields of electricity and electronics. In this chapter, the meaning of Ohm's law is examined, and its use in practical circuit applications is discussed and demonstrated by numerous examples.

In addition to Ohm's law, the concept and definition of power in electric circuits are introduced. Practical considerations are emphasized.

In this chapter you will learn:

- The meaning of Ohm's law.
- The formula for Ohm's law.
- How to use Ohm's law in the determination of voltage, current, and resistance in electric circuits.
- How to work with current, voltage, and resistance in both large and small units.
- The meaning of the linear relationship between current and voltage.
- The definition of power (P) and its unit, the watt (W).
- The difference between power and energy.
- How to determine the amount of power in a resistive circuit.
- How to choose resistors to handle the power.
- The definition of the kilowatthour (kWh) unit of energy.

3

APPLICATION ASSIGNMENT

As a technician in the development lab for a large manufacturer of electronic equipment, you are given three specific assignments during your first week on the job:

1. By specifying the necessary current capacity, choose a 5-V power supply that is capable of providing current to a 75-Ω load.
2. Measure the current through a certain device when 25 V are applied, and determine the resistance value and power rating of a "dummy" load resistor that can be substituted for the device for test purposes.
3. Specify the source voltage necessary for a certain display element that requires 20 mA of current to achieve a desired illumination. The internal resistance of the element is known to be 100 Ω.

After studying this chapter, you should be able to complete this assignment. The Application Note at the end of the chapter gives one solution.

DEMONSTRATION OF OHM'S LAW

3–1

Ohm's law describes how *current, voltage,* and *resistance* are related and provides a formula that you can use to determine any one of the quantities when you know the other two. We will now examine Ohm's law step-by-step using the simple circuit in Figure 3–1 to demonstrate its meaning. Then you will see how to use the formula in practical circuit calculations.

In Figure 3–1 there is a variable voltage source with which the voltage can be adjusted over a range of values. A voltmeter is connected across the terminals of the voltage source, and an ammeter is connected in the current path with the resistor.

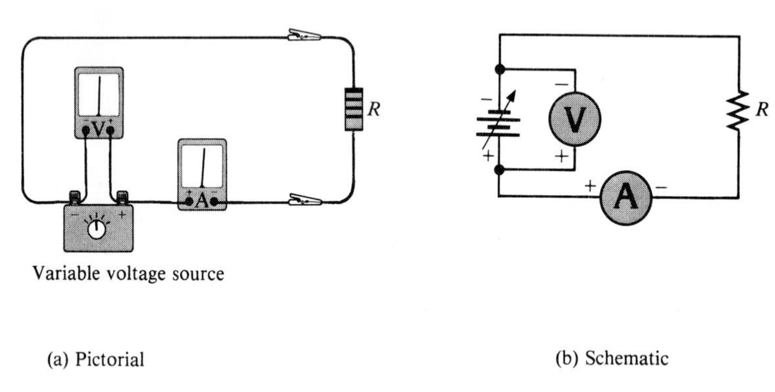

(a) Pictorial (b) Schematic

FIGURE 3–1
Circuit used for the demonstration of Ohm's law.

How Voltage Affects Current

Figure 3–2 demonstrates how a change in the voltage affects the current. Specific values are used for illustration. In Part (a), the voltage is set at 10 V, and the resulting current of 1 A is measured. Notice that the resistance is kept at 10 Ω. In Part (b), the voltage is reduced to 5 V, and a resulting current of 0.5 A is measured. Notice that when the voltage is halved, the current is halved. In Part (c), the voltage is increased to 20 V (double its initial value), and a resulting current of 2 A is measured. Thus, as you can see, doubling the voltage also doubles the current, tripling the voltage triples the current, and so on.

Linearly Proportional Relationship between Current and Voltage

The previous demonstration showed that the voltage and current in a resistive circuit are directly related and *linearly proportional;* that is, they have a straight-line relationship. If the voltage goes up by a certain amount, the current will go up by a proportional amount. If the voltage goes down by a certain amount, the current will go down by a proportional amount. This relationship is illustrated by the straight-line graph of Figure 3–3 for a resistance of 10 Ω. For any constant resistance, a straight-line (linear) relationship between current and voltage will always exist.

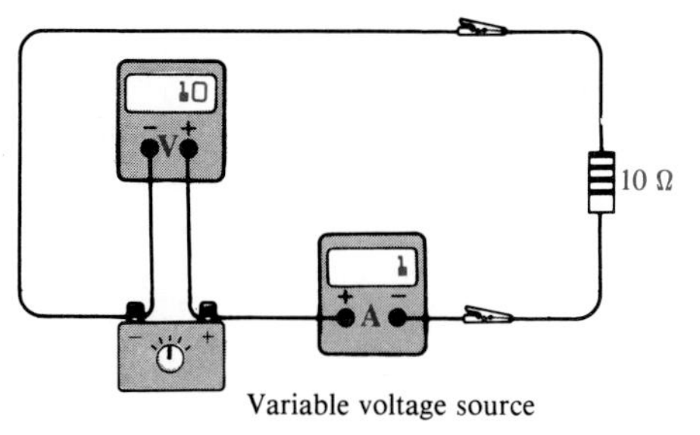

(a) Initial setting, $V = 10$ V

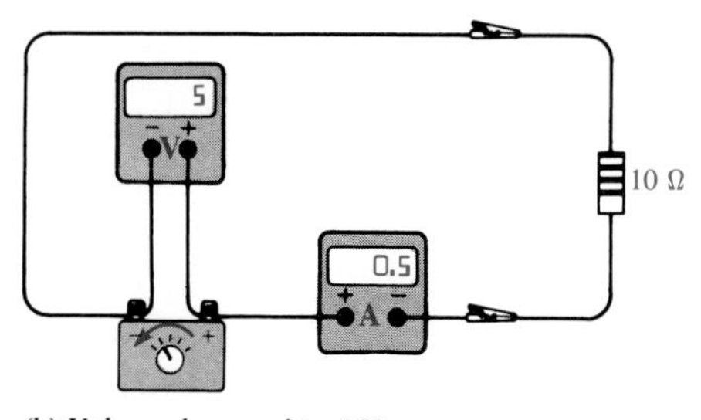

(b) Voltage decreased to 5 V

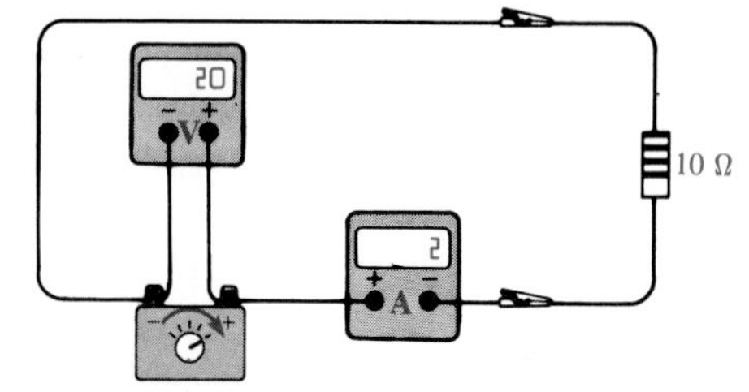

(c) Voltage increased to 20 V

FIGURE 3–2
Demonstration of Ohm's law: How voltage affects current with resistance constant.

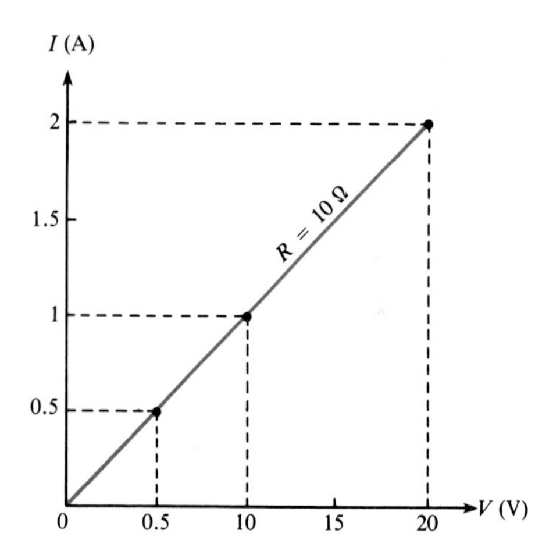

FIGURE 3–3
Graph of current versus voltage for Figure 3–2 with resistance constant.

How Resistance Affects Current

Figure 3–4 demonstrates how a change in resistance affects the current when the voltage is kept at a constant value. Again, specific values are used for illustration. In Part (a), the voltage is set at 10 V and remains there for each step. A 10-Ω resistor is connected in the circuit, and the resulting current is 1 A as indicated. In Part (b), the resistor is changed to 5 Ω, and the resulting current is 2 A. As you can see, halving the resistance doubles the current. In Part (c), the resistor is changed to 20 Ω, and the resulting current is 0.5 A. In this case, doubling the resistance reduces the current to half its original value.

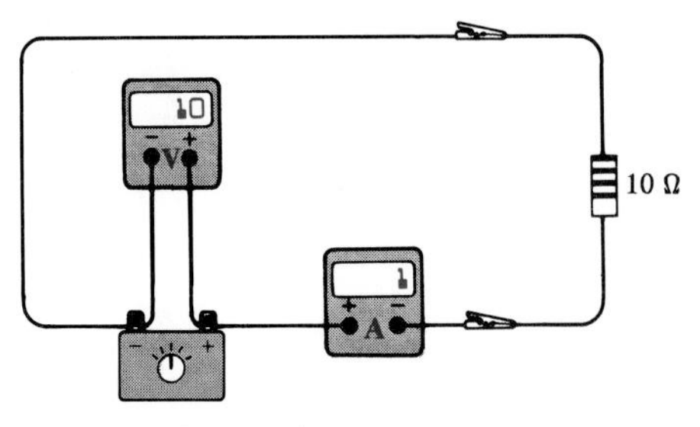

(a) Initial value, $R = 10\ \Omega$

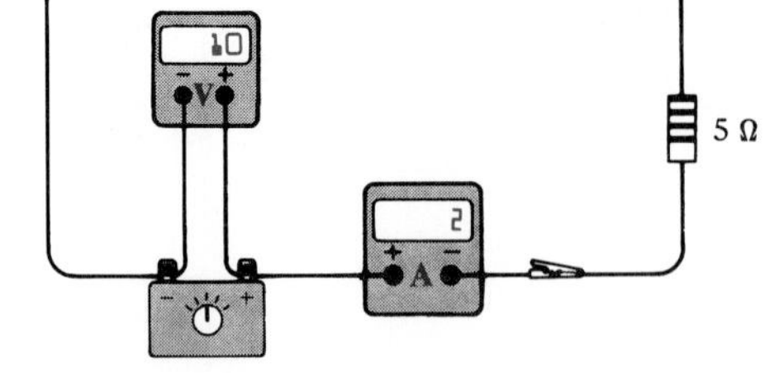

(b) Resistance decreased to 5 Ω

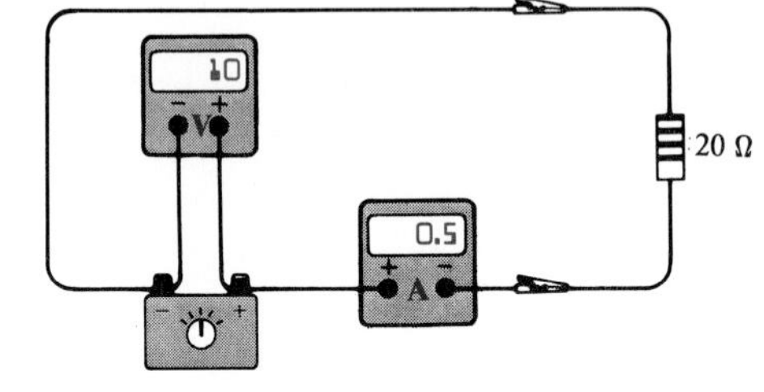

(c) Resistance increased to 20 Ω

FIGURE 3–4
Demonstration of Ohm's law: How resistance affects current with voltage constant.

This demonstration shows that current is *inversely* proportional to resistance. When resistance goes up by a certain amount, the current goes down by a proportional amount. When resistance goes down by a certain amount, the current goes up by a proportional amount.

The Ohm's Law Formula

Figures 3–2 and 3–4 demonstrate Ohm's law, which can be stated by a formula as follows:

$$I = \frac{V}{R} \tag{3–1}$$

This formula describes what was indicated by the demonstration: *For a constant resistance, if the voltage in a circuit is increased, more current will flow, and if the voltage is decreased, less current will flow. Also, for a constant voltage, if the resistance in a circuit is increased, less current will flow, and if the resistance is decreased, more current will flow.*

Ohm's law is illustrated graphically as follows.

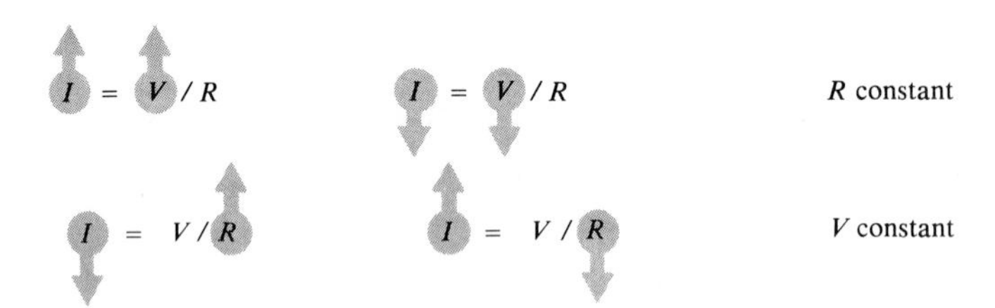

EXAMPLE 3–1 Verify that the Ohm's law formula in Equation (3–1) gives results that agree with the values in the circuits of Figures 3–2 and 3–4.

Solution

In Figure 3–2, $R = 10\ \Omega$. For $V = 10$ V:

$$I = \frac{V}{R} = \frac{10\ \text{V}}{10\ \Omega} = 1\ \text{A}$$

For $V = 5$ V:

$$I = \frac{V}{R} = \frac{5\ \text{V}}{10\ \Omega} = 0.5\ \text{A}$$

For $V = 20$ V:

$$I = \frac{V}{R} = \frac{20\ \text{V}}{10\ \Omega} = 2\ \text{A}$$

The calculator sequence for $V = 10$ V is

[1] [0] [÷] [1] [0] [=]

In Figure 3–4, $V = 10$ V. For $R = 5\ \Omega$:

$$I = \frac{V}{R} = \frac{10\ \text{V}}{5\ \Omega} = 2\ \text{A}$$

For $R = 20\ \Omega$:

$$I = \frac{V}{R} = \frac{10\ \text{V}}{20\ \Omega} = 0.5\ \text{A}$$

Thus, the formula verifies the values of both figures.

Other Forms of the Ohm's Law Formula

As you have seen, you can use the Ohm's law formula in Equation (3–1) to find the current in a circuit when the voltage and resistance are known. By a simple algebraic change, you can use the formula to find the voltage when the current and resistance are known. By multiplying both sides of Equation (3–1) by R, we get:

$$V = IR \tag{3–2}$$

We obtain another simple change in Equation (3–1) by transposing the R and the I:

$$R = \frac{V}{I} \tag{3–3}$$

With this formula, you can find the resistance when the voltage and current are known.

Keep in mind that Equations (3–1), (3–2), and (3–3) are all equivalent expressions of Ohm's law.

SECTION REVIEW 3–1

1. Briefly state Ohm's law in words.
2. If the voltage across a resistor is tripled, does the current increase or decrease? By how much?
3. There is a fixed voltage across a variable resistor, and you measure a current of 10 mA. If you double the resistance, how much current will you measure?
4. Will you observe any change in current in a circuit where both the voltage and the resistance are doubled?

APPLICATION OF OHM'S LAW

3–2 How to Determine I When V and R Are Known

In this section you will learn to determine current values when you know the values of voltage and resistance. In these problems, the formula $I = V/R$ is used. In order to get current in *amperes,* you must express the value of V in *volts* and the value of R in *ohms.*

EXAMPLE 3–2 How many amperes of current are flowing in the circuit of Figure 3–5?

FIGURE 3–5

Solution

Into the formula $I = V/R$, substitute 100 V for V and 20 Ω for R. Divide 20 Ω into 100 V as follows:

$$I = \frac{V}{R} = \frac{100\text{ V}}{20\ \Omega} = 5\text{ A}$$

There are 5 A of current in this circuit.

In electronics work, resistance values of thousands of ohms or even millions of ohms are common. As you learned in Chapter 1, large values of resistance are indicated by the metric system prefixes *kilo* (k) and *mega* (M). Thus, thousands of ohms are expressed in *kilohms* (kΩ), and millions of ohms in *megohms* (MΩ). The following examples illustrate how to use kilohms and megohms when using Ohm's law to find the current.

EXAMPLE 3–3 Calculate the current in Figure 3–6.

FIGURE 3–6

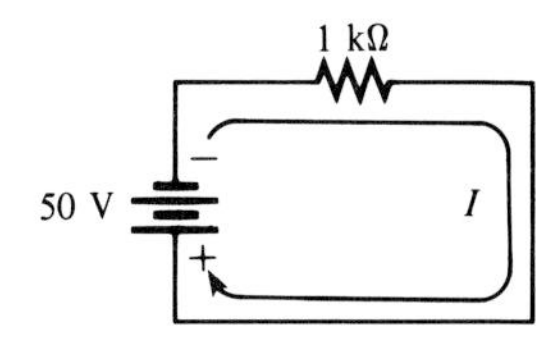

Solution

Remember that 1 kΩ is the same as 1000 Ω. Substituting 50 V for V and 1000 Ω for R gives the current in amperes as follows:

$$I = \frac{V}{R} = \frac{50\ \text{V}}{1000\ \Omega} = 0.05\ \text{A}$$

The calculator sequence is

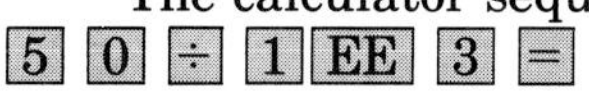

In Example 3–3, 0.05 A can be expressed as 50 milliamperes (50 mA). Thus, when *volts* are divided by *kilohms,* the current will be in *milliamperes.*

If *volts* are applied when resistance values are in *megohms,* the current is in *microamperes* (μA).

EXAMPLE 3–4 Determine the amount of current in the circuit of Figure 3–7.

FIGURE 3–7

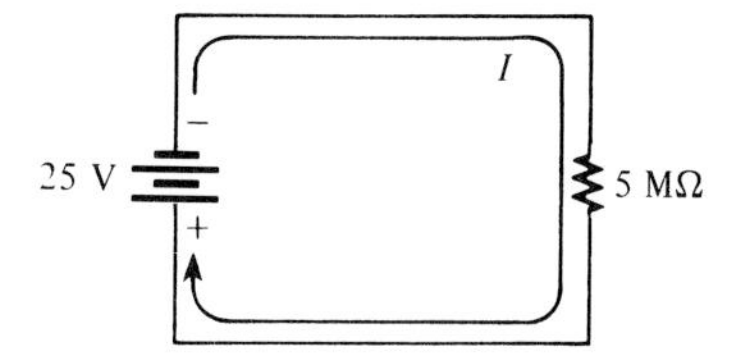

Solution

Recall that 5 MΩ equals 5,000,000 Ω. Substituting 25 V for V and 5,000,000 Ω for R gives the following result:

$$I = \frac{V}{R} = \frac{25\ \text{V}}{5{,}000{,}000\ \Omega} = 0.000005\ \text{A}$$

Notice that 0.000005 A equals 5 microamperes (5 μA), which can also be expressed as 5×10^{-6} A.

The calculator sequence is

[2] [5] [÷] [5] [EE] [6] [=]

Small voltages, usually less than 50 V, are common in electronic circuits. Occasionally, however, large voltages are encountered. For example, the high-voltage supply in a television receiver is around 20,000 V (20 kilovolts, or 20 kV), and transmission voltages generated by the power companies may be as high as 345,000 V (345 kV).

EXAMPLE 3–5 How much current will flow through 100 MΩ when 50 kV are applied?

Solution

In this case, we divide 50 kV by 100 MΩ to get the current. Using 50,000 V for 50 kV and 100,000,000 Ω for 100 MΩ, we obtain the current as follows:

$$I = \frac{V}{R} = \frac{50 \text{ kV}}{100 \text{ M}\Omega} = \frac{50{,}000 \text{ V}}{100{,}000{,}000 \text{ }\Omega}$$
$$= 0.0005 \text{ A} = 0.5 \text{ mA}$$

Remember to move the decimal point three places to the right to convert from amperes to milliamperes.

The calculator sequence is

[5] [0] [EE] [3] [÷] [1] [0] [0] [EE] [6] [=]

How to Determine *V* When *I* and *R* Are Known

In this section you will learn to determine voltage values when the current and resistance are known. In these problems, the formula $V = IR$ is used. To obtain voltage in *volts,* you must express the value of I in *amperes* and the value of R in *ohms*.

EXAMPLE 3–6 In the circuit of Figure 3–8, how much voltage is needed to produce 5 A of current?

FIGURE 3–8

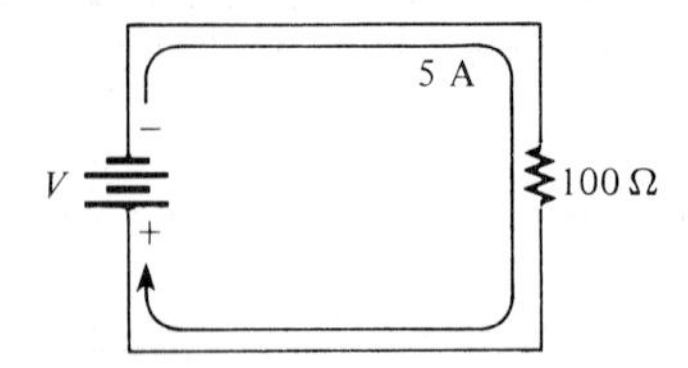

Solution

Substitute 5 A for I and 100 Ω for R into the formula $V = IR$ as follows:

$$V = IR = (5 \text{ A})(100 \text{ }\Omega) = 500 \text{ V}$$

Thus, 500 V are required to produce 5 A of current through a 100-Ω resistor.

EXAMPLE 3–7 How much voltage will be measured across the resistor in Figure 3–9?

FIGURE 3–9

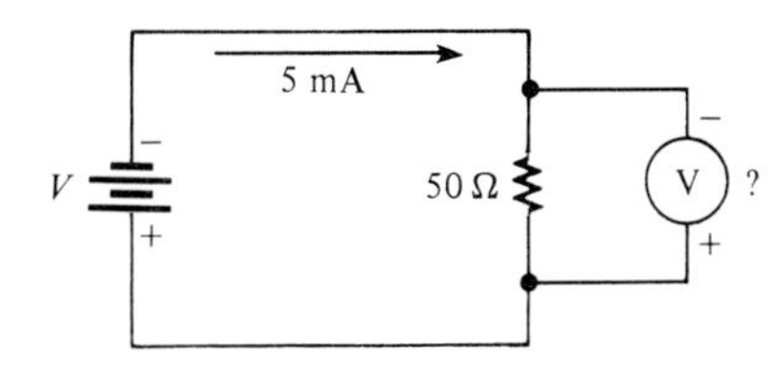

Solution
Note that 5 mA equals 0.005 A. Substituting the values for I and R into the formula $V = IR$, we get the following result:

$$V = IR = (5 \text{ mA})(50\ \Omega)$$
$$= (0.005 \text{ A})(50\ \Omega) = 0.25 \text{ V}$$

When *milliamperes* are multiplied by *ohms*, we get *millivolts*.

The calculator sequences is

5 EE +/- 3 × 5 0 =

EXAMPLE 3–8 The circuit in Figure 3–10 has a current of 10 mA. What is the voltage?

FIGURE 3–10

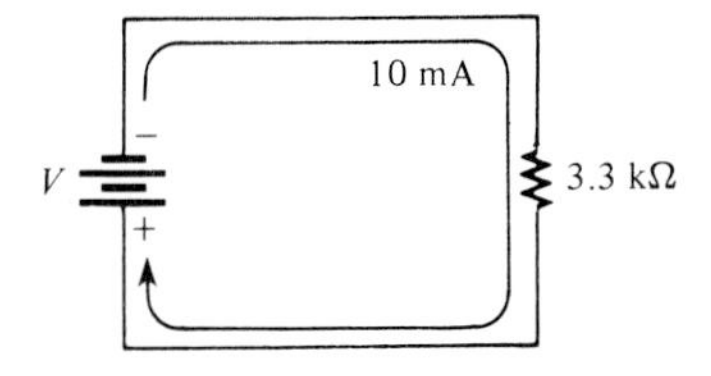

Solution
Note that 10 mA equals 0.010 A and that 3.3 kΩ equals 3300 Ω. Substituting these values into the formula $V = IR$, we get

$$V = IR = (10 \text{ mA})(3.3 \text{ k}\Omega)$$
$$= (0.010 \text{ A})(3300\ \Omega) = 33 \text{ V}$$

Milliamperes cancel *kilohms* when multiplied, and the result is *volts*.

The calculator sequence is

1 0 EE +/- 3 × 3 · 3 EE 3 =

How to Determine *R* When *V* and *I* Are Known

In this section you will learn to determine resistance values when the current and voltage are known. In these problems, the formula $R = V/I$ is used. To find resistance in *ohms*, you must express the value of I in *amperes* and the value of V in *volts*.

EXAMPLE 3–9 In the circuit of Figure 3–11, how much resistance is needed to draw 3 A of current from the battery?

FIGURE 3–11

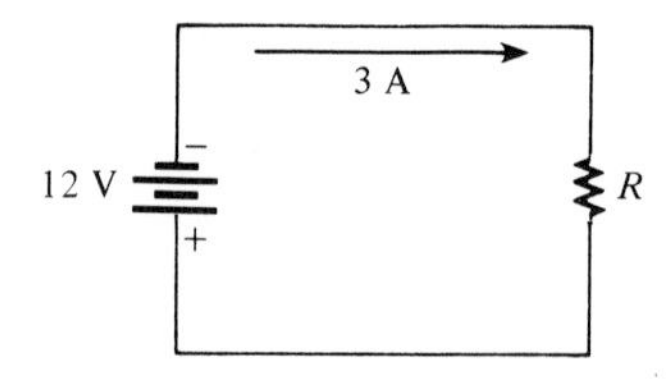

Solution

Substitute 12 V for V and 3 A for I into the formula $R = V/I$:

$$R = \frac{V}{I} = \frac{12\text{ V}}{3\text{ A}} = 4\ \Omega$$

EXAMPLE 3–10 Suppose that the ammeter in Figure 3–12 indicates 5 mA of current and the voltmeter reads 150 V. What is the value of R?

FIGURE 3–12

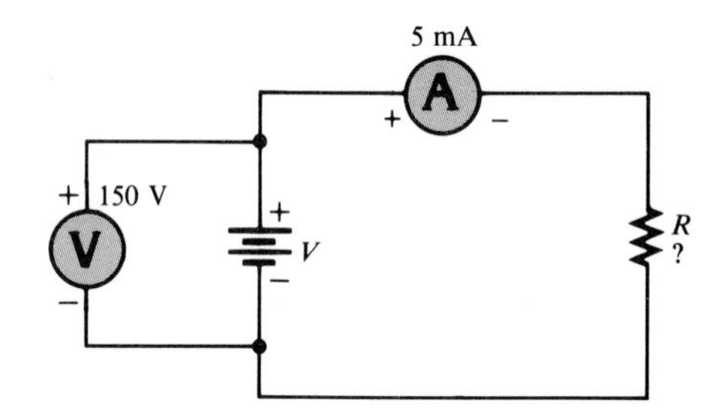

Solution

Note that 5 mA equals 0.005 A. Substituting the voltage and current values into the formula $R = V/I$, we get

$$R = \frac{V}{I} = \frac{150\text{ V}}{5\text{ mA}} = \frac{150\text{ V}}{0.005\text{ A}}$$
$$= 30{,}000\ \Omega = 30\text{ k}\Omega$$

Thus, if *volts* are divided by *milliamperes*, the resistance will be in *kilohms*.

SECTION REVIEW 3–2

1. $V = 10$ V and $R = 5\ \Omega$. Find I.
2. If a 5-MΩ resistor has 20 kV across it, how much current flows?
3. How much current will 10 kV across 2 kΩ produce?
4. $I = 1$ A and $R = 10\ \Omega$. Find V.
5. What voltage do you need to produce 3 mA of current in a 3-kΩ resistance?
6. A battery produces 2 A of current into a 6-Ω resistive load. What is the battery voltage?

7. $V = 10$ V and $I = 2$ A. Find R.
8. You have a resistor across which you measure 25 V, and your ammeter indicates 50 mA of current. What is the resistor's value in kilohms? In ohms?

POWER AND ENERGY

3–3

Power is the rate at which energy is used. In other words, power, symbolized by *P*, is a certain amount of energy used in a certain length of time, expressed as follows:

$$\text{Power} = \frac{\text{energy}}{\text{time}}$$

$$P = \frac{W}{t} \qquad (3\text{–}4)$$

Energy is measured in *joules* (J), time is measured in *seconds* (s), and power is measured in *watts* (W).

Energy in *joules* divided by time in *seconds* gives power in *watts*. For example, if 50 J of energy are used in 2 s, the power is 50 J/2 s = 25 W.

By definition, *one watt is the amount of power when one joule of energy is consumed in one second.* Thus, the number of joules consumed in one second is always equal to the number of watts. For example, if 75 J are used in 1 s, the power is 75 W. The following example illustrates use of these units.

EXAMPLE 3–11 An amount of energy equal to 100 J is used in 5 s. What is the power in watts?

Solution

$$P = \frac{\text{energy}}{\text{time}} = \frac{100\text{ J}}{5\text{ s}} = 20\text{ W}$$

Amounts of power much less than one watt are common in certain areas of electronics. As with small current and voltage values, metric prefixes are used to designate small amounts of power. Thus, *milliwatts* (mW) and *microwatts* (μW) are commonly found in some applications.

In the electrical utilities field, *kilowatts* (kW) and *megawatts* (MW) are common units. Radio and television stations also use large amounts of power to transmit signals.

EXAMPLE 3–12 Express the following powers using appropriate metric prefixes:
(a) 0.045 W **(b)** 0.000012 W **(c)** 3500 W **(d)** 10,000,000 W

Solution

(a) 0.045 W = 45 mW **(b)** 0.000012 W = 12 μW
(c) 3500 W = 3.5 kW **(d)** 10,000,000 W = 10 MW

The Kilowatthour (kWh) Unit of Energy

Since power is the rate of energy usage, power utilized over a period of time represents energy consumption. If we multiply *power* and *time*, we have *energy*, symbolized by W:

$$\text{Energy} = \text{power} \times \text{time}$$
$$W = Pt \qquad (3\text{–}5)$$

Earlier, the *joule* was defined as a unit of energy. However, there is another way of expressing energy. Since power is expressed in *watts* and time in *seconds*, we can use units of energy called the *wattsecond* (Ws), *watthour* (Wh), and *kilowatthour* (kWh).

When you pay your electric bill, you are charged on the basis of the amount of *energy* you use. Because power companies deal in huge amounts of energy, the most practical unit is the *kilowatthour*. *You use a kilowatthour of energy when you use 1000 watts of power for one hour.* For example, a 100-W light bulb burning for ten hours uses 1 kWh of energy.

EXAMPLE 3–13

Determine the number of kilowatthours for each of the following energy consumptions:

(a) 1400 W for 1 h **(b)** 2500 W for 2 h **(c)** 100,000 W for 5 h

Solution

(a) 1400 W = 1.4 kW
Energy = (1.4 kW)(1 h) = 1.4 kWh

(b) 2500 W = 2.5 kW
Energy = (2.5 kW)(2 h) = 5 kWh

(c) 100,000 W = 100 kW
Energy = (100 kW)(5 h) = 500 kWh

SECTION REVIEW 3–3

1. Define *power*.
2. Write the formula for power in terms of energy and time.
3. Define *watt*.
4. Express each of the following values of power in the most appropriate units:
   (a) 68,000 W **(b)** 0.005 W **(c)** 0.000025 W
5. If you use 100 W of power for 10 h, how much energy (in kilowatthours) have you consumed?
6. Convert 2000 Wh to kilowatthours.
7. Convert 360,000 Ws to kilowatthours.

3–4 POWER IN AN ELECTRIC CIRCUIT

When current flows through resistance, the collisions of the electrons give off heat, resulting in a loss of energy as indicated in Figure 3–13. There is always a certain amount of power in an electric circuit, and it is dependent on the amount of *resistance* and on the amount of *current*, expressed as follows:

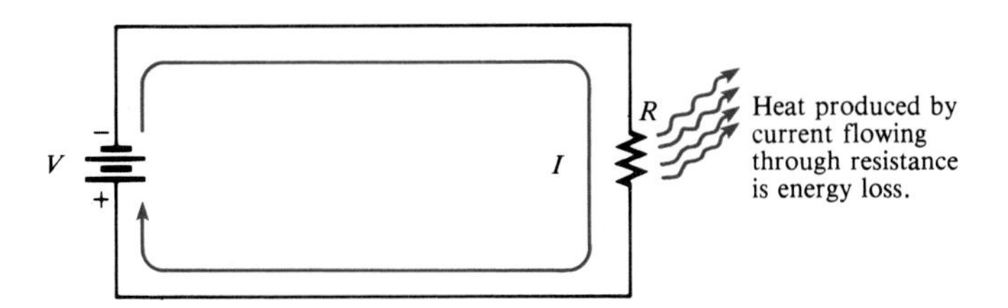

FIGURE 3–13
Power in an electric circuit is seen as heat given off by the resistance.

$$P = I^2R \quad (3\text{–}6)$$

We can produce an *equivalent* expression for power by substituting V for IR (I^2 is $I \times I$).

$$P = I^2R = (I \times I)R = I(IR) = (IR)I$$
$$P = VI \quad (3\text{–}7)$$

We obtain another *equivalent* expression by substituting V/R for I (Ohm's law) as follows:

$$P = VI = V\left(\frac{V}{R}\right)$$
$$P = \frac{V^2}{R} \quad (3\text{–}8)$$

How to Use the Appropriate Power Formula

To calculate the power in a resistance, you can use any one of the three power formulas, depending on what information you have. For example, assume that you know the values of current and voltage. In this case you calculate the power with the formula $P = VI$. If you know I and R, use the formula $P = I^2R$. If you know V and R, use the formula $P = V^2/R$.

EXAMPLE 3–14 Calculate the power in each of the three circuits of Figure 3–14.

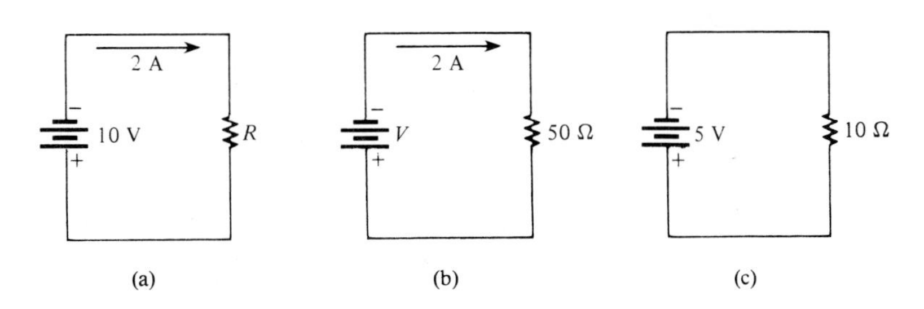

FIGURE 3–14

Solution
In circuit (a), V and I are known. The power is determined as follows:

$$P = VI = (10\text{ V})(2\text{ A}) = 20\text{ W}$$

In circuit (b), I and R are known. The power is determined as follows:

$$P = I^2R = (2\text{ A})^2(50\ \Omega) = 200\text{ W}$$

In circuit (c), V and R are known. The power is determined as follows:

$$P = \frac{V^2}{R} = \frac{(5\text{ V})^2}{10\ \Omega} = 2.5\text{ W}$$

The calculator sequence for circuit (c) is
5 2nd x^2 ÷ 1 0 =

EXAMPLE 3–15 A 100-W light bulb operates on 120 V. How much current does it require?

Solution
Use the formula $P = VI$ and solve for I as follows: Transpose the terms to get I on the left of the equation:

$$VI = P$$

Divide both sides of equation by V to get I by itself:

$$\frac{\cancel{V}I}{\cancel{V}} = \frac{P}{V}$$

The V's cancel on the left, leaving

$$I = \frac{P}{V}$$

Substituting 100 W for P and 120 V for V yields

$$I = \frac{P}{V} = \frac{100\text{ W}}{120\text{ V}} = 0.833\text{ A}$$

SECTION REVIEW 3–4

1. If there are 10 V across a resistor and a current of 3 A flowing through it, what is the power?
2. How much power does the source in Figure 3–15 generate? What is the power in the resistor? Are the two values the same? Why?
3. If a current of 5 A is flowing through a 50-Ω resistor, what is the power?
4. How much power is produced by 20 mA through a 5-kΩ resistor?
5. Five volts are applied to a 10-Ω resistor. What is the power?
6. How much power does a 2-kΩ resistor with 8 V across it produce?
7. What is the resistance of a 75-W bulb that takes 0.5 A?

FIGURE 3–15

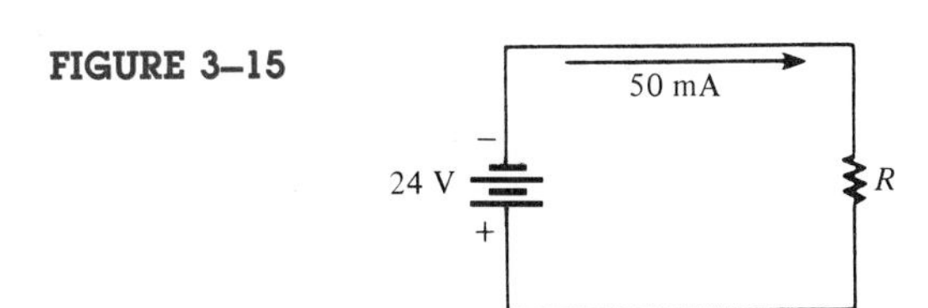

THE POWER RATING OF RESISTORS

3–5

As you know, a resistor gives off heat when current flows through it. There is a limit to the amount of heat that a resistor can give off, which is specified by its power rating.

The *power rating* is the maximum amount of power that a resistor can dissipate without being damaged by excessive heat buildup. The power rating is not related to the ohmic value (resistance) but rather is determined mainly by the physical size and shape of the resistor. The larger the *surface area* of a resistor, the more power it can dissipate. *The surface area of a cylindrically shaped resistor is equal to the length (l) times the circumference (c), as indicated in Figure 3–16.*

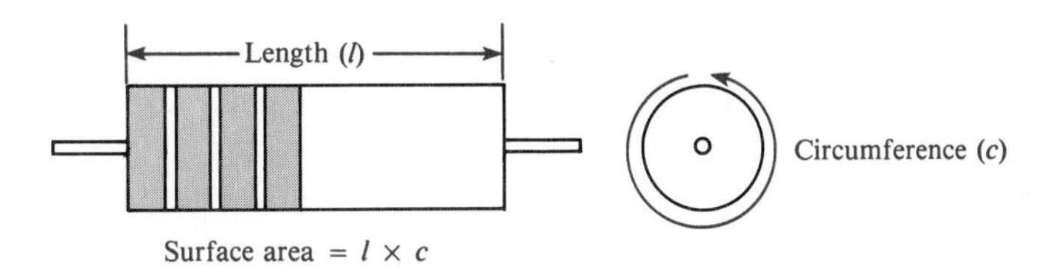

FIGURE 3–16
The power rating of a resistor is directly related to its surface area.

Carbon-composition resistors are available in standard power ratings from 1/8 W to 2 W, as shown in Figure 3–17. Available power ratings for other types of resistors vary. For example, carbon-film and metal-film resistors have ratings up to 10 W, and wirewound resistors have ratings up to 225 W or greater. Figure 3–18 shows some of these resistors.

FIGURE 3–17
Carbon-composition resistors with standard power ratings of 1/8 W, 1/4 W, 1/2 W, 1 W, and 2 W (courtesy of Allen-Bradley Co.).

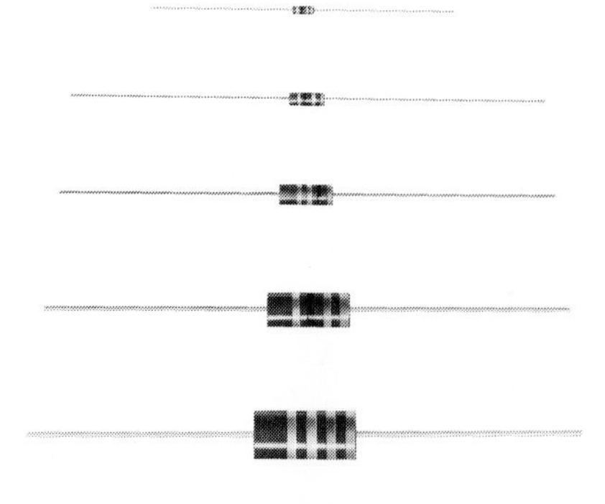

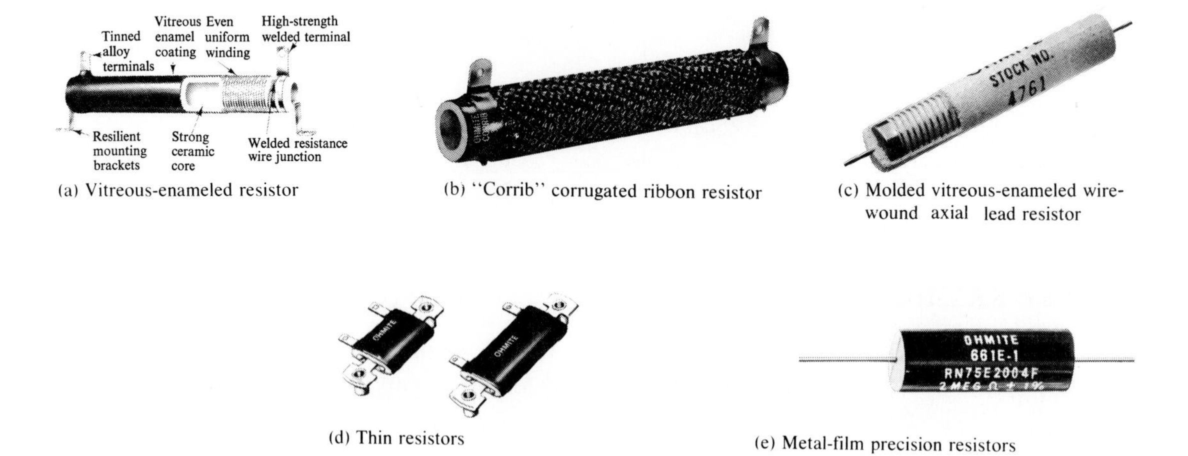

(a) Vitreous-enameled resistor (b) "Corrib" corrugated ribbon resistor (c) Molded vitreous-enameled wire-wound axial lead resistor

(d) Thin resistors (e) Metal-film precision resistors

FIGURE 3–18
Typical resistors with high power ratings (courtesy of Ohmite Manufacturing Co.)

How to Select the Proper Power Rating for an Application

When a resistor is used in a circuit, its power rating must be greater than the maximum power that it will have to handle. For example, if a carbon-composition resistor is to dissipate 0.75 W in a circuit application, its rating should be at least the next higher standard value which is 1 W. It is common practice to use a rating that is approximately double the actual power when possible.

EXAMPLE 3–16 Choose an adequate power rating for each of the carbon-composition resistors in Figure 3–19.

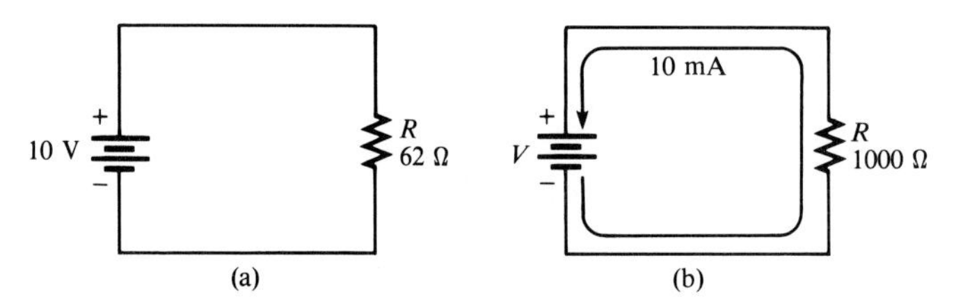

FIGURE 3–19

Solution

(a) The actual power is

$$P = \frac{V^2}{R} = \frac{(10\text{ V})^2}{62\ \Omega} = \frac{100\text{ V}^2}{62\ \Omega} = 1.6\text{ W}$$

Select a resistor with a power rating higher than the actual power. In this case, a 2-W resistor should be used.

(b) The actual power is

$$P = I^2R = (10 \text{ mA})^2(1000 \ \Omega) = 0.1 \text{ W}$$

A 1/8-W (0.125-W) resistor should be used in this case.

Resistor Failures

When the power into a resistor is greater than its rating, the resistor will become excessively hot. As a result, either the resistor will burn open or its resistance value will be greatly altered.

A resistor that has been damaged because of overheating can often be detected by the charred or altered appearance of its surface. If there is no visual evidence, a resistor that is suspected of being damaged can be checked with an ohmmeter for an open or increased resistance value.

EXAMPLE 3–17 Determine whether the resistor in each circuit of Figure 3–20 has possibly been damaged by overheating.

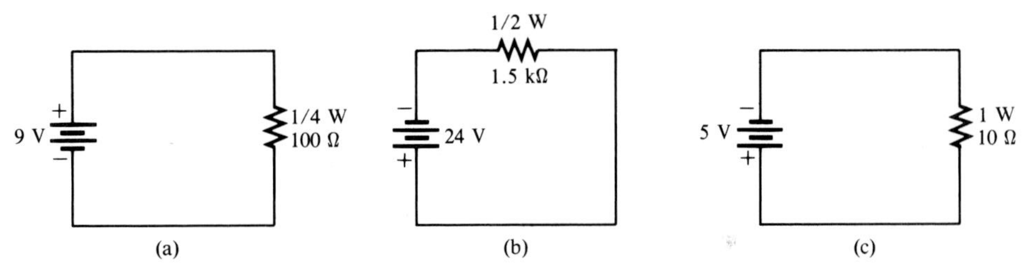

FIGURE 3–20

Solution

(a)

$$P = \frac{V^2}{R} = \frac{(9 \text{ V})^2}{100 \ \Omega} = 0.81 \text{ W}$$

The rating of the resistor is 1/4 W (0.25 W), which is insufficient to handle the power. The resistor may be burned out.

(b)

$$P = \frac{V^2}{R} = \frac{(24 \text{ V})^2}{1.5 \text{ k}\Omega} = 0.384 \text{ W}$$

The rating of the resistor is 1/2 W (0.5 W), which is sufficient to handle the power.

(c)

$$P = \frac{V^2}{R} = \frac{(5 \text{ V})^2}{10 \ \Omega} = 2.5 \text{ W}$$

The rating of the resistor is 1 W, which is insufficient to handle the power. The resistor may be burned out.

How to Check a Resistor with an Ohmmeter

A typical analog multimeter (VOM) and a digital multimeter are shown in Figures 3–21(a) and 3–21(b), respectively. The large switch on the analog meter is called a *range switch.* Notice the resistance (OHMS) settings on both meters. For the analog meter in Part (a), each setting indicates the amount by which the *ohms scale* (top scale) on the meter is to be *multiplied.* For example, if the pointer is at 50 on the ohms scale and the range switch is set at $\times 100$, the resistance being measured is $50 \times 100\ \Omega = 5000\ \Omega$. *If the resistor is open, the pointer will stay at full left scale (∞ means infinite) regardless of the range switch setting.* For the digital meter, you use the range switch to select the appropriate setting for the value of resistance being measured. You do not have to multiply to get the correct reading because you have a direct digital readout of the resistance value.

(a)

(b)

FIGURE 3–21
Typical multimeters (courtesy of Triplett Corp.).

The Basic Ohmmeter

A basic ohmmeter, shown in Figure 3–22(a), contains a battery and a variable resistor in series with the meter movement. To measure resistance, connect the leads across the external resistor to be measured, as shown in Part (b). This connection completes the circuit, allowing the internal battery to produce current through the movement, in turn causing a deflection of the pointer proportional to the value of the external resistance being measured.

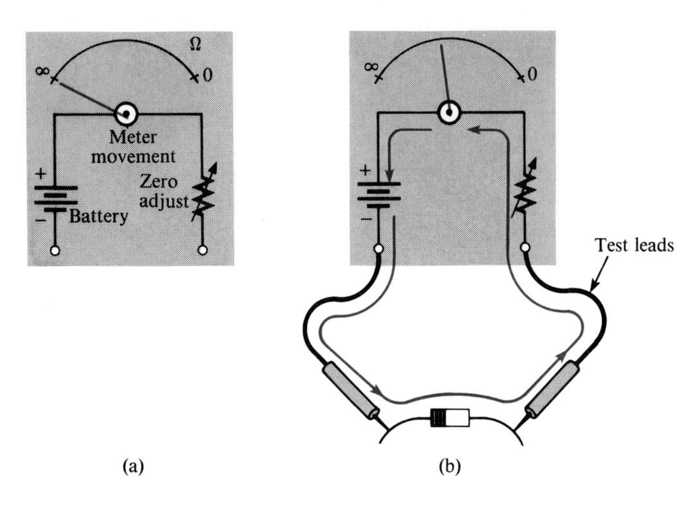

FIGURE 3–22
Basic ohmmeter circuit.

Zero Adjustment

When the ohmmeter leads are open, as in Figure 3–23(a), the pointer is at full left scale, indicating *infinite* (∞) resistance (open circuit). When the leads are *shorted,* as in Figure 3–23(b), the pointer is at full right scale, indicating *zero* resistance.

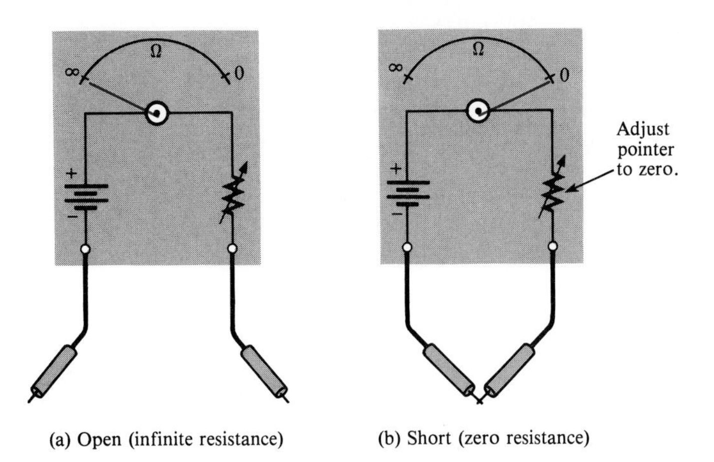

FIGURE 3–23
Zero adjustment.

The purpose of the variable resistor is to adjust the current so that the pointer is at exactly zero when the leads are shorted. It is used to compensate for changes in the internal battery voltage due to aging.

Because of the ohmmeter's internal battery, any other voltage source must be disconnected from a circuit before a resistance measurement is made. If the voltage is left on, there is a risk of damaging the meter.

The following steps should be used in measuring resistance:

1. Disconnect the voltage from the circuit.
2. To prevent other parts of the circuit from altering the measurement, disconnect the resistor to be checked from its circuit.
3. Touch the test leads together, and use the *ohms adjust* knob to set the pointer at 0 Ω (analog meter only).
4. Select the proper range switch setting. On the analog meter, it is best to select a setting that produces a reading on the right half of the scale for better accuracy. On the digital meter, set the switch at the lowest setting that is greater than the value you expect to measure.
5. Connect the meter leads across the resistor. Polarity is not important in resistance measurements. Be careful not to touch the test lead contacts with your fingers because your body resistance will affect the reading.

SECTION REVIEW 3–5

1. Name two important values associated with a resistor.
2. How does the physical size of a resistor determine the amount of power that it can handle?
3. List the standard power ratings of carbon-composition resistors.
4. A resistor must handle 0.3 W. What size carbon resistor should be used to dissipate the energy properly?
5. Why is an internal battery required in an ohmmeter?
6. If the pointer indicates 8 on the ohmmeter scale and the range switch is set at R × 1k, what resistance is being measured?

APPLICATION NOTE

The Application Assignment at the beginning of the chapter could be completed as follows:

1. To choose a 5-V supply that is capable of providing current to a 75-Ω load, calculate the current using Ohm's law:

$$I = \frac{V}{R} = \frac{5\ \text{V}}{75\ \Omega} = 66.67\ \text{mA}$$

   Specify a power supply that can provide more than the amount of current expected.
2. Connect an ammeter between the 25-V source and the device as indicated in Figure 3–24.

FIGURE 3–24

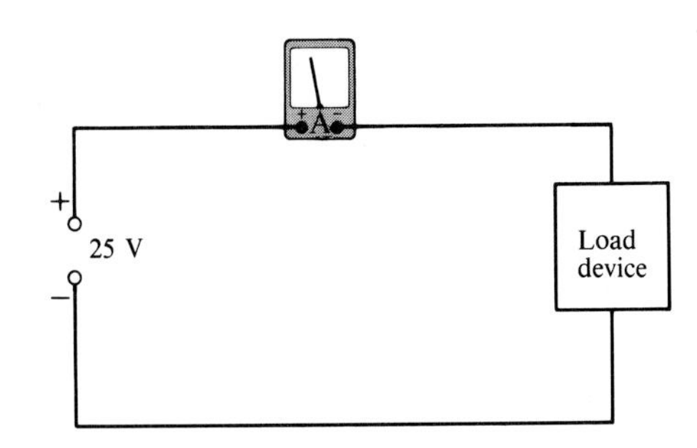

Take the value of I on the meter and determine the resistance and power rating of the "dummy" load as follows:

$$R = \frac{V}{I} = \frac{25\ \text{V}}{I}$$
$$P = VI = (25\ V)(I)$$

3. Use Ohm's law to find the voltage necessary for the 20-mA display element which has an internal resistance of 100 Ω.

$$V = IR = (20\ \text{mA})(100\ \Omega) = 2\ \text{V}$$

SUMMARY

Facts

- Voltage and current are linearly proportional.
- Ohm's law gives the relationship of voltage, current, and resistance.
- Current is directly proportional to voltage.
- Current is inversely proportional to resistance.
- A kilohm (kΩ) is one thousand ohms.
- A megohm (MΩ) is one million ohms:
- A microampere (μA) is one-millionth of an ampere.
- A milliampere (mA) is one-thousandth of an ampere.
- The amount of heat that a resistor can dissipate is specified by its power rating.
- The power rating of a resistor is determined by its physical size and shape.
- Power rating is not related to ohmic value.
- A resistor normally opens when it burns out.
- The symbol for infinity is ∞.

Units

- Watt (W)—the unit of power *(P)*.
- Kilowatthour (kWh)—a unit of energy.

Definitions

- *Ohm's law*—a law which states that current is directly proportional to voltage and inversely proportional to resistance.
- *Power*—the rate at which energy is used.
- *One kilowatthour*—1000 watts of power used for one hour.

Formulas

$$I = \frac{V}{R} \quad \text{Ohm's law} \tag{3–1}$$

$$V = IR \quad \text{Ohm's law} \tag{3–2}$$

$$R = \frac{V}{I} \quad \text{Ohm's law} \tag{3–3}$$

$$P = \frac{W}{t} \quad \text{Power equals energy divided by time.} \tag{3–4}$$

$$W = Pt \quad \text{Energy equals power multiplied by time.} \tag{3–5}$$

$$P = I^2R \quad \text{Power equals current squared times resistance.} \tag{3–6}$$

$$P = VI \quad \text{Power equals voltage times current.} \tag{3–7}$$

$$P = \frac{V^2}{R} \quad \text{Power equals voltage squared divided by resistance.} \tag{3–8}$$

SELF-TEST

1. Explain Ohm's law.
2. In a circuit consisting of a voltage source and a resistor, describe what happens to the current when
   (a) the voltage is tripled.
   (b) the voltage is reduced by 75%.
   (c) the resistance is doubled.
   (d) the resistance is reduced by 35%.
   (e) the voltage is doubled and the resistance is cut in half.
   (f) the voltage is doubled and the resistance is doubled.
3. State the formula used to find I when the values of V and R are known.
4. State the formula used to find V when the values of I and R are known.
5. State the formula used to find R when the values of V and I are known.
6. (a) Volts divided by kΩ gives ________.
   (b) Volts divided by MΩ gives ________.
   (c) Milliamperes times kΩ gives ________.
   (d) Milliamperes times MΩ gives ________.
   (e) Volts divided by mA gives ________.
   (f) Volts divided by μA gives ________.
7. Distinguish between power and energy.
8. (a) How many watts are in a kilowatt?
   (b) How many watts are in a megawatt?
9. Show how to use a voltmeter and an ammeter to measure the power in a resistor when it is connected to a voltage source.
10. (a) State the formula used to find power when V and I are known.
    (b) State the formula used to find power when I and R are known.
    (c) State the formula used to find power when V and R are known.
11. What power rating would you use for a carbon-composition resistor that will be required to handle 1.1 W?
12. What reading would you expect on an analog ohmmeter when the resistor being measured is open?

PROBLEM SET A

Section 3–1

3–1 The current in a circuit is 1 A. Determine what the current will be when
(a) the voltage is tripled.
(b) the voltage is reduced by 80%.
(c) the voltage is increased by 50%.

3–2 The current in a circuit is 100 mA. Determine what the current will be when
(a) the resistance is increased by 100%.
(b) the resistance is reduced by 30%.
(c) the resistance is quadrupled.

3–3 The current in a circuit is 10 mA. What will the current be if the voltage is tripled and the resistance is doubled?

Section 3–2

3–4 Determine the current in each case.
(a) $V = 5\text{ V}, R = 1\ \Omega$ **(b)** $V = 15\text{ V}, R = 10\ \Omega$
(c) $V = 50\text{ V}, R = 100\ \Omega$ **(d)** $V = 30\text{ V}, R = 15\text{ k}\Omega$
(e) $V = 250\text{ V}, R = 5\text{ M}\Omega$

3–5 Determine the current in each case.
(a) $V = 9\text{ V}, R = 2.7\text{ k}\Omega$ **(b)** $V = 5.5\text{ V}, R = 10\text{ k}\Omega$
(c) $V = 40\text{ V}, R = 68\text{ k}\Omega$ **(d)** $V = 1\text{ kV}, R = 2\text{ k}\Omega$
(e) $V = 66\text{ kV}, R = 10\text{ M}\Omega$

3–6 A 10-Ω resistor is connected across a 12-V battery. How much current flows through the resistor?

3–7 A resistor is connected across the terminals of a dc voltage source in each part of Figure 3–25. Determine the current in each resistor. (Full-color photos appear between pages 128 and 129.)

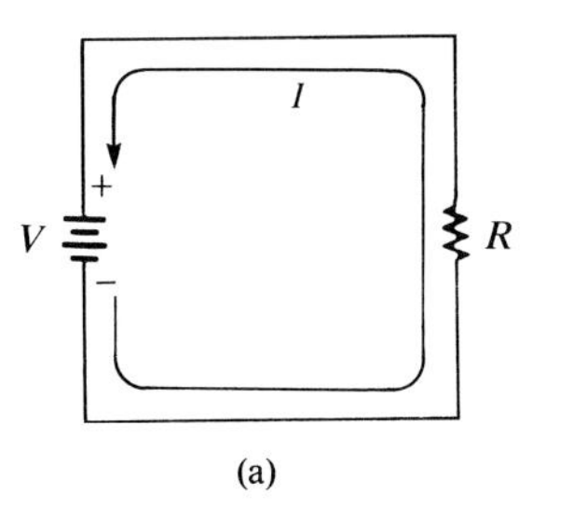

(a)

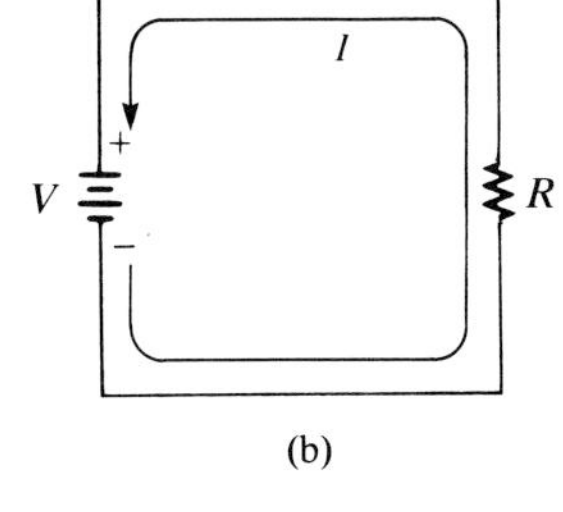

(b)

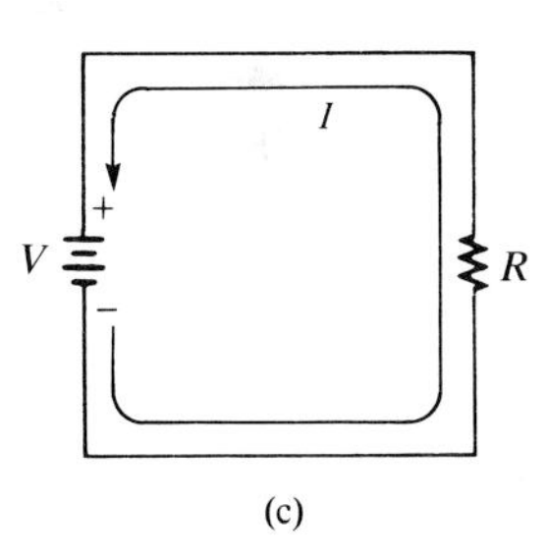

(c)

FIGURE 3–25

3–8 Calculate the voltage for each value of I and R.
(a) $I = 2\text{ A}, R = 18\Omega$ **(b)** $I = 5\text{ A}, R = 50\ \Omega$
(c) $I = 2.5\text{ A}, R = 600\ \Omega$ **(d)** $I = 0.6\text{ A}, R = 47\ \Omega$
(e) $I = 0.1\text{ A}, R = 500\ \Omega$

3–9 Calculate the voltage for each value of I and R.
(a) $I = 1\text{ mA}, R = 10\ \Omega$ **(b)** $I = 50\text{ mA}, R = 33\ \Omega$
(c) $I = 3\text{ A}, R = 5\text{ k}\Omega$ **(d)** $I = 1.6\text{ mA}, R = 2.2\text{ k}\Omega$
(e) $I = 250\ \mu\text{A}, R = 1\text{ k}\Omega$ **(f)** $I = 500\text{ mA}, R = 1.5\text{ M}\Omega$
(g) $I = 850\ \mu\text{A}, R = 10\text{ M}\Omega$ **(h)** $I = 75\ \mu\text{A}, R = 50\ \Omega$

3–10 Three amperes of current are measured through a 27-Ω resistor connected across a voltage source. How much voltage does the source produce?

3–11 Assign a voltage value to each source in the circuits of Figure 3–26 to obtain the indicated amounts of current.

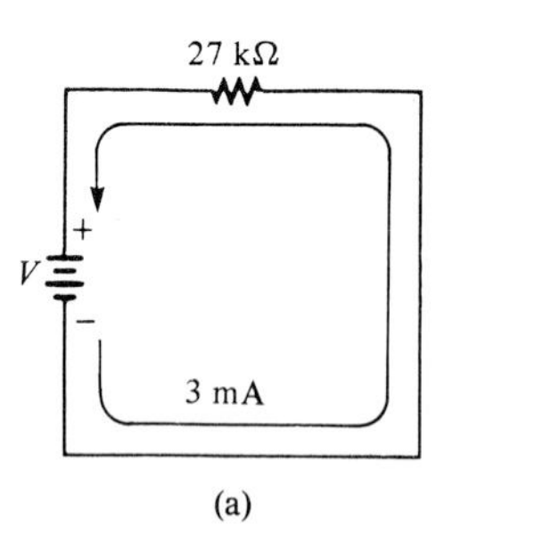

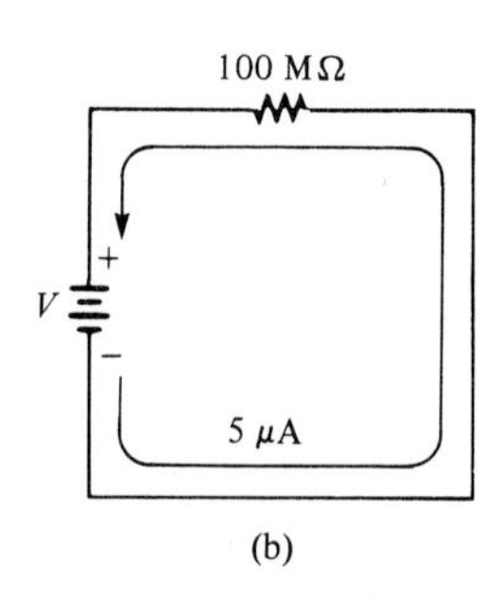

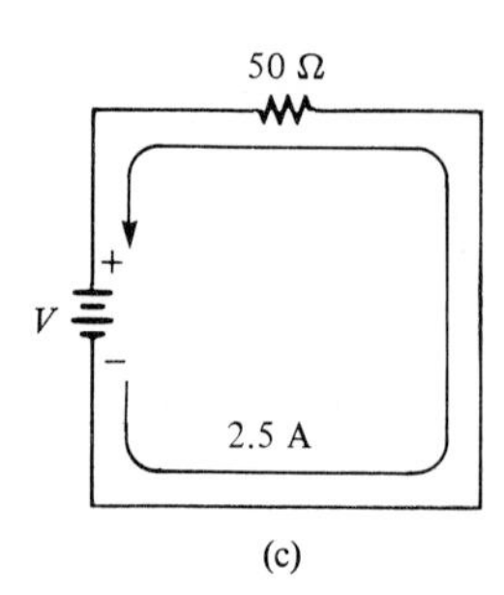

FIGURE 3–26

3–12 Calculate the resistance for each value of V and I.
(a) $V = 10$ V, $I = 2$ A **(b)** $V = 90$ V, $I = 45$ A
(c) $V = 50$ V, $I = 5$ A **(d)** $V = 5.5$ V, $I = 10$ A
(e) $V = 150$ V, $I = 0.5$ A

3–13 Calculate R for each set of V and I values.
(a) $V = 10$ kV, $I = 5$ A **(b)** $V = 7$ V, $I = 2$ mA
(c) $V = 500$ V, $I = 250$ mA **(d)** $V = 50$ V, $I = 500$ μA
(e) $V = 1$ kV, $I = 1$ mA

3–14 Six volts are applied across a resistor. A current of 2 mA is measured. What is the value of the resistor?

3–15 Choose the correct value of resistance to get the current values indicated in each circuit of Figure 3–27.

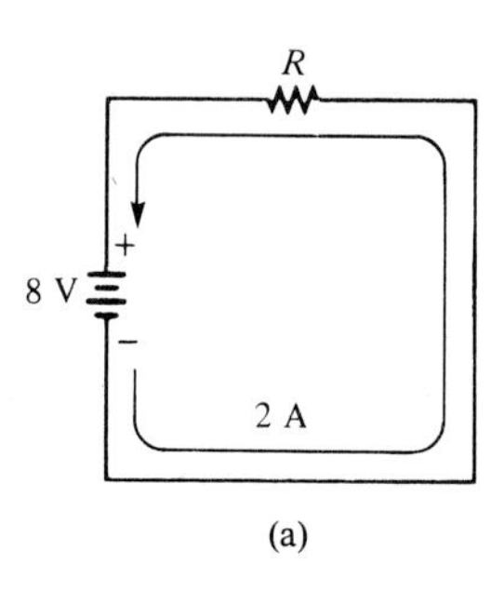

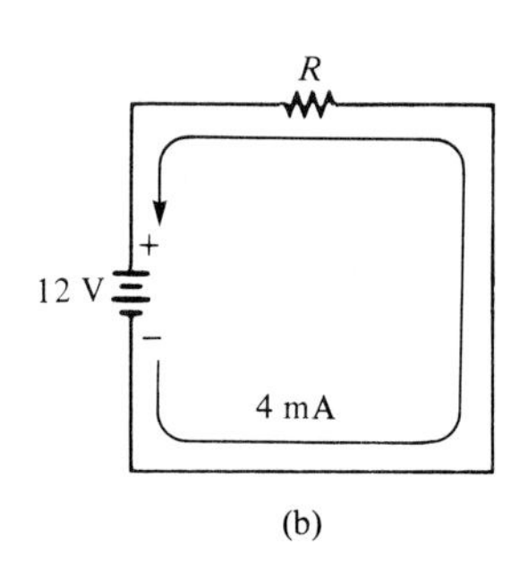

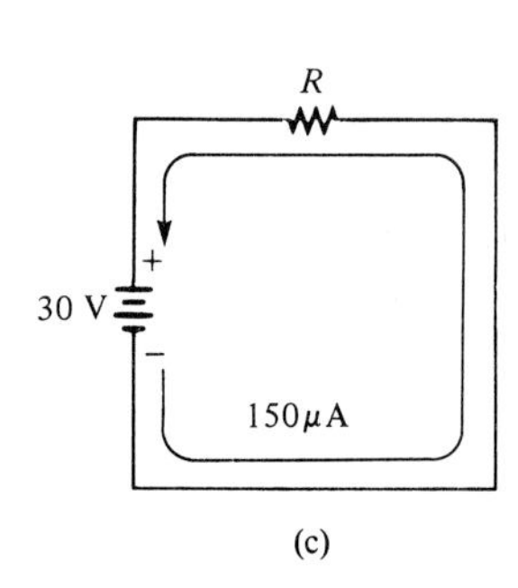

FIGURE 3–27

Section 3–3

3–16 What is the power when energy is consumed at the rate of 350 J/s?

3–17 How many watts are used when 7500 J of energy are consumed in 5 h?

3–18 Convert the following to kilowatts:
(a) 1000 W **(b)** 3750 W **(c)** 160 W **(d)** 50,000 W

3–19 Convert the following to megawatts:
(a) 1,000,000 W **(b)** 3×10^6 W **(c)** 15×10^7 W
(d) 8700 kW

3–20 Convert the following to milliwatts:
(a) 1 W **(b)** 0.4 W **(c)** 0.002 W **(d)** 0.0125 W

3–21 Convert the following to microwatts:
(a) 2 W (b) 0.0005 W (c) 0.25 mW (d) 0.00667 mW

3–22 Convert the following to watts:
(a) 1.5 kW (b) 0.5 MW (c) 350 mW (d) 9000 μW

Section 3–4

3–23 If a resistor has 5.5 V across it and 3 mA flowing through it, what is the power?

3–24 An electric heater works on 115 V and draws 3 A of current. How much power does it use?

3–25 How much power is produced by 500 mA of current through a 4.7-kΩ resistor?

3–26 Calculate the power handled by a 10-kΩ resistor carrying 100 μA.

3–27 If there are 60 V across a 600-Ω resistor, what is the power?

3–28 A 50-Ω resistor is connected across the terminals of a 1.5-V battery. What is the power dissipation in the resistor?

3–29 If a resistor is to carry 2 A of current and handle 100 W of power, how many ohms must it be? Assume that the voltage can be adjusted to any required value.

3–30 Convert 5×10^6 wattminutes to kWh.

3–31 Convert 6700 wattseconds to kWh.

3–32 How many watthours does 50 W used for 12 h equal? How many kilowatthours?

Section 3–5

3–33 A 6.8-kΩ resistor has burned out in a circuit. You must replace it with another resistor with the same ohmic value. If the resistor carries 10 mA, what should its power rating be? Assume that you have available carbon-composition resistors in all the standard power ratings.

3–34 A certain type of power resistor comes in the following ratings: 3 W, 5 W, 8 W, 12 W, 20 W. Your particular application requires a resistor that can handle approximately 8 W. Which rating would you use? Why?

PROBLEM SET B

3–35 The filament of a light bulb in the circuit of Figure 3–28(a) has a certain amount of resistance, represented by an equivalent resistance in Figure 3–28(b). If the bulb operates with 120 V and 0.8 A of current, what is the resistance of its filament?

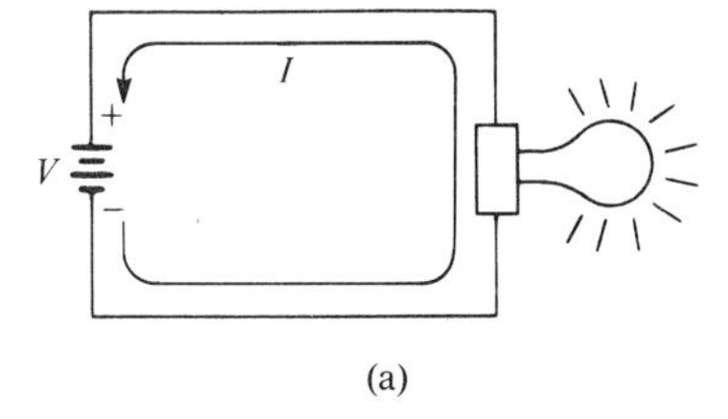

(a)

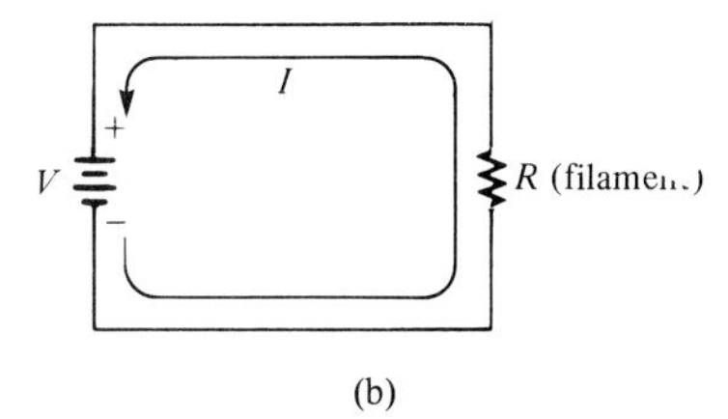

(b)

FIGURE 3–28

3–36 A certain electrical device has an unknown resistance. You have available a 12-V battery and an ammeter. How would you determine the value of the unknown resistance? Draw the necessary circuit connections.

3–37 A variable voltage source is connected to the circuit of Figure 3–29. Start at 0 V and increase the voltage in 10-V steps up to 100 V. Determine the current at

FIGURE 3–29

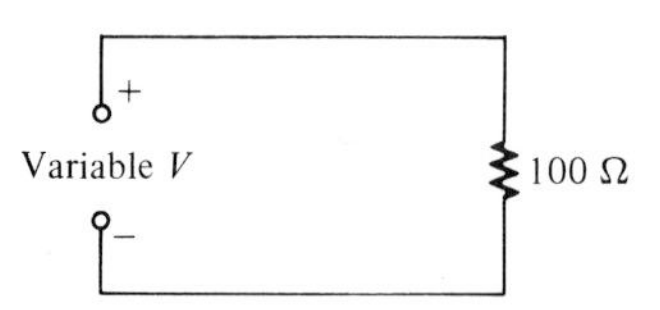

each voltage value, and plot a graph of V versus I. Is the graph a straight line? What does the graph indicate?

3–38 In a certain circuit, $V = 1$ V and $I = 5$ mA. Determine the current for each of the following voltages in the same circuit:
(a) $V = 1.5$ V **(b)** $V = 2$ V **(c)** $V = 3$ V
(d) $V = 4$ V **(e)** $V = 10$ V

3–39 Figure 3–30 is a graph of voltage versus current for three resistance values. Determine R_1, R_2, and R_3.

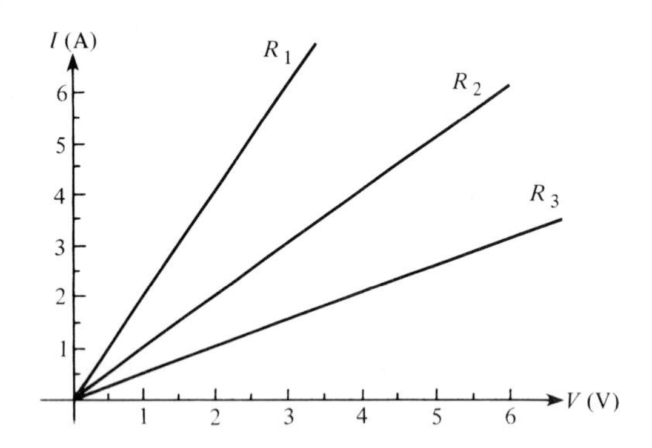

FIGURE 3–30

3–40 You are measuring the current in a circuit that is operated on a 10-V battery. The ammeter reads 50 mA. Later, you notice that the current has dropped to 30 mA. Eliminating the possibility of a resistance change, you must conclude that the voltage has changed. How much has the voltage of the battery changed, and what is its new value?

3–41 If you wish to increase the amount of current in a resistor from 100 mA to 150 mA by changing the 20-V source, by how many volts should you change the source? To what new value should you set it?

3–42 By varying the rheostat (variable resistor) in the circuit of Figure 3–31, you can change the amount of current. The setting of the rheostat is such that the current is 750 mA. What is the ohmic value of this setting? To adjust the current to 1 A, to what ohmic value must you set the rheostat?

FIGURE 3–31

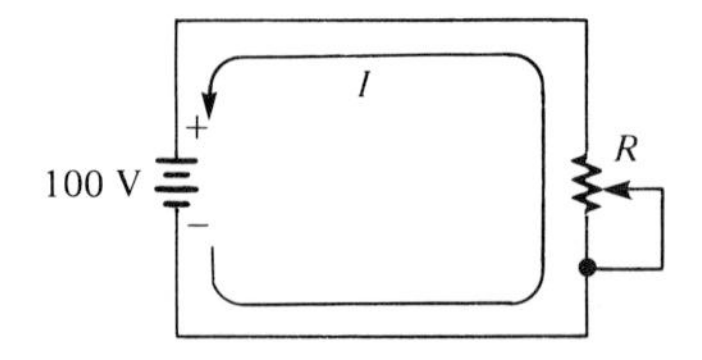

3–43 A certain resistor has the following color code: orange, orange, red, gold. Determine the maximum and minimum currents you should expect to measure when a 12-V source is connected across the resistor.

3–44 A 6-V source is connected to a 100-Ω resistor by two 12-ft lengths of 18-gage copper wire. Refer to the wire table in Appendix A to determine the following:

(a) current. (b) resistor voltage.
(c) voltage across each length of wire.

3–45 Assume that you are measuring the current in a circuit that is operating with 25 V. The ammeter reads 50 mA. Later, you notice that the current has dropped to 40 mA. Assuming that the resistance did not change, you must conclude that the voltage has changed. How much has the voltage changed, and what is its new value?

3–46 A certain appliance uses 300 W. If it is allowed to run continuously for 30 days, how many kilowatthours of energy does it consume?

3–47 At the end of a 31-day period, your utility bill shows that you have used 1500 kWh. What is your average daily power?

3–48 An 8.2-kΩ resistor has burned out in a circuit. You must replace it with another resistor with the same ohmic value. If the resistor carries 5 mA, what should its power rating be? Assume that you have available carbon-composition resistors in all the standard power ratings.

3–49 A certain type of power resistor comes in the following ratings: 3 W, 5 W, 8 W, 12 W, 20 W. Your particular application requires a resistor that can handle approximately 10 W. Which rating would you use? Why?

3–50 A 12-V source is connected across a 10-Ω resistor.
(a) How much energy is used in 2 minutes?
(b) If the resistor is disconnected after 1 minute, does the power increase or decrease?

ANSWERS TO SECTION REVIEWS

Section 3–1

1. Current varies directly with voltage and inversely with resistance.
2. Increases, three times. **3.** 5 mA. **4.** No.

Section 3–2

1. 2 A. **2.** 4 mA. **3.** 5 A.
4. 10 V. **5.** 9 V. **6.** 12 V.
7. 5 Ω. **8.** 0.5 kΩ, 500 Ω.

Section 3–3

1. The rate at which energy is used. **2.** $P = W/t$.
3. The unit of power. One watt is the power when 1 J of energy is used in 1 s.
4. (a) 68 kW; (b) 5 mW; (c) 25 μW.
5. 1 kWh. **6.** 2 kWh. **7.** 0.1 kWh.

Section 3–4

1. 30 W.
2. 1.2 W, 1.2 W. Yes, because all energy produced by the source is dissipated by the resistance.
3. 1250 W. **4.** 2 W. **5.** 2.5 W.
6. 32 mW. **7.** 300 Ω.

Section 3–5

1. Ohmic value, power rating. **2.** A larger surface area dissipates more energy.
3. 0.125 W, 0.25 W, 0.5 W, 1 W, 2 W. **4.** At least 0.5 W.
5. To produce current through the resistance being measured.
6. 8 kΩ.

Series Circuits

Resistive circuits can be of two basic forms: *series* or *parallel*. In this chapter we discuss series circuits. Parallel circuits are presented in Chapter 5, and combinations of series and parallel are examined in Chapter 6. In this chapter you will see how Ohm's law is used in series circuits, and you will study another very important law, called *Kirchhoff's voltage law*. Also, several important applications of series circuits are given. Specifically, you will learn:

- How to identify a series circuit.
- How to determine the current in a series circuit.
- How to determine total resistance.
- How to apply Ohm's law to find the current, the voltages, and the resistances in a series circuit.
- How to connect voltage sources in series to achieve a higher voltage.
- The definition of Kirchhoff's voltage law and how to apply it.
- What voltage dividers are and how they can be used.
- How to determine the total power in a series circuit.
- How to identify some of the common problems in series circuits.

4

APPLICATION ASSIGNMENT

You are the electronic technician for an expedition to investigate volcanic phenomena in the Hawaiian Islands. The investigators will use several specialized electronic instruments to make measurements on and near volcanoes. You are assigned to devise a circuit, housed in a portable case, that will provide a selection of voltages to power the equipment. The following voltages are required and are to be derived from a 24-V battery: 20 V, 16 V, 10 V, and 8 V. You must determine how to build the circuit and must specify all the components. The maximum current drain on the 24 V battery is to be 1 A.

Two plug-type terminals are to be provided for the source battery connection, and two terminals for the output voltage. There must be a means to connect any one of the four voltages to the output terminals. Only one of the voltages is to be available at a time.

After completing this chapter, you should have the basic knowledge to complete this assignment. The Application Note at the end of the chapter gives one solution.

RESISTORS IN SERIES

4–1

Resistors in series are connected end-to-end or in a "string," as shown in Figure 4–1. Part (a) of the figure shows two resistors connected in series between point *A* and point *B*. Part (b) shows three in series, and Part (c) shows four in series. Of course, there can be any number of resistors in a series connection.

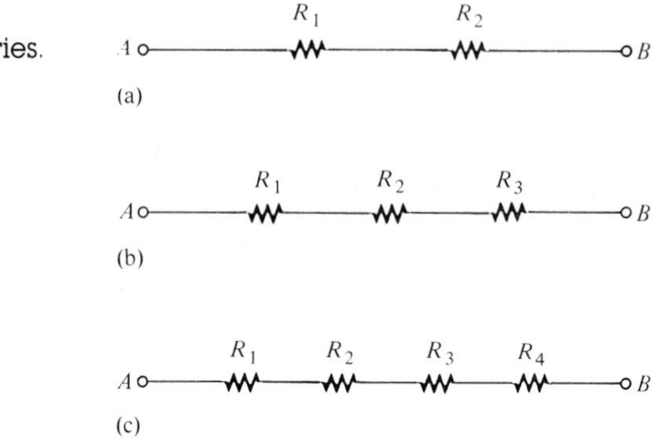

FIGURE 4–1
Resistors in series.

The only way for electrons to get from point *A* to point *B* in any of the connections of Figure 4–1 is to go through *each* of the resistors. The following is an important way to identify a series connection:

> **A series connection provides only one path for current between two points in a circuit so that the same current flows through each series resistor.**

Identifying Series Connections

In an actual circuit diagram, a series connection may not always be as easy to identify as those in Figure 4–1. For example, Figure 4–2 shows series resistors drawn in other ways. Remember, *if there is only one current path between two points, the resistors between those two points are in series,* no matter how they appear in a diagram.

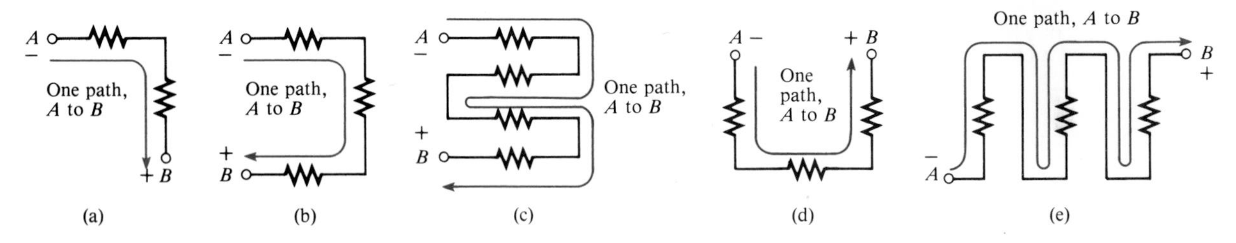

FIGURE 4–2
Some examples of series connections. Notice that the current must be the same at all points.

EXAMPLE 4–1

Five resistors are positioned on a circuit board as shown in Figure 4–3. Wire them together in series so that, starting from the negative (−) terminal, R_1 is first, R_2 is second, R_3 is third, and so on. Draw a schematic diagram showing this connection.

FIGURE 4–3

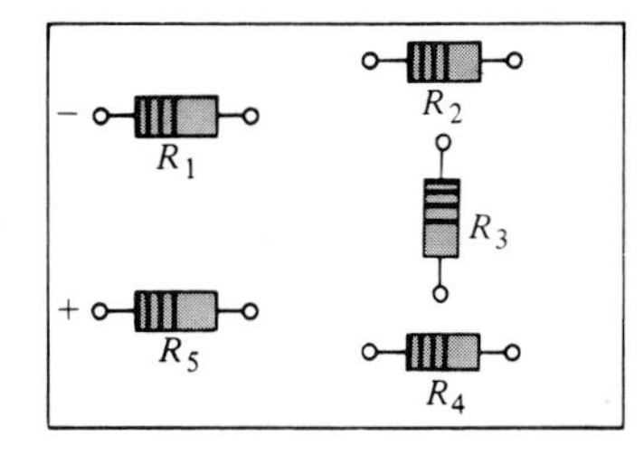

Solution

The wires are connected as shown in Figure 4–4(a), which is the *assembly diagram*. The *schematic diagram* is shown in Figure 4–4(b). Note that the schematic diagram does not necessarily show the actual physical arrangement of the resistors as does the assembly diagram. The purpose of the *schematic* is to show how components are connected *electrically*. The purpose of the *assembly* diagram is to show how components are arranged *physically*.

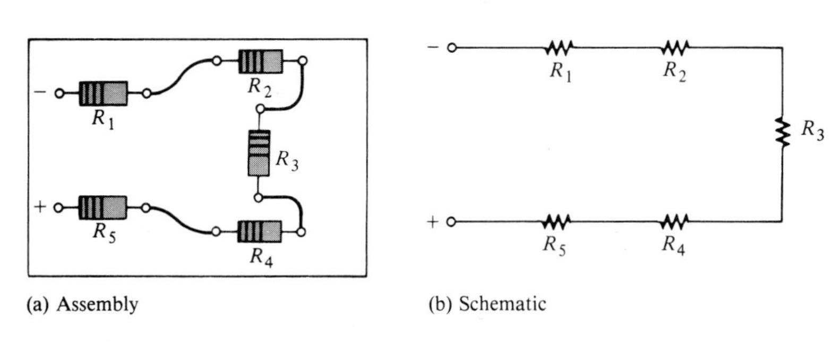

(a) Assembly (b) Schematic

FIGURE 4–4

EXAMPLE 4–2 Describe how the resistors on the printed circuit (PC) board in Figure 4–5 are related electrically.

FIGURE 4–5

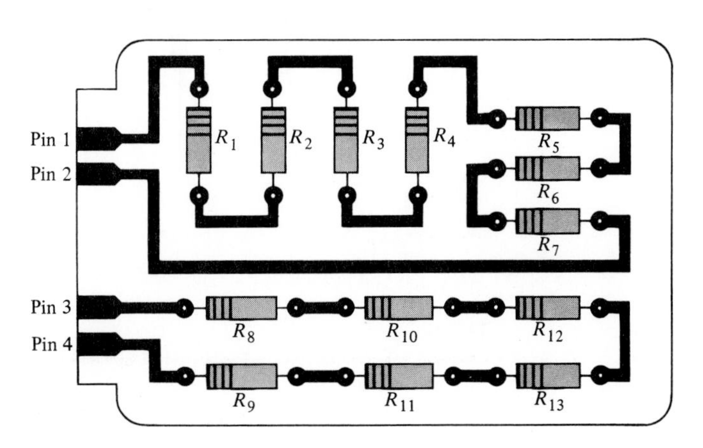

Solution
Resistors R_1 through R_7 are in series with each other. This series combination is connected between pins 1 and 2 on the PC board.

Resistors R_8 through R_{13} are in series with each other. This series combination is connected between pins 3 and 4 on the PC board.

SECTION REVIEW 4–1

1. How can you identify a series connection?
2. Complete the schematic diagrams for the circuits in each part of Figure 4–6 by connecting the resistors in series in numerical order from A to B.
3. Now connect each *group* of series resistors in Figure 4–6 in series.

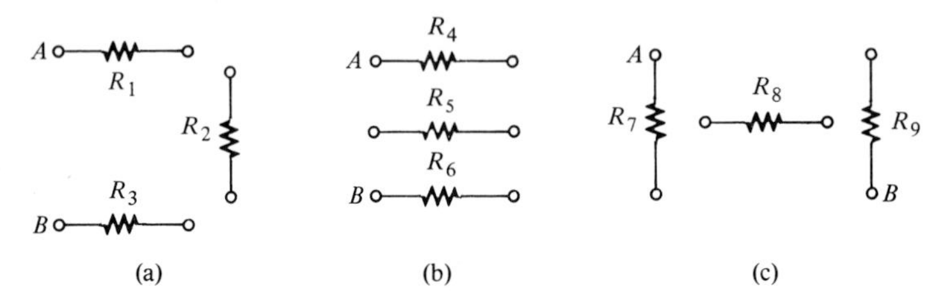

FIGURE 4–6

CURRENT IN A SERIES CIRCUIT

4–2

Equal Current through All Points

Figure 4–7 shows three resistors connected in series to a voltage source. *At any point in this circuit, the current flowing into that point must equal the current flowing out of that point,* as illustrated by the directional arrows. Notice that

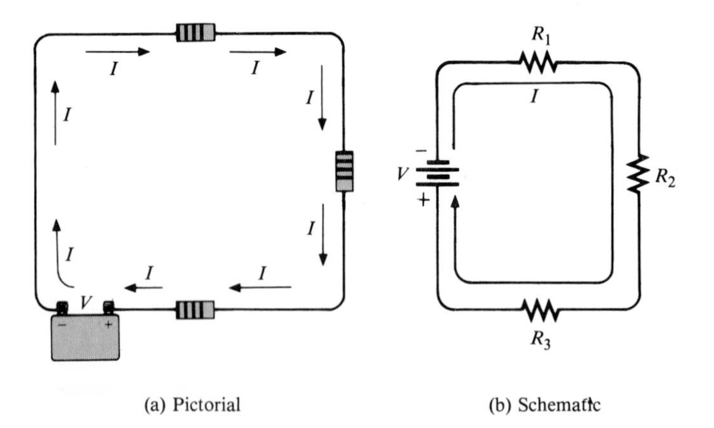

FIGURE 4–7
Current entering any point in a series circuit is the same as the current leaving that point.

the current flowing out of each of the resistors must equal the current flowing in, because it is essentially forced through the resistor and out the other end. Since there are no paths other than the single series path into which the current can branch off, the current in each section of the circuit must be the same as the current in all other sections. Although resistance reduces the current, the reduction is felt at all points equally.

In Figure 4–8, the battery supplies one ampere (1 A) of current to the series resistors. One ampere is flowing out of the negative terminal of the source. When ammeters are connected in series at several points as shown, *each* meter reads one ampere.

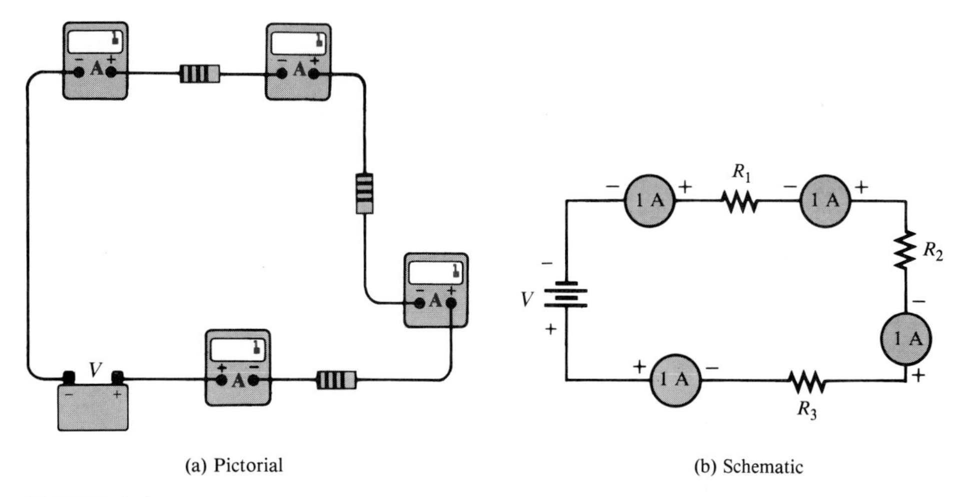

(a) Pictorial (b) Schematic

FIGURE 4–8
Current is the same at all points in a series circuit.

SECTION REVIEW 4–2

1. In a series circuit with a 10-Ω and a 5-Ω resistor in series, 2 A flows through the 10-Ω resistor. How much current flows through the 5-Ω resistor?
2. A milliammeter is connected between points *A* and *B* in Figure 4–9. It measures 50 mA. If you move the meter and connect it between points *C* and *D*, how much current will it indicate? Between *E* and *F*?
3. In Figure 4–10, how much current does ammeter 1 indicate? Ammeter 2?

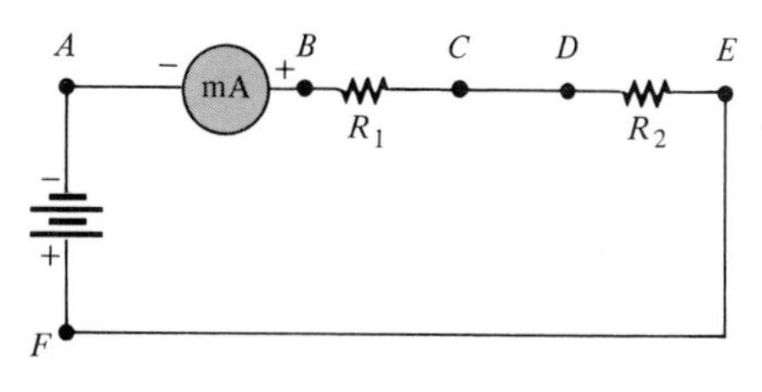

FIGURE 4–9

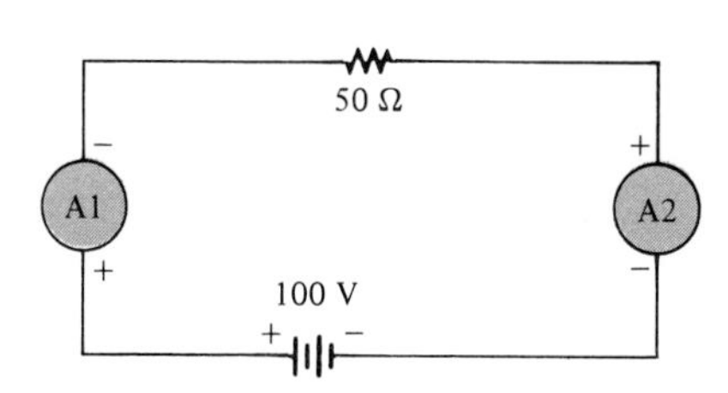

FIGURE 4–10

TOTAL SERIES RESISTANCE

4–3

The total resistance of a series connection is equal to the sum of the resistances of each individual resistor.

This fact is understandable because each of the resistors in series offers opposition to the current in direct proportion to its ohmic value. A greater number of resistors connected in series creates *more opposition* to current. More opposition to current implies a higher ohmic value of resistance. So every time a resistor is added in series, the total resistance increases.

How Series Resistor Values Add

Figure 4–11 illustrates how series resistances add to *increase* the total resistance (R_T). Part (a) of the figure has a single 10-Ω resistor. Part (b) shows another 10-Ω resistor connected in series with the first one, making a total resistance of 20 Ω. If a third 10-Ω resistor is connected in series with the first two, as shown in Part (c), the total resistance becomes 30 Ω.

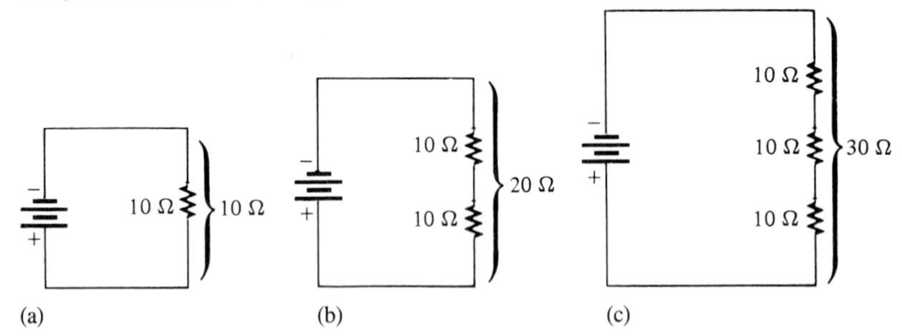

FIGURE 4–11
Total resistance increases with each additional series resistor.

Series Resistance Formula

For *any number* of individual resistors connected in series, the total resistance is the sum of each of the individual values:

$$R_T = R_1 + R_2 + R_3 + \cdots + R_n \qquad \textbf{(4–1)}$$

where R_T is the total resistance and R_n is the last resistor in the series string (n can be any positive integer equal to the number of resistors in series). For example, if there are four resistors in series ($n = 4$), the total resistance formula is

$$R_T = R_1 + R_2 + R_3 + R_4$$

If there are six resistors in series ($n = 6$), the total resistance formula is

$$R_T = R_1 + R_2 + R_3 + R_4 + R_5 + R_6$$

To illustrate the calculation of total series resistance, let us determine R_T of circuit of Figure 4–12, where V_S is the source voltage. This circuit has five

FIGURE 4–12
Example of five resistors in series.

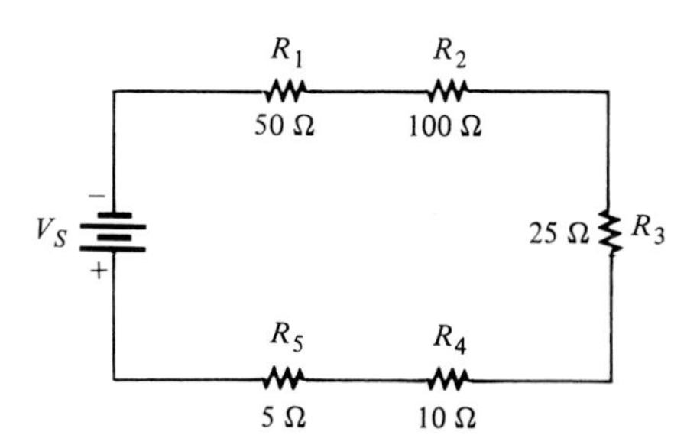

resistors in series. To find the total resistance, we simply add the values as follows:

$$R_T = 50\ \Omega + 100\ \Omega + 25\ \Omega + 10\ \Omega + 5\ \Omega = 190\ \Omega$$

This equation illustrates an important point: In Figure 4–12, the order in which the resistances are added does not matter; we still get the same total. Also, we can physically change the positions of the resistors in the circuit without affecting the total resistance.

EXAMPLE 4–3

Connect the resistors in Figure 4–13 (a full-color photo appears between pages 128 and 129) in series, and determine the total resistance, R_T.

FIGURE 4–13

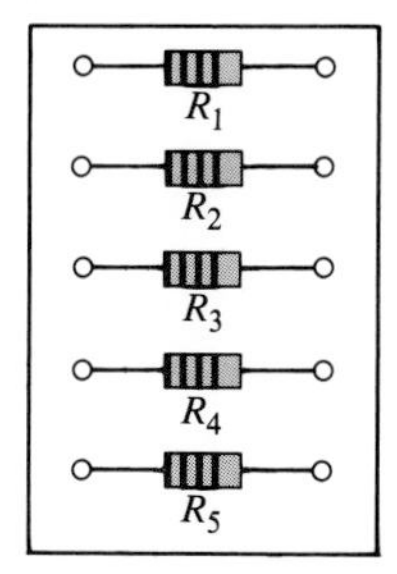

Solution

The resistors are connected as shown in Figure 4–14 (a full-color photo appears between pages 128 and 129).

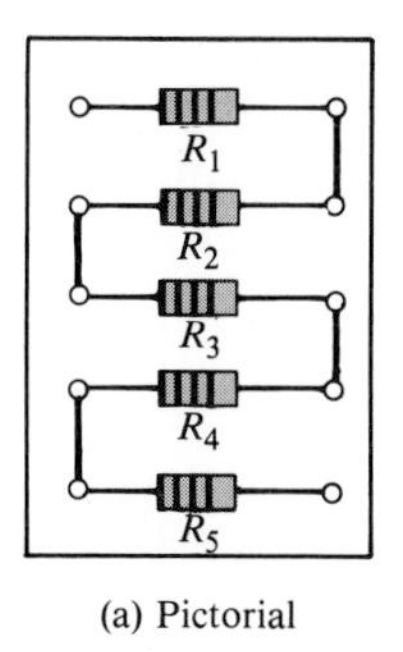

(a) Pictorial

R_1 33 Ω R_2 68 Ω 100 Ω R_3 R_5 10 Ω R_4 47 Ω

(b) Schematic

FIGURE 4–14

We find the total resistance by adding all the values as follows:

$$R_T = 33\ \Omega + 68\ \Omega + 100\ \Omega + 47\ \Omega + 10\ \Omega = 258\ \Omega$$

EXAMPLE 4–4 Calculate R_T for each circuit in Figure 4–15.

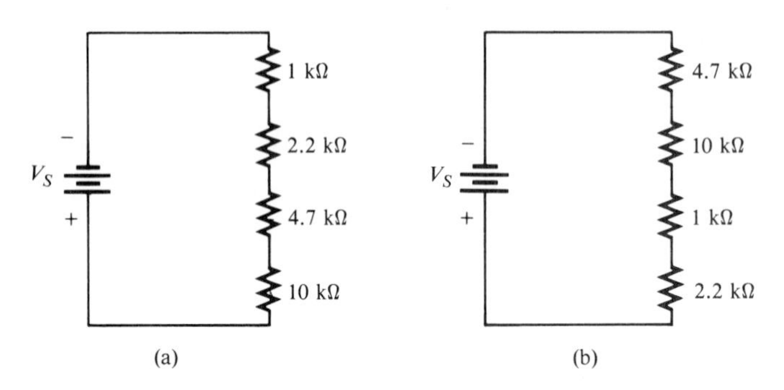

FIGURE 4–15

Solution

(a) Add the resistor values:

$$R_T = 1\ \text{k}\Omega + 2.2\ \text{k}\Omega + 4.7\ \text{k}\Omega + 10\ \text{k}\Omega = 17.9\ \text{k}\Omega$$

(b) Add the resistor values:

$$R_T = 4.7\ \text{k}\Omega + 10\ \text{k}\Omega + 1\ \text{k}\Omega + 2.2\ \text{k}\Omega = 17.9\ \text{k}\Omega$$

Notice that the total resistance does not depend on the position of the resistors. Both circuits are identical in terms of total resistance.

The calculator sequence for (a) is

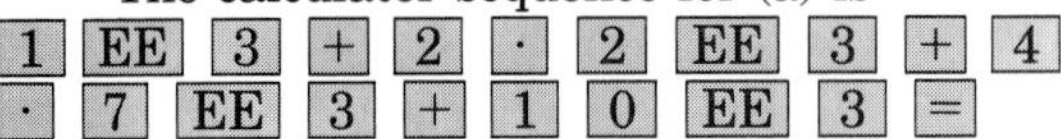

Equal-Value Series Resistors

When a circuit has more than one resistor of the *same* value in series, there is a shortcut method to obtain the total resistance: Simply multiply the *ohmic value* of the resistors having the same value by the *number* of equal-value resistors that are in series. This method is essentially the same as adding the values. For example, five 100-Ω resistors in series have an R_T of 5(100 Ω) = 500 Ω. In general, the formula is expressed as

$$R_T = nR \qquad \textbf{(4–2)}$$

where n is the number of equal-value resistors and R is the value.

EXAMPLE 4–5 Find the R_T of eight 22-Ω resistors in series.

Solution
We find R_T by adding the values as follows:

$$R_T = 22\ \Omega + 22\ \Omega + 22\ \Omega + 22\ \Omega + 22\ \Omega + 22\ \Omega + 22\ \Omega + 22\ \Omega$$
$$= 176\ \Omega$$

However, it is much easier to multiply:

$$R_T = 8(22\ \Omega) = 176\ \Omega$$

SECTION REVIEW 4–3

1. Calculate R_T between points A and B for each circuit in Figure 4–16. (A full-color photo appears between pages 128 and 129.)

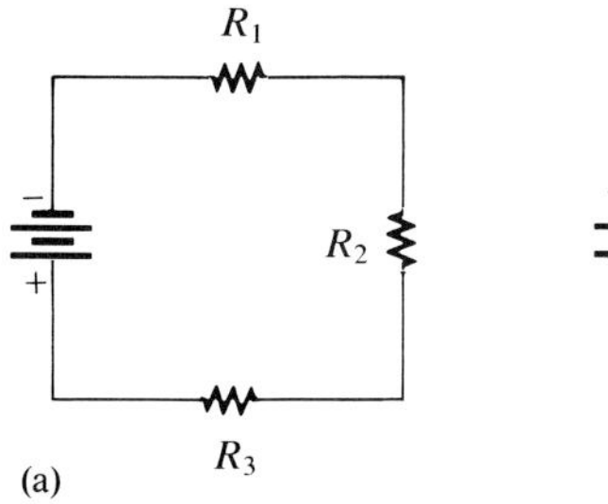

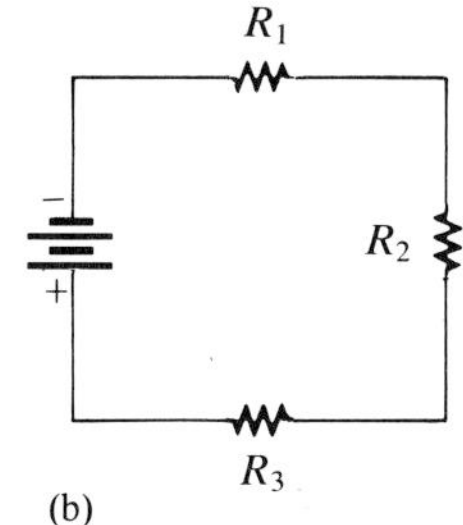

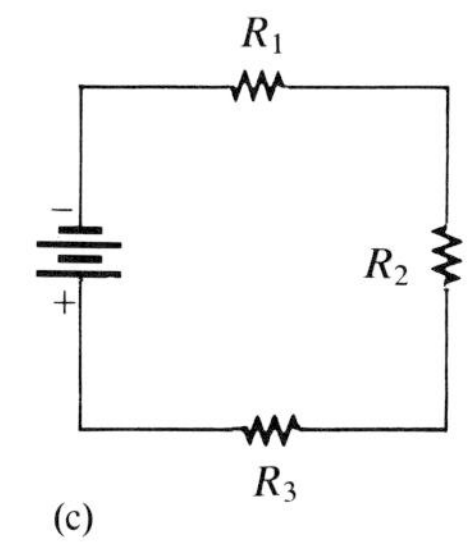

FIGURE 4–16

2. The following resistors are in series: one 100 Ω, two 47 Ω, four 12 Ω, and one 330 Ω. What is the total resistance?
3. Suppose that you have one resistor each of the following values: 1 kΩ, 2.7 kΩ, 3.3 kΩ, and 1.8 kΩ. To get a total resistance of 10 kΩ, you need one more resistor. What should its value be?
4. What is the R_T for twelve 47-Ω resistors in series?

OHM'S LAW IN SERIES CIRCUITS

4–4

In this section, we use Ohm's law and the basic concepts of series circuits to analyze several circuits.

EXAMPLE 4–6 As you have seen, some resistors are not color coded with bands but have the values stamped on the resistor body. When the circuit board shown in Figure 4–17 was assembled, someone mounted the resistors with the labels turned

FIGURE 4–17

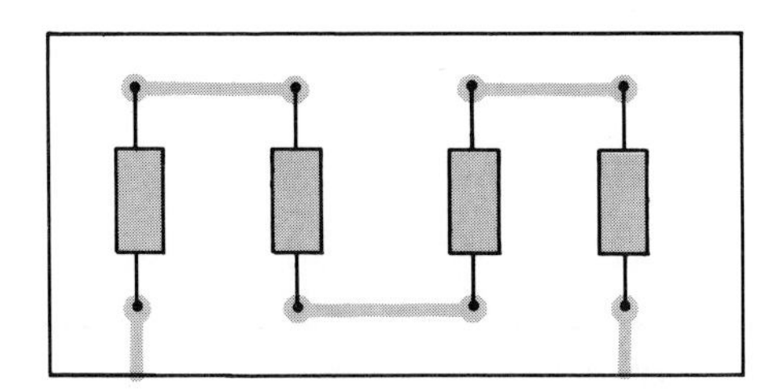

down, and there is no documentation showing the resistor values (inexcusable!). Without removing the resistors from the board, use Ohm's law to determine the resistance of each one.

Solution

The resistors are all in series, so the current is the same through each one. Measure the current by connecting a 12-V source (arbitrary value) and an ammeter as shown in Figure 4–18. Measure the voltage across each resistor by placing the voltmeter across the first resistor. Then repeat this measurement for the other three resistors. For illustration, the values indicated are assumed to be the measured values.

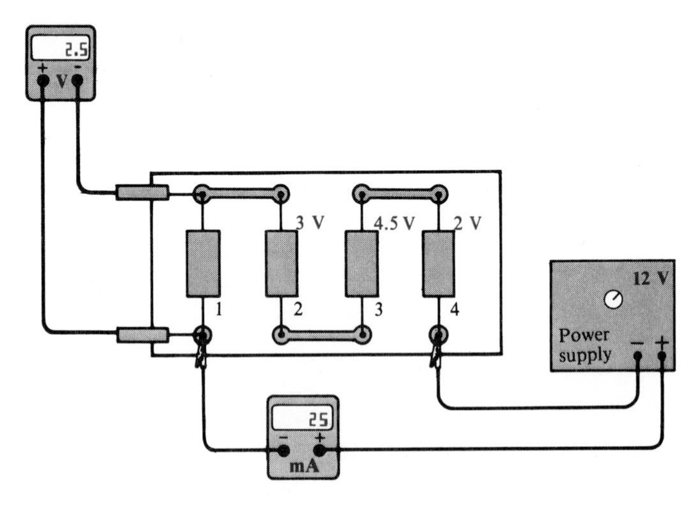

FIGURE 4–18
The voltmeter readings for each resistor are indicated.

Determine the resistance of each resistor by substituting the measured values of current and voltage into the Ohm's law formula as follows:

$$R_1 = \frac{V_1}{I} = \frac{2.5\text{ V}}{25\text{ mA}} = 100\ \Omega$$

$$R_2 = \frac{V_2}{I} = \frac{3\text{ V}}{25\text{ mA}} = 120\ \Omega$$

$$R_3 = \frac{V_3}{I} = \frac{4.5\text{ V}}{25\text{ mA}} = 180\ \Omega$$

$$R_4 = \frac{V_4}{I} = \frac{2\text{ V}}{25\text{ mA}} = 80\ \Omega$$

The calculator sequence for R_1 is

[2] [·] [5] [÷] [2] [5] [EE] [+/−] [3] [=]

EXAMPLE 4–7

Find the current in the circuit of Figure 4–19.

FIGURE 4–19

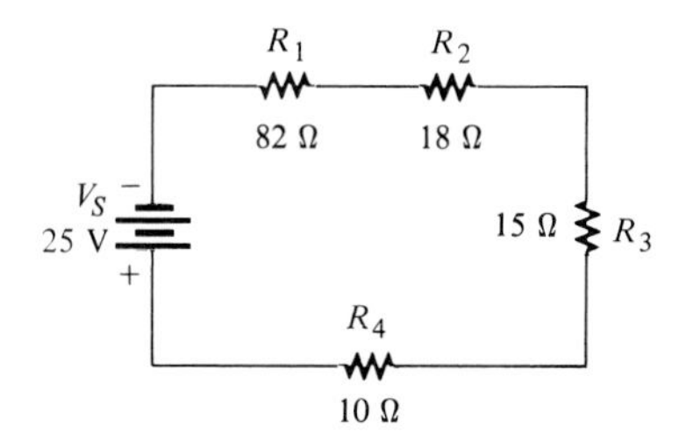

Solution

The current is determined by the voltage and the *total resistance*. First, calculate the total resistance as follows:

$$\begin{aligned} R_T &= R_1 + R_2 + R_3 + R_4 \\ &= 82\ \Omega + 18\ \Omega + 15\ \Omega + 10\ \Omega = 125\ \Omega \end{aligned}$$

Next, using Ohm's law, calculate the current as follows:

$$I = \frac{V_S}{R_T} = \frac{25\ \text{V}}{125\ \Omega} = 0.2\ \text{A}$$

Remember, the *same* current flows at all points in the circuit. Thus, *each* resistor has 0.2 A through it.

EXAMPLE 4–8

In the circuit of Figure 4–20, 1 mA of current flows. For this amount of current to flow, what must the source voltage V_S be?

FIGURE 4–20

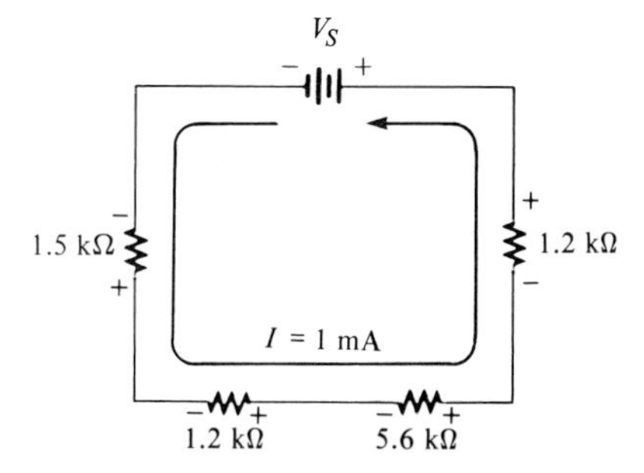

Solution

In order to calculate V_S, we must determine R_T as follows:

$$R_T = 1.2\ \text{k}\Omega + 5.6\ \text{k}\Omega + 1.2\ \text{k}\Omega + 1.5\ \text{k}\Omega = 9.5\ \text{k}\Omega$$

Now use Ohm's law to get V_S:

$$V_S = IR_T = (1\ \text{mA})(9.5\ \text{k}\Omega) = 9.5\ \text{V}$$

EXAMPLE 4–9

Calculate the voltage across each resistor in Figure 4–21, and find the value of V_S.

FIGURE 4–21

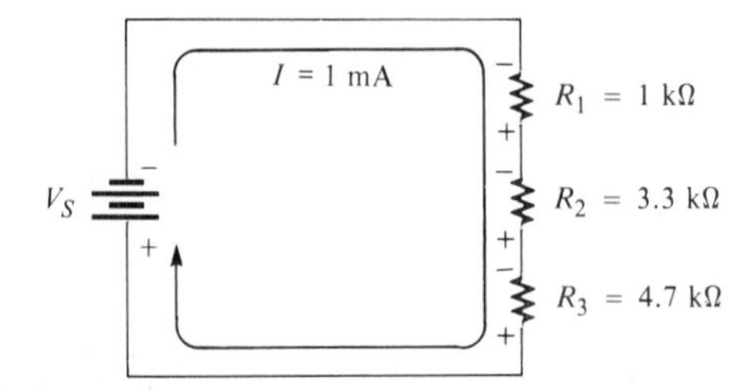

Solution

By Ohm's law, the voltage across each resistor is equal to its resistance multiplied by the current through it. Using the Ohm's law formula $V = IR$, we determine the voltage across each of the resistors. Keep in mind that the same current flows through each series resistor.

The voltage across R_1 is

$$V_1 = IR_1 = (1 \text{ mA})(1 \text{ k}\Omega) = 1 \text{ V}$$

The voltage across R_2 is

$$V_2 = IR_2 = (1 \text{ mA})(3.3 \text{ k}\Omega) = 3.3 \text{ V}$$

The voltage across R_3 is

$$V_3 = IR_3 = (1 \text{ mA})(4.7 \text{ k}\Omega) = 4.7 \text{ V}$$

The source voltage V_S is equal to the current times the *total resistance:*

$$R_T = 1 \text{ k}\Omega + 3.3 \text{ k}\Omega + 4.7 \text{ k}\Omega = 9 \text{ k}\Omega$$

$$V_S = (1 \text{ mA})(9 \text{ k}\Omega) = 9 \text{ V}$$

Notice that if you add the voltage drops of the resistors, they total 9 V, which is the same as the source voltage.

The calculator sequence for V_S is

[1] [EE] [+/−] [3] [×] [9] [EE] [3] [=]

SECTION REVIEW 4–4

1. A 10-V battery is connected across three 100-Ω resistors in series. What is the current through each resistor?
2. How much voltage is required to produce 5 A through the circuit of Figure 4–22?
3. How much voltage is dropped across each resistor in Figure 4–22 with 5 A flowing?
4. Four equal-value resistors are connected in series with a 5-V source. Five milliamperes of current are measured. What is the value of each resistor?

FIGURE 4–22

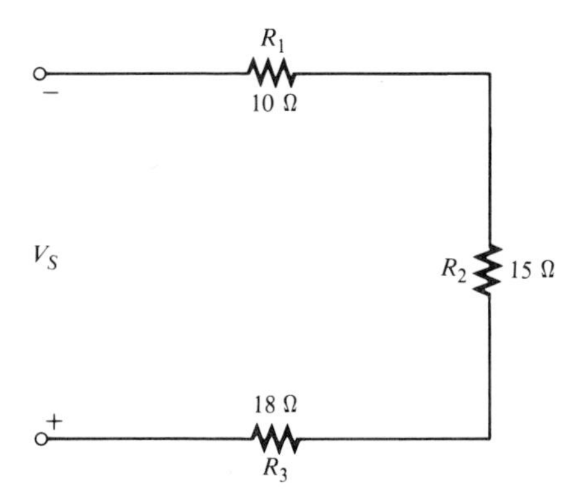

VOLTAGE SOURCES IN SERIES

4–5

When batteries are placed in a flashlight, they are connected in series to produce a larger voltage, as illustrated in Figure 4–23. In this example, three 1.5-V cells are placed in series to produce a total voltage (V_T) as follows:

$$V_T = V_{S1} + V_{S2} + V_{S3} = 1.5\text{ V} + 1.5\text{ V} + 1.5\text{ V} = 4.5\text{ V}$$

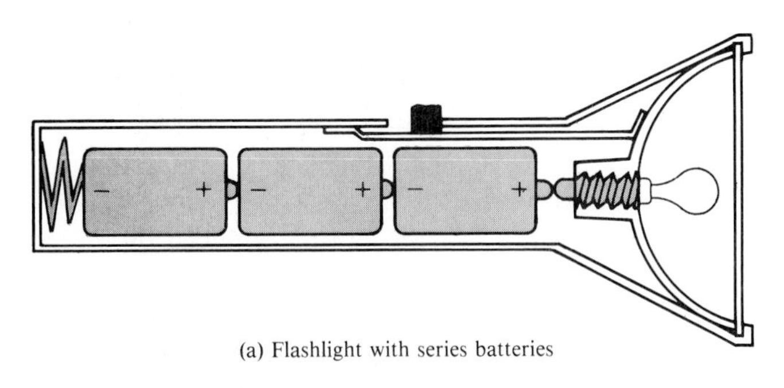

(a) Flashlight with series batteries

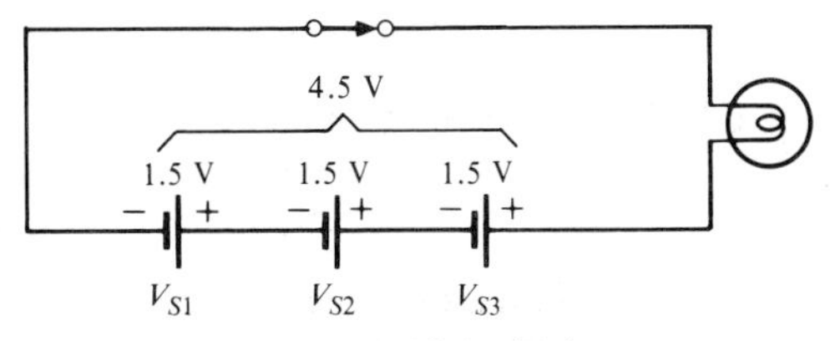

(b) Schematic of flashlight circuit

FIGURE 4–23
Example of series voltage sources.

Series voltage sources are *added* when their polarities are in the same direction and are *subtracted* when their polarities are in opposite directions. For example, if one of the batteries in the flashlight is turned around as indicated in the schematic of Figure 4–24, its voltage subtracts as follows and reduces the total voltage:

$$V_T = V_{S1} - V_{S2} + V_{S3} = 1.5\text{ V} - 1.5\text{ V} + 1.5\text{ V} = 1.5\text{ V}$$

FIGURE 4–24
Opposite polarities subtract.

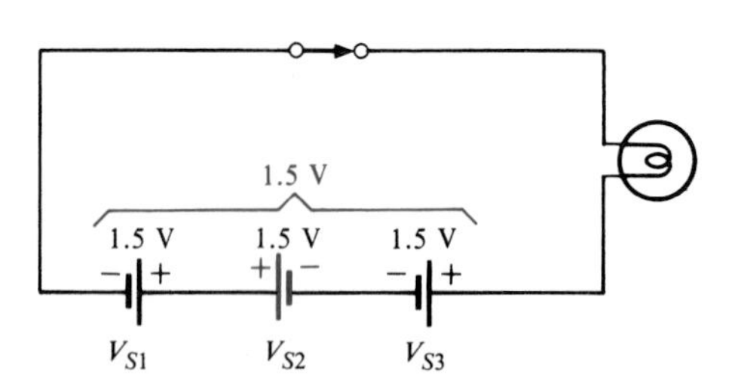

EXAMPLE 4–10 What is the total source voltage (V_T) in Figure 4–25?

FIGURE 4–25

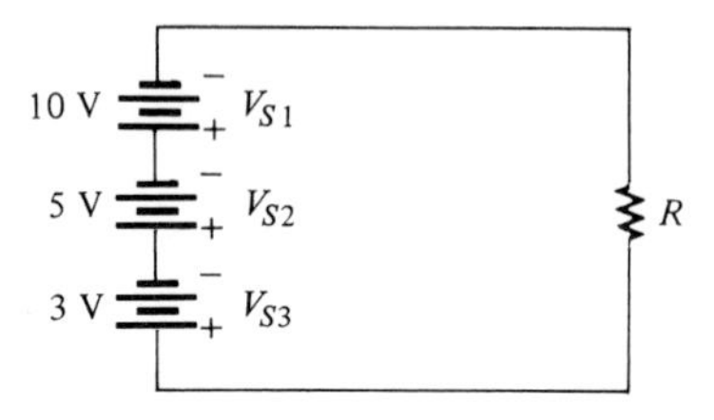

Solution

The polarity of each source is the same (the sources are connected in the same direction in the circuit). So we sum the three voltages to get the total:

$$V_T = V_{S1} + V_{S2} + V_{S3} = 10\text{ V} + 5\text{ V} + 3\text{ V} = 18\text{ V}$$

The three individual sources can be replaced by a single equivalent source of 18 V with its polarity as shown in Figure 4–26.

FIGURE 4–26

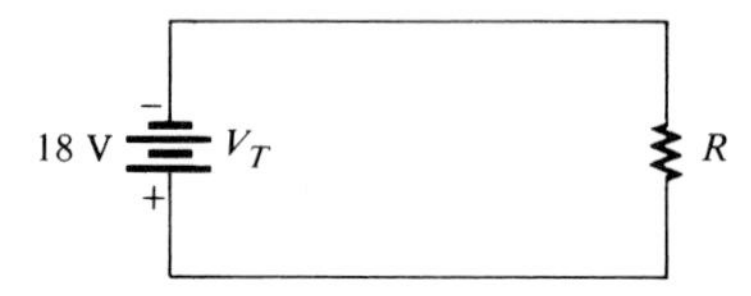

EXAMPLE 4–11 Determine V_T in Figure 4–27.

FIGURE 4–27

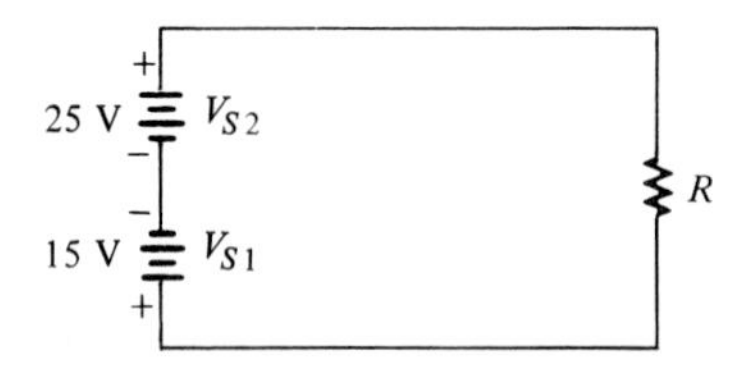

Solution

These sources are connected in *opposing* directions. If you go clockwise around the circuit, you go from plus to minus through V_{S1}, and minus to plus through V_{S2}. The total voltage is the *difference* of the two source voltages. The total voltage has the same polarity as the larger-value source. Here we will choose V_{S2} to be positive:

$$V_T = V_{S2} - V_{S1} = 25\text{ V} - 15\text{ V} = 10\text{ V}$$

The two sources in Figure 4–27 can be replaced by a 10-V equivalent one with polarity as shown in Figure 4–28.

FIGURE 4–28

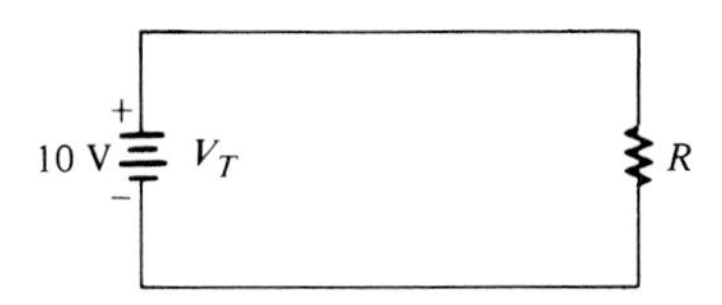

SECTION REVIEW 4–5

1. How many 12-V batteries must be connected in series to produce 60 V? Sketch a schematic that shows the battery connections.
2. The resistive circuit in Figure 4–29 is used to bias a transistor amplifier. Show how to connect two 15-V power supplies in order to get 30 V across the two resistors.
3. Determine the total source voltage in each circuit of Figure 4–30.
4. Sketch the *equivalent single source* circuit for each circuit of Figure 4–30.

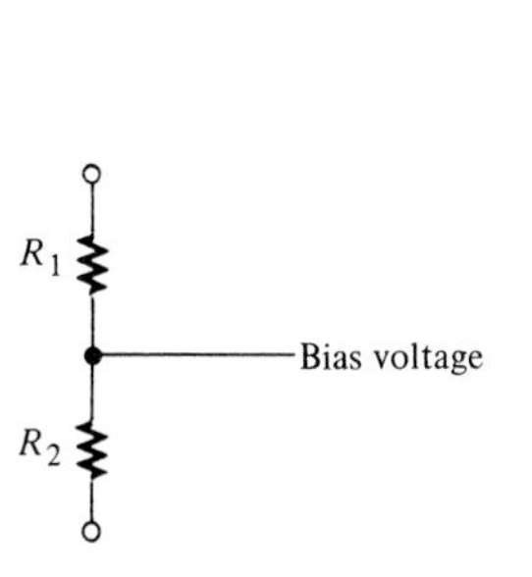

FIGURE 4–29

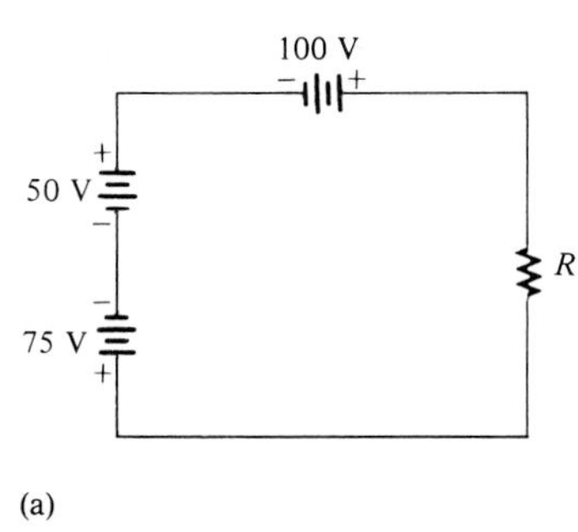

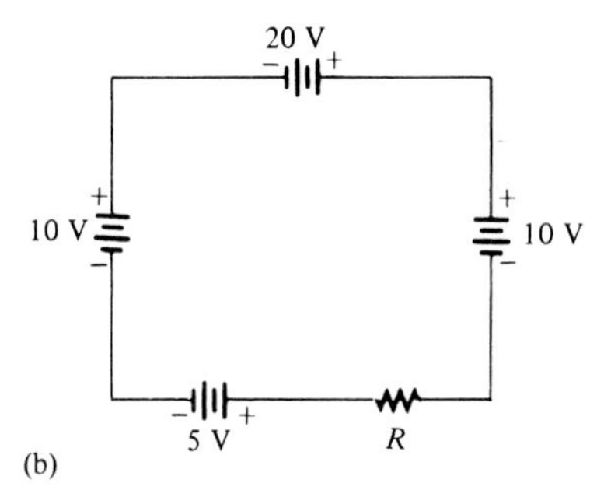

FIGURE 4–30

KIRCHHOFF'S VOLTAGE LAW

4–6

In an electric circuit, the voltages across the resistors (called *voltage drops*) *always* have polarities *opposite* to the source voltage polarity. For example, in Figure 4–31, follow a clockwise loop around the circuit and note that the source polarity is *plus-to-minus* and each voltage drop is *minus-to-plus*.

Also notice in Figure 4–31 that the *current flows out of the negative side of the source* and through the resistors as indicated by the arrows. The *current flows into the negative side of each resistor and out the positive side.* The reason is that as the electrons flow through a resistor, they lose energy and are therefore at a lower energy level when they emerge. The lower energy side is less negative (more positive) than the higher energy side. The drop in energy level across a resistor creates a potential difference, or *voltage drop,* with a minus-to-plus polarity in the direction of the current.

Notice that the voltage from point A to point B in the circuit of Figure 4–31 equals the source voltage, V_S. Also, the voltage from A to B is the *sum* of

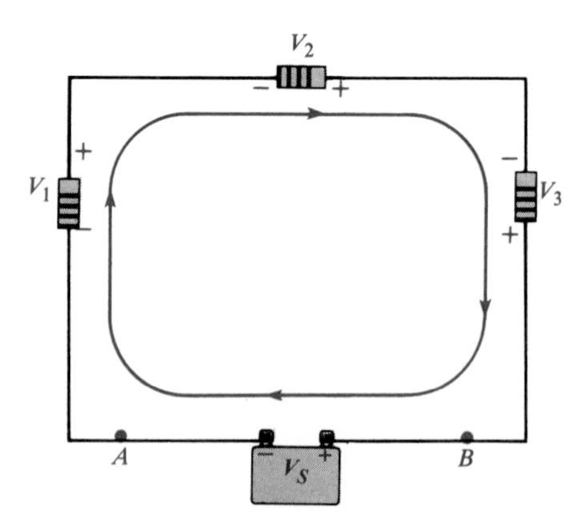

FIGURE 4–31
Kirchhoff's voltage law: The sum of the voltage drops equals the source voltage.

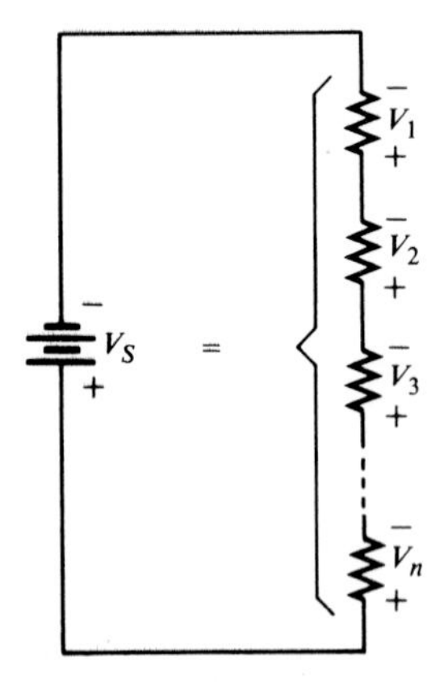

FIGURE 4–32
Sum of n voltage drops equals the source voltage.

the series resistor voltage drops. Therefore, the source voltage is equal to the sum of the three voltage drops.

This discussion is an example of *Kirchhoff's voltage law,* which is generally stated as follows:

The sum of all the voltage drops around a single closed loop in a circuit is equal to the total source voltage in that loop.

The general concept of Kirchhoff's voltage law is illustrated in Figure 4–32 and expressed by Equation (4–3):

$$V_S = V_1 + V_2 + V_3 + \cdots + V_n \qquad (4\text{–}3)$$

where n represents the number of voltage drops.

Another Way to State Kirchhoff's Voltage Law

If all the voltage drops around a closed loop are added and then this total is subtracted from the source voltage, the result is zero. This result occurs because the sum of the voltage drops equals the source voltage.

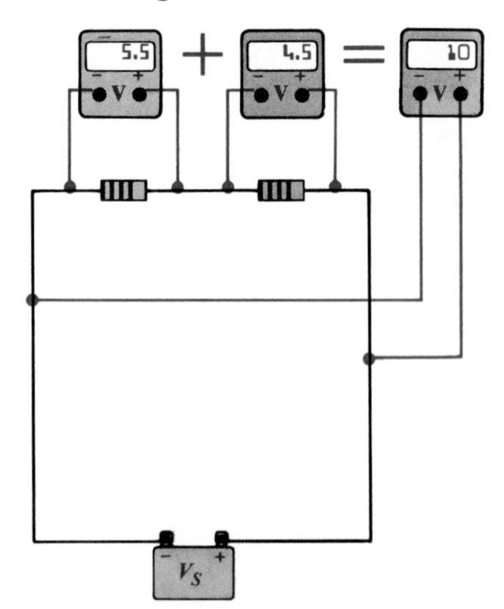

FIGURE 4–33
Experimental verification of Kirchhoff's voltage law.

The sum of all voltages (both source and drops) around a closed path is zero.

You can verify Kirchhoff's voltage law by connecting a circuit and measuring each resistor voltage and the source voltage as illustrated in Figure 4–33. When the resistor voltages are added together, their sum will equal the source voltage. Any number of resistors can be added.

EXAMPLE 4–12 Determine the source voltage V_S in Figure 4–34 where the two voltage drops are given.

FIGURE 4–34

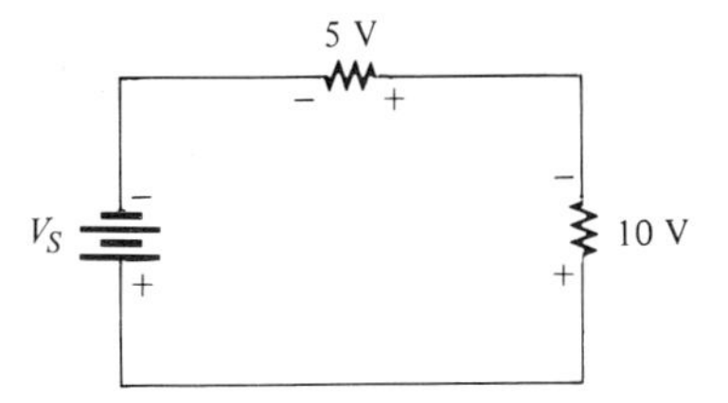

Solution

By Kirchhoff's voltage law, the source voltage (applied voltage) must equal the sum of the voltage drops. Adding the voltage drops gives us the value of the source voltage:

$$V_S = 5\text{ V} + 10\text{ V} = 15\text{ V}$$

EXAMPLE 4–13 Determine the unknown voltage drop, V_3, in Figure 4–35.

FIGURE 4–35

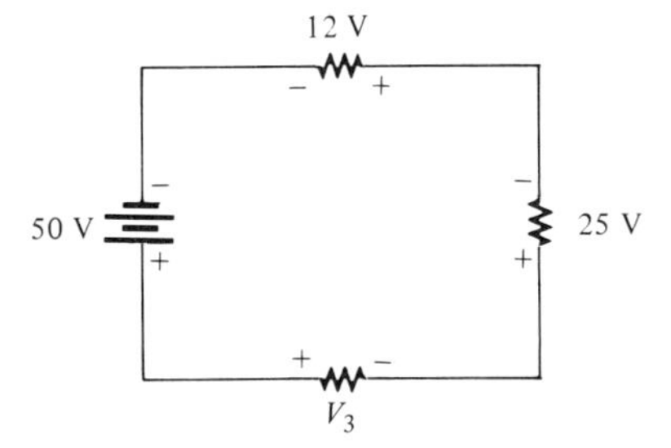

Solution

By Kirchhoff's voltage law, the sum of all the voltages around the circuit is zero (the signs of the voltage drops are opposite the sign of the source).

$$V_S - V_1 - V_2 - V_3 = 0$$

The value of each voltage drop except V_3 is known. Substitute these values into the equation as follows:

$$50\text{ V} - 12\text{ V} - 25\text{ V} - V_3 = 0$$

Next combine the known values:

$$13\text{ V} - V_3 = 0$$

Transpose 13 V to the right side of the equation, and cancel the minus signs:

$$-V_3 = -13\text{ V}$$
$$V_3 = 13\text{ V}$$

The voltage drop across R_3 is 13 V, and its polarity is as shown in Figure 4–35.

EXAMPLE 4–14 Find the value of R_4 in Figure 4–36.

FIGURE 4–36

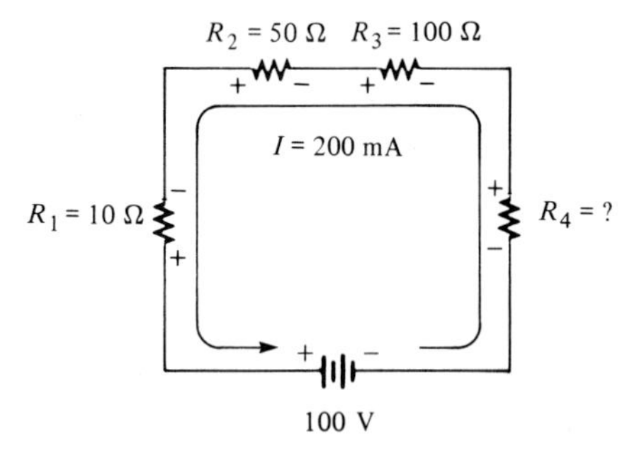

Solution

In this problem we must use both Ohm's law *and* Kirchhoff's voltage law. Follow this procedure carefully.

First use Ohm's law to find the voltage drop across each of the *known* resistors:

$$V_1 = IR_1 = (200\text{ mA})(10\ \Omega) = 2\text{ V}$$
$$V_2 = IR_2 = (200\text{ mA})(50\ \Omega) = 10\text{ V}$$
$$V_3 = IR_3 = (200\text{ mA})(100\ \Omega) = 20\text{ V}$$

Next, use Kirchhoff's voltage law to find V_4, the voltage drop across the *unknown* resistor:

$$V_S - V_1 - V_2 - V_3 - V_4 = 0$$
$$100\text{ V} - 2\text{ V} - 10\text{ V} - 20\text{ V} - V_4 = 0$$
$$68\text{ V} - V_4 = 0$$
$$V_4 = 68\text{ V}$$

Now that we know V_4, we can use Ohm's law to calculate R_4 as follows:

$$R_4 = \frac{V_4}{I} = \frac{68\text{ V}}{200\text{ mA}} = 340\ \Omega$$

SECTION REVIEW 4–6

1. State Kirchhoff's voltage law in two ways.
2. A 50-V source is connected to a series resistive circuit. What is the total of the voltage drops in this circuit?
3. Two equal-value resistors are connected in series across a 10-V battery. What is the voltage drop across each resistor?
4. In a series circuit with a 25-V source, there are three resistors. One voltage drop is 5 V, and the other is 10 V. What is the value of the third voltage drop?

VOLTAGE DIVIDERS

4–7

A series circuit acts as a *voltage divider*. In this section you will see what this term means and why voltage dividers are an important application of series circuits.

To illustrate how a series string of resistors acts as a voltage divider, we will examine Figure 4–37(a), where there are two resistors in series. As you already know, there are two voltage drops: one across R_1 and one across R_2. These voltage drops are V_1 and V_2, respectively, as indicated in the diagram.

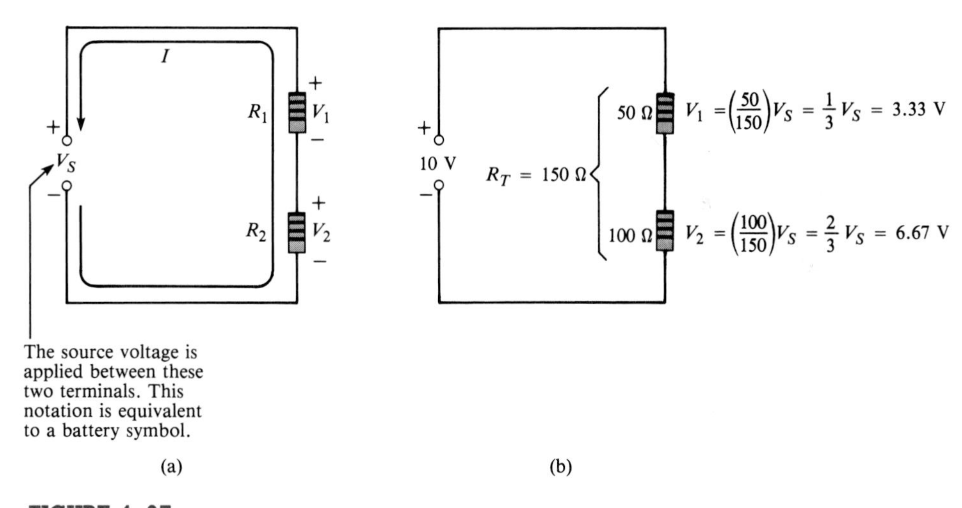

FIGURE 4–37
A two-resistor voltage divider.

Since the same current flows through each resistor, the voltage drops are proportional to the ohmic values of the resistors. For example, if the value of R_2 is twice that of R_1, then the value of V_2 is twice that of V_1. In other words, the total voltage drop *divides* among the series resistors in amounts directly proportional to the resistance values. The smallest resistance has the least voltage, and the largest resistance has the most voltage ($V = IR$). For example, in Figure 4–37(b), if V_S is 10 V, R_1 is 50 Ω, and R_2 is 100 Ω, then V_1 is one-third the total voltage, or 3.33 V, because R_1 is one-third the *total* resistance. Likewise, V_2 is two-thirds V_S, or 6.67 V, because R_2 is two-thirds the total resistance.

Voltage Divider Formula

Although you can use Ohm's law to determine the voltage drop across any resistor in a series circuit, you must first know the current. It is often easier to use resistance ratios to find the voltage across a particular resistor as was demonstrated in Figure 4–37(b). The general voltage divider formula is

$$V_x = \left(\frac{R_x}{R_T}\right)V_S \tag{4–4}$$

where V_x is the voltage across any resistor or combination of resistors, R_x is the resistance across which the voltage is to be determined, and R_T is the total series resistance. For example, in Figure 4–38, $V_1 = (R_1/R_T)V_S$, $V_2 = (R_2/R_T)V_S$, $V_3 = (R_3/R_T)V_S$, and so on.

FIGURE 4–38
Five-resistor voltage divider.

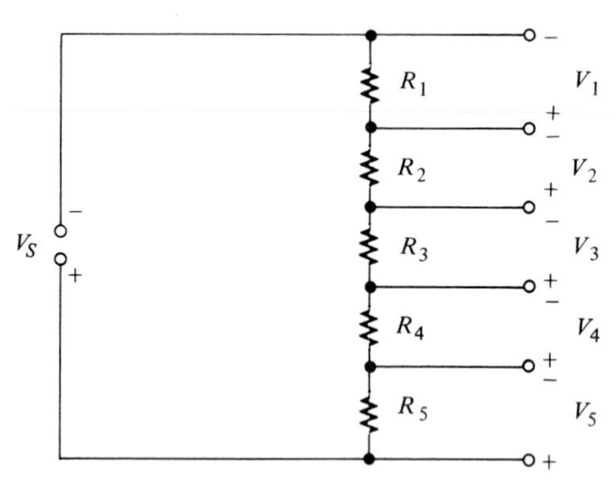

The following general statement about voltage dividers applies to series resistive circuits and describes Equation (4–4):

> **The voltage drop across any resistor or combination of resistors in a series circuit is equal to the ratio of that resistance value to the total resistance, multiplied by the source voltage.**

The following examples illustrate how the voltage divider formula is used.

EXAMPLE 4–15

Determine the voltage across R_1 and the voltage across R_2 in the voltage divider in Figure 4–39.

FIGURE 4–39

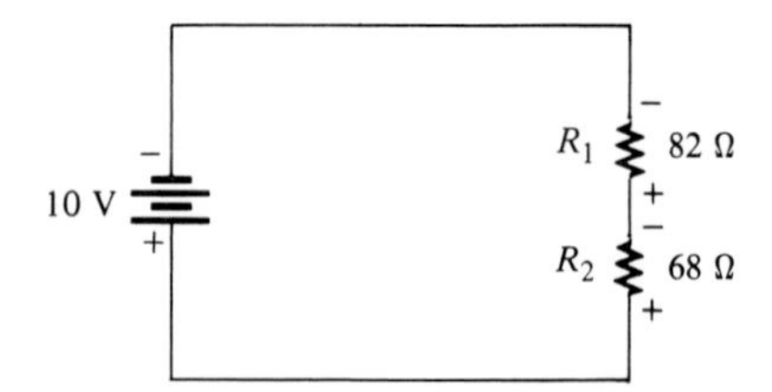

Solution
Use the voltage divider formula, $V_x = (R_x/R_T)V_S$. In this problem we are looking for V_1; so $V_x = V_1$ and $R_x = R_1$. The total resistance is

$$R_T = R_1 + R_2 = 82\ \Omega + 68\ \Omega = 150\ \Omega$$

R_1 is 82 Ω and V_S is 10 V. Substituting these values into the voltage divider formula, we have

$$V_1 = \left(\frac{R_1}{R_T}\right)V_S = \left(\frac{82\ \Omega}{150\ \Omega}\right)10\ \text{V} = 5.47\ \text{V}$$

There are two ways to find the value of V_2 in this problem: Kirchhoff's voltage law or the voltage divider formula.

First, using Kirchhoff's voltage law, we know that $V_S = V_1 + V_2$. By substituting the values for V_S and V_1, we can solve for V_2 as follows:

$$V_2 = 10\ \text{V} - 5.47\ \text{V} = 4.53\ \text{V}$$

A second way is to use the voltage divider formula to find V_2 as follows:

$$V_2 = \left(\frac{R_2}{R_T}\right)V_S = \left(\frac{68\ \Omega}{150\ \Omega}\right)10\ \text{V} = 4.53\ \text{V}$$

We get the same result either way.

The calculator sequence for V_2 is
[6] [8] [÷] [1] [5] [0] [=] [×] [1] [0] [=]

EXAMPLE 4–16 Calculate the voltage drop across each resistor in the voltage divider of Figure 4–40.

FIGURE 4–40

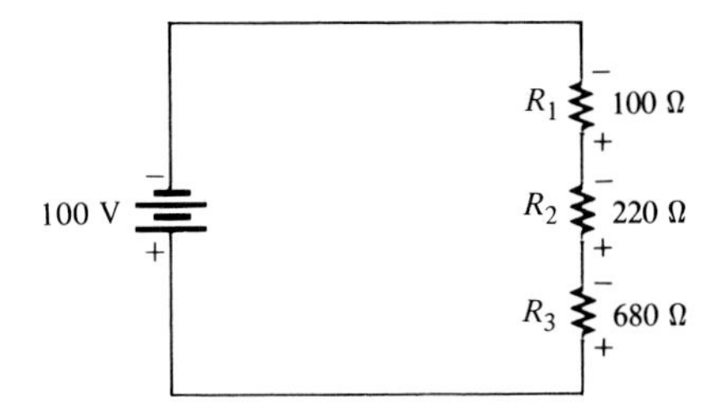

Solution

Look at the circuit for a moment and consider the following: The total resistance is 1000 Ω. We can examine the circuit and determine that 10% of the total voltage is across R_1 because it is 10% of the total resistance (100 Ω is 10% of 1000 Ω). Likewise, we see that 22% of the total voltage is dropped across R_2 because it is 22% of the total resistance (220 Ω is 22% of 1000 Ω). Finally, R_3 drops 68% of the total voltage (680 Ω is 68% of 1000 Ω).

Because of the convenient values in this problem, it is easy to figure the voltages mentally. Such is not always the case, but sometimes a little thinking will produce a result more efficiently and eliminate some calculating.

Although we have already reasoned through this problem, the calculations will verify our results:

$$V_1 = \left(\frac{R_1}{R_T}\right)V_S = \left(\frac{100\ \Omega}{1000\ \Omega}\right)100\ \text{V} = 10\ \text{V}$$

$$V_2 = \left(\frac{R_2}{R_T}\right)V_S = \left(\frac{220\ \Omega}{1000\ \Omega}\right)100\ \text{V} = 22\ \text{V}$$

$$V_3 = \left(\frac{R_3}{R_T}\right)V_S = \left(\frac{680\ \Omega}{1000\ \Omega}\right)100\ \text{V} = 68\ \text{V}$$

Notice that the sum of the voltage drops is equal to the source voltage, in accordance with Kirchhoff's voltage law. This check is a good way to verify your results.

The Potentiometer as an Adjustable Voltage Divider

Recall from Chapter 2 that a potentiometer is a variable resistor with three terminals. A potentiometer connected to a voltage source is shown in Figure 4–41. Notice that the two end terminals are labeled 1 and 2. The adjustable terminal or wiper is labeled 3. The potentiometer acts as a voltage divider. We can illustrate this concept better by separating the total resistance into two parts, as shown in Figure 4–41(c). The resistance between terminal 1 and terminal 3 (R_{13}) is one part, and the resistance between terminal 3 and terminal 2 (R_{32}) is the other part. So this potentiometer actually is a two-resistor voltage divider that can be manually adjusted.

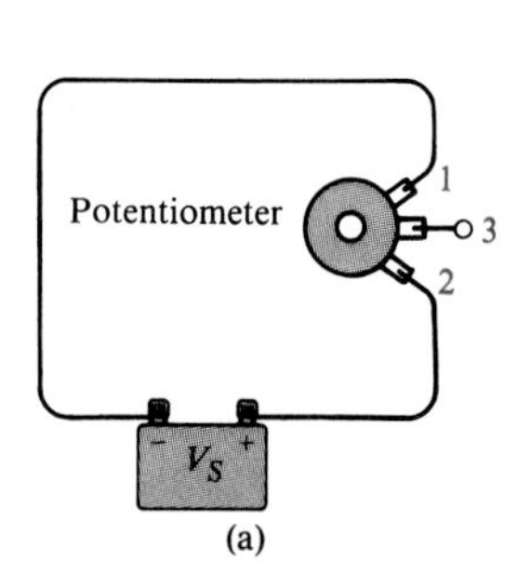

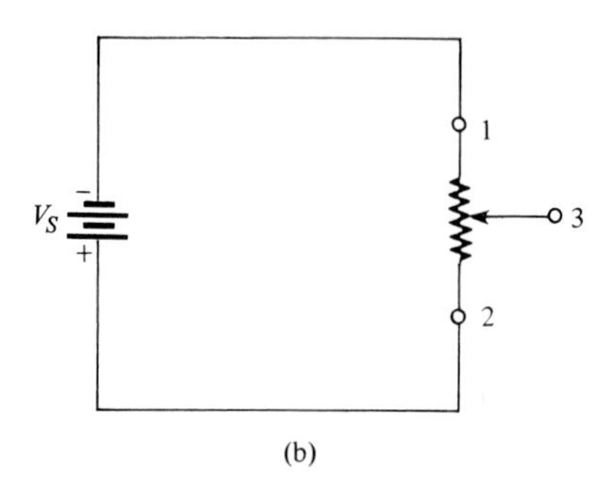

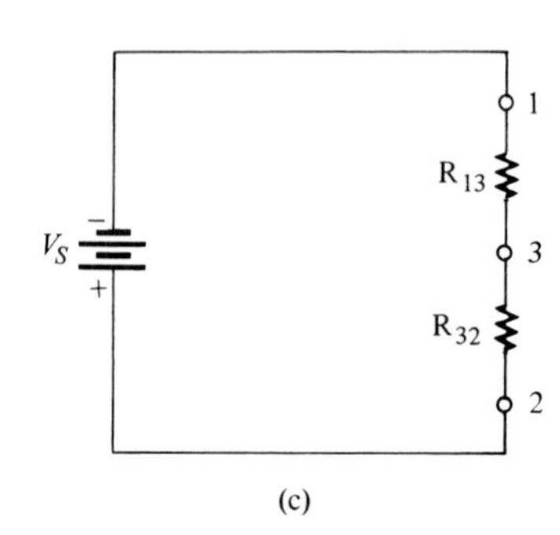

FIGURE 4–41
The potentiometer as a voltage divider.

Figure 4–42 shows what happens when the wiper terminal (3) is moved. In Part (a), the wiper is exactly centered, making the two resistances equal. If we measure the voltage across terminals 3 to 2 as indicated by the voltmeter symbol, we have one-half of the total source voltage. When the wiper is moved up, as in Part (b), the resistance between terminals 3 and 2 increases, and the voltage across it increases proportionally. When the wiper is moved down, as in

Part (c), the resistance between terminals 3 and 2 decreases, and the voltage decreases proportionally.

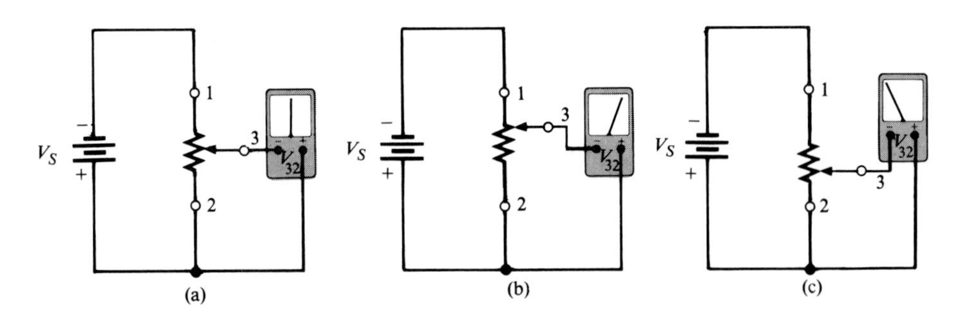

FIGURE 4–42
Adjusting the voltage divider.

Applications of Voltage Dividers

The volume control of radio or TV receivers is a common application of a potentiometer used as a voltage divider. Since the loudness of the sound is dependent on the amount of voltage associated with the audio signal, you can increase or decrease the volume by adjusting the potentiometer, that is, by turning the knob of the volume control on the set. The block diagram in Figure 4–43 shows how a potentiometer can be used for volume control in a typical receiver.

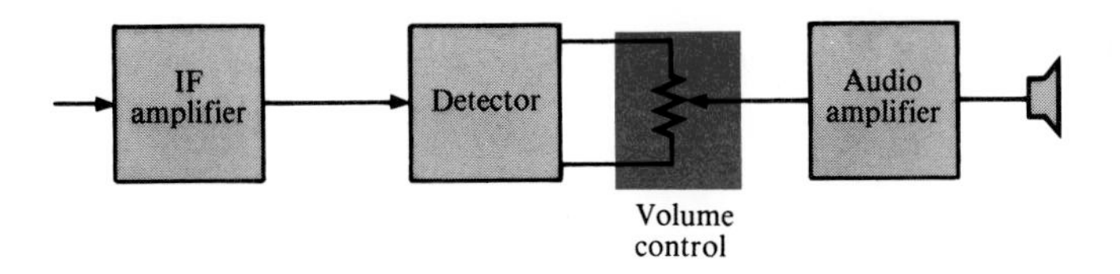

FIGURE 4–43
A voltage divider used for volume control.

Another application of a voltage divider is illustrated in Figure 4–44, which depicts a potentiometer voltage divider as a fuel-level sensor in an automobile gas tank. As shown in Part (a), the float moves up as the tank is filled and moves down as the tank empties. The float is mechanically linked to the wiper arm of a potentiometer, as shown in Part (b). The output voltage varies proportionally with the position of the wiper arm. As the fuel in the tank decreases, the sensor output voltage also decreases. The output voltage goes to the indicator circuitry, which controls the fuel gauge or digital readout to show the fuel level. The schematic of this system is shown in Part (c).

Still another application for voltage dividers is in setting the dc operating voltage *(bias)* in transistor amplifiers. Figure 4–45 shows a voltage divider used for this purpose. You will study transistor amplifiers and biasing later, so it is important that you understand the basics of voltage dividers at this point.

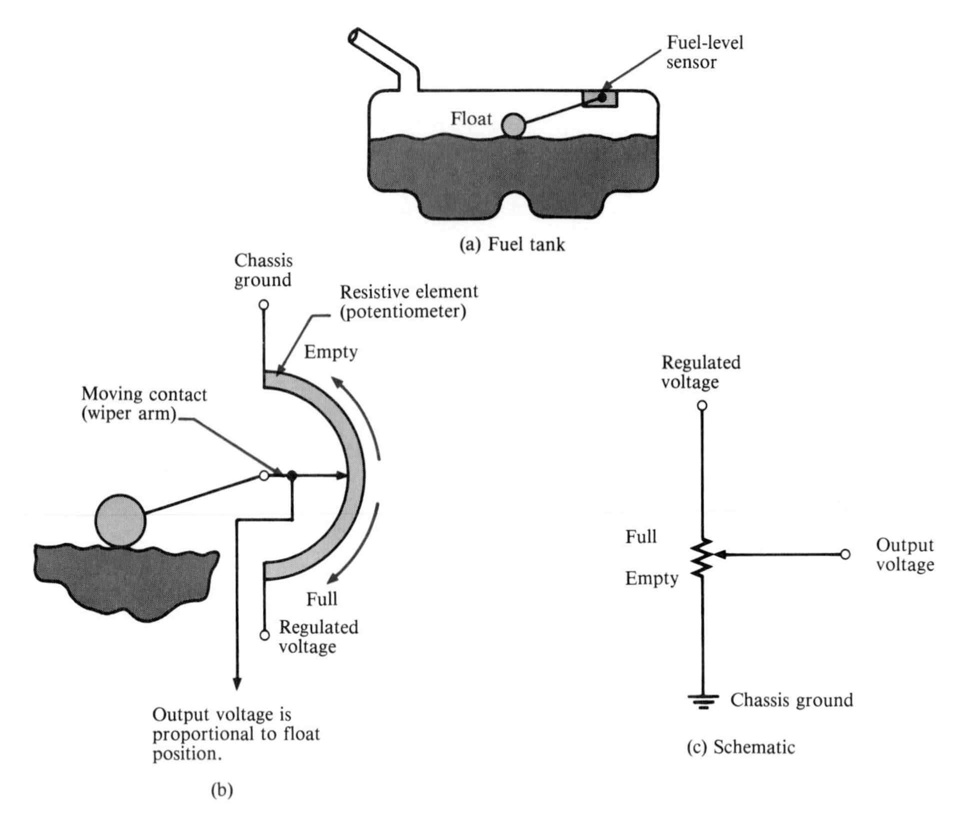

FIGURE 4–44
A potentiometer voltage divider used as an automotive fuel-level sensor.

FIGURE 4–45
The voltage divider as a bias circuit for a transistor amplifier.

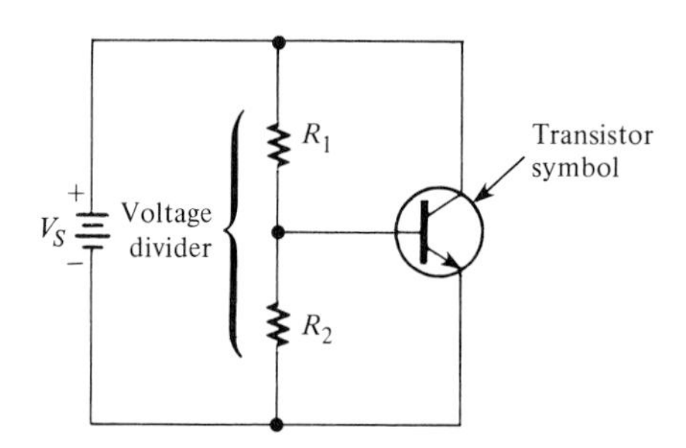

These examples are only three out of an almost endless number of applications of voltage dividers.

SECTION REVIEW 4–7

1. If two series resistors of equal value are connected across a 20-V source, how much voltage is there across each resistor?
2. A 56-Ω resistor and a 82-Ω resistor are connected as a voltage divider. The source voltage is 100 V. Sketch the circuit, and determine the voltage across each of the resistors.

3. The circuit of Figure 4–46 is an adjustable voltage divider. If the potentiometer is linear, where would you set the wiper in order to get 5 V from B to A and 5 V from C to B?

FIGURE 4–46

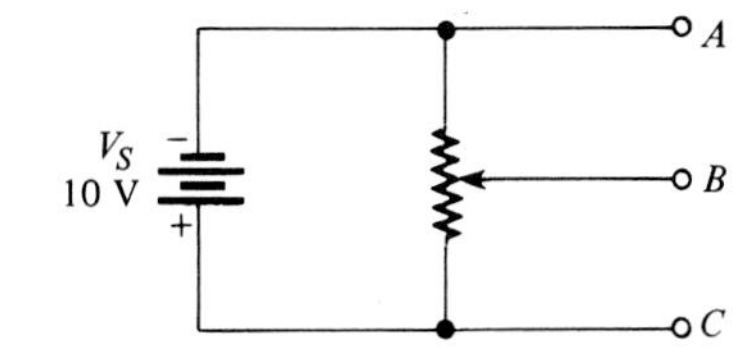

POWER IN A SERIES CIRCUIT

4–8

The total amount of power in a series resistive circuit is equal to the sum of the powers in each resistor in series:

$$P_T = P_1 + P_2 + P_3 + \cdots + P_n \qquad (4\text{–}5)$$

where n is the number of resistors in series, P_T is the total power, and P_n is the power in the last resistor in series. In other words, the powers are additive.

The power formulas that you learned in Chapter 3 are, of course, directly applicable to series circuits. Since the same current flows through each resistor in series, the following formulas are used to calculate the total power:

$$P_T = V_S I$$
$$P_T = I^2 R_T$$
$$P_T = \frac{V_S^2}{R_T}$$

where V_S is the total source voltage across the series circuit and R_T is the total resistance. Example 4–17 illustrates how to calculate total power in a series circuit.

EXAMPLE 4–17 Determine the total amount of power in the series circuit in Figure 4–47.

FIGURE 4–47

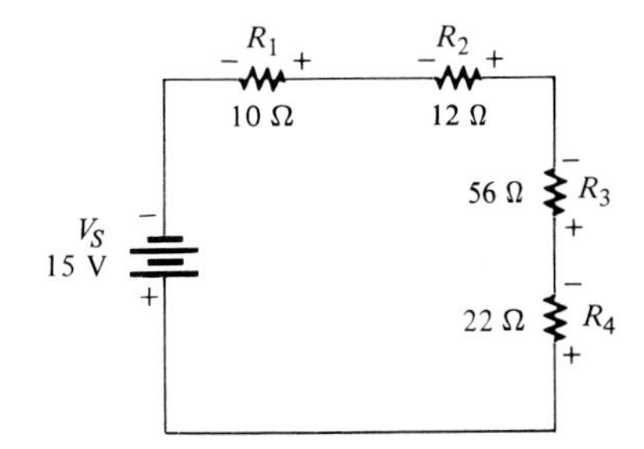

Solution
We know that the source voltage is 15 V. The total resistance is

$$R_T = 10\ \Omega + 12\ \Omega + 56\ \Omega + 22\ \Omega = 100\ \Omega$$

The easiest formula to use is $P_T = V_S^2/R_T$ since we know both V_S and R_T:

$$P_T = \frac{V_S^2}{R_T} = \frac{(15\ \text{V})^2}{100\ \Omega} = \frac{225\ \text{V}^2}{100\ \Omega} = 2.25\ \text{W}$$

If the power of each resistor is determined separately and all of these powers are added, the same result is obtained. We will work through another calculation to illustrate.

First, find the current as follows:

$$I = \frac{V_S}{R_T} = \frac{15\ \text{V}}{100\ \Omega} = 0.15\ \text{A}$$

Next, calculate the power for each resistor using $P = I^2R$:

$$P_1 = (0.15\ \text{A})^2(10\ \Omega) = 0.225\ \text{W}$$
$$P_2 = (0.15\ \text{A})^2(12\ \Omega) = 0.270\ \text{W}$$
$$P_3 = (0.15\ \text{A})^2(56\ \Omega) = 1.260\ \text{W}$$
$$P_4 = (0.15\ \text{A})^2(22\ \Omega) = 0.495\ \text{W}$$

Now, add these powers to get the total power:

$$P_T = 0.225\ \text{W} + 0.270\ \text{W} + 1.260\ \text{W} + 0.495\ \text{W}$$
$$= 2.25\ \text{W}$$

This result shows that the sum of the individual powers is equal to the total power as determined by one of the power formulas.

SECTION REVIEW 4–8

1. If you know the power in each resistor in a series circuit, how can you find the total power?
2. The resistors in a series circuit have the following powers: 2 W, 5 W, 1 W, and 8 W. What is the total power in the circuit?
3. A circuit has a 100-Ω, a 330-Ω, and a 680-Ω resistor in series. A current of 1 A flows through the circuit. What is the total power?

TROUBLES IN SERIES CIRCUITS

4–9 Open Circuit

The most common failure in a series circuit is an *open*. For example, when a resistor or a lamp burns out, it creates a break in the current path, as illustrated in Figure 4–48.

An open in a series circuit prevents any current from flowing.

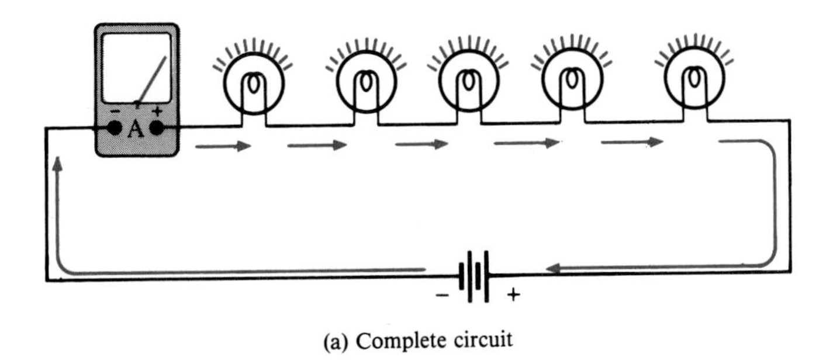

(a) Complete circuit

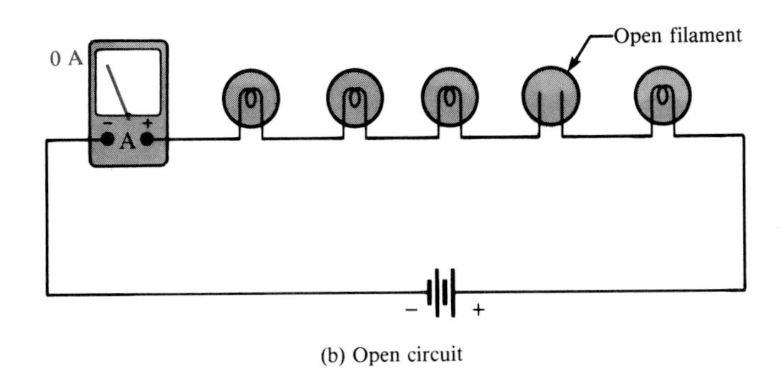

(b) Open circuit

FIGURE 4–48
An open circuit prevents current.

How to Check for an Open Element

Sometimes a visual check will reveal a charred resistor or an open lamp filament. However, it is possible for a resistor to open without showing visible signs of damage. In this situation, a *voltage check* of the series circuit is required. The general procedure is as follows: Measure the voltage across each resistor in series. *The voltage across all of the good resistors will be zero. The voltage across the open resistor will equal the total voltage across the series combination.*

The above condition occurs because an open resistor will prevent current through the series circuit. With no current, there can be no voltage drop across any of the good resistors. Since $IR = 0$, in accordance with *Ohm's law,* the voltage on each side of a good resistor is the same. The total voltage must then appear across the open resistor in accordance with *Kirchhoff's voltage law,* as illustrated in Figure 4–49. To fix the circuit, replace the open resistor.

Short Circuit

Sometimes a short occurs when two conductors touch or a foreign object such as solder or a wire clipping accidentally connects two sections of a circuit together. This situation is particularly common in circuits with a high component density. Several potential causes of short circuits are illustrated on the PC board in Figure 4–50.

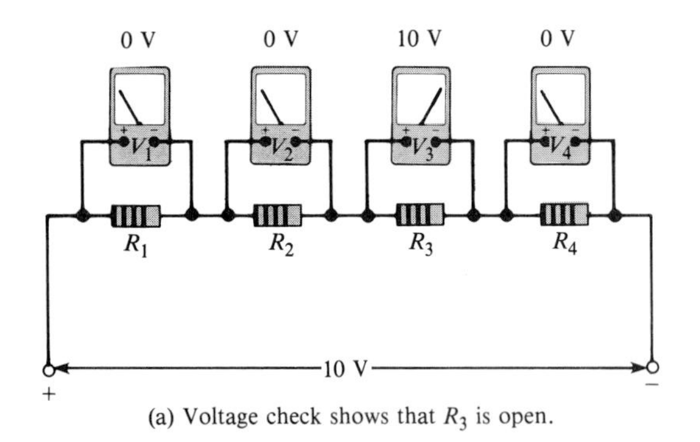

(a) Voltage check shows that R_3 is open.

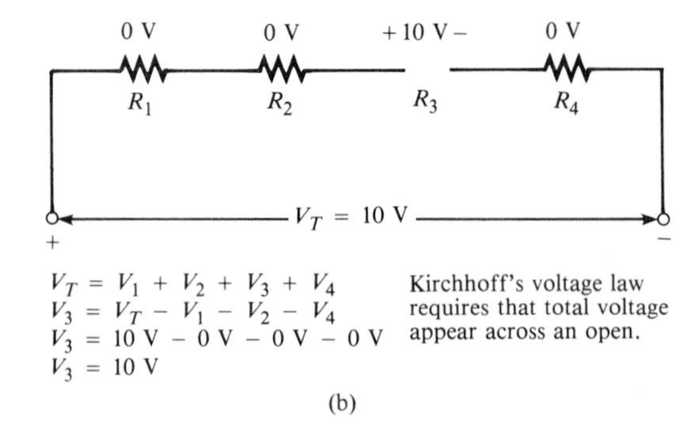

$V_T = V_1 + V_2 + V_3 + V_4$
$V_3 = V_T - V_1 - V_2 - V_4$
$V_3 = 10\text{ V} - 0\text{ V} - 0\text{ V} - 0\text{ V}$
$V_3 = 10\text{ V}$

Kirchhoff's voltage law requires that total voltage appear across an open.

(b)

FIGURE 4–49
Troubleshooting a series circuit for an open element.

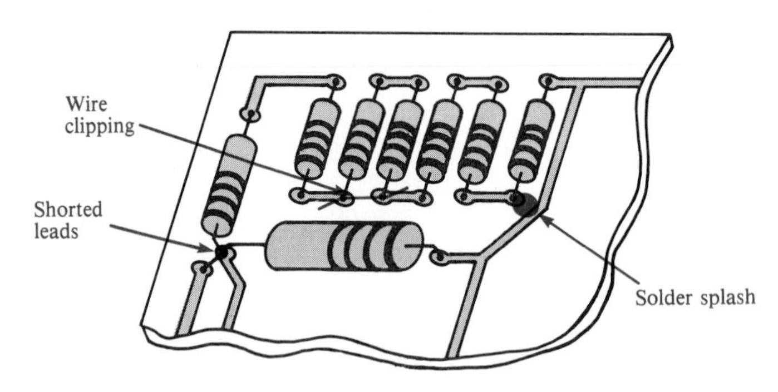

FIGURE 4–50
Examples of shorts on a PC board.

When there is a short, a portion of the series resistance is bypassed (all of the current goes through the short), thus reducing the total resistance as illustrated in Figure 4–51. Notice that the current increases as a result of the short.

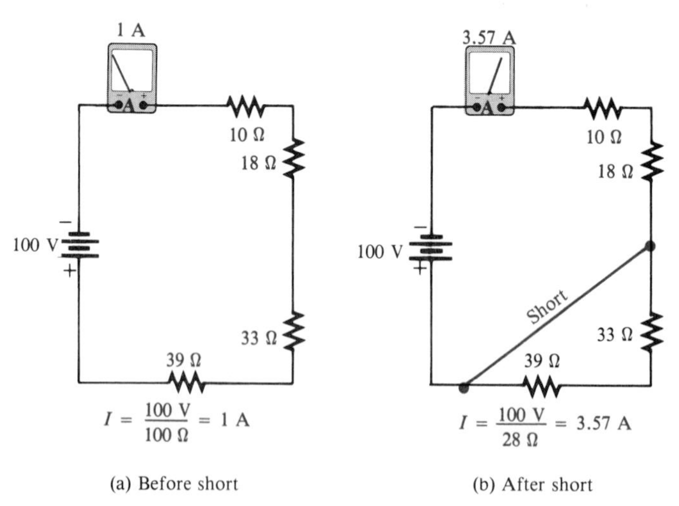

FIGURE 4–51
The effect of a short in a series circuit.

SECTION REVIEW 4–9

1. When a resistor fails, it will normally open (T or F).
2. The total voltage across a string of series resistors is 24 V. If one of the resistors is open, how much voltage is there across it? How much is there across each of the good resistors?

APPLICATION NOTE

The multiple-voltage circuit specified in the Application Assignment at the beginning of the chapter can be implemented with a voltage divider and a rotary switch, as shown in Figure 4–52(a). A suggested case configuration is shown in Part (b). All voltages are with respect to the negative side of the source.

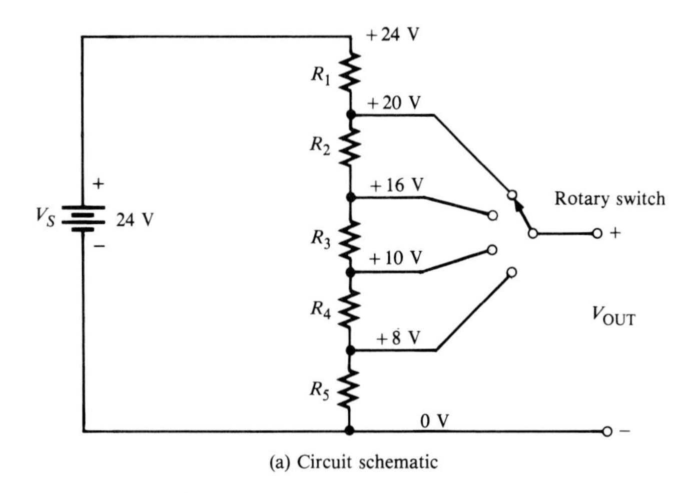

(a) Circuit schematic

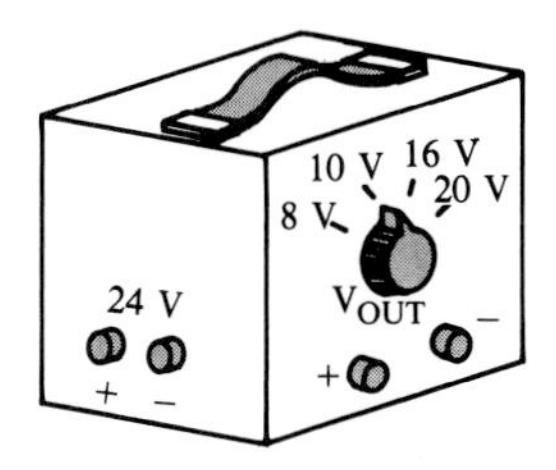

(b) Case configuration

FIGURE 4–52

To keep the current at 1 A, the total resistance must be

$$R_T = \frac{V_S}{I_T} = \frac{24\text{ V}}{1\text{ A}} = 24\ \Omega$$

Determine each resistor value by using the voltage divider formula, Equation (4–4), and noting that the ratio of resistances is equal to the ratio of voltages:

$$\frac{R_x}{R_T} = \frac{V_x}{V_S}$$

From this equation, you can solve for R_x as follows:

$$R_x = \left(\frac{V_x}{V_S}\right)R_T$$

There is 4 V across R_1 (24 V − 20 V):

$$R_1 = \left(\frac{4\ \text{V}}{24\ \text{V}}\right)24\ \Omega = 4\ \Omega$$

There is 4 V across R_2 (20 V − 16 V):

$$R_2 = \left(\frac{4\ \text{V}}{24\ \text{V}}\right)24\ \Omega = 4\ \Omega$$

There is 6 V across R_3 (16 V − 10 V):

$$R_3 = \left(\frac{6\ \text{V}}{24\ \text{V}}\right)24\ \Omega = 6\ \Omega$$

There is 2 V across R_4 (10 V − 8 V):

$$R_4 = \left(\frac{2\ \text{V}}{24\ \text{V}}\right)24\ \Omega = 2\ \Omega$$

There is 8 V across R_5 (8 V − 0 V):

$$R_5 = \left(\frac{8\ \text{V}}{24\ \text{V}}\right)24\ \Omega = 8\ \Omega$$

In order to specify the power rating of each resistor, determine the actual power dissipation as follows:

$$P_1 = P_2 = I^2R_1 = (1\ \text{A})^2(4\ \Omega) = 4\ \text{W}$$
$$P_3 = I^2R_3 = (1\ \text{A})^2(6\ \Omega) = 6\ \text{W}$$
$$P_4 = I^2R_4 = (1\ \text{A})^2(2\ \Omega) = 2\ \text{W}$$
$$P_5 = I^2R_5 = (1\ \text{A})^2(8\ \Omega) = 8\ \text{W}$$

You may have to use wirewound resistors to get the low resistance values and the required power ratings. You may have to place several resistors in series to achieve a given value. For example, use two 1-Ω resistors to get 2 Ω, and so on. The actual power ratings must be greater than the calculated power dissipations.

SUMMARY

Facts

- A series connection provides only one path for current.
- The same amount of current flows at all points in a series circuit.
- The total series resistance is the sum of all resistors in the series circuit.
- Series voltage sources add when their polarities are the same and subtract when their polarities are opposite.

- The voltage across a resistor is called a *voltage drop*.
- The polarity of a voltage drop is opposite to that of the source voltage.
- The voltage across each series resistor equals the ratio of the resistance value to the total resistance, multiplied by the source voltage.
- The highest resistance in series has the greatest voltage drop, and the lowest resistance has the least voltage drop.
- Total power dissipation is the sum of powers dissipated by the individual resistors.
- There is no current in an open circuit.
- All of the voltage appears across an open series element and none across the other elements.
- A short circuit reduces the total resistance.

Definitions

- *Kirchhoff's voltage law (KVL)*—a law which states that (1) the sum of the voltage drops around a closed loop equals the source voltage or (2) the sum of all the voltages (drops and sources) around a closed loop is zero.

Formulas

$$R_T = R_1 + R_2 + R_3 + \cdots + R_n \quad \text{Total resistance of } n \text{ resistors in series} \qquad (4\text{–}1)$$

$$R_T = nR \quad \text{Total resistance of } n \text{ equal-value resistors in series} \qquad (4\text{–}2)$$

$$V_S = V_1 + V_2 + V_3 + \cdots + V_n \quad \text{Kirchhoff's voltage law} \qquad (4\text{–}3)$$

$$V_x = \left(\frac{R_x}{R_T}\right)V_S \quad \text{Voltage divider formula} \qquad (4\text{–}4)$$

$$P_T = P_1 + P_2 + P_3 + \cdots + P_n \quad \text{Total power} \qquad (4\text{–}5)$$

SELF-TEST

1. Sketch five resistors connected in series.
2. Suppose that the five series resistors in Question 1 are connected to a voltage source and a current of 2 A flows into the first resistor. How much current flows out of the third resistor? The fifth resistor?
3. Show how to place an ammeter in the circuit in Question 2 to measure the current.
4. What happens to the total resistance of a given series circuit when additional resistors are connected into the circuit?
5. A certain series circuit has a 100-Ω, a 270-Ω, and a 330-Ω resistor in series. If the 270-Ω resistor is removed, will the current increase or decrease? Why?
6. Show how to connect two 12-V automobile batteries to get 24 V.
7. Suppose you put four 1.5-V batteries in a flashlight, but you accidentally put one of them in backwards. Will the light be brighter or dimmer than it should be? Why?
8. If you measure all the voltage drops and the source voltage in a series circuit and add them together, taking into consideration the polarities, what result will you get? What circuit law does this result verify?
9. There are six resistors in a given series circuit, and each resistor has 5 V dropped across it. What is the source voltage?

10. A series circuit consists of a 4.7-kΩ resistor, a 5.6-kΩ resistor, and a 10-kΩ resistor. Which resistor has the most voltage across it?
11. Which dissipates more power when connected across a 100-V source, one 100-Ω resistor or two 100-Ω resistors in series?
12. The total power in a given series circuit is 10 W. There are five equal-value resistors in the circuit. How much power does each resistor dissipate?
13. When you connect an ammeter in a given series circuit and turn on the source voltage, the meter reads zero. What should you check for, and how will you know when you have found the problem?
14. You are checking out a series circuit and find that the current is higher than it should be. What type of problem should you look for?

PROBLEM SET A

Section 4–1

4–1 Connect each set of resistors in Figure 4–53 in series between points A and B.

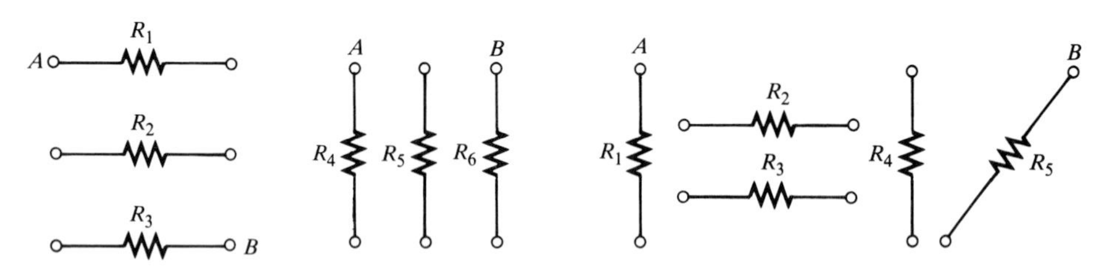

FIGURE 4–53

4–2 Determine which resistors in Figure 4–54 are in series. Show how to interconnect the pins to put all the resistors in series.

FIGURE 4–54

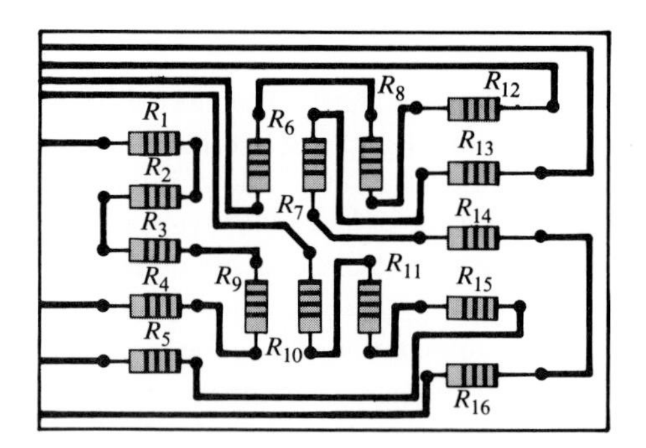

Section 4–2

4–3 What is the current through each of four resistors in a series circuit if the source voltage is 12 V and the total resistance is 120 Ω?

4–4 The current from the source in Figure 4–55 is 5 mA. How much current does each milliammeter in the circuit indicate?

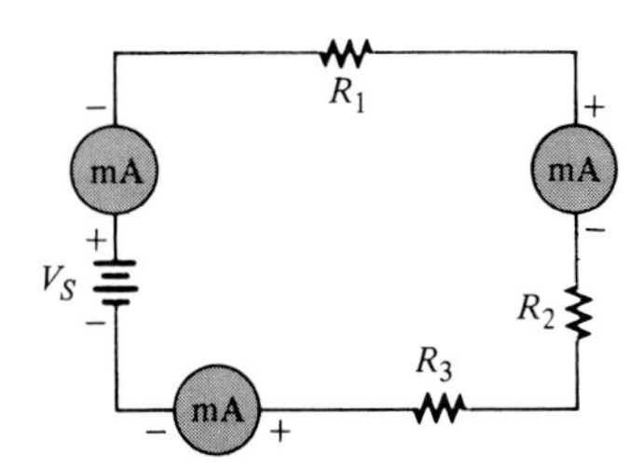

FIGURE 4–55

Section 4–3

4–5 An 82-Ω resistor and a 56-Ω resistor are connected in series. What is the total resistance?

4–6 Find the total resistance of each group of series resistors shown in Figure 4–56 (a full-color photo appears between pages 128 and 129).

4–7 Determine R_T for each circuit in Figure 4–57. Show how to measure R_T with an ohmmeter.

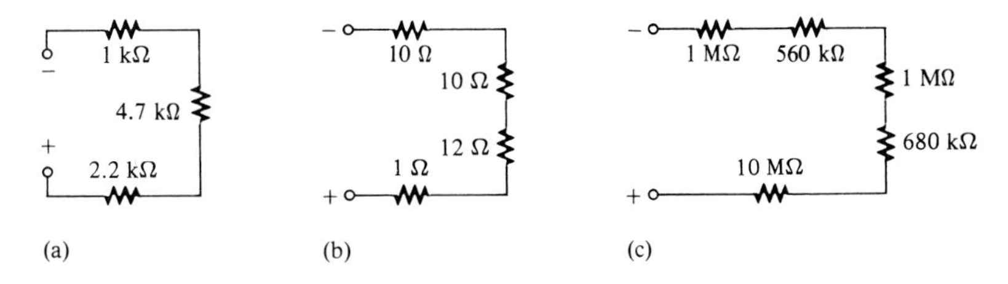

FIGURE 4–57

4–8 What is the total resistance of twelve 5.6-kΩ resistors in series?

4–9 Six 47-Ω resistors, eight 100-Ω resistors, and two 22-Ω resistors are in series. What is the total resistance?

4–10 The total resistance in Figure 4–58 is 20 kΩ. What is the value of R_5?

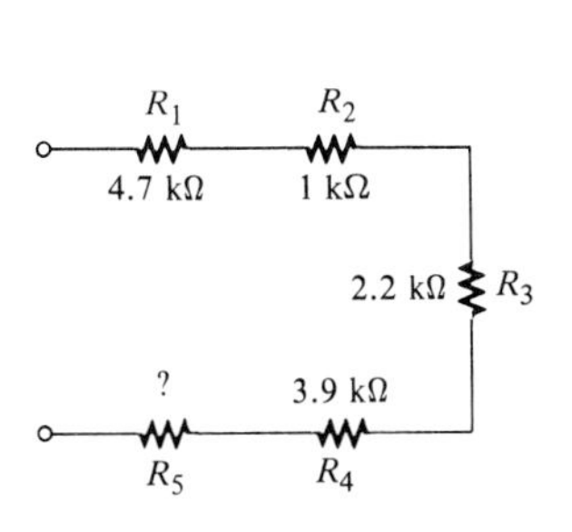

FIGURE 4–58

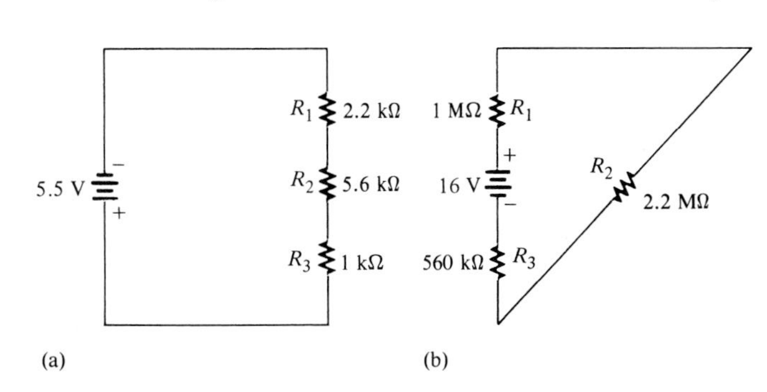

FIGURE 4–59

Section 4–4

4–11 What is the current in each circuit of Figure 4–59? Show how to connect an ammeter in each case.

4–12 Determine the voltage across each resistor in Figure 4–59.

4–13 Three 470-Ω resistors are in series with a 500-V source. How much current is flowing?

4–14 Four equal-value resistors are in series with a 5-V source, and a current of 1 mA is measured. What is the value of each resistor?

Section 4–5

4–15 A 5-V battery and a 9-V battery are connected in series with their polarities in the same direction. What is the total voltage?

4–16 Determine the total source voltage in each circuit of Figure 4–60.

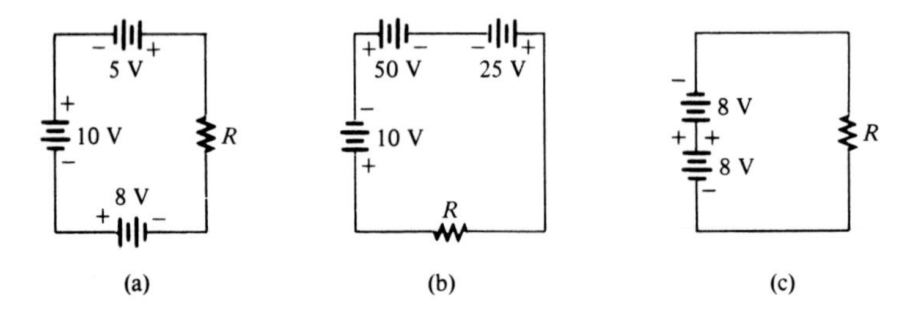

FIGURE 4–60

Section 4–6

4–17 The following voltage drops are measured across each of three resistors in series: 5.5 V, 8.2 V, and 12.3 V. What is the value of the source voltage to which these resistors are connected?

4–18 Five resistors are in series with a 20-V source. The voltage drops across four of the resistors are 1.5 V, 5.5 V, 3 V, and 6 V. How much voltage is across the fifth resistor?

4–19 Determine the unspecified voltage drop(s) in each circuit of Figure 4–61. Show how to connect a voltmeter to measure each unknown voltage drop.

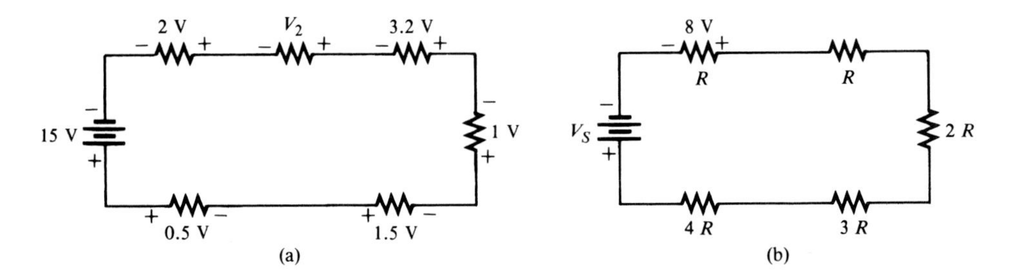

FIGURE 4–61

Section 4–7

4–20 The total resistance of a series circuit is 500 Ω. What percentage of the total voltage appears across a 22-Ω resistor in the series circuit?

4–21 Find the voltage between A and B in each voltage divider of Figure 4–62.

4–22 What is the voltage across each resistor in Figure 4–63? R is the lowest value and all others are multiples of that value as indicated.

4–23 What is the voltage across each resistor in Figure 4–64 (a full-color photo appears between pages 128 and 129)?

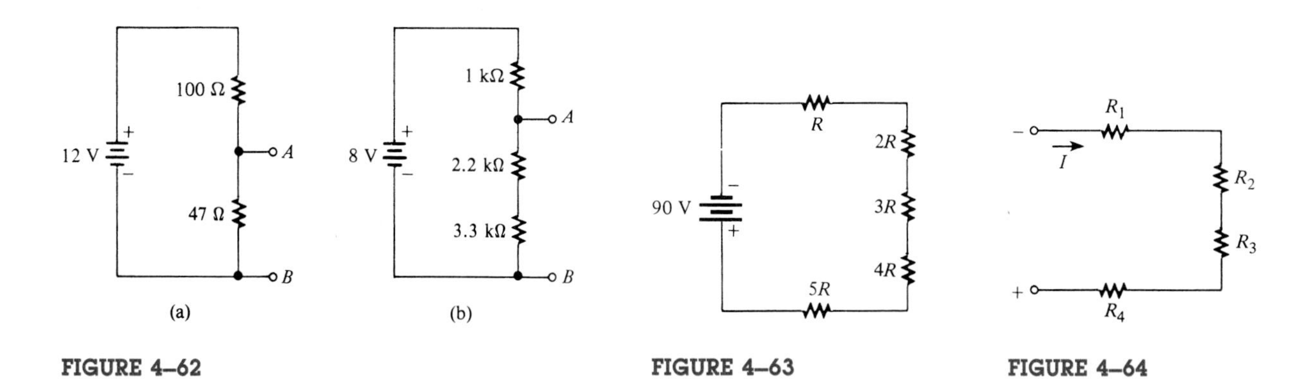

FIGURE 4–62

FIGURE 4–63

FIGURE 4–64

Section 4–8

4–24 Five series resistors each dissipate 50 mW of power. What is the total power?

4–25 Use the results of Problem 4–23 to find the total power in Figure 4–64.

Section 4–9

4–26 By observing the meters in Figure 4–65, determine the types of failures in the circuits and which components have failed.

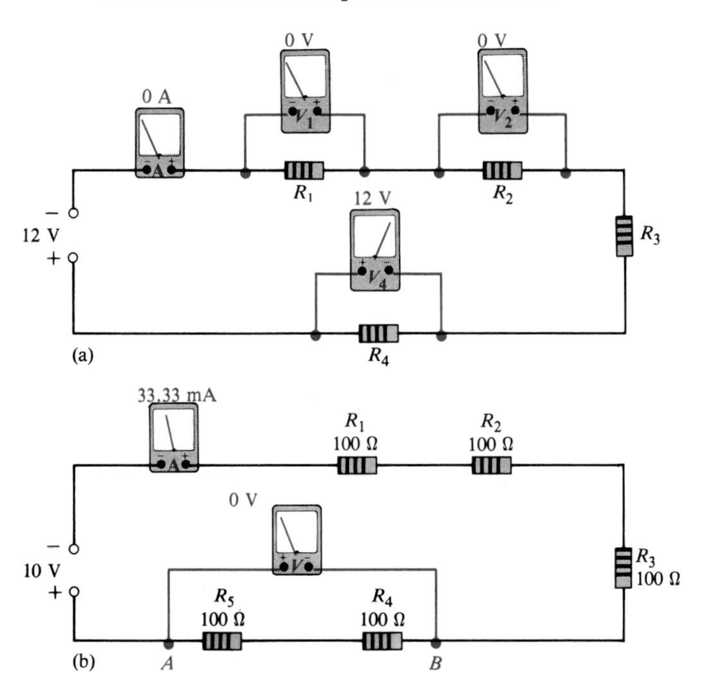

FIGURE 4–65

4–27 Is the ohmmeter reading in Figure 4–66 correct? (A full-color photo appears between pages 128 and 129.) If not, what is wrong?

PROBLEM SET B

4–28 Determine the unknown resistance (R_3) in the circuit of Figure 4–67.

4–29 You have the following resistor values available to you in the lab in unlimited quantities: 10 Ω, 100 Ω, 470 Ω, 560 Ω, 680 Ω, 1 kΩ, 2.2 kΩ, and 5.6 kΩ. All of the other standard values are out of stock. A project that you are working on requires an 18-kΩ resistance. What combination of available values can you use to obtain the needed value?

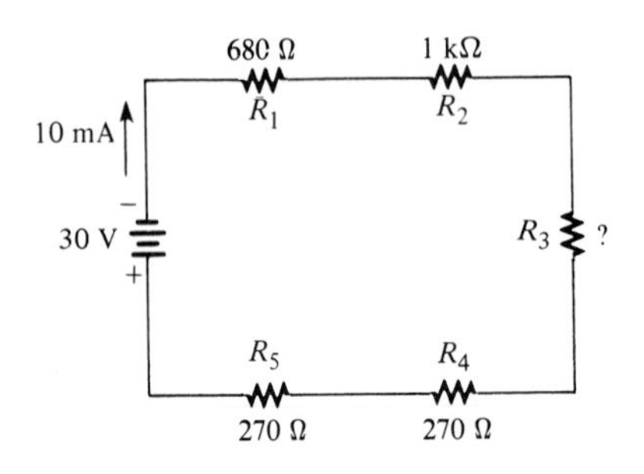

FIGURE 4–67

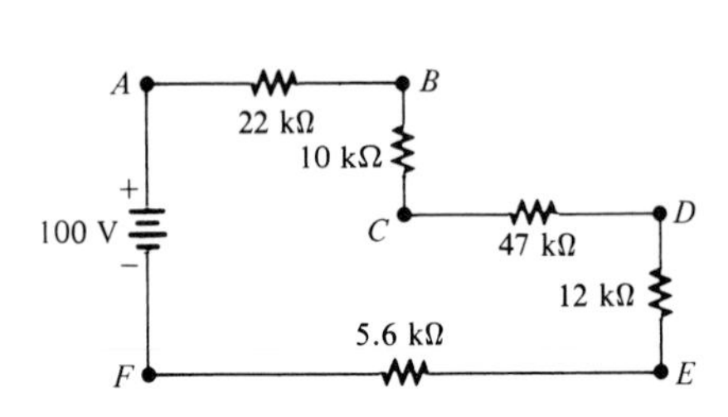

FIGURE 4–68

4–30 Determine the voltage at each point in Figure 4–68 with respect to the negative side of the battery.

4–31 Find all the unknown quantities (in color) in Figure 4–69.

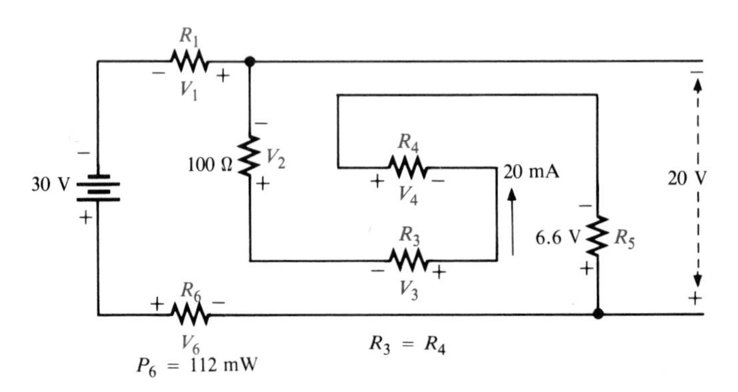

FIGURE 4–69

4–32 There are 250 mA flowing in a series circuit with a total resistance of 1.5 kΩ. The current must be reduced by 25%. Determine how much resistance must be added in order to accomplish this reduction in current.

4–33 Four ½-W resistors are in series: 47 Ω, 68 Ω, 100 Ω, and 120 Ω. To what maximum value can the current be raised before the power rating of one of the resistors is exceeded? Which resistor will burn out first if the current is increased above the maximum?

4–34 A certain series circuit is made up of a ⅛-W resistor, a ¼-W resistor, and a ½-W resistor. The total resistance is 2400 Ω. If each of the resistors is operating at its maximum power level, determine the following:
(a) I. **(b)** V_S. **(c)** the value of each resistor.

4–35 Using 1.5-V batteries, a switch, and three lamps, devise a circuit to apply 4.5 V across one lamp, two lamps in series, or three lamps in series with a single-control switch. Draw the schematic diagram.

4–36 Develop a variable voltage divider to provide output voltages ranging from a minimum of 10 V to a maximum of 100 V using a 120-V source. The maximum voltage must be at the maximum resistance setting of the potentiometer. The minimum voltage must be at the minimum resistance (zero ohms) setting. The maximum current is to be 10 mA.

4–37 Using the standard resistor values given in Appendix B, build a voltage divider to provide the following approximate voltages with respect to the negative terminal of a 30-V source: 8.18 V, 14.73 V, and 24.55 V. The current drain on the source must be limited to no more than 1 mA. The number of resistors, their ohmic values, and their power ratings must be specified. Draw a schematic diagram showing the circuit with all resistor values indicated.

ANSWERS TO SECTION REVIEWS

Section 4–1

1. There is a single current path. **2.** See Figure 4–70. **3.** See Figure 4–71.

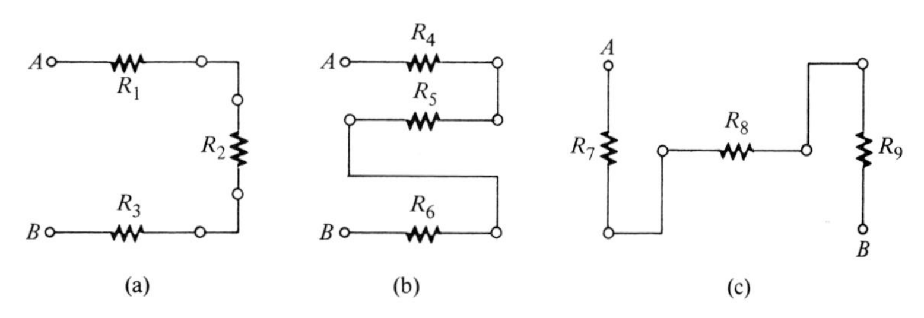

FIGURE 4–70

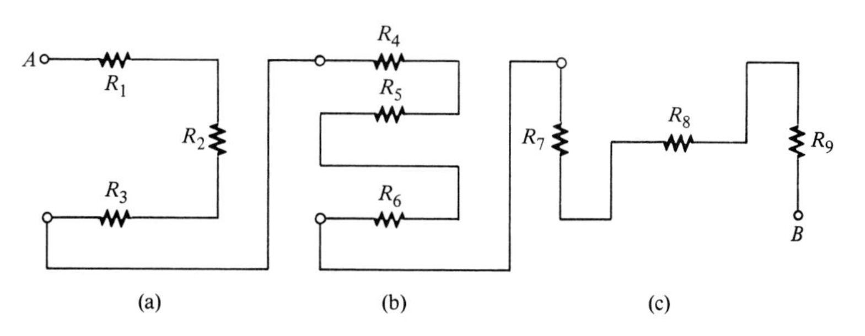

FIGURE 4–71

Section 4–2

1. 2 A. **2.** 50 mA between C and D, 50 mA between E and F. **3.** 2 A, 2 A.

Section 4–3

1. **(a)** 143 Ω; **(b)** 96 Ω; **(c)** 4020 Ω. **2.** 572 Ω. **3.** 1.2 kΩ.
4. 564 Ω.

Section 4–4

1. 0.033 A or 33 mA. 2. 215 V. 3. $V_1 = 50$ V, $V_2 = 75$ V, $V_3 = 90$ V.
4. 250 Ω.

Section 4–5

1. Five; see Figure 4–72. 2. See Figure 4–73.
3. (a) 75 V; (b) 15 V. 4. See Figure 4–74.

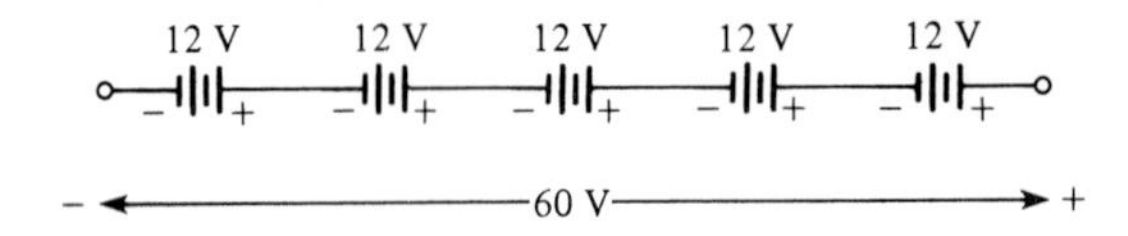

FIGURE 4–72

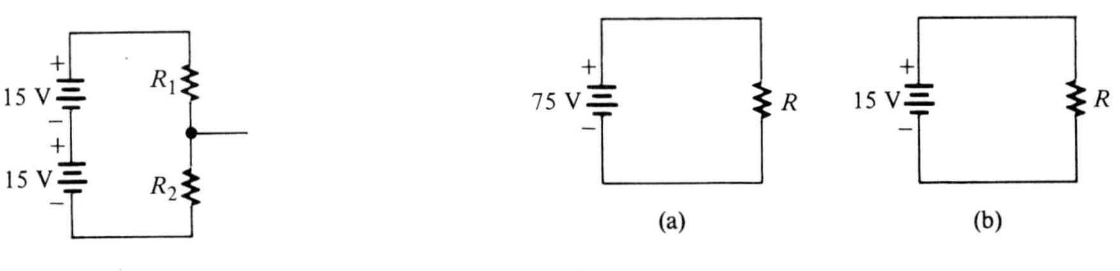

FIGURE 4–73

FIGURE 4–74

Section 4–6

1. (a) The sum of the voltages around a closed path is zero:
   (b) the sum of the voltage drops equals the total source voltage.
2. 50 V. 3. 5 V. 4. 10 V.

Section 4–7

1. 10 V. 2. 40.6 V across the 56-Ω, 59.4 V across the 82-Ω. See Figure 4–75.
3. At the midpoint.

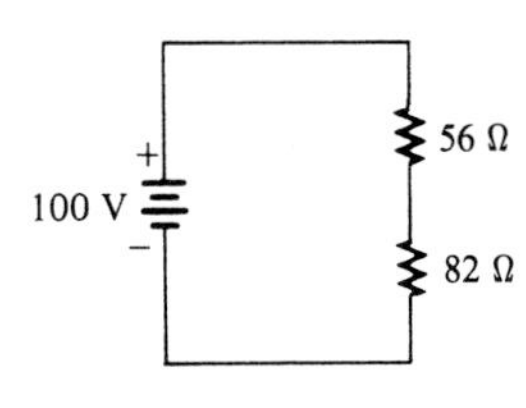

FIGURE 4–75

Section 4–8

1. Add the power in each resistor. 2. 16 W. 3. 1110 W.

Section 4–9

1. T. 2. 24 V, 0 V.

Parallel Circuits

In this chapter, parallel circuits are introduced. Parallel circuits are found in many applications, for example, lighting systems, sound systems, appliances, and most electronic equipment.

Specifically, you will learn:

- How to identify a parallel connection.
- The definition of Kirchhoff's current law and how to apply it.
- How to determine total resistance.
- How to apply Ohm's law to find the currents, voltage, and resistances in a parallel circuit.
- How parallel circuits act as current dividers.
- How to apply the current divider principle.
- How to determine total power in a parallel circuit.
- How to locate an open path in a parallel circuit.

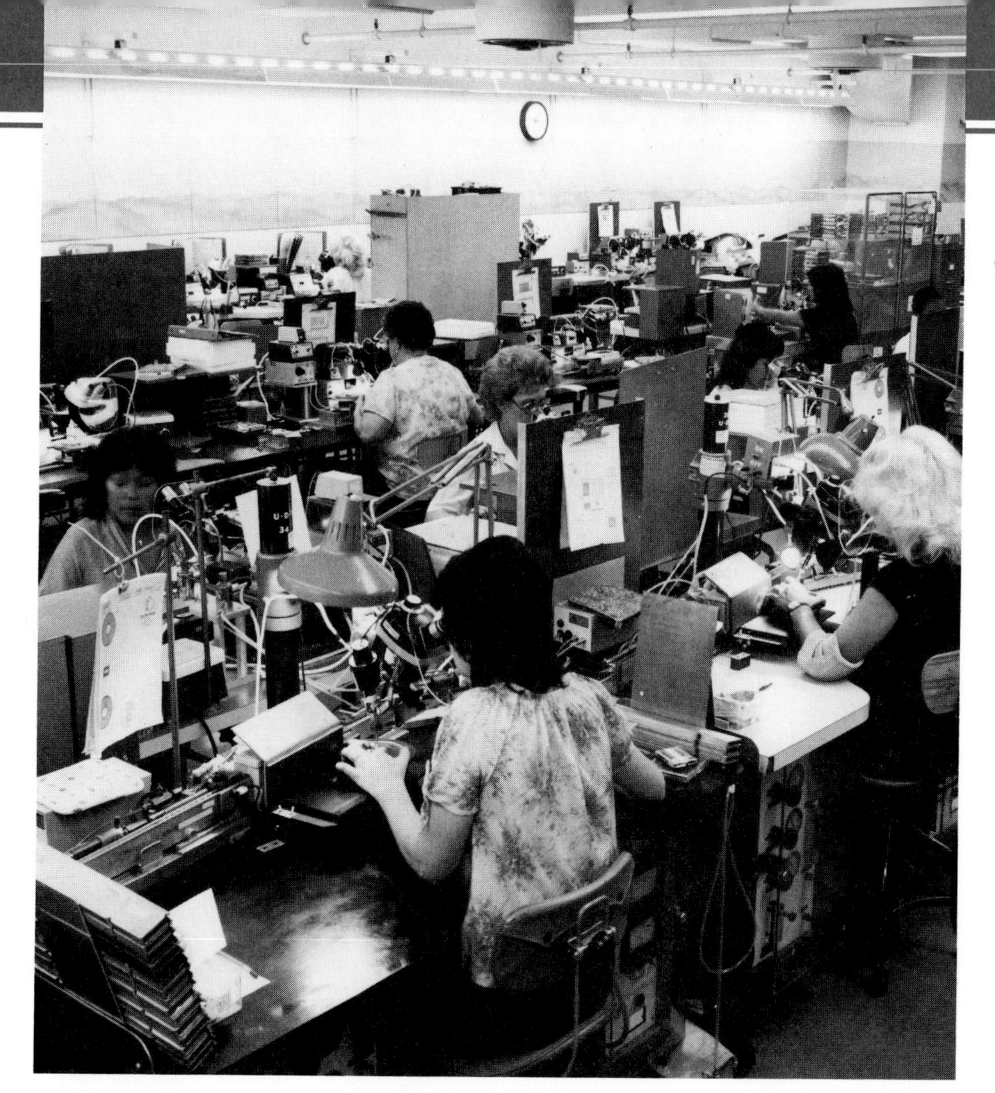

5

APPLICATION ASSIGNMENT

As the technician in charge of testing a particular PC board assembly, you decide to devise a systematic test procedure for checking out each PC board for open resistors as it comes off the assembly line. Each board consists of three groups of parallel resistors. Each group has several 1/2-W resistors in parallel and is connected to pins on the PC board as follows:

- Parallel group 1(R_1 through R_4): 1 kΩ, 1.5 kΩ, 1.8 kΩ, 2.2 kΩ. Pins 1 and 2.
- Parallel group 2 (R_5 through R_8): 10 kΩ, 22 kΩ, 27 kΩ, 39 kΩ. Pins 3 and 4.
- Parallel group 3 (R_9 through R_{12}): 47 kΩ, 56 kΩ, 68 kΩ, 82 kΩ. Pins 5 and 6.

Each parallel group is separate and not interconnected with another group. The PC board layout is similar to the one shown in Figure 5–68 (a full-color photo appears between pages 128 and 129) except for the number of resistors.

The material that you will cover in this chapter will enable you to develop the required test procedure. The Application Note at the end of the chapter gives one solution.

RESISTORS IN PARALLEL

5–1

When two or more components, each forming a separate currrent path, are connected across the same voltage source, they are in parallel. A parallel circuit provides more than one path for current between two given points.

Each parallel path is called a *branch*. Two resistors connected in parallel are shown in Figure 5–1(a). As shown in Part (b), the current flowing out of the source divides when it gets to point A. Part of it goes through R_1 and part through R_2. The two currents come back together at point B. If additional resistors are connected in parallel with the first two, more current paths are provided, as shown in Figure 5–1(c).

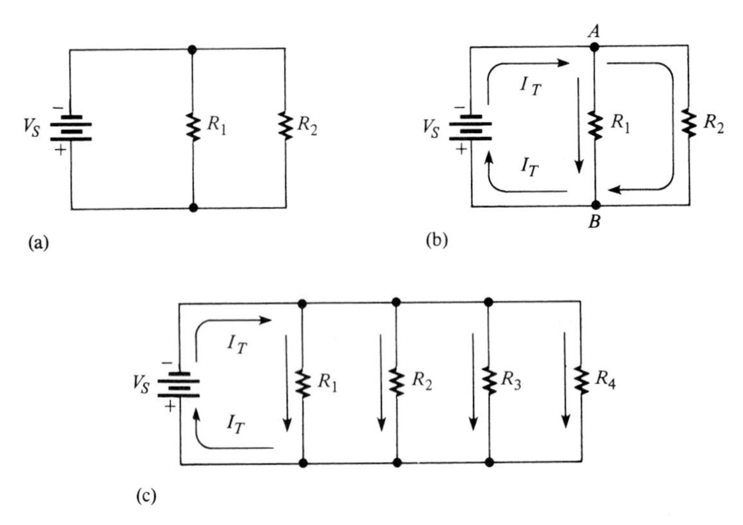

FIGURE 5–1
Resistors in parallel.

Identifying Parallel Connections

It is obvious that the resistors in Figure 5–1 are connected in parallel. However, in actual circuit diagrams, the parallel relationship often is not so clear. It is important that you learn to recognize parallel connections regardless of how they may be drawn.

A rule for identifying parallel circuits is as follows:

If there is more than one current path (branch) between two points, and if the voltage between those two points appears across each of the branches, then there is a parallel circuit between those two points.

Figure 5–2 shows parallel resistors drawn in different ways between two points labeled A and B. Notice that in each case, the current "travels" two paths going from A to B, and the voltage across each branch is the same. Although these figures show only two parallel paths, there can be any number of resistors in parallel.

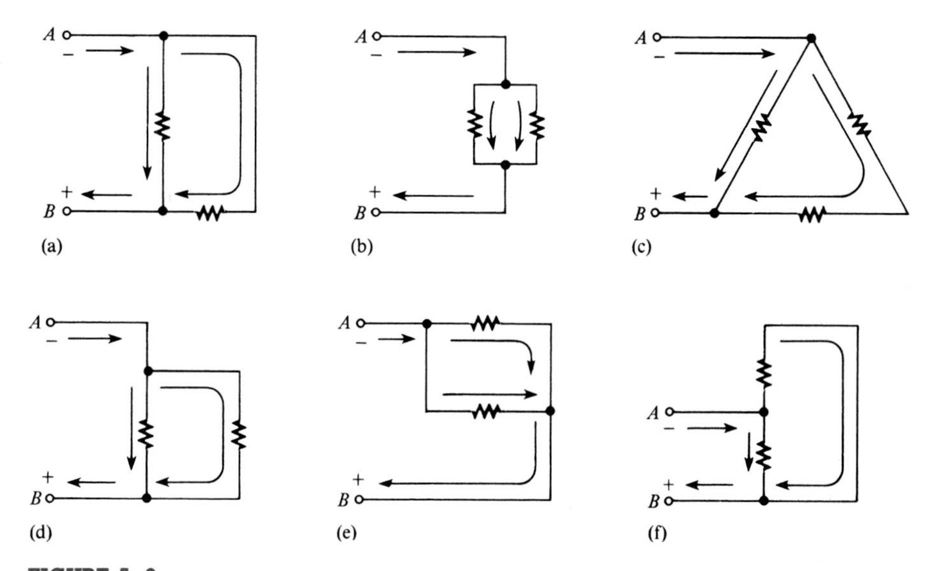

FIGURE 5–2
Examples of circuits with two parallel paths.

EXAMPLE 5–1 Five resistors are positioned on a circuit board as shown in Figure 5–3. Show the wiring required to connect all the resistors in parallel, and draw a schematic diagram showing this connection.

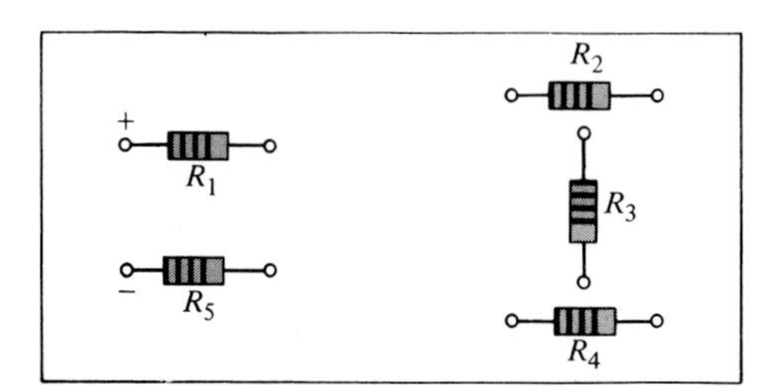

FIGURE 5–3

Solution

Wires are connected as shown in the assembly diagram of Figure 5–4(a). The

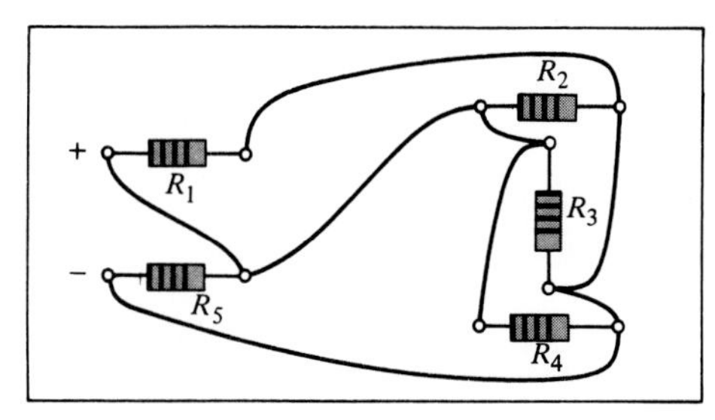

(a) Assembly

(b) Schematic

FIGURE 5–4

schematic diagram is shown in Figure 5–4(b). Again, note that the schematic does not necessarily have to show the actual physical arrangement of the resistors. The purpose of the schematic is to show how components are connected electrically.

EXAMPLE 5–2

Determine the parallel groupings in Figure 5–5.

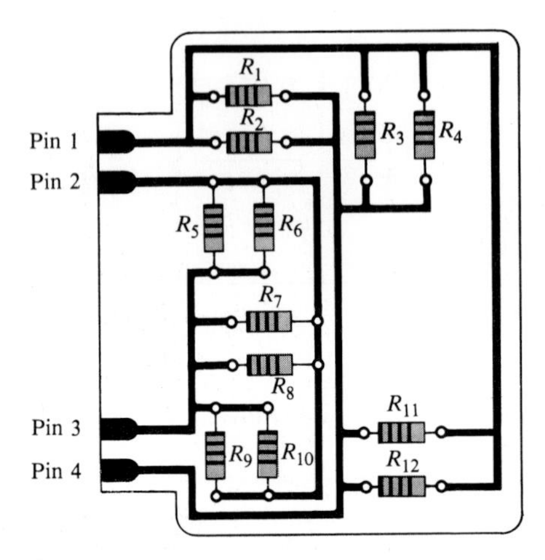

FIGURE 5–5

Solution

Resistors R_1 through R_4 and R_{11} and R_{12} are all in parallel. This parallel combination is connected to pins 1 and 4.

Resistors R_5 through R_{10} are all in parallel. This combination is connected to pins 2 and 3.

SECTION REVIEW 5–1

1. How do you identify a parallel connection?
2. Complete the schematic diagrams for the circuits in each part of Figure 5–6 by connecting the resistors in parallel between points A and B.
3. Now connect each *group* of parallel resistors in Figure 5–6 in parallel with each other.

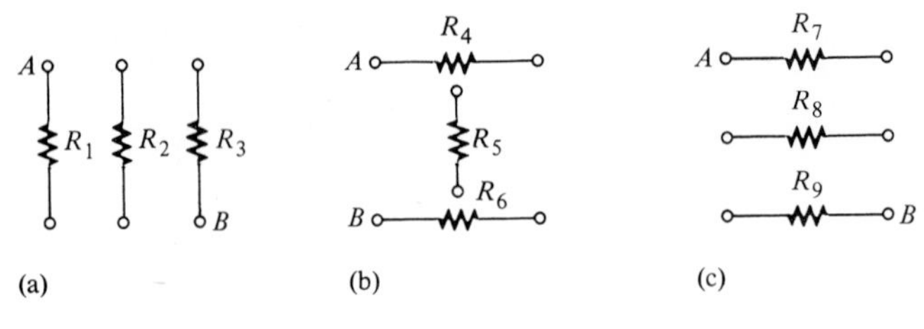

FIGURE 5–6

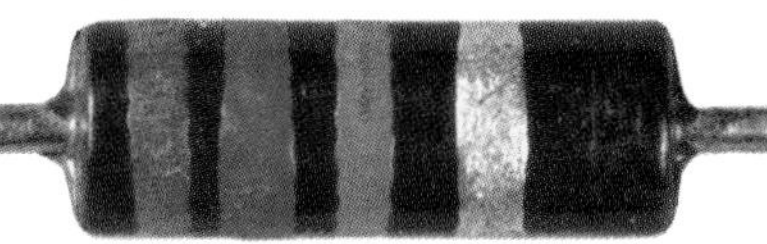

Color Insert: Actual Circuits for Examples, Section Reviews, and Problems

The experimenter boards shown in this full-color section are commonly found in many school laboratories and are used in industry, however, a brief explanation of the board layout may be useful to those unfamiliar with them.

Notice that each row perpendicular to the center slot consists of five connection points. Each of these five points is internally connected to each other to form a common electrical tie point. Each row, however, is electrically separate from all the other rows.

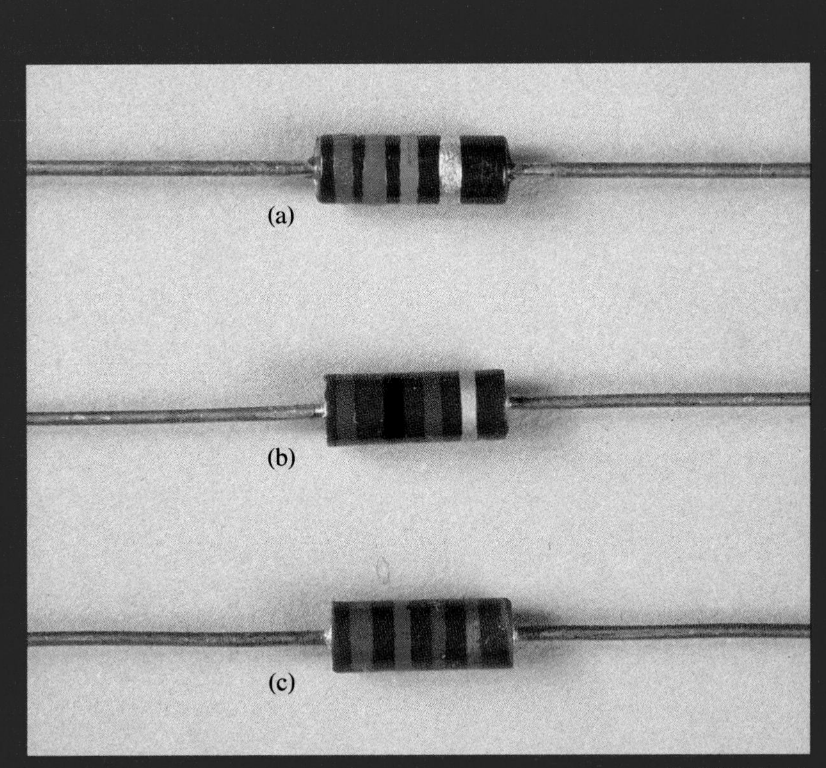

FIGURE 2–19
For Example 2–4

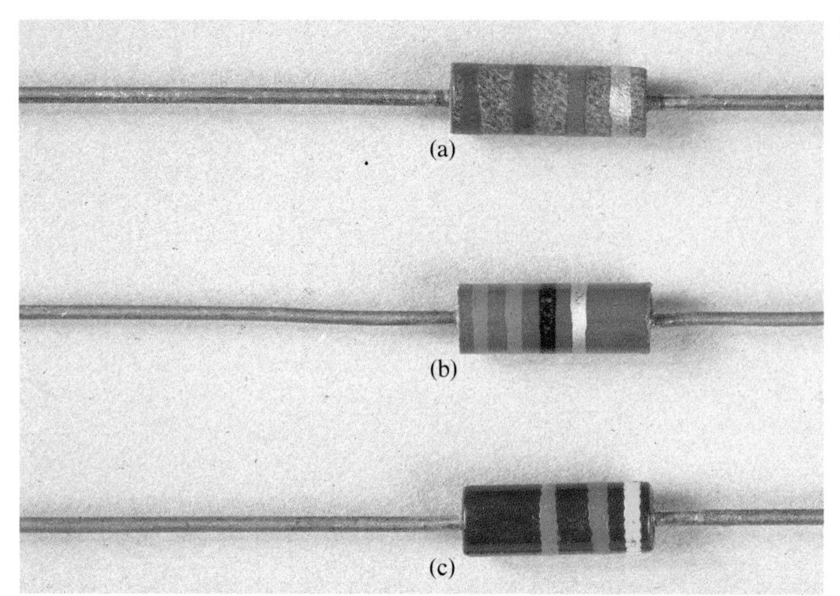

FIGURE 2–37(a)
For Problems 2–11 and 2–12

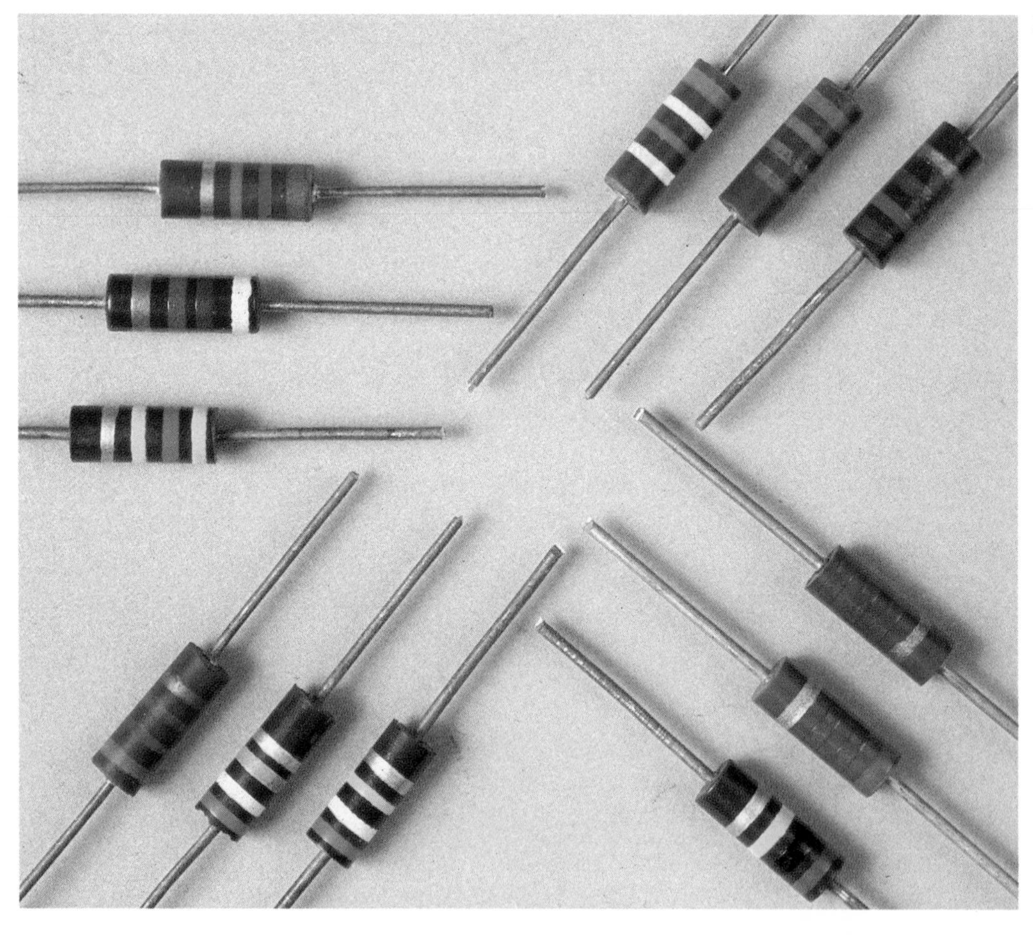

FIGURE 2–37(b)
For Problem 2–13(b)

(a)

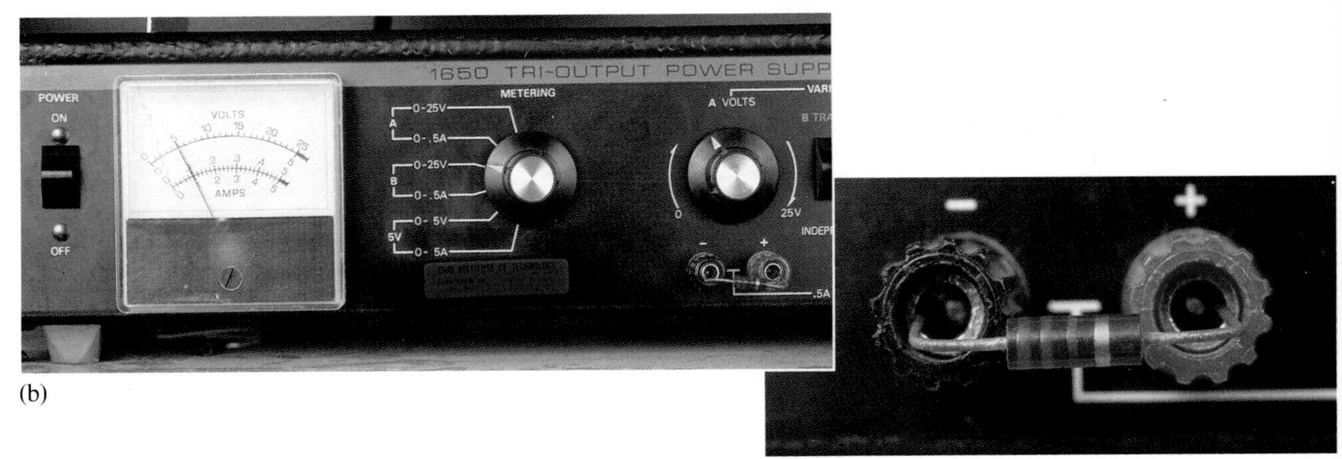

(b)

(c)

FIGURE 3–25
For Problem 3–7

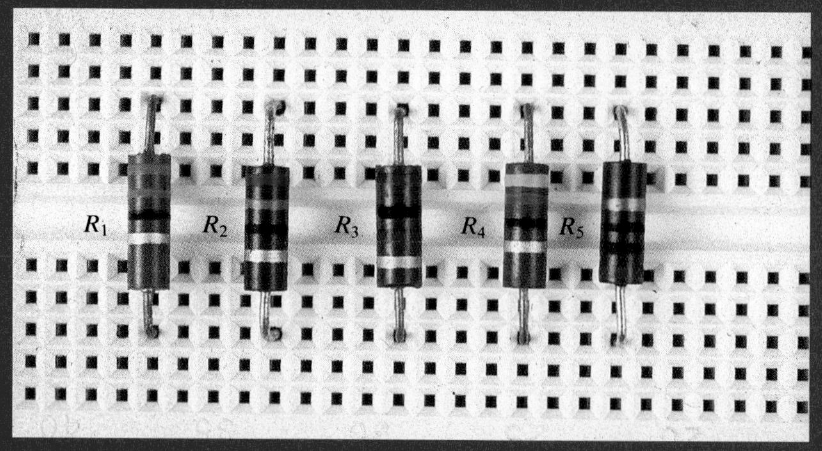

FIGURE 4–13
For Example 4–3

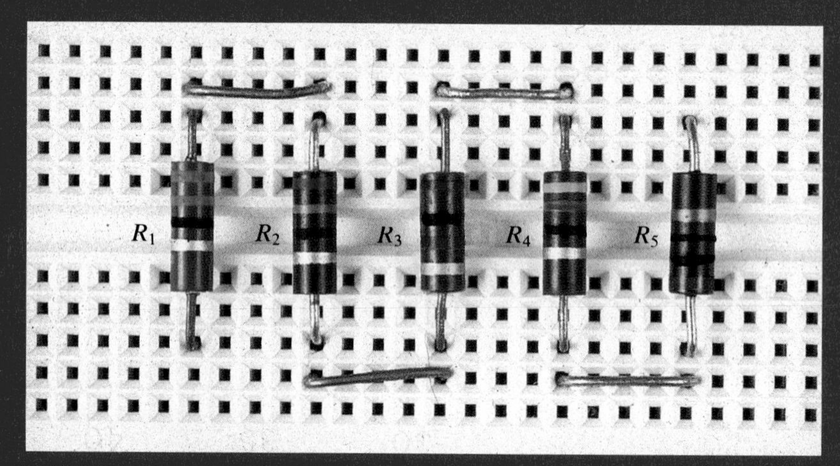

FIGURE 4–14
For Example 4–3

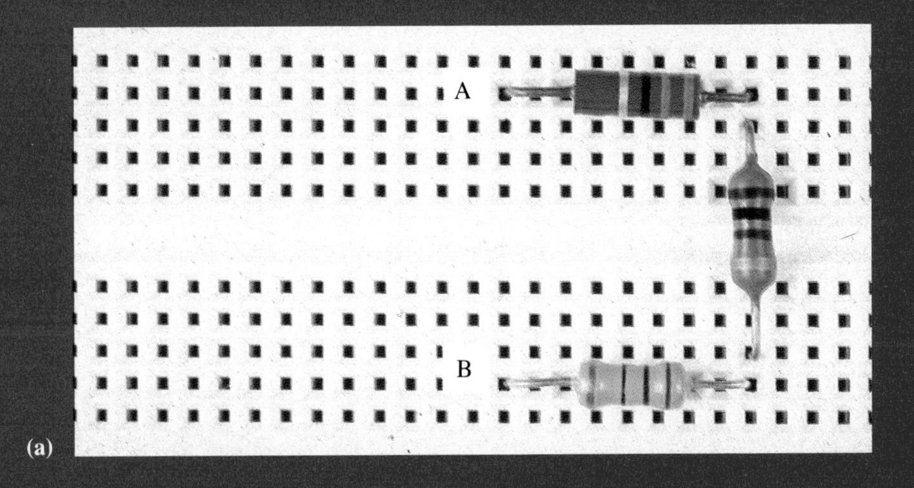

(a)

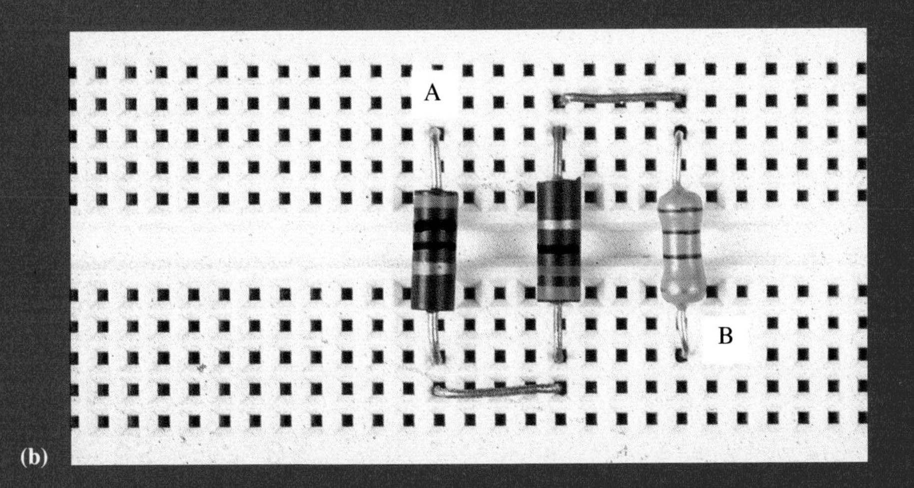

(b)

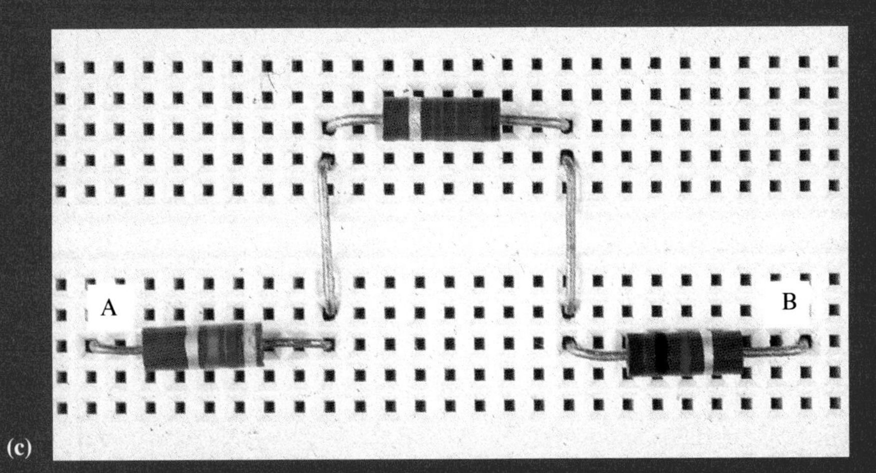

(c)

FIGURE 4–16
For Section Review 4–3, question 1

FIGURE 4–56
For Problem 4–6

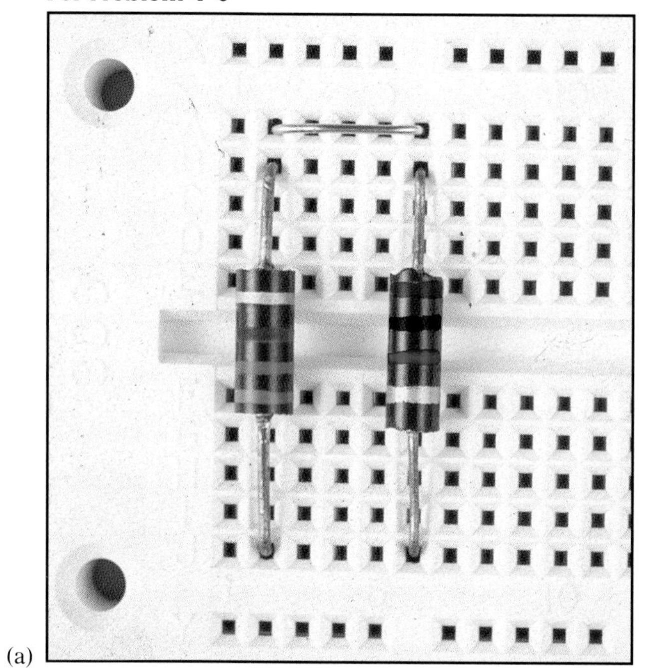
(a)

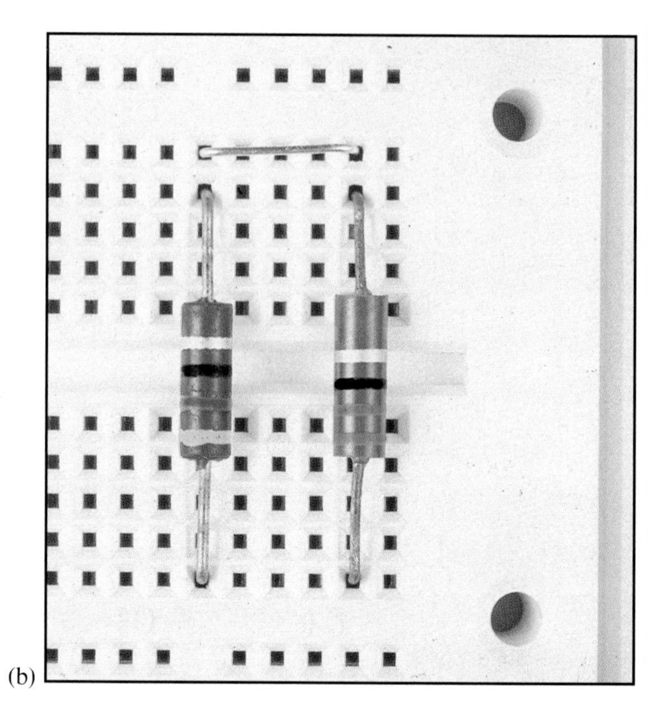
(b)

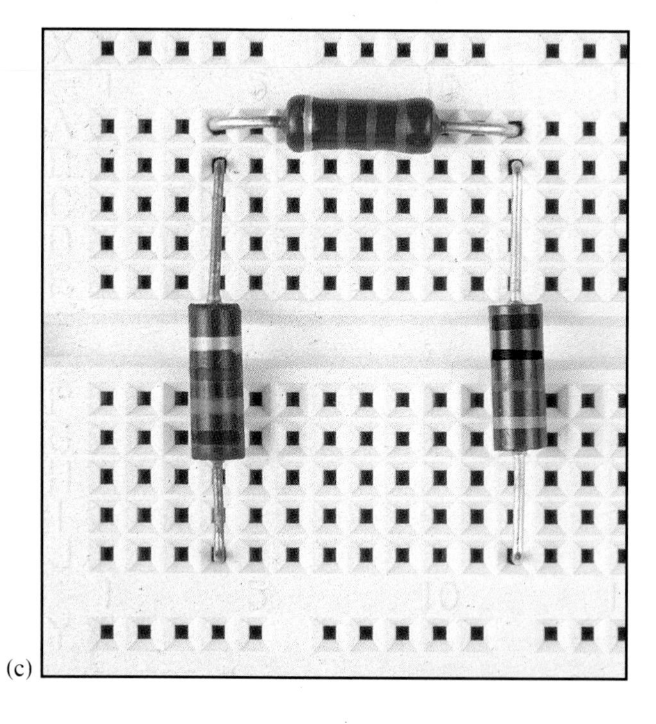
(c)

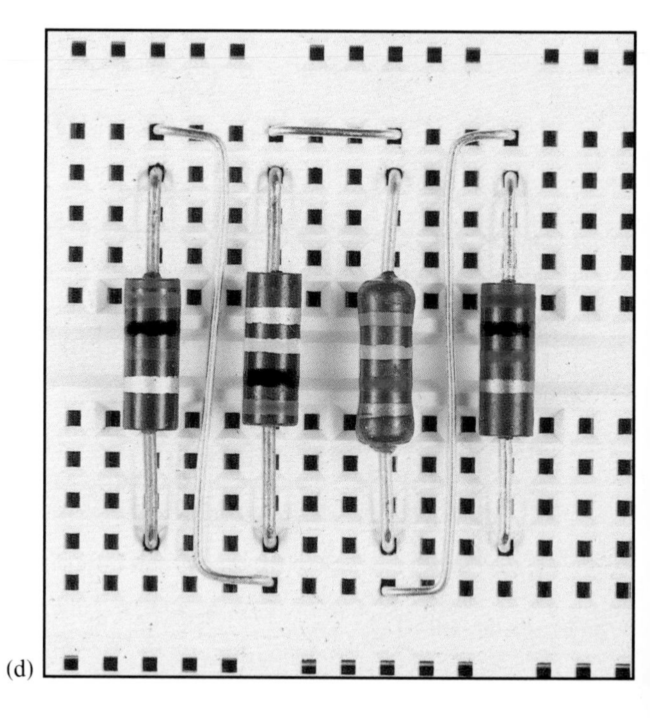
(d)

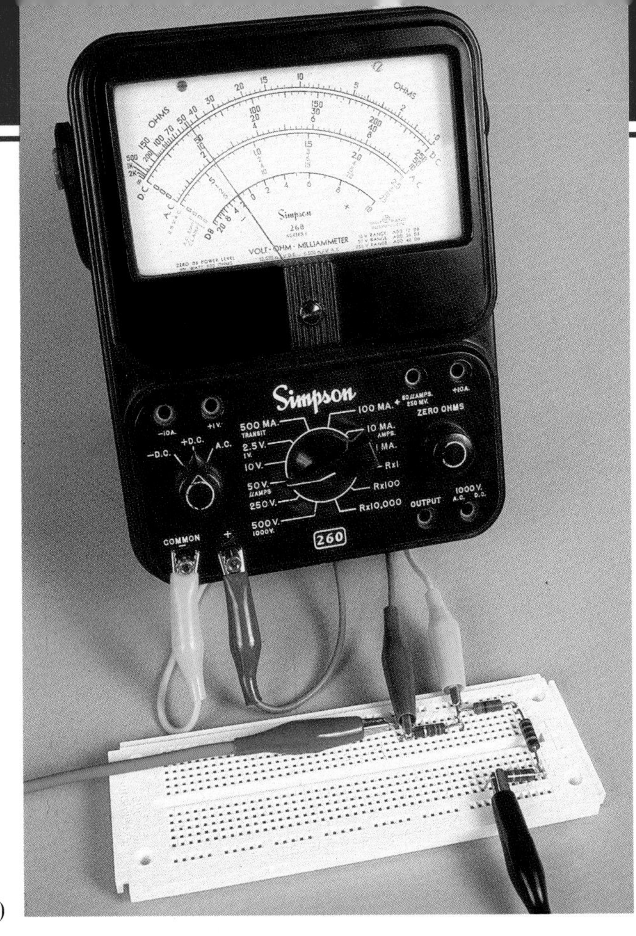

(a)

FIGURE 4–64
For Problem 4–23

(b)

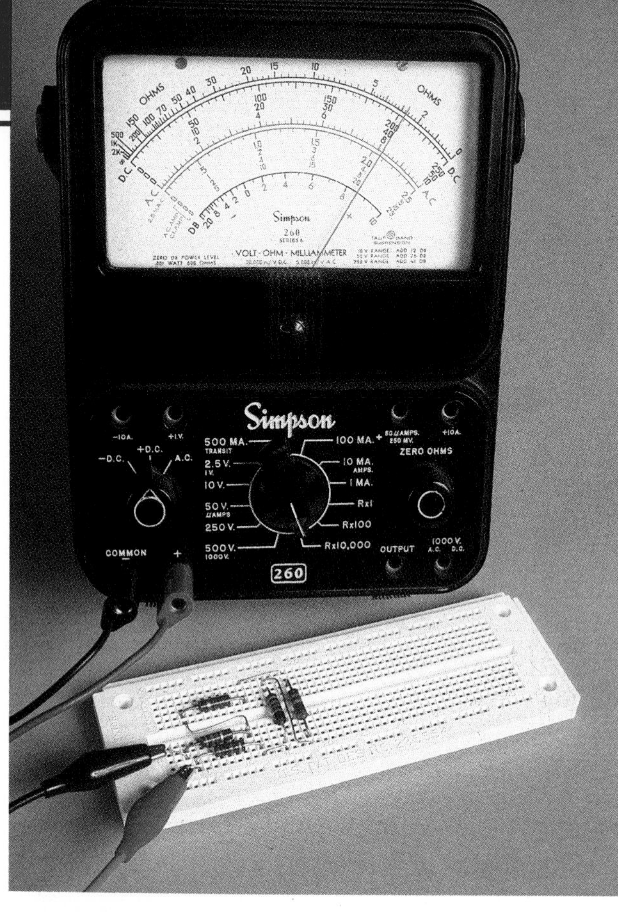

(a)

FIGURE 4–66
For Problem 4–27

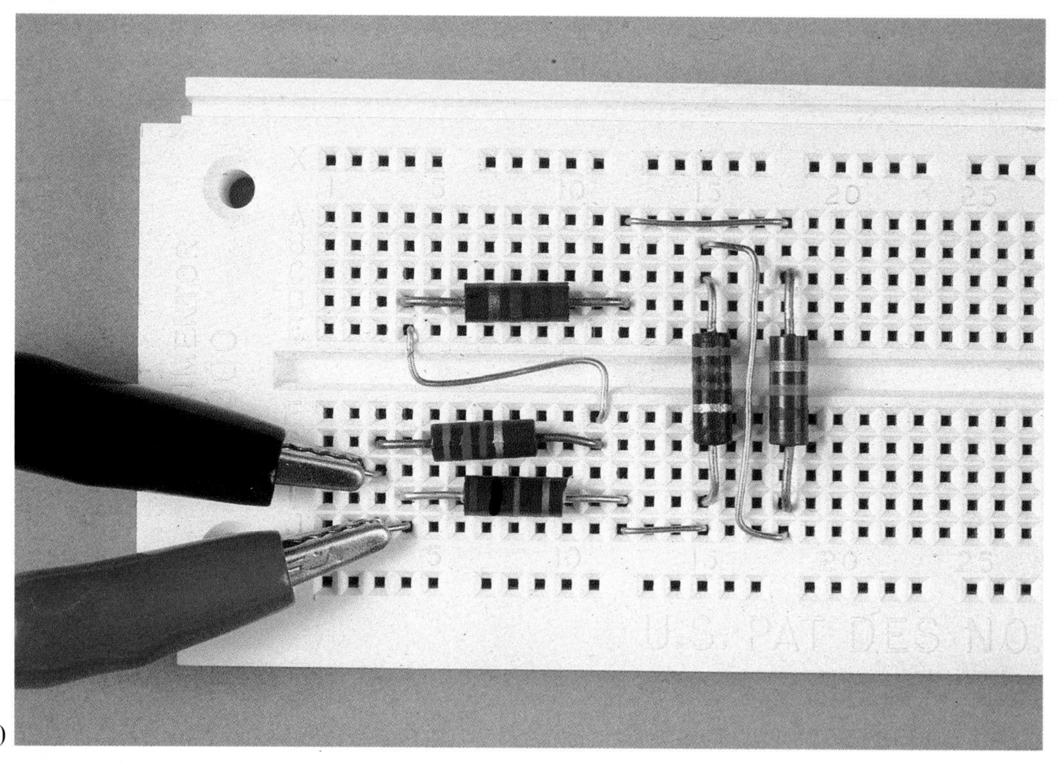

(b)

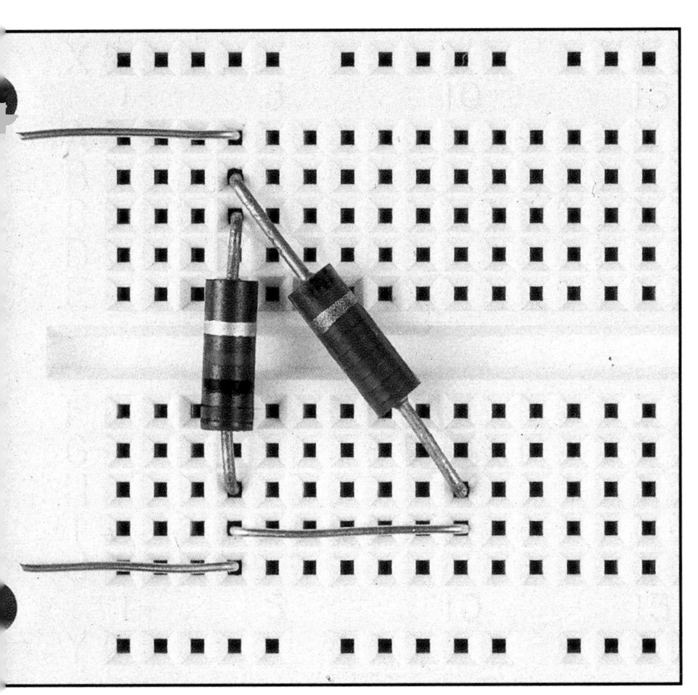

FIGURE 5–25
For Section Review 5–4, question 3

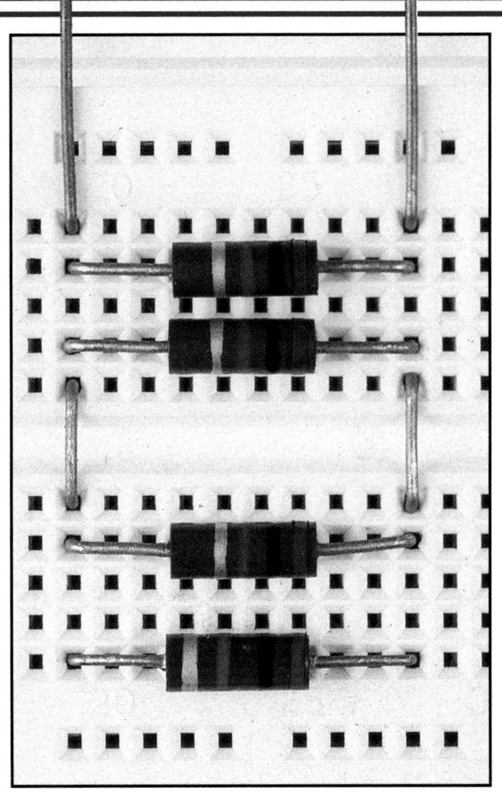

FIGURE 5–26
For Section Review 5–4, question 3

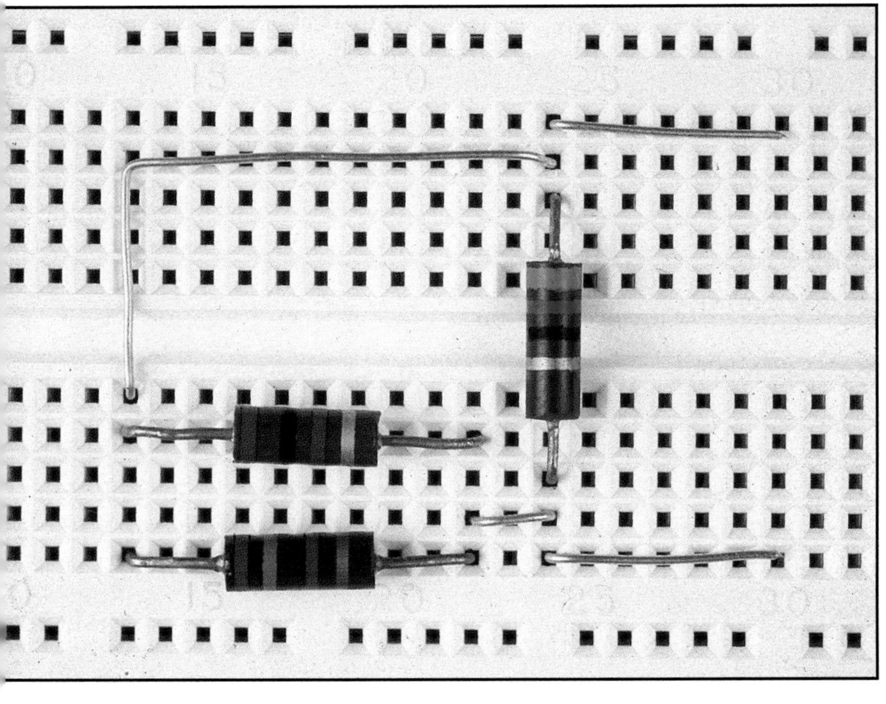

FIGURE 5–27
For Section Review 5–4, question 3

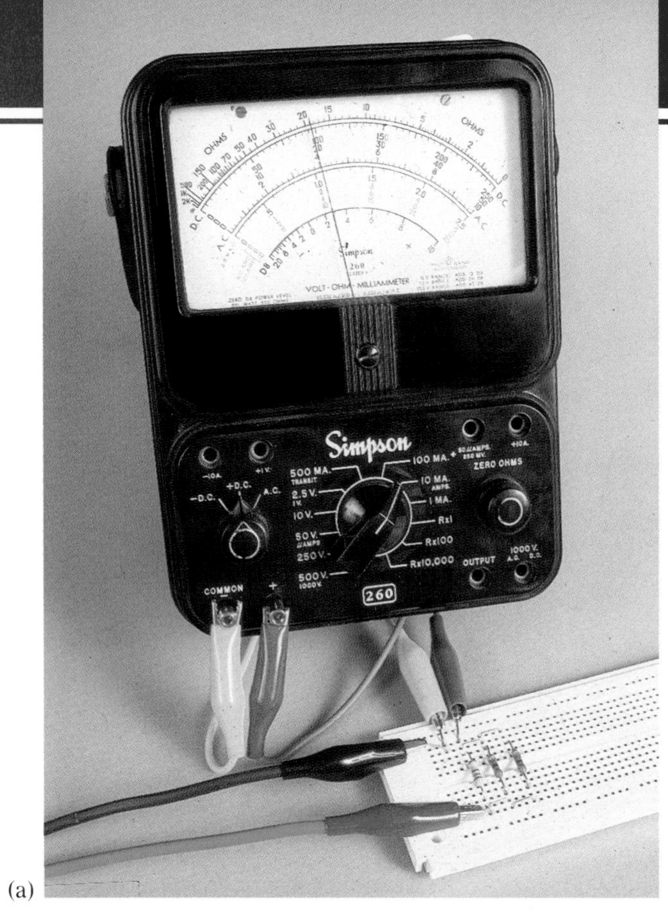

(a)

FIGURE 5–39
For Section Review 5–6, question 2

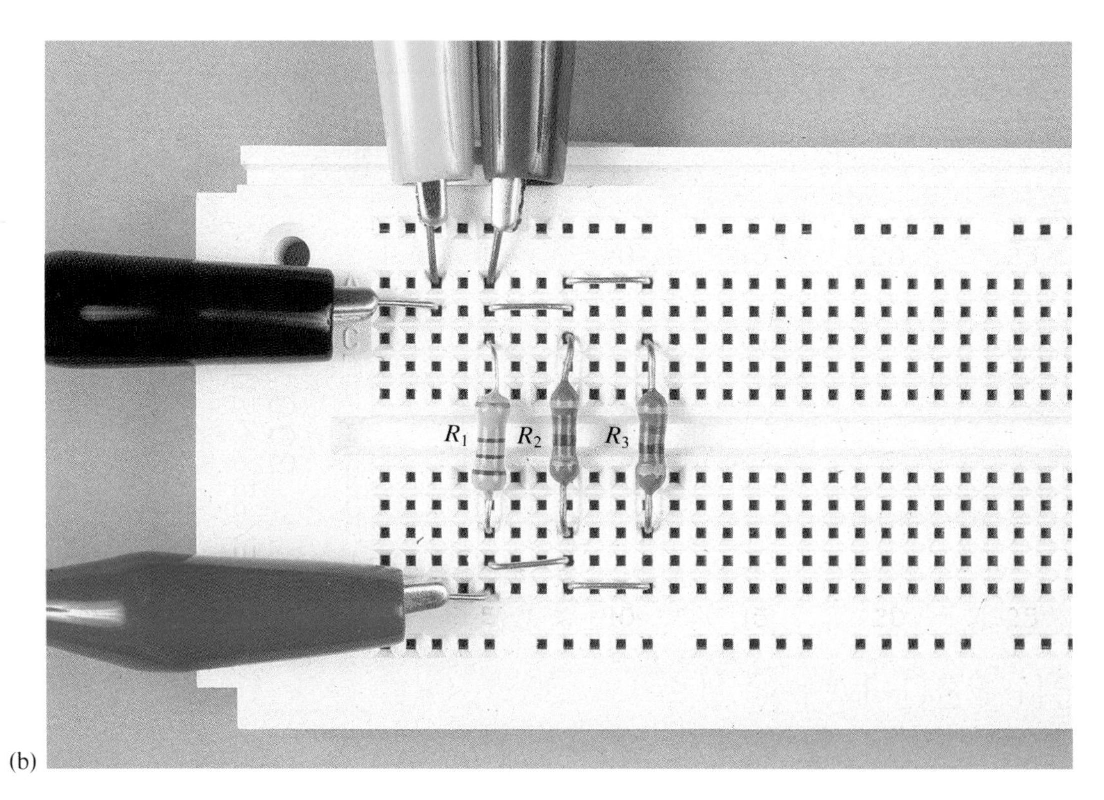

(b)

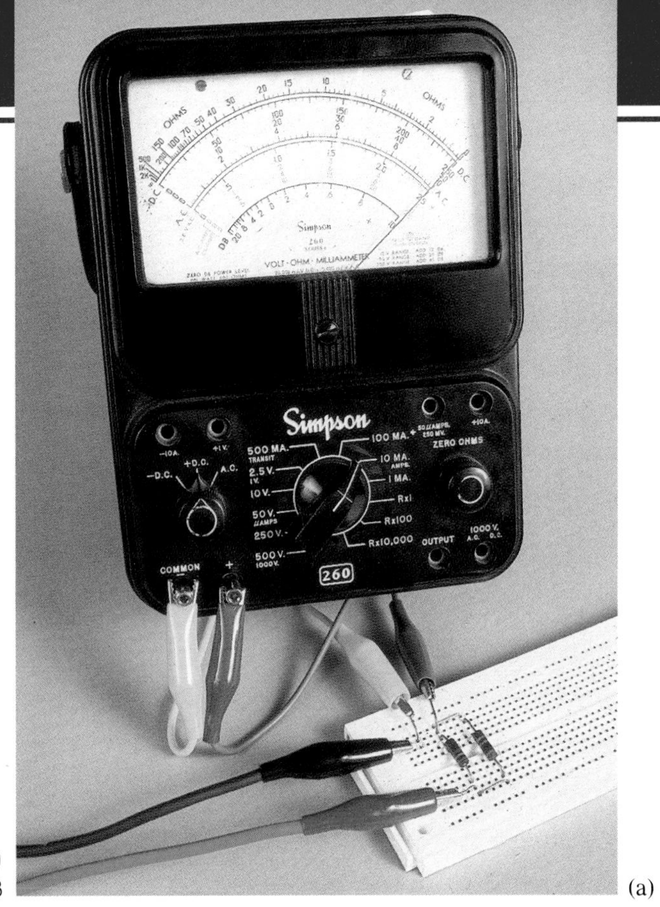

FIGURE 5–40
For Section Review 5–6, question 3

(a)

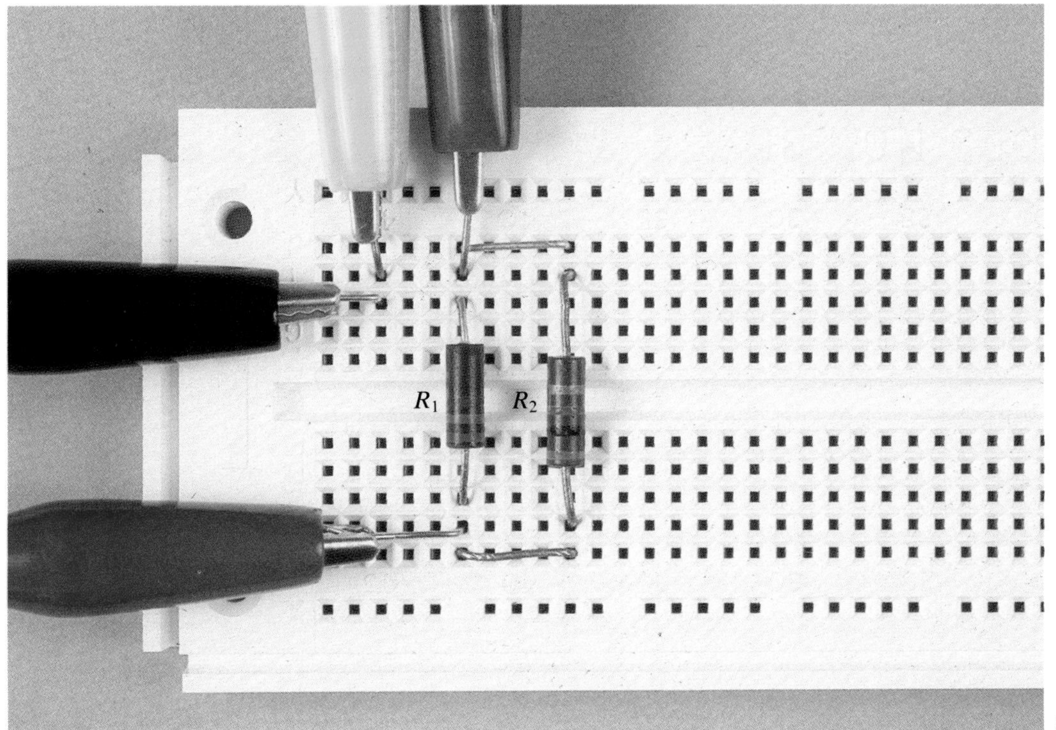

(b)

FIGURE 5–55
For Problem 5–9

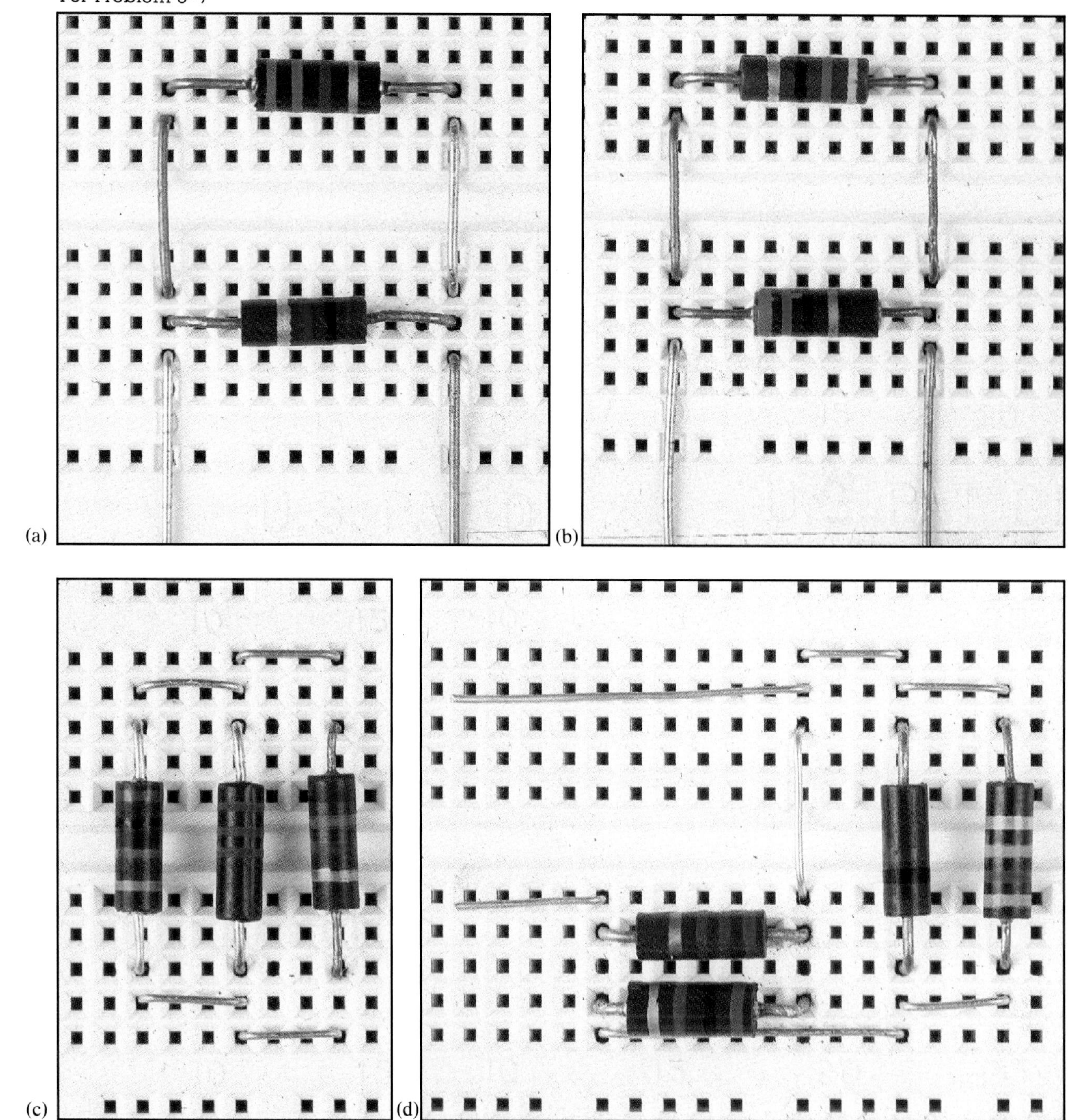

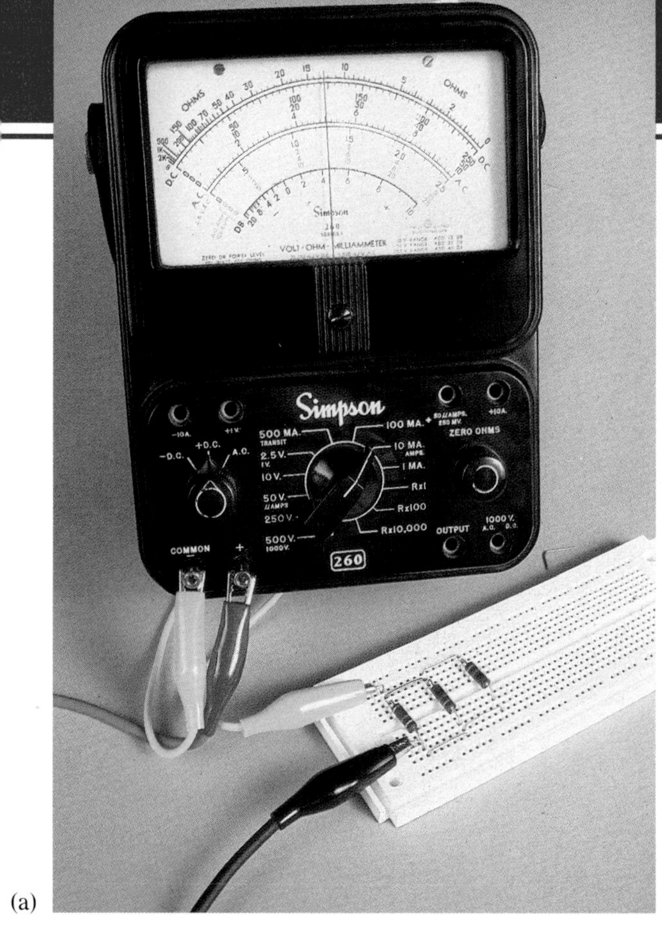

(a)

FIGURE 5–65
For Problem 5–29

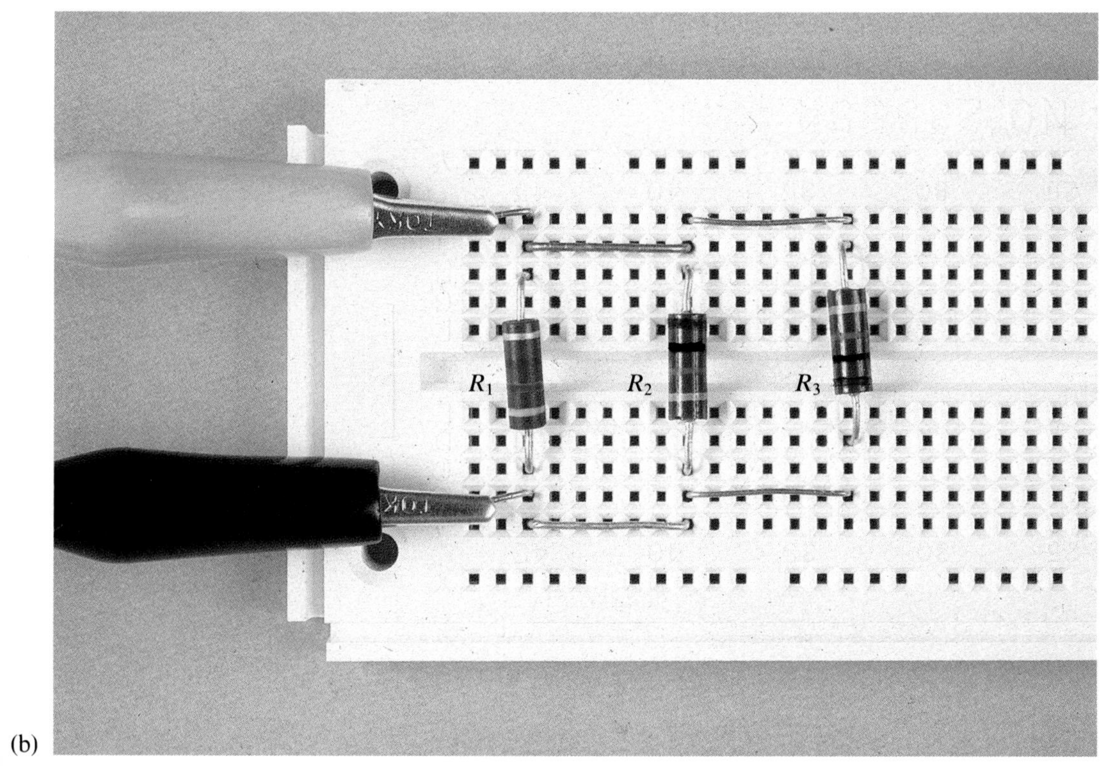

(b)

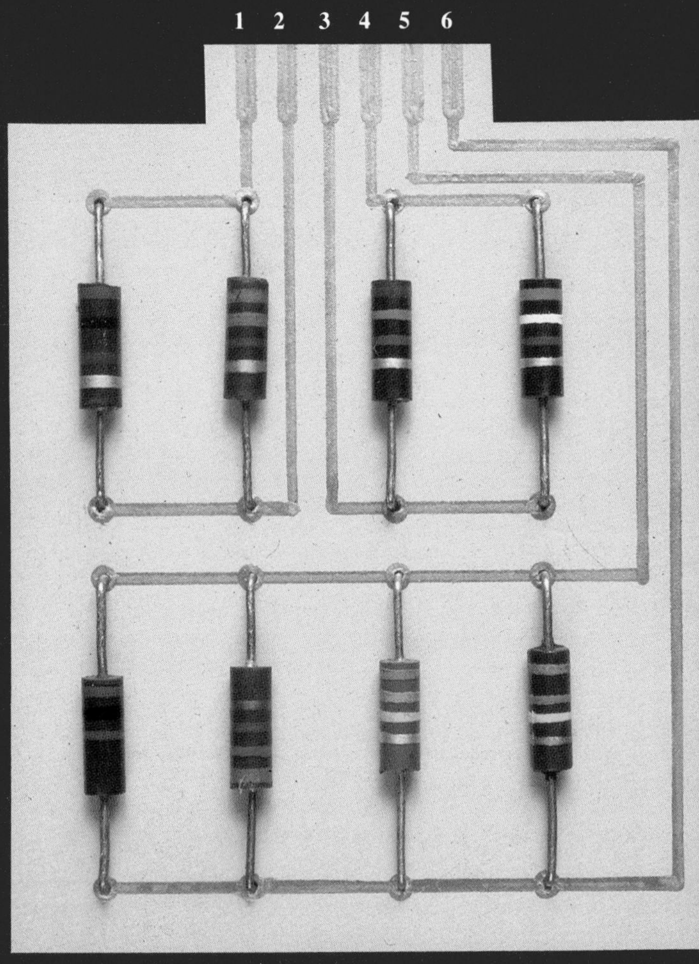

FIGURE 5–68
For Problem 5–33

FIGURE 6–68
For Problem 6–6

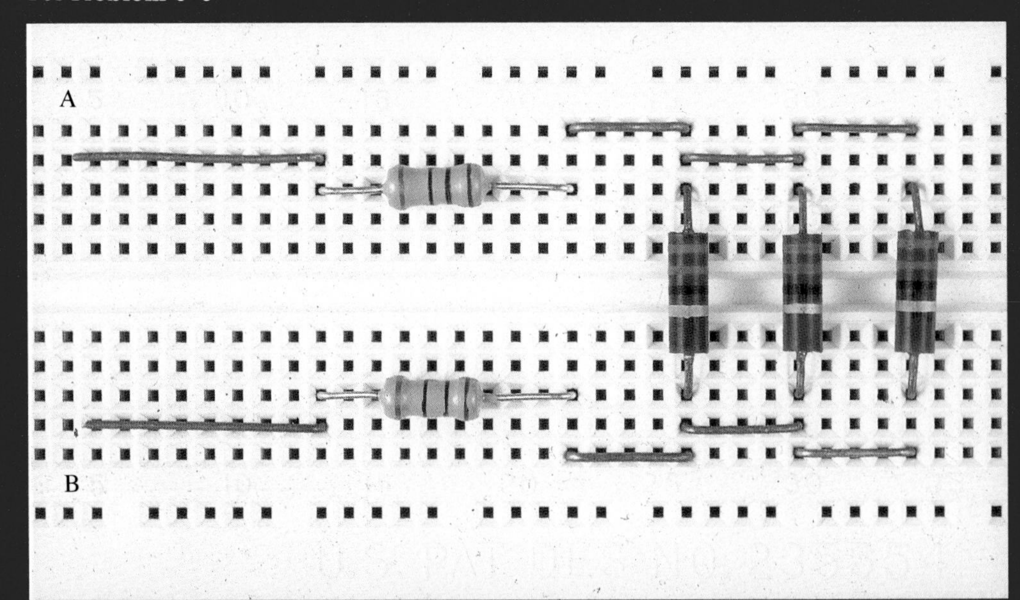

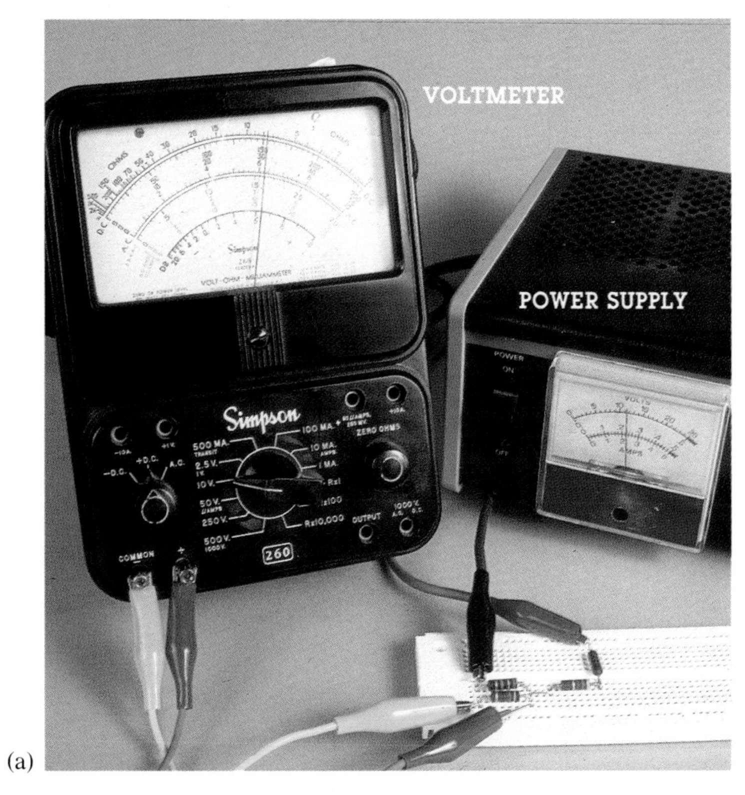

(a)

FIGURE 6–77
For Problem 6–23

(b)

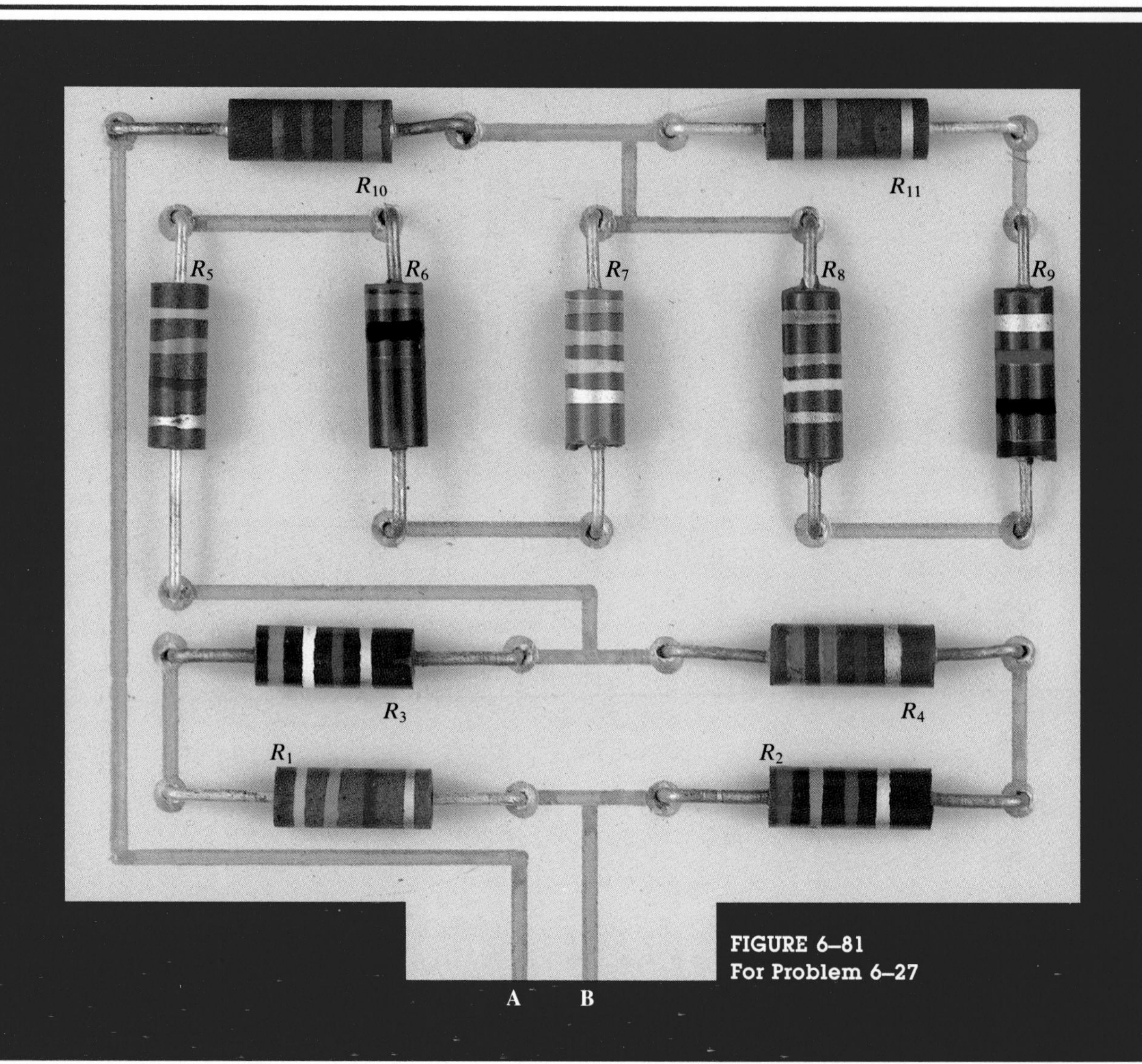

FIGURE 6–81
For Problem 6–27

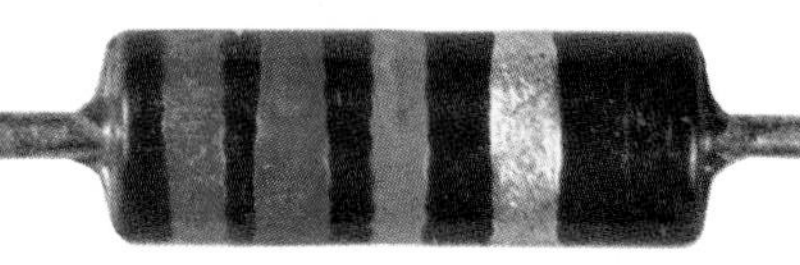

VOLTAGE IN A PARALLEL CIRCUIT

5–2

As mentioned, each path in a parallel circuit is sometimes called a *branch*.

The voltage across any branch of a parallel combination is equal to the voltage across each of the other branches in parallel.

To illustrate voltage drop in a parallel circuit, let us examine Figure 5–7(a). Points *A, B, C,* and *D* along the left side of the parallel circuit are *electrically the same point* because the voltage is the same along this line. You can think of all of these points as being connected by a single wire to the negative terminal of the battery. The points *E, F, G,* and *H* along the right side of the circuit are all at a potential equal to that of the positive terminal of the source. Thus, each voltage across each parallel resistor is the same, and each is equal to the source voltage.

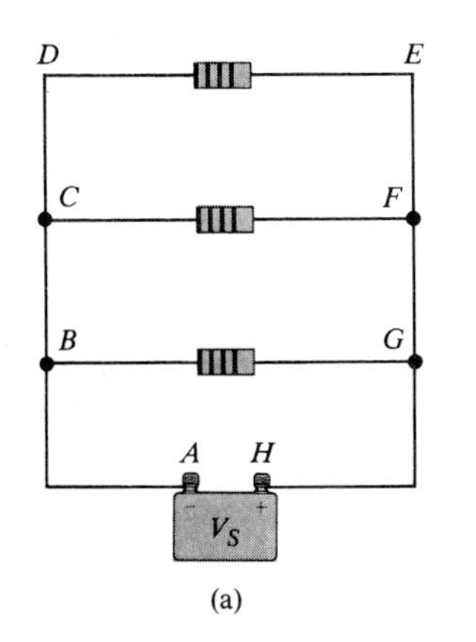

(a)

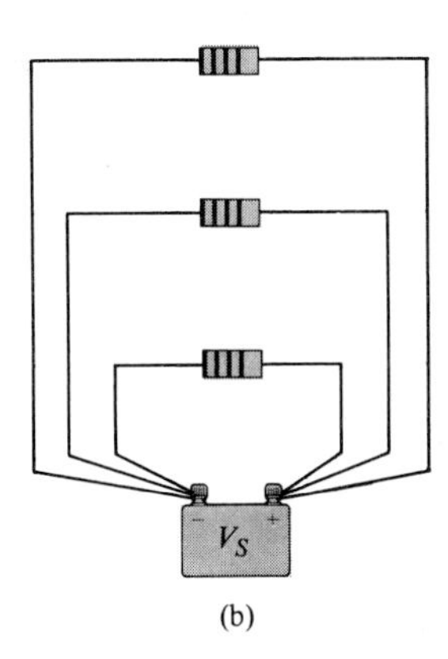

(b)

FIGURE 5–7
Voltage across parallel branches is the same.

Figure 5–7(b) is the same circuit as in Part (a), drawn in a slightly different way. Here the left side of the resistors are connected to a single point, which is the negative battery terminal. The right side of the resistors are all connected to the same point, which is the positive battery terminal. The resistors are still all in parallel across the source.

In Figure 5–8, a 12-V battery is connected across three parallel resistors. When the voltage is measured across the battery and then across each of the resistors, the readings are the same. Thus, as you can see, the same voltage appears across each branch in a parallel circuit.

SECTION REVIEW 5–2

1. A 10-Ω and a 22-Ω resistor are connected in parallel with a 5-V source. What is the voltage across each of the resistors?
2. A voltmeter connected across R_1 in Figure 5–9 measures 118 V. If you move the meter and connect it across R_2, how much voltage will it indicate? What is the source voltage?
3. In Figure 5–10, how much voltage does voltmeter 1 indicate? Voltmeter 2?

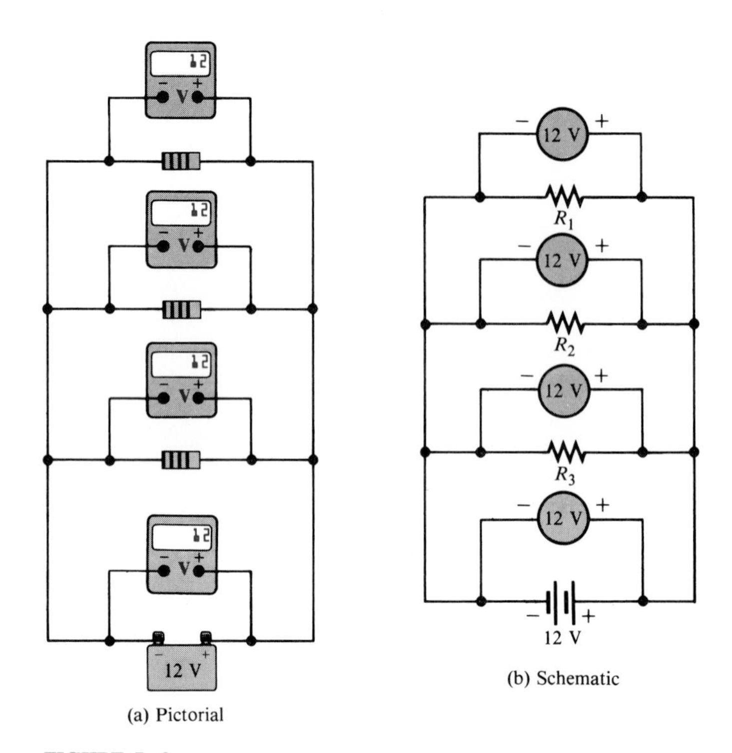

FIGURE 5–8
The same voltage appears across each resistor in parallel.

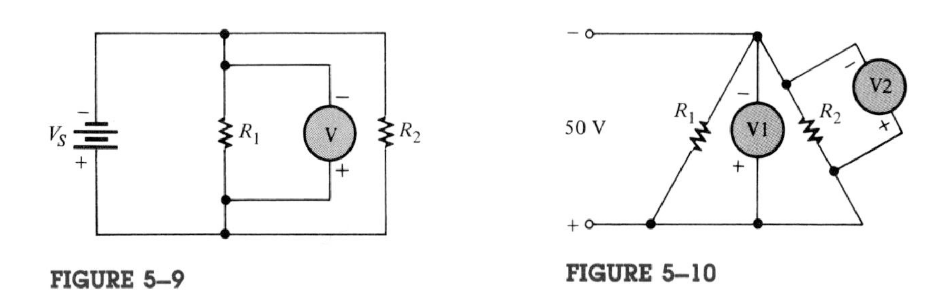

FIGURE 5–9

FIGURE 5–10

KIRCHHOFF'S CURRENT LAW

5–3

Kirchhoff's current law is stated as follows:

> **The sum of the currents into a junction is equal to the sum of the currents out of that junction.**

A *junction* is any point in a circuit where two or more circuit paths come together. In a parallel circuit, a junction is a point where the parallel branches connect together. The following is another way to state Kirchhoff's current law:

> **The total current into a junction is equal to the total current out of that junction.**

For example, in the circuit of Figure 5–11, point A is one junction and point B is another. Let us start at the negative terminal of the source and follow the current. The total current I_T flows from the source and *into* the junction at point A. At this point, the current splits up among the three branches as indicated. Each of the three branch currents (I_1, I_2, and I_3) flows *out of* junction A. Kirchhoff's current law says that the total current into junction A is equal to the total current out of junction A; that is,

$$I_T = I_1 + I_2 + I_3$$

Now, following the currents in Figure 5–11 through the three branches, you see that they come back together at point B. Currents I_1, I_2, and I_3 flow into junction B, and I_T flows out. Kirchhoff's current law formula at this junction is therefore the same as at junction A:

$$I_T = I_1 + I_2 + I_3$$

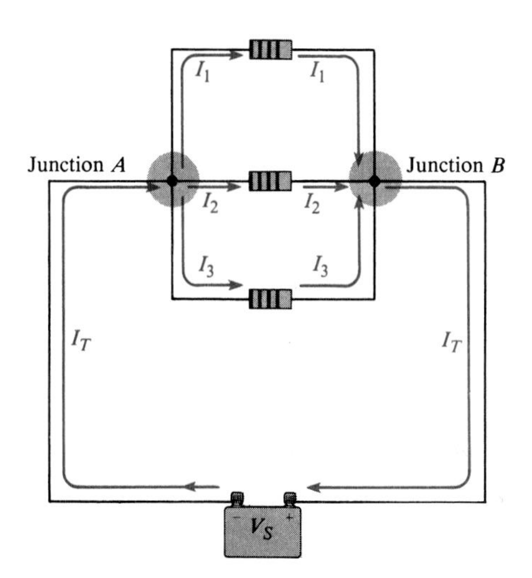

FIGURE 5–11
Kirchhoff's current law: The current into a junction equals the current out of that junction.

General Formula for Kirchhoff's Current Law

The previous discussion was a specific case to illustrate Kirchhoff's current law, often abbreviated KCL. Now let us look at the general case. Figure 5–12 shows a *generalized* circuit junction where a number of branches are connected to a point in the circuit. Currents $I_{IN(1)}$ through $I_{IN(n)}$ flow into the junction (n can be any number). Currents $I_{OUT(1)}$ through $I_{OUT(m)}$ flow out of the junction (m can be any number, but not necessarily equal to n). By Kirchhoff's current law, the sum of the currents into a junction must equal the sum of the currents out of the junction. With reference to Figure 5–12, the general formula for Kirchhoff's current law is

$$I_{IN(1)} + I_{IN(2)} + \cdots + I_{IN(n)} = I_{OUT(1)} + I_{OUT(2)} + \cdots + I_{OUT(m)} \quad (5\text{–}1)$$

If all of the terms on the right side of Equation (5–1) are brought over to the left side, their signs change to negative, and a zero is left on the right side.

Kirchhoff's current law is sometimes stated in this way:

> **The algebraic sum of all the currents entering and leaving a junction is equal to zero.**

This statement is just another, equivalent way of stating what we have just discussed.

You can verify Kirchhoff's current law by connecting a circuit and measuring each branch current and the total current from the source, as illustrated in Figure 5–13. When the branch currents are added together, their sum will equal the total current. This rule applies for any number of branches.

$$I_{IN(1)} + I_{IN(2)} + I_{IN(3)} + \cdots + I_{IN(n)} = I_{OUT(1)} + I_{OUT(2)} + I_{OUT(3)} + \cdots + I_{OUT(m)}$$

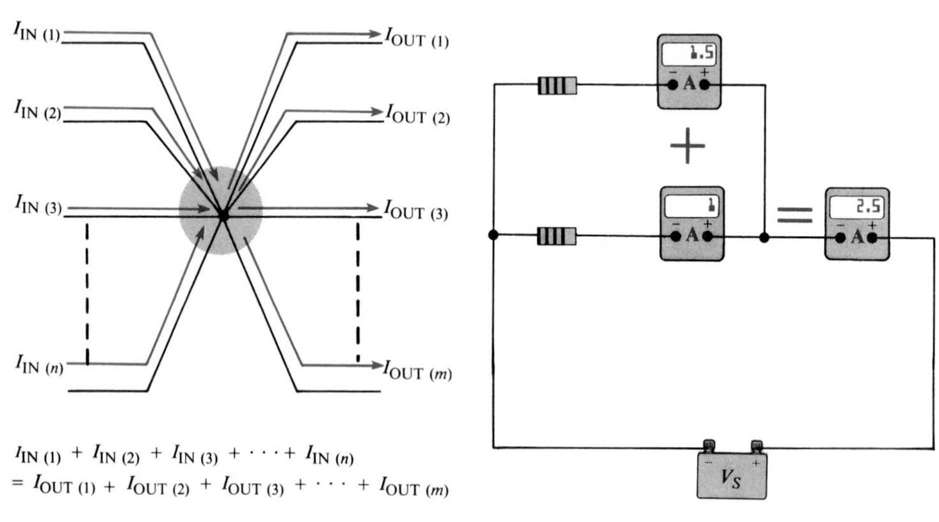

FIGURE 5–12
Generalized circuit junction to illustrate Kirchhoff's current law.

FIGURE 5–13
Experimental verification of Kirchhoff's current law.

The following three examples illustrate use of Kirchhoff's current law.

EXAMPLE 5–3 The branch currents in the circuit of Figure 5–14 are known. Determine the total current entering junction *A* and the total current leaving junction *B*.

FIGURE 5–14

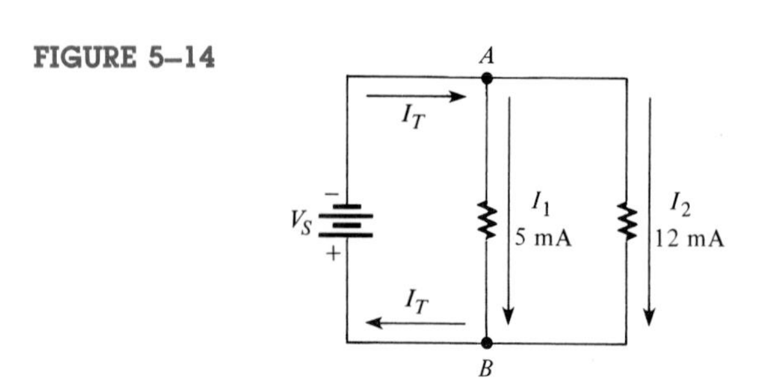

Solution
The total current flowing out of junction A is the sum of the two branch currents. So the total current flowing into A is

$$I_T = I_1 + I_2 = 5\text{ mA} + 12\text{ mA} = 17\text{ mA}$$

The total current entering point B is the sum of the two branch currents. So the total current flowing out of B is

$$I_T = I_1 + I_2 = 5\text{ mA} + 12\text{ mA} = 17\text{ mA}$$

The calculator sequence is
[5] [EE] [+/−] [3] [+] [1] [2] [EE] [+/−] [3] [=]

EXAMPLE 5–4 Determine the current through R_2 in Figure 5–15.

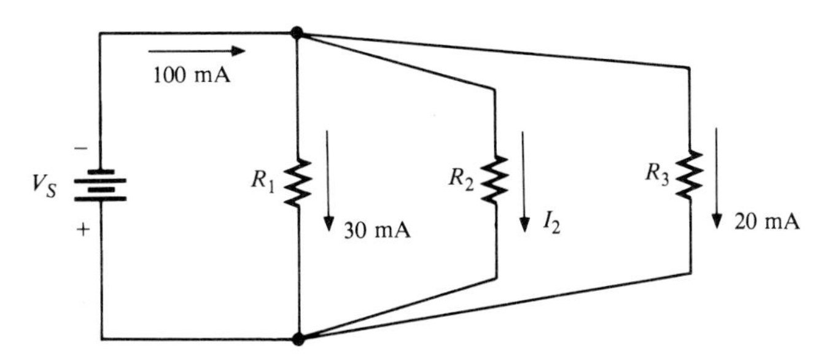

FIGURE 5–15

Solution
The total current flowing into the junction of the three branches is known, and two of the branch currents are known. The current equation at this junction is

$$I_T = I_1 + I_2 + I_3$$

Solving for I_2, we get

$$I_2 = I_T - I_1 - I_3 = 100\text{ mA} - 30\text{ mA} - 20\text{ mA} = 50\text{ mA}$$

EXAMPLE 5–5 Use Kirchhoff's current law to find the current measured by ammeters A_1 and A_2 in Figure 5–16.

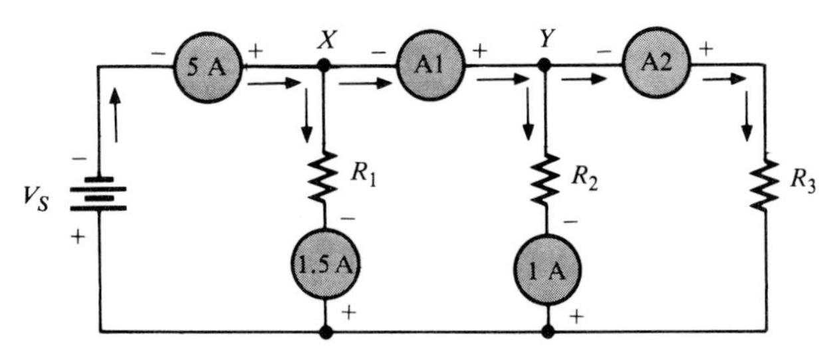

FIGURE 5–16

Solution

The total current into junction X is 5 A. Two currents flow out of junction X: 1.5 A through resistor R_1 and the current through A1. Kirchhoff's current law at junction X is

$$5\text{ A} = 1.5\text{ A} + I_{A1}$$

Solving for I_{A1} yields

$$I_{A1} = 5\text{ A} - 1.5\text{ A} = 3.5\text{ A}$$

The total current into junction Y is $I_{A1} = 3.5$ A. Two currents flow out of junction Y: 1 A through resistor R_2 and the current through A_2 and R_3. Kirchhoff's current law applied at junction Y gives

$$3.5\text{ A} = 1\text{ A} + I_{A2}$$

Solving for I_{A2} yields

$$I_{A2} = 3.5\text{ A} - 1\text{ A} = 2.5\text{ A}$$

SECTION REVIEW 5–3

1. A total current of 2.5 A flows into the junction of three parallel branches. What is the sum of all three branch currents?
2. In Figure 5–17, 100 mA and 300 mA flow into the junction. What is the amount of current flowing out of the junction?
3. Determine I_1 in the circuit of Figure 5–18.
4. Two branch currents enter a junction, and two branch currents leave the same junction. One of the currents entering the junction is 1 A, and one of the currents leaving the junction is 3 A. The total current entering and leaving the junction is 8 A. Determine the value of the unknown current entering the junction and the value of the unknown current leaving the junction.

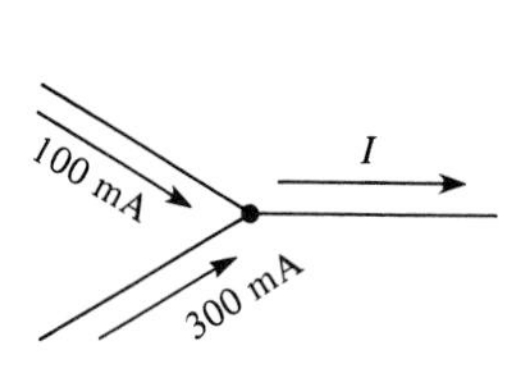

FIGURE 5–17

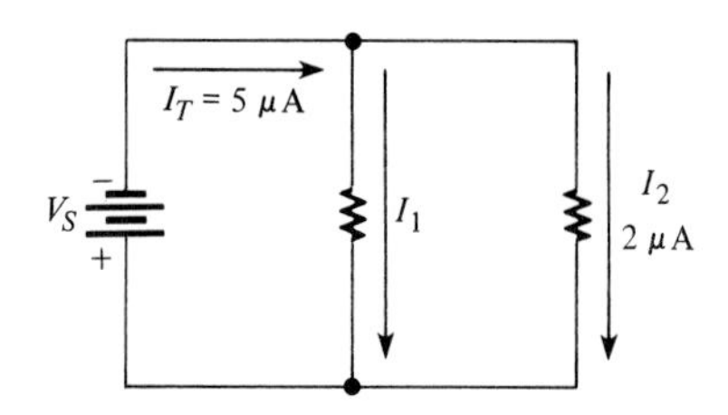

FIGURE 5–18

TOTAL PARALLEL RESISTANCE

5–4

When resistors are connected in parallel, the total resistance of the circuit decreases. The total resistance of a parallel combination is always less than the value of the smallest resistor.

For example, if a 10-Ω resistor and a 100-Ω resistor are connected in parallel, the total resistance is less than 10 Ω. The exact value must be calculated, and you will learn how to do so later in this section.

How the Number of Current Paths Affects Resistance

As you know, when resistors are connected in parallel, the current has more than one path. The number of current paths is equal to the number of parallel branches.

For example, in Figure 5–19(a), there is only one current path since it is a series circuit. A certain amount of current, I_1, flows through R_1. If resistor R_2 is connected in parallel with R_1, as shown in Figure 5–19(b), an *additional* amount of current, I_2, flows through R_2. The total current coming from the source has *increased* with the addition of the parallel branch. An increase in the *total current* from the source means that the total resistance has decreased, in accordance with Ohm's law. Additional resistors connected in parallel will further reduce the resistance and increase the total current.

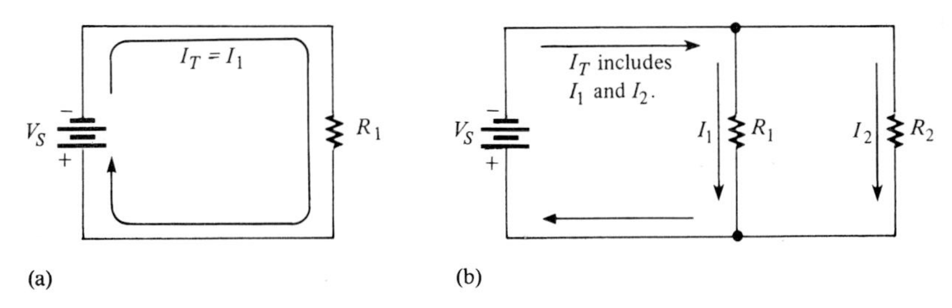

FIGURE 5–19
A greater number of resistors in parallel reduces total resistance and increases total current.

Formula for Total Parallel Resistance

The circuit in Figure 5–20 shows a general case of n resistors in parallel (n can be any number). From Kirchhoff's current law, the current equation is

$$I_T = I_1 + I_2 + I_3 + \cdots + I_n$$

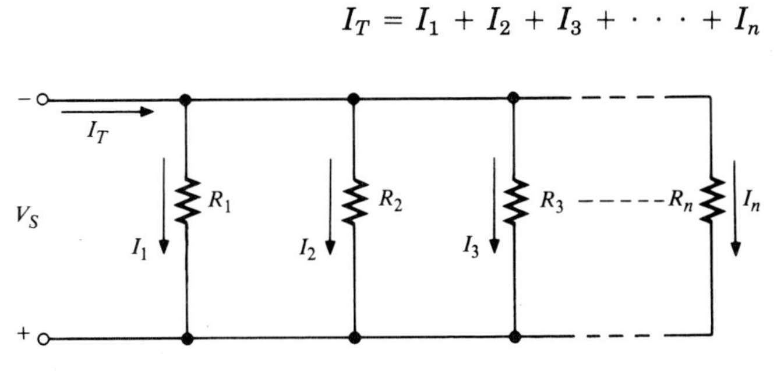

FIGURE 5–20
Circuit with n resistors in parallel.

Since V_S is the voltage across each of the parallel resistors, by Ohm's law, $I_1 = V_S/R_1$, $I_2 = V_S/R_2$, and so on. By substituting into the current equation, we get

$$\frac{V_S}{R_T} = \frac{V_S}{R_1} + \frac{V_S}{R_2} + \frac{V_S}{R_3} + \cdots + \frac{V_S}{R_n}$$

We factor V_S out of the right side of the equation and cancel it with V_S on the left side, leaving only the resistance terms:

$$\frac{1}{R_T} = \frac{1}{R_1} + \frac{1}{R_2} + \frac{1}{R_3} + \cdots + \frac{1}{R_n} \tag{5–2}$$

We get another useful form of Equation (5–2) by taking the reciprocal of (that is, by inverting) both sides of the equation:

$$R_T = \frac{1}{\left(\frac{1}{R_1}\right) + \left(\frac{1}{R_2}\right) + \left(\frac{1}{R_3}\right) + \cdots + \left(\frac{1}{R_n}\right)} \tag{5–3}$$

Equation (5–3) shows that to find the total parallel resistance, add all the $1/R$ terms and then take the reciprocal of the sum. Example 5–6 shows how to use this formula in a specific case, and Example 5–7 shows a simple method of using the calculator in determining parallel resistances. The reciprocal of resistance ($1/R$) is called *conductance* and is symbolized by G. We will, however, continue to use $1/R$.

EXAMPLE 5–6

Calculate the total parallel resistance between points A and B of the circuit in Figure 5–21.

FIGURE 5–21

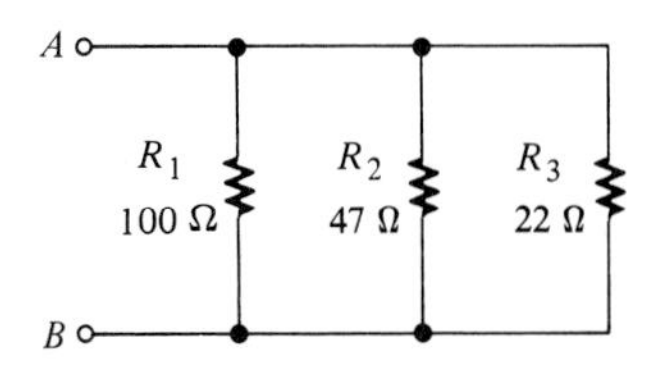

Solution

First, find the reciprocal of each of the three resistors as follows:

$$\frac{1}{R_1} = \frac{1}{100\ \Omega} = 0.01$$

$$\frac{1}{R_2} = \frac{1}{47\ \Omega} = 0.0213$$

$$\frac{1}{R_3} = \frac{1}{22\ \Omega} = 0.0455$$

Next, calculate R_T by adding $1/R_1$, $1/R_2$, and $1/R_3$ and taking the reciprocal of the sum as follows:

$$R_T = \frac{1}{0.01 + 0.0213 + 0.045} = \frac{1}{0.076} = 13.02\ \Omega$$

Notice that the value of R_T (13.02 Ω) is smaller than the smallest value in parallel, which is R_3 (22 Ω).

Calculator Solution

The parallel resistance formula is easily solved on an electronic calculator. The general procedure is to enter the value of R_1 and then take its reciprocal by pressing the [1/x] key. Next press the [+] key; then enter the value of R_2 and take its reciprocal. Repeat this procedure until all of the resistor values have been entered and the reciprocal of each has been added. The final step is to press the [1/x] key to convert $1/R_T$ to R_T. The total parallel resistance is now on the display. This calculator procedure is illustrated in Example 5–7.

EXAMPLE 5–7 Show the steps required for a calculator solution of Example 5–6.

Solution

Step 1: Enter 100. Display shows 100.
Step 2: Press [1/x] key. Display shows 0.01.
Step 3: Press [+] key. Display shows 0.01.
Step 4: Enter 47. Display shows 47.
Step 5: Press [1/x] key. Display shows 0.0212766
Step 6: Press [+] key. Display shows 0.0312766
Step 7: Enter 22. Display shows 22.
Step 8: Press [1/x] key. Display shows 0.0454545
Step 9: Press [=] key. Display shows 0.0767311
Step 10: Press [1/x] key. Display shows 13.032518.

The number displayed in Step 10 is the total *resistance* in ohms.

Two Resistors in Parallel

Equation (5–3) is a general formula for finding the total resistance of any number of resistors in parallel. It is often useful to consider only two resistors in parallel because this situation occurs commonly in practice.

Based on Equation (5–3), the formula for two resistors in parallel is

$$R_T = \frac{R_1 R_2}{R_1 + R_2} \qquad (5\text{–}4)$$

Equation (5–4) states that *the total resistance of two resistors in parallel is equal to the product of the two resistors divided by the sum of the two resistors.* This equation is sometimes referred to as the "product-over-the-sum" formula. Example 5–8 illustrates how to use it.

EXAMPLE 5–8 Calculate the total resistance between the positive and negative terminals of the source of the circuit in Figure 5–22.

FIGURE 5–22

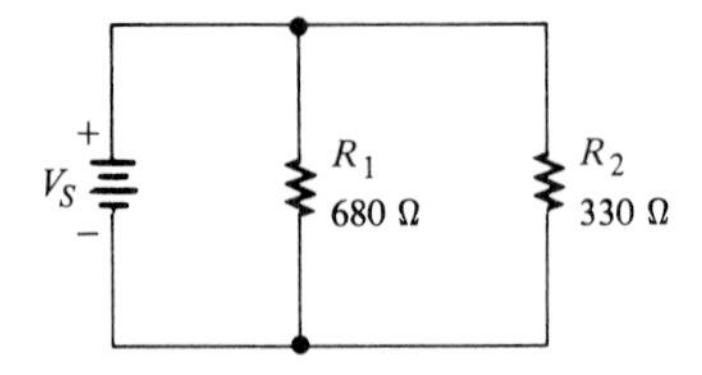

Solution
Use Equation (5–4) as follows:

$$R_T = \frac{R_1 R_2}{R_1 + R_2} = \frac{(680\ \Omega)(330\ \Omega)}{680\ \Omega + 330\ \Omega}$$
$$= \frac{224{,}400\ \Omega^2}{1{,}010\ \Omega} = 222.18\ \Omega$$

The calculator sequence is
6 8 0 × 3 3 0 ÷ (6 8 0 + 3 3 0) =

Resistors of Equal Value in Parallel

Another special case of parallel circuits is the parallel connection of several resistors having the same ohmic value. The following is a shortcut method of calculating R_T when this case occurs:

$$R_T = \frac{R}{n} \tag{5–5}$$

Equation (5–5) says that when any number of resistors (n), all having the same resistance (R), are connected in parallel, R_T is equal to the resistance divided by the number of resistors in parallel. Example 5–9 shows how to use this formula.

EXAMPLE 5–9 Find the total resistance between points A and B in Figure 5–23.

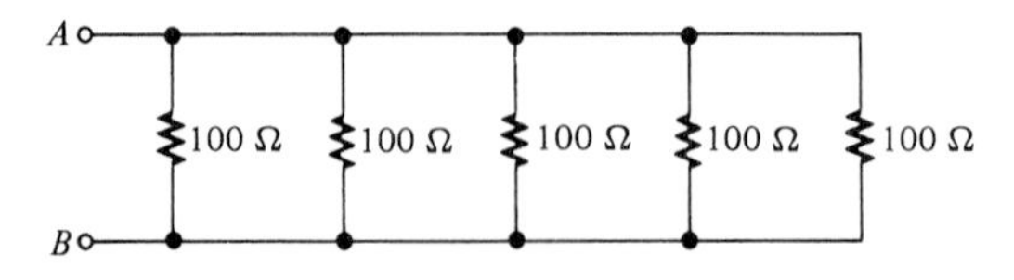

FIGURE 5–23

Solution
There are five 100-Ω resistors in parallel. Use Equation (5–5) as follows:

$$R_T = \frac{R}{n} = \frac{100\ \Omega}{5} = 20\ \Omega$$

EXAMPLE 5–10 A stereo amplifier drives two 8-Ω speakers in parallel as shown in Figure 5–24. What is the total resistance across the output terminals of the amplifier?

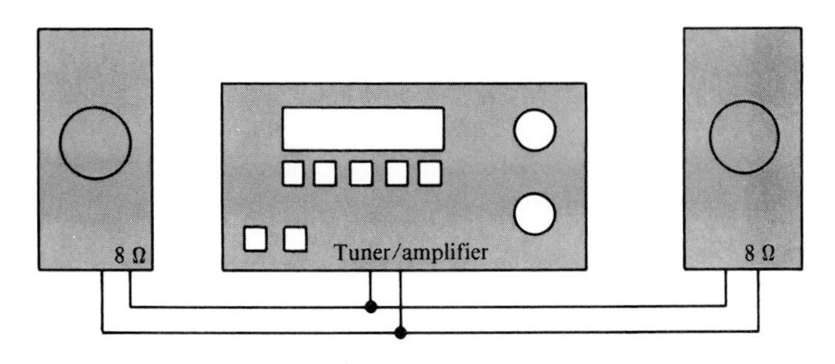

FIGURE 5–24

Solution
The total parallel resistance of the two 8-Ω speakers is

$$R_T = \frac{R}{n} = \frac{8\ \Omega}{2} = 4\ \Omega$$

Notation for Parallel Resistors

Sometimes, for convenience, parallel resistors are designated by two parallel vertical marks. For example, R_1 in parallel with R_2 can be written as $R_1 \| R_2$. Also, when several resistors are in parallel with each other, this notation can be used. For example,

$$R_1 \| R_2 \| R_3 \| R_4 \| R_5$$

indicates that R_1 through R_5 are all in parallel.

This notation is also used with resistance values. For example,

$$10\ \text{k}\Omega \| 5\ \text{k}\Omega$$

means that a 10-kΩ resistor is in parallel with a 5-kΩ resistor.

SECTION REVIEW 5–4

1. Does the total resistance increase or decrease as more resistors are connected in parallel?
2. The total parallel resistance is always less than ________.
3. Determine R_T for each circuit in Figures 5–25, 5–26, and 5–27. (Full-color photos appear between pages 128 and 129.)

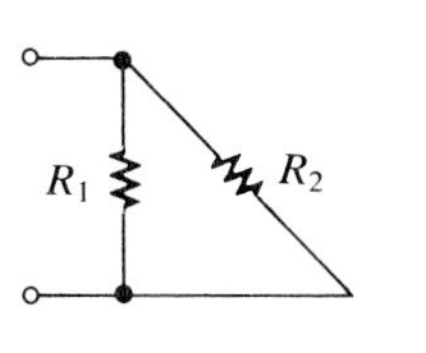

FIGURE 5–25

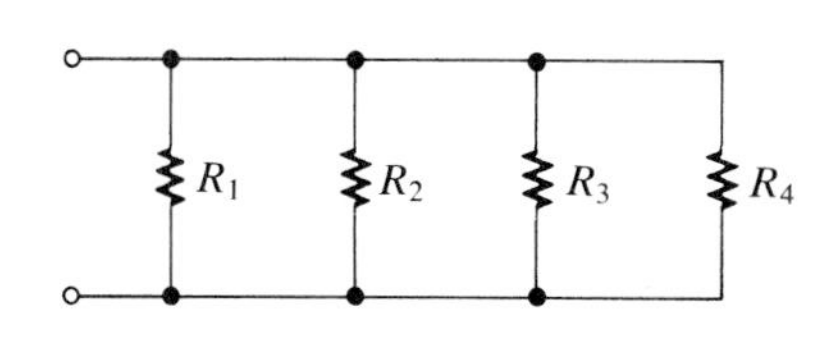

FIGURE 5–26

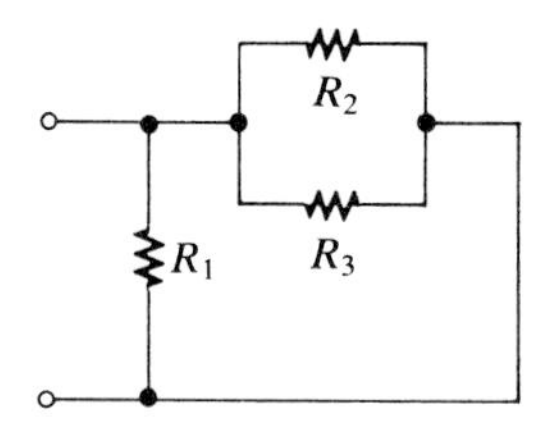

FIGURE 5–27

OHM'S LAW IN PARALLEL CIRCUITS

5–5

In this section, several circuits are analyzed using Ohm's law and the basic concepts of parallel circuits.

EXAMPLE 5–11

The circuit board in Figure 5–28 has three resistors in parallel used for bias modification in an instrumentation amplifier. The values of two of the resistors are known from the color codes, but the third resistor is unmarked (maybe the bands are worn off from handling). Determine the value of the unknown resistor.

FIGURE 5–28

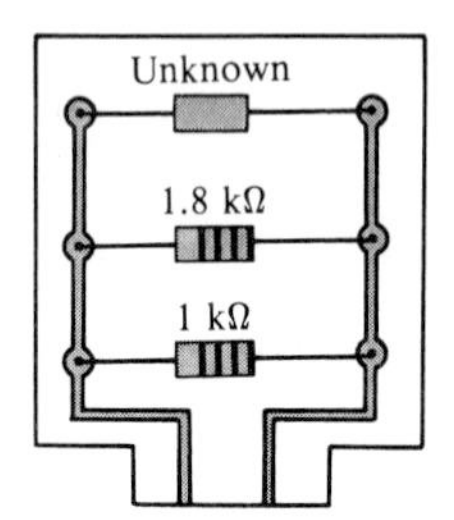

Solution

If we can determine the total resistance of the three resistors in parallel, then we can use the parallel resistance formula to calculate the unknown resistance. We can use Ohm's law to find the total resistance if voltage and total current are known.

In Figure 5–29, a 10-V source (arbitrary value) is connected across the resistors and the total current is measured. Using these measured values, we find that the total resistance is

$$R_T = \frac{V}{I_T} = \frac{10\text{ V}}{20.1\text{ mA}} = 498\ \Omega$$

Use the reciprocal resistance formula to find the unknown resistance as follows:

$$\frac{1}{R_T} = \frac{1}{R_1} + \frac{1}{R_2} + \frac{1}{R_3}$$

$$\frac{1}{R_3} = \frac{1}{R_T} - \frac{1}{R_1} - \frac{1}{R_2} = \frac{1}{498\ \Omega} - \frac{1}{1\ \text{k}\Omega} - \frac{1}{1.8\ \text{k}\Omega} = 0.000453$$

$$R_3 = \frac{1}{0.000453} \cong 2.2\ \text{k}\Omega$$

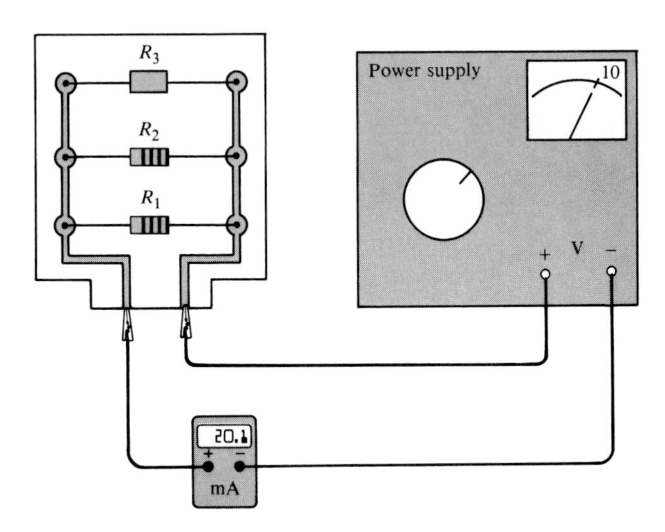

FIGURE 5–29

The calculator sequence for R_3 is

4 9 8 2nd 1/x − 1 EE 3 2nd 1/x − 1 · 8 EE 3 2nd 1/x = 2nd 1/x

---

EXAMPLE 5–12 Find the total current produced by the battery in Figure 5–30.

FIGURE 5–30

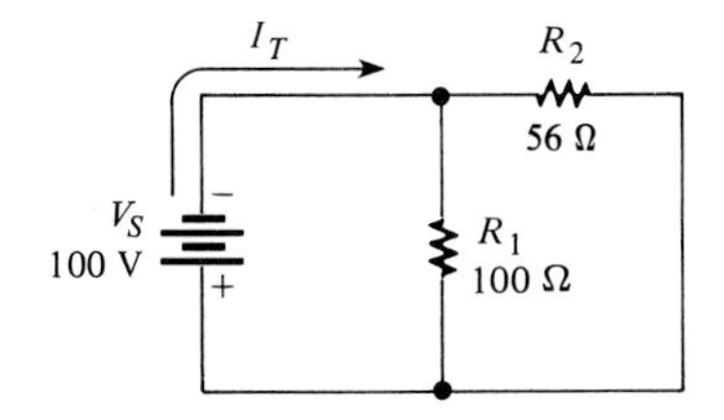

Solution

Step 1: The battery "sees" a total parallel resistance which determines the amount of current that it generates. First, calculate R_T:

$$R_T = \frac{R_1R_2}{R_1 + R_2} = \frac{(100\ \Omega)(56\ \Omega)}{100\ \Omega + 56\ \Omega}$$

$$= \frac{5600\ \Omega^2}{156\ \Omega} = 35.9\ \Omega$$

Step 2: The battery voltage is 100 V. Use Ohm's law to find I_T:

$$I_T = \frac{100 \text{ V}}{35.9 \ \Omega} = 2.79 \text{ A}$$

The calculator sequence for R_T is

1 0 0 × 5 6 ÷ (1 0 0 + 5 6) =

and the calculator sequence for I_T is

1 0 0 ÷ 3 5 . 9 =

EXAMPLE 5–13 Determine the current through each resistor in the parallel circuit of Figure 5–31.

FIGURE 5–31

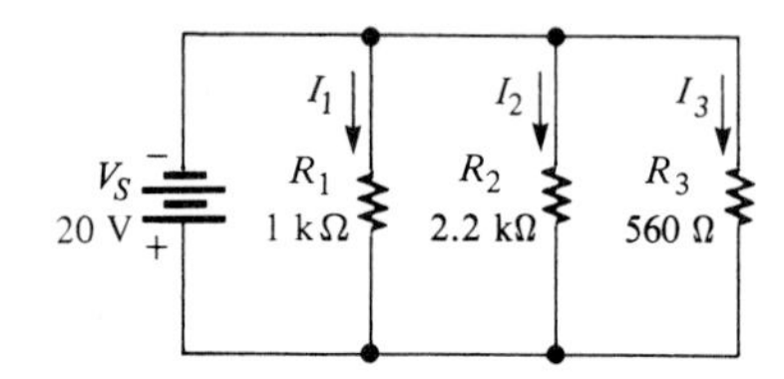

Solution

The voltage across each resistor (branch) is equal to the source voltage. That is, the voltage across R_1 is 20 V, the voltage across R_2 is 20 V, and the voltage across R_3 is 20 V. The current through each resistor is determined as follows:

$$I_1 = \frac{V_S}{R_1} = \frac{20 \text{ V}}{1 \text{ k}\Omega} = 20 \text{ mA}$$

$$I_2 = \frac{V_S}{R_2} = \frac{20 \text{ V}}{2.2 \text{ k}\Omega} = 9.1 \text{ mA}$$

$$I_3 = \frac{V_S}{R_3} = \frac{20 \text{ V}}{560 \ \Omega} = 35.7 \text{ mA}$$

EXAMPLE 5–14 Find the voltage V_S across the parallel circuit in Figure 5–32.

FIGURE 5–32

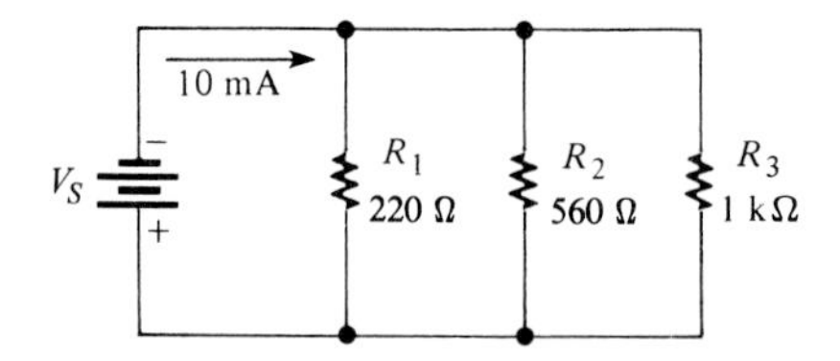

Solution

We know the total current into the parallel circuit. We need to know the total resistance, and then we can apply Ohm's law to get the voltage. The total resistance is

$$R_T = \frac{1}{\left(\frac{1}{R_1}\right) + \left(\frac{1}{R_2}\right) + \left(\frac{1}{R_3}\right)} = \frac{1}{0.00455 + 0.00179 + 0.001}$$

$$= \frac{1}{0.00734} = 136.24\ \Omega$$

$$V_S = I_T R_T = (10\ \text{mA})(136.24\ \Omega) = 1.3624\ \text{V}$$

SECTION REVIEW 5–5

1. A 10-V battery is connected across three 68-Ω resistors that are in parallel. What is the total current from the battery?
2. How much voltage is required to produce 2 A of current through the circuit of Figure 5–33?
3. How much current is there through each resistor of Figure 5–33?
4. There are four equal-value resistors in parallel with a 12-V source, and 6 mA of current from the source. What is the value of each resistor?
5. A 1-kΩ and a 2.2-kΩ resistor are connected in parallel. A total of 100 mA flows through the parallel combination. How much voltage is dropped across the resistors?

FIGURE 5–33

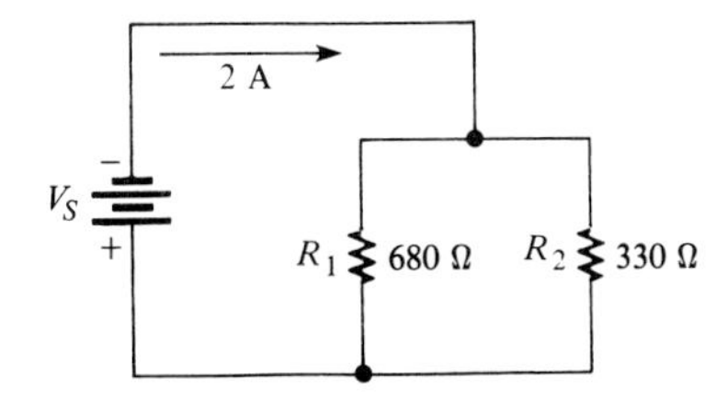

THE CURRENT DIVIDER PRINCIPLE

5–6

In a parallel circuit, when the total current flows into the junction of the parallel branches, it *divides* among the branches. Thus, a parallel circuit acts as a current divider. This current divider principle is illustrated in Figure 5–34 for a two-branch parallel circuit in which part of the total current I_T goes through R_1 and part through R_2.

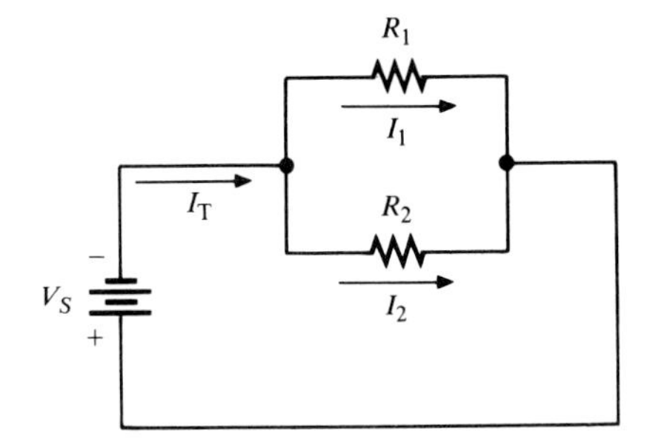

FIGURE 5–34
Total current divides between two branches.

Since the same voltage is across each of the resistors in parallel, the branch currents are inversely proportional to the ohmic values of the resistors. For example, if the value of R_2 is twice that of R_1, then the value of I_2 is one-half that of I_1. In other words,

> **the total current divides among parallel resistors in a manner inversely proportional to the resistance values.**

The branches with higher resistance have less current, and the branches with lower resistance have more current, in accordance with Ohm's law. If all the branches have the same resistance, the branch currents are all equal.

Figure 5–35 shows specific values to demonstrate how the currents divide according to the branch resistances. Notice that in this case the resistance of the upper branch is one-tenth the resistance of the lower branch, but the upper branch current is *ten times* the lower branch current.

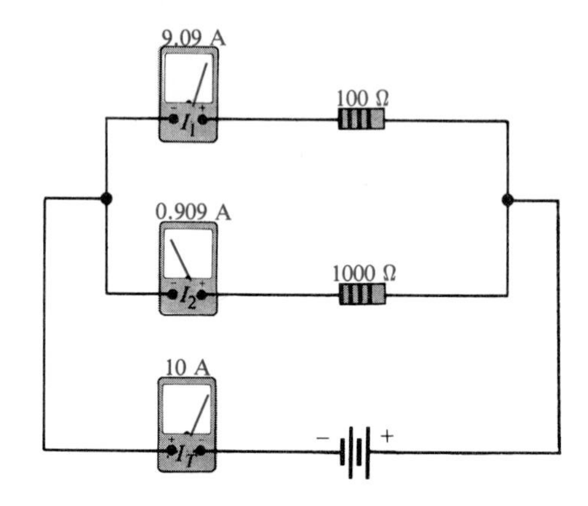

FIGURE 5–35
The branch with the lowest resistance has the most current, and the branch with the highest resistance has the least current.

Formulas for Two Branches

You already know how to use Ohm's law ($I = V/R$) to determine the current in any parallel branch when the voltage and resistance are known. When the voltage is not known but the total current is, you can find both of the branch currents (I_1 and I_2) by using the following formulas:

$$I_1 = \left(\frac{R_2}{R_1 + R_2}\right) I_T \tag{5–6}$$

$$I_2 = \left(\frac{R_1}{R_1 + R_2}\right) I_T \tag{5–7}$$

Note that the current in either branch is equal to the *opposite* branch resistance over the *sum* of the two resistors, multiplied times the total current.

EXAMPLE 5–15 Find I_1 and I_2 in Figure 5–36.

FIGURE 5–36

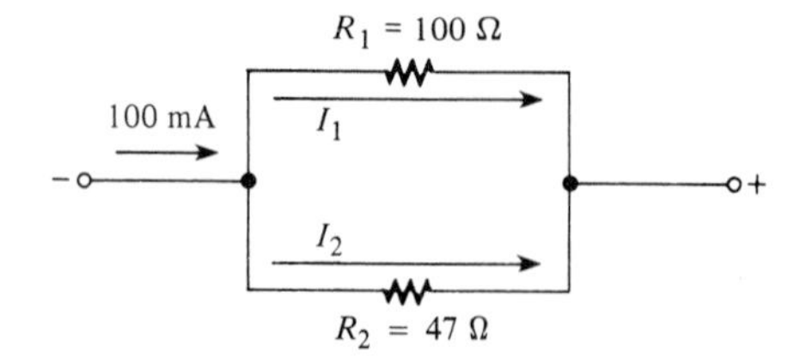

Solution
Using Equation (5–6), we determine I_1 as follows:

$$I_1 = \left(\frac{R_2}{R_1 + R_2}\right)I_T = \left(\frac{47\ \Omega}{147\ \Omega}\right)100\ \text{mA}$$
$$= 31.97\ \text{mA}$$

Using Equation (5–7), we determine I_2 as follows:

$$I_2 = \left(\frac{R_1}{R_1 + R_2}\right)I_T = \left(\frac{100\ \Omega}{1.47\ \Omega}\right)100\ \text{mA}$$
$$= 68.03\ \text{mA}$$

The calculator sequence for I_2 is

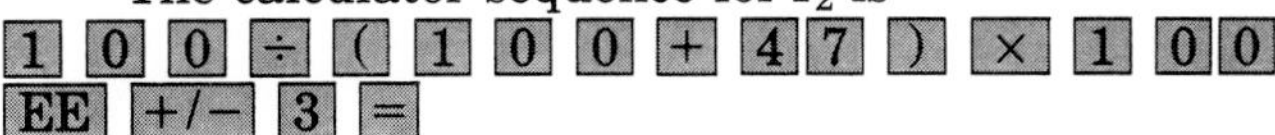

General Formula for Any Number of Branches

A generalized parallel circuit with n branches is shown in Figure 5–37 (n represents any number).

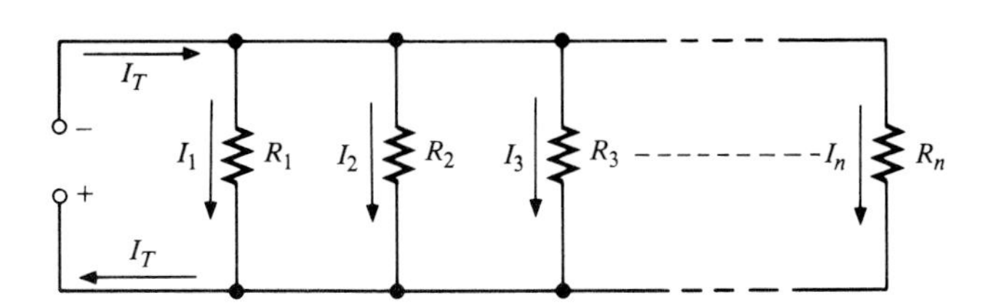

FIGURE 5–37
Generalized parallel circuit with n branches.

The current in any branch can be determined with the following formula:

$$I_x = \left(\frac{R_T}{R_x}\right)I_T \tag{5–8}$$

where I_x represents any branch current (I_1, I_2, and so on) and R_x represents any resistance (R_1, R_2, and so on). For example, the formula for current in the second branch is

$$I_2 = \left(\frac{R_T}{R_2}\right)I_T$$

Equation (5–8) can be used for a parallel circuit with any number of branches. The formulas for the special case of two branches are actually derived from this general formula. Notice that you must determine R_T in order to use Equation (5–8) to find a branch current.

EXAMPLE 5–16 Determine the current through each resistor in the circuit of Figure 5–38.

FIGURE 5–38

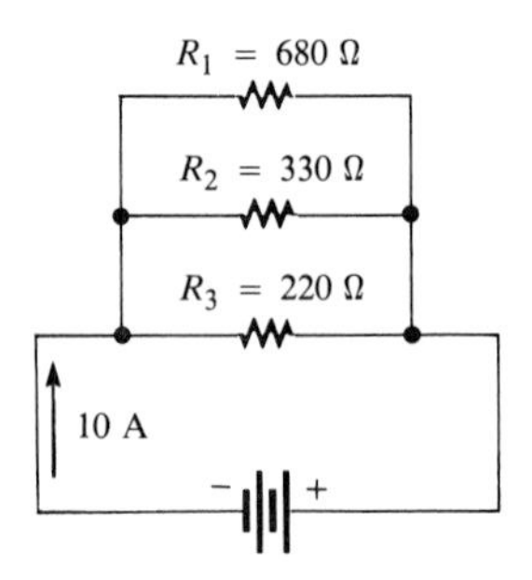

Solution

First calculate the total parallel resistance:

$$R_T = \frac{1}{\left(\frac{1}{R_1}\right) + \left(\frac{1}{R_2}\right) + \left(\frac{1}{R_3}\right)}$$

$$= \frac{1}{\left(\frac{1}{680\ \Omega}\right) + \left(\frac{1}{330\ \Omega}\right) + \left(\frac{1}{220\ \Omega}\right)} = 110.5\ \Omega$$

The total current is 10 A. Using Equation (5–8), we calculate each branch current as follows:

$$I_1 = \left(\frac{R_T}{R_1}\right)I_T = \left(\frac{110.5\ \Omega}{680\ \Omega}\right)10\text{ A} = 1.63\text{ A}$$

$$I_2 = \left(\frac{R_T}{R_2}\right)I_T = \left(\frac{110.5\ \Omega}{330\ \Omega}\right)10\text{ A} = 3.35\text{ A}$$

$$I_3 = \left(\frac{R_T}{R_3}\right)I_T = \left(\frac{110.5\ \Omega}{220\ \Omega}\right)10\text{ A} = 5.02\text{ A}$$

SECTION REVIEW 5–6

1. A parallel circuit has the following resistors in parallel: 220-Ω, 100-Ω, 68-Ω, 56-Ω, and 22-Ω. Which resistor has the most current through it? The least current?
2. Determine the current through the right-most resistor in Figure 5–39. (A full-color photo appears between pages 128 and 129.)
3. Find the currents through each resistor in the circuit of Figure 5–40. (A full-color photo appears between pages 128 and 129.)

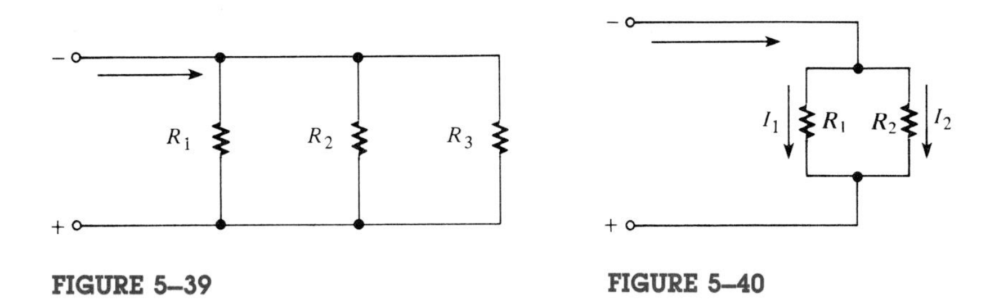

FIGURE 5–39 FIGURE 5–40

POWER IN A PARALLEL CIRCUIT

5–7

The total amount of power in a parallel resistive circuit is equal to the sum of the powers in each resistor in parallel. Equation (5–9) states this formula in a concise way for any number of resistors in parallel:

$$P_T = P_1 + P_2 + P_3 + \cdot \cdot \cdot + P_n \tag{5–9}$$

where P_T is the total power and P_n is the power in the last resistor in parallel. As you can see, the power losses are additive, just as in the series circuit.

EXAMPLE 5–17 The amplifier in the stereo system of Figure 5–41 drives four speakers as shown. If the maximum voltage to the speakers is 15 V, how much power must the amplifier be able to deliver to the speakers?

FIGURE 5–41

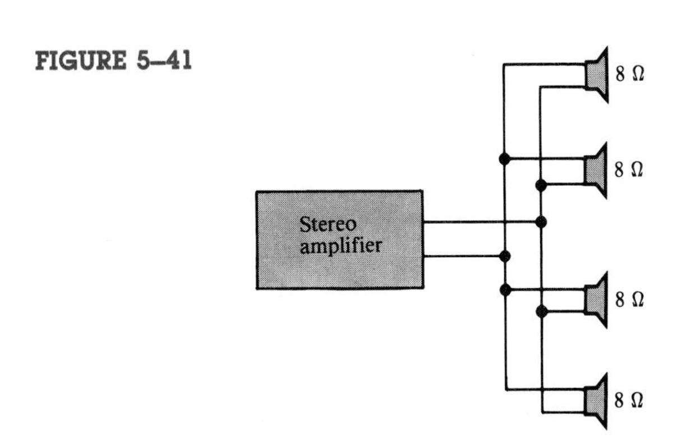

Solution
The speakers are connected in parallel to the amplifier output, so the voltage across each is the same. The maximum power to each speaker is

$$P = \frac{V^2}{R} = \frac{(15\text{ V})^2}{8\ \Omega} = 28.125\text{ W}$$

The total power that the amplifier must be capable of delivering to the speaker system is four times the power in an individual speaker because the total power is the sum of the individual powers:

$$P_T = P + P + P + P = 4P = 112.5\text{ W}$$

SECTION REVIEW 5–7

1. The resistors in a parallel circuit have the following powers: 2 W, 5 W, 1 W, and 8 W. What is the total power in the circuit?
2. A circuit has a 1-kΩ, a 2.2-kΩ, and a 3.9-kΩ resistor in parallel. A total current of 1 A flows into the parallel circuit. What is the total power?

HOW PARALLEL CIRCUITS ARE USED

5–8

Parallel circuits are found in some form in virtually every electronic device. In many of these applications, the parallel relationship of components may not be obvious until you have covered some advanced topics that will come later. Therefore, in this section, some common and familiar applications of parallel circuits are discussed.

Automotive

One advantage of a parallel circuit over a series circuit is that when one branch opens, the other branches are not affected. For example, Figure 5–42 shows a simplified diagram of an automobile lighting system. When one headlight on your car goes out, it does not cause the other lights to go out, because they are all in parallel.

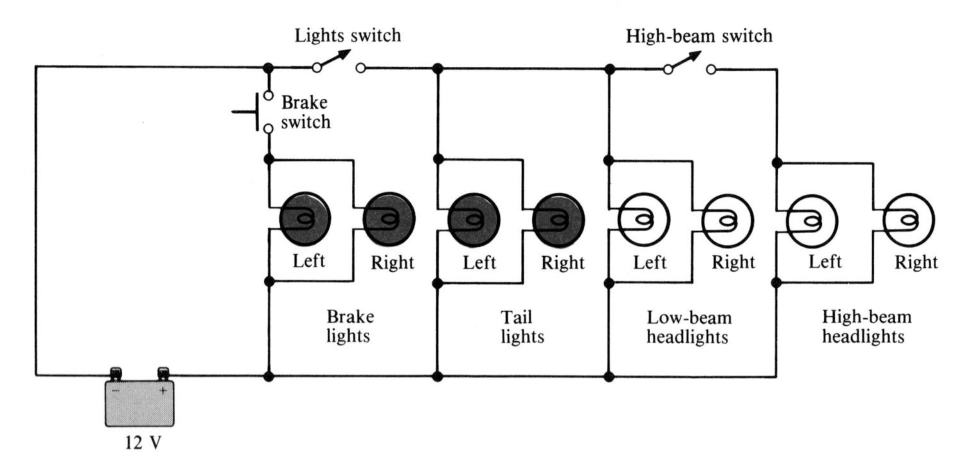

FIGURE 5–42
Simplified diagram of the exterior light system of an automobile.

Notice that the brake lights are switched on independently of the head and tail lights. They come on only when the driver closes the brake light switch by depressing the brake pedal.

When the *lights* switch is closed, both low-beam headlights and both taillights are on. The high-beam headlights are on only when both the *lights* switch and the *high-beam* switch are closed. If any one of the lights burns out (opens), there is still current in each of the other lights.

Residential

Another common use of parallel circuits is in residential electrical systems. All the lights and appliances in a home are wired in parallel. Figure 5–43(a) shows a typical room wiring arrangement with two switch-controlled lights and three wall outlets in parallel.

Figure 5–43(b) shows a simplified parallel arrangement of four heating elements on an electric range. The four-position switches in each branch allow the user to control the amount of current through the heating elements by selecting the appropriate limiting resistor. The lowest resistor value (H setting) allows the highest amount of current for maximum heat. The highest resistor value (L setting) allows the least amount of current for minimum heat; M designates the medium settings.

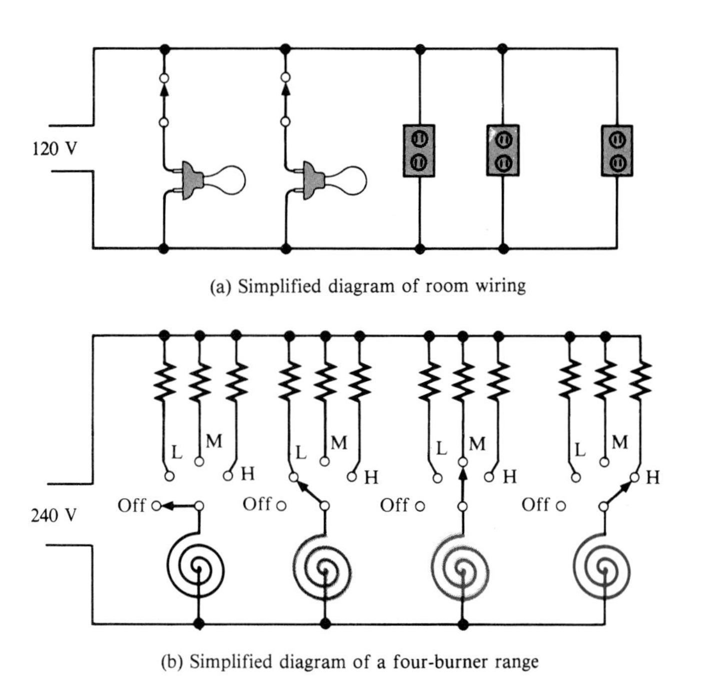

(a) Simplified diagram of room wiring

(b) Simplified diagram of a four-burner range

FIGURE 5–43
Examples of parallel circuits in residential wiring and appliances.

Ammeters

Another example in which parallel circuits are used is the familiar analog (needle-type) ammeter or milliammeter. Parallel circuits are an important part of the operation of the ammeter because they allow the user to select various ranges in order to measure many different current values.

The mechanism in an ammeter that causes the pointer to move in proportion to the current is called the *meter movement,* which is based on a magnetic principle that you will learn later. Right now, all you need to know is that a meter movement has a certain resistance and a maximum current. This maximum current, called the *full-scale deflection* current, causes the pointer to go all the way to the end of the scale. For example, a certain meter movement has a 50-Ω resistance and a full-scale deflection current of 1 mA. A meter with this particular movement can measure currents of 1 mA or less. Currents greater than 1 mA will cause the pointer to "peg" (or stop) at full scale. Figure 5–44 illustrates a 1-mA meter.

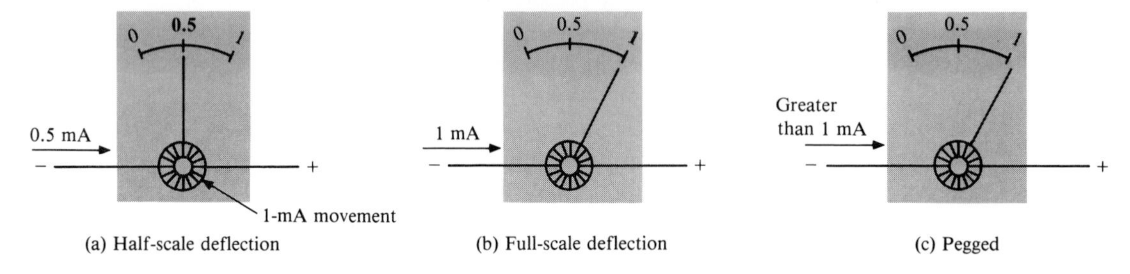

FIGURE 5–44
A 1-mA meter.

Figure 5–45 shows a simple ammeter with a resistor in parallel with the meter movement; this resistor is called a *shunt* resistor. Its purpose is to bypass any current in excess of 1 mA around the meter movement. The figure specifically shows 9 mA through the shunt resistor and 1 mA through the meter movement. Thus, up to 10 mA can be measured. To find the actual current value, simply multiply the reading on the scale by 10.

A practical ammeter has a range switch that permits the selection of several full-scale current settings. In each switch position, a certain amount of current is bypassed through a parallel resistor as determined by the resistance value. In our example, the current through the movement is never greater than 1 mA.

FIGURE 5–45
A 10-mA meter.

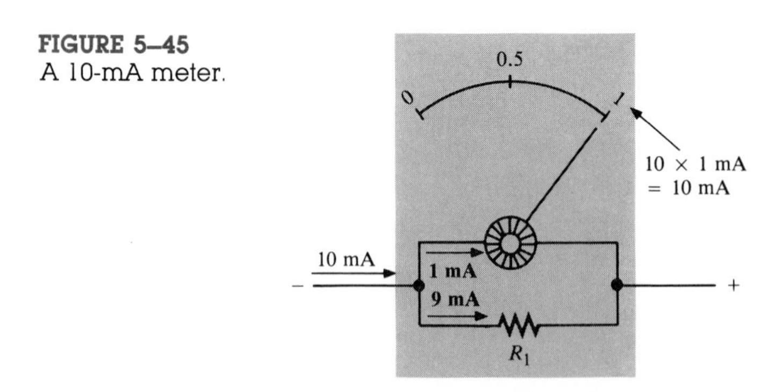

Figure 5–46 illustrates a meter with three ranges: 1 mA, 10 mA, and 100 mA. When the range switch is in the 1-mA position, all of the current coming

into the meter goes through the meter movement. In the 10-mA setting, up to 9 mA goes through R_1 and up to 1 mA through the movement. In the 100-mA setting, up to 99 mA goes through R_2, and the movement can still have only 1 mA for full-scale.

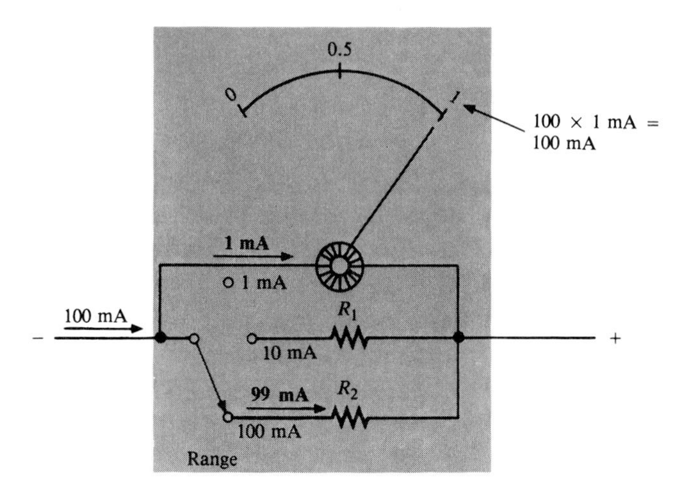

FIGURE 5–46
A milliammeter with three ranges.

For example, in Figure 5–46, if 50 mA of current are being measured, the needle points at the 0.5 mark on the scale; we must multiply 0.5 by 100 to find the current value. In this situation, 0.5 mA flows through the movement (half-scale deflection), and 49.5 mA flows through R_2.

SECTION REVIEW 5–8

1. For the ammeter in Figure 5–46, what is the maximum resistance that the meter will have when connected in a circuit? What is the maximum current that can be measured at the setting?
2. Do the shunt resistors have resistance values considerably less than or more than that of the movement? Why?

TROUBLES IN PARALLEL CIRCUITS

5–9

Recall that an open circuit is one in which the current path is interrupted and no current flows. In this section, the effect of an open branch in a parallel circuit is examined.

When an open occurs in a parallel branch, the total resistance increases, the total current decreases, and the same currents continue to flow in the remaining branches. The decrease in the total current equals the amount of current that was previously flowing in the branch that is now open.

Finding an Open Branch by Measurement

As illustrated in Figure 5–47, when a parallel resistor opens, the voltage across it does not change (unless the source voltage changes) because the voltage across all branches in parallel is the same whether or not they are open. *There is no way of determining the open resistor by measuring the voltage.*

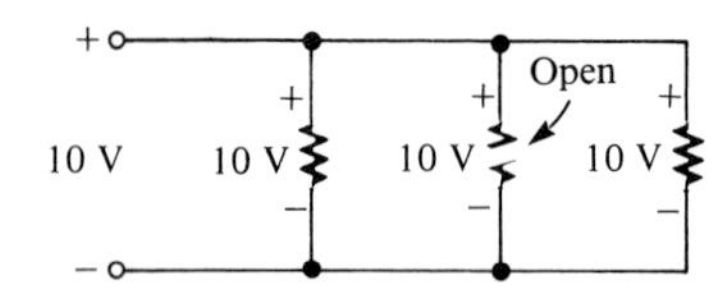

FIGURE 5–47
Parallel branches all have the same voltage.

If a visual inspection does not reveal the open resistor, you must locate it by measuring the total current into the parallel circuit. When one of the parallel branches is open, the total current is less than its normal value, as illustrated in Figure 5–48. Part (a) shows the normal total current for the two resistors in parallel. In Part (b), resistor R_2 is open, and the current through R_1 has not changed from its normal value. The total current, however, has now dropped to the same value as the R_1 current. In Part (c), resistor R_1 is open, and the total current has dropped to the same value as the R_2 current. You can verify the values in Figure 5–48 with Ohm's law.

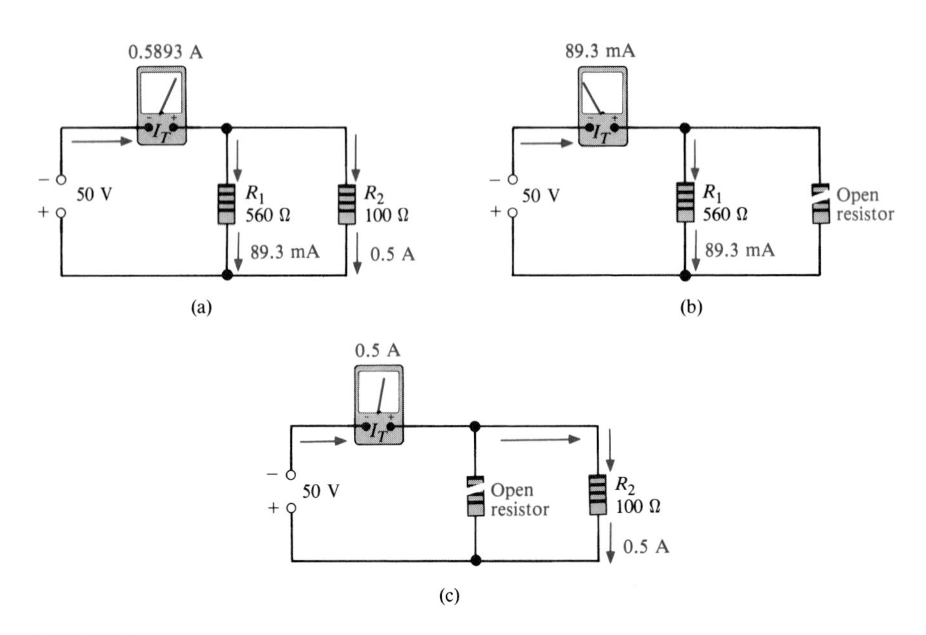

FIGURE 5–48
Effect of open branches on the total current.

The following example illustrates how to locate an open branch when all the branch currents are different.

EXAMPLE 5–18

In Figure 5–49, there is a total current of 31.09 mA, and the voltage across the parallel branches is 20 V. Is there an open resistor, and, if so, which one is it?

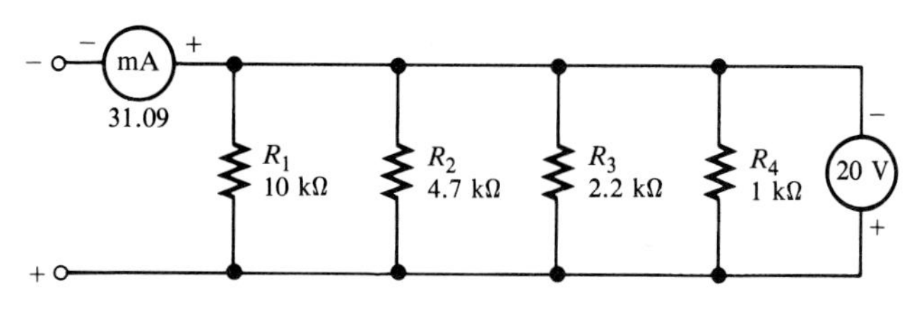

FIGURE 5–49

Solution

Calculate the current in each branch:

$$I_1 = \frac{V}{R_1} = \frac{20\text{ V}}{10\text{ k}\Omega} = 2\text{ mA}$$

$$I_2 = \frac{V}{R_2} = \frac{20\text{ V}}{4.7\text{ k}\Omega} = 4.26\text{ mA}$$

$$I_3 = \frac{V}{R_3} = \frac{20\text{ V}}{2.2\text{ k}\Omega} = 9.09\text{ mA}$$

$$I_4 = \frac{V}{R_4} = \frac{20\text{ V}}{1\text{ k}\Omega} = 20\text{ mA}$$

The total current *should* be

$$I_T = I_1 + I_2 + I_3 + I_4 = 2\text{ mA} + 4.26\text{ mA} + 9.09\text{ mA} + 20\text{ mA}$$
$$= 35.35\text{ mA}$$

The actual *measured* current is 31.09 mA, as stated, which is 4.26 mA less than normal, indicating that the branch carrying 4.26 mA is open. Thus, R_2 must be open.

When two or more branches have the same normal current, the open branch can be isolated only by direct measurements of the branch current or resistance. When a zero current measurement or an infinite resistance measurement occurs, you have found the open branch. Remember, however, that the resistor must be disconnected for a resistance measurement.

SECTION REVIEW 5–9

1. A three-branch circuit normally has the following branch currents 1 A, 2.5 A, and 1.2 A. If the total current measures 3.5 A, which branch is open?
2. How can you tell if a parallel branch is open by measuring the resistance of the parallel circuit?

APPLICATION NOTE

A suggested procedure for testing the PC board for open resistors described in the Application Assignment at the beginning of the chapter is as follows:

Principle

Apply a known voltage to each group of resistors and measure the resulting current. If the ammeter indicates the correct total current, the resistors are all good. If the ammeter indicates an incorrect total current, determine the open resistor(s) by the amount of error in the current reading.

Procedure

Connect a 30-V source in series with an ammeter to each set of pins on the pc board. Make your evaluation based on Table 5–1.

Step 1: Connect the ammeter in series with the 30-V source across designated pins.
Step 2: If the current is correct, all resistors in the group are good.
Step 3: If the current is incorrect, one or more resistors may be open.
Step 4: Isolate the open branch or branches by subtracting the meter reading from the correct current. This step gives the missing branch current(s).

TABLE 5–1

Group	Correct Total Current	Branch Currents (mA)			
		I_1	I_2	I_3	I_4
1 (Pins 1 and 2)	80.31 mA	30	20	16.67	13.64
2 (Pins 3 and 4)	6.24 mA	3	1.36	1.11	0.77
3 (Pins 5 and 6)	1.99 mA	0.64	0.54	0.44	0.37

Example

Suppose that the 30-V source is connected to pins 1 and 2 (group 1) and the ammeter measures 50.31 mA. This reading is incorrect and indicates that a resistor is open. Subtracting 50.31 from 80.31 gives 30 mA, indicating that R_1 is open because I_1 = 30 mA.

Can you think of a better way to test for open resistors?

SUMMARY

Facts

- The same voltage appears across each branch in a parallel circuit.
- The total resistance of a parallel circuit is less than the smallest branch resistance.
- In a parallel circuit, the total current divides into branch currents. According to Ohm's law, each branch current is inversely proportional to the branch resistance.
- The largest branch resistance has the smallest current, and the smallest branch resistance has the largest current.
- The total power in a parallel circuit is the sum of the powers in each resistor.
- When a parallel branch opens, the total resistance increases and the total current decreases. The other branch currents are not affected.

Definitions

- *Branch*—one current path in a parallel circuit.
- *Kirchhoff's current law (KCL)*—a law which states that the total current into a junction equals the total current out of the junction.

Formulas

$$I_{IN(1)} + I_{IN(2)} + I_{IN(3)} + \cdots + I_{IN(n)} = I_{OUT(1)} + I_{OUT(2)} + I_{OUT(3)} + \cdots + I_{OUT(m)}$$ Kirchhoff's current law (5–1)

$$\frac{1}{R_T} = \frac{1}{R_1} + \frac{1}{R_2} + \frac{1}{R_3} + \cdots + \frac{1}{R_n}$$ Reciprocal for total parallel resistance (5–2)

$$R_T = \frac{1}{\left(\frac{1}{R_1}\right) + \left(\frac{1}{R_2}\right) + \left(\frac{1}{R_3}\right) + \cdots + \left(\frac{1}{R_n}\right)}$$ Total parallel resistance (5–3)

$$R_T = \frac{R_1 R_2}{R_1 + R_2}$$ Special case for two resistors in parallel (5–4)

$$R_T = \frac{R}{n}$$ Special case for n equal-value resistors in parallel (5–5)

$$I_1 = \left(\frac{R_2}{R_1 + R_2}\right) I_T$$ Two-branch current divider formula (5–6)

$$I_2 = \left(\frac{R_1}{R_1 + R_2}\right) I_T$$ Two-branch current divider formula (5–7)

$$I_x = \left(\frac{R_T}{R_x}\right) I_T$$ General current divider formula (5–8)

$$P_T = P_1 + P_2 + P_3 + \cdots + P_n$$ Total power (5–9)

SELF-TEST

1. Sketch a parallel circuit with four resistors connected to a 10-V source.
2. Connect each set of resistors in Figure 5–50 in parallel between points A and B.

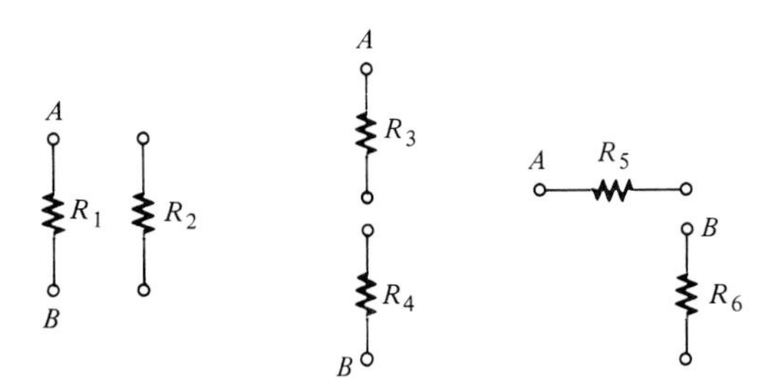

FIGURE 5–50

3. Discuss the difference between a parallel circuit and a series circuit.
4. When a 1.2-kΩ resistor and a 100-Ω resistor are connected in parallel, the total resistance is less than ________.

5. A 330-Ω resistor, a 270-Ω resistor, and a 68-Ω resistor are all in parallel. The total resistance is less than ________.
6. Two 1-kΩ resistors are in parallel. What is the total resistance?
7. If a 9-V battery is connected across the parallel resistors in Problem 5, how much voltage is there across each of the resistors?
8. When additional resistors are connected across an existing parallel circuit, what happens to the total resistance?
9. If one of the resistors in a parallel circuit is removed, what happens to the total resistance?
10. The currents into a junction flow along two paths. One current is 5 A and the other is 3 A. What is the total current flowing out of the junction?
11. The following resistors are in parallel: 390-Ω, 560-Ω, and 820-Ω. Which resistor has the most current? The least current?
12. When the total current into a parallel circuit increases, what does this increase indicate?
13. The power dissipation in each of three parallel branches is 1 W. What is the total power dissipation of the circuit?
14. In a four-branch parallel circuit, 10 mA of current flows in each branch. If one of the branches opens, what is the current in each of the other three branches?

PROBLEM SET A

Section 5–1

5–1 Connect the resistors in Figure 5–51 in parallel across the battery.

5–2 Determine whether or not all the resistors in Figure 5–52 are connected in parallel on the PC board. Sketch the schematic.

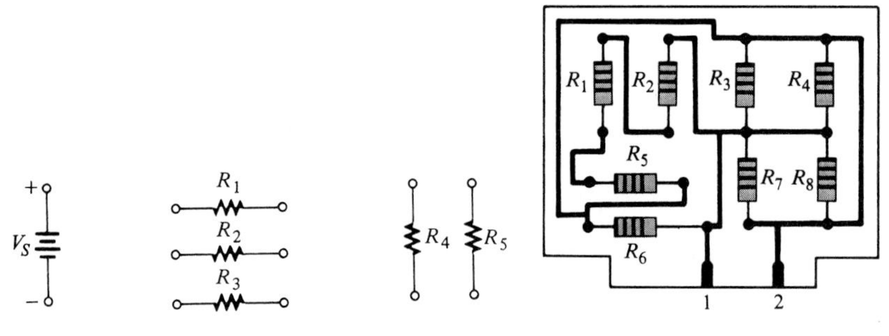

FIGURE 5–51

FIGURE 5–52

Section 5–2

5–3 Determine the voltage across and the current through each parallel resistor if the total voltage is 12 V and the total resistance is 600 Ω. There are four resistors, all of equal value.

5–4 The source voltage in Figure 5–53 is 100 V. How much voltage does each of the meters V1–V3 read?

FIGURE 5–53

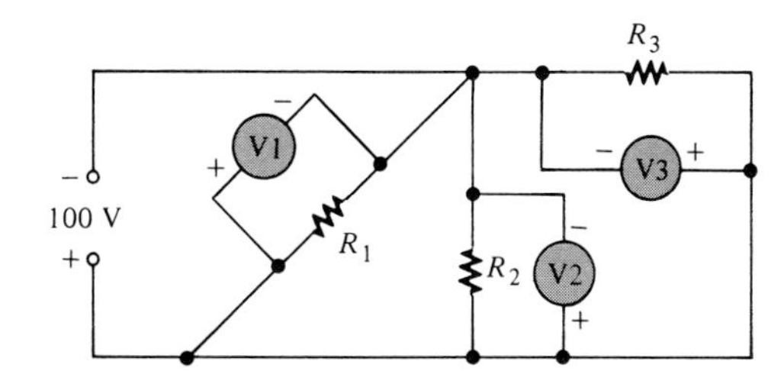

Section 5–3

5–5 The following currents are measured in the same direction in a three-branch parallel circuit: 250-mA, 300-mA, and 800-mA. What is the value of the current into the junction of these three branches?

5–6 Five hundred milliamperes (500 mA) flow into five parallel resistors. The currents through four of the resistors are 50 mA, 150 mA, 25 mA, and 100 mA. How much current flows through the fifth resistor?

5–7 How much current flows through R_2 and R_3 in Figure 5–54 if R_2 and R_3 have the same resistance? Show how to connect ammeters to measure these currents.

FIGURE 5–54

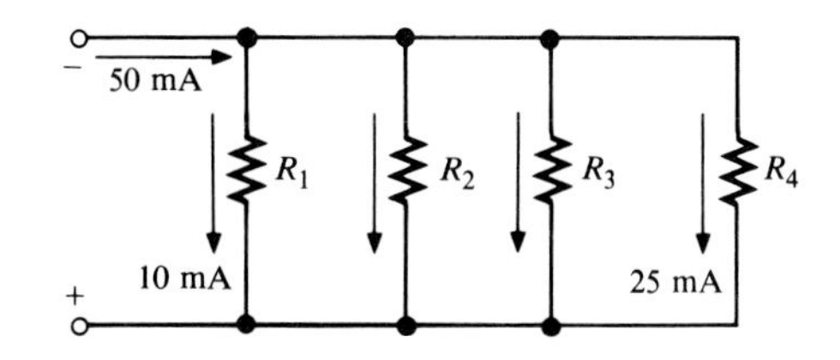

Section 5–4

5–8 The following resistors are connected in parallel: 1-MΩ, 2.2-MΩ, 4.7-MΩ, 12-MΩ, and 22-MΩ. Determine the total resistance.

5–9 Find the total resistance for each group of parallel resistors in Figure 5–55. (A full-color photo appears between pages 128 and 129.)

5–10 Calculate R_T for each circuit in Figure 5–56.

5–11 What is the total resistance of eleven 22-kΩ resistors in parallel?

5–12 Five 15-Ω, ten 100-Ω, and two 10-Ω resistors are all connected in parallel. What is the total resistance?

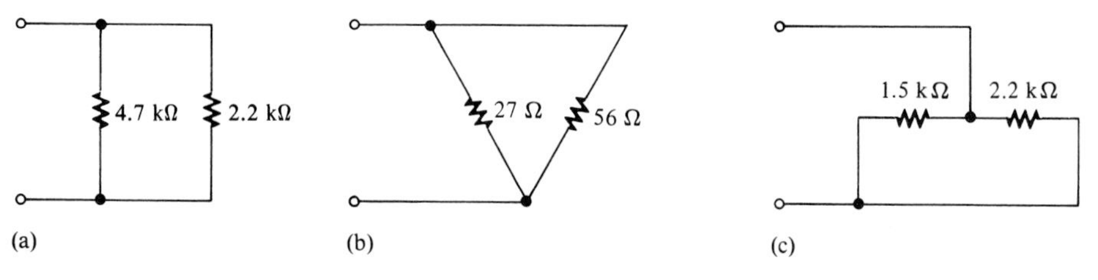

FIGURE 5–56

Section 5–5

5–13 What is the total current I_T in each circuit of Figure 5–57?

5–14 Three 33-Ω resistors are connected in parallel with a 110-V source. How much current is flowing from the source?

5–15 Which circuit of Figure 5–58 has more total current?

5–16 Four equal-value resistors are connected in parallel. Five volts are applied across the parallel circuit, and 2.5 mA are measured from the source. What is the value of each resistor?

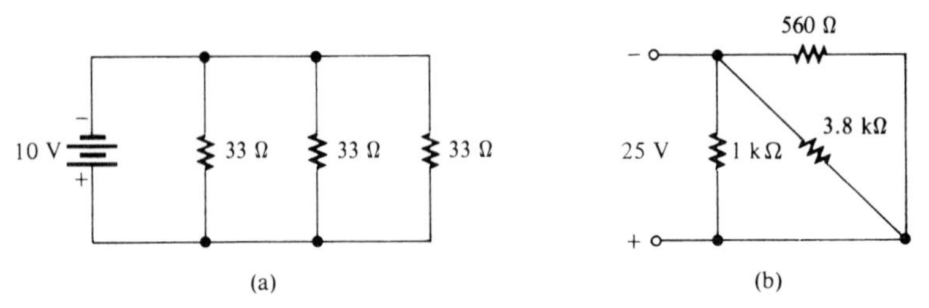

FIGURE 5–57

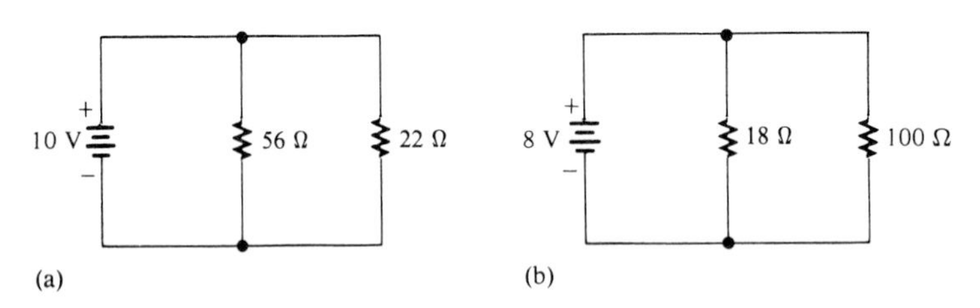

FIGURE 5–58

Section 5–6

5–17 How much branch current should each meter in Figure 5–59 indicate?

5–18 Using the current divider formula, determine the current in each branch of the circuits of Figure 5–60.

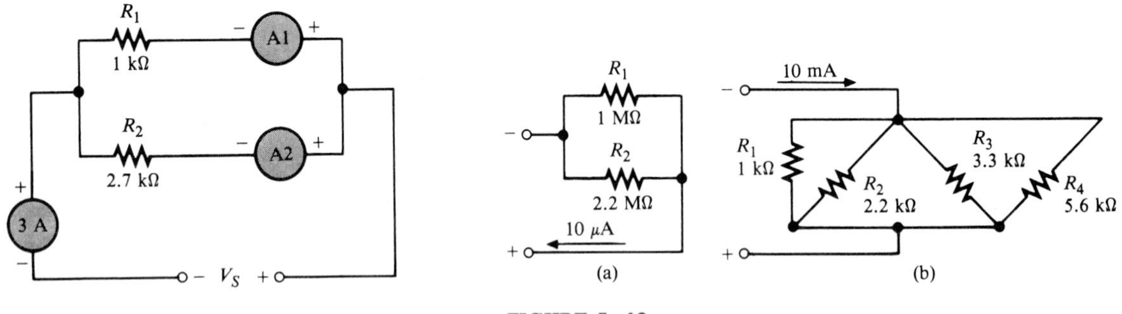

FIGURE 5–59

FIGURE 5–60

Section 5–7

5–19 Five parallel resistors each handle 40 mW. What is the total power?

5–20 Determine the total power in each circuit of Figure 5–60.

5–21 Six light bulbs are connected in parallel across 110 V. Each bulb is rated at 75 W. How much current flows through each bulb, and what is the total current?

Section 5–9

5–22 If one of the bulbs burns out in Problem 5–21, how much current will flow through each of the remaining bulbs? What will the total current be?

5–23 In Figure 5–61, the current and voltage measurements are indicated. Has a resistor opened, and, if so, which one?

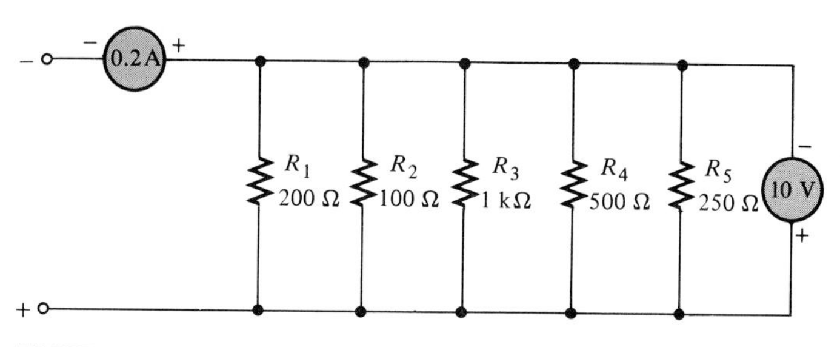

FIGURE 5–61

PROBLEM SET B

5–24 In the circuit of Figure 5–62, determine resistances R_2, R_3, and R_4.

5–25 The total resistance of a parallel circuit is 25 Ω. If the total current is 100 mA, how much current flows through a 220-Ω resistor that makes up part of the parallel circuit?

5–26 What is the current through each resistor in Figure 5–63? R is the lowest-value resistor, and all others are multiples of that value as indicated.

5–27 A certain parallel network consists of only ½-W resistors. The total resistance is 1 kΩ, and the total current is 50 mA. If each resistor is operating at one-half its maximum power level, determine the following:
(a) the number of resistors. **(b)** the value of each resistor.
(c) the current in each branch. **(d)** the applied voltage.

5–28 Find the values of the unspecified labeled quantities in each circuit of Figure 5–64.

5–29 What is wrong with the circuit in Figure 5–65 (a full-color photo appears between pages 128 and 129) if 25V are applied across the red and black leads?

5–30 If the total resistance in Figure 5–66 is 200 Ω, what is the value of R_2?

5–31 Determine the unknown resistances in Figure 5–67.

5–32 A total of 250 mA flows into a parallel circuit with a total resistance of 1.5 kΩ. The current must be increased by 25%. Determine how much resistance to add in parallel to accomplish this increase in current.

5–33 Develop a test procedure to check the circuit in Figure 5–68 (a full-color photo appears between pages 128 and 129) to make sure that there are no open components. You must do this test without removing any component from the board. List the procedure in a detailed step-by-step format.

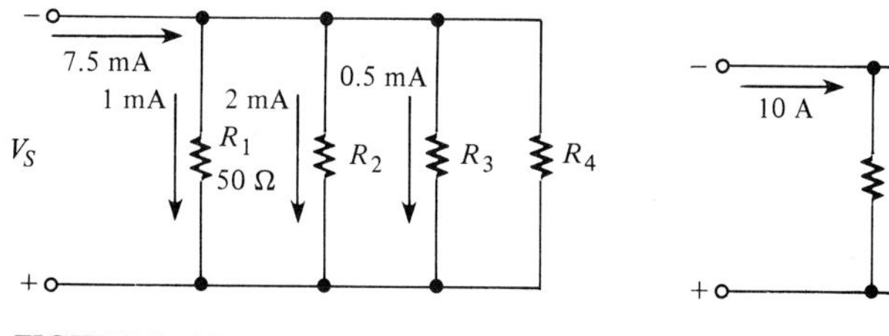

FIGURE 5–62

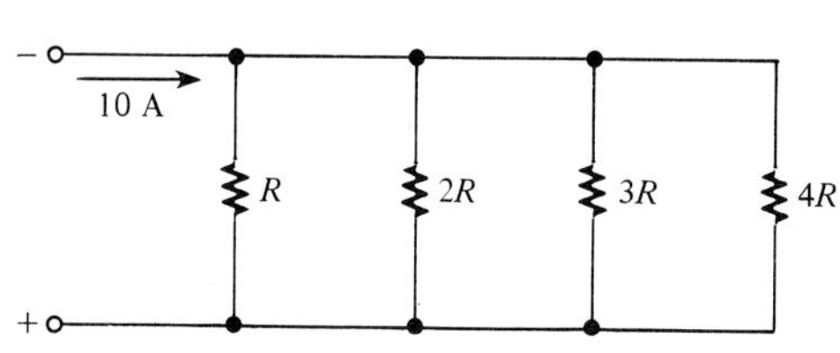

FIGURE 5–63

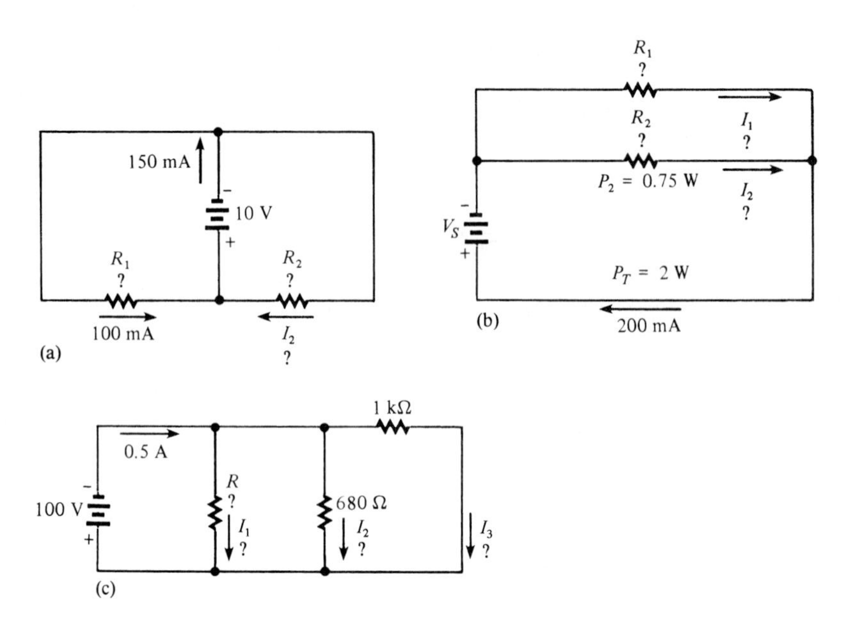

FIGURE 5–64

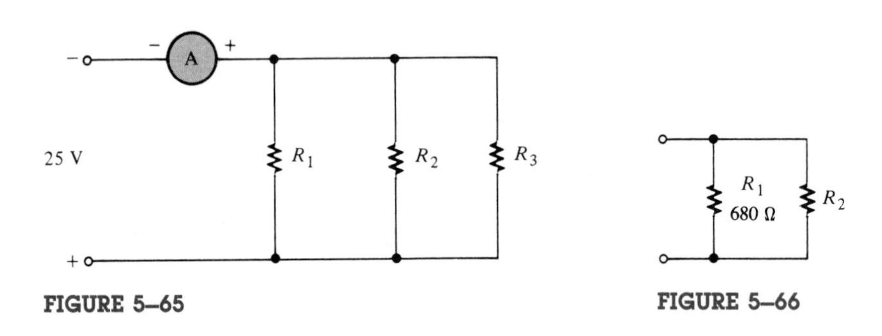

FIGURE 5–65

FIGURE 5–66

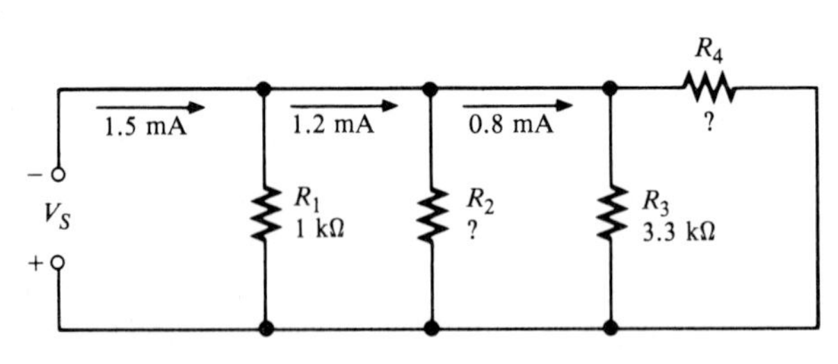

FIGURE 5–67

ANSWERS TO SECTION REVIEWS

Section 5–1

1. A parallel connection has more than one current path between two given points.
2. See Figure 5–69. 4. See Figure 5–70.

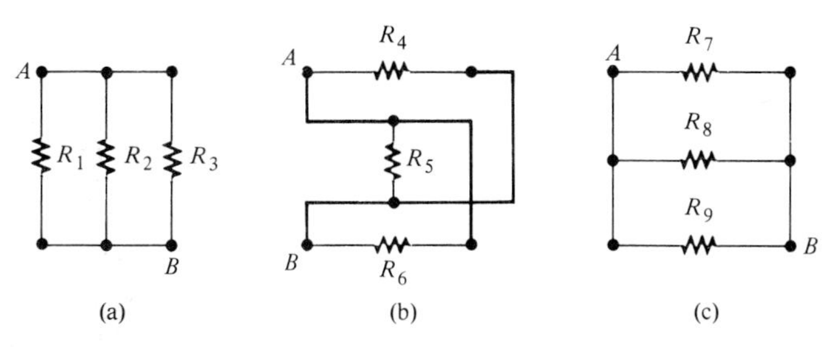

FIGURE 5–69

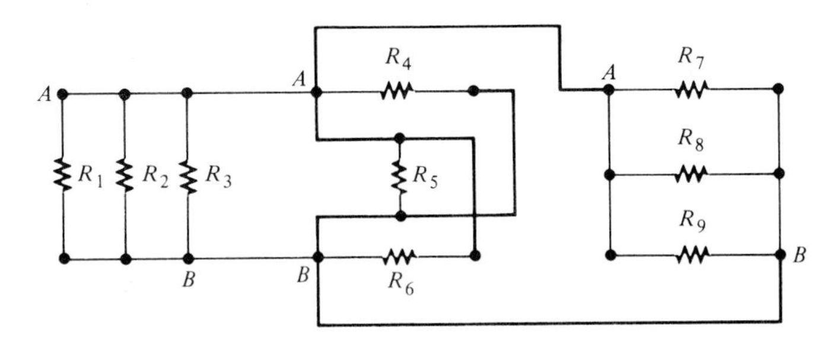

FIGURE 5–70

Section 5–2

1. 5 V. 2. 118 V, 118 V. 3. 50 V, 50 V.

Section 5–3

1. 2.5 A. 2. 400 mA. 3. 3 μA. 4. 7 A entering, 5 A leaving.

Section 5–4

1. Decrease. 2. The smallest resistance value. 3. 688 Ω, 250 Ω, 29.93 Ω.

Section 5–5

1. 0.44 A. 2. 444.36 V
3. 0.653 A through the 680-Ω; 1.35 A through the 330-Ω. 4. 8 kΩ.
5. 68.75 V.

Section 5–6

1. The 22-Ω has most current; the 220-Ω has least current. 2. 0.97 mA.
3. $I_1 = 3.06$ mA, $I_2 = 6.93$ mA.

Section 5–7

1. 16 W. 2. 584.47 W.

Section 5–8

1. 50 Ω, 1 mA.
2. Yes, because the shunt resistors must allow currents much greater than the current through the meter movement.

Section 5–9

1. The 1.2-A branch is open.
2. When the total resistance is greater than it should be, a branch is open.

Series-Parallel Circuits

Various combinations of both series and parallel resistors are often used in electronic circuits. In this chapter, examples of such series-parallel arrangements are examined and analyzed. An important circuit called the *Wheatstone bridge* is introduced, and circuits with more than one voltage source are analyzed in simple steps. Also, you will learn how complex circuits can be simplified using *Thevenin's theorem*. Troubleshooting series-parallel circuits for shorts and opens is also covered.

Specifically, you will learn:

- How to identify series and parallel parts of a series-parallel circuit and how to recognize the relationships of all the resistors.
- How to determine the total resistance of a series-parallel circuit.
- How to determine the currents and voltages in a series-parallel circuit.
- The meaning and use of grounds in a circuit.
- How resistive loads affect voltage divider circuits.
- Under what conditions a voltmeter can affect the value of the voltage that it is measuring.
- What a Wheatstone bridge circuit is and how it can be used.
- The superposition method and how to use it to evaluate circuits with more than one source.
- Thevenin's theorem and how to apply it to simplify complex circuits.
- How to use troubleshooting techniques to identify opens and shorts in a series-parallel circuit.

6

APPLICATION ASSIGNMENT

Recall that the Application Assignment in Chapter 4 required a multiple-output voltage divider to provide dc supply voltages for several electronic instruments to be used in a study of volcanic activity in the Hawaiian Islands.

Although you did a great job on that assignment, it turns out that additional information has become available. As a result, you are asked to review the circuit to make sure that it meets the following new specification: None of the four output voltages from the voltage divider circuit can change more than 5% when the appropriate instrument is connected to it.

The new data are listed below:

Instruments	dc Supply Voltage	Max. Current from dc Supply
Seismometer	20 V	100 mA
Telemeter	16 V	80 mA
Volcanometer	10 V	60 mA
Computer	8 V	60 mA

Your assignment is to analyze the voltage divider that was developed in Chapter 4 under the loading conditions specified above, and determine whether any of the four voltages will not meet the 5% specification.

After studying this chapter, you will be able to complete this assignment. The Application Note at the end of the chapter shows one approach to the required analysis.

IDENTIFYING SERIES-PARALLEL RELATIONSHIPS

6–1

A series-parallel circuit consists of combinations of both series and parallel current paths. It is important to be able to identify how the components in a circuit are arranged in terms of their series and parallel relationships.

Figure 6–1(a) shows a simple series-parallel combination of resistors. Notice that the resistance from point A to point B is R_1. The resistance from point B to point C is R_2 and R_3 in parallel ($R_2 \| R_3$). The resistance from point A to point C is R_1 in series with the parallel combination of R_2 and R_3, as indicated in Figure 6–1(b).

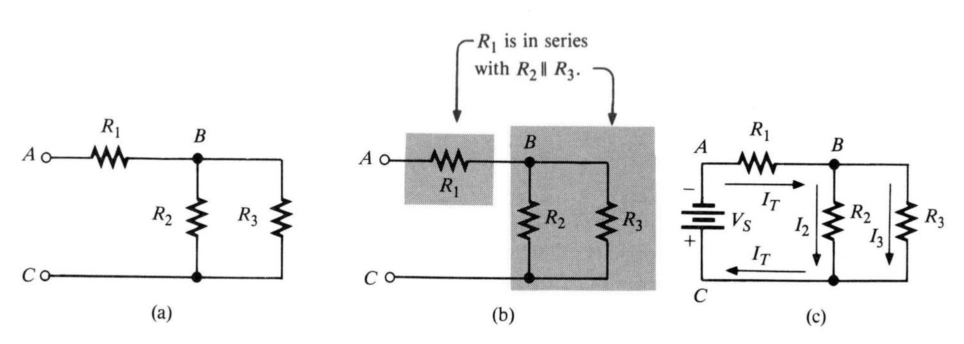

FIGURE 6–1
A simple series-parallel circuit.

When the circuit of Figure 6–1(a) is connected to a voltage source as shown in Part (c), the *total* current flows through R_1 and divides at point B into the two parallel paths. These two branch currents then recombine, and the total current flows into the positive source terminal as shown.

Now, to illustrate series-parallel relationships, we will increase the complexity of the circuit in Figure 6–1(a) step-by-step. In Figure 6–2(a), another resistor (R_4) is connected in series with R_1. The resistance between points A and B is now $R_1 + R_4$, and this combination is in series with the parallel combination of R_2 and R_3, as illustrated in Part (b).

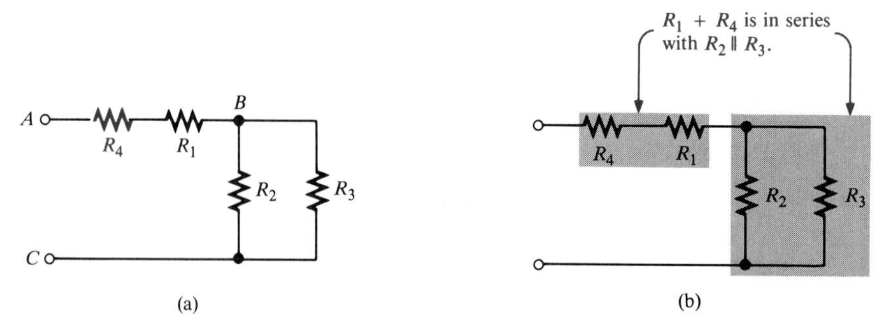

FIGURE 6–2
R_4 is added to the circuit in series with R_1.

In Figure 6–3(a), R_5 is connected in series with R_2. The series combination of R_2 and R_5 is in parallel with R_3. This entire series-parallel combination is in series with the $R_1 + R_4$ combination, as illustrated in Part (b).

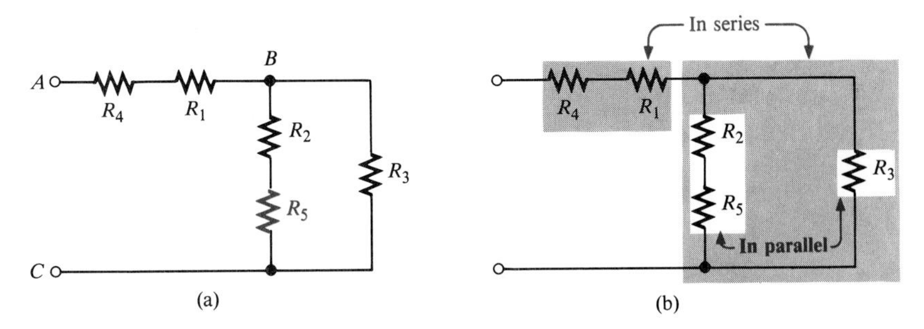

FIGURE 6–3
R_5 is added to the circuit in series with R_2.

In Figure 6–4(a), R_6 is connected in parallel with the series combination of R_1 and R_4. The series-parallel combination of R_1, R_4, and R_6 is in series with the series-parallel combination of R_2, R_3, and R_5, as indicated in Part (b).

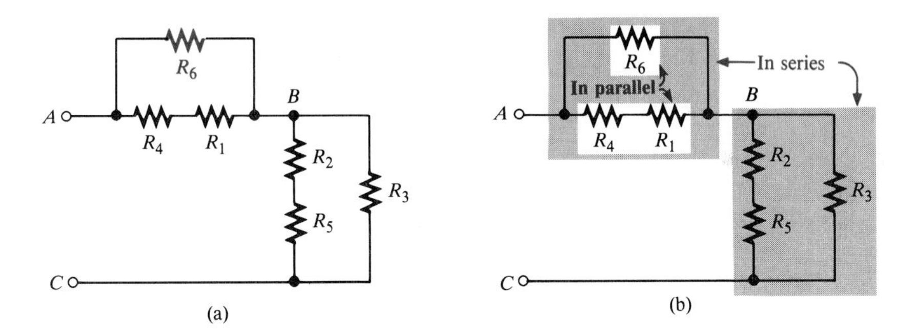

FIGURE 6–4
R_6 is added to the circuit in parallel with the series combination of R_1 and R_4.

EXAMPLE 6–1

Identify the series-parallel relationships in Figure 6–5.

FIGURE 6–5

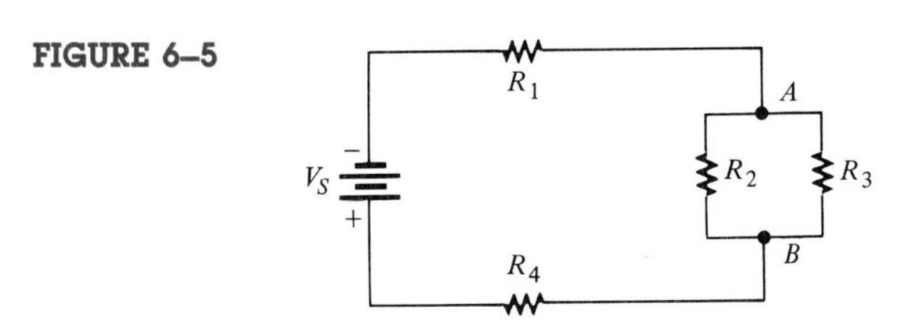

Solution
Starting at the negative terminal of the source, follow the current paths. All of

the current produced by the source must go through R_1, which is in series with the rest of the circuit.

The total current takes two paths when it gets to point A. Part of it flows through R_2, and part of it through R_3. R_2 and R_3 are in parallel with each other, and this parallel combination is in series with R_1.

At point B, the currents through R_2 and R_3 come together again. Thus, the total current flows through R_4. R_4 is in series with R_1 and the parallel combination of R_2 and R_3. The currents are shown in Figure 6–6, where I_T is the total current.

FIGURE 6–6

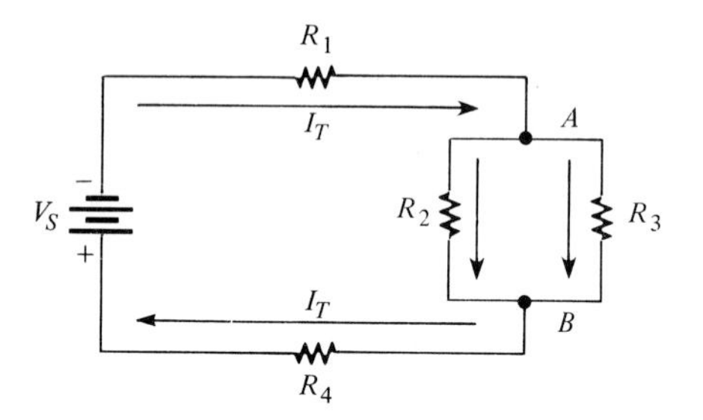

In summary, R_1 and R_4 are in series with the parallel combination of R_2 and R_3.

EXAMPLE 6–2 Describe the series-parallel combination between points A and D in Figure 6–7.

FIGURE 6–7

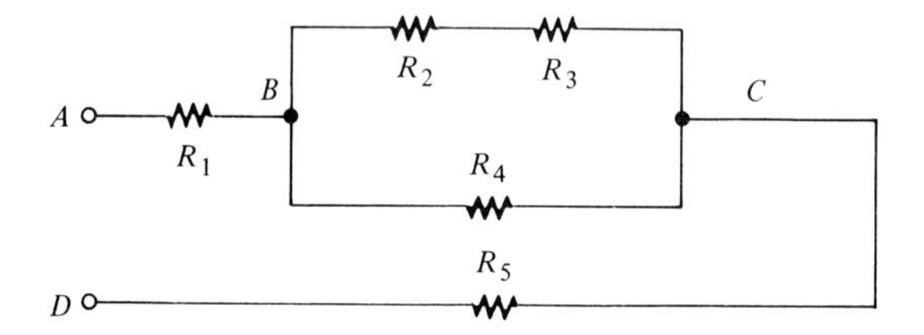

Solution

Between points B and C, there are two parallel paths. The lower path consists of R_4, and the upper path consists of a *series* combination of R_2 and R_3. This parallel combination is in series with both R_1 and R_5.

In summary, R_1 and R_5 are in series with the parallel combination of R_4 and $R_2 + R_3$.

EXAMPLE 6–3 Describe the total resistance between each pair of points in Figure 6–8.

FIGURE 6–8

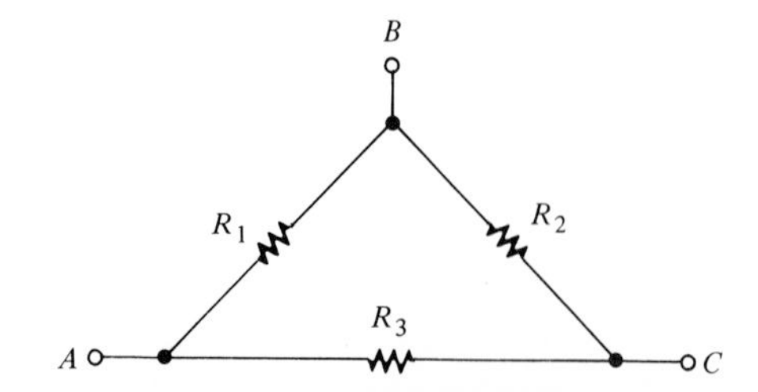

Solution

1. From point A to B: R_1 is in *parallel* with the *series* combination of R_2 and R_3.
2. From point A to C: R_3 is in *parallel* with the *series* combination of R_1 and R_2.
3. From point B to C: R_2 is in *parallel* with the *series* combination of R_1 and R_3.

Redrawing a Schematic to Determine the Series-Parallel Relationships

Sometimes it is difficult to see the series-parallel relationships on a schematic diagram because of the way in which it is drawn. In such a situation, it helps to redraw the diagram so that the relationships become clear. The following example gives a simple illustration.

EXAMPLE 6–4 Identify the series-parallel relationships in Figure 6–9.

FIGURE 6–9

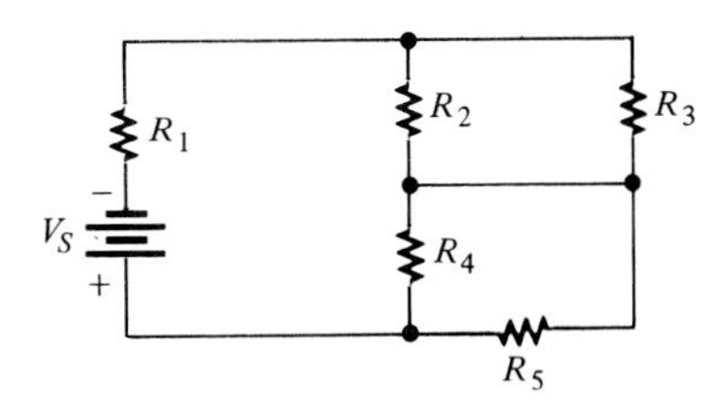

Solution

Sometimes it is easier to see a particular circuit arrangement if it is drawn in a different way. In this example, the circuit schematic is redrawn in Figure 6–10, which better illustrates the series-parallel relationships. Now you can see that R_2 and R_3 are in parallel with each other and also that R_4 and R_5 are in parallel with each other. Both parallel combinations are in series with each other and with R_1.

FIGURE 6–10

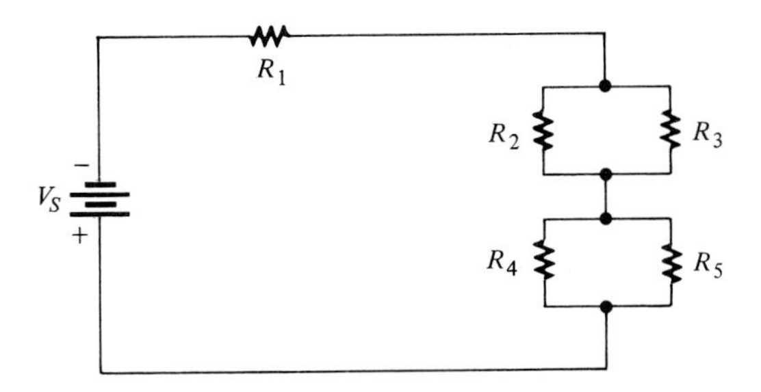

Determining Relationships on a PC Board

Usually, the physical arrangement of components on a PC board bears no resemblance to the actual circuit relationships. By tracing out the circuit on the PC board and rearranging the components on paper into a recognizable form, you can determine the series-parallel relationships. An example will illustrate.

EXAMPLE 6–5

Determine the relationships of the resistors on the PC board in Figure 6–11.

FIGURE 6–11

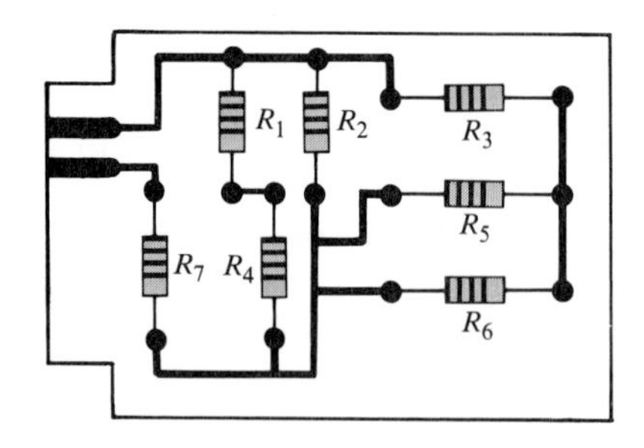

Solution

In Figure 6–12(a), the schematic is drawn in the same arrangement as that of the resistors on the board. In Part (b), the resistors are reoriented so that the series-parallel relationships are obvious. R_1 and R_4 are in series; $R_1 + R_4$ is in parallel with R_2; R_5 and R_6 are in parallel and this combination is in series with R_3. The R_3, R_5, and R_6 series-parallel combination is in parallel with both R_2 and the $R_1 + R_4$ combination. This entire series-parallel combination is in series with R_7, as Figure 6–12(c) illustrates.

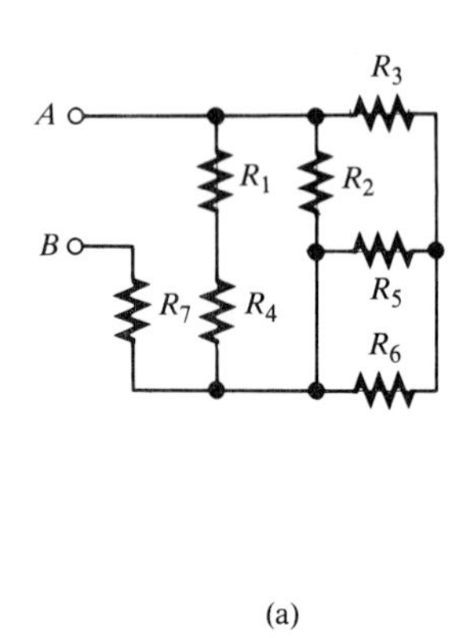

(a)

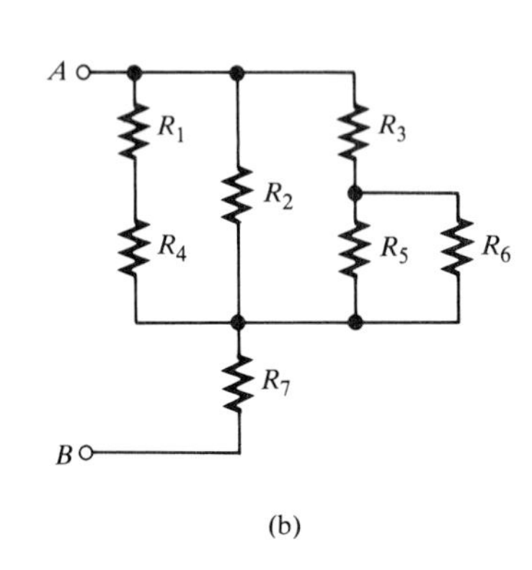

(b)

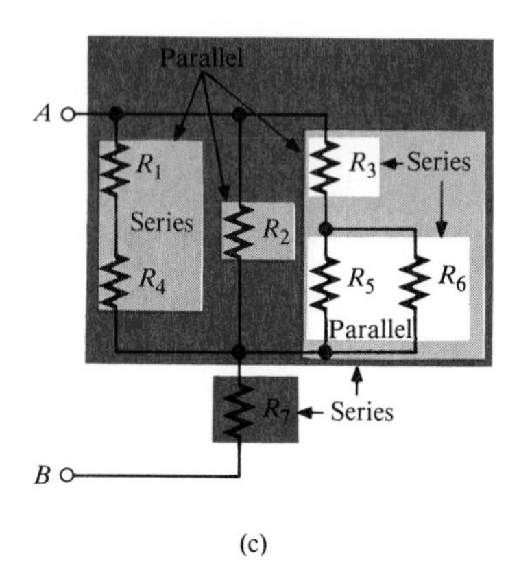

(c)

FIGURE 6–12

SECTION REVIEW 6–1

1. A certain series-parallel circuit is described as follows: R_1 and R_2 are in parallel. This parallel combination is in series with another parallel combination of R_3 and R_4. Sketch the circuit.
2. In the circuit of Figure 6–13, describe the series-parallel relationships of the resistors.

FIGURE 6–13

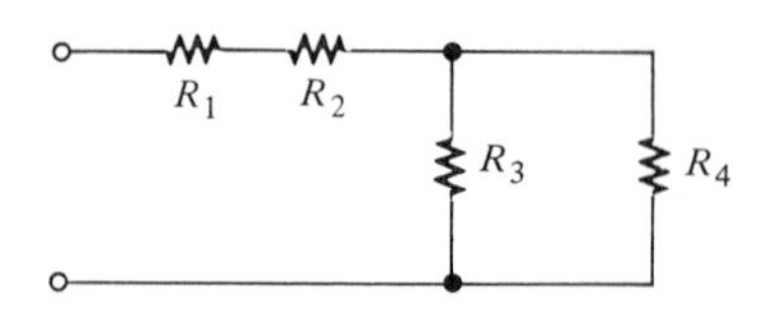

3. Which resistors are in parallel in Figure 6–14?
4. Identify the parallel arrangements in Figure 6–15.
5. Are the parallel combinations in Figure 6–15 in series?

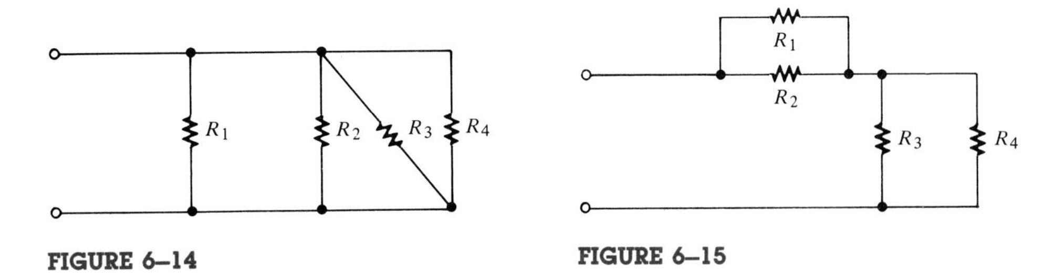

FIGURE 6–14

FIGURE 6–15

ANALYSIS OF SERIES-PARALLEL CIRCUITS

6–2

Several quantities are important in a circuit that is a series-parallel configuration of resistors. In this section you will learn how to find total resistance, total current, branch currents, and the voltage across any portion of a circuit.

Total Resistance

In Chapter 4, you learned how to determine total series resistance. In Chapter 5, you learned how to determine total parallel resistance.

To find the total resistance R_T of a series-parallel combination, first identify the series and parallel relationships, and then apply what you have previously learned. The following two examples illustrate this general approach.

EXAMPLE 6–6 Determine R_T between points A and B of the circuit in Figure 6–16.

FIGURE 6–16

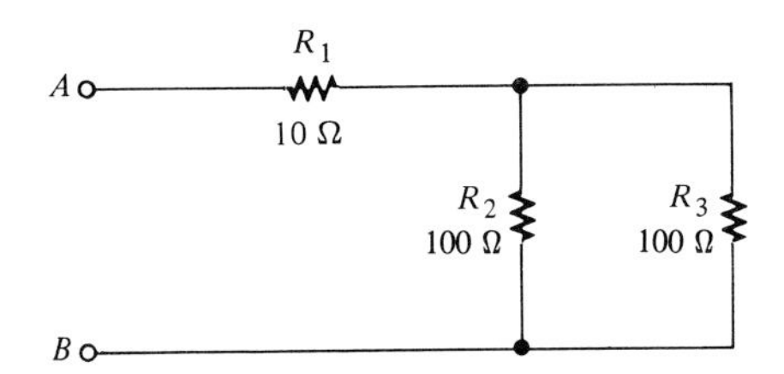

Solution

R_2 and R_3 are in parallel, and this parallel combination is in series with R_1. First find the parallel resistance of R_2 and R_3. Since R_2 and R_3 are equal in value, divide the value by 2:

$$R_{2\text{-}3} = \frac{R}{n} = \frac{100\ \Omega}{2} = 50\ \Omega$$

Now, since R_1 is in series with $R_{2\text{-}3}$, add their values as follows:

$$R_T = R_1 + R_{2\text{-}3} = 10\ \Omega + 50\ \Omega = 60\ \Omega$$

The calculator sequence is
1 0 0 ÷ 2 = + 1 0 =

EXAMPLE 6–7

Find R_T of the circuit in Figure 6–17.

FIGURE 6–17

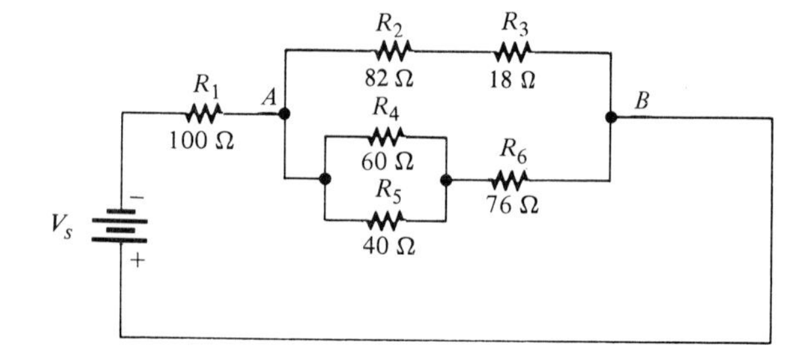

Solution
In the *upper branch,* R_2 is in series with R_3. We will call the series combination $R_{2\text{-}3}$. It is equal to $R_2 + R_3$:

$$R_{2\text{-}3} = R_2 + R_3 = 82\ \Omega + 18\ \Omega = 100\ \Omega$$

In the *lower branch,* R_4 and R_5 are in parallel with each other. We will call this parallel combination $R_{4\text{-}5}$. It is calculated as follows:

$$R_{4\text{-}5} = \frac{R_4 R_5}{R_4 + R_5} = \frac{(60\ \Omega)(40\ \Omega)}{60\ \Omega + 40\ \Omega} = 24\ \Omega$$

Also in the *lower branch,* the parallel combination of R_4 and R_5 is in series with R_6. This series-parallel combination is designated $R_{4\text{-}5\text{-}6}$ and is calculated as follows:

$$R_{4\text{-}5\text{-}6} = R_6 + R_{4\text{-}5} = 76\ \Omega + 24\ \Omega = 100\ \Omega$$

Figure 6–18 shows the original circuit in a simplified *equivalent* form.

FIGURE 6–18

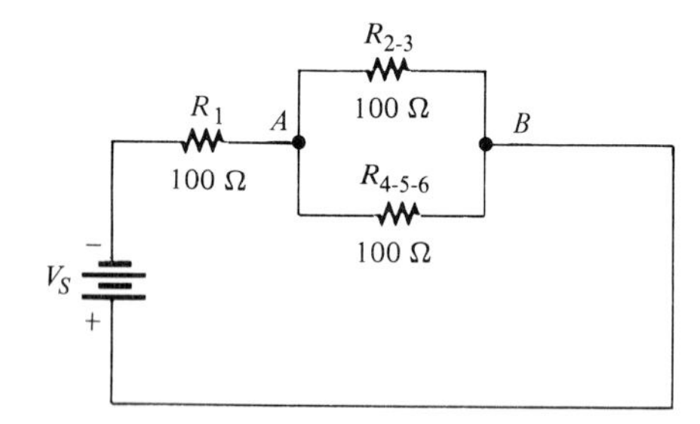

Now we can find the resistance between points A and B. It is $R_{2\text{-}3}$ in parallel with $R_{4\text{-}5\text{-}6}$. Since these resistances are equal, the equivalent resistance is calculated as follows:

$$R_{AB} = \frac{100\ \Omega}{2} = 50\ \Omega$$

Finally, the total resistance is R_1 in series with R_{AB}:

$$R_T = R_1 + R_{AB} = 100\ \Omega + 50\ \Omega = 150\ \Omega$$

Total Current

Once the total resistance and the source voltage are known, you can find total current in a circuit by applying Ohm's law. Total current is the total source voltage divided by the total resistance:

$$I_T = \frac{V_S}{R_T}$$

For example, let us find the total current in the circuit of Example 6–7 (Figure 6–17). Assume that the source voltage is 30 V. The calculation is as follows:

$$I_T = \frac{V_S}{R_T} = \frac{30\ \text{V}}{150\ \Omega} = 0.2\ \text{A}$$

Branch Currents

Using the *current divider formula, Kirchhoff's current law,* or *Ohm's law,* or combinations of these, you can find the current in any branch of a series-parallel circuit. In some cases it may take repeated application of the formula to find a given current.

EXAMPLE 6–8 Determine the current through R_4 in Figure 6–19 if $V_S = 50$ V.

FIGURE 6–19

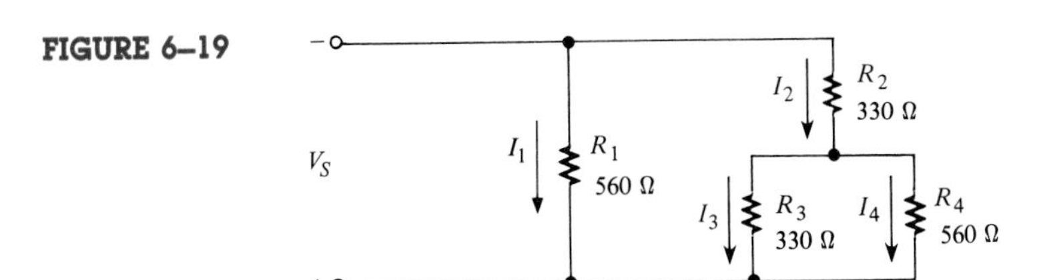

Solution

First the current (I_2) into the junction of R_3 and R_4 must be found. Once we know this current, we can use the current divider formula to find I_4.

Notice that there are two main branches in the circuit. The left-most branch consists of only R_1. The right-most branch has R_2 in series with the parallel combination of R_3 and R_4. The voltage across both of these main branches is the same and equal to 50 V. We can find the current (I_2) into the junction of R_3 and R_4 by calculating the resistance of the right-most main branch and then applying Ohm's law as follows:

$$R_{2\text{-}3\text{-}4} = R_2 + \frac{R_3R_4}{R_3 + R_4} = 330\ \Omega + \frac{(560\ \Omega)(330\ \Omega)}{890\ \Omega}$$
$$= 537.6\ \Omega$$
$$I_2 = \frac{V_S}{R_{2\text{-}3\text{-}4}} = \frac{50\ \text{V}}{537.6\ \Omega}$$
$$= 0.093\ \text{A}$$

Using the current divider formula, calculate I_4 as follows:

$$I_4 = \left(\frac{R_3}{R_3 + R_4}\right)I_2 = \left(\frac{330\ \Omega}{890\ \Omega}\right)0.093\ \text{A}$$
$$= 0.035\ \text{A}$$

You can find the current in R_3 by applying Kirchhoff's law ($I_2 = I_3 + I_4$) and subtracting I_4 from I_2 ($I_3 = I_2 - I_4$).

You can find the current in R_1 by using Ohm's law ($I_1 = V_S/R_1$).

The calculator sequence is

5 6 0 × 3 3 0 ÷ (5 6 0 +
3 3 0) = + 3 3 0 = STO 0
5 0 ÷ RCL 0 = STO 1 3 3 0 ÷
(5 6 0 + 3 3 0) =
× RCL 1 =

Voltage Relationships

The circuit in Figure 6–20 will be used to illustrate voltage relationships in a series-parallel circuit. Voltmeters are connected to measure each of the resistor voltages, and the readings are as indicated.

Some general observations about Figure 6–20 follow:

1. V_1 and V_2 are equal because R_1 and R_2 are in parallel. (Recall that voltages across parallel branches are the same.) V_1 and V_2 are the same as the voltage from A to B.
2. V_3 is equal to $V_4 + V_5$ because R_3 is in parallel with the series combination of R_4 and R_5. (V_3 is the same as the voltage from B to C.)
3. V_4 is about one-third of the voltage from B to C because R_4 is about one-third of the resistance $R_4 + R_5$ (by the voltage divider principle).
4. V_5 is about two-thirds of the voltage from B to C because R_5 is about two-thirds of $R_4 + R_5$.
5. $V_1 + V_3$ equals V_S because, by Kirchhoff's voltage law, the sum of the voltage drops must equal the source voltage.

The following example will verify the meter readings in Figure 6–20.

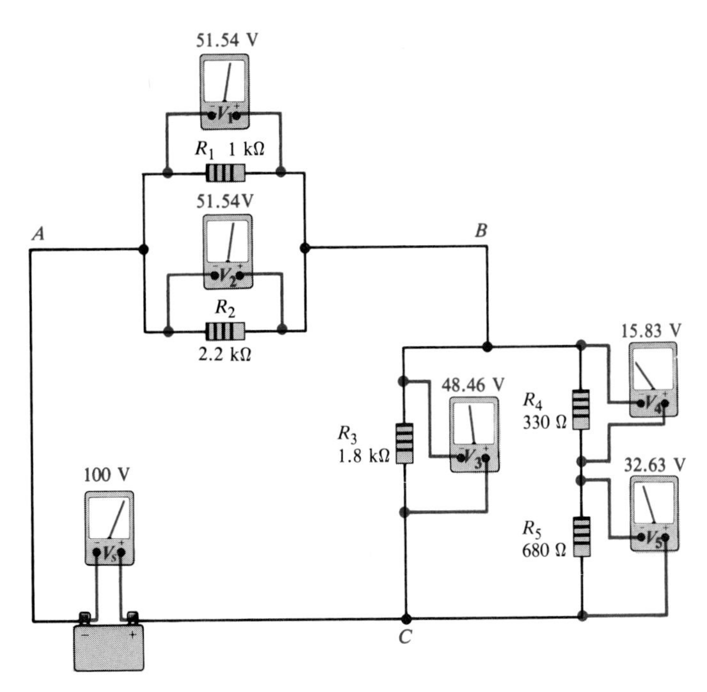

FIGURE 6–20
Illustration of voltage relationships.

EXAMPLE 6–9

Verify that the voltmeter readings in Figure 6–20 are correct. The circuit is redrawn as a schematic in Figure 6–21.

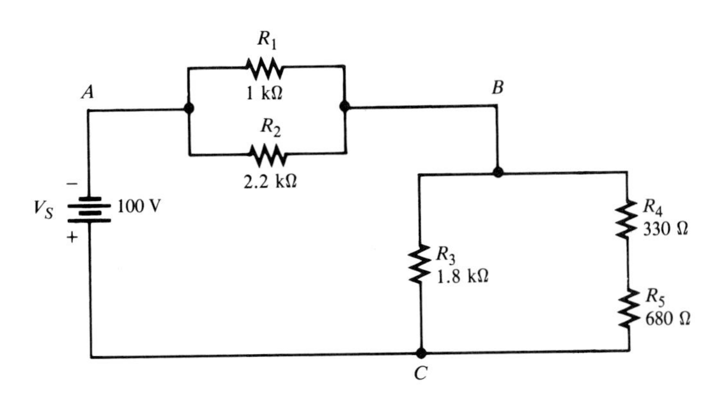

FIGURE 6–21

Solution
The resistance from point A to point B is the parallel combination of R_1 and R_2:

$$R_{AB} = \frac{R_1 R_2}{R_1 + R_2} = \frac{(1\text{ k}\Omega)(2.2\text{ k}\Omega)}{3.2\text{ k}\Omega} = 688\ \Omega$$

The resistance from point B to point C is R_3 in parallel with the series combination of R_4 and R_5:

$$R_4 + R_5 = 330\ \Omega + 680\ \Omega = 1010\ \Omega = 1.01\ \text{k}\Omega$$

$$R_{BC} = \frac{R_3(R_4 + R_5)}{R_3 + R_4 + R_5} = \frac{(1.8\ \text{k}\Omega)(1.01\ \text{k}\Omega)}{2.81\ \text{k}\Omega} = 647\ \Omega$$

The resistance from A to B is in series with the resistance from B to C, so the total circuit resistance is

$$R_T = R_{AB} + R_{BC} = 688\ \Omega + 647\ \Omega = 1335\ \Omega$$

Using the voltage divider principle, we find the voltages as follows:

$$V_{AB} = \left(\frac{R_{AB}}{R_T}\right)V_S = \left(\frac{688\ \Omega}{1335\ \Omega}\right)100\ \text{V} = 51.54\ \text{V}$$

$$V_{BC} = \left(\frac{R_{BC}}{R_T}\right)V_S = \left(\frac{647\ \Omega}{1335\ \Omega}\right)100\ \text{V} = 48.46\ \text{V}$$

$$V_1 = V_2 = V_{AB} = 51.54\ \text{V}$$

$$V_3 = V_{BC} = 48.46\ \text{V}$$

$$V_4 = \left(\frac{R_4}{R_4 + R_5}\right)V_{BC} = \left(\frac{330\ \Omega}{1010\ \Omega}\right)48.46\ \text{V} = 15.83\ \text{V}$$

$$V_5 = \left(\frac{R_5}{R_4 + R_5}\right)V_{BC} = \left(\frac{680\ \Omega}{1010\ \Omega}\right)48.46\ \text{V} = 32.63\ \text{V}$$

SECTION REVIEW 6–2

1. Find the total resistance between A and B in the circuit of Figure 6–22.
2. Find I_3 in Figure 6–22.
3. Find V_2 in Figure 6–22.
4. Determine R_T and I_T in Figure 6–23.

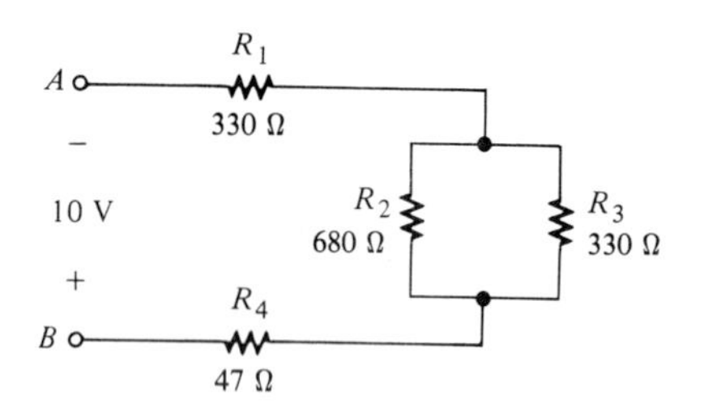

FIGURE 6–22

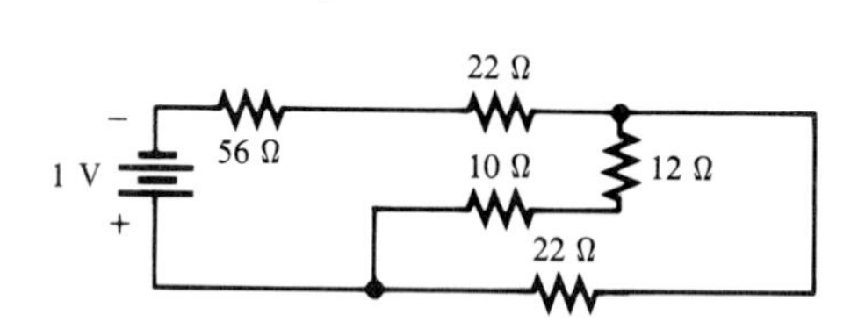

FIGURE 6–23

CIRCUIT GROUND

6–3

Voltage is relative. That is, the voltage at one point in a circuit is always measured relative to another point. For example, if we say that there are +100 V at a certain point in a circuit, we mean that the point is 100 V more positive than some *reference point* in the circuit. This *reference point* in a circuit is usually called *ground,* abbreviated GND.

The term *ground* derives from the method used in ac power lines, in which one side of the line is neutralized by connecting it to a water pipe or a metal rod driven into the ground. This method of grounding is called *earth ground.*

In most electronic equipment, the metal chassis that houses the assembly or a conductive area on a PC board is used as the *common* or *reference point,* called the *chassis ground* or *circuit ground,* as illustrated in Figure 6–24.

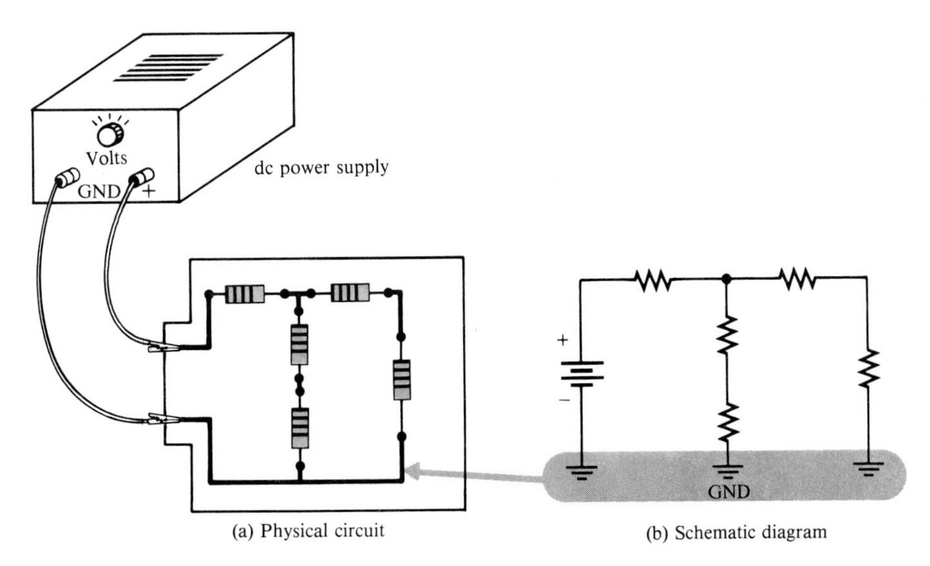

FIGURE 6–24
Illustration of circuit ground.

The ground provides a convenient way of connecting all common points within the circuit back to one side of the battery or other voltage source. The chassis or circuit ground does not necessarily have to be connected to earth ground. However, in many cases it is earth grounded in order to prevent a shock hazard due to a potential difference between chassis and earth ground.

In summary, ground is a reference point in electronic circuits. It has a potential of zero volts (0 V) *with respect to all other points in the circuit that are referenced to it,* as illustrated in Figure 6–25. In Part (a), the negative side of

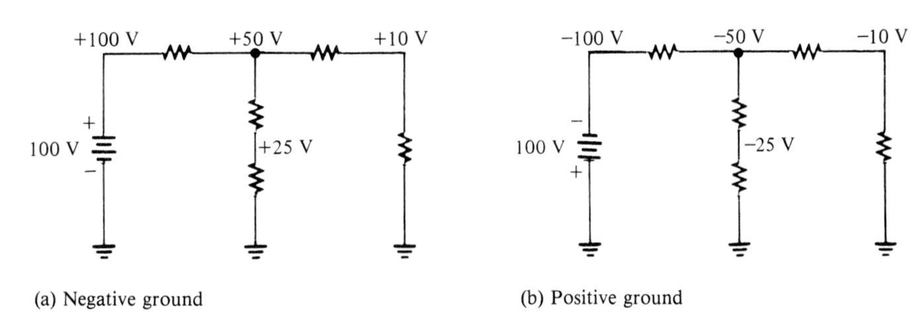

FIGURE 6–25
Example of negative and positive grounds.

the source is grounded, and all voltages indicated are positive with respect to ground. In Part (b), the positive side of the source is ground. The voltages at all other points are therefore negative with respect to ground.

Measurement of Voltages with Respect to Ground

When voltages are measured in a circuit, one meter lead is connected to the circuit ground, and the other to the point at which the voltage is to be measured. In a negative ground circuit, the negative meter terminal is connected to the circuit ground. The positive terminal of the voltmeter is then connected to the positive voltage point. Measurement of positive voltage is illustrated in Figure 6–26, where the meter reads the voltage at point A with respect to ground.

For a circuit with a positive ground, the positive voltmeter lead is connected to ground, and the negative lead is connected to the negative voltage point, as indicated in Figure 6–27. Here the meter reads the voltage at point A with respect to ground.

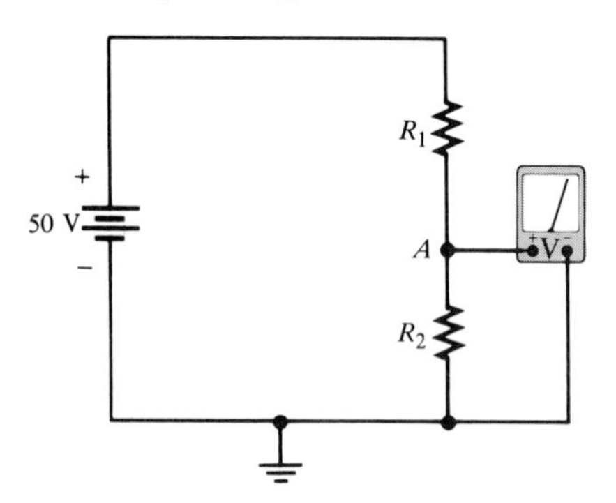

FIGURE 6–26
Measuring a voltage with respect to negative ground.

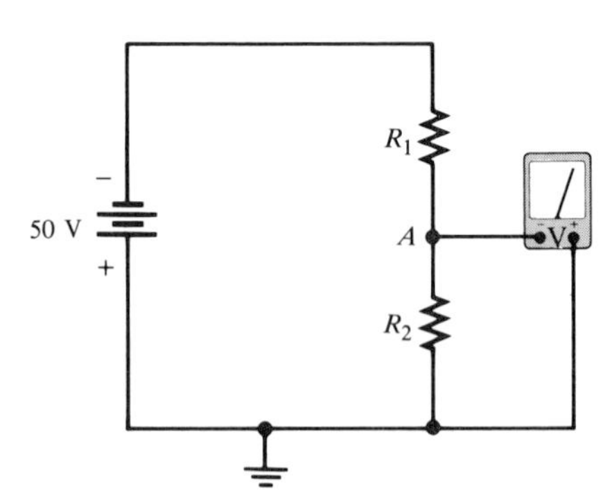

FIGURE 6–27
Measuring a voltage with respect to positive ground.

When voltages must be measured at several points in a circuit, the ground lead can be clipped to ground at one point in the circuit and left there. The other lead is then moved from point to point as the voltages are measured. This method is illustrated in Figure 6–28.

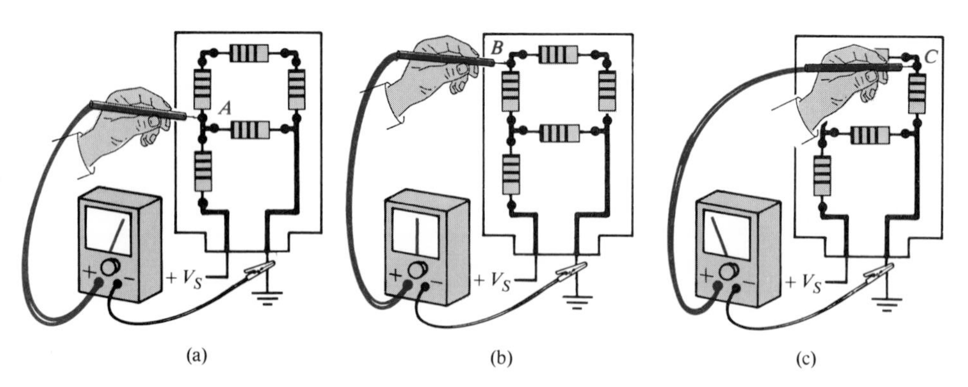

FIGURE 6–28
Measuring voltages with respect to ground at several points in a circuit.

Measurement of Voltage across an Ungrounded Resistor

Voltage can normally be measured across a resistor, as shown in Figure 6–29, even though neither side of the resistor is connected to circuit ground.

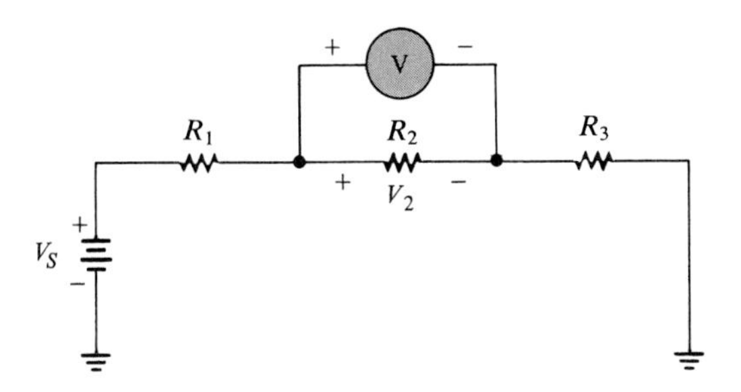

FIGURE 6–29
Measuring voltage across a resistor.

In some cases, when the meter is not isolated from power line ground, the negative lead of the meter will ground one side of the resistor and alter the operation of the circuit. In this situation, another method must be used, as illustrated in Figure 6–30. The voltages on each side of the resistor are measured *with respect to ground.* The *difference* of these two measurements is the voltage drop across the resistor.

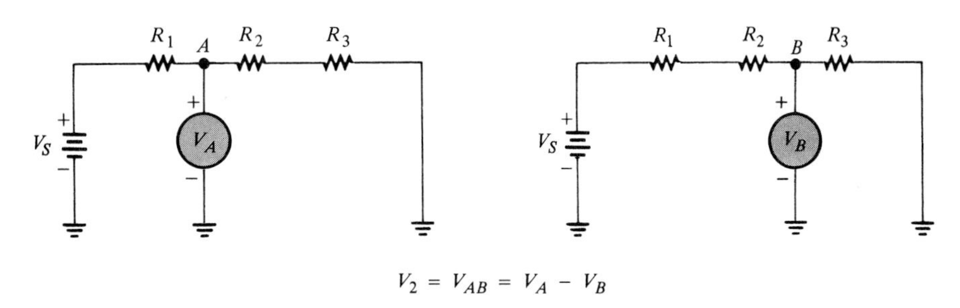

FIGURE 6–30
Measuring voltage across a resistor with two separate measurements to ground.

EXAMPLE 6–10 Determine the voltages with respect to ground at each of the indicated points in each circuit of Figure 6–31. Assume that 25 V are dropped across each resistor.

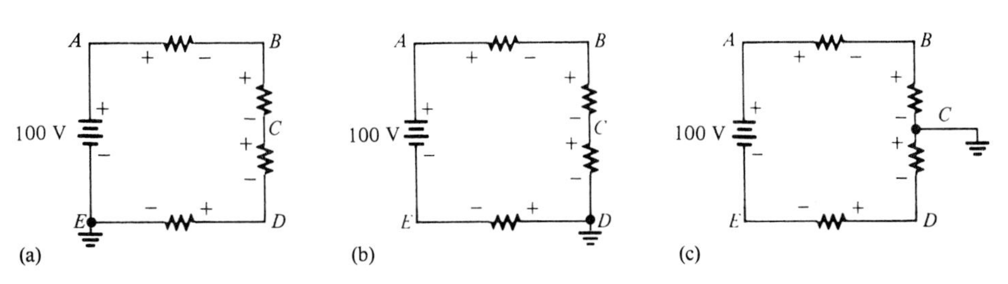

FIGURE 6–31

Solution

In circuit (a), the voltage polarities are as shown. Point E is ground. The voltages with respect to ground are as follows:

$$V_E = 0 \text{ V}$$
$$V_{DE} = +25 \text{ V}$$
$$V_{CE} = +50 \text{ V}$$
$$V_{BE} = +75 \text{ V}$$
$$V_{AE} = +100 \text{ V}$$

In circuit (b), the voltage polarities are as shown. Point D is ground. The voltages with respect to ground are as follows:

$$V_{ED} = -25 \text{ V}$$
$$V_D = 0 \text{ V}$$
$$V_{CD} = +25 \text{ V}$$
$$V_{BD} = +50 \text{ V}$$
$$V_{AD} = +75 \text{ V}$$

In circuit (c), the voltage polarities are as shown. Point C is ground. The voltages with respect to ground are as follows:

$$V_{EC} = -50 \text{ V}$$
$$V_{DC} = -25 \text{ V}$$
$$V_C = 0 \text{ V}$$
$$V_{BC} = +25 \text{ V}$$
$$V_{AC} = +50 \text{ V}$$

SECTION REVIEW 6–3

1. The common point in a circuit is called ________.
2. Most voltages in a circuit are referenced to ground (T or F).
3. The housing or chassis is often used as circuit ground (T or F).
4. What does *earth ground* mean?

LOADED VOLTAGE DIVIDERS

6–4

Voltage dividers were introduced in Chapter 4. In this section we will discuss the effects of resistive loads on voltage dividers. For example, the voltage divider in Figure 6–32(a) produces an output voltage (V_{OUT}) of 5 V because the two resistors are of equal value. This voltage is the *unloaded output voltage.* When a load resistor, R_L, is connected from the output to ground as shown in Figure 6–32(b), the output voltage is reduced by an amount that depends on the value of R_L. The load resistor is in parallel with R_2, reducing the resistance from point A to ground and, as a result, also reducing the voltage across the parallel combination. This is one effect of *loading* a voltage divider. Another effect of a loaded condition is that more current is drawn from the source because the total resistance of the circuit is reduced.

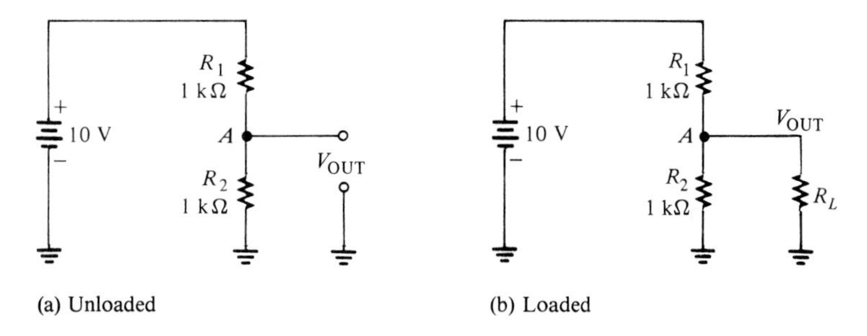

(a) Unloaded (b) Loaded

FIGURE 6–32
A voltage divider with both unloaded and loaded outputs.

The larger R_L is compared to R_2, the less the output voltage is reduced from its unloaded value, as illustrated in Figure 6–33. The reason is that when two resistors are connected in parallel and one of the resistors is much greater than the other, the total resistance is close to the value of the smaller resistance.

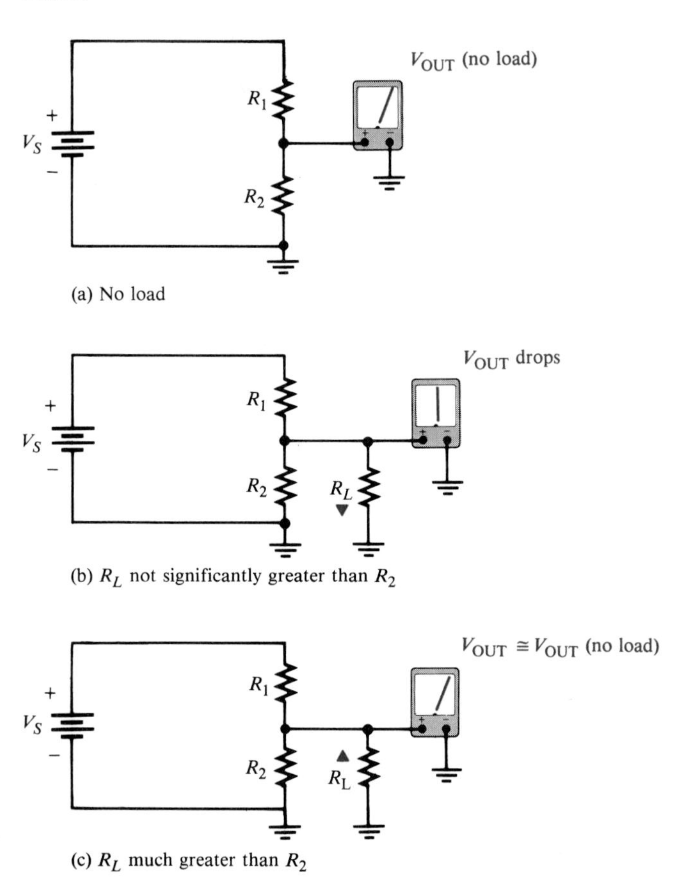

(a) No load

(b) R_L not significantly greater than R_2

(c) R_L much greater than R_2

FIGURE 6–33
The effect of a load resistor.

EXAMPLE 6–11 Determine both the unloaded and the loaded output voltages of the voltage divider in Figure 6–34 for the following two values of load resistance:

(a) $R_L = 10\ \text{k}\Omega$.
(b) $R_L = 100\ \text{k}\Omega$.

FIGURE 6–34

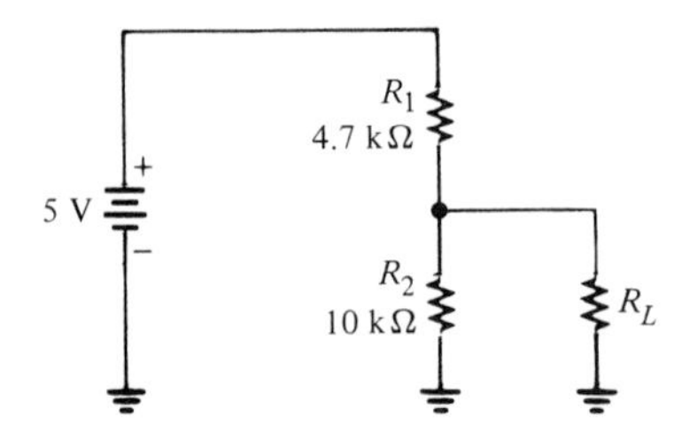

Solution

(a) The *unloaded* output voltage is

$$V_{\text{OUT}} = \left(\frac{10\ \text{k}\Omega}{14.7\ \text{k}\Omega}\right)5\ \text{V} = 3.4\ \text{V}$$

With the 10-kΩ load resistor connected, R_L is in parallel with R_2, which gives 5 kΩ, as shown by the equivalent circuit in Figure 6–35(a).
The *loaded* output voltage is

$$V_{\text{OUT}} = \left(\frac{5\ \text{k}\Omega}{9.7\ \text{k}\Omega}\right)5\ \text{V} = 2.6\ \text{V}$$

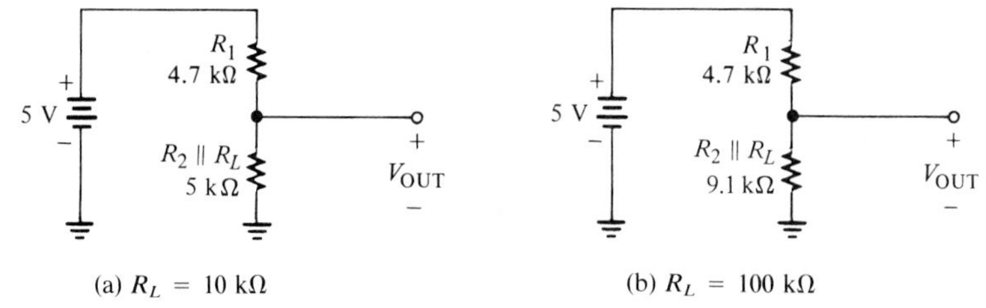

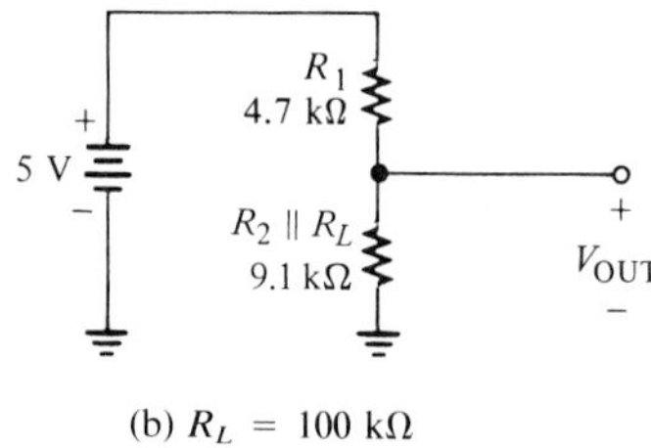

FIGURE 6–35

(b) With the 100-kΩ load, the resistance from output to ground is

$$\frac{R_2 R_L}{R_2 + R_L} = \frac{(10\ \text{k}\Omega)(100\ \text{k}\Omega)}{110\ \text{k}\Omega} = 9.1\ \text{k}\Omega$$

as shown in Figure 6–35(b).
The *loaded* output voltage is

$$V_{\text{OUT}} = \left(\frac{9.1\ \text{k}\Omega}{13.8\ \text{k}\Omega}\right)5\ \text{V} = 3.3\ \text{V}$$

Notice that with the larger value of R_L, the output is reduced from its unloaded value by much less than it is with the smaller R_L. This problem illustrates the loading effect of R_L on the voltage divider.

Load Current and Bleeder Current

In a loaded voltage divider circuit, the total current drawn from the source consists of currents that flow through the load resistors, and current that flows only through the divider resistors. Figure 6–36 shows a voltage divider with two voltage outputs or two *taps*. Notice that part of the total current, I_T, flows through R_1 and R_{L1}; part flows through R_1, R_2, and R_{L2}; and the rest flows through R_3. As you can see, the current through R_3 is not a load current. Called the *bleeder current,* it is what is left of the total current after the load currents branch off.

EXAMPLE 6–12 Determine the load currents and the bleeder current in the two-tap loaded voltage divider in Figure 6–36.

FIGURE 6–36
Currents in a two-tap loaded voltage divider.

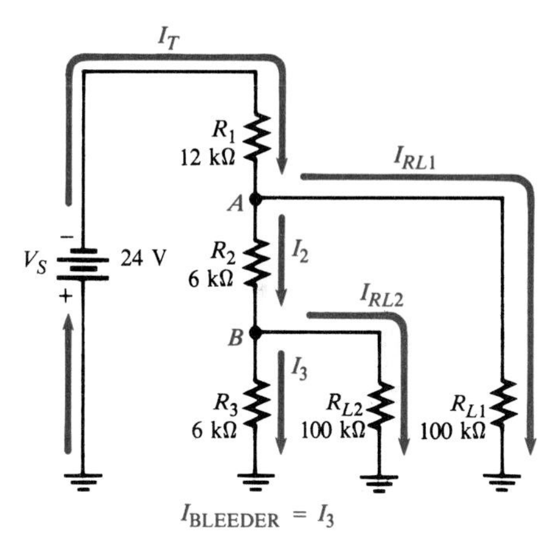

Solution

The equivalent resistance from point A to ground is the 100-kΩ load resistor R_{L1} in parallel with the combination of R_2 in series with the parallel combination of R_3 and R_{L2}. This solution is as follows.

R_3 in parallel with R_{L2} is R_B:

$$R_B = \frac{(6\text{ k}\Omega)(100\text{ k}\Omega)}{106\text{ k}\Omega} = 5.66\text{ k}\Omega$$

R_2 in series with R_B is $R_{2\text{-}B}$:

$$R_{2\text{-}B} = 6\text{ k}\Omega + 5.66\text{ k}\Omega = 11.66\text{ k}\Omega$$

R_{L1} in parallel with $R_{2\text{-}B}$ is R_A:

$$R_A = \frac{(100\ \text{k}\Omega)(11.66\ \text{k}\Omega)}{111.66\ \text{k}\Omega} = 10.44\ \text{k}\Omega$$

$$R_T = R_A + R_1 = 10.44\ \text{k}\Omega + 12\ \text{k}\Omega = 22.44\ \text{k}\Omega$$

R_A is the total resistance from point A to ground.
The voltage across R_{L1} is determined as follows:

$$V_{RL1} = V_A = \left(\frac{R_A}{R_T}\right)V_S = \left(\frac{10.44\ \text{k}\Omega}{22.44\ \text{k}\Omega}\right)24\ \text{V} = 11.17\ \text{V}$$

$$I_{RL1} = \frac{V_{RL1}}{R_{L1}} = \frac{11.17\ \text{V}}{100\ \text{k}\Omega} = 0.112\ \text{mA}$$

The voltage at point B is determined by the parallel combination of R_{L2} and R_3 (R_B), R_2, and the voltage at point A:

$$V_B = \left(\frac{R_B}{R_2 + R_B}\right)V_A = \left(\frac{5.66\ \text{k}\Omega}{11.66\ \text{k}\Omega}\right)11.17\ \text{V} = 5.42\ \text{V}$$

The load current through R_{L2} is

$$I_{RL2} = \frac{V_{RL2}}{R_{L2}} = \frac{V_B}{R_{L2}} = \frac{5.42\ \text{V}}{100\ \text{k}\Omega} = 0.054\ \text{mA}$$

The bleeder current is

$$I_3 = \frac{V_B}{R_3} = \frac{5.42\ \text{V}}{6\ \text{k}\Omega} = 0.903\ \text{mA}$$

The Loading Effect of a Voltmeter

As you have learned, voltmeters are connected in parallel in order to measure voltages across resistors. In effect, a voltmeter is a load on the circuit that is being measured, but until now, we have ignored the loading effect. Why?

The internal resistance of a voltmeter is very high, and normally it has negligible effect on the circuit that it is measuring. However, if the internal resistance of the voltmeter is not sufficiently greater than the circuit resistance across which it is connected, the loading effect will cause the measured voltage to be less than its actual value.

The internal resistance of a voltmeter is specified by the *sensitivity* factor. A common sensitivity value for many voltmeters is 20,000 Ω/V. For example, to use a voltmeter in Figure 6–32, you would set it on the 10-V range in order to measure the 5-V output, and its internal resistance would be

$$20{,}000\ \Omega/\text{V} \times 10\ \text{V} = 200{,}000\ \Omega$$

The loading effect is negligible in this case because 200,000 Ω in parallel with 1 kΩ has little effect on the voltage across the 1-kΩ resistor.

When you are measuring voltages in circuits that have high resistance values, then you must be aware of potential loading effects of the voltmeter.

Bipolar Voltage Dividers

An example of a voltage divider that produces both positive and negative voltages from a single source is shown in Figure 6–37. Notice that neither the positive nor the negative terminal of the source is connected to ground. The voltages at points A and B are *positive with respect to ground,* and the voltages at points C and D are *negative with respect to ground.*

FIGURE 6–37
A bipolar voltage divider. The positive and negative voltages are with respect to ground.

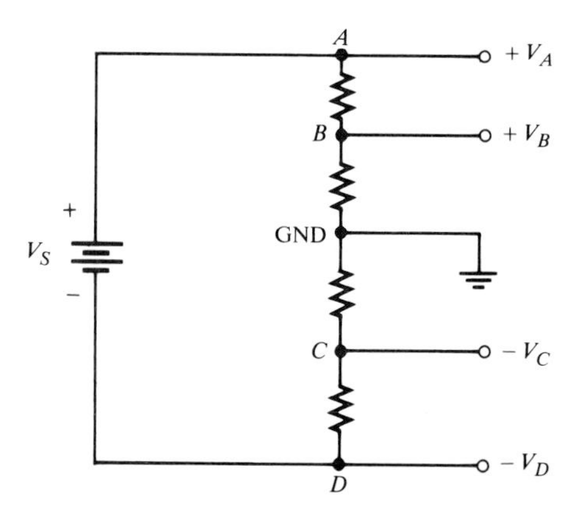

SECTION REVIEW 6–4

1. A load resistor is connected to an output on a voltage divider. What effect does the load resistor have on the output voltage?
2. A larger-value load resistor will cause the output voltage to change less than a smaller-value one will (T or F).
3. For the voltage divider in Figure 6–38, determine the unloaded output voltage. Also determine the output voltage with a 10-kΩ load resistor connected from the output to ground.

FIGURE 6–38

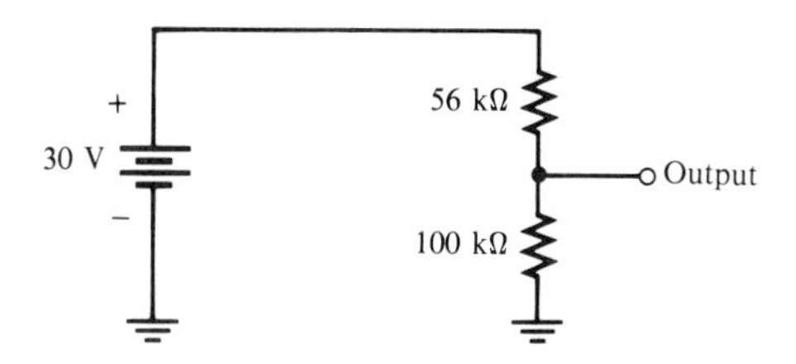

THE WHEATSTONE BRIDGE

6–5

The *bridge circuit* is widely used in measurement devices and other applications that you will learn later. For now, we will consider the *balanced bridge,* which can be used to measure unknown resistance values. This circuit, shown in Figure 6–39(a), is known as a *Wheatstone bridge.* Part (b) is the same circuit electrically, but it is drawn in a different way.

A bridge is said to be *balanced* when the voltage (V_{OUT}) across the output terminals A and B is *zero;* that is, $V_A = V_B$. If V_A equals V_B, then $V_{R1} = V_{R2}$,

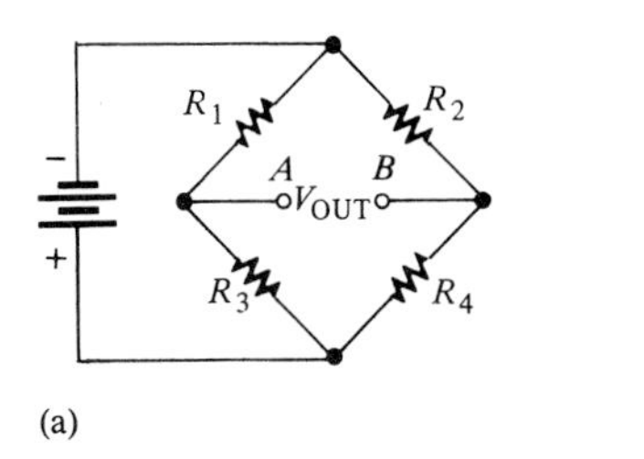

(a) (b)

FIGURE 6–39
Wheatstone bridge.

since the top sides of both R_1 and R_2 are connected to the same point. Also, $V_{R3} = V_{R4}$, since the bottom sides of both R_3 and R_4 connect to the same point. The voltage ratios can be written as

$$\frac{V_1}{V_3} = \frac{V_2}{V_4}$$

Substituting by Ohm's law yields

$$\frac{I_1R_1}{I_1R_3} = \frac{I_2R_2}{I_2R_4}$$

The currents cancel, leaving

$$\frac{R_1}{R_3} = \frac{R_2}{R_4}$$

Solving for R_1, we get

$$R_1 = R_3\left(\frac{R_2}{R_4}\right)$$

How can we use this formula to determine an unknown resistance? First, let us make R_3 a *variable* resistor and call it R_V. Also, we set the ratio R_2/R_4 to a known value. If R_V is adjusted until the bridge is balanced, the product of R_V and the ratio R_2/R_4 is equal to R_1, which is our unknown resistor (R_{UNK}). The formula for unknown resistance is stated as follows:

$$R_{UNK} = R_V\left(\frac{R_2}{R_4}\right) \tag{6–1}$$

The bridge is balanced when the voltage across the output terminals equals zero ($V_A = V_B$). A *galvanometer* (a meter that measures small currents in either direction and is zero at center scale) is connected between the output terminals. Then R_V is adjusted until the galvanometer shows zero current ($V_A = V_B$), indicating a balanced condition. The setting of R_V multiplied by the ratio R_2/R_4 gives the value of R_{UNK}. Figure 6–40 shows this arrangement. For example, if $R_2/R_4 = 1/10$ and $R_V = 680\ \Omega$, then $R_{UNK} = (680\ \Omega)(1/10) = 68\ \Omega$.

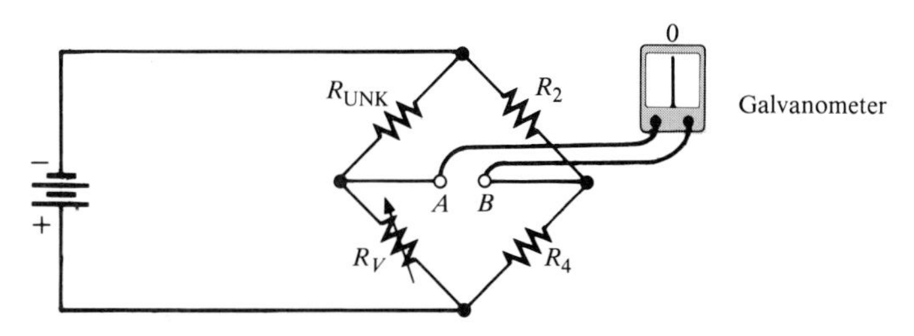

FIGURE 6–40
Balanced Wheatstone bridge.

A Bridge Application

Bridge circuits are used for many measurements other than that of unknown resistance. One application of the Wheatstone bridge is in accurate temperature measurement. A temperature-sensitive element such as a thermistor is connected in a Wheatstone bridge as shown in Figure 6–41. An amplifier is connected across the output from A to B in order to increase the output voltage from the bridge to a usable value. The bridge is calibrated so that it is balanced at a specified reference temperature. As the temperature changes, the resistance of the sensing element changes proportionately, and the bridge becomes unbalanced. As a result, V_{AB} changes and is amplified (increased) and converted to a form for direct temperature readout on a gauge or a digital-type display.

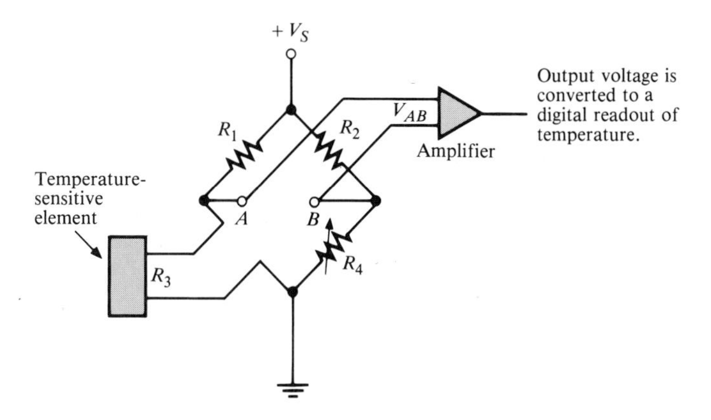

FIGURE 6–41
A simplified circuit for temperature measurement.

SECTION REVIEW 6–5

1. Under what condition is a bridge balanced?
2. What is the unknown resistance in Figure 6–42?

FIGURE 6–42

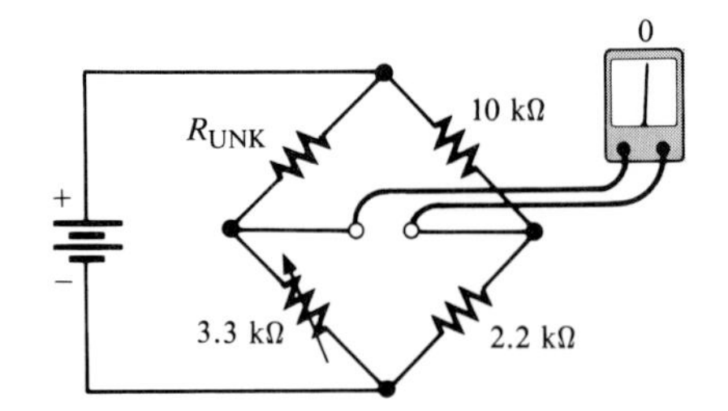

CIRCUITS WITH MORE THAN ONE VOLTAGE SOURCE

6–6

Some circuits require more than one voltage source. For example, certain types of amplifiers require both a positive and a negative voltage source for proper operation.

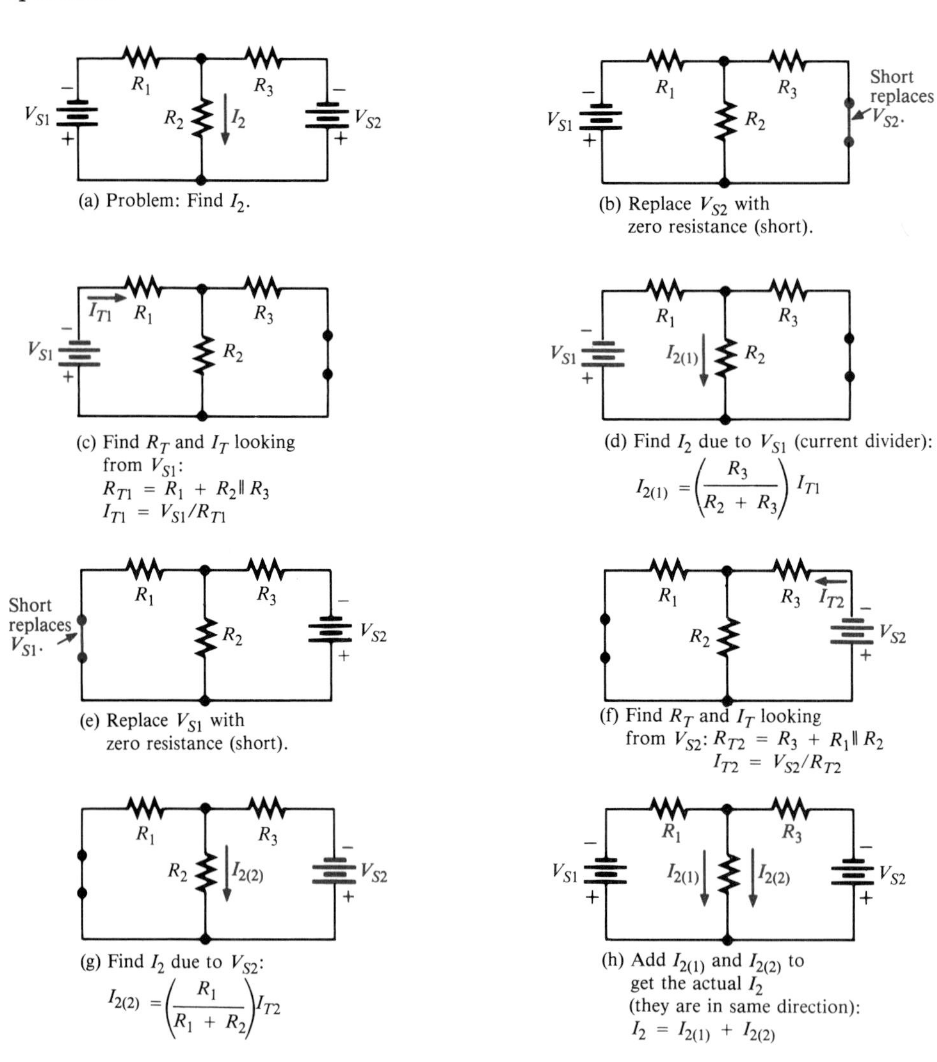

FIGURE 6–43
Demonstration of the superposition method.

In this section, a general method is presented for evaluating series-parallel circuits having multiple voltage sources. The method is illustrated with two-source circuits. However, once you understand the basic principle, you can easily extend this method to circuits with any number of sources.

The Superposition Method

The superposition method is a way to determine currents and voltages in a circuit that has multiple sources by taking *one source at a time*. The other sources are replaced by their internal resistances. Recall from Chapter 4 that the *ideal* voltage source has a *zero* internal resistance. In this section, all voltage sources will be treated as ideal in order to simplify the coverage.

The steps in applying the superposition method are as follows:

Step 1: Take one voltage source at a time and replace each of the other voltage sources with a short (a short represents zero resistance).

Step 2: Determine the current or voltage that you need just as if there were only *one* source in the circuit.

Step 3: Take the next source in the circuit and repeat Steps 1 and 2 for each source.

Step 4: To find the actual current or voltage, add or subtract the currents or voltages due to each individual source. If the currents are in the same direction or the voltages of the same polarity, add them. If the currents are in opposite directions or the voltages of opposite polarities, subtract them.

An example of the approach to superposition is demonstrated in Figure 6–43 for a series-parallel circuit with two voltage sources. Study the steps in this figure.

EXAMPLE 6–13 Find the total current through R_3 in Figure 6–44.

FIGURE 6–44

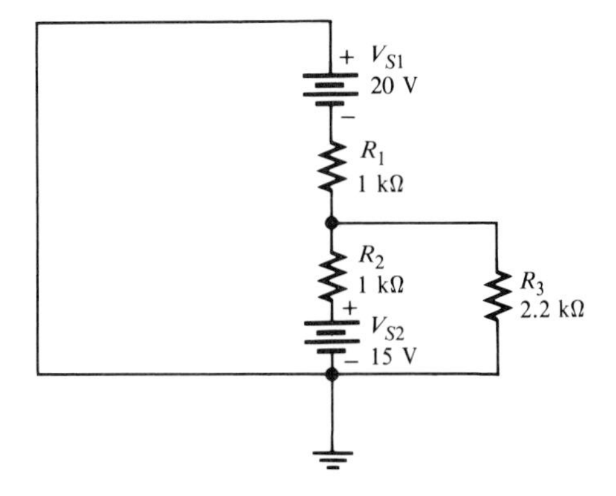

Solution

Step 1: Find the current through R_3 due to source V_{S1} by replacing source V_{S2} with a short, as shown in Figure 6–45(a).

Looking from V_{S1},

$$R_{T1} = R_1 + \frac{R_2R_3}{R_2 + R_3}$$

$$= 1\ \text{k}\Omega + \frac{(1\ \text{k}\Omega)(2.2\ \text{k}\Omega)}{3.2\ \text{k}\Omega} = 1.69\ \text{k}\Omega$$

$$I_{T1} = \frac{V_{S1}}{R_{T1}} = \frac{20\ \text{V}}{1.69\ \text{k}\Omega} = 11.83\ \text{mA}$$

Now apply the current divider formula to find the current through R_3 due to source V_{S1} as follows:

$$I_{3(1)} = \left(\frac{R_2}{R_2 + R_3}\right)I_{T1} = \left(\frac{1\ \text{k}\Omega}{3.2\ \text{k}\Omega}\right)11.83\ \text{mA} = 3.7\ \text{mA}$$

Notice that this current flows *downward* through R_3.

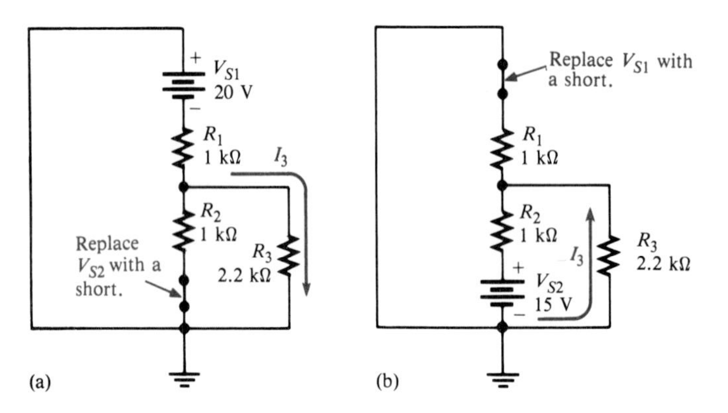

FIGURE 6–45

Step 2: Next find I_3 due to source V_{S2} by replacing source V_{S1} with a short, as shown in Figure 6–45(b).

Looking from V_{S2},

$$R_{T2} = R_2 + \frac{R_1R_3}{R_1 + R_3}$$

$$= 1\ \text{k}\Omega + \frac{(2.2\ \text{k}\Omega)(1\ \text{k}\Omega)}{3.2\ \text{k}\Omega} = 1.69\ \text{k}\Omega$$

$$I_{T2} = \frac{V_{S2}}{R_{T2}} = \frac{15\ \text{V}}{1.69\ \text{k}\Omega} = 8.88\ \text{mA}$$

Now apply the current divider formula to find the current through R_3 due to source V_{S2} as follows:

$$I_{3(2)} = \left(\frac{R_1}{R_1 + R_3}\right)I_T = \left(\frac{1\ \text{k}\Omega}{3.2\ \text{k}\Omega}\right)8.88\ \text{mA} = 2.78\ \text{mA}$$

Notice that this current flows *upward* through R_3.

Step 3: Calculation of the total current through R_3 is as follows:

$$I_{3(T)} = I_{3(1)} - I_{3(2)} = 3.7 \text{ mA} - 2.78 \text{ mA} = 0.92 \text{ mA}$$

This current flows downward through R_3.

SECTION REVIEW 6–6

1. Using the superposition theorem, find the current through R_1 in Figure 6–46.
2. If two currents are in opposing directions through a branch of a circuit, in what direction does the net current flow?

FIGURE 6–46

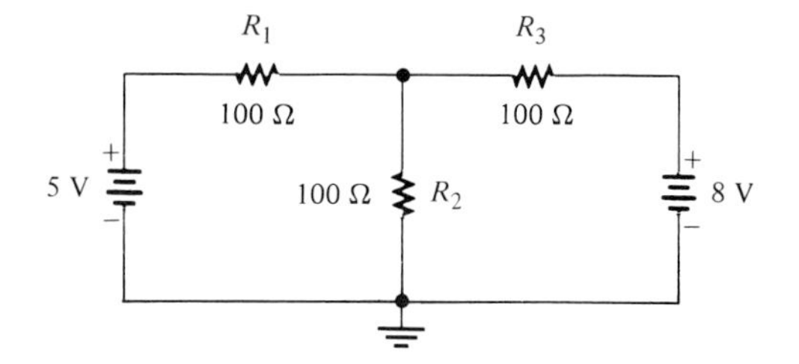

THEVENIN'S THEOREM

6–7

Thevenin's theorem gives us a method for simplifying a circuit to a standard equivalent form. The Thevenin equivalent form of any resistive circuit consists of an *equivalent voltage source* (V_{TH}) and an *equivalent resistance* (R_{TH}), arranged as shown in Figure 6–47. The values of the equivalent voltage and resistance depend on the values in the original circuit. Any resistive circuit can be simplified regardless of its complexity.

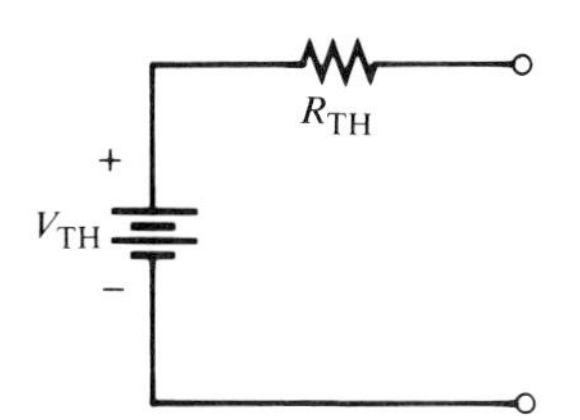

FIGURE 6–47 The general form of a Thevenin equivalent circuit. Any resistive circuit can be reduced to this form.

The Meaning of Equivalency in Thevenin's Theorem

Although a Thevenin equivalent circuit is not the same as its original circuit, it *acts* the same in terms of the output voltage and current. For example, as shown in Figure 6–48, place a resistive circuit of any complexity in a box with only the output terminals exposed. Then place the Thevenin equivalent of that circuit in an *identical* box with, again, only the output terminals exposed. Connect identical load resistors across the output terminals of each box. Next connect a voltmeter and an ammeter to measure the voltage and current for each load as shown in the figure. The measured values will be identical (neglecting tolerance variations), and you will not be able to determine which box contains the original circuit and which contains the Thevenin equivalent. That is, in terms of your observations, both circuits are the same. This condition is sometimes known as *terminal equivalency,* because both circuits look the same from the "viewpoint" of the two output terminals.

How to Find the Thevenin Equivalent of a Circuit

To find the Thevenin equivalent of any circuit, determine the equivalent voltage, V_{TH}, and the equivalent resistance, R_{TH}. For example, in Figure 6–49, the

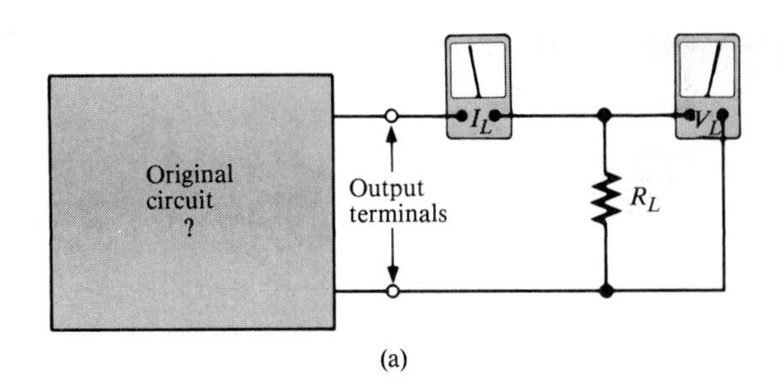

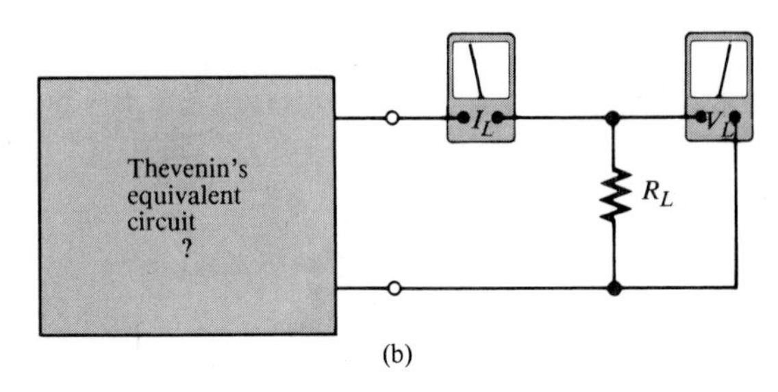

FIGURE 6–48
Which box contains the original circuit and which contains the Thevenin equivalent circuit?

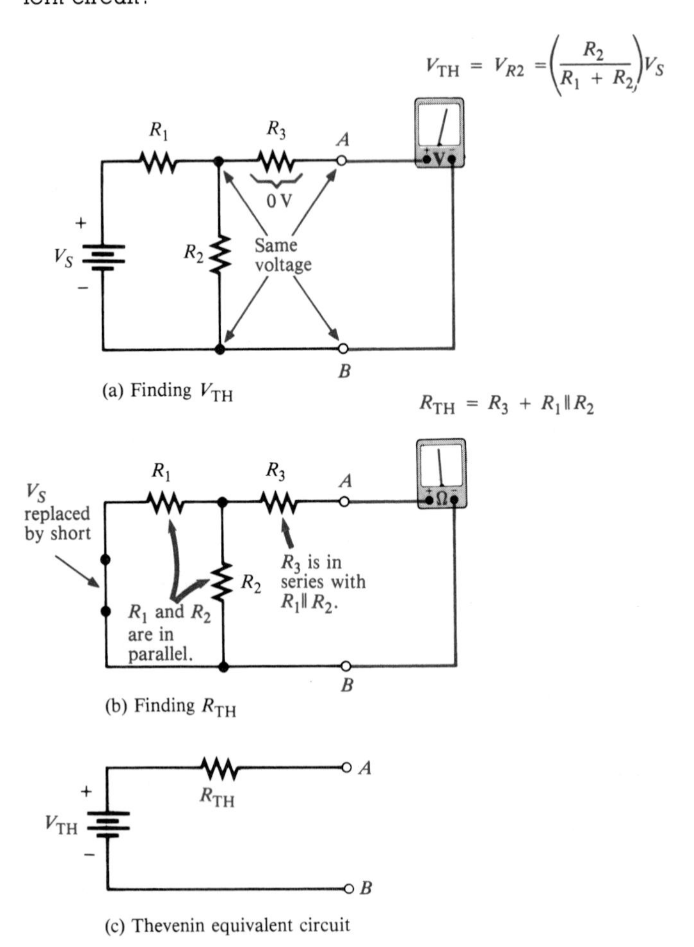

FIGURE 6–49
Example of the simplification of a circuit by Thevenin's theorem.

Thevenin equivalent for the circuit between points A and B is found as follows:

In Part (a), the voltage across the designated points A and B is the Thevenin equivalent voltage. In this particular circuit, the voltage from A to B is the same as the voltage across R_2 because there is no current through R_3 and, therefore, no voltage drop across it. V_{TH} is expressed as follows for this particular example:

$$V_{TH} = \left(\frac{R_2}{R_1 + R_2}\right)V_S$$

In Part (b), the resistance between points A and B with the source replaced by a short (zero internal resistance) is the Thevenin equivalent resistance. In this particular circuit, the resistance from A to B is R_3 in series with the parallel combination of R_1 and R_2. Therefore, R_{TH} is expressed as follows:

$$R_{TH} = R_3 + \frac{R_1R_2}{R_1 + R_2}$$

The Thevenin equivalent circuit is shown in Part (c).

EXAMPLE 6–14 Find the Thevenin equivalent between the output terminals of the circuit in Figure 6–50.

FIGURE 6–50

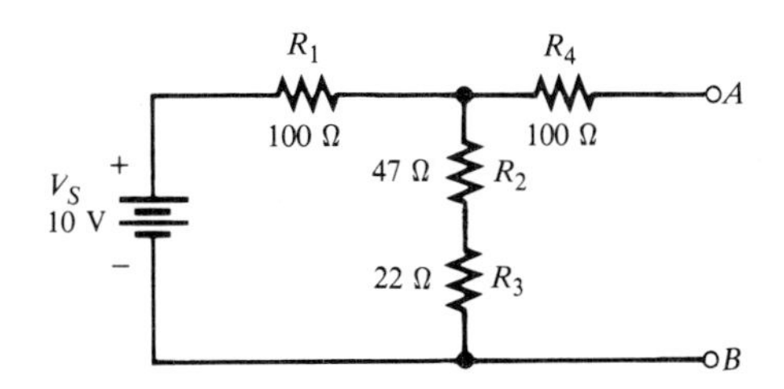

Solution

V_{TH} equals the voltage across $R_2 + R_3$ as shown in Figure 6–51(a). Use the voltage divider principle to find V_{TH}:

$$V_{TH} = \left(\frac{R_2 + R_3}{R_1 + R_2 + R_3}\right)V_S = \left(\frac{69\ \Omega}{169\ \Omega}\right)10\text{ V} = 4.08\text{ V}$$

To find R_{TH}, first replace the source with a short to simulate a zero internal resistance. Then R_1 appears in parallel with $R_2 + R_3$, and R_4 is in series with the series-parallel combination of R_1, R_2, and R_3 as indicated in Figure 6–51(b):

$$R_{TH} = R_4 + \frac{R_1(R_2 + R_3)}{R_1 + R_2 + R_3} = 100\ \Omega + \frac{(100\ \Omega)(69\ \Omega)}{169\ \Omega} = 140.8\ \Omega$$

The resulting Thevenin equivalent circuit is shown in Part (c).

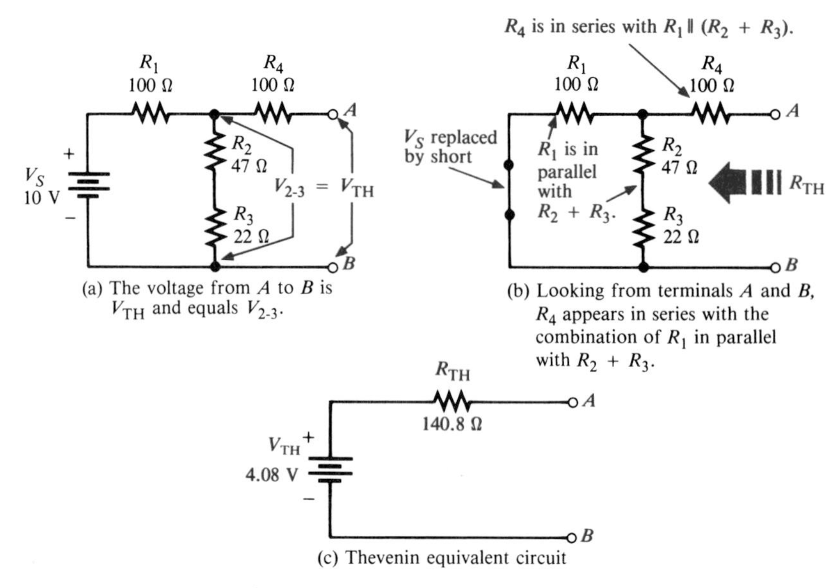

(a) The voltage from A to B is V_{TH} and equals $V_{2\text{-}3}$.

(b) Looking from terminals A and B, R_4 appears in series with the combination of R_1 in parallel with $R_2 + R_3$.

(c) Thevenin equivalent circuit

FIGURE 6–51

How the Thevenin Equivalency Depends on the Viewpoint

The Thevenin equivalent for any circuit depends on the location of the two points from between which the circuit is "viewed." In Figure 6–50, we viewed the circuit from between the two points labeled A and B. Any given circuit can have more than one Thevenin equivalent, depending on how the viewpoints are designated. For example, if you view the circuit in Figure 6–52 from between points A and C, you obtain a completely different result than if you viewed it from between points A and B or from between points B and C.

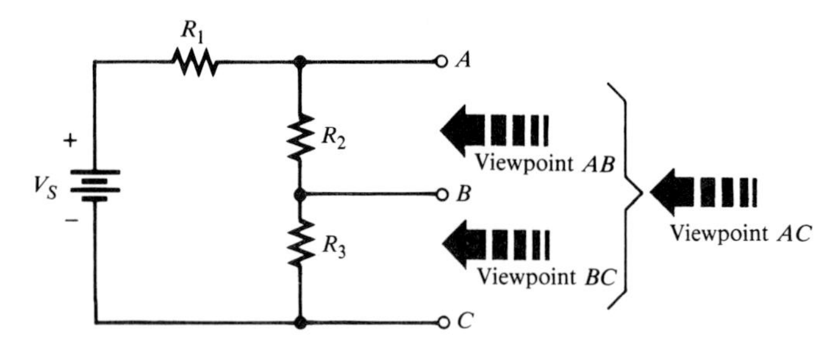

FIGURE 6–52
Thevenin's equivalent depends on viewpoint.

In Figure 6–53(a), when viewed from points A and C, V_{TH} is the voltage across $R_2 + R_3$ and can be expressed using the voltage divider formula as follows:

$$V_{TH} = \left(\frac{R_2 + R_3}{R_1 + R_2 + R_3}\right)V_S$$

Also, as shown in Part (b), the resistance between points A and C is $R_2 + R_3$ in parallel with R_1 (the source is replaced by a short) and can be expressed as follows:

$$R_{TH} = \frac{R_1(R_2 + R_3)}{R_1 + R_2 + R_3}$$

The resulting Thevenin equivalent circuit is shown in Part (c).

When viewed from points B and C as indicated in Figure 6–53(d), V_{TH} is the voltage across R_3 and can be expressed as follows:

$$V_{TH} = \left(\frac{R_3}{R_1 + R_2 + R_3}\right)V_S$$

As shown in Part (e), the resistance between points B and C is R_3 in parallel with the series combination of R_1 and R_2:

$$R_{TH} = \frac{R_3(R_1 + R_2)}{R_1 + R_2 + R_3}$$

The resulting Thevenin equivalent is shown in Part (f).

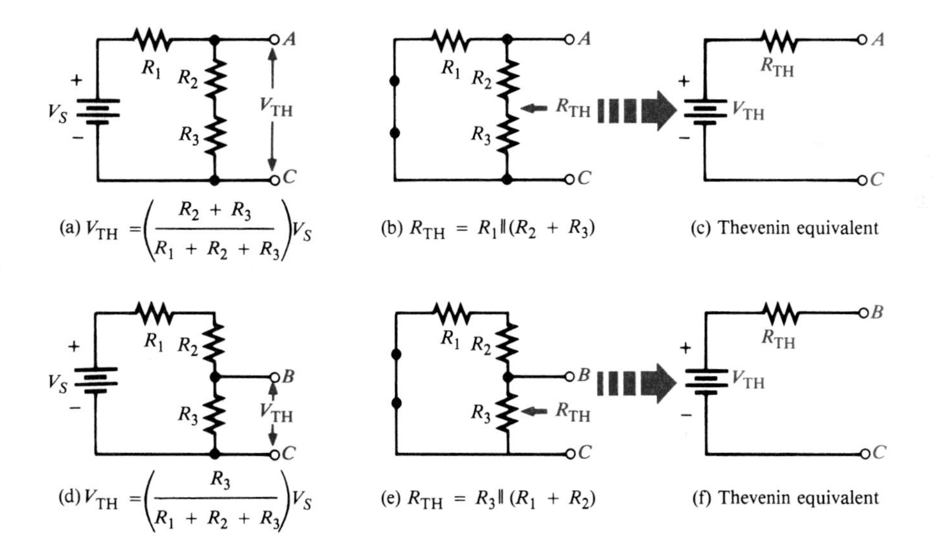

FIGURE 6–53
Example of circuit Thevenized from two viewpoints, resulting in two different equivalent circuits. (The V_{TH} and R_{TH} values are different.)

Thevenizing a Portion of a Circuit

In many cases it helps to Thevenize only a portion of a circuit. For example, when we need to know the equivalent circuit as viewed by one particular resistor in the circuit, we remove that resistor and apply Thevenin's theorem to the remaining part of the circuit as viewed from the points between which that resistor was connected. Figure 6–54 illustrates the Thevenizing of part of a circuit.

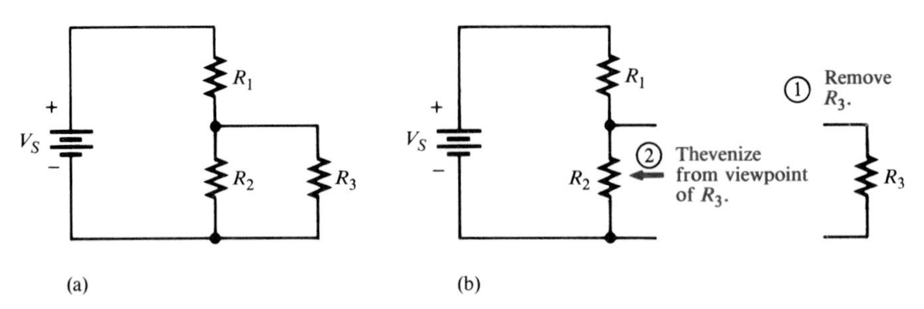

FIGURE 6–54
Example of Thevenizing a portion of a circuit. In this case, the circuit is Thevenized from the viewpoint of R_3.

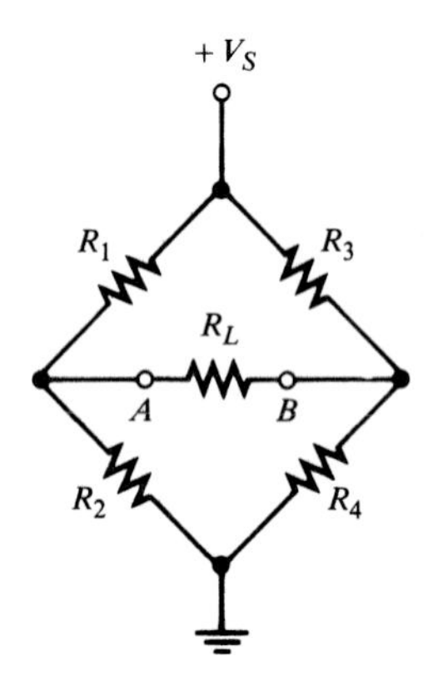

FIGURE 6–55
Wheatstone bridge with load resistor is not a series-parallel circuit.

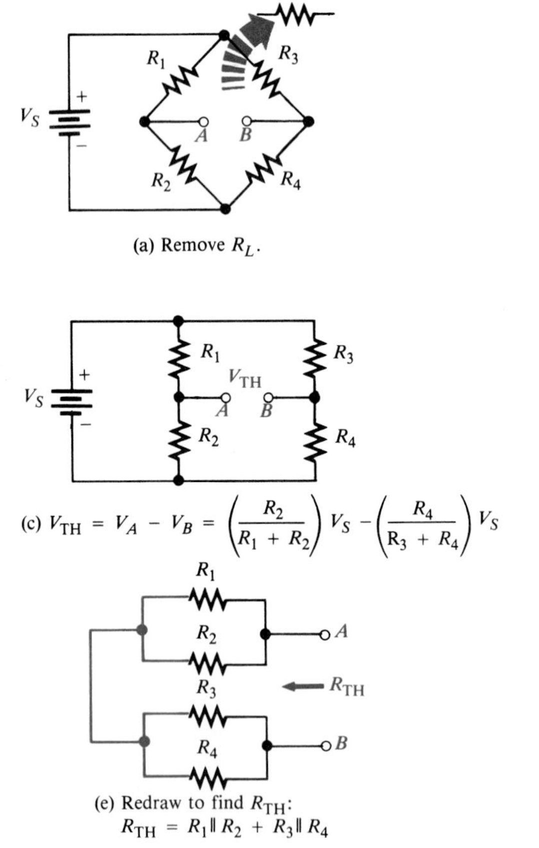

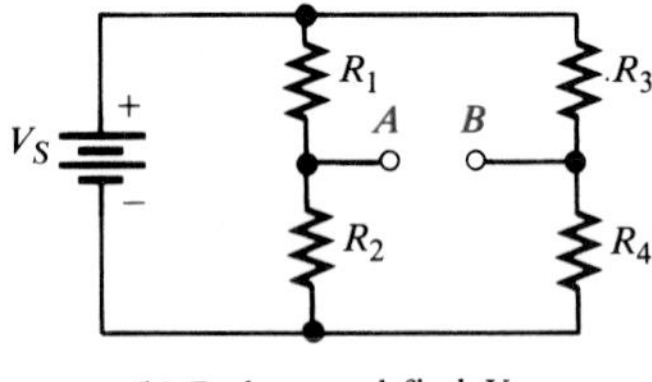

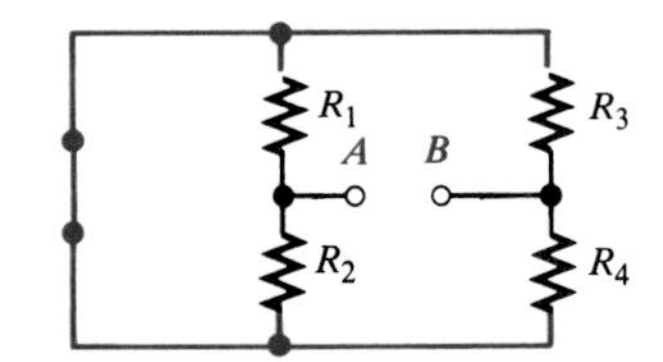

(d) Replace V_S with a short. *Note:* The colored lines are the same electrical point as the colored lines in Part (e).

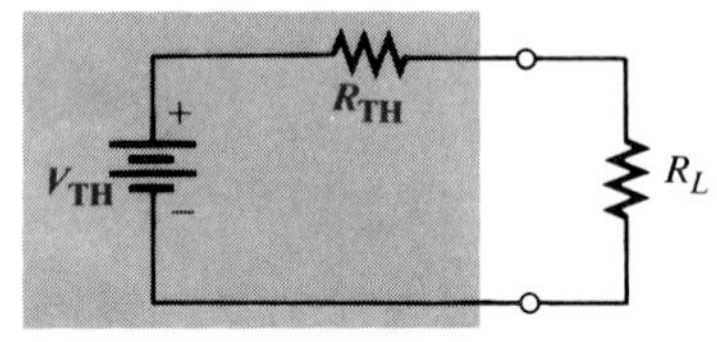

(f) Thevenin's equivalent with R_L reconnected

FIGURE 6–56
Simplifying a Wheatstone bridge with Thevenin's theorem.

Using this type of approach, you can easily find the voltage and current for a specified resistor for any number of resistor values using only Ohm's law. This method eliminates the necessity of reanalyzing the original circuit for each different resistance value.

Thevenizing a Bridge Circuit

The usefulness of Thevenin's theorem is perhaps best illustrated when it is applied to a Wheatstone bridge circuit. For example, when a load resistor is connected to the output terminals of a Wheatstone bridge, as shown in Figure 6–55, the circuit is very difficult to analyze because it is not a straightforward series-parallel arrangement. If you doubt that this analysis is difficult, try to identify which resistors are in parallel and which are in series.

Using Thevenin's theorem, we can simplify the bridge circuit to an equivalent circuit viewed from the load resistor as shown step-by-step in Figure 6–56. Study carefully the steps in this figure. Once the equivalent circuit for the bridge is found, the voltage and current for any value of load resistor can easily be determined.

EXAMPLE 6–15 Determine the voltage and current for the load resistor, R_L, in the bridge circuit of Figure 6–57.

FIGURE 6–57

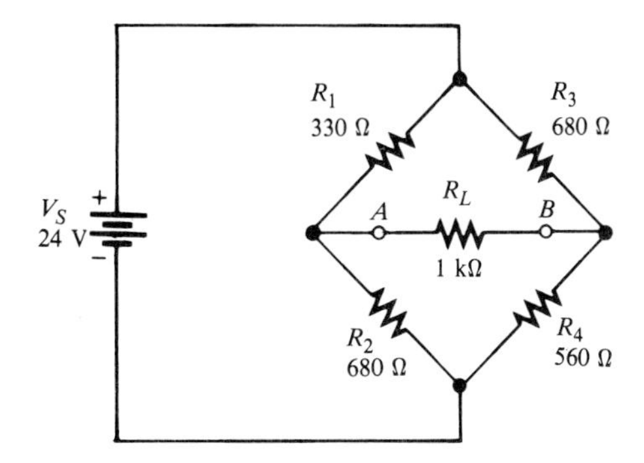

Solution

Step 1: Remove R_L.

Step 2: Thevenize the bridge as viewed from between points A and B, as was shown in Figure 6–56.

$$V_{TH} = V_A - V_B = \left(\frac{R_2}{R_1 + R_2}\right)V_S - \left(\frac{R_4}{R_3 + R_4}\right)V_S$$

$$= \left(\frac{680\ \Omega}{1010\ \Omega}\right)24\text{ V} - \left(\frac{560\ \Omega}{1240\ \Omega}\right)24\text{ V}$$

$$= 16.16\text{ V} - 10.84\text{ V} = 5.32\text{ V}$$

$$R_{TH} = \frac{R_1R_2}{R_1 + R_2} + \frac{R_3R_4}{R_3 + R_4} = \frac{(330\ \Omega)(680\ \Omega)}{1010\ \Omega} + \frac{(680\ \Omega)(560\ \Omega)}{1240\ \Omega}$$

$$= 222.18\ \Omega + 307.1\ \Omega = 529.28\ \Omega$$

Step 3: Connect the load resistor from points A to B of the equivalent circuit, and determine the load voltage and current as illustrated in Figure 6–58.

FIGURE 6–58

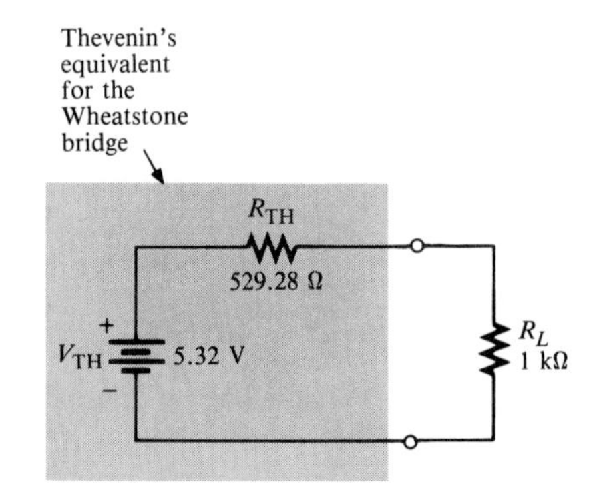

$$V_{RL} = \left(\frac{R_L}{R_L + R_{TH}}\right)V_{TH} = \left(\frac{1\text{ k}\Omega}{1.52928\ \Omega}\right)5.32\text{ V} = 3.48\text{ V}$$

$$I_{RL} = \frac{V_{RL}}{R_L} = \frac{3.48\text{ V}}{1\text{ k}\Omega} = 3.48\text{ mA}$$

SECTION REVIEW 6–7

1. What are the two components of a Thevenin equivalent circuit?
2. Draw the general form of a Thevenin equivalent circuit.
3. How is V_{TH} determined?
4. How is R_{TH} determined?
5. For the original circuit in Figure 6–59, draw the Thevenin equivalent circuit as viewed by R_L.

FIGURE 6–59

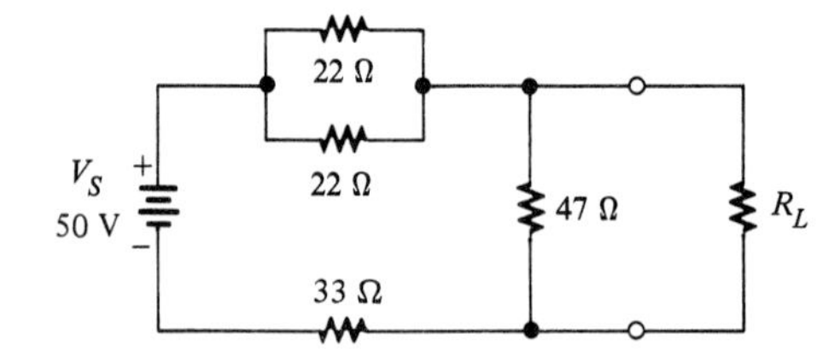

TROUBLES IN SERIES-PARALLEL CIRCUITS

6–8

Troubleshooting is the process of identifying and locating a failure or problem in a circuit. Some troubleshooting techniques have already been discussed in relation to both series circuits and parallel circuits. Now we will extend these methods to the series-parallel networks.

Opens and *shorts* are typical problems that occur in electric circuits. As mentioned before, if a resistor burns out, it will normally produce an *open circuit.* Bad solder connections, broken wires, and poor contacts can also be causes of open paths. Pieces of foreign material, such as solder splashes, broken insulation on wires, and so on, can often lead to shorts in a circuit. A *short* is a zero resistance path between two points.

In addition to complete opens or shorts, partial opens or partial shorts can develop in a circuit. A partial open would be a much higher than normal resistance, but not infinitely large. A partial short would be a much lower than normal resistance, but not zero.

The following examples illustrate troubleshooting in series-parallel circuits.

EXAMPLE 6–16

From the indicated voltmeter reading in Figure 6–60, determine if there is a fault. If there is a fault, identify it as either a short or an open.

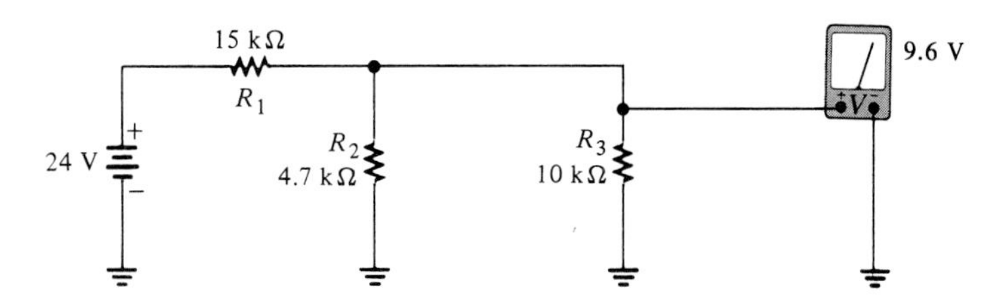

FIGURE 6–60

Solution

First determine what the voltmeter *should* be indicating. Since R_2 and R_3 are in parallel, their combined resistance is

$$R_{2\text{-}3} = \frac{R_2R_3}{R_2 + R_3} = \frac{(4.7\ \text{k}\Omega)(10\ \text{k}\Omega)}{14.7\ \text{k}\Omega} = 3.2\ \text{k}\Omega$$

The voltage across the parallel combination is determined by the voltage divider formula as follows:

$$V_{2\text{-}3} = \left(\frac{R_{2\text{-}3}}{R_1 + R_{2\text{-}3}}\right)V_S = \left(\frac{3.2\ \text{k}\Omega}{18.2\ \text{k}\Omega}\right)24\ \text{V} = 4.22\ \text{V}$$

Thus, 4.22 V is the voltage reading that you *should* get on the meter. But the meter reads 9.6 V instead. This value is incorrect, and, because it is higher than it should be, R_2 or R_3 is probably open. Why? Because if either of these two resistors is open, the resistance across which the meter is connected is larger than expected. A higher resistance will drop a higher voltage in this circuit, which is, in effect, a voltage divider.

Let us start by assuming that R_2 is open. If it is, the voltage across R_3 is as follows:

$$V_3 = \left(\frac{R_3}{R_1 + R_3}\right)V_S = \left(\frac{10\ \text{k}\Omega}{25\ \text{k}\Omega}\right)24\ \text{V} = 9.6\ \text{V}$$

This calculation shows that R_2 is open. Replace R_2 with a new resistor.

EXAMPLE 6–17 Suppose that you measure 24 V with the voltmeter in Figure 6–61. Determine if there is a fault, and, if there is, identify it.

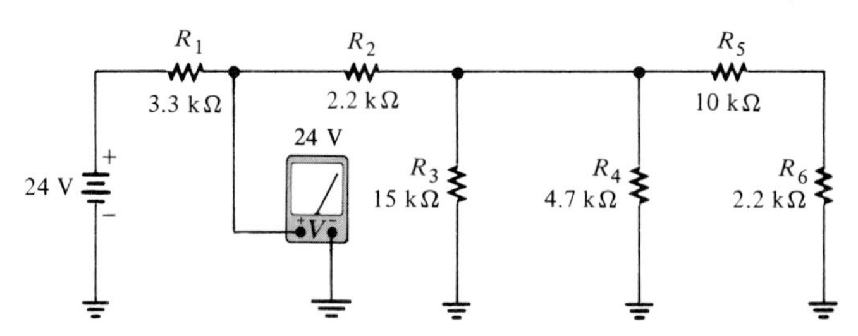

FIGURE 6–61

Solution

There is no voltage drop across R_1 because both sides of the resistor are at +24 V. Either no current is flowing through R_1 from the source, which tells us that R_2 is open in the circuit, or R_1 is shorted.

If R_1 were open, the meter would not read 24 V. The most logical failure is an open R_2. If it *is* open, then no current will flow from the source. To verify this, measure across R_2 with the voltmeter as shown in Figure 6–62. If R_2 is open, the meter will indicate 24 V. The right side of R_2 will be at zero volts because no current is flowing through any of the other resistors to cause a voltage drop.

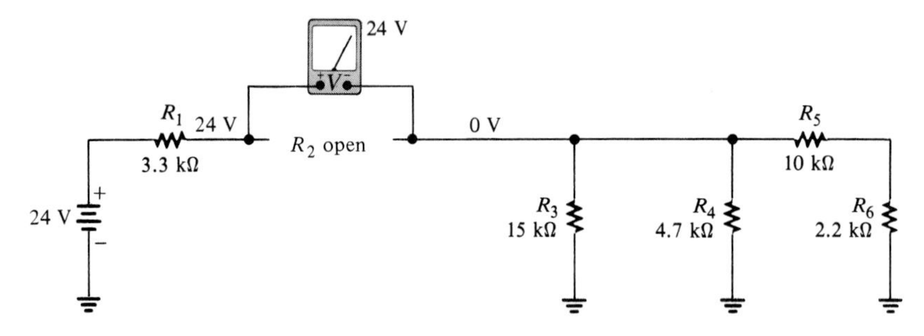

FIGURE 6–62

EXAMPLE 6–18 The two voltmeters in Figure 6–63 indicate the voltages shown. Determine if there are any opens or shorts in the circuit and, if so, where they are located.

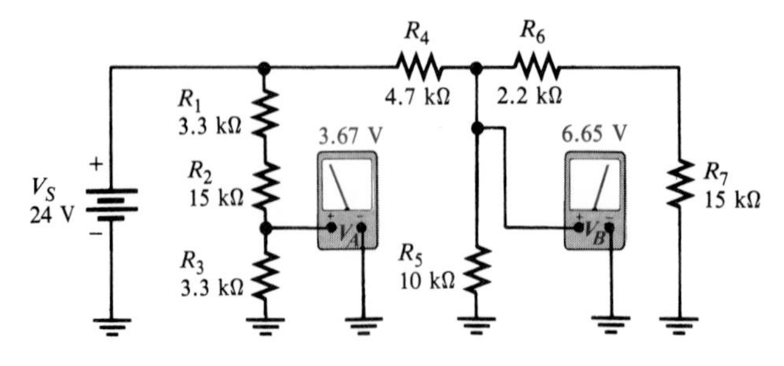

FIGURE 6–63

Solution

First let us see if the voltmeter readings are correct. R_1, R_2, and R_3 act as a voltage divider. The voltage across R_3 (V_A) is calculated as follows:

$$V_A = \left(\frac{R_3}{R_1 + R_2 + R_3}\right)V_S$$
$$= \left(\frac{3.3 \text{ k}\Omega}{21.6 \text{ k}\Omega}\right)24 \text{ V} = 3.67 \text{ V}$$

The voltmeter A reading (V_A) is correct.

Now let us see if the voltmeter B reading (V_B) is correct. $R_6 + R_7$ is in parallel with R_5. The series-parallel combination of R_5, R_6, and R_7 is in series with R_4. The resistance of the R_5, R_6, and R_7 combination is calculated as follows:

$$R_{5\text{-}6\text{-}7} = \frac{(R_6 + R_7)R_5}{R_5 + R_6 + R_7}$$
$$= \frac{(17.2 \text{ k}\Omega)(10 \text{ k}\Omega)}{27.2 \text{ k}\Omega} = 6.3 \text{ k}\Omega$$

$R_{5\text{-}6\text{-}7}$ and R_4 form a voltage divider, and voltmeter B is measuring the voltage across $R_{5\text{-}6\text{-}7}$. Is it correct? We check as follows:

$$V_B = \left(\frac{R_{5\text{-}6\text{-}7}}{R_4 + R_{5\text{-}6\text{-}7}}\right)V_S$$
$$= \left(\frac{6.3 \text{ k}\Omega}{11 \text{ k}\Omega}\right)24 \text{ V} = 13.75 \text{ V}$$

Thus, the actual measured voltage (6.65 V) at this point is incorrect. Some further thought will help to isolate the problem.

We know that R_4 is not open, because if it were, the meter would read 0 V. If there were a short across it, the meter would read 24 V. Since the actual voltage is much less than it should be, $R_{5\text{-}6\text{-}7}$ must be less than the calculated value of 6.3 kΩ. The most likely problem is a short across R_7. If there is a short from the top of R_7 to ground, R_6 is effectively in parallel with R_5. In this case,

$$R_5 \| R_6 = \frac{R_5 R_6}{R_5 + R_5}$$
$$= \frac{(2.2 \text{ k}\Omega)(10 \text{ k}\Omega)}{12.2 \text{ k}\Omega} = 1.8 \text{ k}\Omega$$

Then V_B is

$$V_B = \left(\frac{1.8 \text{ k}\Omega}{6.5 \text{ k}\Omega}\right)24 \text{ V} = 6.65 \text{ V}$$

This value for V_B agrees with the voltmeter B reading. So there is a short across R_7. If this were an actual circuit, you would try to find the physical cause of the short.

SECTION REVIEW 6–8

1. Name two types of common circuit faults.
2. For the following faults in Figure 6–64, determine what voltage would be measured at point A:
   (a) no faults. (b) R_1 open. (c) short across R_5. (d) R_3 and R_4 open.

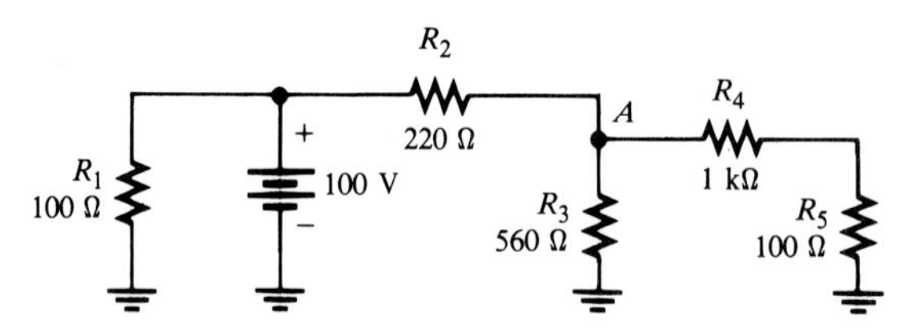

FIGURE 6–64

APPLICATION NOTE

An analytic approach to the Application Assignment at the beginning of the chapter follows:

The seismometer draws 100 mA from the 20-V output. Thus, the effective resistance R_S of this instrument is

$$R_S = \frac{20 \text{ V}}{100 \text{ mA}} = 200 \ \Omega$$

When connected, the seismometer presents a 200-Ω load across R_2, R_3, R_4, and R_5, as shown in Figure 6–65.

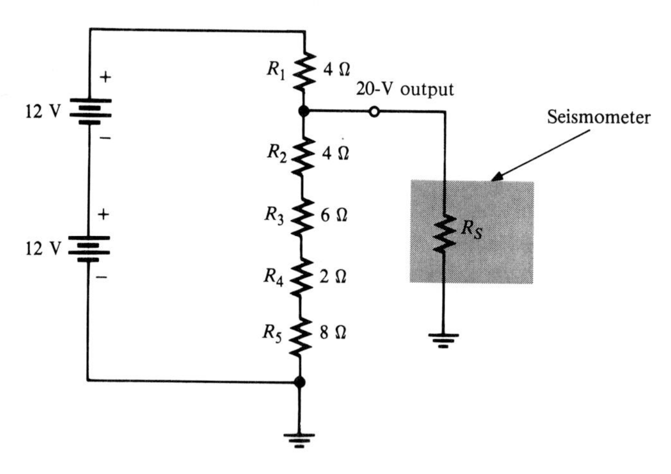

FIGURE 6–65

The total resistance from the 20-V output to ground is

$$R_{20} = \frac{(20 \ \Omega)(200 \ \Omega)}{220 \ \Omega} = 18.18 \ \Omega$$

The 20-V output voltage under the loaded condition is

$$V_{OUT} = \left(\frac{18.18\ \Omega}{22.18\ \Omega}\right)24\ V = 19.67\ V$$

The percentage change in the output when the load is connected is

$$100 - \frac{19.67\ V}{20\ V}(100) = 1.65\%$$

This change is less than the 5% allowed under the specification. Analyze all the remaining instrument loads in the same way to see if you are within the specification.

SUMMARY

Facts

- A series-parallel circuit is a combination of both series components and parallel components.
- When a load resistor is connected across a voltage divider output, the output voltage decreases. This effect is called *loading*.
- To minimize the loading effect, the load resistance should be much larger than the resistance across which it is connected.
- Ground is considered at zero volts with respect to other points in a circuit.
- A Wheatstone bridge is used to measure an unknown resistance.
- When a Wheatstone bridge is *balanced,* the output voltage is zero.
- To find any current or voltage in a circuit with two or more voltage sources, take the sources one at a time using the superposition method.
- Any complex resistive circuit can be replaced by its Thevenin equivalent.
- The Thevenin equivalent circuit is made up of an equivalent resistance (R_{TH}) in series with an equivalent voltage source (V_{TH}).

Definitions

- *Ground*—the common or reference point in a circuit.
- *Negative ground*—the term used when the negative side of the source is grounded.
- *Positive ground*—the term used when the positive side of the source is grounded.
- *Terminal equivalency*—a condition that occurs when two circuits produce the same load voltage and load current for the same value of load resistance.
- *Troubleshooting*—the process of identifying and locating a fault in a circuit.

Formula

$R_{UNK} = R_V\left(\frac{R_2}{R_4}\right)$ Unknown resistance in a Wheatstone bridge **(6–1)**

SELF-TEST

1. Draw the schematic diagram for a series-parallel circuit that is described as follows: R_1 and R_2 are in series with each other, and this series combination is in parallel with a series combination of R_3, R_4, and R_5.
2. How do you find the total resistance for the circuit described in Question 1?
3. If all the resistors in Question 1 are the same value, through which resistors does the most current flow when voltage is applied?

4. Two 1-kΩ resistors are in series, and this series combination is in parallel with a 2.2-kΩ resistor. The voltage across one of the 1-kΩ resistors is 6 V. How much voltage is across the 2.2 k-Ω resistor?
5. A 330-Ω resistor is in series with the parallel combination of four 1-kΩ resistors. A 100-V source is connected to the circuit. Which resistor has the most current through it? Which one has the most voltage across it?
6. In Question 5, what percentage of the total current flows through each of the 1-kΩ resistors?
7. In a certain circuit, the voltage at point A is 5 V with respect to ground. The voltage at point B is 8 V with respect to ground. The voltage at point C is 12 V with respect to ground. Determine the voltages at points B and C with respect to point A.
8. The output of a certain voltage divider is 9 V with no load. When a load is connected, will the voltage increase or decrease?
9. A certain voltage divider consists of two 10-kΩ resistors in series. Which will have more effect on the output voltage, a 1-MΩ load or a 100-kΩ load?
10. Would you expect more or less current to be drawn from the source when a load resistor is connected to the output of a voltage divider?
11. What is the output voltage of a balanced Wheatstone bridge?
12. What is a galvanometer?
13. Explain the basic steps used to analyze a circuit with two or more voltage sources.
14. In a two-source circuit, one source produces 10 mA through a given branch. The other source produces 8 mA in the opposite direction through the same branch. What is the actual current through the branch?
15. The Thevenin values for a certain circuit are found to be 8.5 V and 280 Ω. Draw the Thevenin equivalent circuit.
16. You are measuring the voltage at a given point in a circuit that has very high resistance values. The measured value is somewhat lower than it should be. What could be causing this discrepancy?

PROBLEM SET A

Section 6–1

6–1 Identify the series-parallel relationships in Figure 6–66 as seen from the source terminals.

FIGURE 6–66

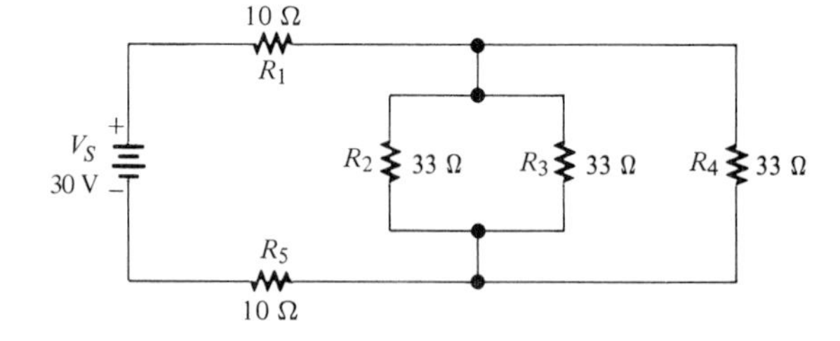

6–2 Visualize and sketch the following series-parallel combinations:
(a) R_1 in series with the parallel combination of R_2 and R_3.
(b) R_1 in parallel with the series combination of R_2 and R_3.
(c) R_1 in parallel with a branch containing R_2 in series with a parallel combination of four other resistors.

6–3 Visualize and sketch the following series-parallel circuits:

(a) a parallel combination of three branches, each containing two series resistors.

(b) a series combination of three parallel circuits, each containing two resistors.

6–4 In each circuit of Figure 6–67, identify the series and parallel relationships of the resistors viewed from the source.

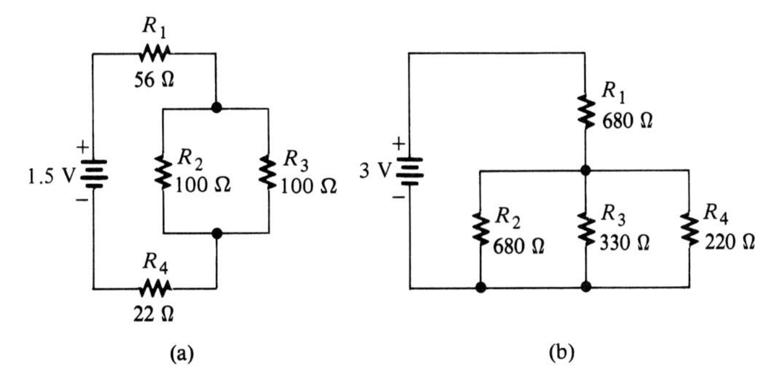

FIGURE 6–67

Section 6–2

6–5 A certain circuit is composed of two parallel resistors. The total resistance is 667 Ω. One of the resistors is 1 kΩ. What is the other resistor?

6–6 For the circuit in Figure 6–68, determine the total resistance between points A and B. (A full-color photo appears between pages 128 and 129.)

6–7 Determine the total resistance for each circuit in Figure 6–67.

6–8 Determine the current through each resistor in Figure 6–66; then calculate each voltage drop.

6–9 Determine the current through each resistor in both circuits of Figure 6–67; then calculate each voltage drop.

6–10 In Figure 6–69, find the following:

(a) total resistance between terminals A and B.

(b) total current drawn from a 6-V source connected from A to B.

(c) current through R_5.

(d) voltage across R_2.

FIGURE 6–69

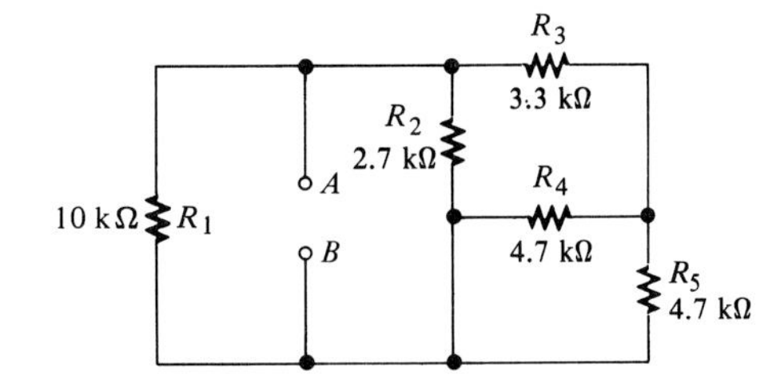

Section 6–3

6–11 Determine the voltages with respect to ground in Figure 6–70.

6–12 Determine the voltage at each point with respect to ground in Figure 6–71.

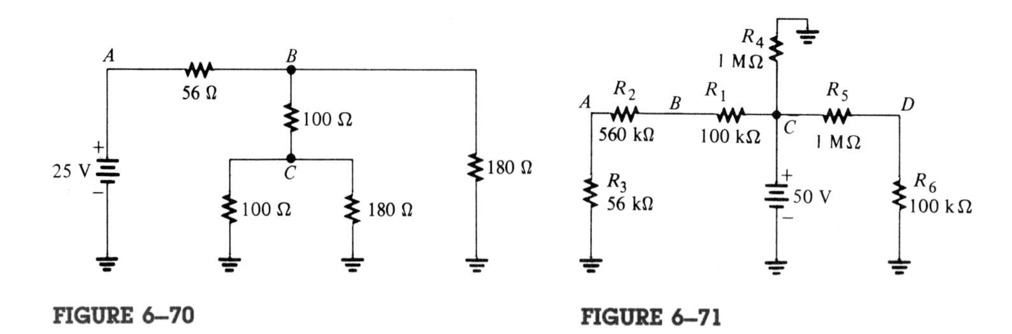

FIGURE 6–70

FIGURE 6–71

Section 6–4

6–13 A voltage divider consists of two 56-kΩ resistors and a 15-V source. Calculate the unloaded output voltage. What will the output voltage be if a load resistor of 1 MΩ is connected to the output?

6–14 A 12-V battery output is divided down to obtain two output voltages. Three 3.3-kΩ resistors are used to provide the two outputs. Determine the output voltages when both load resistances are 10 kΩ.

6–15 Which will cause a smaller decrease in output voltage for a given voltage divider, a 10-kΩ load or a 56-kΩ load?

6–16 In Figure 6–72, determine the current drain on the battery with no load across the output terminals. With a 10-kΩ load, what is the battery current?

FIGURE 6–72

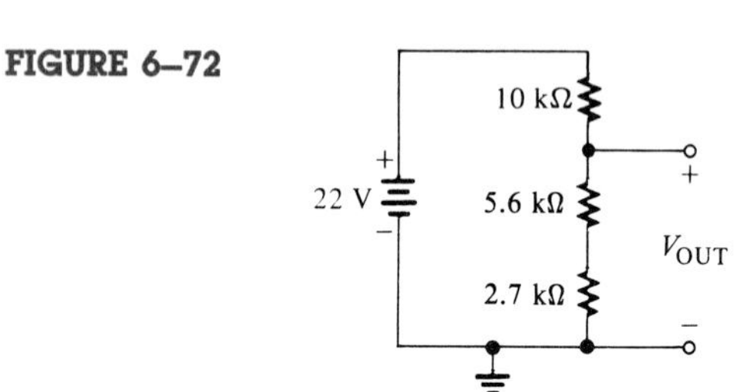

Section 6–5

6–17 A resistor of unknown value is connected to a Wheatstone bridge circuit. The bridge parameters are set as follows: $R_V = 18$ kΩ and $R_2/R_4 = 0.02$. What is R_{UNK}?

6–18 A bridge network is shown in Figure 6–73. To what value must R_V be set in order to balance the bridge?

FIGURE 6–73

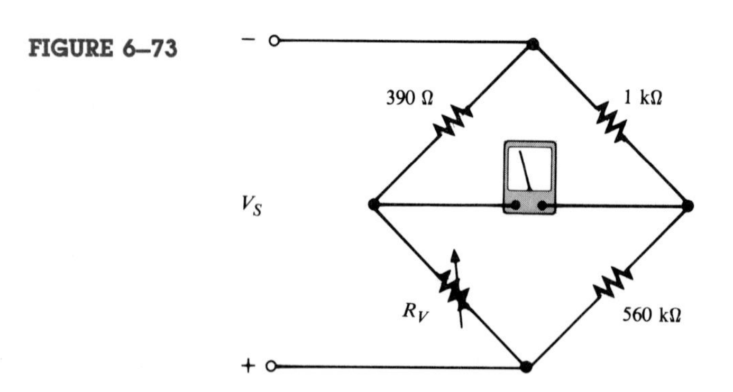

Section 6–6

6–19 In Figure 6–74, use the superposition method to find the total current in R_3.

6–20 In Figure 6–74, what is the total current flowing through R_2?

FIGURE 6–74

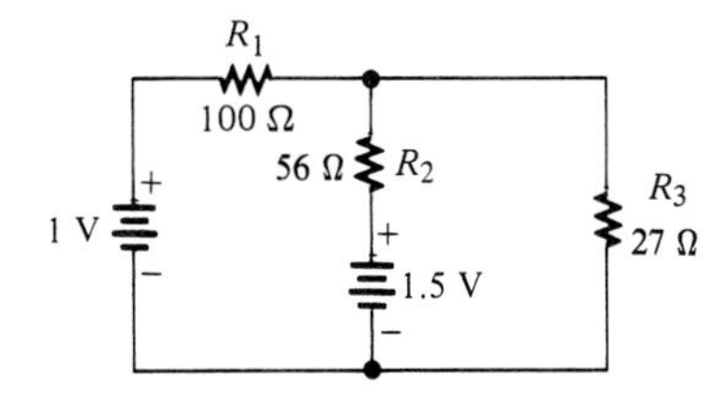

Section 6–7

6–21 Reduce the circuit in Figure 6–75 to its Thevenin equivalent as viewed from terminals A and B.

6–22 For each circuit in Figure 6–76, determine the Thevenin equivalent as seen by R_L.

FIGURE 6–75

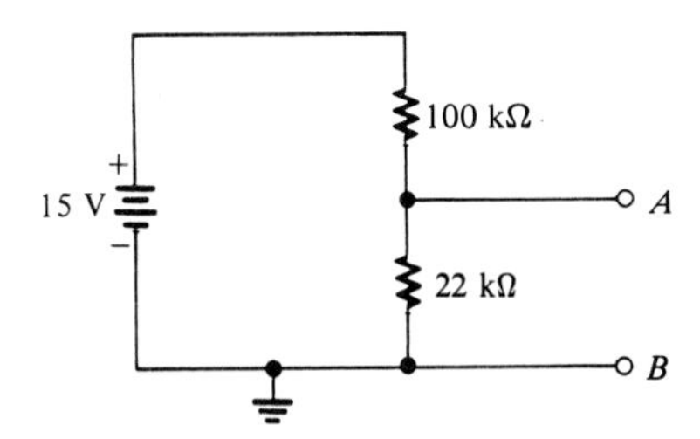

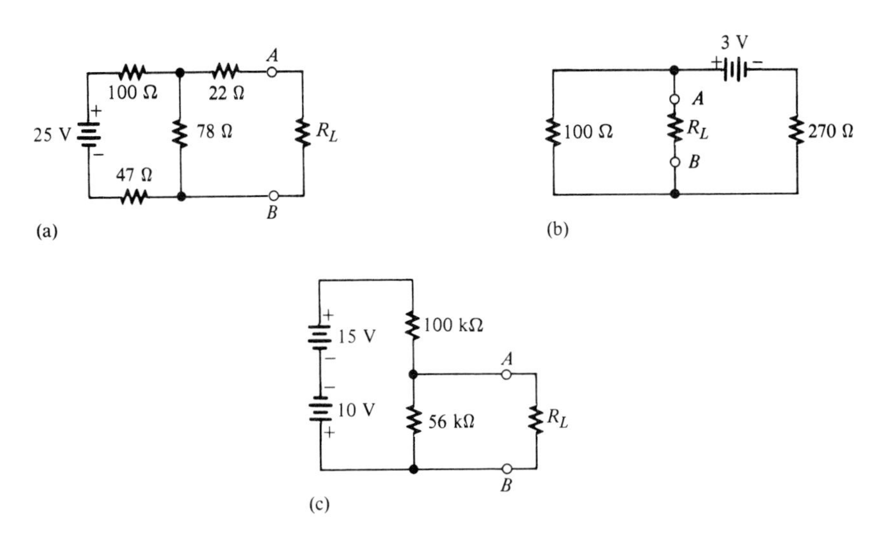

FIGURE 6–76

Section 6–8

6–23 Is the voltmeter reading in Figure 6–77 correct? (A full-color photo appears between pages 128 and 129.)

6–24 If R_2 in Figure 6–78 opens, what voltages will be read at points A, B, and C?

6–25 Check the meter readings in Figure 6–79 and locate any fault that may exist.

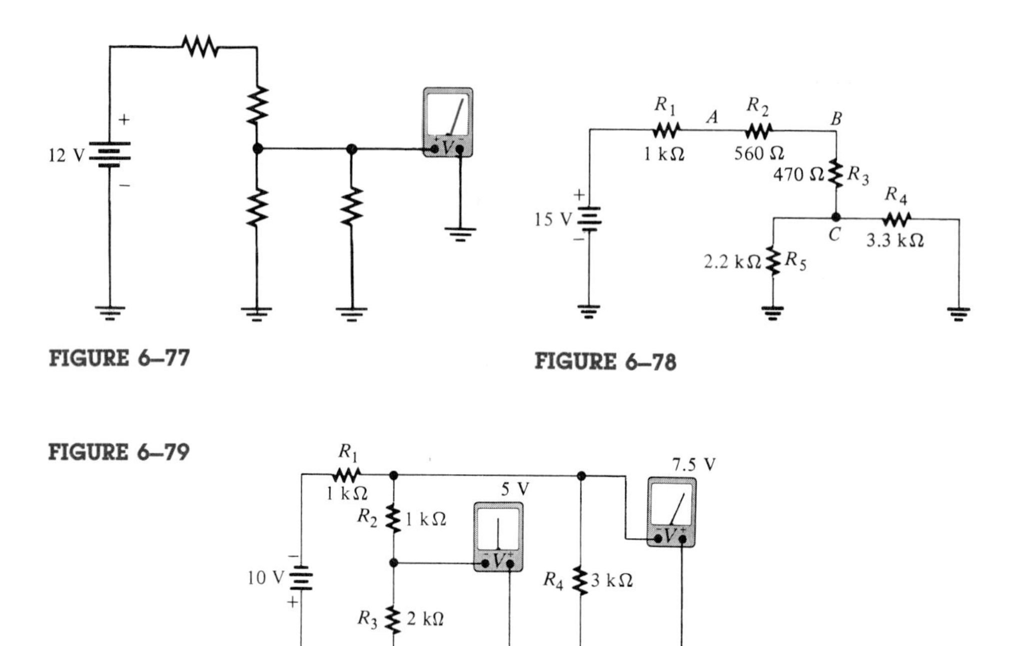

FIGURE 6–77

FIGURE 6–78

FIGURE 6–79

PROBLEM SET B

6–26 In each circuit of Figure 6–80, identify the series and parallel relationships of the resistors viewed from the source.

6–27 Draw the schematic of the PC board layout in Figure 6–81 showing resistor values (a full-color photo appears between pages 128 and 129) and identify the series-parallel relationships. Which resistors can be removed with no effect on R_T?

6–28 For the circuit shown in Figure 6–82, calculate the following:
(a) total resistance across the source.
(b) total current from the source.
(c) current through the 900-Ω resistor.
(d) voltage from point A to point B.

6–29 Determine the total resistance and the voltage at points A, B, and C in the circuit of Figure 6–83.

6–30 Determine the total resistance between terminals A and B of the circuit in Figure 6–84. Also calculate the current in each branch with 10 V between A and B.

6–31 What is the voltage across each resistor in Figure 6–84?

6–32 Determine the voltage, V_{AB}, in Figure 6–85.

6–33 Find the value of R_2 in Figure 6–86.

6–34 Find I_T and V_{OUT} in Figure 6–87.

6–35 Design a voltage divider to provide a 6-V output with no load and a minimum of 5.5 V across a 1-kΩ load. The source voltage is 24 V, and the unloaded current is not to exceed 100 mA.

6–36 Determine the resistance values for a voltage divider that must meet the following specifications: The current under an unloaded condition is not to exceed 5 mA. The source voltage is to be 10 V. A 5-V output and a 2.5-V output are required.

Sketch the circuit. Determine the effect on the output voltages if a 1-kΩ load is connected to each output.

6–37 Using the superposition method, calculate the current in the right-most branch of Figure 6–88.

6–38 Find the current through R_L in Figure 6–89.

6–39 Using Thevenin's theorem, find the voltage across R_4 in Figure 6–90.

6–40 Look at the meters in Figure 6–91 and determine if there is a fault in the circuit. If there is a fault, identify it.

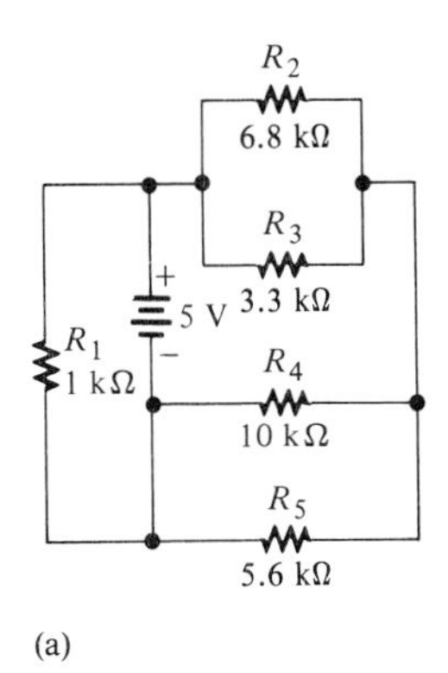

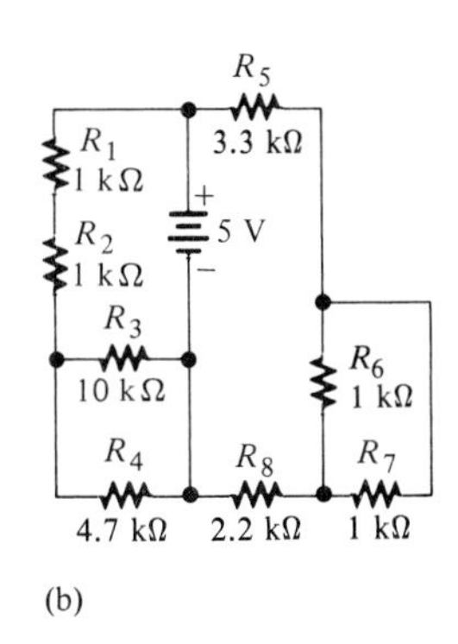

FIGURE 6–80

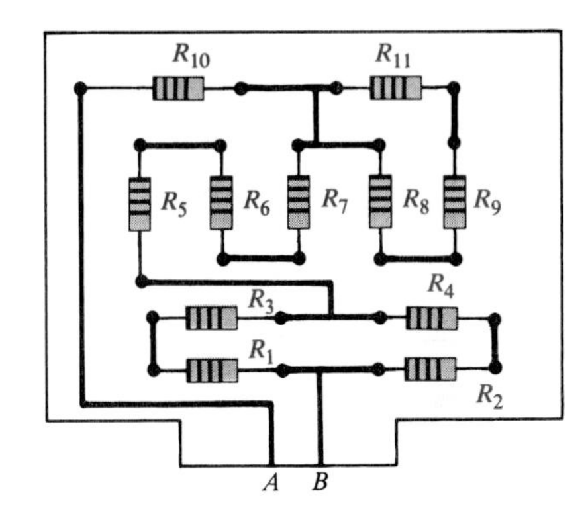

FIGURE 6–81

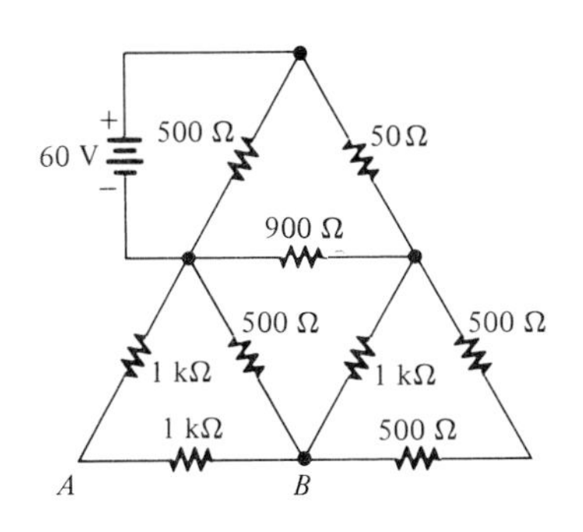

FIGURE 6–82

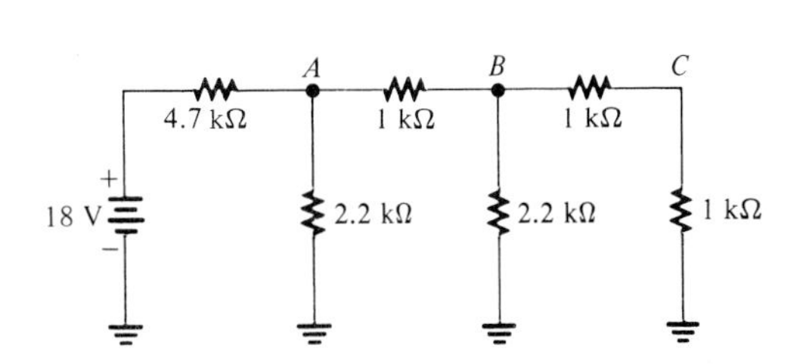

FIGURE 6–83

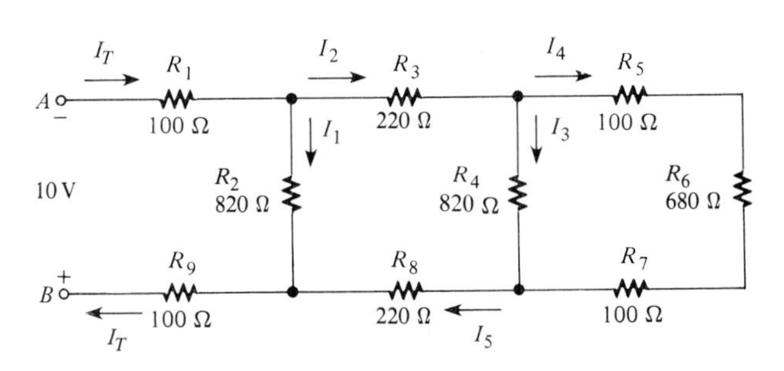

FIGURE 6–84

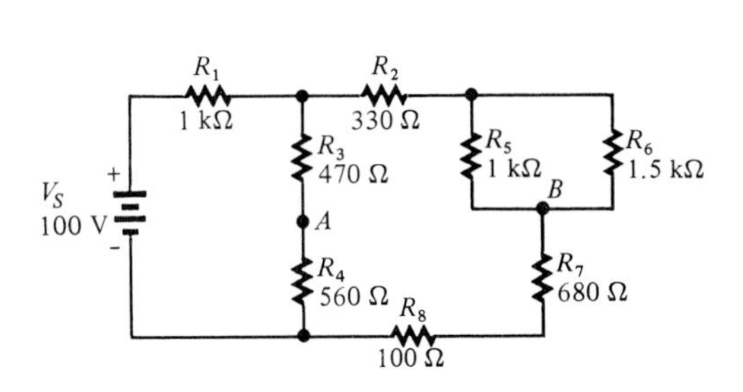

FIGURE 6–85

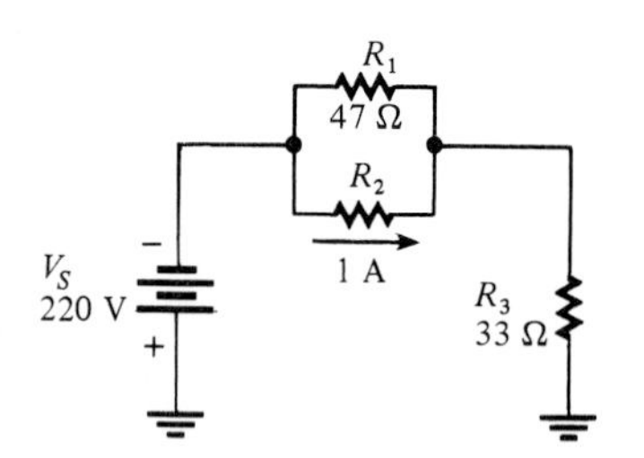

FIGURE 6–86

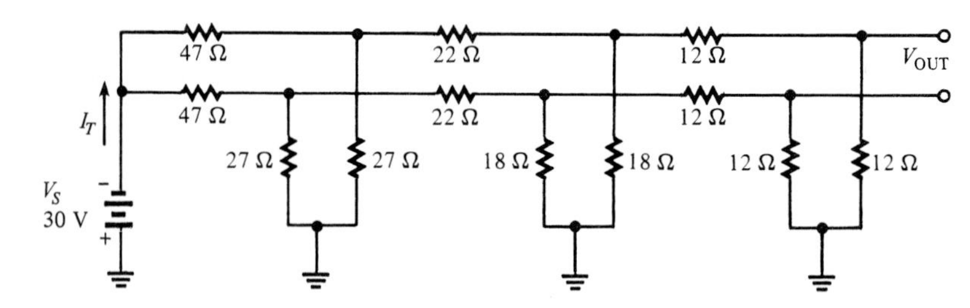

FIGURE 6–87

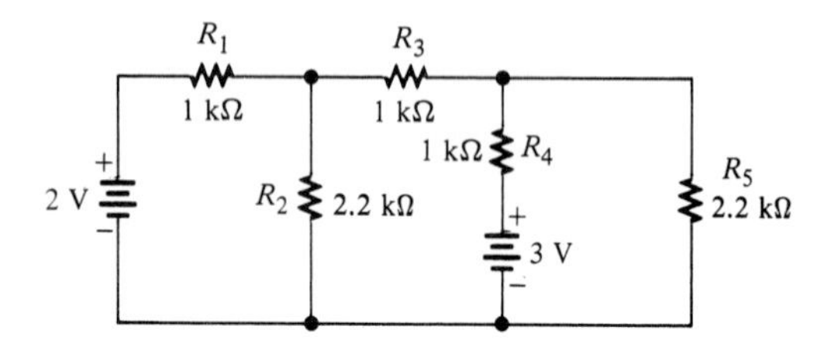

FIGURE 6–88

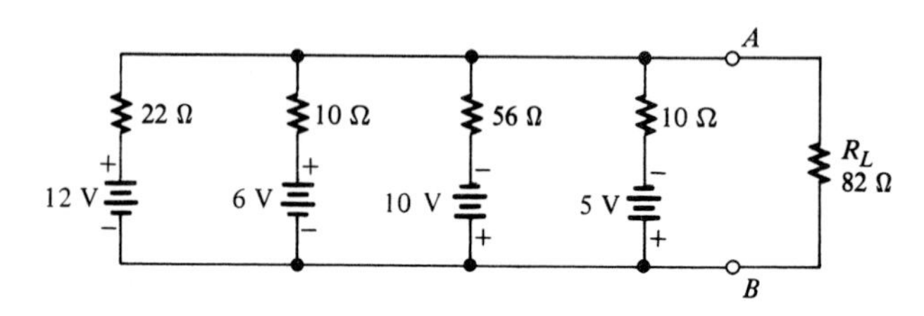

FIGURE 6–89

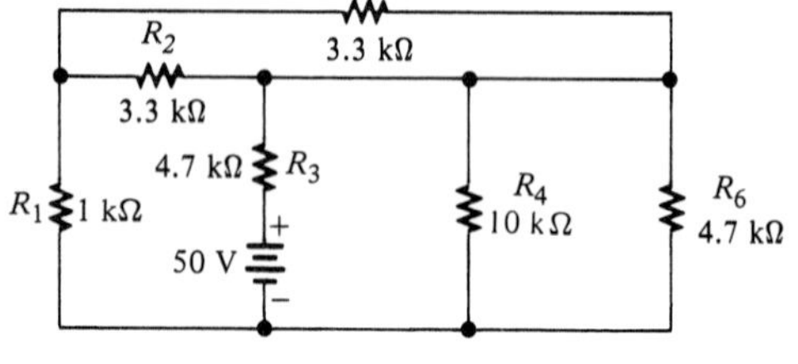
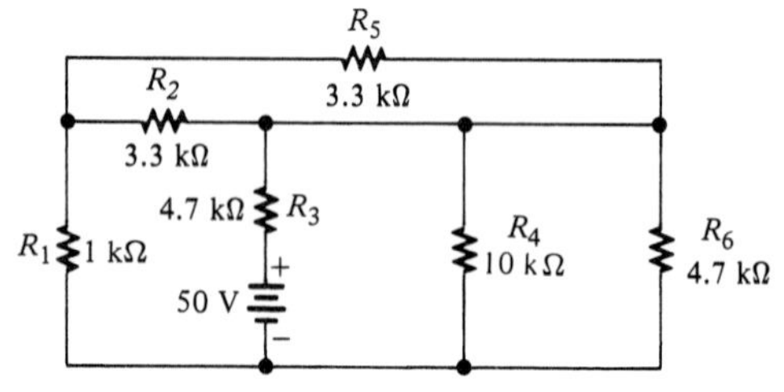

FIGURE 6–90

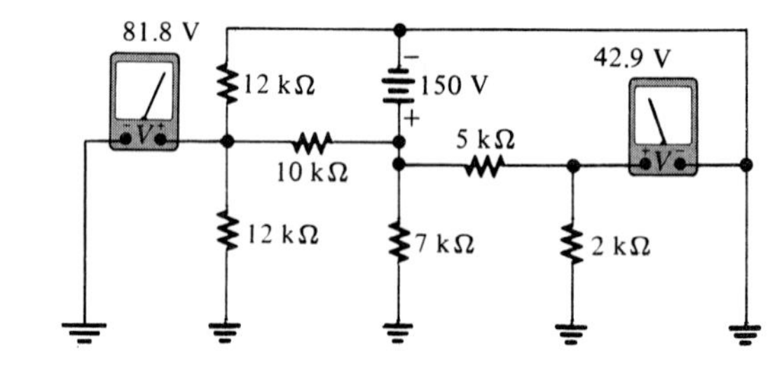

FIGURE 6–91

ANSWERS TO SECTION REVIEWS

Section 6–1

1. See Figure 6–92.
2. R_1 and R_2 are in series with the parallel combination of R_3 and R_4.
3. All resistors.
4. R_1 and R_2 are in parallel; R_3 and R_4 are in parallel. **5.** Yes.

Section 6–2

1. 599 Ω. **2.** 11.2 mA. **3.** 3.7 V. **4.** 89 Ω, 11.2 mA.

Section 6–3

1. Ground. **2.** T. **3.** T.

4. A connection to earth through a metal rod or a water pipe.

Section 6–4

1. It decreases the output voltage. **2.** T. **3.** 19.2 V, 4.19 V.

Section 6–5

1. Zero output. **2.** 15 kΩ.

Section 6–6

1. 6.67 mA. 2. In the direction of the larger.

Section 6–7

1. V_{TH} and R_{TH}. 2. See Figure 6–93.
3. V_{TH} is the open circuit voltage between two terminals in a circuit.
4. R_{TH} is the resistance as viewed from two terminals in a circuit, with all sources replaced by their internal resistances.
5. See Figure 6–94.

Section 6–8

1. Opens and shorts.
2. (a) 62.8 V; (b) 62.8 V; (c) 62 V; (d) 100 V.

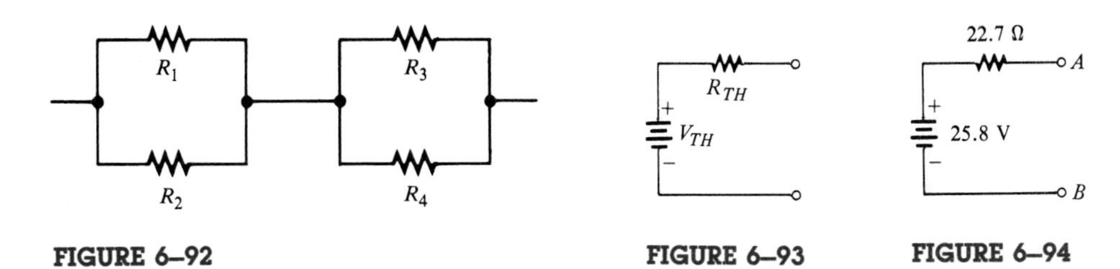

FIGURE 6–92

FIGURE 6–93

FIGURE 6–94

Magnetism and Electromagnetism

This chapter is somewhat of a departure from the previous six chapters because two new concepts are introduced: magnetism and electromagnetism. The operation of many types of electrical devices is based partially on magnetic or electromagnetic principles.

In this chapter you will learn:

- The principles of magnetic fields.
- How materials are magnetized.
- How a magnetic switch operates.
- How magnetic switches can be applied in alarm systems.
- The principles of electromagnetism.
- The important characteristics of electromagnetic fields.
- How electromagnetism is used to record on magnetic tape.
- The principles of operation of solenoids, relays, speakers, and dc generators.
- How voltage is created by a conductor moving through a magnetic field.
- The principles of Faraday's law.

7

APPLICATION ASSIGNMENT

In this chapter, you will become familiar with several devices that can be used in security systems. Using only magnetic and electromagnetic devices, devise a system that will detect illegal entry into your house or apartment through any window or door. When someone breaks into a room, the system will turn on a light in that room and open the door of a cage which is inside the house and in which there is a vicious guard dog.

The Application Note at the end of the chapter gives one solution.

THE MAGNETIC FIELD

7–1

A permanent magnet, such as the bar magnet shown in Figure 7–1, has a magnetic field surrounding it. The magnetic field consists of *lines of force* that radiate from the north pole (N) to the south pole (S) and back to the north pole through the magnetic material. For clarity, only a few lines of force are shown in the figure. Imagine, however, that many lines surround the magnet in three dimensions.

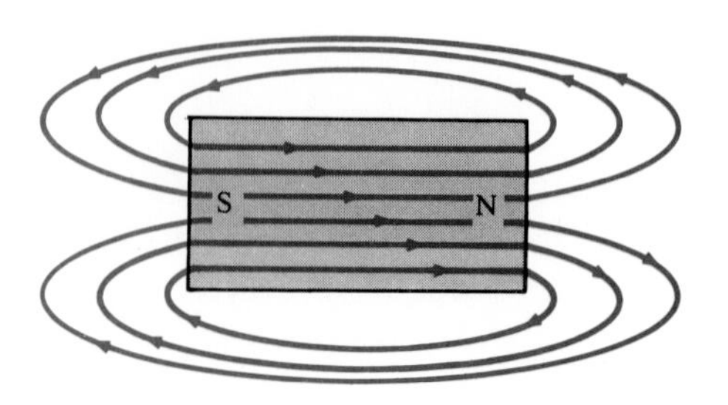

Colored lines represent magnetic lines of force.

FIGURE 7–1
Magnetic lines of force around a bar magnet.

Attraction and Repulsion of Magnetic Poles

When *unlike* poles of two permanent magnets are placed close together, an attractive force is produced by the magnetic fields, as indicated in Figure 7–2(a). When two *like* poles are brought close together, they repel each other, as shown in Part (b).

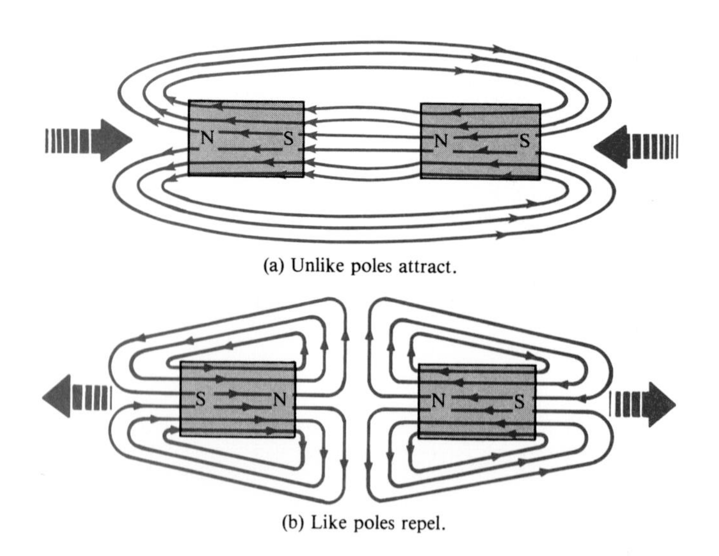

FIGURE 7–2
Magnetic attraction and repulsion.

Altering a Magnetic Field

When a nonmagnetic material, such as paper, glass, wood, or plastic, is placed in a magnetic field, the lines of force are unaltered, as shown in Figure 7–3(a). However, when a *magnetic* material such as iron is placed in the magnetic field, the lines of force tend to change course and pass through the iron rather than through the surrounding air. They do so because the iron provides a magnetic path that is more easily established than that of air. Figure 7–3(b) illustrates this principle.

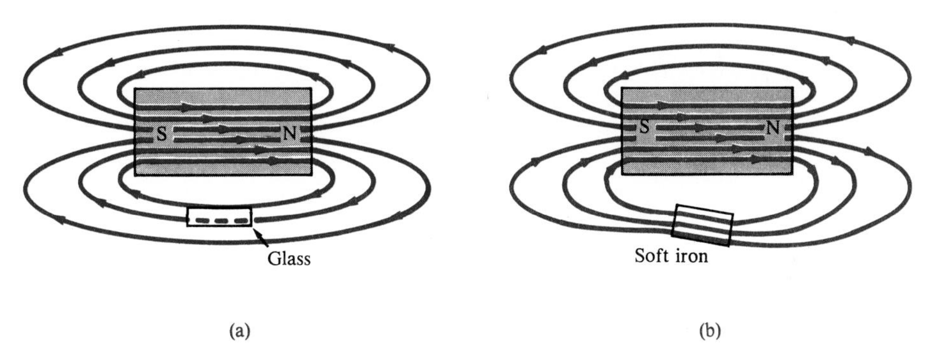

FIGURE 7–3
Effect of (a) nonmagnetic and (b) magnetic materials on a magnetic field.

Magnetic Flux

The group of force lines going from the north pole to the south pole of a magnet is called the *magnetic flux,* symbolized by ϕ (the lower-case Greek letter phi). The number of lines of force in a magnetic field determines the value of the flux. The more lines of force, the greater the flux and the stronger the magnetic field.

The unit of magnetic flux is the *weber* (Wb). One weber equals 10^8 lines. In most practical situations, the weber is a very large unit; thus, the microweber (μWb) is more common. One microweber equals 100 lines of magnetic flux.

Magnetic Flux Density

The *flux density* is the amount of flux per unit area in the magnetic field. Its symbol is B, and its unit is the tesla (T). One tesla equals one weber per square meter (Wb/m^2). The following formula expresses the flux density:

$$B = \frac{\phi}{A} \tag{7–1}$$

where ϕ is the flux and A is the cross-sectional area of the magnetic field.

EXAMPLE 7–1 Find the flux density in a magnetic field in which the flux in 0.1 square meter is 800 μWb.

Solution

$$B = \frac{\phi}{A} = \frac{800\ \mu\text{Wb}}{0.1\ \text{m}^2} = 8000 \times 10^{-6}\ \text{T}$$

How Materials Become Magnetized

Ferromagnetic materials such as iron, nickel, and cobalt become magnetized when placed in the magnetic field of a magnet. We have all seen a permanent magnet pick up paper clips, nails, iron filings, and so on. In these cases, the object becomes magnetized (that is, it actually becomes a magnet itself) under the influence of the permanent magnetic field and becomes attracted to the magnet. When removed from the magnetic field, the object tends to lose its magnetism.

Ferromagnetic materials have minute *magnetic domains* created within their atomic structure. These domains can be viewed as very small bar magnets with north and south poles. When the material is not exposed to an external magnetic field, the magnetic domains are randomly oriented, as shown in Figure 7–4(a). When the material is placed in a magnetic field, the domains align themselves as shown in Part (b). Thus, the object itself effectively becomes a magnet.

FIGURE 7–4
Magnetic domains in (a) an unmagnetized and (b) a magnetized material.

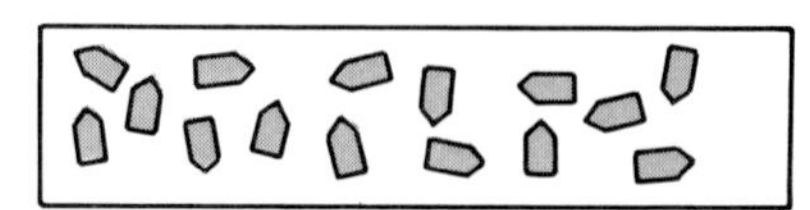

(a) The magnetic domains (N ◁▭ S) are randomly oriented in the unmagnetized material.

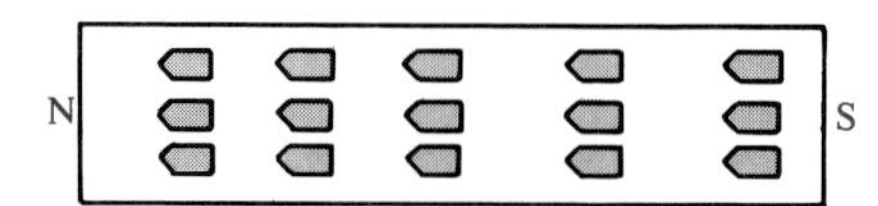

(b) The magnetic domains become aligned when the material is magnetized.

An Application

Permanent magnets have almost endless applications, one of which is presented here as an illustration. Figure 7–5 shows a typical magnetically operated, normally closed (NC) switch. When the magnet is near the switch mechanism, the metallic arm is held in its NC position. When the magnet is moved away, the spring pulls the arm up, breaking the contact as shown in Figure 7–6.

Switches of this type are commonly used in perimeter alarm systems to detect entry into a building through windows or doors. As Figure 7–7 shows, several openings can be protected by magnetic switches wired to a common transmitter. When any one of the switches opens, the transmitter is activated and sends a signal to a central receiver and alarm unit.

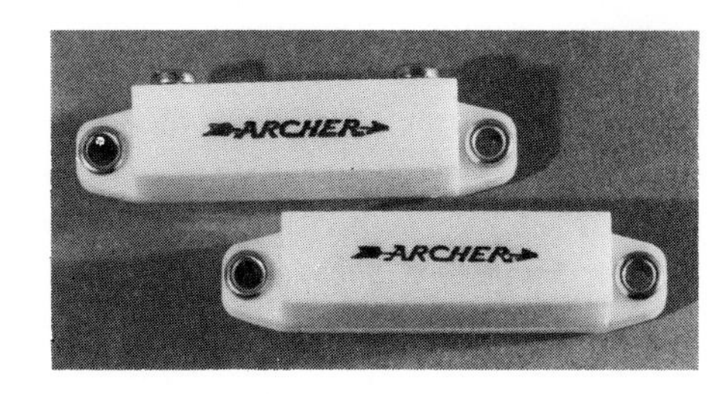

FIGURE 7–5
Magnet and switch set (courtesy of Tandy Corp.).

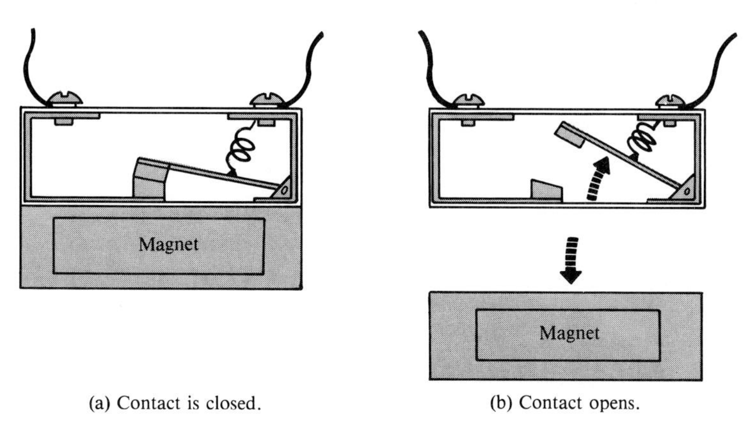

(a) Contact is closed. (b) Contact opens.

FIGURE 7–6
Operation of a magnetic switch.

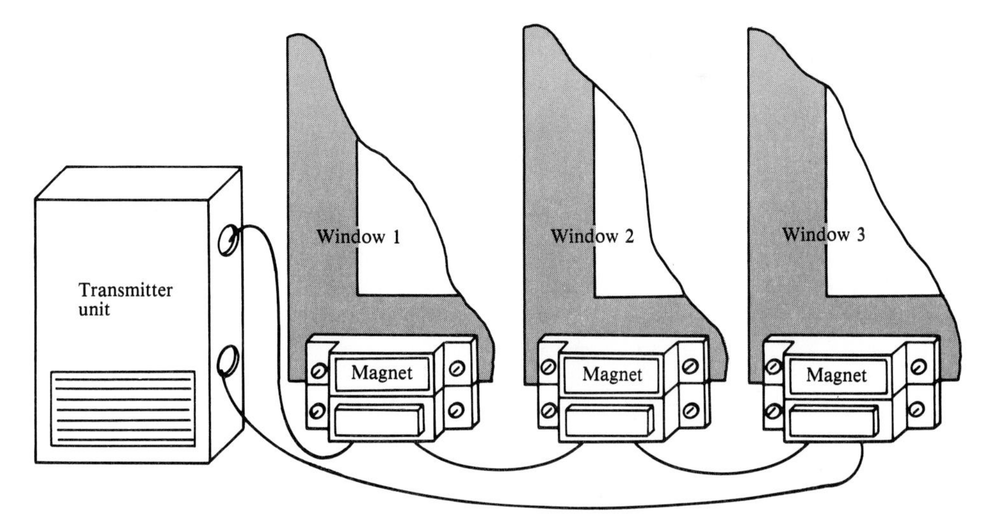

FIGURE 7–7
Connection of a typical perimeter alarm system.

SECTION REVIEW 7–1

1. When the north poles of two magnets are placed close together, do they repel or attract each other?
2. What is magnetic flux?

ELECTROMAGNETISM

7–2

Current produces a magnetic field, called an *electromagnetic field,* around a conductor, as illustrated in Figure 7–8. The invisible lines of force of the magnetic field form a concentric circular pattern around the conductor and are continuous along its length.

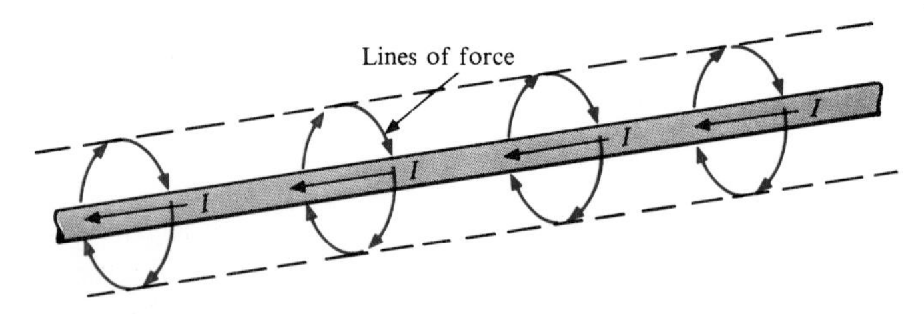

FIGURE 7–8
Magnetic field around a current-carrying conductor.

Although the magnetic field cannot be seen, it is capable of producing visible effects. For example, if a current-carrying wire is inserted through a sheet of paper in a perpendicular direction, iron filings placed on the surface of the paper arrange themselves along the magnetic lines of force in concentric rings, as illustrated in Figure 7–9(a). Part (b) of the figure illustrates that the north pole of a compass placed in the electromagnetic field will point in the direction of the lines of force. The field is stronger closer to the conductor and becomes weaker with increasing distance from the conductor.

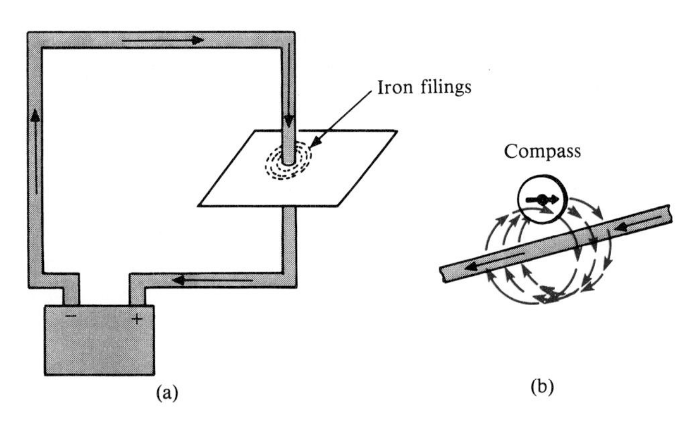

FIGURE 7–9
Visible effects of an electromagnetic field.

Direction of the Lines of Force

The direction of the lines of force surrounding the conductor are indicated in Figure 7–10. When the direction of current is right to left, as in Part (a), the lines are in a *clockwise* direction. When current is left to right, as in Part (b), the lines are in a *counterclockwise* direction.

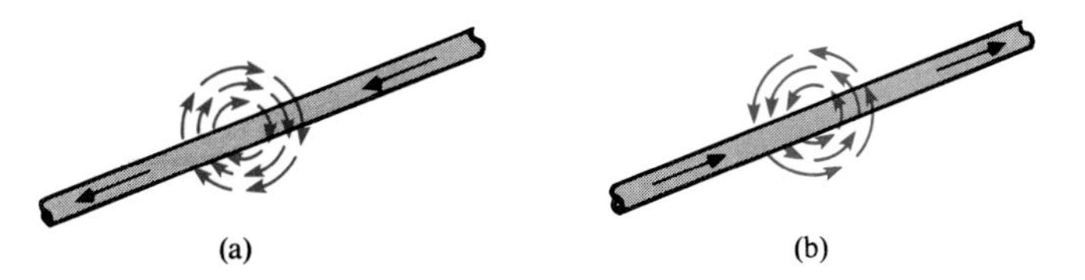

FIGURE 7–10
Magnetic lines of force around a current-carrying conductor.

Left-Hand Rule

An aid to remembering the direction of the lines of force is illustrated in Figure 7–11. Imagine that you are grasping the conductor with your left hand, with your thumb pointing in the direction of current. Your fingers indicate the direction of the magnetic lines of force.

FIGURE 7–11
Illustration of left-hand rule.

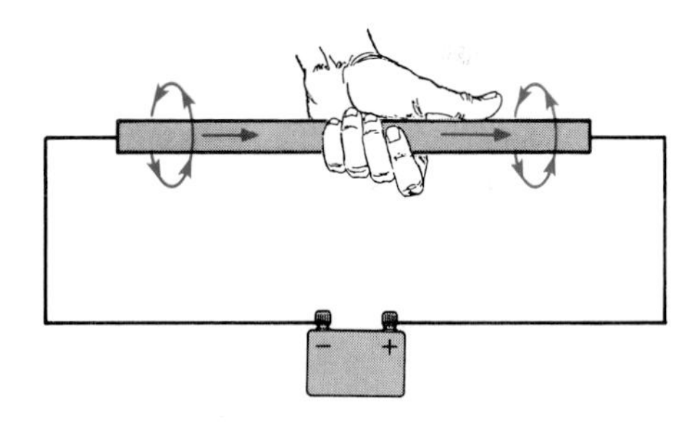

Electromagnetic Properties

Several important properties relating to electromagnetic fields are now presented.

Magnetomotive Force (mmf)

As you have learned, current in a conductor produces a magnetic field. The force that produces the magnetic field is called the *magnetomotive force* (mmf). The unit of mmf, the ampere-turn (At), is established on the basis of the current in a single loop (turn) of wire. The formula for mmf is as follows:

$$F_m = NI \tag{7–2}$$

where F_m is the magnetomotive force, N is the number of turns of wire, and I is the current.

Reluctance

Reluctance ($\mathcal{R}$) is the opposition to the establishment of a magnetic field in an electromagnetic circuit. It is the ratio of the mmf required to establish a given flux to the amount of flux, and its units are ampere-turns per weber. The formula for reluctance is

$$\mathcal{R} = \frac{F_m}{\phi} \tag{7–3}$$

This equation is sometimes known as "Ohm's law for magnetic circuits," because the reluctance is analogous to the resistance in electrical circuits.

Permeability

The ease with which a magnetic field can be established in a given material is measured by the *permeability* of that material. The higher the permeability, the more easily a magnetic field can be established.

The symbol of permeability is μ, and the formula is as follows:

$$\mu = \frac{l}{\mathcal{R}A} \tag{7–4}$$

where $\mathcal{R}$ is the reluctance in ampere-turns per weber, l is the length of the material in meters, and A is the cross-sectional area in square meters. For reference, the permeability of a vacuum is $4\pi \times 10^{-7}$ Wb/At·m.

Magnetic Field Intensity

The amount of magnetomotive force (mmf) per unit length (l) of magnetic material is defined as the *field intensity (H):*

$$H = \frac{F_m}{l} \tag{7–5}$$

The unit of H is ampere-turns per meter. For example, when a wire is wound around a cylindrical core of magnetic material, the intensity of the magnetic field that is established when current flows through the looped wire depends on the amount of current, the number of loops in the winding, and the length of the cylindrical core.

EXAMPLE 7–2

There are two amperes of current through a wire with 5 turns.

(a) What is the mmf?

(b) What is the reluctance of the circuit if the flux is 250 μWb?

Solution

(a) $N = 5$ and $I = 2$ A

$F_m = NI = (5)(2\text{ A}) = 10\text{ At}$

(b) $\mathcal{R} = \dfrac{F_m}{\phi} = \dfrac{10\text{ At}}{250\ \mu\text{Wb}} = 0.04 \times 10^6\text{ At/Wb}$

The Electromagnet

An electromagnet is based on the properties that you have just learned. A basic electromagnet is simply a coil of wire wound around a core material that can be easily magnetized.

The shape of the electromagnet can be designed for various applications. For example, Figure 7–12 shows a U-shaped magnetic core. When the coil of wire is connected to a battery and current flows, as shown in Part (a), a magnetic field is established as indicated. If the current is reversed, as shown in Part (b), the direction of the magnetic field is also reversed. The closer the north and south poles are brought together, the smaller the air gap between them becomes, and the easier it becomes to establish a magnetic field, because the reluctance is lessened.

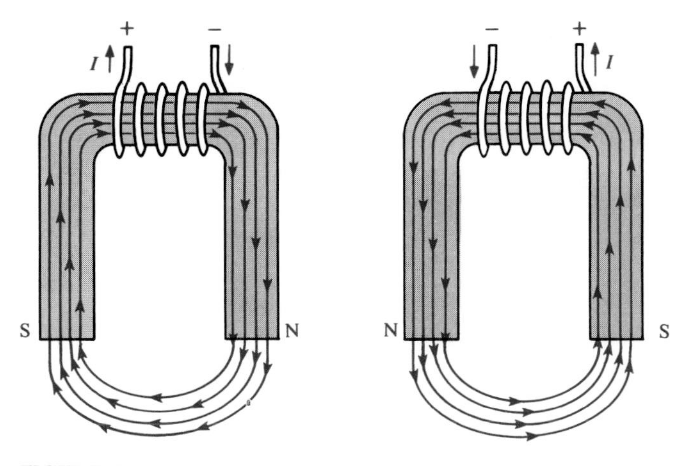

FIGURE 7–12
Reversing the current in the coil causes the electromagnetic field to reverse.

A good example of one application of an electromagnet is the process of recording on magnetic tape. In this situation, the *recording head* is an electromagnet with a narrow air gap, as shown in Figure 7–13. Current sets up a magnetic field across the air gap, and as the recording head passes over the magnetic tape, the tape is permanently magnetized. In digital recording, for example, the tape is magnetized in one direction for a binary 1 and in the other direction for a binary 0, as illustrated in the figure. This magnetization in different directions is accomplished by reversing the coil current in the recording head.

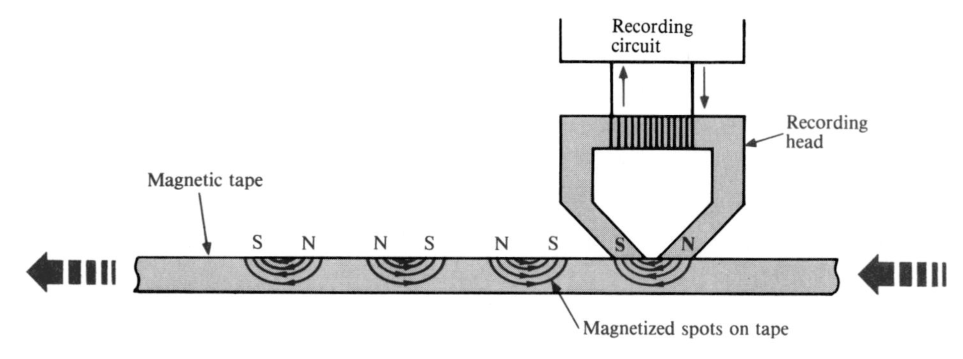

FIGURE 7–13
A recording head recording information on magnetic tape by magnetizing the tape as it passes by.

Electromagnetic Devices

The recording head is one example of an electromagnetic device. Several others are now presented.

The Solenoid

Generally, the solenoid is a type of electromagnet that has a *movable* iron core whose movement depends on both an electromagnetic field and a mechanical spring force. The basic structure is shown in Figure 7–14(a). When current flows through the coil, the electromagnetic field magnetizes the core so that, effectively, there are two magnetic fields. The repulsive force of like poles and the attractive force of unlike poles cause the core to move outward against the spring tension, as shown in Figure 7–14(b). When the coil current stops, the core is pulled back in by the spring.

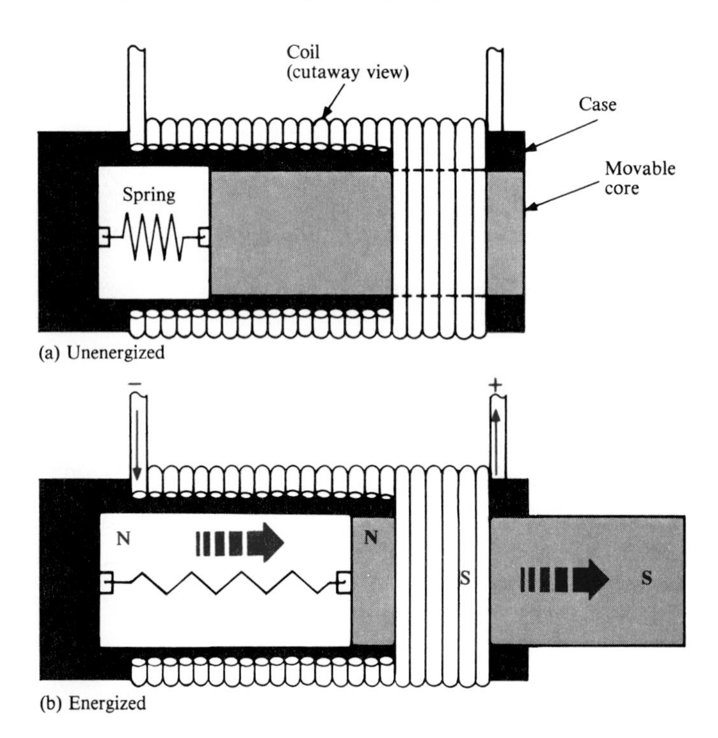

FIGURE 7–14
A basic solenoid structure with a cutaway view through the windings.

This mechanical movement in a solenoid is used for many applications, such as opening and closing valves, automobile door locks, and so on. Typical solenoids are shown in Figure 7–15.

The Relay

Relays differ from solenoids in that the electromagnetic action is used to open or close electrical contacts rather than to provide mechanical movement. Figure 7–16 shows the basic operation of a relay with one normally open (NO) contact

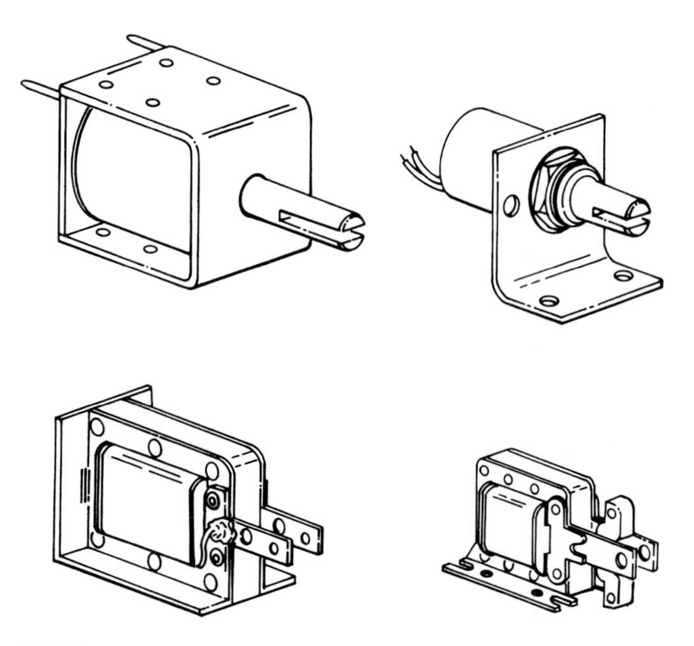

FIGURE 7–15
Typical solenoids.

FIGURE 7–16
Basic structure of a single-pole–double-throw relay.

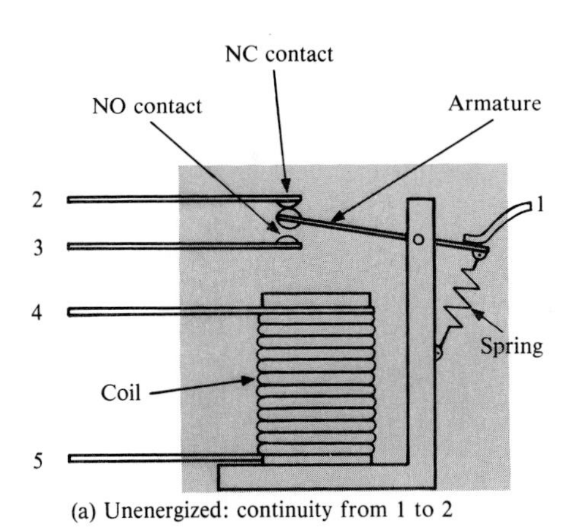

(a) Unenergized: continuity from 1 to 2

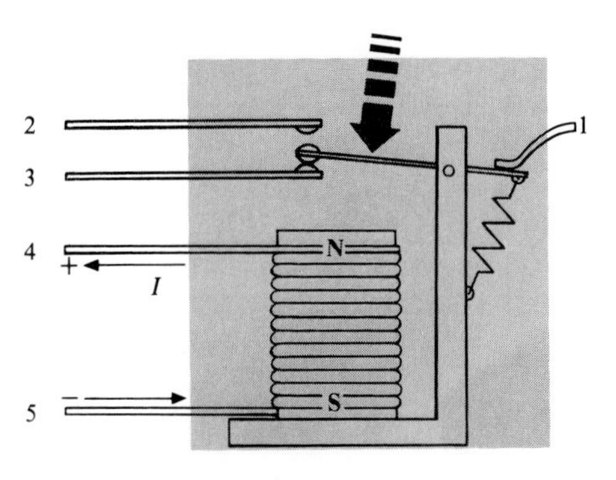

(b) Energized: continuity from 1 to 3

and one normally closed (NC) contact (single-pole–double-throw). When there is no coil current, the armature is held against the upper contact by the spring, thus providing continuity from terminal 1 to terminal 2, as shown in Part (a) of the figure. When energized with coil current, the armature is pulled down by the attractive force of the electromagnetic field and makes connection with the lower contact to provide continuity from terminal 1 to terminal 3, as shown in Part (b).

A typical relay and its schematic symbol are shown in Figure 7–17.

(a)

(b)

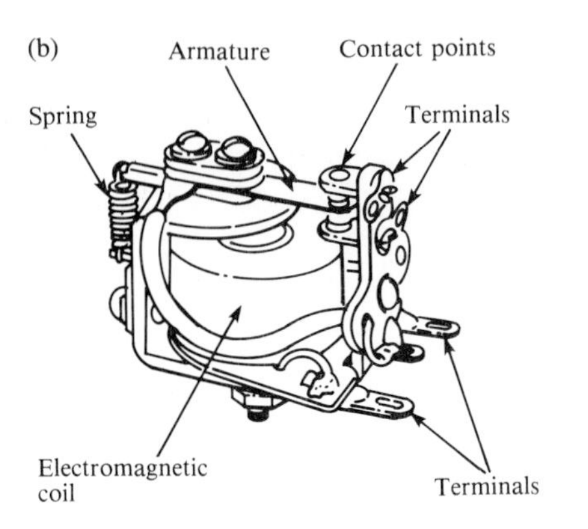

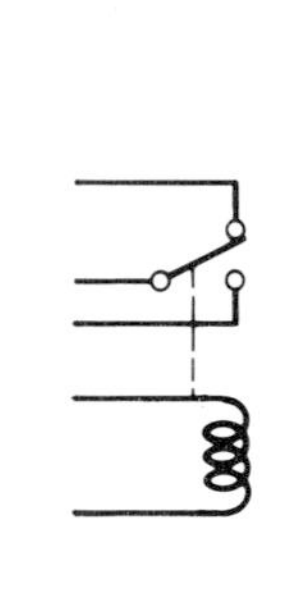

FIGURE 7–17
A typical relay.

The Speaker

Permanent-magnet speakers are commonly used in stereos, radios, and TV, and their operation is based on the principle of electromagnetism. A typical speaker is constructed with a permanent magnet and an electromagnet, as shown in Figure 7–18(a). The cone of the speaker consists of a paper-like diaphragm to which is attached a hollow cylinder with a coil around it, forming an electromagnet. One of the poles of the permanent magnet is positioned within the cylindrical coil. When current flows through the coil in one direction, the interaction of the permanent magnetic field with the electromagnetic field causes the cylinder to move to the right, as indicated in Figure 7–18(b). Current through the coil in the other direction causes the cylinder to move to the left, as shown in Part (c).

The movement of the coil cylinder causes the flexible diaphragm also to move in or out, depending on the direction of the coil current. The amount of coil current determines the intensity of the magnetic field, which controls the amount that the diaphragm moves.

As shown in Figure 7–19, when an audio signal (voice or music) is applied to the coil, the current varies in both direction and amount. In response, the diaphragm will *vibrate* in and out by varying amounts and at varying rates. Vibration in the diaphragm causes the air that is in contact with it to vibrate in the same manner. These air vibrations move through the air as sound waves.

SECTION REVIEW 7–2

1. Explain the difference between magnetism and electromagnetism.
2. What happens to the magnetic field in an electromagnet when the current through the coil is reversed?

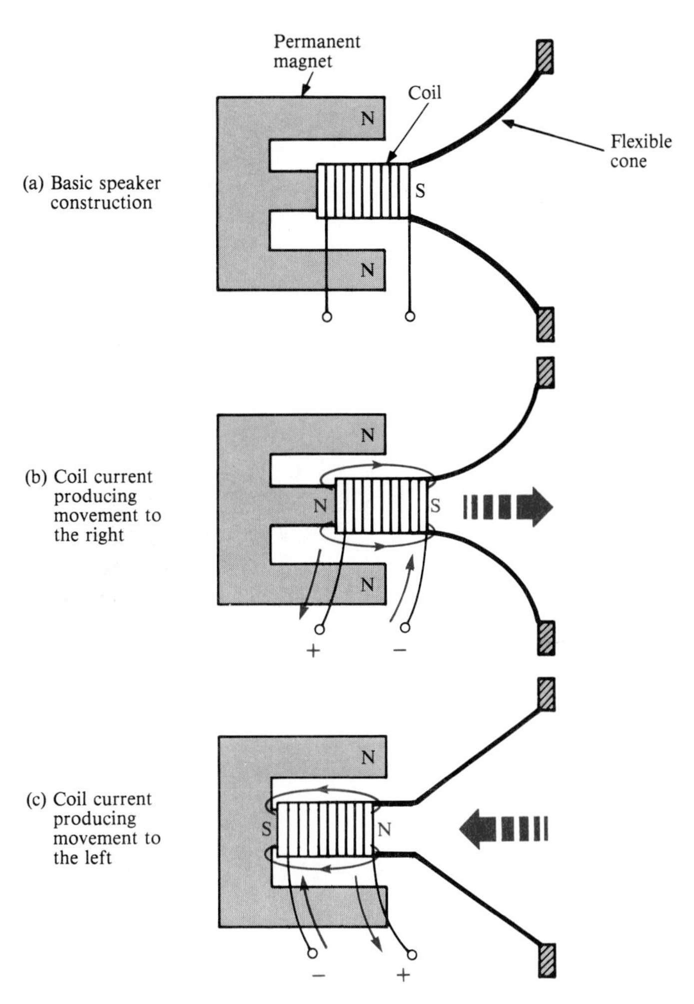

FIGURE 7–18
Basic speaker operation.

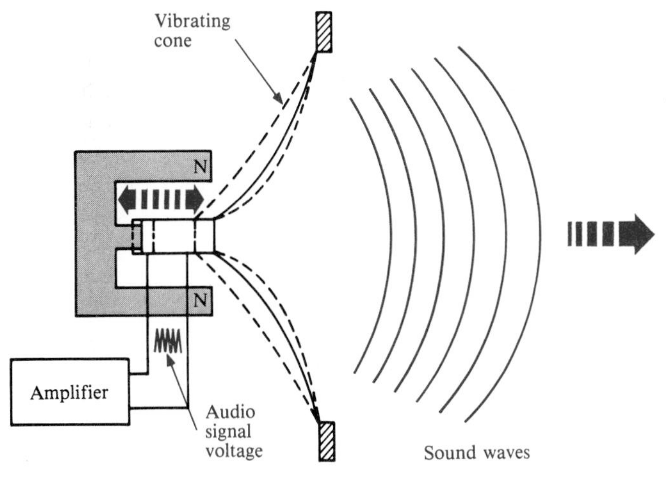

FIGURE 7–19
The speaker converts audio signal voltages into sound waves.

ELECTROMAGNETIC INDUCTION

7–3

When a conductor is moved through a magnetic field, a voltage is produced across the conductor. This principle is known as *electromagnetic induction,* and the resulting voltage is an *induced voltage.*

The principle of electromagnetic induction is widely applied in electrical circuits, as you will learn in this chapter and later in the study of transformers. The operation of electrical motors and generators is also based on this principle.

Relative Motion

When a wire is moved across a magnetic field, there is a relative motion between the wire and the magnetic field. Likewise, when a magnetic field is moved past a stationary wire, there is also relative motion. In either case, there is an *induced voltage* in the wire as a result of this motion, as Figure 7–20 indicates.

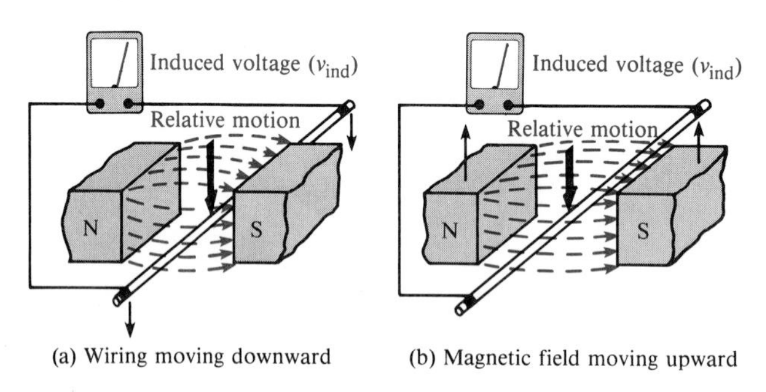

FIGURE 7–20
Relative motion between a wire and a magnetic field.

The amount of the induced voltage depends on the *rate* at which the wire and the magnetic field move with respect to each other: The faster the relative speed, the greater the induced voltage.

Polarity of the Induced Voltage

If the conductor in Figure 7–20 is moved first one way and then another in the magnetic field, a reversal of the polarity of the induced voltage will be observed. As the wire is moved downward, a voltage is induced with the polarity indicated in Figure 7–21(a). As the wire is moved upward, the polarity is as indicated in Part (b) of the figure. The lower-case v stands for instantaneous voltage.

Induced Current

When a load resistor is connected to the wire in Figure 7–21, the voltage induced by the relative motion in the magnetic field will cause a current in the load, as shown in Figure 7–22. This current is called the *induced current.*

The principle of producing a voltage and a current in a load by moving a conductor across a magnetic field is the basis for electrical generators. The concept of a conductor existing in a moving magnetic field is fundamental to inductance in an electrical circuit. The lower-case i stands for instantaneous current.

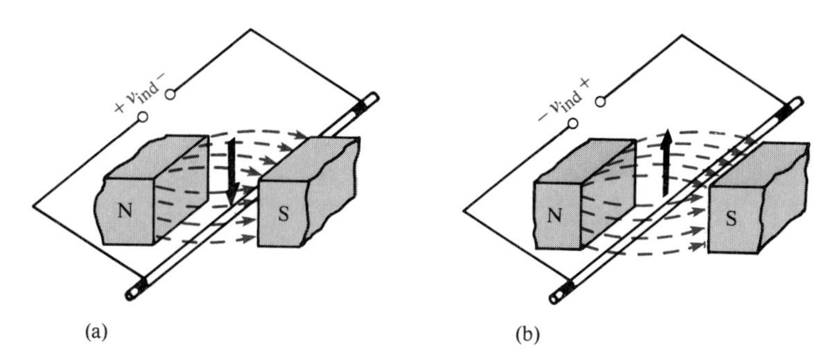

FIGURE 7–21
Polarity of induced voltage depends on direction of motion.

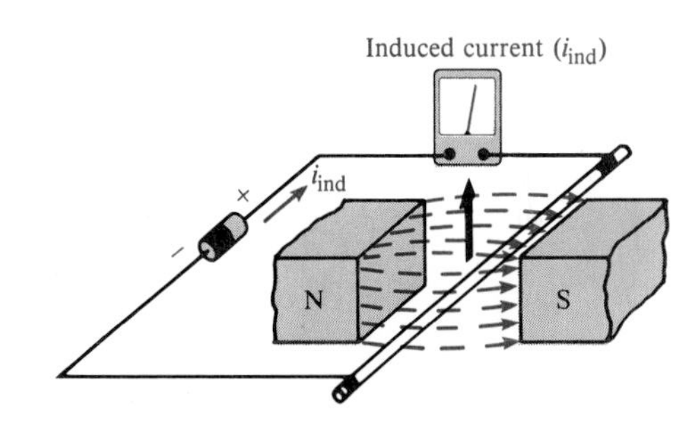

FIGURE 7–22
Induced current in a load as the wire moves through the magnetic field.

Forces on a Current-Carrying Conductor in a Magnetic Field

Figure 7–23(a) shows current outward through a wire in a magnetic field. The electromagnetic field set up by the current interacts with the permanent magnetic field; as a result, the permanent lines of force *above* the wire tend to be deflected down under the wire, because they are opposite in direction to the electromagnetic lines of force. Therefore, the flux density above is reduced, and the magnetic field is weakened. The flux density below the conductor is increased, and the magnetic field is strengthened. An upward force on the conductor results, and the conductor tends to move toward the weaker magnetic field.

Figure 7–23(b) shows the current flowing inward, resulting in a force on the conductor in the downward direction. This principle is the basis for electrical motors.

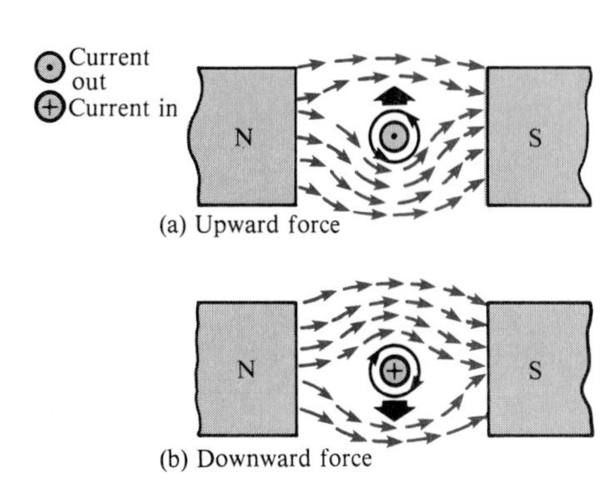

FIGURE 7–23
Forces on a current-carrying conductor in a magnetic field.

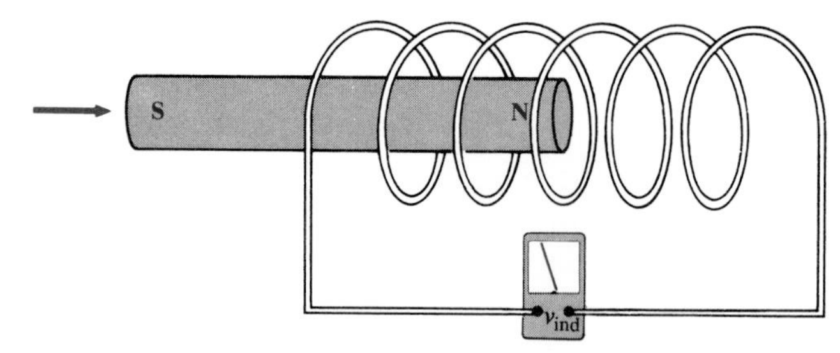

(a) As magnet moves to the right, magnetic field is changing with respect to coil, and a voltage is induced.

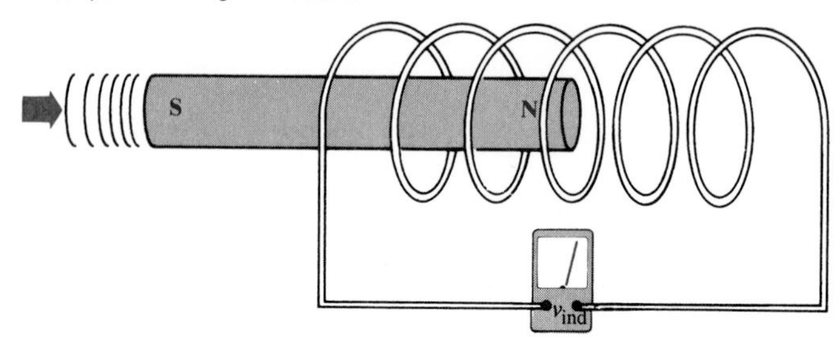

(b) As magnet moves more rapidly to the right, magnetic field is changing more rapidly with respect to coil, and a greater voltage is induced.

FIGURE 7–24
A demonstration of the first part of Faraday's law: The amount of induced voltage is directly proportional to the rate of change of the magnetic field with respect to the coil.

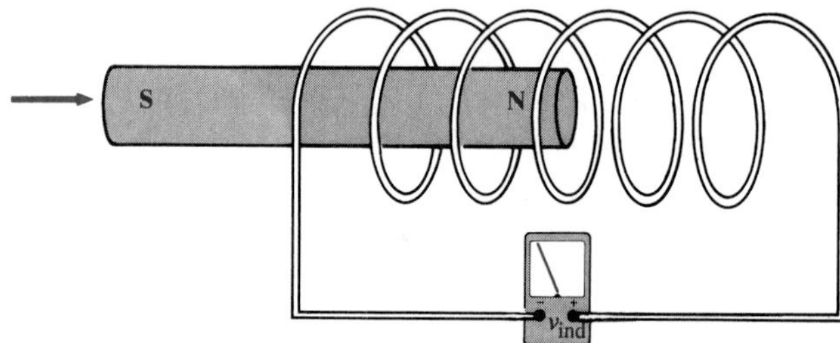

(a) Magnet moves through coil and induces a voltage.

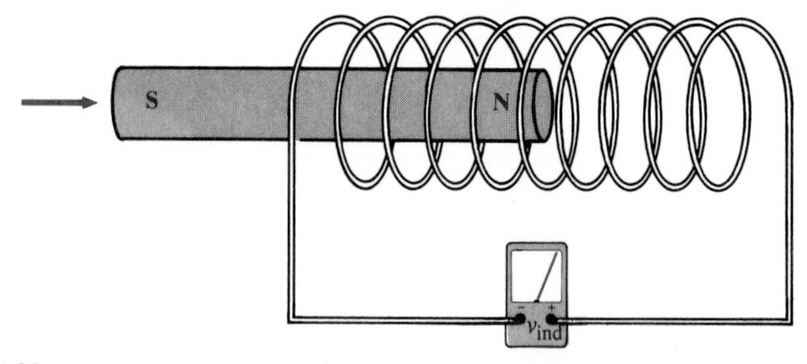

(b) Magnet moves at same rate through a coil with more turns (loops) and induces a greater voltage.

FIGURE 7–25
A demonstration of the second part of Faraday's law: The amount of induced voltage is directly proportional to the number of turns in the coil.

Faraday's Law

Michael Faraday discovered the principle of electromagnetic induction in 1831. He found that moving a magnet through a coil of wire induced a voltage across the coil, and that when a complete path was provided, the induced voltage caused an induced current, as you have seen. Faraday's observations are as follows:

1. The amount of voltage induced in a coil is directly proportional to the rate of change of the magnetic field with respect to the coil.
2. The amount of voltage induced in a coil is directly proportional to the number of turns of wire in the coil.
3. The voltage induced across a coil equals the number of turns in the coil times the rate of change of the magnetic flux.

Part 1 of Faraday's law is demonstrated in Figure 7–24, where a bar magnet is moved through a coil, thus creating a changing magnetic field. In Part (a) of the figure, the magnet is moved at a certain rate, and a certain induced voltage is produced as indicated. In Part (b), the magnet is moved at a faster rate through the coil, creating a greater induced voltage.

Part 2 of Faraday's law is demonstrated in Figure 7–25. In Part (a), the magnet is moved through the coil and a voltage is induced as shown. In Part (b), the magnet is moved at the same speed through a coil that has a greater number of turns. The greater number of turns creates a greater induced voltage.

SECTION REVIEW 7–3

1. What is the induced voltage across a stationary conductor in a stationary magnetic field?
2. When the speed at which a conductor is moved through a magnetic field is increased, does the induced voltage increase, decrease, or remain the same?
3. When there is current through a conductor in a magnetic field, what happens?

APPLICATIONS OF ELECTROMAGNETIC INDUCTION

7–4

In this section, two interesting applications of electromagnetic induction are discussed: an automotive crankshaft position sensor, and a dc generator. Although there are many varied applications, these two are representative.

Automotive Crankshaft Position Sensor

An interesting automotive application is a type of engine sensor that detects the crankshaft position directly using electromagnetic induction. The electronic engine controller in many automobiles uses the position of the crankshaft to set ignition timing and, sometimes, to adjust the fuel control system. Figure 7–26 shows the basic concept. A steel disk is attached to the engine's crankshaft by an extension rod; the protruding tabs on the disk represent specific crankshaft positions.

As illustrated in Figure 7–26, as the disk rotates with the crankshaft, the tabs periodically pass through the air gap of the permanent magnet. Since steel has a much lower reluctance than does air (a magnetic field can be established in steel much more easily than in air), the magnetic flux suddenly increases as a tab comes into the air gap, causing a voltage to be induced across the coil. This process is illustrated in Figure 7–27. The electronic engine control circuit uses the induced voltage as an indicator of the crankshaft position.

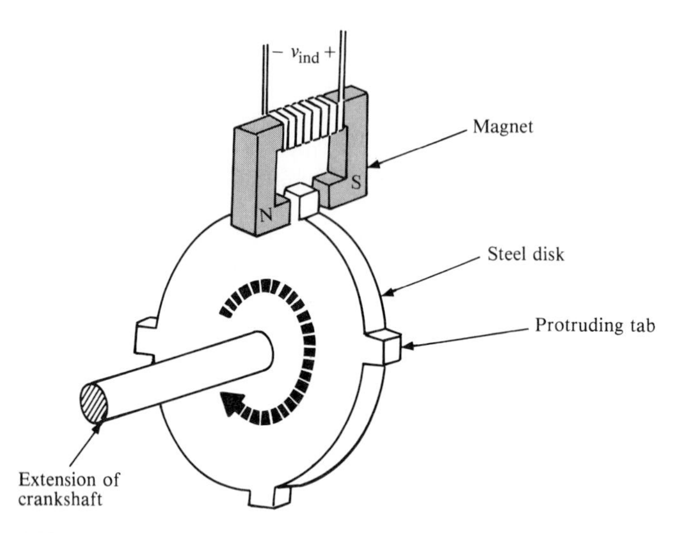

FIGURE 7–26
A crankshaft position sensor that produces a voltage when a tab passes through the air gap of the magnet.

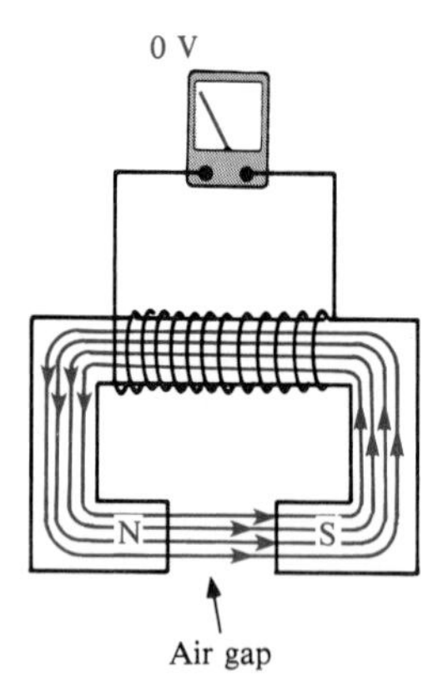

(a) There is no changing magnetic field, so there is no induced voltage.

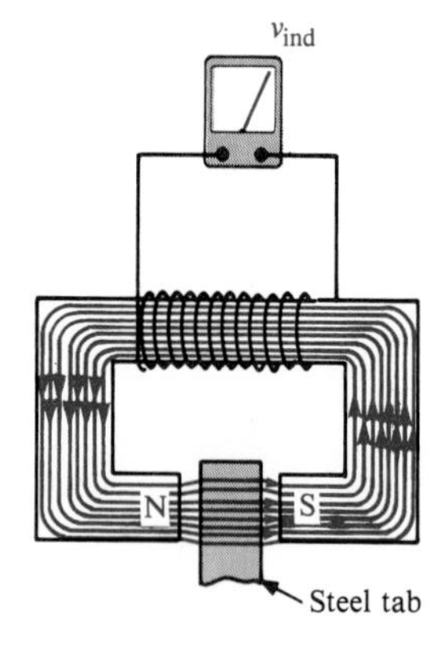

(b) Insertion of the steel tap reduces the reluctance of the air gap, causing the magnetic flux to increase and thus inducing a voltage.

FIGURE 7–27
As the tab passes through the air gap of the magnet, the coil senses a change in the magnetic field, and a voltage is induced.

A dc Generator

Figure 7–28 shows a simplified dc generator consisting of a single loop of wire in a permanent magnetic field. Notice that each end of the loop is connected to a *split-ring* arrangement. This conductive metal ring is called a *commutator*. As the loop is rotated in the magnetic field, the split commutator ring also rotates. Each half of the split ring rubs against the fixed contacts, called *brushes,* and connects the loop to an external circuit.

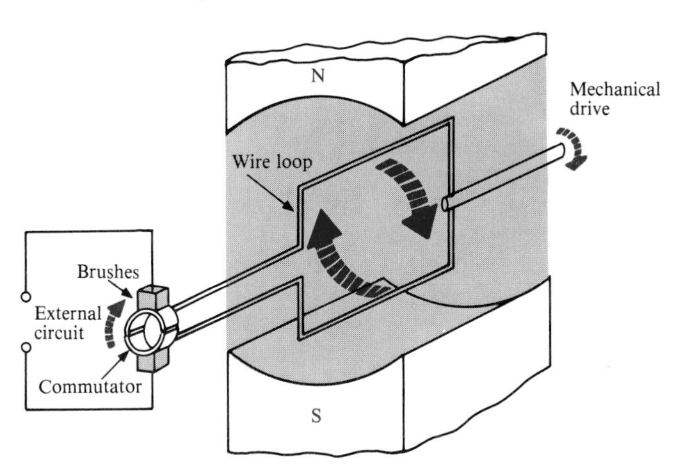

FIGURE 7–28
A basic dc generator.

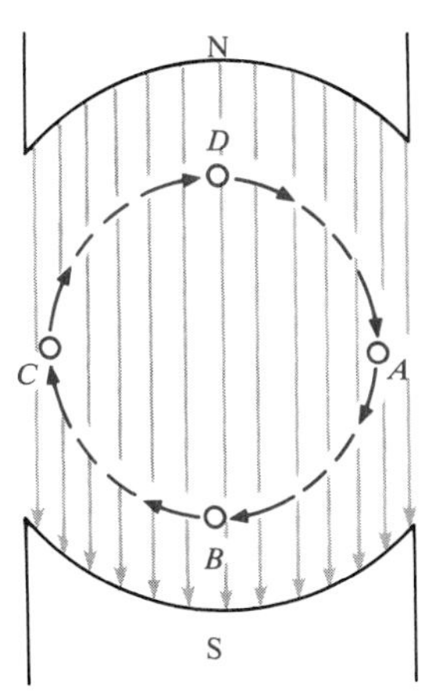

FIGURE 7–29
End view of loop cutting through the magnetic field.

As the loop rotates through the magnetic field, it "cuts" through the flux lines at varying angles, as illustrated in Figure 7–29. At position *A* in its rotation, the loop of wire is effectively moving *parallel* with the magnetic field. Therefore, at this instant, the rate at which it is cutting through the magnetic flux lines is *zero.*

As the loop moves from position *A* to position *B,* it cuts through the flux lines at an increasing rate. At position *B,* it is moving effectively *perpendicular* to the magnetic field and thus is cutting through a maximum number of lines. As the loop rotates from position *B* to position *C,* the rate at which it cuts the flux lines decreases to minimum (zero) at *C*. From position *C* to position *D,* the rate at which the loop cuts the flux lines increases to a maximum at *D* and then back to a minimum again at *A*.

As you previously learned, when a wire moves through a magnetic field, a voltage is induced, and by Faraday's law, the amount of induced voltage is proportional to the number of loops (turns) in the wire and the rate at which it is moving with respect to the magnetic field. Now you know that the *angle* at which the wire moves with respect to the magnetic flux lines determines the amount of induced voltage, because the rate at which the wire cuts through the flux lines depends on the angle of motion.

Figure 7–30 illustrates how a voltage is induced in the external circuit as the single loop rotates in the magnetic field. Assume that the loop is in its instantaneous horizontal position, so the induced voltage is zero. As the loop

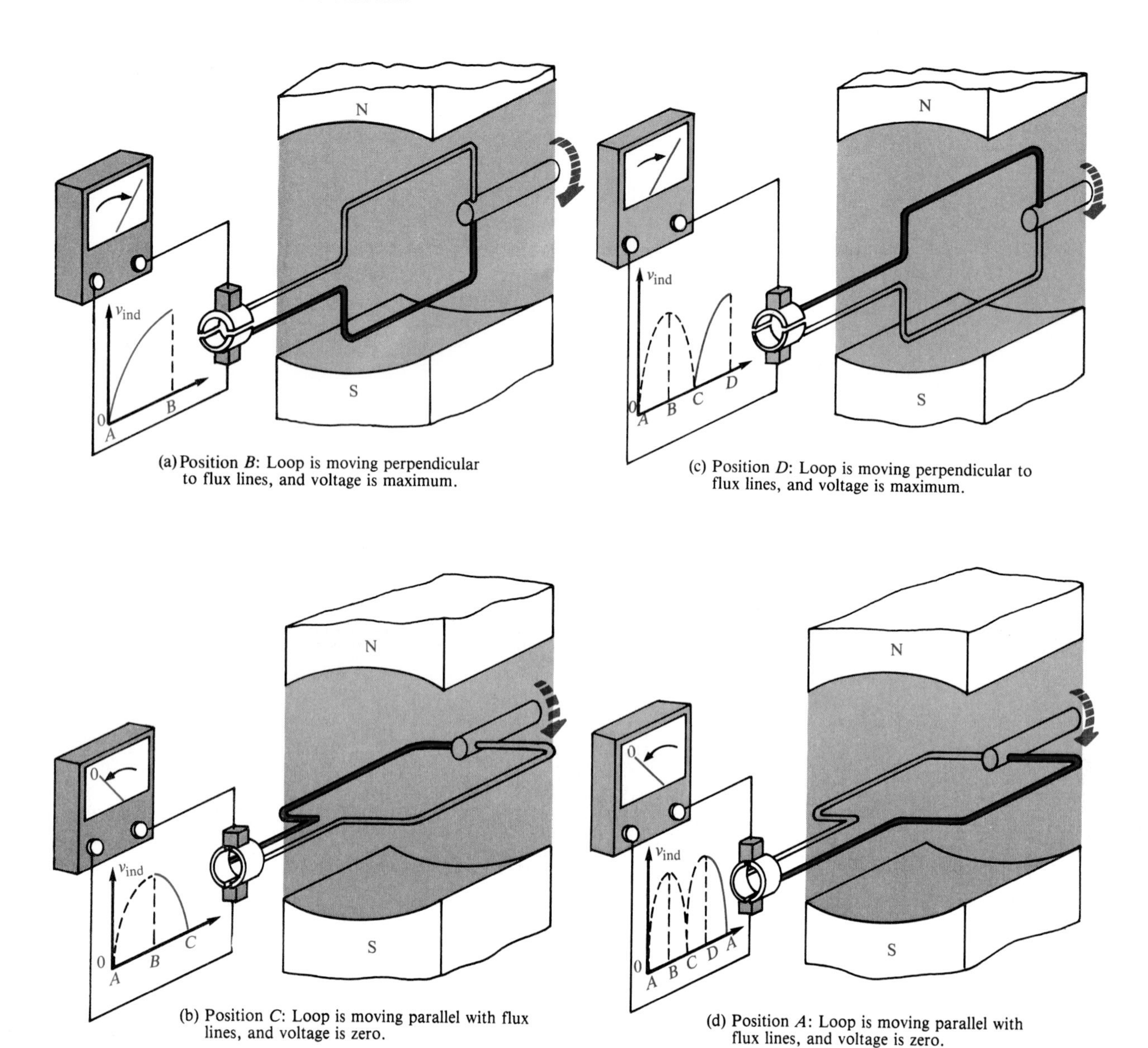

(a) Position B: Loop is moving perpendicular to flux lines, and voltage is maximum.

(b) Position C: Loop is moving parallel with flux lines, and voltage is zero.

(c) Position D: Loop is moving perpendicular to flux lines, and voltage is maximum.

(d) Position A: Loop is moving parallel with flux lines, and voltage is zero.

FIGURE 7–30
Operation of a basic dc generator.

continues in its rotation, the induced voltage builds up to a maximum at position B, as shown in Part (a) of the figure. Then, as the loop continues from B to C, the voltage decreases to zero at C, as shown in Part (b).

During the second half of the revolution, shown in Parts (c) and (d), the brushes switch to opposite commutator sections, so the polarity of the voltage remains the same across the output. Thus, as the loop rotates from position C to position D and then back to A, the voltage increases from zero at C to a maximum at D and back to zero at A.

Figure 7–31 shows how the induced voltage varies as the loop goes through several rotations (three in this case). This voltage is a dc voltage be-

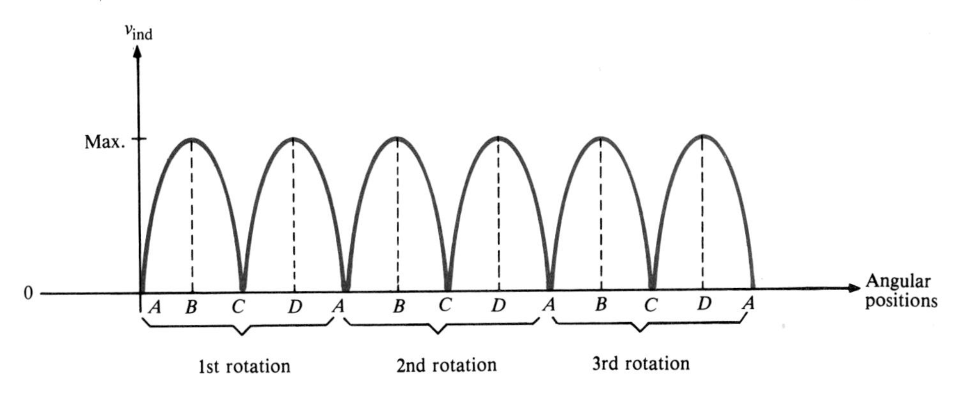

FIGURE 7–31
Induced voltage over three rotations of the loop.

cause its polarities do not change. However, the voltage is pulsating between zero and its maximum value.

When more loops are added, the voltages induced across each loop are combined across the output. Since the voltages are offset from each other, they do not reach their maximum or zero values at the same time. A smoother dc voltage results, as shown in Figure 7–32 for two loops. The variations can be further smoothed out by *filters* to achieve a nearly constant dc voltage. (Filters are discussed in a later chapter.)

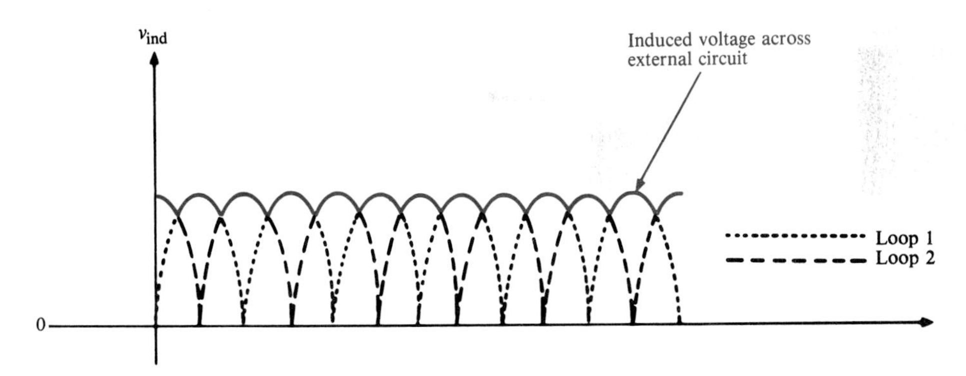

FIGURE 7–32
The induced voltage for a two-loop generator. There is much less variation in the induced voltage.

SECTION REVIEW 7–4

1. If the steel disk in the crankshaft position sensor has stopped, with a tab in the magnet's air gap, what is the induced voltage?
2. What happens to the induced voltage if the loop in the basic dc generator suddenly begins rotating at a faster speed?

APPLICATION NOTE

A suggested approach to the security system problem in the Application Assignment at the beginning of the chapter is as follows:

Use normally-closed (NC) magnetic switches at each entry point in a room, and connect them all in series so that they provide a path for current to energize a relay coil. If the path is interrupted, the relay will be deenergized. In the deenergized position, the relay contacts will turn on a lamp. Also, the relay contacts can be used to interrupt a circuit and deenergize a solenoid that pulls a latch on the door of the dog cage, allowing the door to fall open.

Draw a schematic either for this system or for one that you have created.

SUMMARY

Facts

- Unlike magnetic poles attract each other, and like poles repel each other.
- Materials that can be magnetized are called *ferromagnetic*.
- When current flows through a conductor, it produces an electromagnetic field around the conductor.
- You can use the left-hand rule to establish the direction of the electromagnetic lines of force around a conductor.
- An electromagnet is basically a coil of wire around a magnetic core.
- When a conductor moves within a magnetic field, or when a magnetic field moves relative to a conductor, a voltage is induced across the conductor.
- The faster the relative motion between a conductor and a magnetic field, the greater is the induced voltage.

Units

- Weber (Wb)—the unit of magnetic flux.
- Tesla (T)—the unit of flux density.
- Ampere-turn (At)—the unit of magnetomotive force (mmf).

Definitions

- *Magnetic flux*—the lines of force between the north pole and the south pole of a magnet.
- *Flux density*—the number of lines of force per unit area in a magnetic field.
- *Electromagnetic field*—a magnetic field created by current in a conductor.
- *Magnetomotive force (mmf)*—the force that produces an electromagnetic field; it is the product of current and number of turns, and its unit is ampere-turns.
- *Reluctance*—the opposition to the establishment of a magnetic field.
- *Permeability*—the measure of the ease with which a magnetic field can be established in a material.
- *Magnetic field intensity*—the amount of mmf per unit length of magnetic material.
- *Solenoid*—an electromagnetic device characterized by a movable iron core that is actuated by current in a coil of wire.
- *Relay*—an electromagnetic device in which electrical contacts are controlled by current in a coil.
- *Electromagnetic induction*—the process of producing a voltage by the relative motion between a magnetic field and a conductor.
- *Faraday's law*—the voltage induced across a coil equals the number of turns in the coil times the rate of change of the magnetic flux.

Formulas

$$B = \frac{\phi}{A} \quad \text{Magnetic flux density} \qquad (7\text{–}1)$$

Figure 8–2 illustrates how the interaction of magnetic fields produces rotation of the coil assembly. The current flows inward at the "cross" and outward at the "dot" in the single winding shown. The inward current produces a counterclockwise electromagnetic field that reinforces the permanent magnetic field below it. The result is an upward force on the left side of the coil as shown. A downward force is developed on the right side of the coil, where the current is outward. These forces produce a clockwise rotation of the coil assembly.

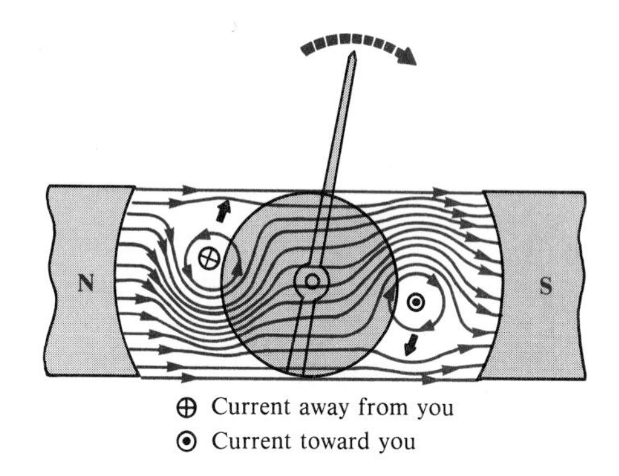

FIGURE 8–2
When the electromagnetic field interacts with the permanent magnetic field, forces are exerted on the rotating coil assembly, causing it to move clockwise and thus deflecting the pointer.

In addition to the d'Arsonval movement, there are two other types of movements that are used in analog meters: the *iron-vane* and the *electrodynamometer,* which are described briefly in the following paragraphs.

Iron-Vane Movement

This type of movement consists basically of two iron bars (vanes) that are located side-by-side within a coil, as shown in Figure 8–3(a). One vane is stationary, and the other movable. The electromagnetic field produced by the coil current magnetizes the vanes, making them like bar magnets with the same

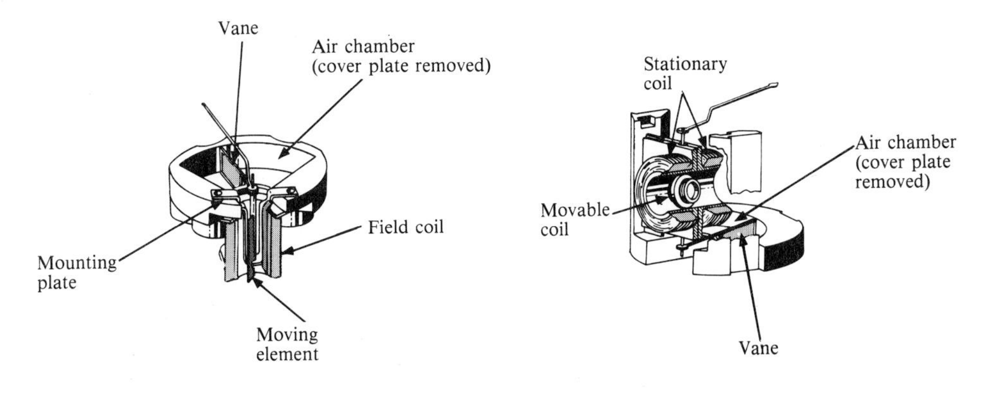

FIGURE 8–3
Construction views of iron-vane and electrodynamometer movements (courtesy of Triplett Corp.).

orientation of their north and south poles. The like poles repel each other, and the movable vane is pushed away from the stationary vane. A pointer attached to the movable vane is thus deflected an amount proportional to the coil current.

The iron-vane movement is very rugged and dependable, but it has the drawback of having a very nonlinear scale. That is, the measurement increments are very crowded on the low end of the scale and are more spread out on the upper end. The most common use for this type of movement is in ac (alternating current) ammeters.

Electrodynamometer Movement

This type of movement differs from the d'Arsonval in that it uses an electromagnet rather than a permanent magnet. As shown in Figure 8–3(b), it also has a moving coil to which the pointer is attached, as in the d'Arsonval. The electrodynamometer movement can be used to measure both dc and ac quantities without additional internal circuitry, whereas the d'Arsonval movement requires extra circuitry for measuring ac. The most common use for this movement is in wattmeters for measuring power.

Current Sensitivity and Resistance of the Meter Movement

The *current sensitivity* of a meter movement is the amount of current required to deflect the pointer full scale (all the way to its right-most position). For example, a 1-mA sensitivity means that when there is 1 mA through the meter coil, the needle is at its maximum deflection. If 0.5 mA flows through the coil, the needle is at the halfway point of its full deflection.

The movement *resistance* is simply the resistance of the coil of wire used in the movement.

SECTION REVIEW 8–1

1. Define *current sensitivity* of a meter movement.
2. Describe the basic difference between an electrodynamometer movement and a d'Arsonval movement.

BASIC ANALOG dc METER CIRCUITRY

8–2

The Ammeter

A typical d'Arsonval movement might have a current sensitivity of 1 mA and a resistance of 50 Ω. In order to measure more than 1 mA, additional circuitry must be used with the basic meter movement. Figure 8–4(a) shows a simple ammeter with a *shunt* (parallel) resistor (R_{SH}) across the movement. The purpose of the shunt resistor is to bypass current in excess of 1 mA around the meter movement. For example, let us assume that this meter must measure currents of up to 10 mA. Thus, for *full-scale* deflection, the movement must carry 1 mA, and the shunt resistor must carry 9 mA, as indicated in Figure 8–4(b).

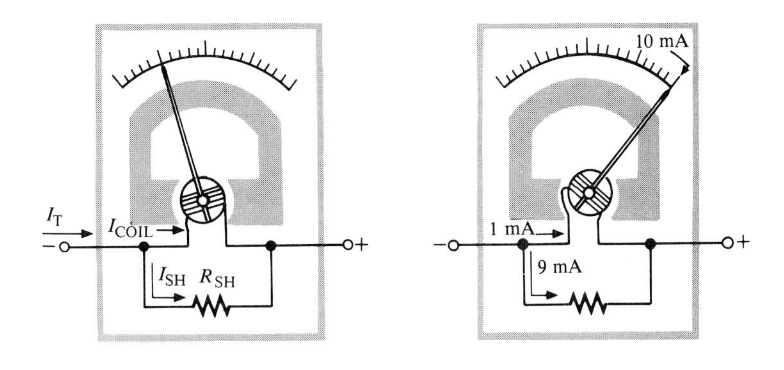

(a) Total current is $I_{COIL} + I_{SH}$.

(b) Meter indicates 10 mA (1 mA + 9 mA).

FIGURE 8–4
Shunt resistance (R_{SH}) in an ammeter.

Multiple-Range Ammeters

The simple ammeter in Figure 8–4 has only one range. As you saw, it can measure currents from 0 to 10 mA and no higher. Most practical ammeters have several ranges. Each range in a multiple-range ammeter has its own shunt resistance which is selected with a multiple-position switch. For example, Figure 8–5 shows a four-range ammeter with a 0.1-mA (100-μA) meter movement. When the switch is in the 0.1-mA position, all of the current being measured

FIGURE 8–5
Example of a multiple-range ammeter.

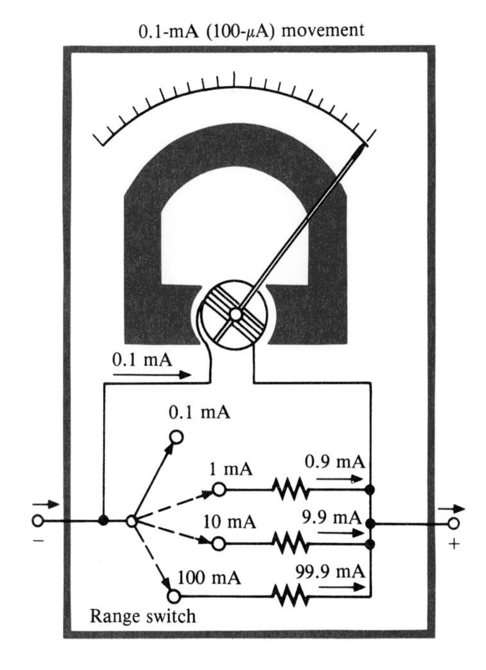

flows through the coil. In the other positions, some of the current flows through the coil, but most of it flows through the shunt resistor. In the 1-mA position, 100 μA (0.1 mA) flows through the coil and 0.9 mA through the shunt resistor for *full-scale deflection.* In the 10-mA position, 100 μA flows through the coil and 9.9 mA through the shunt resistor for full-scale deflection. In the 100-mA position, 100 μA flows through the coil and 99.9 mA through the shunt resistor for full-scale deflection. Of course, for less than full-scale deflection, the current values are less.

Effect of the Ammeter on a Circuit

As you know from the presentation in Chapter 2 and other earlier coverage, an ammeter is connected in *series* to measure the current in a circuit. Ideally, the meter should not alter the current that it is intended to measure. In practice, however, the meter unavoidably has some effect on the circuit, because its *internal resistance* is connected in series with the circuit resistance. However, in most cases, the meter's internal resistance is so small compared to the circuit resistance that it can be neglected.

For example, if the meter has a 50-Ω movement and a 100-μA full-scale current, the voltage dropped across the coil is

$$V_{\text{COIL}} = I_{\text{COIL}}R_{\text{COIL}} = (100\ \mu\text{A})(50\ \Omega) = 5\ \text{mV}$$

The shunt resistance for the 10-mA range, for example, is

$$R_{\text{SH}} = \frac{V_{\text{COIL}}}{I_{\text{SH}}} = \frac{5\ \text{mV}}{9.9\ \text{mA}} = 0.505\ \Omega$$

As you can see, the total resistance of the ammeter on the 10-mA range is the coil resistance in parallel with the shunt resistance:

$$R_{\text{COIL}} \| R_{\text{SH}} = 50\ \Omega \| 0.505\ \Omega = 0.5\ \Omega$$

EXAMPLE 8–1

How much does an ammeter with a 100-μA, 50-Ω movement affect the current in the circuit of Figure 8–6?

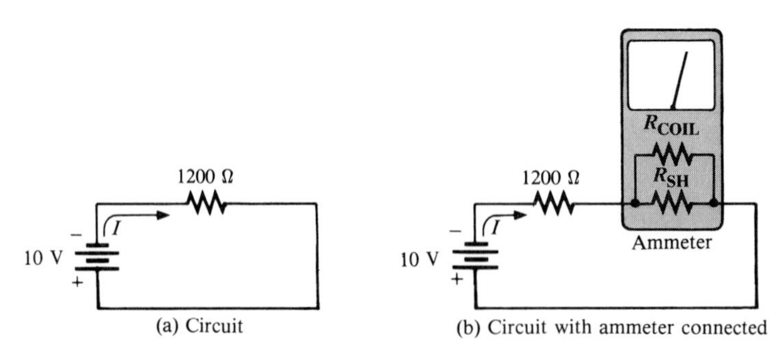

FIGURE 8–6

Solution
The true current flowing in the circuit is

$$I = \frac{10\ \text{V}}{1200\ \Omega} = 8.3333\ \text{mA}$$

The meter is set on the 10-mA range in order to measure this particular amount of current. It was found that the meter's resistance on the 10-mA range is 0.5 Ω. When the meter is connected in the circuit, its resistance is in series with the 1200-Ω resistor. Thus, there is a total of 1200.5 Ω.

The current actually measured by the ammeter is

$$I = \frac{10\ \text{V}}{1200.5\ \Omega} = 8.3299\ \text{mA}$$

This current differs from the true circuit current by only 0.04%. Therefore, the meter does not significantly alter the current value, a situation which, of course, is necessary because the measuring instrument should not change the quantity that is to be measured accurately.

The Voltmeter

The voltmeter utilizes the same type of movement as the ammeter. Different external circuitry is added so that the movement will function to measure voltage in a circuit.

As you have seen, the voltage drop across the meter coil is dependent on the current and the coil resistance. For example, a 50-μA, 1000-Ω movement has a full-scale voltage drop of (50 μA)(1000 Ω) = 50 mV. To use the meter to indicate voltages greater than 50 mV, we must add a *series resistance* to drop any additional voltage beyond that which the movement requires for full-scale deflection. This resistance is called the *multiplier resistance* and is designated R_M.

A basic voltmeter is shown in Figure 8–7(a) with a single multiplier resistor for one range. For this meter to measure 1 V full scale, the multiplier resistor (R_M) must drop 0.95 V because the coil drops only 50 mV (0.95 V + 50 mV = 1 V), as shown in Figure 8–7(b).

Multiple-Range Voltmeters

The simple voltmeter in Figure 8–7 has only one range. It can measure voltages from 0 to 1 V and no higher. Most practical voltmeters have several ranges, each of which has its own multiplier resistor which can be selected by a switch. For example, Figure 8–8 shows a four-range voltmeter with a 50-μA, 1000-Ω meter movement. When the switch is in the 50-mV position, all of the voltage is dropped across the coil. In the other positions, some of the total voltage is dropped across the coil, but most is dropped across the multiplier resistor. In the 1-V position, 50 mV is dropped across the coil and 0.95 V across R_{M1} for *full-scale deflection*. In the 10-V position, 50 mV is dropped across the coil and 9.95 V across $R_{M1} + R_{M2}$ for full-scale deflection. In the 100-V position, 50 mV is dropped across the coil and 99.95 V across $R_{M1} + R_{M2} + R_{M3}$ for full-scale

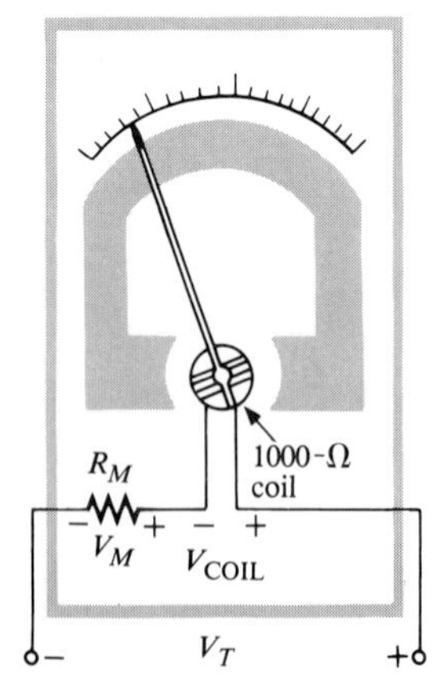

(a) Total voltage is $V_M + V_{COIL}$.

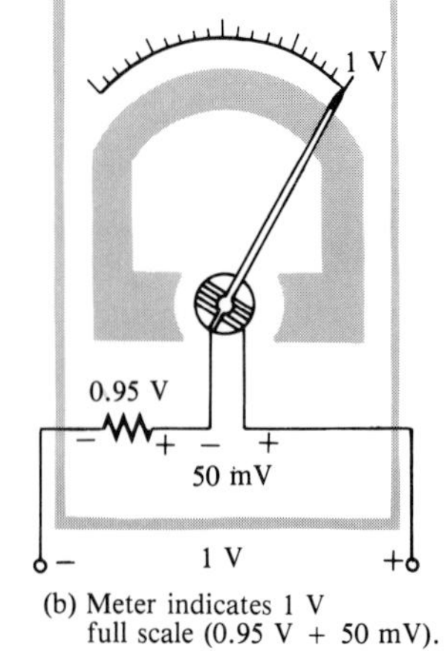

(b) Meter indicates 1 V full scale (0.95 V + 50 mV).

FIGURE 8–7
Multiplier resistance (R_M) in a voltmeter.

FIGURE 8–8
Example of a multiple-range voltmeter.

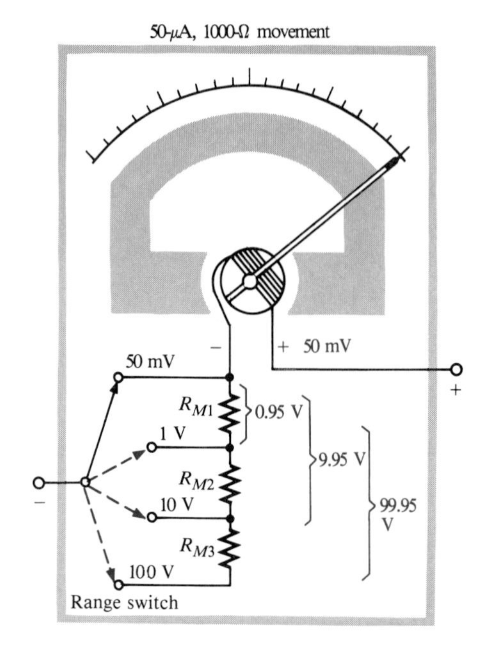

deflection. Of course, for less than full-scale deflection, the voltage values are less.

Loading Effect of a Voltmeter

As you know from Chapter 2 and other earlier coverage, a voltmeter is always connected in parallel with the circuit component across which the voltage is to be measured. Thus, it is much easier to measure voltage than current, because you must break a circuit to insert an ammeter in series. You simply connect a voltmeter across the circuit without disrupting the circuit or breaking a connection.

Since some current is required through the voltmeter to operate the movement, the voltmeter has some effect on the circuit to which it is connected. This effect is called *loading*. However, as long as the meter resistance is much greater than the resistance of the circuit across which it is connected, the loading effect is negligible. This characteristic is necessary because we do not want the measuring instrument to change the voltage that it is measuring.

The meter movement just discussed has a full-scale coil current of 50-μA. This current flows through the multiplier resistors since they are in series with the coil. For example, on the 10-V range, the total multiplier resistance is

$$R_M = \frac{9.95\ \text{V}}{50\ \mu\text{A}} = 199\ \text{k}\Omega$$

The internal resistance of the voltmeter is the multiplier resistance in series with the coil resistance:

$$R_M + R_{\text{COIL}} = 199\ \text{k}\Omega + 1\ \text{k}\Omega = 200\ \text{k}\Omega$$

EXAMPLE 8–2

How much does a voltmeter with a 50-μA, 1000-Ω movement affect the voltage being measured in Figure 8–9?

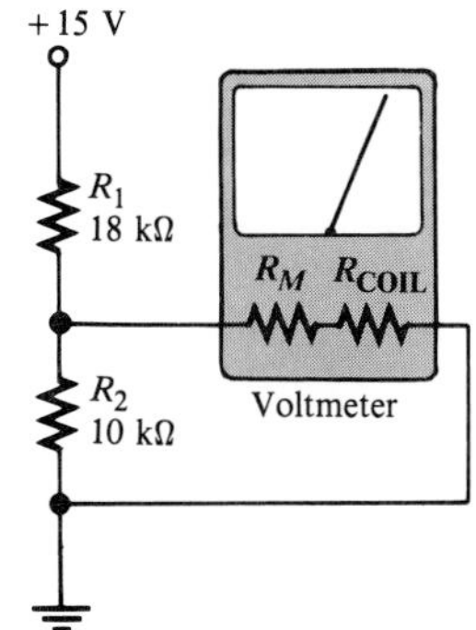

+15 V

R_1 18 kΩ

R_M R_{COIL}

R_2 10 kΩ

Voltmeter

FIGURE 8–9

Solution

The true voltage across R_2 is

$$V_{R2} = \left(\frac{10\ \text{k}\Omega}{28\ \text{k}\Omega}\right)15\ \text{V} = 5.357\ \text{V}$$

The meter is set on the 10-V range in order to measure this particular amount of voltage. The meter's resistance on the 10-V range is 200 kΩ. When the meter is connected across the circuit, its resistance is in *parallel* with R_2 of the circuit, giving a total of

$$\frac{R_2(R_M + R_{\text{COIL}})}{R_2 + R_M + R_{\text{COIL}}} = \frac{(10\ \text{k}\Omega)(200\ \text{k}\Omega)}{210\ \text{k}\Omega} = 9.52\ \text{k}\Omega$$

The voltage actually measured by the voltmeter is

$$V_{R2} = \left(\frac{9.52\ \text{k}\Omega}{27.52\ \text{k}\Omega}\right)15\ \text{V} = 5.189\ \text{V}$$

This differs from the true voltage across R_2 by only 3.1%. Therefore, the meter does not significantly alter the voltage value because its internal resistance is much greater than the circuit resistance across which it is connected.

The Ohmmeter

The meter movement used for the ammeter and the voltmeter can also be adapted for use in an ohmmeter. The ohmmeter is used to measure resistance values.

A basic one-range ohmmeter is shown in Figure 8–10(a). It contains a battery and a variable resistor in series with the movement. To measure resistance, the leads are connected across the external resistor to be measured, as shown in Part (b). This connection completes the circuit, allowing the internal battery to produce current through the movement coil, causing a deflection of the pointer proportional to the value of the external resistance being measured.

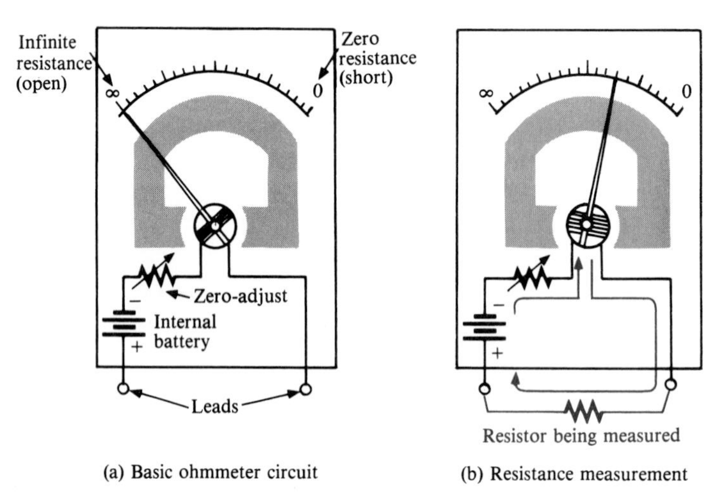

FIGURE 8–10
A basic one-range ohmmeter.

Zero Adjustment

When the ohmmeter leads are open, the pointer is at full left scale, indicating *infinite* (∞) resistance (open circuit). When the leads are *shorted*, the pointer is at full right scale, indicating *zero* resistance.

The purpose of the variable resistor is to adjust the current so that the pointer is at exactly zero when the leads are shorted. It is used to compensate for changes in the internal battery voltage due to aging.

Multiple-Range Ohmmeter

An ohmmeter usually has several ranges. These typically are labeled R×1, R×10, R×100, R×1k, R×10k, R×100k, and R×1M, although some ohmmeters may not have all of the ranges mentioned. These range settings are interpreted differently from those of the ammeter or voltmeter: *The reading on the ohmmeter scale is multiplied by the factor indicated by the range setting.* For example, if the pointer is at 20 on the scale and the range switch is set at R×100, the actual resistance measurement is 20 × 100, or 2 kΩ.

When the meter's leads are open, there is no coil current and the pointer is at infinity. In order to deflect the pointer toward the zero end of the scale when a resistance is measured, coil current must flow, and the smaller the measured resistance, the greater the coil current must be.

Shunt (parallel) resistors are used to provide multiple ranges so that the meter can measure resistance values from very small to very large. For each range, a different value of shunt resistance is switched in. The shunt resistance increases for the higher ohm ranges and is always equal to the *center scale* reading on any range. An example of a multiple-range ohmmeter circuit is shown in Figure 8–11.

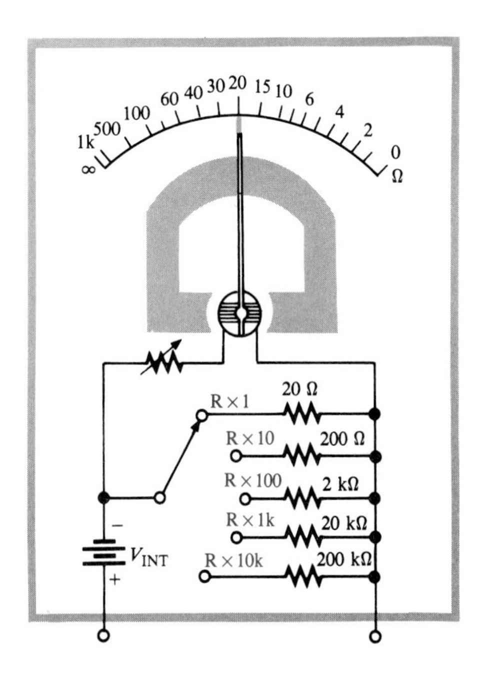

FIGURE 8–11
Example of a multiple-range ohmmeter circuit.

Remember, an ohmmeter must not be connected to a circuit when the circuit's power is on. Always turn the power off before connecting the meter.

SECTION REVIEW 8–2

1. Explain why an ammeter does not significantly alter the current that it is intended to measure when connected in a circuit.
2. Explain why a voltmeter does not significantly alter the voltage that it is intended to measure. What should you be aware of when choosing a voltmeter for a given application?
3. Why does an ohmmeter require an internal battery?
4. What precaution should you take when using an ohmmeter?

READING ANALOG MULTIMETERS

8–3

A typical analog multimeter is shown in Figure 8–12. This particular instrument can be used to measure both direct current (dc) and alternating current (ac) quantities as well as resistance values. It has four selectable functions: dc volts (DC VOLTS), dc milliamperes (DC MA), ac volts (AC VOLTS), and OHMS. Most analog multimeters are similar to this one.

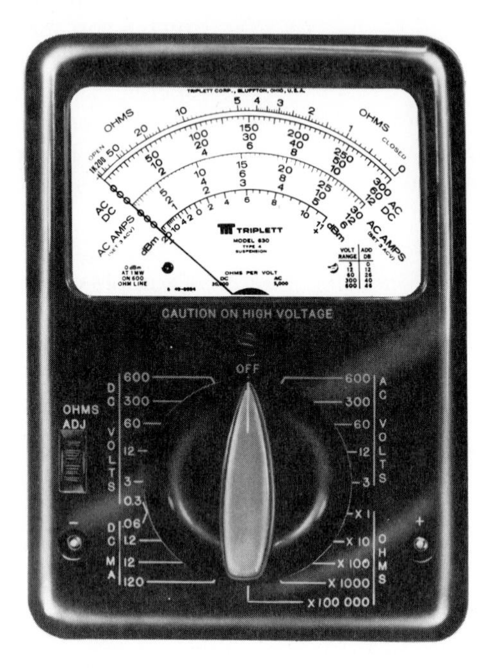

FIGURE 8–12
A typical analog multimeter (courtesy of Triplett Corp.).

Within each function there are several ranges, as indicated by the brackets around the selector switch. For example, the DC VOLTS function has 0.3-V, 3-V, 12-V, 60-V, 300-V, and 600-V ranges. Thus, dc voltages from 0.3 V full-scale to 600 V full-scale can be measured. On the DC MA function, direct currents from 0.06 mA full-scale to 120 mA full-scale can be measured. On the ohm scale, the settings are $\times 1$, $\times 10$, $\times 100$, $\times 1000$, and $\times 100000$.

The Ohm Scale

Ohms are read on the top scale of the meter. This scale is *nonlinear;* that is, the values represented by each division (large or small) vary as you go across the scale. In Figure 8–12, notice how the scale becomes more compressed as you go from right to left.

To read the actual value in ohms, multiply the number on the scale as indicated by the pointer by the factor selected by the switch. For example, when the switch is set at $\times 100$ and the pointer is at 20, the reading is $20 \times 100 = 2000\ \Omega$.

As another example, assume that the switch is at $\times 10$ and the pointer is at the seventh small division between the 1 and 2 marks, indicating 17 Ω (1.7×10). Now, if the meter remains connected to the same resistance and the switch setting is changed to $\times 1$, the pointer will move to the second small division between the 15 and 20 marks. This, of course, is also a 17-Ω reading, illustrating that a given resistance value can often be read at more than one switch setting.

The ac-dc Scales

The second, third, and fourth scales from the top, labeled "AC" and "DC," are used in conjunction with the DC VOLTS, DC MA, and AC VOLTS functions. The upper ac-dc scale ends at the 300 mark and is used with the range settings that are multiples of three, such as 0.3, 3, and 300. For example, when the switch is at 3 on the DC VOLTS function, the 300 scale has a full-scale value of 3 V. At the range setting of 300, the full-scale value is 300 V, and so on.

The middle ac-dc scale ends at 60. This scale is used in conjunction with range settings that are multiples of 6, such as 0.06, 60, and 600. For example, when the switch is at 60 on the DC VOLTS function, the full-scale value is 60 V.

The lower ac-dc scale ends at 12 and is used in conjunction with switch settings that are multiples of 12, such as 1.2, 12, and 120.

The remaining scales are for ac current and for decibels, which are discussed later in the book.

EXAMPLE 8–3 In Figure 8–13, determine the quantity that is being measured and its value.

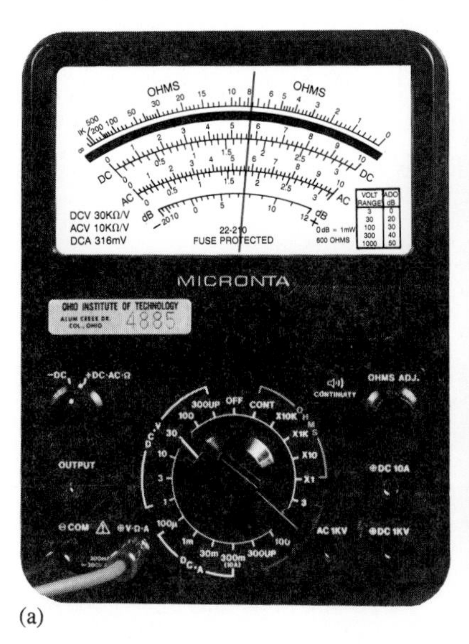

(a)

(b)

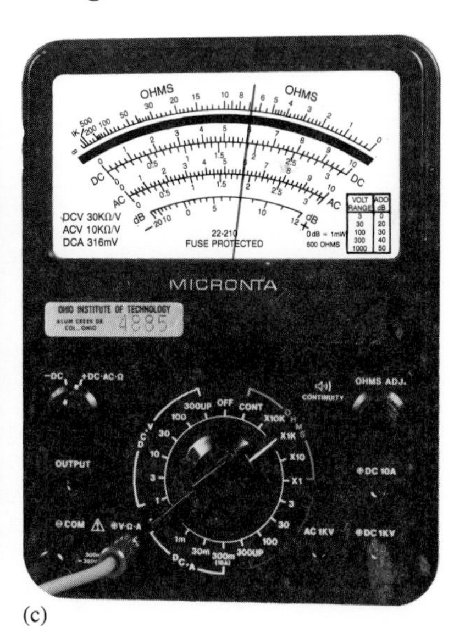

(c)

FIGURE 8–13

Solution

(a) The switch is set on the dc volts (DC-V) function and the 30-V range. The reading taken from the lower dc scale is 18 V.

(b) The switch is set on the dc ampere (DC-A) function and the 100-μA range. The reading taken from the upper dc scale is 42 μA.

(c) The switch is set on the ohm (OHMS) function and the $\times$1K range. The reading taken from the ohm scale (top scale) is 7 kΩ.

SECTION REVIEW 8–3

1. The multimeter in Figure 8–12 is set on the 3-V range to measure dc voltage. The pointer is at 150 on the upper ac-dc scale. What voltage is being measured?
2. How do you set up the meter to measure 275 V dc, and on what scale do you read the voltage?
3. If you expect to measure a resistance in excess of 20 kΩ, where do you set the switch?

DIGITAL MULTIMETERS (DMMs)

8–4

DMMs are perhaps the most widely used type of electronic measuring instrument. Generally, DMMs provide more functions, better accuracy, greater ease of reading, and greater reliability than do many analog meters. Analog meters have at least one advantage over DMMs, however: They can track short-term variations and trends in a measured quantity that many DMMs are too slow to respond to. Several typical DMMs are shown in Figure 8–14.

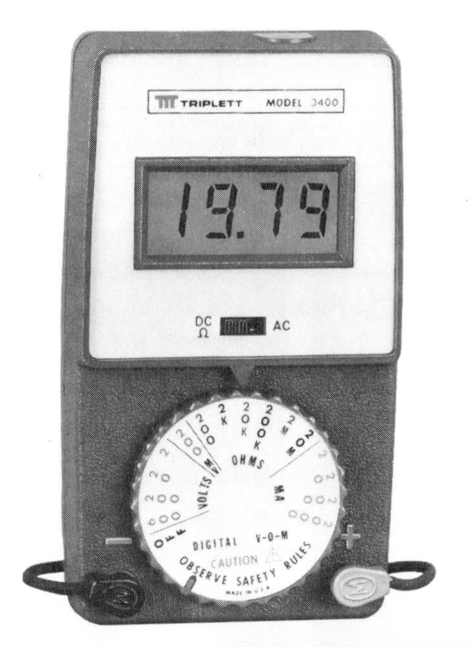

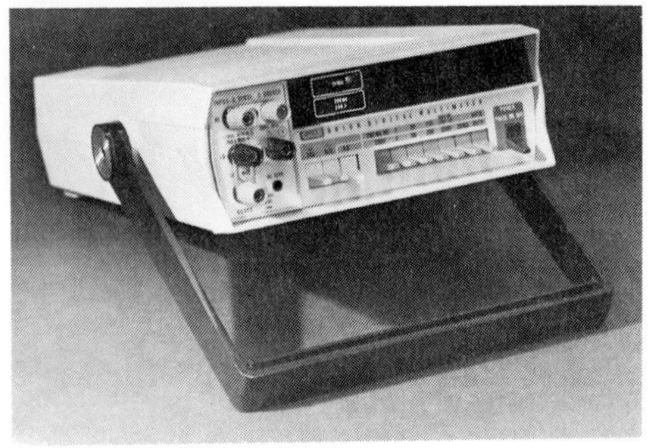

FIGURE 8–14
Typical digital multimeters (DMMs) [(top) courtesy of Triplett Corp. (bottom) courtesy of John Fluke Manufacturing Co.].

DMM Functions

The basic functions found on most DMMs include the following:

- ohms
- dc voltage and current
- ac voltage and current

Some DMMs provide special functions such as transistor or diode tests, power measurement, and decibel measurement for audio amplifier tests.

DMM Displays

DMMs are available with either LCD (liquid-crystal display) or LED (light-emitting diode) readouts. The LCD is the most commonly used readout in battery-powered instruments because it requires only very small amounts of current. A typical battery-powered DMM with an LCD readout operates on a 9-V battery that will last from a few hundred hours to 2000 hours and more. The disadvantages of LCD readouts are that (a) they are difficult or impossible to see in low-light conditions and (b) they are relatively slow to respond to measurement changes. LEDs, on the other hand, can be seen in the dark and respond quickly to changes in measured values. LED displays require much more current than LCDs, and, therefore, battery life is shortened when they are used in portable equipment.

Both LCD and LED DMM displays are in a *seven-segment* format. Each digit in the display consists of seven separate segments, as shown in Figure 8–15(a). Each of the ten decimal digits is formed by activation of appropriate segments, as illustrated in Figure 8–15(b). In addition to the seven segments, there is also a decimal point.

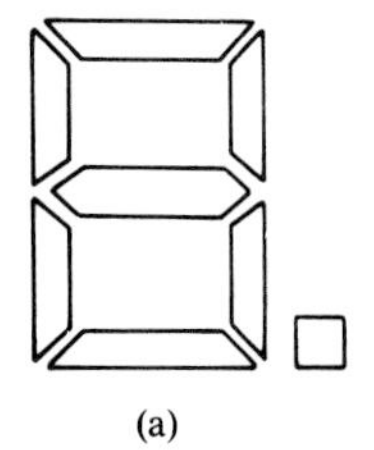

(a)

(b)

FIGURE 8–15
Seven-segment display.

Resolution

The resolution of a meter is the smallest increment of a quantity that the meter can measure. The smaller the increment, the better the resolution. One factor that determines the resolution of a meter is the number of digits in the display.

Because many meters have 3½ digits in their display, we will use this case for illustration. A 3½-digit multimeter has three digit positions that can indicate from 0 through 9, and one digit position that can indicate only a value of 1. This latter digit, called the *half-digit,* is always the most significant digit in the display. For example, suppose that a DMM is reading 0.999 volt, as shown in Figure 8–16(a). If the voltage increases by 0.001 V to 1 V, the display correctly shows 1.000 V, as shown in Part (b). The "1" is the half-digit. Thus, with 3½ digits, a variation of 0.001 V, which is the resolution, can be observed.

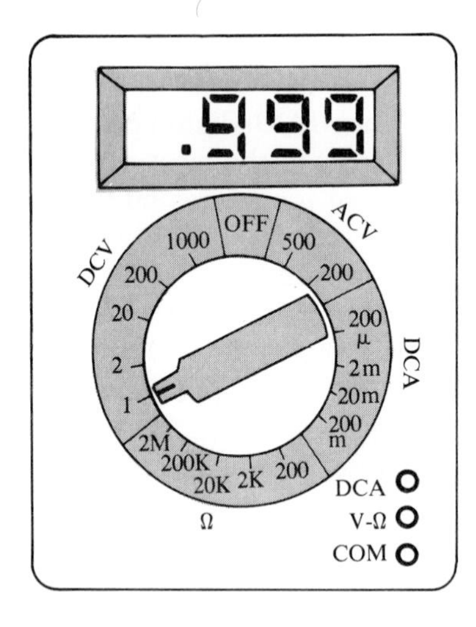

(a) Resolution: 0.001 V

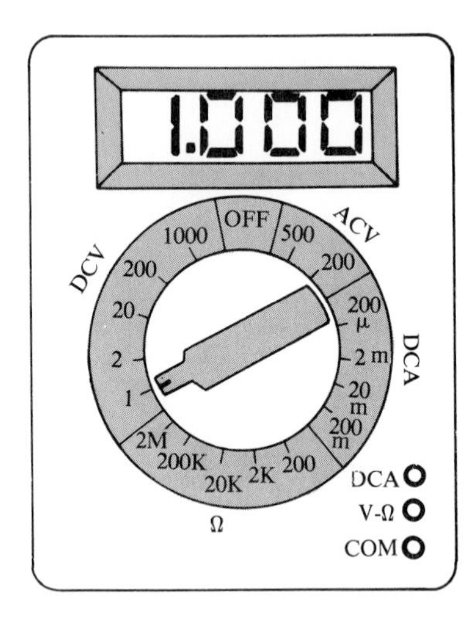

(b) Resolution: 0.001 V

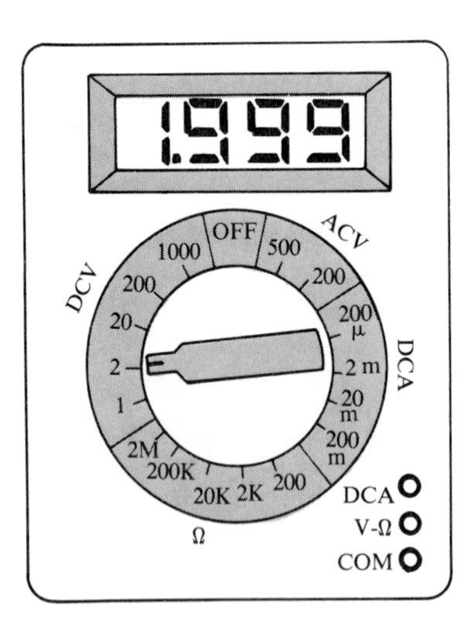

(c) Resolution: 0.001 V

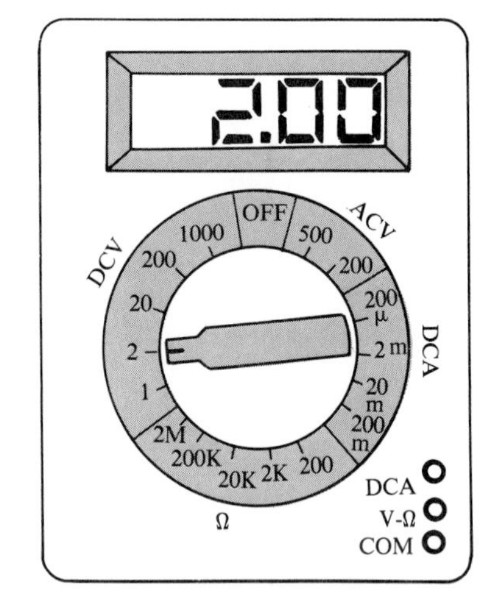

(d) Resolution: 0.01 V

FIGURE 8–16
A 3½-digit DMM illustrates how the resolution changes with the number of digits in use.

Now, suppose that the voltage increases to 1.999 V. This value is indicated on the meter as shown in Part (c) of the figure. If the voltage increases by 0.001 V to 2 V, the half-digit cannot display the "2," so the display shows 2.00. The half-digit is blanked and only three digits are active, as indicated in Part (d). With only three digits active, the resolution is 0.01 V rather than 0.001 V as it is with 3½ active digits. The resolution remains 0.01 V up to 19.99 V. The resolution goes to 0.1 V for readings of 20.0 V to 199.9 V. At 200 V, the resolution goes to 1 V, and so on.

The resolution capability of a DMM is also determined by the internal circuitry and the rate at which the measured quantity is *sampled.* DMMs with displays of 4½ through 8½ digits are also available.

Accuracy

The accuracy of a DMM is established strictly by its internal circuitry. For typical meters, accuracies range from 0.01% to 0.5%, with some precision laboratory-grade meters going to 0.002%.

SECTION REVIEW 8–4

1. List two common types of DMM displays, and discuss the advantages and disadvantages of each.
2. Define *resolution* in a DMM.

THE OSCILLOSCOPE

8–5

The oscilloscope, or *scope* for short, is one of the most widely used and versatile test instruments. It displays on a screen the actual shape of a voltage that is changing with time so that various measurements can be made.

Figure 8–17 shows two typical oscilloscopes. The one in Part (a) is a simpler and less expensive model. The one in Part (b) has better performance characteristics and plug-in modules that provide a variety of specialized functions.

(a)

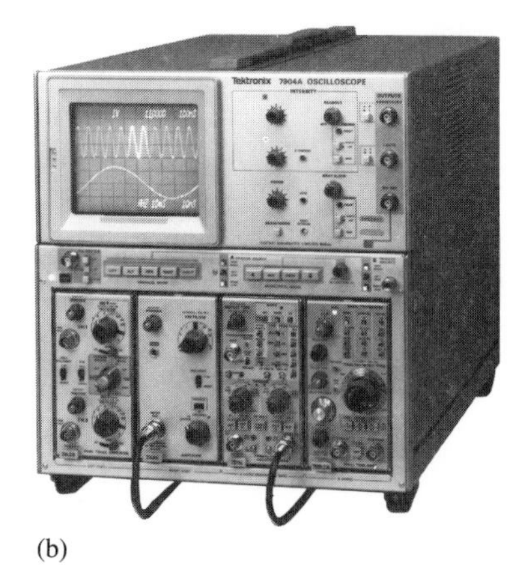

(b)

FIGURE 8–17
Oscilloscopes ((a) courtesy of B&K Precision/ Dynascan Corp. (b) courtesy of Tektronix, Inc.).

Cathode-Ray Tube (CRT)

The oscilloscope is built around the cathode-ray tube (CRT), which is the device that displays the waveforms. The screen of the scope is the front of the CRT.

The CRT is a vacuum tube device containing an *electron gun* that emits a narrow, focused *beam* of electrons. A phosphorescent coating on the face of the tube forms the *screen*. The beam is electronically focused and accelerated so that it strikes the screen, causing light to be emitted at the point of impact.

Figure 8–18 shows the basic construction of a CRT. The *electron gun* assembly contains a *heater,* a *cathode,* a *control grid,* and *accelerating* and *focusing grids*. The heater carries current that indirectly heats the cathode. The heated cathode emits electrons. The amount of voltage on the control grid determines the flow of electrons and thus the *intensity* of the beam. The electrons are accelerated by the accelerating grid and are focused by the focusing grid into a narrow beam that converges at the screen. The beam is further accelerated to a high speed after it leaves the electron gun by a high voltage on the *anode* surfaces of the CRT.

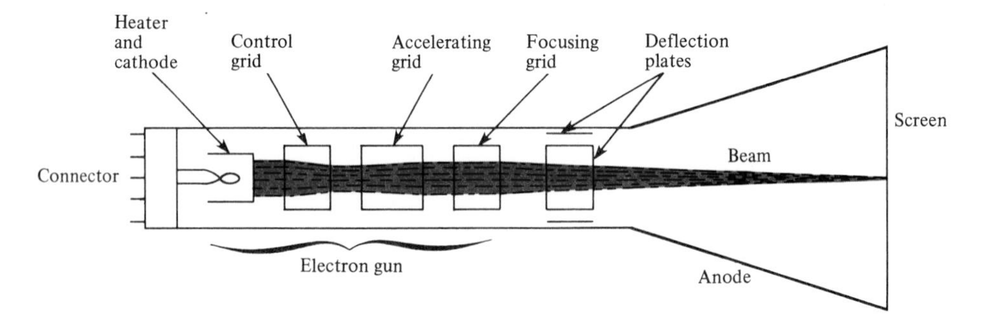

FIGURE 8–18
Basic construction of a CRT.

Deflection of the Beam

The purpose of the *deflection plates* in the CRT is to produce a "bending" or deflection of the electron beam. This deflection allows the position of the point of impact on the screen to be varied. There are two sets of deflection plates: one set for *vertical deflection,* and the other set for *horizontal deflection.*

Figure 8–19 shows a front view of the CRT's deflection plates. One plate from each set normally is grounded as shown. If there is *no voltage* on the other

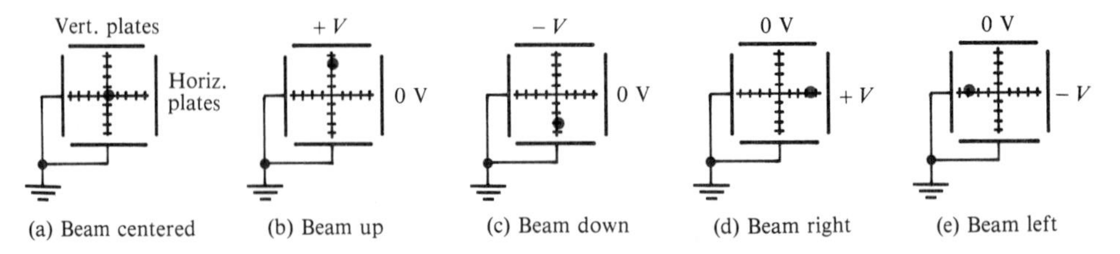

FIGURE 8–19
Deflection of an electron beam in a CRT.

plates, as in Figure 8–19(a), the beam is not deflected and hits the center of the screen. If a *positive* voltage is on the vertical plate, the beam is attracted upward, as indicated in Part (b) of the figure. Remember that opposite charges attract. If a *negative* voltage is applied, the beam is deflected downward because like charges repel, as shown in Part (c).

Likewise, a positive or a negative voltage on the horizontal plate deflects the beam right or left, respectively, as shown in Figure 8–19(d) and (e). The amount of deflection is proportional to the amount of voltage on the plates.

Sweeping the Beam Horizontally

In normal oscilloscope operation, the beam is horizontally deflected from left to right across the screen at a certain rate. This *sweeping action* produces a horizontal line or *trace* across the screen, as shown in Figure 8–20.

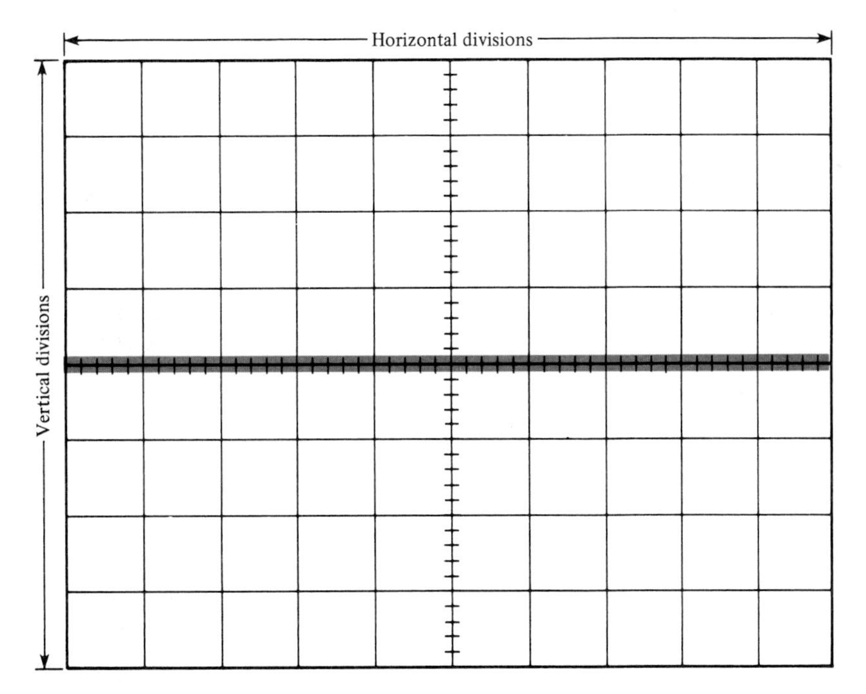

FIGURE 8–20
Scope screen with horizontal trace in color (8 cm × 10 cm).

The *rate* at which the beam is swept across the screen establishes a *time base*. The scope screen is divided into horizontal and vertical divisions, as shown in Figure 8–20. For a given time base, each horizontal division represents a fixed interval of time. For example, if the beam takes 1 second for a full left-to-right sweep, then each division represents 0.1 second. All scopes have provisions for selecting various sweep rates, which will be discussed later.

The actual sweeping of the beam is accomplished by application of a *sawtooth voltage* across the horizontal plates. The basic idea is illustrated in Figure 8–21. When the sawtooth is at its maximum negative peak, the beam is de-

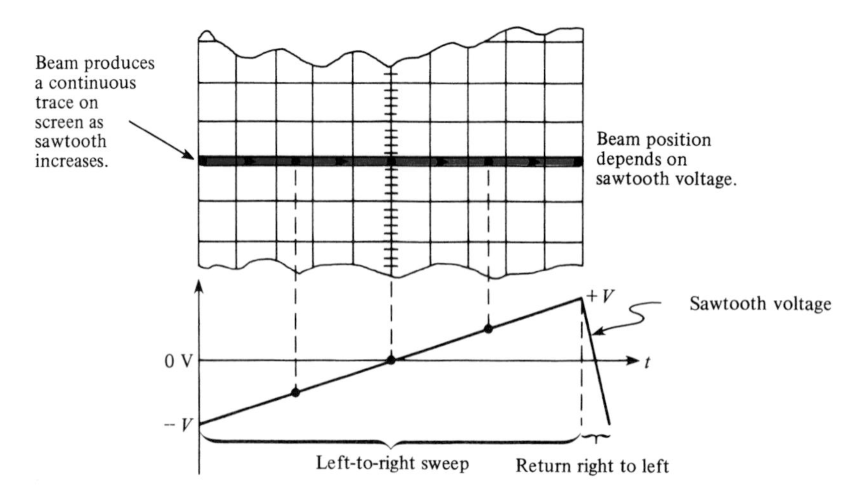

FIGURE 8–21
Sweeping the beam across the screen with a sawtooth voltage.

flected to its left-most screen position. This deflection is due to maximum repulsion from the right deflection plate.

As the sawtooth voltage *increases,* the beam moves toward the center of the screen. When the sawtooth voltage is *zero,* the beam is at the center of the screen, because there is no repulsion or attraction from the plate. As the voltage increases *positively,* the plate attracts the beam, causing it to move toward the right side of the screen. At the positive peak of the sawtooth, the beam is at its right-most screen position.

The rate at which the sawtooth goes from negative to positive is determined by its frequency. This rate in turn establishes the *sweep rate* of the beam. When the sawtooth makes the abrupt change from positive back to negative, the beam is rapidly returned to the left side of the screen, ready for another sweep. During this "flyback" time, the beam is *blanked* out and thus does not produce a trace on the screen.

How a Voltage Pattern Is Produced

The main purpose of the scope is to display the shape (waveform) of a voltage under test. To do so, we apply the voltage under test across the *vertical plates* through a vertical amplifier circuit. As you have seen, a voltage across the vertical plates causes a vertical deflection of the beam. A negative voltage causes the beam to go below the center of the screen, and a positive voltage makes it go above center.

Assume, for example, that a varying dc voltage (such as the output of a dc generator) is applied across the vertical plates. As a result, the beam will move up and down on the screen. The amount that the beam goes above center depends on the maximum value of the voltage.

At the same time that the beam is being deflected vertically, it is also sweeping horizontally, causing the voltage variation to be traced out across the screen, as illustrated in Figure 8–22. All scopes provide for the calibrated adjustment of the vertical deflection, so each vertical division represents a known amount of voltage.

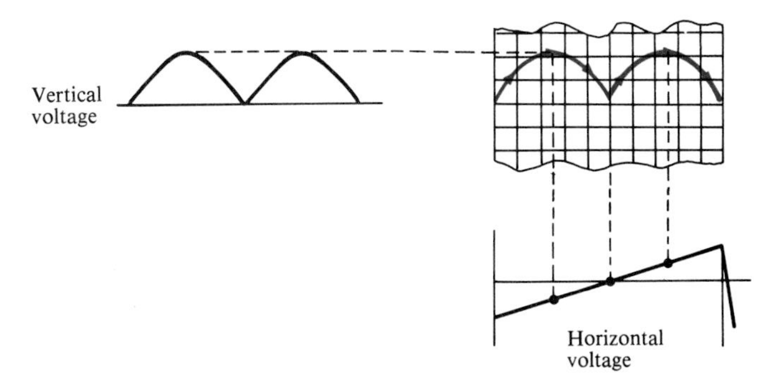

FIGURE 8–22
Vertical and horizontal deflection combined to produce a pattern on the screen.

Oscilloscope Controls

A wide variety of oscilloscopes are available, ranging from relatively simple instruments with limited capabilities to much more sophisticated models that provide a variety of optional functions and precision measurements. Regardless of their complexity, however, all scopes have certain operational features in common. In this section, we examine the most common front panel controls. Each control and its basic function are described. Figure 8–23 shows a representative front panel of an oscilloscope.

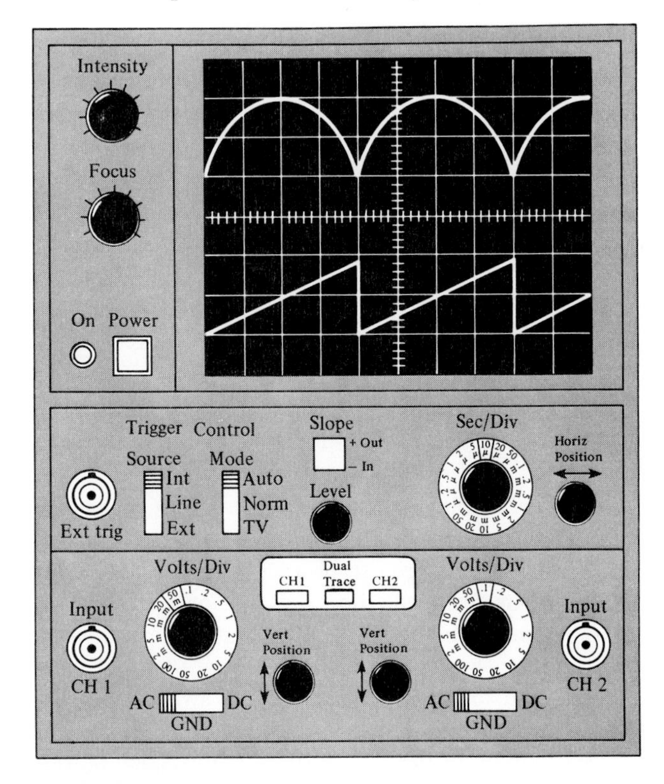

FIGURE 8–23
Representative front panel of a dual-trace oscilloscope.

Screen

In the upper portion of Figure 8–23 is the CRT screen. There are 8 vertical divisions and 10 horizontal divisions indicated with grid lines or graticules. A standard screen size is 8 cm × 10 cm. The screen is coated with phosphor that emits light when struck by the electron beam.

Power Switch and Light

This switch turns the power to the scope on and off. The light indicates when the power is on.

Intensity

The intensity control knob varies the brightness of the trace on the screen. Caution should be used so that the intensity is not left too high for an extended period of time, especially when the beam forms a motionless dot on the screen. Damage to the screen can result from excessive intensity.

Focus

This control focuses the beam so that it converges to a tiny point at the screen. An out-of-focus condition results in a fuzzy trace.

Horizontal Position

This control knob adjusts the neutral horizontal position of the beam. It is used to reposition horizontally a waveform display for more convenient viewing or measurement.

Seconds/Division

This selector switch sets the horizontal sweep rate. It is the *time base control.* The switch selects the time interval that is to be represented by each horizontal division in seconds, milliseconds, or microseconds. The setting in Figure 8–23 is at 10 μs. Thus, each of the ten horizontal divisions represents 10 μs, so there are 100 μs from the extreme left of the screen to the extreme right.

Trigger Control

The trigger controls allow the beam to be triggered from various selected sources. The triggering of the beam causes it to begin its sweep across the screen. It can be triggered from an internally generated signal derived from an input signal, or from the line voltage, or from an externally applied trigger signal. The *modes* of triggering are auto, normal, and TV. In the auto mode, sweep occurs in the absence of an adequate trigger signal. In the normal mode, a trigger signal must be present for the sweep to occur. The TV mode provides triggering on the TV field or TV line signals. The *slope* switch allows the triggering to occur on either the positive-going slope or the negative-going slope of the trigger waveform. The *level* control selects the voltage level on the trigger signal at which the triggering occurs.

Basically, the trigger controls provide for synchronization of the horizontal sweep waveform and the input signal waveform. As a result, the display of the input signal is stable on the screen, rather than appearing to drift across the screen.

Volts/Division

The example scope in Figure 8–23 is a *dual-trace* type, which allows two waveforms to be displayed simultaneously. Many scopes have only single-trace capability. Notice that there are two identical volts/div selectors. There is a set of controls for each of the two *input channels*. Only one is described here, but the same applies to the other as well.

The volts/div selector switch sets the number of volts to be represented by each division on the *vertical* scale. For example, the upper waveform is applied to channel 1 and covers two vertical divisions from maximum to minimum. The volts/div switch for channel 1 is set at 50 mV, which means that each vertical division represents 50 mV. Therefore, the peak-to-peak value of the sine wave is (2 div)(50 mV/div) = 100 mV. If a lower setting were selected, the displayed wave would cover more vertical divisions. If a higher setting were selected, the displayed wave would cover fewer vertical divisions.

Notice that there is a set of three switches for selecting channel 1 (CH 1), channel 2 (CH 2), or dual trace. Either input signal can be displayed separately, or both can be displayed as illustrated.

Vertical Position

The two vertical position controls move the traces up or down for easier measurement or observation.

AC-GND-DC Switch

This switch, located below the volts/div control, allows the input signal to be ac coupled, dc coupled, or grounded. The ac coupling eliminates any dc component on the input signal. The dc coupling permits dc values to be displayed. The ground position allows a 0-V reference to be established on the screen.

Input

The signals to be displayed are connected into the channel 1 and channel 2 input connectors. This connection is normally done with a special *probe* that minimizes the loading effect of the scope's input resistance on the circuit being measured.

SECTION REVIEW 8–5

1. What does *CRT* stand for?
2. On an oscilloscope, voltage is measured (horizontally, vertically) on the screen and time is measured (horizontally, vertically).
3. What can an oscilloscope do that a multimeter cannot?

APPLICATION NOTE

Each of the meters in the Application Assignment at the beginning of the chapter is set up as follows:

- Meter for TP1: Function—DCV, range—20
- Meter for TP2: Function—ACV, range—500
- Meter for TP3: Function—DCV, range—200
- Meter for TP4: Function—DCA, range—20 m
- Meter for TP5: Function—DCA, range—2 m

For the meters set on the DCV or ACV functions, the lead to the test point must be connected into the V-Ω plug on the meter. For the meters set on the DCA function, the lead to the test point must be connected into the DCA plug on the meter. The common (COM) lead must be connected to the ground point on the system under test.

SUMMARY

Facts

- The d'Arsonval meter movement consists of a moving coil placed within a permanent magnetic field.
- Current through the coil of a d'Arsonval movement causes the coil to rotate an amount proportional to the amount of current.
- The iron-vane movement consists of two iron bars placed within an electromagnet.
- The electrodynamometer movement consists of a moving coil placed within an electromagnetic field.
- An ammeter shunt is a parallel resistor that diverts a part of the total current around the coil in an analog ammeter.
- A multiple-range analog ammeter has a shunt resistor for each range.
- A multiplier resistor is in series with the coil in an analog voltmeter and drops a part of the total voltage.
- Multiple-range analog voltmeters have a multiplier resistor for each range.
- An ohmmeter uses an internal battery to produce current through a resistance being measured.
- The half-digit in a DMM can display only a 1.
- Resolution of a DMM depends, in part, on the number of display digits.
- An oscilloscope can display the shape of a varying voltage, allowing both voltage and time measurements.

Definitions

- *Current sensitivity* of a meter—the amount of current required to deflect the pointer full scale.
- *Movement resistance*—the resistance of the coil.
- *Resolution* of a meter—the smallest increment of a quantity that the meter can measure.
- *Cathode-ray tube (CRT)*—the device within the oscilloscope that displays the waveforms.

SELF-TEST

1. Which type of meter movement, d'Arsonval or electrodynamometer, is applicable to both dc and ac measurements?

2. What is the purpose of a shunt resistor in an ammeter?
3. What is the purpose of a multiplier resistor in a voltmeter?
4. What is the purpose of the internal battery in an ohmmeter?
5. What does the term "4½ digits" mean when applied to DMMs?
6. What type of DMM display is most appropriate for low-light conditions?
7. In a CRT, what is used to trace a pattern across the screen?

PROBLEM SET A

Section 8–1

8–1 If there are 25 μA through a 50-μA meter movement, how much will the pointer be deflected?

8–2 A certain milliammeter has a 10-mA sensitivity. At what scale position is the pointer with 10 mA through the movement? With 5 mA through the movement?

8–3 How much current is required to produce a full-scale deflection in a meter with a sensitivity of 100 μA?

Section 8–2

8–4 A 50-μA, 1000-Ω movement is used in an ammeter. Determine the full-scale shunt current (I_{SH}) on each of the following ranges:
(a) 100 μA **(b)** 1 mA **(c)** 10 mA **(d)** 100 mA **(e)** 1 A

8–5 Repeat Problem 8–4 for a 1-mA, 50-Ω movement.

8–6 Calculate the shunt resistor value (R_{SH}) for each range in Problem 8–4.

8–7 A voltmeter has a sensitivity of 20,000 Ω/V. What is its internal resistance on the 100-V range?

8–8 What multiplier resistance is used in the meter in Problem 8–7 on the 100-V range? The resistance of the movement is 1 kΩ.

8–9 An ohmmeter uses a 50-μA, 1000-Ω movement. It has a 3-V internal battery. What is the value of resistance that it is measuring when the pointer is at center scale, assuming no internal shunt?

Section 8–3

8–10 What are the voltage readings in Figures 8–24(a) and 8–24(b)?

8–11 How much resistance is the ohmmeter in Figure 8–25 measuring?

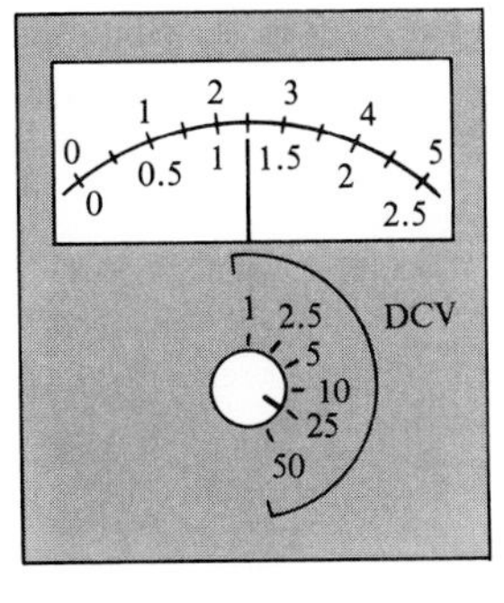

(a)

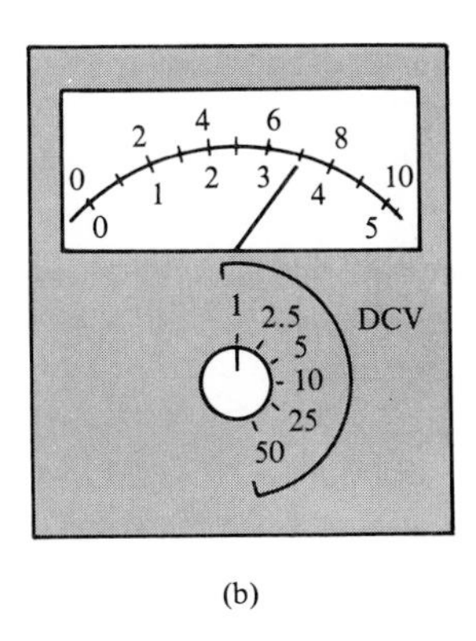

(b)

FIGURE 8–24

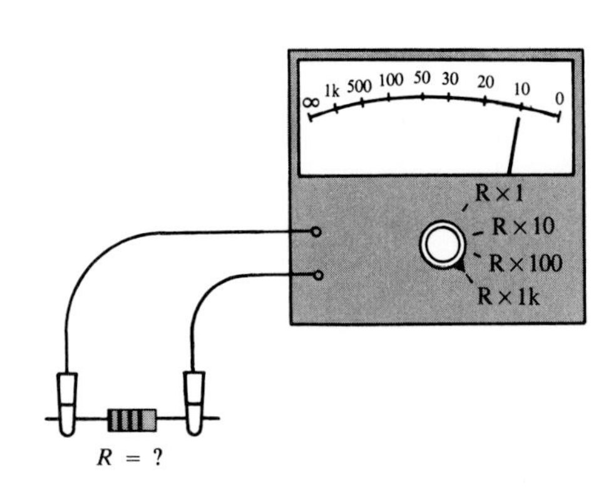

FIGURE 8–25

8–12 Determine the resistance indicated by each of the following ohmmeter readings and range settings:
(a) pointer at 2, range setting at R×100.
(b) pointer at 15, range setting at R×10M.
(c) pointer at 45, range setting at R×100.

8–13 A multimeter has the following ranges: 1 mA, 10 mA, 100 mA; 100 mV, 1 V, 10 V; R×1, R×10, R×100. Indicate schematically how you would connect the multimeter in Figure 8–26 for the following:
(a) I_1
(b) V_1
(c) R_1
In each case indicate the *function* on which you would set the meter and the *range* that you would use.

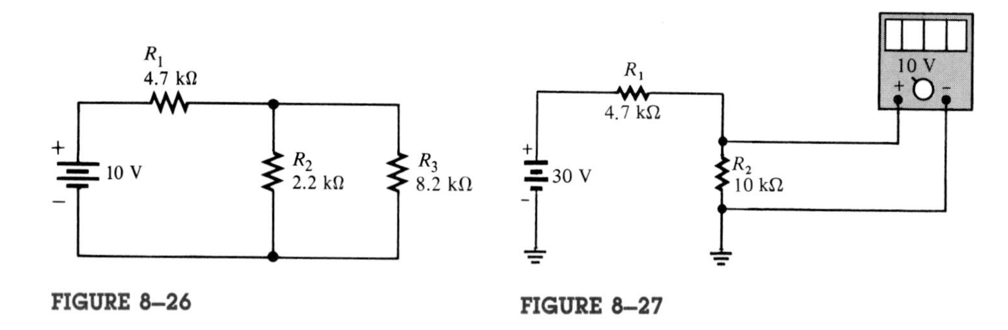

FIGURE 8–26 **FIGURE 8–27**

Section 8–4

8–14 What is the maximum resolution of a 4½-digit DMM?

8–15 What voltage reading would you expect on the 3½-digit multimeter in Figure 8–27?

Section 8–5

8–16 Determine from the oscilloscope in Figure 8–23 the number of times that the waveforms repeat each second.

8–17 What is the voltage from minimum to maximum of the sawtooth-shaped pattern (CH 2) in Figure 8–23?

PROBLEM SET B

8–18 The voltmeter in Figure 8–28 has a sensitivity of 20,000 Ω/V. The range switch is set to the 1-V position in order to measure the voltage across R_2. What voltage is indicated by the meter? What is the actual voltage across R_2 with the voltmeter disconnected? Explain the difference.

FIGURE 8–28

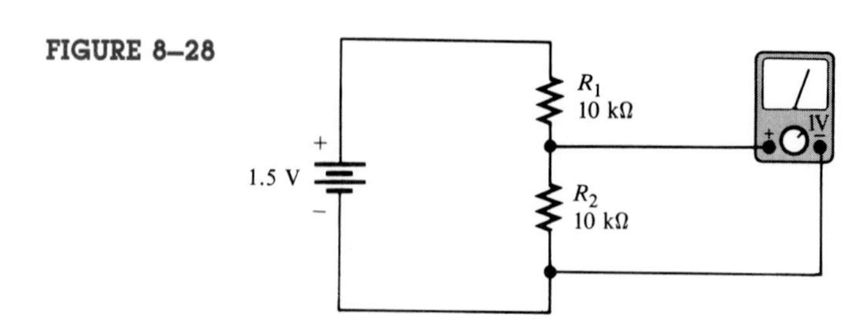

8–19 Are the voltage readings in Figure 8–29 correct? If not, what has malfunctioned in the circuit?

8–20 A 20,000-Ω/V voltmeter is set on the 10-V range scale. Determine the percent errors for the reading with respect to ground taken at each test point in Figure 8–30 due to the loading effect of the voltmeter.

8–21 Using only a voltmeter, show how you would determine by measurement the current, power, and voltage for each resistor on the PC board in Figure 8–29.

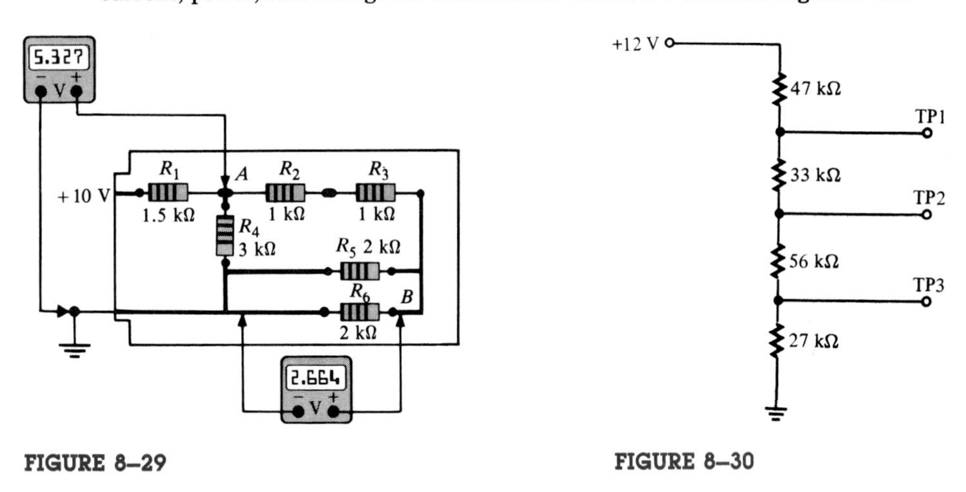

FIGURE 8–29

FIGURE 8–30

ANSWERS TO SECTION REVIEWS

Section 8–1

1. Current sensitivity is the amount of current that deflects the pointer full-scale.
2. The d'Arsonval movement uses a permanent magnet, and the electrodynamometer uses an electromagnet.

Section 8–2

1. Because the internal resistance is very low compared to the external circuit resistance.
2. Because the internal resistance is very high compared to the external circuit resistance. You should be sure that the internal resistance of the voltmeter is sufficiently large so as not to load the circuit.
3. To produce current through the resistance being measured.
4. Never connect it to an external voltage.

Section 8–3

1. 1.5 V. **2.** Set the range switch to 300 and read on the upper ac-dc scale.
3. ×1000

Section 8–4

1. The LCD requires little current, but it is difficult to see in low light and is slow to respond.
   The LED can be seen in the dark, and it responds quickly. However, it requires much more current than does the LCD.
2. The smallest increment of a quantity that the meter can measure.

Section 8–5

1. Cathode ray tube. **2.** Vertically, horizontally.
3. It can display time-varying quantities.

PART TWO AC CIRCUITS

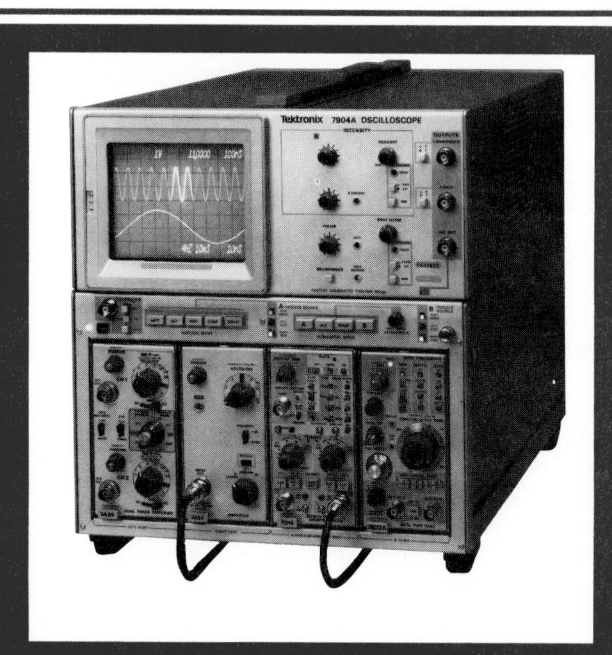

Introduction to Alternating Current and Voltage

This chapter serves as an introduction to alternating current (ac) circuits. Alternating voltages and currents fluctuate with time and periodically change polarity and direction according to certain patterns called *waveforms.* Particular emphasis is given to the sine wave because of its basic importance in ac circuits. Other types of waveforms are also introduced, including pulse, triangular, and sawtooth.

The concept of phasors is introduced to aid your understanding of ac circuits. Phasors provide a concise format for representing sine wave voltages and currents and their relationships with each other.

Specifically, you will learn:

- How to identify sine waves and measure their characteristics.
- How the frequency and period of a sine wave are related.
- The various ways to measure the following voltage or current values of a sine wave: instantaneous, peak, peak-to-peak, rms, and average.
- The fundamentals of ac generators and other sine wave voltage sources.
- How to measure points on a sine wave in terms of angular units.
- How to determine the phase angle between two sine waves.
- The meaning of *phase lead* and *phase lag.*
- How to express sine waves with a mathematical formula.
- What a phasor is and how it can be used to represent sine waves.
- How Ohm's law and Kirchhoff's laws apply to ac as well as dc circuits.
- The characteristics of pulse, triangular, and sawtooth waveforms.
- What harmonics are and how nonsinusoidal waveforms are made up of various harmonic components.

APPLICATION ASSIGNMENT

As an electronic technician working for a company that designs and manufactures laboratory instruments, you have been assigned to a team that is responsible for the design and testing of a new signal generator that produces various types of time-varying voltages with adjustable frequency and amplitude ranges. The project engineer has assigned you to test the prototype model for the proper output voltages.

You must measure the frequency, period, and amplitude for all output waveforms to assure compliance with the design specifications. In addition, you must measure the rise time, fall time, and duty cycle for the specified ranges of adjustment for the pulse output. The specifications are as follows:

1. *Sine wave output* Frequency adjustable from 30 Hz to 30 kHz; amplitude adjustable from 0.5 V peak to 25 V peak.
2. *Pulse output* Frequency adjustable from 1 kHz to 50 kHz; amplitude adjustable from 0.1 V to 15 V; duty cycle adjustable from 10% to 75%; rise and fall times to be no greater than 0.1 μs.
3. *Sawtooth output* Frequency adjustable from 100 Hz to 10 kHz; amplitude adjustable from 1 V to 30 V.

The Application Note at the end of the chapter gives one solution to this assignment.

THE SINE WAVE

9–1

The sine wave is one very common type of alternating current (ac) and alternating voltage. It is also referred to as a *sinusoidal wave* or, simply, *sinusoid.* The electrical service provided by the power companies is in the form of sinusoidal voltage and current. In addition, other types of waveforms are composites of many individual sine waves called *harmonics,* as you will see later.

Figure 9–1 shows the general shape of a sine wave, which can be either current or voltage. Notice how the voltage (or current) varies with time. Starting at zero, it *increases* to a positive maximum (peak), returns to zero, and then increases to a *negative* maximum (peak) before returning again to zero.

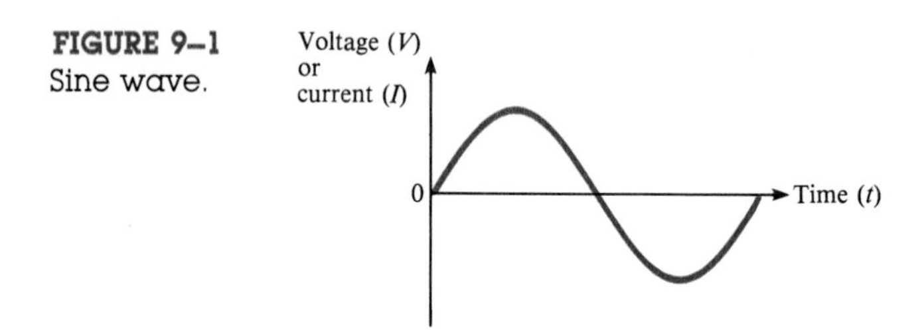

FIGURE 9–1
Sine wave.

The Polarity of a Sine Wave

As you have seen, a sine wave changes *polarity* at its zero value; that is, it *alternates* between positive and negative values. When a sine wave voltage is applied to a resistive circuit, as in Figure 9–2, an alternating sine wave current

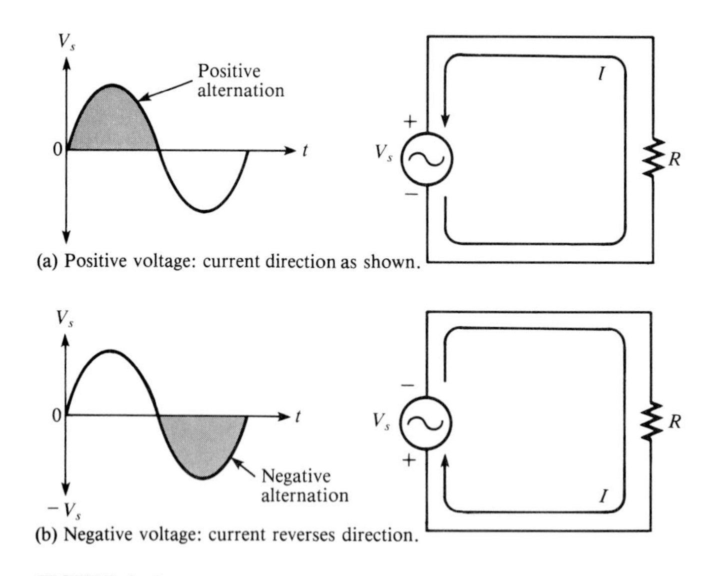

FIGURE 9–2
Alternating current and voltage. Notice the symbol for a sine wave voltage source in the circuit diagrams.

results. When the voltage changes polarity, the current correspondingly changes direction as indicated.

During the *positive alternation* of the applied voltage V_s, the current is in the direction shown in Figure 9–2(a). During a *negative alternation* of the applied voltage, the current is in the opposite direction, as shown in Part (b). The combined positive and negative alternations make up one *cycle* of a sine wave.

The Period of a Sine Wave

As you have seen, a sine wave varies with time in a definable manner. Time is designated by *t*. The time required for a sine wave to complete one full cycle is called the *period (T)*, as indicated in Figure 9–3(a). Typically, a sine wave continues to repeat itself in identical cycles, as shown in Part (b). Since all cycles of a repetitive sine wave are the same, the period is always a fixed value for a given sine wave.

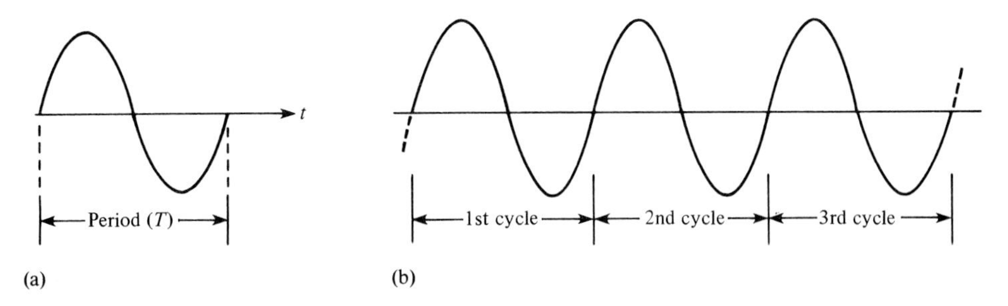

FIGURE 9–3
The period of a sine wave is the same for each cycle.

The period of a sine wave does not necessarily have to be measured between the zero crossings at the beginning and end of a cycle. It can be measured from any point in a given cycle to the *corresponding* point in the next cycle.

EXAMPLE 9–1 What is the period of the sine wave in Figure 9–4?

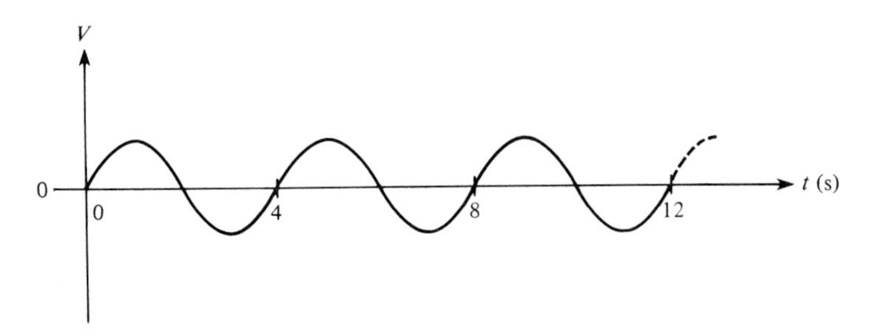

FIGURE 9–4

Solution

As you can see, it takes four seconds (4 s) to complete each cycle. Therefore, the period is 4 s.

$$T = 4 \text{ s}$$

EXAMPLE 9–2 Show three possible ways to measure the period of the sine wave in Figure 9–5. How many cycles are shown?

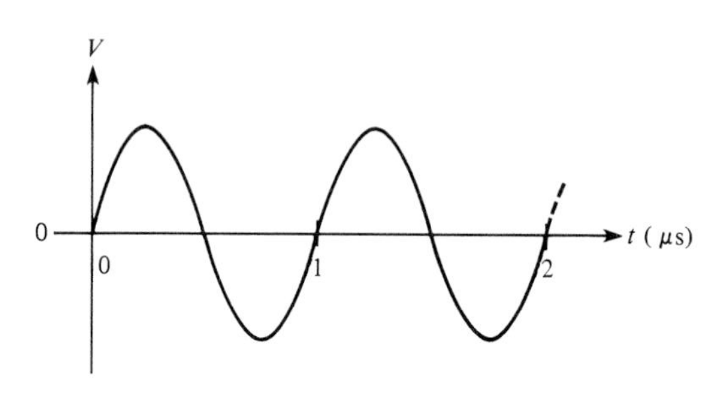

FIGURE 9–5

Solution

Method 1: The period can be measured from one zero crossing to the corresponding zero crossing in the next cycle.

Method 2: The period can be measured from the positive peak in one cycle to the positive peak in the next cycle.

Method 3: The period can be measured from the negative peak in one cycle to the negative peak in the next cycle.

These measurements are indicated in Figure 9–6, where two cycles of the sine wave are shown. Keep in mind that you obtain the same value for the period no matter which of these points on the waveform you use.

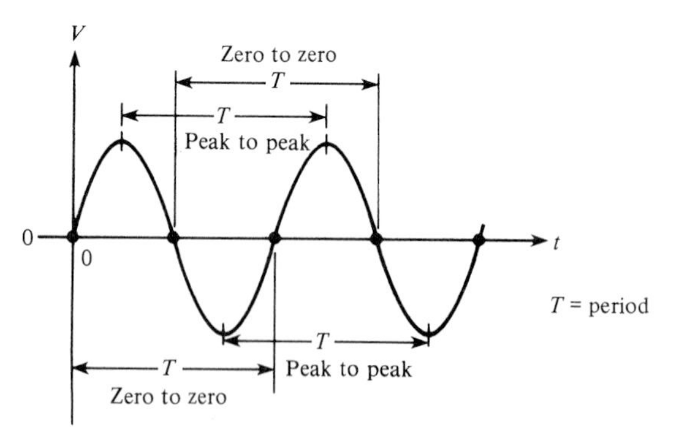

FIGURE 9–6
Measurement of the period.

The Frequency of a Sine Wave

Frequency is the number of cycles that a sine wave completes in one second.

The more cycles completed in one second, the higher the frequency.

Figure 9–7 shows two sine waves. The sine wave in Part (a) completes two full cycles in one second. The one in Part (b) completes four cycles in one second. Therefore, the sine wave in Part (b) has twice the frequency of the one in Part (a).

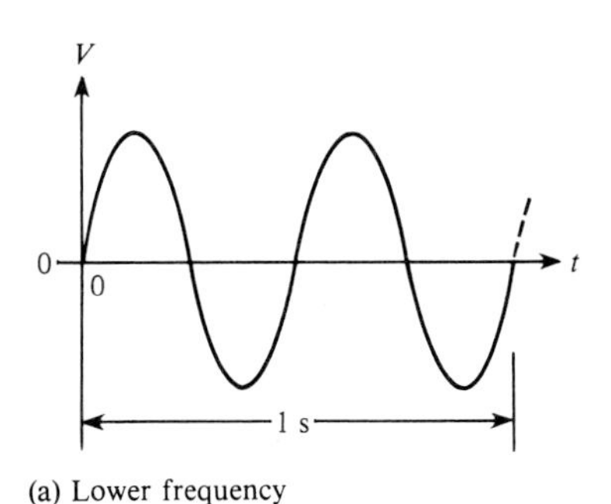

(a) Lower frequency

(b) Higher frequency

FIGURE 9–7
Illustration of frequency.

The Unit of Frequency

Frequency is measured in units of *hertz,* abbreviated Hz. One hertz is equivalent to one cycle per second; 60 Hz is 60 cycles per second; and so on. The symbol for frequency is f.

Relationship of Frequency and Period

The relationship between frequency and period is very important. The formulas for this relationship are as follows:

$$f = \frac{1}{T} \qquad (9\text{–}1)$$

$$T = \frac{1}{f} \qquad (9\text{–}2)$$

There is a *reciprocal* relationship between f and T. Knowing one, you can calculate the other with the 1/x key on your calculator. This relationship is logical because a sine wave with a longer period goes through fewer cycles in one second than one with a shorter period.

EXAMPLE 9–3 Which sine wave in Figure 9–8 has the higher frequency? Determine the period and the frequency of both waveforms.

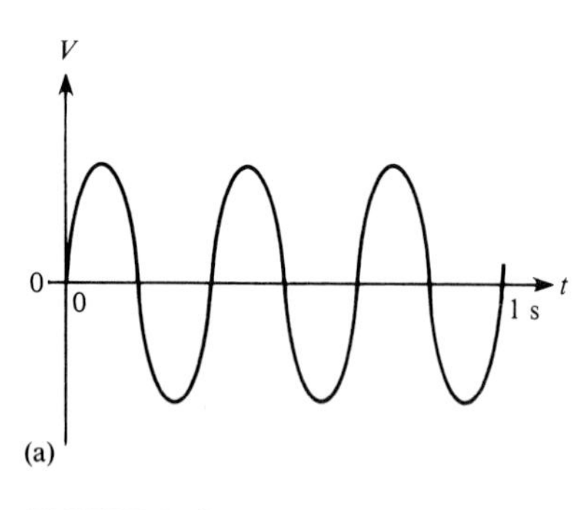

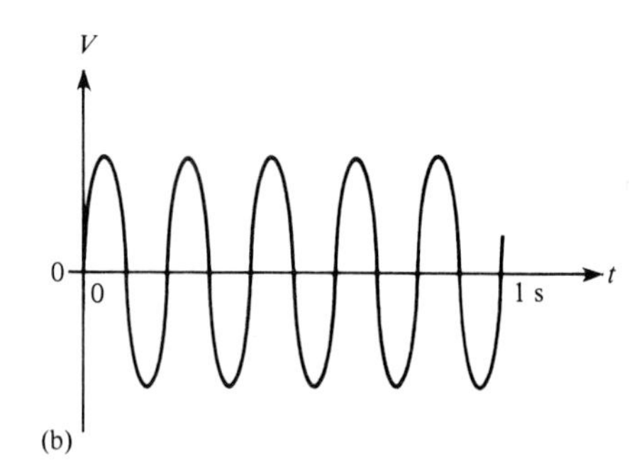

FIGURE 9–8

Solution
In Part (a) of the figure, three cycles take 1 s. Therefore, one cycle takes 0.333 s (one-third second), and this is the period.

$$T = 0.333 \text{ s}$$

$$f = \frac{1}{T} = \frac{1}{0.333 \text{ s}} = 3 \text{ Hz}$$

In Part (b) of the figure, five cycles take 1 s. Therefore, one cycle takes 0.2 s (one-fifth second), and this is the period.

$$T = 0.2 \text{ s}$$

$$f = \frac{1}{T} = \frac{1}{0.2 \text{ s}} = 5 \text{ Hz}$$

EXAMPLE 9–4

The period of a certain sine wave is 10 milliseconds (10 ms). What is the frequency?

Solution
Using Equation (9–1), we obtain the following result:

$$f = \frac{1}{T} = \frac{1}{10 \text{ ms}} = \frac{1}{10 \times 10^{-3} \text{ s}}$$
$$= 0.1 \times 10^{3} \text{ Hz} = 100 \text{ Hz}$$

EXAMPLE 9–5

The frequency of a sine wave is 60 Hz. What is the period?

Solution
Using Equation (9–2), we determine the period as follows:

$$T = \frac{1}{f} = \frac{1}{60 \text{ Hz}} = 0.01667 \text{ s} = 16.67 \text{ ms}$$

SECTION REVIEW 9–1

1. Describe one cycle of a sine wave.
2. At what point does a sine wave change polarity?
3. How many maximum points does a sine wave have during one cycle?
4. How is the period of a sine wave measured?
5. Define *frequency*, and state its unit.
6. Determine f when $T = 5\ \mu s$.
7. Determine T when $f = 120$ Hz.

VOLTAGE AND CURRENT VALUES OF A SINE WAVE

9–2

There are several ways to measure the value of a sine wave in terms of its voltage or its current magnitude. These are *instantaneous, peak, peak-to-peak, rms, and average* values.

Instantaneous Value

Figure 9–9 illustrates that at any point in time on a sine wave, the voltage (or current) has an *instantaneous* value. This instantaneous value is different at different points along the curve. Instantaneous values are positive during the positive alternation and negative during the negative alternation. For example, in Figure 9–9, the instantaneous voltage is 3.1 V at 1 μs, 7.07 V at 2.5 μs, 10 V at 5 μs, 0 V at 10 μs, -3.1 V at 11 μs, and so on. Instantaneous values of voltage and current are symbolized by lower-case v and i, respectively.

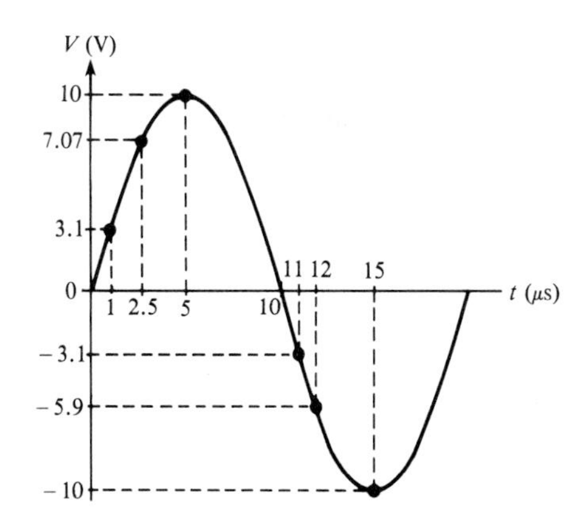

FIGURE 9–9
Example of instantaneous values of a sine wave voltage.

Peak Value

The peak value of a sine wave is the value of voltage (or current) at the positive or the negative maximum (peaks) with respect to zero. Since the peaks are equal in magnitude, a sine wave is characterized by a single peak value, as is illustrated in Figure 9–10. For a given sine wave, the peak value is constant and is represented by V_p or I_p. In this case, the peak value is 8 V.

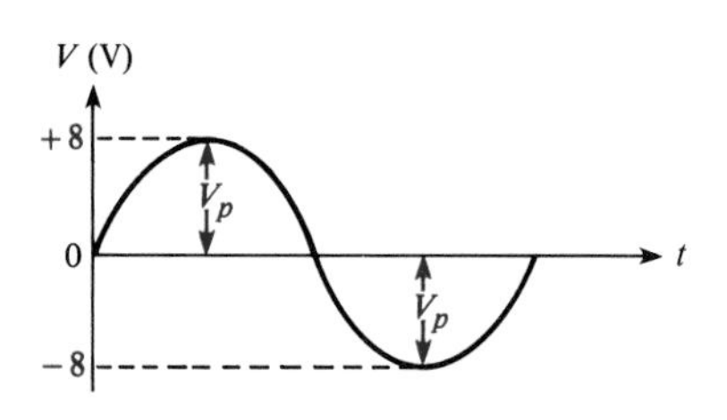

FIGURE 9–10
Example of the peak value of a sine wave voltage.

Peak-to-Peak Value

The peak-to-peak value of a sine wave, as illustrated in Figure 9–11, is the voltage (or current) from the positive peak to the negative peak. Of course, it is always *twice the peak value* as expressed in the following equations:

$$V_{pp} = 2V_p \tag{9–3}$$

$$I_{pp} = 2I_p \tag{9–4}$$

where peak-to-peak values are represented by the symbols V_{pp} or I_{pp}. In Figure 9–11, the peak-to-peak value is 16 V.

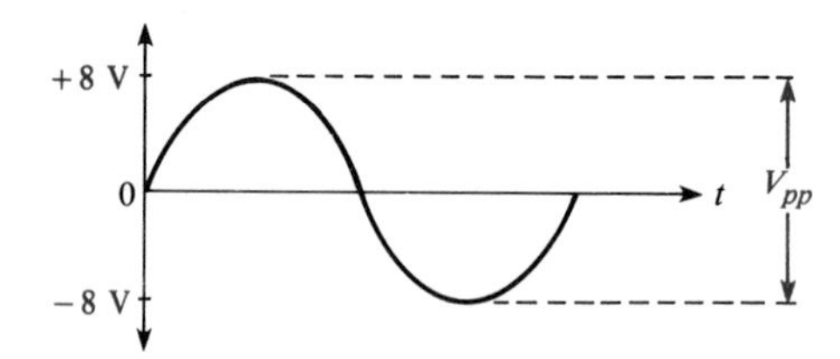

FIGURE 9–11
Example of the peak-to-peak value of a sine wave voltage.

rms Value

The term *rms* stands for *root mean square.* It refers to the mathematical process by which this value is derived. The rms value is also referred to as the *effective value.* Most ac voltmeters display rms voltage. The 120 volts at your wall outlet is an rms value.

The rms value of a sine wave is actually a measure of the *heating effect* of the sine wave. For example, when a resistor is connected across an ac (sine wave) voltage source, as shown in Figure 9–12(a), a certain amount of heat is generated by the power in the resistor. Part (b) shows the *same* resistor connected across a dc voltage source. The value of the ac voltage can be adjusted so that the resistor gives off the same amount of heat as it does when connected to the dc source.

The rms value of a sine wave is equal to the dc voltage that produces the same amount of heat as the sinusoidal voltage.

The peak value of a sine wave can be converted to the corresponding rms value using the following relationships for either voltage or current:

$$V_{rms} = \sqrt{0.5}\,V_p \cong 0.707V_p \tag{9–5}$$

$$I_{rms} = \sqrt{0.5}\,I_p \cong 0.707I_p \tag{9–6}$$

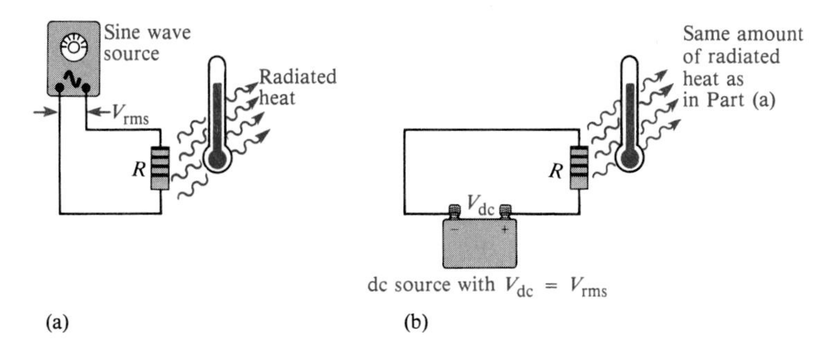

FIGURE 9–12
When the same amount of heat is being produced in both cases, the sine wave has an rms value equal to the dc voltage.

Using these formulas, we can also determine the peak value knowing the rms value as follows:

$$V_p = \left(\frac{1}{0.707}\right)V_{rms}$$
$$V_p = \sqrt{2}\, V_{rms} \cong 1.414V_{rms} \tag{9–7}$$

Similarly,

$$I_p = \sqrt{2}\, I_{rms} \cong 1.414I_{rms} \tag{9–8}$$

To find the peak-to-peak value, simply double the peak value:

$$V_{pp} = 2.828V_{rms} \tag{9–9}$$

and

$$I_{pp} = 2.828I_{rms} \tag{9–10}$$

Average Value

The average value of a sine wave taken over *one complete cycle* is always zero, because the positive values (above the zero crossing) offset the negative values (below the zero crossing).

To be useful for comparison purposes, the average value of a sine wave is defined over a *half-cycle* rather than over a full cycle. The average value is the total area under the half-cycle curve divided by the distance of the curve along the horizontal axis. The result is expressed in terms of the peak value as follows for both voltage and current:

$$V_{avg} = \left(\frac{2}{\pi}\right)V_p \cong 0.637V_p \tag{9–11}$$

$$I_{avg} = \left(\frac{2}{\pi}\right)I_p \cong 0.637I_p \tag{9–12}$$

The average value of a sine wave voltage is illustrated in Figure 9–13.

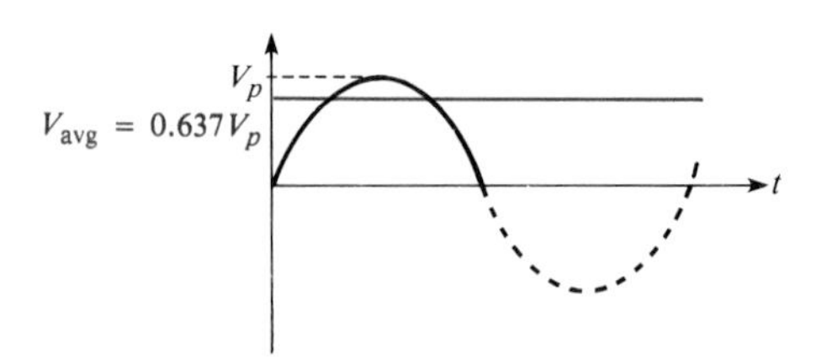

FIGURE 9–13
Average value.

EXAMPLE 9–6 Determine V_p, V_{pp}, V_{rms}, and V_{avg} for the sine wave in Figure 9–14.

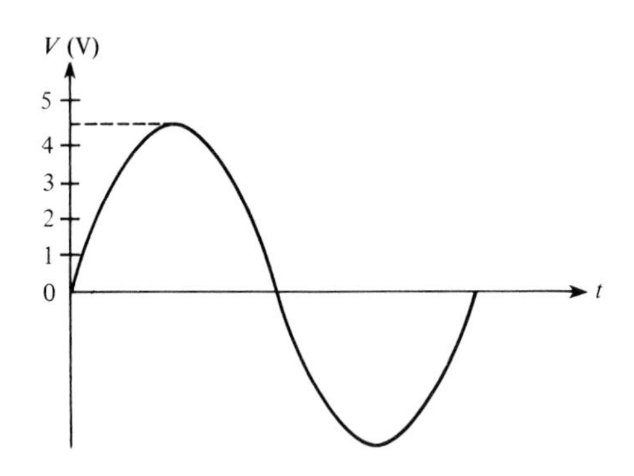

FIGURE 9–14

Solution
As taken directly from the graph, $V_p = 4.5$ V. From this value we calculate the other values:

$$V_{pp} = 2V_p = 2(4.5 \text{ V}) = 9 \text{ V}$$
$$V_{rms} = 0.707V_p = 0.707(4.5 \text{ V}) = 3.182 \text{ V}$$
$$V_{avg} = 0.637V_p = 0.637(4.5 \text{ V}) = 2.867 \text{ V}$$

SECTION REVIEW 9–2

1. Determine V_{pp} when
   (a) $V_p = 1$ V. **(b)** $V_{rms} = 1.414$ V. **(c)** $V_{avg} = 3$ V.
2. Determine V_{rms} when
   (a) $V_p = 2.5$ V. **(b)** $V_{pp} = 10$ V. **(c)** $V_{avg} = 1.5$ V.

SINE WAVE VOLTAGE SOURCES

9–3

There are two basic methods of generating sine wave voltages: *electromagnetic* and *electronic*. Sine waves are produced electromagnetically by ac generators and electronically by oscillator circuits.

An ac Generator

Figure 9–15 shows a basic ac generator consisting of a single loop of wire in a permanent magnetic field. Notice that each end of the loop is connected to a

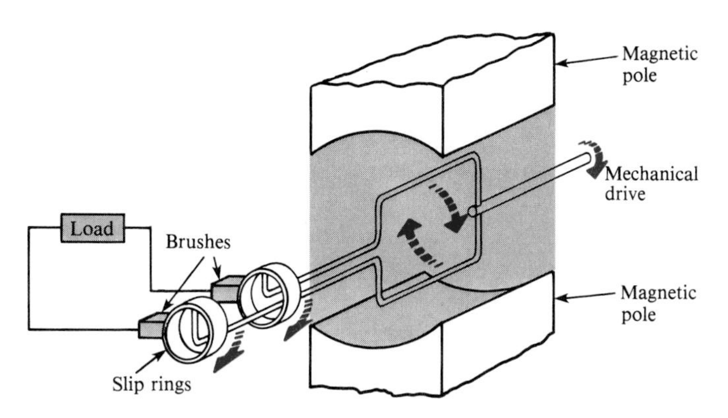

FIGURE 9–15
A basic ac generator.

separate solid conductive ring called a *slip ring*. As the loop rotates in the magnetic field, the slip rings also rotate and rub against the brushes which connect the loop to an external load. Compare this generator to the basic dc generator in Chapter 7, and note the difference in the ring and brush arrangements.

As you learned in Chapter 7, when a conductor moves through a magnetic field, a voltage is induced. Figure 9–16 illustrates how a sine-wave voltage is

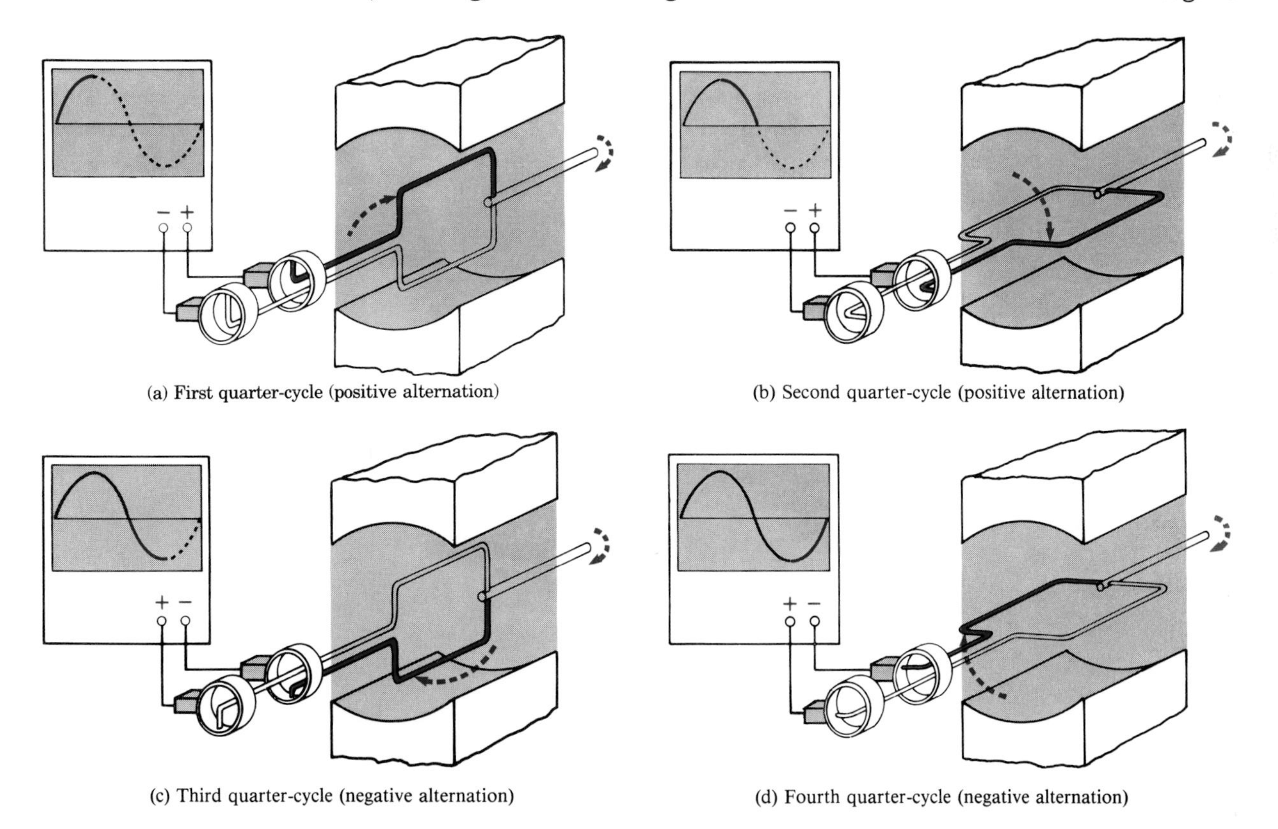

(a) First quarter-cycle (positive alternation)

(b) Second quarter-cycle (positive alternation)

(c) Third quarter-cycle (negative alternation)

(d) Fourth quarter-cycle (negative alternation)

FIGURE 9–16
One revolution of the loop generates one cycle of the sine wave voltage.

produced by the basic ac generator as the loop rotates. An oscilloscope is used to display the voltage waveform.

To begin, Figure 9–16(a) shows the loop rotating through the first quarter of a revolution. It goes from an instantaneous horizontal position, where the induced voltage is zero, to an instantaneous vertical position, where the induced voltage is maximum. As shown on the display, this part of the rotation produces the first quarter of the sine wave cycle as the voltage builds up from zero to its positive maximum. Part (b) of the figure shows the loop completing the first half of a revolution. During this part of the rotation, the voltage decreases from its positive maximum back to zero.

During the second half of the revolution, illustrated in Figures 9–16(c) and 9–16(d), the loop is cutting through the magnetic field in the opposite direction, so the voltage produced has a polarity opposite to that produced during the first half of the revolution. After one complete revolution of the loop, one full cycle of the sine wave voltage has been produced. As the loop continues to rotate, repetitive cycles of the sine wave are generated.

Factors That Affect Frequency in ac Generators

You have seen that one revolution of the conductor through the magnetic field in the basic ac generator (also called an *alternator*) produces one cycle of induced sinusoidal voltage. It is obvious that the rate at which the conductor is rotated determines the time for completion of one cycle. For example, if the conductor completes 60 revolutions in one second, the period of the resulting sine wave is 1/60 second, corresponding to a frequency of 60 Hz. Thus, the faster the conductor rotates, the higher the resulting frequency of the induced voltage, as illustrated in Figure 9–17.

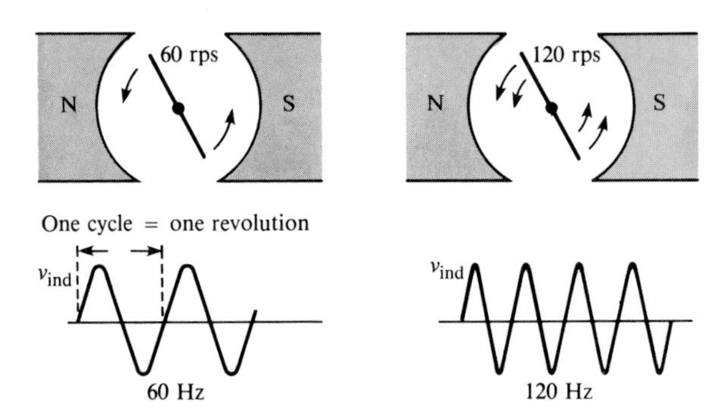

FIGURE 9–17
Frequency is directly proportional to the rate of rotation of the loop.

Another way of achieving a higher frequency is to increase the number of magnetic poles. In the previous discussion, two magnetic poles were used to illustrate the ac generator principle. During one revolution, the conductor passes under a north pole and a south pole, thus producing one cycle of a sine wave. When four magnetic poles are used instead of two, as shown in Figure 9–18, one cycle is generated during *one-half* a revolution. This doubles the frequency for the same rate of rotation.

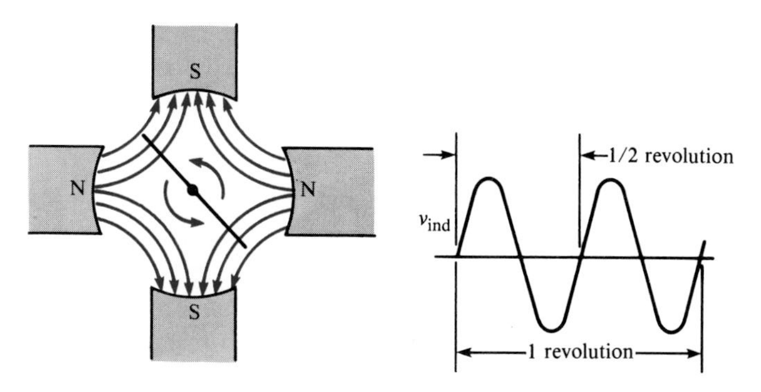

FIGURE 9–18
Four poles achieve a higher frequency than two for the same rps.

An expression for frequency in terms of the number of poles and the number of revolutions per second (rps) is as follows:

$$f = (\text{number of pole pairs})(\text{rps}) \tag{9–13}$$

EXAMPLE 9–7

A four-pole generator has a rotation speed of 100 rps. Determine the frequency of the output voltage.

Solution

$$f = (\text{number of pole pairs})(\text{rps}) = 2(100) = 200 \text{ Hz}$$

Factors That Affect Voltage Amplitude in ac Generators

Recall from Chapter 7 that the voltage induced in a conductor depends on the number of turns *(N)* and the rate of change with respect to the magnetic field. Therefore, when the speed of rotation of the conductor is increased, not only the frequency of the induced voltage increases—so also does the amplitude. Since the frequency value normally is fixed, the most practical method of increasing the amount of induced voltage is to increase the number of loops.

Signal Generators

The signal generator is an instrument that electronically produces sine waves for use in testing or controlling electronic circuits and systems. There are a variety of signal generators, ranging from special-purpose instruments that produce only one type of waveform in a limited frequency range, to programmable instruments that produce a wide range of frequencies and a variety of waveforms. All signal generators consist basically of an oscillator, which is an electronic circuit that produces sine wave voltages whose amplitude and frequency can be adjusted.

Audio Frequency Generators

The audible range of frequencies is typically from about 20 Hz to about 20 kHz. As a minimum requirement, an audio frequency (af) generator must provide sine wave voltages within this range. Many af generators or *audio oscillators,* however, have frequency ranges much greater than the actual audible range. A typical laboratory af generator produces frequencies from less than 1 Hz to greater than 1 MHz.

Figure 9–19 shows one type of audio generator. Notice that there is a range switch for selecting the desired frequency range. The frequency control sets the exact frequency within the selected range. The amplitude control adjusts the output voltage. Once it is set, it should remain essentially constant over the frequency range of the instrument. The maximum specified output voltage occurs when there is no load connected across the output terminals of the generator.

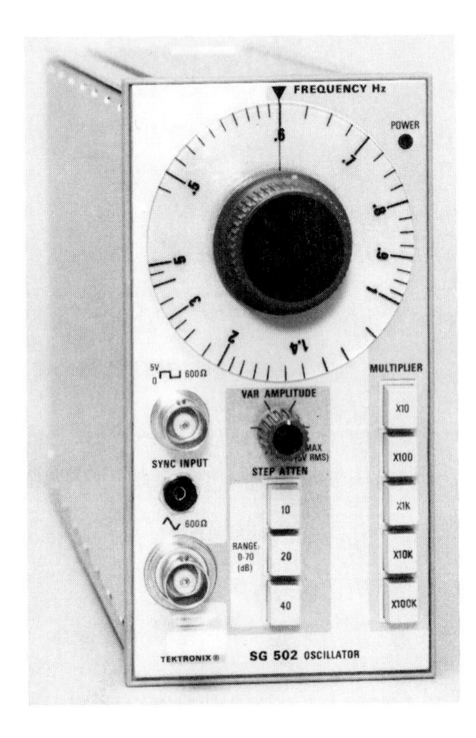

FIGURE 9–19
Audio oscillator (courtesy of Tektronix, Inc.).

Radio Frequency Generators

Many radio frequency (rf) generators cover frequencies ranging from 30 kHz to 3000 MHz. The lower end of the range, of course, overlaps with the range of many af generators. Most rf generators produce at least two types of output signal: *continuous-wave* and *modulated.*

A continuous-wave or CW signal is a single-frequency sine wave with a steady amplitude. A modulated signal is a sine wave with an amplitude that varies sinusoidally at a much lower frequency, called *amplitude modulation* (AM). The two types of signals are illustrated in Figure 9–20.

There are many types of rf generators. All have provisions for adjusting the frequency over the specified range and for setting the output voltage amplitude. The percentage of modulation can also be set, as illustrated in Figure

9–21. Some rf generators also have provisions for frequency modulation (FM). Figure 9–22 shows two typical rf generators.

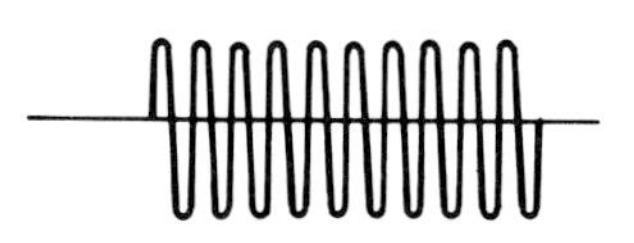

(a) Continuous wave (CW)

(b) Amplitude modulation (AM)

FIGURE 9–20
CW and amplitude-modulated rf signals.

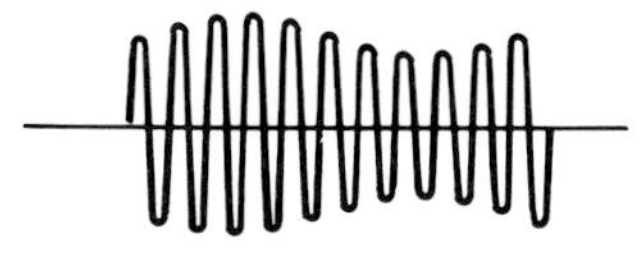

(a) Lower percentage of modulation

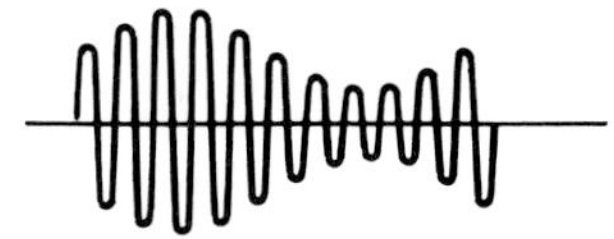

(b) Higher percentage of modulation

FIGURE 9–21
Variations in amplitude modulation.

(a)

(b)

FIGURE 9–22
Typical rf signal generators [(a) courtesy of B&K Precision Test Instruments, Dynascan Corp. (b) courtesy of Hewlett-Packard Co.]

SECTION REVIEW 9–3

1. How are sine waves generated electromagnetically?
2. List two classifications of sine wave frequency ranges.

9–4 ANGULAR RELATIONSHIPS OF A SINE WAVE

As you have seen, sine waves can be measured along the horizontal axis on a time basis. However, since the time for completion of one full cycle or any portion of a cycle is frequency-dependent, it is often useful to specify points on the sine wave in terms of an *angular measurement* expressed in degrees or radians. Angular measurement is independent of frequency.

As previously discussed, sine wave voltage can be produced by rotating electromechanical machines. As the rotor of the ac generator goes through a full 360° of rotation, the resulting voltage output is one full cycle of a sine wave. Thus, the angular measurement of a sine wave can be related to the angular rotation of a generator, as shown in Figure 9–23.

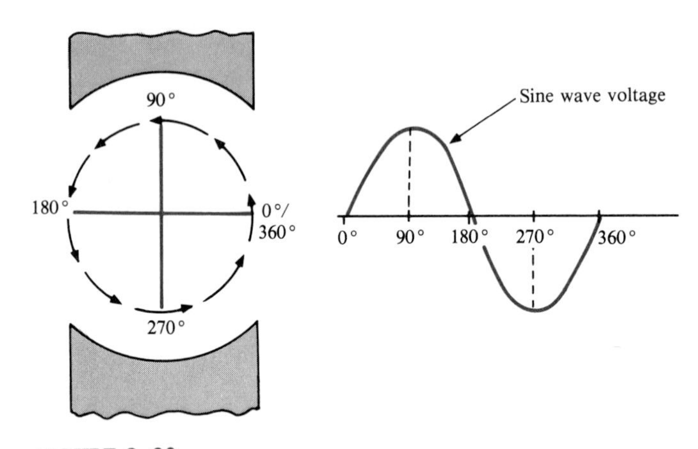

FIGURE 9–23

Angular Measurement

A *radian* (rad) is defined as the angular distance along the circumference of a circle equal to the radius of the circle. One radian is equivalent to 57.3°, as illustrated in Figure 9–24. In a 360° revolution, there are 2π radians.

FIGURE 9–24
Angular measurement showing relationship of radian to degrees.

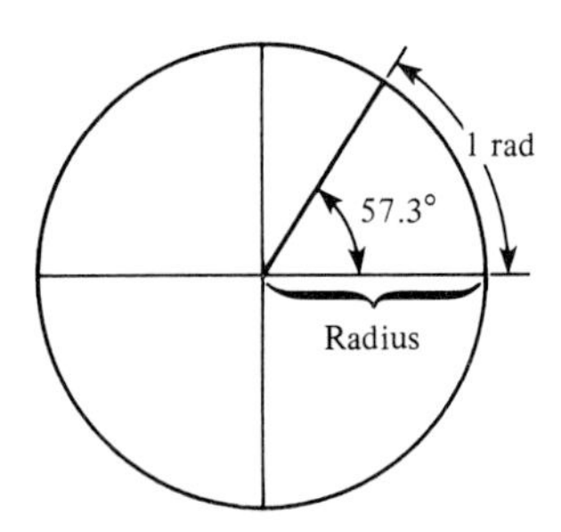

π is the ratio of the circumference of any circle to its diameter and has a constant value of 3.1416.

Most calculators have a π key so that the actual numerical value does not have to be entered.

Table 9–1 lists several values of degrees and the corresponding radian values. These angular measurements are illustrated in Figure 9–25.

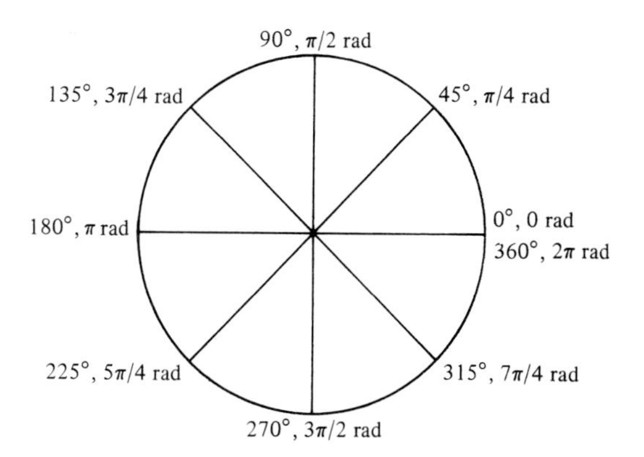

FIGURE 9–25
Angular measurements.

TABLE 9–1

Degrees	Radians (rad)
0	0
45	$\pi/4$
90	$\pi/2$
135	$3\pi/4$
180	π
225	$5\pi/4$
270	$3\pi/2$
315	$7\pi/4$
360	2π

Sine Wave Angles

The angular measurement of a sine wave is based on 360° or 2π radians for a complete cycle. A half-cycle is 180° or π radians; a quarter-cycle is 90° or $\pi/2$ radians; and so on. Figure 9–26(a) shows angles in degrees for a full cycle of a sine wave; Part (b) shows the same points in radians.

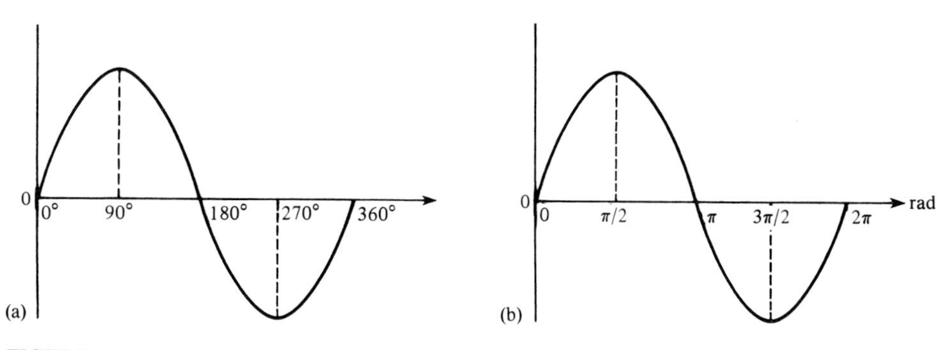

FIGURE 9–26
Sine wave angles.

Phase of a Sine Wave

The phase of a sine wave is an angular measurement that specifies the position of that sine wave *relative* to a reference. Figure 9–27 shows one cycle of a sine wave to be used as the *reference*. Note that the first positive-going crossing of the horizontal axis (zero crossing) is at 0° (0 rad), and the positive peak is at 90° ($\pi/2$ rad). The negative-going zero crossing is at 180° (π rad), and the negative peak is at 270° ($3\pi/2$ rad). The cycle is completed at 360° (2π rad). When the sine wave is shifted left or right with respect to this reference, there is a *phase shift.*

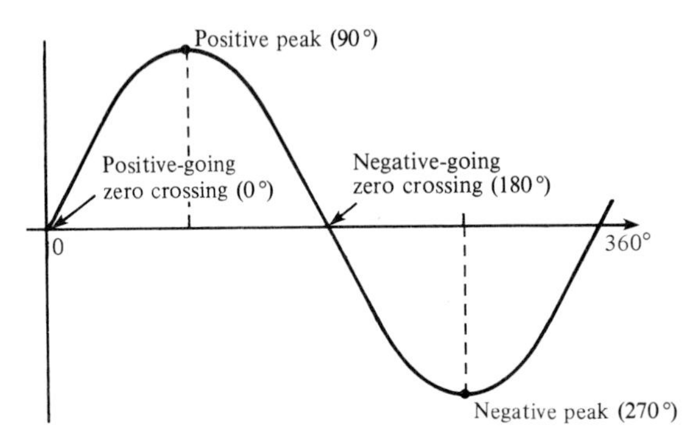

FIGURE 9–27
Phase reference.

Figure 9–28 illustrates phase shifts of a sine wave. In Part (a) of the figure, sine wave B is shifted to the right by 90° ($\pi/2$ rad). Thus, there is a *phase angle* of 90° between sine wave A and sine wave B. In terms of time, the positive peak of sine wave B occurs *later* than the positive peak of sine wave A, because time increases to the right along the horizontal axis. In this case, sine wave B is said to *lag* sine wave A by 90° or $\pi/2$ radians. In other words, sine wave A *leads* sine wave B by 90°.

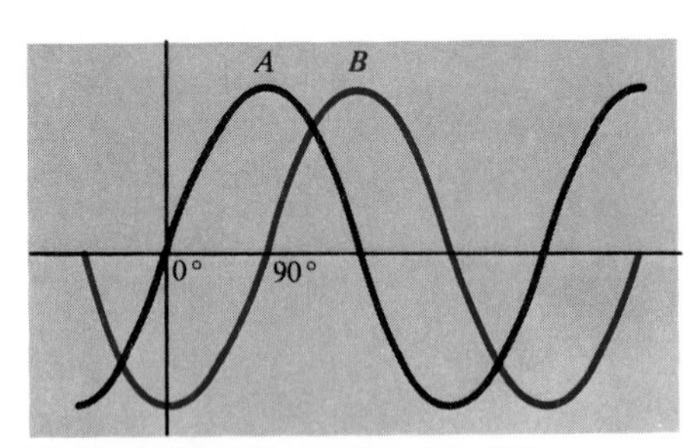

(a) A leads B by 90°, or B lags A by 90°.

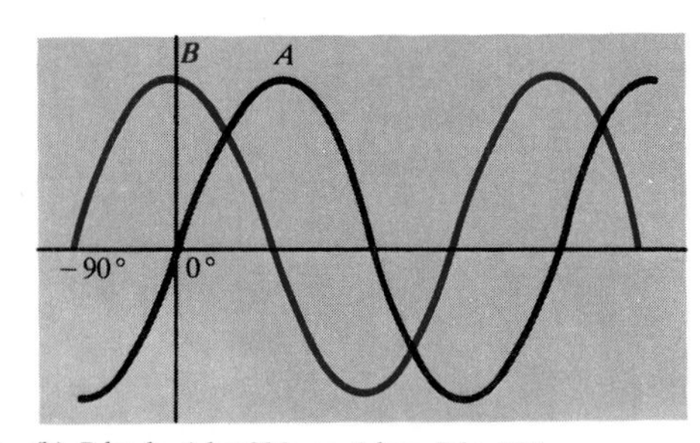

(b) B leads A by 90°, or A lags B by 90°.

FIGURE 9–28
Illustration of a phase shift.

In Figure 9–28(b), sine wave B is shown shifted left by 90°. Thus, again there is a phase angle of 90° between sine wave A and sine wave B. In this case, the positive peak of sine wave B occurs earlier in time than that of sine wave A; therefore, sine wave B is said to *lead* by 90°. In both cases there is a 90° *phase angle* between the two waveforms.

EXAMPLE 9–8

What are the phase angles between the two sine waves in Figures 9–29(a) and 9–29(b)?

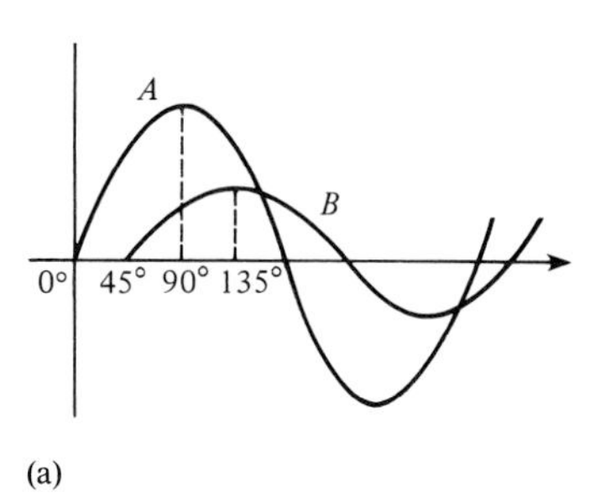

(a) (b)

FIGURE 9–29

Solution

(a) The zero crossing of sine wave A is at 0°, and the corresponding zero crossing of sine wave B is at 45°. There is a 45° phase angle between the two waveforms (45° − 0° = 45°) with sine wave A leading.

(b) The zero crossing of sine wave B is at 30°, and the corresponding zero crossing of sine wave A is at 60 °. There is a 30° phase angle between the two waveforms (60° − 30° = 30°) with sine wave B leading.

SECTION REVIEW 9–4

1. When the positive-going zero crossing of a sine wave occurs at 0°, at what angle does each of the following points occur?
   (a) positive peak. **(b)** negative-going zero crossing.
   (c) negative peak. **(d)** end of first complete cycle.
2. A half-cycle is completed in ______ degrees or ______ radians.
3. A full cycle is completed in ______ degrees or ______ radians.
4. Determine the phase angle between the two sine waves in Figure 9–30.

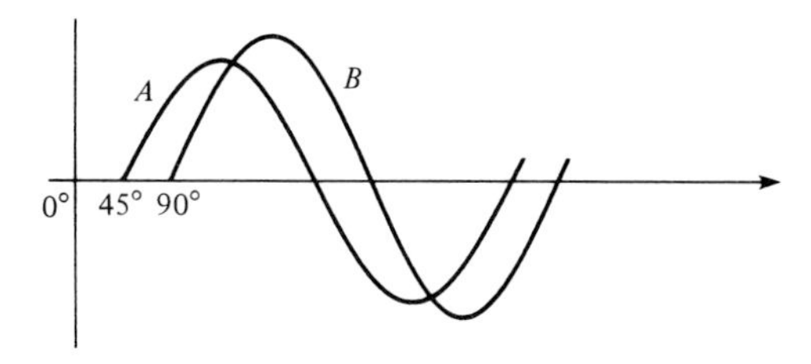

FIGURE 9–30

THE SINE WAVE EQUATION

9–5

As you have seen, a sine wave can be graphically represented by voltage (or current) values on the vertical axis and by angular measurement (in degrees or radians) along the horizontal axis. A generalized graph of one cycle of a sine wave is shown in Figure 9–31. The *amplitude, A,* is the maximum value of the voltage or current on the vertical axis, and angular values run along the horizontal axis. The variable y is an instantaneous value representing either voltage or current at a given angle, θ.

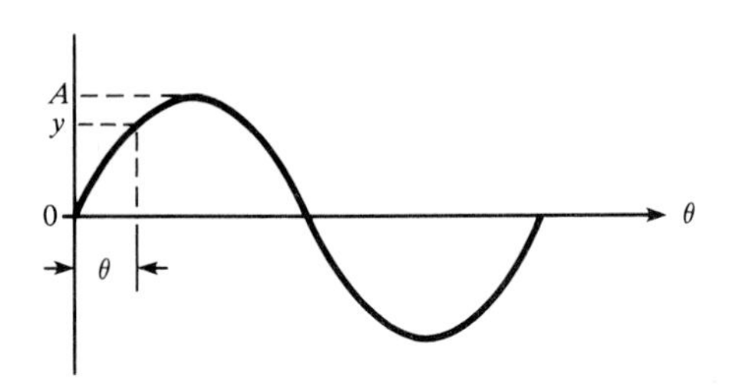

FIGURE 9–31
One cycle of a generalized sine wave showing amplitude and phase.

All sine wave curves follow a specific mathematical formula. The general expression for the sine wave curve in Figure 9–31 is

$$y = A \sin \theta \tag{9–14}$$

This expression states that any point on the sine wave, represented by an instantaneous value y, is equal to the maximum value times the sine (sin) of the angle θ at that point. For example, if a certain sine wave voltage has a peak value of 10 V, the instantaneous voltage at a point 60° along the horizontal axis can be calculated as follows:

$$v = V_p \sin \theta = 10 \sin 60° = 10(0.866) = 8.66 \text{ V}$$

Figure 9–32 shows this instantaneous value on the curve. You can find the sine of any angle on your calculator by first entering the value of the angle and then pressing the sin key.

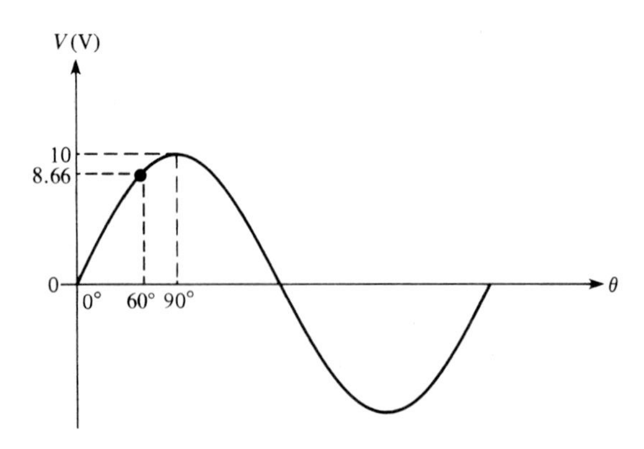

FIGURE 9–32
Illustration of the instantaneous value at $\theta = 60°$.

Expressions for Shifted Sine Waves

When a sine wave is shifted to the right of the reference (lagging) by a certain angle, ϕ, as illustrated in Figure 9–33(a), the general expression is

$$y = A \sin(\theta - \phi) \tag{9–15}$$

When a sine wave is shifted to the left of the reference (leading) by a certain angle, ϕ, as shown in Figure 9–33(b), the general expression is

$$y = A \sin(\theta + \phi) \tag{9–16}$$

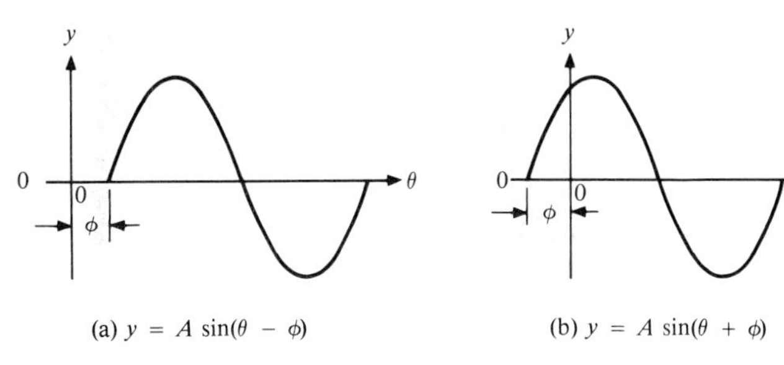

FIGURE 9–33
Shifted sine waves.

EXAMPLE 9–9

Determine the instantaneous value at the 90° reference point on the horizontal axis for each sine wave voltage in Figure 9–34.

FIGURE 9–34

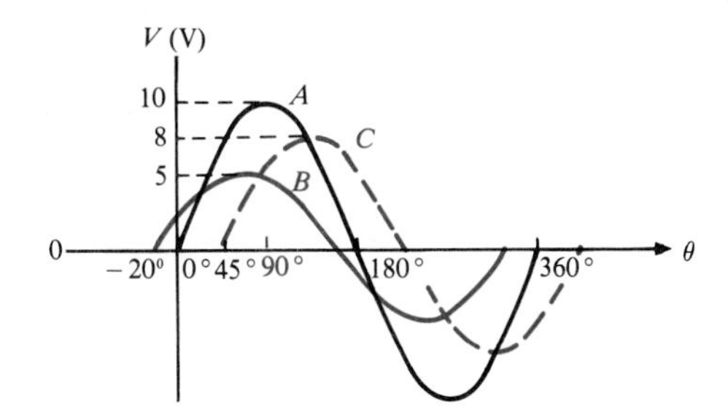

Solution

Sine wave A is the reference. Sine wave B is shifted left 20° with respect to A, so it leads. Sine wave C is shifted right 45° with respect to A, so it lags.

$$\begin{aligned}
v_A &= V_p \sin \theta \\
&= 10 \sin 90° = 10(1) = 10 \text{ V} \\
v_B &= V_p \sin(\theta + \phi_B) \\
&= 5 \sin(90° + 20°) = 5 \sin 110° = 5(0.9397) = 4.7 \text{ V} \\
v_C &= V_p \sin(\theta - \phi_C) \\
&= 8 \sin(90° - 45°) = 8 \sin 45° = 8(0.7071) = 5.66 \text{ V}
\end{aligned}$$

SECTION REVIEW 9–5

1. Calculate the instantaneous value at 120° for the sine wave in Figure 9–32.
2. Determine the instantaneous value at the 45° point on the reference axis of a sine wave shifted 10° to the left of the zero reference ($V_p = 10$ V).

INTRODUCTION TO PHASORS

9–6

Phasors can be used to represent time-varying quantities, such as sine waves, in terms of their magnitude and angular position (phase angle). Examples of phasors are shown in Figure 9–35. The *length* of the phasor "arrow" represents the magnitude. The angle, θ (relative to 0°), represents the angular position, as shown in Part (a). The specific phasor example in Part (b) has a magnitude of 2 and a phase angle of 60°. The phasor in Part (c) has a magnitude of 3 and a phase angle of 180°. The phasor in Part (d) has a magnitude of 1 and a phase angle of −45°.

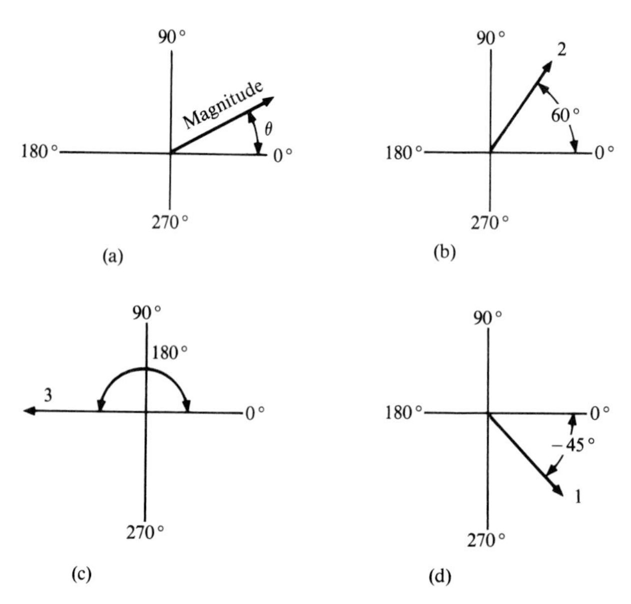

FIGURE 9–35
Examples of phasors.

Phasor Representation of a Sine Wave

A full cycle of a sine wave can be represented by rotation of a phasor through 360°.

> **The instantaneous value of the sine wave at any point is equal to the vertical distance from the tip of the phasor to the horizontal axis.**

Figure 9–36 shows how the phasor "traces out" the sine wave as it goes from 0° to 360°. You can relate this concept to the rotation in an ac generator.

Notice in Figure 9–36 that the length of the phasor is equal to the *peak* value of the sine wave (observe the 90° and the 270° points). The angle of the phasor measured from 0° is the corresponding angular point on the sine wave.

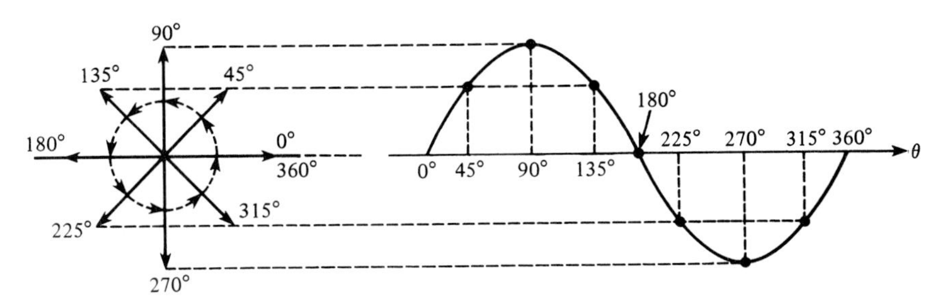

FIGURE 9–36
Sine wave represented by rotational phasor motion.

Derivation of the Sine Wave Formula

Let's examine a phasor representation at one specific angle. Figure 9–37 shows a voltage phasor at an angular position of 45° and the corresponding point on the sine wave. The instantaneous value of the sine wave at this point is related to both the position and the length of the phasor. As previously mentioned, the vertical distance from the phasor tip down to the horizontal axis represents the instantaneous value of the sine wave at that point.

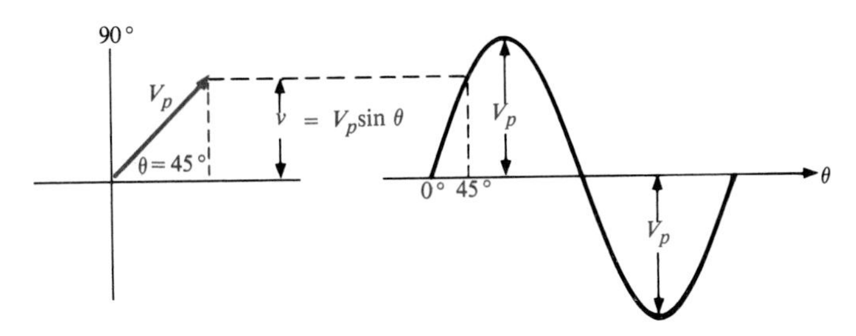

FIGURE 9–37
Right triangle derivation of sine wave formula.

Notice that when a vertical line is drawn from the phasor tip down to the horizontal axis, a *right triangle* is formed as shown in the figure. The length of the phasor is the *hypotenuse* of the triangle, and the vertical projection is the *opposite side*. From trigonometry, *the opposite side of a right triangle is equal to the hypotenuse times the sine of the angle θ*. In this case, the length of the phasor is the *peak* value of the sine wave voltage, V_p. Thus, the opposite side of the triangle, which is the instantaneous value, can be expressed as $v = V_p \sin \theta$. Recall that this formula is the one stated earlier for calculating instantaneous sine wave values. Of course, this formula also applies to a current sine wave.

Positive and Negative Phasor Angles

The position of a phasor at any instant can be expressed as a *positive* angle, as you have seen, or as an *equivalent negative angle*. Positive angles are measured counterclockwise from 0°. Negative angles are measured clockwise from 0°. For a given positive angle θ, the corresponding negative angle is $\theta - 360°$, as illustrated in Figure 9–38(a). In Part (b), a specific example is shown. The angle of the phasor in this case can be expressed as $+225°$ or $-135°$.

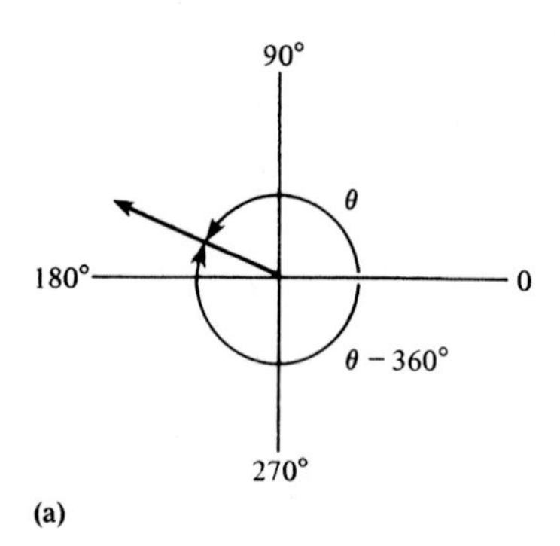

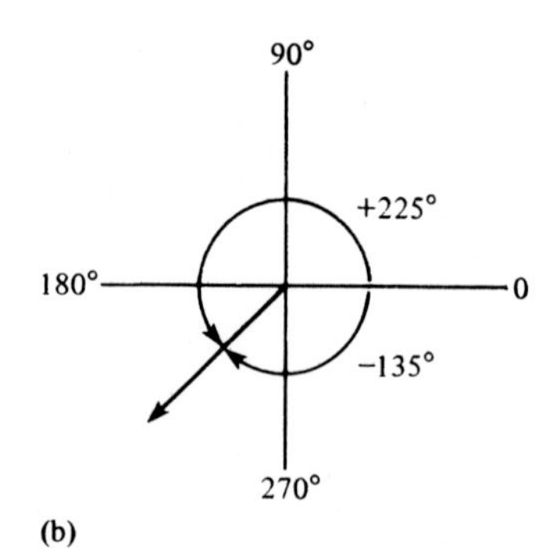

FIGURE 9–38
Positive and negative phasor angles.

EXAMPLE 9–10

For each phasor in Figure 9–39, determine the instantaneous sine wave value. Also express each positive angle shown as an equivalent negative angle. The length of each phasor represents the peak value of the sine wave.

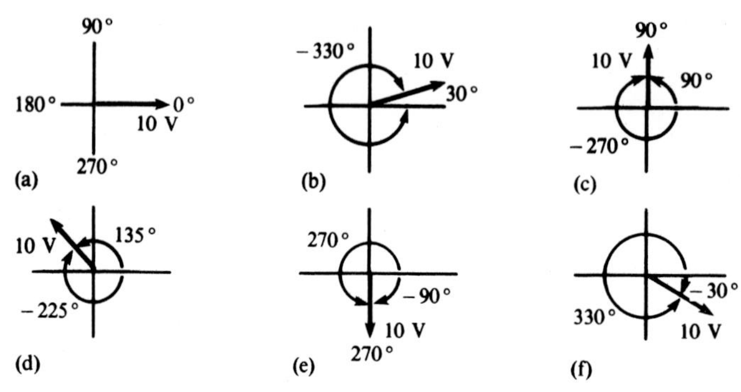

FIGURE 9–39

Solution:

(a) $\theta = 0°$

$$v = 10 \sin 0° = 10(0) = 0 \text{ V}$$

(b) $\theta = 30° = -330°$

$$v = 10 \sin 30° = 10(0.5) = 5 \text{ V}$$

(c) $\theta = 90° = -270°$

$$v = 10 \sin 90° = 10(1) = 10 \text{ V}$$

(d) $\theta = 135° = -225°$

$$v = 10 \sin 135° = 10(0.707) = 7.07 \text{ V}$$

(e) $\theta = 270° = -90°$

$v = 10 \sin 270° = 10(-1) = -10\text{ V}$

(f) $\theta = 330° = -30°$

$v = 10 \sin 330° = 10(-0.5) = -5\text{ V}$

The equivalent negative angles are shown in Figure 9–39.

Phasor Diagrams

A phasor diagram shows the *relative relationship* of two or more sine waves of the same frequency. A phasor in a *fixed* position represents a *complete* sine wave, because once the phase angle between two or more sine waves of the same frequency is established, it remains constant throughout the cycles. For example, the two sine waves in Figure 9–40(a) can be represented by a phasor diagram, as shown in Part (b). As you can see, sine wave *B* leads sine wave *A* by 30°. The length of the phasors can be used to represent peak, rms, or average values as long as the representation is consistent.

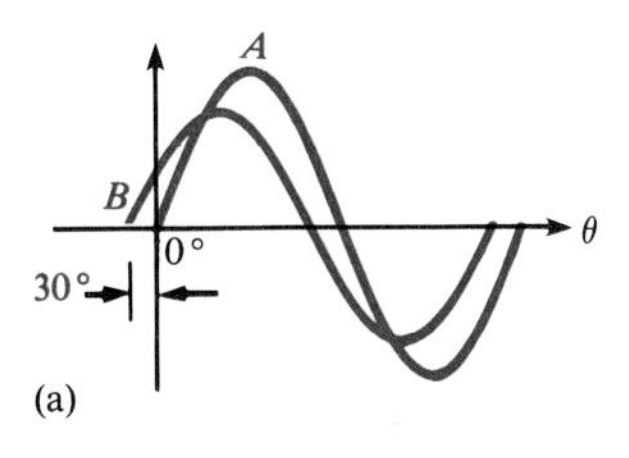

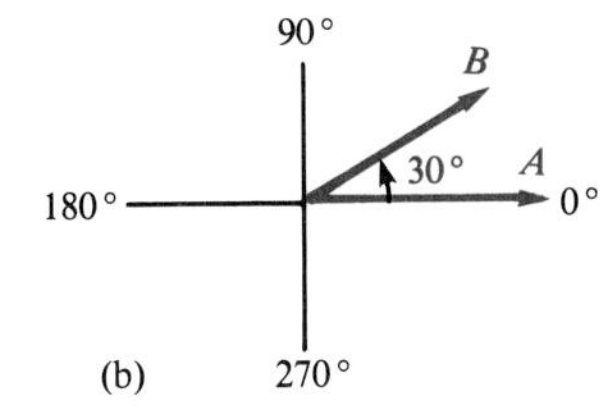

FIGURE 9–40
Example of a phasor diagram.

EXAMPLE 9–11 Use a phasor diagram to represent the sine waves in Figure 9–41.

FIGURE 9–41

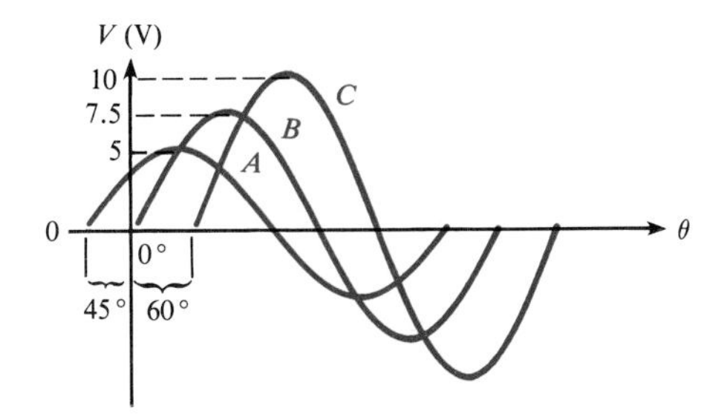

Solution

The phasor diagram representing the sine waves is shown in Figure 9–42. In this case, the length of each phasor represents the peak value of the sine wave.

FIGURE 9–42

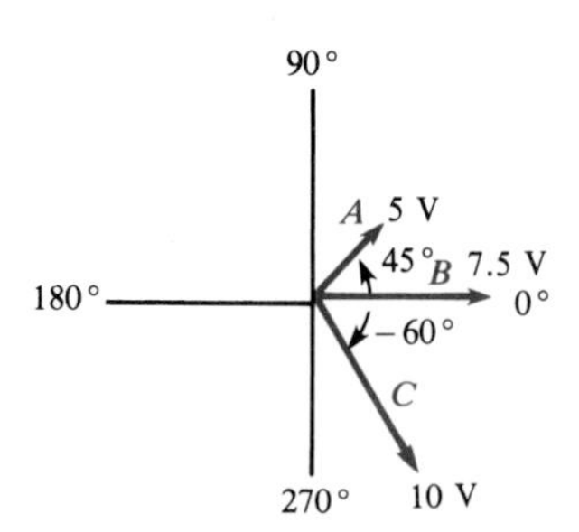

SECTION REVIEW 9–6

1. What is a phasor?
2. Sketch a phasor diagram to represent the two sine waves in Figure 9–43. Use peak values.

FIGURE 9–43

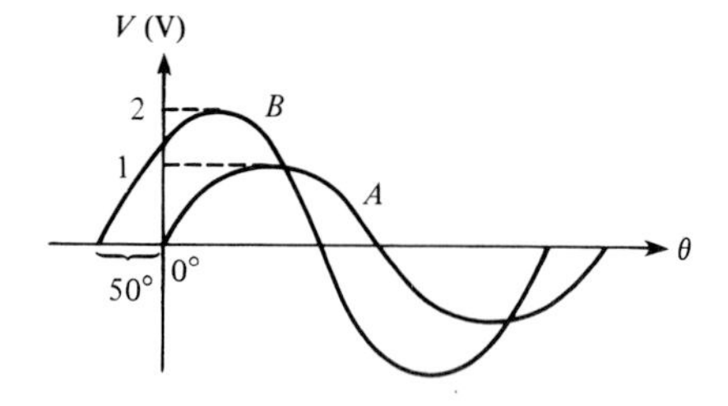

OHM'S LAW AND KIRCHHOFF'S LAWS

9–7

When time-varying ac voltages such as the sine wave are applied to resistive circuits, the circuit laws that you studied earlier still apply. Ohm's law applies to resistive ac circuits in the same way that it applies to dc circuits. If a sine wave voltage is applied across a resistor as shown in Figure 9–44, a sine wave current flows. The current is zero when the voltage is zero and is maximum when the voltage is maximum. When the voltage changes polarity, the current reverses direction. As a result, the voltage and current are said to be *in phase* with each other.

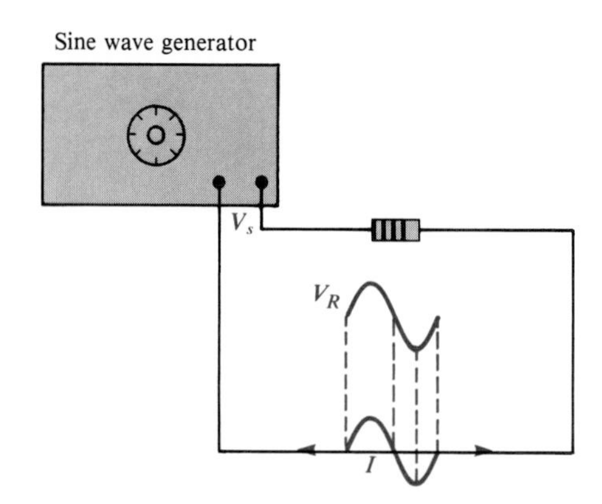

FIGURE 9–44
A sine wave voltage produces a sine wave current.

When using Ohm's law in ac circuits, remember that both the voltage and the current must be expressed consistently, that is, both as peak values, both as rms values, both as average values, and so on.

EXAMPLE 9–12 Determine the rms voltage across each resistor and the rms current in Figure 9–45. The source voltage is given as an rms value.

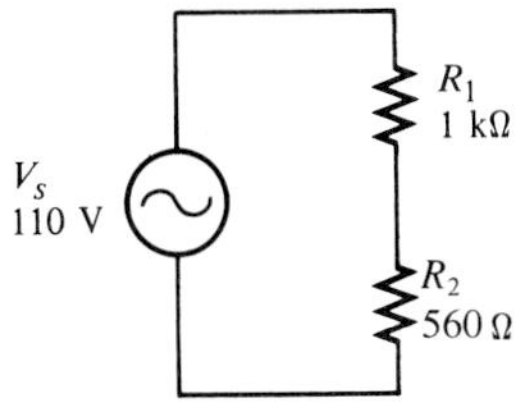

V_s 110 V, R_1 1 kΩ, R_2 560 Ω

FIGURE 9–45

Solution

The total resistance of the circuit is

$$R_T = R_1 + R_2 = 1\text{ k}\Omega + 560\ \Omega = 1.56\text{ k}\Omega$$

Use Ohm's law to find the rms current:

$$I_{\text{rms}} = \frac{V_{s(\text{rms})}}{R_T} = \frac{110\text{ V}}{1.56\text{ k}\Omega} = 70.5\text{ mA}$$

The rms voltage drop across each resistor is

$$V_{1(\text{rms})} = I_{\text{rms}} R_1 = (70.5\text{ mA})(1\text{ k}\Omega) = 70.5\text{ V}$$

$$V_{2(\text{rms})} = I_{\text{rms}} R_2 = (70.5\text{ mA})(560\ \Omega) = 39.5\text{ V}$$

Kirchhoff's voltage and current laws apply to ac circuits as well as to dc circuits. Figure 9–46 illustrates Kirchhoff's voltage law in a resistive circuit that has a sine wave voltage source. As you can see, the source voltage is the sum of all the voltage drops across the resistors, just as in a dc circuit.

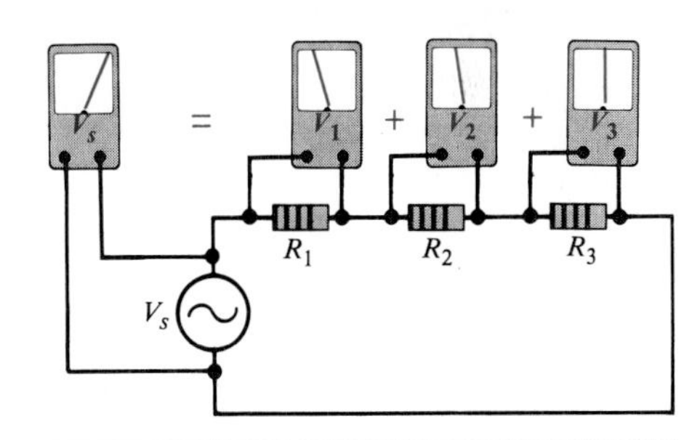

FIGURE 9–46
Illustration of Kirchhoff's voltage law in an ac circuit.

EXAMPLE 9–13 (a) Find the unknown peak voltage drop in Figure 9–47(a). All values are given in rms.

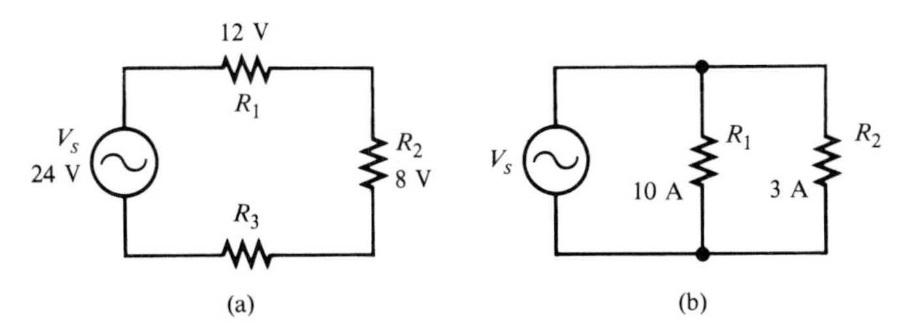

FIGURE 9–47

(b) Find the total rms current in Figure 9–47(b). All values are given in rms.

Solution

(a) Use Kirchhoff's voltage law to find V_3:

$$V_s = V_1 + V_2 + V_3$$

Solve for V_3:

$$\begin{aligned} V_{3(\text{rms})} &= V_{s(\text{rms})} - V_{1(\text{rms})} - V_{2(\text{rms})} \\ &= 24\text{ V} - 12\text{ V} - 8\text{ V} = 4\text{ V} \end{aligned}$$

Convert rms to peak:

$$V_{3(\text{peak})} = 1.414V_{3(\text{rms})} = 1.414(4\text{ V}) = 5.66\text{ V}$$

(b) Use Kirchhoff's current law to find I_T:

$$\begin{aligned} I_{T(\text{rms})} &= I_{1(\text{rms})} + I_{2(\text{rms})} \\ &= 10\text{ A} + 3\text{ A} = 13\text{ A} \end{aligned}$$

SECTION REVIEW 9–7

1. A sine wave voltage with an average value of 12.5 V is applied to a circuit with a resistance of 330 Ω. What is the average current in the circuit?
2. The peak voltage drops in a series resistive circuit are 6.2 V, 11.3 V, and 7.8 V. What is the rms value of the source voltage?

NONSINUSOIDAL WAVEFORMS

9–8

Sine waves are very important in electronics, but they are by no means the only type of ac or time-varying waveform. Two other categories are discussed next: the *pulse waveform* and the *triangular waveform.*

Pulse Waveforms

Basically, a pulse can be described as a very rapid transition from one voltage or current level (baseline) to another, and then, after an interval of time, a very rapid transition back to the original baseline level. The transitions in level are called *steps*. An *ideal* pulse consists of two opposite-going steps of equal ampli-

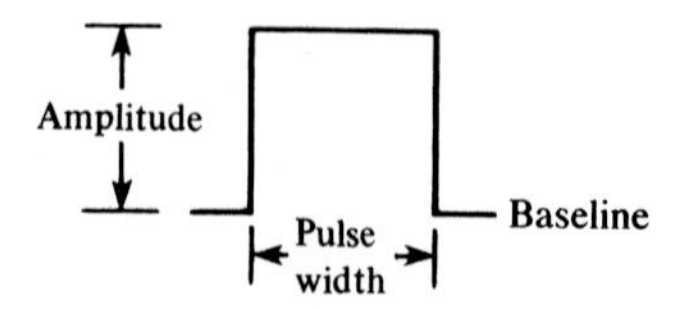

(a) Positive-going pulse

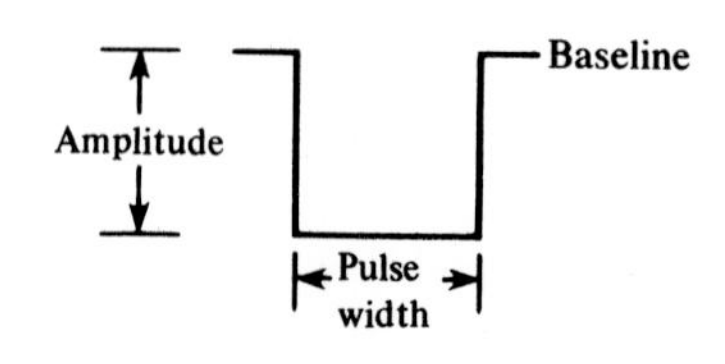

(b) Negative-going pulse

FIGURE 9–48
Ideal pulses.

tude. Figure 9–48(a) shows an ideal positive-going pulse consisting of two equal but opposite instantaneous steps separated by an interval of time called the *pulse width*. Part (b) of the figure shows an ideal negative-going pulse. The height of the pulse measured from the baseline is its amplitude.

In many applications, analysis is simplified by treating all pulses as ideal (composed of instantaneous steps and perfectly rectangular in shape). Actual pulses, however, are never ideal. All pulses possess certain characteristics that cause them to be different from the ideal.

In practice, pulses cannot change from one level to another instantaneously. Time is always required for a transition (step), as illustrated in Figure 9–49(a). As you can see, there is an interval of time during which the pulse is rising from its lower value to its higher value. This interval is called the *rise time*, t_r.

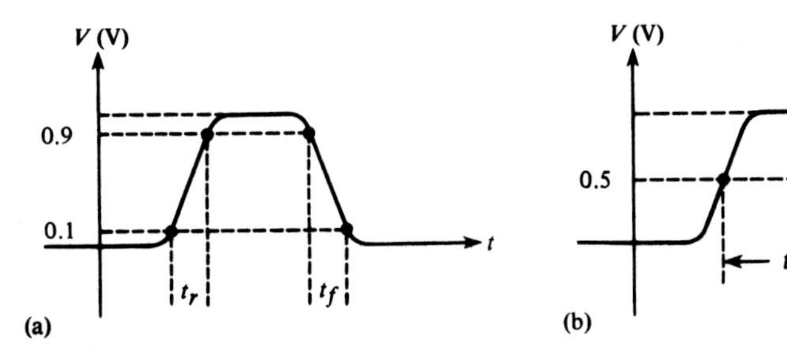

FIGURE 9–49
Nonideal pulse.

The most widely accepted definition of rise time is the time required for the pulse to go from 10% of its full amplitude to 90% of its full amplitude.

The interval of time during which the pulse is falling from its higher value to its lower value is called the *fall time*, t_f.

The accepted definition of fall time is the time required for the pulse to go from 90% of its full amplitude to 10% of its full amplitude.

Pulse width also requires a precise definition for the nonideal pulse, because the rising and falling edges are not vertical.

A widely accepted definition of pulse width (t_w) is the time between the point on the rising edge, where the value is 50% of full amplitude, to the point on the falling edge, where the value is 50% of full amplitude.

Pulse width is shown in Figure 9–49(b).

Repetitive Pulses

Any waveform that repeats itself at fixed intervals is *periodic*. Some examples of periodic pulse waveforms are shown in Figure 9–50. Notice that in each case, the pulses repeat at regular intervals. The rate at which the pulses repeat is

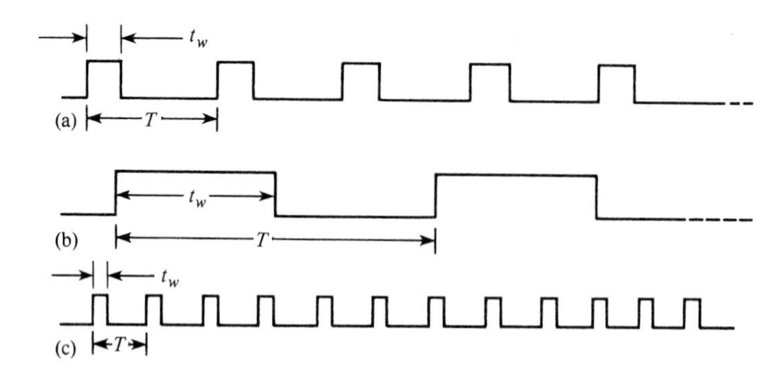

FIGURE 9–50
Examples of repetitive pulse waveforms.

the pulse repetition rate, which is the *fundamental frequency* of the waveform. The frequency can be expressed in *hertz* or in *pulses per second.* The time from one pulse to the corresponding point on the next pulse is the period T.

Another very important characteristic of periodic pulse waveforms is the *duty cycle.*

The duty cycle is defined as the ratio of the pulse width (t_w) to the period T and is usually expressed as a percentage:

$$\text{Percent duty cycle} = \left(\frac{t_w}{T}\right)100\% \qquad \textbf{(9–17)}$$

EXAMPLE 9–14 Determine the period, frequency, and duty cycle for the pulse waveform in Figure 9–51.

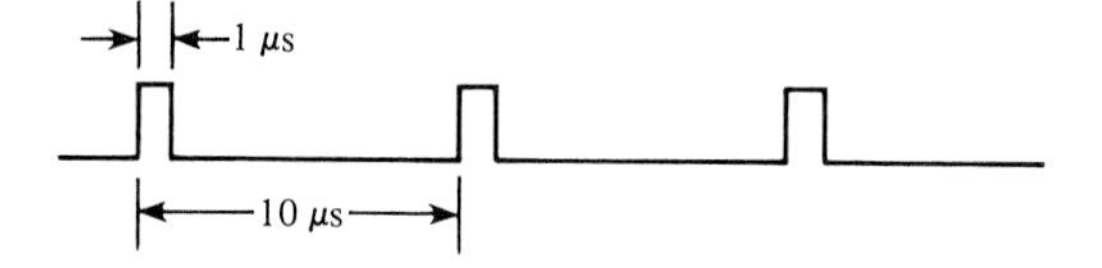

FIGURE 9–51

Solution

$$T = 10\ \mu\text{s}$$

$$f = \frac{1}{T} = \frac{1}{10\ \mu\text{s}} = 0.1\ \text{MHz}$$

$$\text{Percent duty cycle} = \left(\frac{1\ \mu\text{s}}{10\ \mu\text{s}}\right)100\% = 10\%$$

Square Waves

A square wave is a pulse waveform with a duty cycle of 50%. Thus, the pulse width is equal to one-half of the period. A square wave is shown in Figure 9–52.

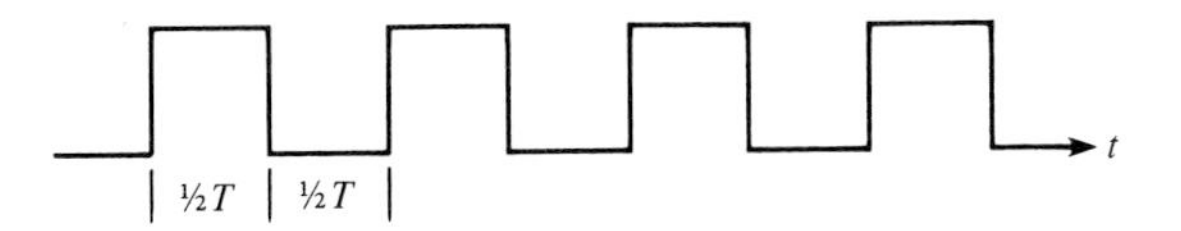

FIGURE 9–52
A square wave has a duty cycle of 50%.

The Average Value of a Pulse Waveform

An alternating square wave that has equal positive and negative amplitudes has a zero average value. If its positive amplitude is greater than its negative amplitude, it has a positive nonzero average value. If its negative amplitude is greater, it has a negative nonzero average value. For periodic pulse waveforms that are not square waves, the duty cycle also affects the average value, as we will demonstrate.

The average value (V_{avg}) of a pulse waveform is equal to its baseline value plus its duty cycle times its amplitude.

The lower level of the waveform is taken as the baseline. The formula is as follows:

$$V_{avg} = \text{baseline} + (\text{duty cycle})(\text{amplitude}) \quad \textbf{(9–18)}$$

The following example illustrates the calculation of the average value.

EXAMPLE 9–15 Determine the average value of each of the waveforms in Figure 9–53.

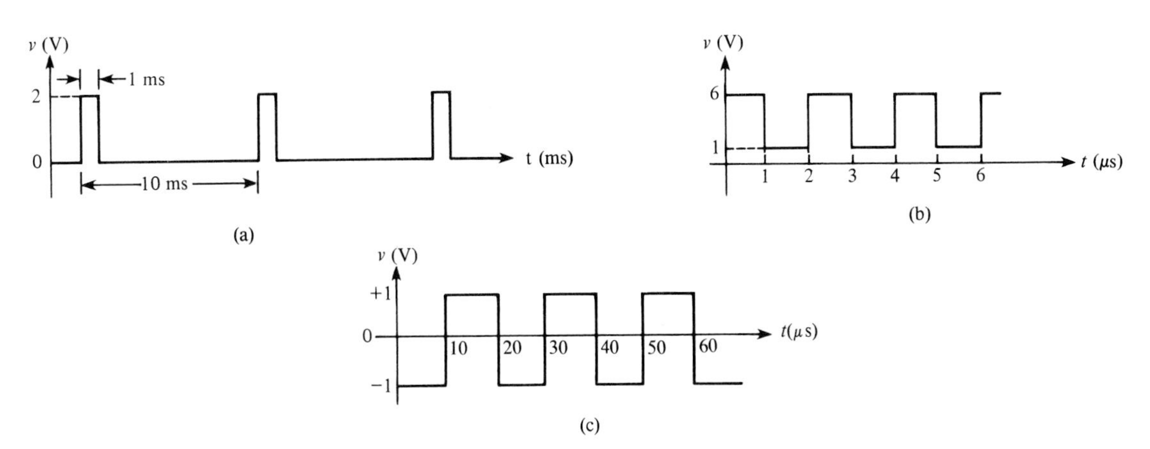

FIGURE 9–53

Solution

(a) Here the baseline is at 0 V, the amplitude is 2 V, and the duty cycle is 10%. The average value is as follows:

$$V_{avg} = \text{baseline} + (\text{duty cycle})(\text{amplitude})$$
$$= 0 \text{ V} + (0.1)(2 \text{ V}) = 0.2 \text{ V}$$

(b) This waveform has a baseline of +1 V, an amplitude of 5 V, and a duty cycle of 50%. The average value is

$$V_{avg} = \text{baseline} + (\text{duty cycle})(\text{amplitude})$$
$$= 1\text{ V} + (0.5)(5\text{ V}) = 1\text{ V} + 2.5\text{ V} = 3.5\text{ V}$$

(c) This waveform is a square wave with a baseline of −1 V and an amplitude of 2 V. The average value is

$$V_{avg} = \text{baseline} + (\text{duty cycle})(\text{amplitude})$$
$$= -1\text{ V} + (0.5)(2\text{ V}) = -1\text{ V} + 1\text{ V} = 0\text{ V}$$

This is an *alternating* square wave, and, like an alternating sine wave, it has an average of zero.

Triangular and Sawtooth Waveforms

Triangular and sawtooth waveforms are formed by voltage or current *ramps*. A ramp is a *linear* increase or decrease in the voltage or current. Figure 9–54 shows both positive- and negative-going ramps. In Part (a) of the figure, the ramp has a positive slope; in Part (b), the ramp has a negative slope. The *slope* of a voltage ramp is $\pm V/t$ and is expressed in units of V/s. The slope of a current ramp is $\pm I/t$ and is expressed in units of A/s.

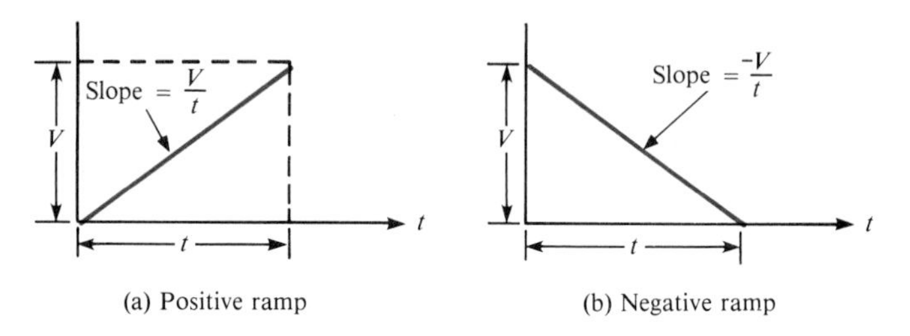

FIGURE 9–54
Ramps.

EXAMPLE 9–16 What are the slopes of the voltage ramps in Figure 9–55?

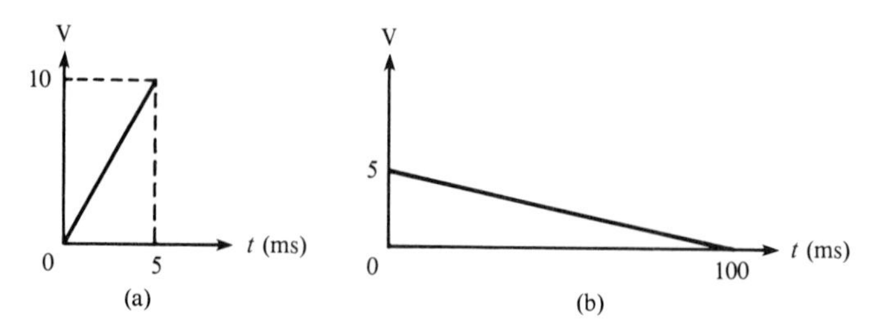

FIGURE 9–55

Solution

(a) The voltage increases from 0 V to +10 V in 5 ms. Thus, $V = 10$ V and $t = 5$ ms. The slope is

$$\frac{V}{t} = \frac{10\text{ V}}{5\text{ ms}} = 2\text{ V/ms} = 2\text{ kV/s}$$

(b) The voltage decreases from +5 V to 0 V in 100 ms. Thus, $V = -5$ V and $t = 100$ ms. The slope is

$$\frac{V}{t} = \frac{-5\text{ V}}{100\text{ ms}} = -0.05\text{ V/ms} = -50\text{ V/s}$$

Triangular Waveforms

Figure 9–56 shows that a triangular waveform is composed of positive- and negative-going ramps having equal slopes. The period of this waveform is measured from one peak to the next corresponding peak, as illustrated. This particular triangular waveform is alternating and has an average value of zero.

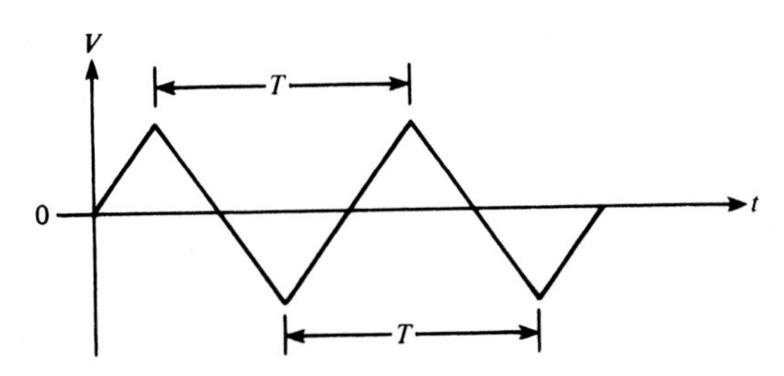

FIGURE 9–56
Alternating triangular waveform.

Figure 9–57 depicts a triangular waveform with a nonzero average value. The frequency for triangular waves is determined in the same way as for sine waves, that is, $f = 1/T$.

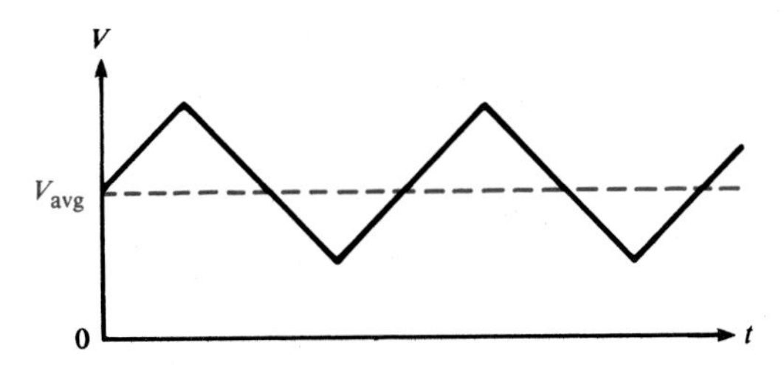

FIGURE 9–57
Nonalternating triangular waveform.

Sawtooth Waveforms

The sawtooth wave is actually a special case of the triangular wave consisting of two ramps, one of much longer duration than the other. Sawtooth waveforms are commonly used in many electronic systems, as you have seen in the case of the oscilloscope. Another example is the electron beam that sweeps across the screen of your TV receiver, creating the picture; it also is controlled by sawtooth voltages and currents. One sawtooth wave produces the horizontal beam movement, and the other produces the vertical beam movement. The sawtooth is sometimes called a *sweep*.

Figure 9–58 is an example of a sawtooth wave. Notice that is consists of a positive-going ramp of relatively long duration, followed by a negative-going ramp of relatively short duration.

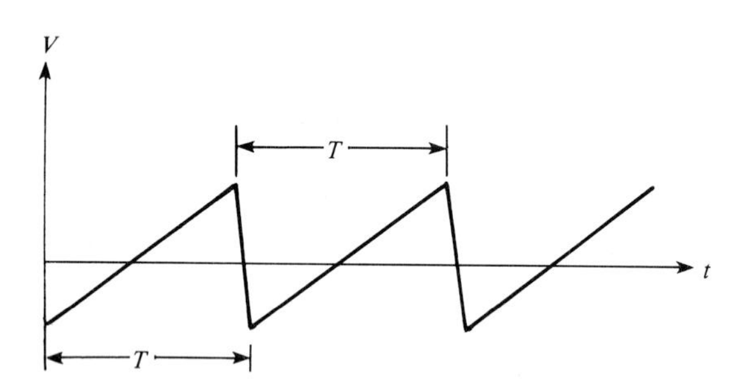

FIGURE 9–58
Alternating sawtooth waveform.

Harmonics

A repetitive nonsinusoidal waveform is composed of a *fundamental frequency* and *harmonic frequencies*. The fundamental frequency is the repetition rate of the waveform, and the harmonics are higher-frequency sine waves that are multiples of the fundamental.

Odd Harmonics

Odd harmonics are frequencies that are *odd multiples* of the fundamental frequency of a waveform. For example, a 1-kHz square wave consists of a fundamental of 1 kHz and odd harmonics of 3 kHz, 5 kHz, 7 kHz, and so on. The 3-kHz frequency in this case is called the *third harmonic;* the 5-kHz frequency is the *fifth harmonic;* and so on.

Even Harmonics

Even harmonics are frequencies that are *even multiples* of the fundamental frequency. For example, if a certain wave has a fundamental of 200 Hz, the second harmonic is 400 Hz, the fourth harmonic is 800 Hz, the sixth harmonic is 1200 Hz, and so on. These are even harmonics.

Composite Waveform

Any variation from a pure sine wave produces harmonics. A nonsinusoidal wave is a composite of the fundamental and the harmonics. Some types of waveforms have only odd harmonics, some have only even harmonics, and some contain both. The shape of the wave is determined by its harmonic content. Generally, only the fundamental and the first few harmonics are of significant importance in determining the waveshape.

A *square wave* is an example of a waveform that consists of a fundamental and only odd harmonics. When the instantaneous values of the fundamental and each odd harmonic are added algebraically at each point, the resulting curve will have the shape of a square wave, as illustrated in Figure 9–59. In Part (a) of the figure, the fundamental and the third harmonic produce a waveshape that begins to resemble a square wave. In Part (b), the fundamental, third, and fifth harmonics produce a closer resemblance. When the seventh harmonic is included, as in Part (c), the resulting waveshape becomes even more like a square wave. As more harmonics are included, a square wave is approached.

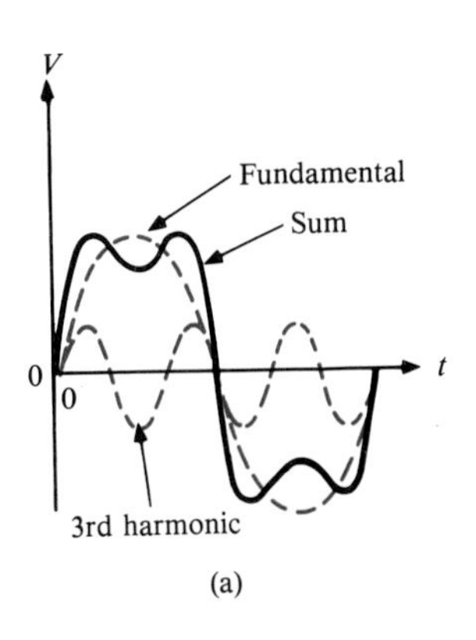

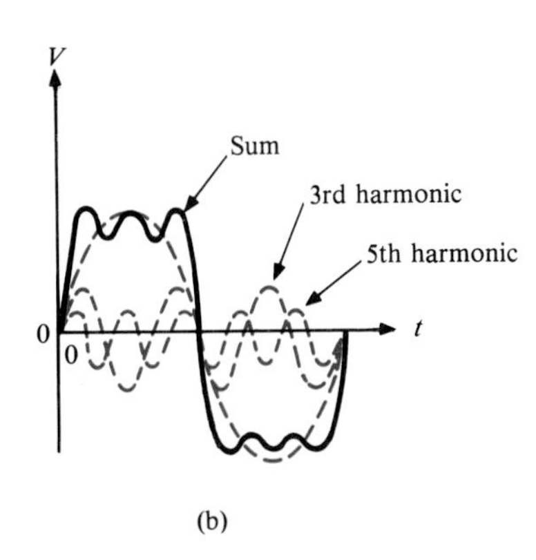

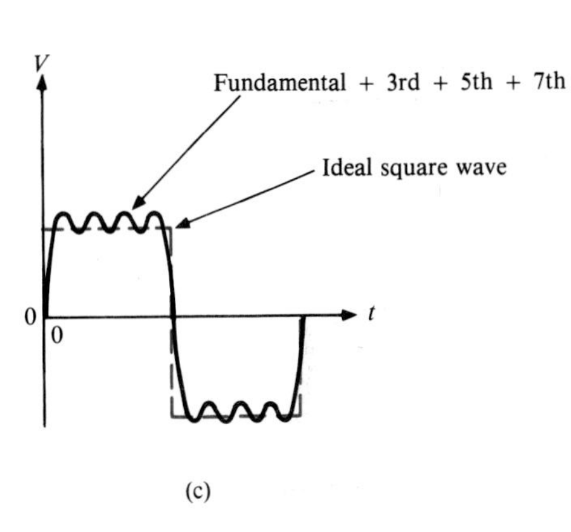

FIGURE 9–59
Odd harmonics produce a square wave.

SECTION REVIEW 9–8

1. Define the following parameters:
   (a) rise time. **(b)** fall time. **(c)** pulse width.
2. In a certain repetitive pulse waveform, the pulses are 200 μs wide and occur once every millisecond. What is the frequency of this waveform?
3. Determine the duty cycle, amplitude, and average value of the waveform in Problem 2 if the pulses go from −2 V to +6 V.
4. What is the second harmonic of a fundamental frequency of 1 kHz?

APPLICATION NOTE

The Application Assignment at the beginning of the chapter required you to check the output voltage parameters of a signal generator. Figure 9–60 shows channel 1 (CH 1) of the oscilloscope being used to display each waveform. The volts/div control and the sec/div control must be properly set for each measurement. As a rule of thumb, when measuring both amplitude and period, use the lowest settings that result in at least one full cycle and the display of the full waveform amplitude on the screen.

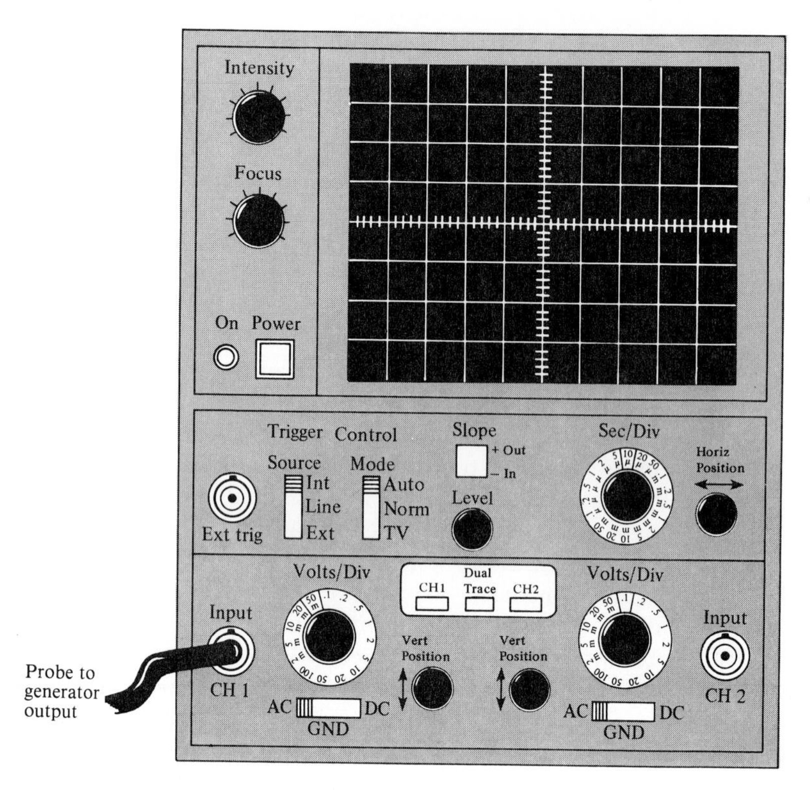

FIGURE 9–60

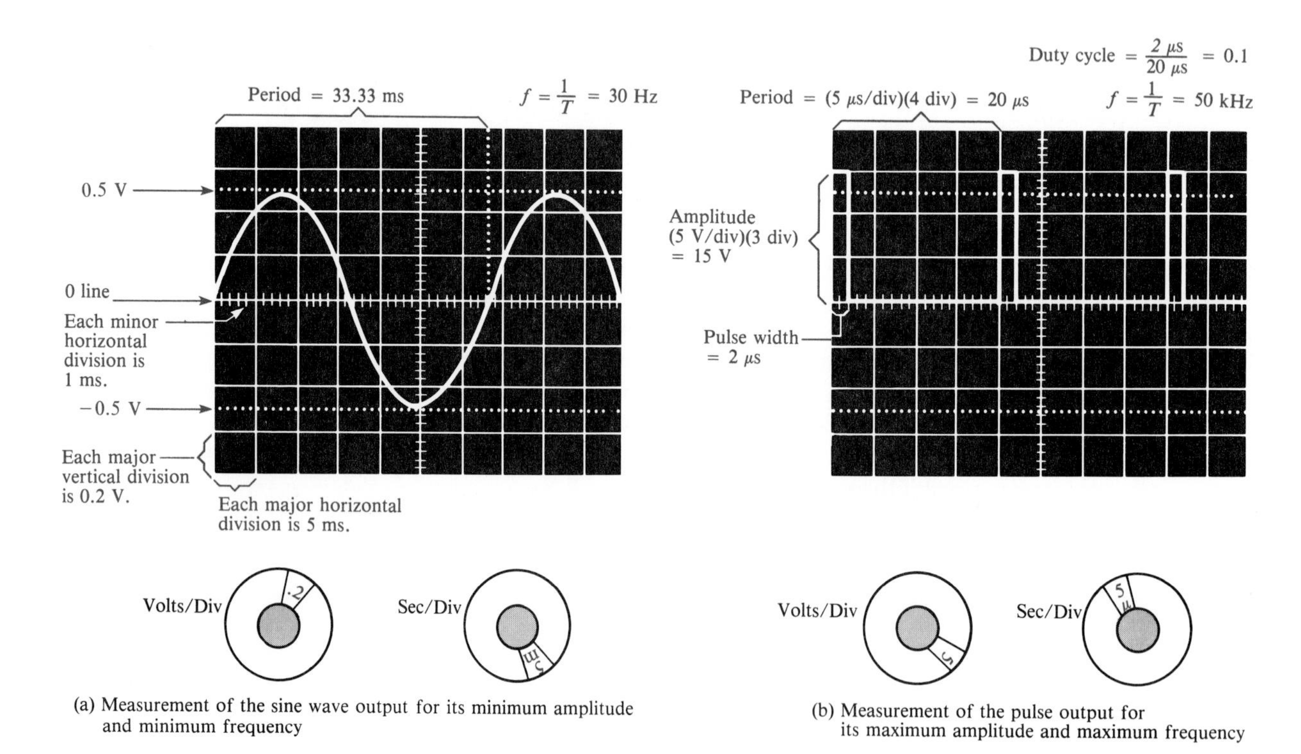

(a) Measurement of the sine wave output for its minimum amplitude and minimum frequency

(b) Measurement of the pulse output for its maximum amplitude and maximum frequency

FIGURE 9–61

Figure 9–61 shows the settings and the displays for two of the required measurements. Peak amplitude and period are measured directly on the screen. Frequency and duty cycle must be calculated from period and pulse width measurements.

SUMMARY

Facts

- The sine wave is a time-varying, periodic waveform.
- The sine wave is a common form of ac (alternating current).
- Alternating current changes direction in response to changes in the polarity of the source voltage.
- One cycle of an alternating sine wave consists of a positive alternation and a negative alternation.
- Two common sources of sine waves are the electromagnetic ac generator and the electronic oscillator circuit.
- A full cycle of a sine wave is 360° or 2π radians. A half-cycle is 180° or π radian. A quarter-cycle is 90° or $\pi/2$ radians.
- A sine wave voltage can be generated by a conductor rotating in a magnetic field.
- Phase angle is the difference in degrees or radians between two sine waves.
- The angular position of a phasor represents the angle of a sine wave, and the length of the phasor represents the amplitude.
- A pulse consists of a transition from a baseline level to an amplitude level, followed by a transition back to the baseline level.
- A triangle or sawtooth wave consists of positive- and negative-going ramps.
- Harmonic frequencies are odd or even multiples of the repetition rate of a nonsinusoidal waveform.

Units

- Hertz (Hz)—the unit of frequency (f). One hertz is one cycle per second.
- Second (s)—the unit of time (t).
- Degree (°)—a unit of angular measure.
- Radian (rad)—a unit of angular measure.

Definitions

- *Cycle*—one repetition of a periodic waveform.
- *Frequency*—the number of cycles completed in one second.
- *Period*—the time required for a periodic waveform to complete one cycle.
- *Instantaneous value*—the voltage or current value of a waveform at a given instant in time.
- *Peak value*—the voltage or current value of a waveform at its maximum point.
- *Peak-to-peak value*—the voltage or current value of a waveform from its minimum to its maximum.
- *rms value*—the root mean square or effective value of a sine wave; it is a measure of the heating effect and is equal to 0.707 times the peak value.
- *Average value*—the average of a sine wave over one half-cycle. It is 0.637 times the peak value.
- *Phasor*—a representation of a time-varying quantity in terms of both magnitude and direction.
- *Radian*—an angular unit equal to 57.3°.
- *Duty cycle*—the ratio of pulse width to period.

- *Pulse width* (t_w)—the time between the 50% points on the rising and falling edges of a pulse.
- *Rise time* (t_r)—the time required for a pulse to change from 10% to 90% of its full amplitude.
- *Fall time* (t_f)—the time required for a pulse to change from 90% to 10% of its full amplitude.

Formulas

$f = \frac{1}{T}$ Frequency (9–1)

$T = \frac{1}{f}$ Period (9–2)

$V_{pp} = 2V_p$ Peak-to-peak voltage (sine wave) (9–3)

$V_{rms} \cong 0.707V_p$ Root-mean-square voltage (sine wave) (9–5)

$V_p \cong 1.414V_{rms}$ Peak voltage (sine wave) (9–7)

$V_{pp} = 2.828V_{rms}$ Peak-to-peak voltage (sine wave) (9–9)

$V_{avg} \cong 0.637V_p$ Average voltage (sine wave) (9–11)

$y = A \sin \theta$ General equation for a sine wave (9–14)

$\text{Percent duty cycle} = \left(\frac{t_w}{T}\right)100$ (9–17)

$V_{avg} = \text{baseline} + (\text{duty cycle})(\text{amplitude})$ Average value of pulse waveform (9–18)

Symbol

FIGURE 9–62
Sine wave voltage source.

SELF-TEST

1. How does alternating current (ac) differ from direct current (dc)?
2. How many times does a sine wave reach a peak value during each cycle?
3. Define *cycle* in relation to a sine wave.
4. What is the difference between a periodic and a nonperiodic waveform?
5. One sine wave has a frequency of 12 kHz, and another has a frequency of 20 kHz. Which sine wave is changing at a faster rate?
6. One sine wave has a period of 2 ms, and another has a period of 5 ms. Which sine wave is changing at a faster rate?
7. How many cycles does a sine wave go through in 10 seconds when its frequency is 60 Hz?

8. If the peak value of a certain sine wave voltage is 10 V, what is the peak-to-peak value? The rms value?
9. Explain the significance of the rms value of a sine wave.
10. What is the average of an ac sine wave over a full cycle? Why?
11. How is the average value of a sine wave determined?
12. What type of sine wave voltage sources are available in your laboratory?
13. What determines the frequency of the output of an electromagnetic ac generator?
14. What do the terms "AM" and "FM" mean?
15. How many degrees are in a half-cycle of a sine wave?
16. How many degrees are there in two radians?
17. One sine wave has a positive-going zero crossing at 10°, and another sine wave has a positive-going zero crossing at 45°. What is the phase angle between the two waveforms?
18. What is the instantaneous value of a sine wave current at 32° from its positive-going zero crossing if its peak value is 15 A?
19. The rms current through a 10-kΩ resistor is 5 mA. What is the rms voltage drop across the resistor?
20. Two series resistors are connected to an ac source. There is 6.5 V rms across one resistor and 3.2 V rms across the other. What is the rms source voltage?
21. One 10-kHz pulse waveform consists of pulses that are 10 μs wide, and another consists of pulses that are 50 μs wide. Which 10-kHz waveform has a higher duty cycle?
22. What is the duty cycle of a square wave?
23. Explain why a sawtooth is a special case of a triangular waveform.

PROBLEM SET A

Section 9–1

9–1 Calculate the frequency for each of the following values of period:
(a) 1 s **(b)** 0.2 s **(c)** 50 ms
(d) 1 ms **(e)** 500 μs **(f)** 10 μs

9–2 Calculate the period for each of the following values of frequency:
(a) 1 Hz **(b)** 60 Hz **(c)** 500 Hz
(d) 1 kHz **(e)** 200 kHz **(f)** 5 MHz

9–3 A sine wave goes through 5 cycles in 10 μs. What is its period?

9–4 A sine wave has a frequency of 50 kHz. How many cycles does it complete in 10 ms?

Section 9–2

9–5 A sine wave has a peak value of 12 V. Determine the following values:
(a) rms. **(b)** peak-to-peak. **(c)** average.

9–6 A sinusoidal current has an rms value of 5 mA. Determine the following values:
(a) peak. **(b)** average. **(c)** peak-to-peak.

9–7 For the sine wave in Figure 9–63, determine the peak, peak-to-peak, rms, and average values.

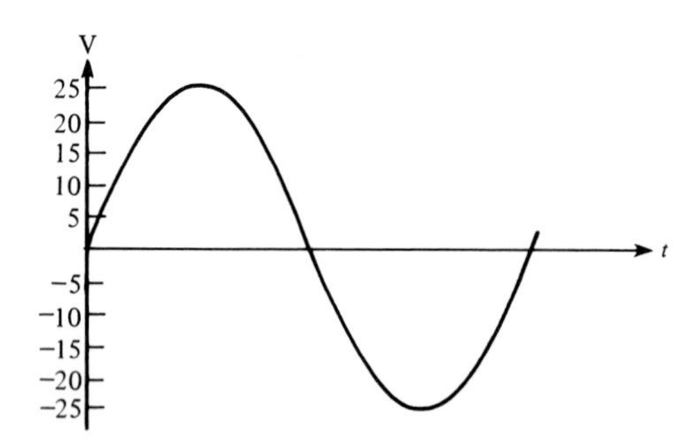

FIGURE 9–63

Section 9–3

9–8 The conductive loop on the rotor of a simple two-pole, single-phase generator rotates at a rate of 250 rps. What is the frequency of the induced output voltage?

9–9 A certain four-pole generator has a speed of rotation of 3600 rpm. What is the frequency of the voltage produced by this generator?

Section 9–4

9–10 Sine wave A has a positive-going zero crossing at 30°. Sine wave B has a positive-going zero crossing at 45°. Determine the phase angle between the two signals. Which signal leads?

9–11 One sine wave has a positive peak at 75°, and another has a positive peak at 100°. How much is each sine wave shifted in phase from the 0° reference? What is the phase angle between them?

9–12 Make a sketch of two sine waves as follows: Sine wave A is the reference, and sine wave B lags A by 90°. Both have equal amplitudes.

Section 9–5

9–13 A certain sine wave has a positive-going zero crossing at 0° and an rms value of 20 V. Calculate its instantaneous value at each of the following angles:

(a) 15° **(b)** 33° **(c)** 50°
(d) 110° **(e)** 70° **(f)** 145°
(g) 250° **(h)** 325°

9–14 For a particular 0° reference sinusoidal current, the peak value is 100 mA. Determine the instantaneous value at each of the following points:

(a) 35° **(b)** 95° **(c)** 190°
(d) 215° **(e)** 275° **(f)** 360°

9–15 Sine wave A lags sine wave B by 30°. Both have peak values of 15 V. Sine wave A is the reference with a positive-going crossing at 0°. Determine the instantaneous value of sine wave B at 30°, 45°, 90°, 180°, 200°, and 300°.

9–16 Repeat Problem 9–15 for the case when sine wave A *leads* sine wave B by 30°.

Section 9–6

9–17 Draw a phasor diagram to represent the sine waves in Figure 9–64.

9–18 Sketch the sine waves represented by the phasor diagram in Figure 9–65. The phasor lengths represent peak values.

FIGURE 9–64

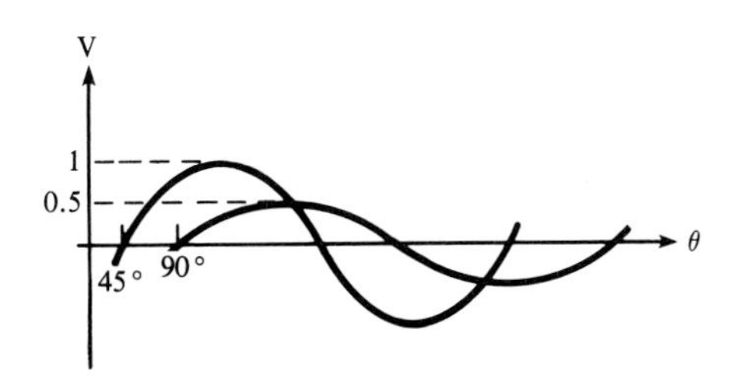

FIGURE 9–65

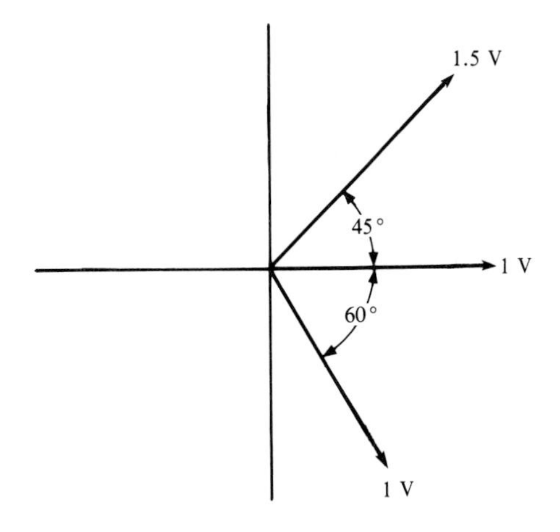

Section 9–7

9–19 A sinusoidal voltage is applied to the resistive circuit in Figure 9–66. Determine the following:

(a) I_{rms} **(b)** I_{avg} **(c)** I_p **(d)** I_{pp} **(e)** i at the positive peak

9–20 Find the average values of the voltages across R_1 and R_2 in Figure 9–67. Values shown are rms.

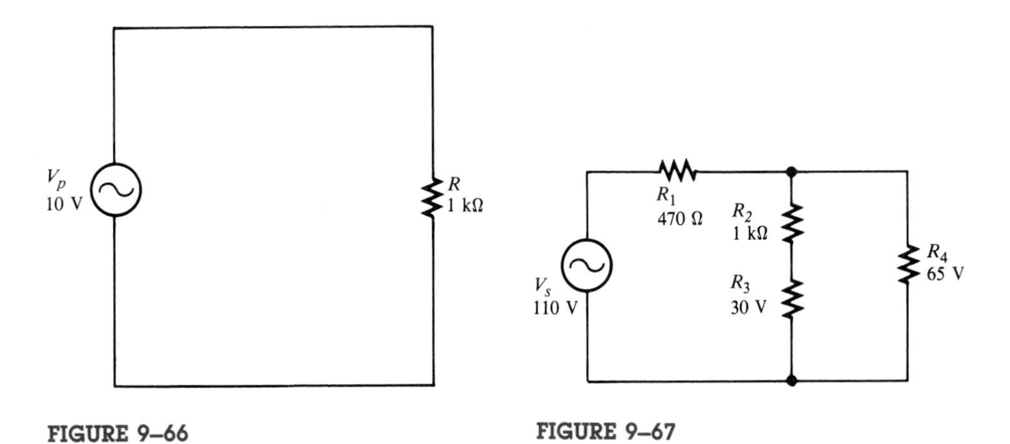

FIGURE 9–66

FIGURE 9–67

Section 9–8

9–21 From the graph in Figure 9–68, determine the approximate values of t_r, t_f, t_w, and amplitude.

9–22 Determine the duty cycle for each pulse waveform in Figure 9–69.

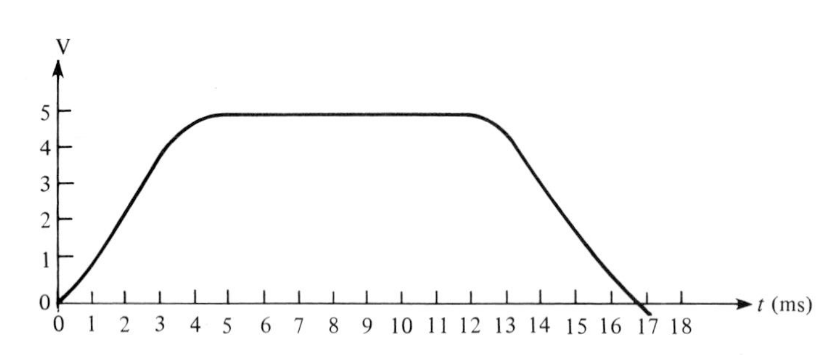

FIGURE 9–68

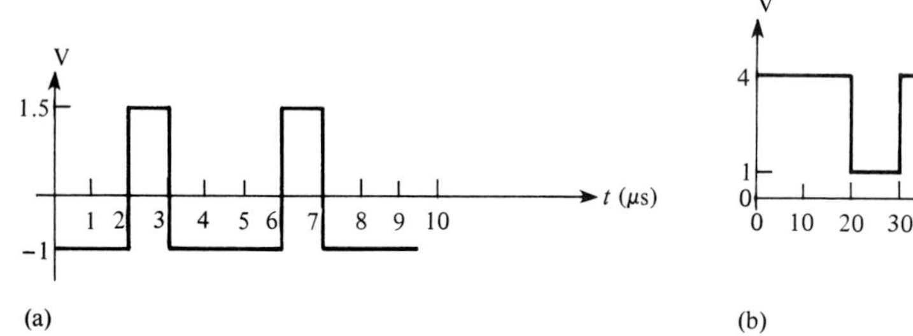

FIGURE 9–69

9–23 Find the average value of each pulse waveform in Figure 9–69.

9–24 What is the frequency of each waveform in Figure 9–69?

9–25 What is the frequency of each sawtooth waveform in Figure 9–70?

9–26 A square wave has a period of 40 μs. List the first six odd harmonics.

9–27 What is the fundamental frequency of the square wave mentioned in Problem 9–26?

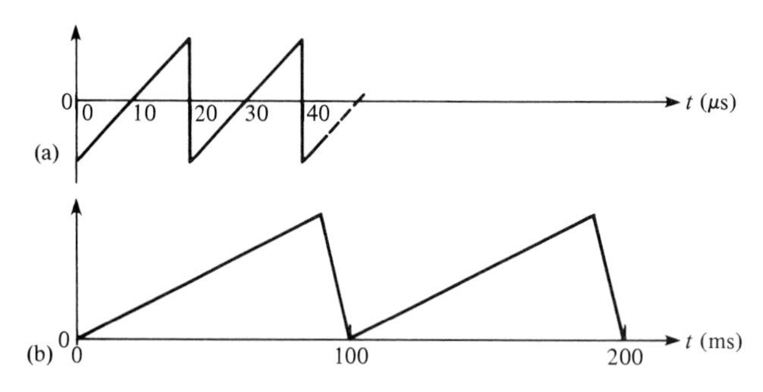

FIGURE 9–70

PROBLEM SET B

9–28 A certain sine wave has a frequency of 2.2 kHz and an rms value of 25 V. Assuming a given cycle begins (zero crossing) at $t = 0$ s, what is the change in voltage from 0.12 ms to 0.2 ms?

9–29 Figure 9–71 shows a sinusoidal voltage source in series with a dc source. Effectively, the two voltages are superimposed. Sketch the voltage across R_L. Determine the maximum current through R_L and the average voltage across R_L.

9–30 A nonsinusoidal waveform called a *stairstep* is shown in Figure 9–72. Determine its average value.

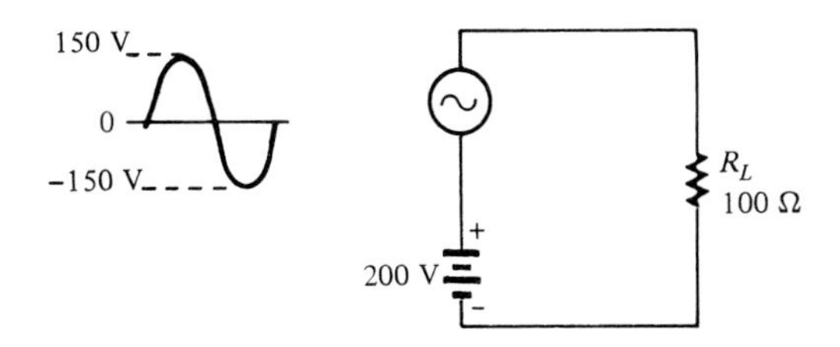

FIGURE 9–71

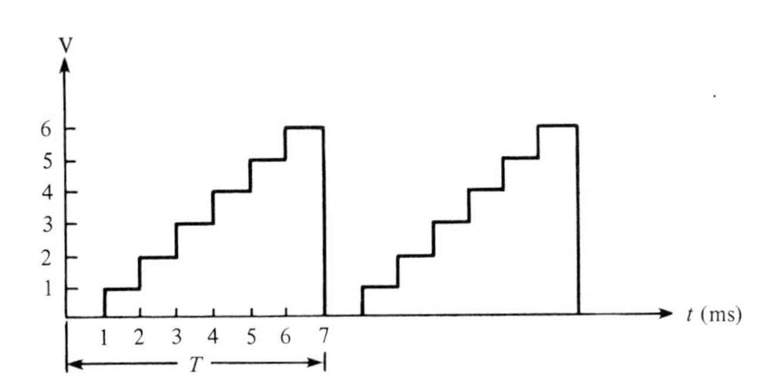

FIGURE 9–72

ANSWERS TO SECTION REVIEWS

Section 9–1

1. From the zero crossing through a positive peak, then through zero to a negative peak and back to the zero crossing.
2. At the zero crossings. **3.** 2.
4. From one zero crossing to the next *corresponding* zero crossing, or from one peak to the next *corresponding* peak.
5. The number of cycles completed in one second, hertz.
6. 200 kHz. **7.** 8.33 ms.

Section 9–2

1. **(a)** 2 V; **(b)** 4 V; **(c)** 9.42 V.
2. **(a)** 1.77 V; **(b)** 3.54 V; **(c)** 1.66 V.

Section 9–3

1. By rotating a conductor in a magnetic field.
2. Audio frequency (af), radio frequency (rf).

Section 9–4

1. **(a)** 90°; **(b)** 180°; **(c)** 270°; **(d)** 360°. **2.** 180, π.
3. 360, 2π. **4.** 45°.

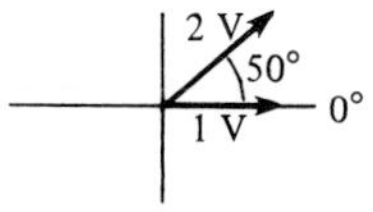

FIGURE 9–73

Section 9–5

1. 8.66 V. 2. 8.19 V.

Section 9–6

1. A graphic representation of the magnitude and angular position of a time-varying quantity.
2. See Figure 9–73.

Section 9–7

1. 37.9 mA. 2. 17.89 V.

Section 9–8

1. (a) The time interval from 10% to 90% of the rising pulse edge;
   (b) the time interval from 90% to 10% of the falling pulse edge;
   (c) the time interval from 50% of the leading pulse edge to 50% of the trailing pulse edge.
2. 1 kHz. 3. 20%, 8 V, −0.4 V. 4. 2 kHz.

Capacitors

The capacitor is an electrical device that can store electrical charge, thereby creating an electric field which in turn stores energy; the measure of that energy-storing ability is called *capacitance*. In this chapter, the basic capacitor is introduced and its characteristics are studied. The physical construction and electric properties of various types of capacitors are discussed.

The basic behavior of capacitors in both dc and ac circuits is studied, and series and parallel combinations are analyzed. Representative applications and methods of testing capacitors also are discussed.

Specifically, in this chapter, you will learn:

- The basic structure of a capacitor.
- What a capacitor is and what it does.
- The definition of *capacitance* and how it is measured.
- How a capacitor stores energy.
- The definition of *Coulomb's law* and how it relates to an electric field and the storage of energy.
- How a capacitor charges and discharges.
- How various physical parameters determine capacitance value.
- Several common capacitor classifications according to dielectric material.
- What happens when capacitors are connected in series.
- What happens when capacitors are connected in parallel.
- The meaning of *time constant*.
- How the time constant affects the charging and discharging of a capacitor.
- Why a capacitor blocks dc.
- How a capacitor introduces a phase shift between current and voltage.
- The definition of *capacitive reactance* and how to determine its value in a circuit.
- Why there is ideally no energy loss in a capacitor.
- The significance of reactive power.
- Several common capacitor applications.
- How to check a capacitor with an ohmmeter.

10

APPLICATION ASSIGNMENT

A defective stereo amplifier is returned for repair. Your supervisor thinks that a coupling or a bypass capacitor in the amplifier circuitry may be faulty. Your assignment is to check all points in the circuit to which a capacitor is connected for the proper ac and/or dc voltage levels, which are specified on the schematic.

After checking the voltage levels, you must remove from the circuit board any capacitor that you suspect of being bad and test it for proper functioning and value.

After completing this chapter, you will be able to carry out this assignment. The solution is given in the Application Note at the end of the chapter.

THE CAPACITOR

10–1

Basic Construction

In its simplest form, a capacitor is an electrical device constructed of two parallel conductive plates separated by an insulating material called the *dielectric*. Connecting leads are attached to the parallel plates. A basic capacitor is shown in Figure 10–1(a), and the schematic symbol is shown in Part (b).

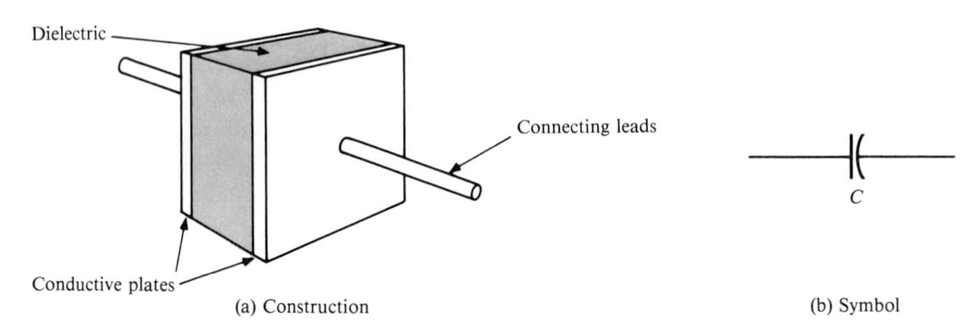

FIGURE 10–1
The basic capacitor.

How a Capacitor Stores Charge

In the neutral state, both plates of a capacitor have an equal number of free electrons, as indicated in Figure 10–2(a). When the capacitor is connected to a voltage source through a resistor, as shown in Part (b), electrons (negative charge) are removed from plate A, and an equal number are deposited on plate B. As plate A loses electrons and plate B gains electrons, plate A becomes positive with respect to plate B. During this charging process, electrons flow only through the connecting leads and the source. No electrons flow through the dielectric of the capacitor because it is an insulator. The movement of electrons ceases when the voltage across the capacitor equals the source voltage, as indicated in Part (c). If the capacitor is disconnected from the source, it retains the stored charge for a long period of time (the length of time depends on the type of capacitor) and still has the voltage across it, as shown in Part (d). Actually, the charged capacitor can be considered as a temporary battery.

Capacitance

The amount of charge per unit of voltage that a capacitor can store is its *capacitance,* designated C. That is, capacitance is a measure of a capacitor's ability to store charge. The more charge per unit of voltage that a capacitor can store, the greater its capacitance, as expressed by the following formula:

$$C = \frac{Q}{V} \qquad (10\text{–}1)$$

where C is capacitance, Q is charge, and V is voltage.

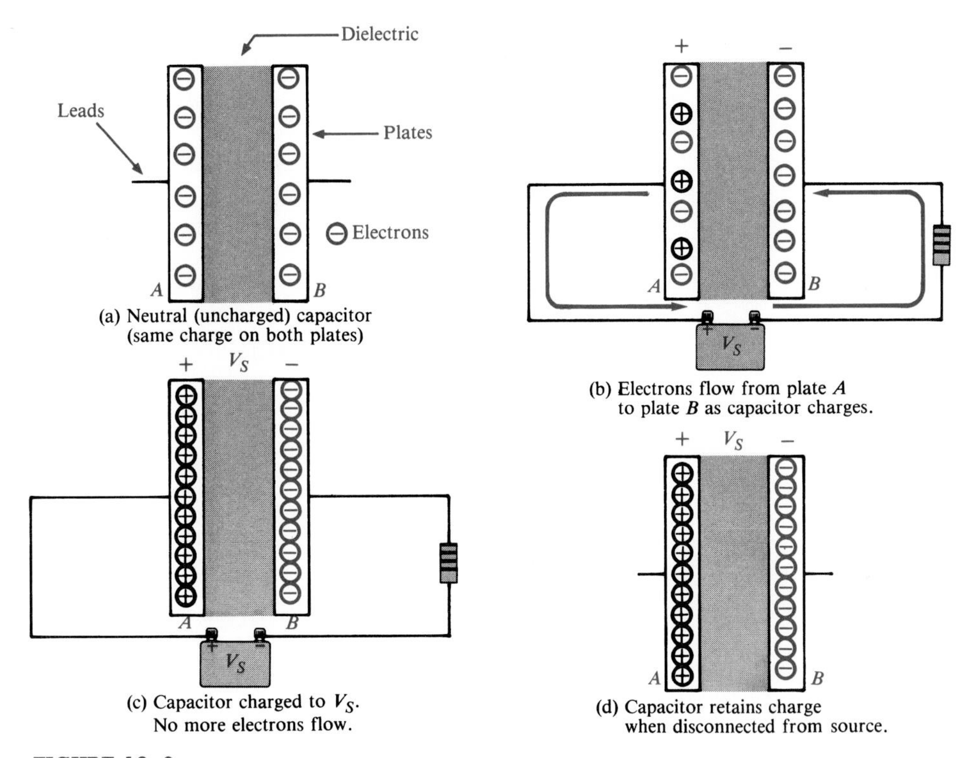

(a) Neutral (uncharged) capacitor (same charge on both plates)

(b) Electrons flow from plate A to plate B as capacitor charges.

(c) Capacitor charged to V_S. No more electrons flow.

(d) Capacitor retains charge when disconnected from source.

FIGURE 10–2
Illustration of a capacitor storing charge.

The Unit of Capacitance

The *farad* (F) is the basic unit of capacitance. By definition, *one farad* is the amount of capacitance when *one coulomb* of charge is stored with *one volt* across the plates.

Most capacitors that you will use in electronics work have capacitance values in microfarads (μF) and picofarads (pF). A microfarad is one-millionth of a farad ($1\ \mu\text{F} = 1 \times 10^{-6}$ F), and a picofarad is one-trillionth of a farad ($1\ \text{pF} = 1 \times 10^{-12}$ F).

EXAMPLE 10–1

(a) A certain capacitor stores 50 microcoulombs (50 μC) when 10 V are applied across its plates. What is its capacitance?

(b) A 2-μF capacitor has 100 V across its plates. How much charge does it store?

(c) Determine the voltage across a 100-pF capacitor that is storing 2 μC of charge.

Solution

(a) Use Equation (10–1):

$$C = \frac{Q}{V} = \frac{50\ \mu\text{C}}{10\ \text{V}} = 5\ \mu\text{F}$$

The calculator sequence is

5 0 EE +/− 6 ÷ 1 0 =

(b) Solve Equation (10–1) for Q as follows:

$$\frac{Q}{V} = C$$

$$\frac{Q}{\cancel{V}}(\cancel{V}) = C(V)$$

$$Q = CV = (2\ \mu\text{F})(100\ \text{V}) = 200\ \mu\text{C}$$

(c) Solve Equation (10–1) for V as follows:

$$\cancel{C}\left(\frac{V}{\cancel{C}}\right) = \frac{Q}{\cancel{V}}\left(\frac{\cancel{V}}{C}\right)$$

$$V = \frac{Q}{C} = \frac{2\ \mu\text{C}}{100\ \text{pF}} = 2 \times 10^4\ \text{V} = 20\ \text{kV}$$

How a Capacitor Stores Energy

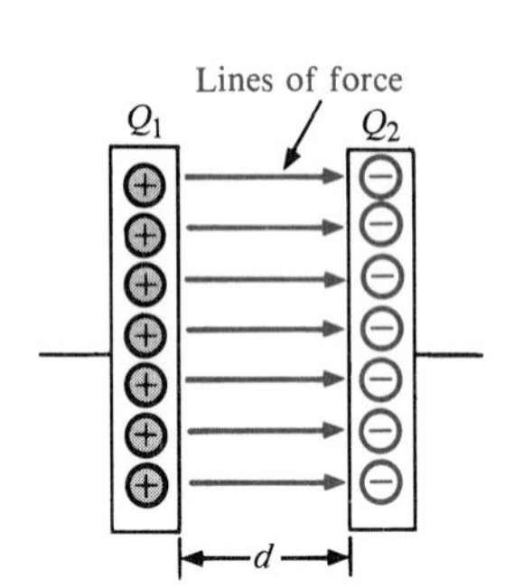

FIGURE 10–3
An electric field exists between the plates of a charged capacitor.

A capacitor stores energy in the form of an *electric field* that is established by the opposite charges on the two plates. The electric field is represented by *lines of force* between the positive and negative charges and concentrated within the dielectric, as shown in Figure 10–3.

Coulomb's Law

A force exists between two charged bodies that is directly proportional to the product of the two charges and inversely proportional to the square of the distance between the bodies.

This relationship is expressed as

$$F = \frac{kQ_1Q_2}{d^2} \qquad (10\text{–}2)$$

where F is the force in newtons, Q_1 and Q_2 are the charges in coulombs, d is the distance between the charges in meters, and k is a proportionality constant equal to 9×10^9. Figure 10–4 illustrates the line of force between a positive and a negative charge. Many opposite charges on the plates of a capacitor create many lines of force, which form an *electric field* that stores energy within the dielectric.

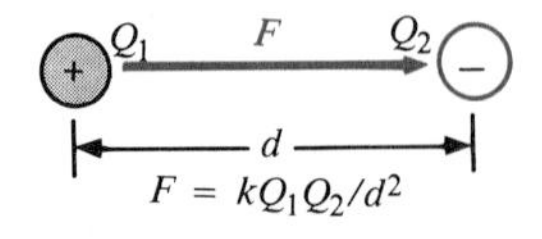

FIGURE 10–4
A force exists between two charged bodies.

The greater the forces between the charges on the plates of a capacitor, the more energy is stored. The amount of energy stored therefore is directly proportional to the capacitance, because, from Coulomb's law, the more charge stored, the greater the force.

Also, from the equation $Q = CV$, the amount of charge stored is directly related to the voltage as well as the capacitance. Therefore, the amount of energy stored also is dependent on the square of the voltage across the plates of the capacitor. The formula for the energy stored by a capacitor is as follows:

$$W = \frac{1}{2}CV^2 \quad (10\text{–}3)$$

where the energy, W, is in joules when C is in farads and V is in volts.

Voltage Rating

Every capacitor has a limit on the amount of voltage that it can withstand across its plates. The *voltage rating* specifies the maximum dc voltage that can be applied without risk of damage to the device. If this maximum voltage, commonly called the *breakdown voltage* or *working voltage,* is exceeded, permanent damage to the capacitor can result.

Both the capacitance and the voltage rating must be taken into consideration before a capacitor is used in a circuit application. The choice of capacitance value is based on particular circuit requirements (and on factors that are studied later). The voltage rating should always be well above the maximum voltage expected in a particular application.

Dielectric Strength

The breakdown voltage of a capacitor is determined by the dielectric strength of the dielectric material used. The dielectric strength is expressed in volts/mil (1 mil = 0.001 in). Table 10–1 lists typical values for several materials. Exact values vary depending on the specific composition of the material.

TABLE 10–1
Some common dielectric materials and their dielectric strengths.

Material	Dielectric Strength (volts/mil)
Air	80
Oil	375
Ceramic	1000
Paper (paraffined)	1200
Teflon®	1500
Mica	1500
Glass	2000

The dielectric strength can best be explained by an example. Assume that a certain capacitor has a plate separation of 1 mil and that the dielectric material is ceramic. This particular capacitor can withstand a maximum voltage of 1000 V, because its dielectric strength is 1000 V/mil. If the maximum voltage is exceeded, the dielectric may break down and conduct current, causing permanent damage to the capacitor. If the ceramic capacitor has a plate separation of 2 mils, its breakdown voltage is 2000 V.

Temperature Coefficient

The temperature coefficient indicates the amount and direction of a change in capacitance value with temperature. A *positive* temperature coefficient means that the capacitance increases with an increase in temperature or decreases with a decrease in temperature. A *negative* coefficient means that the capacitance decreases with an increase in temperature or increases with a decrease in temperature.

Temperature coefficients typically are specified in *parts per million per degree Celsius* (ppm/°C). For example, a negative temperature coefficient of 150 ppm/°C for a 1-μF capacitor means that for every degree rise in temperature, the capacitance decreases by 150 pF (there are one million picofarads in one microfarad).

Leakage

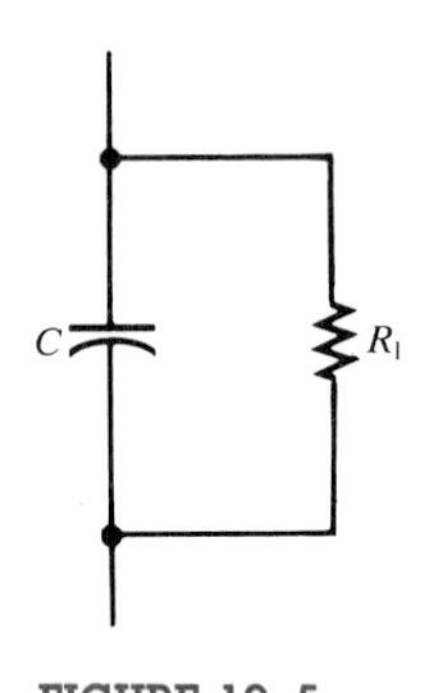

FIGURE 10–5
Equivalent circuit for a nonideal capacitor.

No insulating material is perfect. The dielectric of any capacitor will conduct some very small amount of current. Thus, the charge on a capacitor will eventually leak off. Some types of capacitors have higher leakages than others. An equivalent circuit for a nonideal capacitor is shown in Figure 10–5. The parallel resistor represents the extremely high resistance of the dielectric material through which leakage current flows.

Physical Characteristics of a Capacitor

The following parameters are important in establishing the capacitance and the voltage rating of a capacitor: plate area, plate separation, and dielectric constant.

Plate Area

Capacitance is directly proportional to the physical size of the plates as determined by the plate area, A. A larger plate area produces a larger capacitance, and vice versa. Figure 10–6(a) shows that the plate area of a parallel plate capacitor is the area of one of the plates. If the plates are moved in relation to each other, as shown in Part (b), the *overlapping area* determines the effective plate area. This variation in effective plate area is the basis for a certain type of variable capacitor.

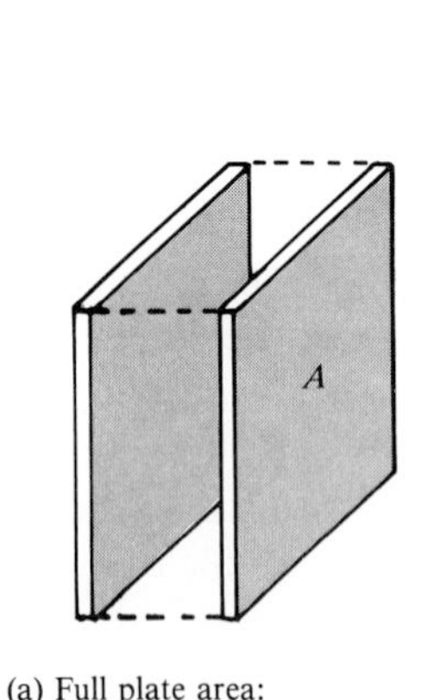

(a) Full plate area: more capacitance

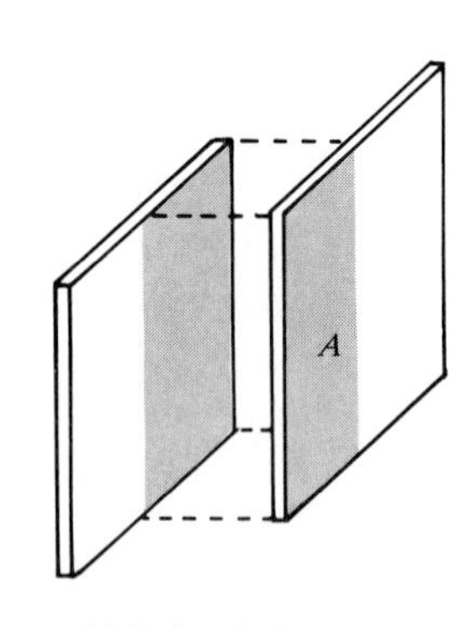

(b) Reduced plate area: less capacitance

FIGURE 10–6
Capacitance is directly proportional to plate area (A).

Plate Separation

Capacitance is inversely proportional to the distance between the plates. The plate separation is designated *d,* as shown in Figure 10–7. A greater separation of the plates produces a smaller capacitance, as illustrated in the figure. The breakdown voltage is directly proportional to the plate separation. The further the plates are separated, the greater the breakdown voltage.

FIGURE 10–7
Capacitance is inversely proportional to the distance *d* between the plates.

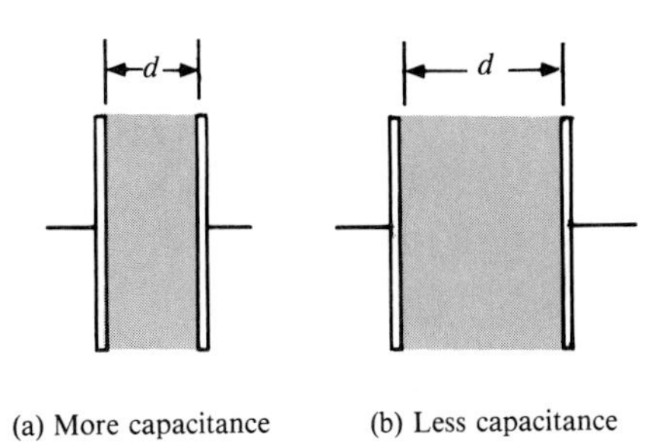

(a) More capacitance (b) Less capacitance

Dielectric Constant

As you know, the insulating material between the plates of a capacitor is called the *dielectric.* Every dielectric material has the ability to concentrate the lines of force of the electric field existing between the oppositely charged plates of a capacitor and thus increase the capacity for energy storage. The measure of a material's ability to establish an electric field is called the *dielectric constant* or *relative permittivity,* symbolized by ϵ_r (the Greek letter epsilon).

Capacitance is directly proportional to the dielectric constant. The dielectric constant of a vacuum is defined as 1, and that of air is very close to 1. These values are used as a reference, and all other materials have values of ϵ_r specified with respect to that of a vacuum or air. For example, a material with $\epsilon_r = 8$ can have a capacitance eight times greater than that of air, with all other factors being equal.

The dielectric constant (relative permittivity) is dimensionless, because it is a relative measure and is a ratio of the *absolute permittivity,* ϵ, of a material to the *absolute permittivity,* ϵ_0, of a vacuum, as expressed by the following formula:

$$\epsilon_r = \frac{\epsilon}{\epsilon_0} \tag{10–4}$$

The value of ϵ_0 is 8.85×10^{-12} F/m (farads per meter).

Formula for Capacitance in Terms of Physical Parameters

You have seen how capacitance is directly related to plate area A and the dielectric constant ϵ_r, and inversely related to plate separation d. An exact formula for calculating the capacitance in terms of these three quantities is as follows:

$$C = \frac{A\epsilon_r(8.85 \times 10^{-12}\ \text{F/m})}{d} \tag{10–5}$$

where A is in square meters (m^2), d is in meters (m), and C is in farads (F). Recall that 8.85×10^{-12} F/m is the absolute permittivity, ϵ_0, of a vacuum and that $\epsilon_r(8.85 \times 10^{-12})$ F/m is the absolute permittivity of a dielectric, as derived from Equation (10–4).

EXAMPLE 10–2 Determine the capacitance of a parallel plate capacitor having a plate area of $0.01\ m^2$ and a plate separation of 0.02 m. The dielectric is mica which has a dielectric constant of 5.0.

Solution
Use Equation (10–5):

$$C = \frac{A\epsilon_r(8.85 \times 10^{-12}\ \text{F/m})}{d}$$
$$= \frac{(0.01\ \text{m}^2)(5.0)(8.85 \times 10^{-12}\ \text{F/m})}{0.02\ \text{m}} = 22.13\ \text{pF}$$

The calculator sequence is

SECTION REVIEW 10–1

1. Define *capacitance*.
2. (a) How many microfarads in a farad?
   (b) How many picofarads in a farad?
   (c) How many picofarads in a microfarad?
3. (a) When the plate area of a capacitor is increased, does the capacitance increase or decrease?
   (b) When the distance between the plates is increased, does the capacitance increase or decrease?

TYPES OF CAPACITORS

10–2

Capacitors normally are classified according to the type of dielectric material. The most common types of dielectric materials are mica, ceramic, paper/plastic, and electrolytic (aluminum oxide and tantalum oxide). In this section, the characteristics and construction of each of these types of capacitors and the variable capacitors are examined.

Mica Capacitors

There are two types of mica capacitors: stacked-foil and silver-mica. The basic construction of the stacked-foil type is shown in Figure 10–8. It consists of alternate layers of metal foil and thin sheets of mica. The metal foil forms the plate, with alternate foil sheets connected together to increase the plate area. More layers are used to increase the plate area, thus increasing the capacitance.

The mica/foil stack is encapsulated in an insulating material such as Bakelite®, as shown in Part (b) of the figure. The silver-mica capacitor is formed in a similar way by stacking mica sheets with silver electrode material screened on them.

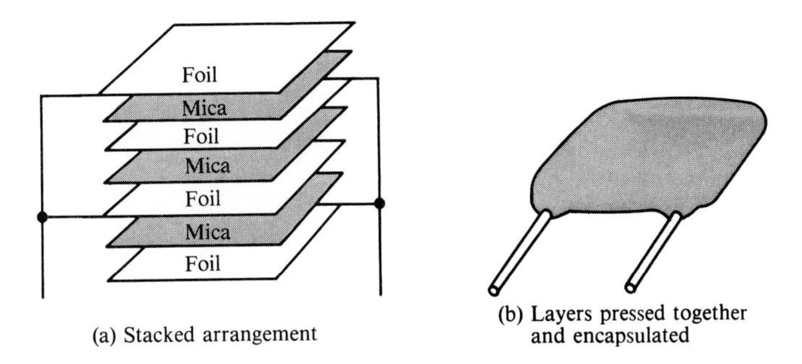

FIGURE 10–8
Construction of a typical mica capacitor.

Mica capacitors are available with capacitance values ranging from 1 pF to 0.1 μF and voltage ratings from 100 to 2500 V dc. Temperature coefficients from −20 to +100 PPM/°C are common. Mica has a typical dielectric constant of 5.

Ceramic Capacitors

Ceramic dielectrics provide very high dielectric constants (1200 is typical). As a result, comparatively high capacitance values can be achieved in a small physical size. Ceramic capacitors are available in either ceramic disk, as shown in Figure 10–9, or in a multilayer configuration, as shown in Figure 10–10.

FIGURE 10–9
A ceramic disk capacitor and its basic construction.

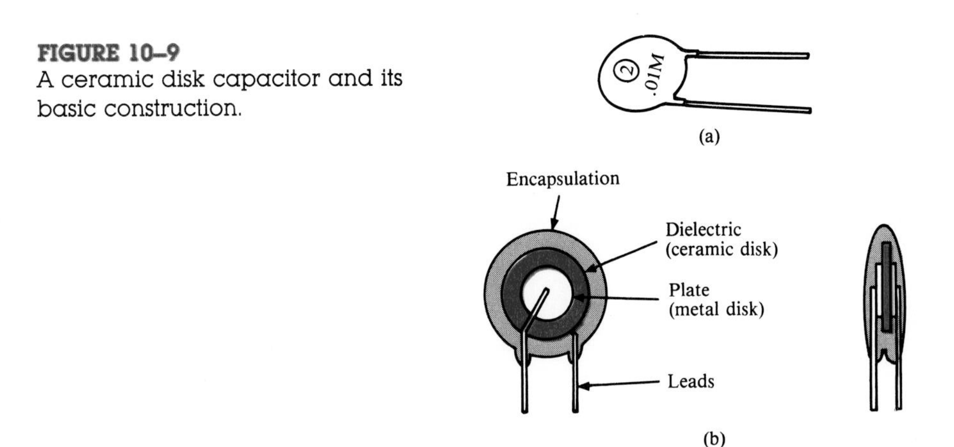

Ceramic capacitors typically are available in capacitance values ranging from 1 pF to 2.2 μF with voltage ratings up to 6 kV. A typical temperature coefficient for ceramic capacitors is 200,000 PPM/°C.

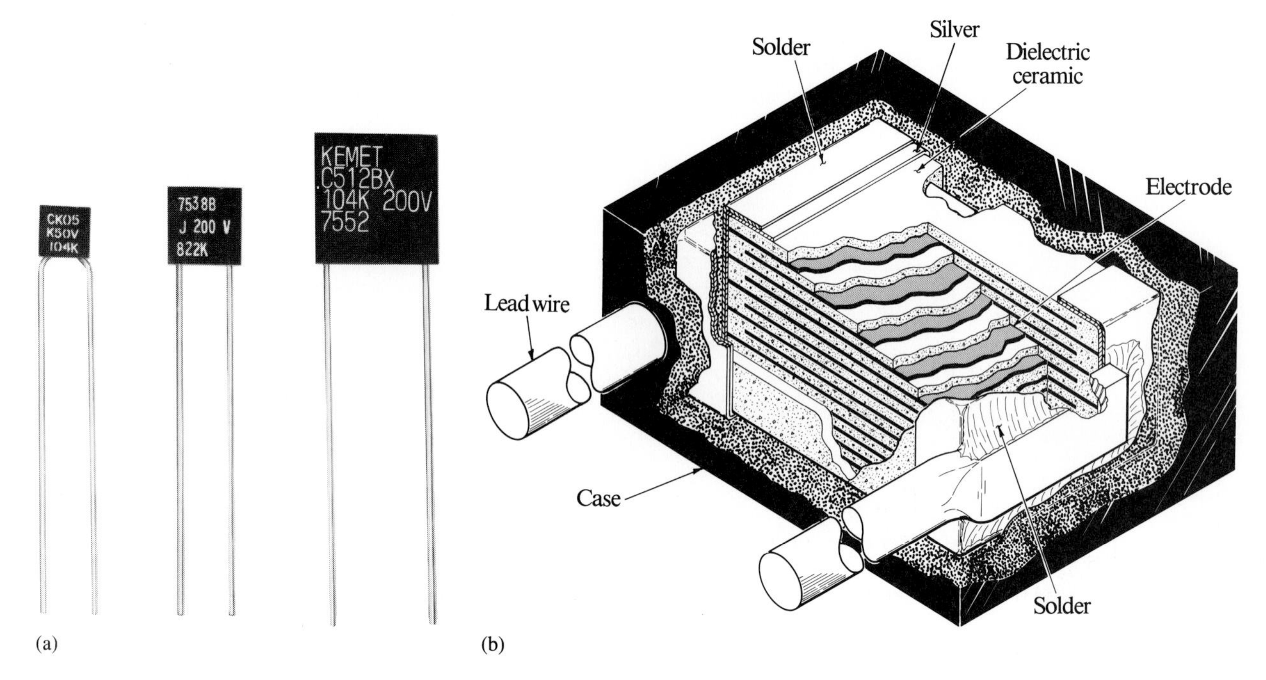

FIGURE 10–10
Ceramic capacitors. (a) Typical capacitors. (b) Construction view (courtesy of Union Carbide Electronics Division).

Paper/Plastic Capacitors

There are several types of plastic-film capacitors and the older paper dielectric capacitors. Polycarbonate, parylene, polyester, polystyrene, polypropylene, mylar, and paper are some of the more common dielectric materials used. Some of these types have capacitance values up to 100 μF.

Figure 10–11 shows a common basic construction used in many plastic-film and paper capacitors. A thin strip of plastic-film dielectric is sandwiched between two thin metal strips which act as plates. One lead is connected to the inner plate and one to the outer plate as indicated. The strips are then rolled in

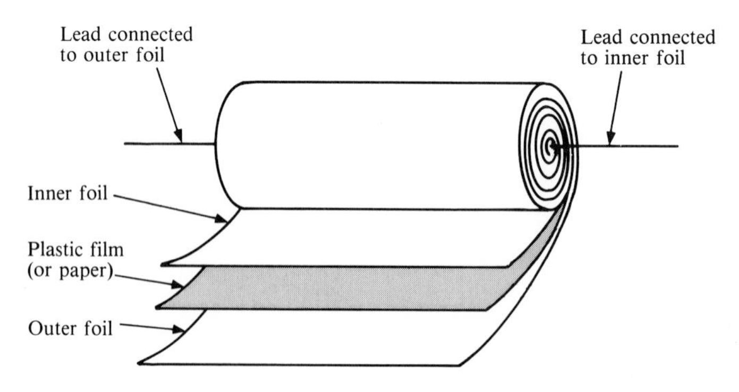

FIGURE 10–11
Basic construction of tubular paper/plastic dielectric capacitors.

a spiral configuration and encapsulated in a molded case. Thus, a large plate area can be packaged in a relatively small physical size, thereby achieving large capacitance values. Figure 10–12 shows a construction view for one type of plastic-film capacitor.

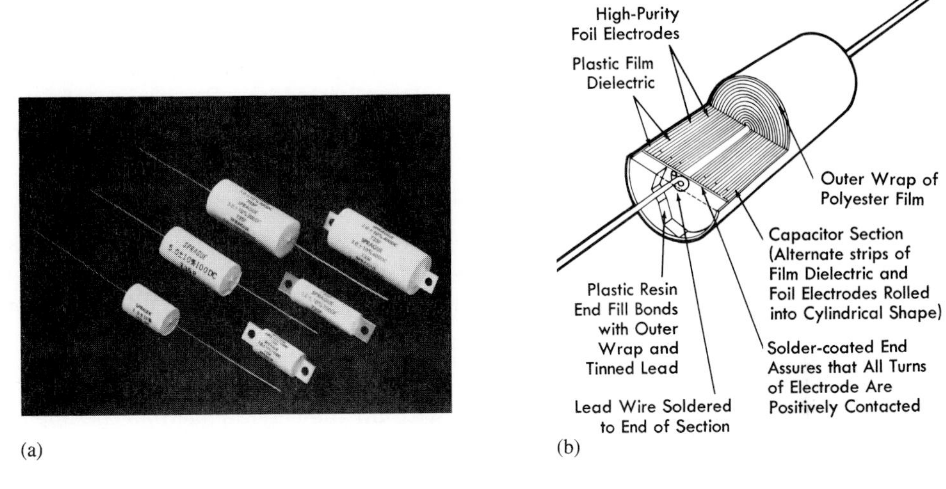

FIGURE 10–12
Film capacitors. (a) Typical example (courtesy of Siemens Corp.); (b) Construction view of plastic film capacitor (courtesy of Union Carbide Electronics Division).

Electrolytic Capacitors

Electrolytic capacitors are *polarized* so that one plate is positive and the other negative. These capacitors are used for high capacitance values up to over 200,000 μF, but they have relatively low breakdown voltages (350 V is a typical maximum) and high amounts of leakage.

Electrolytic capacitors are available in two types: *aluminum* and *tantalum*. The basic construction of an electrolytic capacitor is shown in Figure 10–13(a).

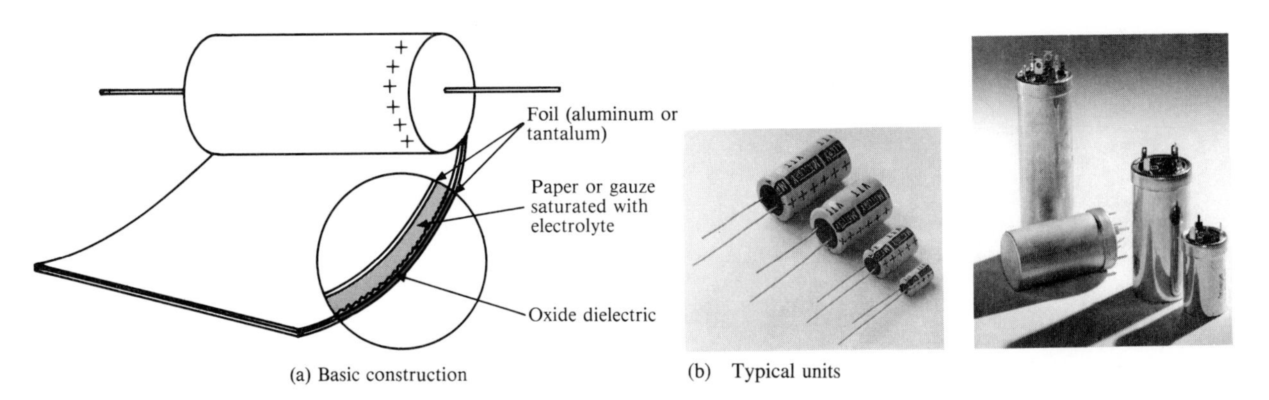

FIGURE 10–13
Electrolytic capacitors.

The capacitor consists of two strips of either aluminum or tantalum foil separated by a paper or gauze strip saturated with an electrolyte. During manufacturing, an electrochemical reaction is induced which causes an oxide layer (either aluminum oxide or tantalum oxide) to form on the inner surface of the positive plate. This oxide layer acts as the dielectric. Part (b) of Figure 10–13 shows several typical electrolytic capacitors.

Since an electrolytic capacitor is polarized, *the positive plate must always be connected to the positive side of a circuit.* The positive end is indicated by plus signs or some other obvious marking. Be very careful to make the correct connection and to install the capacitor only in a dc, not ac, circuit.

Variable Capacitors

Variable capacitors are used in a circuit when there is a need to adjust the capacitance value either manually or automatically, for example, in radio or TV tuners. The major types of variable or adjustable capacitors are now introduced.

Air Capacitor

Variable capacitors with air dielectrics, such as the one shown in Figure 10–14, are sometimes used as tuning capacitors in applications requiring frequency selection. This type of capacitor is constructed of several plates that mesh together. One set of plates can be moved relative to the other, thus changing the effective plate area and the capacitance. The movable plates are linked together mechanically so that they all move when a shaft is rotated. The schematic symbol for a variable capacitor is shown in Figure 10–15.

FIGURE 10–14
A typical variable air capacitor.

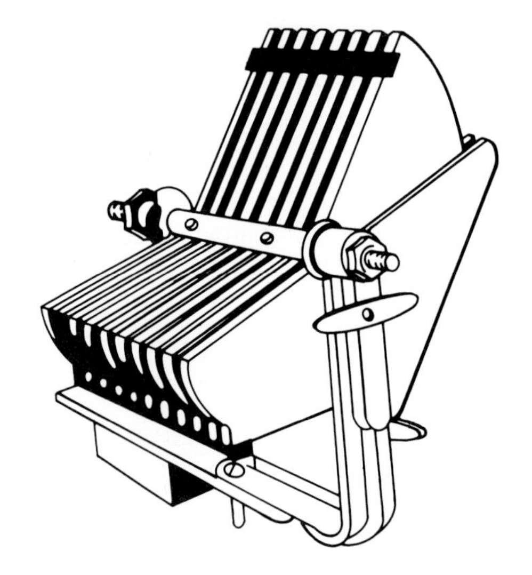

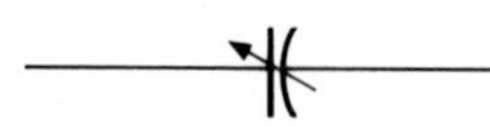

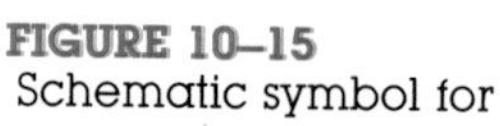

FIGURE 10–15
Schematic symbol for a variable capacitor.

Trimmers and Padders

These adjustable capacitors normally have screwdriver adjustments and are used for very fine adjustments in a circuit. Ceramic or mica is a common dielectric in these types of capacitors, and the capacitance usually is changed by adjusting the plate separation. Figure 10–16 shows some typical devices.

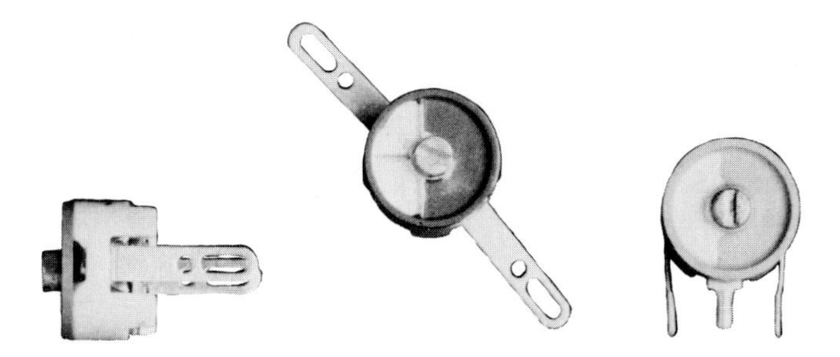

FIGURE 10–16
Trimmer capacitors (courtesy of Murata Erie North America, Inc.).

Varactors

The varactor is a semiconductor device that exhibits a capacitance characteristic which is varied by changing the voltage across its terminals. This device usually is covered in detail in a course on electronic devices.

Capacitor Labeling

Capacitor values are indicated on the body of the capacitor either by *typographical labels* or by *color codes*. Typographical labels consist of letters and numbers that indicate various parameters such as capacitance, voltage rating, tolerance, and others.

Some capacitors carry no unit designation for capacitance. In these cases, the units are implied by the value indicated. For example, a ceramic capacitor marked .001 or .01 has units of microfarads because picofarad values that small are not available. As another example, a ceramic capacitor labeled 50 or 330 has units of picofarads because microfarad units that large normally are not available in this type.

In some instances, the units are labeled as pF or μF; often the microfarad unit is labeled as MF or MFD. Voltage rating appears on some types of capacitors and is omitted on others. When it is omitted, the voltage rating can be determined from information supplied by the manufacturer. The tolerance of the capacitor is usually labeled as a percentage, such as $\pm 10\%$. The temperature coefficient is indicated by a *parts per million* marking. This type of label consists of a *P* or *N* followed by a number. For example, N750 means a negative temperature coefficient of 750 ppm/°C, and P30 means a positive temperature coefficient of 30 ppm/°C.

SECTION REVIEW 10–2

1. How are capacitors commonly classified?
2. What is the difference between a fixed and a variable capacitor?
3. What type of capacitor normally is polarized?
4. What precautions must be taken when a polarized capacitor is installed in a circuit?

10–3 SERIES CAPACITORS

When capacitors are connected in series, *the total capacitance is less than the smallest capacitance value,* because the *effective* plate separation increases. The calculation of total series capacitance is analogous to the calculation of total resistance of parallel resistors.

To start, we will use two capacitors in series to show how the total capacitance is determined. Figure 10–17 shows two capacitors, which initially are uncharged, connected in series with a dc voltage source. When the switch is closed, as shown in Part (a), current begins to flow.

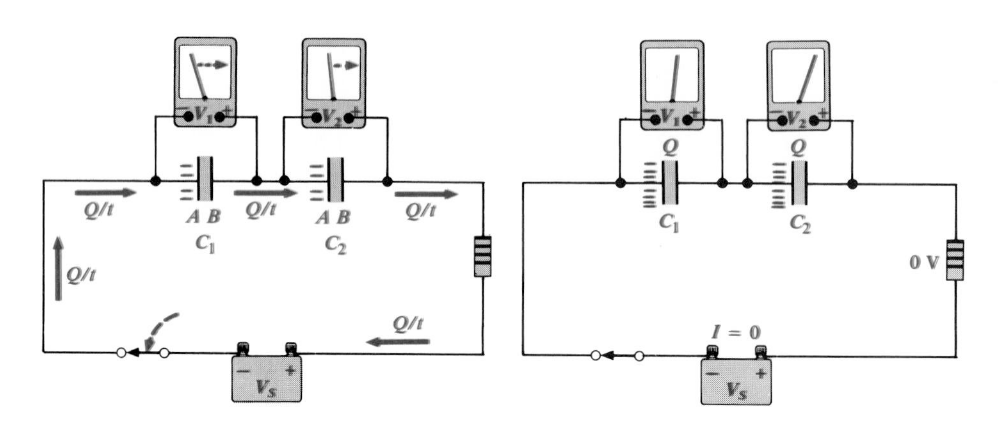

(a) $I = Q/t$ is the same at all points.

(b) Both capacitors store the same amount of charge ($Q = Q_T = Q_1 = Q_2$).

FIGURE 10–17
Capacitors in series produce a total capacitance that is less than the smallest value.

Recall that (1) current is the same at all points in a series circuit and (2) current is defined as the rate of flow of charge ($I = Q/t$). In a certain period of time, a certain amount of charge moves through the circuit. Since current is the same everywhere in the circuit of Figure 10–17(a), the same amount of charge is moved from the negative side of the source to plate A of C_1, and from plate B of C_1 to plate A of C_2, and from plate B of C_2 to the positive side of the source. As a result, of course, the same amount of charge is deposited on the plates of both capacitors in a given period of time, and the total charge (Q_T) moved through the circuit in that period of time equals the charge stored by C_1 and also equals the charge stored by C_2:

$$Q_T = Q_1 = Q_2$$

As the capacitors charge, the voltage across each one increases as indicated.

Figure 10–17(b) shows the capacitors after they have been completely charged and the current has ceased to flow. Both capacitors store an equal amount of charge (Q), and the voltage across each one depends on its capacitance value ($V = Q/C$). By Kirchhoff's voltage law, which applies to capacitive circuits as well as to resistive circuits, the sum of the capacitor voltages equals the source voltage:

$$V_S = V_1 + V_2$$

Using the fact that $V = Q/C$, we can substitute into the formula for Kirchhoff's law and get the following relationship (where $Q = Q_T = Q_1 = Q_2$):

$$\frac{Q}{C_T} = \frac{Q}{C_1} + \frac{Q}{C_2}$$

The Q can be factored out of the right side of the equation and canceled with the Q on the left side as follows:

$$\frac{\not{Q}}{C_T} = \not{Q}\left(\frac{1}{C_1} + \frac{1}{C_2}\right)$$

Thus we have the following relationship for two capacitors in series:

$$\frac{1}{C_T} = \frac{1}{C_1} + \frac{1}{C_2} \quad (10\text{–}6)$$

Taking the reciprocal of both sides of Equation (10–6) gives the formula for the total capacitance:

$$C_T = \frac{1}{(1/C_1) + (1/C_2)} \quad (10\text{–}7)$$

EXAMPLE 10–3 Find C_T in Figure 10–18.

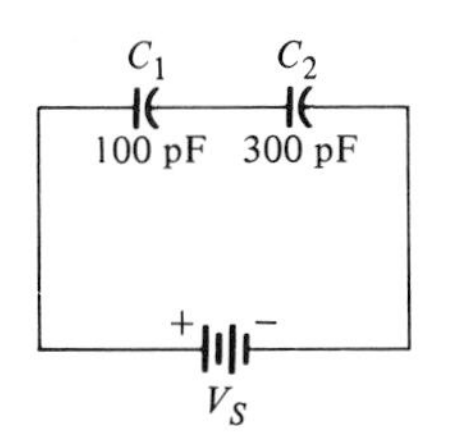

FIGURE 10–18

Solution

$$C_T = \frac{1}{(1/C_1) + (1/C_2)}$$

$$= \frac{1}{(1/100\text{ pF}) + (1/300\text{ pF})} = 75\text{ pF}$$

The calculator sequence is

[1] [0] [0] [EE] [+/−] [1] [2] [2nd] [1/x] [+] [3] [0] [0]
[EE] [+/−] [1] [2] [2nd] [1/x] [=] [2nd] [1/x] [=]

General Formula for Series Capacitance

Equations (10–6) and (10–7) can be extended to any number of capacitors in series, as shown in Figure 10–19. The expanded formulas are as follows, where the subscript n can be any number:

$$\frac{1}{C_T} = \frac{1}{C_1} + \frac{1}{C_2} + \frac{1}{C_3} + \cdots + \frac{1}{C_n} \quad (10\text{–}8)$$

$$C_T = \frac{1}{(1/C_1) + (1/C_2) + (1/C_3) + \cdots + (1/C_n)} \quad (10\text{–}9)$$

FIGURE 10–19
General series circuit with n capacitors.

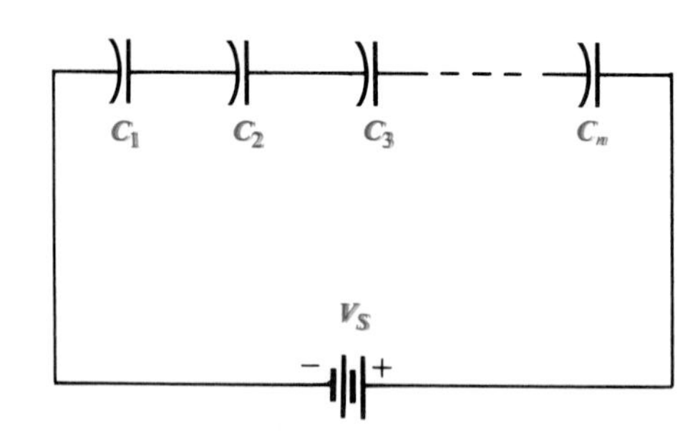

EXAMPLE 10–4 Determine the total capacitance in Figure 10–20.

FIGURE 10–20

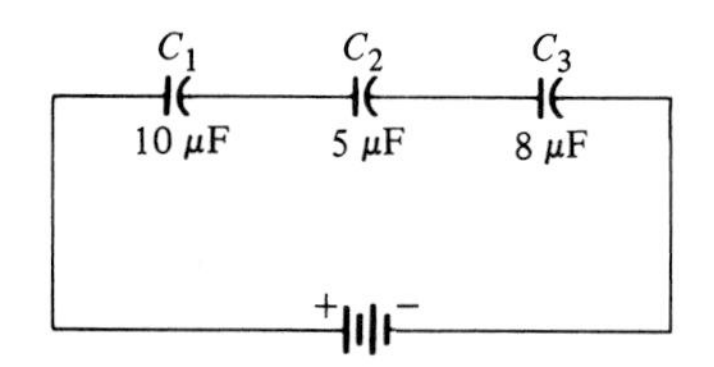

Solution

$$\frac{1}{C_T} = \frac{1}{C_1} + \frac{1}{C_2} + \frac{1}{C_3} = \frac{1}{10\ \mu\text{F}} + \frac{1}{5\ \mu\text{F}} + \frac{1}{8\ \mu\text{F}}$$

Taking the reciprocal of both sides yields

$$C_T = \frac{1}{(1/10\ \mu\text{F}) + (1/5\ \mu\text{F}) + (1/8\ \mu\text{F})}$$

$$= \frac{1}{0.425}\ \mu\text{F} = 2.35\ \mu\text{F}$$

How Series Capacitors Divide the Voltage

The voltage across each capacitor in a series connection depends on its capacitance value according to the formula $V = Q/C$. The largest-value capacitor will have the smallest voltage because of the reciprocal relationship. Likewise, the smallest capacitance value will have the largest voltage. The voltage across any individual capacitor in a series connection can be determined using the following formula:

$$V_x = \left(\frac{C_T}{C_x}\right)V_S \qquad \textbf{(10–10)}$$

where C_x is any capacitor such as C_1, C_2, C_3, and so on.

EXAMPLE 10–5 Find the voltage across each capacitor in Figure 10–21.

FIGURE 10–21

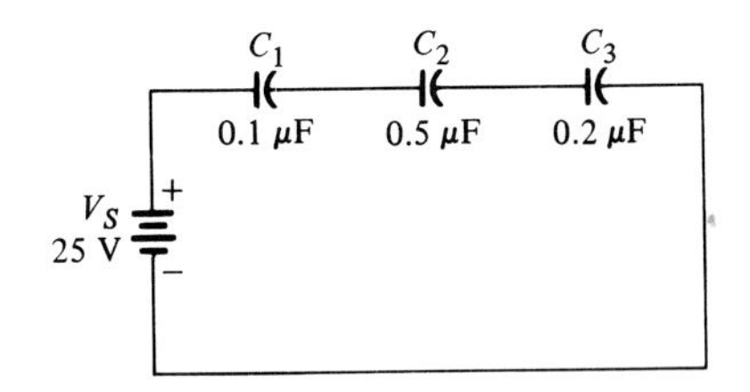

Solution

$$\frac{1}{C_T} = \frac{1}{C_1} + \frac{1}{C_2} + \frac{1}{C_3}$$

$$= \frac{1}{0.1\ \mu\text{F}} + \frac{1}{0.5\ \mu\text{F}} + \frac{1}{0.2\ \mu\text{F}}$$

$$C_T = \frac{1}{17}\ \mu\text{F} = 0.0588\ \mu\text{F}$$

$$V_1 = \left(\frac{C_T}{C_1}\right)V = \left(\frac{0.0588\ \mu\text{F}}{0.1\ \mu\text{F}}\right)25\ \text{V} = 14.71\ \text{V}$$

$$V_2 = \left(\frac{C_T}{C_2}\right)V = \left(\frac{0.0588\ \mu\text{F}}{0.5\ \mu\text{F}}\right)25\ \text{V} = 2.94\ \text{V}$$

$$V_3 = \left(\frac{C_T}{C_3}\right)V = \left(\frac{0.0588\ \mu\text{F}}{0.2\ \mu\text{F}}\right)25\ \text{V} = 7.35\ \text{V}$$

The calculator sequence for V_1 is

[·] [0] [5] [8] [8] [EE] [+/−] [6] [÷] [·] [1] [EE] [+/−] [6] [×] [2] [5] [=]

SECTION REVIEW 10–3

1. Is the total capacitance of a series connection less than or greater than the value of the smallest capacitor?
2. The following capacitors are in series: 100 pF, 250 pF, and 500 pF. What is the total capacitance?
3. A 0.01-μF and a 0.015-μF capacitor are in series. Determine the total capacitance.
4. Determine the voltage across the 0.01-μF capacitor in Problem 3 if 10 V are connected across the two series capacitors.

10–4 PARALLEL CAPACITORS

When capacitors are connected in parallel, *the total capacitance is the sum of the individual capacitances,* because the *effective* plate area increases. The calculation of total parallel capacitance is analogous to the calculation of total series resistance.

Figure 10–22 shows two parallel capacitors connected to a dc voltage source. When the switch is closed, as shown in Part (a), current begins to flow. A total amount of charge (Q_T) moves through the circuit in a certain period of time. Part of the total charge is stored by C_1 and part by C_2. The portion of the total charge that is stored by each capacitor depends on its capacitance value according to the relationship $Q = CV$.

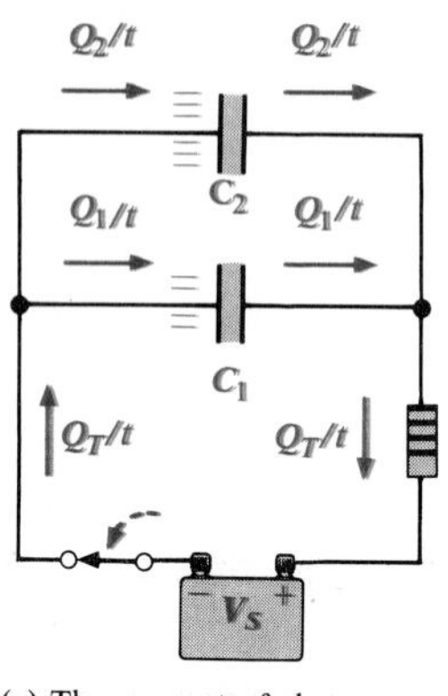

(a) The amount of charge on each capacitor is directly proportional to its capacitance value.

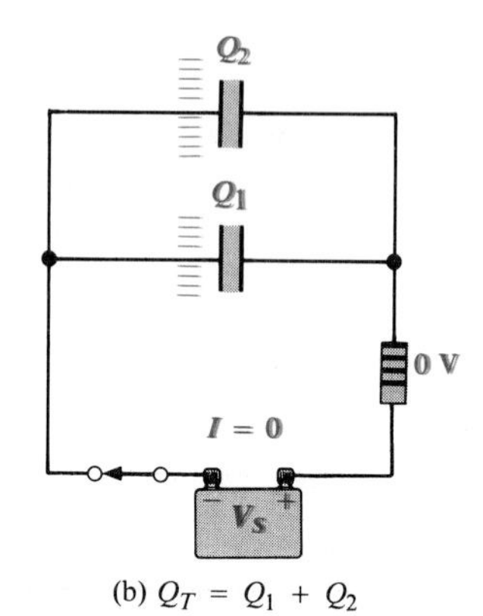

(b) $Q_T = Q_1 + Q_2$

FIGURE 10–22
Capacitors in parallel produce a total capacitance that is the sum of the individual capacitances.

Figure 10–22(b) shows the capacitors after they have been completely charged and the current has stopped. Since the voltage across both capacitors is the same, the larger capacitor stores more charge. If the capacitors are equal in value, they store an equal amount of charge. The charge stored by both of the capacitors together equals the total charge that was delivered from the source:

$$Q_T = Q_1 + Q_2$$

Using the fact that $Q = CV$, we can substitute into the above formula and get the following relationship:

$$C_T V_S = C_1 V_S + C_2 V_S$$

Because all the V_S terms are equal, they can be canceled, leaving

$$C_T = C_1 + C_2 \qquad \textbf{(10–11)}$$

EXAMPLE 10–6 What is the total capacitance in Figure 10–23?

FIGURE 10–23

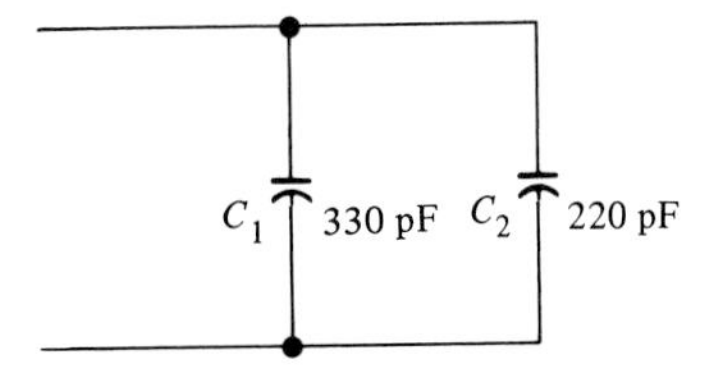

Solution

$$C_T = C_1 + C_2 = 330 \text{ pF} + 220 \text{ pF} = 550 \text{ pF}$$

General Formula for Parallel Capacitance

Equation (10–11) can be extended to any number of capacitors in parallel, as shown in Figure 10–24. The expanded formula is as follows, where the subscript n can be any number:

$$C_T = C_1 + C_2 + C_3 + \cdots + C_n \tag{10–12}$$

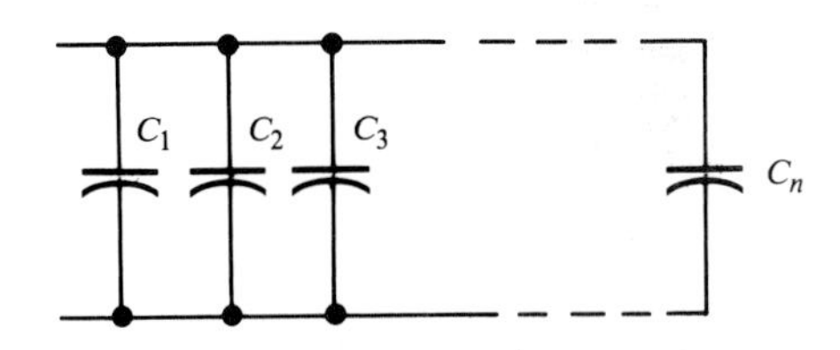

FIGURE 10–24
General parallel circuit with n capacitors.

EXAMPLE 10–7 Determine C_T in Figure 10–25.

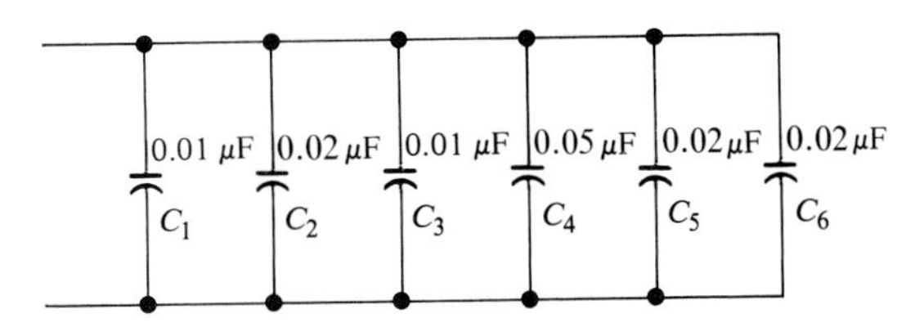

FIGURE 10–25

Solution

$$\begin{aligned} C_T &= C_1 + C_2 + C_3 + C_4 + C_5 + C_6 \\ &= 0.01\ \mu\text{F} + 0.02\ \mu\text{F} + 0.01\ \mu\text{F} + 0.05\ \mu\text{F} + 0.02\ \mu\text{F} + 0.02\ \mu\text{F} \\ &= 0.13\ \mu\text{F} \end{aligned}$$

SECTION REVIEW 10–4

1. How is total parallel capacitance determined?
2. In a certain application, you need 0.05 μF. The only values available are 0.01 μF, which are available in large quantities. How can you get the total capacitance that you need?
3. The following capacitors are in parallel: 10 pF, 5 pF, 33 pF, and 50 pF. What is C_T?

CAPACITORS IN dc CIRCUITS

10–5

In this section, the response of a simple capacitive circuit during charging and discharging is examined. Figure 10–26(a) shows a capacitor connected in series with a resistor and a switch to a dc voltage source. Initially, the switch is open and the capacitor is uncharged with zero volts across its plates. At the instant the switch is closed, the current jumps to its maximum value and the capacitor begins to charge. The current is maximum initially because the capacitor has zero volts across it and, therefore, appears as a *short;* thus, the current is limited only by the resistance. As time passes and the capacitor charges, the current decreases and the voltage V_C across the capacitor increases. The resistor voltage is proportional to the current during this charging period.

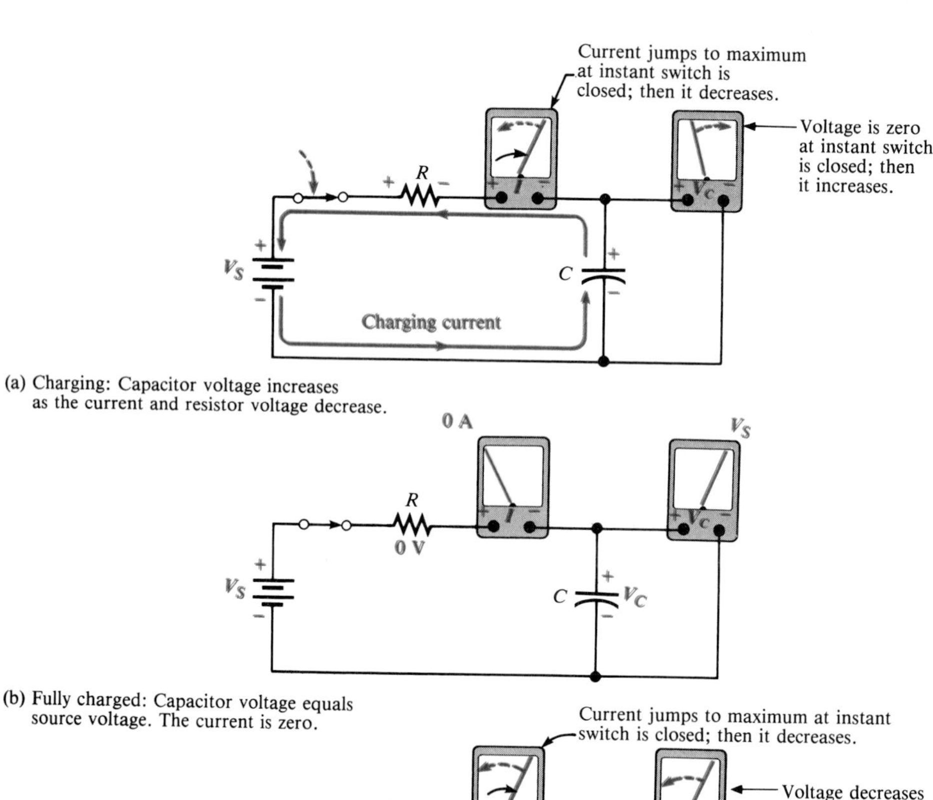

(a) Charging: Capacitor voltage increases as the current and resistor voltage decrease.

(b) Fully charged: Capacitor voltage equals source voltage. The current is zero.

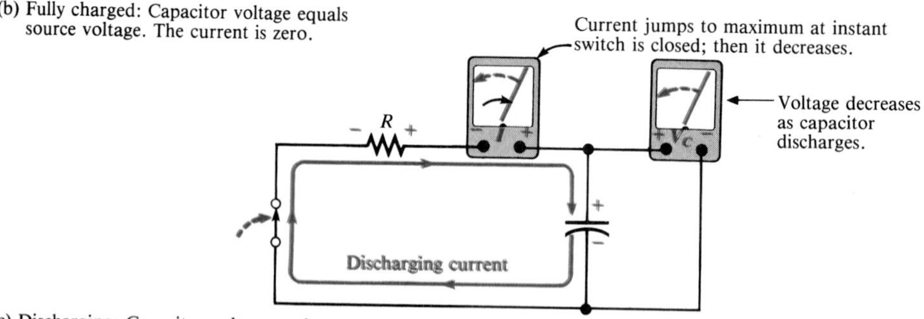

(c) Discharging: Capacitor voltage, resistor voltage, and the current decrease from initial maximums. Note that the discharge current is opposite to the charge current.

FIGURE 10–26
Charging and discharging of a capacitor.

After a certain period of time, the capacitor reaches full charge. At this point, the current is zero and the capacitor voltage is equal to the dc source voltage, as shown in Figure 10–26(b). If the switch were opened now, the capacitor would retain its full charge (neglecting any leakage).

In Figure 10–26(c), the voltage source has been removed. When the switch is closed, the capacitor begins to discharge. Initially, the current jumps to a maximum but in a direction opposite to its direction during charging. As time passes, the current and capacitor voltage decrease. The resistor voltage is always proportional to the current. When the capacitor has fully discharged, the current and the capacitor voltage are zero.

Note the following about capacitors in dc circuits:

1. Voltage across a capacitor cannot change instantaneously.
2. Current in a capacitive circuit can change instantaneously.
3. A fully charged capacitor appears as an *open* to nonchanging current.
4. An uncharged capacitor appears as a *short* to an instantaneous change in current.

Now we will examine in more detail how the voltage and current change with time in a capacitive circuit.

The Time Constant

As you have seen when a capacitor charges or discharges through a resistance, a certain time is required for the capacitor to charge fully or discharge fully. *The voltage across a capacitor cannot change instantaneously,* because a finite time is required to move charge from one point to another. The rate at which the capacitor charges or discharges is determined by the *time constant* of the circuit. *The time constant of a series RC circuit is a time interval that equals the product of the resistance and the capacitance.* The time constant is symbolized by τ (Greek letter tau), and the formula is as follows:

$$\tau = RC \tag{10–13}$$

Recall that $I = Q/t$. The current is the amount of charge moved in a given time. When the resistance is increased, the charging current is reduced, thus increasing the charging time of the capacitor. When the capacitance is increased, the amount of charge increases; thus, for the same current, more time is required to charge the capacitor.

EXAMPLE 10–8 A series RC circuit has a resistance of 1 MΩ and a capacitance of 5 μF. What is the time constant?

Solution

$$\tau = RC = (1 \times 10^6\ \Omega)(5 \times 10^{-6}\ \text{F}) = 5 \text{ seconds}$$

During one time constant interval, the charge on a capacitor changes approximately 63%. Therefore, an uncharged capacitor charges to 63% of its fully charged voltage in one time constant. When a capacitor is discharging, its voltage drops to approximately 37% (100% − 63%) of its initial value in one time constant. This change also corresponds to a 63% change.

The Charging and Discharging Curves

A capacitor charges and discharges following a nonlinear curve, as shown in Figure 10–27. In these graphs, the percentage of full charge is shown at each time constant interval. This type of curve follows a precise mathematical formula and is called an *exponential curve.* The charging curve is an *increasing exponential,* and the discharging curve is a *decreasing exponential.* As you can see, it takes *five time constants* to approximately reach the final value. Five time constants is *accepted* as the time to fully charge or discharge a capacitor.

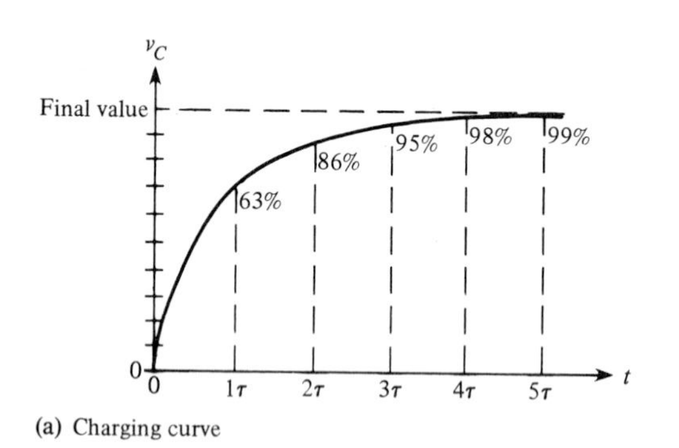

(a) Charging curve

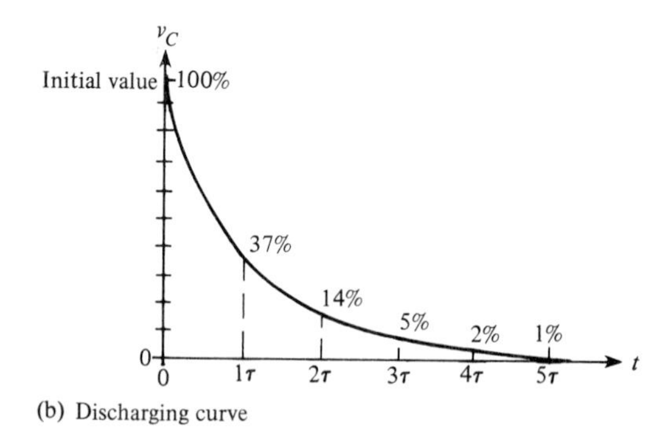

(b) Discharging curve

FIGURE 10–27
Charging and discharging exponential curves for an RC circuit. These curves apply to both voltage and current.

General Formula

The general expressions for either increasing or decreasing exponential curves are given in the following equations for both voltage and current:

$$v = V_F + (V_i - V_F)e^{-t/\tau} \qquad \textbf{(10–14)}$$

$$i = I_F + (I_i - I_F)e^{-t/\tau} \qquad \textbf{(10–15)}$$

where V_F and I_F are the *final* values, and V_i and I_i are the *initial* values. The lower-case letters v and i are the instantaneous values of the capacitor voltage and current at time t, and e is the base of natural logarithms with a value of 2.718. The e^x key or the INV and ln x keys on your calculator make it easy to work with these formulas.

The Charging Curve

The formula for the special case in which an increasing exponential voltage curve begins at zero ($V_i = 0$) is given in Equation (10–16). It is developed as follows, starting with the general formula:

$$v = V_F + (V_i - V_F)e^{-t/\tau}$$
$$= V_F + (0 - V_F)e^{-t/RC}$$
$$v = V_F(1 - e^{-t/RC}) \qquad \textbf{(10–16)}$$

Using Equation (10–16), we can calculate the value of the charging voltage of a capacitor at any instant of time. The same is true for an increasing current.

EXAMPLE 10–9 In Figure 10–28, determine the capacitor voltage 50 microseconds (μs) after the switch is closed if the capacitor initially is uncharged. Sketch the charging curve.

FIGURE 10–28

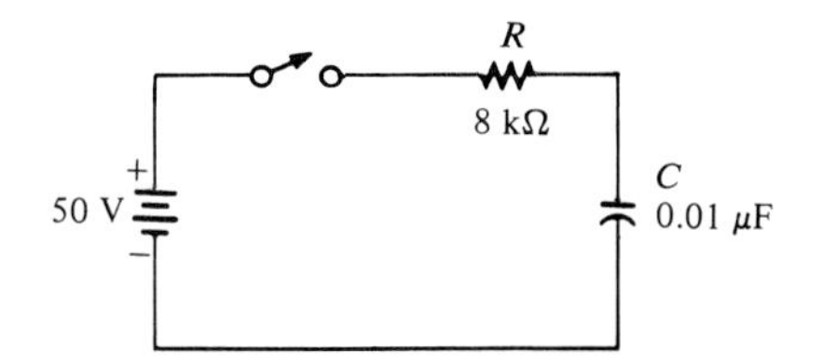

Solution

The time constant is $RC = (8\ \text{k}\Omega)(0.01\ \mu\text{F}) = 80\ \mu\text{s}$. The voltage to which the capacitor will fully charge is 50 V (this is V_F). The initial voltage is zero. Notice that 50 μs is less than one time constant; so the capacitor will charge less than 63% of the full voltage in that time.

$$v_C = V_F(1 - e^{-t/RC}) = 50\ \text{V}(1 - e^{-50\,\mu\text{s}/80\,\mu\text{s}})$$
$$= 50\ \text{V}(1 - e^{-0.625}) = 50\ \text{V}(1 - 0.535) = 23.2\ \text{V}$$

We determine the value of $e^{-0.625}$ on the calculator by entering -0.625 and then pressing the [e^x] key (or [INV] and then [ln x]).

The charging curve for the capacitor is shown in Figure 10–29.

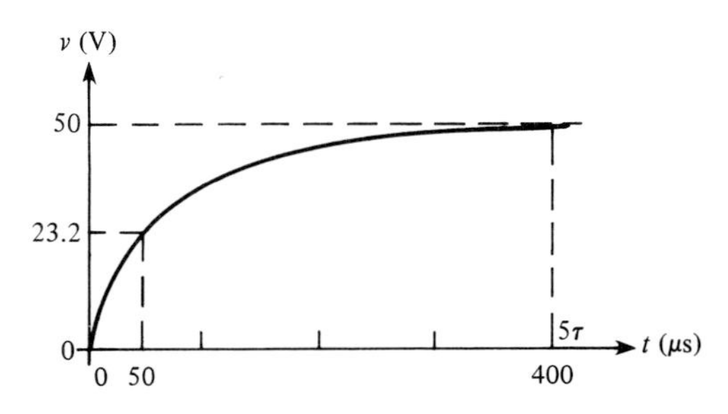

FIGURE 10–29

The calculator sequence is

[5] [0] [÷] [8] [0] [=] [+/−] [INV] [ln x] [+/−] [+] [1] [=] [×] [5] [0] [=]

The Discharging Curve

The formula for the special case in which a decreasing exponential voltage curve ends at zero is derived from the general formula as follows:

$$\begin{aligned} v &= V_F + (V_i - V_F)e^{-t/\tau} \\ &= 0 + (V_i - 0)e^{-t/RC} \\ v &= V_i e^{-t/RC} \end{aligned} \qquad \textbf{(10–17)}$$

where V_i is the voltage at the beginning of the discharge. We can use this formula to calculate the discharging voltage at any instant, as Example 10–10 illustrates.

EXAMPLE 10–10 Determine the capacitor voltage in Figure 10–30 at a point in time 6 milliseconds (ms) after the switch is closed. Sketch the discharging curve.

FIGURE 10–30

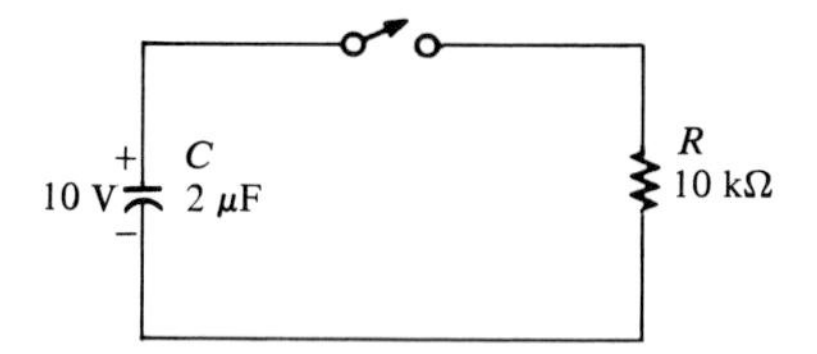

Solution

The discharge time constant is $RC = (10\ \text{k}\Omega)(2\ \mu\text{F}) = 20$ ms. The initial capacitor voltage is 10 V. Notice that 6 ms is less than one time constant, so the capacitor will discharge less than 63%. Therefore, it will have a voltage greater than 37% of the initial voltage at 6 ms.

$$\begin{aligned} v_C &= V_i e^{-t/RC} = 10e^{-6\ \text{ms}/20\ \text{ms}} \\ &= 10e^{-0.3} = 10(0.741) = 7.41\ \text{V} \end{aligned}$$

Again, the value of $e^{-0.3}$ can be determined with a calculator.

The discharging curve for the capacitor is shown in Figure 10–31.

FIGURE 10–31

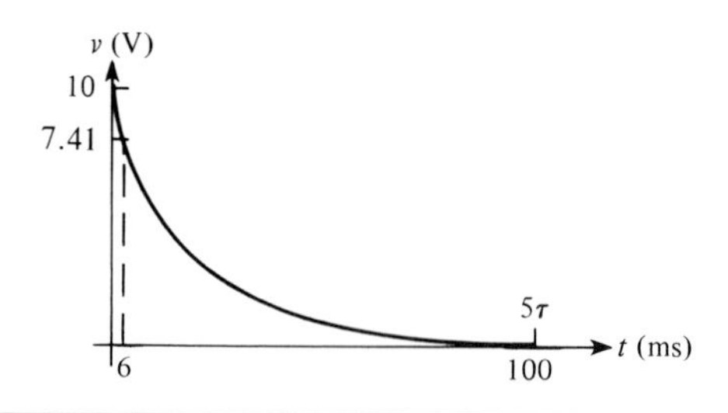

Universal Exponential Curves

The universal curves in Figure 10–32 provide a graphical solution of the charge and discharge of capacitors. Example 10–11 illustrates this graphical method.

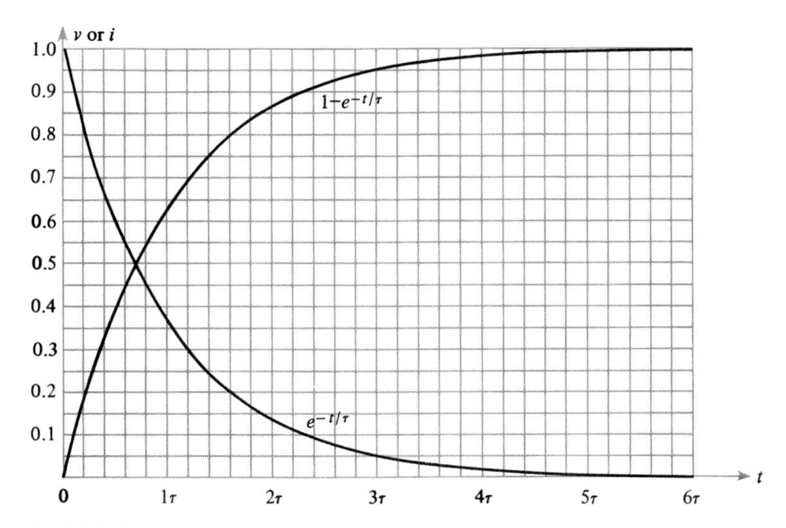

FIGURE 10–32
Universal exponential curves.

EXAMPLE 10–11 How long will it take the capacitor in Figure 10–33 to charge to 75 V? What is the capacitor voltage 2 ms after the switch is closed? Use the universal curves in Figure 10–32 to determine the answers.

FIGURE 10–33

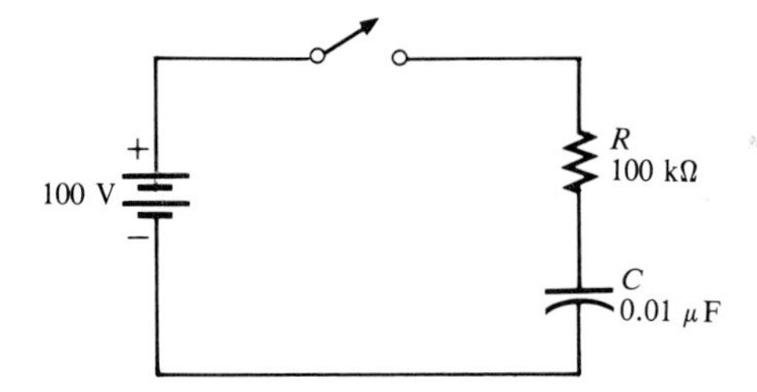

Solution

The full charge voltage is 100 V, which is the 100% level on the graph. Since 75 V is 75% of the maximum, you can see that this value occurs at 1.4 time constants. One time constant is 1 ms. Therefore, the capacitor voltage reaches 75 V at 1.4 ms after the switch is closed.

The capacitor is at approximately 87 V in 2 ms. These graphical solutions are shown in Figure 10–34 on page 340.

SECTION REVIEW 10–5

1. Determine the time constant when $R = 1.2\ \text{k}\Omega$ and $C = 1000\ \text{pF}$.
2. If the circuit in Problem 1 is charged with a 5-V source, how long will it take the capacitor to reach full charge? At full charge, what is the capacitor voltage?
3. A certain circuit has a time constant of 1 ms. If it is charged with a 10-V battery, what will the capacitor voltage be at each of the following intervals: 2 ms, 3 ms, 4 ms, and 5 ms?
4. A capacitor is charged to 100 V. If it is discharged through a resistor, what is the capacitor voltage at one time constant?

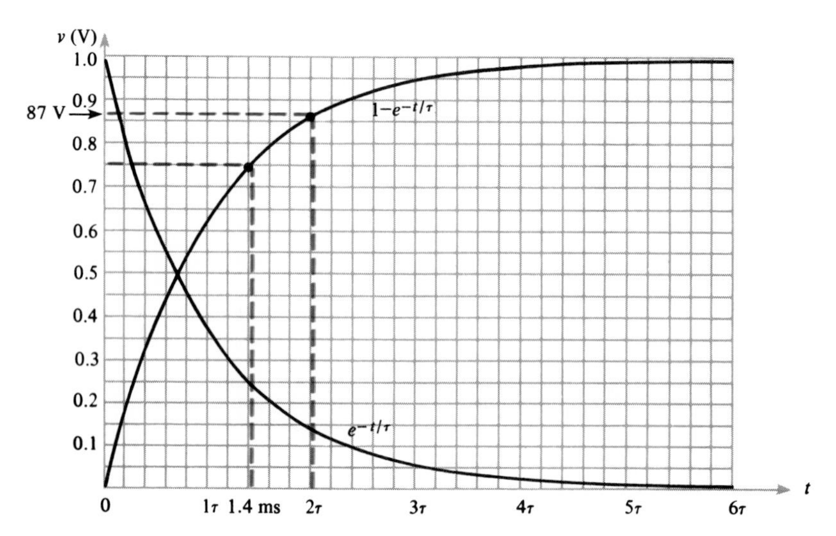

FIGURE 10–34

CAPACITORS IN ac CIRCUITS

10–6 The Phase Relationship of Current and Voltage

We will illustrate the relationship of current and capacitor voltage in an ac circuit by connecting a capacitor in series with a resistor to a sine wave voltage source and then observing the capacitor voltage and the resistor voltage with an oscilloscope. The connection is shown in Figure 10–35. Recall that the resistor voltage and the current are always in phase with each other. Therefore, by observing the resistor voltage, we are also indirectly observing the current in terms of its phase relationship to the capacitor voltage. Notice that the frequency of the voltage and of the current are always the same in an ac circuit.

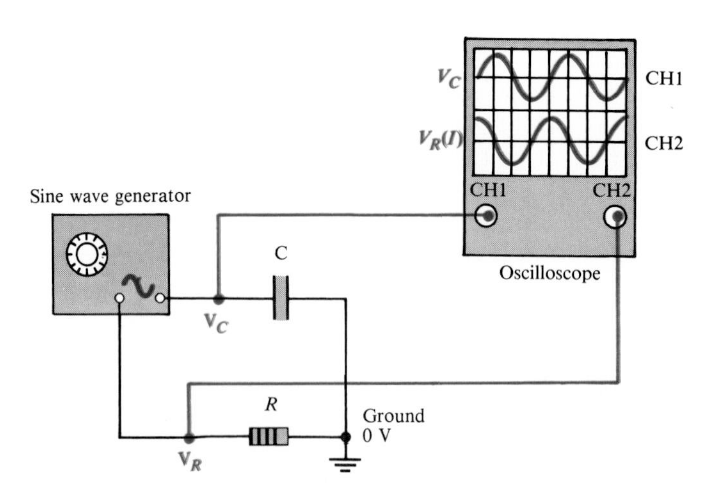

FIGURE 10–35
Oscilloscope display showing that the current leads the capacitor voltage by 90°. The current is in phase with the resistor voltage.

Why the Current Leads the Capacitor Voltage by 90°

A sine wave voltage is shown in Figure 10–36. Notice that the rate at which the voltage is changing varies along the sine wave curve as indicated by the "steepness" of the curve. At the zero crossings, the curve is changing at a faster rate than anywhere else along the curve. At the peaks, the curve has a zero rate of change because it has just reached its maximum and is at the point of changing direction.

FIGURE 10–36
The rates of change of a sine wave.

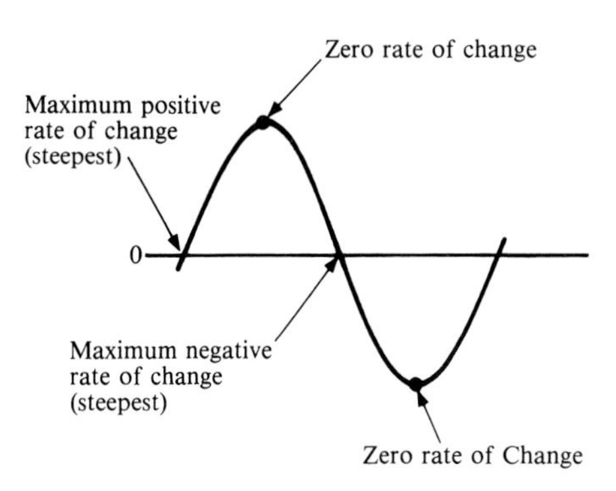

The amount of charge stored by a capacitor determines the voltage across it. Therefore, the rate at which the charge is moved (Q/t = current) from one plate to the other determines the rate at which the voltage changes. When the current is changing at its maximum rate (at the zero crossings), the voltage is at its maximum value (peak). When the current is changing at its minimum rate (zero at the peaks), the voltage is at its minimum value (zero). This relationship is illustrated in Figure 10–37. As you can see, the current peaks occur a quarter of a cycle before the voltage peaks. Thus, the current *leads* the voltage by 90°.

FIGURE 10–37
Current is always leading the capacitor voltage by 90°.

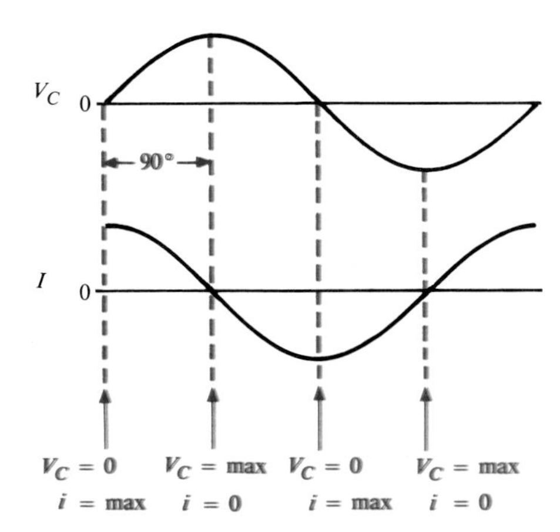

Capacitive Reactance, X_C

In Figure 10–38, a capacitor is connected to a sine wave voltage source. Note that when the source voltage is held at a constant amplitude value and its

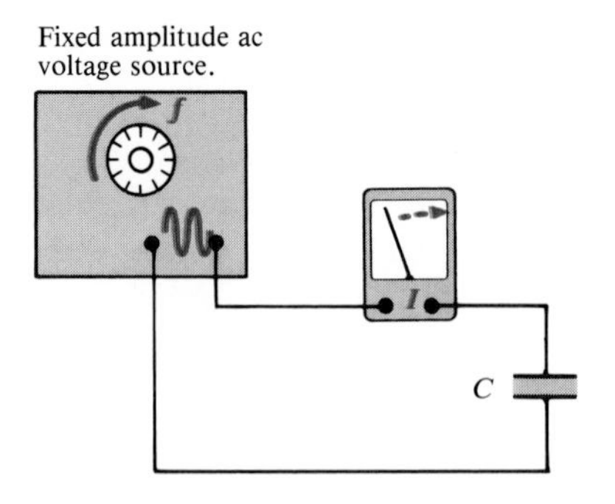

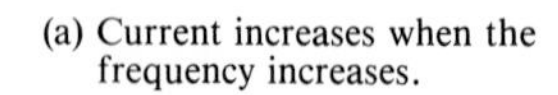
(a) Current increases when the frequency increases.

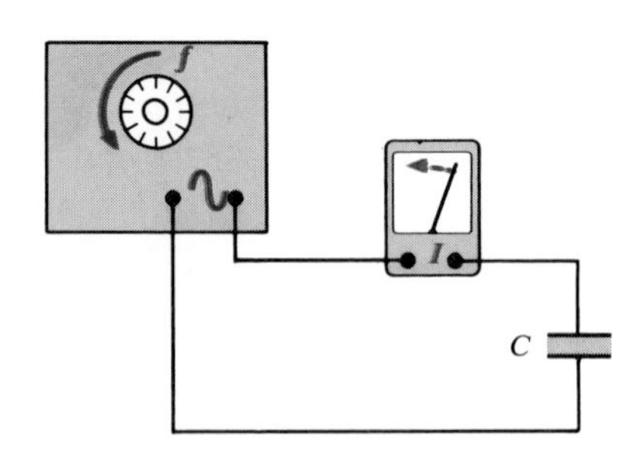

(b) Current decreases when the frequency decreases.

FIGURE 10–38
The current in a capacitive circuit varies directly with the frequency of the applied voltage.

frequency is increased, the amplitude of the current increases. Also, when the frequency of the source is decreased, the current amplitude decreases. The reason is as follows: When the frequency of the voltage increases, its rate of change also increases. This relationship is illustrated in Figure 10–39, where the frequency is doubled.

FIGURE 10–39
Rate of change increases with frequency.

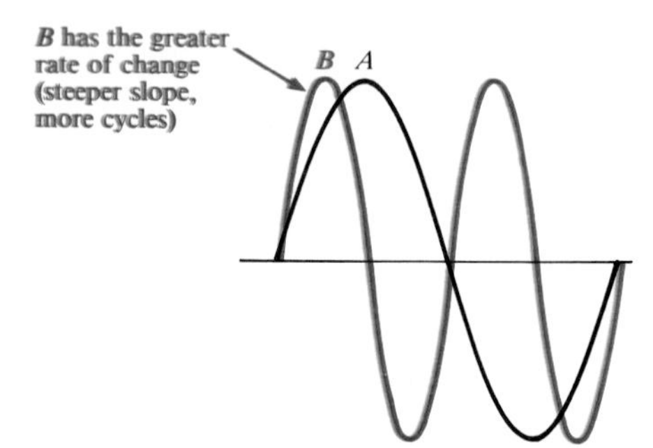

Now, if the rate at which the voltage is changing increases, the amount of charge moving through the circuit in a given period of time must also increase. More charge in a given period of time means more current. For example, a tenfold increase in frequency means that the capacitor is charging and discharging 10 times as much in a given time interval. Therefore, since the rate of charge movement has increased 10 times, the current must increase by 10 ($I = Q/t$).

An *increase* in the amount of current for a fixed amount of voltage indicates that *opposition* to the current has *decreased.* Therefore, the capacitor offers opposition to current, which varies *inversely* with frequency. The opposition to sinusoidal current is called *capacitive reactance.* The symbol for capacitive reactance is X_C, and its unit is the ohm (Ω).

You have just seen how frequency affects the opposition to current (capacitive reactance) in a capacitor. Now let's see how the capacitance affects the reactance. Figure 10–40(a) shows that when a sine wave voltage with a fixed amplitude and frequency is applied to a 1-μF capacitor, a certain amount of current flows. When the capacitance value is increased to 2 μF, the current increases, as shown in Part (b). Thus, when the capacitance increases, the opposition to current (capacitive reactance) decreases. Therefore, not only is the capacitive reactance inversely proportional to frequency, but it is also inversely proportional to capacitance:

$$X_C \text{ is proportional to } \frac{1}{fC}$$

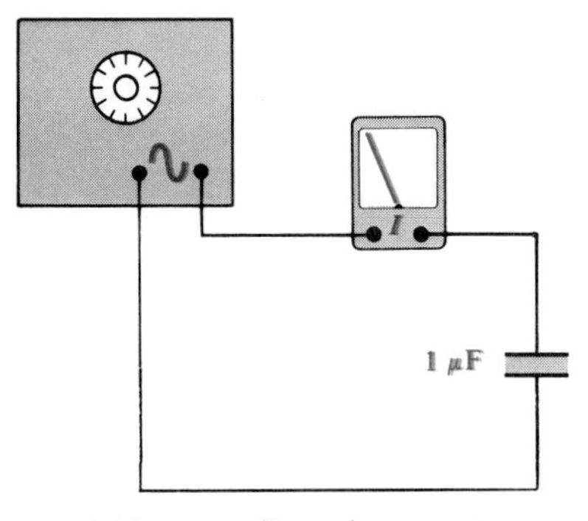

(a) Less capacitance, less current

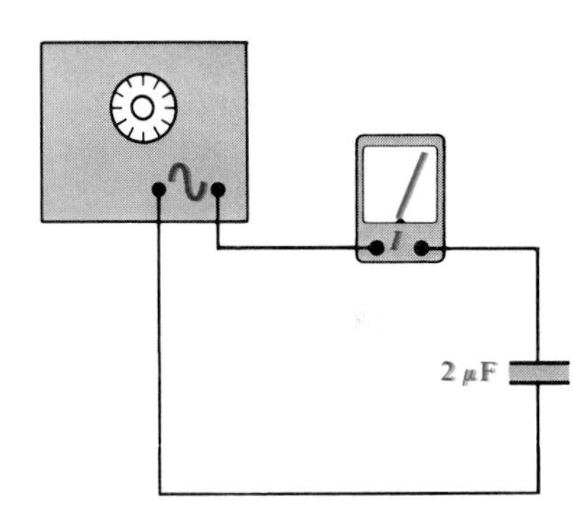

(b) More capacitance, more current

FIGURE 10–40
For a fixed voltage and frequency, the current varies directly with capacitance.

It can be proven that the constant of proportionality is $1/2\pi$. Therefore, the formula for X_C is

$$X_C = \frac{1}{2\pi fC} \tag{10–18}$$

The 2π term comes from the fact that a sine wave can be described in terms of rotational motion, and one revolution contains 2π radians.

EXAMPLE 10–12 A sinusoidal voltage is applied to a capacitor, as shown in Figure 10–41. The frequency of the sine wave is 1 kHz. Determine the capacitive reactance.

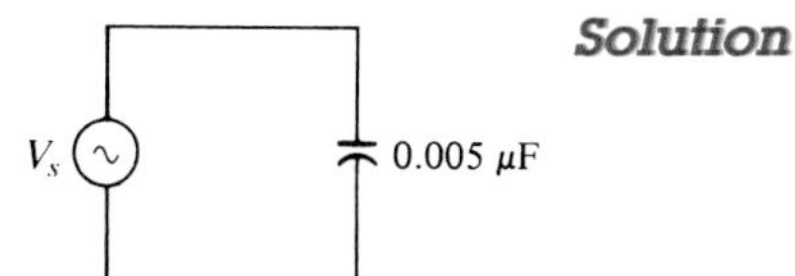

FIGURE 10–41

Solution

$$X_C = \frac{1}{2\pi fC} = \frac{1}{2\pi(1 \times 10^3\ \text{Hz})(0.005 \times 10^{-6}\ \text{F})}$$
$$= 31.83\ \text{k}\Omega$$

The calculator sequence is

[2] [×] [π] [×] [1] [EE] [3] [×] [·] [0] [0] [5] [EE] [+/−] [6] [=] [2nd] [1/x]

Ohm's Law in Capacitive Circuits

The reactance of a capacitor is analogous to the resistance of a resistor. In fact, both are expressed in ohms. Since both are forms of opposition to current, Ohm's law applies to capacitive circuits as well as to resistive circuits and is stated as follows for capacitive reactance:

$$V = IX_C \tag{10–19}$$

When applying ohm's law in ac circuits, you must express both the current and the voltage in the same way, that is, both in rms, both in peak, and so on.

EXAMPLE 10–13 Determine the rms current in Figure 10–42.

FIGURE 10–42

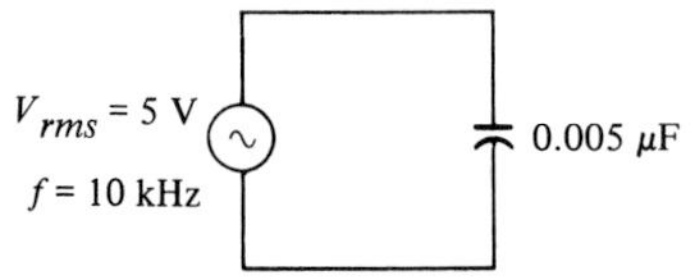

$V_{rms} = 5$ V
$f = 10$ kHz
0.005 μF

Solution

$$X_C = \frac{1}{2\pi fC} = \frac{1}{2\pi(10 \times 10^3\ \text{Hz})(0.005 \times 10^{-6}\ \text{F})}$$
$$= 3.18\ \text{k}\Omega$$

$$I_{\text{rms}} = \frac{V_{\text{rms}}}{X_C} = \frac{5\ \text{V}}{3.18\ \text{k}\Omega} = 1.57\ \text{mA}$$

Power in a Capacitor

As discussed earlier in this chapter, a charged capacitor stores energy in the electric field within the dielectric. An ideal capacitor does not dissipate energy; it only stores it. When an ac voltage is applied to a capacitor, energy is stored by the capacitor during a portion of the voltage cycle then the stored energy is *returned to the source* during another portion of the cycle. *There is no net energy loss.* Figure 10–43 shows the power curve that results from one cycle of capacitor voltage and current.

Instantaneous Power (p)

The product of v and i gives instantaneous power, p. At points where v or i is zero, p is also zero. When both v and i are positive, p is also positive. When either v or i is positive and the other negative, p is negative. When both v and i are negative, p is positive. As you can see, the power follows a sinusoidal-type curve. Positive values of power indicate that energy is stored by the capacitor. Negative values of power indicate that energy is returned from the capacitor to the source. Note that the power fluctuates at a frequency twice that of the voltage or current as energy is alternately stored and returned to the source.

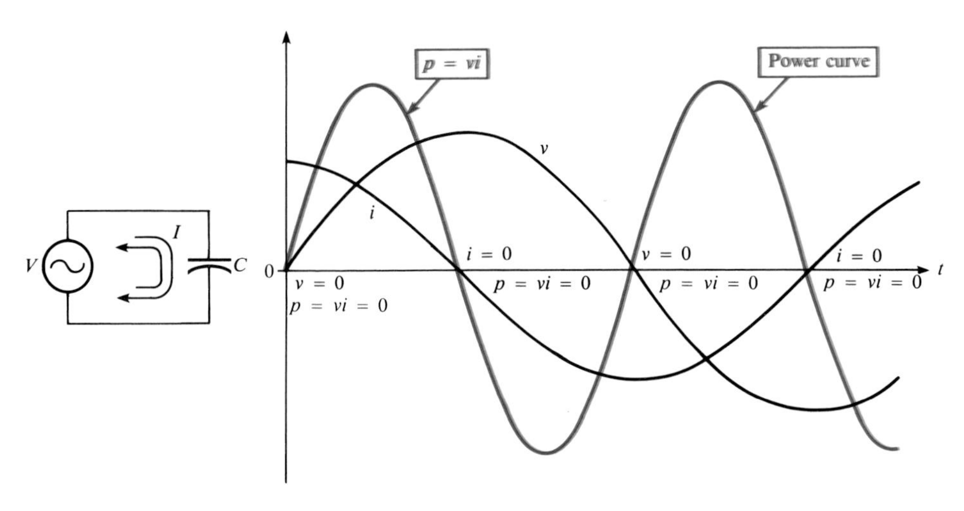

FIGURE 10–43
Power curve.

True Power (P_{true})

Ideally, all of the energy stored by a capacitor during the positive portion of the power cycle is returned to the source during the negative portion. *No net energy is consumed in the capacitor, so the true power is zero.* Actually, because of leakage and foil resistance in a practical capacitor, a small percentage of the total power is dissipated.

Reactive Power (P_r)

The rate at which a capacitor stores or returns energy is called its *reactive power, P_r*. The reactive power is a nonzero quantity, because at any instant in time, the capacitor is actually taking energy from the source or returning energy to it. Reactive power does not represent an energy loss. The following formulas apply:

$$P_r = V_{rms}I_{rms} \tag{10–20}$$

$$P_r = \frac{V_{rms}^2}{X_C} \tag{10–21}$$

$$P_r = I_{rms}^2 X_C \tag{10–22}$$

Notice that these equations are of the same form as those for true power in a resistor. The voltage and current are expressed in rms. The unit of reactive power is *volt-amperes reactive* (VAR).

EXAMPLE 10–14 Determine the true power and the reactive power in Figure 10–44.

FIGURE 10–44

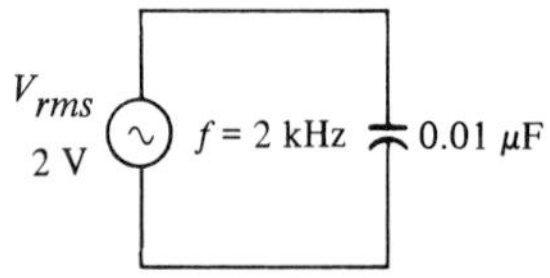

Solution
The true power P_{true} is *always zero for a capacitor.* The reactive power is as follows:

$$X_C = \frac{1}{2\pi fC} = \frac{1}{2\pi(2 \times 10^3 \text{ Hz})(0.01 \times 10^{-6} \text{ F})}$$
$$= 7.958 \text{ k}\Omega$$
$$P_r = \frac{V_{rms}^2}{X_C} = \frac{(2 \text{ V})^2}{7.958 \text{ k}\Omega}$$
$$= 0.503 \times 10^{-3} \text{ VAR} = 0.503 \text{ mVAR}$$

SECTION REVIEW 10–6

1. State the phase relationship between current and voltage in a capacitor.
2. Calculate X_C for $f = 5$ kHz and $C = 50$ pF.
3. At what frequency is the reactance of a 0.1-μF capacitor equal to 2 kΩ?
4. Calculate the rms current in Figure 10–45.
5. A 1-μF capacitor is connected to an ac voltage source of 12 V rms. What is the true power?
6. In Problem 5, determine reactive power at a frequency of 500 Hz.

FIGURE 10–45

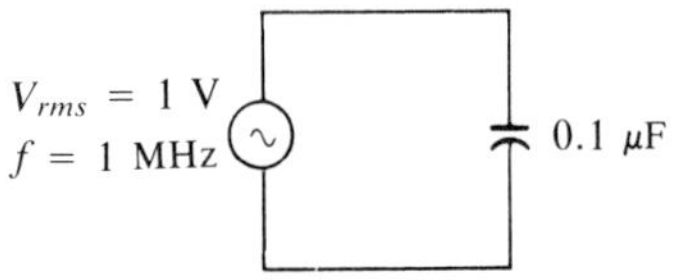

CAPACITOR APPLICATIONS

10–7

Capacitors are very widely used in electrical and electronic applications. A few typical applications are discussed here to illustrate the usefulness of this component.

Power Supply Filter

A device that converts the 60 Hz sine wave voltage from your wall outlet to a pulsating dc voltage is called a *full-wave rectifier.* The basic concept of a power supply with a full-wave rectifier is shown in the block diagram of Figure 10–46. In order to be useful in powering most systems such as the radio, TV, or computer, the pulsating dc voltage must be converted to a nearly constant level. This conversion is accomplished with a power supply filter as indicated.

A basic power supply filter is implemented with a capacitor as shown in Figure 10–47(a). As shown in Part (b), the basic operation is as follows: The capacitor charges as the full-wave voltage increases. When the peak is reached

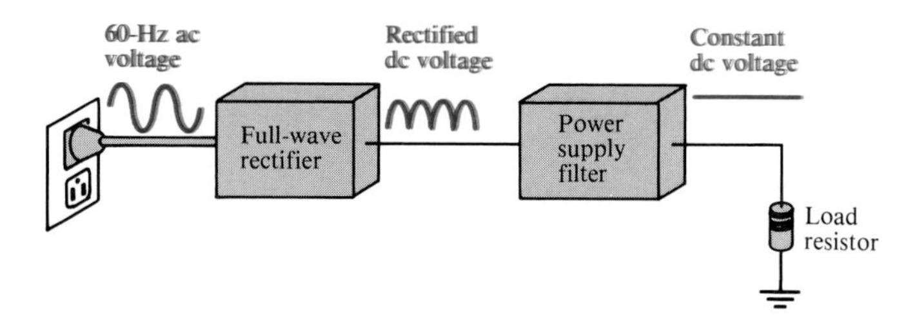

FIGURE 10–46
Basic concept of a power supply which converts ac to dc.

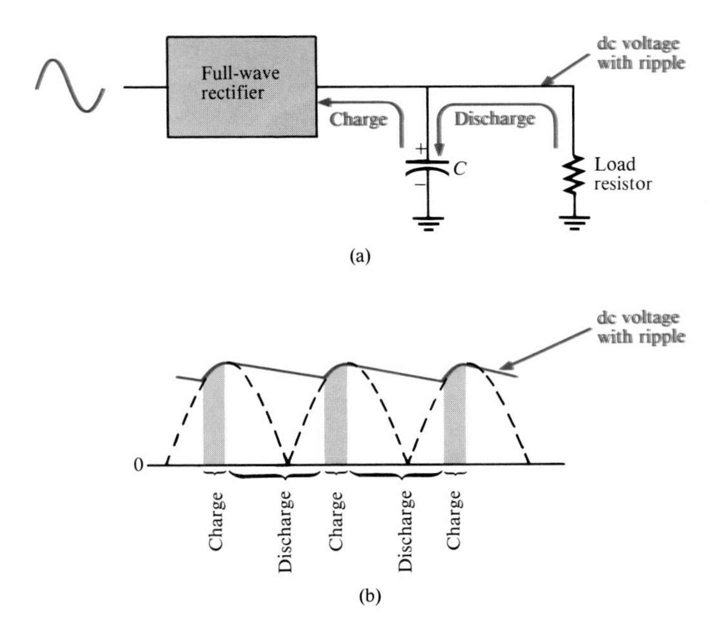

FIGURE 10–47
Basic operation of a capacitor power supply filter.

and the full-wave voltage starts to decrease, the capacitor begins to discharge through the load resistance. The value of the capacitance is selected so that the time constant is very long compared to the period of the full-wave voltage. The rectifier circuitry allows current only in a direction to charge the capacitor and prevents discharge current. By the time the full-wave voltage nears its next peak, the capacitor has discharged only a small amount and requires only a small amount of recharging to get it back to the peak. This action results in an almost constant dc voltage as shown. The small fluctuation is caused by the slight discharging and recharging of the capacitor. This fluctuation is called *ripple voltage*.

Coupling and Bypass Capacitors

Many applications, such as transistor amplifiers, require that an ac voltage be superimposed on a dc voltage at a certain point in the circuit, while at other points, the ac voltage must be removed without affecting the dc voltage.

The first situation is illustrated in Figure 10–48(a), where a capacitor is used to *couple* an ac voltage from the source to a point on a voltage divider that has a dc voltage. Since a capacitor blocks dc, the ac source is unaffected by the dc level, but the ac signal is passed through and superimposed on the dc level.

The second situation is illustrated in Figure 10–48(b), where a capacitor is used to *bypass* the ac voltage to ground, leaving only the dc voltage. Details of electronic amplifiers with ac coupling and bypass circuits are covered in a later course.

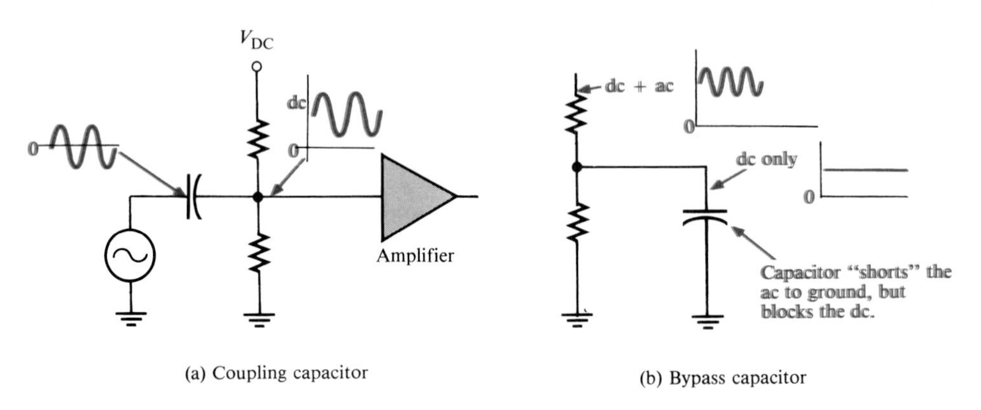

FIGURE 10–48
Coupling and bypass capacitors.

Tuned Circuits

Capacitors are used in conjunction with other components such as the inductor (to be covered in the next chapter) to provide *frequency selection* in communications systems. These *tuned circuits* allow a narrow band of frequencies to be selected while all other frequencies are rejected. A tuned circuit is one form of *filter*. The tuners in your TV and radio receivers are based on this principle and permit you to select one channel or station out of the many that are available.

Frequency selectivity is based on the fact the reactance of a capacitor depends on the frequency. The basic concept of a tuned circuit is shown in Figure 10–49. This topic will be covered in detail in Chapter 16.

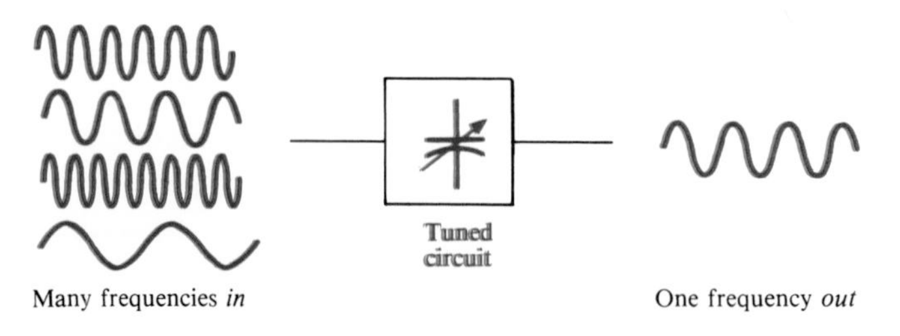

FIGURE 10–49
Basic concept of a tuned circuit.

Computer Memories

Many computer memories utilize capacitors as the *storage element* for binary data, which consist of arrangements of only two types of digits, 1s and 0s. A charged capacitor can represent a 1, and a discharged capacitor can represent a 0. Patterns of 1s and 0s can be stored in a memory that consists of an array of capacitors with associated circuitry. You will study this topic later in a computer or digital fundamentals course.

SECTION REVIEW 10–7

1. Explain how pulsating dc voltage is smoothed out by a capacitor in a power supply filter.
2. How can you use a capacitor to remove ac voltage from a given point in a circuit?

TESTING CAPACITORS

10–8

Capacitors are generally very reliable devices. Their useful life can be extended significantly by operating well within the voltage rating and at moderate temperatures.

Failures can be categorized into two areas: *catastrophic* and *degradation.* The catastrophic failures are usually a short circuit caused by dielectric breakdown or an open circuit caused by connection failure. Degradation usually results in a gradual decrease in leakage resistance, and hence an increase in leakage current.

Ohmmeter Check

When there is a suspected problem, the capacitor can be removed from the circuit and checked with an ohmmeter. First, to be sure that the capacitor is discharged, short its leads, as indicated in Figure 10–50(a). Connect the meter, set on a high ohms range such as R × 1M, to the capacitor, as shown in Figure 10–50(b), and observe the needle. It should initially indicate near zero ohms. Then it should begin to move toward the high-resistance end of the scale as the capacitor charges from the ohmmeter's battery, as shown in Figure 10–50(c). When the capacitor is fully charged, the meter will indicate an extremely high resistance, as shown in Figure 10–50(d).

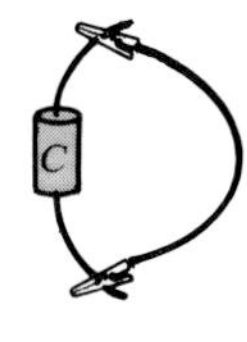

(a) Discharging

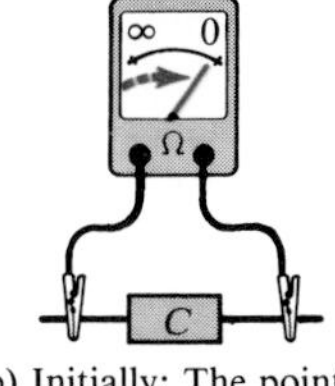

(b) Initially: The pointer jumps to zero.

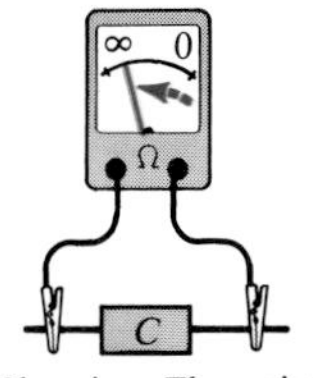

(c) Charging: The pointer slowly moves back.

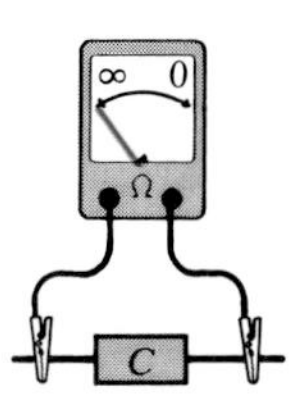

(d) Fully charged

FIGURE 10–50
Checking a capacitor with an ohmmeter. This check shows a good capacitor.

As mentioned, the capacitor charges from the internal battery of the ohmmeter, and the meter responds to the charging current. The larger the capacitance value, the more slowly the capacitor will charge, as indicated by the needle movement. For very small pF values, the meter response may be insufficient to indicate the fast charging action.

If the capacitor is internally shorted, the meter will go to zero and stay there. If it is leaky, the final meter reading will be much less than normal. Most capacitors have a resistance of several hundred megohms. The exception is the electrolytic, which may normally have less than one megohm of leakage resistance. If the capacitor is open, no charging action will be observed, and the meter will indicate an infinite resistance. The actual capacitance value can be measured on an instrument such as the *LCR* meter (inductance-capacitance-resistance). An example of this type of instrument is shown in Figure 10–51.

FIGURE 10–51
LCR meter for measuring capacitance (courtesy of Hewlett-Packard Co.).

SECTION REVIEW 10–8

1. How can a capacitor be discharged after removal from the circuit?
2. Describe how the needle of an ohmmeter responds when a good capacitor is checked.

APPLICATION NOTE

For the Application Assignment at the beginning of the chapter, you can check the capacitor for an open, short, or leakage condition by using the ohmmeter as described in Section 10–8. If the capacitor passes these checks, you can measure its capacitance value by using a capacitance or *LCR* meter.

SUMMARY

Facts

- A capacitor blocks dc and passes ac.
- A capacitor is composed of two parallel conductive plates separated by an insulating material called the *dielectric*.
- A capacitor stores electrical charge on its plates.
- Energy is stored in a capacitor by the electric field created between the charged plates.

- Capacitance is measured in units of farads (F).
- Capacitance is directly proportional to the plate area and the dielectric constant and inversely proportional to the distance between the plates (the dielectric thickness).
- The dielectric constant is an indication of the ability of a material to establish an electric field.
- The dielectric strength is one factor that determines the breakdown voltage of a capacitor.
- Capacitors are commonly classified according to the dielectric material. Typical materials are mica, ceramic, paper/plastic, and electrolytic (aluminum oxide and tantalum oxide).
- The total capacitance of series capacitors is less than the smallest capacitance.
- The total capacitance of parallel capacitors is the sum of all the capacitances.
- The time constant determines the charging and discharging time of a capacitor with resistance in series.
- In a circuit with capacitance and resistance, the voltage and current during charging and discharging make a 63% change during each time constant interval.
- Five time constants are required for a capacitor to fully charge or discharge.
- Current in a capacitor leads the voltage by 90°.
- Capacitive reactance (X_C) is inversely proportional to frequency and to capacitance.
- Ideally, there is no energy loss in a capacitor and, thus, the true power (watts) is zero. However, in most capacitors there is some energy loss due to leakage.

Units

- Farad (F)—the unit of capacitance (C).
- Volts/mil—the unit of dielectric strength (1 mil = 0.001 in.).
- Volt-ampere reactive (VAR)—the unit of reactive power (P_r).

Definitions

- *Capacitance*—the measure of a capacitor's ability to store charge.
- *One farad*—the amount of capacitance when one coulomb of charge is stored with one volt across the plates.
- *Breakdown voltage*—the amount of voltage per mil (0.001 in.) of dielectric thickness that a capacitor can withstand without damage.
- *Temperature coefficient*—the amount of change in capacitance per degree change in temperature.
- *Time constant*—the time required for a capacitor to charge or discharge by 63%. The capacitance multiplied by the resistance.
- *Capacitive reactance*—the opposition to sinusoidal current expressed in ohms.

Formulas

$$C = \frac{Q}{V} \quad \text{Capacitance in terms of charge and voltage} \quad (10\text{–}1)$$

$$F = \frac{kQ_1Q_2}{d^2} \quad \text{Coulomb's law (force between two charged bodies)} \quad (10\text{–}2)$$

$$W = \frac{1}{2}CV^2 \quad \text{Energy stored by a capacitor} \quad (10\text{–}3)$$

$$\epsilon_r = \frac{\epsilon}{\epsilon_0} \quad \text{Dielectric constant (relative permittivity)} \quad (10\text{–}4)$$

$$C = \frac{A\epsilon_r(8.85 \times 10^{-12}\text{F/m})}{d}$$ Capacitance in terms of physical parameters (10–5)

$$\frac{1}{C_T} = \frac{1}{C_1} + \frac{1}{C_2}$$ Reciprocal of total series capacitance (two capacitors) (10–6)

$$C_T = \frac{1}{(1/C_1) + (1/C_2)}$$ Total series capacitance (two capacitors) (10–7)

$$\frac{1}{C_T} = \frac{1}{C_1} + \frac{1}{C_2} + \frac{1}{C_3} + \cdots + \frac{1}{C_n}$$ Reciprocal of total series capacitance (general) (10–8)

$$C_T = \frac{1}{(1/C_1) + (1/C_2) + (1/C_3) + \cdots + (1/C_n)}$$ Total series capacitance (general) (10–9)

$$V_x = \left(\frac{C_T}{C_x}\right)V_S$$ Voltage across series capacitor (10–10)

$$C_T = C_1 + C_2$$ Two capacitors in parallel (10–11)

$$C_T = C_1 + C_2 + \cdots + C_n$$ n capacitors in parallel (10–12)

$$\tau = RC$$ Time constant (10–13)

$$v = V_F + (V_i - V_F)e^{-t/\tau}$$ Exponential voltage (10–14)

$$i = I_F + (I_i - I_F)e^{-t/\tau}$$ Exponential current (10–15)

$$v = V_F(1 - e^{-t/RC})$$ Increasing exponential voltage (10–16)

$$v = V_ie^{-t/RC}$$ Decreasing exponential voltage (10–17)

$$X_C = \frac{1}{2\pi fC}$$ Capacitive reactance (10–18)

$$V = IX_C$$ Ohm's law (10–19)

$$P_r = V_{rms}I_{rms}$$ Reactive power (10–20)

$$P_r = \frac{V_{rms}^2}{X_C}$$ Reactive power (10–21)

$$P_r = I_{rms}^2X_C$$ Reactive power (10–22)

Symbols

Capacitor symbols are shown in Figure 10–52.

FIGURE 10–52
Capacitor symbols.

SELF-TEST

1. Indicate true or false for each of the following statements:
   (a) The plates of a capacitor are conductive.
   (b) The dielectric is the insulating material that separates the plates.
   (c) Constant dc flows through a fully charged capacitor.
   (d) A practical capacitor stores charge indefinitely when it is disconnected from the source.

2. Indicate true or false for each of the following statements:
   (a) There is current through the dielectric of a charging capacitor.
   (b) When a capacitor is connected to a dc voltage source, it will charge to the value of the source voltage.
   (c) You can discharge an ideal capacitor by simply disconnecting it from the voltage source.
   (d) When a capacitor is completely discharged, the voltage across its plates is zero.
3. Which is the larger capacitance, 0.01 μF or 0.00001 F?
4. Which is the smaller capacitance, 1000 pF or 0.0001 μF?
5. If the voltage across a given capacitor is increased, does the amount of stored charge increase or decrease?
6. If the voltage across a given capacitor is doubled, how much does the stored energy increase?
7. How can the voltage rating of a capacitor be increased?
8. Select the physical parameter changes that result in an increase in the capacitance value:
   (a) Reduce plate area. (b) Move plates further apart.
   (c) Move plates closer. (d) Increase plate area.
   (e) Increase the thickness of the dielectric.
   (f) Decrease the thickness of the dielectric.
9. A 1-μF, a 2.2-μF, and a 0.05-μF capacitor are connected in series. The total capacitance is (greater, less than) ______ μF.
10. Four 0.02-μF capacitors are connected in parallel. What is the total capacitance?
11. An uncharged capacitor and a resistor are placed in series with a switch and are connected to a dc voltage source.
    (a) At the instant the switch is closed, what is the voltage across the capacitor?
    (b) When will the capacitor reach full charge?
    (c) What is the voltage across the capacitor when it is fully charged?
    (d) What is the current in the circuit when the capacitor is fully charged?
12. A sine wave voltage is applied across a capacitor. When the frequency of the voltage is increased, does the current increase or decrease? Why?
13. A capacitor and a resistor are connected in series with a sine wave voltage source. The frequency is set so that the capacitive reactance is equal to the resistance, and an equal amount of voltage is dropped across each component. If the frequency is decreased, which component will have the greater voltage across it?
14. An ohmmeter is connected across a discharged capacitor, and the pointer stabilizes at a low resistance value. What is your opinion of this capacitor?

PROBLEM SET A

Section 10–1

10–1 (a) Find the capacitance when $Q = 50\ \mu C$ and $V = 10$ V.
(b) Find the charge when $C = 0.001\ \mu F$ and $V = 1$ kV.
(c) Find the voltage when $Q = 2$ mC and $C = 200\ \mu F$.

10–2 Convert the following values from microfarads to picofarads:
(a) 0.1 μF (b) 0.0025 μF (c) 5 μF

10–3 Convert the following values from picofarads to microfarads:
(a) 1000 pF (b) 3500 pF (c) 250 pF

10–4 Convert the following values from farads to microfarads:
(a) 0.0000001 F (b) 0.0022 F (c) 0.0000000015 F

10–5 What size capacitor is capable of storing 10 mJ of energy with 100 V across its plates?

10–6 A mica capacitor has a plate area of 0.04 m^2 and a dielectric thickness of 0.008 m. What is its capacitance?

10–7 An air capacitor has 0.1-m square plates. The plates are separated by 0.01 m. Calculate the capacitance.

10–8 At ambient temperature (25°C), a certain capacitor is specified to be 1000 pF. It has a negative temperature coefficient of 200 ppm/°C. What is its capacitance at 75°C?

10–9 A 0.001-μF capacitor has a positive temperature coefficient of 500 ppm/°C. How much change in capacitance will a 25°C increase in temperature cause?

Section 10–2

10–10 In the construction of a stacked-foil mica capacitor, how is the plate area increased?

10–11 What type of capacitor has the higher dielectric constant, mica or ceramic?

10–12 Show how to connect an electrolytic capacitor between points A and B in Figure 10–53.

10–13 Determine the value of the typographically labeled ceramic disk capacitors in Figure 10–54.

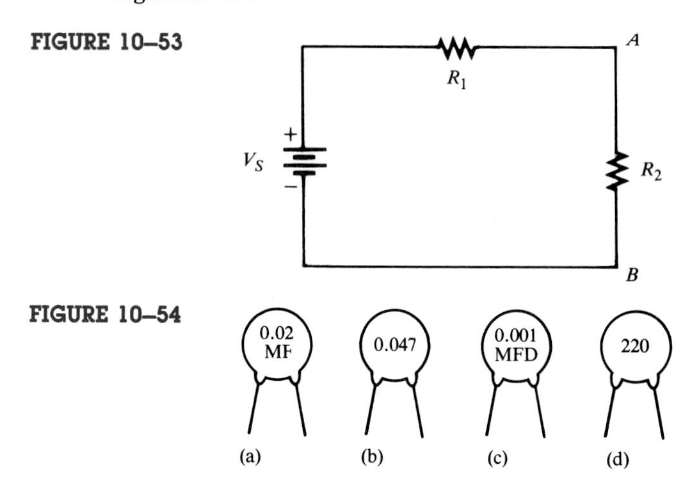

FIGURE 10–53

FIGURE 10–54

Section 10–3

10–14 Five 1000-pF capacitors are in series. What is the total capacitance?

10–15 Find the total capacitance for each circuit in Figure 10–55.

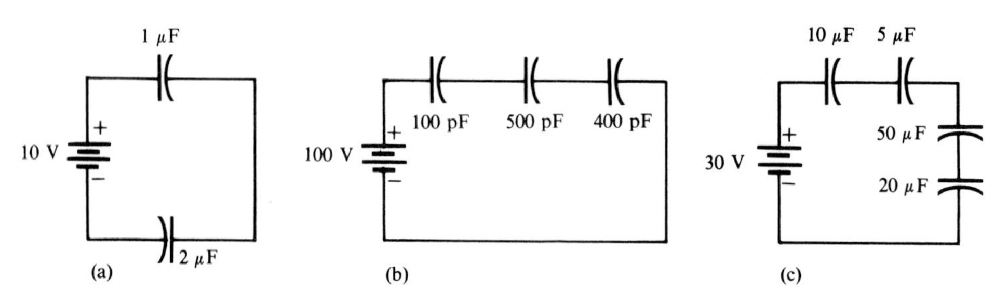

FIGURE 10–55

10–16 For each circuit in Figure 10–55, determine the voltage across each capacitor.

Section 10–4

10–17 Determine C_T for each circuit in Figure 10–56.

10–18 Determine C_T for each circuit in Figure 10–57.

10–19 What is the voltage between points A and B in each circuit in Figure 10–57?

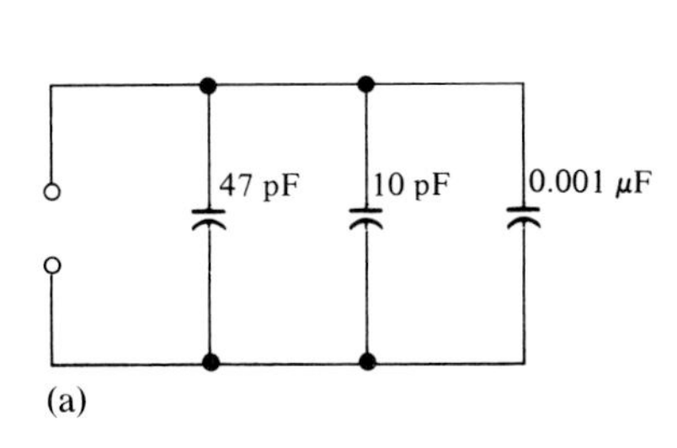

(a)

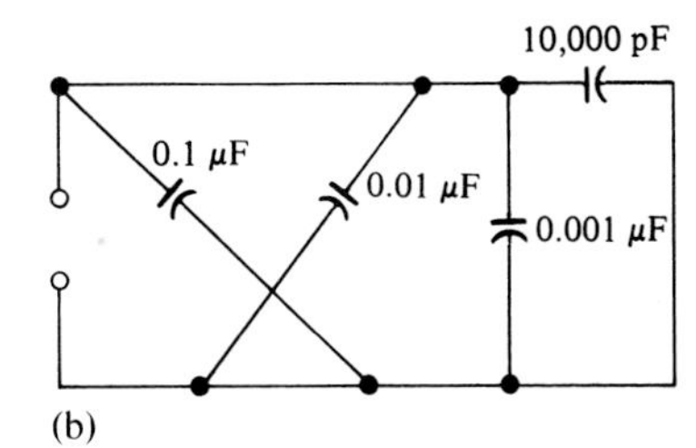

(b)

FIGURE 10–56

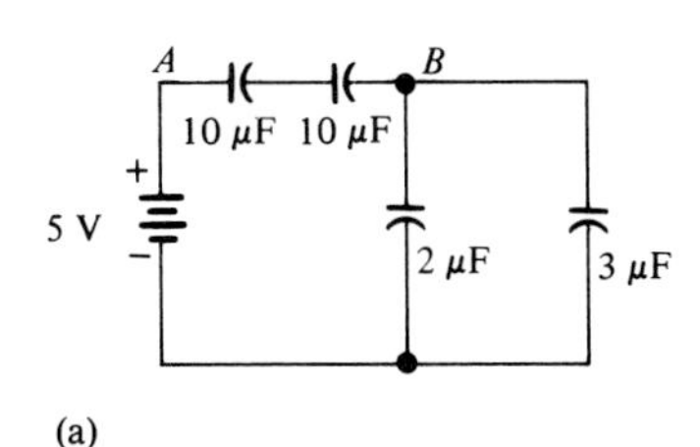

(a)

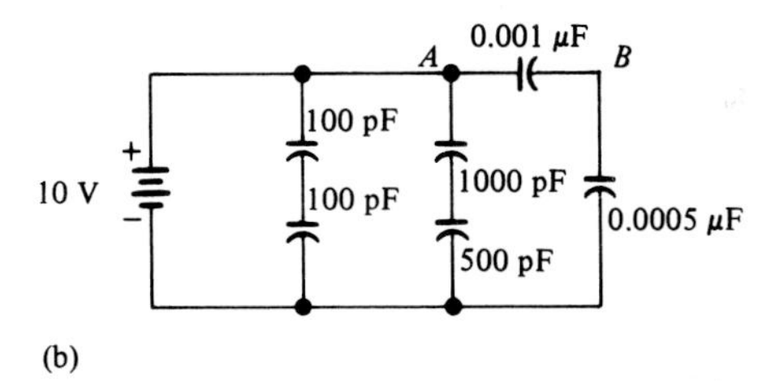

(b)

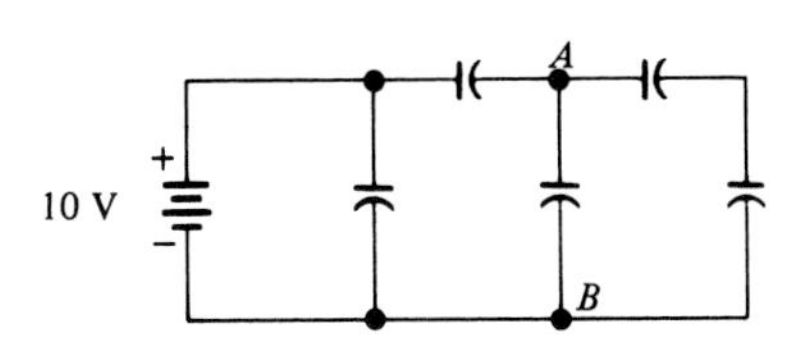

(c) $C = 1\ \mu\text{F}$ for each capacitor

FIGURE 10–57

Section 10–5

10–20 Determine the time constant for each of the following series RC combinations:

(a) $R = 100\ \Omega$, $C = 1\ \mu\text{F}$ **(b)** $R = 10\ \text{M}\Omega$, $C = 50\ \text{pF}$
(c) $R = 4.7\ \text{k}\Omega$, $C = 0.005\ \mu\text{F}$ **(d)** $R = 1.5\ \text{M}\Omega$, $C = 0.01\ \mu\text{F}$

10–21 Determine how long it takes the capacitor to reach full charge for each of the following combinations:

(a) $R = 50\ \Omega$, $C = 50\ \mu\text{F}$ **(b)** $R = 3300\ \Omega$, $C = 0.015\ \mu\text{F}$
(c) $R = 22\ \text{k}\Omega$, $C = 100\ \text{pF}$ **(d)** $R = 5\ \text{M}\Omega$, $C = 10\ \text{pF}$

10–22 In the circuit of Figure 10–58, the capacitor initially is uncharged. Determine the capacitor voltage at the following times after the switch is closed:

(a) 10 μs **(b)** 20 μs **(c)** 30 μs **(d)** 40 μs **(e)** 50 μs

10–23 In Figure 10–59, the capacitor is charged to 25 V. Find the capacitor voltage after the following times when the switch is closed:

(a) 1.5 ms **(b)** 4.5 ms **(c)** 6 ms **(d)** 7.5 ms

10–24 Repeat Problem 10–22 for the following time intervals:

(a) 2 μs **(b)** 5 μs **(c)** 15 μs

10–25 Repeat Problem 10–23 for the following times:
(a) 0.5 ms **(b)** 1 ms **(c)** 2 ms

FIGURE 10–58

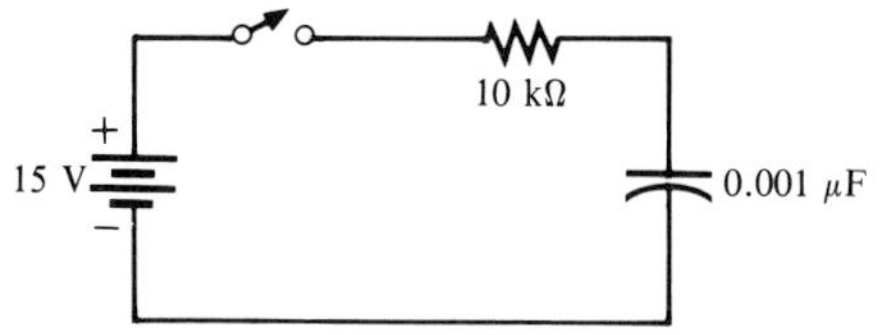

FIGURE 10–59

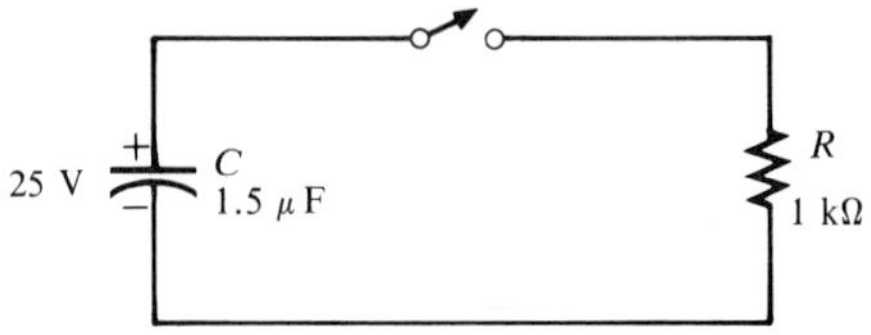

Section 10–6

10–26 What is the value of the total capacitive reactance in each circuit in Figure 10–60?

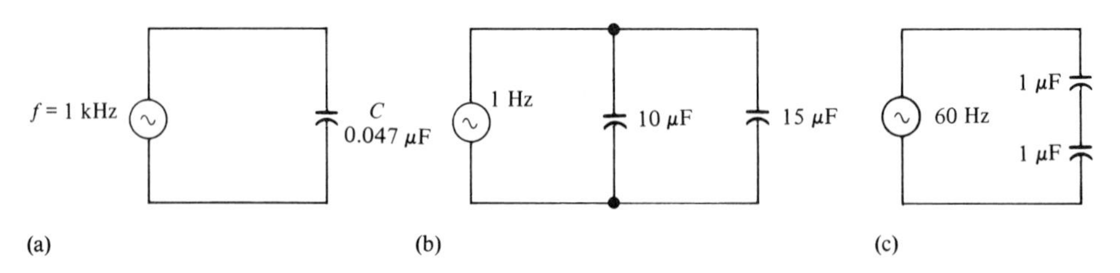

FIGURE 10–60

10–27 In Figure 10–57, each dc voltage source is replaced by a 10-V rms, 2-kHz ac source. Determine the reactance in each case.

10–28 In each circuit of Figure 10–60, what frequency is required to produce an X_C of 100 Ω? An X_C of 1 kΩ?

10–29 A sinusoidal voltage of 20 V rms produces an rms current of 100 mA when connected to a certain capacitor. What is the reactance?

10–30 A 10-kHz voltage is applied to a 0.0047-μF capacitor, and 1 mA of rms current is measured. What is the value of the voltage?

10–31 Determine the true power and the reactive power in Problem 10–30.

Section 10–7

10–32 If another capacitor is connected in parallel with the existing capacitor in the power supply filter of Figure 10–47, how is the ripple voltage affected?

10–33 Ideally, what should the reactance of a bypass capacitor be in order to eliminate a 10-kHz ac voltage at a given point in an amplifier circuit?

Section 10–8

10–34 Assume that you are checking a capacitor with an ohmmeter, and when you connect the leads across the capacitor, the pointer does not move from its left-end scale position. What is the problem?

10–35 In checking a capacitor with the ohmmeter, you find that the pointer goes all the way to the right end of the scale and stays there. What is the problem?

PROBLEM SET B

10–36 Two series capacitors (one 1-μF, the other of unknown value) are charged from a 12-V source. The 1-μF capacitor is charged to 8 V, and the other to 4 V. What is the value of the unknown capacitor?

10–37 How long does it take C to discharge to 3 V in Figure 10–59?

10–38 How long does it take C to charge to 8 V in Figure 10–58?

10–39 Determine the time constant for the circuit in Figure 10–61.

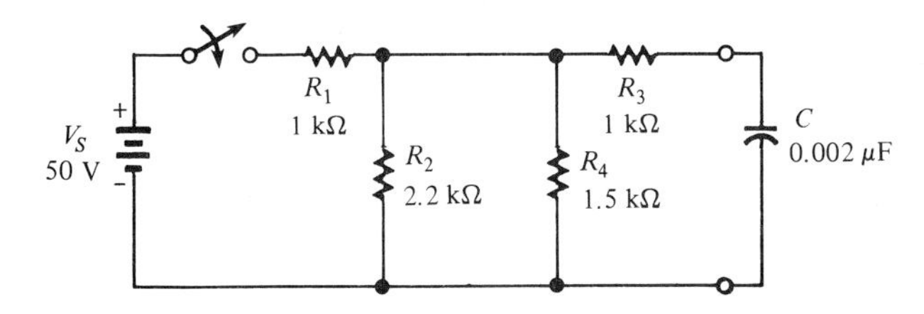

FIGURE 10–61

10–40 In Figure 10–62, the capacitor initially is uncharged. At $t = 10\ \mu$s after the switch is closed, the instantaneous capacitor voltage is 7.2 V. Determine the value of R.

FIGURE 10–62

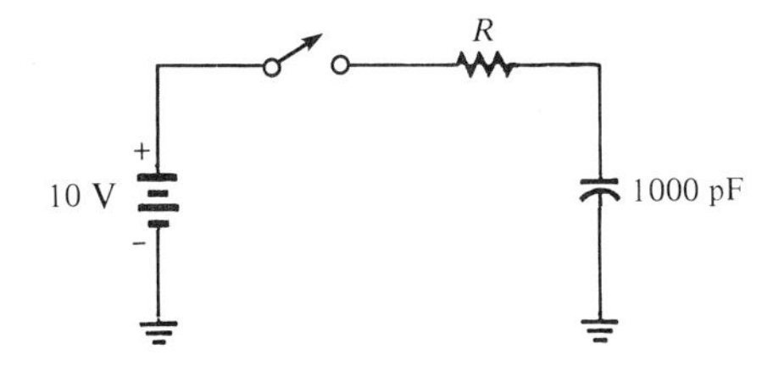

10–41 **(a)** The capacitor in Figure 10–63 is uncharged when the switch is thrown into position 1. The switch remains in position 1 for 10 ms and then is thrown into position 2, where it remains indefinitely. Sketch the complete waveform for the capacitor voltage.

(b) If the switch is thrown back to position 1 after 5 ms in position 2, and then is left in position 1, how will the waveform appear?

FIGURE 10–63

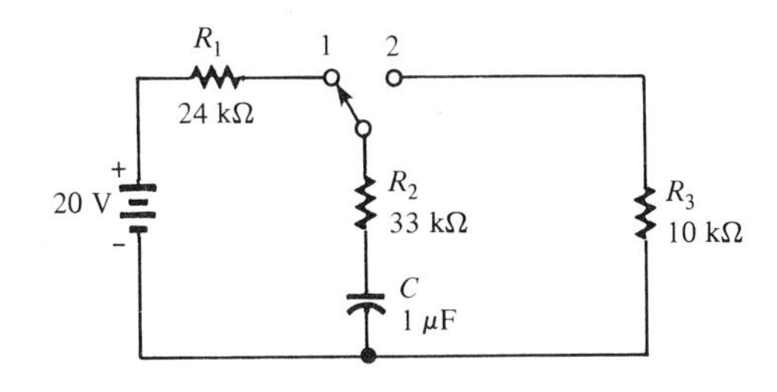

10–42 Determine the ac voltage across each capacitor and the current in each branch of the circuit in Figure 10–64.

FIGURE 10–64

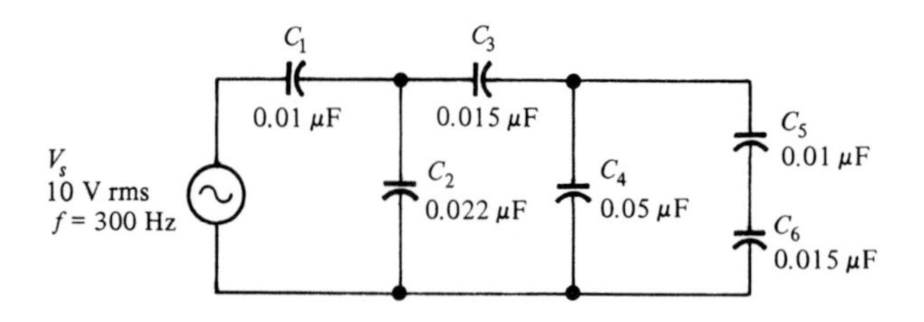

10–43 Find the value of C_1 in Figure 10–65.

FIGURE 10–65

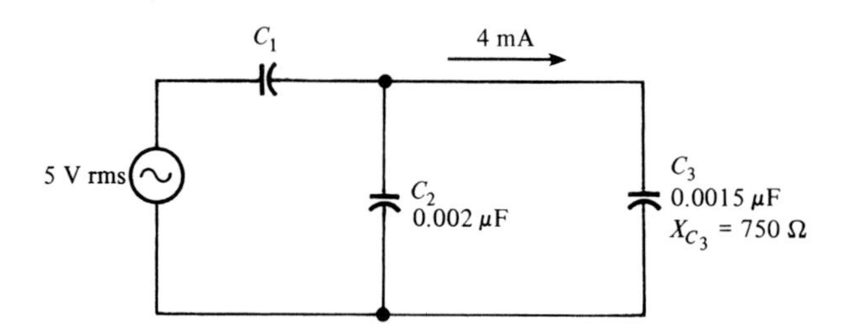

ANSWERS TO SECTION REVIEWS

Section 10–1

1. The ability (capacity) to store electrical charge.
2. (a) 10^6; (b) 10^{12}; (c) 10^6. **3.** (a) increase; (b) decrease.

Section 10–2

1. By the dielectric material.
2. A fixed capacitance cannot be changed; a variable can. **3.** Electrolytic.
4. Be sure that the voltage rating is sufficient. Connect the positive end to the positive side of the circuit.

Section 10–3

1. Less. **2.** 62.5 pF. **3.** 0.006 μF. **4.** 6 V.

Section 10–4

1. The individual capacitors are added.
2. By using five 0.01-μF capacitors in parallel. **3.** 98 pF.

Section 10–5

1. 1.2 μs. **2.** 6 μs, 5 V. **3.** 8.6 V, 9.5 V, 9.8 V, 9.9 V. **4.** 37 V.

Section 10–6

1. Current leads voltage by 90°. **2.** 637 kΩ. **3.** 796 Hz. **4.** 628 mA.
5. 0. **6.** 0.453 VAR.

Section 10–7

1. Once the capacitor charges to the peak voltage, it discharges very little before the next peak.

2. By selecting a capacitor that has a reactance of almost zero at the frequency of the ac voltage and connecting that capacitor from that point to ground.

Section 10–8

1. Short its leads.
2. Initially, the needle jumps to zero; then it moves to the high-resistance end of the scale.

Inductors

Inductance is the property of a coil of wire that opposes a change in current. The basis for inductance is the electromagnetic field that surrounds any conductor when there is current through it. The electrical component designed to have the property of inductance is called an *inductor, coil,* or *choke.* All of these terms refer to essentially the same type of device.

In this chapter, the basic inductor is introduced and its characteristics are studied. Various types of inductors are covered in terms of their physical construction and their electrical properties. The basic behavior of inductors in both dc and ac circuits is studied, and series and parallel combinations are analyzed. A method of testing inductors is discussed.

Specifically in this chapter, you will learn:

- What an inductor is and what it does in a circuit.
- How an inductor stores energy.
- The unit of inductance.
- How physical characteristics determine inductance.
- Why inductors exhibit resistance and capacitance.
- How Lenz's law and Faraday's law apply to inductors.
- Various types of inductors.
- What happens when inductors are connected in series.
- What happens when inductors are connected in parallel.
- The meaning of the *time constant* in an inductive circuit.
- How an inductor introduces phase shift between current and voltage.
- The definition of *inductive reactance* and how to determine its value in a circuit.
- The meaning of *reactive power* in an inductive circuit.
- Several common applications of inductors.
- How to check an inductor with an ohmmeter.

APPLICATION ASSIGNMENT

A defective power supply is returned for repair. A quick check of the output voltage reveals that there is excessive ripple voltage. Your assignment is to check out both the capacitor and the inductor in the power supply filter to see if either or both are defective.

After completing this chapter, you will be able to carry out this assignment. The solution is given in the Application Note at the end of the chapter.

THE INDUCTOR

11–1

When a length of wire is formed into a coil, as shown in Figure 11–1(a), it becomes a basic *inductor*. Current flowing through the coil produces a magnetic field, as illustrated in Figure 11–1(b). The magnetic lines of force around each loop (turn) in the coil effectively add to the lines of force around the adjoining loops, forming a strong magnetic field within and around the coil, as shown. The net direction of the total magnetic field creates a north and a south pole, as shown in Part (b).

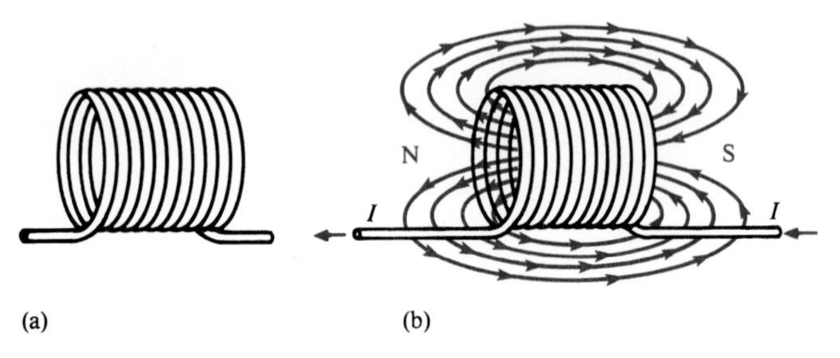

FIGURE 11–1
A coil of wire forms an inductor. When current flows through it, a three-dimensional electromagnetic field is created, surrounding the coil in all directions.

To understand the formation of the total magnetic field in a coil, let's discuss the interaction of the magnetic fields around two adjacent loops. The magnetic lines of force around adjacent loops are deflected into an outer path when the loops are brought close together. This effect occurs because the magnetic lines of force are in *opposing* directions between adjacent loops, as illustrated in Figure 11–2(a). The total magnetic field for the two loops is depicted in Part (b) of the figure. For simplicity, only single lines of force are shown. This effect is additive for many closely adjacent loops in a coil; that is, each additional loop adds to the strength of the electromagnetic field.

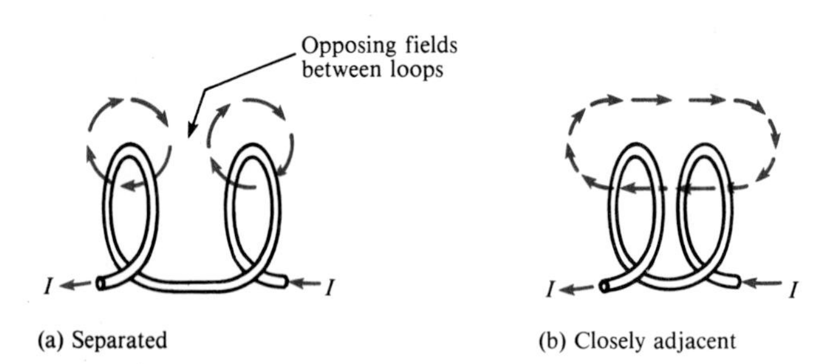

FIGURE 11–2
Interaction of magnetic lines of force in two adjacent loops of a coil.

Self-Inductance

When there is current through an inductor, an electromagnetic field is established. When the current changes, the electromagnetic field also changes. An increase in current expands the field, and a decrease in current reduces it. Therefore, a changing current produces a changing electromagnetic field around the inductor (coil). In turn, the changing electromagnetic field induces a voltage across the coil because of a property called *self-inductance.*

Self-inductance is a measure of a coil's ability to establish an induced voltage as a result of a change in its current. Self-inductance is usually referred to as simply *inductance.* Inductance is symbolized by L.

The Unit of Inductance

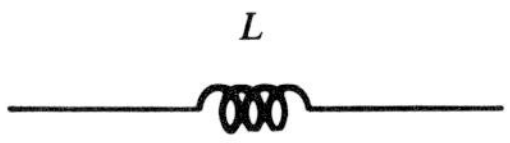

L

FIGURE 11–3
A symbol for the inductor.

The *henry,* symbolized by H, is the basic unit of inductance. By definition, the inductance is *one henry* when current through the coil, changing at the rate of *one ampere per second,* induces *one volt* across the coil. In many practical applications, *millihenries* (mH) and microhenries (μH) are the more common units. A common schematic symbol for the inductor is shown in Figure 11–3.

Energy Storage

An inductor stores energy in the magnetic field created by the current. The energy stored is expressed as follows:

$$W = \frac{1}{2} LI^2 \qquad (11\text{–}1)$$

As you can see, the energy stored is proportional to the inductance and the square of the current. When I is in amperes and L is in henries, the energy is in joules.

Physical Characteristics

The following characteristics are important in establishing the inductance of a coil: the core material and the parameters of the number of turns of wire, the length, and the cross-sectional area.

Core Material

As discussed earlier, an inductor is basically a coil of wire. The material around which the coil is formed is called the *core.* Coils are wound on either nonmagnetic or magnetic materials. Examples of nonmagnetic materials are air, wood, copper, plastic, and glass. The permeabilities of these materials are the same as for a vacuum. Examples of magnetic materials are iron, nickel, steel, cobalt, or alloys. These materials have permeabilities that are hundreds or thousands of times greater than that of a vacuum and are classified as *ferromagnetic.* A ferromagnetic core provides a better path for the magnetic lines of force and thus permits a stronger magnetic field.

As you learned in chapter 7, the permeability (μ) of the core material determines how easily a magnetic field can be established. *The inductance is directly proportional to the permeability of the core material.*

Parameters

As indicated in Figure 11–4, the number of turns of wire, the length, and the cross-sectional area of the core are factors in setting the value of inductance. The inductance is inversely proportional to the length of the core and directly proportional to the cross-sectional area. Also, the inductance is directly related to the number of turns squared. This relationship is as follows:

$$L = \frac{N^2 \mu A}{l} \tag{11–2}$$

where L is the inductance in henries, N is the number of turns, μ is the permeability, A is the cross-sectional area in meters squared, and l is the core length in meters.

FIGURE 11–4
Factors that determine the inductance of a coil.

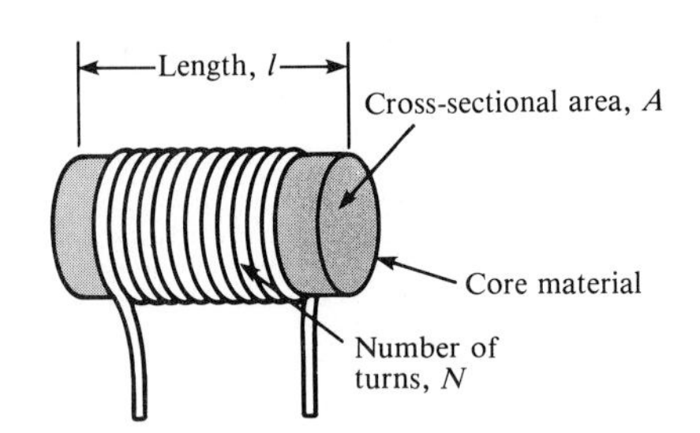

EXAMPLE 11–1 Determine the inductance of the coil in Figure 11–5. The permeability of the core is 0.25×10^{-3}.

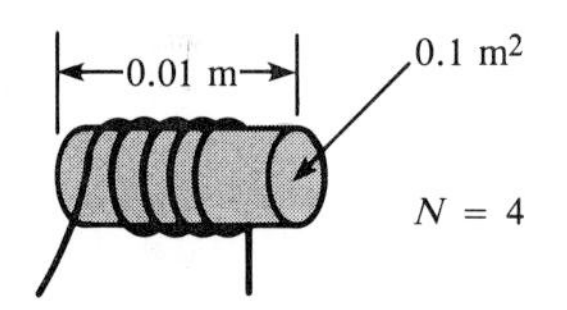

FIGURE 11–5

Solution

$$L = \frac{N^2 \mu A}{l} = \frac{(4)^2(0.25 \times 10^{-3})(0.1)}{0.01} = 40 \text{ mH}$$

The calculator sequence is
[4] [2nd] [x²] [×] [·] [2] [5] [EE] [+/−] [3] [×] [·] [1] [÷] [·] [0] [1] [=]

Winding Resistance

When a coil is made of a certain material, for example, insulated copper wire, that wire has a certain *resistance* per unit of length. When many turns of wire are used to construct a coil, the total resistance may be significant. This inherent resistance is called the *dc resistance* or the *winding resistance* (R_W). Although this resistance is distributed along the length of the wire, it effectively appears in series with the inductance of the coil, as shown in Figure 11–6. In many applications, the winding resistance can be ignored and the coil considered as an ideal inductor. In other cases, the resistance must be considered.

FIGURE 11–6
Winding resistance of a coil.

(a) The wire has resistance. (b) Equivalent circuit

Winding Capacitance

When two conductors are placed side by side, there is always some capacitance between them. Thus, when many turns of wire are placed close together in a coil, a certain amount of *stray* capacitance is a natural side effect. In many applications, this stray capacitance is very small and has no significant effect. In other cases, particularly at high frequencies, it may become quite important.

The equivalent circuit for an inductor with both its winding resistance (R_W) and its winding capacitance (C_W) is shown in Figure 11–7. The capacitance effectively acts in parallel.

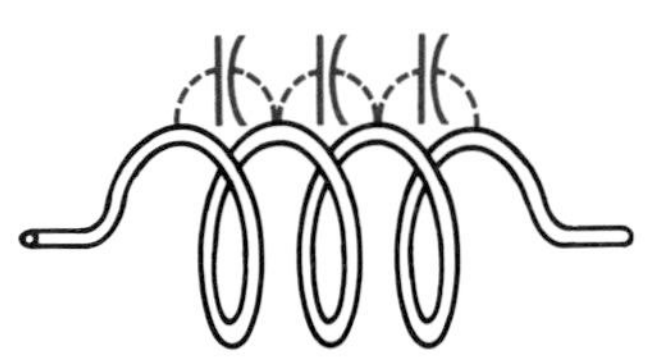

(a) Stray capacitance between each loop appears as a total parallel capacitance.

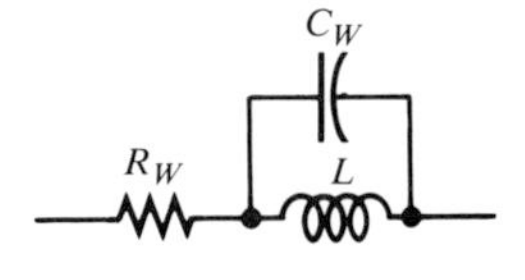

(b) Equivalent circuit

FIGURE 11–7
Winding capacitance of a coil.

Lenz's Law

In Chapter 7, you learned that a changing magnetic field induces a voltage in a coil that is directly proportional to the rate of change of the magnetic field and the number of turns in the coil. This principle was called Faraday's law. Lenz's law adds to this by defining the direction of induced voltage:

> **When the current through a coil changes and an induced voltage is created as a result of the changing magnetic field, the direction of the induced voltage is such that it always opposes the change in current.**

In Figure 11–8(a), the current is constant and is limited by R_1. There is no induced voltage because the magnetic field is unchanging. In Part (b), the switch suddenly is closed, placing R_2 in parallel with R_1 and thus reducing the resistance. Naturally, the current tries to increase and the magnetic field begins to expand, but the induced voltage opposes this attempted increase in current for an instant.

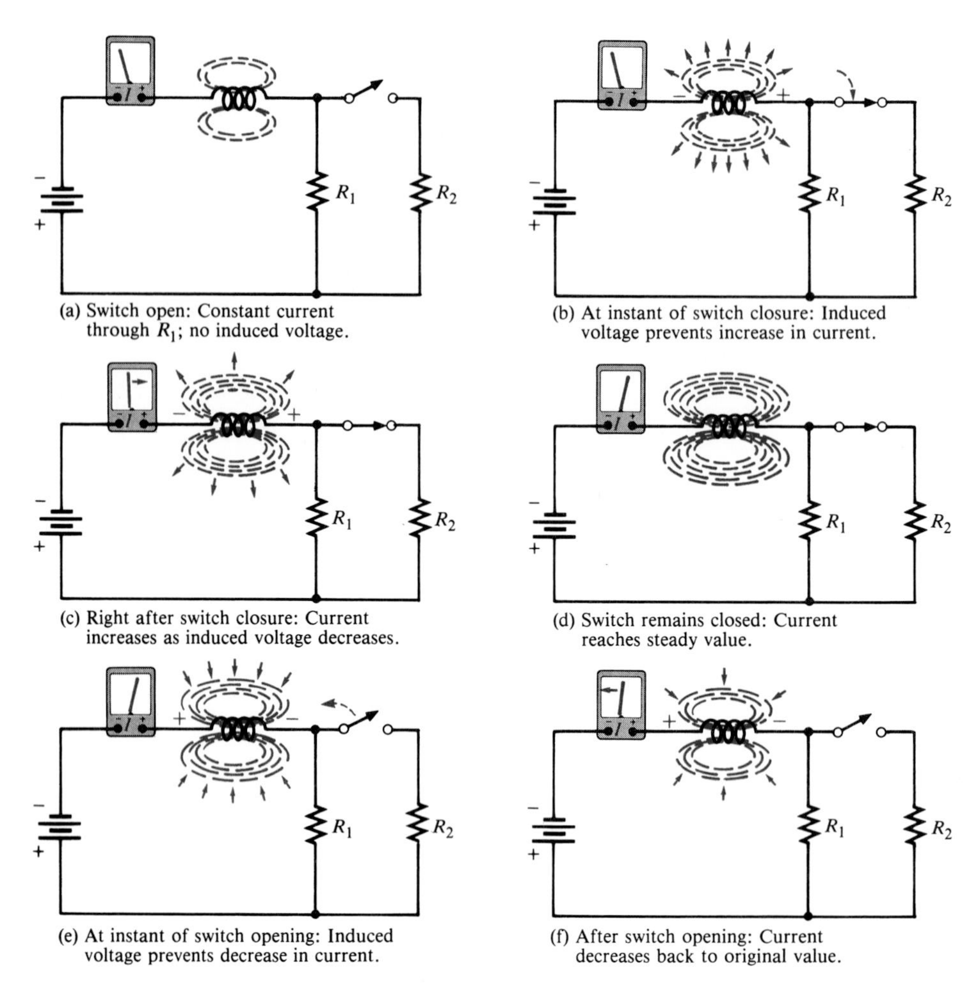

(a) Switch open: Constant current through R_1; no induced voltage.

(b) At instant of switch closure: Induced voltage prevents increase in current.

(c) Right after switch closure: Current increases as induced voltage decreases.

(d) Switch remains closed: Current reaches steady value.

(e) At instant of switch opening: Induced voltage prevents decrease in current.

(f) After switch opening: Current decreases back to original value.

FIGURE 11–8
Demonstration of Lenz's law: When the current tries to change suddenly, the electromagnetic field changes and induces a voltage in a direction that opposes that change in current.

In Part (c), the induced voltage gradually decreases, allowing the current to increase. In Part (d), the current has reached a constant value as determined by the parallel resistors, and the induced voltage is zero. In Part (e), the switch has been suddenly opened, and, for an instant, the induced voltage prevents any decrease in current. In Part (f), the induced voltage gradually decreases, allowing the current to decrease back to a value determined by R_1. Notice that the induced voltage has a polarity that opposes any current change. The polarity of the induced voltage is opposite that of the battery voltage for an increase in current and aids the battery voltage for a decrease in current.

SECTION REVIEW 11–1

1. Describe what happens to L when
   (a) N is increased.
   (b) the core length is increased.

(c) the cross-sectional area of the core is decreased.
(d) a ferromagnetic core is replaced by an air core.

2. Explain why inductors inevitably have some winding resistance.

TYPES OF INDUCTORS

11–2

Inductors are made in a variety of shapes and sizes. Basically, they fall into two general categories: *fixed* and *variable*. The standard schematic symbols are shown in Figure 11–9.

FIGURE 11–9
Symbols for (a) fixed and (b) variable inductors.

(a) Fixed (b) Variable

Both fixed and variable inductors can be classified according to the type of core material. Three common types are the air core, the iron core, and the ferrite core. Each has a unique symbol, as shown in Figure 11–10.

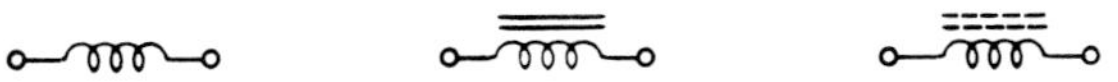

(a) Air core (b) Iron core (c) Ferrite core

FIGURE 11–10
Inductor symbols.

Adjustable (variable) inductors usually have a screw-type adjustment that moves a sliding core in and out, thus changing the inductance. A wide variety of inductors exists, some of which are shown in Figure 11–11.

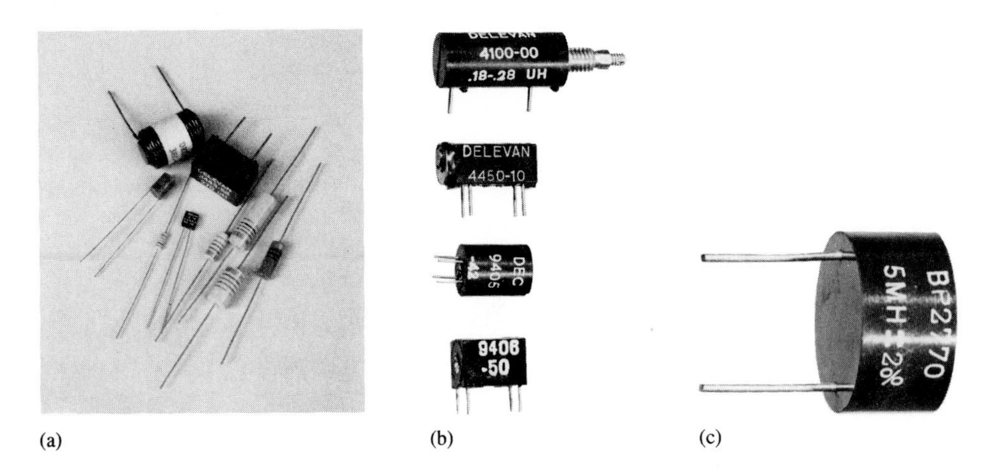

FIGURE 11–11
Typical inductors. (a) Fixed molded inductors; (b) variable coils; (c) toroid inductor [(a) and (b) courtesy of Delevan/American Precision. (c) courtesy of Dale Electronics].

SECTION REVIEW 11–2

1. Name two general categories of inductors.
2. Identify the inductor symbols in Figure 11–12.

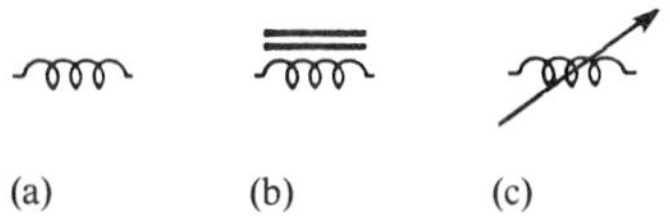

FIGURE 11–12

(a) (b) (c)

SERIES INDUCTORS

11–3

When inductors are connected in series, as in Figure 11–13, *the total inductance, L_T, is the sum of the individual inductances.* The formula for L_T is expressed in the following equation for the general case of n inductors in series:

$$L_T = L_1 + L_2 + L_3 + \cdots + L_n \quad (11\text{–}3)$$

Notice that inductance in series is similar to resistance in series.

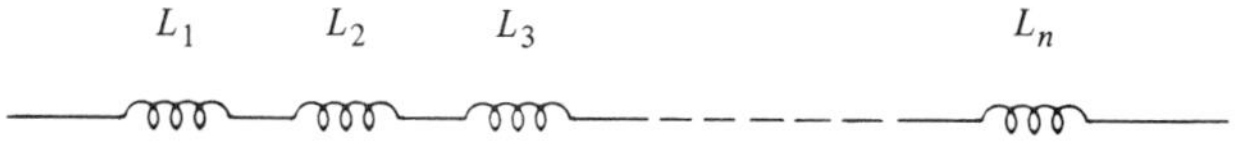

L_1 L_2 L_3 L_n

FIGURE 11–13
Inductors in series.

EXAMPLE 11–2

Determine the total inductance for each of the series connections in Figure 11–14.

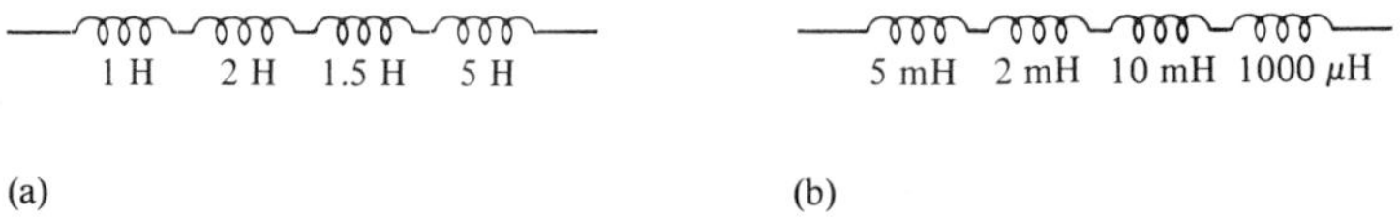

1 H 2 H 1.5 H 5 H

5 mH 2 mH 10 mH 1000 μH

(a) (b)

FIGURE 11–14

Solution

(a) $L_T = 1\text{ H} + 2\text{ H} + 1.5\text{ H} + 5\text{ H} = 9.5\text{ H}$

(b) $L_T = 5\text{ mH} + 2\text{ mH} + 10\text{ mH} + 1\text{ mH} = 18\text{ mH}$

Note: $1000\ \mu\text{H} = 1\text{ mH}$

SECTION REVIEW 11–3

1. State the rule for combining inductors in series.
2. What is L_T for a series connection of 100 μH, 500 μH, and 2 mH?

PARALLEL INDUCTORS

11–4

When inductors are connected in parallel, as in Figure 11–15, *the total inductance is less than the smallest inductance.* The formula for total inductance in parallel is similar to that for total parallel resistance or total series capacitance:

$$\frac{1}{L_T} = \frac{1}{L_1} + \frac{1}{L_2} + \frac{1}{L_3} + \cdots + \frac{1}{L_n} \tag{11–4}$$

FIGURE 11–15
Inductors in parallel.

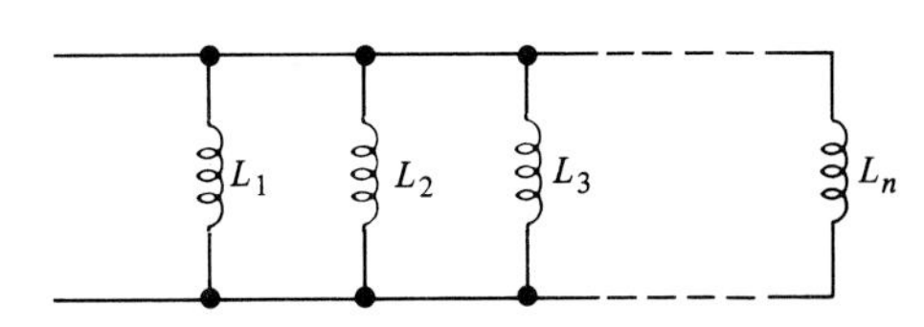

This general formula states that the reciprocal of the total inductance is equal to the sum of the reciprocals of the individual inductances. L_T can be found by taking the reciprocal of both sides of Equation (11–4):

$$L_T = \frac{1}{(1/L_1) + (1/L_2) + (1/L_3) + \cdots + (1/L_n)} \tag{11–5}$$

EXAMPLE 11–3 Determine L_T in Figure 11–16.

FIGURE 11–16

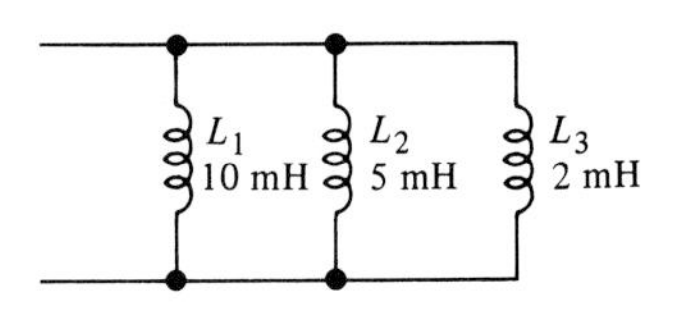

Solution

$$L_T = \frac{1}{(1/10 \text{ mH}) + (1/5 \text{ mH}) + (1/2 \text{ mH})}$$
$$= \frac{1}{0.8 \text{ mH}} = 1.25 \text{ mH}$$

The calculator sequence is

[1] [0] [EE] [+/−] [3] [2nd] [1/x] [+] [5] [EE] [+/−] [3] [2nd] [1/x] [+] [2] [EE] [+/−] [3] [2nd] [1/x] [=] [2nd] [1/x]

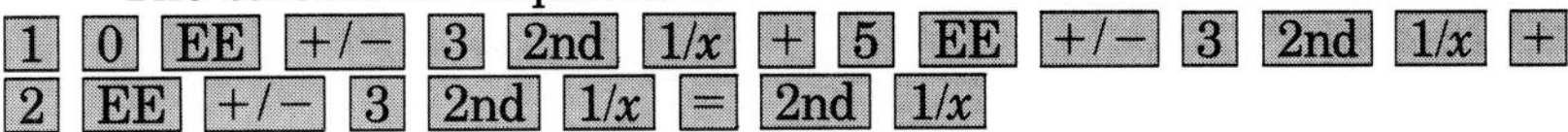

SECTION REVIEW 11–4

1. Compare the total inductance in parallel with the smallest-value individual inductor.
2. The calculation of total parallel inductance is similar to that for parallel resistance (T or F).
3. Determine L_T for each parallel combination:
   (a) 100 mH, 50 mH, and 10 mH
   (b) 40 μH and 60 μH
   (c) Ten 1-H coils

INDUCTORS IN dc CIRCUITS

11–5

When constant direct current flows in an inductor, there is no induced voltage. There is, however, a voltage drop due to the winding resistance of the coil. The inductance itself appears as a *short* to dc. Energy is stored in the magnetic field according to the formula $W = \frac{1}{2}LI^2$. The only energy loss occurs in the winding resistance ($P = I^2R_W$). This condition is illustrated in Figure 11–17.

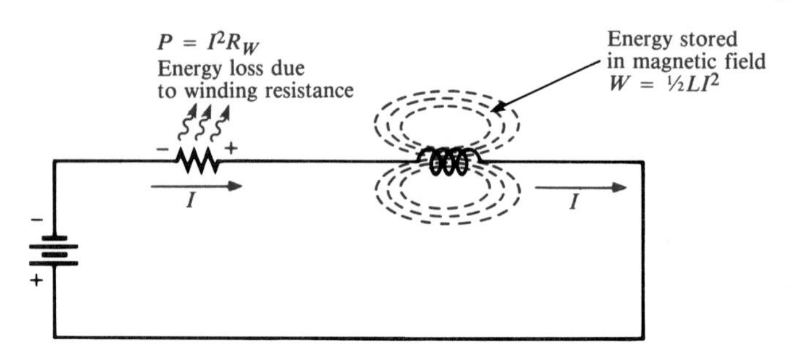

FIGURE 11–17
Energy storage and loss in an inductor. The only dc voltage drop across the coil is due to the winding resistance.

Time Constant

Because the inductor's basic action is to *oppose a change in its current,* it follows that *current cannot change instantaneously in an inductor.* A certain time is required for the current to make a change from one value to another. The rate at which the current changes is determined by the *time constant.* The time constant for a series *RL* circuit is

$$\tau = \frac{L}{R} \tag{11–6}$$

where τ is in seconds when L is in henries and R is in ohms.

EXAMPLE 11–4 A series *RL* circuit has a resistance of 1 kΩ and an inductance of 1 mH. What is the time constant?

Solution

$$\tau = \frac{L}{R} = \frac{1 \text{ mH}}{1 \text{ k}\Omega} = \frac{1 \times 10^{-3} \text{ H}}{1 \times 10^{3} \ \Omega}$$
$$= 1 \times 10^{-6} \text{ s} = 1 \ \mu\text{s}$$

Energizing Current in an Inductor

In a series *RL* circuit, the current will increase to 63% of its full value in one time constant interval after the switch is closed. This buildup of current is analogous to the buildup of capacitor voltage during the charging in an *RC* circuit; they both follow an exponential curve and reach the approximate percentages of final value as indicated in Figure 11–18.

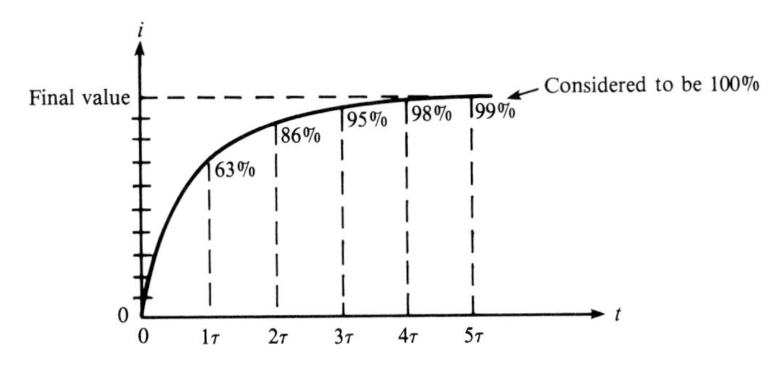

FIGURE 11–18
Energizing current in an inductor.

The change in current over five time constant intervals is illustrated in Figure 11–19. When the current reaches its final value at approximately 5τ, it ceases to change. At this time, *the inductor acts as a short (except for winding resistance) to the constant current.* The final value of the current is $V_S/R_W = 10\ \text{V}/10\ \Omega = 1\ \text{A}$.

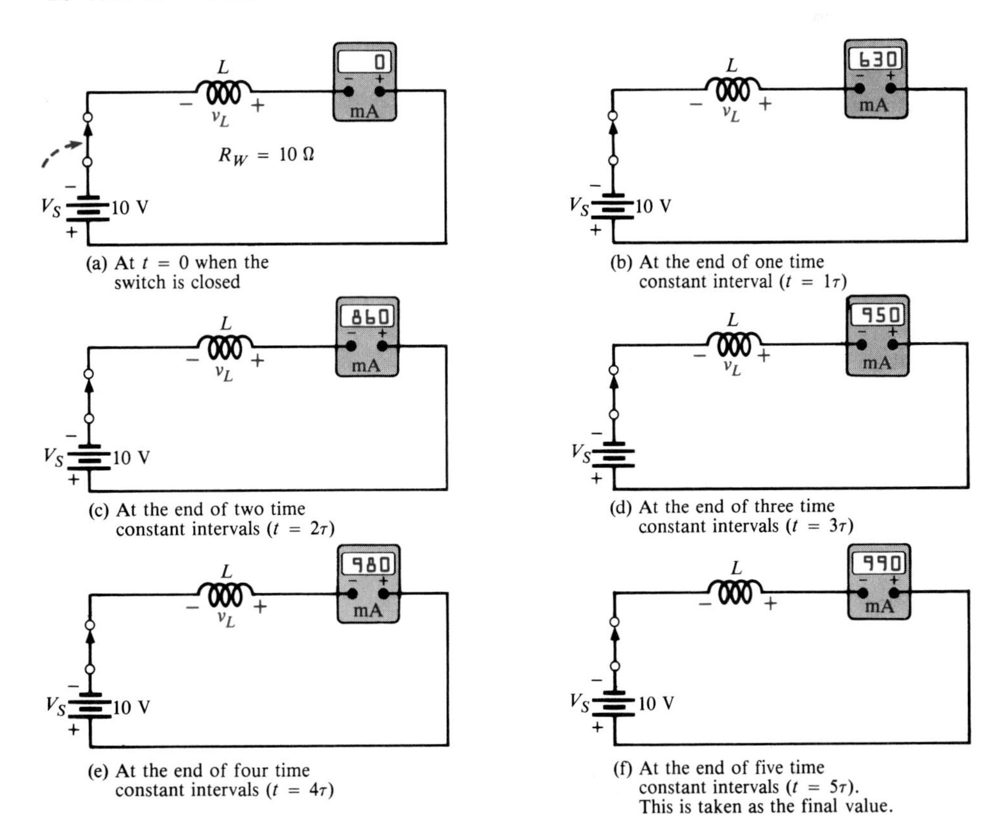

(a) At $t = 0$ when the switch is closed

(b) At the end of one time constant interval ($t = 1\tau$)

(c) At the end of two time constant intervals ($t = 2\tau$)

(d) At the end of three time constant intervals ($t = 3\tau$)

(e) At the end of four time constant intervals ($t = 4\tau$)

(f) At the end of five time constant intervals ($t = 5\tau$). This is taken as the final value.

FIGURE 11–19
Illustration of the exponential build-up of current in an inductor. The current increases another 63% during each time constant interval. A winding resistance of 10 Ω is assumed. A voltage (v_L) is induced in the coil that tends to oppose the increase in current.

EXAMPLE 11–5 Calculate the time constant for Figure 11–20. Then determine the current and the time at each time constant interval, measured from the instant the switch is closed.

FIGURE 11–20

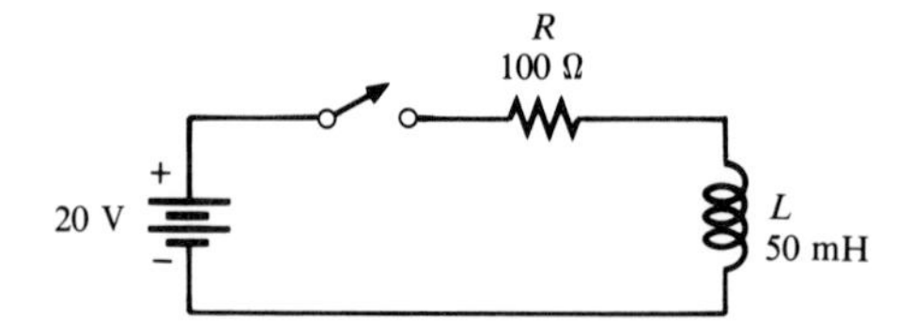

Solution

$$I_{\text{final}} = \frac{V_S}{R} = \frac{20\text{ V}}{100\ \Omega} = 0.2\text{ A}$$

$$\tau = \frac{L}{R} = \frac{50\text{ mH}}{100\ \Omega} = 0.5\text{ ms}$$

$$\begin{aligned}
\text{At } 1\tau &= 0.5\text{ ms:} & i &= 0.63(0.2\text{ A}) = 0.126\text{ A}\\
\text{At } 2\tau &= 1\text{ ms:} & i &= 0.86(0.2\text{ A}) = 0.172\text{ A}\\
\text{At } 3\tau &= 1.5\text{ ms:} & i &= 0.95(0.2\text{ A}) = 0.190\text{ A}\\
\text{At } 4\tau &= 2\text{ ms:} & i &= 0.98(0.2\text{ A}) = 0.196\text{ A}\\
\text{At } 5\tau &= 2.5\text{ ms:} & i &= 0.99(0.2\text{ A}) = 0.198\text{ A}\\
& & &\cong 0.2\text{ A}
\end{aligned}$$

Deenergizing Current in an Inductor

Current in an inductor decreases exponentially according to the approximate percentage values shown on the curve in Figure 11–21.

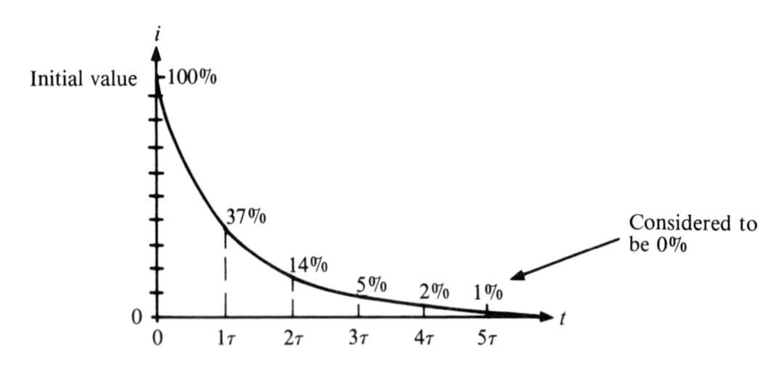

FIGURE 11–21
Deenergizing current in an inductor.

Figure 11–22(a) shows a constant current of 1 A (1000 mA) through an inductor. Switch 1 (SW1) is opened and switch 2 (SW2) is closed simultaneously, and for an instant the induced voltage keeps the 1 A flowing. During the first time constant interval, the current decreases by 63% down to 370 mA (37% of its initial value), as indicated in Part (b). During the second time constant interval, the current decreases by another 63% to 140 mA (14% of its initial value), as shown in Part (c). The continued decrease in the current is illustrated by the remaining parts of Figure 11–22. Part (f) shows that only 1% of the initial current is left at the end of five time constants. Traditionally, this value is accepted as the final value and is approximated as zero current. Notice that until after the five time constants have elapsed, there is an induced voltage across the coil which is trying to keep the current flowing. This voltage follows a decreasing exponential curve, as will be shown later.

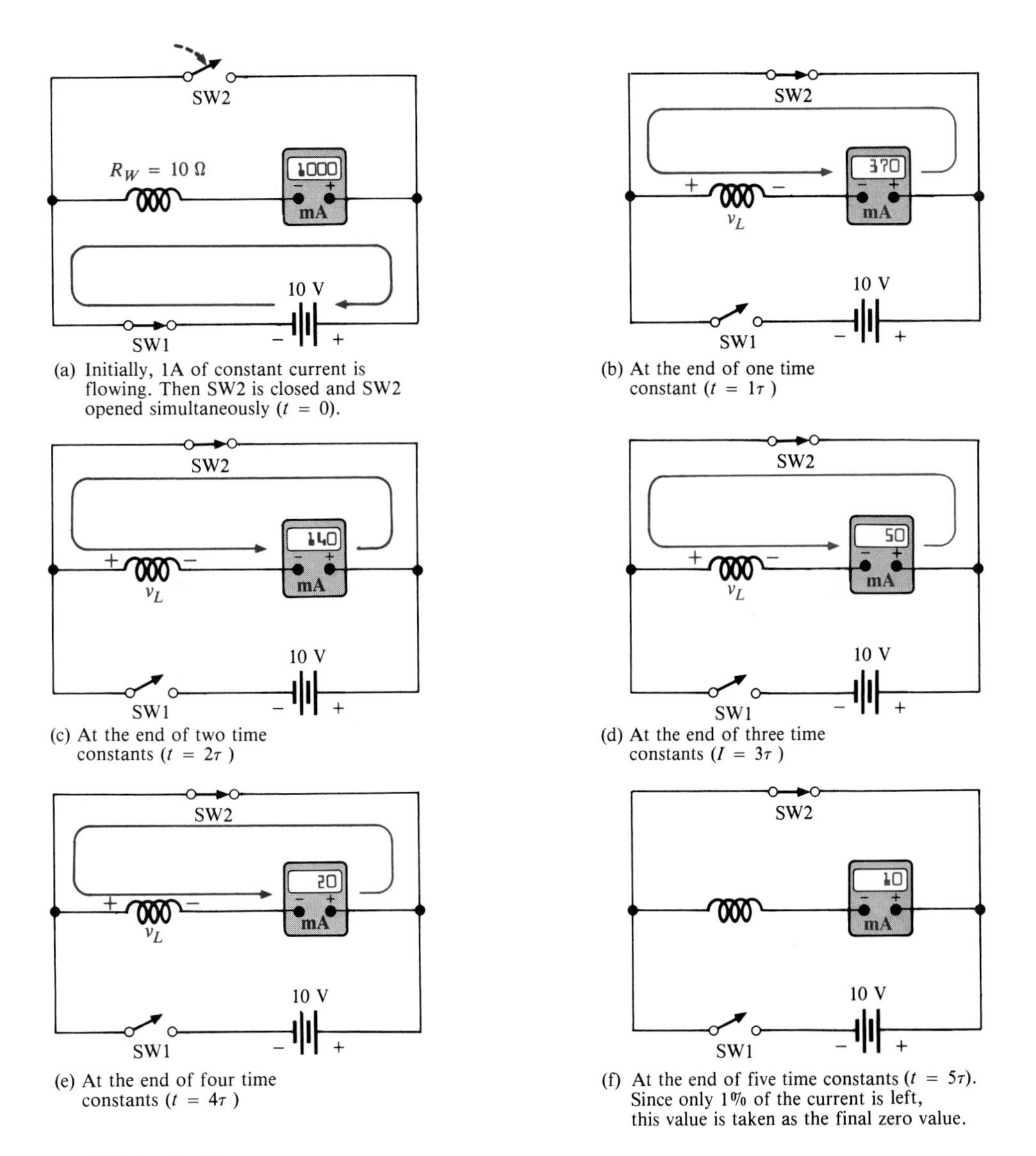

(a) Initially, 1A of constant current is flowing. Then SW2 is closed and SW2 opened simultaneously (t = 0).

(b) At the end of one time constant ($t = 1\tau$)

(c) At the end of two time constants ($t = 2\tau$)

(d) At the end of three time constants ($I = 3\tau$)

(e) At the end of four time constants ($t = 4\tau$)

(f) At the end of five time constants ($t = 5\tau$). Since only 1% of the current is left, this value is taken as the final zero value.

FIGURE 11–22
Illustration of the exponential decrease of current in an inductor. The current decreases another 63% during each time constant interval.

EXAMPLE 11–6 In Figure 11–23, switch 1 (SW1) is opened at the instant that switch 2 (SW2) is closed.

(a) What is the time constant?
(b) What is the initial coil current at the instant of switching?
(c) What is the coil current at 1τ?

Assume steady state current through the coil prior to switch change.

FIGURE 11–23

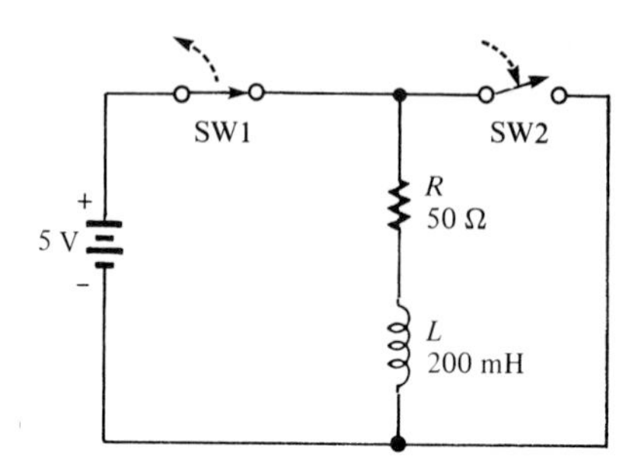

Solution

(a) $\tau = L/R = 200\ \mu\text{H}/10\ \Omega = 20\ \mu\text{s}$.

(b) Current cannot change instantaneously in an inductor. Therefore, the current at the instant of the switch change is the same as the steady state current:

$$I = \frac{5\ \text{V}}{10\ \Omega} = 0.5\ \text{A}$$

(c) At 1τ, the current has decreased to 37% of its initial value:

$$i = 0.37(0.5\ \text{A}) = 0.185\ \text{A}$$

Induced Voltage in the Series *RL* Circuit

As you know, when current changes in an inductor, a voltage is induced. We now examine what happens to the voltages across the resistor and the coil in a series circuit when a change in current occurs.

Look at the circuit in Figure 11–24(a). When the switch is open, there is no current, and the resistor voltage and the coil voltage are both zero. At the instant the switch is closed, as indicated in Part (b), v_R is zero and v_L is 10 V. The reason for this change is that the induced voltage across the coil is equal and opposite to the applied voltage to prevent the current from changing instantaneously. Therefore, at the instant of switch closure, *L effectively acts as an open with all the applied voltage across it.*

During the first five time constants, the current is building up exponentially, and the induced coil voltage is decreasing. The resistor voltage increases with the current, as Part (c) illustrates.

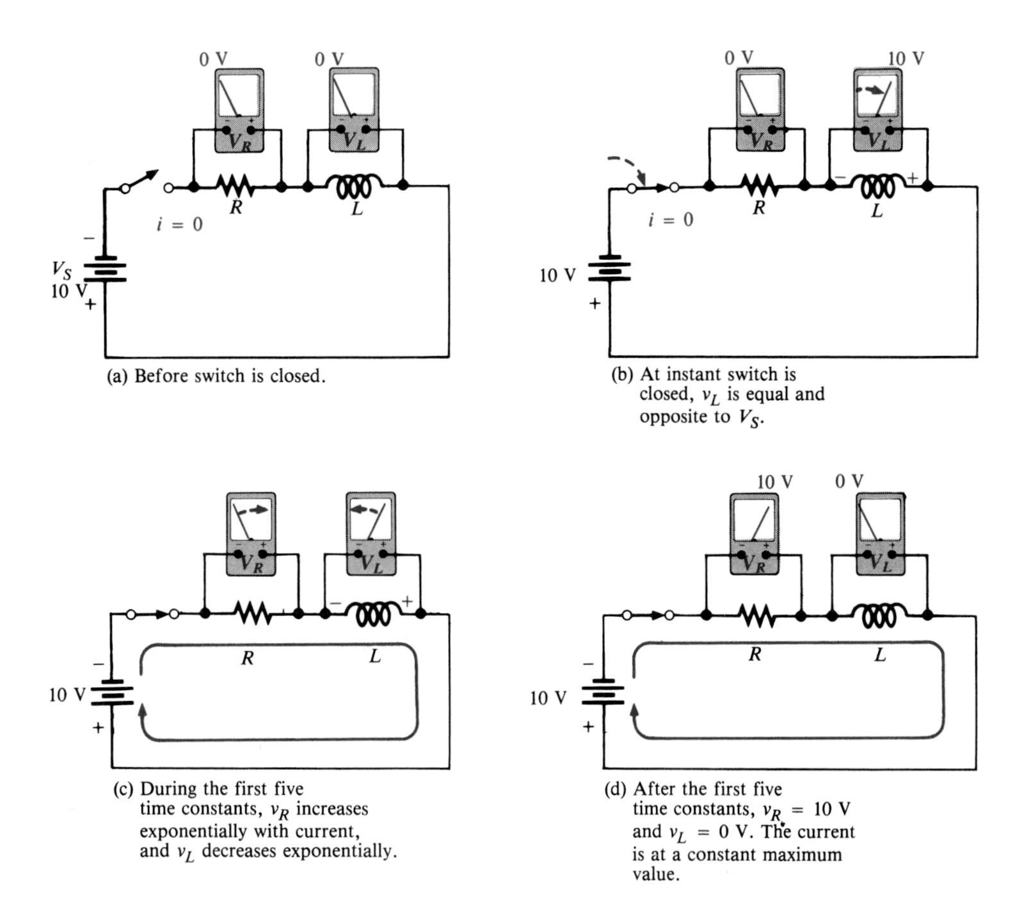

FIGURE 11–24
Illustration of how the coil voltage and the resistor voltage change in response to an increase in current.

After five time constants have elapsed, the current has reached its final value, V_S/R. At this time, all of the applied voltage is dropped across the resistor and none across the coil. Thus, L effectively acts as a *short* to nonchanging current, as Part (d) illustrates. Keep in mind that the inductor always reacts to a change in current by creating an induced voltage in order to counteract that change.

Now let us examine the case illustrated in Figure 11–25, where the current is switched out, and the inductor discharges through another path. Part (a) shows the steady state condition, and Part (b) illustrates the instant at which the source is removed by opening SW1 and the discharge path is connected by the closure of SW2. There was 1 A through L prior to this. Notice that 10 V are induced in L in the direction to aid the 1 A in an effort to keep it from changing. Then, as shown in Part (c), the current decays exponentially, and so do v_R and v_L. After 5τ, as shown in Part (d), all of the energy stored in the magnetic field of L is dissipated, and all values are zero.

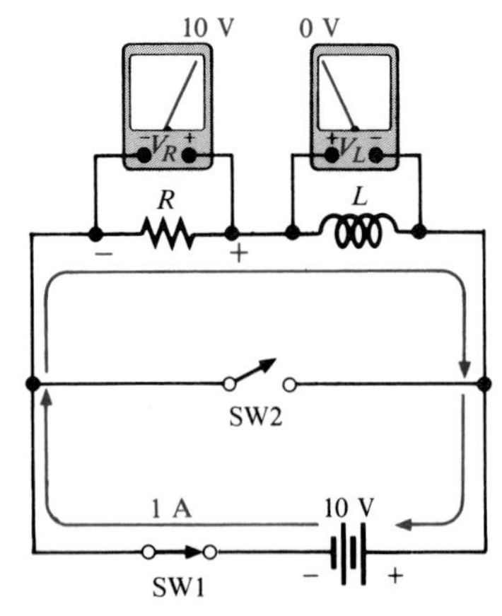

(a) Initially, a constant 1 A flows and the voltage is dropped across R.

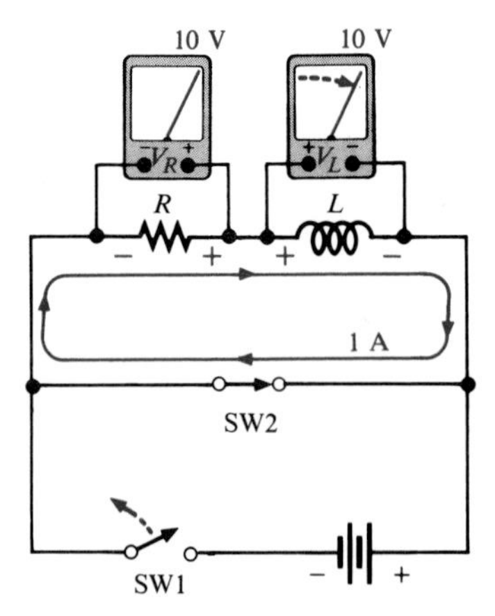

(b) At the instant that SW1 is opened and SW2 closed, 10 V is induced across L.

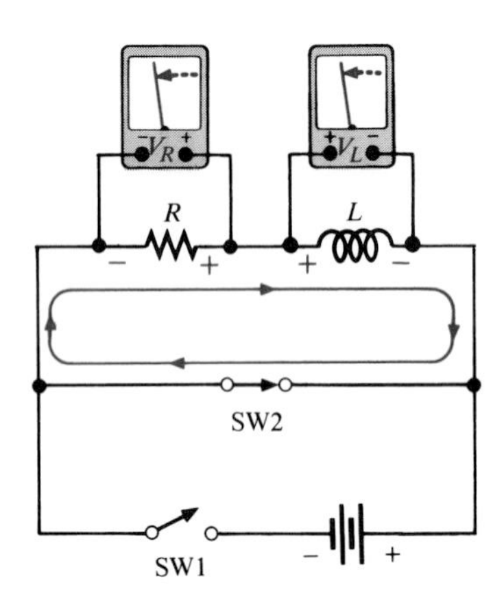

(c) During the five time constant interval, v_R and v_L decrease exponentially with the current.

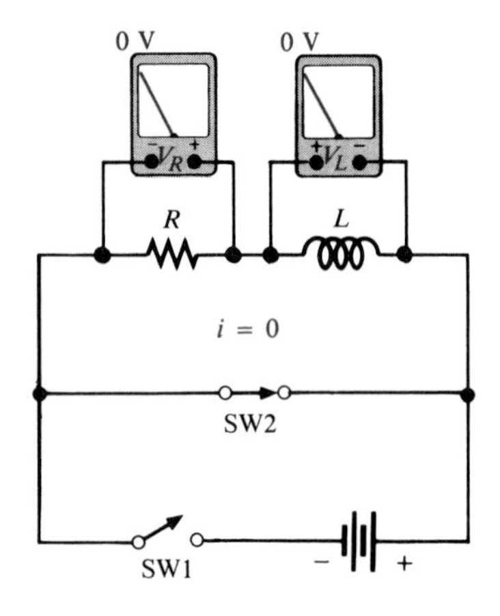

(d) After five time constants, V_R, V_L, and i are all zero.

FIGURE 11–25
Illustration of how the coil voltage and the resistor voltage change in response to a decrease in current.

EXAMPLE 11–7 **(a)** In Figure 11–26(a), what is v_L at the instant the switch is closed? What is v_L after 5τ?

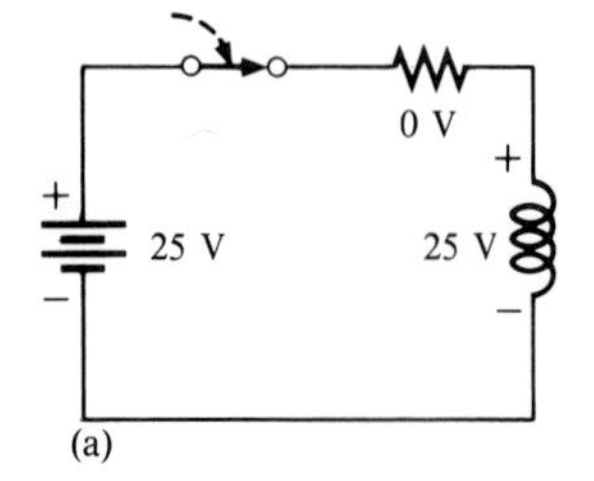

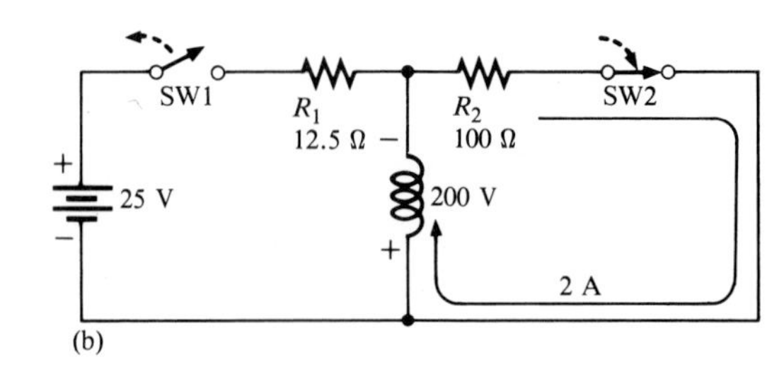

FIGURE 11–26

(b) In Figure 11–26(b), what is v_L at the instant SW1 opens and SW2 closes? What is v_L after 5τ?

Solution

(a) At the instant the switch is closed, all of the voltage is across L. Thus, $v_L = 25$ V, with the polarity as shown. After 5τ, L acts as a short, so $v_L = 0$ V.

(b) With S_1 closed and S_2 open, the steady state current is

$$\frac{25 \text{ V}}{12.5\ \Omega} = 2 \text{ A}$$

When the switches are thrown, an induced voltage is created across L sufficient to keep this 2-A current for an instant. In this case, it takes $v_L = IR_2 = (2 \text{ A})(100\ \Omega) = 200$ V. After 5τ, the inductor voltage is zero. These results are indicated in the circuit diagrams.

The Exponential Formulas

The formulas for the exponential current and voltage in an RL circuit are similar to those used in the last chapter for the RC circuit, and the universal exponential curves in Figure 10–32 apply to inductors as well as capacitors. The general formulas for RL circuits are stated as follows:

$$v = V_F + (V_i - V_F)e^{-Rt/L} \quad (11\text{–}7)$$

$$i = I_F + (I_i - I_F)e^{-Rt/L} \quad (11\text{–}8)$$

where V_F and I_F are the final values, V_i and I_i are the initial values, and v and i are the instantaneous values of the inductor voltage or current at time t.

Increasing Current

The formula for the special case in which an increasing exponential current curve begins at zero ($I_i = 0$) is

$$i = I_F(1 - e^{-Rt/L}) \quad (11\text{–}9)$$

Using Equation (11–9), we can calculate the value of the increasing inductor current at any instant of time. The same is true for voltage.

EXAMPLE 11–8 In Figure 11–27, determine the inductor current 30 μs after the switch is closed.

FIGURE 11–27

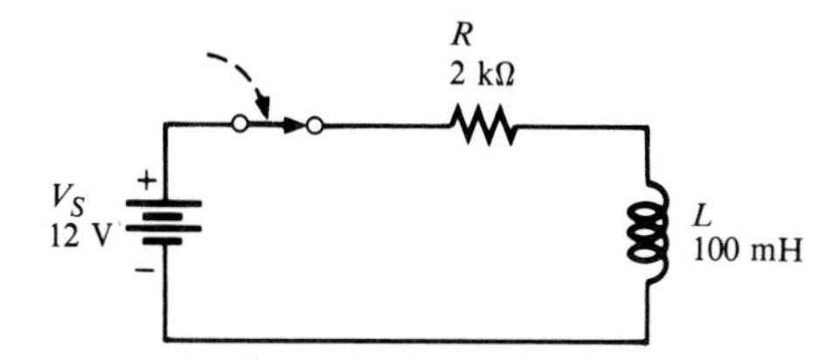

Solution

The time constant is $L/R = 100\text{ mH}/2\text{ k}\Omega = 50\ \mu\text{s}$. The final current is $V_S/R = 12\text{ V}/2\text{ k}\Omega = 6\text{ mA}$. The initial current is zero. Notice that 30 μs is less than one time constant, so the current will reach less than 63% of its final value in that time:

$$i_L = I_F(1 - e^{-Rt/L}) = 6\text{ mA}(1 - e^{-0.6})$$
$$= 6\text{ mA}(1 - 0.549) = 2.71\text{ mA}$$

The calculator sequence is

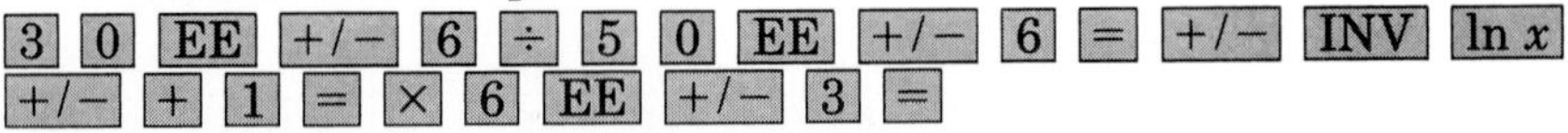

Decreasing Current

The formula for the special case in which a decreasing exponential current has a final value of zero is as follows:

$$i = I_i e^{-Rt/L} \qquad (11\text{–}10)$$

This formula can be used to calculate the deenergizing current at any instant, as the following example shows.

EXAMPLE 11–9 Determine the inductor current in Figure 11–28 at a point in time 2 ms after the switches are thrown (SW1 opened and SW2 closed).

FIGURE 11–28

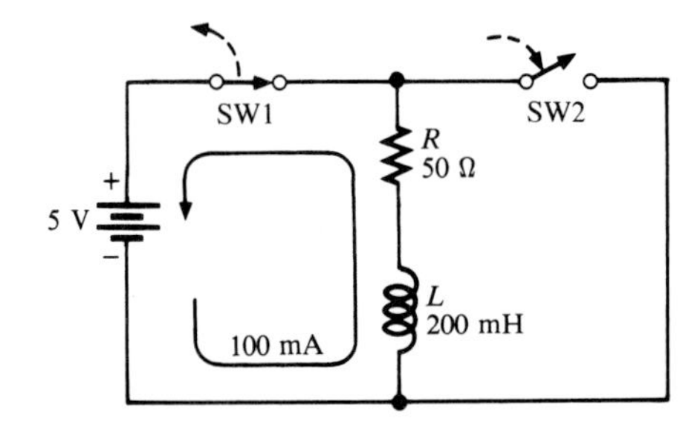

Solution

The deenergizing time constant is $L/R = 200\text{ mH}/50\ \Omega = 4\text{ ms}$. The initial current in the inductor is 100 mA. Notice that 2 ms is less than one time constant, so the current will show a decrease less than 63%. Therefore, the current will be greater than 37% of its initial value at 2 ms after the switches are thrown:

$$i = I_i e^{-Rt/L} = (100\text{ mA})e^{-0.5} = 60.65\text{ mA}$$

The calculator sequence is

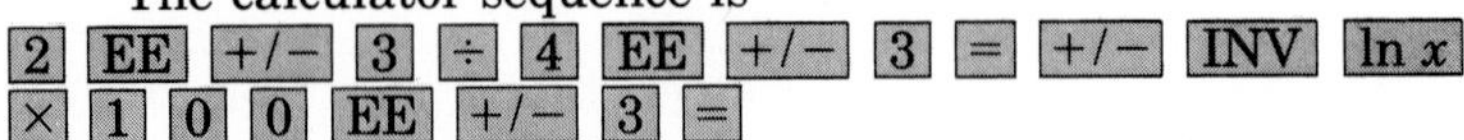

SECTION REVIEW 11–5

1. A 15-mH inductor with a winding resistance of 10 Ω has a constant direct current of 10 mA through it. What is the voltage drop across the inductor?
2. A 20-V dc source is connected to a series *RL* circuit with a switch. At the instant of switch closure, what are the values of v_R and v_L?
3. In the same circuit, after a time interval equal to 5τ from switch closure, what are v_R and v_L?
4. In a series *RL* circuit where $R = 1\ \text{k}\Omega$ and $L = 500\ \mu\text{H}$, what is the time constant? Determine the current 0.25 μs after a switch connects 10 V across the circuit.

INDUCTORS IN ac CIRCUITS

11–6

We will illustrate the relationship of current and inductor voltage in an ac circuit by connecting an inductor in series with a resistor to a sine wave voltage source and then observing the inductor voltage and the resistor voltage with an oscilloscope, as shown in Figure 11–29. Recall that the resistor voltage and the current are always in phase with each other, so if you are looking at the resistor voltage, you are also effectively looking at the current in terms of its phase relationship to the inductor voltage. In this case, it is assumed that the winding resistance is negligible. Notice that the frequency of the voltage and current are always the same in an ac circuit.

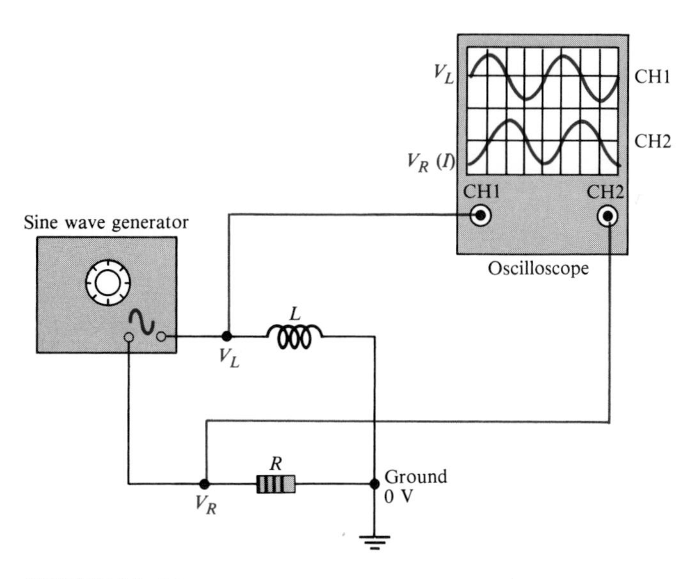

FIGURE 11–29
Oscilloscope display showing that the current lags the inductor voltage by 90°. The current is in phase with the resistor voltage.

Why the Current Lags the Inductor Voltage by 90°

As you know, a sine wave voltage has a maximum rate of change at its zero crossings and a zero rate of change at the peaks. From Faraday's law (Chapter 7) we know that the amount of voltage induced across a coil is directly proportional to the rate at which the current is changing. Therefore, the coil voltage is maximum at the zero crossings of the current where the rate of change of the current is the greatest. Also, the amount of voltage is zero at the peaks of the current where its rate of change is zero. This relationship is illustrated in Figure 11–30. As you can see, the current peaks occur a quarter cycle after the voltage peaks. Thus, the current *lags* the voltage by 90°.

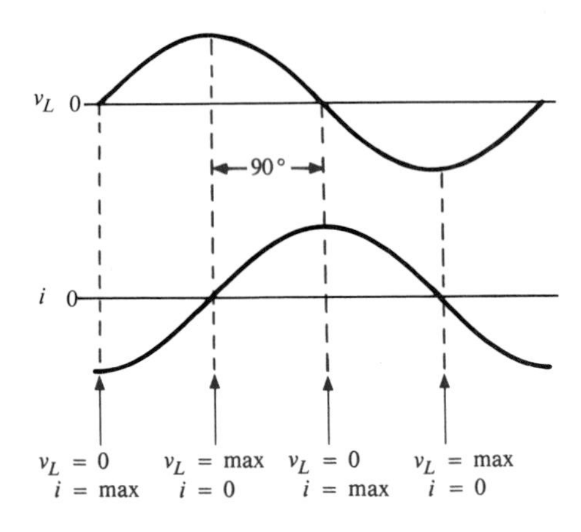

FIGURE 11–30
Current is always lagging the inductor voltage by 90°.

Inductive Reactance, X_L

In Figure 11–31, an inductor is connected to a sine wave source. Note that when the source voltage is held at a constant amplitude value and its frequency is increased, the amplitude of the current decreases. Also, when the frequency of the source is decreased, the current amplitude increases. The reason is as follows: When the frequency of the applied voltage increases, its rate of change also increases, as you already know. Now, if the frequency of the applied voltage is increased, the frequency of the current also increases. According to Faraday's and Lenz's laws, this increase in frequency induces more voltage across the inductor in a direction to oppose the current and cause it to decrease in amplitude. Similarly, a decrease in frequency will cause an increase in current.

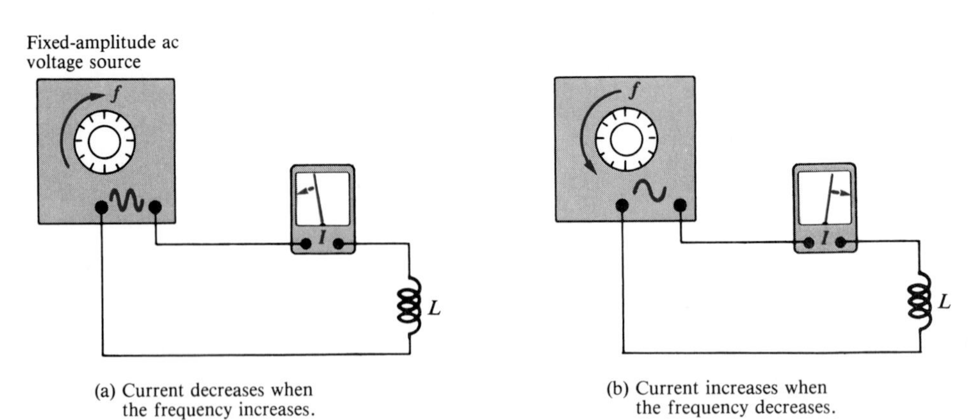

(a) Current decreases when the frequency increases.

(b) Current increases when the frequency decreases.

FIGURE 11–31
The current in an inductive circuit varies inversely with the frequency of the applied voltage.

A *decrease* in the amount of current for a fixed amount of voltage indicates that *opposition* to the current has *increased.* Thus, the inductor offers opposition to current which varies *directly* with frequency. The opposition to sinusoidal current is called *inductive reactance.* The symbol for inductive reactance is X_L, and its unit is the ohm (Ω).

You have just seen how frequency affects the opposition to current (inductive reactance) in an inductor. Now let's see how the *inductance, L,* affects the reactance. Figure 11–32(a) shows that when a sine wave voltage with a fixed amplitude and frequency is applied to a 1-mH inductor, a certain amount of current flows. When the inductance value is increased to 2 mH, the current decreases, as shown in Part (b). Thus, when the inductance increases, the opposition to current (inductive reactance) increases. So not only is the inductive reactance directly proportional to frequency, but it is also directly proportional to inductance:

$$X_L \text{ is proportional to } fL$$

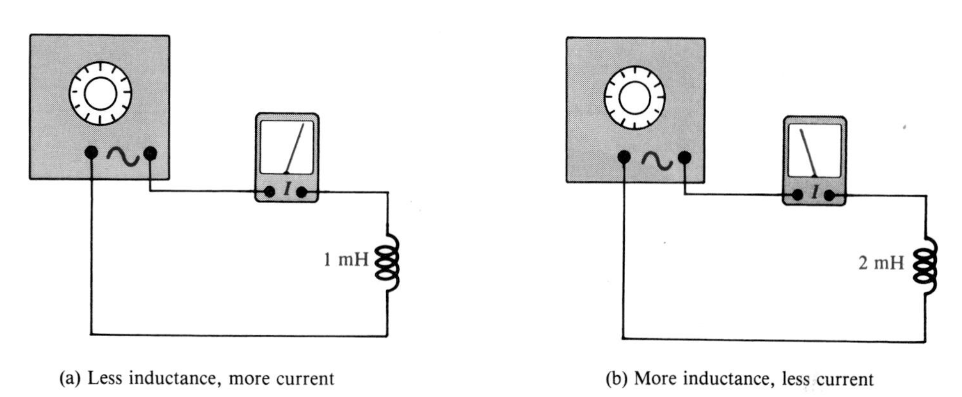

(a) Less inductance, more current

(b) More inductance, less current

FIGURE 11–32
For a fixed voltage and frequency, the current varies inversely with inductance.

It can be proven that the constant of proportionality is 2π, so the formula for X_L is

$$X_L = 2\pi fL \tag{11–11}$$

As with capacitive reactance, the 2π term comes from the relationship of the sine wave to rotational motion.

EXAMPLE 11–10 A sinusoidal voltage is applied to the circuit in Figure 11–33. The frequency is 1 kHz. Determine the inductive reactance.

FIGURE 11–33

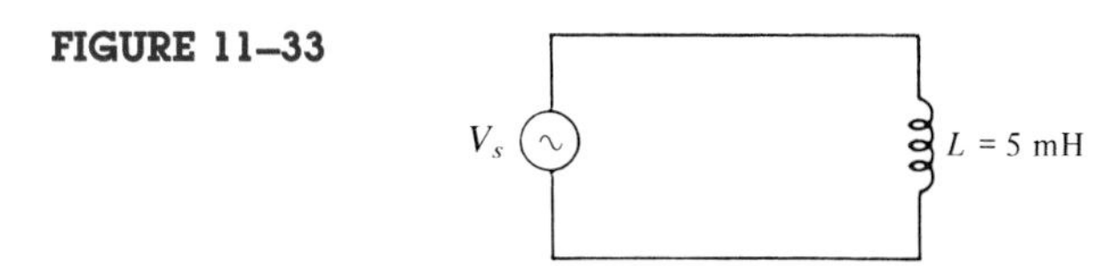

Solution

$$1 \text{ kHz} = 1 \times 10^3 \text{ Hz}$$
$$5 \text{ mH} = 5 \times 10^{-3} \text{ H}$$
$$X_L = 2\pi fL = 2\pi(1 \times 10^3 \text{ Hz})(5 \times 10^{-3} \text{ H}) = 31.4 \ \Omega$$

Ohm's Law in Inductive Circuits

The reactance of an inductor is analogous to the resistance of a resistor. In fact, X_L, just like X_C and R, is expressed ohms. Since inductive reactance is a form of opposition to current, Ohm's law applies to inductive circuits as well as to resistive circuits and capacitive circuits, and it is stated as follows:

$$V = IX_L \tag{11–12}$$

When applying Ohm's law in ac circuits, you must express both the current and the voltage in the same way, that is, both in rms, both in peak, and so on.

EXAMPLE 11–11 Determine the rms current in Figure 11–34.

FIGURE 11–34

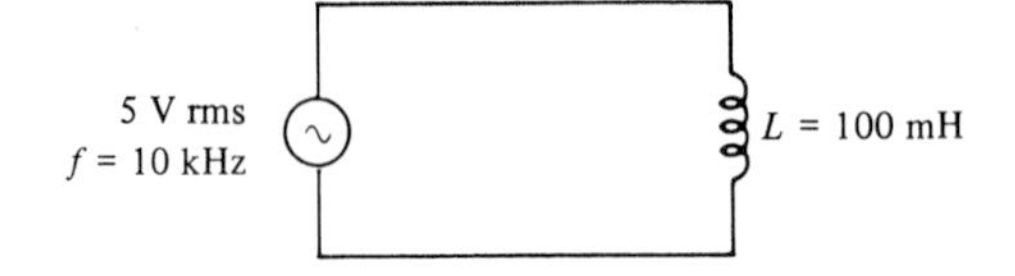

Solution

$$10 \text{ kHz} = 10 \times 10^3 \text{ Hz}$$
$$100 \text{ mH} = 100 \times 10^{-3} \text{ H}$$

First calculate X_L:

$$X_L = 2\pi fL = 2\pi(10 \times 10^3 \text{ Hz})(100 \times 10^{-3} \text{ H}) = 6283 \ \Omega$$

Using Ohm's law, we obtain

$$I_{\text{rms}} = \frac{V_{\text{rms}}}{X_L} = \frac{5 \text{ V}}{6283 \ \Omega} = 795.8 \ \mu\text{A}$$

Power in an Inductor

As discussed earlier, an inductor stores energy in its magnetic field when there is current through it. An ideal inductor (assuming no winding resistance) does not dissipate energy; it only stores it. When an ac voltage is applied to an inductor, energy is stored by the inductor during a portion of the cycle; then the stored energy is returned to the source during another portion of the cycle.

There is no net energy loss. Figure 11–35 shows the power curve that results from one cycle of inductor current and voltage.

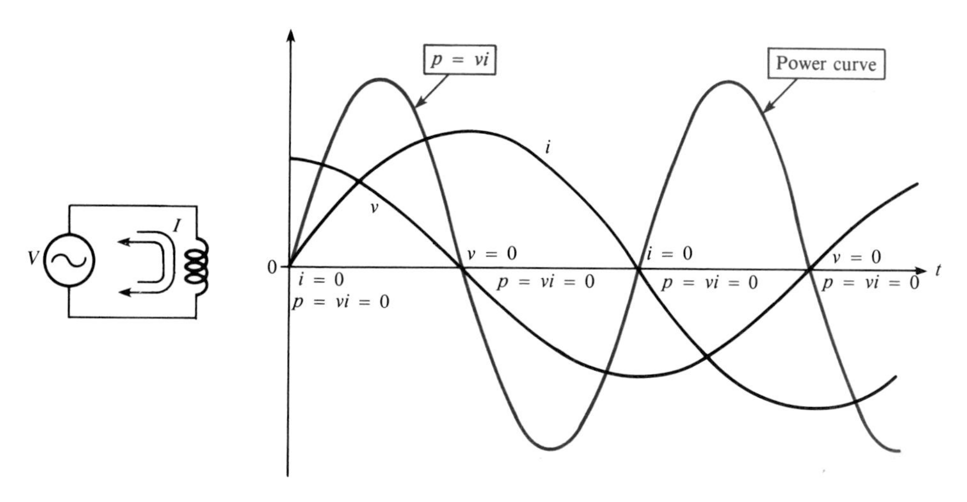

FIGURE 11–35
Power curve.

Instantaneous Power (p)

The product of v and i gives instantaneous power, p. At points where v or i is zero, p is also zero. When both v and i are positive, p is also positive. When either v or i is positive and the other negative, p is negative. When both v and i are negative, p is positive. As you can see in Figure 11–35, the power follows a sinusoidal-type curve. Positive values of power indicate that energy is stored by the inductor. Negative values of power indicate that energy is returned from the inductor to the source. Note that the power fluctuates at a frequency twice that of the voltage or current as energy is alternately stored and returned to the source.

True Power (P_{true})

Ideally, all of the energy stored by an inductor during the positive portion of the power cycle is returned to the source during the negative portion. *No net energy is consumed in the inductance,* so the power is zero. Actually, because of winding resistance in a practical inductor, some power is always dissipated:

$$P_{\text{true}} = I_{\text{rms}}^2 R_W \tag{11–13}$$

Reactive Power (P_r)

The rate at which an inductor stores or returns energy is called its *reactive power, P_r*. The reactive power is a nonzero quantity, because at any instant in time, the inductor is actually taking energy from the source or returning energy to it. Reactive power does not represent an energy loss. The following formulas apply:

$$P_r = V_{rms}I_{rms} \tag{11–14}$$

$$P_r = \frac{V_{rms}^2}{X_L} \tag{11–15}$$

$$P_r = I_{rms}^2 X_L \tag{11–16}$$

EXAMPLE 11–12 A 10-V rms signal with a frequency of 1 kHz is applied to a 10-mH coil with a winding resistance of 5 Ω. Determine the reactive power (P_r).

Solution

$$X_L = 2\pi fL = 2\pi(1\text{ kHz})(10\text{ mH}) = 62.8\ \Omega$$

$$I = \frac{V_S}{X_L} = \frac{10\text{ V}}{63\ \Omega} = 0.159\text{ A}$$

$$P_r = I^2X_L = (0.159\text{ A})^2(62.8\ \Omega) = 1.59\text{ VAR}$$

SECTION REVIEW 11–6

1. State the phase relationship between current and voltage in an inductor.
2. Calculate X_L for $f = 5$ kHz and $L = 100$ mH.
3. At what frequency is the reactance of a 50-μH inductor equal to 800 Ω?
4. Calculate the rms current in Figure 11–36.
5. A 50-mH inductor is connected to a 12-V rms source. What is the true power? What is the reactive power at a frequency of 1 kHz?

FIGURE 11–36

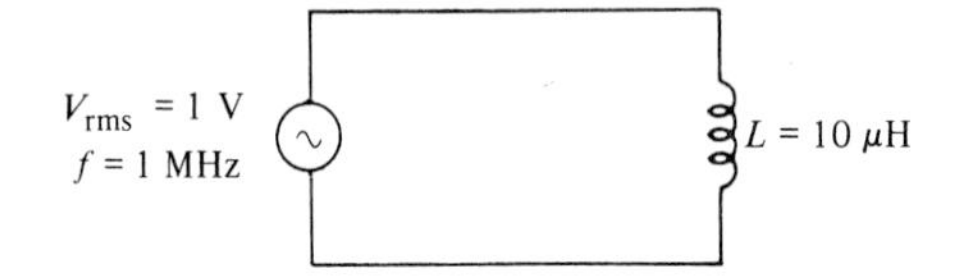

INDUCTOR APPLICATIONS

11–7

Inductors are not as versatile as capacitors and tend to be more limited in their applications. However, there are many practical uses for inductors (coils), such as those discussed in Chapter 7. Recall that relay and solenoid coils, recording and pick-up heads, and sensing elements were introduced as electromagnetic applications of coils. In this section, some additional uses of inductors are presented.

Power Supply Filter

In Chapter 10, you saw that a capacitor is used to filter the pulsating dc in a power supply. The final output voltage was a dc voltage with a small amount of *ripple*. In many cases, an inductor is used in the filter, as shown in Figure 11–37(a), to smooth out the ripple voltage. The inductor, placed in series with the load as shown, tends to oppose the current fluctuations caused by the ripple voltage, and thus the voltage developed across the load is more constant, as shown in Figure 11–37(b).

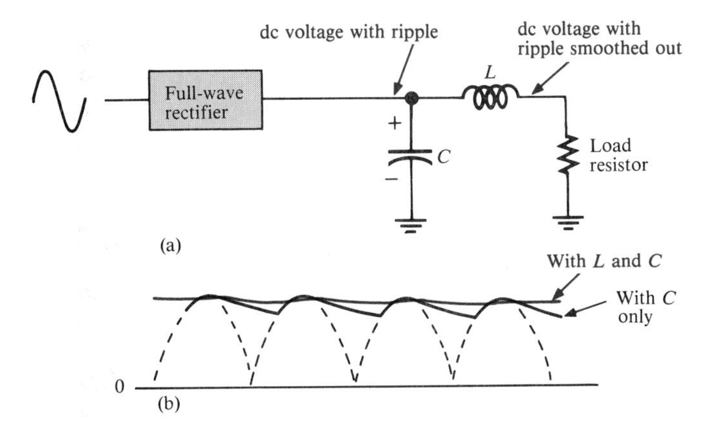

FIGURE 11–37
Basic capacitor power supply filter with a series inductor.

rf Choke

Certain types of inductors are used in applications where radio frequencies (rf) must be prevented from getting into parts of a system, such as the power supply or the audio section of a receiver. In these situations, an inductor is used as a series filter and "chokes" off any unwanted rf signals that may be picked up on a line. This filtering action is based on the fact that the reactance of a coil increases with frequency. When the frequency of the current is sufficiently high, the reactance of the coil becomes extremely large and essentially blocks the current. A basic illustration of an inductor used as an rf choke is shown in Figure 11–38.

FIGURE 11–38
An inductor used as an rf choke to minimize interfering signals on the power supply line.

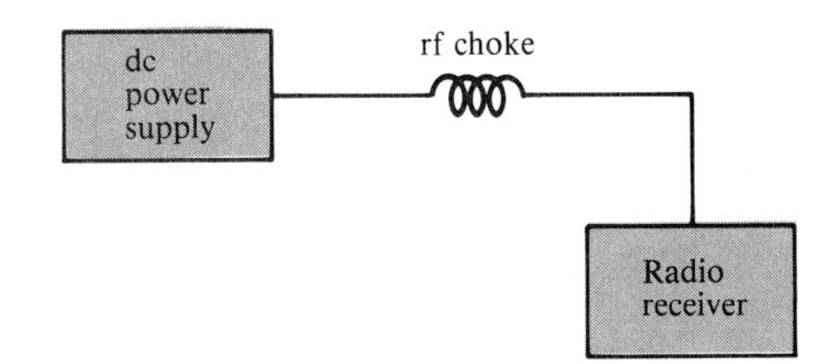

Tuned Circuits

As mentioned in Chapter 10, inductors are used in conjunction with capacitors to provide *frequency selection* in communications systems. These *tuned circuits* allow a narrow band of frequencies to be selected while all other frequencies are rejected. The tuners in your TV and radio receivers are based on this principle and permit you to select one channel or station out of the many that are available.

Frequency selectivity is based on the fact that the reactances of both capacitors and inductors depend on the frequency and on the interaction of these two components when connected in series or parallel. Since the capacitor and the inductor produce opposite phase shifts, their combined opposition to current can be used to obtain a desired response at a selected frequency. Tuned *LC* circuits are covered in detail in a later chapter.

SECTION REVIEW 11–7

1. Explain how the ripple voltage from a power supply filter can be reduced through use of an inductor.
2. How does an inductor connected in series act as an rf choke?

TESTING INDUCTORS

11–8

The most common failure in an inductor is an open. To check for an open, remove the coil from the circuit. If there is an open, an ohmmeter check will indicate infinite resistance, as shown in Figure 11–39(a). If the coil is good, the ohmmeter will show the winding resistance. The value of the winding resistance depends on the wire size and length of the coil. It can be anywhere from one ohm to several hundred ohms. Part (b) shows a good reading.

Occasionally, when an inductor is overheated with excessive current, the wire insulation will melt, and some coils will short together. This produces a reduction in the inductance by reducing the effective number of turns and a corresponding reduction in winding resistance, as shown in Part (c).

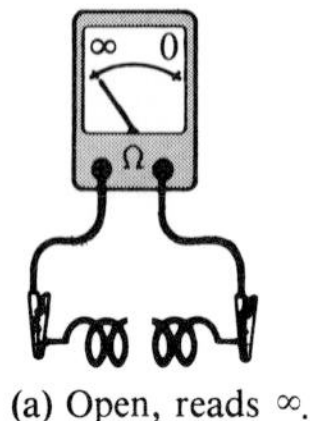

(a) Open, reads ∞.

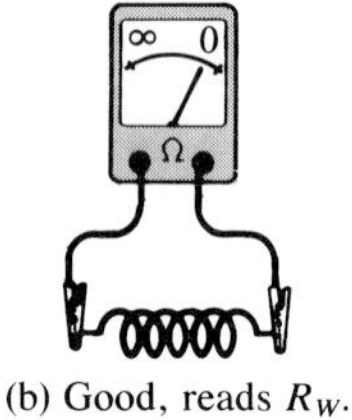

(b) Good, reads R_W.

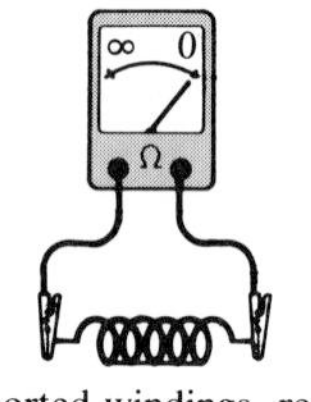

(c) Shorted windings, reads low R_W or near zero.

FIGURE 11–39
Checking a coil by measuring the resistance.

SECTION REVIEW 11–8

1. When a coil is checked, a reading of infinity on the ohmmeter indicates a partial short (T or F).
2. An ohmmeter check of a good coil will indicate the value of the inductance (T or F).

APPLICATION NOTE

For the Application Assignment at the beginning of this chapter, you can check the capacitor and inductor for an open, short, or leakage condition (capacitor only) by using the ohmmeter as described in Sections 10–8 and 11–8. If the components pass these checks, you can measure their values by using an *LCR* meter.

SUMMARY

Facts

- An inductor passes dc and opposes ac.
- An inductor is also called a *coil* and, in some applications, a *choke*.

- An inductor consists of a coil of wire.
- Energy is stored in an inductor by the electromagnetic field created when current flows through it.
- Inductance is measured in units of henries (H).
- The inductance of a coil is directly proportional to the number of turns squared and to the permeability and cross-sectional area of the core, and it is inversely proportional to the length of the core.
- The total inductance of series inductors is the sum of all the inductances.
- The total inductance of parallel inductors is less than the smallest inductance.
- In an RL circuit, the voltage and current in an energizing or deenergizing inductor make a 63% change during each time constant interval.
- Current in an inductor lags the voltage by 90°.
- Inductive reactance X_L is directly proportional to frequency and to inductance.
- Ideally, there is no energy loss in an inductor, and thus the true power is zero. However, in all inductors there is some energy loss due to the winding resistance.

Units

- Henry (H)—the unit of inductance (L).

Definitions

- *Inductance*—the property of a coil that opposes a change in current.
- *One henry*—the amount of inductance that creates an induced voltage of one volt when the current is changing at one ampere per second.
- *Time constant*—the time required for the inductor current to change by 63%. It equals the inductance divided by the resistance.
- *Winding resistance*—the inherent resistance of the coil of wire. It depends on the size and length of the wire.
- *Inductive reactance*—the opposition to sinusoidal current, expressed in ohms.

Formulas

$$W = \frac{1}{2}LI^2 \quad \text{Energy stored by an inductor} \tag{11–1}$$

$$L = \frac{N^2\mu A}{l} \quad \text{Inductance in terms of physical parameters} \tag{11–2}$$

$$L_T = L_1 + L_2 + L_3 + \cdots + L_n \quad \text{Series inductance} \tag{11–3}$$

$$\frac{1}{L_T} = \frac{1}{L_1} + \frac{1}{L_2} + \frac{1}{L_3} + \cdots + \frac{1}{L_n} \quad \text{Reciprocal of total parallel inductance} \tag{11–4}$$

$$L_T = \frac{1}{(1/L_1) + (1/L_2) + (1/L_3) + \cdots + (1/L_n)} \quad \text{Total parallel inductance} \tag{11–5}$$

$$\tau = \frac{L}{R} \quad \text{Time constant} \tag{11–6}$$

$$v = V_F + (V_i - V_F)e^{-Rt/L} \quad \text{Exponential voltage} \tag{11–7}$$

$$i = I_F + (I_i - I_F)e^{-Rt/L} \quad \text{Exponential current} \tag{11–8}$$

$$i = I_F(1 - e^{-Rt/L}) \quad \text{Increasing exponential current} \tag{11–9}$$

$$i = I_ie^{-Rt/L} \quad \text{Decreasing exponential current} \tag{11–10}$$

$$X_L = 2\pi fL \quad \text{Inductive reactance} \tag{11–11}$$

$V = IX_L$ Ohm's law (11–12)

$P_{true} = I_{rms}^2 R_W$ True power (11–13)

$P_r = V_{rms} I_{rms}$ Reactive power (11–14)

$P_r = \frac{V_{rms}^2}{X_L}$ Reactive power (11–15)

$P_r = I_{rms}^2 X_L$ Reactive power (11–16)

Symbols

The symbols for fixed and variable inductors are shown in Figure 11–40.

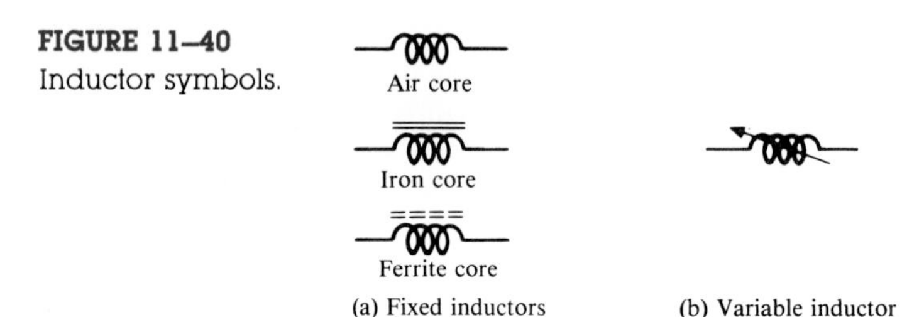

FIGURE 11–40
Inductor symbols.

SELF-TEST

1. Which is the larger inductance, 0.05 μH or 0.000005 H?
2. Which is the smaller inductance, 0.33 mH or 33 μH?
3. If the current through an inductor is increased, does the amount of energy stored in the electromagnetic field increase or decrease?
4. If the current through an inductor is doubled, how much does the stored energy increase?
5. How can the winding resistance of a coil be decreased?
6. Select the changes in physical parameters that result in an increase in the inductance value:
   - **(a)** Decrease the number of turns.
   - **(b)** Change from an air core to an iron core.
   - **(c)** Increase the number of turns.
   - **(d)** Increase the length of the core.
   - **(e)** Decrease the length of the core.
   - **(f)** Use a larger-size wire.
7. A 1-mH, a 3.3-mH, and a 0.1-mH inductor are connected in parallel. The total inductance is (greater, less than) ________ mH.
8. Five 100-μH inductors are connected in parallel. What is the total inductance?
9. An inductor and a resistor are placed in series with a switch and are connected to a dc voltage source.
   - **(a)** At the instant the switch is closed, what is the voltage across the inductor?
   - **(b)** At the instant the switch is closed, what is the current?
   - **(c)** How long will it take the current to reach its maximum value?
   - **(d)** What is the voltage across the inductor when the current has reached its maximum value?
   - **(e)** What is the voltage across the series resistor at the instant the switch is closed and after the current has reached its maximum value?

10. A sine wave voltage is applied across an inductor. When the frequency of the voltage is increased, does the current increase or decrease? Why?
11. An inductor and a resistor are connected in series with a sine wave voltage source. The frequency is set so that the inductive reactance is equal to the resistance and an equal amount of voltage is dropped across each component. If the frequency is decreased, which component will have the greater voltage across it?
12. An ohmmeter is connected across an inductor, and the pointer indicates an infinite value. What is your opinion of this inductor?

PROBLEM SET A

Section 11–1

11–1 Convert the following to millihenries:
(a) 1 H **(b)** 250 μH **(c)** 10 μH **(d)** 0.0005 H

11–2 Convert the following to microhenries:
(a) 300 mH **(b)** 0.08 H **(c)** 5 mH **(d)** 0.00045 mH

11–3 How many turns are required to produce 30 mH with a coil wound on a cylindrical core having a cross-sectional area of 10×10^{-5} m^2 and a length of 0.05 m? The core has a permeability of 1.2×10^{-6}.

11–4 A 12-V battery is connected across a coil with a winding resistance of 12 Ω. How much current is there in the coil?

11–5 How much energy is stored by a 100-mH inductor with a current of 1 A?

11–6 The current through a 100-mH coil is changing at a rate of 200 mA/s. How much voltage is induced across the coil?

Section 11–3

11–7 Five inductors are connected in series. The lowest value is 5 μH. If the value of each inductor is twice that of the preceding one, and if the inductors are connected in order of ascending values, what is the total inductance?

11–8 Suppose that you require a total inductance of 50 mH. You have available a 10-mH coil and a 22-mH coil. How much additional inductance do you need?

Section 11–4

11–9 Determine the total parallel inductance for the following coils in parallel: 75 μH, 50 μH, 25 μH, and 15 μH.

11–10 You have a 12-mH inductor, and it is your smallest value. You need an inductance of 8 mH. What value can you use in parallel with the 12-mH to obtain 8 mH?

11–11 Determine the total inductance of each circuit in Figure 11–41.

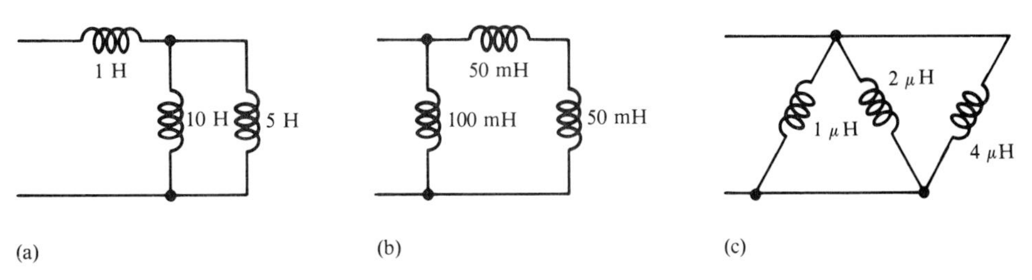

FIGURE 11–41

11–12 Determine the total inductance of each circuit in Figure 11–42.

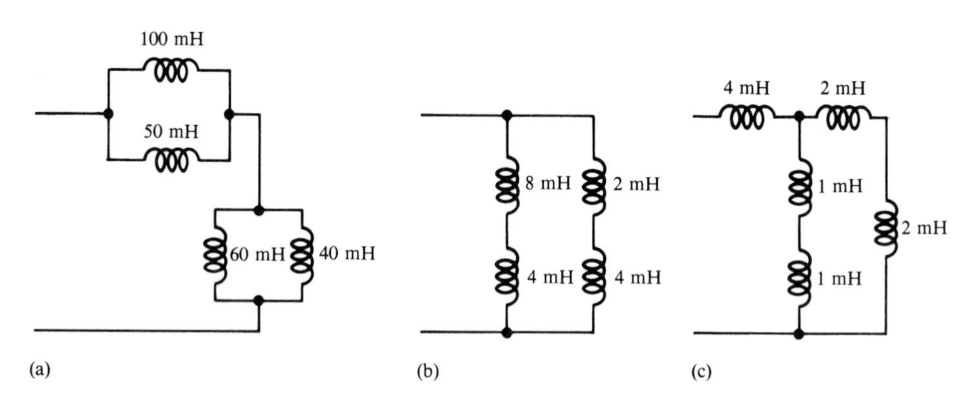

FIGURE 11–42

Section 11–5

11–13 Determine the time constant for each of the following series RL combinations:
(a) $R = 100\ \Omega, L = 100\ \mu\text{H}$ **(b)** $R = 4.7\ \text{k}\Omega, L = 10\ \text{mH}$
(c) $R = 1.5\ \text{M}\Omega, L = 3\ \text{H}$

11–14 In a series RL circuit, determine how long it takes the current to build up to its full value for each of the following:
(a) $R = 50\ \Omega, L = 50\ \mu\text{H}$ **(b)** $R = 3300\ \Omega, L = 15\ \text{mH}$
(c) $R = 22\ \text{k}\Omega, L = 100\ \text{mH}$

11–15 In the circuit of Figure 11–43, there is initially no current. Determine the inductor voltage at the following times after the switch is closed:
(a) 10 μs **(b)** 20 μs **(c)** 30 μs **(d)** 40 μs **(e)** 50 μs

FIGURE 11–43

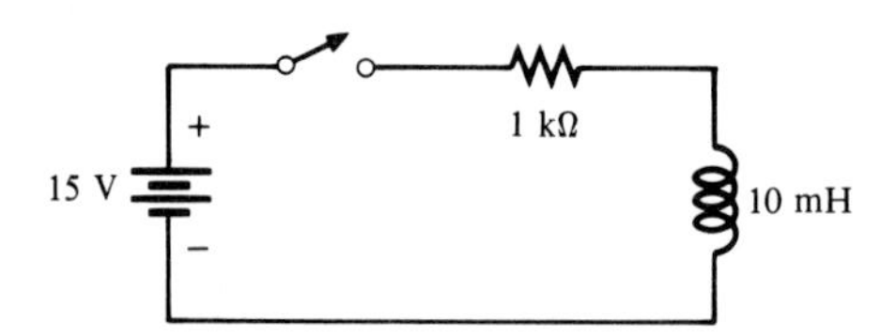

11–16 In Figure 11–44, there are 100 mA through the coil. When SW1 is opened and SW2 simultaneously closed, find the inductor voltage at the following times:
(a) initially **(b)** 1.5 ms **(c)** 4.5 ms **(d)** 6 ms

FIGURE 11–44

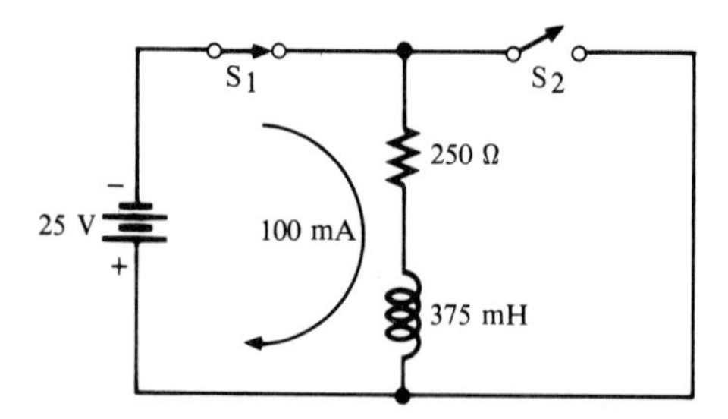

Section 11–6

11–17 Find the total reactance for each circuit in Figure 11–41 when a voltage with a frequency of 5 kHz is applied across the terminals.

11–18 Find the total reactance for each circuit in Figure 11–42 when a 400-Hz voltage is applied.

11–19 Determine the total rms current in Figure 11–45. What are the currents through L_2 and L_3?

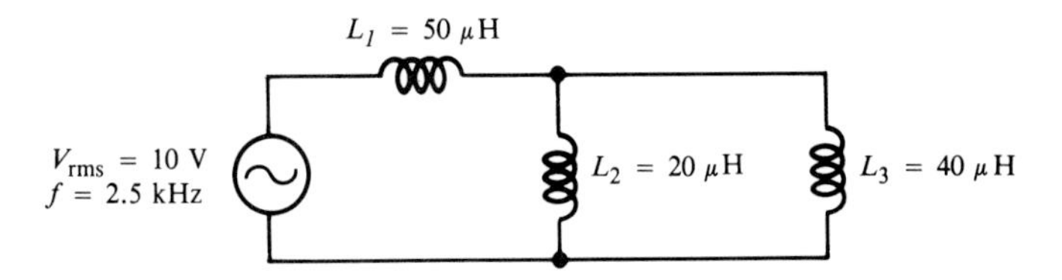

FIGURE 11–45

11–20 What frequency will produce 500-mA total rms current in each circuit of Figure 11–42 with an rms input voltage of 10 V?

11–21 Determine the reactive power in Figure 11–45 neglecting the winding resistance.

Section 11–7

11–22 A certain coil that is supposed to have a 5-Ω winding resistance is measured with an ohmmeter. The meter indicates 2.8 Ω. What is the problem with the coil?

11–23 Determine the indication corresponding to each of the following failures in a coil: **(a)** open. **(b)** shorted. **(c)** partially shorted.

PROBLEM SET B

11–24 Determine the time constant for the circuit in Figure 11–46.

11–25 Find the inductor current 10 μs after the switch is thrown from position 1 to position 2 in Figure 11–47. For simplicity, assume that the switch makes contact with position 2 at the same instant it breaks contact with position 1.

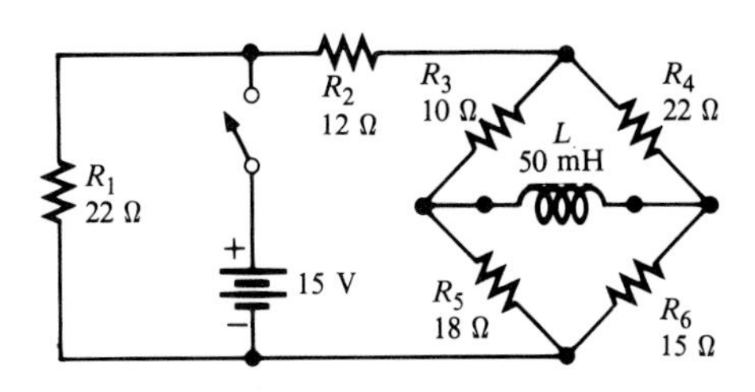

FIGURE 11–46

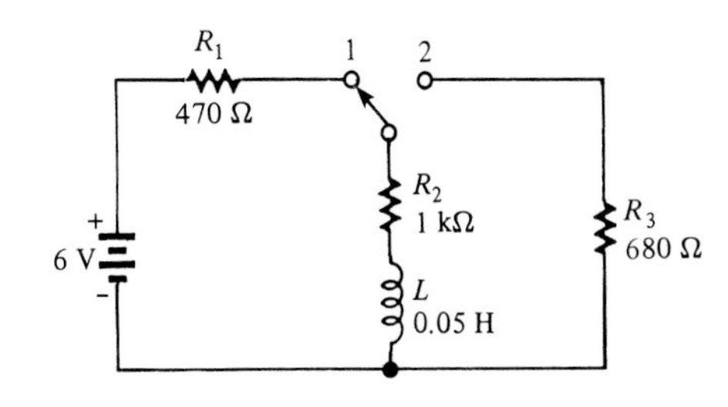

FIGURE 11–47

11–26 In Figure 11–48, SW1 is opened and SW2 is closed at the same instant (t_0). What is the instantaneous voltage across R_2 at t_0?

FIGURE 11–48

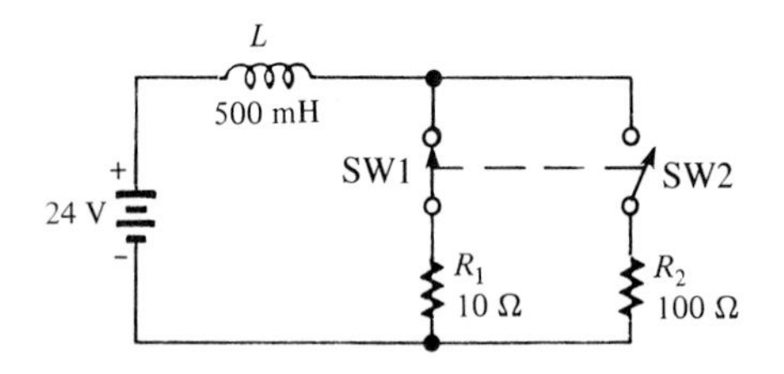

11–27 Determine the value of I_{L2} in Figure 11–49.

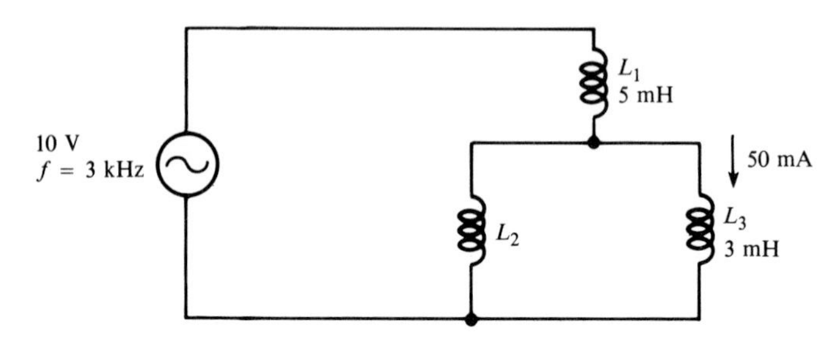

FIGURE 11–49

ANSWERS TO SECTION REVIEWS

Section 11–1

1. (a) L increases; (b) L decreases; (c) L decreases; (d) L decreases.
2. All wire has some resistance, and since inductors are made from turns of wire, there is always resistance.

Section 11–2

1. Fixed and variable. 2. (a) Air core; (b) iron core; (c) variable.

Section 11–3

1. Inductances are added in series. 2. 2600 μH.

Section 11–4

1. The total parallel inductance is smaller than that of the smallest individual inductor in parallel.
2. T.
3. (a) 7.69 mH; (b) 24 μH; (c) 0.1 H.

Section 11–5

1. 0.1 V. 2. $v_R = 0$ V, $v_L = 20$ V. 3. $v_R = 20$ V, $v_L = 0$ V.
4. 0.5 μs, 3.93 mA.

Section 11–6

1. Voltage leads current by 90°. 2. 3.14 kΩ. 3. 2.55 MHz. 4. 15.9 mA.
5. 0, 458.4 mVAR.

Section 11–7

1. The inductor tends to level out the ripple because of its opposition to changes in current.

2. The inductive reactance is very high at radio frequencies and blocks these frequencies.

Section 11–8

1. F. 2. F.

Transformers

In the last chapter, self-inductance was studied. In this chapter, another inductive effect, called *mutual inductance*, is introduced. Mutual inductance occurs when two or more coils are very close together. It is based on the principle of magnetic coupling. A basic transformer is formed by two coils that are magnetically coupled but have no electrical connection between them. The transfer of energy from one coil to the other occurs entirely through a changing magnetic field. There are many applications for transformers, and a basic understanding of their operation is essential.

Specifically, in this chapter you will learn:

- The principle of mutual inductance.
- How a basic transformer operates.
- How the physical characteristics of a transformer, such as the core construction and the turns ratio, affect the performance.
- How the windings determine the voltage polarities.
- How a transformer steps a voltage up to a higher value or down to a lower value.
- The meaning of the terms *primary* and *secondary*.
- How a transformer can change the effective value of a load by reflecting it into the primary circuit.
- The maximum power transfer theorem.
- How a transformer can be used for matching a source resistance to a load resistance to achieve a maximum transfer of power.
- How a transformer can be used to electrically isolate a circuit from the power line for safety.
- How a transformer can be used to prevent dc voltage from passing from one point to another.
- How an actual transformer differs from the ideal model.
- What tapped transformers, multiple-winding transformers, and autotransformers are.
- How transformers are applied in several common situations.
- How to check a transformer for common failures.

12

APPLICATION ASSIGNMENT

You are in charge of modifying the sound system for a sports arena. The audio amplifier has a 50-Ω internal resistance (output resistance) and must drive four 8-Ω speakers (in parallel). You have been requested to get the maximum power possible delivered to the speakers.

After completing this chapter, you will be able to carry out this assignment. The solution is given in the Application Note at the end of this chapter.

THE BASIC TRANSFORMER

12–1

When two coils are placed close together and a sinusoidal current is applied to one coil, the changing magnetic field produced by the current causes an induced voltage in the other coil, in accordance with Faraday's law (Chapter 7). This situation basically describes how a *transformer* operates. Thus, two coils that are *magnetically linked* to each other form a transformer.

Basic Transformer Operation

Recall that the electromagnetic field surrounding a coil expands and collapses as the current increases and decreases, as shown in Figure 12–1.

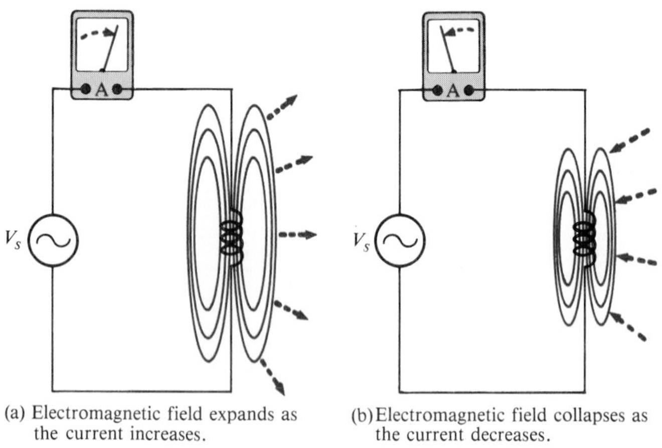

(a) Electromagnetic field expands as the current increases.

(b) Electromagnetic field collapses as the current decreases.

FIGURE 12–1
The electromagnetic field changes with the current.

When a second coil is placed very close to the first coil so that the changing magnetic flux lines cut through the second coil, a voltage is induced, as indicated in Figure 12–2. If the current in the first coil is a sine wave, the voltage induced in the second coil is also a sine wave. The amount of voltage induced in the second coil as a result of the current in the first coil is dependent on the *mutual inductance* between the two coils. *The mutual inductance is established by the inductance of each coil and by the amount of coupling between the two coils.*

A schematic diagram of a transformer is shown in Figure 12–3(a). One coil is called the *primary* winding and one the *secondary* winding as indicated. In standard operation, the source voltage is applied to the primary, and a load is connected to the secondary, as shown in Part (b).

Basic Transformer Construction

The windings of a transformer are formed around the *core*. The purpose of the core is to provide both a physical structure for placement of the windings and a magnetic path so that the magnetic flux lines are concentrated close to the coils.

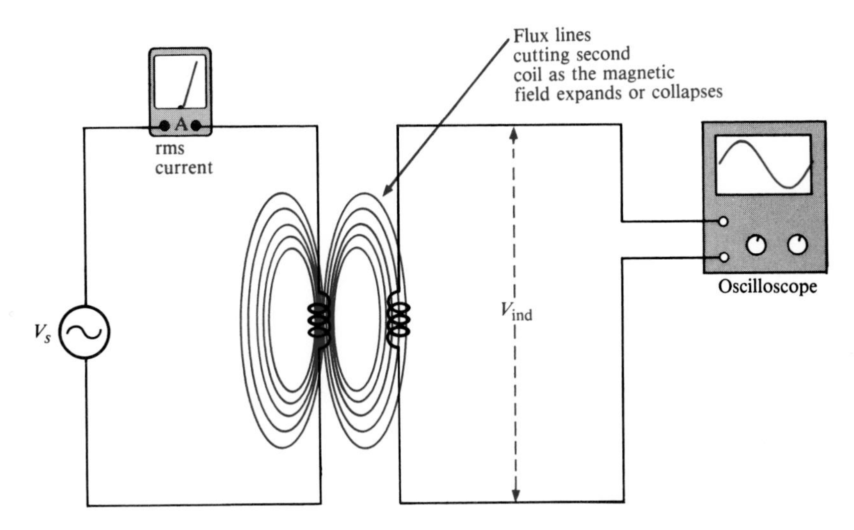

FIGURE 12–2
A voltage is induced in the second coil as a result of the changing current in the first coil.

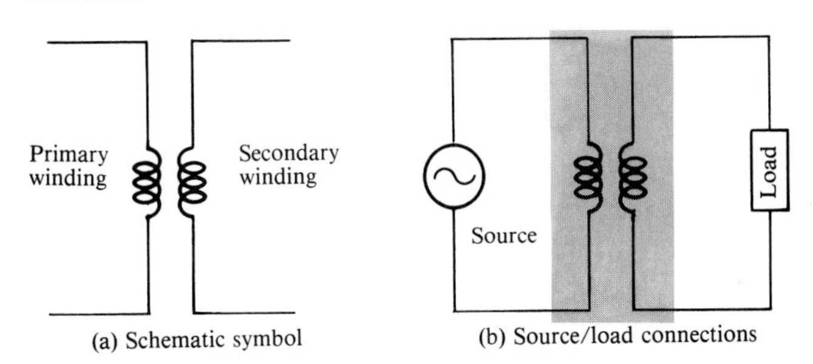

FIGURE 12–3
The basic transformer.

There are three general categories of core material: air, ferrite, and iron. Schematic symbols for each type are shown in Figure 12–4.

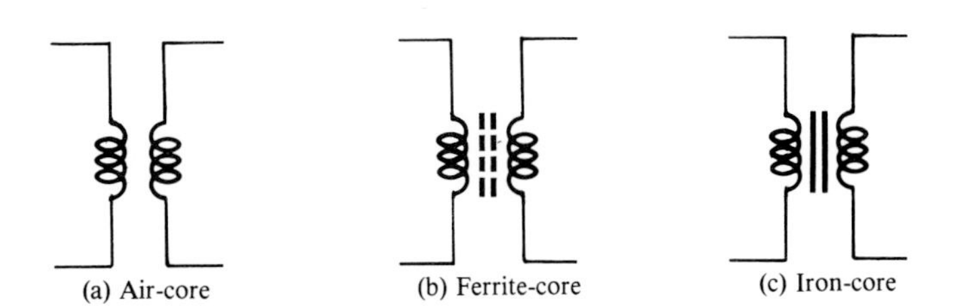

FIGURE 12–4
Schematic symbols based on type of core.

Air-core and ferrite-core transformers generally are used for high-frequency applications and consist of windings on an insulating shell which is hollow (air) or constructed of ferrite, such as depicted in Figure 12–5. The wire is covered by an insulated sheathing or coating to prevent the windings from shorting together. The amount of magnetic coupling between the primary and the secondary is set by the type of core material and by the relative positions of the windings. In Part (a) of the figure, the windings are loosely coupled because

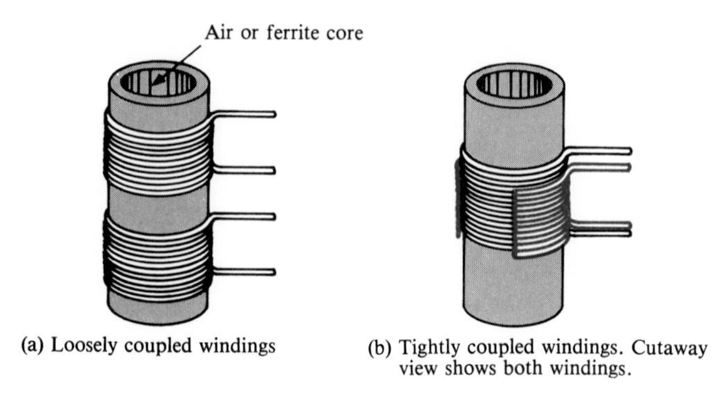

(a) Loosely coupled windings

(b) Tightly coupled windings. Cutaway view shows both windings.

FIGURE 12–5
Transformers with cylindrical-shaped cores.

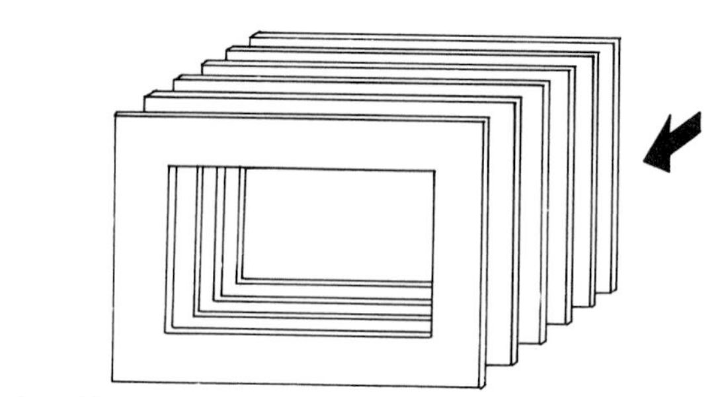

(a) In laminated iron-core construction, the layers are insulated from each other.

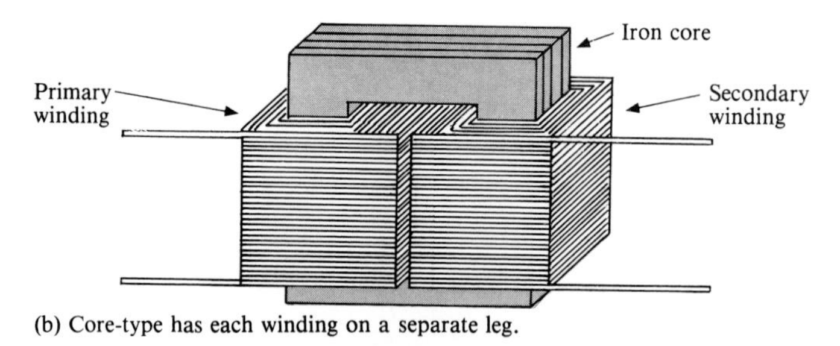

(b) Core-type has each winding on a separate leg.

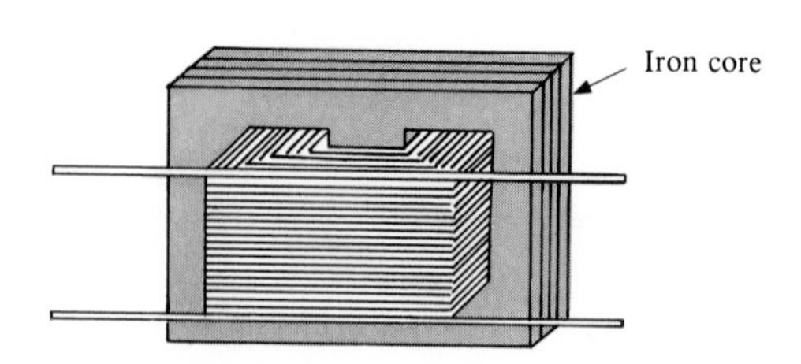

(c) Shell-type has both windings on the same leg.

FIGURE 12–6
Iron-core transformer construction with multilayer windings.

they are separated, and in Part (b) they are tightly coupled because they are close together. The tighter the coupling, the greater the induced voltage in the secondary for a given current in the primary.

Iron-core transformers generally are used for audio frequency (af) and power applications. These transformers consist of windings on a core constructed from laminated sheets of ferromagnetic material insulated from each other, as shown in Figure 12–6(a). This construction provides an easy path for the magnetic flux and increases the amount of coupling between the windings. Figures 12–6(b) and 12–6(c) show the basic construction of two major configurations of iron-core transformer. In the core-type construction, shown in Part (b), the windings are on separate legs of the laminated core. In the shell-type construction, shown in Part (c), both windings are on the same leg.

A variety of transformers are shown in Figure 12–7.

FIGURE 12–7
Some common types of transformers (courtesy of Litton Triad-Utrad).

Turns Ratio

An important parameter of a transformer is its *turns ratio. The turns ratio (n) is the ratio of the number of turns in the secondary winding (N_s) to the number of turns in the primary winding (N_p):*

$$n = \frac{N_s}{N_p} \qquad (12\text{–}1)$$

In the following sections, you will see how the turns ratio affects the voltages and currents in a transformer.

EXAMPLE 12–1 A certain transformer used in a terrain-following radar system has a primary winding with 100 turns and a secondary winding with 400 turns. What is the turns ratio?

Solution

$$N_s = 400 \quad \text{and} \quad N_p = 100$$

$$n = \frac{N_s}{N_p} = \frac{400}{100} = \frac{4}{1} = 4$$

A turns ratio of 4 is usually expressed as 1:4 on a schematic.

Direction of Windings

Another important transformer parameter is the direction in which the windings are placed around the core. As illustrated in Figure 12–8, the direction of the windings determines the polarity of the secondary voltage with respect to the primary voltage. *Phase dots* are used on the schematic symbols to indicate polarities, as shown in Figure 12–9.

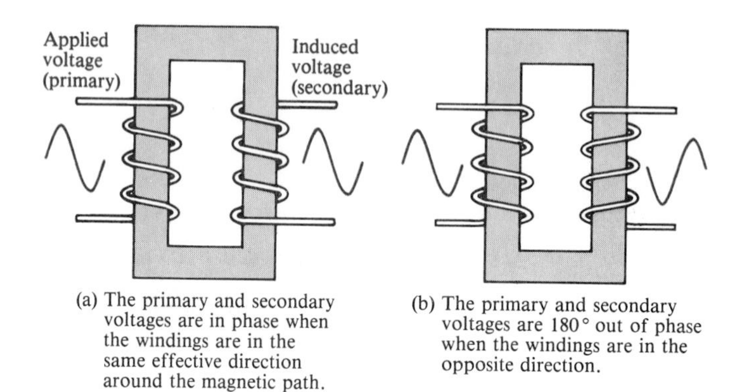

(a) The primary and secondary voltages are in phase when the windings are in the same effective direction around the magnetic path.

(b) The primary and secondary voltages are 180° out of phase when the windings are in the opposite direction.

FIGURE 12–8
Relative polarities of the voltages are determined by the direction of the windings.

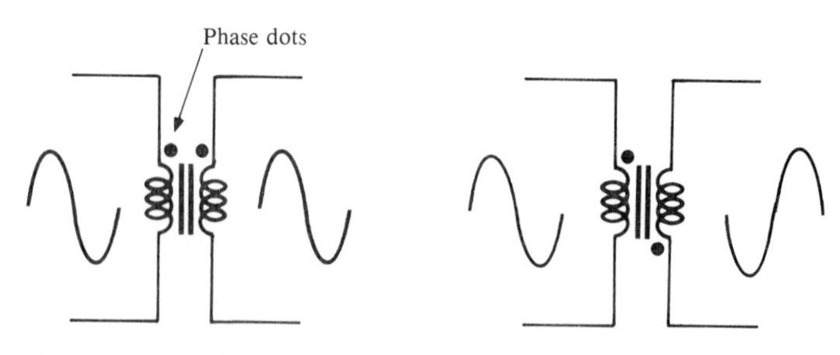

(a) Voltages are in phase.

(b) Voltages are out of phase.

FIGURE 12–9
Phase dots indicate relative polarities of primary and secondary voltages.

SECTION REVIEW 12–1

1. Upon what principle is the operation of a transformer based?
2. Define *mutual inductance.*
3. Define *turns ratio.*
4. Why are the directions of the windings of a transformer important?

STEP-UP TRANSFORMERS

12–2

A transformer in which the secondary voltage is greater than the primary voltage is called a step-up transformer. The amount that the voltage is stepped up depends on the turns ratio.

The ratio of secondary voltage to primary voltage is equal to the ratio of the number of secondary turns to the number of primary turns:

$$\frac{V_s}{V_p} = \frac{N_s}{N_p}$$

where V_s is the secondary voltage and V_p is the primary voltage.

From this relationship, we get

$$V_s = \left(\frac{N_s}{N_p}\right)V_p \quad (12\text{–}2)$$

This equation shows that the secondary voltage is equal to the turns ratio times the primary voltage. This condition assumes that *all* the magnetic flux lines produced in the primary pass through the secondary.

The turns ratio for a step-up transformer is always *greater* than 1.

EXAMPLE 12–2 The transformer in Figure 12–10 has a 200-turn primary winding and a 600-turn secondary winding. What is the voltage across the secondary?

FIGURE 12–10

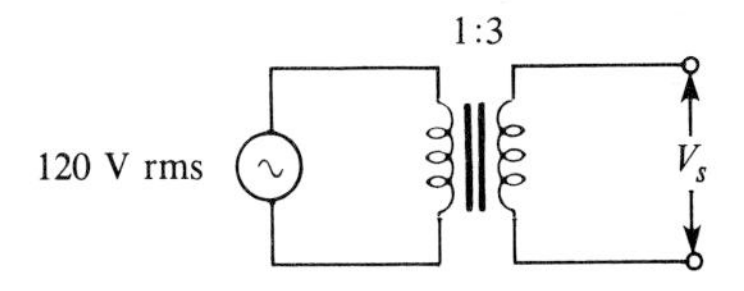

Solution

$$N_p = 200 \quad \text{and} \quad N_s = 600$$

$$V_s = \left(\frac{N_s}{N_p}\right)V_p = \left(\frac{600}{200}\right)V_p = 3V_p = 3(120 \text{ V}) = 360 \text{ V}$$

Note that the turns ratio of 3 is indicated on the schematic as 1:3, meaning that there are 3 secondary turns for each primary turn.

The calculator sequence is

[6] [0] [0] [÷] [2] [0] [0] [×] [1] [2] [0] [=]

SECTION REVIEW 12–2

1. What does a step-up transformer do?
2. If the turns ratio is 5, how much greater is the secondary voltage than the primary voltage?
3. When 240 V ac are applied to a transformer with a turns ratio of 10, what is the secondary voltage?

STEP-DOWN TRANSFORMERS

12–3

A transformer in which the secondary voltage is less than the primary voltage is called a step-down transformer. The amount by which the voltage is stepped down depends on the turns ratio. Equation (12–2) also applies to a step-down transformer; the turns ratio, however, is always *less* than 1.

EXAMPLE 12–3 The transformer in Figure 12–11 is part of a laboratory power supply and has 50 turns in the primary and 10 turns in the secondary. What is the secondary voltage?

FIGURE 12–11

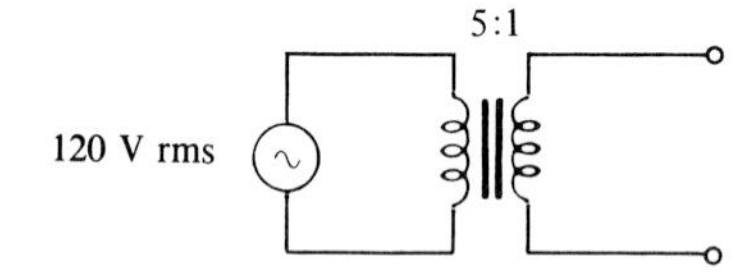

Solution

$$N_p = 50 \quad \text{and} \quad N_s = 10$$

$$\text{Turns ratio} = \frac{N_s}{N_p} = \frac{10}{50} = 0.2$$

$$V_s = 0.2\ V_p = 0.2(120\ \text{V}) = 24\ \text{V}$$

SECTION REVIEW 12–3

1. What does a step-down transformer do?
2. One hundred twenty volts are applied to the primary of a transformer with a turns ratio of 0.5. What is the secondary voltage?
3. A primary voltage of 120 V is reduced to 12 V. What is the turns ratio?

TRANSFORMER POWER

12–4

When a load is connected to the secondary of a transformer, the power transferred to the load can never be greater than the power that is put into the primary. For an ideal transformer, the power in the secondary equals the power in the primary. When losses are considered, the secondary power is always less.

Power is dependent on voltage and current, and there can be no increase in power. Therefore, if the voltage is stepped up, the current is stepped down, and vice versa. We will now examine these conditions more closely.

Loading the Secondary

When a load resistor (R_L) is connected to the secondary, as shown in Figure 12–12, secondary current will flow because of the voltage induced in the secondary coil. It can be shown that the ratio of the primary current I_p to the secondary current I_s is equal to the turns ratio, as expressed in the following equation:

$$\frac{I_p}{I_s} = \frac{N_s}{N_p}$$

Inverting both sides of the equation yields

$$\frac{I_s}{I_p} = \frac{N_p}{N_s}$$

Solving for I_s, we have

$$I_s = \left(\frac{N_p}{N_s}\right)I_p \tag{12–3}$$

Note that N_p/N_s is the *reciprocal* of the turns ratio. Thus, as shown in Figure 12–12(a), for a step-up transformer, in which N_s/N_p is greater than 1, the secondary current is less than the primary current. For a step-down transformer, shown in Part (b), N_s/N_p is less than 1, and I_s is greater than I_p.

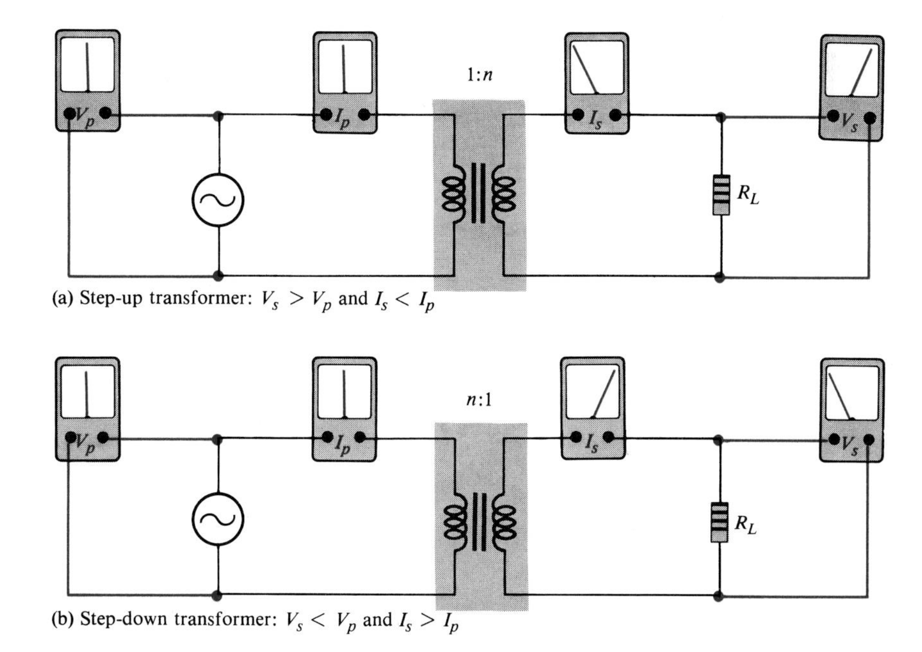

(a) Step-up transformer: $V_s > V_p$ and $I_s < I_p$

(b) Step-down transformer: $V_s < V_p$ and $I_s > I_p$

FIGURE 12–12
Illustration of voltages and currents in transformers with loaded secondaries.

EXAMPLE 12–4 The transformers in Figures 12–13(a) and 12–13(b) have loaded secondaries. If the primary current is 100 mA in each case, how much current flows through the load?

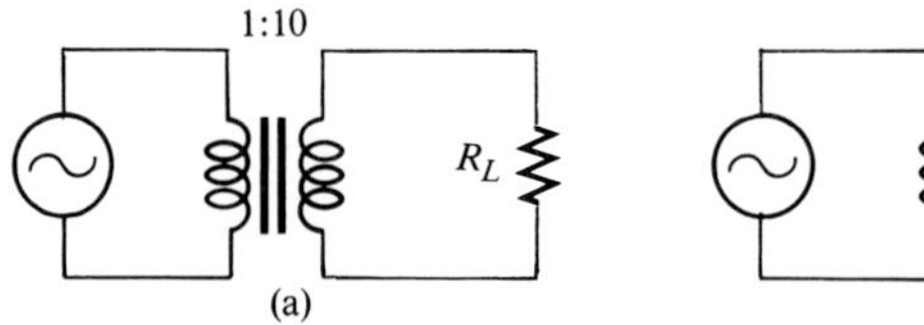

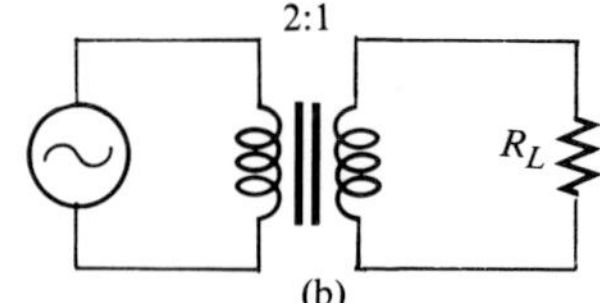

FIGURE 12–13

Solution

(a) $I_s = \left(\frac{N_p}{N_s}\right)I_p = 0.1(100\text{ mA}) = 10\text{ mA}$

(b) $I_s = \left(\frac{N_p}{N_s}\right)I_p = 2(100\text{ mA}) = 200\text{ mA}$

Primary Power Equals Secondary Power

In an ideal transformer, the secondary power is the same as the primary power regardless of the turns ratio, as shown in the following steps:

$$P_p = V_pI_p \quad \text{and} \quad P_s = V_sI_s$$

$$I_s = \left(\frac{N_p}{N_s}\right)I_p \quad \text{and} \quad V_s = \left(\frac{N_s}{N_p}\right)V_p$$

By substitution, we obtain

$$P_s = \left(\frac{\cancel{N_p}}{\cancel{N_s}}\right)\left(\frac{\cancel{N_s}}{\cancel{N_p}}\right)V_pI_p$$

Canceling yields

$$P_s = V_pI_p = P_p$$

SECTION REVIEW 12–4

1. If the turns ratio of a transformer is 2, is the secondary current greater than or less than the primary current? By how much?
2. A transformer has 100 primary turns and 25 secondary turns, and I_p is 0.5 A. What is the value of I_s?

REFLECTED LOAD IN A TRANSFORMER

12–5

From the viewpoint of the primary, a load connected across the secondary of a transformer appears to have an ohmic value that is not necessarily equal to the actual ohmic value of the load, but rather depends on the turns ratio. This load that is reflected into the primary is what the source effectively sees, and it determines the amount of primary current. The concept of the *reflected load* is illustrated in Figure 12–14.

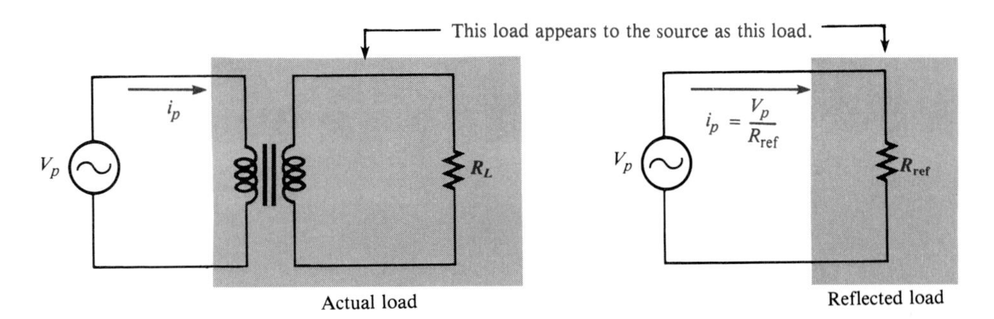

FIGURE 12–14
The load in the secondary of a transformer is reflected into the primary by transformer action. It appears to the source to be a resistance with a value determined by the turns ratio and the actual value of the load resistance.

From the relationships of the currents, voltages, and the turns ratio in a transformer, an equation for the reflected load resistance can be derived. The result is given in the following formula:

$$R_{ref} = \left(\frac{1}{n}\right)^2 R_L \tag{12–4}$$

where $1/n$ is the reciprocal of the turns ratio (N_p/N_s).

EXAMPLE 12–5

Figure 12–15 shows a source that is transformer-coupled to a load resistor of 100 Ω. The transformer has a turns ratio of 4. What is the reflected resistance seen by the source?

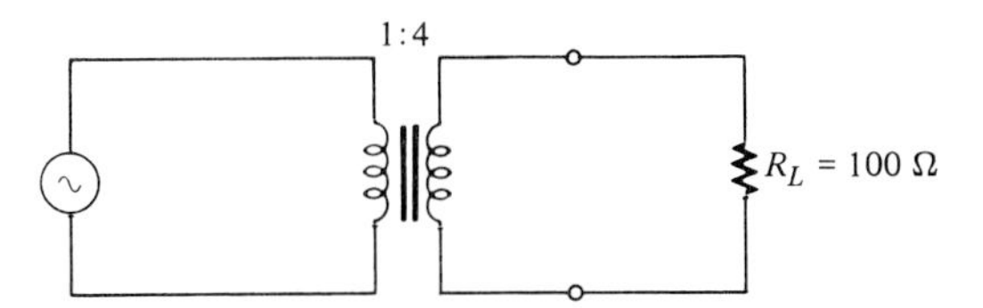

FIGURE 12–15

Solution
The reflected resistance is determined as follows:

$$R_{ref} = \left(\frac{1}{4}\right)^2 R_L = \left(\frac{1}{16}\right)(100\ \Omega) = 6.25\ \Omega$$

The source sees a resistance of 6.25 Ω just as if it were connected directly, as shown in the equivalent circuit of Figure 12–16.

FIGURE 12–16

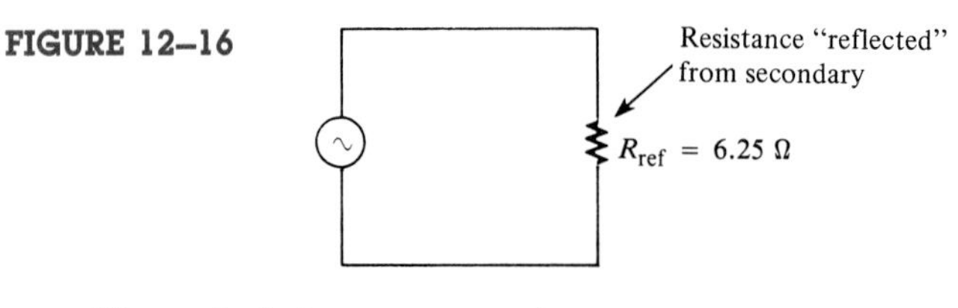

The calculator sequence is

4 2nd 1/x 2nd x^2 × 1 0 0 =

EXAMPLE 12–6 In Figure 12–15, if a transformer is used having 40 primary turns and 10 secondary turns, what is the reflected resistance?

Solution

$$\text{Turns ratio} = 0.25$$

$$R_p = \left(\frac{1}{0.25}\right)^2(100\ \Omega) = (4)^2(100\ \Omega) = 1600\ \Omega$$

Examples 12–5 and 12–6 illustrate that in a step-up transformer ($n > 1$), the reflected resistance is less than the actual load resistance; and in a step-down transformer ($n < 1$), the reflected resistance is greater than the load resistance.

Matching the Load Resistance to the Source Resistance with a Transformer

One application of transformers is in *impedance matching* a load to a source in order to achieve maximum transfer of power from the source to the load. The term *impedance* will become very familiar to you in the next chapter. Basically, impedance is a general term for the opposition to current, including the effects of both resistance and reactance combined. However, for the time being, we will confine our usage to resistance only.

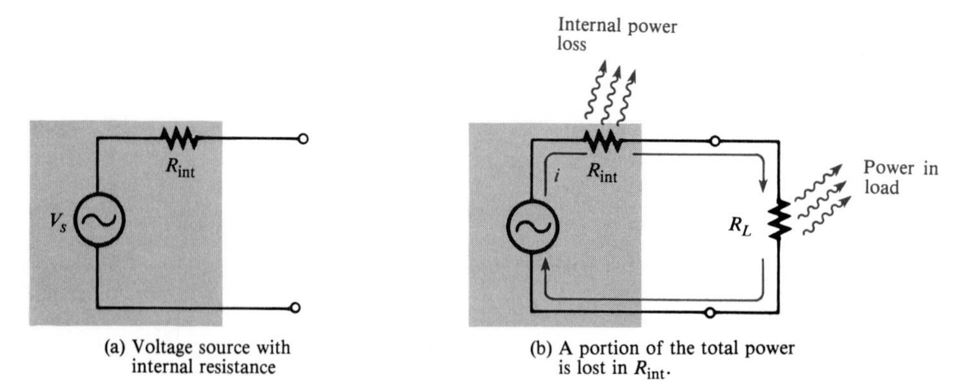

(a) Voltage source with internal resistance

(b) A portion of the total power is lost in R_{int}.

FIGURE 12–17
Power transfer from a nonideal voltage source to a load.

The concept of impedance matching is illustrated in the basic circuit of Figure 12–17. Part (a) shows an ac voltage source with a series resistance representing its internal resistance. Some internal resistance is inherent in all sources due to their internal circuitry or physical makeup. When the source is connected directly to a load, as shown in Part (b), often the objective is to transfer as much of the power produced by the source to the load as possible. However, a certain amount of the power produced by the source is lost in its internal resistance, and the remaining power goes to the load.

Maximum Power Transfer Theorem

When a source is connected to a load, maximum power is delivered to the load when the load resistance is equal to the source resistance.

We will demonstrate this theorem by finding the load power for various values of load resistance and a specific source resistance in Example 12–7.

EXAMPLE 12–7

The source in Figure 12–18 has an internal resistance of 75 Ω. Determine the power in each of the following values of load resistor:
(a) 25 Ω
(b) 50 Ω
(c) 75 Ω
(d) 100 Ω
(e) 125 Ω
Draw a graph showing the load power versus the load resistance.

FIGURE 12–18

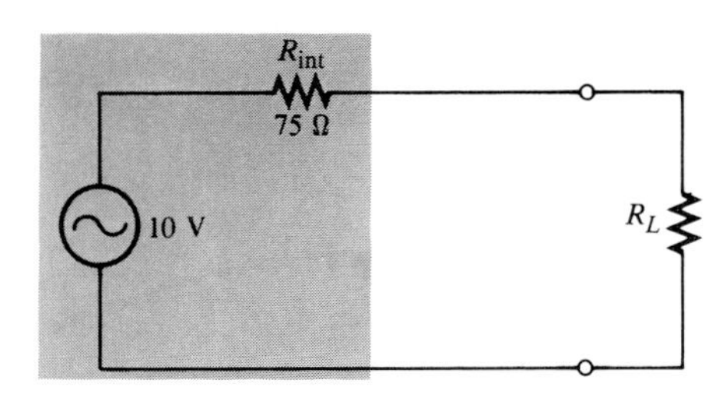

Solution
We will use Ohm's law ($I = V/R$) and the power formula ($P = I^2R$) to find the load power for each value of load resistance.
(a) $R_L = 25\ \Omega$:

$$I = \frac{V_s}{R_{int} + R_L} = \frac{10\text{ V}}{100\ \Omega} = 0.1\text{ A}$$
$$P_L = I^2R_L = (0.1\text{ A})^2(25\ \Omega) = 0.25\text{ W}$$

(b) $R_L = 50\ \Omega$:

$$I = \frac{V_s}{R_{int} + R_L} = \frac{10\text{ V}}{125\ \Omega} = 0.08\text{ A}$$
$$P_L = I^2R_L = (0.08\text{ A})^2(50\ \Omega) = 0.32\text{ W}$$

(c) $R_L = 75\ \Omega$:

$$I = \frac{V_s}{R_{int} + R_L} = \frac{10\text{ V}}{150\ \Omega} = 0.067\text{ A}$$

$$P_L = I^2R_L = (0.067)^2(75\ \Omega) = 0.337\text{ W}$$

(d) $R_L = 100\ \Omega$:

$$I = \frac{V_s}{R_{int} + R_L} = \frac{10\text{ V}}{175\ \Omega} = 0.057\text{ A}$$

$$P_L = I^2R_L = (0.057\text{ A})^2(100\ \Omega) = 0.33\text{ W}$$

(e) $R_L = 125\ \Omega$:

$$I = \frac{V_s}{R_{int} + R_L} = \frac{10\text{ V}}{200\ \Omega} = 0.05\text{ A}$$

$$P_L = I^2R_L = (0.05\text{ A})^2(125\ \Omega) = 0.31\text{ W}$$

Notice that the load power is greatest when $R_L = 75\ \Omega$, which is the same as the internal source resistance. When the load resistance is less than or greater than this value, the power drops off, as the curve in Figure 12–19 graphically illustrates.

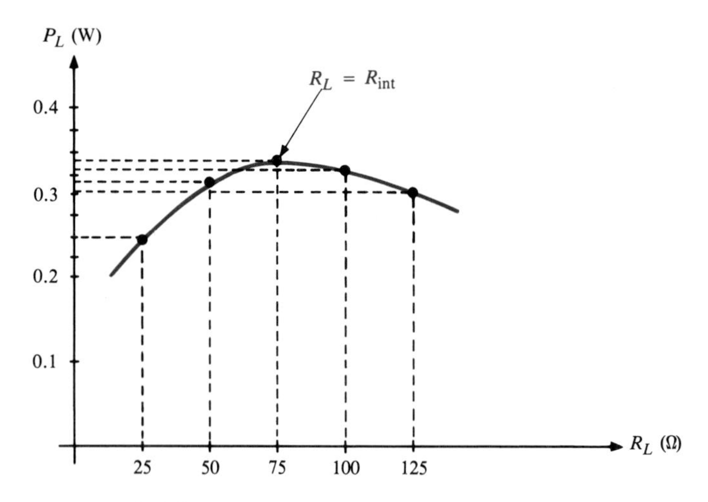

FIGURE 12–19
Curve showing that the load power is maximum when $R_L = R_{int}$.

The calculator sequence for (a) is
1 0 ÷ (7 5 + 2 5) = 2nd x^2 × 2 5 =

A Practical Application

In most practical situations, the internal source resistance of various types of sources is fixed. Also, in many cases, the resistance of a device that acts as a load is fixed and cannot be altered. If you need to connect a given source to a given load, remember that only by chance will their resistances match. In this situation a transformer comes in handy. You can use the reflected-resistance characteristic of a transformer to make the load resistance appear to have the same value as the source resistance, thereby "fooling" the source into "thinking" that there is a match.

Let's take a practical, everyday situation to illustrate. The typical resistance of the input to a TV receiver is 300 Ω. An antenna must be connected to this input by a lead-in cable in order to receive TV signals. In this situation, the antenna and the lead-in act as the source, and the input resistance of the TV receiver is the load, as illustrated in Figure 12–20.

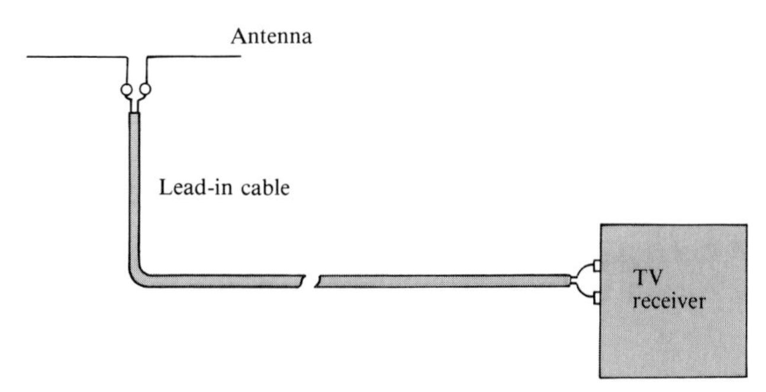

(a) The antenna/lead-in is the source; the TV input is the load.

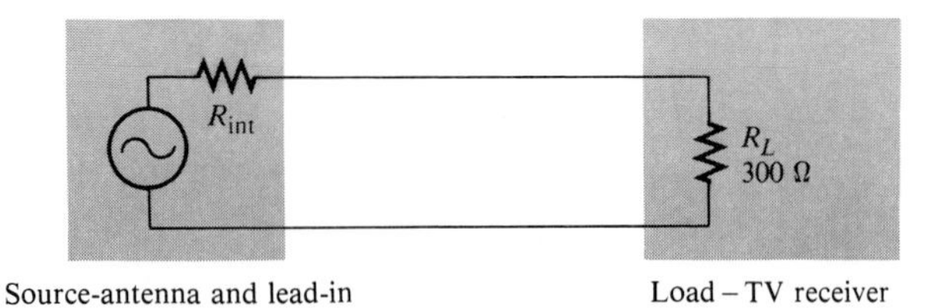

(b) Circuit equivalent of antenna and TV receiver system

FIGURE 12–20
An antenna directly coupled to a TV receiver.

It is common for an antenna system to have a *characteristic resistance* of 75 Ω. Thus, if the 75-Ω source (antenna and lead-in) is connected directly to the 300-Ω TV input, maximum power will not be delivered to the input to the TV, and you will have poor signal reception. The solution is to use a *matching transformer* connected as indicated in Figure 12–21, in order to match the 300-Ω load resistance to the 75-Ω source resistance.

To match the resistances, we must select a proper value of turns ratio (n). We want the 300-Ω load to look like 75 Ω to the source. We solve equation (12–4) for n, using 300 Ω for R_L and 75 Ω for R_{ref}:

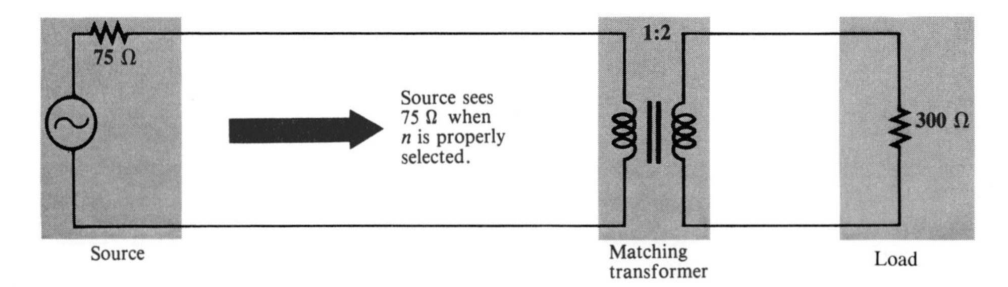

FIGURE 12–21
Example of a load matched to a source by transformer coupling for maximum power transfer.

$$R_{ref} = \left(\frac{1}{n}\right)^2 R_L$$

Transposing terms, we get

$$\left(\frac{1}{n}\right)^2 = \frac{R_{ref}}{R_L}$$

Taking the square root of both sides, we have

$$\frac{1}{n} = \sqrt{\frac{R_{ref}}{R_L}}$$

Inverting both sides and solving for n yields

$$n = \sqrt{\frac{R_L}{R_{ref}}} = \sqrt{\frac{300\ \Omega}{75\ \Omega}} = \sqrt{4} = 2$$

Therefore, a matching transformer with a turns ratio of 2 must be used in this application.

EXAMPLE 12–8 An amplifier has an 800-Ω internal resistance. In order to provide maximum power to an 8-Ω speaker, what turns ratio must be used in the coupling transformer?

Solution
The reflected resistance must equal 800 Ω. Thus,

$$n = \sqrt{\frac{R_L}{R_{ref}}} = \sqrt{\frac{8\ \Omega}{800\ \Omega}} = \sqrt{0.01} = 0.1$$

There must be ten primary turns for each secondary turn. The diagram and its equivalent reflected circuit are shown in Figure 12–22.

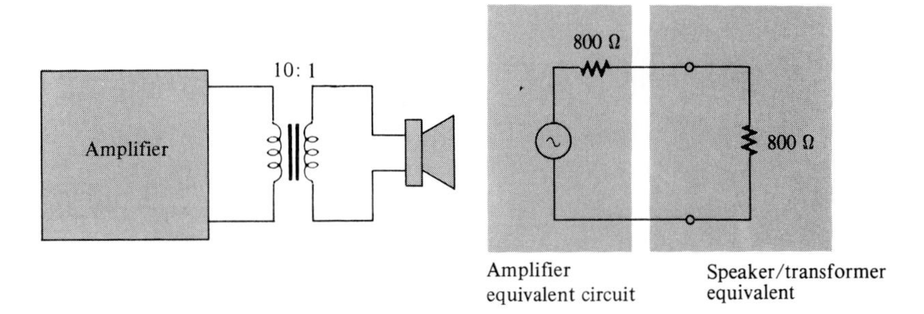

FIGURE 12–22

The calculator sequence is

SECTION REVIEW 12–5

1. Define *reflected impedance.*
2. What transformer characteristic determines the reflected impedance?
3. A given transformer has a turns ratio of 10, and the load is 50 Ω. How much impedance is reflected into the primary?
4. What does *impedance matching* mean?
5. What is the advantage of matching the load impedance to the impedance of a source?
6. A transformer has 100 primary turns and 50 secondary turns. What is the reflected impedance with 100 Ω across the secondary?

THE TRANSFORMER AS AN ISOLATION DEVICE

12–6

Transformers are useful in providing *electrical isolation* between the primary and the secondary, since there is no electrical connection between the two windings. As you know, energy is transferred entirely by magnetic coupling.

dc Isolation

If a nonchanging direct current flows through the primary of a transformer, nothing happens in the secondary, as indicated in Figure 12–23(a). The reason is that a changing primary current is necessary to create a changing magnetic field, which will cause voltage to be induced in the secondary, as indicated in Part (b) of the figure. Therefore, the transformer *isolates* the secondary from any dc in the primary.

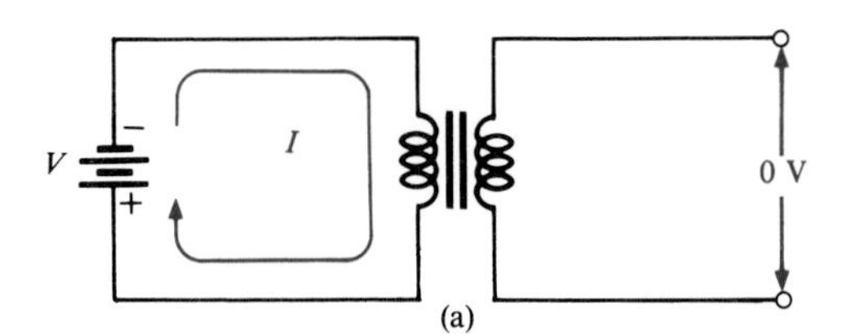

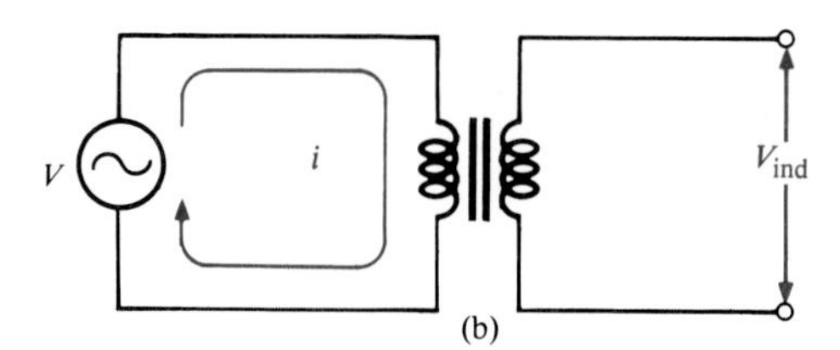

FIGURE 12–23
dc isolation and ac coupling.

For example, in some amplifiers, transformers are used to prevent dc voltages at certain points in the circuit, while allowing the signal (ac) voltage to get through. Figure 12–24 is a simplified illustration of transformer coupling in a two-stage audio amplifier.

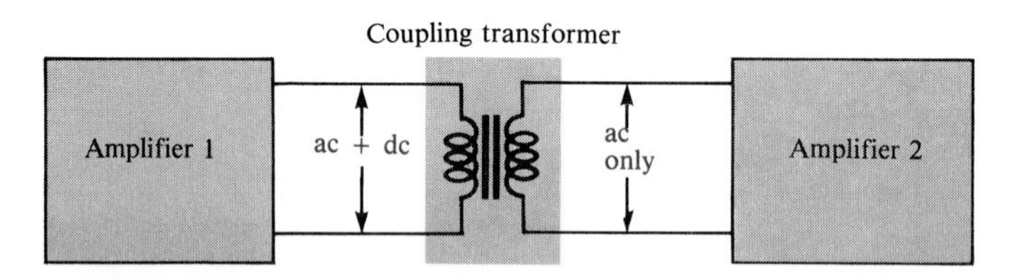

FIGURE 12–24
Audio amplifier stages with transformer coupling for dc isolation.

Power Line Isolation

Transformers are often used to electrically isolate electronic equipment from the 60-Hz, 120-V ac power line. The reason for using an isolation transformer to couple the 60-Hz ac to an instrument is to prevent a possible shock hazard if the 120-V line is connected to the metal chassis of the equipment. This condition is possible if the line cord plug can be inserted into an outlet either way. Incidentally, to prevent this situation, many plugs have keyed prongs so that they can be plugged in only one way.

Figure 12–25 illustrates how a transformer can prevent the metal chassis from being connected to the 120-V line rather than to neutral (ground), no matter how the cord is plugged into the outlet. When an isolation transformer is used, the secondary is said to be "floating" because it is not referenced to the

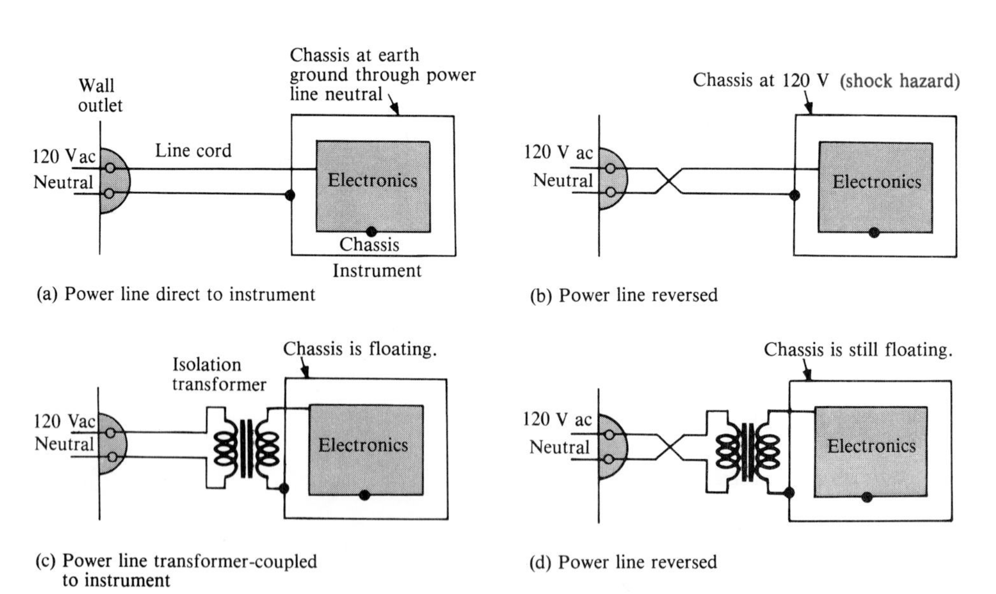

FIGURE 12–25
The use of an isolation transformer to prevent a shock hazard.

power line ground. Should a person come in contact with the secondary voltage, there is no complete current path back to ground, and therefore there is no shock hazard. Current must flow through your body in order for you to receive an electrical shock.

Some TV sets, for example, do not have isolation transformers for reasons of economy. When working on such a unit, be careful to use an external isolation transformer or be sure that the outlet connection does not put the chassis at 120 V.

SECTION REVIEW 12–6

1. What does the term *electrical isolation* mean?
2. Can a dc voltage be coupled by a transformer?

NONIDEAL TRANSFORMER CHARACTERISTICS

12–7

Up to this point, transformer operation has been presented from an ideal point of view, and this approach is valid when you are learning new concepts. However, you should be aware of the nonideal characteristics of practical transformers and how they affect performance.

Winding Resistance

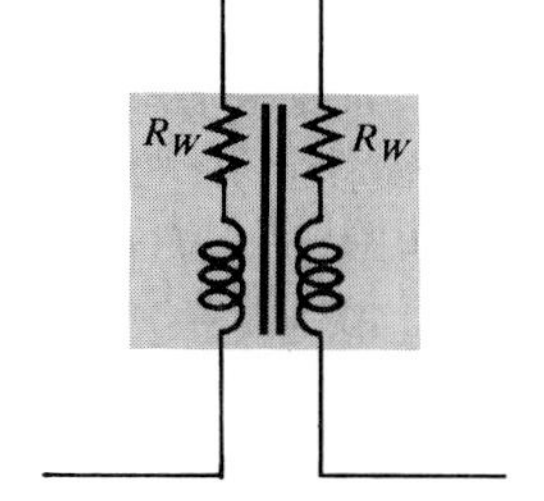

FIGURE 12–26 Winding resistance in a nonideal transformer.

Both the primary and the secondary windings of a transformer have winding resistance. (You learned about winding resistance in Chapter 11 on inductors.) The winding resistances of a transformer are represented as resistors in series with the windings as shown in Figure 12–26.

Winding resistance in a transformer results in less voltage across a secondary load. Voltage drops due to the winding resistance effectively subtract from the primary and secondary voltages and result in load voltage that is less than that predicted by the relationship $V_s = (N_2/N_1)V_p$. In most cases the effect is relatively small and is not considered.

Losses in the Core

There is always some energy loss in the core material of a nonideal transformer. This loss is seen as a heating of ferrite and iron cores, but it does not occur in air cores. Part of this energy is consumed in the continuous reversal of the magnetic field due to the changing direction of the primary current; this energy loss is called *hysteresis loss.* The rest of the energy loss is caused by *eddy currents* induced in the core material by the changing magnetic flux. The eddy-current loss is greatly reduced by the use of laminated construction of iron cores. The thin layers of ferromagnetic material are insulated from each other to minimize the build-up of eddy currents by confining them to a small area and to keep core losses to a minimum.

Magnetic Flux Leakage

In an ideal transformer, all of the magnetic flux produced by the primary current is assumed to pass through the core to the secondary winding, and vice versa. In an actual transformer, some of the magnetic flux lines break out of the core and pass through the surrounding air back to the other end of the winding, as illustrated in Figure 12–27 for the magnetic field produced by the primary current. Magnetic flux leakage results in a reduced secondary voltage.

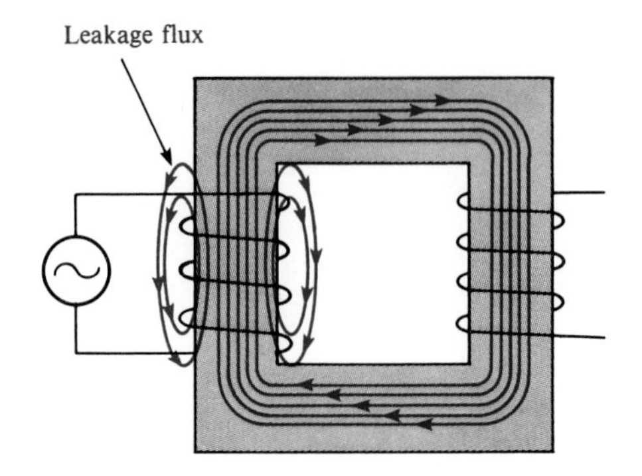

FIGURE 12–27
Flux leakage in a nonideal transformer.

The percentage of magnetic flux that actually reaches the secondary determines the *coefficient of coupling* of the transformer. For example, if nine out of ten flux lines remain inside the core, the coefficient of coupling is 0.90 or 90%. Most iron-core transformers have very high coefficients of coupling (greater than 0.99), while ferrite-core and air-core devices have lower values.

Winding Capacitance

As you learned in Chapter 11, there is always some stray capacitance between adjacent turns of a winding. These stray capacitances result in an effective capacitance in parallel with each winding of a transformer, as indicated in Figure 12–28.

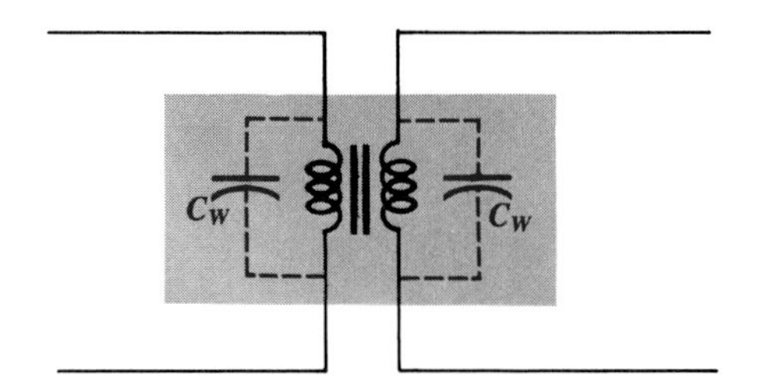

FIGURE 12–28
Winding capacitance in a nonideal transformer.

These stray capacitances have very little effect on the transformer's operation at low frequencies, because the reactances (X_C) are very high. However, at higher frequencies, the reactances decrease and begin to produce a bypassing effect across the primary winding and across the secondary load. As a result, less of the total primary current flows through the primary winding, and less of the total secondary current flows through the load. This effect reduces the load voltage as the frequency goes up.

Transformer Power Rating

A transformer is typically rated in volt-amperes (VA), primary/secondary voltage, and operating frequency. For example, a given transformer rating may be specified as 2 kVA, 500/50, 60 Hz. The 2-kVA value is the *apparent power rating*. The 500 and the 50 can be either secondary or primary voltages.

Let's assume, for example, that 50 V is the secondary voltage. In this case the load current is $I_L = P_s/V_s = 2\text{ kVA}/50\text{ V} = 40\text{ A}$. On the other hand, if 500 V is the secondary voltage, then $I_L = P_s/V_s = 2\text{ kVA}/500\text{ V} = 4\text{ A}$. These are the maximum currents that the secondary can handle in either case.

The reason that the power rating is in volt-amperes (VA) rather than in watts (true power) is as follows: If the transformer load is purely capacitive or purely inductive, the true power (watts) delivered to the load is zero. However, the current for $V_s = 500$ V and $X_C = 100\ \Omega$ at 60 Hz, for example, is 5 A. This current exceeds the maximum that the 2-kVA secondary can handle, and the transformer may be damaged. So it is meaningless to specify power in watts.

Transformer Efficiency

Recall that the secondary power is equal to the primary power in an ideal transformer. Because the nonideal characteristics just discussed result in a power loss in the transformer, the secondary (output) power is always less than the primary (input) power. The *efficiency* (η) of a transformer is a measure of the percentage of the input power that is delivered to the output:

$$\eta = \left(\frac{P_{\text{out}}}{P_{\text{in}}}\right)100\% \qquad (12\text{–}5)$$

Most power transformers have efficiencies in excess of 95%.

EXAMPLE 12–9

A certain type of transformer has a primary current of 5 A and a primary voltage of 4800 V. The secondary current is 90 A and the secondary voltage is 240 V. Determine the efficiency of this transformer.

Solution

The input power is

$$P_{\text{in}} = V_pI_p = (4800\text{ V})(5\text{ A}) = 24\text{ kVA}$$

The output power is

$$P_{\text{out}} = V_sI_s = (240\text{ V})(90\text{ A}) = 21.6\text{ kVA}$$

The efficiency is

$$\eta = \left(\frac{P_{\text{out}}}{P_{\text{in}}}\right)100\% = \left(\frac{21.6\text{ kVA}}{24\text{ kVA}}\right)100\% = 90\%$$

SECTION REVIEW 12–7

1. Explain how an actual transformer differs from the ideal model.
2. The coefficient of coupling of a certain transformer is 0.85. What does this mean?
3. A certain transformer has a rating of 10 kVA. If the secondary voltage is 250 V, how much load current can the transformer handle?

OTHER TYPES OF TRANSFORMERS

12–8

There are several important variations of the basic transformer that you have studied so far. They include

- tapped transformers,
- multiple-winding transformers, and
- autotransformers.

Tapped Transformers

A schematic diagram of a transformer with a *center-tapped* secondary winding is shown in Figure 12–29(a). The center tap (CT) is equivalent to two secondary windings with half the total voltage across each.

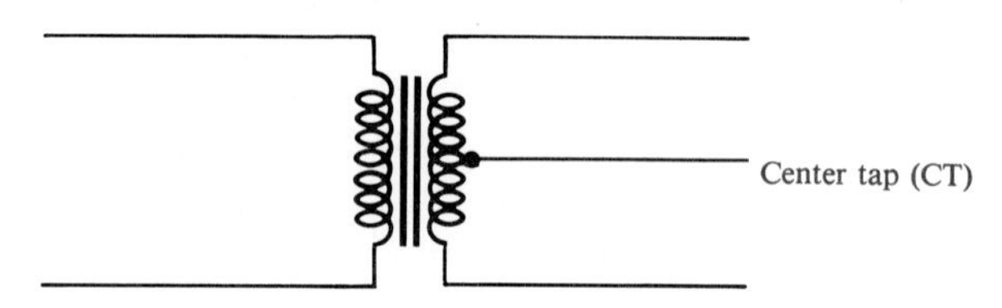

(a) Center-tapped transformer

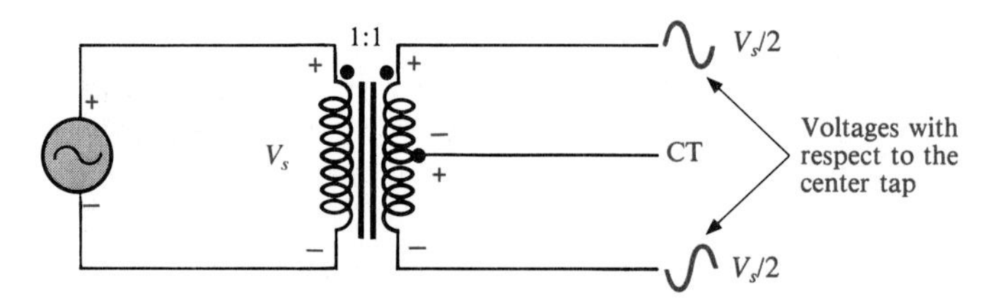

(b) Output voltages with respect to the center tap are 180° out of phase with each other and are one-half the magnitude of the secondary voltage.

FIGURE 12–29
Operation of a center-tapped transformer.

The voltages between either end of the secondary and the center tap are, at any instant, equal in magnitude but opposite in polarity, as illustrated in Figure 12–29(b). Here, for example, at some instant on the sine wave voltage, the polarity across the entire secondary is as shown (top end +, bottom −). At the center tap, the voltage is less positive than the top end but more positive than the bottom end of the secondary. Therefore, measured *with respect to the center tap,* the top end of the secondary is positive, and the bottom end is negative. This center-tapped feature is used in power supply rectifiers in which the ac voltage is converted to dc, as illustrated in Figure 12–30.

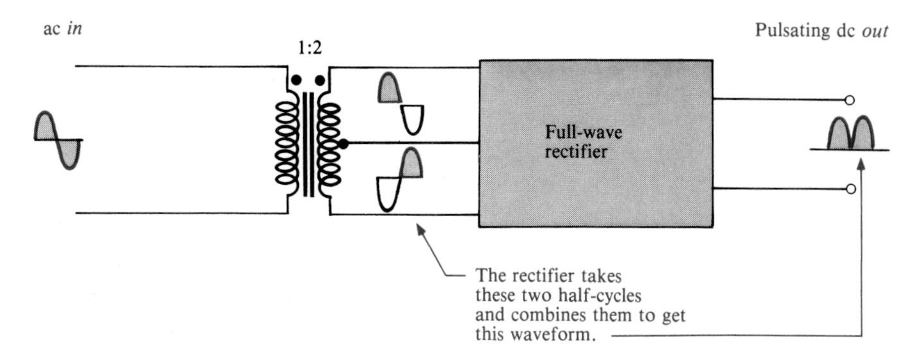

FIGURE 12–30
Application of a center-tapped transformer in ac-to-dc conversion.

Some tapped transformers have taps on the secondary winding at points other than the electrical center. Also, multiple primary and secondary taps are sometimes used in certain applications. Examples of these types of transformers are shown in Figure 12–31.

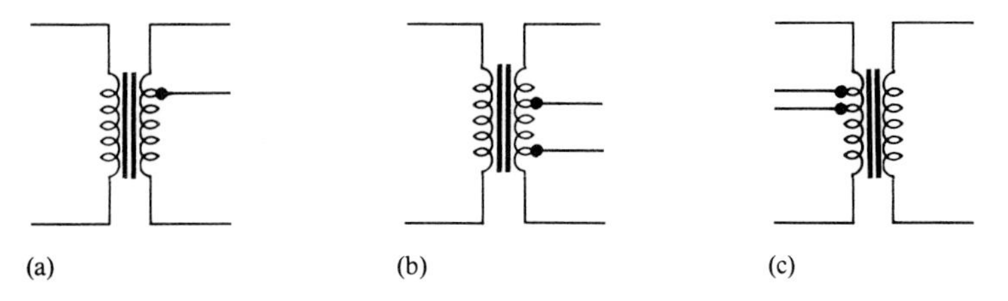

FIGURE 12–31
Tapped transformers.

One example of a transformer with a multiple-tap primary and a center-tapped secondary is the utility-pole transformer used by power companies to step down the high voltage from the power line to 120 V/240 V service for residential and commercial customers, as shown in Figure 12–32. The multiple taps on the primary are used for minor adjustments in the turns ratio in order to overcome line voltages that are slightly too high or too low.

Multiple-Winding Transformers

Some transformers are designed to operate from either 120-V ac or 240-V ac lines. These transformers usually have two primary windings, each of which is designed for 120 V ac. When the two are connected in series, the transformer can be used for 240-V ac operation, as illustrated in Figure 12–33.

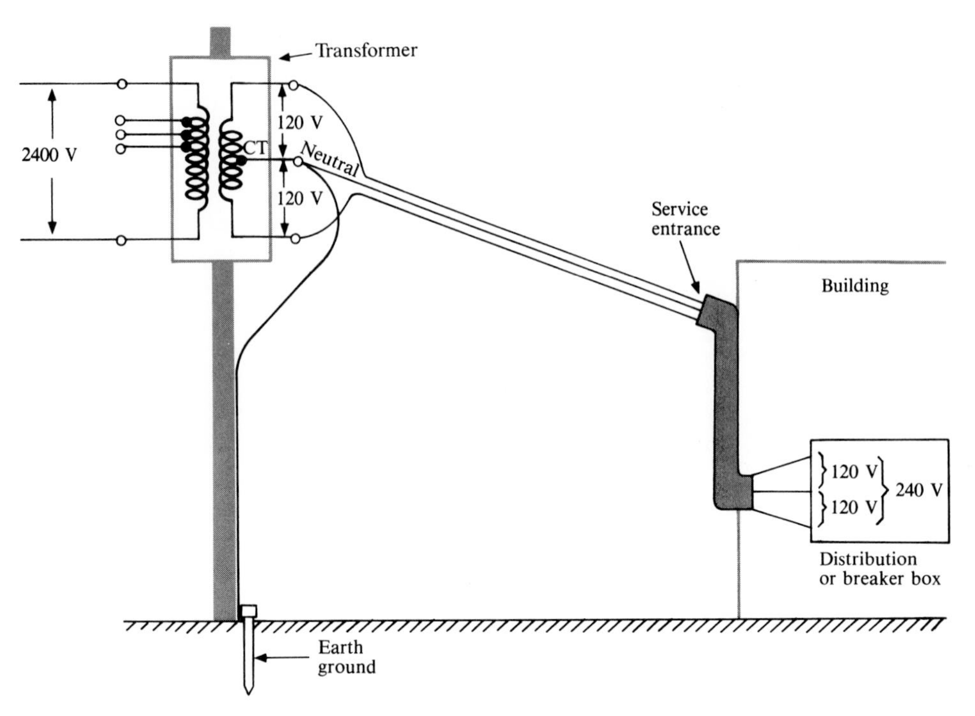

FIGURE 12–32
Utility-pole transformer in a typical power distribution system.

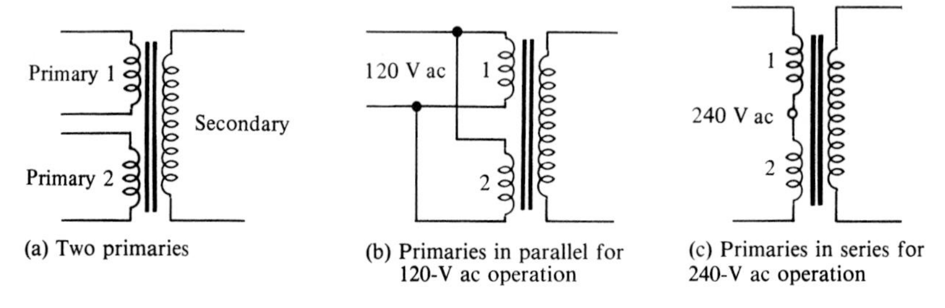

FIGURE 12–33
Multiple-primary transformer.

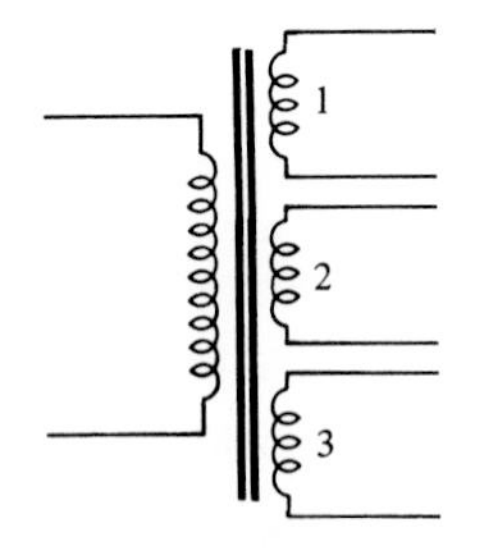

FIGURE 12–34
Multiple-secondary transformer.

More than one secondary can be wound on a common core. Transformers with several secondaries are often used to achieve several voltages by either stepping up or stepping down the primary voltage. These types are commonly used in power supply applications in which several voltage levels are required for the operation of an electronic instrument.

A typical schematic of a multiple-secondary transformer is shown in Figure 12–34; this transformer has three secondaries. Sometimes you will find combinations of multiple-primary, multiple-secondary, and tapped transformers all in one unit.

EXAMPLE 12–10 The transformer shown in Figure 12–35 has the numbers of turns (denoted as T) indicated. One of the secondaries is also center-tapped. If 120 V ac are connected to the primary, determine each secondary voltage and the voltages with respect to CT on the middle secondary.

FIGURE 12–35

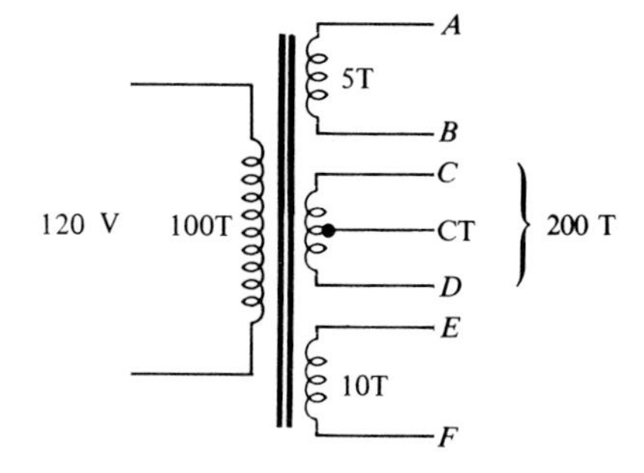

Solution

$$V_{AB} = \left(\frac{5}{100}\right)120\text{ V} = 6\text{ V}$$

$$V_{CD} = \left(\frac{200}{100}\right)120\text{ V} = 240\text{ V}$$

$$V_{CTC} = V_{CTD} = \frac{240\text{ V}}{2} = 120\text{ V}$$

$$V_{EF} = \left(\frac{10}{100}\right)120\text{ V} = 12\text{ V}$$

Autotransformers

In an autotransformer, *one winding* serves as both the primary and the secondary. The winding is tapped at the proper points to achieve the desired turns ratio for stepping up or stepping down the voltage.

Autotransformers differ from conventional transformers in that there is no electrical isolation between the primary and the secondary because both are on

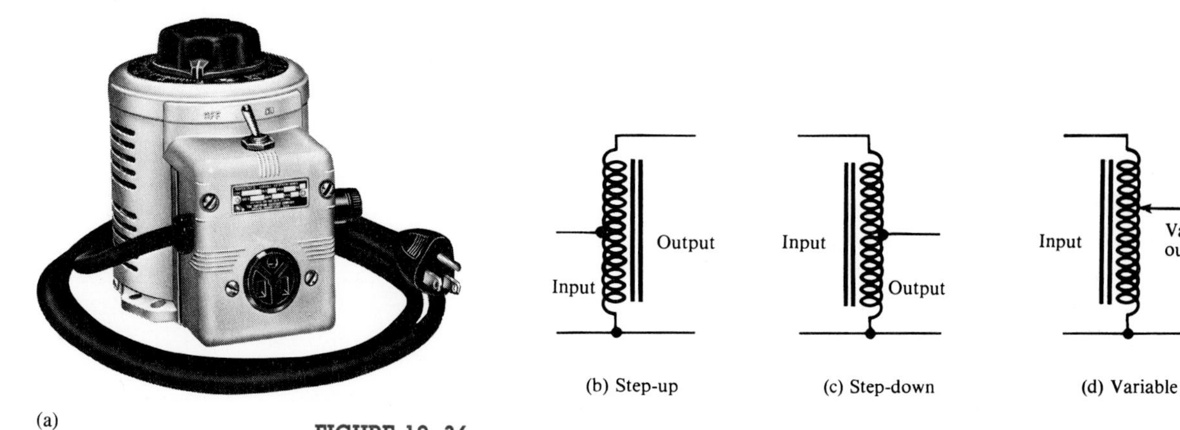

FIGURE 12–36
The autotransformer (courtesy of Superior Electric Co.).

one winding. Autotransformers normally are smaller and lighter than equivalent conventional transformers because they require a much lower kVA rating for a given load. Many autotransformers provide an adjustable tap using a sliding contact mechanism so that the output voltage can be varied (these are often called *variacs*). Figure 12–36 shows a typical autotransformer and several schematic symbols.

The following example illustrates why an autotransformer has a kVA requirement that is less than the input or output kVA.

EXAMPLE 12–11 A certain autotransformer is used to change a source voltage of 240 V to a load voltage of 160 V across an 8-Ω load resistance. Determine the input and output power in kVA, and show that the actual kVA requirement is less than this value. Assume that this transformer is ideal.

Solution

The circuit is shown in Figure 12–37 with the voltages and currents indicated. The current directions have been assigned arbitrarily for convenience.

FIGURE 12–37

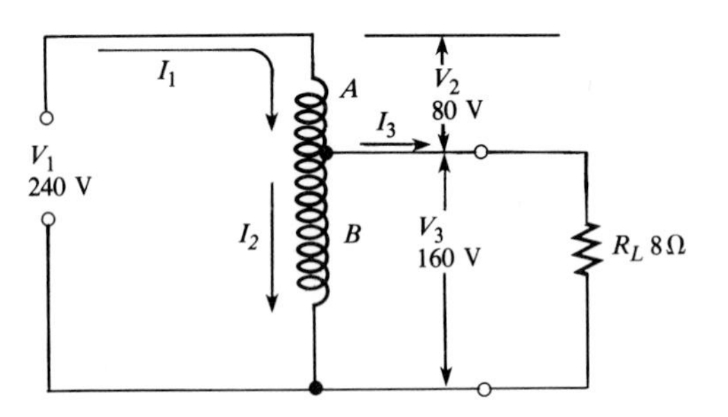

The load current I_3 is determined as follows:

$$I_3 = \frac{V_3}{R_L} = \frac{160\text{ V}}{8\ \Omega} = 20\text{ A}$$

The input power is the total source voltage (V_1) times the total current from the source (I_1):

$$P_{\text{in}} = V_1 I_1$$

The output power is the load voltage V_3 times the load current I_3:

$$P_{\text{out}} = V_3 I_3$$

For an ideal transformer, $P_{\text{in}} = \text{P}_{\text{out}}$; thus,

$$V_1 I_1 = V_3 I_3$$

Solving for I_1 yields

$$I_1 = \frac{V_3 I_3}{V_1} = \frac{(160\text{ V})(20\text{ A})}{240\text{ V}} = 13.33\text{ A}$$

Applying Kirchhoff's current law at the tap junction, we get

$$I_1 = I_2 + I_3$$

Solving for I_2 yields

$$I_2 = I_1 - I_3 = 13.33 \text{ A} - 20 \text{ A} = -6.67 \text{ A}$$

The minus sign can be dropped because the current directions are arbitrary. The input and output power are

$$P_{in} = P_{out} = V_3 I_3 = (160 \text{ V})(20 \text{ A}) = 3.2 \text{ kVA}$$

The power in winding A is

$$V_2 I_1 = (80 \text{ V})(13.33 \text{ A}) = 1.07 \text{ kVA}$$

The power in winding B is

$$V_3 I_2 = (160 \text{ V})(6.67 \text{ A}) = 1.07 \text{ kVA}$$

Thus, the power rating required for each winding is less than the power that is delivered to the load.

SECTION REVIEW 12–8

1. A certain transformer has two secondaries. The turns ratio from the primary to the first secondary is 10. The turns ratio from the primary to the other secondary is 0.2. If 240 V ac are applied to the primary, what are the secondary voltages?
2. Name one advantage and one disadvantage of an autotransformer over a conventional transformer.

TROUBLES IN TRANSFORMERS

12–9

The common failures in transformers are opens, shorts, or partial shorts in either the primary or the secondary windings. One cause of such failures is the operation of the device under conditions that exceed its ratings. A few transformer failures and the typical symptoms are discussed in this section.

Open Primary Winding

When there is an open primary winding, there is no primary current and, therefore, no induced voltage or current in the secondary. This condition is illustrated in Figure 12–38(a), and the method of checking with an ohmmeter is shown in Part (b).

Open Secondary Winding

When there is an open secondary winding, there is no current in the secondary and, as a result, no voltage across the load. Also, an open secondary causes the primary current to be very small (only a small magnetizing current flows). In fact, the primary current may be practically zero. This condition is illustrated in Figure 12–39(a), and the ohmmeter check is shown in Part (b).

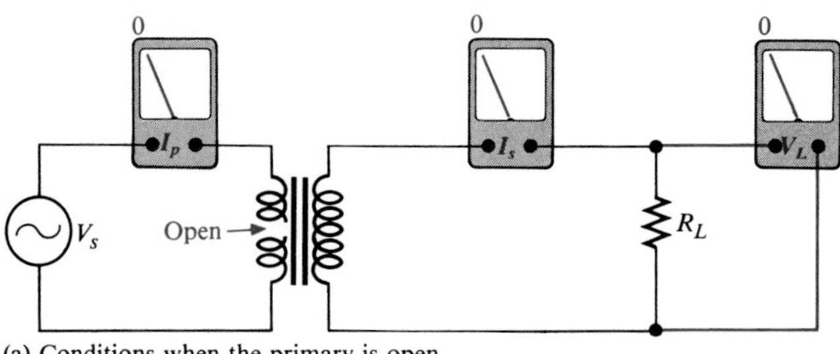

(a) Conditions when the primary is open

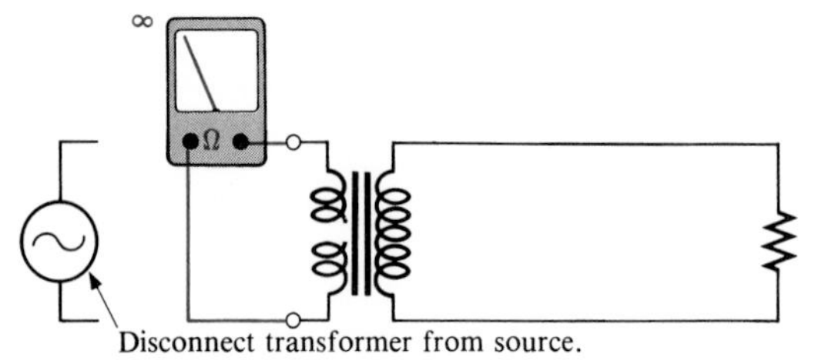

(b) Checking the primary with an ohmmeter

FIGURE 12–38
Open primary winding.

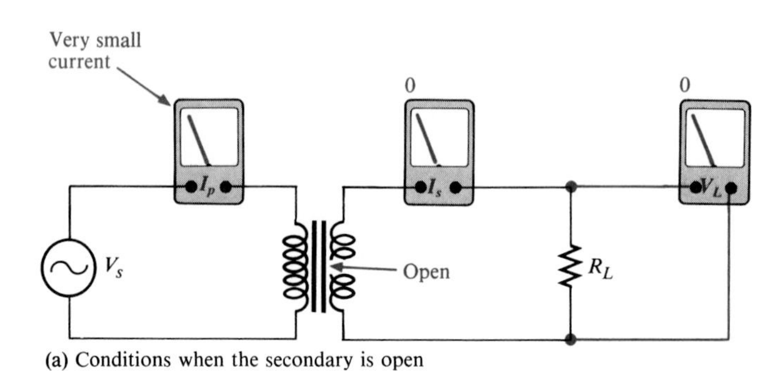

(a) Conditions when the secondary is open

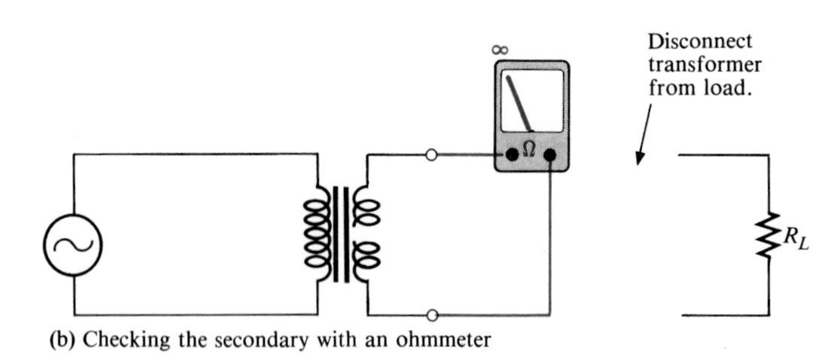

(b) Checking the secondary with an ohmmeter

FIGURE 12–39
Open secondary winding.

Shorted or Partially Shorted Secondary

In this case, there is an excessive primary current because of the low reflected impedance due to the short. Often, this excessive current will burn out the primary and result in an open. The short-circuit current in the secondary causes the load current to be zero (full short) or smaller than normal (partial short), as demonstrated in Figures 12–40(a) and 12–40(b). The ohmmeter check for this condition is shown in Part (c).

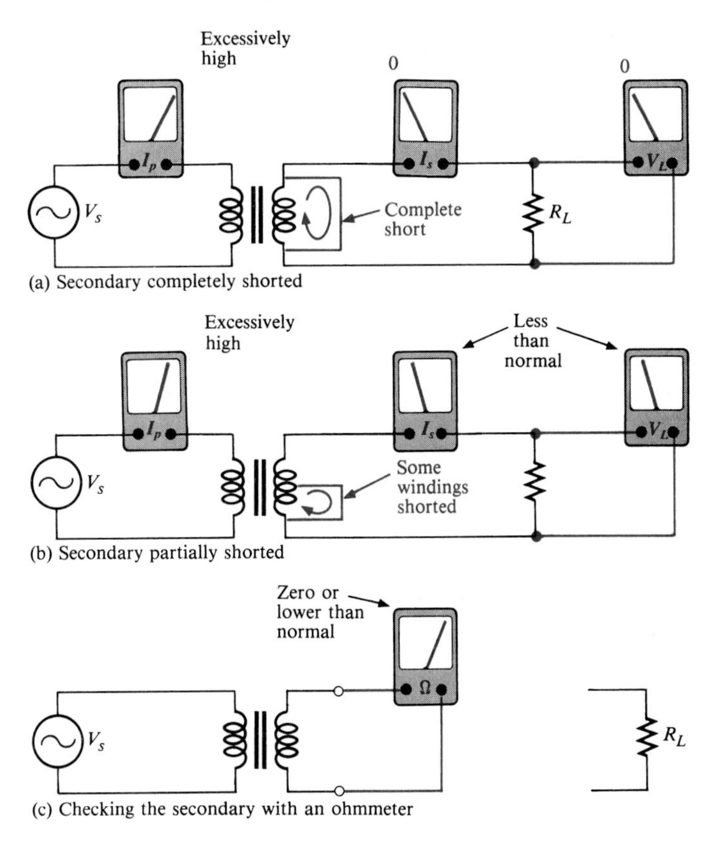

FIGURE 12–40
Shorted secondary winding.

Normally, when a transformer fails, it is very difficult to repair, and therefore the simplest procedure is to replace it.

SECTION REVIEW 12–9

1. Name two possible failures in a transformer.
2. What is often the cause of transformer failure?

APPLICATION NOTE

For the Application Assignment at the beginning of the chapter, you have four 8-Ω speakers in parallel, creating a total load resistance of 2 Ω. The 50-Ω output amplifier must be matched to the 2-Ω load. The matching transformer must be a step-down transformer with a turns ratio of

$$n = \sqrt{\frac{R_L}{R_{ref}}} = \sqrt{\frac{2\ \Omega}{50\ \Omega}} = 0.2$$

SUMMARY

Facts

- A transformer consists of two or more coils that are magnetically coupled on a common core.
- There is mutual inductance between two magnetically coupled coils.
- When current in one coil changes, voltage is induced in the other coil.
- The primary is the winding connected to the source, and the secondary is the winding connected to the load.
- The number of turns in the primary and the number of turns in the secondary determine the *turns ratio.*
- The relative polarities of the primary and secondary voltages are determined by the direction of the windings around the core.
- A step-up transformer has a turns ratio greater than 1.
- A step-down transformer has a turns ratio less than 1.
- A transformer cannot increase power.
- In an ideal transformer, the power from the source (input power) is equal to the power delivered to the load (output power).
- If the voltage is stepped up, the current is stepped down, and vice versa.
- A load in the secondary of a transformer appears to the source as a *reflected* load having a value dependent on the reciprocal of the turns ratio squared.
- A transformer can match a load resistance to a source resistance to achieve maximum power transfer to the load by selecting the proper turns ratio.
- A transformer does not respond to dc.
- Energy losses in an actual transformer result from winding resistances, hysteresis loss in the core, eddy currents in the core, and flux leakage.

Definitions

- *Mutual inductance*—the amount of inductance between two coils.
- *Turns ratio*—the number of secondary turns over the number of primary turns.
- *Core*—the physical structure around which the windings are placed.
- *Step-up transformer*—a transformer in which the secondary voltage is greater than the primary voltage.
- *Step-down transformer*—a transformer in which the secondary voltage is less than the primary voltage.
- *Reflected load*—the load as it appears to the source in the primary.
- *Maximum power transfer theorem*—a theorem stating that maximum power is transferred from a source to a load when the load resistance equals the source resistance.
- *Apparent power*—the product of the voltage times the current, expressed in volt-amperes (VA). In a purely resistive circuit it is the same as the true power (watts).
- *Autotransformer*—a transformer in which the primary and secondary are in a single winding.

Formulas

$$n = \frac{N_s}{N_p} \quad \text{Turns ratio} \qquad \textbf{(12–1)}$$

$$V_s = \left(\frac{N_s}{N_p}\right)V_p \quad \text{Secondary voltage} \qquad \textbf{(12–2)}$$

$$I_s = \left(\frac{N_p}{N_s}\right)I_p \quad \text{Secondary current} \tag{12–3}$$

$$R_{\text{ref}} = \left(\frac{1}{n}\right)^2 R_L \quad \text{Reflected resistance} \tag{12–4}$$

$$\eta = \left(\frac{P_{\text{out}}}{P_{\text{in}}}\right)100\% \quad \text{Transformer efficiency} \tag{12–5}$$

Symbols

Transformer symbols are shown in Figure 12–41.

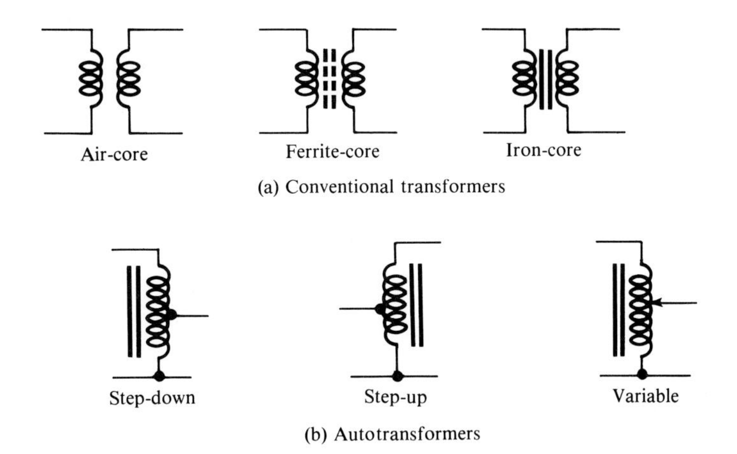

FIGURE 12–41
Transformer symbols.

SELF-TEST

1. Explain the basic operation of a transformer.
2. How does the turns ratio affect the operation of a transformer?
3. If the windings of a certain transformer are in opposite directions around the core, is the secondary voltage in phase or out of phase with the primary voltage?
4. Name two functions of a transformer's core.
5. The turns ratio of a certain transformer is 10. Is the secondary voltage greater than or less than the primary voltage?
6. The turns ratio of a transformer is 0.5. Is it a step-up or a step-down type? If the primary voltage is 100 V, what is the secondary voltage?
7. Suppose that 10 W of power are applied to the primary of an ideal step-up transformer that has a turns ratio of 5. What is the power delivered to the load?
8. For an ideal step-down transformer, the secondary voltage is one-third the primary voltage. What is the relationship between the primary and secondary currents?
9. A 1000-Ω load resistor is connected across the secondary winding of a transformer with a turns ratio of 1. What load value does the source see?
10. In a transformer with a turns ratio of 2, is the reflected resistance greater than or less than the actual load resistance?
11. How is a transformer used to match a load resistance to a source resistance?
12. When is maximum power transferred from a source to a load?

13. A 12-V dc battery is connected across the primary of a transformer with a turns ratio of 4. What is the secondary voltage?
14. Why is the output voltage of an actual transformer usually less than what you expect based on the turns ratio?
15. The coefficient of coupling of a certain transformer is 0.95. What does this mean?

PROBLEM SET A

Section 12–1

12–1 What is the turns ratio of a transformer having 12 primary turns and 36 secondary turns?

12–2 What is the turns ratio of a transformer having 250 primary turns and 1000 secondary turns? What is the turns ratio when the primary has 400 turns and the secondary has 100 turns?

12–3 Determine the polarity of the secondary voltage for each transformer in Figure 12–42.

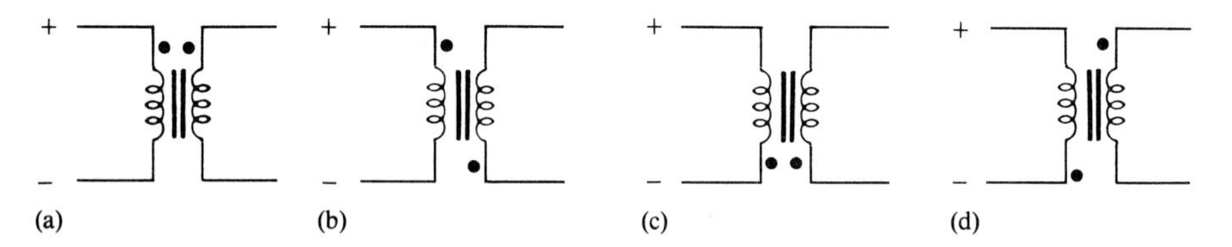

FIGURE 12–42

Section 12–2

12–4 If 120 V ac are connected to the primary of a transformer with 10 primary turns and 15 secondary turns, what is the secondary voltage?

12–5 A certain transformer has 25 turns in its primary. In order to double the voltage, how many turns must be in the secondary?

12–6 How many primary volts must be applied to a transformer with a turns ratio of 10 to obtain a secondary voltage of 60 V ac?

12–7 For each transformer in Figure 12–43, sketch the secondary voltage showing its relationship to the primary voltage. Also indicate the amplitude.

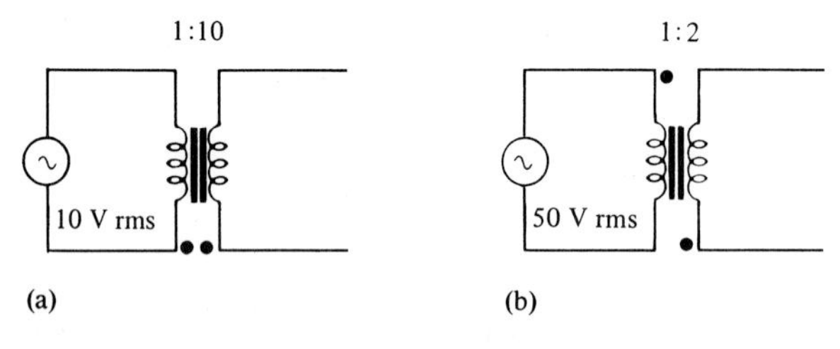

FIGURE 12–43

Section 12–3

12–8 To step 120 V down to 30 V, what must the turns ratio be?

12–9 The primary of a transformer has 1200 V across it. What is the secondary voltage if the turns ratio is 0.2?

12–10 How many primary volts must be applied to a transformer with a turns ratio of 0.1 to obtain a secondary voltage of 6 V ac?

Section 12–4

12–11 Determine I_s in Figure 12–44.

FIGURE 12–44

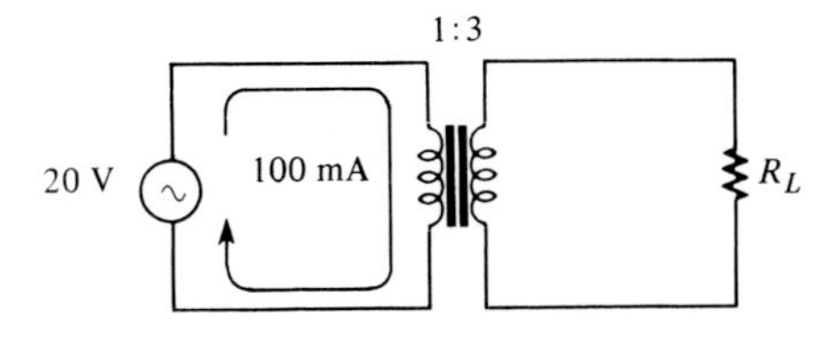

12–12 Determine the following quantities in Figure 12–45:
(a) primary current. **(b)** secondary current.
(c) secondary voltage. **(d)** power in the load.

FIGURE 12–45

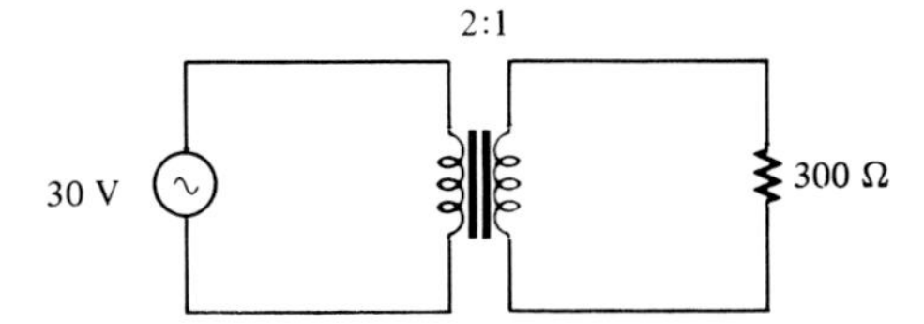

Section 12–5

12–13 What is the load resistance as seen by the source in Figure 12–46?

FIGURE 12–46

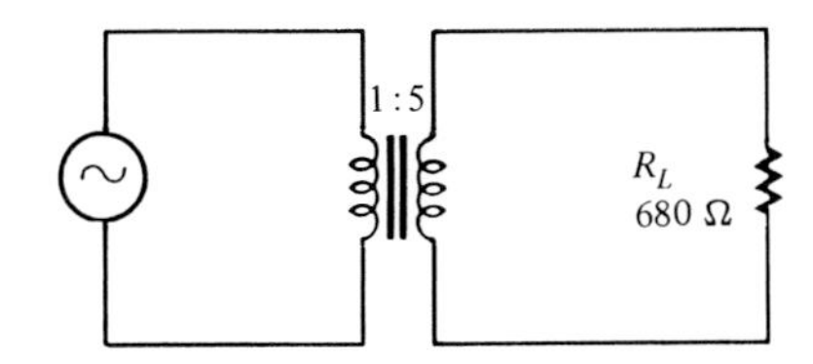

12–14 What must the turns ratio be in Figure 12–47 in order to reflect 300 Ω into the primary?

FIGURE 12–47

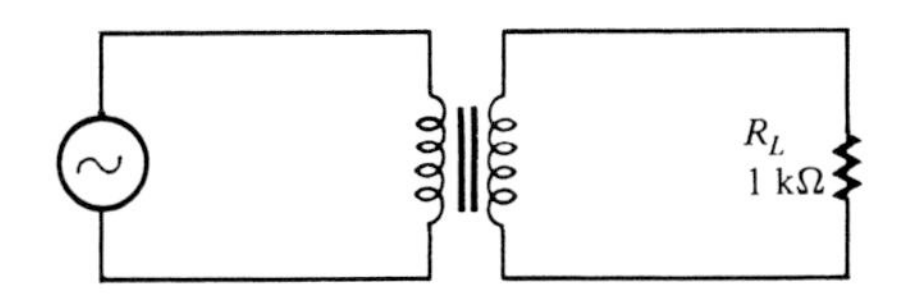

12–15 For the circuit in Figure 12–48, find the turns ratio required to deliver maximum power to the 4-Ω speaker.

FIGURE 12–48

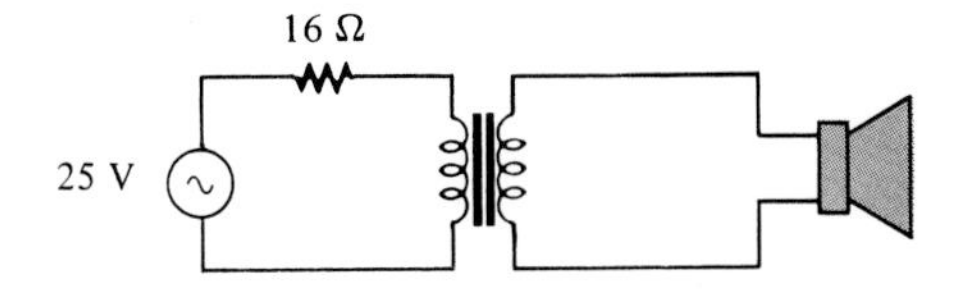

12–16 In Figure 12–48, what is the maximum power delivered to the speaker?

Section 12–6

12–17 What is the voltage across the load in each circuit of Figure 12–49?

12–18 If the bottom of each secondary in Figure 12–49 were grounded, would the values of the load voltages be changed?

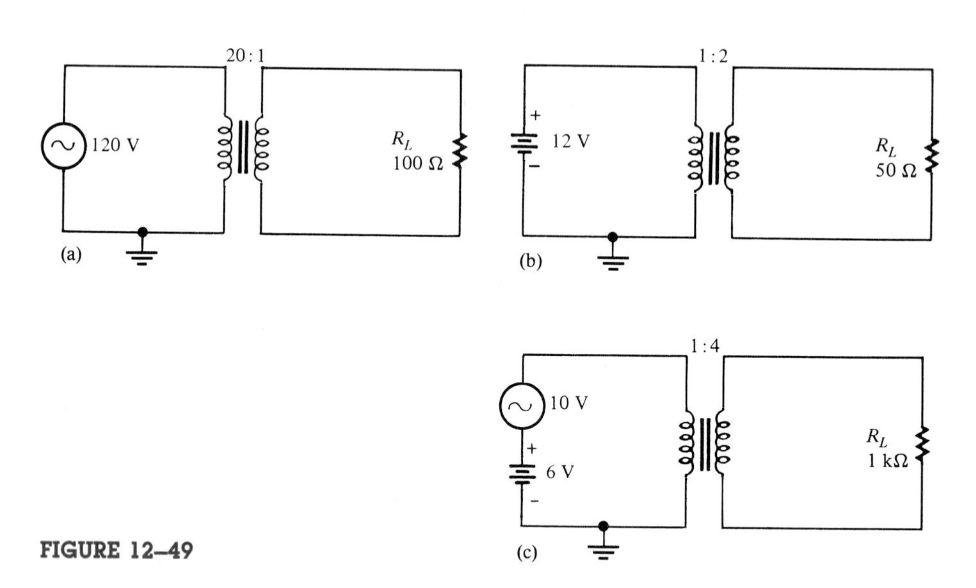

FIGURE 12–49

Section 12–7

12–19 In a certain transformer, the input power to the primary is 100 W. If 5.5 W are lost in the winding resistances, what is the output power to the load, neglecting any other losses?

12–20 What is the efficiency of the transformer in Problem 12–19?

12–21 Determine the coefficient of coupling for a transformer in which 2% of the total flux generated in the primary does not pass through the secondary.

12–22 A certain transformer is rated at 1 kVA. It operates on 60 Hz, 120 V ac. The secondary voltage is 600 V.

(a) What is the maximum load current?
(b) What is the smallest R_L that you can drive?
(c) What is the largest capacitor that can be connected as a load?

12–23 What kVA rating is required for a transformer that must handle a maximum load current of 10 A with a secondary voltage of 2.5 kV?

Section 12–8

12–24 Determine each unknown voltage indicated in Figure 12–50.

FIGURE 12–50

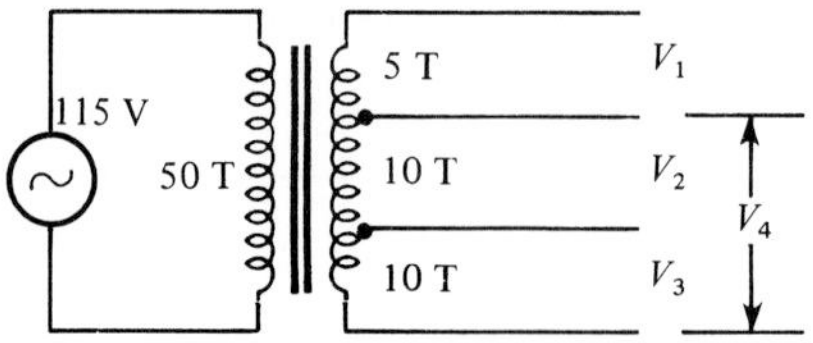

12–25 Using the indicated secondary voltages in Figure 12–51, determine the turns ratio of the primary to each tapped section.

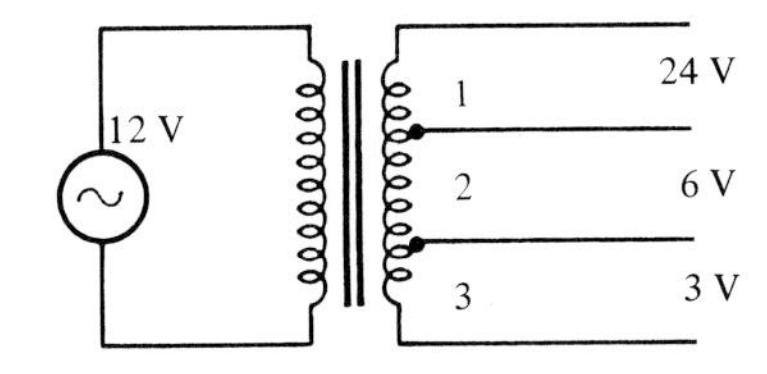

FIGURE 12–51

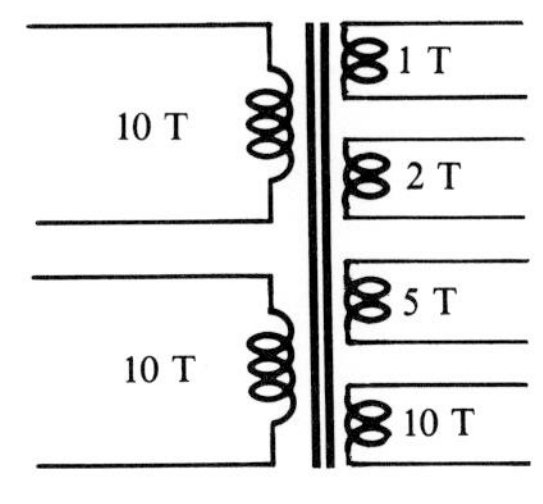

FIGURE 12–52

12–26 In Figure 12–52, each primary can accommodate 120 V ac. How should the primaries be connected for 240-V ac operation? Determine each secondary voltage.

12–27 Find the secondary voltage for each autotransformer in Figure 12–53.

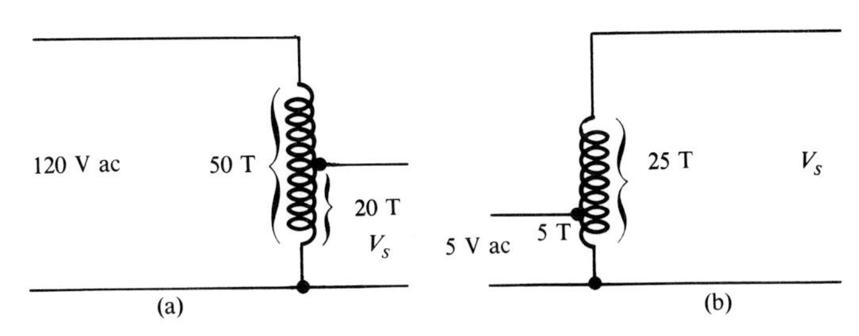

FIGURE 12–53

PROBLEM SET B

12–28 For the loaded, tapped-secondary transformer in Figure 12–54, determine the following:

(a) all load voltages and currents.
(b) the impedance looking into the primary.

FIGURE 12–54

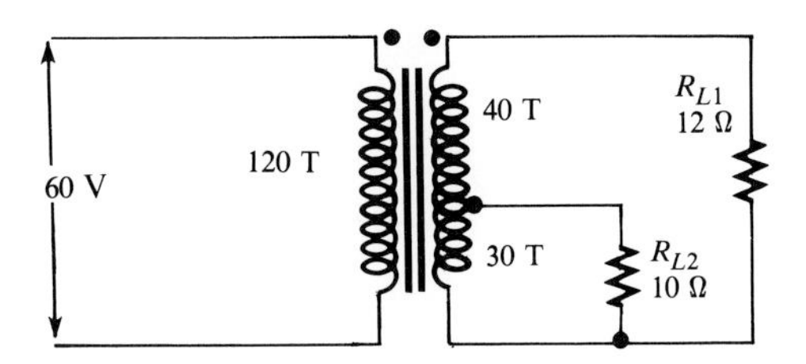

12–29 A certain transformer is rated at 5 kVA, 2400/120 V, at 60 Hz.

(a) What is the turns ratio if the 120 V is the secondary voltage?
(b) What is the current rating of the secondary if 2400 V is the primary voltage?
(c) What is the current rating of the primary if 2400 V is the primary voltage?

12–30 Determine the voltage measured by each voltmeter in Figure 12–55. The bench-type meters have one terminal that connects to ground as indicated.

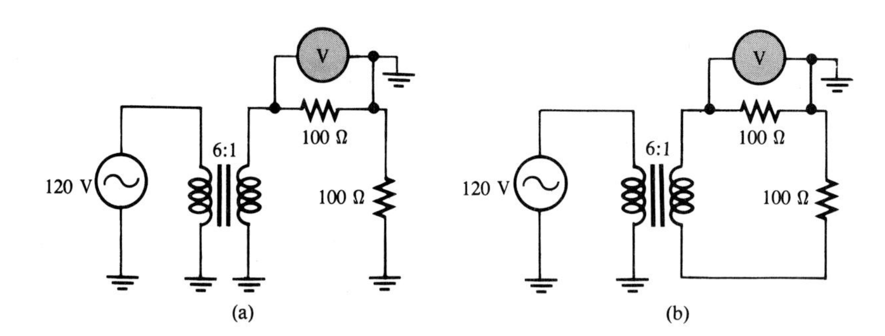

FIGURE 12–55

ANSWERS TO SECTION REVIEWS

Section 12–1

1. Mutual inductance.
2. The inductance between two coils, established by the amount of coupling between the coils.
3. The ratio of secondary turns to primary turns.
4. The directions determine the relative polarities of the voltages.

Section 12–2

1. Increases primary voltage. 2. Five times greater. 3. 2400 V.

Section 12–3

1. Decreases primary voltage. 2. 60 V. 3. 0.1.

Section 12–4

1. Less, half. 2. 2 A.

Section 12–5

1. The impedance in the secondary reflected into the primary.
2. The turns ratio. 3. 0.5 Ω.
4. Making the load impedance equal the source impedance.
5. Maximum power is delivered to the load. 6. 400 Ω.

Section 12–6

1. There is no electrical connection between primary and secondary. 2. No.

Section 12–7

1. In an actual transformer, energy loss reduces the efficiency. An ideal model has an efficiency of 100%.
2. 85% of the magnetic flux generated in the primary passes through the secondary.
3. 40 A.

Section 12–8

1. 2400 V, 48 V. 2. Smaller and lighter for same rating, no electrical isolation.

Section 12–9

1. Open primary, open secondary. 2. Operating above rated values.

Frequency Response of *RC* Circuits

An *RC* circuit contains both resistance and capacitance. It is one of the basic types of *reactive* circuits that we will study. In this chapter, basic series and parallel *RC* circuits and their responses to sine wave voltages are covered. Series-parallel combinations are also examined. Power considerations in *RC* circuits are introduced, and practical aspects of power ratings are discussed. Several basic areas of application are presented to give you an idea of how simple combinations of resistors and capacitors can be applied.

In this chapter you will learn:

- How an *RC* circuit responds to a sine wave input voltage.
- What *impedance* is.
- How to determine the impedance of both series and parallel *RC* circuits.
- Why there is a phase difference between current and voltage in an *RC* circuit, and how to determine its value.
- How to convert a parallel circuit to an equivalent series circuit.
- How to determine the effects of frequency on an *RC* circuit in terms of changes in impedance and phase angle.
- Definitions of *conductance, capacitive susceptance*, and *admittance.*
- How to apply Ohm's law to *RC* circuits to find currents and voltages.
- How to determine the true power, reactive power, and apparent power.
- The meaning and significance of the *power factor.*
- How *RC* circuits can be used as phase-shifting circuits (lead and lag networks).
- How *RC* circuits can be used as frequency-selective circuits (filters).
- How *RC* circuits are used in amplifiers for signal coupling.

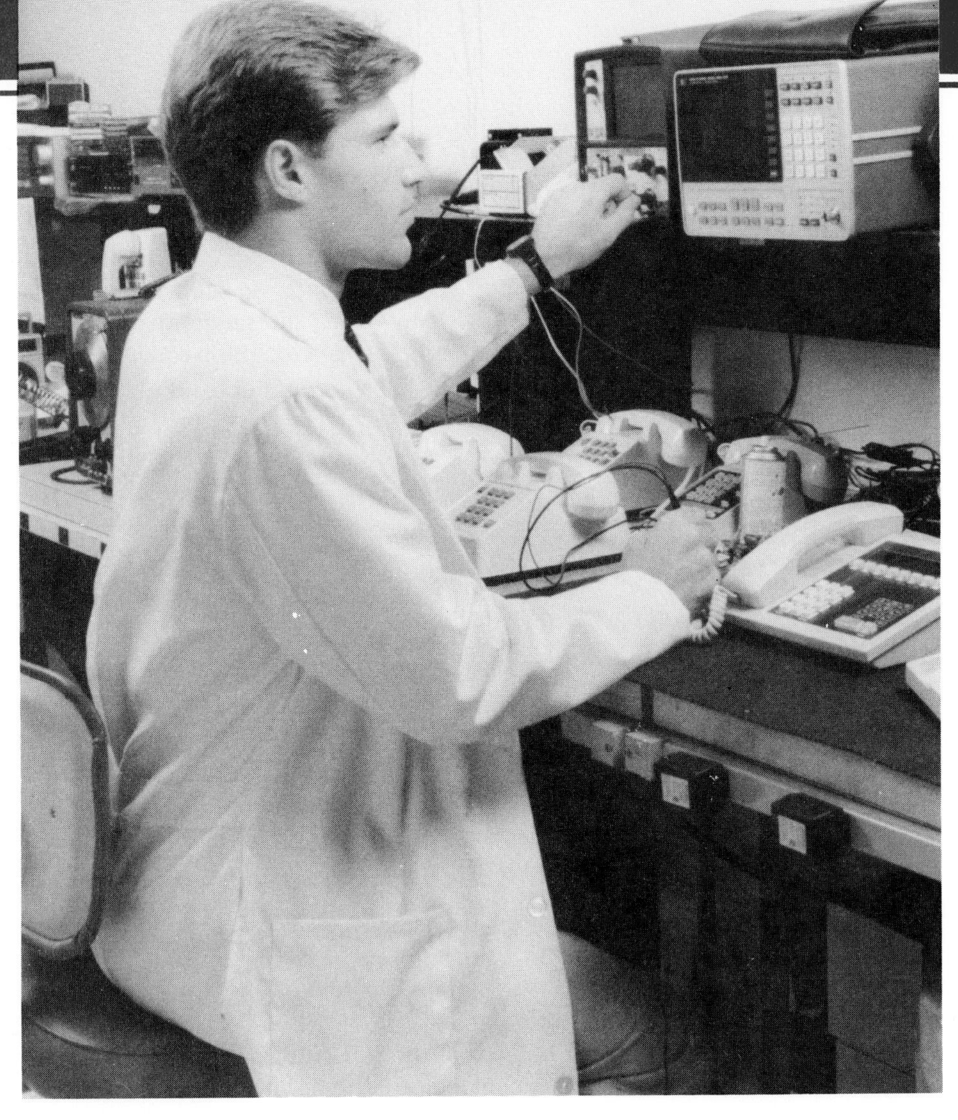

APPLICATION ASSIGNMENT

In a certain communications system, a section of audio amplifiers is being upgraded to accommodate a wider bandwidth so that a lower range of audio frequencies can be amplified without significant loss due to coupling capacitors. As lead technician, you have been asked to examine the coupling networks between each amplifier stage to ensure that there is no significant signal loss at the lower frequencies. If there is, you are to make the appropriate modifications.

After studying this chapter, you will be able to accomplish this assigned task. The solution is given in the Application Note at the end of the chapter.

SINUSOIDAL RESPONSE OF *RC* CIRCUITS

13–1

When a sinusoidal voltage is applied to an *RC* circuit, *each resulting voltage drop and the current in the circuit is also sinusoidal with the same frequency as that of the applied voltage.* As shown in Figure 13–1, the resistor voltage, the capacitor voltage, and the current are all sine waves with the frequency of the source.

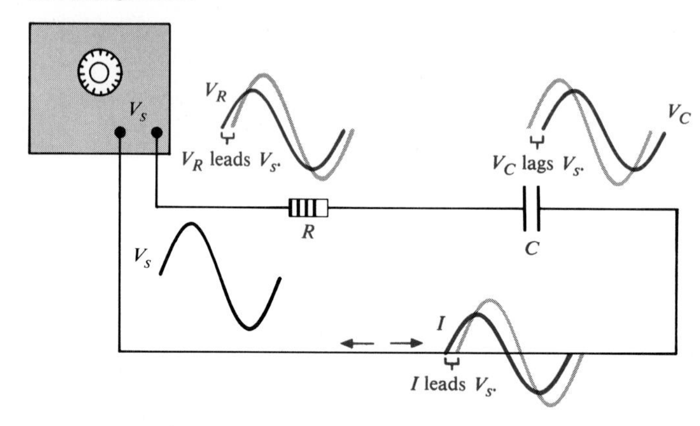

FIGURE 13–1
Illustration of sinusoidal response with general phase relationships of V_R, V_C, and I relative to the source voltage. V_R and I are in phase; V_R leads V_s; V_C lags V_s; and V_R and V_C are 90° out of phase.

Phase shifts are introduced because of the capacitance. As you will learn, the resistor voltage and current are in phase with each other and lead the source voltage in phase. The capacitor voltage lags the source voltage. The phase angle between the current and the capacitor voltage is always 90°. These generalized phase relationships are indicated in the figure.

The amplitudes and the phase relationships of the voltages and current depend on the ohmic values of the resistance and the capacitive reactance (defined in Section 10–6). When a circuit is purely resistive, the phase angle between the applied (source) voltage and the total current is zero. When a circuit is purely capacitive, the phase angle between the applied voltage and the total current is 90°, with the current *leading* the voltage. When there is a combination of both resistance and capacitive reactance in a circuit, the phase angle between the applied voltage and the total current is somewhere between zero and 90°, depending on the relative values of the resistance and the reactance.

Signal Generators

When a circuit is hooked up for a laboratory experiment or for troubleshooting, a signal generator similar to those shown in Figure 13–2 is used to provide the source voltage. These instruments, depending on their capability, are classified as *sine wave generators,* which produce only sine waves; *sine/square generators,*

(a)

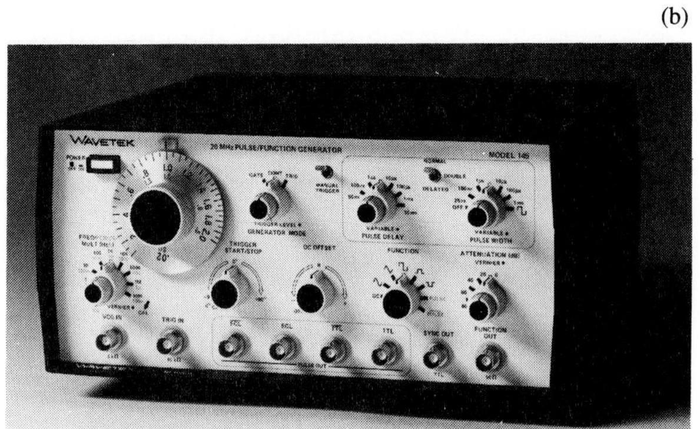

(b)

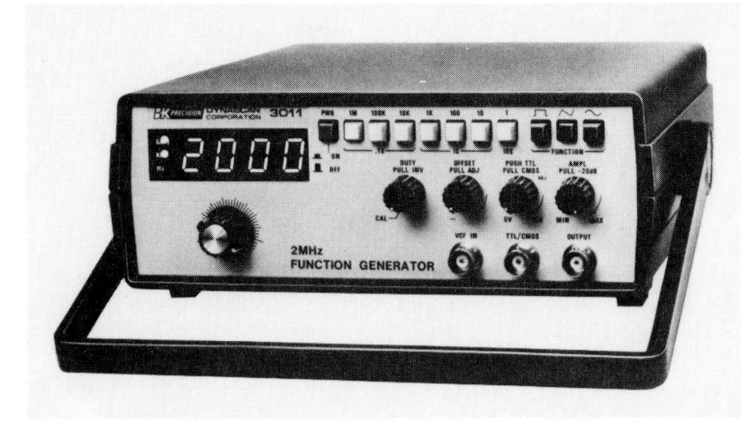

(c)

FIGURE 13–2
Typical signal (function) generators used in circuit testing and troubleshooting [(a) courtesy of Hewlett-Packard Company. (b) courtesy of Wavetek. (c) courtesy of B&K Precision/Dynascan Corp.].

which produce both sine waves and square waves; or *function generators,* which produce sine waves, pulse waveforms, and triangular (ramp) waveforms.

SECTION REVIEW 13–1

1. A 60-Hz sinusoidal voltage is applied to an *RC* circuit. What is the frequency of the capacitor voltage? The current?
2. When the resistance in an *RC* circuit is greater than the capacitive reactance, is the phase angle between the applied voltage and the total current closer to 0° or to 90°?

13–2 IMPEDANCE AND PHASE ANGLE OF A SERIES *RC* CIRCUIT

Impedance is the total opposition to sinusoidal current and is expressed in *ohms.* In a purely resistive circuit, the impedance is simply equal to the total resistance. In a purely capacitive circuit, the impedance is the total capacitive reactance. The impedance of a series RC circuit is determined by *both the resistance and the capacitive reactance.* These cases are illustrated in Figure 13–3. The magnitude of the impedance is symbolized by Z.

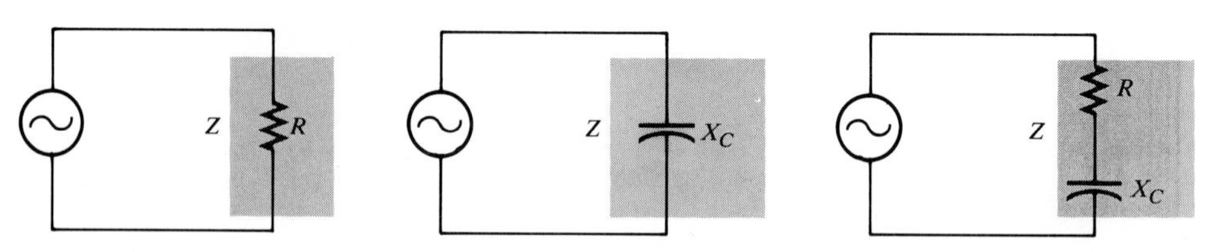

FIGURE 13–3
Three cases of impedance.

The Impedance Triangle

In ac analysis, both R and X_C are treated as *phasor quantities,* as shown in the phasor diagram of Figure 13–4(a), with X_C appearing at a $-90°$ angle with respect to R. This relationship comes from the fact that the capacitor voltage in a series RC circuit *lags* the current, and thus the resistor voltage, by 90°.

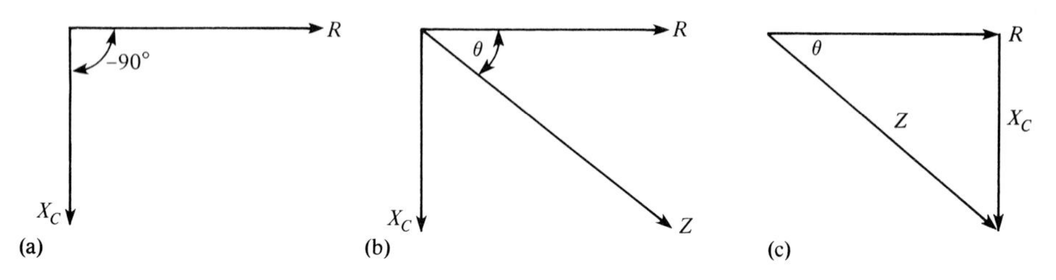

FIGURE 13–4
Development of the impedance triangle for a series RC circuit.

Since Z is the phasor sum of R and X_C, its phasor representation is as shown in Figure 13–4(b). A repositioning of the phasors, as shown in Part (c), forms a right triangle, called the *impedance triangle.* The length of each phasor represents the *magnitude* in ohms, and the angle θ (the Greek letter theta) is the phase angle of the RC circuit and represents the phase difference between the applied (source) voltage and the current.

From right-angle trigonometry (Pythagorean theorem), the magnitude of the impedance can be expressed in terms of the resistance and reactance as

$$Z = \sqrt{R^2 + X_C^2} \qquad \textbf{(13–1)}$$

This is the *magnitude* of Z, as shown in Figure 13–5, and is expressed in ohms.

The phase angle, θ, is expressed as

$$\theta = \arctan\left(\frac{X_C}{R}\right) \qquad \textbf{(13–2)}$$

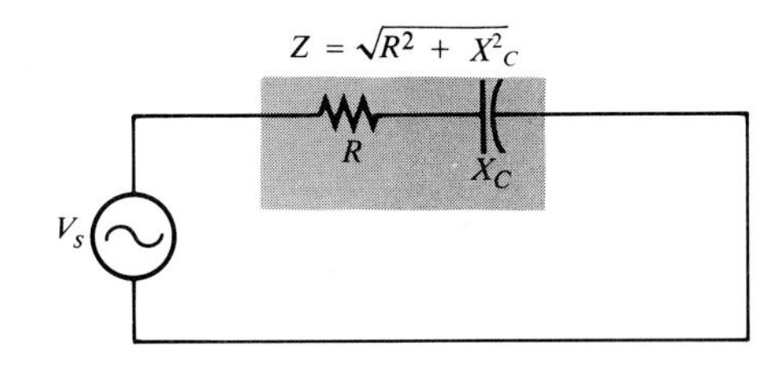

FIGURE 13–5
Impedance of a series *RC* circuit.

Arctan can be found on most calculators by pressing INV and then tan. The symbol $\tan^{-1}$ also means the same thing as arctan and may replace the two key strokes (INV and tan) on some calculators.

EXAMPLE 13–1

Determine the impedance and the phase angle of the circuit in Figure 13–6.

FIGURE 13–6

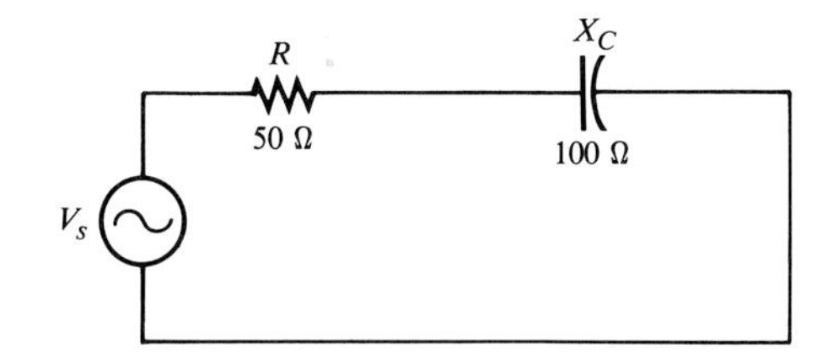

Solution
The impedance is

$$Z = \sqrt{R^2 + X_C^2} = \sqrt{(50\ \Omega)^2 + (100\ \Omega)^2} = 111.8\ \Omega$$

The phase angle is

$$\theta = \arctan\left(\frac{X_C}{R}\right) = \arctan\left(\frac{100}{50}\right) = \arctan(2) = 63.4°$$

The applied voltage *lags* the current by 63.4°.

The calculator sequences are as follows:

Z: 5 0 2nd x^2 + 1 0 0 2nd x^2 = $\sqrt{x}$
θ: 1 0 0 ÷ 5 0 = INV tan

SECTION REVIEW 13–2

1. Define *impedance*.
2. Does the source voltage lead or lag the current in a series *RC* circuit?
3. What causes the phase angle in the *RC* circuit?
4. A series *RC* circuit has a resistance of 33 kΩ and a capacitive reactance of 50 kΩ. What is the value of the impedance? What is the phase angle?

ANALYSIS OF SERIES *RC* CIRCUITS

13–3

In the previous section, you learned how to express the impedance and the phase angle of a series *RC* circuit. In this section, Ohm's law and Kirchhoff's voltage law are utilized in the analysis of *RC* circuits.

Ohm's Law

The application of Ohm's law to series *RC* circuits involves the use of the quantities of Z, V, and I. The three equivalent forms of Ohm's law are as follows:

$$V = IZ \tag{13–3}$$

$$I = \frac{V}{Z} \tag{13–4}$$

$$Z = \frac{V}{I} \tag{13–5}$$

The following two examples illustrate the use of Ohm's law.

EXAMPLE 13–2 If the current in Figure 13–7 is 0.2 mA, determine the source voltage and the phase angle.

FIGURE 13–7

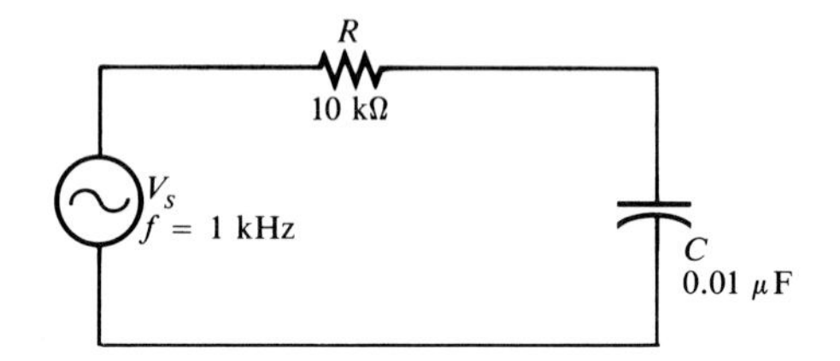

Solution
The capacitive reactance is

$$X_C = \frac{1}{2\pi fC} = \frac{1}{2\pi(1000\text{ Hz})(0.01\ \mu\text{F})} = 15.9\text{ k}\Omega$$

The impedance is

$$Z = \sqrt{R^2 + X_C^2} = \sqrt{(10\text{ k}\Omega)^2 + (15.9\text{ k}\Omega)^2} = 18.78\text{ k}\Omega$$

Applying Ohm's law, we obtain

$$V_s = IZ = (0.2\text{ mA})(18.78\text{ k}\Omega) = 3.76\text{ V}$$

The phase angle is

$$\theta = \arctan\left(\frac{X_C}{R}\right) = \arctan\left(\frac{15.9\text{ k}\Omega}{10\text{ k}\Omega}\right) = 57.83°$$

The source voltage has a magnitude of 3.76 V and lags the current by 57.83°.

The calculator sequence for V_s is

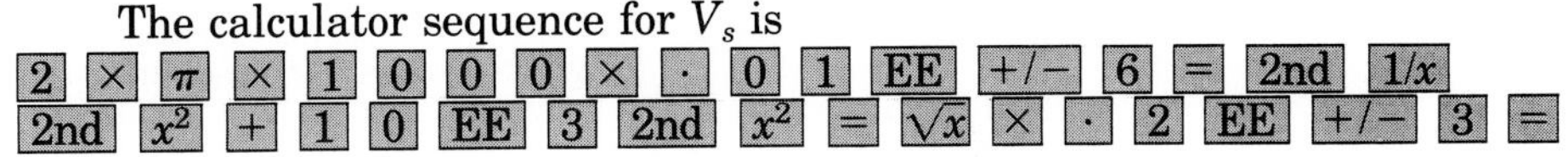

EXAMPLE 13–3 Determine the current in the circuit of Figure 13–8.

FIGURE 13–8

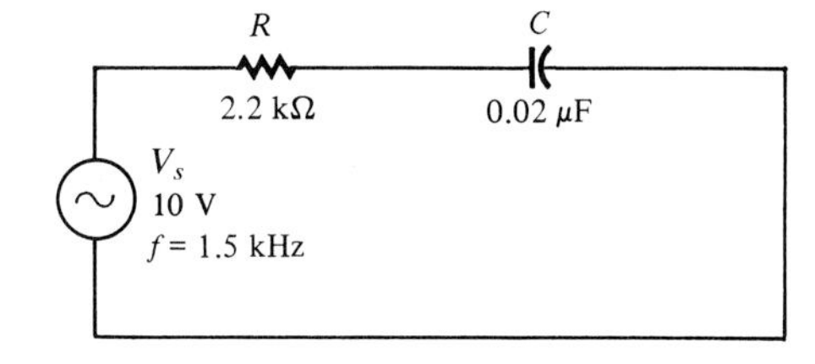

Solution

$$X_C = \frac{1}{2\pi fC} = \frac{1}{2\pi(1.5\text{ kHz})(0.02\ \mu\text{F})} = 5.3\text{ k}\Omega$$

The impedance is

$$Z = \sqrt{R^2 + X_C^2} = \sqrt{(2.2\text{ k}\Omega)^2 + (5.3\text{ k}\Omega)^2}$$
$$= 5.74\text{ k}\Omega$$

Applying Ohm's law, we obtain

$$I = \frac{V}{Z} = \frac{10\text{ V}}{5.74\text{ k}\Omega} = 1.74\text{ mA}$$

Relationships of the Current and Voltages in a Series *RC* Circuit

In a series circuit, the current is the same through both the resistor and the capacitor. Thus, *the resistor voltage is in phase with the current, and the capacitor voltage lags the current by 90°*. Therefore, there is a phase difference of 90° between the resistor voltage, V_R, and the capacitor voltage, V_C, as shown in the waveform diagram of Figure 13–9.

We know from Kirchhoff's voltage law that the sum of the voltage drops must equal the applied voltage. However, since V_R and V_C are not in phase with each other, they must be added as phasor quantities, with V_C lagging V_R by 90°, as shown in Figure 13–10(a). As shown in Part (b), V_s is the phasor sum of V_R and V_C, as expressed in the following equation:

$$V_s = \sqrt{V_R^2 + V_C^2} \tag{13–6}$$

The phase angle between the resistor voltage and the source voltage is

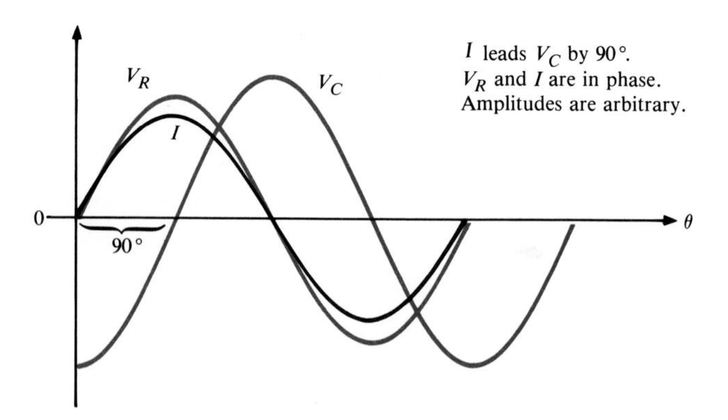

FIGURE 13–9
Phase relation of voltages and current in a series *RC* circuit.

FIGURE 13–10
Voltage phasor diagram for a series *RC* circuit.

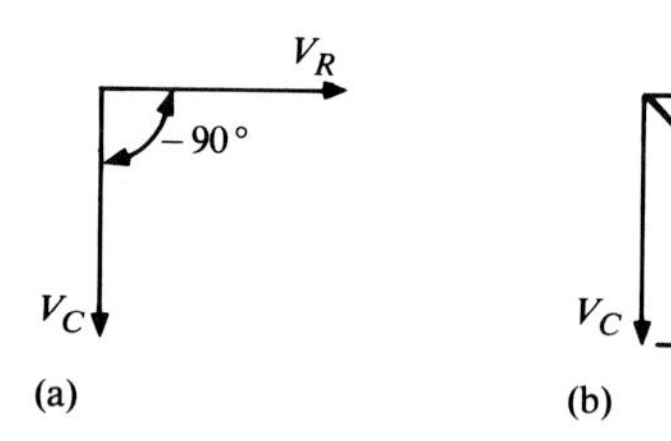

$$\theta = \arctan\left(\frac{V_C}{V_R}\right) \tag{13–7}$$

Since the resistor voltage and the current are in phase, θ also represents the phase angle between the source voltage and the current. Figure 13–11 shows a complete voltage and current phasor diagram representing the waveform diagram of Figure 13–9.

FIGURE 13–11
Voltage and current phasor diagram for the waveforms in Figure 13–9.

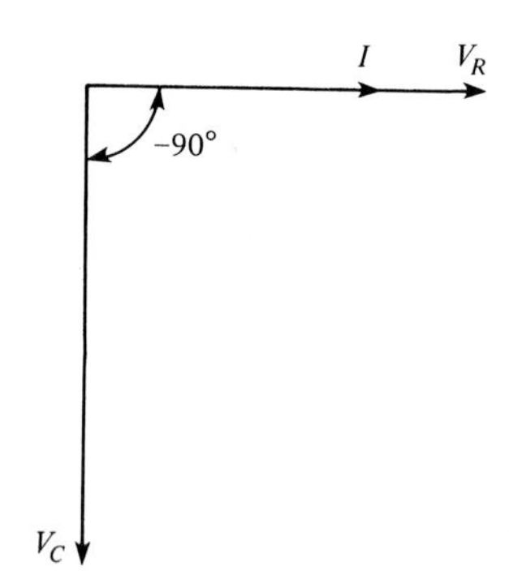

EXAMPLE 13–4 Determine the source voltage and the phase angle in Figure 13–12.

FIGURE 13–12

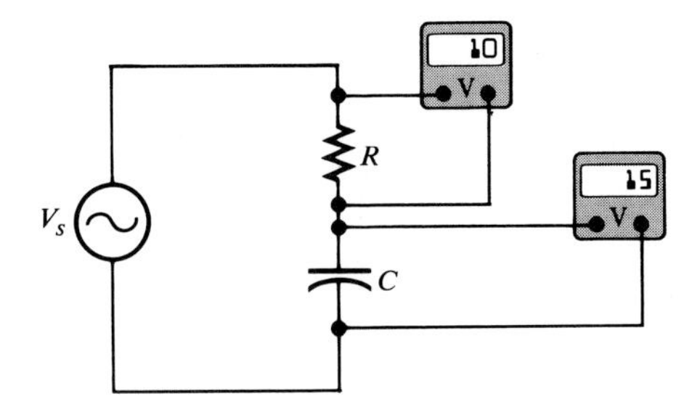

Solution

Since V_R and V_C are 90° out of phase, they cannot be added directly. The source voltage is the *phasor sum* of V_R and V_C:

$$V_s = \sqrt{V_R^2 + V_C^2} = \sqrt{(10\text{ V})^2 + (15\text{ V})^2} = 18\text{ V}$$

The angle between the current and source voltage is

$$\theta = \arctan\left(\frac{V_C}{V_R}\right) = \arctan\left(\frac{15\text{ V}}{10\text{ V}}\right) = 56.3°$$

The calculator sequences are as follows:

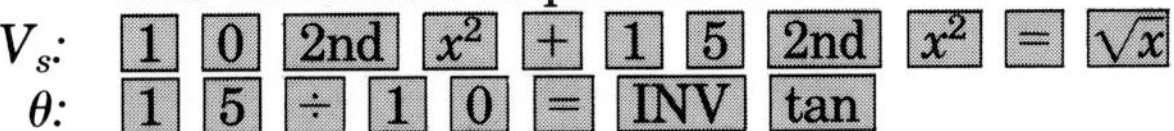

Variation of Voltage and Current Magnitudes with Frequency

As you know, capacitive reactance varies *inversely* with frequency. Since $Z = \sqrt{R^2 + X_C^2}$, you can see that when X_C increases, the entire term under the square root sign increases and thus the total impedance also increases; and when X_C decreases, the total impedance also decreases. Therefore, *in an RC circuit Z is inversely dependent on frequency.*

Figure 13–13 illustrates how the voltages and current in a series *RC* circuit vary as the frequency increases or decreases, with the source voltage held at a constant value. In Part (a), as the frequency is increased, X_C decreases, so less voltage is dropped across the capacitor. Also, Z decreases as X_C decreases, causing the current to increase. An increase in the current causes more voltage across R.

In Part (b) of the figure, as the frequency is decreased X_C increases, so more voltage is dropped across the capacitor. Also, Z increases as X_C increases, causing the current to decrease. A decrease in the current causes less voltage across R.

Changes in Z and X_C can be observed as shown in Figure 13–14. As the frequency increases, the voltage across Z remains constant because V_s is constant. Also, the voltage across C decreases. The increasing current indicates that Z is decreasing. It does so because of the inverse relationship stated in

Ohm's law ($Z = V_Z/I$). The increasing current also indicates that X_C is decreasing ($X_C = V_C/I$). The decrease in V_C corresponds to the decrease in X_C.

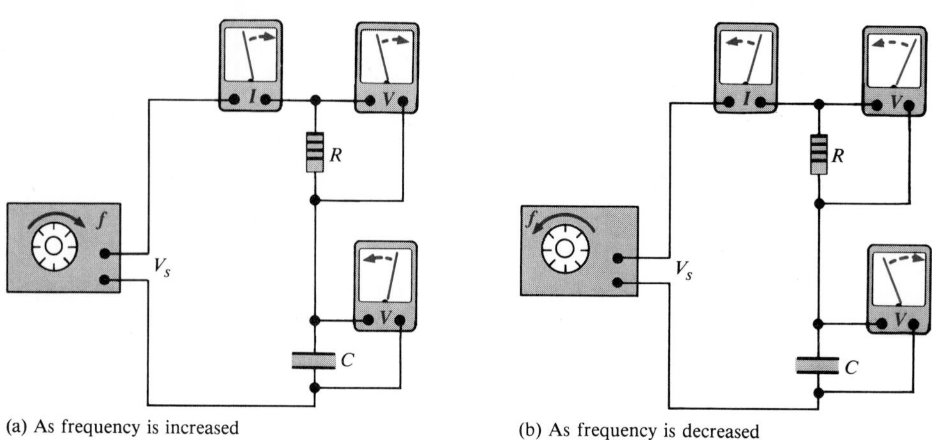

FIGURE 13–13
An illustration of how the voltages and current change as the source frequency is varied. The source voltage is held at a constant amplitude.

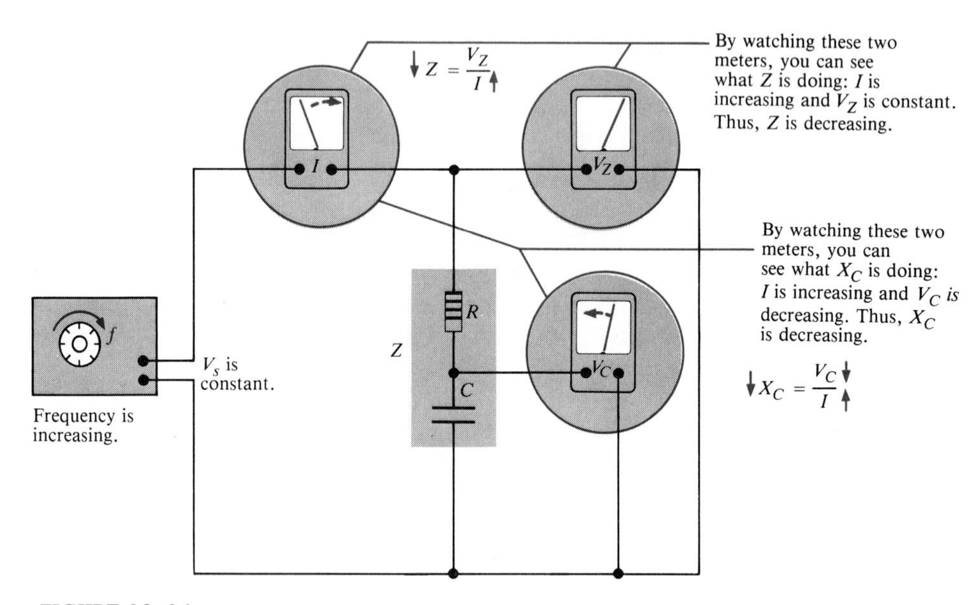

FIGURE 13–14
An illustration of how Z and X_C change with frequency.

Variation of the Phase Angle with Frequency

Since X_C is the factor that introduces the phase angle in a series RC circuit, a change in X_C produces a change in the phase angle. As the frequency is increased, X_C becomes smaller, and thus the phase angle decreases. As the frequency is decreased, X_C becomes larger, and thus the phase angle increases. This variation is illustrated in Figure 13–15 where a phase meter is connected to V_s and to V_R, giving the angle between V_s and V_R. This angle is the phase angle of the circuit because I is in phase with V_R. By measuring the phase of V_R, we are effectively measuring the phase of I. We can use the phase-meter portion of a vector voltmeter such as the one shown in Part (c) of the figure to measure the phase of one voltage relative to another. You can also see the variation of phase angle with frequency by using the impedance triangle as shown in Figure 13–16.

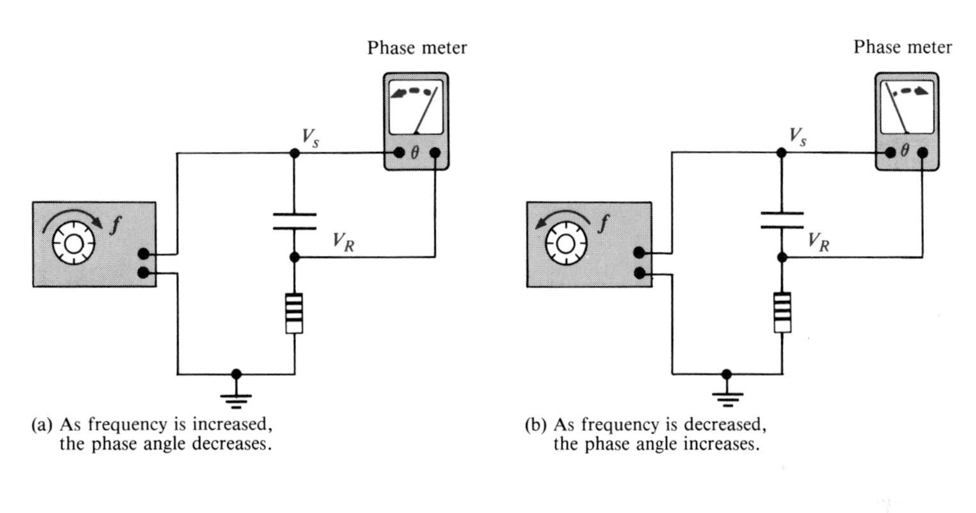

(a) As frequency is increased, the phase angle decreases.

(b) As frequency is decreased, the phase angle increases.

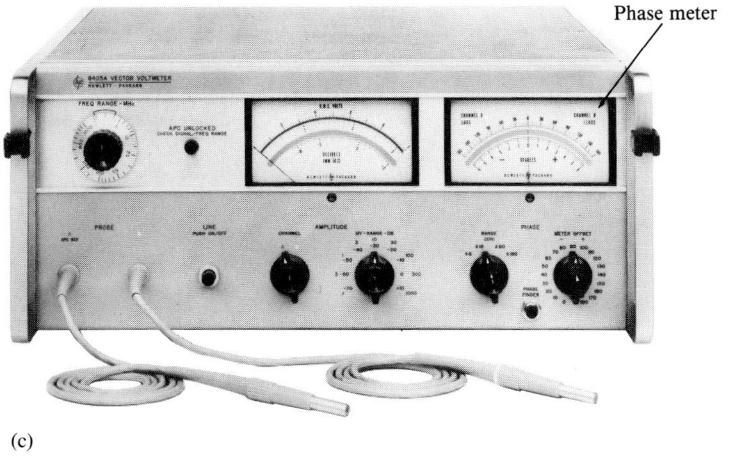

(c)

FIGURE 13–15

A phase meter shows the effect of frequency on the phase angle of a circuit. The phase meter indicates the phase angle between two voltages, V_s and V_R. Since V_R and I are in phase, this angle is the same as the angle between V_s and I (photo courtesy of Hewlett-Packard Co.).

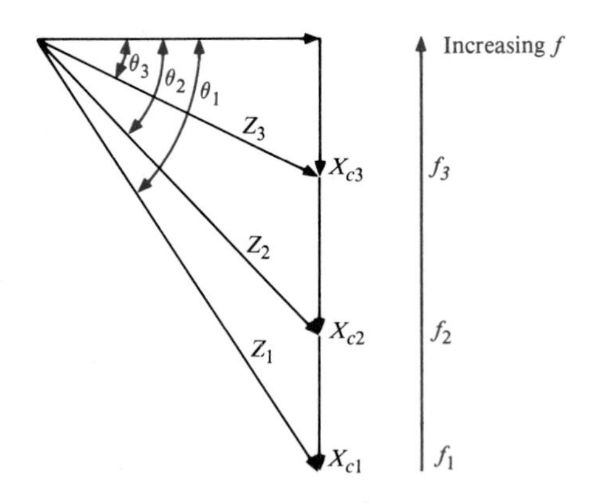

FIGURE 13–16
As the frequency increases, X_C decreases, Z decreases, and θ decreases. Each value of frequency can be visualized as forming a different impedance triangle.

EXAMPLE 13–5 For the series RC circuit in Figure 13–17, determine the impedance and phase angle for each of the following values of frequency:
(a) 10 kHz **(b)** 20 kHz **(c)** 30 kHz

FIGURE 13–17

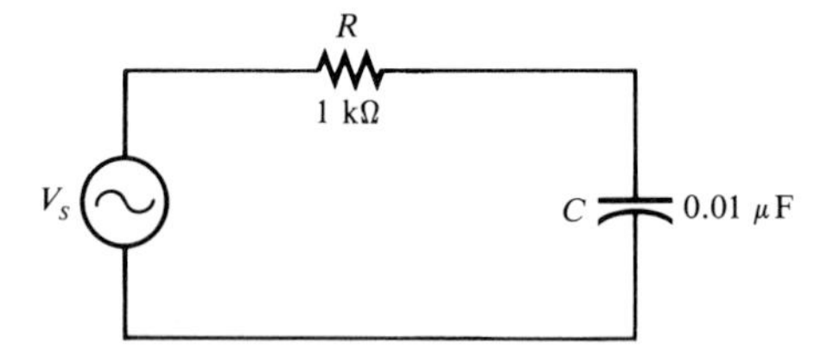

Solution
(a) For $f = 10$ kHz:

$$X_C = \frac{1}{2\pi fC} = \frac{1}{2\pi(10\text{ kHz})(0.01\ \mu\text{F})} = 1592\ \Omega$$

$$Z = \sqrt{R^2 + X_C^2} = \sqrt{(1000\ \Omega)^2 + (1592\ \Omega)^2} = 1880\ \Omega$$

$$\theta = \arctan\left(\frac{X_C}{R}\right) = \arctan\left(\frac{1592\ \Omega}{1000\ \Omega}\right) = 57.87°$$

(b) For $f = 20$ kHz:

$$X_C = \frac{1}{2\pi(20\text{ kHz})(0.01\ \mu\text{F})} = 796\ \Omega$$

$$Z = \sqrt{(1000\ \Omega)^2 + (796\ \Omega)^2} = 1278\ \Omega$$

$$\theta = \arctan\left(\frac{796\ \Omega}{1000\ \Omega}\right) = 38.52°$$

(c) For $f = 30$ kHz:

$$X_C = \frac{1}{2\pi(30\text{ kHz})(0.01\ \mu\text{F})} = 531\ \Omega$$

$$Z = \sqrt{(1000\ \Omega)^2 + (531\ \Omega)^2} = 1132\ \Omega$$

$$\theta = \arctan\left(\frac{531\ \Omega}{1000\ \Omega}\right) = 27.97^\circ$$

Notice that as the frequency increases, X_C, Z, and θ decrease.

SECTION REVIEW 13–3

1. In a certain series *RC* circuit, $V_R = 4$ V and $V_C = 6$ V. What is the magnitude of the total voltage?
2. In Problem 1, what is the phase angle?
3. What is the phase difference between the capacitor voltage and the resistor voltage in a series *RC* circuit?
4. When the frequency of the applied voltage in a series *RC* circuit is increased, what happens to
   (a) the capacitive reactance? **(b)** the impedance? **(c)** the phase angle?

13–4 IMPEDANCE AND PHASE ANGLE OF A PARALLEL *RC* CIRCUIT

A basic parallel *RC* circuit is shown in Figure 13–18. The expression for the impedance is

$$Z = \frac{RX_C}{\sqrt{R^2 + X_C^2}} \tag{13–8}$$

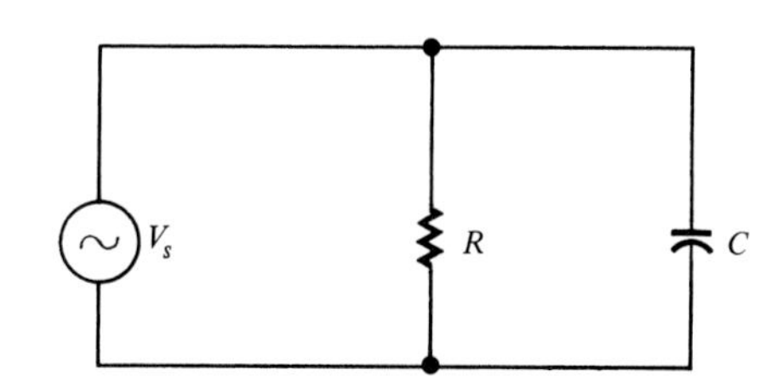

FIGURE 13–18
Parallel *RC* circuit.

The phase angle between the applied voltage and the total current is

$$\theta = \arctan\left(\frac{R}{X_C}\right) \tag{13–9}$$

EXAMPLE 13–6

For each circuit in Figure 13–19, determine the impedance and the phase angle.

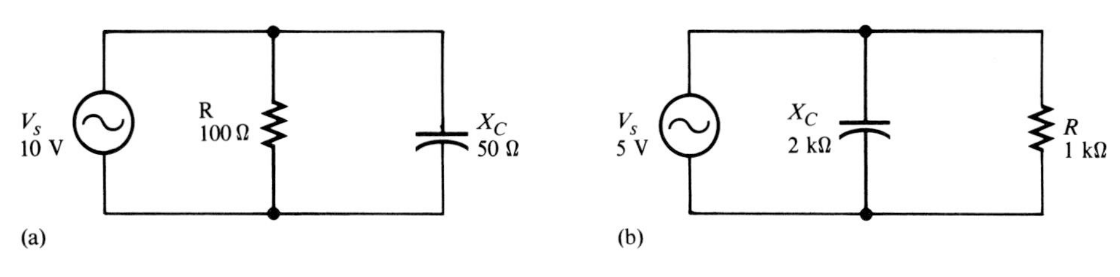

FIGURE 13–19

Solution

$$\textbf{(a)}\quad Z = \frac{RX_c}{\sqrt{R^2 + X_C^2}} = \frac{(100\ \Omega)(50\ \Omega)}{\sqrt{(100\ \Omega)^2 + (50\ \Omega)^2}} = 44.72\ \Omega$$

$$\theta = \arctan\left(\frac{R}{X_C}\right) = \arctan\left(\frac{100\ \Omega}{50\ \Omega}\right) = 63.43^\circ$$

$$\textbf{(b)}\quad Z = \frac{(1\ \text{k}\Omega)(2\ \text{k}\Omega)}{\sqrt{(1\ \text{k}\Omega)^2 + (2\ \text{k}\Omega)^2}} = 894.4\ \Omega$$

$$\theta = \arctan\left(\frac{1\ \text{k}\Omega}{2\ \text{k}\Omega}\right) = 26.57^\circ$$

The calculator sequences for Part (a) are as follows:

Z: [1] [0] [0] [×] [5] [0] [÷] [(] [(] [1] [0] [0] [2nd] [x^2] [+] [5] [0] [2nd] [x^2] [)] [$\sqrt{x}$] [)] [=]

θ: [1] [0] [0] [÷] [5] [0] [=] [INV] [tan]

Conductance, Susceptance, and Admittance

Conductance (G) is the reciprocal of resistance and is expressed as

$$G = \frac{1}{R} \tag{13–10}$$

Two other terms are convenient for use in parallel *RC* circuits. *Capacitive susceptance (B_C)* is the reciprocal of capacitive reactance and is expressed as

$$B_C = \frac{1}{X_C} \tag{13–11}$$

Admittance (Y) is the reciprocal of impedance and is expressed as

$$Y = \frac{1}{Z} \tag{13–12}$$

The unit of each of these quantities is the *siemen* (S), which is the reciprocal of the ohm.

In working with parallel circuits, you will often find it easier to use G, B_C, and Y rather than R, X_C, and Z, as we now discuss. In a parallel *RC* circuit, as shown in Figure 13–20, the total admittance is simply the phasor sum of the conductance and the susceptance:

$$Y = \sqrt{G^2 + B_C^2} \tag{13–13}$$

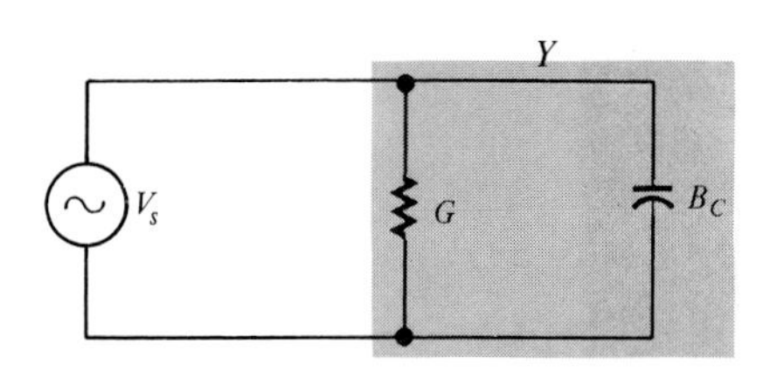

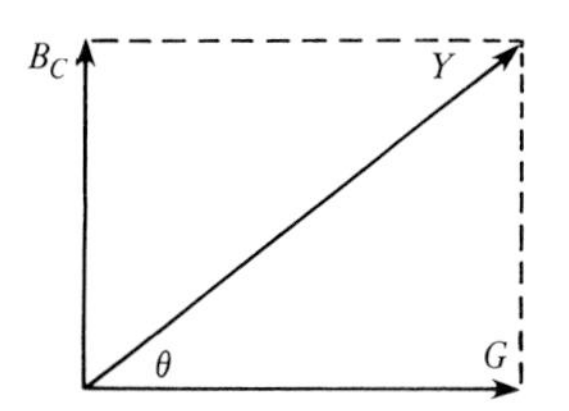

FIGURE 13–20
Admittance is the reciprocal of impedance in a parallel *RC* circuit.

EXAMPLE 13–7 Determine the admittance in Figure 13–21, and then convert it to impedance.

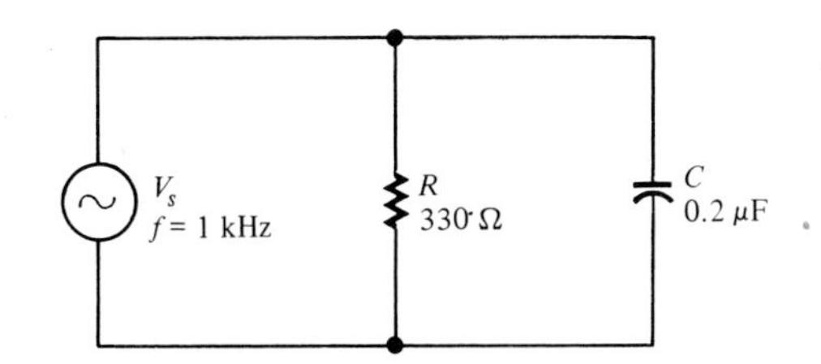

FIGURE 13–21

Solution
Since $R = 330\ \Omega$, then $G = 1/R = 1/330\ \Omega = 0.003$ S.

$$X_C = \frac{1}{2\pi(1000\ \text{Hz})(0.2\ \mu\text{F})} = 796\ \Omega$$

$$B_C = \frac{1}{X_C} = \frac{1}{796\ \Omega} = 0.00126\ \text{S}$$

$$Y = \sqrt{G^2 + B_C^2} = \sqrt{(0.003\ \text{S})^2 + (0.00126\ \text{S})^2} = 0.00325\ \text{S}$$

Now, this can be converted to impedance as follows:

$$Z = \frac{1}{Y} = \frac{1}{0.00325\ \text{S}} = 307.69\ \Omega$$

The calculator sequence for Y is

SECTION REVIEW 13–4

1. Define *conductance, capacitive susceptance,* and *admittance.*
2. If $Z = 100\ \Omega$, what is the value of Y?
3. In a certain parallel *RC* circuit, $R = 50\ \Omega$ and $X_C = 75\ \Omega$. Determine Y.

ANALYSIS OF PARALLEL *RC* CIRCUITS

13–5

For convenience in the analysis of parallel circuits, the Ohm's law formulas using impedance, previously stated, can be rewritten for admittance using the relation $Y = 1/Z$:

$$V = \frac{I}{Y} \tag{13–14}$$

$$I = VY \tag{13–15}$$

$$Y = \frac{I}{V} \tag{13–16}$$

EXAMPLE 13–8 Determine the total current and the phase angle in Figure 13–22.

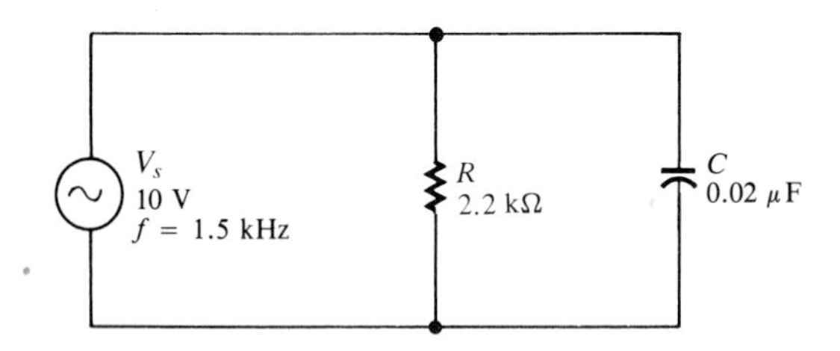

FIGURE 13–22

Solution

$$X_C = \frac{1}{2\pi(1.5 \text{ kHz})(0.02 \ \mu\text{F})} = 5.3 \text{ k}\Omega$$

The susceptance is

$$B_C = \frac{1}{X_C} = \frac{1}{5.3 \text{ k}\Omega} = 0.189 \text{ mS}$$

The conductance is

$$G = \frac{1}{R} = \frac{1}{2.2 \text{ k}\Omega} = 0.455 \text{ mS}$$

The admittance is

$$Y = \sqrt{G^2 + B_C^2} = \sqrt{(0.455 \text{ mS})^2 + (0.189 \text{ mS})^2}$$
$$= 0.493 \text{ mS}$$

Applying Ohm's law, we obtain

$$I_T = VY = (10 \text{ V})(0.493 \text{ mS}) = 4.93 \text{ mA}$$

The phase angle is

$$\theta = \arctan\left(\frac{R}{X_C}\right) = \arctan\left(\frac{2.2\text{ k}\Omega}{5.3\text{ k}\Omega}\right) = 22.56°$$

The total current is 4.93 mA, and it *leads* the applied voltage by 22.56°.

Relationships of the Currents and Voltages in a Parallel *RC* Circuit

Figure 13–23(a) shows all the currents and voltages in a basic parallel *RC* circuit. As you can see, the applied voltage, V_s, appears across both the resistive and the capacitive branches, so V_s, V_R, and V_C are all in phase and of the same magnitude. The total current, I_T, divides at the junction into the two branch currents, I_R and I_C.

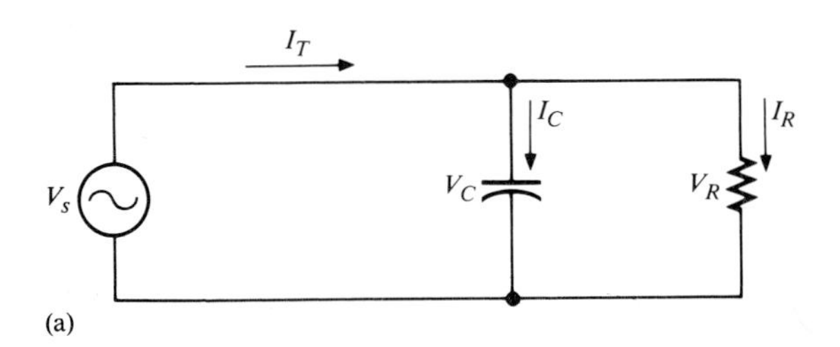

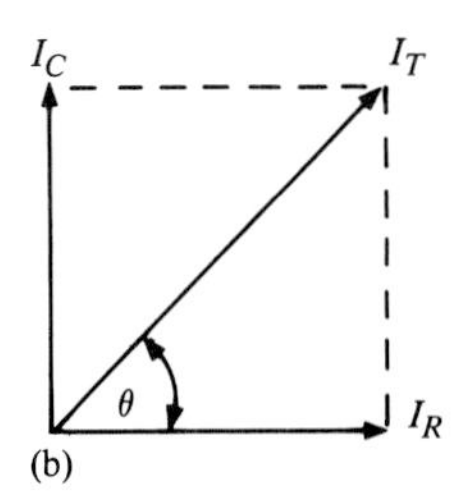

FIGURE 13–23
Currents and voltages in a parallel *RC* circuit. The current directions shown in (a) are instantaneous.

The current through the resistor is in phase with the voltage. The current through the capacitor leads the voltage, and thus the resistive current, by 90°. By Kirchhoff's current law, the total current is the phasor sum of the two branch currents, as shown by the phasor diagram in Figure 13–23(b). The total current is expressed as

$$I_T = \sqrt{I_R^2 + I_C^2} \tag{13–17}$$

The phase angle can be found from the current values as follows:

$$\theta = \arctan\left(\frac{I_C}{I_R}\right) \tag{13–18}$$

which is equivalent to the previous expression, $\theta = \arctan(R/X_C)$.

Figure 13–24 shows a complete current and voltage phasor diagram. Notice that I_C leads I_R by 90° and that I_R is in phase with the voltage ($V_s = V_R = V_C$).

FIGURE 13–24
Current and voltage phasor diagram for a parallel RC circuit (amplitudes are arbitrary).

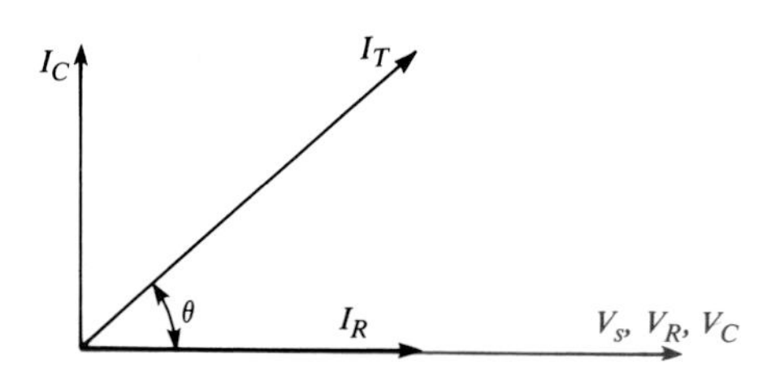

EXAMPLE 13–9 Determine the value of each current in Figure 13–25, and describe the phase relationship of each with the applied voltage.

FIGURE 13–25

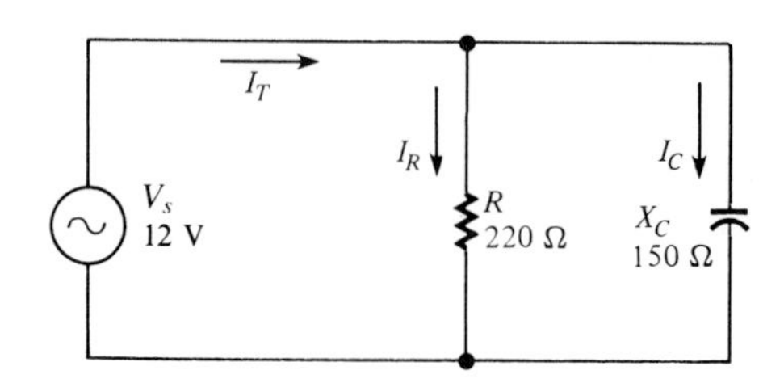

Solution

$$I_R = \frac{V_s}{R} = \frac{12\text{ V}}{220\ \Omega} = 54.55\text{ mA}$$

$$I_C = \frac{V_s}{X_C} = \frac{12\text{ V}}{150\ \Omega} = 80\text{ mA}$$

$$I_T = \sqrt{I_R^2 + I_C^2} = \sqrt{(54.55\text{ mA})^2 + (80\text{ mA})^2} = 96.83\text{ mA}$$

$$\theta = \arctan\left(\frac{I_C}{I_R}\right) = \arctan\left(\frac{80\text{ mA}}{54.55\text{ mA}}\right) = 55.7^\circ$$

I_R is in phase with the voltage, I_C leads the voltage by 90°, and I_T leads the voltage by 55.7°.

Conversion from Parallel to Series Form

For every parallel RC circuit, there is an equivalent series RC circuit. Two circuits are equivalent when they both present an equal impedance and phase angle at their terminals. To obtain the equivalent series circuit for a given parallel RC circuit, first find the impedance and phase angle of the parallel circuit. Then use the values of Z and θ to construct an impedance diagram as shown in Figure 13–26. The unknown sides of the triangle (black) represent the equivalent series resistance and capacitive reactance as indicated. These values can be found using the following trigonometric relationships:

$$R_{eq} = Z\cos\theta \qquad (13\text{–}19)$$

$$X_{C(eq)} = Z\sin\theta \qquad (13\text{–}20)$$

The cosine (cos) and the sine (sin) functions are available on your calculator.

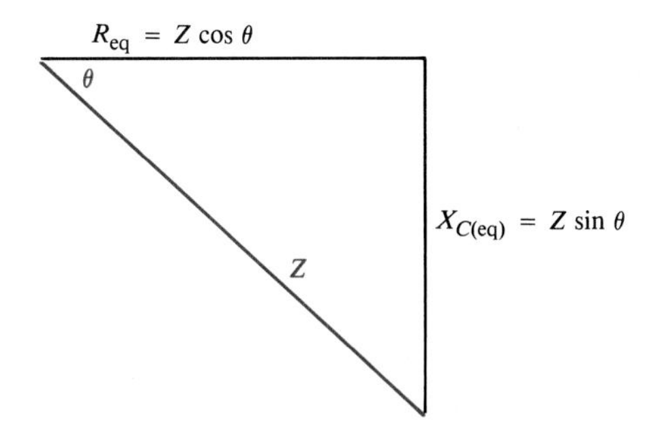

FIGURE 13–26
Impedance triangle for the series equivalent of a parallel *RC* circuit. Z and θ are the known values for the parallel circuit. R_{eq} and $X_{C(eq)}$ are the series equivalent values.

EXAMPLE 13–10

Convert the parallel circuit in Figure 13–27 to an equivalent series form.

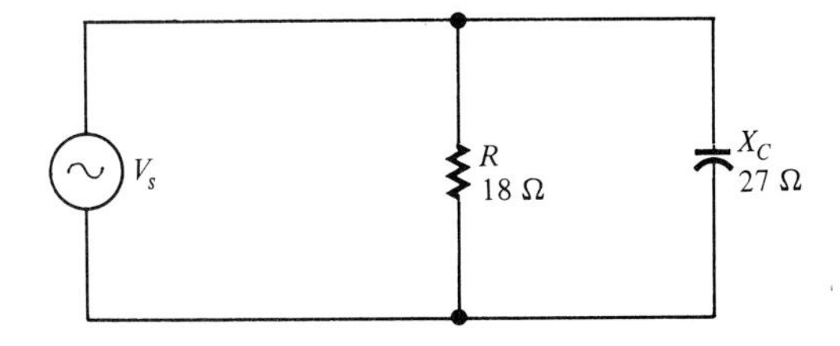

FIGURE 13–27

Solution
First, find the impedance of the parallel circuit:

$$G = \frac{1}{R} = \frac{1}{18\ \Omega} = 0.055\ \text{S}$$

$$B_C = \frac{1}{X_C} = \frac{1}{27\ \Omega} = 0.037\ \text{S}$$

$$Y = \sqrt{G^2 + B_C^2} = \sqrt{(0.055\ \text{S})^2 + (0.037\ \text{S})^2} = 0.066\ \text{S}$$

The total impedance is

$$Z = \frac{1}{Y} = \frac{1}{0.066\ \text{S}} = 15.15\ \Omega$$

The phase angle is

$$\theta = \arctan\left(\frac{R}{X_C}\right) = \arctan\left(\frac{18\ \Omega}{27\ \Omega}\right) = 33.69^\circ$$

The equivalent series values are

$$R_{eq} = Z\cos\theta = 15.15\cos(33.69^\circ) = 12.6\ \Omega$$

$$X_{C(eq)} = Z\sin\theta = 15.15\sin(33.69^\circ) = 8.4\ \Omega$$

The series equivalent circuit is shown in Figure 13–28. Note that the value of C can be determined if the frequency is known.

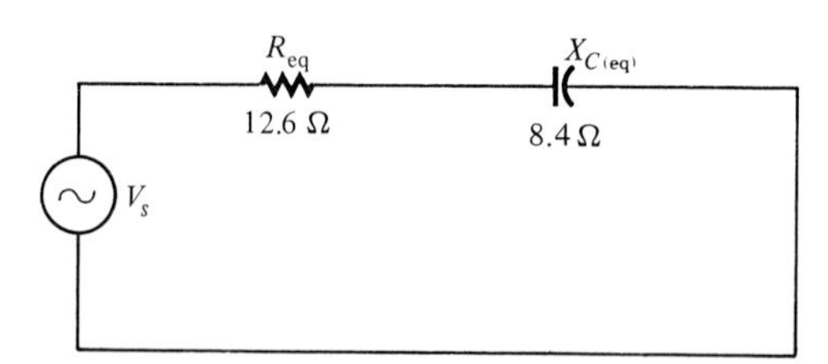

FIGURE 13–28

The calculator sequence for R_{eq} is

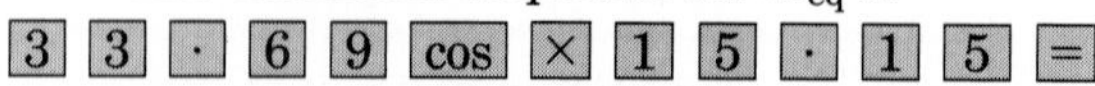

It is interesting to observe that a parallel *RC* circuit becomes *less* reactive when X_C is increased. That is, the circuit phase angle becomes smaller. The reason for this effect is that when X_C is increased relative to R, less of the total current flows through the capacitive branch, and thus the in-phase or resistive current becomes greater.

SECTION REVIEW 13–5

1. The admittance of an *RC* circuit is 0.0035 S, and the applied voltage is 6 V. What is the total current?
2. In a certain parallel *RC* circuit, the resistor current is 10 mA, and the capacitor current is 15 mA. Determine the phase angle and the total current.
3. What is the phase angle between the capacitor current and the applied voltage in a parallel *RC* circuit?

SERIES-PARALLEL ANALYSIS

13–6

In this section, we use the concepts studied in the previous sections to analyze circuits with combinations of both series and parallel *R* and *C* elements. The following example demonstrates the procedures.

EXAMPLE 13–11 In the circuit of Figure 13–29, determine the following:
(a) total impedance. **(b)** total current. **(c)** phase angle.

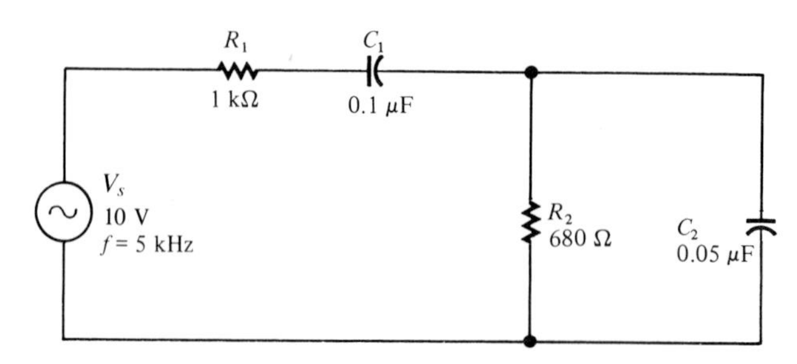

FIGURE 13–29

Solution

(a) First, calculate the magnitudes of capacitive reactances:

$$X_{C1} = \frac{1}{2\pi(5\text{ kHz})(0.1\ \mu\text{F})} = 318.3\ \Omega$$
$$X_{C2} = \frac{1}{2\pi(5\text{ kHz})(0.05\ \mu\text{F})} = 636.6\ \Omega$$

One approach is to find the series equivalent resistance and capacitive reactance for the parallel portion of the circuit, and then add the resistances to get total resistance and add the reactances to get total reactance. From this, the total impedance can be determined.

We find the impedance of the parallel portion (Z_2) by first finding the admittance:

$$G_2 = \frac{1}{R_2} = \frac{1}{680\ \Omega} = 0.00147\text{ S}$$
$$B_{C2} = \frac{1}{X_{C2}} = \frac{1}{636.6\ \Omega} = 0.00157\text{ S}$$
$$Y_2 = \sqrt{G_2^2 + B_{C2}^2} = \sqrt{(0.00147\text{ S})^2 + (0.00157\text{ S})^2} = 2.15\text{ mS}$$
$$Z_2 = \frac{1}{Y_2} = \frac{1}{0.00215\text{ S}} = 465.12\ \Omega$$

The phase angle associated with the parallel portion of the circuit is

$$\theta_p = \arctan\left(\frac{R_2}{X_{C2}}\right) = \arctan\left(\frac{680\ \Omega}{636.6\ \Omega}\right) = 46.89°$$

The series equivalent values for the parallel portion are

$$R_{\text{eq}} = Z_2 \cos\theta_p = 465.12 \cos(46.89°) = 317.9\ \Omega$$
$$X_{C(\text{eq})} = Z_2 \sin\theta_p = 465.12 \sin(46.89°) = 339.6\ \Omega$$

The total resistance is

$$R_T = R_1 + R_{\text{eq}} = 1000\ \Omega + 317.9\ \Omega = 1317.9\ \Omega$$

The total reactance is

$$X_{CT} = X_{C1} + X_{C(\text{eq})} = 318.3\ \Omega + 339.6\ \Omega = 657.9\ \Omega$$

The total impedance is

$$Z_T = \sqrt{R_T^2 + X_{CT}^2} = \sqrt{(1317.9\ \Omega)^2 + (657.9\ \Omega)^2} = 1473\ \Omega$$

(b) We find the total current by using Ohm's law:

$$I_T = \frac{V_s}{Z_T} = \frac{10 \text{ V}}{1473 \ \Omega} = 6.79 \text{ mA}$$

(c) To get the phase angle, view the circuit as a series combination of R_T and X_{CT}:

$$\theta = \arctan\left(\frac{X_{CT}}{R_T}\right) = \arctan\left(\frac{657.9 \ \Omega}{1317.9 \ \Omega}\right) = 26.53°$$

The total current, of course, *leads* the source voltage by 26.53°.

The calculator sequence for Z_T is

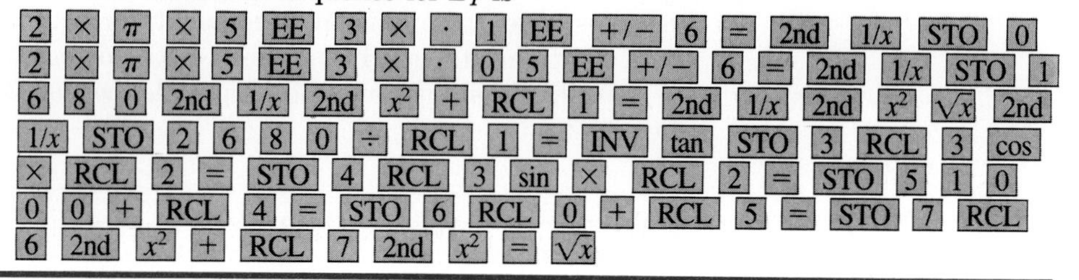

Measurement of Z_T and θ

Now, let's see how the values of Z_T and θ for the circuit in Example 13–11 can be determined by measurement. First, the total impedance is measured as outlined in the following steps and as illustrated in Figure 13–30 (other ways are also possible):

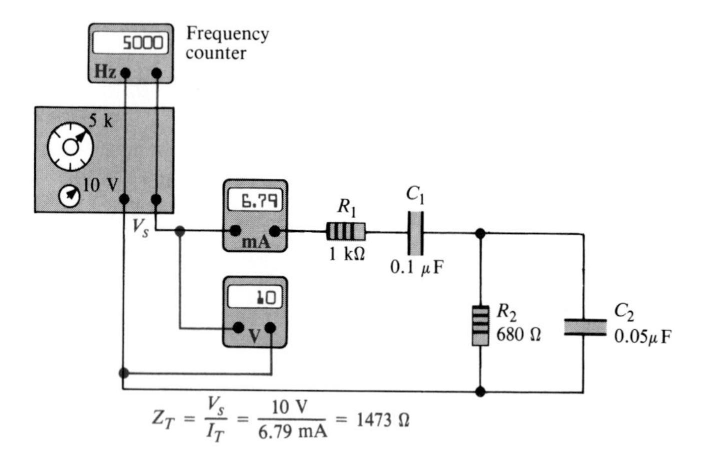

FIGURE 13–30
Determining Z_T by measurement of V_s and I_T.

1. Using a sine wave generator, set the source voltage to a known value (10 V) and the frequency to 5 kHz. It is advisable to check the voltage with an ac voltmeter and the frequency with a frequency counter rather than relying on the marked values on the generator controls.

2. Connect an ac ammeter as shown in the figure, and measure the total current.
3. Calculate the total impedance by using Ohm's law.

Although we could use a phase meter to measure the phase angle, we will use an oscilloscope in this illustration because it is more commonly available. To measure the phase angle, we must have the source voltage and the total current displayed on the screen in the proper time relationship. Two basic types of scope probes are available to measure the quantities with an oscilloscope: the voltage probe and the current probe. These probes are shown in Figure 13–31. Although the current probe is a convenient device, it is often not as readily available as a voltage probe. For this reason, we will confine our phase measurement technique to the use of voltage probes in conjunction with the oscilloscope. A typical oscilloscope voltage probe has two points that are connected to the circuit: the probe tip and the ground lead. Thus, all voltage measurements must be referenced to ground.

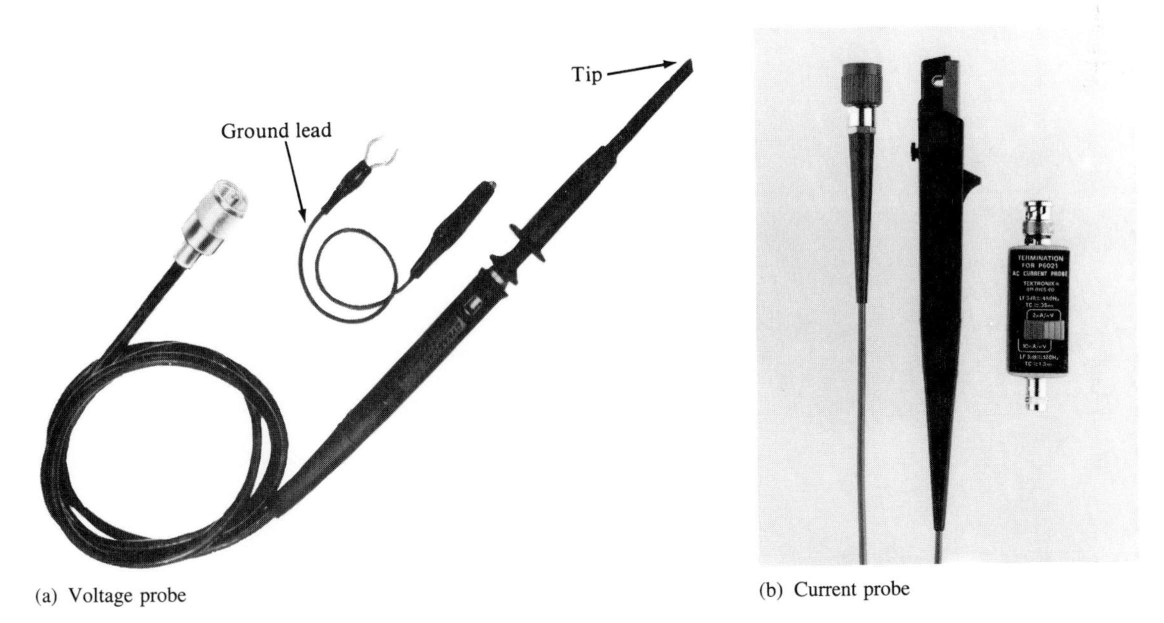

(a) Voltage probe

(b) Current probe

FIGURE 13–31
Typical oscilloscope probes ((a) courtesy of Hewlett-Packard Co. (b) courtesy of Tektronix, Inc.).

Since only voltage probes are to be used, the total current cannot be measured directly. However, for phase measurement, the voltage across R_1 is in phase with the total current and can be used to establish the phase angle. In setting up this circuit, we take the lower side of the source as circuit ground, as shown in Figure 13–32(a).

Before proceeding with the actual phase measurement, note that there is a problem with displaying V_{R1}. If the scope probe is connected across the resistor, as indicated in Figure 13–32(b), the ground lead of the scope will short point

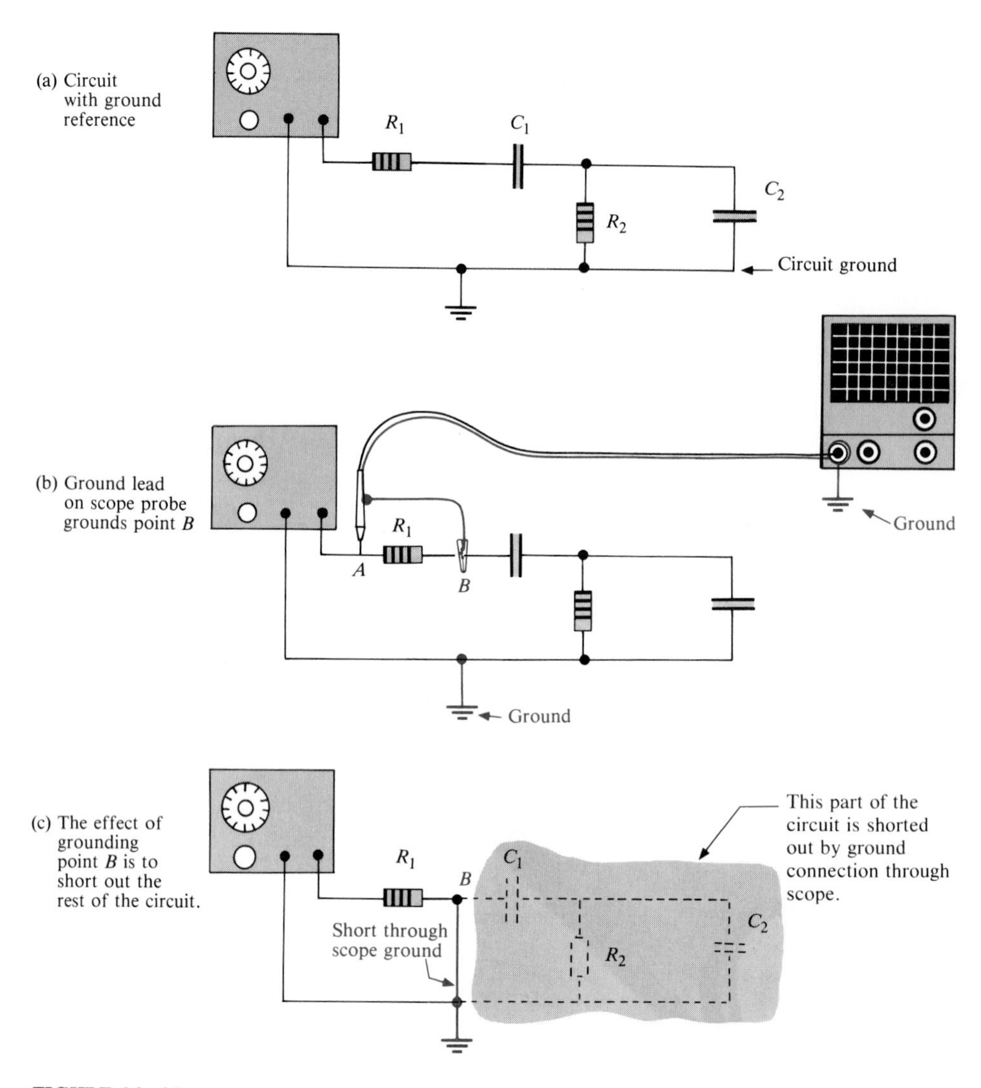

FIGURE 13–32
Effects of measuring directly across a component when the instrument and the circuit are grounded.

B to ground, thus bypassing the rest of the components and effectively removing them from the circuit electrically, as illustrated in Part (c) (assuming that the scope is not isolated from power line ground).

To avoid this problem, we can reposition R_1 in the circuit (when possible) so that one end of it is connected to ground, as shown in Figure 13–33(a). This connection does not alter the circuit electrically, because R_1 still has the same series relationship with the rest of the circuit. Now the scope can be connected across it to display V_{R1}, as indicated in Part (b) of the figure. The other probe is connected across the voltage source to display V_s as indicated. Now channel 1 of the scope has V_s as an input, and channel 2 has V_{R1}. The *trigger source* switch on the scope should be on internal so that each trace on the screen will be triggered by one of the inputs and the other will then be shown in the proper time relationship to it. Since amplitudes are not important, the *volts/div* settings are arbitrary. The *sec/div* settings should be adjusted so that one half-cycle of the waveforms appears on the screen.

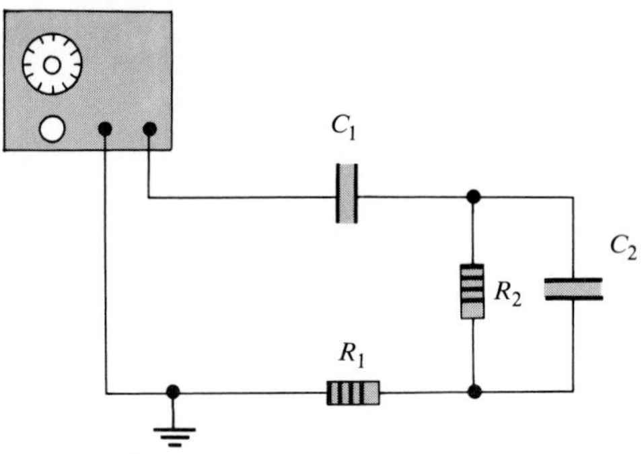

(a) R_1 repositioned so that one end is grounded

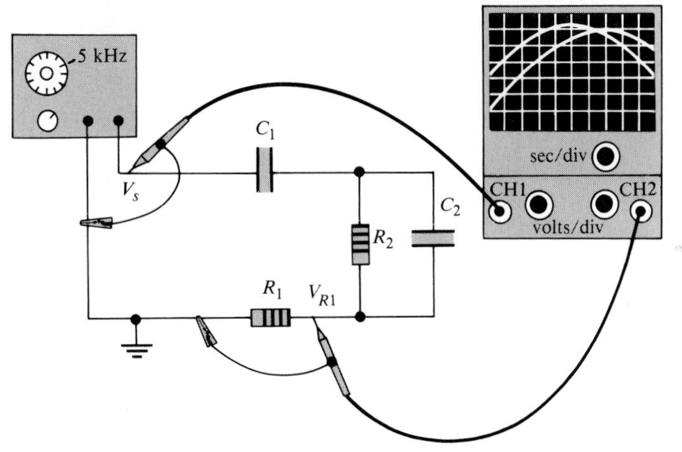

(b) The scope displays a half-cycle of V_{R1} and V_s. V_{R1} represents the phase of the total current.

FIGURE 13–33
Repositioning R_1 so that a direct voltage measurement can be made with respect to ground.

Before connecting the probes to the circuit, we must align the two horizontal lines (traces) so that they appear as a single line across the center of the screen. To do so, ground the probe tips and adjust the *vertical position* knobs to move the traces toward the center line of the screen until they are superimposed. The reason for this procedure is to ensure that both waveforms have the same zero crossing so that an accurate phase measurement can be made.

The resulting oscilloscope display is shown in Figure 13–34. Since there is 180° in one half-cycle, each of the ten horizontal divisions across the screen represents 18°. Thus, the horizontal distance between the corresponding points of the two waveforms is the phase angle in degrees as indicated.

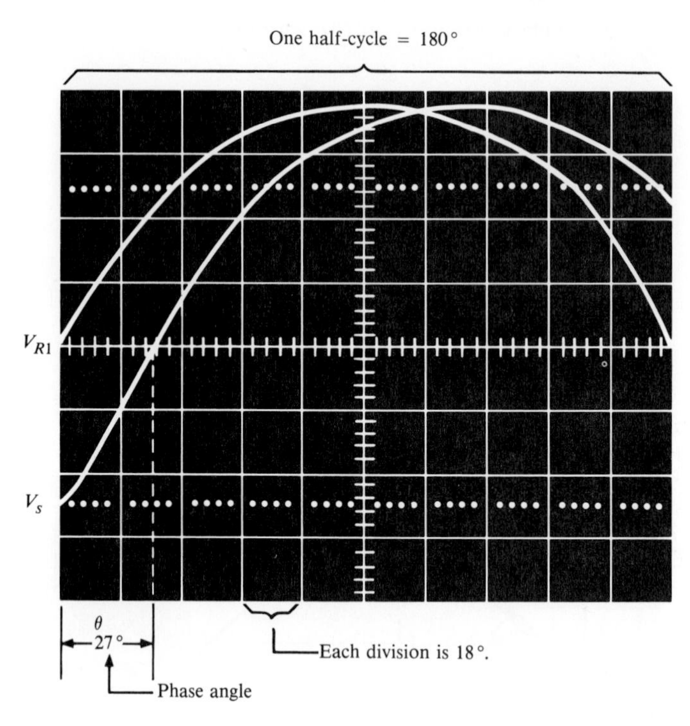

FIGURE 13–34
Measurement of the phase angle on the oscilloscope.

SECTION REVIEW 13–6

1. Determine the series equivalent circuit for the series-parallel circuit in Figure 13–29.
2. What is the voltage across R_1 in Figure 13–29?

POWER IN *RC* CIRCUITS

13–7

In a purely resistive ac circuit, all of the energy delivered by the source is dissipated in the form of heat by the resistance. In a purely capacitive ac circuit, all of the energy delivered by the source is stored by the capacitor during a portion of the voltage cycle and then is returned to the source during another portion of the cycle so that there is no net energy transferred. What happens when both resistance and capacitance exist in a circuit? *Some of the energy is alternately stored and returned by the capacitance, and some is dissipated by the resistance.* The amount of energy dissipated is determined by the relative values of the resistance and the capacitive reactance.

It is reasonable to assume that when the resistance is greater than the reactance, more of the total energy delivered by the source is dissipated by the resistance than is stored by the capacitance. Likewise, when the reactance is greater than the resistance, more of the total energy is stored and returned than is dissipated.

The formulas for power in a resistor, sometimes called *true power* (P_{true}), and the power in a capacitor, called *reactive power* (P_r), are restated here. The unit of true power is the *watt* (W), and the unit of reactive power is the *volt-ampere reactive* (VAR):

$$P_{true} = I_T^2 R \tag{13–21}$$

$$P_r = I_T^2 X_C \tag{13–22}$$

The Power Triangle

The generalized impedance phasor diagram is shown in Figure 13–35(a). A phasor relationship for power can also be represented by a similar diagram, because the respective magnitudes of the powers, P_{true} and P_r, differ from R and X_C by a factor of I_T^2, as shown in Figure 13–35(b).

The resultant power phasor, $I_T^2 Z$, represents the *apparent power*, P_a. At any instant in time, P_a is the total power that *appears* to be transferred between the source and the *RC* circuit. Part of the apparent power is true power, and part of it is reactive power. The unit of apparent power is the *volt-ampere* (VA). The expression for apparent power is

$$P_a = I_T^2 Z \tag{13–23}$$

The power phasor diagram in Figure 13–35(b) can be rearranged in the form of a right triangle as shown in Part (c), which is called the *power triangle*. Using the rules of trigonometry, P_{true} can be expressed as

$$P_{true} = P_a \cos\theta$$

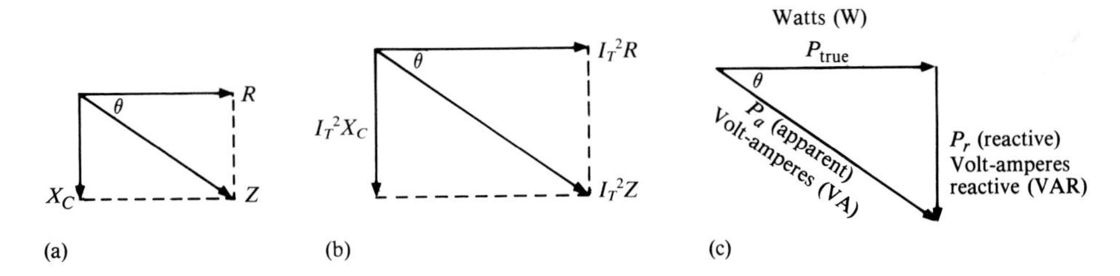

FIGURE 13–35
Development of the power triangle for an *RC* circuit.

Since P_a equals $I_T^2 Z$ or $V_s I_T$, the equation for the true power can be written as

$$P_{true} = V_s I_T \cos\theta \tag{13–24}$$

where V_s is the applied voltage and I_T is the total current.

For the case of a purely resistive circuit, $\theta = 0°$ and $\cos 0° = 1$, so P_{true} equals $V_s I_T$. For the case of a purely capacitive circuit, $\theta = 90°$ and $\cos 90° =$

0, so P_{true} is zero. As you already know, there is no power dissipated in an ideal capacitor.

Power Factor

The term cos θ is called the *power factor* and is stated as follows:

$$PF = \cos\theta \tag{13–25}$$

As the phase angle between applied voltage and total current increases, the power factor decreases, indicating an increasingly reactive circuit. The smaller the power factor, the smaller the power dissipation.

The power factor can vary from 0 for a purely reactive circuit to 1 for a purely resistive circuit. In an RC circuit, the power factor is referred to as a *leading* power factor because the current leads the voltage.

EXAMPLE 13–12 Determine the power factor and the true power in the circuit of Figure 13–36.

FIGURE 13–36

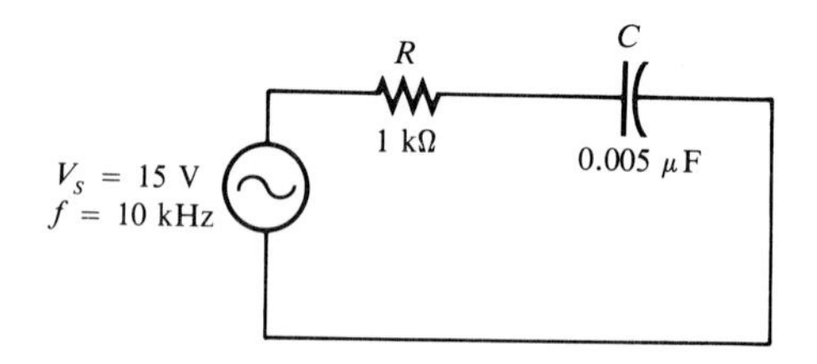

Solution

$$X_C = \frac{1}{2\pi fC} = \frac{1}{2\pi(10\text{ kHz})(0.005\ \mu\text{F})} = 3183\ \Omega$$

$$\theta = \arctan\left(\frac{X_c}{R}\right) = \arctan\left(\frac{3183\ \Omega}{1000\ \Omega}\right) = 72.56^\circ$$

$$PF = \cos\theta = \cos(72.56^\circ) = 0.2997$$

$$Z = \sqrt{R^2 + X_c^2} = \sqrt{(1000\ \Omega)^2 + (3183\ \Omega)^2} = 3336.4\ \Omega$$

$$I = \frac{V_s}{Z} = \frac{15\text{ V}}{3336.4\ \Omega} = 4.496\text{ mA}$$

The true power is

$$P_{\text{true}} = V_s I\cos\theta = (15\text{ V})(4.496\text{ mA})(0.2997) = 20.21\text{ mW}$$

EXAMPLE 13–13 For the circuit in Figure 13–37, find the true power, the reactive power, and the apparent power. X_C has been determined to be 2 kΩ.

FIGURE 13–37

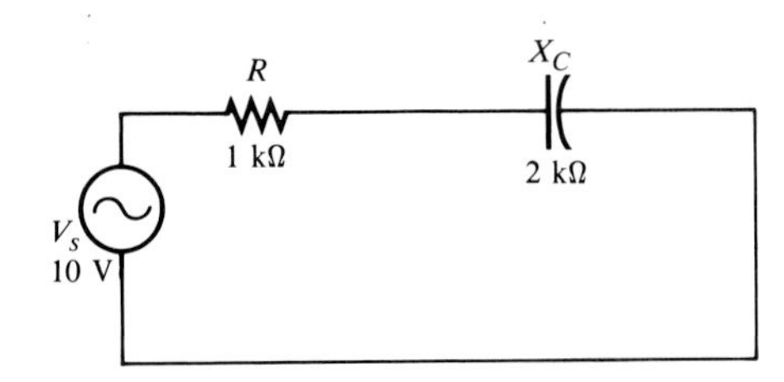

Solution
We first find the total impedance so that the current can be calculated.

$$Z = \sqrt{R^2 + X_C^2} = \sqrt{(1000\ \Omega)^2 + (2000\ \Omega)^2} = 2236\ \Omega$$

$$I = \frac{V_s}{Z} = \frac{10\ \text{V}}{2236\ \Omega} = 4.47\ \text{mA}$$

The phase angle, θ, is

$$\theta = \arctan\left(\frac{2\ \text{k}\Omega}{1\ \text{k}\Omega}\right) = 63.44^\circ$$

The true power is

$$P_{\text{true}} = V_s I \cos\theta = (10\ \text{V})(4.47\ \text{mA})\cos(63.44^\circ) = 19.99\ \text{mW}$$

(The same result is realized using the formula $P_{\text{true}} = I^2R$.)
The reactive power is

$$P_r = I^2X_C = (4.47\ \text{mA})^2\ (2\ \text{k}\Omega) = 39.96\ \text{mVAR}$$

The apparent power is

$$P_a = I^2Z = (4.47\ \text{mA})^2\ (2236\ \Omega) = 44.68\ \text{mVA}$$

The apparent power is also the phasor sum of P_{true} and P_r:

$$P_a = \sqrt{P_{\text{true}}^2 + P_r^2} = 44.68\ \text{mVA}$$

The Significance of Apparent Power

As mentioned, apparent power is the power that *appears* to be transferred between the source and the load, and it consists of two components: a true power component and a reactive power component.

In all electrical and electronic systems, it is the true power that does the work. The reactive power is simply shuttled back and forth between the source and load. Ideally, in terms of performing useful work, all of the power transferred to the load should be true power and none of it reactive power. However, in most practical situations the load must have some reactance associated with it, and therefore we must deal with both power components.

In the last chapter, the use of apparent power was discussed in relation to transformers. For any reactive load, there are two components of the total current: the resistive component and the reactive component. If we consider only the true power (watts) in a load, we are dealing with only a portion of the total current that the load demands from a source. In order to have a realistic picture of the actual current that a load will draw, we must consider apparent power (in VA).

A source such as an ac generator can provide current to a load up to some maximum value. *If the load draws more than this maximum value, the source can be damaged.* Figure 13–38(a) shows a 120-V generator that can deliver a maximum current of 5 A to a load. Assume that the generator is rated at 600 W and is connected to a purely resistive load of 24 Ω (power factor of 1). The ammeter shows that the current is 5 A, and the wattmeter indicates that the power is 600 W. The generator has no problem under these conditions, although it is operating at maximum current and power.

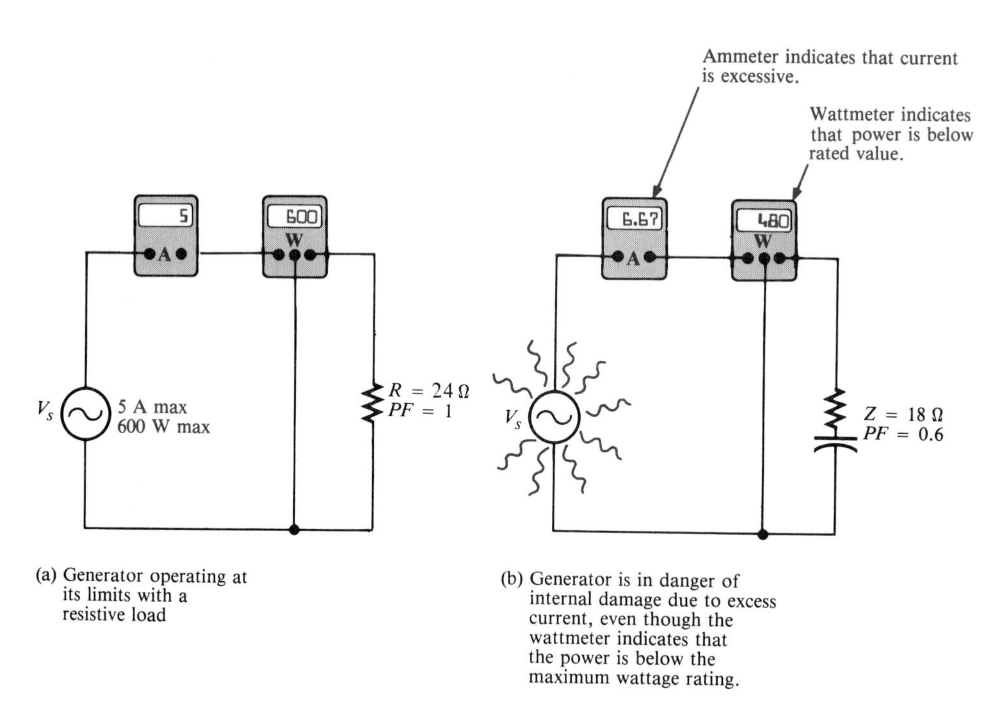

(a) Generator operating at its limits with a resistive load

(b) Generator is in danger of internal damage due to excess current, even though the wattmeter indicates that the power is below the maximum wattage rating.

FIGURE 13–38
The wattage rating of a source is inappropriate when the load is reactive. The rating should be in VA rather than in watts.

Now, consider what happens if the load is changed to a reactive one with an impedance of 18 Ω and a power factor of 0.6, as indicated in Figure 13–38(b). The current is 120 V/18 Ω = 6.67 A, which *exceeds* the maximum. Even though the wattmeter reads 480 W, which is less than the power rating of the generator, the excessive current probably will cause damage. This example shows that a true power rating can be deceiving and is inappropriate for ac sources. The ac generator should be rated at 600 VA, a rating that manufacturers generally use, rather than 600 W.

SECTION REVIEW 13–7

1. To which component in an *RC* circuit is the power dissipation due?
2. If the phase angle, θ, is 45°, what is the power factor?
3. A certain series *RC* circuit has the following parameter values: $R = 300\ \Omega$, $X_C = 460\ \Omega$, and $I = 2$ A. Determine the true power, the reactive power, and the apparent power.

BASIC APPLICATIONS OF *RC* CIRCUITS

13–8

In this section, several basic types of *RC* circuits that have applications in a broad range of areas are introduced. The basic circuits to be discussed are *phase-shift networks, filters,* and *signal-coupling networks.*

The *RC* Lag Network

The *RC* lag network is a phase-shift circuit in which the output voltage *lags* the input voltage by a specified angle. Phase-shift circuits are commonly used in electronic communication systems and in other application areas.

A basic *RC* lag network is shown in Figure 13–39(a). Notice that the output voltage is taken across the *capacitor,* and the input voltage is the total voltage applied across the circuit. The relationship of the voltages is shown in the phasor diagram in Figure 13–39(b). Keep in mind that the circuit phase angle is the angle between V_{in} and the current. Notice that V_C, which is the output voltage, *lags* V_{in} by an angle that is the difference between 90° and the circuit phase angle θ.

Since $\theta = \arctan(X_C/R)$, the phase lag, ϕ, can be expressed as follows:

$$\phi = 90° - \arctan\left(\frac{X_C}{R}\right) \tag{13–26}$$

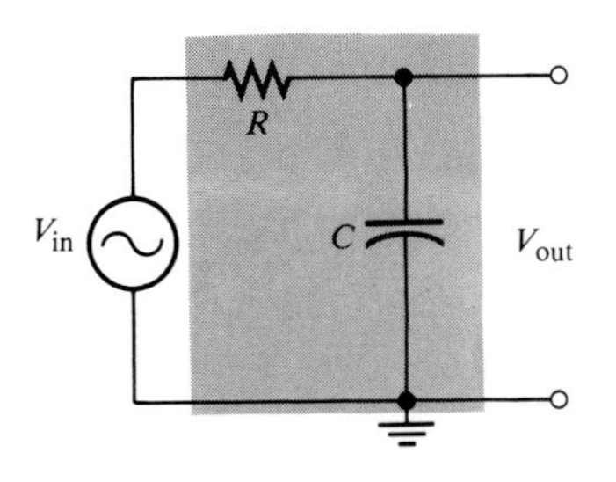

(a) A basic *RC* lag network

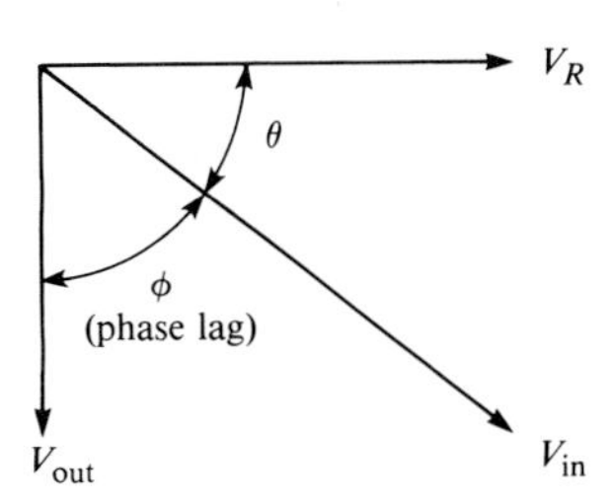

(b) Phasor voltage diagram showing the phase lag between V_{in} and V_{out}

FIGURE 13–39
The *RC* lag network ($V_{out} = V_C$).

When the input and output waveforms are displayed on an oscilloscope, a relationship similar to that in Figure 13–40 is observed. The exact amount of phase lag between the input and the output depends on the values of the resistance and the capacitive reactance. The magnitude of the output voltage depends on these values also.

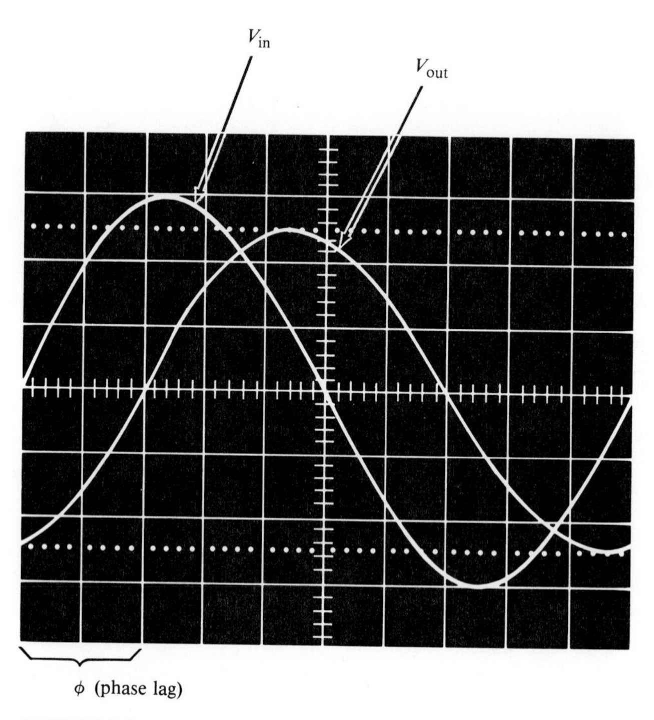

FIGURE 13–40
General oscilloscope display of the input and output waveforms of a lag network (V_{out} lags V_{in}).

EXAMPLE 13–14 Determine the amount of phase lag from input to output in the lag network in Figure 13–41.

FIGURE 13–41

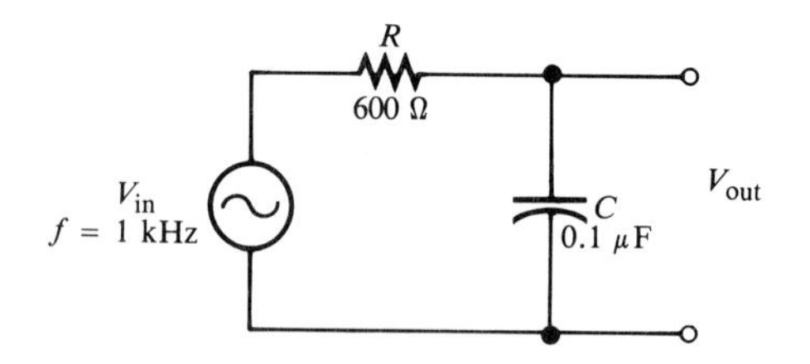

Solution

$$X_C = \frac{1}{2\pi fC} = \frac{1}{2\pi(1 \text{ kHz})(0.1 \ \mu\text{F})} = 1592 \ \Omega$$

$$\phi = 90° - \arctan\left(\frac{X_C}{R}\right) = 90° - \arctan\left(\frac{1592\ \Omega}{600\ \Omega}\right) = 20.65°$$

The output lags the input by 20.65°.

The phase-lag network can be viewed as a voltage divider with a portion of the input voltage dropped across R and a portion across C. The output voltage can be determined with the following formula:

$$V_{out} = \left(\frac{X_C}{\sqrt{R^2 + X_C^2}}\right)V_{in} \qquad (13\text{–}27)$$

EXAMPLE 13–15 For the lag network in Figure 13–41 of Example 13–14, determine the output voltage in phasor form when the input voltage has an rms value of 10 V. Sketch the input and output waveforms showing the proper relationships. ϕ was found in Example 13–14 to be 20.65°.

Solution

$$V_{out} = \left(\frac{X_C}{\sqrt{R^2 + X_C^2}}\right)V_{in}$$

$$= \left(\frac{1592\ \Omega}{\sqrt{(600\ \Omega)^2 + (1592\ \Omega)^2}}\right)10\text{V} = 9.36\text{ V}$$

The waveforms are shown in Figure 13–42.

FIGURE 13–42

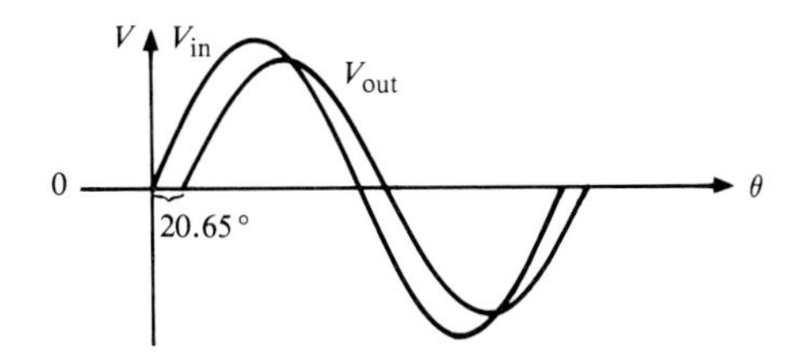

Effects of Frequency on the Phase Shift and Output Voltage of the Lag Network

Since the circuit phase angle, θ, decreases as frequency increases, the phase lag between the input and the output voltages *increases*. You can see this relationship by examining Equation (13–26). Also, the magnitude of V_{out} decreases as the frequency increases, because X_C becomes smaller and less of the total input voltage is dropped across the capacitor. Figure 13–43 illustrates the changes in phase lag and output voltage magnitude with frequency.

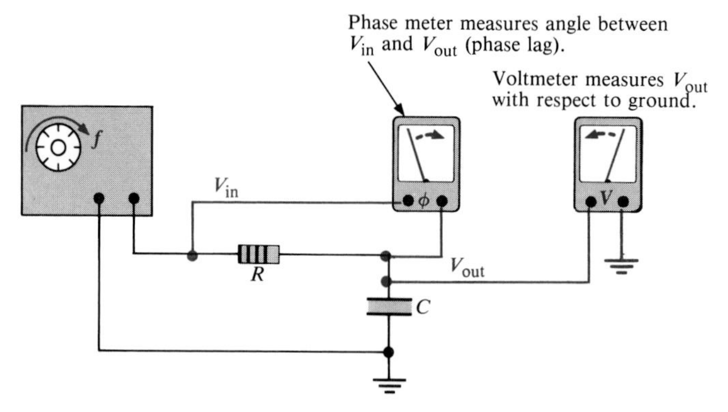

(a) When f increases, the phase lag ϕ increases and V_{out} decreases.

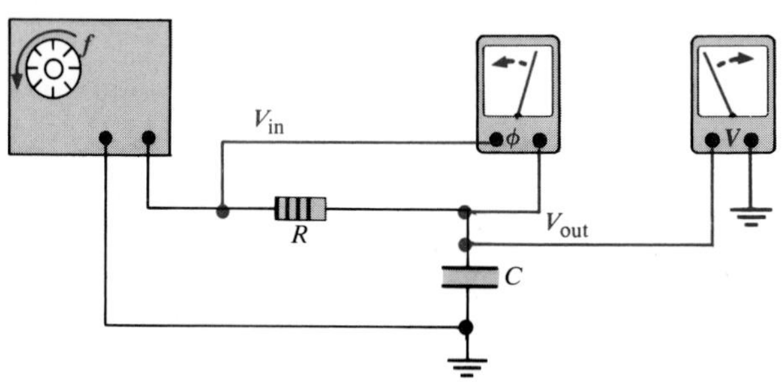

(b) When f decreases, ϕ decreases and V_{out} increases.

FIGURE 13–43
Illustration of how the frequency affects the phase lag and the output voltage in an RC lag network, with the amplitude of V_{in} held constant.

The *RC* Lead Network

The RC lead network is a phase-shift circuit in which the output voltage *leads* the input voltage by a specified angle. A basic RC lead network is shown in Figure 13–44(a). Notice how it differs from the lag network. Here the output voltage is taken across the resistor. The relationship of the voltages is given in the phasor diagram in Figure 13–44(b). The output voltage, V_R, *leads* V_{in} by an angle that is the same as the circuit phase angle, because V_R and I are in phase with each other.

When the input and output waveforms are displayed on an oscilloscope, a relationship similar to that in Figure 13–45 is observed. Of course, the exact amount of phase lead and the output voltage magnitude depend on the values of R and X_C. The phase lead, ϕ, is expressed as follows:

$$\phi = \arctan\left(\frac{X_C}{R}\right) \tag{13–28}$$

The output voltage is expressed as

$$V_{out} = \left(\frac{R}{\sqrt{R^2 + X_C^2}}\right)V_{in} \tag{13–29}$$

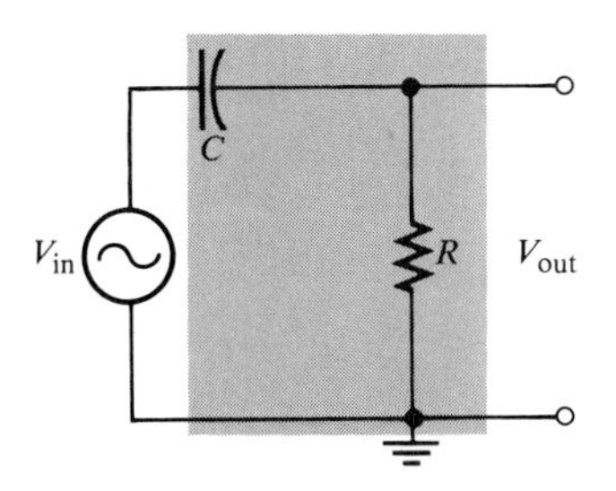

(a) A basic *RC* lead network

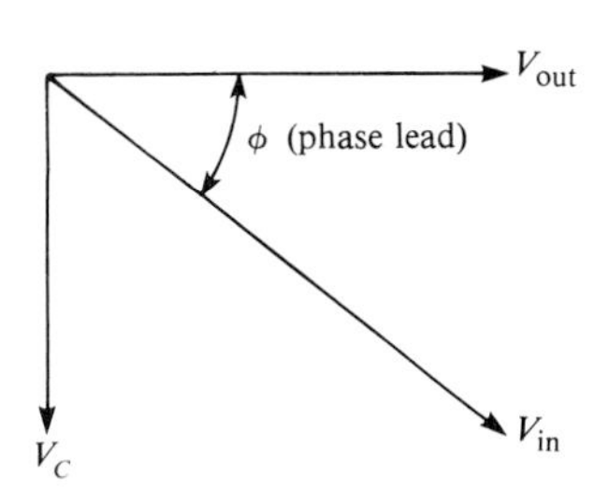

(b) Phasor voltage diagram showing the phase lead between V_{in} and V_{out}

FIGURE 13–44
The *RC* lead network ($V_{out} = V_R$).

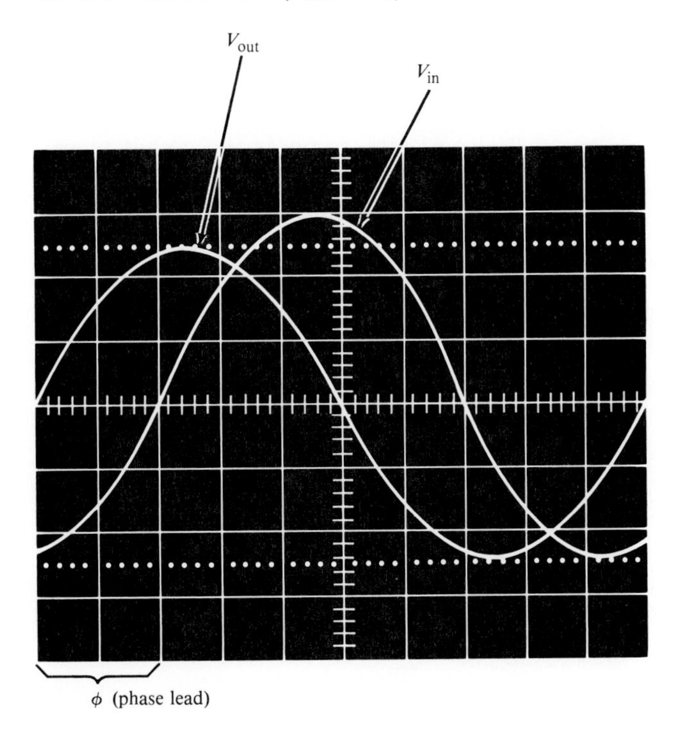

FIGURE 13–45
General oscilloscope display of the input and output waveforms of a lead network (V_{out} leads V_{in}).

EXAMPLE 13–16 Calculate the phase lead and the output voltage for the circuit in Figure 13–46.

FIGURE 13–46

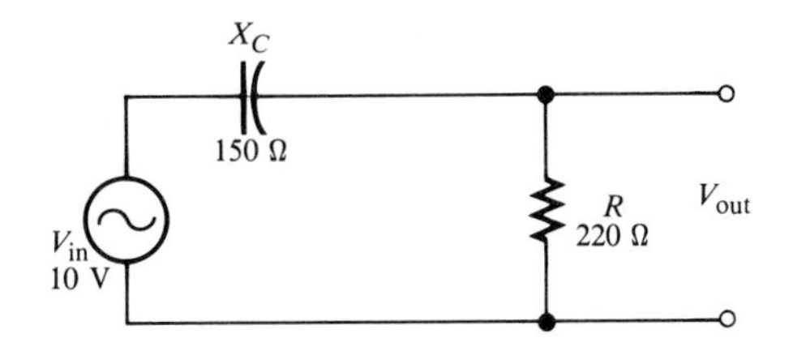

Solution

$$\phi = \arctan\left(\frac{X_C}{R}\right) = \arctan\left(\frac{150\ \Omega}{220\ \Omega}\right) = 34.29°$$

The output leads the input by 34.29°.

$$V_{out} = \left(\frac{R}{\sqrt{R^2 + X_C^2}}\right)V_{in} = \left(\frac{220\ \Omega}{\sqrt{(220\ \Omega)^2 + (150\ \Omega)^2}}\right)10\ \text{V}$$
$$= 8.26\ \text{V}$$

The calculator sequence for V_{out} is

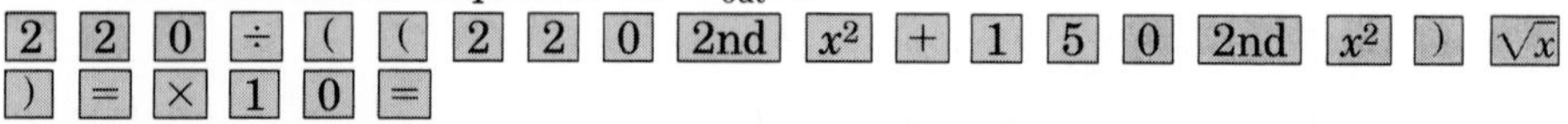

Effects of Frequency on the Phase Shift and Output Voltage of the Lead Network

Since the phase lead is the same as the circuit phase angle θ, it *decreases* as frequency increases. The output voltage increases with frequency because as X_C becomes smaller, more of the input voltage is dropped across the resistor. Figure 13–47 illustrates this relationship.

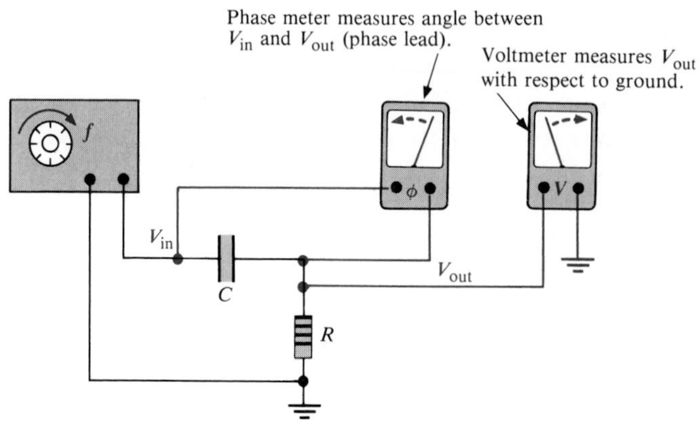

(a) When f increases, the phase lead ϕ decreases and V_{out} increases.

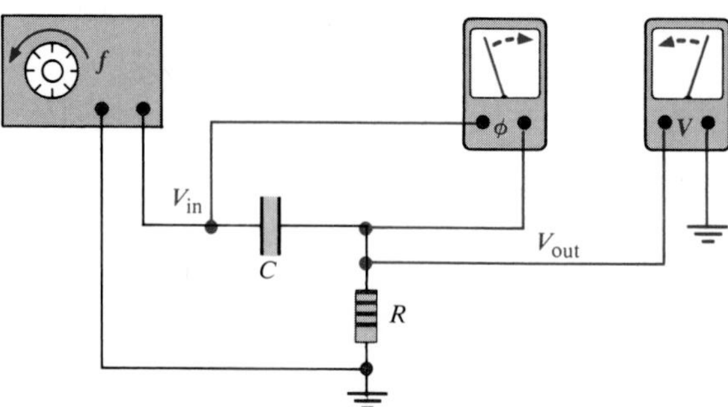

(b) When f decreases, ϕ increases and V_{out} decreases.

FIGURE 13–47
Illustration of how the frequency affects the phase lead and output voltage in an *RC* lead network with the amplitude of V_{in} held constant.

The *RC* Circuit as a Filter

Filters are frequency-selective circuits that permit signals of certain frequencies to pass from the input to the output while blocking all others. That is, all frequencies but the selected ones are *filtered* out.

Series *RC* circuits exhibit a frequency-selective characteristic and therefore act as basic filters. There are two types. The first one that we examine, called a *low-pass filter,* is realized by taking the output across the capacitor, just as in a lag network. The second type, called a *high-pass filter,* is implemented by taking the output across the resistor, as in a lead network.

The *RC* Low-Pass Filter

You have seen what happens to the phase and the output voltage in the lag network. In terms of its filtering action, we are interested primarily in how the output voltage varies with frequency.

To illustrate low-pass filter action, Figure 13–48 shows a specific series of measurements in which the frequency starts at zero (dc) and is increased in increments up to 20 kHz. At each value of frequency, the output voltage is measured. As you can see, the capacitive reactance decreases as frequency goes up, thus dropping less voltage across the capacitor while the input voltage is held at a constant 10 V throughout each step.

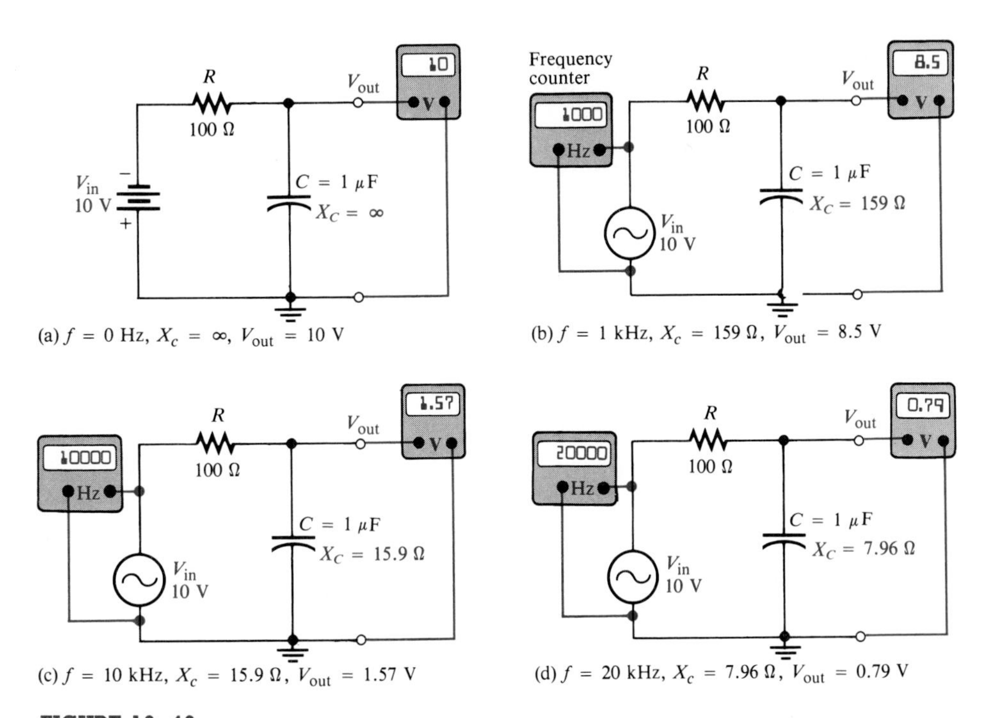

FIGURE 13–48
Example of low-pass filter action.

In Figure 13–49, the measured values are plotted on a graph of V_{out} versus frequency, and a smooth curve is drawn connecting the points. This graph, called a *response curve,* shows that the output voltage is greater at the lower frequencies and decreases as the frequency increases.

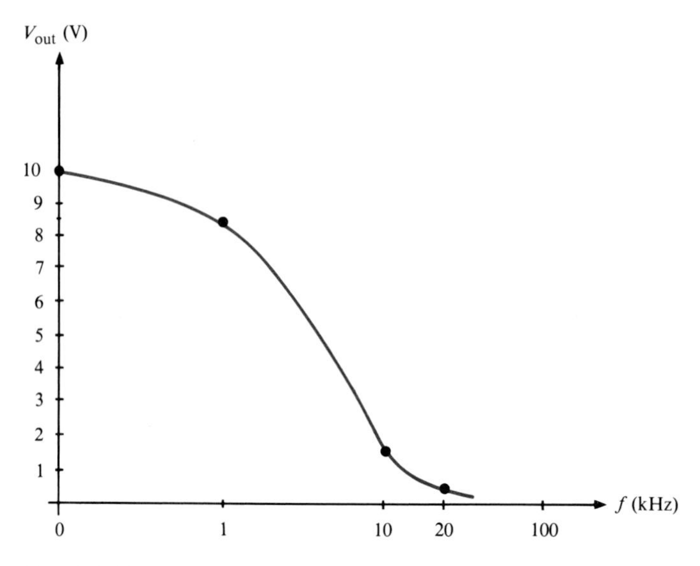

FIGURE 13–49
Frequency response curve for the low-pass filter in Figure 13–48.

The *RC* High-Pass Filter

To illustrate high-pass filter action, Figure 13–50 shows a series of specific measurements. Again, the frequency starts at zero (dc) and is increased in increments up to 10 kHz. As you can see, the capacitive reactance decreases as the frequency goes up, thus causing more of the total input voltage to be dropped across the resistor.

In Figure 13–51, the measured values have been plotted to produce a response curve for this circuit. As you can see, the output voltage is greater at the higher frequencies and decreases as the frequency is reduced.

The Cutoff Frequency and the Bandwidth of a Filter

The frequency at which the capacitive reactance equals the resistance in a low-pass or high-pass *RC* filter is called the *cutoff frequency* and is designated f_c. This condition is expressed as $1/(2\pi f_c C) = R$. Solving for f_c, we get the following formula:

$$f_c = \frac{1}{2\pi RC} \qquad \textbf{(13–30)}$$

At f_c, the output voltage of the filter is 70.7% of its maximum value. It is standard practice to consider the cutoff frequency as the limit of a filter's performance in terms of passing or rejecting frequencies. For example, in a low-pass filter, all frequencies above f_c are considered to be passed by the filter, and all those below f_c are considered to be rejected. The reverse is true for a high-pass filter.

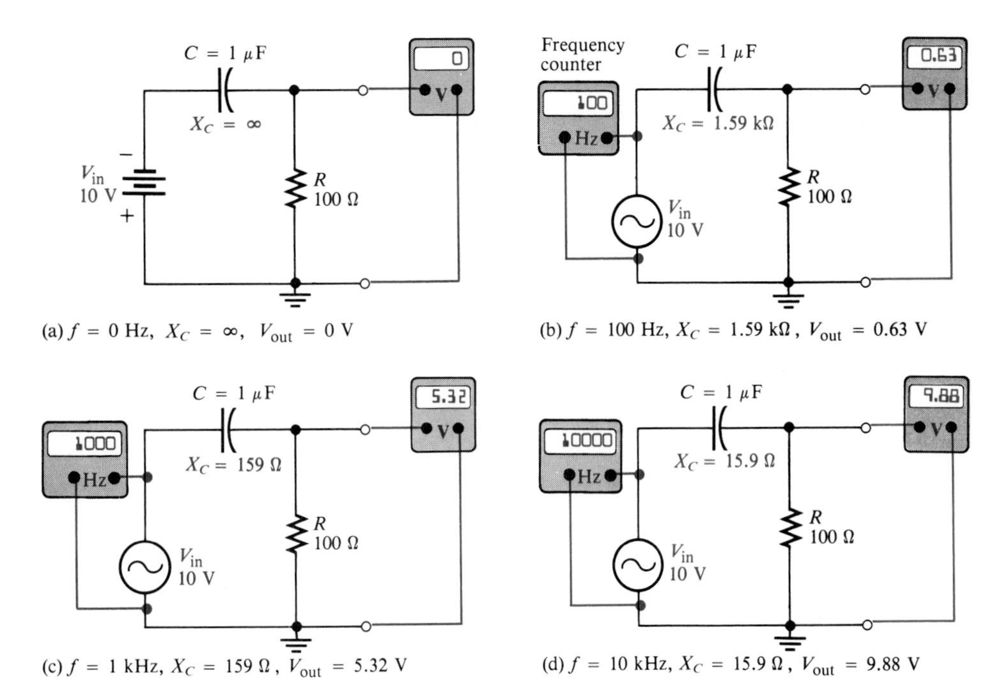

FIGURE 13–50
Example of high-pass filter action.

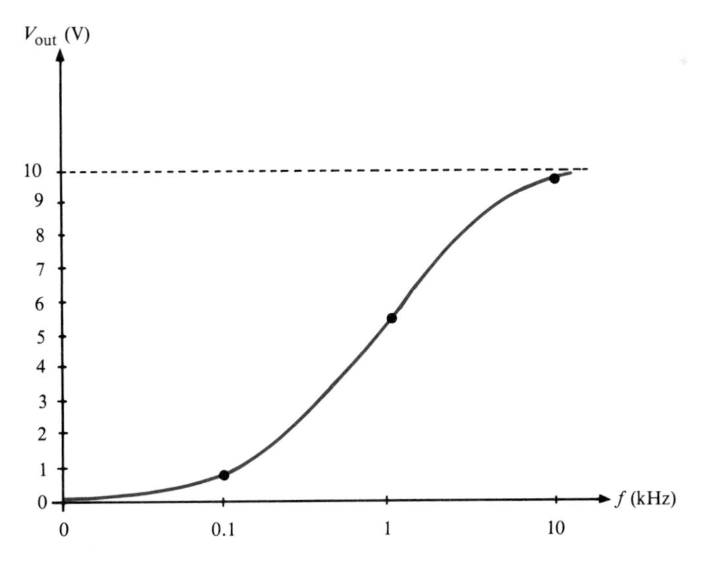

FIGURE 13–51
Frequency response curve for the high-pass filter in Figure 13–50.

The range of frequencies that are considered to be passed by a filter is called the *bandwidth*. Figure 13–52 illustrates the bandwidth and the cutoff frequency for a low-pass filter.

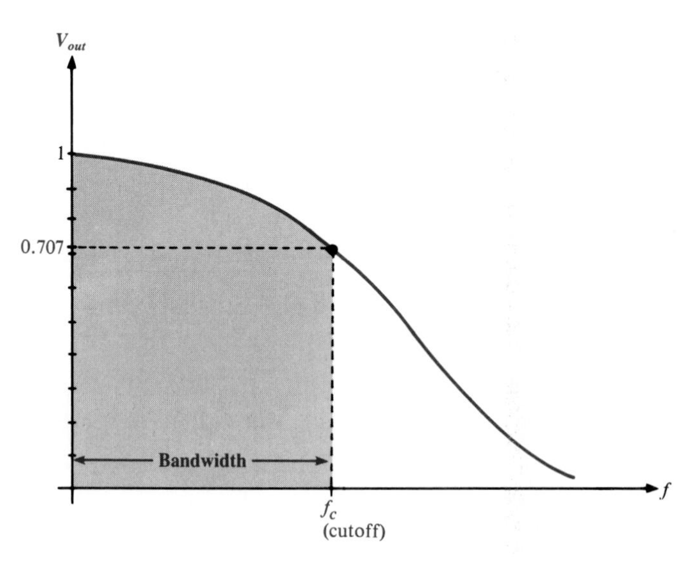

FIGURE 13–52
Normalized general response curve of a low-pass filter showing the cutoff frequency and the bandwidth.

Coupling an ac Signal into a dc Bias Network

Figure 13–53 shows an RC network that is used to create a dc voltage level with an ac voltage superimposed on it. This type of circuit is commonly found in amplifiers in which the dc voltage is required to *bias* the amplifier to the proper operating point and the signal voltage to be amplified is coupled through a capacitor and superimposed on the dc level. The capacitor prevents the low internal resistance of the signal source from affecting the dc bias voltage.

FIGURE 13–53
Amplifier bias and signal-coupling circuit.

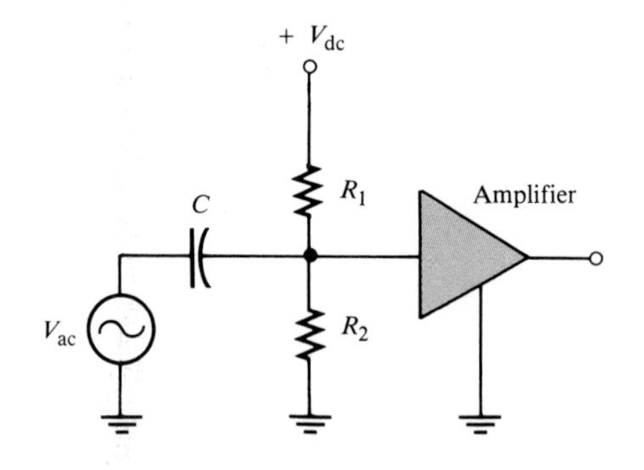

In this type of application, a relatively high value of capacitance is selected so that for the frequencies to be amplified, the reactance is very small compared to the resistance of the bias network. When the reactance is very small (ideally zero), there is practically no phase shift or signal voltage dropped across the capacitor. Therefore, all of the signal voltage passes from the source to the input to the amplifier.

Figure 13–54 illustrates the application of the superposition principle to circuits. In Part (a), we have effectively removed the ac source from the circuit by replacing it with a short to represent its ideal internal resistance. Since C is open to dc, the voltage at point A is determined by the voltage divider action of R_1 and R_2 and the dc voltage source.

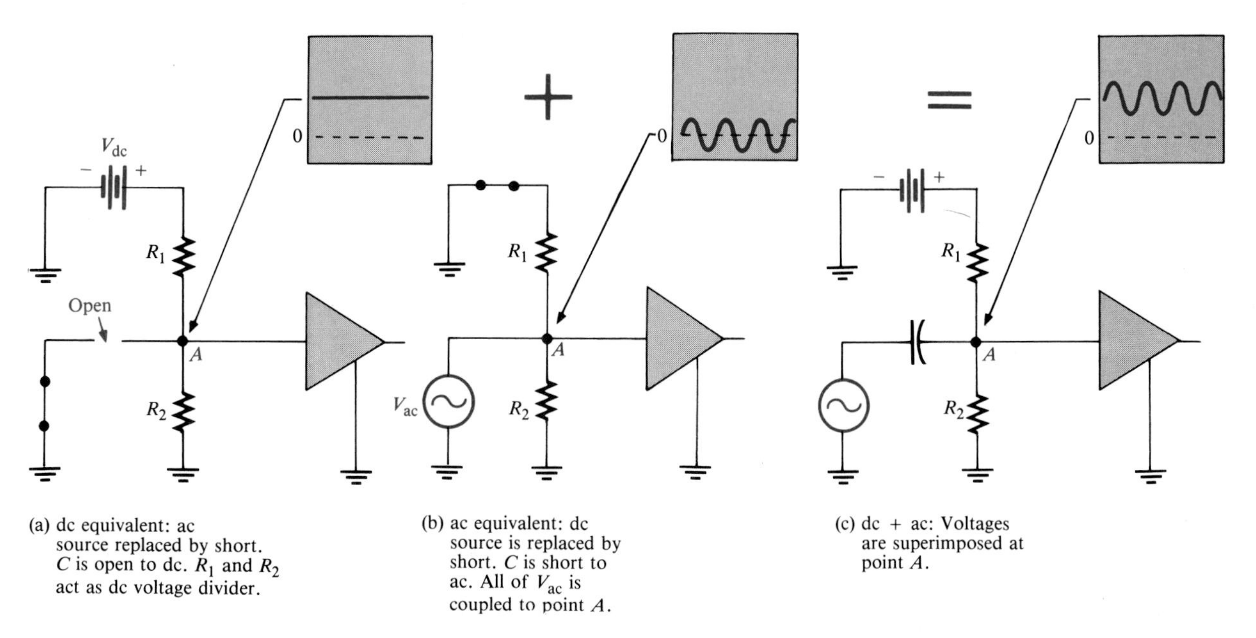

FIGURE 13–54
The superposition of dc and ac voltages in an RC bias and coupling circuit.

In Part (b), we have effectively removed the dc source from the circuit by replacing it with a short to represent its ideal internal resistance. Since C appears as a short at the frequency of the ac, the signal voltage is coupled directly to point A and appears across the parallel combination of R_1 and R_2. Part (c) illustrates that the combined effect of the superposition of the dc and the ac voltages results in the signal voltage "riding" on the dc level.

Troubles in *RC* Networks

An Open Component in a Basic Low-Pass Filter

If the *capacitor* opens, the input signal passes through the resistor to the output and is unaffected by the open capacitor. Thus, you will observe a complete loss of normal filter action. If the *resistor* opens, there is no output signal because the path between the input and output is broken.

A Shorted or Leaky Capacitor in a Basic Low-Pass Filter

If the capacitor shorts out, the output signal voltage is zero due to the direct short through the capacitor to ground.

Another possible condition is that the capacitor becomes excessively leaky, effectively placing a resistance in parallel with the capacitor to ground, which, in effect, combines with the series resistor to form a voltage divider. As a result, the normal output signal voltage is reduced.

An Open Component in a Basic High-Pass Filter

An open resistor to ground causes a reduction in the cutoff frequency to a value dependent on the load resistance (if any) or the resistance of the measuring instrument that is connected to the output. An open capacitor breaks the path between input and output and results in zero output voltage.

A Shorted or Leaky Capacitor in a Basic High-Pass Filter

A shorted capacitor causes the input signal to pass directly through to the output, so you will observe a complete loss of filter action.

If the capacitor becomes leaky, you will observe an output voltage all the way down to dc because the leakage resistance is, in effect, in parallel with the capacitor, thus permitting even a dc voltage to pass through to the output.

SECTION REVIEW 13–8

1. A certain *RC* lag network consists of a 4.7-kΩ resistor and a 0.022-μF capacitor. Determine the phase lag between input and output at a frequency of 3 kHz.
2. An *RC* lead network has the same component values as the lag network in Problem 1. What is the magnitude of the output voltage at 3 kHz when the input is 10 V rms?
3. When an *RC* circuit is used as a low-pass filter, across which component is the output taken?

APPLICATION NOTE

For the Application Assignment at the beginning of the chapter, you were to review the audio amplifier section of a communication system to see if it will accommodate lower frequencies without significant coupling loss. As you know, when frequency decreases, the reactance of a capacitor increases. Therefore, the coupling capacitors between each audio amplifier will drop more voltage at the lower frequencies. You should check how much signal voltage is dropped across each coupling capacitor at the lowest frequency in the band. If there is a noticeable difference in voltage from one side of the capacitor to the other, increase the capacitance value to reduce the drop in signal voltage.

SUMMARY

Facts

- When a sine wave voltage is applied to an *RC* circuit, the current and all the voltage drops are also sine waves.
- Total current in an *RC* circuit always leads the source voltage.
- The resistor voltage is always in phase with the current.
- The capacitor voltage always lags the current by 90°.
- In an *RC* circuit, the impedance is determined by both the resistance and the capacitive reactance combined.
- Impedance is expressed in units of ohms.

- The circuit phase angle is the angle between the total current and the applied (source) voltage.
- The impedance of an *RC* circuit varies inversely with frequency.
- The phase angle (θ) of a series *RC* circuit varies inversely with frequency.
- For each parallel *RC* circuit, there is an equivalent series circuit.
- The impedance of a circuit can be determined by measuring the applied voltage and the total current and then applying Ohm's law.
- In an *RC* circuit, part of the power is resistive and part reactive.
- The phasor combination of resistive power (true power) and reactive power is called *apparent power.*
- Apparent power is expressed in volt-amperes (VA).
- The power factor *(PF)* indicates how much of the apparent power is true power.
- A power factor of 1 indicates a purely resistive circuit, and a power factor of 0 indicates a purely reactive circuit.
- In a lag network, the output voltage lags the input voltage in phase.
- In a lead network, the output voltage leads the input voltage.
- A filter passes certain frequencies and rejects others.

Units

- Volt-amperes (VA)—the unit of apparent power (P_a).
- Volt-amperes reactive (VAR)—the unit of reactive power (P_r).

Definitions

- *Impedance* (Z)—the total opposition to sinusoidal current expressed in ohms.
- *Conductance* (G)—the reciprocal of resistance.
- *Capacitive susceptance* (B_C)—the reciprocal of capacitive reactance.
- *Admittance* (Y)—the reciprocal of impedance.
- *Filter*—a frequency-selective circuit.
- *Cutoff frequency* (f_c)—the frequency at which the output voltage of a filter is 70.7% of the maximum output voltage.
- *Bandwidth*—the range of frequencies that are considered to be passed by a filter.

Formulas

$$Z = \sqrt{R^2 + X_C^2} \quad \text{Series } RC \text{ impedance} \tag{13–1}$$

$$\theta = \arctan\left(\frac{X_C}{R}\right) \quad \text{Series } RC \text{ phase angle} \tag{13–2}$$

$$V = IZ \quad \text{Ohm's law} \tag{13–3}$$

$$I = \frac{V}{Z} \quad \text{Ohm's law} \tag{13–4}$$

$$Z = \frac{V}{I} \quad \text{Ohm's law} \tag{13–5}$$

$$V_s = \sqrt{V_R^2 + V_C^2} \quad \text{Total voltage in series } RC \tag{13–6}$$

$$\theta = \arctan\left(\frac{V_C}{V_R}\right) \quad \text{Series } RC \text{ phase angle} \tag{13–7}$$

$$Z = \frac{RX_C}{\sqrt{R^2 + X_C^2}} \quad \text{Parallel } RC \text{ impedance} \tag{13–8}$$

$$\theta = \arctan\left(\frac{R}{X_C}\right)$$ Parallel RC phase angle (13–9)

$$G = \frac{1}{R}$$ Conductance (13–10)

$$B_C = \frac{1}{X_C}$$ Capacitive susceptance (13–11)

$$Y = \frac{1}{Z}$$ Admittance (13–12)

$$Y = \sqrt{G^2 + B_C^2}$$ Admittance (13–13)

$$V = \frac{I}{Y}$$ Ohm's law (13–14)

$$I = VY$$ Ohm's law (13–15)

$$Y = \frac{I}{V}$$ Ohm's law (13–16)

$$I_T = \sqrt{I_R^2 + I_C^2}$$ Total current in parallel RC circuits (13–17)

$$\theta = \arctan\left(\frac{I_C}{I_R}\right)$$ Parallel RC phase angle (13–18)

$$R_{eq} = Z \cos \theta$$ Equivalent series resistance (13–19)

$$X_{C(eq)} = Z \sin \theta$$ Equivalent series reactance (13–20)

$$P_{true} = I_T^2 R$$ True power (W) (13–21)

$$P_r = I_T^2 X_C$$ Reactive power (VAR) (13–22)

$$P_a = I_T^2 Z$$ Apparent power (VA) (13–23)

$$P_{true} = V_s I_T \cos \theta$$ True power (13–24)

$$PF = \cos \theta$$ Power factor (13–25)

$$\phi = 90° - \arctan\left(\frac{X_C}{R}\right)$$ Phase lag (13–26)

$$V_{out} = \left(\frac{X_C}{\sqrt{R^2 + X_C^2}}\right)V_{in}$$ Output of lag network (13–27)

$$\phi = \arctan\left(\frac{X_C}{R}\right)$$ Phase lead (13–28)

$$V_{out} = \left(\frac{R}{\sqrt{R^2 + X_C^2}}\right)V_{in}$$ Output of lead network (13–29)

$$f_c = \frac{1}{2\pi RC}$$ Cutoff frequency of an RC filter (13–30)

SELF-TEST

1. In a series RC circuit, explain how each voltage drop differs from the applied voltage.
2. Describe the phase relationships between the current in a series RC circuit and **(a)** the resistor voltage. **(b)** the capacitor voltage.

3. If the frequency of the voltage applied to an *RC* circuit is increased, what happens to the impedance? To the phase angle?
4. If the frequency is doubled and the resistance is doubled, how much does the value of the impedance change?
5. To reduce the current in a series *RC* circuit, should you increase or decrease the frequency?
6. In a series *RC* circuit, 10 V rms is measured across the resistor and 10 V rms across the capacitor. Is the rms source voltage
   (a) 20 V, **(b)** 14.14 V, **(c)** 28.28 V, or **(d)** 10 V?
7. The voltages in Question 6 are measured at a certain frequency. To make the resistor voltage greater than the capacitor voltage, must the frequency be increased or reduced?
8. When the resistor voltage in Question 7 becomes greater than the capacitor voltage, does the phase angle increase or decrease?
9. Does the impedance of a parallel *RC* circuit increase or decrease with frequency?
10. In a parallel *RC* circuit, 1 A rms flows through the resistive branch and 1 A rms through the capacitive branch. Is the total rms current
    (a) 1 A, **(b)** 2 A, **(c)** 1.414 A, or **(d)** 2.28 A?
11. Refer to Figure 13–34. The V_{R1} waveform is as shown, but the V_s waveform crosses the horizontal axis at the third division from the left of the screen. What is the phase angle indicated in this case?
12. What does a power factor close to 1 indicate?
13. If a load is purely resistive and the true power is 5 W, what is the apparent power?
14. For a certain load, the true power is 100 W and the reactive power is 100 VAR. Is the apparent power
    (a) 200 VA, **(b)** 100 VA, or **(c)** 141.4 VA?
15. Why is it important to rate sources according to volt-amperes rather than watts?
16. If the bandwidth of a certain low-pass filter is 1 kHz, what is its cutoff frequency?

PROBLEM SET A

Section 13–1

13–1 A 8-kHz sinusoidal voltage is applied to a series *RC* circuit. What is the frequency of the voltage across the resistor? The capacitor?

13–2 What is the waveshape of the current in the circuit of Problem 13–1?

Section 13–2

13–3 Find the total impedance of each circuit in Figure 13–55.

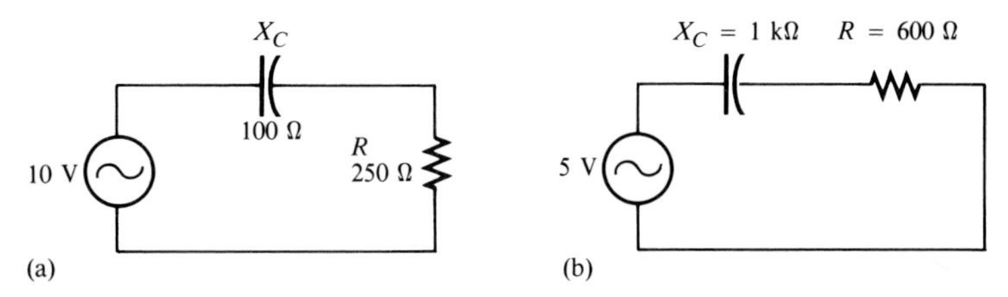

FIGURE 13–55

13–4 Determine the impedance and the phase angle in each circuit in Figure 13–56.

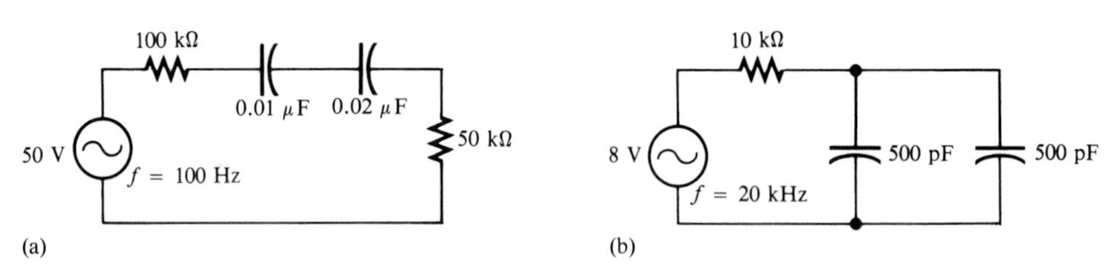

FIGURE 13–56

Section 13–3

13–5 Calculate the total current in each circuit of Figure 13–55.

13–6 Repeat Problem 13–5 for the circuits in Figure 13–56.

13–7 For the circuit in Figure 13–57, draw the phasor diagram showing all voltages and the total current. Indicate the phase angles.

FIGURE 13–57

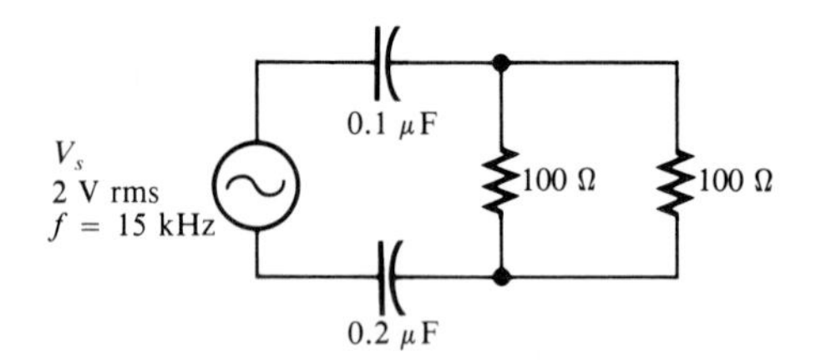

13–8 For the circuit in Figure 13–58, determine the following:
(a) Z **(b)** I **(c)** V_R **(d)** V_C

FIGURE 13–58

Section 13–4

13–9 Determine the impedance for the circuit in Figure 13–59.

FIGURE 13–59

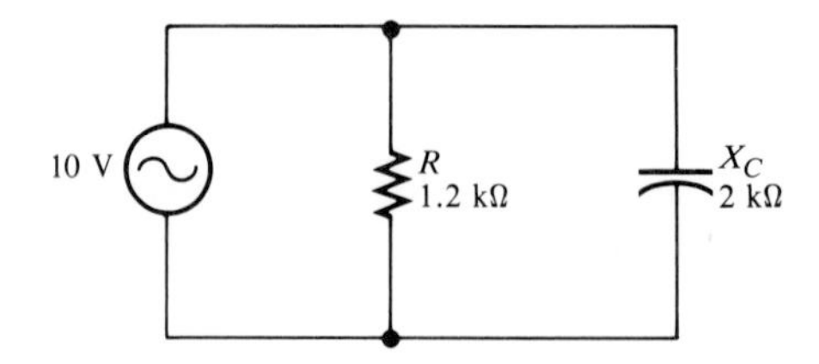

13–10 Determine the impedance and the phase angle in Figure 13–60.

FIGURE 13–60

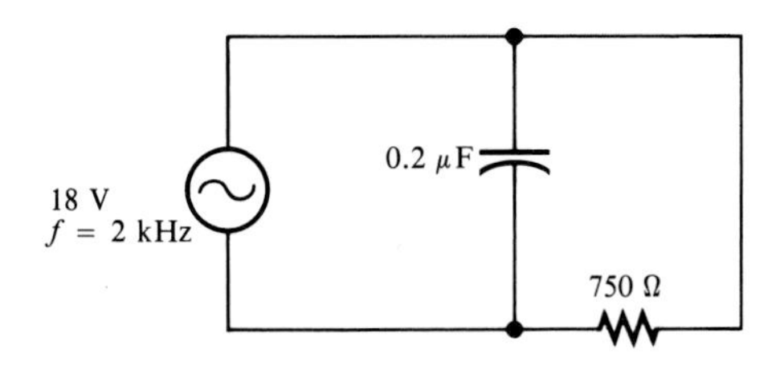

13–11 Repeat Problem 13–10 for the following frequencies:
(a) 1.5 kHz **(b)** 3 kHz **(c)** 5 kHz **(d)** 10 kHz

Section 13–5

13–12 For the circuit in Figure 13–61, find all the currents and voltages.

FIGURE 13–61

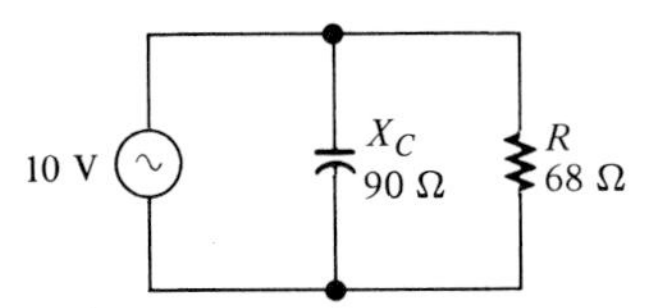

13–13 For the parallel circuit in Figure 13–62, find the magnitude of each branch current and the total current. What is the phase angle between the applied voltage and the total current?

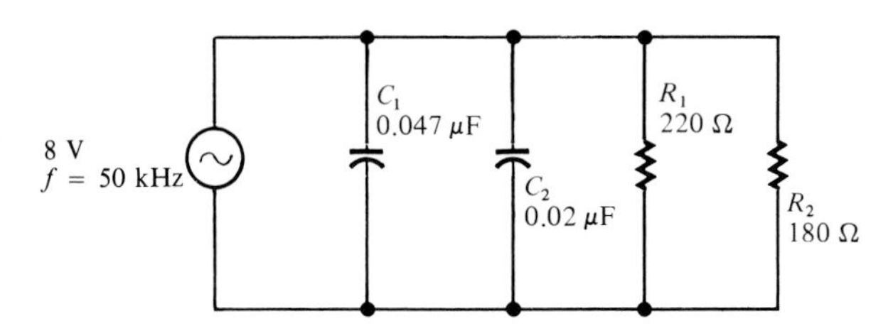

FIGURE 13–62

13–14 For the circuit in Figure 13–63, determine the following:
(a) Z **(b)** I_R **(c)** I_C **(d)** I_T **(e)** θ

FIGURE 13–63

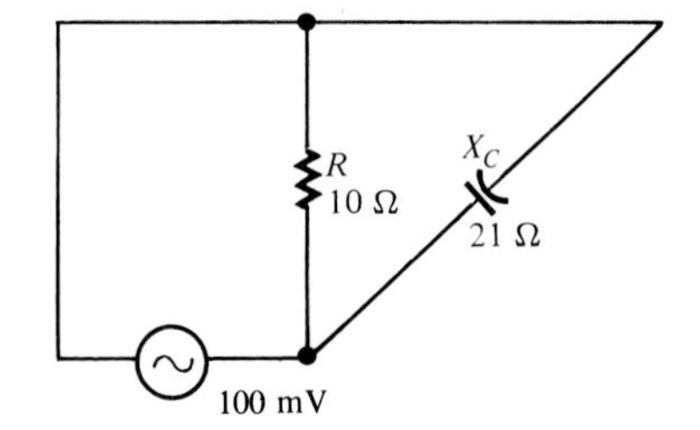

13–15 Repeat Problem 13–14 for $R = 5\ \text{k}\Omega$, $C = 0.05\ \mu\text{F}$, and $f = 500$ Hz.

13–16 Convert the circuit in Figure 13–64 to an equivalent series form.

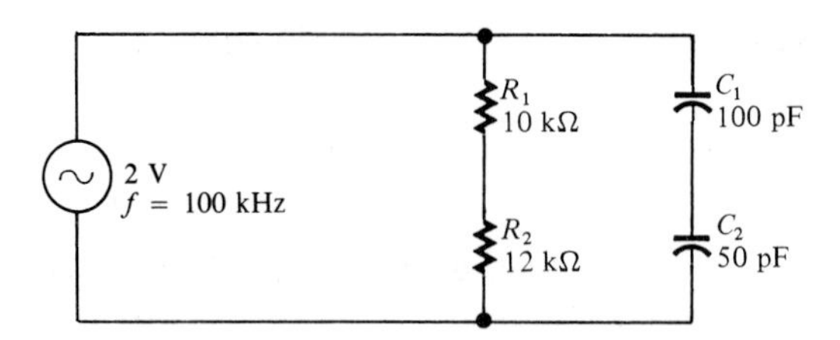

FIGURE 13–64

Section 13–6

13–17 Determine the voltages across each element in Figure 13–65. Sketch the voltage phasor diagram.

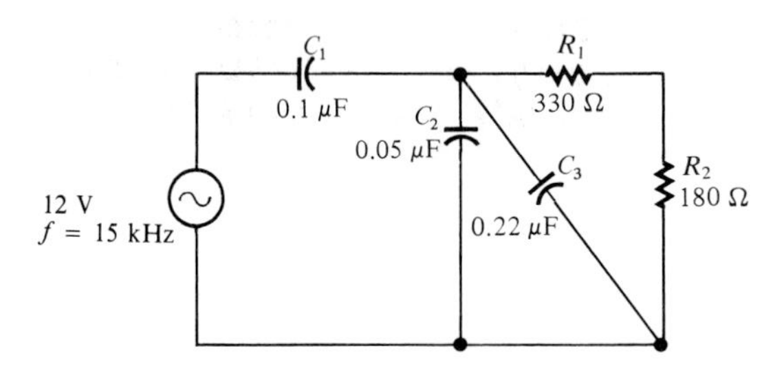

FIGURE 13–65

13–18 Is the circuit in Figure 13–65 predominantly resistive or predominantly capacitive?

13–19 Find the current through each branch and the total current in Figure 13–65. Sketch the current phasor diagram.

13–20 For the circuit in Figure 13–66, determine the following:
(a) I_T **(b)** θ **(c)** V_{R1} **(d)** V_{R2} **(e)** V_{R3} **(f)** V_C

FIGURE 13–66

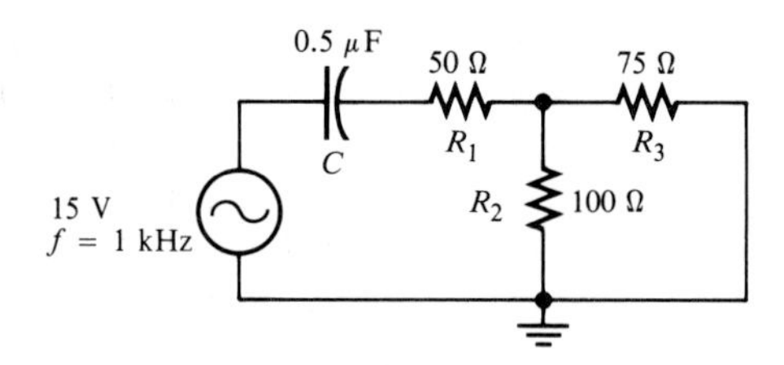

Section 13–7

13–21 In a certain series RC circuit, the true power is 2 W, and the reactive power is 3.5 VAR. Determine the apparent power.

13–22 In Figure 13–58, what is the true power and the reactive power?

13–23 What is the power factor for the circuit of Figure 13–64?

13–24 Determine P_{true}, P_r, P_a, and PF for the circuit in Figure 13–66. Sketch the power triangle.

Section 13–8

13–25 For the lag network in Figure 13–67, determine the phase lag between the input voltage and the output voltage for each of the following frequencies:
(a) 1 Hz **(b)** 100 Hz **(c)** 1 kHz **(d)** 10 kHz

FIGURE 13–67

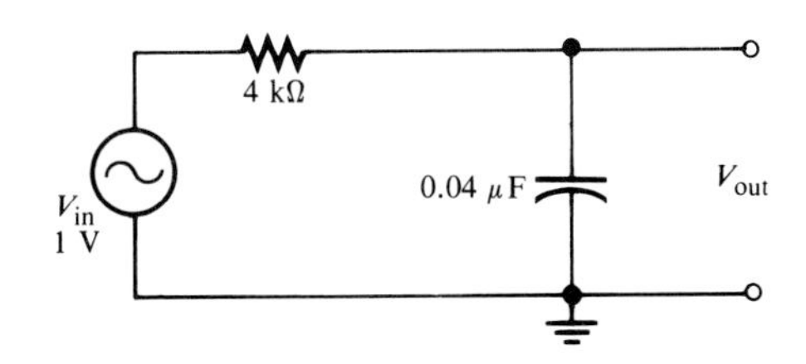

13–26 The lag network in Figure 13–67 also acts as a low-pass filter. Draw a response curve for this circuit by plotting the output voltage versus frequency for 0 Hz to 10 kHz in 1-kHz increments.

13–27 Repeat Problem 13–25 for the lead network in Figure 13–68.

FIGURE 13–68

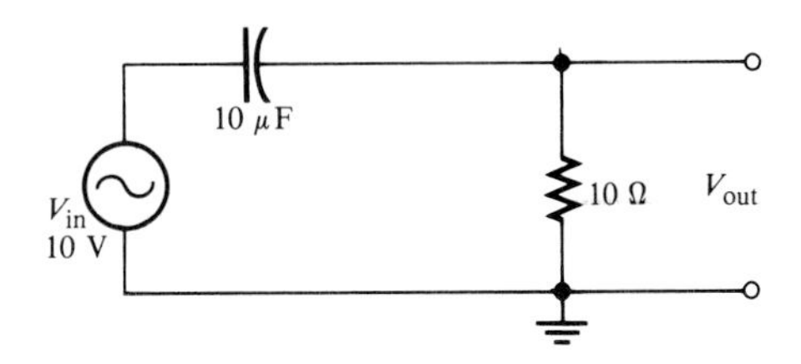

13–28 Plot the frequency response curve for the lead network in Figure 13–68 for a frequency range of 0 Hz to 10 kHz in 1-kHz increments.

13–29 Draw the voltage phasor diagram for each circuit in Figures 13–67 and 13–68 for a frequency of 5 kHz with $V_{in} = 1$ V rms.

13–30 The rms value of the signal voltage out of amplifier *A* in Figure 13–69 is 50 mV. If the input resistance to amplifier *B* is 10 kΩ, how much of the signal is lost due to the coupling capacitor when the frequency is 3 kHz?

FIGURE 13–69

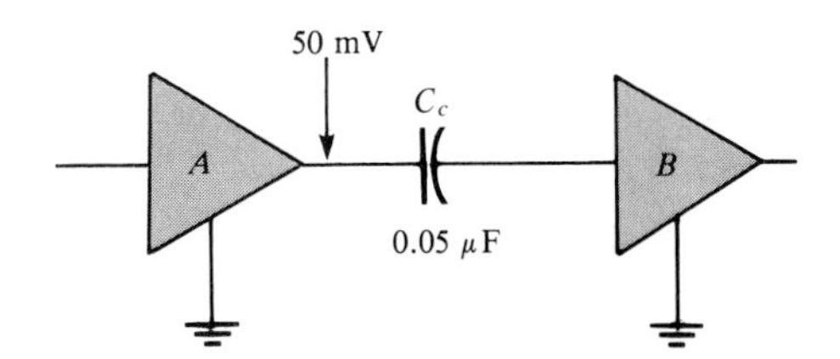

13–31 Determine the cutoff frequency for each circuit in Figures 13–67 and 13–68.

13–32 Determine the bandwidth of the circuit in Figure 13–67.

PROBLEM SET B

13–33 A single 240-V, 60-Hz source drives two loads. Load *A* has an impedance of 50 Ω and a power factor of 0.85. Load *B* has an impedance of 72 Ω and a power factor of 0.95.

(a) How much current does each load draw?
(b) What is the reactive power in each load?
(c) What is the true power in each load?
(d) What is the apparent power in each load?
(e) Which load has more voltage drop along the lines connecting it to the source?

13–34 A certain load dissipates 1.5 kW of power with an impedance of 12 Ω and a power factor of 0.75. What is its reactive power? What is its apparent power?

13–35 Determine the series element or elements that are in the block of Figure 13–70 to meet the following overall circuit requirements:
(a) $P_{true} = 400$ W **(b)** leading power factor (I_T leads V_s).

FIGURE 13–70

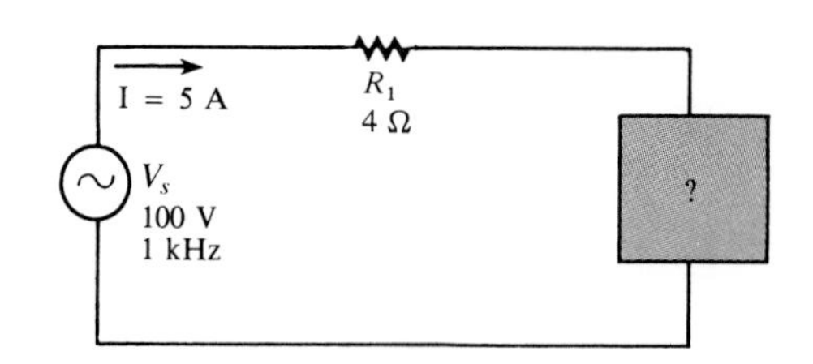

13–36 Determine the value of C_2 in Figure 13–71 when $V_A = V_B$.

FIGURE 13–71

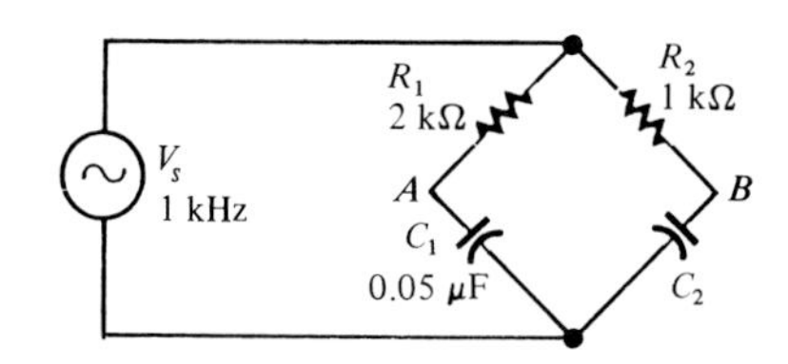

13–37 What minimum value of coupling capacitor is required in Figure 13–69 so that the signal voltage at the input of amplifier B is at least 70.7% of the signal voltage at the output of amplifier A when the frequency is 20 Hz? Neglect the input resistance of the amplifier.

ANSWERS TO SECTION REVIEWS

Section 13–1
1. 60 Hz, 60 Hz. **2.** Closer to 0°.

Section 13–2
1. The opposition to sinusoidal current flow. **2.** It lags.
3. The capacitive reactance. **4.** 59.91 Ω, 56.6°.

Section 13–3
1. 7.2 V. **2.** 56.3°. **3.** 90°.
4. **(a)** X_C decreases; **(b)** Z decreases; **(c)** θ decreases.

Section 13–4
1. *Conductance* is the reciprocal of resistance; *capacitive susceptance* is the reciprocal of capacitive reactance; and *admittance* is the reciprocal of impedance.
2. $Y = 0.01$ S. **3.** $Y = 0.024$ S.

Section 13–5
1. 21 mA. **2.** 56.3°, 18 mA. **3.** 90°.

Section 13–6
1. See Figure 13–72. **2.** 6.79 V.

FIGURE 13–72

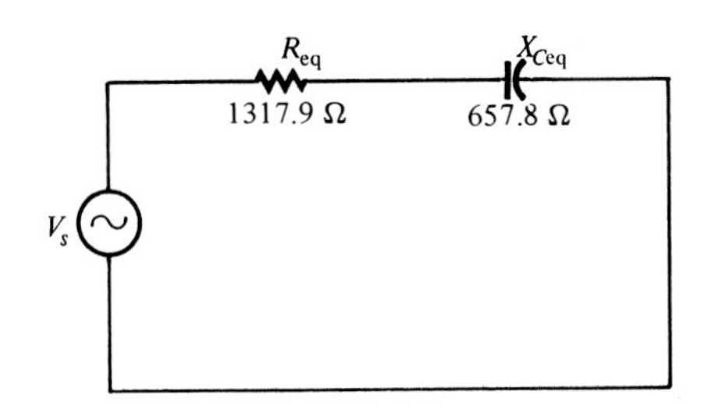

Section 13–7

1. Resistance. **2.** 0.707.

3. $P_{true} = 1200$ W, $P_r = 1840$ VAR, $P_a = 2196.73$ VA.

Section 13–8

1. 62.84°. **2.** 8.9 V rms. **3.** Capacitor.

Frequency Response of *RL* Circuits

An *RL* circuit contains both resistance and inductance. In this chapter, basic series and parallel *RL* circuits and their responses to sine wave voltages are covered. In addition, series-parallel combinations are examined. Power considerations in *RL* circuits are introduced, and practical aspects of the power factor are discussed. A method of improving the power factor is presented, and basic areas of application are covered. As you study this chapter, note the difference in response of *RL* circuits compared to *RC* circuits.

In this chapter, you will learn:

- How an *RL* circuit responds to a sine wave input voltage.
- How to determine the impedance of both series and parallel *RL* circuits.
- Why there is a phase difference between current and voltage in an *RL* circuit and how to determine its value.
- How the phase angle in an *RL* circuit differs from that in an *RC* circuit.
- How to determine the effects of frequency on an *RL* circuit in terms of changes in impedance and phase angle.
- The definition of *inductive susceptance.*
- How to apply Ohm's law to *RL* circuit to find currents and voltages.
- How to determine the true power, reactive power, and apparent power.
- More about the significance of the power factor.
- How to increase the power factor of an inductive load.
- How *RL* circuits can be used as phase-shifting circuits.
- How *RL* circuits can be used as filters.

14

APPLICATION ASSIGNMENT

As a technician, you are assigned to set up a system of servomotors that drive an array of antennas. The purpose of the system is to gather information for a wildlife management study by tracking the movements of wild animals to which transmitters have been attached.

Each of the motors represents an inductive load that is effectively an *RL* circuit with defined values of *R* and *L*. Since the system must be portable, the size of the source that drives each motor must be kept to a minimum. Your project director has asked you to reduce the amount of current that each motor draws from the source and to keep the size of the interconnecting cables to a minimum. The director suggests that reducing the power factor of each motor is a way to accomplish this objective.

After completing this chapter, you will be able to carry out this assignment. The solution is given in the Application Note at the end of the chapter.

SINUSOIDAL RESPONSE OF *RL* CIRCUITS

14–1

As with the *RC* circuit, all currents and voltages in an *RL* circuit are sinusoidal when the input is sinusoidal. Phase shifts are introduced because of the inductance. As you will learn, the resistor voltage and current are in phase with each other but lag the source voltage, and the inductor voltage leads the source voltage. The phase angle between the current and the inductor voltage is always 90°. These generalized phase relationships are indicated in Figure 14–1. Notice that they are opposite from those of the *RC* circuit, as discussed in Chapter 13.

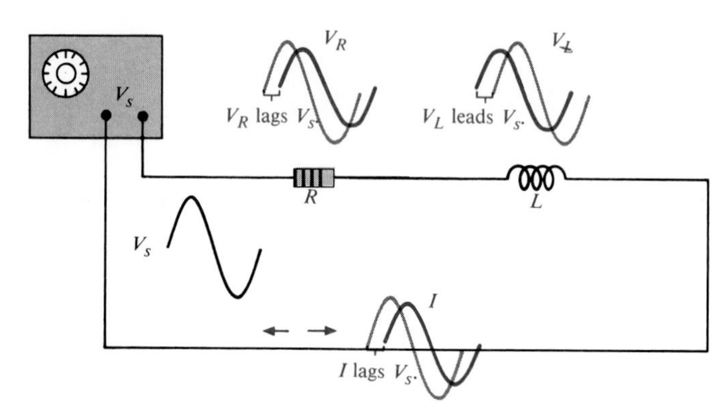

FIGURE 14–1
Illustration of sinusoidal response with general phase relationships of V_R, V_L, and I relative to V_s. V_R and I are in phase, V_R lags V_s, and V_L leads V_s. V_R and V_L are 90° out of phase with each other.

The amplitudes and the phase relationships of the voltages and current depend on the ohmic values of the resistance and the inductive reactance. When a circuit is purely inductive, the phase angle between the applied voltage and the total current is 90°, with the current *lagging* the voltage. When there is a combination of both resistance and inductive reactance in a circuit, the phase angle is somewhere between zero and 90°, depending on the relative values of R and X_L.

SECTION REVIEW 14–1

1. A 1-kHz sinusoidal voltage is applied to an *RL* circuit. What is the frequency of the resulting current?
2. When the resistance in an *RL* circuit is greater than the inductive reactance, is the phase angle between the applied voltage and the total current closer to 0° or to 90°?

IMPEDANCE AND PHASE ANGLE OF SERIES *RL* CIRCUITS

14–2

As you know, impedance is the total opposition to sinusoidal current in a circuit and is expressed in ohms. The impedance of a series *RL* circuit is determined by the resistance and the inductive reactance, as indicated in Figure 14–2.

FIGURE 14–2
Impedance of a series RL circuit.

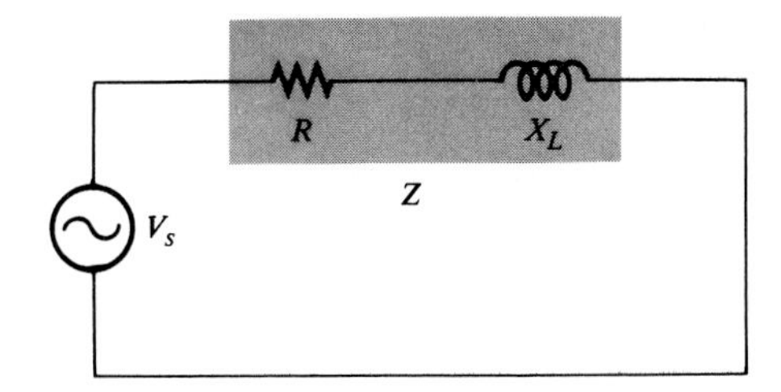

The Impedance Triangle

In ac analysis, both R and X_L are treated as *phasor quantities,* as shown in the phasor diagram of Figure 14–3(a), with X_L appearing at a $+90°$ angle with respect to R. This relationship comes from the fact that the inductor voltage leads the current, and thus the resistor voltage, by 90°. Since Z is the phasor sum of R and X_L, its phasor representation is as shown in Figure 14–3(b). A repositioning of the phasors, as shown in Part (c), forms a right triangle. This formation, as you have learned, is called the *impedance triangle.* The length of each phasor represents the magnitude of the quantity, and θ is the phase angle between the applied voltage and the current in the RL circuit.

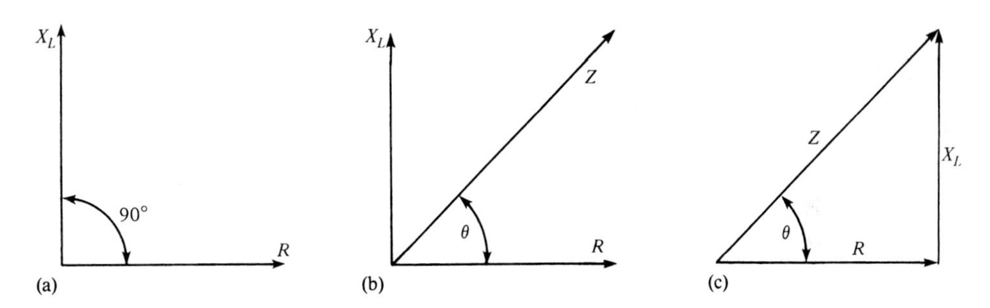

FIGURE 14–3
Development of the impedance triangle for a series RL circuit.

The impedance Z of the series RL circuit can be expressed in terms of the resistance and reactance as

$$Z = \sqrt{R^2 + X_L^2} \tag{14–1}$$

where Z is expressed in ohms. The phase angle θ is expressed as

$$\theta = \arctan\left(\frac{X_L}{R}\right) \tag{14–2}$$

EXAMPLE 14–1 Determine the impedance and phase angle of the circuit in Figure 14–4.

FIGURE 14–4

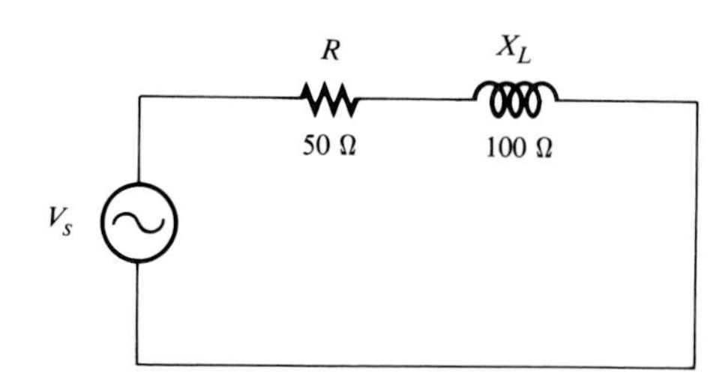

Solution

The impedance is

$$Z = \sqrt{R^2 + X_L^2} = \sqrt{(50\ \Omega)^2 + (100\ \Omega)^2} = 111.8\ \Omega$$

The phase angle is

$$\theta = \arctan\left(\frac{X_L}{R}\right) = \arctan\left(\frac{100\ \Omega}{50\ \Omega}\right) = \arctan 2 = 63.4°$$

The applied voltage *leads* the current by 63.4°.

The calculator sequences are as follows:

Z: 5 0 2nd x^2 + 1 0 0 2nd x^2 = $\sqrt{x}$

θ: 1 0 0 ÷ 5 0 = INV tan

SECTION REVIEW 14–2

1. Does the source voltage lead or lag the current in a series RL circuit?
2. What causes the phase angle in the RL circuit?
3. How does the response of an RL circuit differ from that of an RC circuit?
4. A series RL circuit has a resistance of 33 kΩ and an inductive reactance of 50 kΩ. Determine Z and θ.

ANALYSIS OF SERIES *RL* CIRCUITS

14–3

Ohm's Law

The application of Ohm's law to series RL circuits involves the use of the quantities of Z, V, and I. The three equivalent forms of Ohm's law were stated in Chapter 13 for RC circuits. They apply also to RL circuits and are restated here for convenience:

$$V = IZ, \quad I = \frac{V}{Z}, \quad Z = \frac{V}{I}$$

The following example illustrates the use of Ohm's law.

EXAMPLE 14–2 The current in Figure 14–5 is 0.2 mA. Determine the source voltage.

FIGURE 14–5

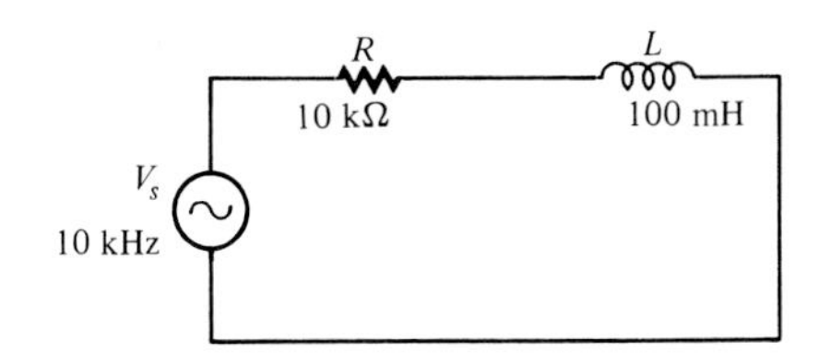

Solution
From Equation (11–11), the inductive reactance is

$$X_L = 2\pi fL = 2\pi(10 \text{ kHz})(100 \text{ mH}) = 6.28 \text{ k}\Omega$$

The impedance is

$$Z = \sqrt{R^2 + X_L^2} = \sqrt{(10 \text{ k}\Omega)^2 + (6.28 \text{ k}\Omega)^2} = 11.81 \text{ k}\Omega$$

Applying Ohm's law, we have

$$V_s = IZ = (0.2 \text{ mA})(11.81 \text{ k}\Omega) = 2.36 \text{ V}$$

Relationships of the Current and Voltages in a Series *RL* Circuit

In a series *RL* circuit, the current is the same through both the resistor and the inductor. Thus, the resistor voltage is in phase with the current, and *the inductor voltage leads the current by 90°*. Therefore, there is a phase difference of 90° between the resistor voltage, V_R, and the inductor voltage, V_L, as shown in the waveform diagram of Figure 14–6.

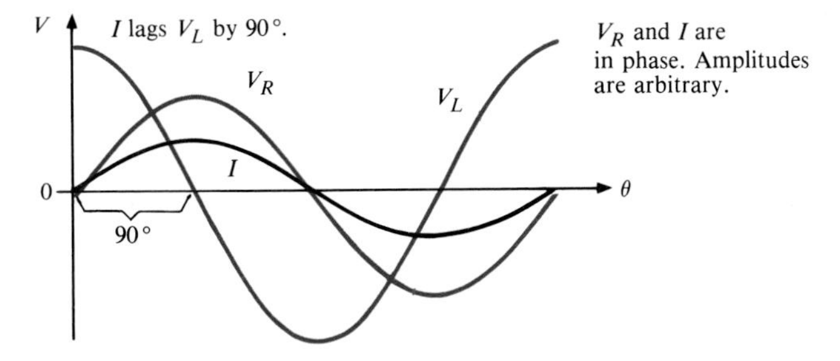

FIGURE 14–6
Phase relation of current and voltages in a series *RL* circuit.

From Kirchhoff's voltage law, the sum of the voltage drops must equal the applied voltage. However, since V_R and V_L are not in phase with each other, they must be added as phasor quantities with V_L leading V_R by 90°, as shown in Figure 14–7(a). As shown in Part (b), V_s is the phasor sum of V_R and V_L. This equation can be expressed as

$$V_s = \sqrt{V_R^2 + V_L^2} \tag{14–3}$$

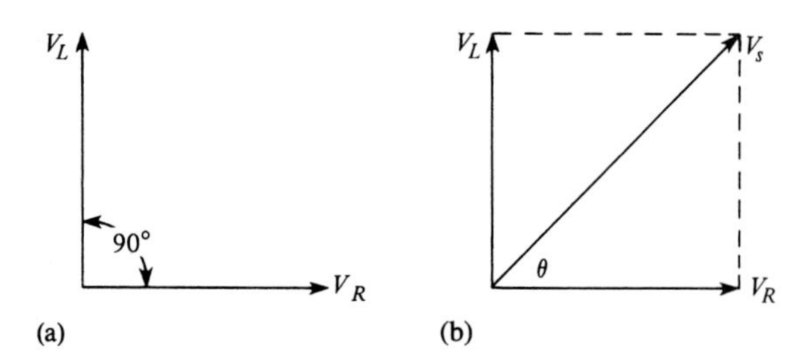

FIGURE 14–7
Voltage phasor diagram for a series *RL* circuit.

The phase angle between the resistor voltage and the source voltage is

$$\theta = \arctan\left(\frac{V_L}{V_R}\right) \tag{14–4}$$

θ is also the phase angle between the source voltage and the current. Figure 14–8 shows a voltage and current phasor diagram that represents the waveform diagram of Figure 14–6.

FIGURE 14–8
Voltage and current phasor diagram for the waveforms in Figure 14–6.

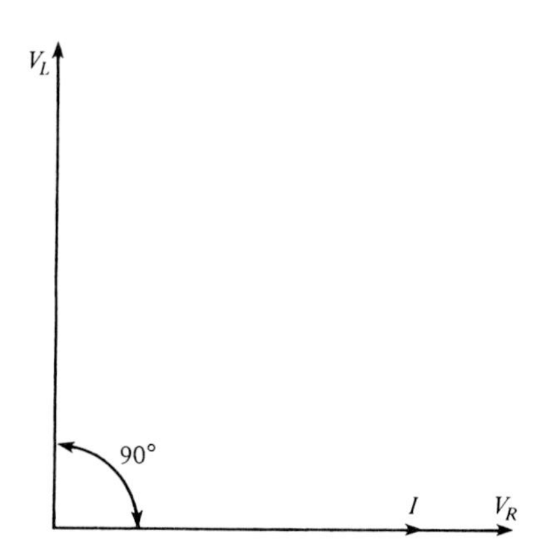

EXAMPLE 14–3 Determine the source voltage and the phase angle in Figure 14–9.

FIGURE 14–9

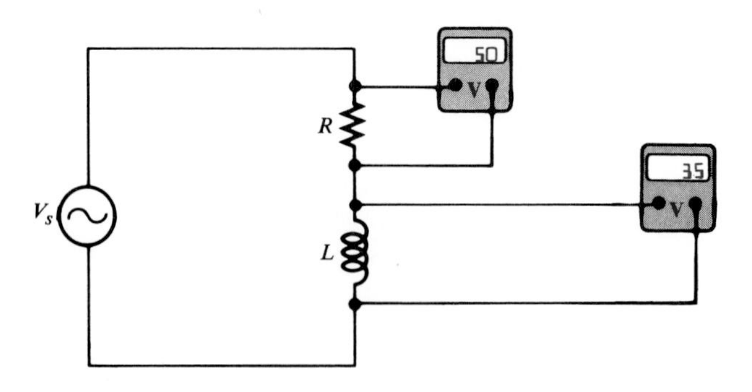

Solution
Since V_R and V_L are 90° out of phase, they cannot be added directly and must be added as phasor quantities:

$$V_s = \sqrt{V_R^2 + V_L^2} = \sqrt{(50\ \text{V})^2 + (35\ \text{V})^2} = 61\ \text{V}$$

The angle between the current and the source voltage is

$$\theta = \arctan\left(\frac{V_L}{V_R}\right) = \arctan\left(\frac{35\ \text{V}}{50\ \text{V}}\right) = 34.99°$$

Variation of Impedance and Phase Angle with Frequency

As you know, inductive reactance varies *directly* with frequency. When X_L increases, the total impedance also increases; and when X_L decreases, the total impedance decreases. Thus, *Z is directly dependent on frequency.*

Figure 14–10 illustrates how the voltages and current in a series *RL* circuit vary as the frequency increases or decreases, with the source voltage held at a constant value. In Part (a), as the frequency is increased, X_L increases, so more of the total voltage is dropped across the inductor. Also, Z increases as X_L increases, causing the current to decrease. A decrease in current causes less voltage across R.

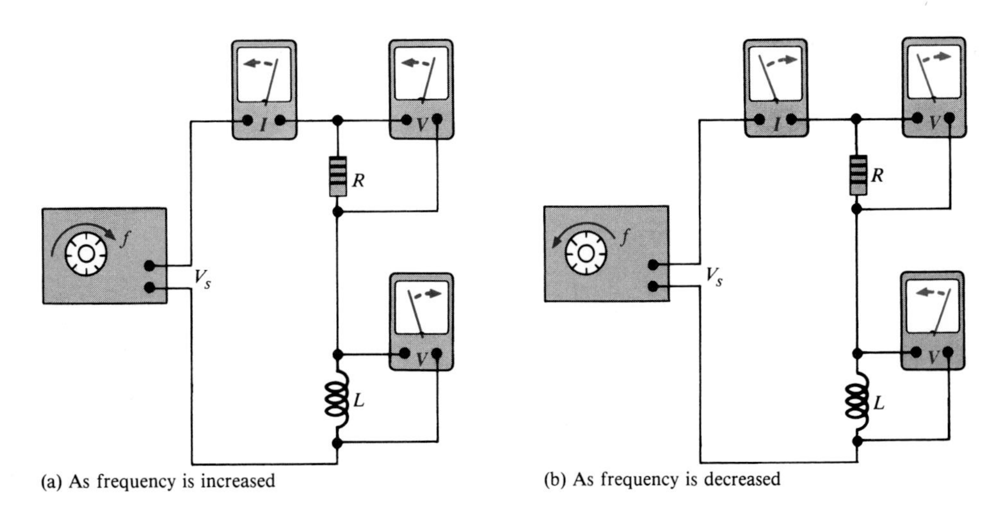

FIGURE 14–10
An illustration of how the voltages and current change as the source frequency is varied. The source voltage is held at a constant amplitude.

In Part (b) of the figure, as the frequency is decreased, X_L decreases, so less voltage is dropped across the inductor. Also, Z decreases as X_L decreases, causing the current to increase. An increase in current causes more voltage across R.

Changes in Z and X_L can be observed as shown in Figure 14–11. As the frequency increases, the voltage across Z remains constant because V_s is constant, but the voltage across L increases. The decreasing current indicates that Z is increasing. It does so because of the inverse relationship stated in Ohm's

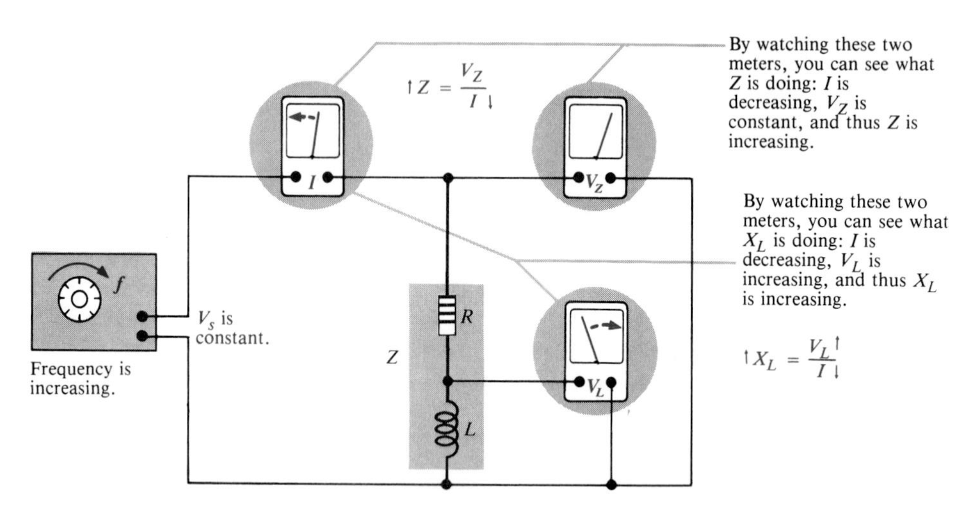

FIGURE 14–11
Observing changes in Z and X_L with frequency by watching the meters and recalling Ohm's law.

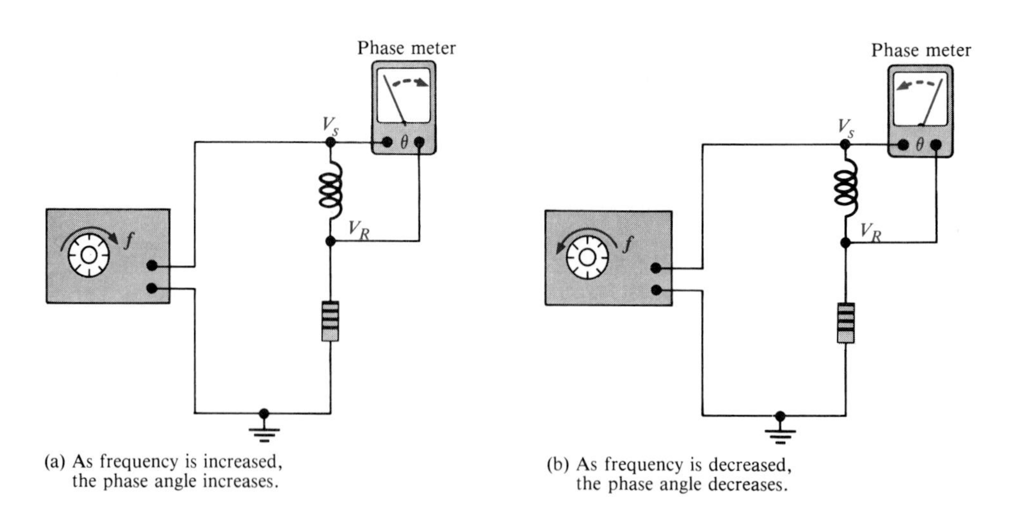

FIGURE 14–12
A phase meter indicates that the phase angle between V_s and V_R varies directly with frequency. V_R is the same phase as I.

law ($Z = V_Z/I$). The decreasing current also indicates that X_L is increasing. The increase in V_L corresponds to the increase in X_L.

Variation of the Phase Angle with Frequency

Since X_L is the factor that introduces the phase angle in a series RL circuit, a change in X_L produces a change in the phase angle. As the frequency is increased, X_L becomes greater, and thus the phase angle increases. As the frequency is decreased, X_L becomes smaller, and thus the phase angle decreases. This relationship is illustrated in Figure 14–12 where a phase meter is connected to V_s and to V_R. As you have seen before, this connection gives the angle between V_s and V_R, which is the phase angle of the circuit because I is in phase with V_R. By measuring the phase of V_R, we are effectively measuring the phase of I.

The variations of phase angle with frequency are illustrated with the impedance triangle as shown in Figure 14–13.

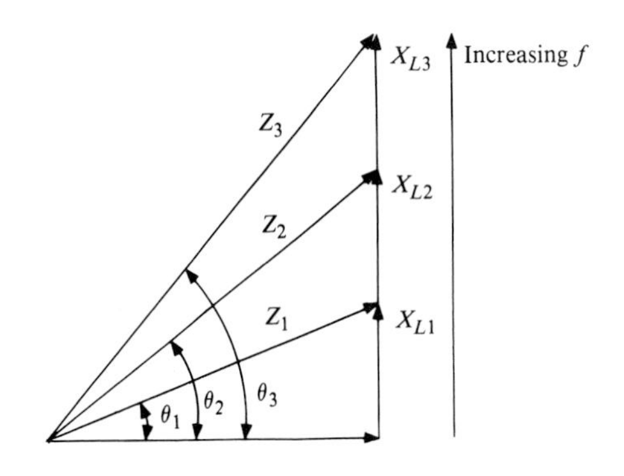

FIGURE 14–13
As the frequency increases, the phase angle θ increases.

EXAMPLE 14–4 For the series RL circuit in Figure 14–14, determine the impedance and the phase angle for each of the following frequencies:
(a) 10 kHz **(b)** 20 kHz **(c)** 30 kHz

FIGURE 14–14

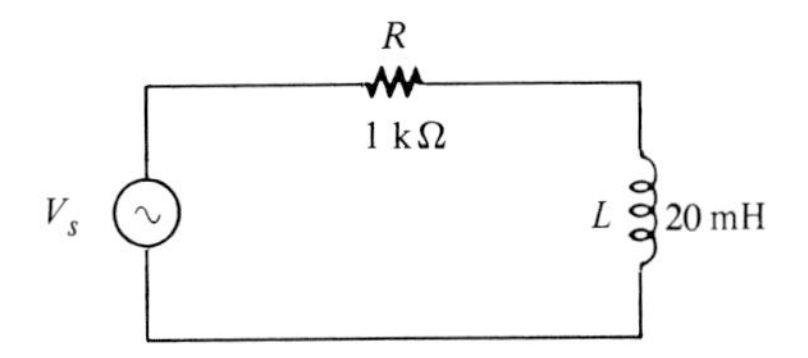

Solution
(a) For $f = 10$ kHz:

$$X_L = 2\pi fL = 2\pi(10 \text{ kHz})(20 \text{ mH}) = 1.26 \text{ k}\Omega$$
$$Z = \sqrt{R^2 + X_L^2} = \sqrt{(1 \text{ k}\Omega)^2 + (1.26 \text{ k}\Omega)^2} = 1.61 \text{ k}\Omega$$

$$\theta = \arctan\left(\frac{X_L}{R}\right) = \arctan\left(\frac{1.26\ \text{k}\Omega}{1\ \text{k}\Omega}\right) = 51.56°$$

(b) For $f = 20$ kHz:

$$X_L = 2\pi(20\ \text{kHz})(20\ \text{mH}) = 2.52\ \text{k}\Omega$$
$$Z = \sqrt{(1\ \text{k}\Omega)^2 + (2.52\ \text{k}\Omega)^2} = 2.71\ \text{k}\Omega$$
$$\theta = \arctan\left(\frac{2.52\ \text{k}\Omega}{1\ \text{k}\Omega}\right) = 68.36°$$

(c) For $f = 30$ kHz:

$$X_L = 2\pi(30\ \text{kHz})(20\ \text{mH}) = 3.77\ \text{k}\Omega$$
$$Z = \sqrt{(1\ \text{k}\Omega)^2 + (3.77\ \text{k}\Omega)^2} = 3.9\ \text{k}\Omega$$
$$\theta = \arctan\left(\frac{3.77\ \text{k}\Omega}{1\ \text{k}\Omega}\right) = 75.14°$$

Notice that as the frequency increases, X_L, Z, and θ also increase.

SECTION REVIEW 14–3

1. In a certain series *RL* circuit, $V_R = 2$ V and $V_L = 3$ V. What is the magnitude of the total voltage?
2. In Problem 1, what is the phase angle?
3. When the frequency of the applied voltage in a series *RL* circuit is increased, what happens to the inductive reactance? To the impedance? To the phase angle?

IMPEDANCE AND PHASE ANGLE OF A PARALLEL *RL* CIRCUIT

14–4

A basic parallel *RL* circuit is shown in Figure 14–15. The expression for the impedance is

$$Z = \frac{RX_L}{\sqrt{R^2 + X_L^2}} \tag{14–5}$$

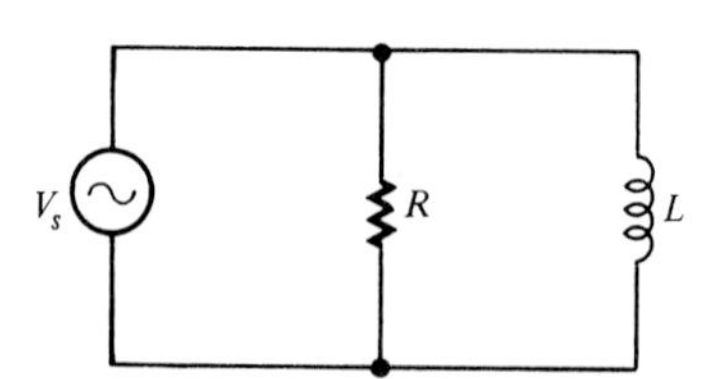

FIGURE 14–15
Parallel *RL* circuit.

The phase angle between the applied voltage and the total current is

$$\theta = \arctan\left(\frac{R}{X_L}\right) \tag{14–6}$$

EXAMPLE 14–5 For each circuit in Figure 14–16, determine the impedance and the phase angle.

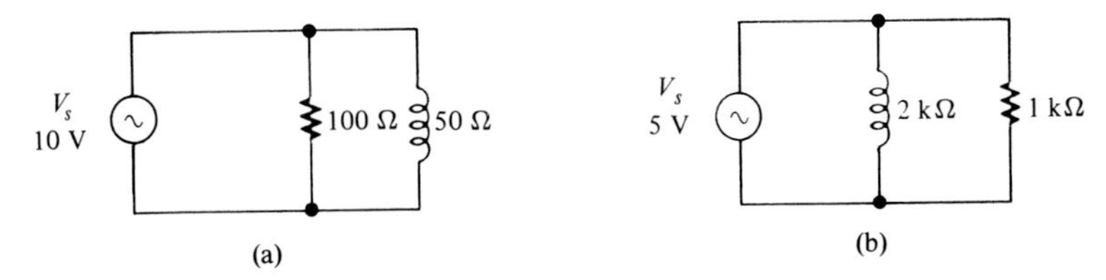

FIGURE 14–16

Solution

$$\textbf{(a)}\quad Z = \frac{RX_L}{\sqrt{R^2 + X_L^2}} = \frac{(100\ \Omega)(50\ \Omega)}{\sqrt{(100\ \Omega)^2 + (50\ \Omega)^2}}$$

$$= 44.72\ \Omega$$

$$\theta = \arctan\left(\frac{R}{X_L}\right) = \arctan\left(\frac{100\ \Omega}{50\ \Omega}\right) = 63.43°$$

$$\textbf{(b)}\quad Z = \frac{(1\ \text{k}\Omega)(2\ \text{k}\Omega)}{\sqrt{(1\ \text{k}\Omega)^2 + (2\ \text{k}\Omega)^2}} = 894.4\ \Omega$$

$$\theta = \arctan\left(\frac{1\ \text{k}\Omega}{2\ \text{k}\Omega}\right) = 26.57°$$

The voltage leads the current, as opposed to the *RC* case where the voltage lags the current.

The calculator sequences for (a) are as follows:

Z: [1] [0] [0] [×] [5] [0] [÷] [(] [(] [1] [0] [0] [2nd] [x^2] [+] [5] [0] [2nd] [x^2] [)] [$\sqrt{x}$] [)] [=]

θ: [1] [0] [0] [÷] [5] [0] [=] [INV] [tan]

Susceptance and Admittance

As you know from Chapter 13, Section 13–4, conductance is the reciprocal of resistance, susceptance is the reciprocal of reactance, and admittance is the reciprocal of impedance.

For parallel *RL* circuits, inductive susceptance is expressed as

$$B_L = \frac{1}{X_L} \tag{14–7}$$

and admittance as

$$Y = \frac{1}{Z} \tag{14–8}$$

As with the *RC* circuit, the unit for G, B_L, and Y is the siemen (S). In the basic parallel *RL* circuit shown in Figure 14–17, the total admittance is the phasor sum of the conductance and the susceptance and is expressed as follows:

$$Y = \sqrt{G^2 + B_L^2} \tag{14–9}$$

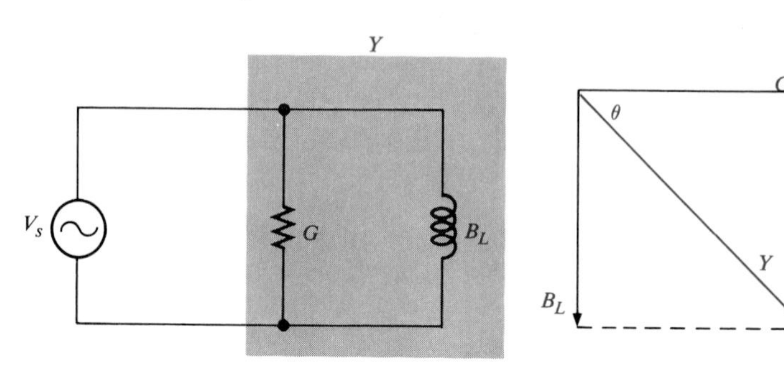

FIGURE 14–17
Admittance is the reciprocal of impedance in a parallel *RL* circuit.

EXAMPLE 14–6 Determine the admittance in Figure 14–18; then convert it to impedance.

FIGURE 14–18

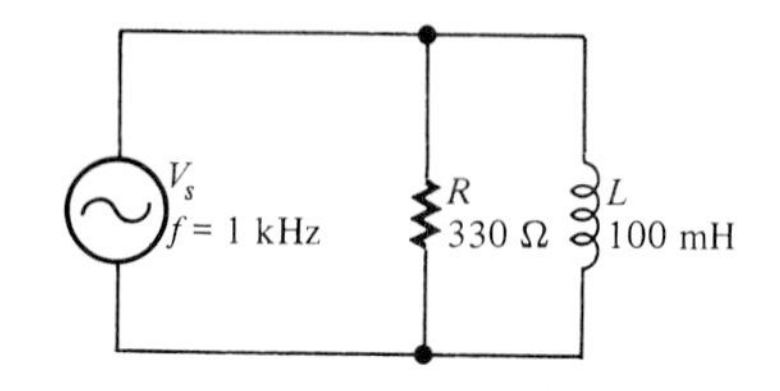

Solution
Since $R = 330\ \Omega$, then

$$G = \frac{1}{R} = \frac{1}{330\ \Omega} = 0.003\ \text{S}$$

$$X_L = 2\pi(1000\ \text{Hz})(100\ \text{mH}) = 628.3\ \Omega$$

$$B_L = \frac{1}{X_L} = 0.00159\ \text{S}$$

$$Y = \sqrt{G^2 + B_L^2} = \sqrt{(0.003\ \text{S})^2 + (0.00159\ \text{S})^2} = 0.0034\ \text{S}$$

Converting to impedance, we get

$$Z = \frac{1}{Y} = \frac{1}{0.0034\ \text{S}} = 294.12\ \Omega$$

SECTION REVIEW 14–4

1. If $Y = 0.05$ S, what is the value of Z?
2. In a certain parallel *RL* circuit, $R = 50\ \Omega$ and $X_L = 75\ \Omega$. Determine the admittance.

3. In the circuit of Problem 2, does the total current lead or lag the applied voltage and by what phase angle?

ANALYSIS OF PARALLEL *RL* CIRCUITS

14–5

The following example applies Ohm's law to the analysis of a parallel *RL* circuit.

EXAMPLE 14–7 Determine the total current and the phase angle in the circuit of Figure 14–19.

FIGURE 14–19

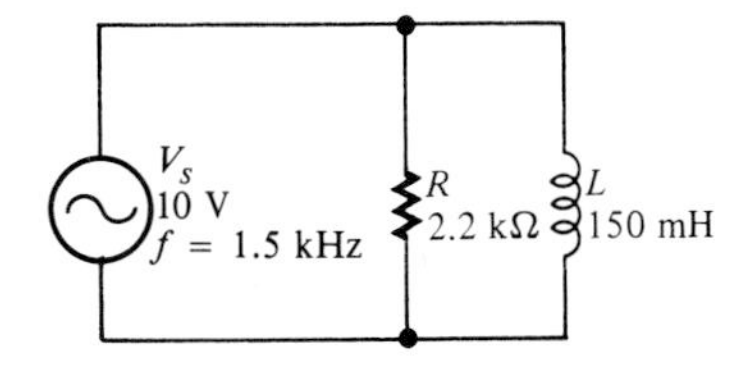

Solution

$$X_L = 2\pi(1.5\text{ kHz})(150\text{ mH}) = 1.41\text{ k}\Omega$$

$$B_L = \frac{1}{X_L} = \frac{1}{1.41\text{ k}\Omega} = 0.709\text{ mS}$$

$$G = \frac{1}{R} = \frac{1}{2.2\text{ k}\Omega} = 0.455\text{ mS}$$

$$Y = \sqrt{G^2 + B_L^2} = \sqrt{(0.455\text{ mS})^2 + (0.709\text{ mS})^2}$$

$$= 0.842\text{ mS}$$

Apply Ohm's law to get the total current:

$$I_T = VY = (10\text{ V})(0.842\text{ mS}) = 8.42\text{ mA}$$

$$\theta = \arctan\left(\frac{R}{X_L}\right) = \arctan\left(\frac{2.2\text{ k}\Omega}{1.41\text{ k}\Omega}\right) = 57.34°$$

The total current lags the source voltage by 57.34°.

Relationships of the Currents and Voltages in a Parallel *RL* Circuit

Figure 14–20(a) shows all the currents and voltages in a basic parallel *RL* circuit. As you can see, the applied voltage, V_s, appears across both the resistive and the inductive branches, so V_s, V_R, and V_L are all in phase and of the same magnitude. The total current, I_T, divides at the junction into the two branch currents, I_R and I_L.

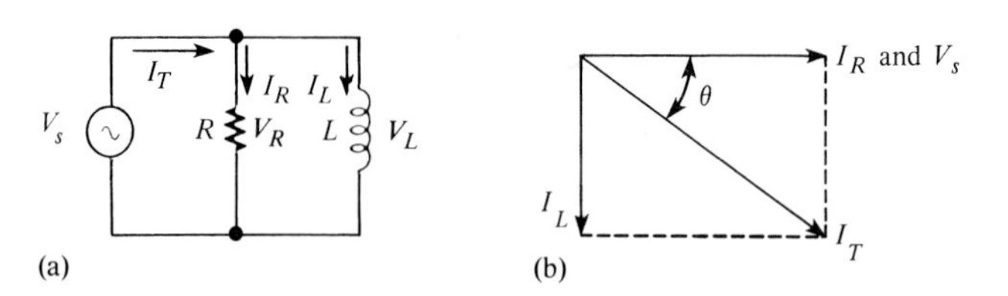

FIGURE 14–20
Currents and voltages in a parallel *RL* circuit. The current directions shown are instantaneous.

The current through the resistor is in phase with the voltage. The current through the inductor lags the voltage and the resistor current by 90°. By Kirchhoff's current law, the total current is the phasor sum of the two branch currents, as shown by the phasor diagram in Figure 14–20(b). The total current is expressed as

$$I_T = \sqrt{I_R^2 + I_L^2} \tag{14–10}$$

and the phase angle between the resistor current and the total current is

$$\theta = \arctan\left(\frac{I_L}{I_R}\right) \tag{14–11}$$

Figure 14–21 shows a complete current and voltage phasor diagram.

FIGURE 14–21
Current and voltage phasor diagram for a parallel *RL* circuit (amplitudes are arbitrary).

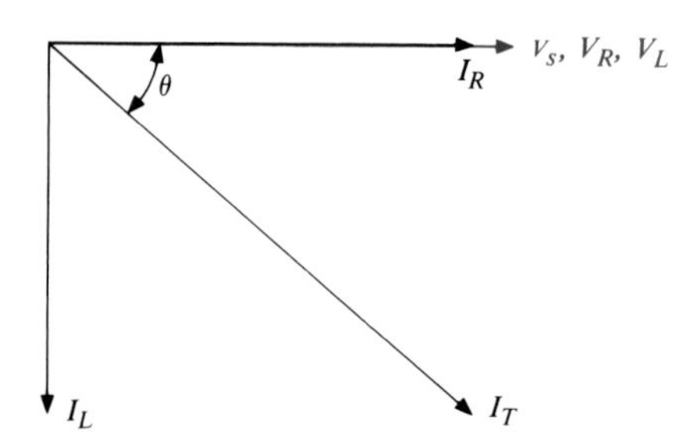

EXAMPLE 14–8 Determine the value of each current in Figure 14–22, and describe the phase relationship of each with the applied voltage.

FIGURE 14–22

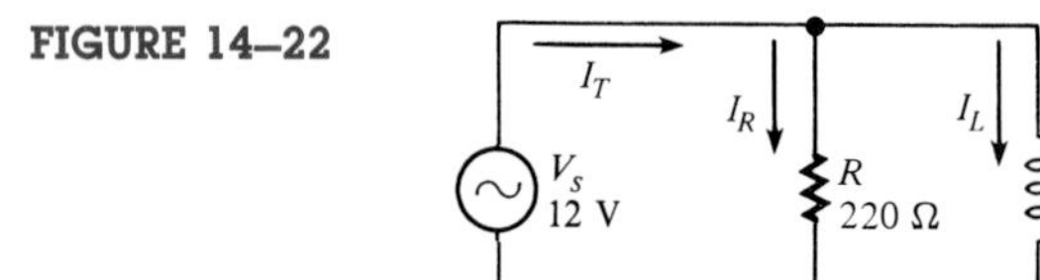

Solution

$$I_R = \frac{V_s}{R} = \frac{12\text{ V}}{220\ \Omega} = 54.55\text{ mA}$$
$$I_L = \frac{V_s}{X_L} = \frac{12\text{ V}}{150\ \Omega} = 80\text{ mA}$$
$$I_T = \sqrt{I_R^2 + I_L^2} = \sqrt{(54.55\text{ mA})^2 + (80\text{ mA})^2} = 96.83\text{ mA}$$
$$\theta = \arctan\left(\frac{R}{X_L}\right) = \arctan\left(\frac{220\ \Omega}{150\ \Omega}\right) = 55.71°$$

The resistor current is 54.55 mA and is in phase with the applied voltage. The inductor current is 80 mA and lags the applied voltage by 90°. The total current is 96.83 mA and lags the voltage by 55.71°.

SECTION REVIEW 14–5

1. The admittance of an *RL* circuit is 0.004 S, and the applied voltage is 8 V. What is the total current?
2. In a certain parallel *RL* circuit, the resistor current is 12 mA, and the inductor current is 20 mA. Determine the phase angle and the total current.
3. What is the phase angle between the inductor current and the applied voltage in a parallel *RL* circuit?

SERIES-PARALLEL ANALYSIS

14–6

In this section, we analyze *RL* circuits with combinations of both series and parallel *R* and *L* elements. Examples illustrate the procedures.

EXAMPLE 14–9 In the circuit of Figure 14–23, determine the values of the following:
(a) Z_T **(b)** I_T **(c)** θ

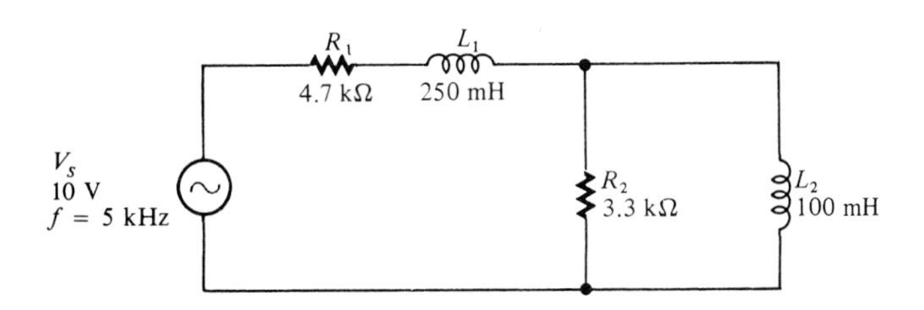

FIGURE 14–23

Solution
(a) First, the inductive reactances are calculated:

$$X_{L1} = 2\pi(5\text{ kHz})(250\text{ mH}) = 7.85\text{ k}\Omega$$
$$X_{L2} = 2\pi(5\text{ kHz})(100\text{ mH}) = 3.14\text{ k}\Omega$$

The approach we will use is to find the series equivalent resistance and inductive reactance for the parallel portion of the circuit. Then we will add the resistances to get the total resistance, add the reactances to get the total reactance, and then determine the total impedance.

The impedance of the parallel portion (Z_2) is determined as follows:

$$G_2 = \frac{1}{R_2} = \frac{1}{3.3\ \text{k}\Omega} = 303\ \mu\text{S}$$

$$B_{L2} = \frac{1}{X_{L2}} = \frac{1}{3.14\ \text{k}\Omega} = 318.5\ \mu\text{S}$$

$$Y_2 = \sqrt{G_2^2 + B_L^2} = \sqrt{(303\ \mu\text{S})^2 + (318.5\ \mu\text{S})^2} = 439.6\ \mu\text{S}$$

Then

$$Z_2 = \frac{1}{Y_2} = \frac{1}{439.6\ \mu\text{S}} = 2.28\ \text{k}\Omega$$

The phase angle associated with the parallel portion of the circuit is

$$\theta_p = \arctan\left(\frac{R_2}{X_{L2}}\right) = \arctan\left(\frac{3.3\ \text{k}\Omega}{3.14\ \text{k}\Omega}\right) = 46.42°$$

The series equivalent values for the parallel portion are

$$R_{\text{eq}} = Z_2\cos\theta_p = (2.28\ \text{k}\Omega)\cos(46.42°) = 1.57\ \text{k}\Omega$$

$$X_{L(\text{eq})} = Z_2\sin\theta_p = (2.28\ \text{k}\Omega)\sin(46.42°) = 1.65\ \text{k}\Omega$$

The total circuit resistance is

$$R_T = R_1 + R_{\text{eq}} = 4.7\ \text{k}\Omega + 1.57\ \text{k}\Omega = 6.27\ \text{k}\Omega$$

The total circuit reactance is

$$X_{LT} = X_{L1} + X_{L(\text{eq})} = 7.85\ \text{k}\Omega + 1.65\ \text{k}\Omega = 9.5\ \text{k}\Omega$$

The total circuit impedance is

$$Z_T = \sqrt{R_T^2 + X_{LT}^2} = \sqrt{(6.27\ \text{k}\Omega)^2 + (9.5\ \text{k}\Omega)^2} = 11.38\ \text{k}\Omega$$

(b) Use Ohm's law to find the total current:

$$I_T = \frac{V_s}{Z_T} = \frac{10\ \text{V}}{11.38\ \text{k}\Omega} = 0.879\ \text{mA}$$

(c) To find the phase angle, view the circuit as a series combination of R_T and X_{LT}:

$$\theta = \arctan\left(\frac{X_{LT}}{R_T}\right) = \arctan\left(\frac{9.5\ \text{k}\Omega}{6.27\ \text{k}\Omega}\right) = 56.58°$$

The total current *lags* the source voltage by 56.58°.

EXAMPLE 14–10 Determine the voltage across each element in Figure 14–24. Sketch a voltage phasor diagram.

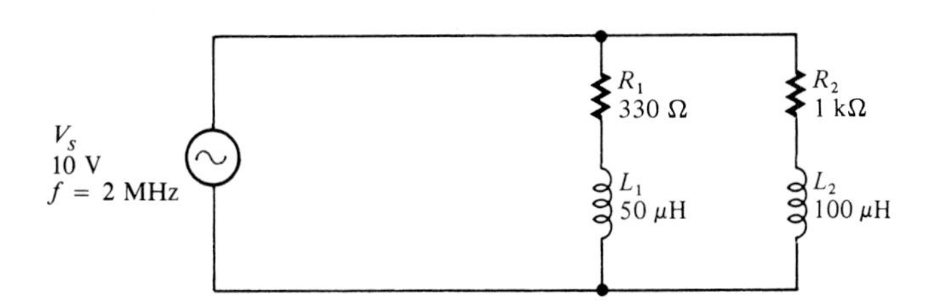

FIGURE 14–24

Solution

First calculate X_{L1} and X_{L2}:

$$X_{L1} = 2\pi f L_1 = 2\pi(2 \text{ MHz})(50\ \mu\text{H}) = 628.3\ \Omega$$
$$X_{L2} = 2\pi f L_2 = 2\pi(2 \text{ MHz})(100\ \mu\text{H}) = 1.257\ \text{k}\Omega$$

Now, determine the impedance of each branch:

$$Z_1 = \sqrt{R_1^2 + X_{L1}^2} = \sqrt{(330\ \Omega)^2 + (628.3\ \Omega)^2} = 709.69\ \Omega$$
$$Z_2 = \sqrt{R_2^2 + X_{L2}^2} = \sqrt{(1\ \text{k}\Omega)^2 + (1.257\ \text{k}\Omega)^2} = 1.61\ \text{k}\Omega$$

Calculate each branch current:

$$I_1 = \frac{V_s}{Z_1} = \frac{10\ \text{V}}{709.69\ \Omega} = 14.09\ \text{mA}$$
$$I_2 = \frac{V_s}{Z_2} = \frac{10\ \text{V}}{1.61\ \text{k}\Omega} = 6.21\ \text{mA}$$

Now, use Ohm's law to find the voltage across each element:

$$V_{R1} = I_1 R_1 = (14.09\ \text{mA})(330\ \Omega) = 4.65\ \text{V}$$
$$V_{L1} = I_1 X_{L1} = (14.09\ \text{mA})(628.3\ \Omega) = 8.85\ \text{V}$$
$$V_{R2} = I_2 R_2 = (6.21\ \text{mA})(1\ \text{k}\Omega) = 6.21\ \text{V}$$
$$V_{L2} = I_2 X_{L2} = (6.21\ \text{mA})(1.257\ \text{k}\Omega) = 7.81\ \text{V}$$

The angles associated with each parallel branch are now determined:

$$\theta_1 = \arctan\left(\frac{X_{L1}}{R_1}\right) = \arctan\left(\frac{628.3\ \Omega}{330\ \Omega}\right) = 62.29°$$
$$\theta_2 = \arctan\left(\frac{X_{L2}}{R_2}\right) = \arctan\left(\frac{1.257\ \text{k}\Omega}{1\ \text{k}\Omega}\right) = 51.5°$$

Thus, I_1 lags V_s by 62.29°, and I_2 lags V_s by 51.5°, as indicated in Figure 14–25(a), where the negative signs indicate lagging angles.

The phase relationships of the voltages are determined as follows:

- V_{R1} is in phase with I_1 and therefore lags V_s by 62.2°.
- V_{L1} leads I_1 by 90°, so its angle is $90° - 62.2° = 27.8°$.
- V_{R2} is in phase with I_2 and therefore lags V_s by 51.5°.
- V_{L2} leads I_2 by 90°, so its angle is $90° - 51.5° = 38.5°$.

These phase relationships are shown in Figure 14–25(b).

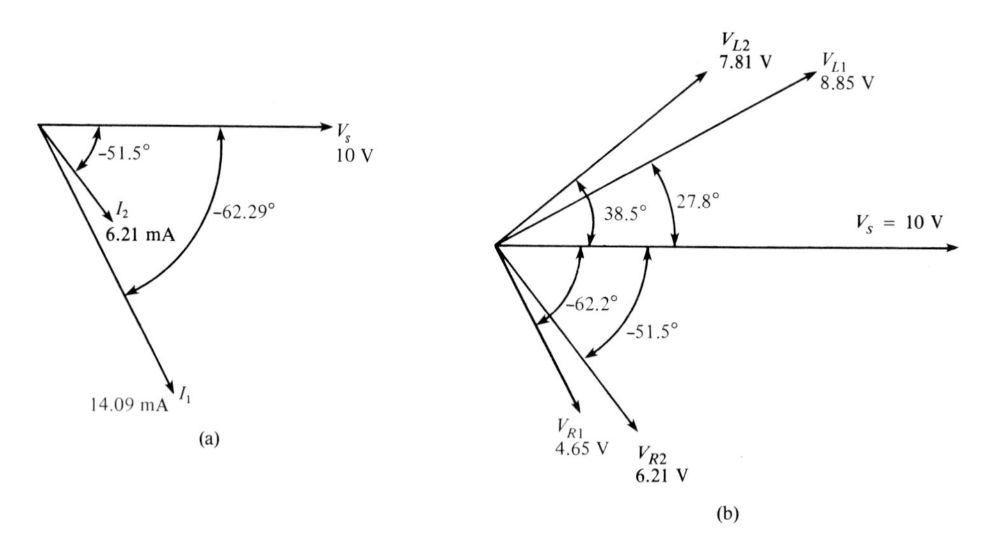

FIGURE 14–25

The two previous examples should give you a feel for how complex *RL* networks respond to sine wave inputs.

SECTION REVIEW 14–6

1. What is the total impedance of the circuit in Figure 14–24?
2. Determine the total current for the circuit in Figure 14–24.

POWER IN *RL* CIRCUITS

14–7

As you know, in a purely resistive ac circuit, all of the energy that is delivered by the source is dissipated in the form of heat by the resistance. In a purely inductive ac circuit, all of the energy delivered by the source is stored by the inductor in its magnetic field during a portion of the voltage cycle and then is returned to the source during another portion of the cycle so that no net energy is transferred.

When there is both resistance and inductance in a circuit, some of the energy is alternately stored and returned by the inductance, and some is dissipated by the resistance. The amount of energy dissipated is determined by the relative values of resistance and inductive reactance.

When the resistance is greater than the inductive reactance, more of the total energy delivered by the source is dissipated by the resistance than is stored by the inductor, and when the reactance is greater than the resistance, more of the total energy is stored and returned than is dissipated.

As you know, the power in a resistor is called the *true power*. The power in an inductor is *reactive power* and is expressed as

$$P_r = I^2X_L \tag{14–12}$$

The Power Triangle

The generalized power triangle for the *RL* circuit is shown in Figure 14–26. The apparent power, P_a, is the resultant of the true power, P_{true}, and the reactive power, P_r.

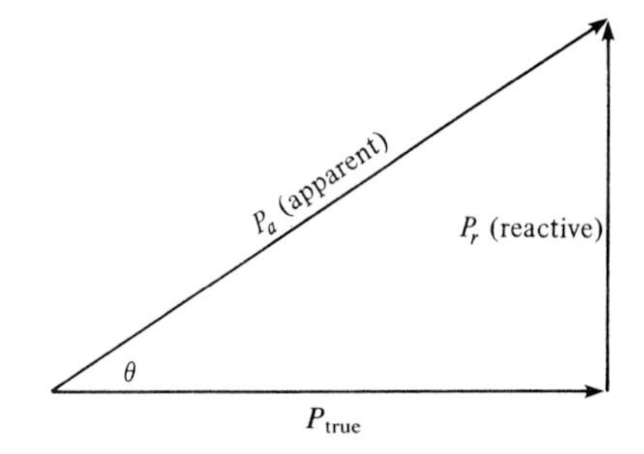

FIGURE 14–26
Generalized power triangle for an *RL* circuit.

Recall that the power factor equals the cosine of θ ($PF = \cos \theta$). As the phase angle between the applied voltage and the total current increases, the power factor decreases, indicating an increasingly reactive circuit. The smaller the power factor, the smaller the true power is compared to the reactive power. The power factor of inductive loads is called a *lagging power factor* because the current lags the source voltage.

EXAMPLE 14–11 Determine the power factor, the true power, the reactive power, and the apparent power in Figure 14–27.

FIGURE 14–27

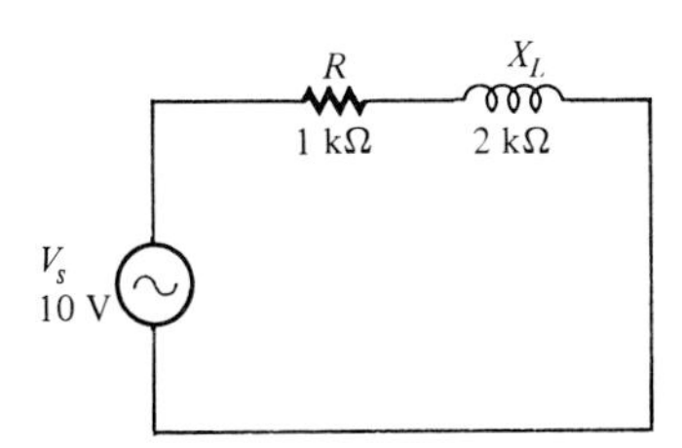

Solution
The impedance of the circuit is

$$Z = \sqrt{R^2 + X_L^2} = \sqrt{(1\ \text{k}\Omega)^2 + (2\ \text{k}\Omega)^2} = 2236\ \Omega$$

The current is

$$I = \frac{V_s}{Z} = \frac{10\ \text{V}}{2236\ \Omega} = 4.47\ \text{mA}$$

The phase angle is

$$\theta = \arctan\left(\frac{X_L}{R}\right) = \arctan\left(\frac{2\ \text{k}\Omega}{1\ \text{k}\Omega}\right) = 63.4°$$

Therefore, the power factor is

$$PF = \cos\theta = \cos(63.4°) = 0.448$$

The true power is

$$P_{\text{true}} = V_s I \cos\theta = (10\ \text{V})(4.47\ \text{mA})(0.448) = 20\ \text{mW}$$

The reactive power is

$$P_r = I^2 X_L = (4.47\ \text{mA})^2(2\ \text{k}\Omega) = 39.96\ \text{mVAR}$$

The apparent power is

$$P_a = I^2 Z = (4.47\ \text{mA})^2(2236\ \Omega) = 44.68\ \text{mVA}$$

Significance of the Power Factor

As you learned in Chapter 13, the power factor *(PF)* is very important in determining how much useful power (true power) is transferred to a load. The highest power factor is 1, which indicates that all of the current to a load is in-phase with the voltage (resistive). When the power factor is 0, all of the current to a load is 90° out-of-phase with the voltage (reactive).

Generally, a power factor as close to 1 as possible is desirable, because then most of the power transferred from the source to the load is useful or true power. True power goes only one way—from source to load—and performs work on the load in terms of energy dissipation. Reactive power simply goes back and forth between the source and the load with no net work being done. Energy must be used in order for work to be done.

Many practical loads have inductance as a result of their particular function, and it is essential for their proper operation. Examples are transformers, electric motors, and speakers, to name a few. Therefore, inductive (and capacitive) loads are a fact of life and we must live with them.

To see the effect of the power factor on system requirements, refer to Figure 14–28 which shows a representation of a typical inductive load consisting effectively of inductance and resistance in parallel. Part (a) shows a load with a relatively low power factor (0.75), and Part (b) shows a load with a relatively high power factor (0.95). Both loads dissipate equal amounts of power as indicated by the wattmeters. Thus, an equal amount of work is done on both loads.

Although both loads are equivalent in terms of the amount of work done (true power), the low–power factor load in Part (a) draws more current from the source than does the high–power factor load in Part (b), as indicated by the ammeters. Therefore, the source in Part (a) must have a higher VA rating than the one in Part (b). Also, the lines connecting the source to the load must be a larger wire gage than those in Part (b), a condition that becomes significant when very long transmission lines are required, such as in power distribution.

Figure 14–28 has demonstrated that a higher power factor is an advantage in delivering power more efficiently to a load.

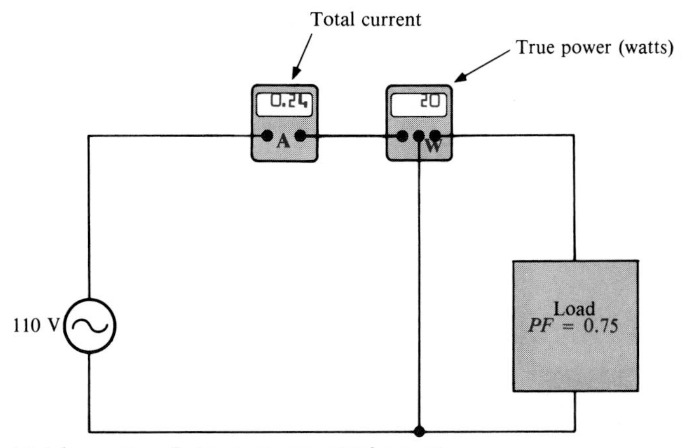

(a) A lower power factor means more total current for a given power dissipation (watts). A larger source is required to deliver the watts.

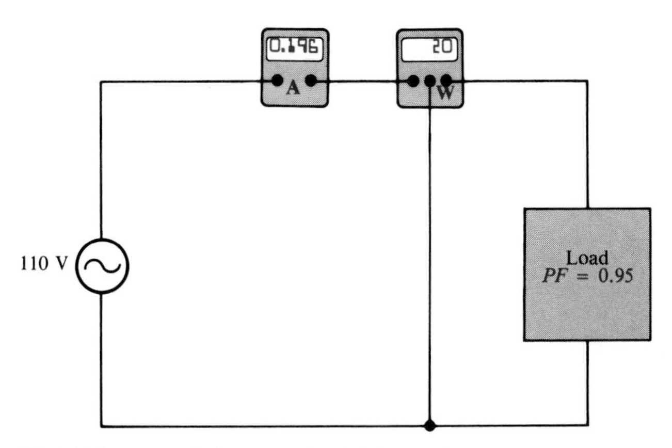

(b) A higher power factor means less total current for a given power dissipation. A smaller source can deliver the same true power (watts).

FIGURE 14–28

Illustration of the effect of the power factor on system requirements such as source rating (VA) and conductor size.

Power Factor Correction

The power factor of an inductive load can be increased by the addition of a capacitor in parallel, as shown in Figure 14–29. The capacitor compensates for the phase lag of the total current by creating a capacitive component of current that is 180° out of phase with the inductive component. This has a canceling effect and reduces the phase angle (and power factor) as well as the total current, as illustrated in the figure.

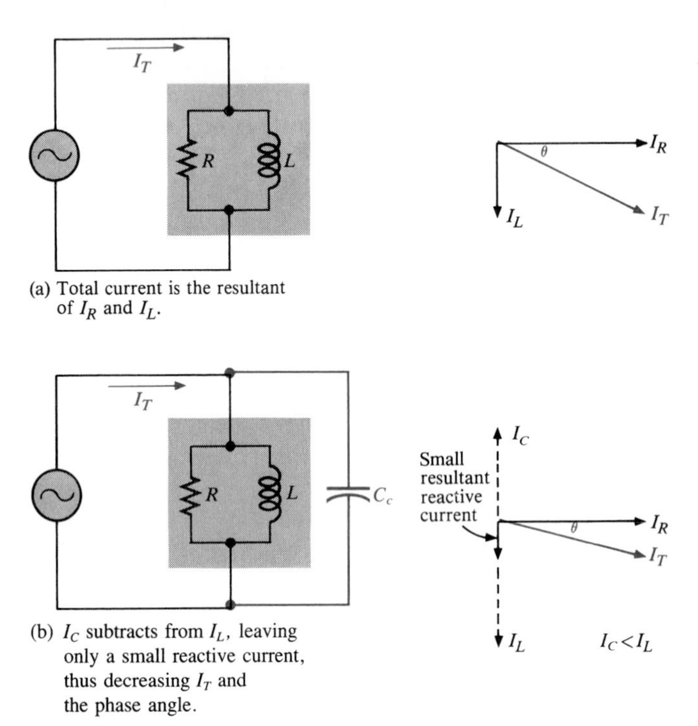

FIGURE 14–29
Example of how the power factor can be increased by the addition of a compensating capacitor.

SECTION REVIEW 14–7

1. To which component in an *RL* circuit is the power dissipation due?
2. Calculate the power factor when $\theta = 50°$.
3. A certain series *RL* circuit consists of a 470-Ω resistor and an inductive reactance of 620 Ω at the operating frequency. Determine P_{true}, P_r, and P_a when $I = 100$ mA.

BASIC APPLICATIONS OF *RL* CIRCUITS

14–8

Like *RC* circuits, *RL* circuits can also be applied in phase-shifting and filter applications.

The *RL* Lag Network

The *RL* lag network is a phase-shift circuit in which the output voltage *lags* the input voltage by a specified angle, ϕ. A basic *RL* lag network is shown in Figure 14–30(a). Notice that the output voltage is taken across the *resistor* and the input voltage is the total voltage applied across the circuit. The relationship of the voltages is shown in the phasor diagram in Figure 14–30(b), and a waveform diagram is shown in Part (c). Notice that V_R, which is the output voltage, *lags* V_{in} by an angle that is the same as the circuit phase angle. The angles are equal, of course, because V_R and I are in phase with each other.

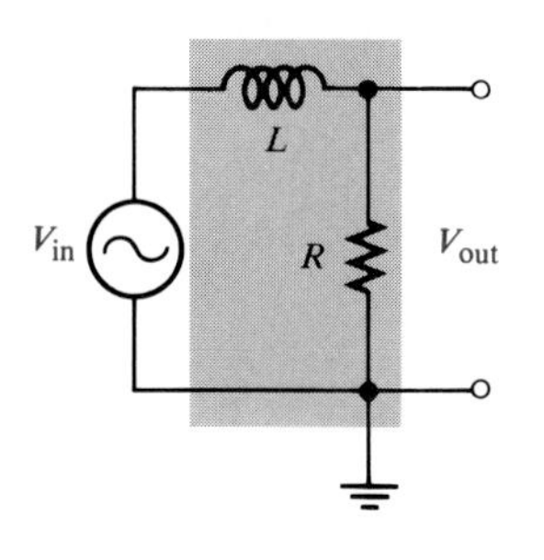

(a) A basic *RL* lag network

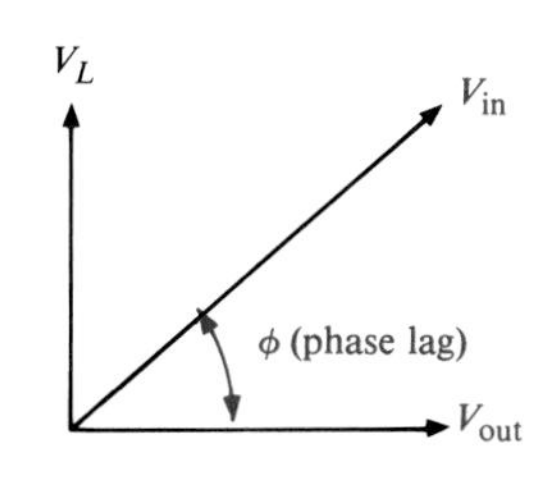

(b) Phasor voltage diagram showing phase lag between V_{in} and V_{out}

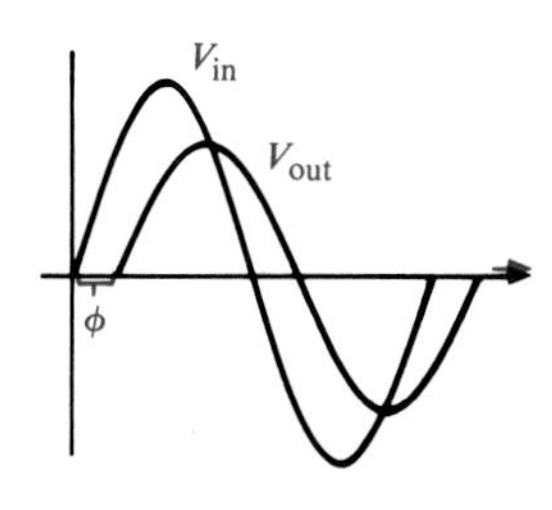

(c) Input and output waveforms

FIGURE 14–30
The *RL* lag network ($V_{out} = V_R$).

The phase lag, ϕ, can be expressed as

$$\phi = \arctan\left(\frac{X_L}{R}\right) \quad (14\text{–}13)$$

EXAMPLE 14–12 Calculate the phase lag for each circuit in Figure 14–31.

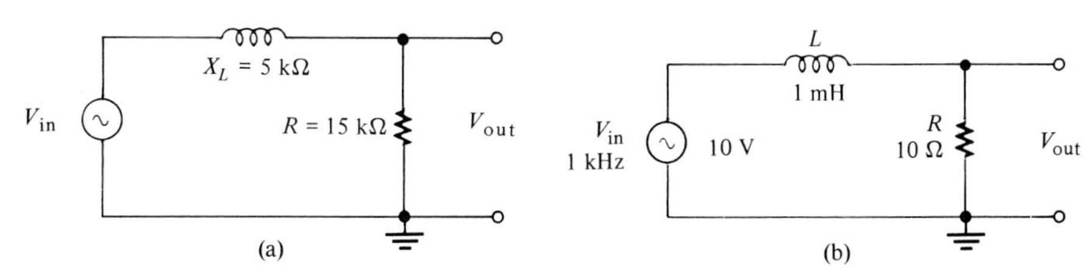

FIGURE 14–31

Solution

(a) $\phi = \arctan\left(\frac{X_L}{R}\right) = \arctan\left(\frac{5\text{ k}\Omega}{15\text{ k}\Omega}\right) = 18.43°$

The output lags the input by 18.43°.

(b) $X_L = 2\pi fL = 2\pi(1\text{ kHz})(1\text{ mH}) = 6.28\ \Omega$

$\phi = \arctan\left(\frac{X_L}{R}\right) = \arctan\left(\frac{6.28\ \Omega}{10\ \Omega}\right) = 32.13°$

The output lags the input by 32.13°.

The phase-lag network can be considered as a voltage divider with a portion of the input voltage dropped across L and a portion across R. The output voltage can be determined with the following formula:

$$V_{out} = \left(\frac{R}{\sqrt{R^2 + X_L^2}}\right)V_{in} \tag{14–14}$$

EXAMPLE 14–13 The input voltage in Figure 14–31(b) of Example 14–12 has an rms value of 10 V. Determine the output voltage. Sketch the waveform relationships for the input and output voltages.

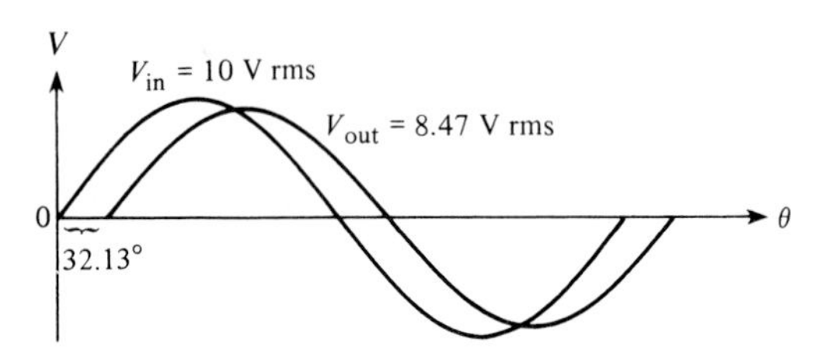

FIGURE 14–32

Solution
The phase lag was found to be 32.13° in Example 14–12.

$$V_{out} = \left(\frac{R}{\sqrt{R^2 + X_L^2}}\right)V_{in} = \left(\frac{10\ \Omega}{11.81\ \Omega}\right)10\ \text{V} = 8.47\ \text{V rms}$$

The waveforms are shown in Figure 14–32.

The calculator sequence is

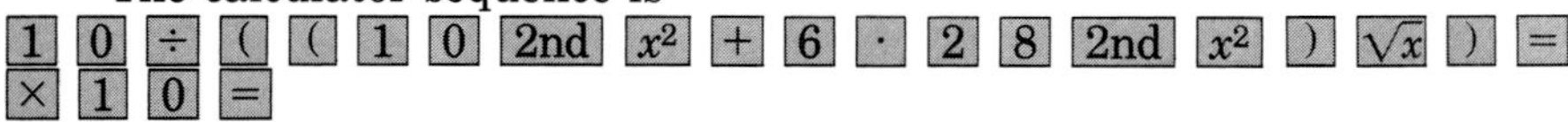

Effects of Frequency on the Phase Shift and Output Voltage of the Lag Network

Since the circuit phase angle and the phase lag are the same, an increase in frequency causes an increase in phase lag. Also, an increase in frequency causes a decrease in the magnitude of the output voltage, because X_L becomes greater and more of the total voltage is dropped across the inductor and less across the resistor. Figure 14–33 illustrates the changes in phase lag and output voltage magnitude with frequency.

The *RL* Lead Network

The *RL* lead network is a phase-shift circuit in which the output voltage *leads* the input voltage by a specified angle, ϕ. A basic *RL* lead network is shown in Figure 14–34(a). Notice how this network differs from the lag network. Here the output voltage is taken across the inductor rather than across the resistor. The relationship of the voltages is shown in the phasor diagram of Figure 14–34(b) and in the waveform plot of Part (c).

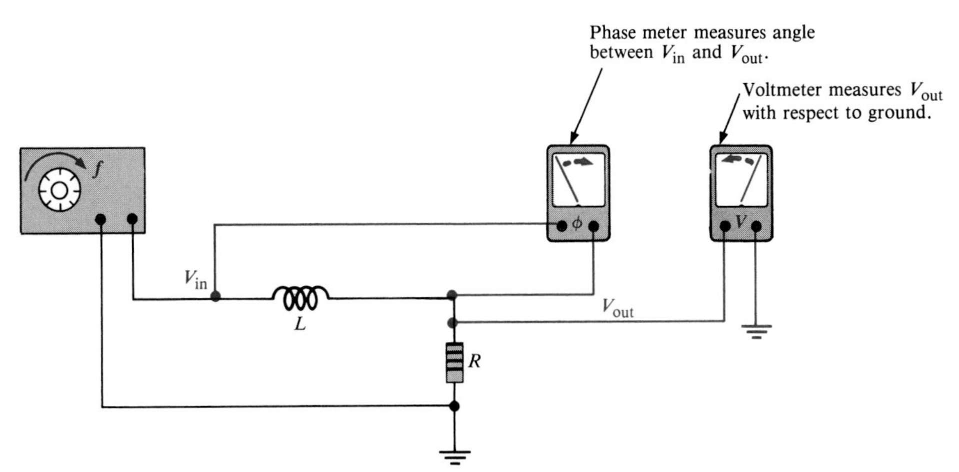

(a) When f increases, the phase lag ϕ increases and V_{out} decreases.

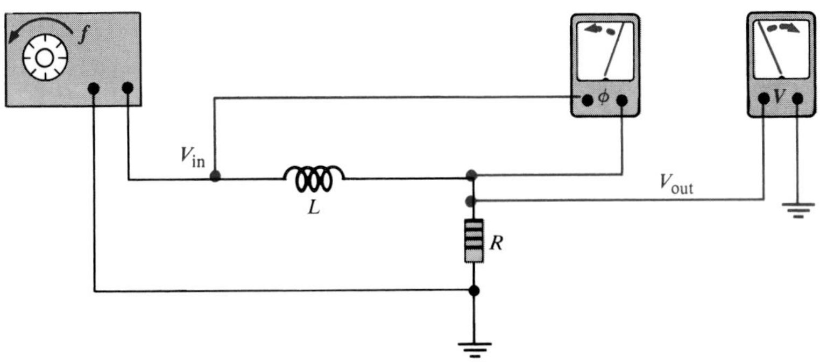

(b) When f decreases, ϕ decreases and V_{out} increases.

FIGURE 14–33
Illustration of how the frequency affects the phase lag and the output voltage in an *RL* lag network with the amplitude of V_{in} held constant.

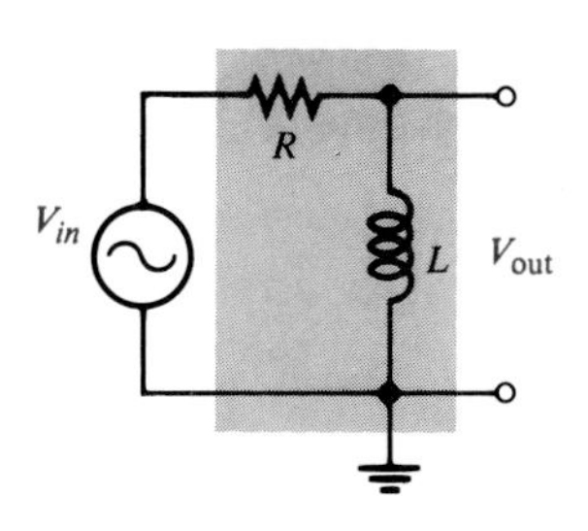

(a) A basic *RL* lead network

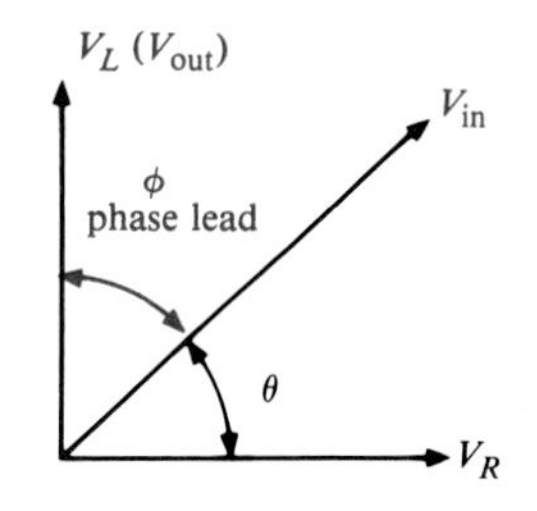

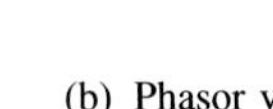

(b) Phasor voltage diagram showing phase lead between V_{in} and V_{out}

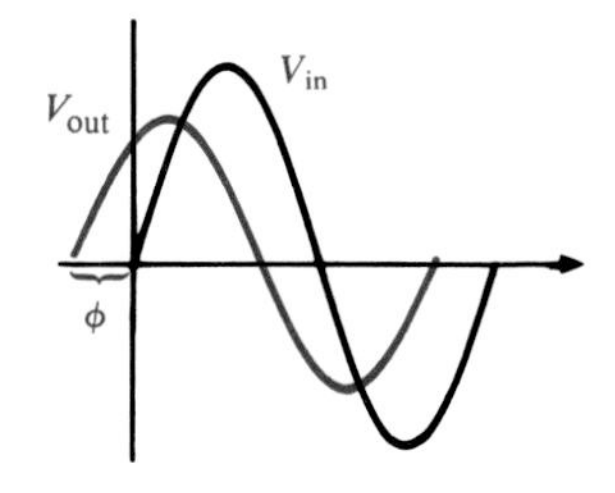

(c) Input and output waveforms

FIGURE 14–34
The *RL* lead network ($V_{out} = V_L$).

The output voltage V_L leads V_{in} by an angle (phase lead) that is the difference between 90° and the circuit phase angle θ, as you can see in the figure [$\theta = \arctan(X_L/R)$]. The amount of phase lead is expressed as follows:

$$\phi = 90° - \arctan\left(\frac{X_L}{R}\right) \tag{14–15}$$

Again, the phase-lead network can be considered as a voltage divider with the voltage across L being the output. The expression for the output voltage is

$$V_{out} = \left(\frac{X_L}{\sqrt{R^2 + X_L^2}}\right)V_{in} \tag{14–16}$$

EXAMPLE 14–14 Determine the amount of phase lead and output voltage in the lead network in Figure 14–35.

FIGURE 14–35

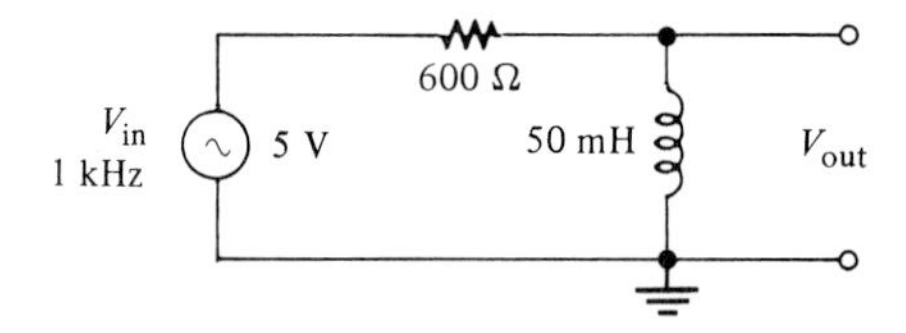

Solution

$$X_L = 2\pi fL = 2\pi(1\text{ kHz})(50\text{ mH}) = 314.2\ \Omega$$

$$\phi = 90° - \arctan\left(\frac{X_L}{R}\right) = 90° - \arctan\left(\frac{314.2}{600}\right) = 62.36°$$

The output leads the input by 62.36°.

$$V_{out} = \left(\frac{X_L}{\sqrt{R^2 + X_L^2}}\right)V_{in} = \left(\frac{314.2\ \Omega}{\sqrt{(600\ \Omega)^2 + (314.2\ \Omega)^2}}\right)5\text{ V}$$
$$= 2.32\text{ V}$$

Effects of Frequency on the Phase Shift and Output Voltage of the Lead Network

Since the circuit phase angle, θ, increases as frequency increases, the phase lead between the input and the output voltages *decreases*. You can see this relationship by examining Equation (14–15). Also, the amplitude of the output voltage increases as the frequency increases because X_L becomes greater and more of the total input voltage is dropped across the inductor. Figure 14–36 illustrates the changes in the phase lead and output voltage with frequency.

The *RL* Circuit as a Filter

Like *RC* circuits, series *RL* circuits also exhibit a frequency-selective characteristic and therefore act as basic filters.

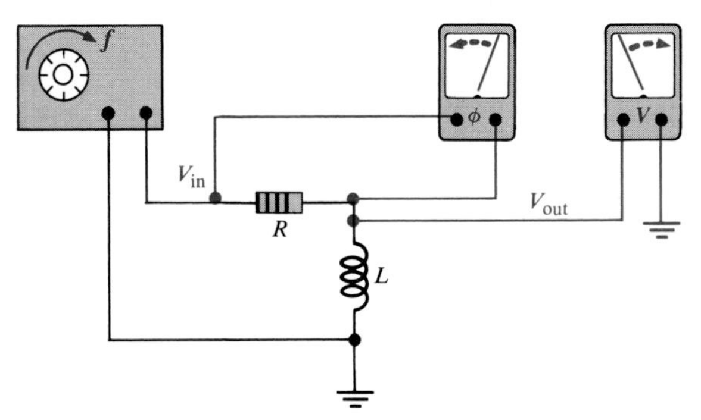

(a) When f increases, the phase lead ϕ decreases and V_{out} increases.

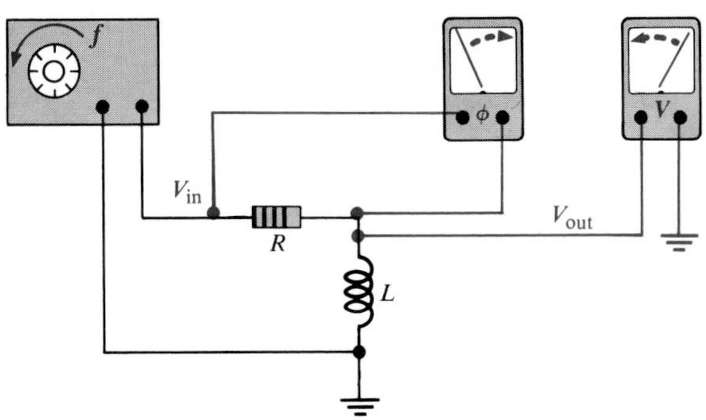

(b) When f decreases, ϕ increases and V_{out} decreases.

FIGURE 14–36
Illustration of how the frequency affects the phase lead and output voltage in an *RL* lead network with the amplitude of V_{in} held constant.

The *RL* Low-Pass Filter

You have seen what happens to the phase and the output voltage in the lag network. In terms of its filtering action, the change in the amplitude of the output voltage with frequency is the primary consideration.

To illustrate low-pass filter action, Figure 14–37 shows a specific series of measurements in which the frequency starts at zero (dc) and is increased in increments up to 20 kHz. At each value of frequency the output voltage is measured. As you can see, the inductive reactance increases as frequency goes up, thus causing less voltage to be dropped across the resistor while the input voltage is held at a constant 10 V throughout each step. The response curve for these particular values would appear the same as the response curve in Figure 13–49 for the *RC* low-pass filter.

The *RL* High-Pass Filter

To illustrate *RL* high-pass filter action, Figure 14–38 shows a series of specific measurements. Again, the frequency starts at zero (dc) and is increased in increments up to 10 kHz. As you can see, the inductive reactance increases as the frequency goes up, thus causing more voltage to be dropped across the inductor.

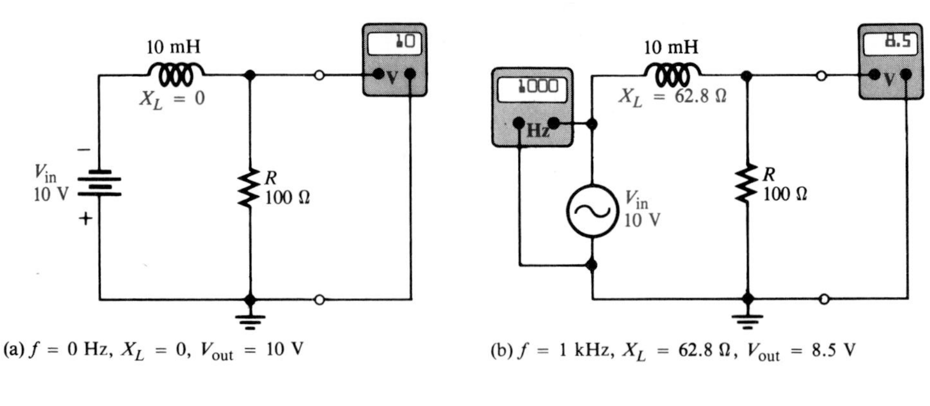

(a) $f = 0$ Hz, $X_L = 0$, $V_{out} = 10$ V

(b) $f = 1$ kHz, $X_L = 62.8\ \Omega$, $V_{out} = 8.5$ V

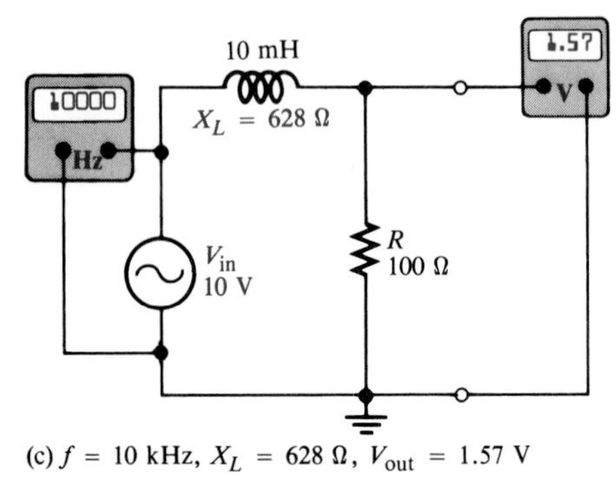

(c) $f = 10$ kHz, $X_L = 628\ \Omega$, $V_{out} = 1.57$ V

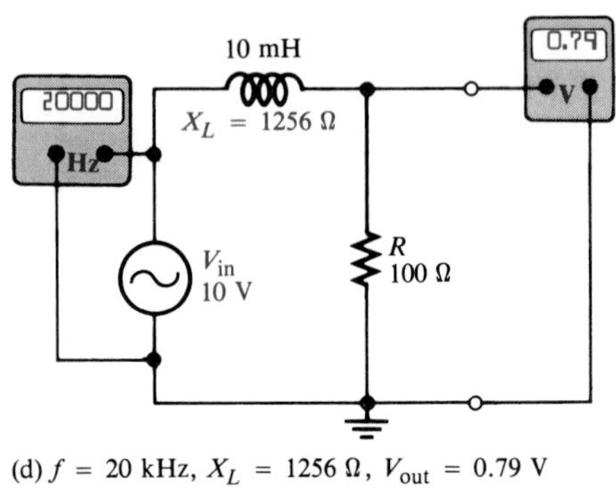

(d) $f = 20$ kHz, $X_L = 1256\ \Omega$, $V_{out} = 0.79$ V

FIGURE 14–37
Example of low-pass filter action. Winding resistance has been neglected.

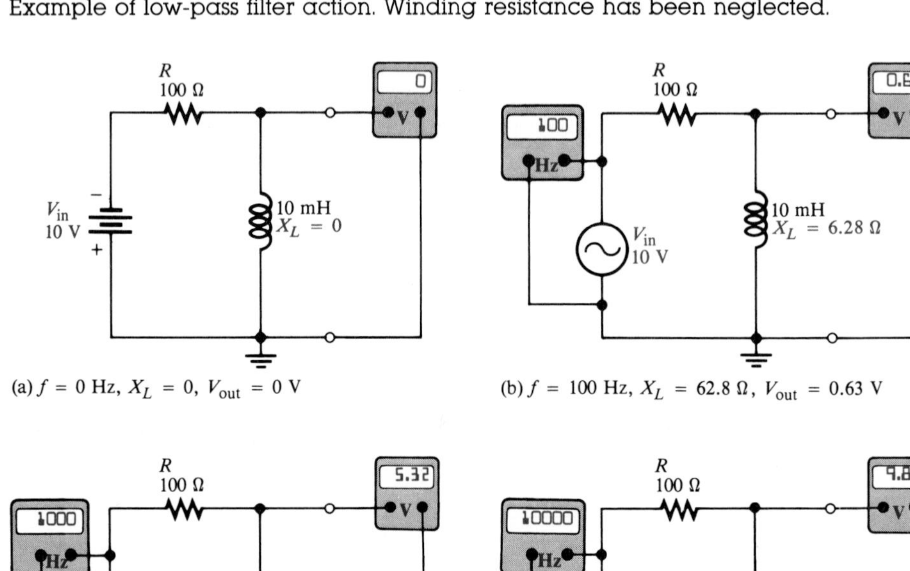

(a) $f = 0$ Hz, $X_L = 0$, $V_{out} = 0$ V

(b) $f = 100$ Hz, $X_L = 62.8\ \Omega$, $V_{out} = 0.63$ V

(c) $f = 1$ kHz, $X_L = 62.8\ \Omega$, $V_{out} = 5.32$ V

(d) $f = 10$ kHz, $X_L = 628\ \Omega$, $V_{out} = 9.88$ V

FIGURE 14–38
Example of high-pass filter action. Winding resistance has been neglected.

Again, when the values are plotted, the response curve is the same as the one for the *RC* high-pass filter as shown in Figure 13–51.

The Cutoff Frequency of an *RL* Filter

The frequency at which the inductive reactance equals the resistance in a low-pass or high-pass *RL* filter is called the *cutoff frequency* and is designated f_c. This condition is expressed as $2\pi f_c L = R$. Solving for f_c, we get the following formula:

$$f_c = \frac{1}{2\pi(L/R)} \tag{14–17}$$

As with the *RC* filter, the output voltage is 70.7% of its maximum value at f_c. In a high-pass filter, all frequencies above f_c are considered to be passed by the filter, and all those below f_c are considered to be rejected. The reverse, of course, is true for a low-pass filter. *Filter bandwidth,* defined in Chapter 13, applies to both *RC* and *RL* filter circuits.

Troubles in *RL* Networks

An Open Component in a Basic Low-Pass Filter

If the *resistor* opens, the input signal passes through the inductor to the output and is unaffected by the open resistor. Thus, you will observe an increase in cutoff frequency to a value dependent on the load resistance (if any) or the resistance of the measuring instrument that is connected to the output. If the *inductor* opens, there is no output signal because the path between input and output is broken.

A Shorted Inductor in a Basic Low-Pass Filter

If the inductor shorts out, the input signal voltage is connected directly to the output, and there is a complete loss of filter action.

Another possible condition is that the inductor may become partially shorted, thereby reducing the inductance and creating an increase in the cutoff frequency.

An Open Component in a Basic High-Pass Filter

An open inductor to ground causes a loss of filter action. An open resistor breaks the path between input and output and results in zero output voltage.

A Shorted Inductor in a Basic High-Pass Filter

A shorted inductor results in zero output voltage due to the direct short through the inductor to ground. A partially shorted inductor will cause an increase in the critical frequency.

SECTION REVIEW 14–8

1. A certain *RL* lead network consists of a 3.3-kΩ resistor and a 15-mH inductor. Determine the phase lead between input and output at a frequency of 5 kHz.
2. An *RL* lag network has the same component value as the lead network in Problem 1. What is the magnitude of the output voltage at 5 kHz when the input is 10 V rms?

3. When an RL circuit is used as a low-pass filter, across which component is the output taken?

APPLICATION NOTE

You can decrease the power factor of the motor from the Application Assignment by adding a capacitor across the motor terminals. The value selected will depend on the inductance of the motor. Decreasing the power factor reduces the current drawn from the source, and thus the system will require a lower-rated source and smaller interconnecting cables.

SUMMARY

Facts

- When a sine wave voltage is applied to an RL circuit, the current and all the voltage drops are also sine waves.
- Total current in an RL circuit always lags the source voltage.
- The resistor voltage is always in phase with the current.
- The inductor voltage always leads the current by 90°.
- In an RL circuit, the impedance is determined by both the resistance and the inductive reactance combined.
- Impedance is expressed in units of ohms.
- The impedance of an RL circuit varies directly with frequency.
- The phase angle (θ) of a series RL circuit varies directly with frequency.
- You can determine the impedance of a circuit by measuring the applied voltage and the total current and then applying Ohm's law.
- In an RL circuit, part of the power is resistive and part reactive.
- The phasor combination of resistive power (true power) and reactive power is called *apparent power*.
- The power factor indicates how much of the apparent power is true power.
- A power factor of 1 indicates a purely resistive circuit, and a power factor of 0 indicates a purely reactive circuit.
- In a lag network, the output voltage lags the input voltage in phase.
- In a lead network, the output voltage leads the input voltage in phase.
- A filter passes certain frequencies and rejects others.

Definitions

- *Impedance* (Z)—the total opposition to sinusoidal current expressed in ohms.
- *Inductive susceptance* (B_L)—the reciprocal of inductive reactance.
- *Filter*—a frequency-selective circuit.
- *Cutoff frequency* (f_c)—the frequency at which the output voltage of a filter is 70.7% of the maximum output voltage.
- *Bandwidth*—the range of frequencies that are considered to be passed by a filter.

Formulas

$$Z = \sqrt{R^2 + X_L^2} \quad \text{Series } RL \text{ impedance} \qquad (14\text{–}1)$$

$$\theta = \arctan\left(\frac{X_L}{R}\right) \quad \text{Series } RL \text{ phase angle} \qquad (14\text{–}2)$$

$$V_s = \sqrt{V_R^2 + V_L^2} \quad \text{Total voltage in a series } RL \text{ circuit} \qquad (14\text{–}3)$$

$$\theta = \arctan\left(\frac{V_L}{V_R}\right) \quad \text{Series } RL \text{ phase angle} \qquad (14\text{–}4)$$

$$Z = \frac{RX_L}{\sqrt{R^2 + X_L^2}} \quad \text{Parallel } RL \text{ impedance} \qquad (14\text{–}5)$$

$$\theta = \arctan\left(\frac{R}{X_L}\right) \quad \text{Parallel } RL \text{ phase angle} \qquad (14\text{–}6)$$

$$B_L = \frac{1}{X_L} \quad \text{Inductive susceptance} \qquad (14\text{–}7)$$

$$Y = \frac{1}{Z} \quad \text{Admittance} \qquad (14\text{–}8)$$

$$Y = \sqrt{G^2 + B_L^2} \quad \text{Admittance} \qquad (14\text{–}9)$$

$$I_T = \sqrt{I_R^2 + I_L^2} \quad \text{Total current in parallel } RL \text{ circuit} \qquad (14\text{–}10)$$

$$\theta = \arctan\left(\frac{I_L}{I_R}\right) \quad \text{Parallel } RL \text{ phase angle} \qquad (14\text{–}11)$$

$$P_r = I^2 X_L \quad \text{Reactive power} \qquad (14\text{–}12)$$

$$\phi = \arctan\left(\frac{X_L}{R}\right) \quad \text{Phase lag} \qquad (14\text{–}13)$$

$$V_{out} = \left(\frac{R}{\sqrt{R^2 + X_L^2}}\right)V_{in} \quad \text{Output of lag network} \qquad (14\text{–}14)$$

$$\phi = 90^\circ - \arctan\left(\frac{X_L}{R}\right) \quad \text{Phase lead} \qquad (14\text{–}15)$$

$$V_{out} = \left(\frac{X_L}{\sqrt{R^2 + X_L^2}}\right)V_{in} \quad \text{Output of lead network} \qquad (14\text{–}16)$$

$$f_c = \frac{1}{2\pi(L/R)} \quad \text{Cutoff frequency of } RL \text{ filter} \qquad (14\text{–}17)$$

SELF-TEST

1. In a series *RL* circuit, in what way does each voltage drop differ from the applied voltage?
2. Describe the phase relationships between the current in a series *RL* circuit and
   (a) the resistor voltage.
   (b) the inductor voltage.
3. If the frequency of the voltage applied to an *RL* circuit is increased, what happens to the impedance? To the phase angle?
4. If the frequency is doubled and the resistance is halved, how much does the value of the impedance change?
5. To reduce the current in a series *RL* circuit, should you increase or decrease the frequency?
6. In a series *RL* circuit, 10 V rms is measured across the resistor, and 10 V rms is measured across the inductor. What is the peak value of the source voltage?
7. The voltages in Question 6 are measured at a certain frequency. To make the resistor voltage greater than the inductor voltage, should you increase or decrease the frequency?

8. When the resistor voltage in Question 7 becomes greater than the inductor voltage, does the phase angle increase or decrease?
9. Does the impedance of a parallel *RL* circuit increase or decrease with frequency?
10. In a parallel *RL* circuit, 2 A rms flows through the resistive branch, and 2 A rms flows through the inductive branch. What is the total rms current?
11. You are observing two voltage waveforms on an oscilloscope. The time base of the scope is adjusted so that one half-cycle of the waveforms covers the ten divisions on the horizontal axis of the screen. The positive-going zero crossing of one waveform is at the left-most division, and the positive-going zero crossing of the other is three divisions to the right. What is the phase shift between these two waveforms?
12. In general, which is more desirable, a power factor of 0.9 or a power factor of 0.78?
13. If a load is purely inductive and the reactive power is 10 VAR, what is the apparent power?
14. For a certain load, the true power is 10 W and the reactive power is 10 VAR. Which of the following does the apparent power equal:
    (a) 5 VA **(b)** 20 VA **(c)** 14.14 VA **(d)** 100 VA
15. The cutoff frequency of a certain low-pass *RL* filter is 20 kHz. What is the filter's bandwidth?

PROBLEM SET A

Section 14–1

14–1 A 15-kHz sinusoidal voltage is applied to a series *RL* circuit. Determine the frequency of I, V_R, and V_L.

14–2 What are the waveshapes of I, V_R, and V_L in Problem 14–1?

Section 14–2

14–3 Find the impedance of each circuit in Figure 14–39.

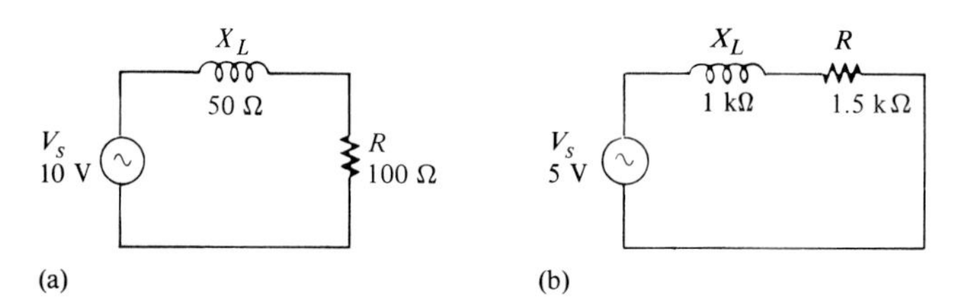

FIGURE 14–39

14–4 Determine the impedance and phase angle in each circuit in Figure 14–40.

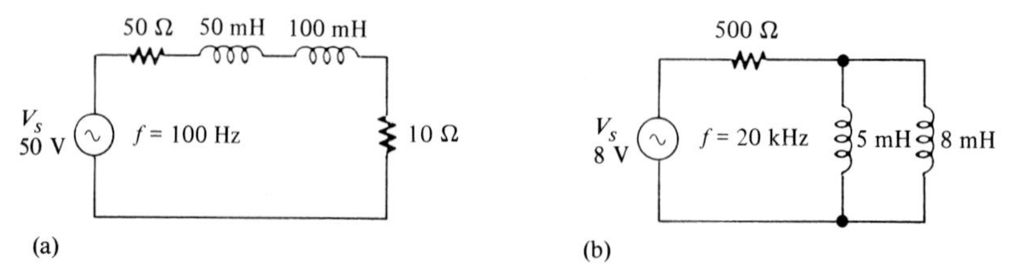

FIGURE 14–40

14–5 In Figure 14–41, determine the impedance at each of the following frequencies:
(a) 100 Hz **(b)** 500 Hz **(c)** 1 kHz **(d)** 2 kHz

FIGURE 14–41

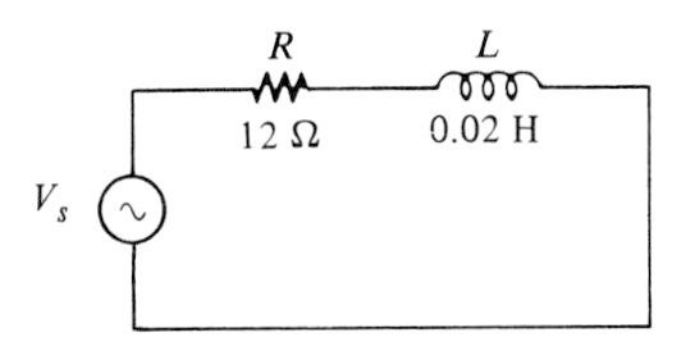

14–6 Determine the values of R and X_L in a series RL circuit for the following values of impedance and phase angle:
(a) $Z = 20\ \Omega, \theta = 45°$ **(b)** $Z = 500\ \Omega, \theta = 35°$
(c) $Z = 2.5\ k\Omega, \theta = 72.5°$ **(d)** $Z = 998\ \Omega, \theta = 45°$

Section 14–3

14–7 Find the current for each circuit of Figure 14–39.

14–8 Calculate the total current in each circuit of Figure 14–40.

14–9 Determine θ for the circuit in Figure 14–42.

14–10 If the inductance in Figure 14–42 is doubled, does θ increase or decrease, and by how many degrees?

14–11 Sketch the waveforms for V_s, V_R, and V_L in Figure 14–42. Show the proper phase relationships.

FIGURE 14–42

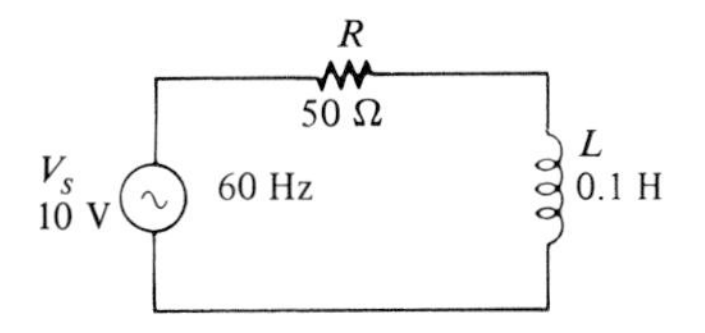

14–12 For the circuit in Figure 14–43, find the rms values for V_R and V_L for each of the following frequencies:
(a) 60 Hz **(b)** 200 Hz **(c)** 500 Hz **(d)** 1 kHz

FIGURE 14–43

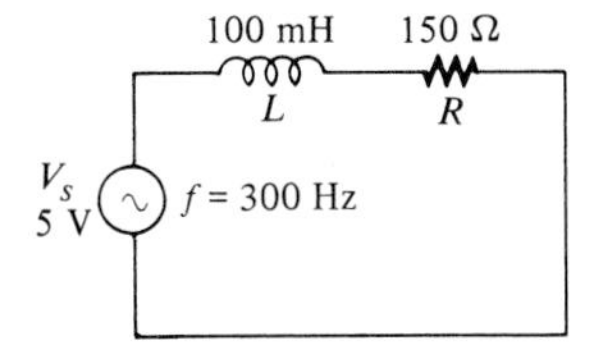

Section 14–4

14–13 What is the impedance for the circuit in Figure 14–44?

FIGURE 14–44

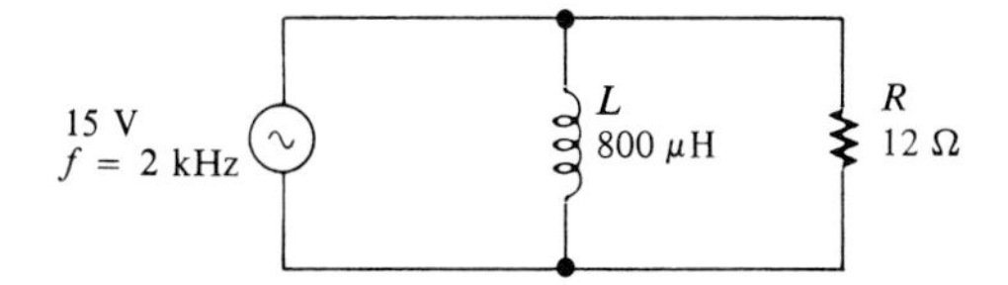

14–14 Repeat Problem 14–13 for the following frequencies:
(a) 1.5 kHz (b) 3 kHz (c) 5 kHz (d) 10 kHz

14–15 At what frequency does X_L equal R in Figure 14–44?

Section 14–5

14–16 Find the total current and each branch current in Figure 14–45.

FIGURE 14–45

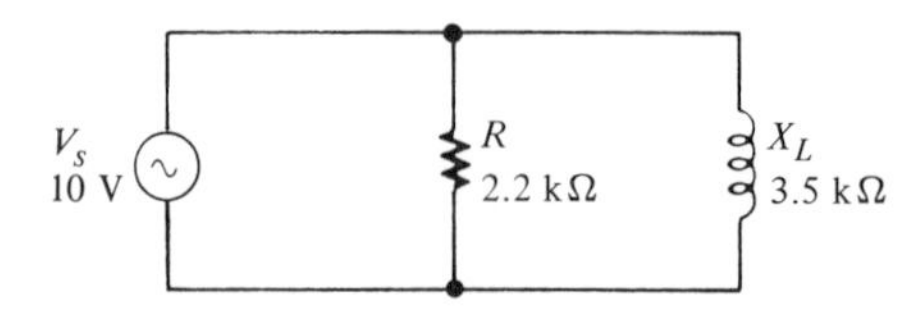

14–17 Determine the following quantities in Figure 14–46:
(a) Z (b) I_R (c) I_L (d) I_T (e) θ

FIGURE 14–46

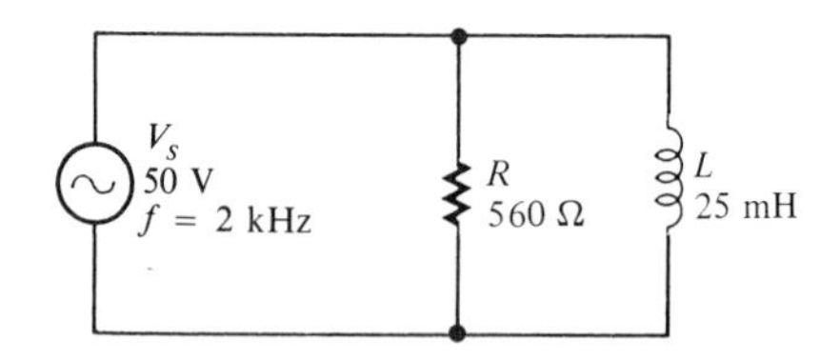

14–18 Convert the circuit in Figure 14–47 to an equivalent series form.

FIGURE 14–47

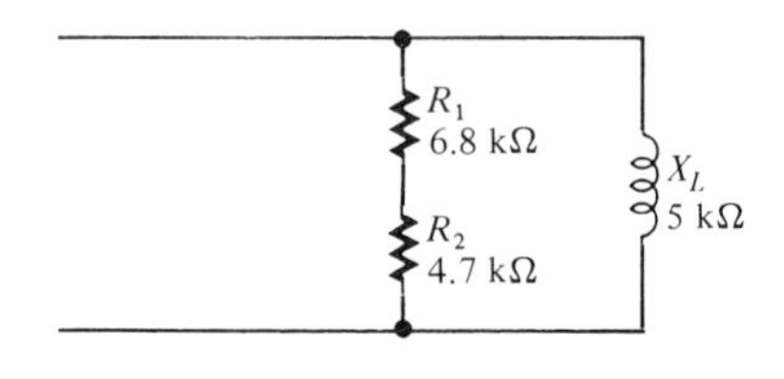

Section 14–6

14–19 Determine the voltage across each element in Figure 14–48.

FIGURE 14–48

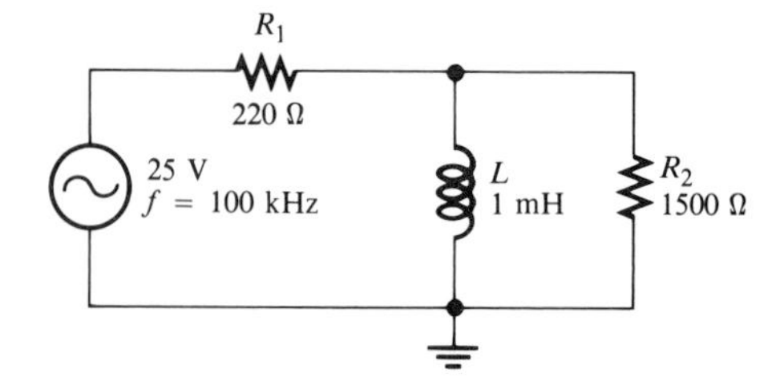

14–20 Is the circuit in Figure 14–48 predominantly resistive or predominantly inductive?

14–21 Find the current in each branch and the total current in Figure 14–48.

Section 14–7

14–22 In a certain *RL* circuit, the true power is 100 mW, and the reactive power is 340 mVAR. What is the apparent power?

14–23 Determine the true power and the reactive power in Figure 14–42.

14–24 What is the power factor in Figure 14–45?

14–25 Determine P_{true}, P_r, P_a, and *PF* for the circuit in Figure 14–48. Sketch the power triangle.

Section 14–8

14–26 For the lag network in Figure 14–49, determine the phase lag of the output voltage with respect to the input for the following frequencies:
(a) 1 Hz **(b)** 100 Hz **(c)** 1 kHz **(d)** 10 kHz

FIGURE 14–49

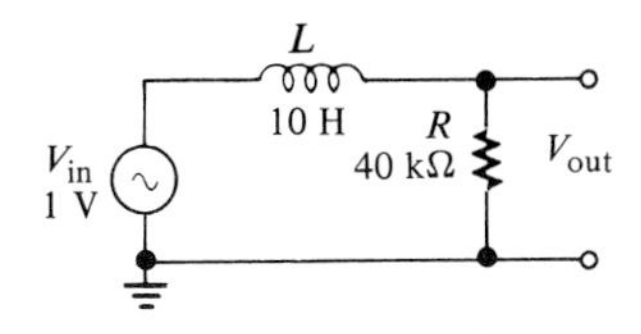

14–27 Plot the response curve for the circuit in Figure 14–49. Show the output voltage versus frequency in 1-kHz increments from 0 Hz to 5 kHz.

14–28 Repeat Problem 14–26 for the lead network in Figure 14–50.

FIGURE 14–50

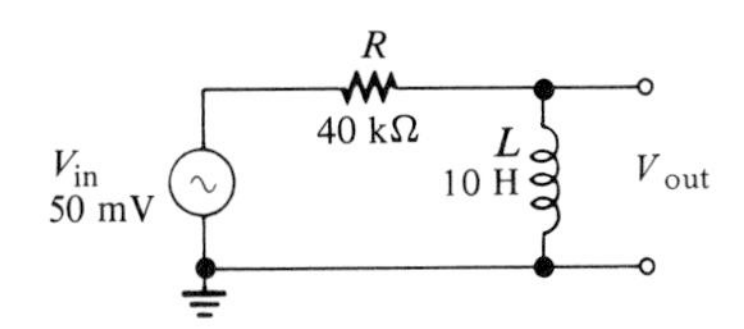

14–29 Using the same procedure as in Problem 14–27, plot the response curve for Figure 14–50.

14–30 Sketch the voltage phasor diagram for each circuit in Figures 14–49 and 14–50 for a frequency of 8 kHz.

PROBLEM SET B

14–31 Determine the voltage across the inductors in Figure 14–51.

FIGURE 14–51

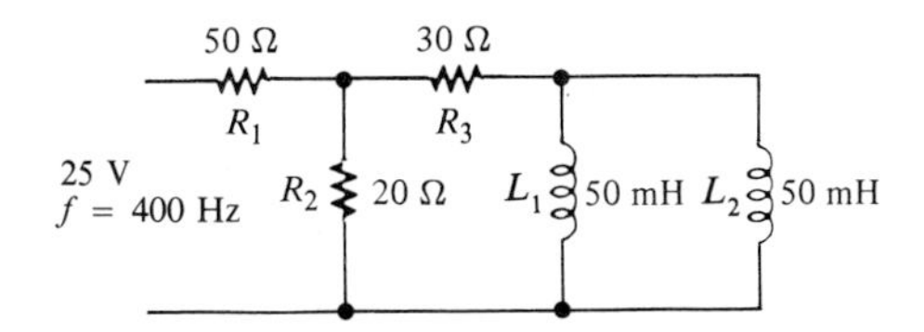

14–32 Is the circuit in Figure 14–51 predominantly resistive or predominantly inductive?

14–33 Find the total current in Figure 14–51.

14–34 For the circuit in Figure 14–52, determine the following:
(a) Z_T (b) I_T (c) θ (d) V_L
(e) V_{R3}

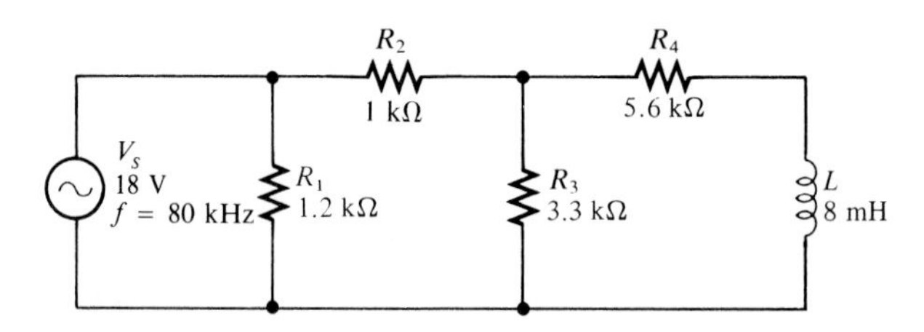

FIGURE 14–52

14–35 For the circuit in Figure 14–53, determine the following:
(a) I_{R1} (b) I_{L1} (c) I_{L2} (d) I_{R2}

FIGURE 14–53

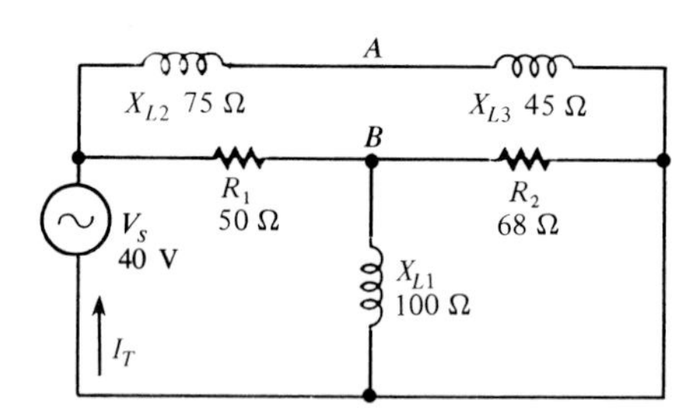

14–36 Determine the phase shift and attenuation (ratio of V_{out} to V_{in}) from the input to the output for the network in Figure 14–54.

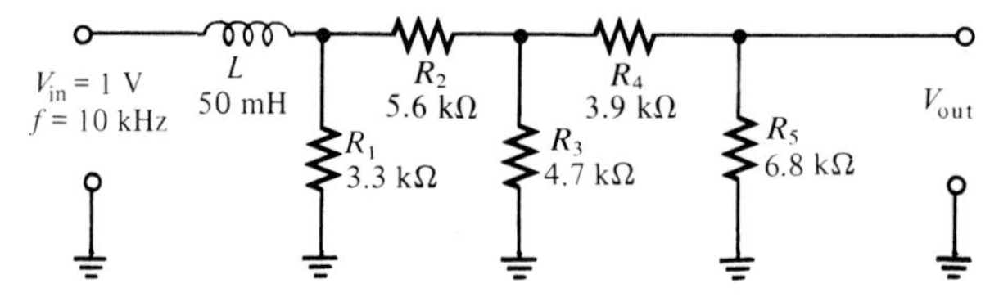

FIGURE 14–54

ANSWERS TO SECTION REVIEWS

Section 14–1

1. 1 kHz. 2. Closer to 0°.

Section 14–2

1. Leads. 2. The inductive reactance. 3. Opposite phase angles.
4. 59.9 kΩ, 56.58°.

Section 14–3

1. 3.61 V. 2. 56.31°. 3. X_L increases, Z increases, and θ increases.

Section 14–4

1. 20 Ω. 2. 0.024 S. 3. Lags by 33.69°.

Section 14–5

1. 32 mA. **2.** 59.04°, 23.32 mA. **3.** 90°.

Section 14–6

1. $Z = 494.6\ \Omega$. **2.** $I_T = 20.22$ mA.

Section 14–7

1. Resistor. **2.** 0.643. **3.** $P_{true} = 4.7$ W, $P_r = 6.2$ VAR, $P_a = 7.78$ VA.

Section 14–8

1. 81.87°. **2.** 1.41 V. **3.** Resistor.

Resonant Circuits

In this chapter, we introduce frequency response of *RLC* circuits—that is, circuits with combinations of resistance, capacitance, and inductance. We will discuss both series and parallel *RLC* circuits, including the concepts of series and parallel resonance.

Because it is the basis for frequency selectivity, resonance in electrical circuits is very important to the operation of many types of electronic systems, particularly in the area of communications. For example, the ability of a radio or television receiver to select a certain frequency transmitted by a certain station and to eliminate frequencies from other stations is based on the principle of resonance.

The operation of both band-pass and band-stop filters is based on resonance of circuits containing inductance and capacitance, and these filters are discussed in this chapter. Also, certain system applications are presented.

Specifically, you will learn:

- How to determine the impedance and phase angle in series and parallel *RLC* circuits.
- How to analyze *RLC* circuits for voltage and current values.
- The characteristics of the series and parallel resonant conditions.
- How to determine the frequency at which resonance occurs.
- The differences between series resonance and parallel resonance.
- How to analyze a nonideal parallel resonant circuit that has resistance in its inductive branch.
- The effects of parallel resistance on a parallel resonant circuit.
- How to determine the *Q* of both series and parallel resonant circuits.
- The response characteristics of band-pass and band-stop filters, both series resonant and parallel resonant types.
- How the quality factor *(Q)* affects the bandwidth *(BW)* of a filter.
- How the *Q* affects the resonant frequency and impedance of a parallel resonant filter.
- The meaning of *decibel* (dB).
- How resonant filters are applied in several specific areas.

APPLICATION ASSIGNMENT

As a service technician in an electronics shop, you are working on a television receiver that has dark bars across the picture screen. Being an inexperienced technician, you are not sure what is causing this problem. Your supervisor tells you that the dark bars are *sound bars* caused by the audio signal getting through to the picture tube and interfering with the video signal.

After studying this chapter, you will know the most likely cause of the problem. A possible cause is given in the Application Note at the end of the chapter.

IMPEDANCE AND PHASE ANGLE OF SERIES *RLC* CIRCUITS

15–1

A series *RLC* circuit is shown in Figure 15–1. It contains resistance, inductance, and capacitance. As you know, X_L causes the total current to lag the applied voltage. X_C has the opposite effect: It causes the current to lead the voltage. Thus, X_L and X_C tend to offset each other. When they are equal, they cancel, and the total reactance is zero. In any case, the total reactance in the series circuit is

$$X_T = |X_L - X_C| \tag{15–1}$$

The term $|X_L - X_C|$ means the *absolute value* of the difference of the two reactances. That is, the sign of the result is considered positive no matter which reactance is greater. For example, $3 - 7 = -4$, but the absolute value is

$$|3 - 7| = 4$$

When $X_L > X_C$, the circuit is predominantly inductive, and when $X_C > X_L$, the circuit is predominantly capacitive.

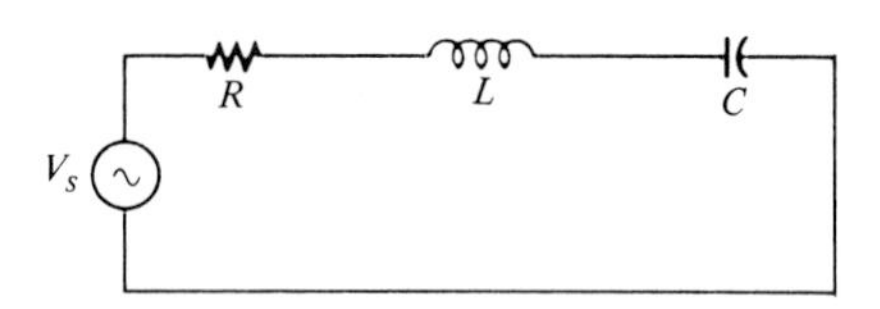

FIGURE 15–1
Series *RLC* circuit.

The total impedance for the series *RLC* circuit is given by

$$Z = \sqrt{R^2 + (X_L - X_C)^2} \tag{15–2}$$

and the phase angle is

$$\theta = \arctan\left(\frac{X_T}{R}\right) \tag{15–3}$$

EXAMPLE 15–1 Determine the total impedance and the phase angle in Figure 15–2.

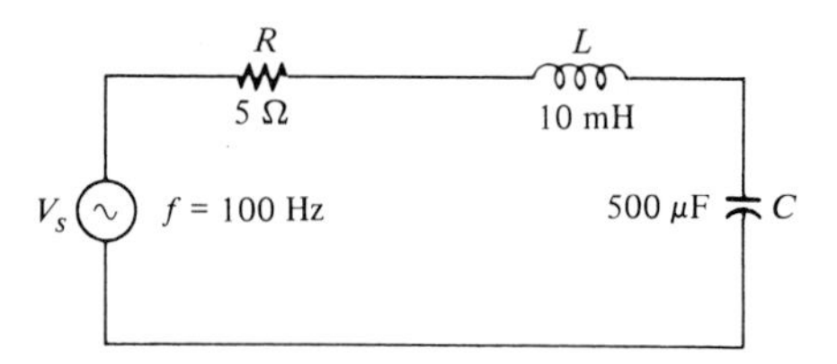

FIGURE 15–2

Solution
First find X_C and X_L:

$$X_C = \frac{1}{2\pi fC} = \frac{1}{2\pi(100 \text{ Hz})(500 \ \mu\text{F})} = 3.18 \ \Omega$$
$$X_L = 2\pi fL = 2\pi(100 \text{ Hz})(10 \text{ mH}) = 6.28 \ \Omega$$

In this case, X_L is greater than X_C, and thus the circuit is more inductive than capacitive. The magnitude of the total reactance is

$$X_T = |X_L - X_C| = |6.28\ \Omega - 3.18\ \Omega| = 3.1\ \Omega \qquad \text{(inductive)}$$

The total circuit impedance is

$$Z = \sqrt{R^2 + X_T^2} = \sqrt{(5\ \Omega)^2 + (3.1\ \Omega)^2} = 5.88\ \Omega$$

The phase angle (between I and V_s) is

$$\theta = \arctan\left(\frac{X_T}{R}\right) = \arctan\left(\frac{3.1\ \Omega}{5\ \Omega}\right)$$
$$= 31.8° \qquad \text{(current lagging } V_s\text{)}$$

The calculator sequences are as follows:

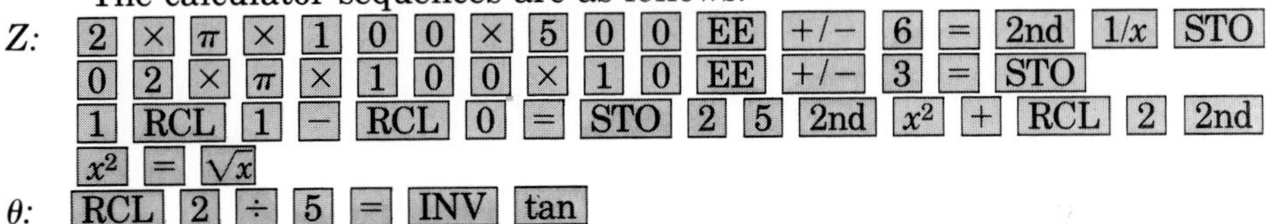

As you have seen, when the inductive reactance is greater than the capacitive reactance, the circuit appears inductive, and the current lags the applied voltage. When the capacitive reactance is greater, the circuit appears capacitive, and the current leads the applied voltage.

SECTION REVIEW 15–1

1. In a given series RLC circuit, X_C is 150 Ω and X_L is 80 Ω. What is the total reactance in ohms? Is it inductive or capacitive?
2. Determine the impedance for the circuit in Problem 1 when $R = 45\ \Omega$. What is the phase angle? Is the current leading or lagging the applied voltage?

ANALYSIS OF SERIES *RLC* CIRCUITS

15–2

Recall that capacitive reactance varies inversely with frequency and inductive reactance varies directly. As shown in Figure 15–3, for a typical series RLC circuit the total reactance behaves as follows: Starting at a very low frequency,

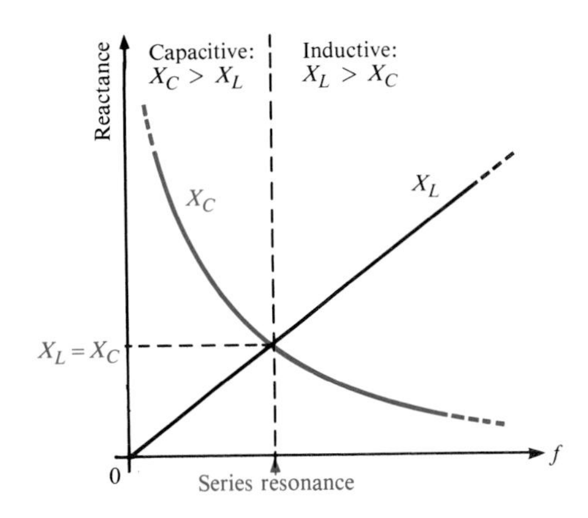

FIGURE 15–3
How X_C and X_L vary with frequency.

X_C is high, X_L is low, and the circuit is predominantly capacitive. As the frequency is increased, X_C decreases and X_L increases until a value is reached where $X_C = X_L$ and the two reactances cancel, making the circuit purely resistive. This condition is *series resonance* and will be studied in the next section. As the frequency is increased further, X_L becomes greater than X_C, and the circuit is predominantly inductive. Example 15–2 illustrates how the impedance and phase angle change as the source frequency is varied.

EXAMPLE 15–2 For each of the following frequencies of the source voltage, find the impedance and the phase angle for the circuit in Figure 15–4:
(a) $f = 1$ kHz **(b)** $f = 3.5$ kHz **(c)** $f = 5$ kHz
Note how the impedance and the phase angle change.

FIGURE 15–4

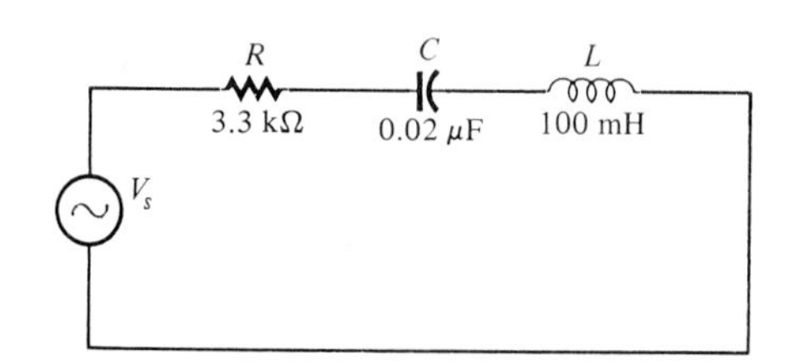

Solution

(a) At $f = 1$ kHz:

$$X_C = \frac{1}{2\pi(1 \text{ kHz})(0.02\ \mu\text{F})} = 7.96 \text{ k}\Omega$$
$$X_L = 2\pi(1 \text{ kHz})(100 \text{ mH}) = 628.3\ \Omega$$

The circuit is highly capacitive because X_C is much larger than X_L. The impedance is

$$Z = \sqrt{R^2 + (X_L - X_C)^2} = \sqrt{(3.3 \text{ k}\Omega)^2 + (628.3\ \Omega - 7.96 \text{ k}\Omega)^2}$$
$$= 8.04 \text{ k}\Omega$$

The phase angle is

$$\theta = \arctan\left(\frac{X_T}{R}\right) = \arctan\left(\frac{7.33 \text{ k}\Omega}{3.3 \text{ k}\Omega}\right) = 65.76°$$

I leads V_s by 65.76°.

(b) At $f = 3.5$ kHz:

$$X_C = \frac{1}{2\pi(3.5 \text{ kHz})(0.02\ \mu\text{F})} = 2.27 \text{ k}\Omega$$
$$X_L = 2\pi(3.5 \text{ kHz})(100 \text{ mH}) = 2.2 \text{ k}\Omega$$

The circuit is very close to being purely resistive but is still slightly capacitive because X_C is slightly larger than X_L.

$$Z = \sqrt{(3.3 \text{ k}\Omega)^2 + (2.2 \text{ k}\Omega - 2.27 \text{ k}\Omega)^2} = 3.3 \text{ k}\Omega$$
$$\theta = \arctan\left(\frac{X_T}{R}\right) = \arctan\left(\frac{0.07 \text{ k}\Omega}{3.3 \text{ k}\Omega}\right) = 1.2°$$

I leads V_s by 1.2°.

(c) At $f = 5$ kHz:

$$X_C = \frac{1}{2\pi(5\text{ kHz})(0.02\ \mu\text{F})} = 1.59\text{ k}\Omega$$

$$X_L = 2\pi(5\text{ kHz})(100\text{ mH}) = 3.14\text{ k}\Omega$$

The circuit is now predominantly inductive because $X_L > X_C$.

$$Z = \sqrt{(3.3\text{ k}\Omega)^2 + (3.14\text{ k}\Omega - 1.59\text{ k}\Omega)^2} = 3.64\text{ k}\Omega$$

$$\theta = \arctan\left(\frac{X_T}{R}\right) = \arctan\left(\frac{1.55\text{ k}\Omega}{3.3\text{ k}\Omega}\right) = 25.16°$$

I lags V_s by 25.16°.

Notice how the circuit changed from capacitive to inductive as the frequency increased. The phase condition changed from the leading current to the lagging current. It is interesting to note that both the impedance and the phase angle decreased to a minimum and then began increasing again as the frequency went up.

In a series *RLC* circuit, the capacitor voltage and the inductor voltage are *always* 180° out of phase with each other. For this reason, V_C and V_L subtract from each other, and thus the voltage across L and C combined is always less than the larger individual voltage across either element, as illustrated in Figure 15–5 and in the waveform diagram of Figure 15–6.

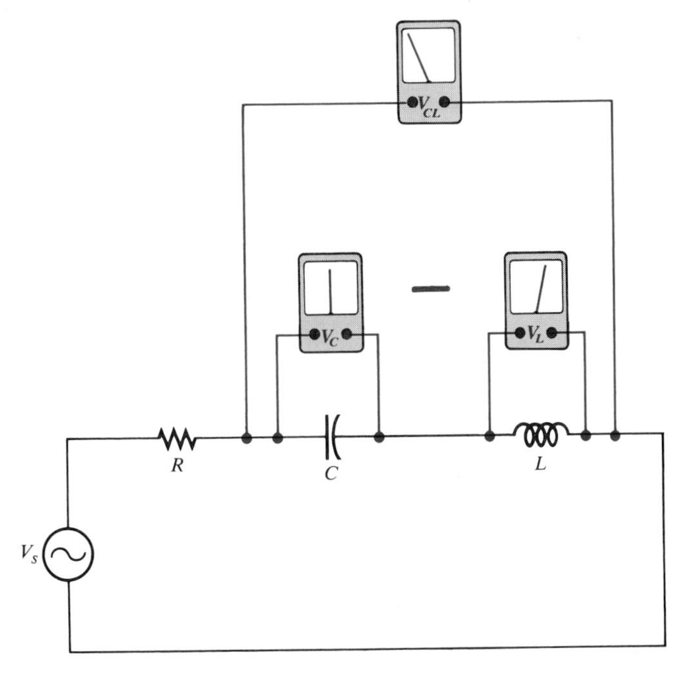

FIGURE 15–5
The voltage across the series combination of C and L is always less than the larger individual voltage.

FIGURE 15–6
V_L and V_C effectively subtract.

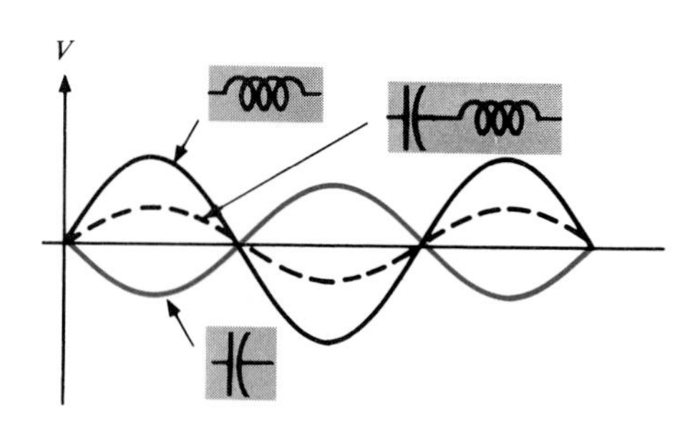

In the next example, we use Ohm's law to find the voltages in a series *RLC* circuit.

EXAMPLE 15–3 Find the voltages across each element in Figure 15–7, and draw a complete voltage phasor diagram.

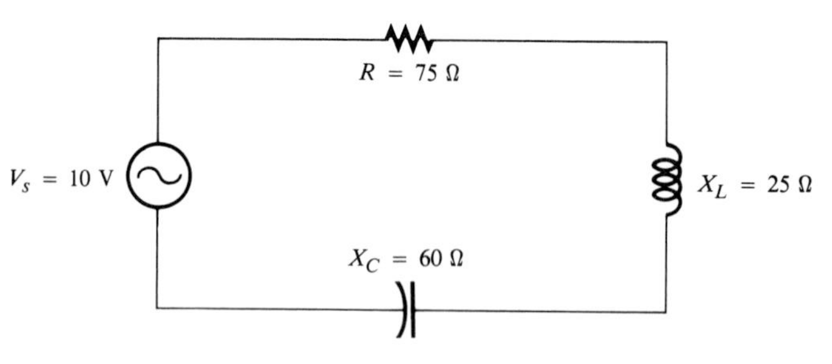

FIGURE 15–7

Solution
First find the total impedance:

$$Z = \sqrt{R^2 + (X_L - X_C)^2} = \sqrt{(75\ \Omega)^2 + (25\ \Omega - 60\ \Omega)^2}$$
$$= \sqrt{(75\ \Omega)^2 + (35\ \Omega)^2} = 82.76\ \Omega$$

Apply Ohm's law to find the current:

$$I = \frac{V_s}{Z} = \frac{10\ \text{V}}{82.76\ \Omega} = 0.121\ \text{A}$$

Now apply Ohm's law to find the voltages across *R*, *L*, and *C*:

$$V_R = IR = (0.121\ \text{A})(75\ \Omega) = 9.08\ \text{V}$$
$$V_L = IX_L = (0.121\ \text{A})(25\ \Omega) = 3.03\ \text{V}$$
$$V_C = IX_C = (0.121\ \text{A})(60\ \Omega) = 7.26\ \text{V}$$

The voltage across *L* and *C* combined is

$$V_{LC} = V_C - V_L = 7.26\ \text{V} - 3.03\ \text{V} = 4.23\ \text{V}$$

The circuit phase angle is

$$\theta = \arctan\left(\frac{X_T}{R}\right) = \arctan\left(\frac{35\ \Omega}{75\ \Omega}\right) = 25°$$

Since the circuit is capacitive ($X_C > X_L$), the current leads the source voltage by 25°.

The phasor diagram is shown in Figure 15–8. Notice that V_L is leading V_R by 90°, and V_C is lagging V_R by 90°. Also, there is a 180° phase difference between V_L and V_C. If the current phasor were shown, it would be at the same angle as V_R.

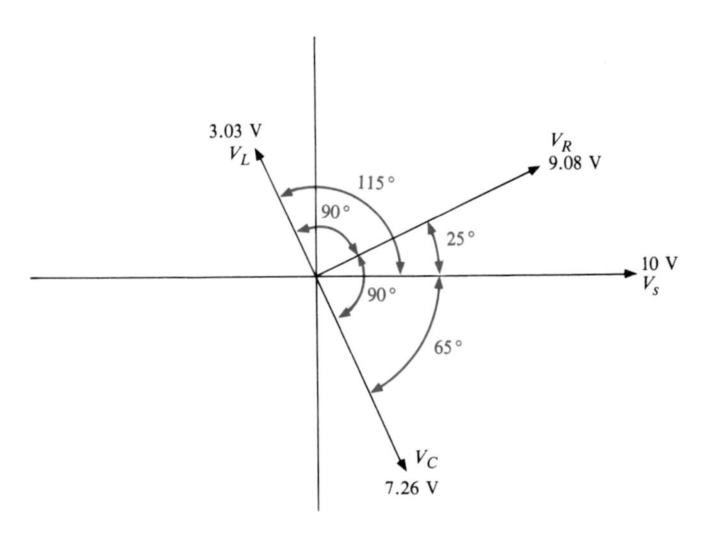

FIGURE 15–8

SECTION REVIEW 15–2

1. The following voltages occur in a certain series *RLC* circuit:

$$V_R = 24\text{ V} \quad V_L = 15\text{ V} \quad V_C = 45\text{ V}$$

Determine the source voltage.
2. When $R = 10\ \Omega$, $X_C = 18\ \Omega$, and $X_L = 12\ \Omega$, does the current lead or lag the applied voltage?
3. Determine the total reactance in Problem 2.

SERIES RESONANCE

15–3

In a series *RLC* circuit, *series resonance* occurs when $X_L = X_C$. The frequency at which resonance occurs is called the *resonant frequency*, f_r. Figure 15–9 illustrates the series resonant condition. Since $X_L = X_C$, the reactances effectively cancel, and the *impedance is purely resistive.* These resonant conditions are stated in the following equations:

$$X_L = X_C \tag{15–4}$$

$$Z_r = R \tag{15–5}$$

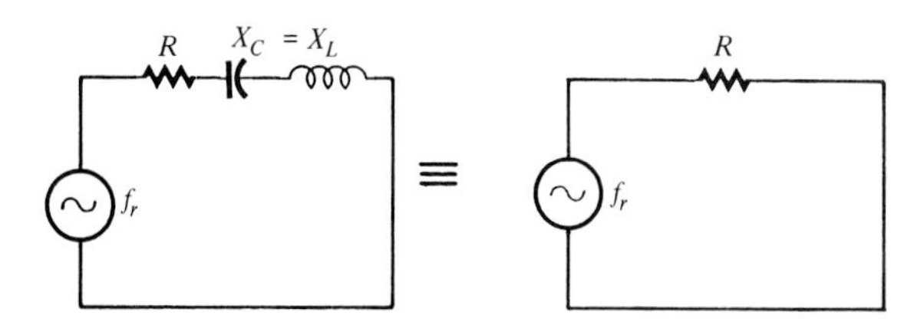

FIGURE 15–9
At the resonant frequency f_r, the reactances effectively cancel, leaving $Z = R$.

EXAMPLE 15–4 For the series *RLC* circuit in Figure 15–10, determine X_C and Z at resonance.

FIGURE 15–10

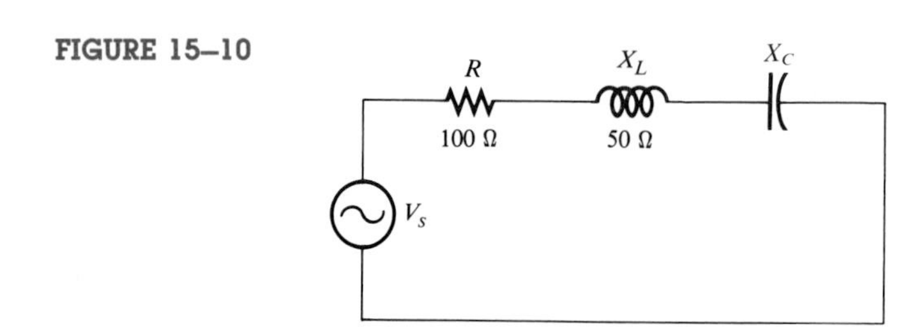

Solution

$$X_C = X_L = 50\ \Omega \quad \text{at the resonant frequency}$$
$$Z_r = R = 100\ \Omega$$

Why X_L and X_C Effectively Cancel at Resonance

At the series resonant frequency, the voltages across C and L are equal in magnitude because the reactances are equal and because the same current flows through both since they are in series ($IX_C = IX_L$). Also, V_L and V_C are always 180° out of phase with each other.

During any given cycle, the polarities of the voltages across C and L are opposite, as shown in Figures 15–11(a) and 15–11(b). The equal and opposite voltages across C and L cancel, leaving zero volts from point A to point B as shown in the figure. Since there is no voltage drop from A to B but there is still current, the total reactance must be zero, as indicated in Part (c) of the figure. Also, the voltage phasor diagram in Part (d) shows that V_C and V_L are equal in magnitude and 180° out of phase with each other.

The Series Resonant Frequency

For a given series *RLC* circuit, resonance occurs at only one specific frequency. A formula for this resonant frequency is developed as follows:

$$X_L = X_C$$

Substituting the reactance formulas, we have

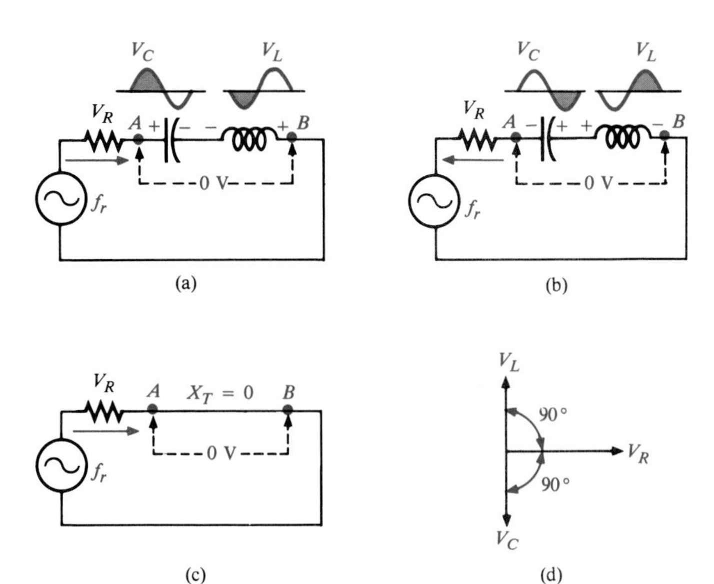

FIGURE 15–11
At the resonant frequency, f_r, the voltages across C and L are equal in magnitude. Since they are 180° out of phase with each other, they cancel, leaving 0 V across the LC combination (point A to point B). The section of the circuit from A to B effectively looks like a short at resonance.

$$2\pi f_r L = \frac{1}{2\pi f_r C}$$

Solving for the resonant frequency f_r yields

$$(2\pi f_r L)(2\pi f_r C) = 1$$
$$4\pi^2 f_r^2 LC = 1$$
$$f_r^2 = \frac{1}{4\pi^2 LC}$$

Taking the square root of both sides, we obtain

$$f_r = \frac{1}{2\pi\sqrt{LC}} \qquad (15\text{–}6)$$

EXAMPLE 15–5 Find the series resonant frequency for the circuit in Figure 15–12.

FIGURE 15–12

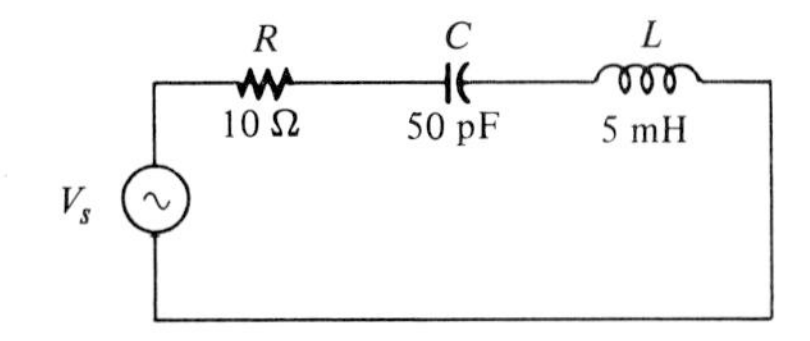

Solution

$$f_r = \frac{1}{2\pi\sqrt{LC}} = \frac{1}{2\pi\sqrt{(5\text{ mH})(50\text{ pF})}} = 318\text{ kHz}$$

The calculator sequence is

[5] [EE] [+/−] [3] [×] [5] [0] [EE] [+/−] [1] [2] [=] [$\sqrt{x}$] [×] [2] [×] [π] [=] [2nd] [1/x]

Voltages and Current in a Series Resonant Circuit

It is interesting to see how the current and the voltages in a series *RLC* circuit vary as the frequency is increased from below the resonant frequency, through resonance, and then above resonance. Figure 15–13 illustrates the general response of a circuit in terms of changes in the current and in the voltage drops.

Below the Resonant Frequency

Figure 15–13(a) indicates the response as the source frequency is increased from zero toward f_r. At $f = 0$ Hz (dc), the capacitor appears open and blocks the current. Thus, there is no voltage across R or L, and the entire source voltage appears across C. The impedance of the circuit is infinitely large at 0 Hz because X_C is infinite (C is open). As the frequency begins to increase, X_C decreases and X_L increases, causing the total reactance, $X_C - X_L$, to decrease. As a result, the impedance decreases and the current increases. As the current increases, V_R also increases, and both V_C and V_L increase. The voltage across C and L combined decreases from its maximum value of V_s because as V_C and V_L approach the same value, their difference becomes less.

At the Resonant Frequency

When the frequency reaches its resonant value, f_r, as shown in Figure 15–13(b), V_C and V_L cancel because they are equal in magnitude but opposite in phase. At this point the total impedance is equal to R and is at its minimum value because the total reactance is zero. Thus, the current is at its maximum value, V_s/R, and V_R is at its maximum value equal to the source voltage.

Above the Resonant Frequency

As the frequency is increased above resonance, as indicated in Figure 15–13(c), X_L continues to increase and X_C continues to decrease, causing the total reactance, $X_L - X_C$, to increase. As a result, there is an increase in impedance and a decrease in current. As the current decreases, V_R also decreases and V_C and V_L decrease. As V_C and V_L decrease, their difference becomes greater, so V_{CL} increases. As the frequency becomes very high, the current approaches zero, both V_R and V_C approach zero, and V_L approaches V_s.

The responses of voltage and current to increasing frequency are summarized as follows: As frequency is increased, the current increases below resonance, peaks at the resonant frequency, and then decreases above resonance. The resistor voltage responds in the same way as the current. The voltage

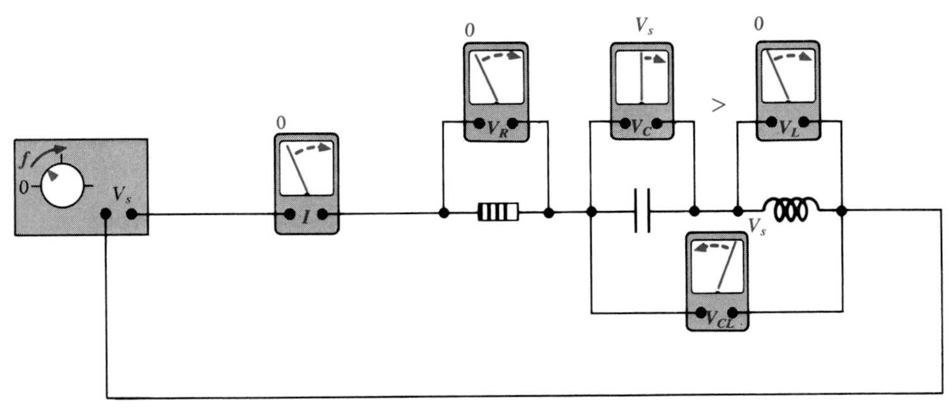

(a) Below resonance ($X_C > X_L$)

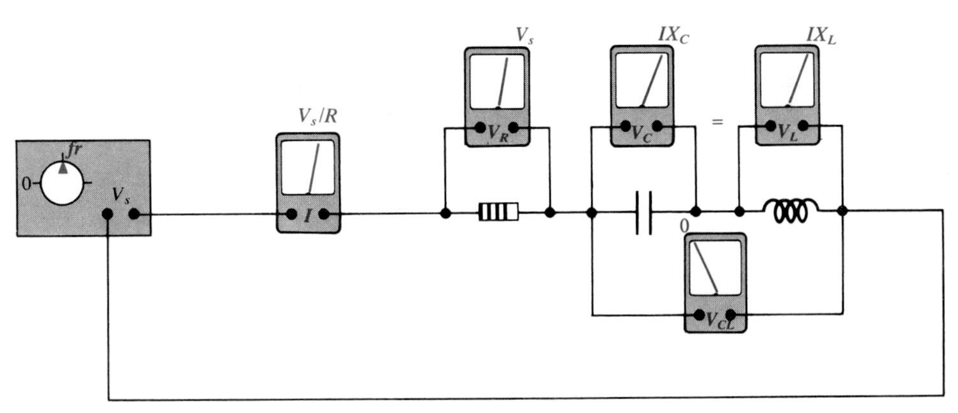

(b) At resonance ($X_C = X_L$)

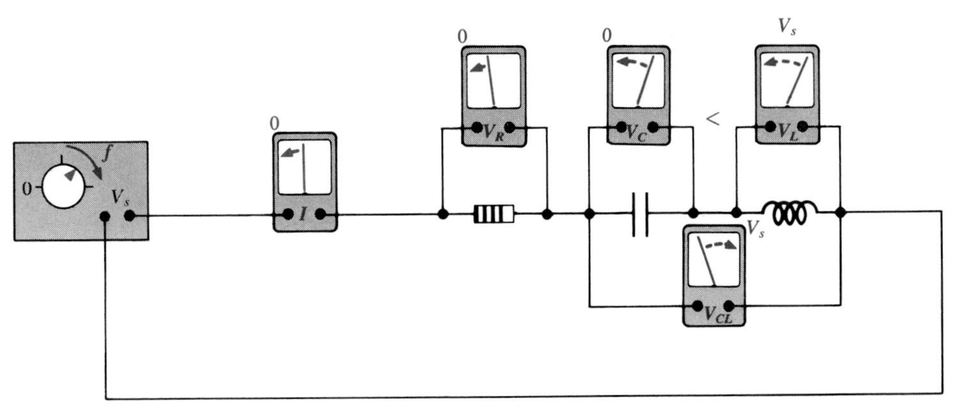

(c) Above resonance ($X_L > X_C$)

FIGURE 15–13
An illustration of how the voltages and the current respond in a series *RLC* circuit as the frequency is increased from below to above its resonant value. The source voltage is held at a constant amplitude.

across the C and L combination decreases as the frequency increases below resonance, reaching a minimum of zero at the resonant frequency; then it increases above resonance. The general response curves in Figure 15–14 illustrate this action graphically.

(a) Current (b) Resistor voltage

(c) Capacitor voltage (d) Inductor voltage

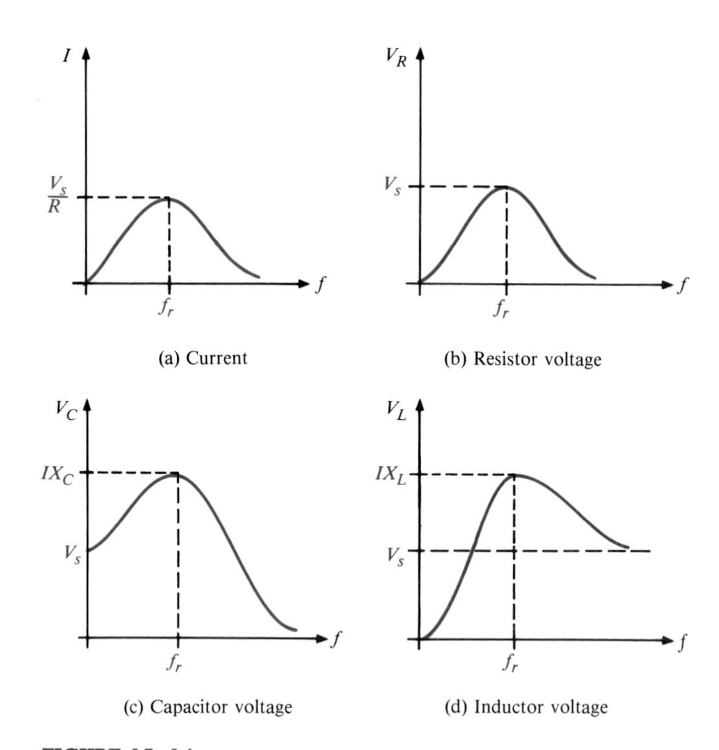

FIGURE 15–14
Current and voltages as functions of frequency in a series *RLC* circuit.

EXAMPLE 15–6 Find I, V_R, V_L, and V_C at resonance in Figure 15–15.

FIGURE 15–15

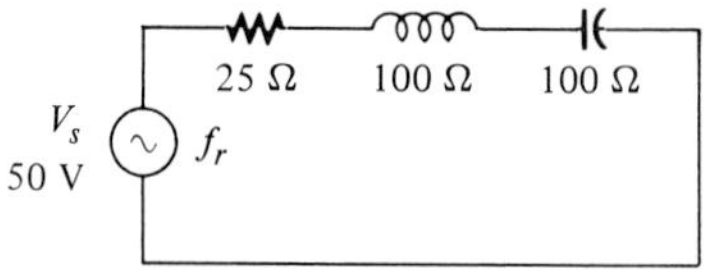

Solution
At resonance, I is maximum and equal to V_s/R:

$$I = \frac{V_s}{R} = \frac{50\ \text{V}}{25\ \Omega} = 2\ \text{A}$$

Applying Ohm's law, we obtain the following voltages:

$$V_R = IR = (2\ \text{A})(25\ \Omega) = 50\ \text{V}$$

$$V_L = IX_L = (2\text{ A})(100\ \Omega) = 200\text{ V}$$
$$V_C = IX_C = (2\text{ A})(100\ \Omega) = 200\text{ V}$$

The voltages are maximum at resonance but drop off above and below f_r. The voltages across L and C at resonance are exactly *equal in magnitude but 180° out of phase, so they cancel.* Thus, the total voltage across both L and C is zero, and $V_R = V_s$ at resonance. Individually, V_L and V_C can be much greater than the source voltage, as you see. Keep in mind that V_L and V_C are always opposite in polarity regardless of the frequency, but only at resonance are their magnitudes equal.

The Impedance of a Series Resonant Circuit

Figure 15–16 shows a general graph of impedance versus frequency superimposed on the curves for X_C and X_L. At zero frequency, both X_C and Z are infinitely large and X_L is zero, because the capacitor looks like an open at 0 Hz and the inductor looks like a short. As the frequency increases, X_C decreases and X_L increases. Since X_C is larger than X_L at frequencies below f_r, Z decreases along with X_C. At f_r, $X_C = X_L$ and $Z = R$. At frequencies above f_r, X_L becomes increasingly larger than X_C, causing Z to increase.

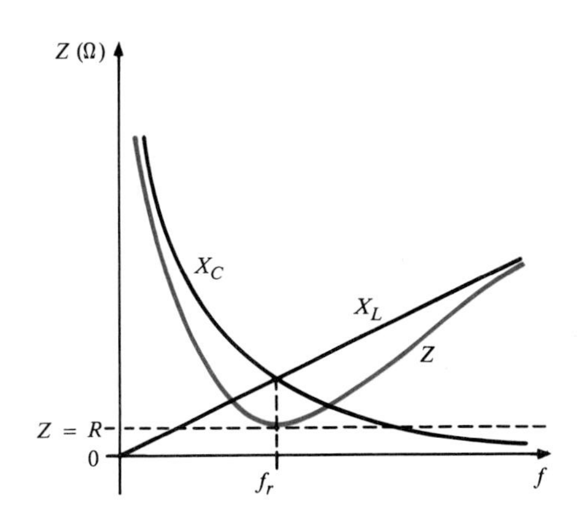

FIGURE 15–16
Series RLC impedance as a function of frequency.

EXAMPLE 15–7 For the circuit in Figure 15–17, determine the impedance at the following frequencies:
(a) f_r **(b)** 1000 Hz below f_r **(c)** 1000 Hz above f_r

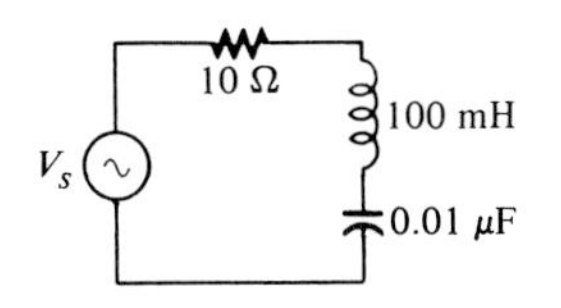

FIGURE 15–17

Solution

$$f_r = \frac{1}{2\pi\sqrt{LC}} = \frac{1}{2\pi\sqrt{(100\text{ mH})(0.01\ \mu\text{F})}} = 5.03\text{ kHz}$$

At 1000 Hz below f_r,

$$f_r - 1\text{ kHz} = 4.03\text{ kHz}$$

At 1000 Hz above f_r,

$$f_r + 1\text{ kHz} = 6.03\text{ kHz}$$

(a) The impedance at resonance is equal to R:

$$Z = 10\ \Omega$$

(b) The impedance at $f_r - 1$ kHz is

$$Z = \sqrt{R^2 + (X_L - X_C)^2} = \sqrt{(10\ \Omega)^2 + (2.53\text{ k}\Omega - 3.95\text{ k}\Omega)^2}$$
$$= 1.42\text{ k}\Omega$$

Notice that X_C is greater than X_L, so Z is capacitive.

(c) The impedance at $f_r + 1$ kHz is

$$Z = \sqrt{R^2 + (X_L - X_C)^2} = \sqrt{(10\ \Omega)^2 + (3.79\text{ k}\Omega - 2.64\text{ k}\Omega)^2}$$
$$= 1.15\text{ k}\Omega$$

X_L is greater than X_C, so Z is inductive.

The Phase Angle of a Series Resonant Circuit

At frequencies below resonance, $X_C > X_L$, and the current *leads* the source voltage, as indicated in Figure 15–18(a). The phase angle decreases as the fre-

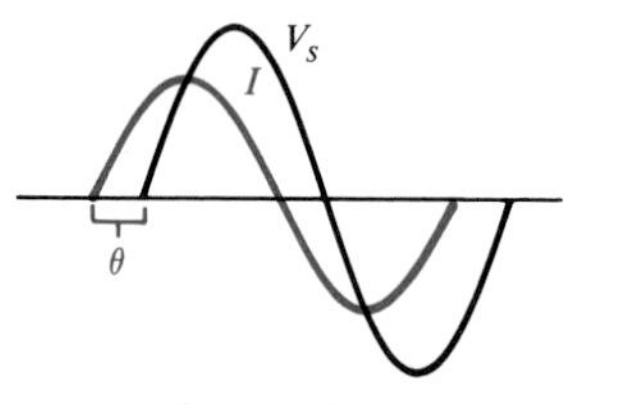

(a) Below f_r, I leads V_s.

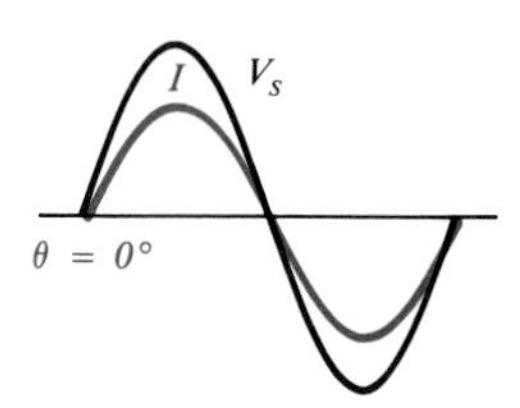

(b) At f_r, I is in phase with V_s.

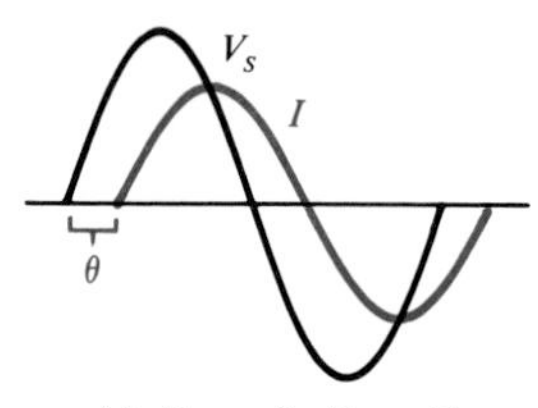

(c) Above f_r, I lags V_s.

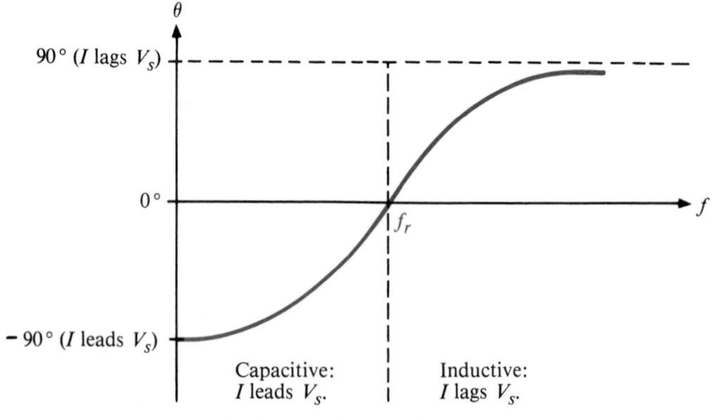

(d) Phase angle versus frequency

FIGURE 15–18
The phase angle as a function of frequency in a series *RLC* circuit.

quency approaches the resonant value and is 0° at resonance, as indicated in Part (b). At frequencies above resonance, $X_L > X_C$, and the current *lags* the source voltage, as indicated in Part (c). As the frequency goes higher, the phase angle approaches 90°. A plot of phase angle versus frequency is shown in Part (d) of the figure.

SECTION REVIEW 15–3

1. What is the condition for series resonance?
2. Why is the current maximum at the resonant frequency?
3. Calculate the resonant frequency for $C = 1000$ pF and $L = 1000\ \mu$H.
4. In Problem 3, is the circuit inductive or capacitive at 50 kHz?

SERIES RESONANT FILTERS

15–4

A common use of series *RLC* circuits is in filter applications. In this section, you will learn the basic configurations for *band-pass* and *band-stop* filters and several important filter characteristics.

The Band-Pass Filter

A basic series resonant band-pass filter is shown in Figure 15–19. Notice that the series *LC* portion is placed between the input and the output and that the output is taken across the resistor.

FIGURE 15–19
A series resonant band-pass filter.

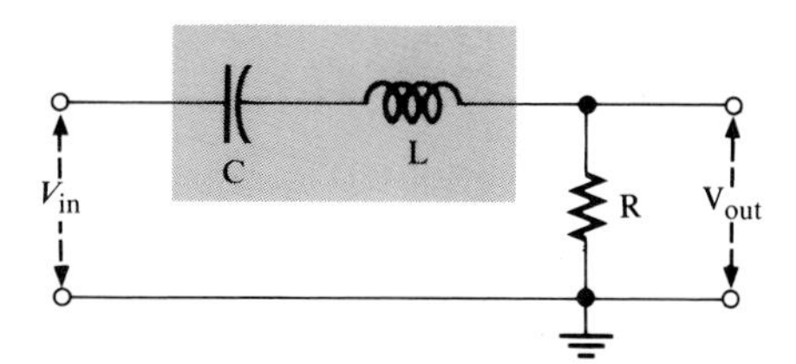

This filter allows signals with the resonant frequency and within a band of frequencies below and above the resonant value to pass from input to output without a significant reduction in amplitude. Signals with frequencies lying outside this specified band (called the *pass band*) are reduced in amplitude to below a certain level and are considered to be rejected by the filter.

The filtering action is the result of the impedance characteristic of the filter. As you learned in the previous section, the impedance is minimum at resonance and has increasingly higher values below and above the resonant frequency. At very low frequencies, the impedance is very high and tends to block the current. As the frequency increases, the impedance drops, allowing more current to flow and thus more voltage across the output resistor. At the resonant frequency, the impedance is very low and equal to the winding resistance of the coil. At this point a maximum current is flowing and the resulting output voltage is maximum. As the frequency goes above resonance, the impedance again increases, causing the current and the resulting output voltage to drop.

Bandwidth of the Pass Band

The bandwidth of a band-pass filter is *the range of frequencies for which the current (or output voltage) is equal to or greater than 70.7% of its value at the resonant frequency.* As you know, bandwidth is often abbreviated as *BW*.

Figure 15–20 illustrates the general response of a series resonant band-pass filter, and Figure 15–21 shows the bandwidth on the response curve for the filter. Notice that frequency f_1, which is below f_r, is the frequency at which I (or V_{out}) is 70.7% of the resonant value (I_{max}); f_1 is commonly called the *lower cutoff frequency*. At frequency f_2, above f_r, the current (or V_{out}) is again 70.7% of its maximum; f_2 is called the *upper cutoff frequency*. Other names for f_1 and f_2 are *−3 dB frequencies, band frequencies,* and *half-power frequencies*. (The term *decibel*, abbreviated dB, is defined on page 540).

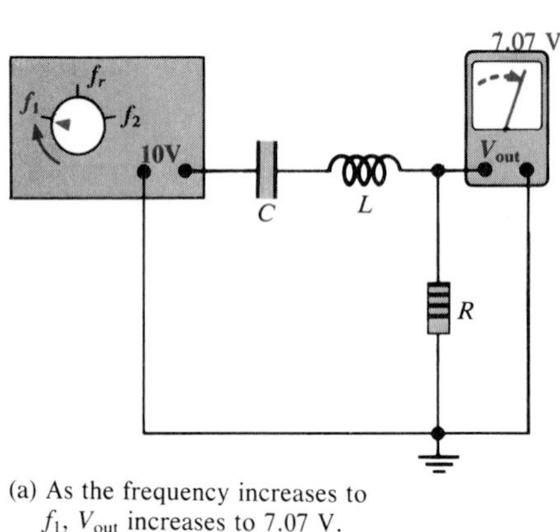

(a) As the frequency increases to f_1, V_{out} increases to 7.07 V.

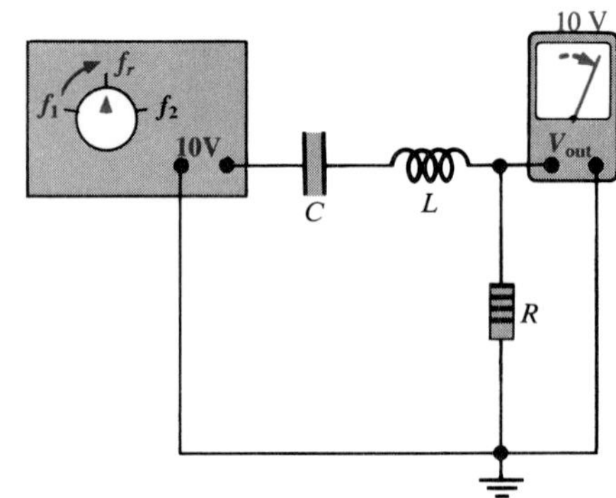

(b) As the frequency increases from f_1 to f_r, V_{out} increases from 7.07 V to 10 V.

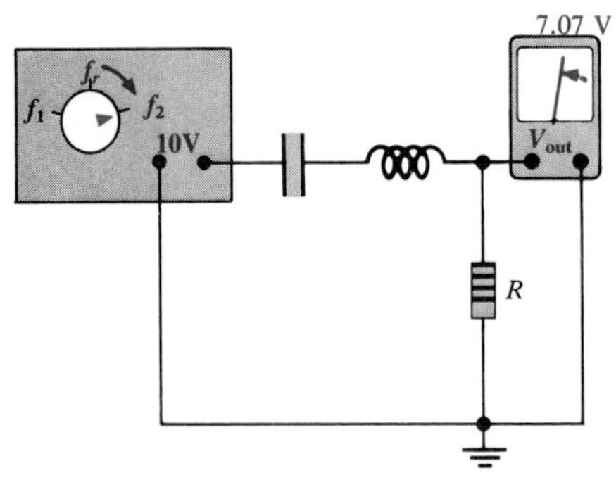

(c) As the frequency increases from f_r to f_2, V_{out} decreases from 10 V to 7.07 V.

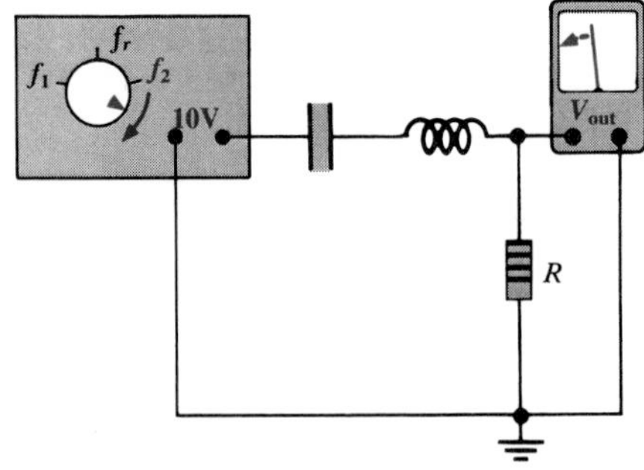

(d) As the frequency increases above f_2, V_{out} decreases below 7.07 V.

FIGURE 15–20
Example of the response of a series resonant band-pass filter with the input voltage at a constant 10 V rms. The winding resistance of the coil is neglected.

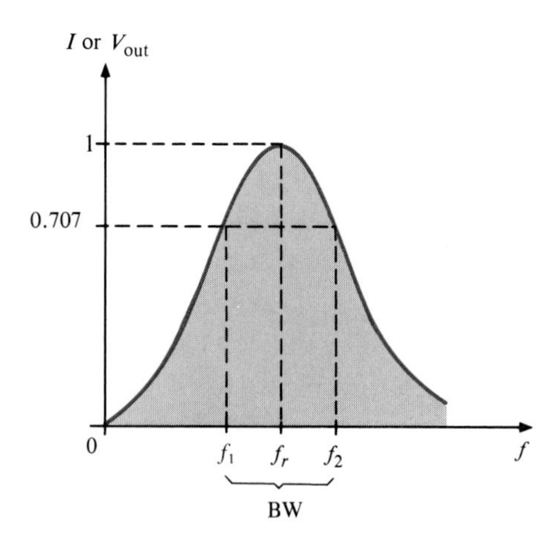

FIGURE 15–21
Generalized response curve of a series resonant band-pass filter.

The formula for calculating the bandwidth is as follows:

$$BW = f_2 - f_1 \tag{15–7}$$

The unit of bandwidth is the hertz (Hz), the same as for frequency.

EXAMPLE 15–8 A certain series resonant band-pass filter has a maximum current of 100 mA at the resonant frequency. What is the value of the current at the cutoff frequencies?

Solution
Current at the cutoff frequencies is 70.7% of maximum:

$$I_{f1} = I_{f2} = 0.707I_{max} = 0.707(100 \text{ mA}) = 70.7 \text{ mA}$$

EXAMPLE 15–9 A resonant circuit has a lower cutoff frequency of 8 kHz and an upper cutoff frequency of 12 kHz. Determine the bandwidth.

Solution

$$BW = f_2 - f_1 = 12 \text{ kHz} - 8 \text{ kHz} = 4 \text{ kHz}$$

Half-Power Points of the Filter Response

As previously mentioned, the upper and lower cutoff frequencies are sometimes called the *half-power frequencies*. This term is derived from the fact that the true power from the source at these frequencies is one-half the power delivered at the resonant frequency. The following steps show that this relationship is true for a series resonant circuit.

At resonance,

$$P_{max} = I_{max}^2 R$$

The power at f_1 or f_2 is

$$P_{f1} = I_{f1}^2R = (0.707I_{max})^2R = (0.707)^2I_{max}^2R$$
$$= 0.5I_{max}^2R = 0.5P_{max}$$

Decibel (dB) Measurement

As previously mentioned, another common term for the upper and lower cutoff frequencies is *−3 dB frequencies*. The decibel (dB) is a logarithmic measurement of voltage or power ratios which can be used to express the input-to-output relationship of a filter. The following equation expresses a *voltage* ratio in decibels:

$$\text{dB} = 20\log\left(\frac{V_{out}}{V_{in}}\right) \tag{15–8}$$

The following equation is the decibel formula for a *power* ratio:

$$\text{dB} = 10\log\left(\frac{P_{out}}{P_{in}}\right) \tag{15–9}$$

EXAMPLE 15–10 The output voltage of a filter is 5 V and the input is 10 V. Express the voltage ratio in decibels.

Solution

$$20\log\left(\frac{V_{out}}{V_{in}}\right) = 20\log\left(\frac{5\text{ V}}{10\text{ V}}\right) = 20\log(0.5) = -6.02\text{ dB}$$

The calculator sequence is

[5] [÷] [1] [0] [=] [log] [×] [2] [0] [=]

The −3 dB Frequencies

The output of a filter is said to be *down 3 dB* at the cutoff frequencies. Actually, this frequency is the point at which the output voltage is 70.7% of the maximum voltage at resonance. We can show that the 70.7% point is the same as 3 dB below maximum (or −3 dB) as follows. The maximum voltage is the zero dB reference:

$$20\log\left(\frac{0.707V_{max}}{V_{max}}\right) = 20\log(0.707) = -3\text{ dB}$$

Selectivity of a Band-Pass Filter

The response curve in Figure 15–21 is also called a *selectivity curve.* Selectivity defines how well a resonant circuit responds to a certain frequency and discriminates against all others. *The narrower the bandwidth, the greater the selectivity.*

We normally assume that a resonant circuit accepts frequencies within its bandwidth and completely eliminates frequencies outside the bandwidth. Such is not actually the case, however, because signals with frequencies outside the bandwidth are not completely eliminated. Their magnitudes, however, are greatly reduced. The further the frequencies are from the cutoff frequencies, the greater is the reduction, as illustrated in Figure 15–22(a). An ideal selectivity curve is shown in Figure 15–22(b).

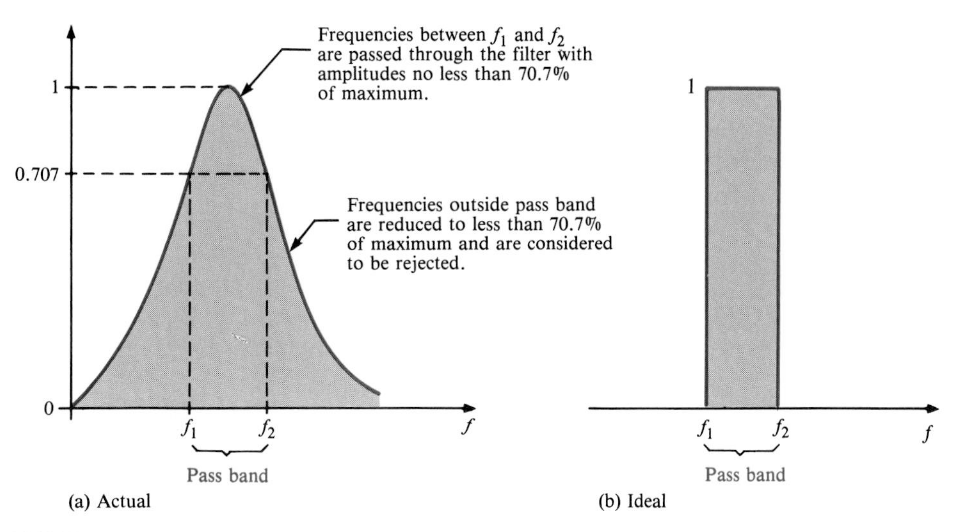

FIGURE 15–22
Generalized selectivity curve of a band-pass filter.

As you can see in Figure 15–23, another factor that influences selectivity is the steepness of the slopes of the curve. The faster the curve drops off at the cutoff frequencies, the more selective the circuit is, because the frequencies outside the pass band are more quickly reduced (attenuated).

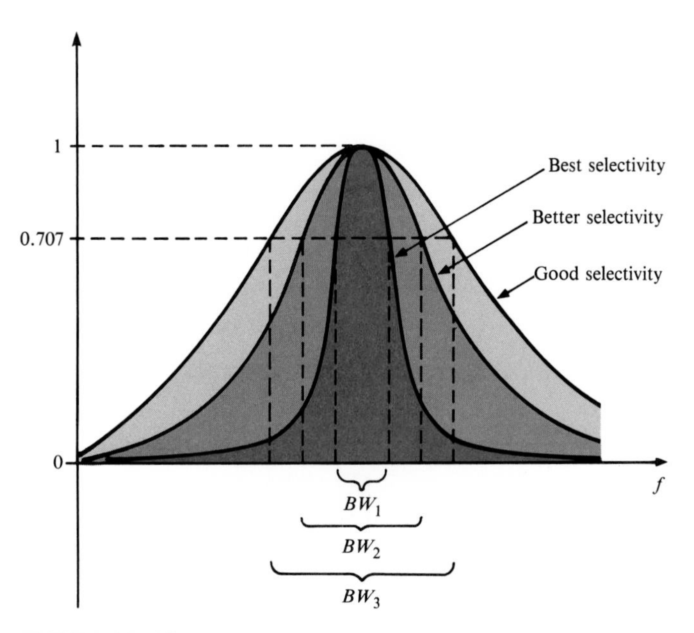

FIGURE 15–23
Comparative selectivity curves.

The Quality Factor (*Q*) of a Resonant Circuit and Its Effect on Bandwidth

The quality factor (Q) is the ratio of the reactive power in the inductor to the true power in the winding resistance of the coil or the resistance in series with the coil. It is a ratio of the power in L to the power in R. The quality factor is very important in resonant circuits. A formula for Q is developed as follows:

$$Q = \frac{\text{energy stored}}{\text{energy dissipated}} = \frac{\text{reactive power}}{\text{true power}} = \frac{I^2X_L}{I^2R}$$

In a series circuit, I is the same in L and R; thus, the I^2 terms cancel, leaving

$$Q = \frac{X_L}{R} \tag{15–10}$$

When the resistance is just the winding resistance of the coil, the circuit Q and the coil Q are the same. Since Q varies with frequency because X_L varies, we are interested mainly in Q at *resonance*. Note that Q is a ratio of like units and, therefore, has no unit itself.

EXAMPLE 15–11 Determine Q at resonance for the circuit in Figure 15–24.

FIGURE 15–24

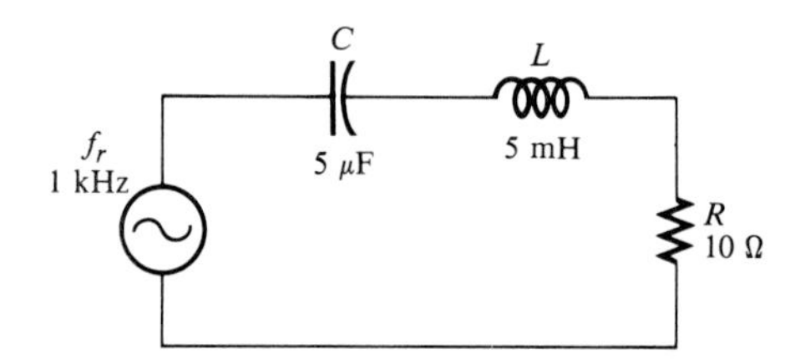

Solution

$$X_L = 2\pi f_r L = 31.4\ \Omega$$

$$Q = \frac{X_L}{R} = \frac{31.4\ \Omega}{10\ \Omega} = 3.14$$

How Q Affects Bandwidth

A higher value of circuit Q results in a smaller bandwidth. A lower value of Q causes a larger bandwidth. A formula for the bandwidth of a resonant circuit in terms of Q is stated in the following equation:

$$BW = \frac{f_r}{Q} \tag{15–11}$$

EXAMPLE 15–12 What is the bandwidth of the filter in Figure 15–25?

FIGURE 15–25

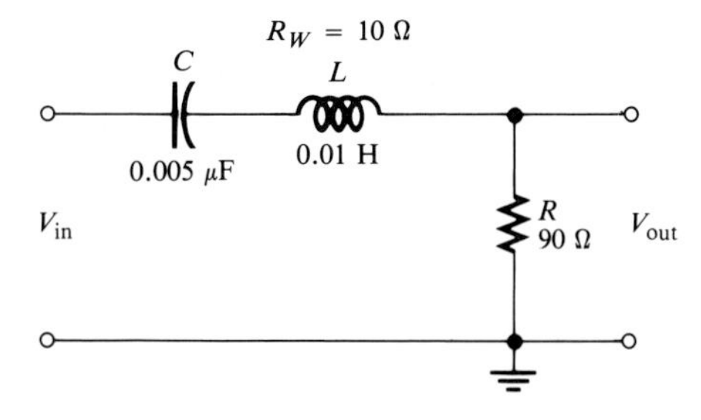

Solution

The total resistance is

$$R_T = R + R_W = 90\ \Omega + 10\ \Omega = 100\ \Omega$$

The bandwidth is found as follows:

$$f_r = \frac{1}{2\pi\sqrt{LC}} = 22.5\text{ kHz}$$

$$Q = \frac{X_L}{R} = \frac{1.41\text{ k}\Omega}{100\ \Omega} = 14.1$$

$$BW = \frac{f_r}{Q} = \frac{22.5\text{ kHz}}{14.1} = 1.6\text{ kHz}$$

The Band-Stop Filter

A basic series resonant band-stop filter is shown in Figure 15–26. Notice that the output voltage is taken across the *LC* portion of the circuit. This filter is still a series *RLC* circuit, just as the band-pass filter is. The difference is that in this case, the output voltage is taken across the combination of *L* and *C* rather than across *R*.

FIGURE 15–26
A series resonant band-stop filter.

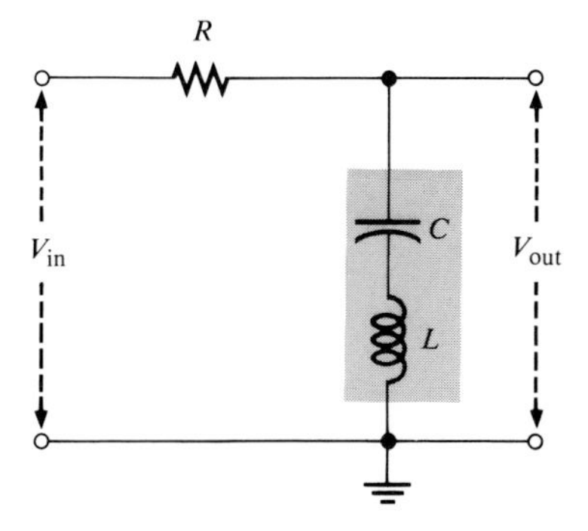

This filter rejects signals with frequencies between the upper and lower cutoff frequencies and passes those signals with frequencies below and above the cutoff values, as shown in the response curve of Figure 15–27. The range of

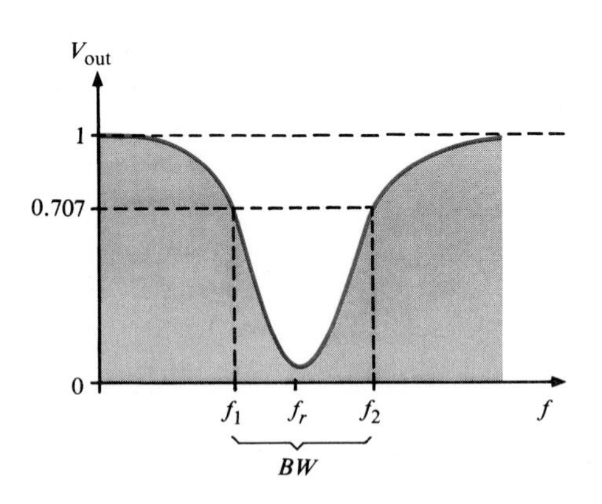

FIGURE 15–27
Generalized response curve for a band-stop filter.

frequencies between the lower and upper cutoff points is called the *stop band.* This type of filter is also referred to as a *band-elimination filter, band-reject filter,* or a *notch filter.*

At very low frequencies, the *LC* combination appears as a near open due to the high X_C, thus allowing most of the input voltage to pass through to the

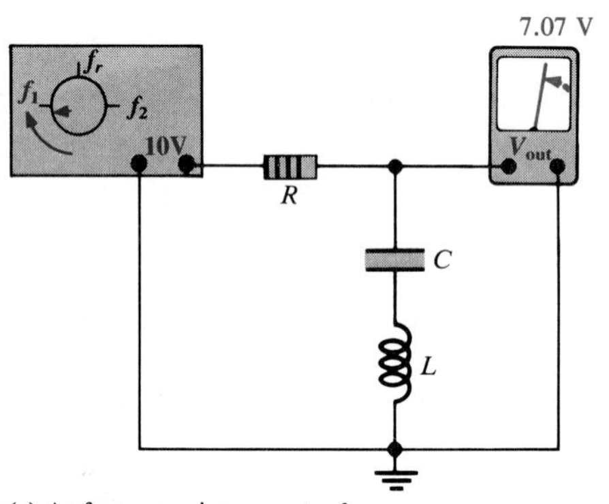

(a) As frequency increases to f_1, V_{out} decreases from 10 V to 7.07 V.

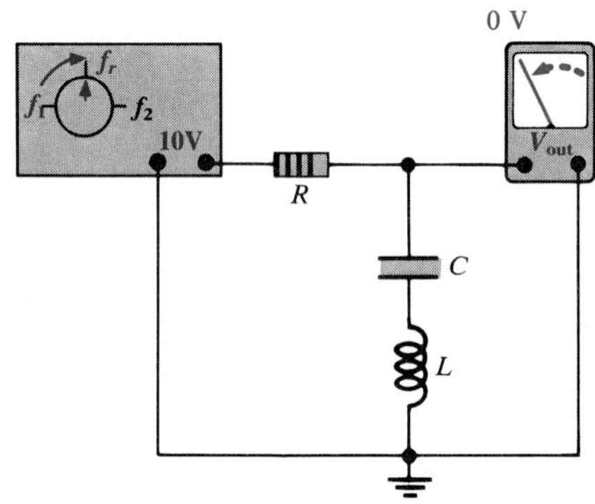

(b) As frequency increases from f_1 to f_r, V_{out} decreases from 7.07 V to 0 V.

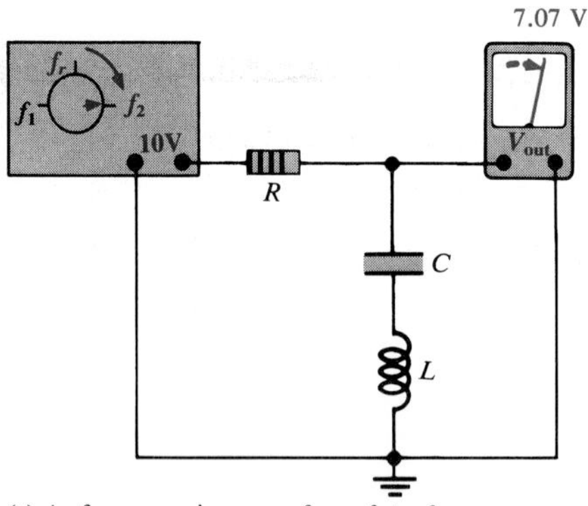

(c) As frequency increases from f_r to f_2, V_{out} increases from 0 V to 7.07 V.

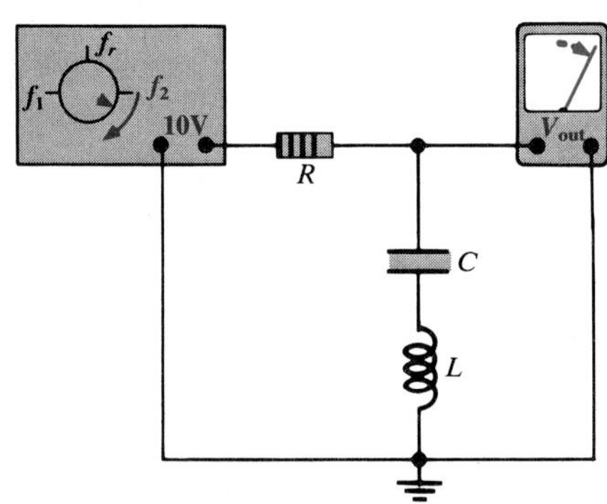

(d) As frequency increases above f_2, V_{out} increases toward 10 V.

FIGURE 15–28
Example of the response of a series resonant band-stop filter with V_{in} at a constant 10 V rms. The winding resistance is neglected.

output. As the frequency increases, the impedance of the LC combination decreases until, at resonance, it is zero (ideally). Thus, the input signal is shorted to ground, and there is very little output voltage. As the frequency goes above its resonant value, the LC impedance increases, allowing an increasing amount of voltage to be dropped across it. The general response of a series resonant band-stop filter is illustrated in Figure 15–28.

Characteristics of the Band-Stop Filter

All of the characteristics that have been discussed in relation to the band-pass filter (current response, impedance characteristic, bandwidth, selectivity, and Q) apply equally to the band-stop filter, with the exception that the response curve of the output voltage is opposite. For the band-pass filter, V_{out} is maximum at resonance. For the band-stop filter, V_{out} is minimum at resonance.

EXAMPLE 15–13 Find the output voltage at f_r and the bandwidth in Figure 15–29.

FIGURE 15–29

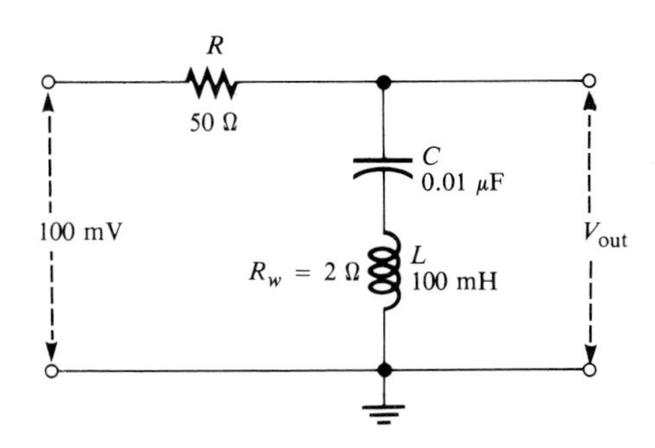

Solution

Since $X_L = X_C$ at resonance, then

$$V_{out} = \left(\frac{R_W}{R + R_W}\right)V_{in} = \left(\frac{2\ \Omega}{52\ \Omega}\right)100\text{ mV} = 3.85\text{ mV}$$

$$f_r = \frac{1}{2\pi\sqrt{LC}} = \frac{1}{2\pi\sqrt{(100\text{ mH})(0.01\ \mu\text{F})}} = 5.03\text{ kHz}$$

$$Q = \frac{X_L}{R} = \frac{3160\ \Omega}{52\ \Omega} = 60.8$$

$$BW = \frac{f_r}{Q} = \frac{5.03\text{ kHz}}{60.8} = 82.7\text{ Hz}$$

SECTION REVIEW 15–4

1. The output voltage of a certain band-pass filter is 15 V at the resonant frequency. What is its value at the cutoff frequencies?
2. For a certain band-pass filter, $f_r = 120$ kHz and $Q = 12$. What is the bandwidth of the filter?
3. In a band-stop filter, is the current minimum or maximum at resonance? Is the output voltage minimum or maximum at resonance?

PARALLEL *RLC* CIRCUITS

15–5

In this section, a basic parallel *RLC* circuit, shown in Figure 15–30, is examined in terms of its impedance, phase angle, and current relationships.

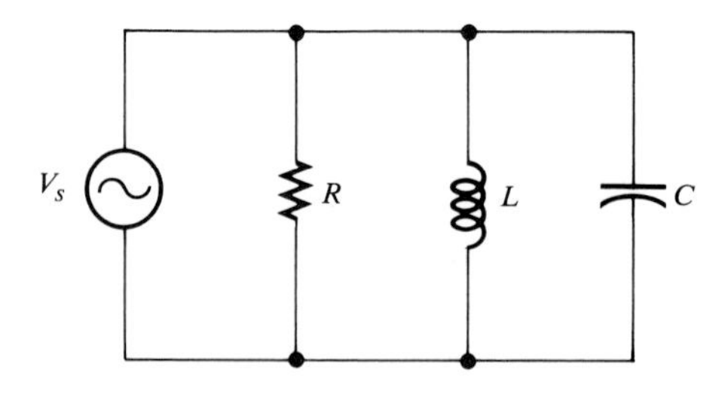

FIGURE 15–30
Parallel *RLC* circuit.

Impedance and Phase Angle

The circuit in Figure 15–30 consists of the parallel combination of *R, L,* and *C.* To find the admittance, add the conductance, *G,* and the total susceptance, B_T, as phasor quantities. B_T is the difference of the capacitive susceptance and the inductive susceptance. Thus,

$$Y = \sqrt{G^2 + (B_C - B_L)^2} \quad \textbf{(15–12)}$$

The total impedance is the reciprocal of the admittance:

$$Z = \frac{1}{Y}$$

The phase angle of the circuit is given by the following formula:

$$\theta = \arctan\left(\frac{B_T}{G}\right) \quad \textbf{(15–13)}$$

When the frequency is above its resonant value ($X_C < X_L$), the impedance of the circuit in Figure 15–30 is predominantly capacitive, and the total current *leads* the source voltage. When the frequency is below its resonant value ($X_L < X_C$), the impedance of the circuit is predominantly inductive, and the total current *lags* the source voltage.

EXAMPLE 15–14 Determine the total impedance and the phase angle in Figure 15–31.

FIGURE 15–31

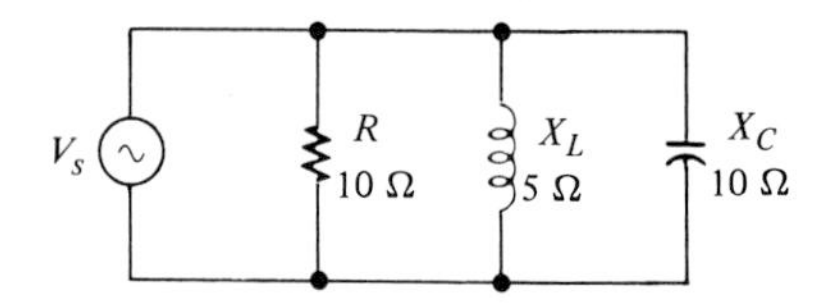

Solution

$$G = \frac{1}{R} = \frac{1}{10\ \Omega} = 0.1\ \text{S}$$

$$B_C = \frac{1}{X_C} = \frac{1}{10\ \Omega} = 0.1\ \text{S}$$

$$B_L = \frac{1}{X_L} = \frac{1}{5\ \Omega} = 0.2\ \text{S}$$

$$Y = \sqrt{G^2 + (B_C - B_L)^2} = \sqrt{(0.1\ \text{S})^2 + (0.1\ \text{S} - 0.2\ \text{S})^2}$$
$$= \sqrt{(0.1\ \text{S})^2 + (-0.1\ \text{S})^2} = 0.1414\ \text{S}$$

From Y, we can get Z:

$$Z = \frac{1}{Y} = \frac{1}{0.1414\ \text{S}} = 7.07\ \Omega$$

The phase angle is

$$\theta = \arctan\left(\frac{B_T}{G}\right) = \arctan\left(\frac{0.1\ \text{S}}{0.1\ \text{S}}\right) = 45^\circ$$

The total current *lags* V_s by 45° because the circuit is predominantly inductive ($X_L < X_C$).

Current Relationships

In a parallel *RLC* circuit, the current in the capacitive branch and the current in the inductive branch are *always* 180° out of phase with each other (neglecting any coil resistance). For this reason, I_C and I_L subtract from each other, and thus the total current into the parallel branches of L and C is always less than the largest individual branch current, as illustrated in Figure 15–32 and in the waveform diagram of Figure 15–33. Of course, the current in the resistive branch is always 90° out of phase with both reactive currents, as shown in the current phasor diagram of Figure 15–34.

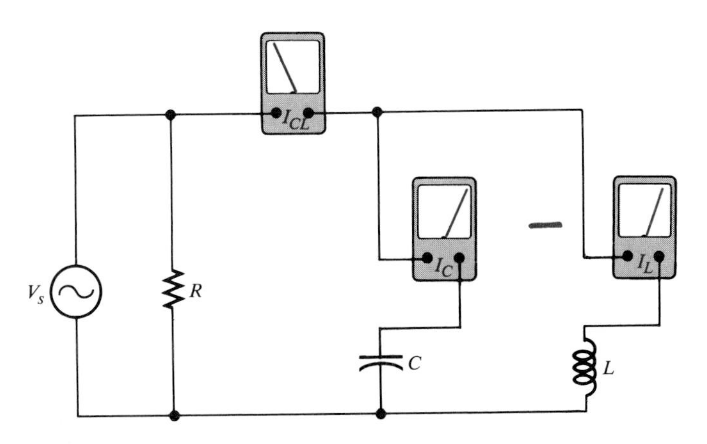

FIGURE 15–32
The total current into the parallel combination of C and L is the difference of the two branch currents.

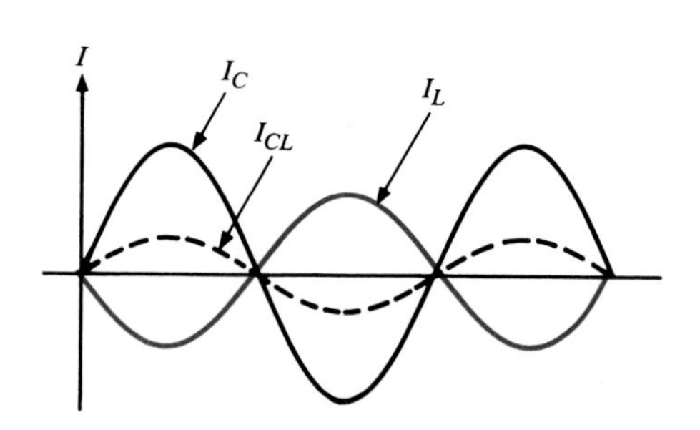

FIGURE 15–33
I_C and I_L effectively subtract.

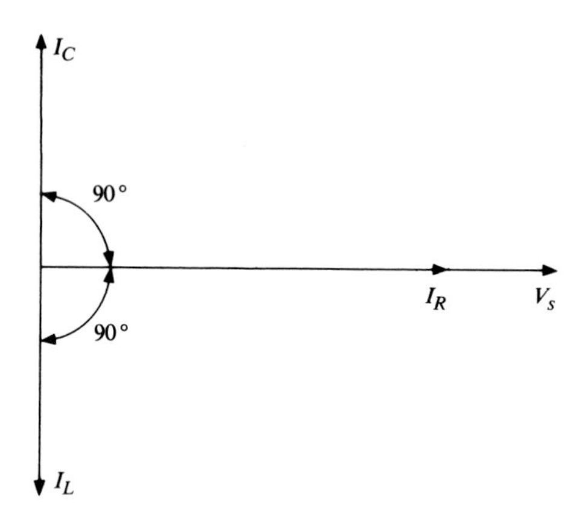

FIGURE 15–34
Current phasor diagram for a parallel *RLC* circuit.

The total current can be expressed as

$$I_T = \sqrt{I_R^2 + (I_C - I_L)^2} \tag{15–14}$$

The phase angle can also be expressed in terms of the branch currents as

$$\theta = \arctan\left(\frac{I_{CL}}{I_R}\right) \tag{15–15}$$

where I_{CL} is $|I_C - I_L|$.

EXAMPLE 15–15 Find each branch current and the total current in Figure 15–35, and show their relationship.

FIGURE 15–35

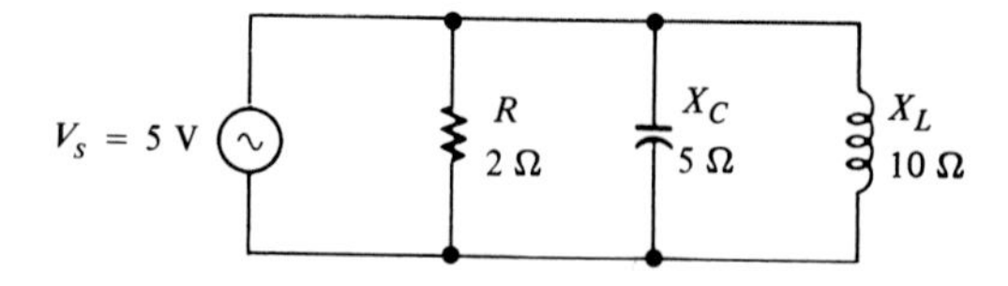

Solution
Each branch current can be found using Ohm's law:

$$I_R = \frac{V_s}{R} = \frac{5\text{ V}}{2\ \Omega} = 2.5\text{ A}$$

$$I_C = \frac{V_s}{X_C} = \frac{5\text{ V}}{5\ \Omega} = 1\text{ A}$$

$$I_L = \frac{V_s}{X_L} = \frac{5\text{ V}}{10\ \Omega} = 0.5\text{ A}$$

The total current is the phasor sum of the branch currents:

$$\begin{aligned} I_T &= \sqrt{I_R^2 + (I_C - I_L)^2} = \sqrt{(2.5\text{ A})^2 + (1\text{ A} - 0.5\text{ A})^2} \\ &= \sqrt{(2.5\text{ A})^2 + (0.5\text{ A})^2} = 2.55\text{ A} \end{aligned}$$

The phase angle is

$$\theta = \arctan\left(\frac{I_{CL}}{I_R}\right) = \arctan\left(\frac{0.5\text{ A}}{2.5\text{ A}}\right) = 11.31°$$

The total current is 2.55 A, *leading* V_s by 11.31°. Figure 15–36 is the current phasor diagram for the circuit.

FIGURE 15–36

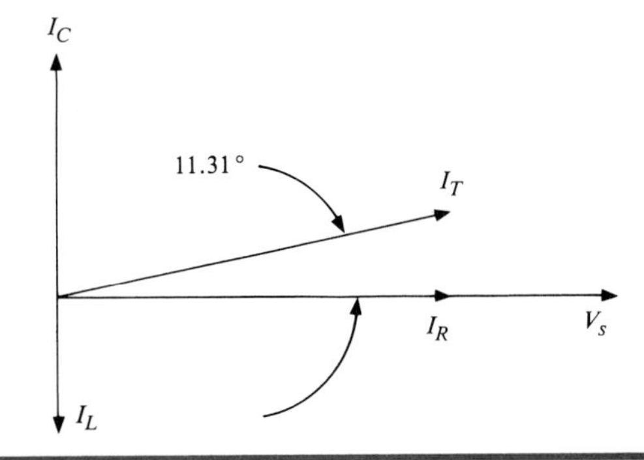

Conversion of Series-Parallel to Parallel

The series-parallel configuration shown in Figure 15–37 is important because it represents a circuit having parallel L and C branches, with the winding resistance of the coil taken into account as a series resistance in the L branch.

It is helpful to view the series-parallel circuit in Figure 15–37 in an *equivalent* purely parallel form, as indicated in Figure 15–38. This form will simplify our analysis of parallel resonant characteristics in the next section.

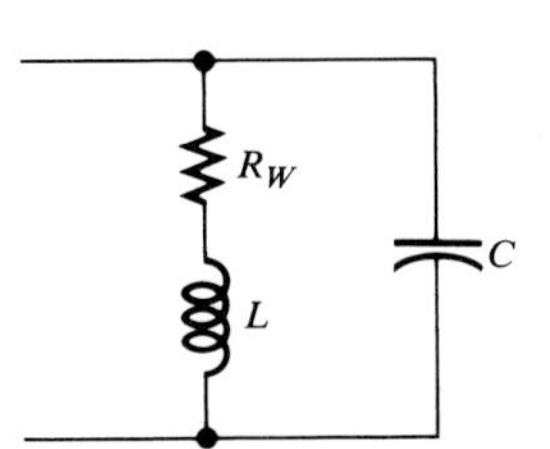

FIGURE 15–37
An important series-parallel RLC circuit ($Q = X_L/R_W$).

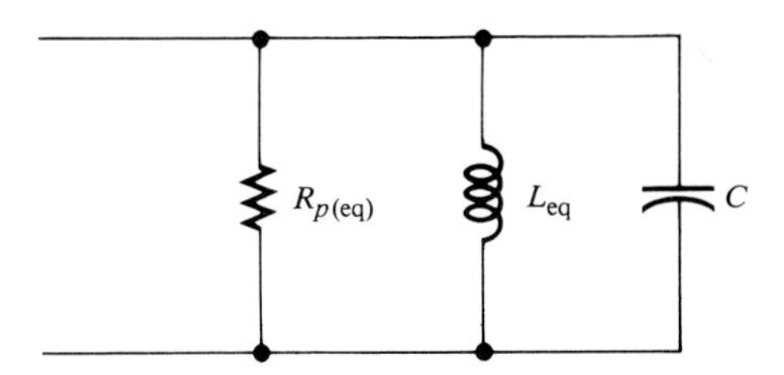

FIGURE 15–38
Parallel equivalent form of the circuit in Figure 15–37.

The equivalent inductance L_{eq} and the equivalent parallel resistance $R_{p(\text{eq})}$ are given by the following formulas:

$$L_{eq} = L\left(\frac{Q^2 + 1}{Q^2}\right) \quad (15\text{–}16)$$

$$R_{p(eq)} = R_W(Q^2 + 1) \quad (15\text{–}17)$$

where Q is the quality factor of the coil, X_L/R_W. Derivations of these formulas are quite involved and thus are not given here. Notice in the equations that for a $Q \geq 10$, the value of L_{eq} is approximately the same as the original value of L. For example, if $L = 10$ mH, then

$$L_{eq} = 10\ \text{mH}\left(\frac{10^2 + 1}{10^2}\right) = 10\ \text{mH}(1.01) = 10.1\ \text{mH}$$

The equivalency of the two circuits means that at a given frequency, when the same value of voltage is applied to both circuits, the same total current flows in both circuits and the phase angles are the same. Basically, an equivalent circuit simply makes circuit analysis more convenient.

EXAMPLE 15–16 Convert the series-parallel circuit in Figure 15–39 to an equivalent parallel form at the given frequency.

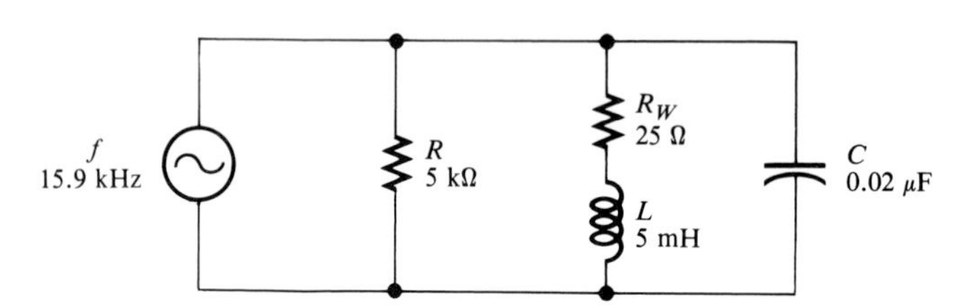

FIGURE 15–39

Solution

$$X_L = 2\pi fL = 2\pi(15.9\ \text{kHz})(5\ \text{mH}) = 500\ \Omega$$

The Q of the coil is

$$Q = \frac{X_L}{R_W} = \frac{500\ \Omega}{25\ \Omega} = 20$$

Since $Q > 10$, then $L_{eq} \cong L = 5$ mH.

The equivalent parallel resistance is

$$R_{p(eq)} = R_W(Q^2 + 1) = (25\ \Omega)(20^2 + 1) = 10.025\ \text{k}\Omega$$

This equivalent resistance appears in parallel with R as shown in Figure 15–40(a). When combined, they give a total parallel resistance (R_{pT}) of 3.37 kΩ, as indicated in Figure 15–40(b).

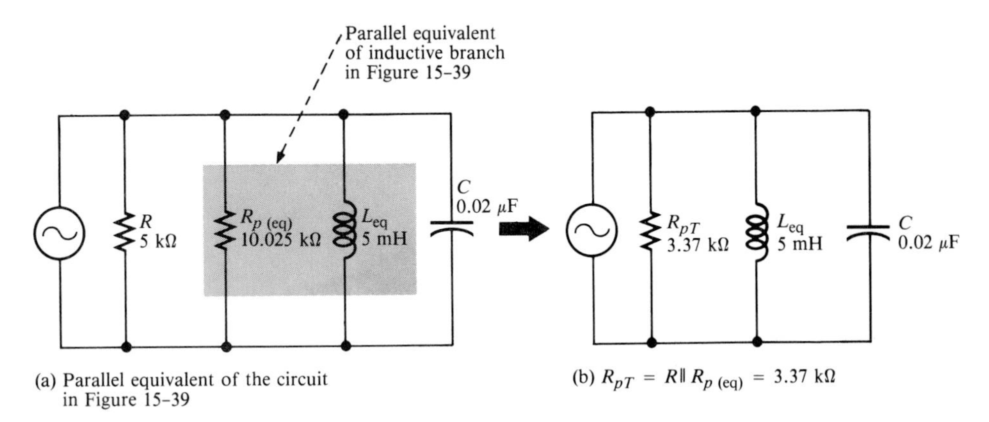

(a) Parallel equivalent of the circuit in Figure 15–39

(b) $R_{pT} = R \| R_{p\,(eq)} = 3.37\ \text{k}\Omega$

FIGURE 15–40

SECTION REVIEW 15–5

1. In a three-branch parallel circuit, $R = 150\ \Omega$, $X_C = 100\ \Omega$, and $X_L = 50\ \Omega$. Determine the current in each branch when $V_s = 12$ V.
2. Is the circuit in Problem 1 capacitive or inductive? Why?
3. Find the equivalent parallel inductance and resistance for a 20-mH coil with a winding resistance of 10 Ω at a frequency of 1 kHz.

PARALLEL RESONANCE

15–6

First we will look at the resonant condition in an *ideal* parallel *LC* circuit. Then we will progress to the more realistic case where the resistance of the coil is taken into account.

Ideally, parallel resonance occurs when $X_C = X_L$. The frequency at which resonance occurs is called the *resonant frequency,* just as in the series case. When $X_C = X_L$, the two branch currents, I_C and I_L, are equal in magnitude, and, of course, they are always 180° out of phase with each other. Thus, the two currents cancel and the total current is zero, as presented in Figure 15–41.

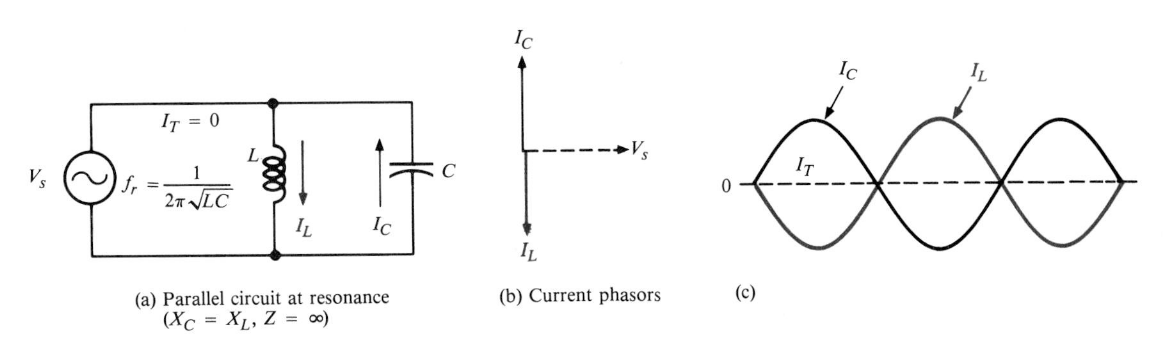

(a) Parallel circuit at resonance ($X_C = X_L$, $Z = \infty$)

(b) Current phasors

(c)

FIGURE 15–41
An ideal parallel *LC* circuit at resonance.

Since the total current is zero, the impedance of the parallel *LC* circuit is infinitely large (∞). These ideal resonant conditions are stated in the following equations:

$$X_L = X_C$$

$$Z_r = \infty$$

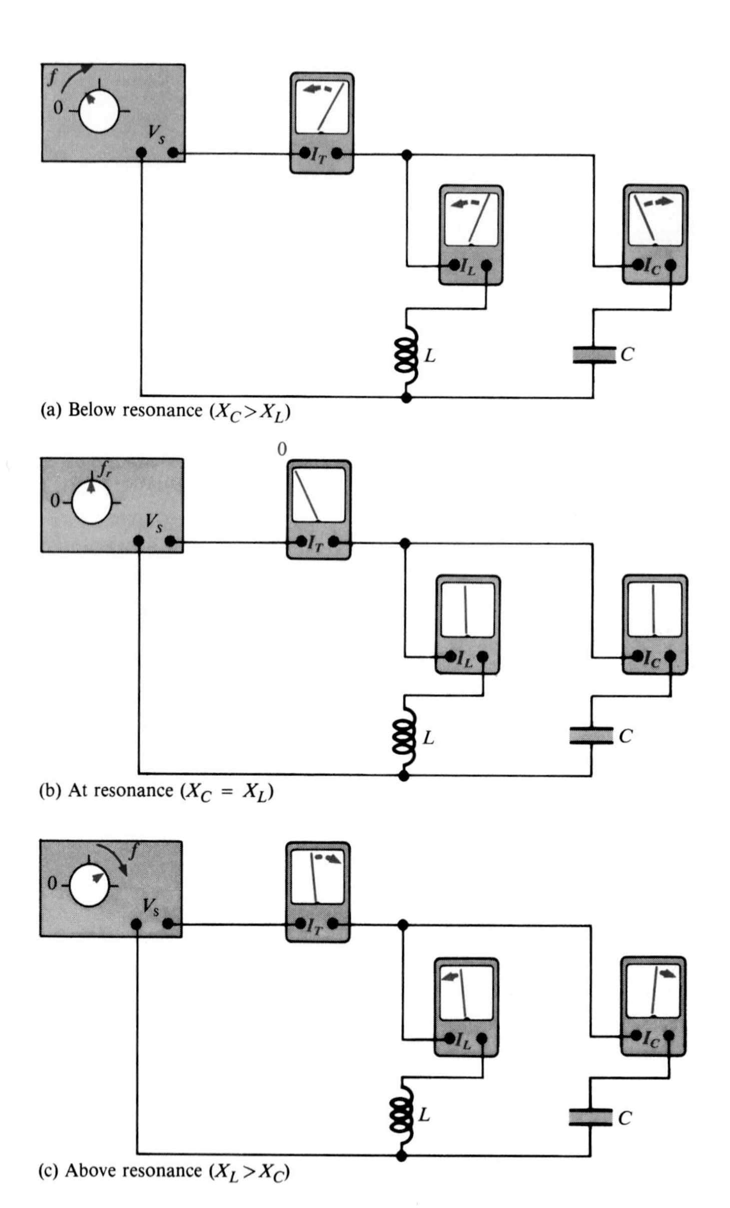

FIGURE 15–42
An illustration of how the currents respond in a parallel *LC* circuit as the frequency is varied from below resonance to above resonance. The source voltage amplitude is constant.

The Parallel Resonant Frequency

For an ideal parallel resonant circuit, the frequency at which resonance occurs is determined by the same formula as in series resonant circuits:

$$f_r = \frac{1}{2\pi\sqrt{LC}}$$

Currents in a Parallel Resonant Circuit

It is interesting to see how the currents in a parallel LC circuit vary as the frequency is increased from below the resonant value, through resonance, and then above the resonant value. Figure 15–42 illustrates the general response of an ideal circuit in terms of changes in the currents. Part (a) indicates the response as the source frequency is increased from zero toward f_r. At very low frequencies, X_C is very high and X_L is very low, so most of the current is through L. As the frequency increases, the current through L decreases and the current through C increases, causing the total current to decrease. At all times, I_L and I_C are 180° out of phase with each other, and thus the total current is the difference of the two branch currents. During this time, the impedance is increasing, as indicated by the decrease in total current.

When the frequency reaches its resonant value, f_r, as shown in Figure 15–42(b), X_C and X_L are equal, so I_C and I_L cancel because they are equal in magnitude but opposite in phase. At this point the total current is at its minimum value of zero. Since I_T is zero, Z is infinite. Thus, the ideal parallel LC circuit appears as an open at the resonant frequency.

As the frequency is increased above resonance, as indicated in Figure 15–42(c), X_C continues to decrease and X_L continues to increase, causing the branch currents again to be unequal, with I_C being larger. As a result, total current increases and impedance decreases. As the frequency becomes very high, the impedance becomes very small due to the dominance of X_C.

In summary, the current dips to a minimum and the impedance peaks to a maximum at parallel resonance. The expression for the total current into the L and C branches is

$$I_T = |I_C - I_L| \tag{15–18}$$

EXAMPLE 15–17 Find the resonant frequency and the branch currents in the ideal parallel LC circuit of Figure 15–43.

FIGURE 15–43

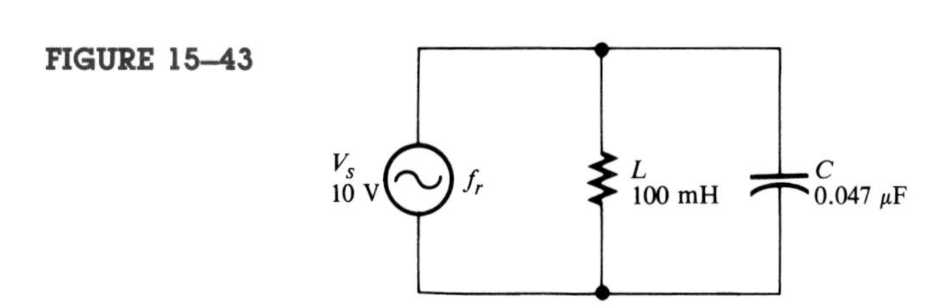

Solution

$$f_r = \frac{1}{2\pi\sqrt{LC}} = \frac{1}{2\pi\sqrt{(100 \text{ mH})(0.047 \ \mu\text{F})}} = 2322 \text{ Hz}$$

$$X_L = 2\pi f_r L = 2\pi(2322 \text{ Hz})(100 \text{ mH}) = 1459 \ \Omega$$

$$X_C = X_L = 1459 \ \Omega$$

$$I_L = \frac{V_s}{X_L} = \frac{10 \text{ V}}{1459 \ \Omega} = 6.85 \text{ mA}$$

$$I_C = I_L = 6.85 \text{ mA}$$

$$I_{LC} = 0$$

Why the Parallel Resonant *LC* Circuit is Often Called a *Tank* Circuit

The term *tank circuit* refers to the fact that the parallel resonant circuit stores energy in the magnetic field of the coil and in the electric field of the capacitor. The stored energy is transferred back and forth between the capacitor and the coil on alternate half-cycles as the current goes first one way and then the other when the inductor deenergizes and the capacitor charges, and vice versa, as illustrated in Figure 15–44.

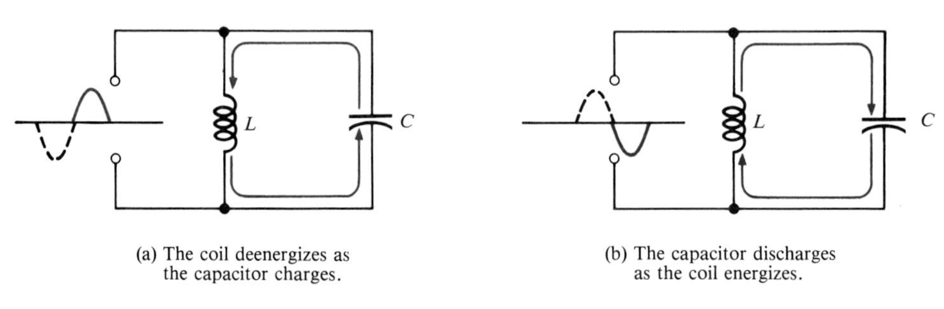

(a) The coil deenergizes as the capacitor charges.

(b) The capacitor discharges as the coil energizes.

FIGURE 15–44
Energy storage in an ideal parallel resonant tank circuit.

Parallel Resonant Conditions in a Nonideal Circuit

So far, the resonance of an ideal parallel *LC* circuit has been examined. Now we will consider resonance in a tank circuit with the resistance of the coil taken into account. Figure 15–45 shows a nonideal tank circuit and its parallel *RLC* equivalent.

The Q of the circuit at resonance is simply the Q of the coil:

$$Q = \frac{X_L}{R_W}$$

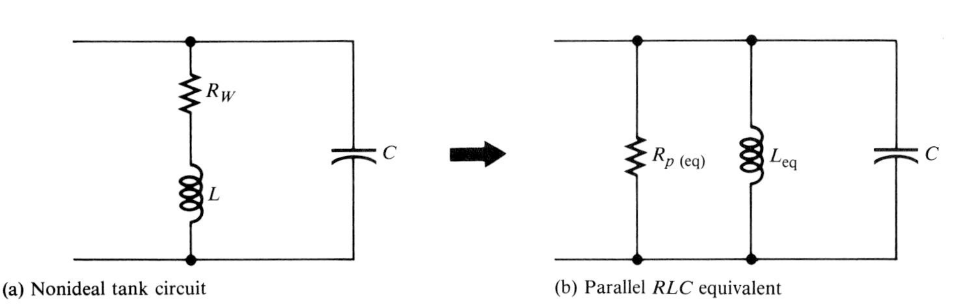

FIGURE 15–45
A practical treatment of parallel resonant circuits must include the coil resistance.

The expressions for the equivalent parallel resistance and the equivalent inductance were given in the previous section as

$$R_{p(eq)} = R_W(Q^2 + 1)$$
$$L_{eq} = L\left(\frac{Q^2 + 1}{Q^2}\right)$$

Recall that for $Q \geq 10$, $L_{eq} \cong L$.

At parallel resonance,

$$X_{L(eq)} = X_C$$

In the parallel equivalent circuit, we have $R_{p(eq)}$ in parallel with an ideal coil and a capacitor, so the L and C branches act as an ideal tank circuit which has an infinite impedance at resonance as shown in Figure 15–46. Therefore, the total impedance of the nonideal tank circuit at resonance can be expressed as simply the equivalent parallel resistance:

$$Z_r = R_W(Q^2 + 1) \tag{15–19}$$

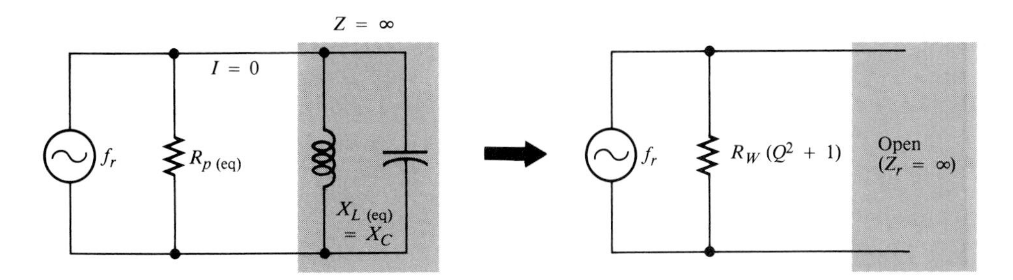

FIGURE 15–46
At resonance, the parallel LC portion appears open and the source sees only $R_{p(eq)}$.

EXAMPLE 15–18 Determine the impedance of the circuit in Figure 15–47 at the resonant frequency ($f_r \cong 17{,}794$ Hz).

FIGURE 15–47

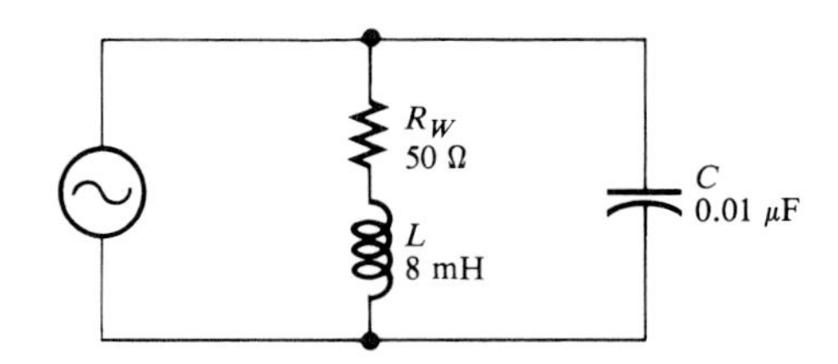

Solution

$$X_L = 2\pi f_r L = 2\pi(17{,}794 \text{ Hz})(8 \text{ mH}) = 894 \ \Omega$$

$$Q = \frac{X_L}{R_W} = \frac{894 \ \Omega}{50 \ \Omega} = 17.9$$

$$Z_r = R_W(Q^2 + 1) = 50 \ \Omega(17.9^2 + 1) = 16.1 \text{ k}\Omega$$

The calculator sequence is

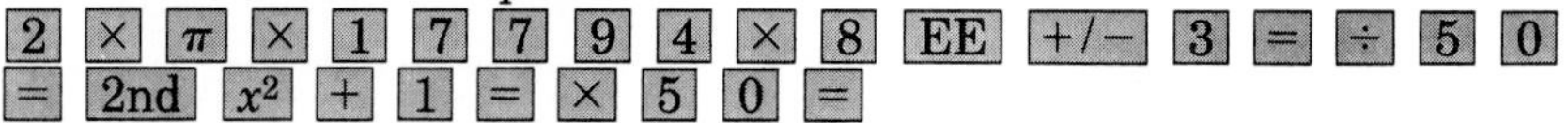

Variation of the Impedance with Frequency

The impedance of a parallel resonant circuit is maximum at the resonant frequency and decreases at lower and higher frequencies, as indicated by the curve in Figure 15–48.

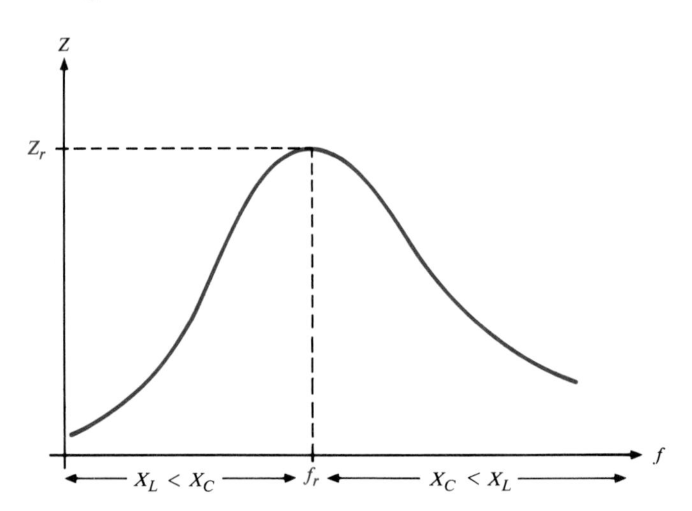

FIGURE 15–48
Generalized impedance curve for a parallel resonant circuit. The circuit is inductive below f_r, resistive at f_r, and capacitive above f_r.

At very low frequencies, X_L is very small and X_C is very high, so the total impedance is essentially equal to that of the inductive branch. As the frequency goes up, the impedance also increases, and the inductive reactance dominates (because it is less than X_C) until the resonant frequency is reached. At this point, of course, $X_L \cong X_C$ (for $Q > 10$) and the impedance is at its maximum. As the frequency goes above resonance, the capacitive reactance dominates (because it is less than X_L) and the impedance decreases.

Current and Phase Angle at Resonance

In the ideal tank circuit, the total current from the source at resonance is zero because the impedance is infinite. In the nonideal case, there is some total current at the resonant frequency, and it is determined by the impedance at resonance:

$$I_T = \frac{V_s}{Z_r} \tag{15–20}$$

The phase angle of the parallel resonant circuit is 0° because the impedance is purely resistive at the resonant frequency.

Parallel Resonant Frequency in a Nonideal Circuit

As you know, when the coil resistance is considered, the resonant condition is

$$X_{L(\text{eq})} = X_C$$

which can be expressed as

$$2\pi f_r L\left(\frac{Q^2 + 1}{Q^2}\right) = \frac{1}{2\pi f_r C}$$

Solving for f_r, we get

$$f_r = \frac{1}{2\pi\sqrt{LC}}\sqrt{\frac{Q^2}{Q^2 + 1}} \tag{15–21}$$

When $Q \geq 10$, the term with the Q factors is approximately 1:

$$\sqrt{\frac{Q^2}{Q^2 + 1}} = \sqrt{\frac{100}{101}} = 0.995 \cong 1$$

Therefore, the parallel resonant frequency is approximately the same as the series resonant frequency as long as Q is equal to or greater than 10:

$$f_r \cong \frac{1}{2\pi\sqrt{LC}} \qquad \text{for } Q \geq 10$$

Equation (15–21) was given only to show the effect of Q on the resonant frequency. You cannot use it to calculate the f_r of a given circuit, because you must first know the value of Q. Since $Q = X_L/R_W$, you must know X_L at the resonant frequency. In order to get X_L, you must know f_r. Since f_r is what we are looking for in the first place, there is no way to proceed with only Equation (15–21) at our disposal.

A more precise expression for f_r in terms of the circuit component values is as follows:

$$f_r = \frac{\sqrt{1 - (R_W^2 C/L)}}{2\pi\sqrt{LC}} \tag{15–22}$$

Now, f_r can be found from the component values alone.

EXAMPLE 15–19 Find the frequency, impedance, and total current at resonance for the circuit in Figure 15–49.

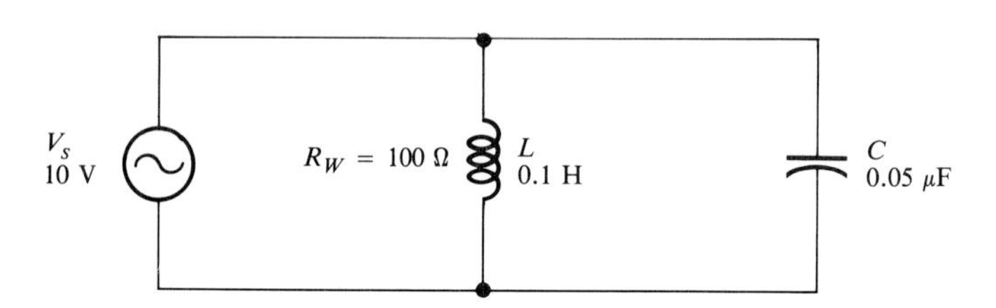

FIGURE 15–49

Solution

$$f_r = \frac{\sqrt{1 - (R_W^2 C/L)}}{2\pi\sqrt{LC}} = \frac{\sqrt{1 - [(100\ \Omega)^2(0.05\ \mu\text{F})/0.1\ \text{H}]}}{2\pi\sqrt{(0.05\ \mu\text{F})(0.1\ \text{H})}} = 2.25\ \text{kHz}$$

$$X_L = 2\pi f_r L = 2\pi(2.25\ \text{kHz})(0.1\ \text{H}) = 1.4\ \text{k}\Omega$$

$$Q = \frac{X_L}{R_W} = \frac{1.4\ \text{k}\Omega}{100\ \Omega} = 14$$

$$Z_r = R_W(Q^2 + 1) = 100\ \Omega(14^2 + 1) = 19.7\ \text{k}\Omega$$

$$I_T = \frac{V_s}{Z_r} = \frac{10\ \text{V}}{19.7\ \text{k}\Omega} = 0.51\ \text{mA}$$

Note that since $Q > 10$, the approximate formula, $f_r \cong 1/2\pi\sqrt{LC}$, could be used.

The calculator sequence for f_r is

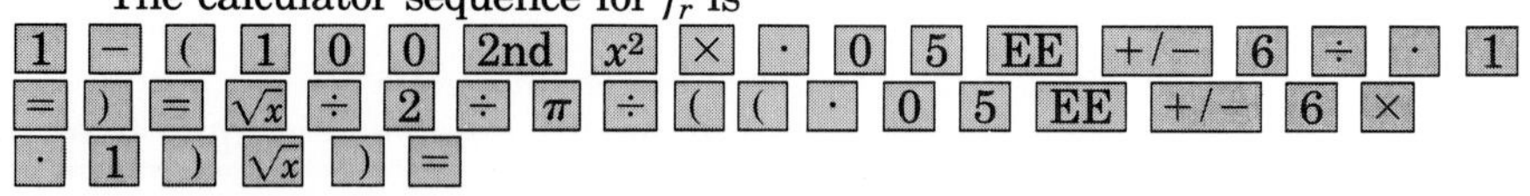

How an External Parallel Load Resistance Affects a Tank Circuit

There are many practical situations in which an external load resistance appears in parallel with a tank circuit as shown in Figure 15–50(a). Obviously, the external resistor (R_L) will dissipate more of the energy delivered by the source and thus will lower the *overall* Q of the circuit. The external resistor effectively appears in parallel with the equivalent parallel resistance of the coil,

$R_{p(eq)}$, and both are combined to determine a total parallel resistance, R_{pT}, as indicated in Figure 15–50(b):

$$R_{pT} = R_L \| R_{p(eq)}$$

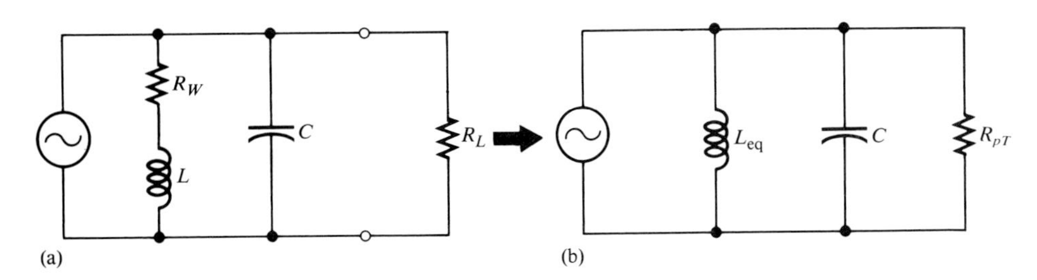

FIGURE 15–50
Tank circuit with a parallel load resistor and its equivalent circuit.

The overall parallel *RLC* circuit Q is expressed differently from the Q of a series circuit:

$$Q_O = \frac{R_{pT}}{X_{L(eq)}} \qquad (15\text{–}23)$$

As you can see, the effect of loading the tank circuit is to reduce its overall Q (which is equal to the coil Q when unloaded).

SECTION REVIEW 15–6

1. Is the impedance minimum or maximum at parallel resonance?
2. Is the current minimum or maximum at parallel resonance?
3. If $Q = 25$, $L = 50$ mH, and $C = 1000$ pF, what is f_r?
4. In Question 3, if $Q = 2.5$, what is f_r?
5. In a certain tank circuit, the coil resistance is 20 Ω. What is the total impedance at resonance if $Q = 20$?

PARALLEL RESONANT FILTERS

15–7

Parallel resonant circuits are commonly applied in band-pass and band-stop filters. In this section, we will examine these applications.

The Band-Pass Filter

A basic parallel resonant band-pass filter is shown in Figure 15–51. Notice that the output is taken across the tank circuit in this application.

The filtering action is as follows: At very low frequencies, the impedance of the tank circuit is very low, and therefore only a small amount of voltage is dropped across it. As the frequency increases, the impedance of the tank circuit increases, and, as a result, the output voltage increases. When the frequency reaches its resonant value, the impedance is at its maximum and so is the output voltage. As the frequency goes above resonance, the impedance begins to

FIGURE 15–51
A basic parallel resonant band-pass filter.

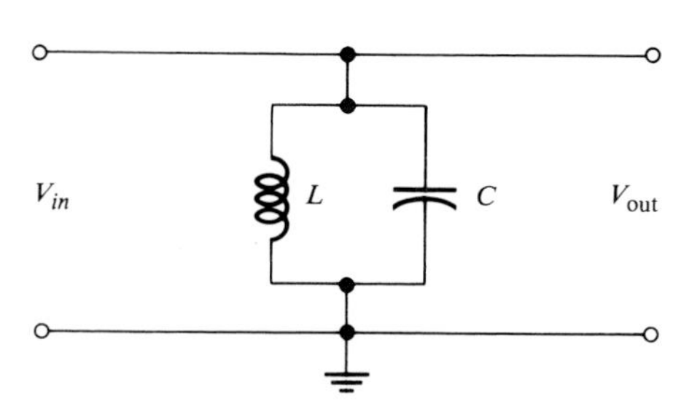

decrease, causing the output voltage to decrease. The bandwidth and cutoff frequencies for a parallel resonant band-pass filter are defined in the same way as for the series resonant circuit, and the formulas given in Section 15–4 still apply. General band-pass response curves showing both V_{out} and I_T versus frequency are given in Figures 15–52(a) and 15–52(b), respectively.

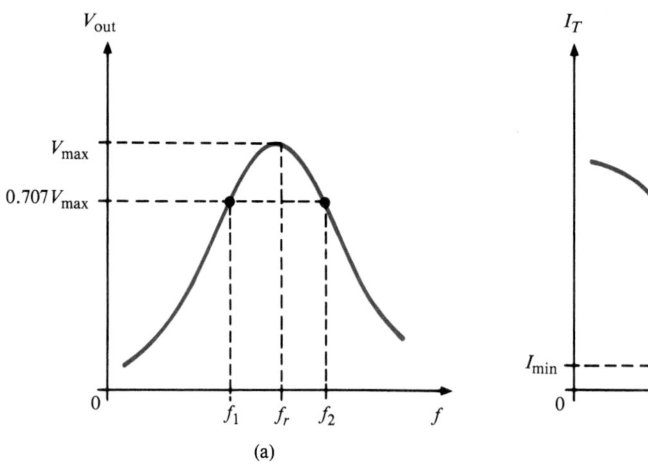

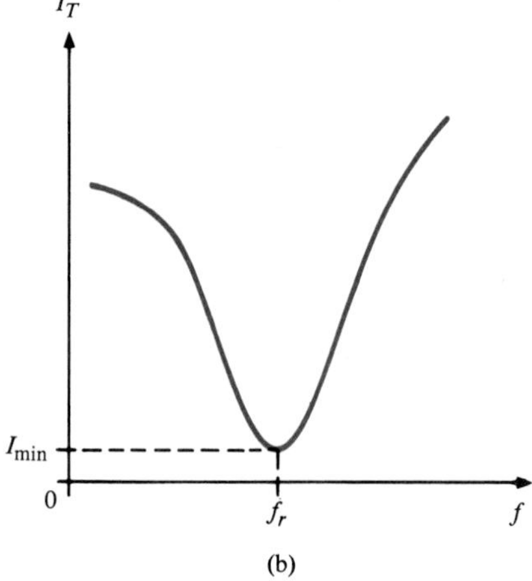

FIGURE 15–52
Generalized response curves for a parallel resonant band-pass filter.

The general response of a parallel resonant band-pass filter is illustrated in Figure 15–53. This illustration pictorially shows how the current and the output voltage change with frequency.

EXAMPLE 15–20 A certain parallel resonant band-pass filter has a maximum output voltage of 4 V at f_r. What is the value of V_{out} at the cutoff frequencies?

Solution
V_{out} at the cutoff frequencies is 70.7% of maximum:

$$V_{out(1)} = V_{out(2)} = 0.707V_{out(max)} = 0.707(4\text{ V}) = 2.828\text{ V}$$

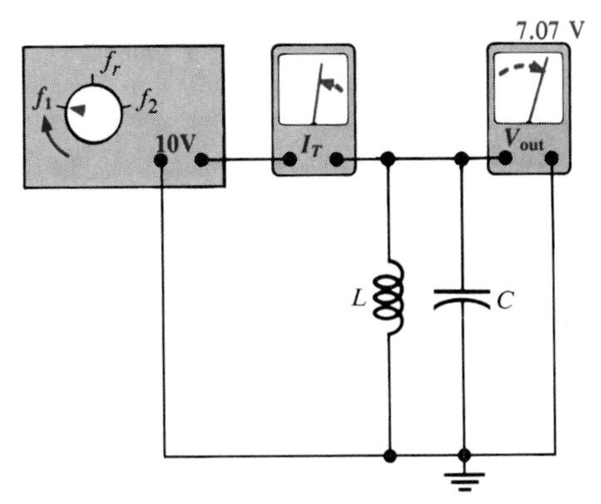

(a) As the frequency increases to f_1, V_{out} increases to 7.07 V, and I_T decreases.

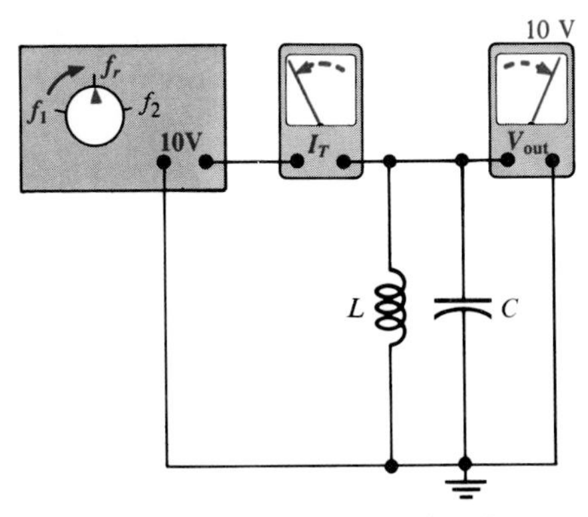

(b) As the frequency increases from f_1 to f_r, V_{out} increases from 7.07 V to 10 V, and I_T decreases to its minimum value.

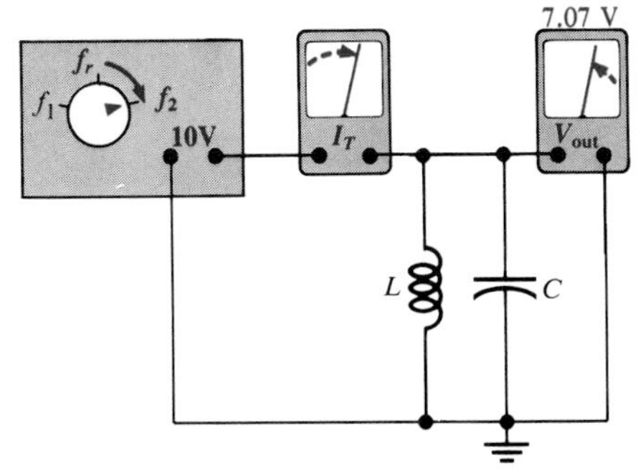

(c) As the frequency increases from f_r to f_2, V_{out} decreases from 10 V to 7.07 V, and I_T increases from its minimum.

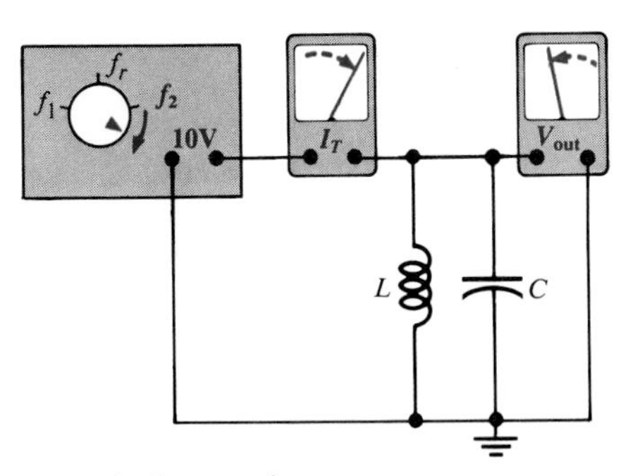

(d) As the frequency increases above f_2, V_{out} decreases below 7.07 V, and I_T continues to increase.

FIGURE 15–53
Example of the response of a parallel resonant band-pass filter with the input voltage at a constant 10 V rms.

EXAMPLE 15–21 A parallel resonant circuit has a lower cutoff frequency of 3.5 kHz and an upper cutoff frequency of 6 kHz. What is the bandwidth?

Solution

$$BW = f_2 - f_1 = 6\text{ kHz} - 3.5\text{ kHz} = 2.5\text{ kHz}$$

EXAMPLE 15–22 A certain parallel resonant band-pass filter has a resonant frequency of 12 kHz and a Q of 10. What is its bandwidth?

Solution

$$BW = \frac{f_r}{Q} = \frac{12\text{ kHz}}{10} = 1.2\text{ kHz}$$

How Loading Affects the Selectivity of a Parallel Resonant Band-Pass Filter

When a resistive load is connected across the output of a filter as shown in Figure 15–54(a), the Q of the filter is reduced. Since $BW = f_r/Q$, the bandwidth is increased, thus reducing the selectivity. Also, the impedance of the filter at resonance is decreased because R_L effectively appears in parallel with $R_{p(eq)}$. Thus, the maximum output voltage is reduced by the voltage divider effect of R_{pT} and the internal source resistance R_s, as illustrated in Figure 15–54(b). Part (c) of the figure shows the general effect on the filter response curve.

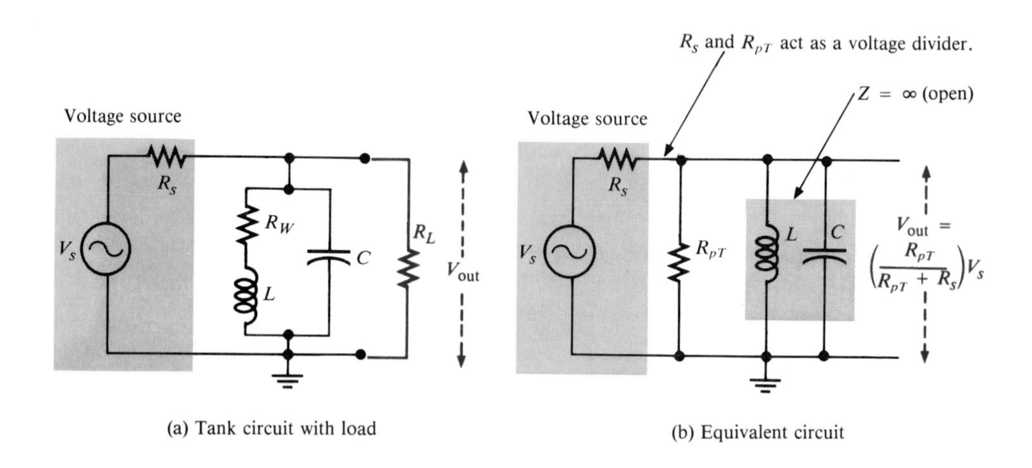

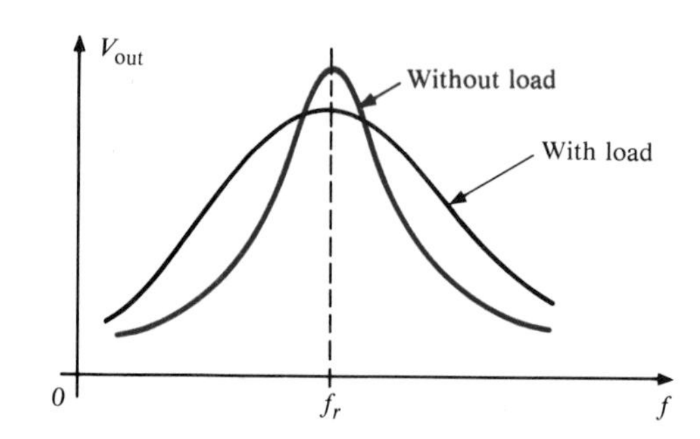

(c) Loading widens the bandwidth and reduces the output.

FIGURE 15–54
Effects of loading on a parallel resonant band-pass filter.

EXAMPLE 15–23

(a) Determine f_r, BW, and V_{out} at resonance for the unloaded filter in Figure 15–55. The source resistance is 600 Ω.

(b) Repeat Part (a) when the filter is loaded with a 50-kΩ resistance, and compare the results.

FIGURE 15–55

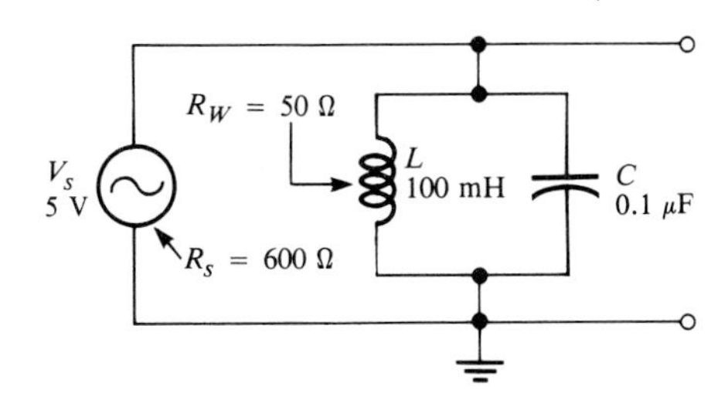

Solution

(a) $f_r = \dfrac{\sqrt{1 - (R_W^2 C/L)}}{2\pi\sqrt{LC}} = 1.59 \text{ kHz}$

At resonance,

$$X_L = 2\pi f_r L = 2\pi(1.59 \text{ kHz})(100 \text{ mH}) = 999\ \Omega$$

$$Q = \frac{X_L}{R_W} = \frac{999\ \Omega}{50\ \Omega} = 19.98$$

$$BW = \frac{f_r}{Q} = \frac{1.59 \text{ kHz}}{19.98} = 79.58 \text{ Hz}$$

$$R_{p(\text{eq})} = R_W(Q^2 + 1) = 50\ \Omega(19.98^2 + 1) = 20.01 \text{ k}\Omega$$

$$V_{\text{out}} = \left(\frac{R_{p(\text{eq})}}{R_{p(\text{eq})} + R_s}\right)V_s = \left(\frac{20.01 \text{ k}\Omega}{20.61 \text{ k}\Omega}\right)5 \text{ V} = 4.85 \text{ V}$$

(b) When a 50-kΩ load resistance is connected, the following values are obtained:

$$R_{pT} = R_{p(\text{eq})} \| R_L = 20.01 \text{ k}\Omega \| 50 \text{ k}\Omega = 14.34 \text{ k}\Omega$$

$$Q_O = \frac{R_{pT}}{X_L} = \frac{14.34 \text{ k}\Omega}{999\ \Omega} = 14.35$$

$$BW = \frac{f_r}{Q_O} = \frac{1.59 \text{ kHz}}{14.35} = 110.8 \text{ Hz}$$

$$V_{\text{out}} = \left(\frac{R_{pT}}{R_{pT} + R_s}\right)V_s = \left(\frac{14.34 \text{ k}\Omega}{14.94 \text{ k}\Omega}\right)5 \text{ V} = 4.799 \text{ V}$$

The result of adding the load resistance is an increased bandwidth and a decreased output voltage at resonance.

The Band-Stop Filter

A basic parallel resonant band-stop filter is shown in Figure 15–56. The output is taken across a load resistor in series with the tank circuit.

The variation of the tank circuit impedance with frequency produces the familiar current response that has been previously discussed; that is, the cur-

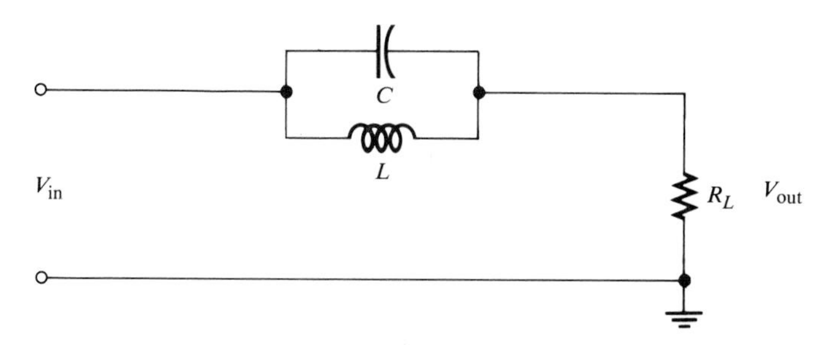

FIGURE 15–56
A basic parallel resonant band-stop filter.

rent is minimum at resonance and increases on both sides of resonance. Since the output voltage is across the series load resistor, it follows the current, thus creating the band-stop response characteristic as indicated in Figure 15–57.

Actually, the band-stop filter in Figure 15–56 can be viewed as a voltage divider created by Z_r of the tank and the load resistance. Thus, the output voltage at f_r is

$$V_{out} = \left(\frac{R_L}{R_L + Z_r}\right)V_{in}$$

FIGURE 15–57
Band-stop filter response.

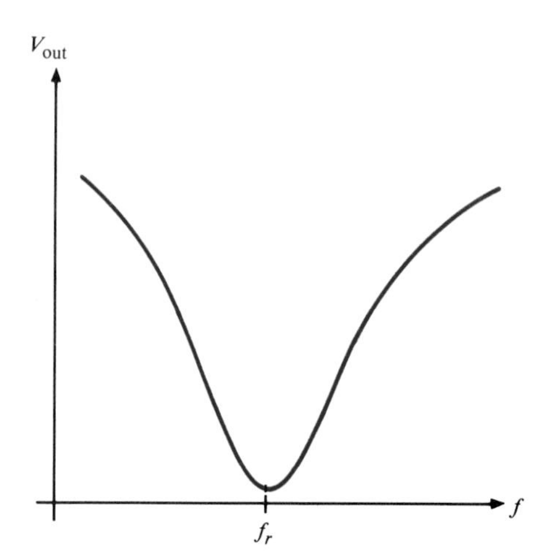

EXAMPLE 15–24 Find f_r and the output voltage at resonance for the band-stop filter in Figure 15–58.

Solution

$$f_r = \frac{\sqrt{1 - (R_W^2 C/L)}}{2\pi\sqrt{LC}} = 249.4 \text{ Hz}$$

$$X_L = 2\pi f_r L = 517\ \Omega$$

$$Q = \frac{X_L}{R_W} = \frac{517\ \Omega}{250\ \Omega} = 2.07$$

$$Z_r = R_W(Q^2 + 1) = 1.32\ \text{k}\Omega$$

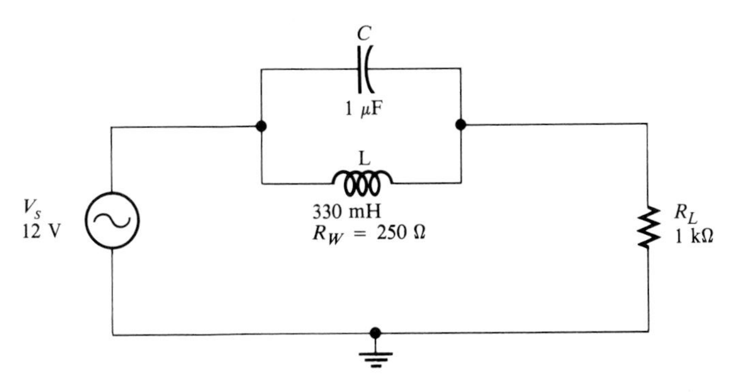

FIGURE 15–58

At resonance,

$$V_{out} = \left(\frac{R_L}{R_L + Z_r}\right)V_s = \left(\frac{1\ \text{k}\Omega}{2.32\ \text{k}\Omega}\right)12\ \text{V} = 5.17\ \text{V}$$

Basic filter configurations have been introduced in this section and in Section 15–4. These basic configurations are sometimes used in combination to increase filter selectivity. For example, a series resonant and a parallel resonant band-pass filter can be used together, with the series circuit connected between the input and output and the parallel circuit connected across the output.

Variable capacitors or inductors are sometimes used in certain filter applications so that the resonant circuits can be *tuned* over a range of resonant frequencies.

SECTION REVIEW 15–7

1. How can the bandwidth of a parallel resonant filter be increased?
2. The resonant frequency of a certain high-Q ($Q > 10$) filter is 5 kHz. If the Q is lowered to 2, does f_r change and, if so, to what value?
3. What is the impedance of a tank circuit at its resonant frequency if $R_W = 75\ \Omega$ and $Q = 25$?

SYSTEM APPLICATIONS

15–8

Resonant filters are used in a wide variety of applications, particularly in communication systems. In this section, we will look briefly at a few common communication system applications. The purpose of this coverage is not to explain the systems that are used as examples but rather to show how important filters are in electronic communication.

Tuned Amplifiers

A tuned amplifier is a circuit that amplifies signals within a specified band. Typically, a parallel resonant circuit is used in conjunction with an amplifier to achieve the selectivity. In terms of the general operation, input signals with frequencies that range over a wide band are accepted on the amplifier's input and are amplified. The function of the resonant circuit is to allow only a relatively narrow band of those frequencies to be passed on. The variable capacitor allows tuning over the range of input frequencies so that a desired frequency can be selected, as indicated in Figure 15–59.

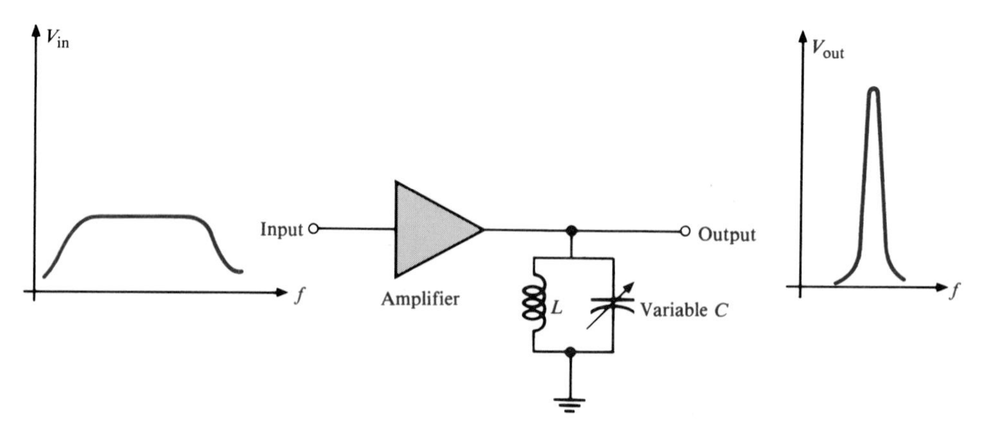

FIGURE 15–59
A basic tuned band-pass amplifier.

Antenna Input to a Receiver

Radio signals are sent out from a transmitter via electromagnetic waves which propagate through the atmosphere. When the electromagnetic waves cut across the receiving antenna, small voltages are induced. Out of all the wide range of electromagnetic frequencies, only one frequency or a limited band of frequencies must be extracted. Figure 15–60 shows a typical arrangement of an antenna coupled to the receiver input by a transformer. A variable capacitor is connected across the transformer secondary to form a parallel resonant circuit.

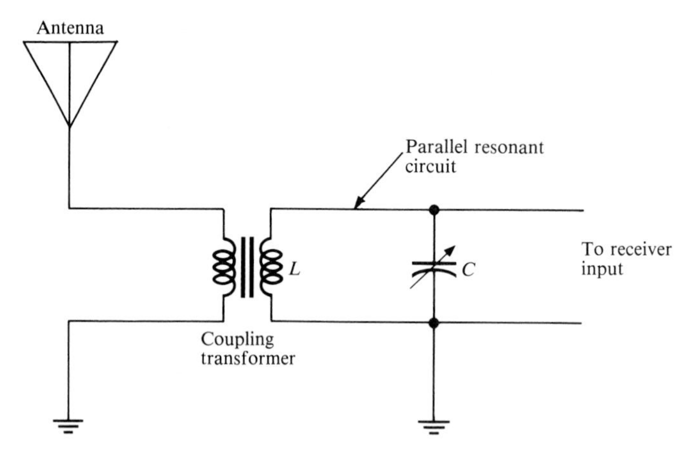

FIGURE 15–60
Resonant coupling from an antenna.

Double-Tuned Transformer Coupling in a Receiver

In some types of communication receivers, tuned amplifiers are transformer-coupled together to increase the amplification. Capacitors can be placed in parallel with the primary and secondary of the transformer, effectively creating two parallel resonant band-pass filters that are coupled together. This technique, illustrated in Figure 15–61, can result in a wider bandwidth and steeper slopes on the response curve, thus increasing the selectivity for a desired band of frequencies.

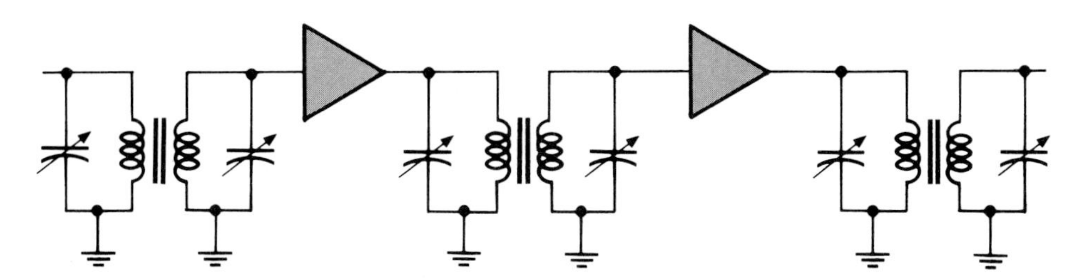

FIGURE 15–61
Double-tuned amplifiers.

Signal Reception and Separation in a TV Receiver

A television receiver must handle both video (picture) signals and audio (sound) signals. Each TV transmitting station is allotted a 6-MHz bandwidth. Channel 2 is allotted a band from 54 MHz through 59 MHz, channel 3 is allotted a band from 60 MHz through 65 MHz, on up to Channel 13 which has a band from 210 MHz through 215 MHz. You can tune the front end of the TV receiver to select any one of these channels by using tuned amplifiers. The signal output of the front end of the receiver has a bandwidth from 41 MHz through 46 MHz, regardless of the channel that is tuned in. This band, called the *intermediate frequency* (IF) band, contains both video and audio. Amplifiers tuned to the IF band boost the signal and feed it to the video amplifier.

Before the output of the video amplifier is applied to the picture tube, the audio signal is removed by a 4.5-MHz band-stop filter (called a *wave trap*), as shown in Figure 15–62. This trap keeps the sound signal from interfering with the picture. The video amplifier output is also applied to band-pass circuits that are tuned to the sound carrier frequency of 4.5 MHz. The sound signal is then processed and applied to the speaker as indicated in Figure 15–62.

Superheterodyne Receiver

Another good example of filter applications is in the common AM (amplitude modulation) receiver. The AM broadcast band ranges from 535 kHz to 1605 kHz. Each AM station is assigned a certain narrow bandwidth within that range. A simplified block diagram of a superheterodyne AM receiver is shown in Figure 15–63.

There are basically three parallel resonant band-pass filters in the front end of the receiver. Each of these filters is gang-tuned by capacitors; that is, the capacitors are mechanically or electronically linked together so that they change together as the tuning knob is turned. The front end is tuned to receive a desired station, for example, one that transmits at 600 kHz. The input filter from the antenna and the RF (radio frequency) amplifier filter select only a frequency of 600 kHz out of all the frequencies crossing the antenna. The actual

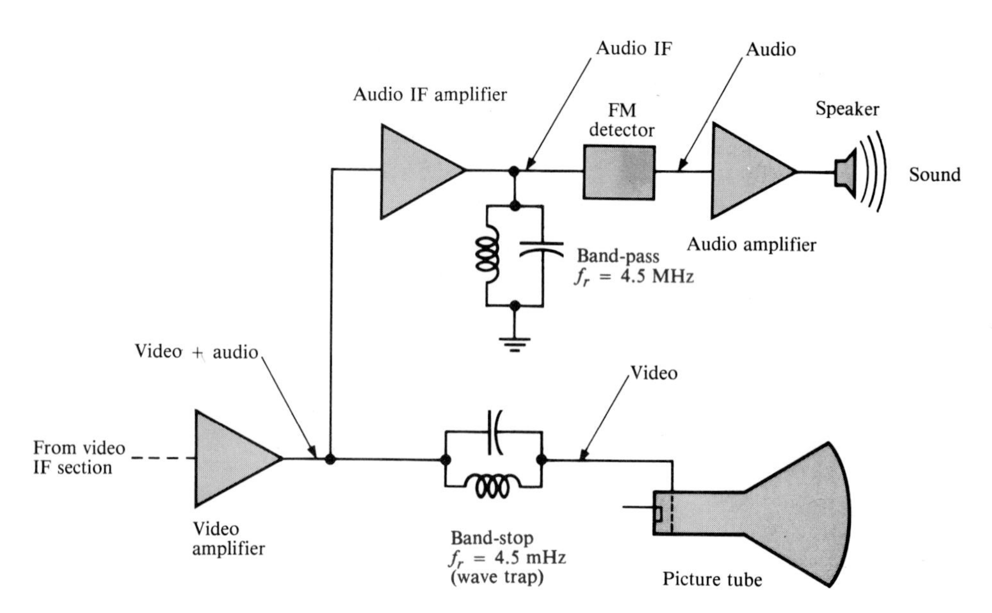

FIGURE 15–62
A simplified portion of a TV receiver showing filter usage.

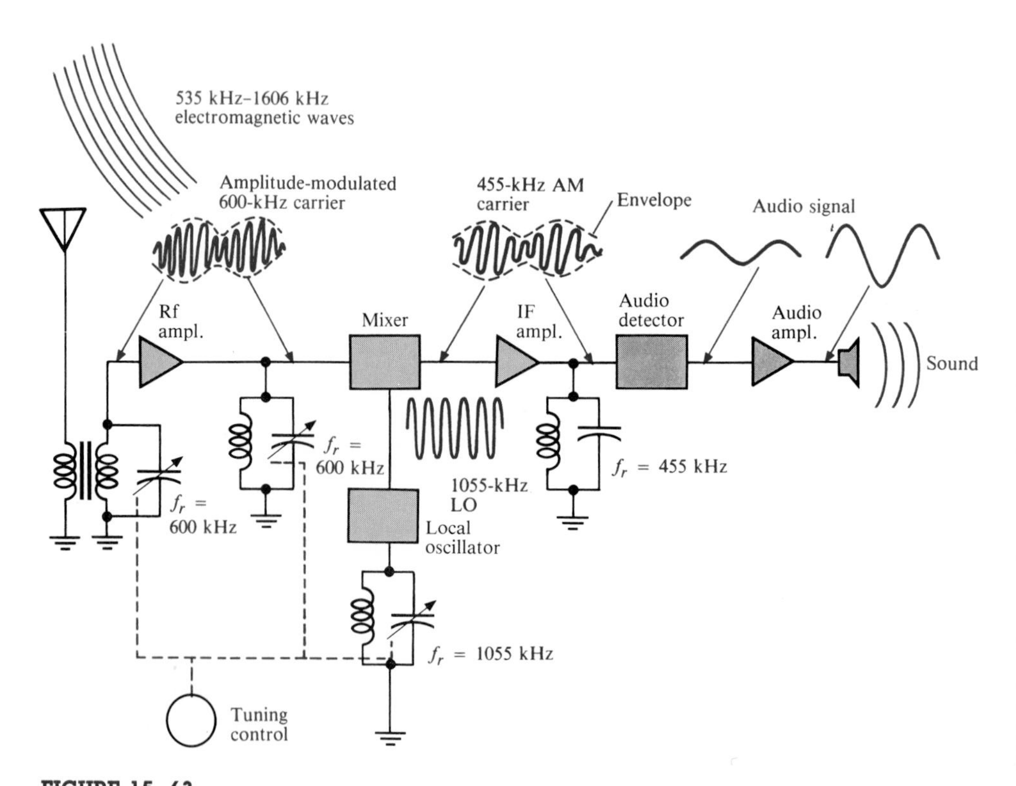

FIGURE 15–63
A simplified diagram of a superheterodyne AM radio broadcast receiver showing the application of tuned resonant circuits.

audio (sound) signal is carried by the 600-kHz carrier frequency by modulating the amplitude of the carrier so that it follows the audio signal as indicated. The variation in the amplitude of the carrier corresponding to the audio signal is called the *envelope*. The 600 kHz is then applied to a circuit called the *mixer*. The *local oscillator* (LO) is tuned to a frequency that is 455 kHz above the selected frequency (1055 kHz, in this case). By a process called *heterodyning* or *beating,* the AM signal and the local oscillator signal are mixed together, and the 600-kHz AM signal is converted to a 455-kHz AM signal (1055 kHz − 600 kHz = 455 kHz). The 455 kHz is the intermediate frequency (IF) for standard AM receivers. No matter which station within the broadcast band is selected, its frequency is always converted to the 455-kHz IF. The amplitude-modulated IF is applied to an *audio detector* which removes the IF, leaving only the envelope or audio signal. The audio signal is then amplified and applied to the speaker.

SECTION REVIEW 15–8

1. Generally, why is a tuned filter necessary when a signal is coupled from an antenna to the input of a receiver?
2. What is a *wave trap?*
3. What is meant by *ganged tuning?*

APPLICATION NOTE

For the Application Assignment at the beginning of the chapter, you were to determine the cause of the sound bars on a TV screen. When such sound bars appear, the most likely culprit is the 4.5-MHz band-stop filter (wave trap) that blocks the passage of the audio signal to the picture tube. You should check this filter for a shorted or faulty component.

SUMMARY

Facts

At series resonance:

- The reactances are equal.
- The impedance is minimum and equal to the resistance.
- The current is maximum.
- The phase angle is zero.
- The voltages across L and C are equal in magnitude and, as always, 180° out of phase with each other and thus they cancel.

At parallel resonance:

- The reactances are approximately equal for $Q \geq 10$.
- The impedance is maximum.
- The current is minimum and, ideally, equal to zero.
- The phase angle is zero.
- The currents in the L and C branches are equal in magnitude and, as always, 180° out of phase with each other.

Definitions

- *Q of a series resonant circuit*—the ratio of reactance to series resistance.
- *Q of a parallel resonant circuit*—the ratio of parallel resistance to reactance. If there is no parallel resistance, the Q is the Q of the coil.

- *Bandwidth of a band-pass or a band-stop filter*—the upper cutoff frequency minus the lower cutoff frequency.
- *Cutoff frequencies*—the frequencies at which the filter response is 70.7% of its maximum.
- *Half-power frequencies and −3 dB frequencies*—other names for the cutoff frequencies.

Formulas

$X_T = |X_L - X_C|$ Total series reactance (absolute value) (15–1)

$Z = \sqrt{R^2 + (X_L - X_C)^2}$ Series *RLC* impedance (15–2)

$\theta = \arctan\left(\frac{X_T}{R}\right)$ Series *RLC* phase angle (15–3)

$X_L = X_C$ Condition for series resonance (15–4)

$Z_r = R$ Series resonant impedance (15–5)

$f_r = \frac{1}{2\pi\sqrt{LC}}$ Series resonant frequency (15–6)

$BW = f_2 - f_1$ Bandwidth (15–7)

$\text{dB} = 20 \log\left(\frac{V_{\text{out}}}{V_{\text{in}}}\right)$ Decibel formula for voltage ratio (15–8)

$\text{dB} = 10 \log\left(\frac{P_{\text{out}}}{P_{\text{in}}}\right)$ Decibel formula for power ratio (15–9)

$Q = \frac{X_L}{R}$ Series resonant Q (15–10)

$BW = \frac{f_r}{Q}$ Bandwidth (15–11)

$Y = \sqrt{G^2 + (B_C - B_L)^2}$ Parallel *RLC* admittance (15–12)

$\theta = \arctan\left(\frac{B_T}{G}\right)$ Parallel *RLC* phase angle (15–13)

$I_T = \sqrt{I_R^2 + (I_C - I_L)^2}$ Total parallel *RLC* current (15–14)

$\theta = \arctan\left(\frac{I_{CL}}{I_R}\right)$ Parallel *RLC* phase angle (15–15)

$L_{\text{eq}} = L\left(\frac{Q^2 + 1}{Q^2}\right)$ Equivalent parallel inductance (15–16)

$R_{p(\text{eq})} = R_W(Q^2 + 1)$ Equivalent parallel resistance (15–17)

$I_T = |I_C - I_L|$ Total parallel *LC* current (absolute value) (15–18)

$Z_r = R_W(Q^2 + 1)$ Impedance at parallel resonance (15–19)

$I_T = \frac{V_s}{Z_r}$ Total current at parallel resonance (15–20)

$f_r = \frac{1}{2\pi\sqrt{LC}}\sqrt{\frac{Q^2}{Q^2 + 1}}$ Parallel resonant frequency (exact) (15–21)

$f_r = \frac{\sqrt{1 - (R_W^2 C/L)}}{2\pi\sqrt{LC}}$ Parallel resonant frequency (exact) (15–22)

$Q_O = \frac{R_{pT}}{X_{L(\text{eq})}}$ Parallel *RLC* Q (15–23)

SELF-TEST

1. What is the total reactance of a series *RLC* circuit at resonance?
2. What is the phase angle of a series *RLC* circuit at resonance?
3. What is the impedance of a series *LC* circuit at the resonant frequency? The component values are $L = 15$ mH, $C = 0.015\ \mu$F, and $R_W = 80\ \Omega$.
4. In a series *RLC* circuit operating below the resonant frequency, does the current lead or lag the applied voltage? Why?
5. In a series *RLC* circuit, if *C* is increased, what happens to the resonant frequency?
6. In a certain series resonant circuit, $V_C = 150$ V, $V_L = 150$ V, and $V_R = 50$ V. What is the value of the source voltage? What is the voltage across *L* and *C* combined?
7. A certain series resonant band-pass filter has a bandwidth of 1000 Hz. If the existing coil is replaced by a coil with a lower *Q*, what happens to the bandwidth?
8. In a parallel *RLC* circuit, why does the current lag the source voltage at frequencies below resonance?
9. Ideally, what is the total current into the *L* and *C* branches at resonance?
10. In order to tune a parallel resonant circuit to a lower frequency, should you increase or decrease the capacitance?
11. Under what condition is the parallel resonant frequency the same as for series resonance?
12. What happens to the bandwidth of a parallel resonant filter if the parallel resistance is reduced?

PROBLEM SET A

Section 15–1

15–1 A certain series *RLC* circuit has the following values: $R = 10\ \Omega$, $C = 0.05\ \mu$F, and $L = 5$ mH. Determine the impedance and phase angle. What is the total reactance? $f = 5$ kHz.

15–2 Find the impedance in Figure 15–64.

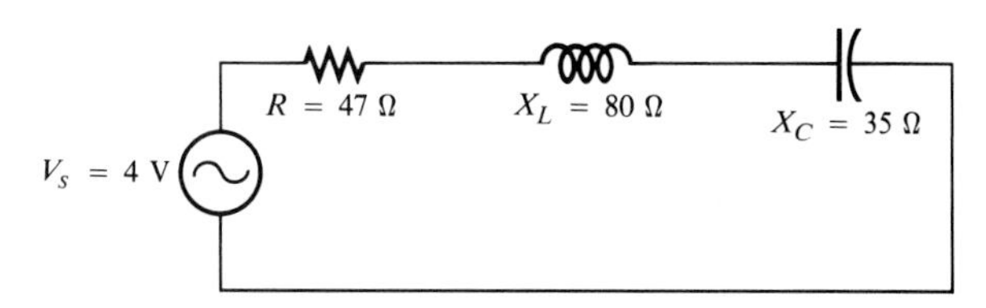

FIGURE 15–64

15–3 If the frequency of the source voltage in Figure 15–64 is doubled from the value that produces the indicated reactances, how does the impedance change?

Section 15–2

15–4 For the circuit in Figure 15–64, find I_T, V_R, V_L, and V_C.

15–5 Sketch the voltage phasor diagram for the circuit in Figure 15–64.

Section 15–3

15–6 Find X_L, X_C, Z, and I at the resonant frequency in Figure 15–65.

15–7 A certain series resonant circuit has a maximum current of 50 mA and a V_L of 100 V. The applied voltage is 10 V. What is *Z*? What are X_L and X_C?

FIGURE 15–65

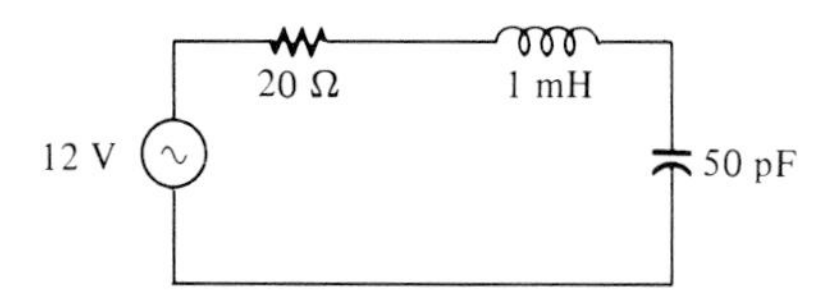

15–8 For the *RLC* circuit in Figure 15–66, determine the resonant frequency and the cutoff frequencies.

15–9 What is the value of the current at the half-power points in Figure 15–66?

FIGURE 15–66

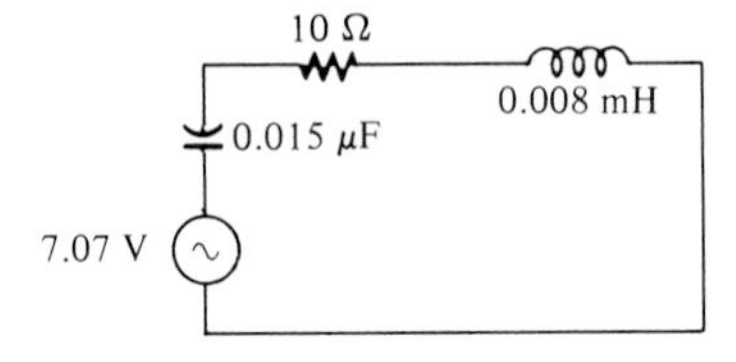

Section 15–4

15–10 Determine the resonant frequency for each filter in Figure 15–67. Are these filters band-pass or band-stop types?

FIGURE 15–67

15–11 Assuming that the coils in Figure 15–67 have a winding resistance of 10 Ω, find the bandwidth for each filter.

15–12 Determine the resonant frequency and bandwidth for each filter in Figure 15–68.

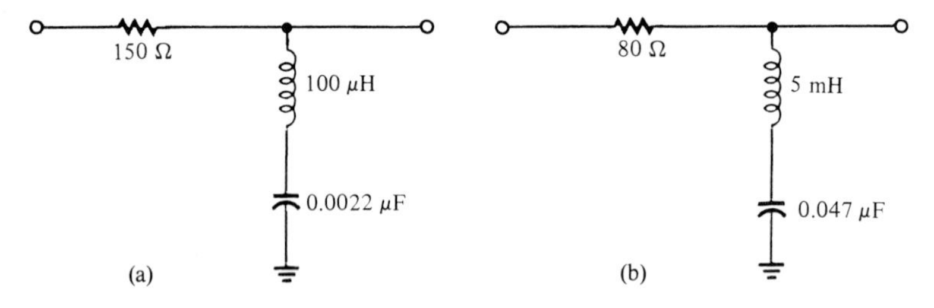

FIGURE 15–68

Section 15–5

15–13 Find the total impedance of the circuit in Figure 15–69.

15–14 Is the circuit in Figure 15–69 capacitive or inductive? Explain.

15–15 For the circuit in Figure 15–69, find all the currents and voltages.

FIGURE 15–69

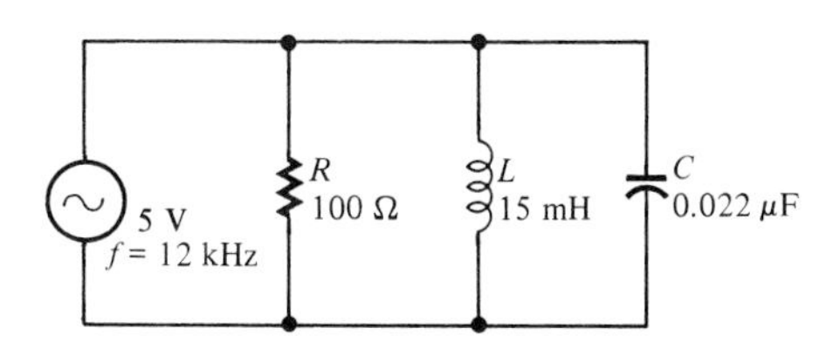

15–16 Find the total impedance for the circuit in Figure 15–70.

FIGURE 15–70

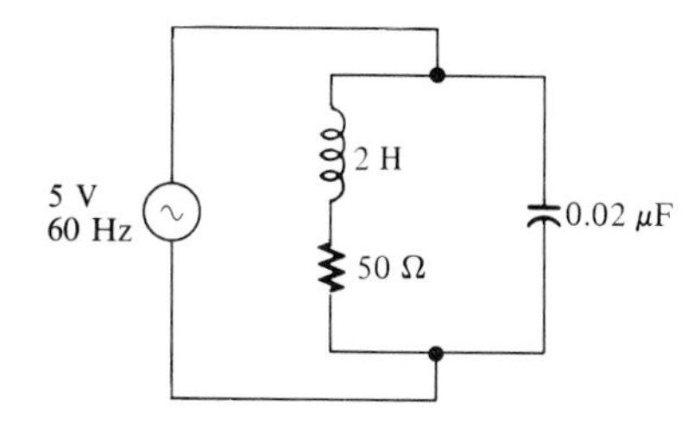

Section 15–6

15–17 What is the impedance of an ideal parallel resonant circuit (no resistance in either branch)?

15–18 Find Z at resonance and f_r for the tank circuit in Figure 15–71.

FIGURE 15–71

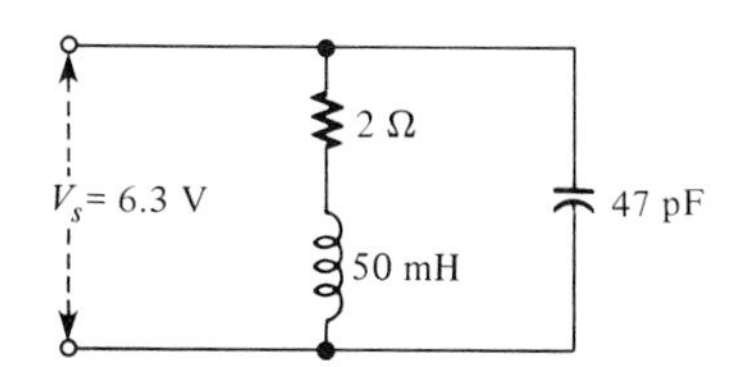

15–19 How much current is drawn from the source in Figure 15–71 at resonance? What are the inductive current and the capacitive current at the resonant frequency?

Section 15–7

15–20 At resonance, $X_L = 2\text{ k}\Omega$ and $R_W = 25\ \Omega$ in a parallel resonant band-pass filter. The resonant frequency is 5 kHz. Determine the bandwidth.

15–21 If the lower cutoff frequency is 2400 Hz and the upper cutoff frequency is 2800 Hz, what is the bandwidth?

15–22 In a certain resonant circuit, the power at resonance is 2.75 W. What is the power at the lower and upper cutoff frequencies?

15–23 What values of L and C should be used in a tank circuit to obtain a resonant frequency of 8 kHz? The bandwidth must be 800 Hz. The winding resistance of the coil is 10 Ω.

15–24 A parallel resonant circuit has a Q of 50 and a BW of 400 Hz. If Q is doubled, what is the bandwidth for the same f_r?

PROBLEM SET B

15–25 For each following case, express the voltage ratio in dB:

(a) $V_{in} = 1$ V, $V_{out} = 1$ V **(b)** $V_{in} = 5$ V, $V_{out} = 3$ V
(c) $V_{in} = 10$ V, $V_{out} = 7.07$ V **(d)** $V_{in} = 25$ V, $V_{out} = 5$ V

15–26 Find the current through each component in Figure 15–72. Find the voltage across each component.

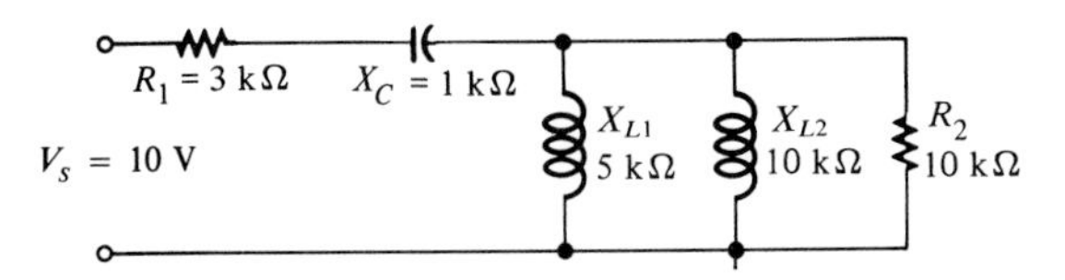

FIGURE 15–72

15–27 Determine whether there is a value of C that will make $V_{ab} = 0$ V in Figure 15–73. If not, explain.

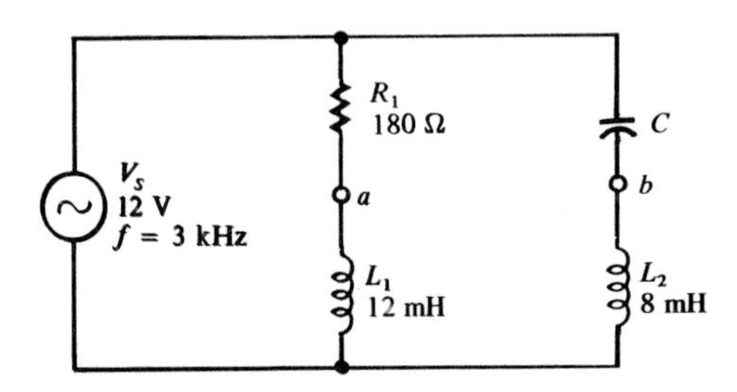

FIGURE 15–73

15–28 If the value of C is 0.2 μF, how much current flows through each branch in Figure 15–73? What is the total current?

15–29 Determine the resonant frequencies in Figure 15–74.

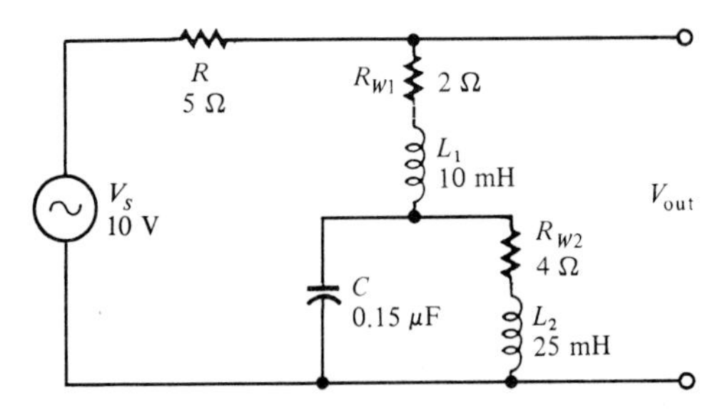

FIGURE 15–74

15–30 Design a band-pass filter using a parallel resonant circuit to meet the following specifications:
(a) $BW = 500$ Hz (b) $Q = 40$ (c) $I_{C(max)} = 20$ mA, $V_{C(max)} = 2.5$ V

ANSWERS TO SECTION REVIEWS

Section 15–1

1. 70 Ω, capacitive. 2. 83.22 Ω, 57.26°, leading.

Section 15–2

1. 38.42 V. 2. Leads. 3. 6 Ω.

Section 15–3

1. $X_L = X_C$. 2. The impedance is minimum. 3. 159 kHz. 4. Capacitive.

Section 15–4

1. 10.61 V. 2. 10 kHz. 3. Maximum, minimum.

Section 15–5

1. $I_R = 80$ mA, $I_C = 120$ mA, $I_L = 240$ mA. 2. Inductive ($X_L < X_C$).
3. 20.13 mH, 1589 Ω.

Section 15–6

1. Maximum. 2. Minimum. 3. 22.51 kHz. 4. 20.9 kHz. 5. 8020 Ω.

Section 15–7

1. By reducing the parallel resistance. 2. Yes, 4.47 kHz. 3. 46.95 kΩ.

Section 15–8

1. To select a narrow band of frequencies. 2. A band-stop filter.
3. Several variable capacitors (or inductors) whose values can be varied simultaneously with a common control.

Pulse Response of *RC* and *RL* Circuits

In Chapter 15, the frequency response of *RC* and *RL* circuits was covered. In this chapter, the response of *RC* and *RL* filters to pulse inputs is examined. With pulse inputs, the *time* response of the circuits is of primary importance. In the areas of pulse and digital circuits, the technician is often concerned with how a circuit responds over an interval of time to rapid changes in voltage or current. The relationship of the circuit *time constant* to the input pulse characteristics, such as pulse width and period, determines the shapes of the circuit voltages.

Before starting this chapter, you should review the material in Sections 10–5 and 11–5. Understanding exponential changes in voltages and currents in inductors and capacitors is crucial to the study of pulse response. Exponential formulas that were given in Chapter 10 are used throughout this chapter.

Integrator and *differentiator*, terms used throughout this coverage, refer to mathematical functions that are approximated by these circuits under certain conditions. Integration is an averaging process, and differentiation is a process for establishing a rate of change of a function. We will not deal with the purely mathematical aspects at this level of coverage.

In this chapter you will learn:

- How to recognize an *RC* and an *RL* integrating circuit.
- How to recognize an *RC* and an *RL* differentiating circuit.
- The meaning of *transient time.*
- The meaning of the term *steady state.*
- How the output voltage of integrators and differentiators changes with changes in input pulse width and/or frequency.
- How the output voltage of integrators and differentiators is a function of the circuit time constant.

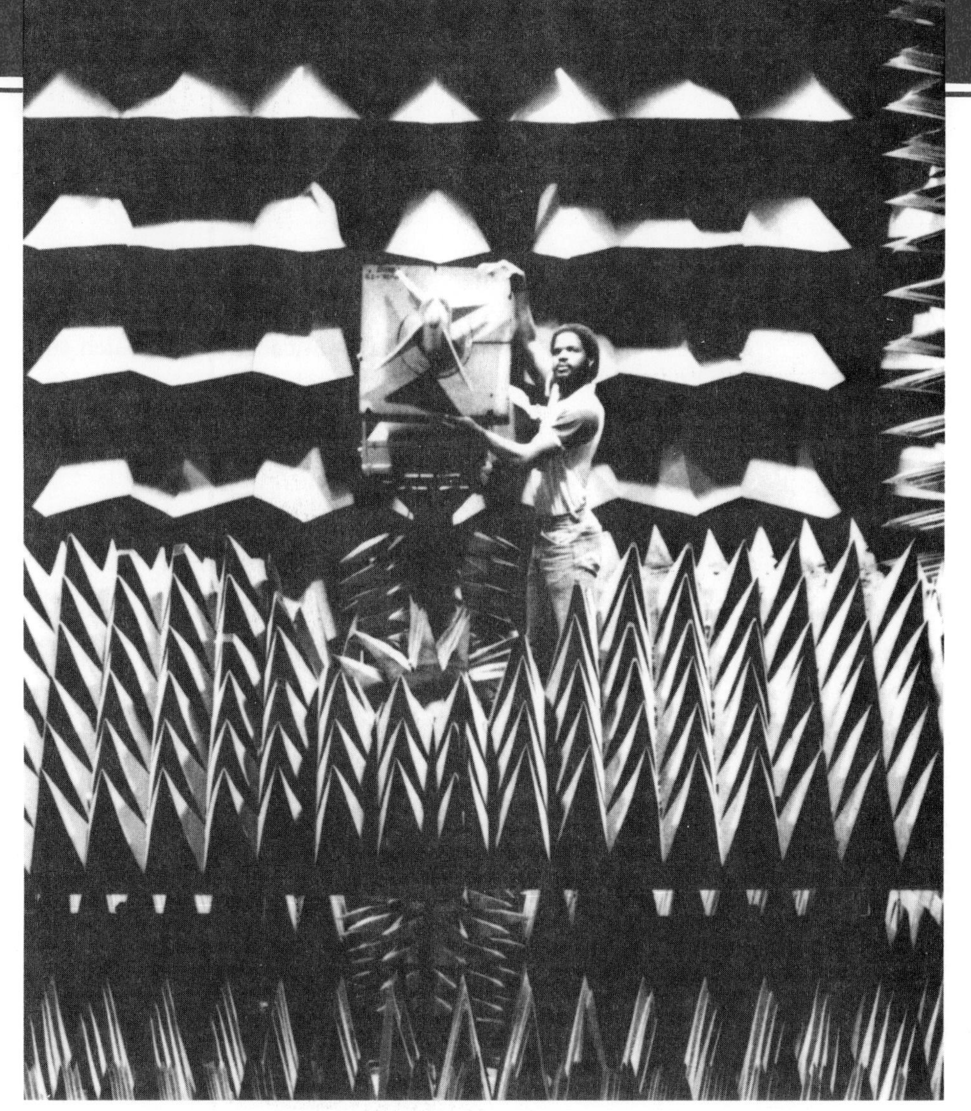

16

APPLICATION ASSIGNMENT

You are the lead technician on an engineering development project for a digital radar system. The design engineer has asked you to devise a simple circuit that will produce an alternating series of positive and negative voltage spikes (pulses of very short duration) derived from an available square wave source of variable amplitude and frequency. The spikes are to occur at each transition of the square wave and are to be used as timing pulses for testing a portion of the digital radar system. The spikes are to be no wider than 0.1 μs at their 37% amplitude points. The amplitude of each spike should be approximately the same as that of the square wave.

This chapter will prepare you to approach this assignment. A solution is given in the Application Note at the end of the chapter.

THE *RC* INTEGRATOR

16–1

Figure 16–1 shows a series *RC* circuit with a pulse input. When the output is taken across the capacitor, this circuit is known as an *integrator* in terms of its pulse response. Recall that in terms of frequency response, it is a basic *low-pass* filter. The term *integrator* is derived from a mathematical function which this circuit approximates under certain conditions.

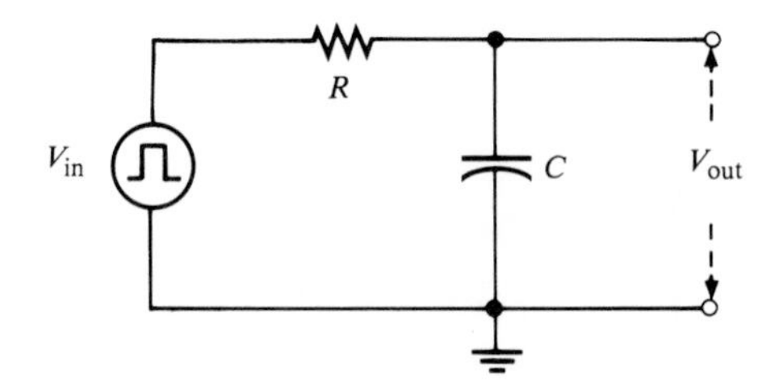

FIGURE 16–1
An *RC* integrating circuit.

How the Capacitor Charges and Discharges with a Pulse Input

When a pulse generator is connected to the input, as symbolized in Figure 16–1, the capacitor will charge and discharge in response to the pulses. When the input goes from its low level to its high level, the capacitor *charges* toward the high level of the pulse through the resistor. This action is analogous to connecting a battery through a switch to the *RC* network, as illustrated in Figure 16–2(a). When the pulse goes from its high level back to its low level, the capacitor *discharges* back through the source. The resistance of the source is assumed to be negligible compared to *R*. This action is analogous to replacing the source with a closed switch, as illustrated in Figure 16–2(b).

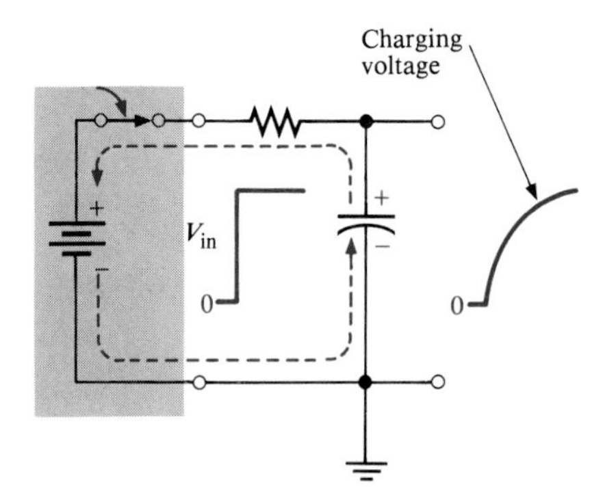

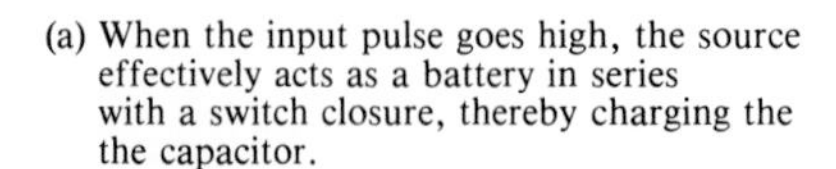
(a) When the input pulse goes high, the source effectively acts as a battery in series with a switch closure, thereby charging the the capacitor.

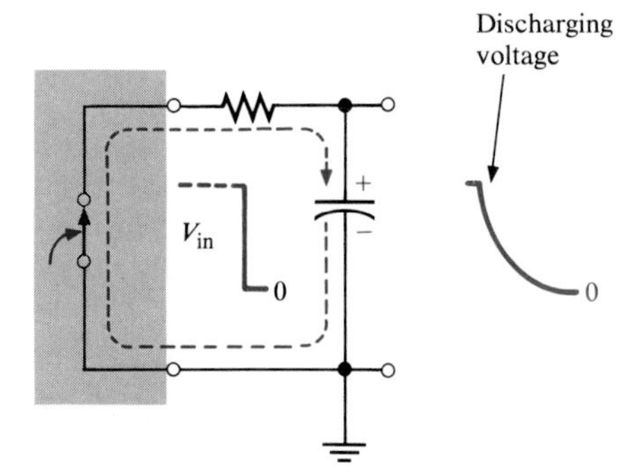

(b) When the input pulse goes back low, the source effectively acts as a closed switch, providing a discharge path for the capacitor.

FIGURE 16–2
The equivalent action when a pulse source charges and discharges the capacitor.

The capacitor will charge and discharge following an *exponential curve*. Its rate of charging and discharging, of course, depends on the RC time constant ($\tau = RC$).

For an ideal pulse, both edges are considered to occur instantaneously. Two basic rules of capacitor behavior aid in understanding RC circuit responses to pulse inputs:

1. The capacitor appears as a *short* to an instantaneous change in current and as an open to dc.
2. The voltage across the capacitor cannot change instantaneously—it can change only exponentially.

The Capacitor Voltage

In an RC integrator, *the output is the capacitor voltage*. The capacitor charges during the time that the pulse is high. If the pulse is at its high level long enough, the capacitor will fully charge to the voltage amplitude of the pulse, as illustrated in Figure 16–3. The capacitor discharges during the time that the pulse is low. If the low time between pulses is long enough, the capacitor will fully discharge to zero, as shown in the figure. Then when the next pulse occurs, it will charge again.

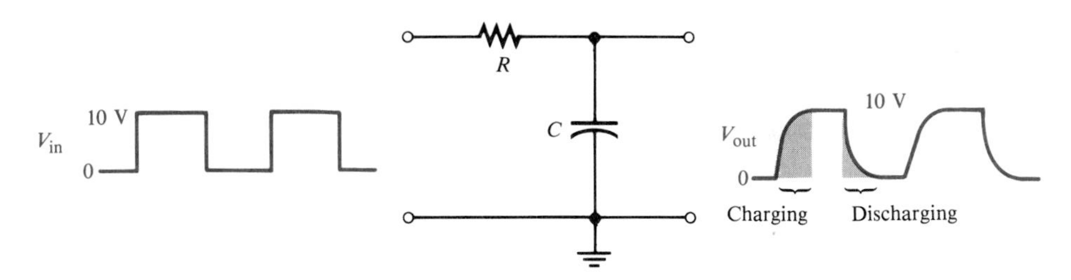

FIGURE 16–3
Illustration of a capacitor fully charging and discharging in response to a pulse input.

SECTION REVIEW 16–1

1. Define the term *integrator* in relation to an RC circuit.
2. What causes a capacitor in an RC circuit to charge and discharge?

RESPONSE OF AN *RC* INTEGRATOR TO A SINGLE PULSE

16–2

From the previous section, you have a general idea of how an RC integrator responds to a pulse input. In this section, we examine the response to a single pulse in detail. Then in the next section, we will expand this basic knowledge to include the general case of a series of pulses.

Two conditions of pulse response must be considered:

1. when the input pulse width (t_W) is equal to or greater than five time constants ($t_W \geq 5\tau$)
2. when the input pulse width is less than five time constants ($t_W < 5\tau$)

Recall that five time constants is accepted as the time for a capacitor to fully charge or fully discharge; this time is often called the *transient time.*

When the Pulse Width Is Equal to or Greater than Five Time Constants

The capacitor will *fully charge* if the pulse width is equal to or greater than five time constants (5τ). This condition is expressed as $t_W \geq 5\tau$. At the end of the pulse, the capacitor *fully discharges* back through the source.

Figure 16–4 illustrates the output waveforms for various values of time constant and a fixed input pulse width. Notice that the shape of the output pulse approaches that of the input as the transient time is made small compared to the pulse width. In each case the output reaches the full amplitude of the input.

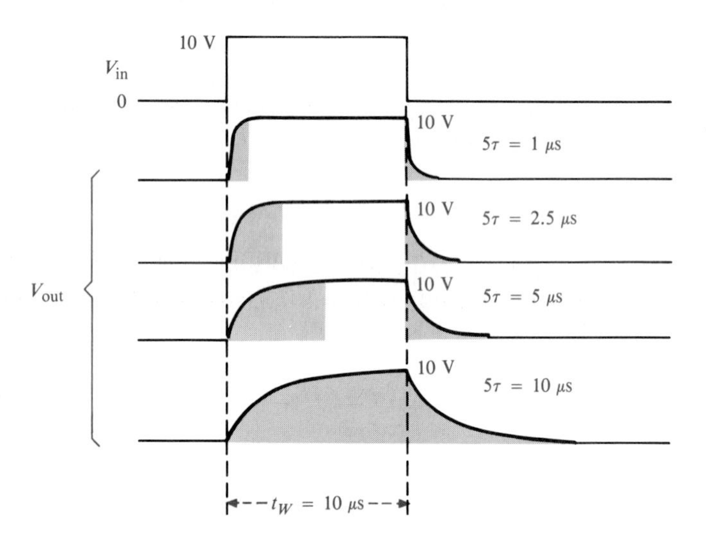

FIGURE 16–4
Variation of an integrator's output pulse shape with time constant. The shaded areas indicate when the capacitor is charging and discharging.

Figure 16–5 shows how a fixed time constant and a variable input pulse width affect the integrator output. Notice that as the pulse width is increased, the shape of the output pulse approaches that of the input. Again, this means that the transient time is short compared to the pulse width.

When the Pulse Width Is Less than Five Time Constants

Now let us examine the case in which the width of the input pulse is less than five time constants of the integrator. This condition is expressed as $t_W < 5\tau$.

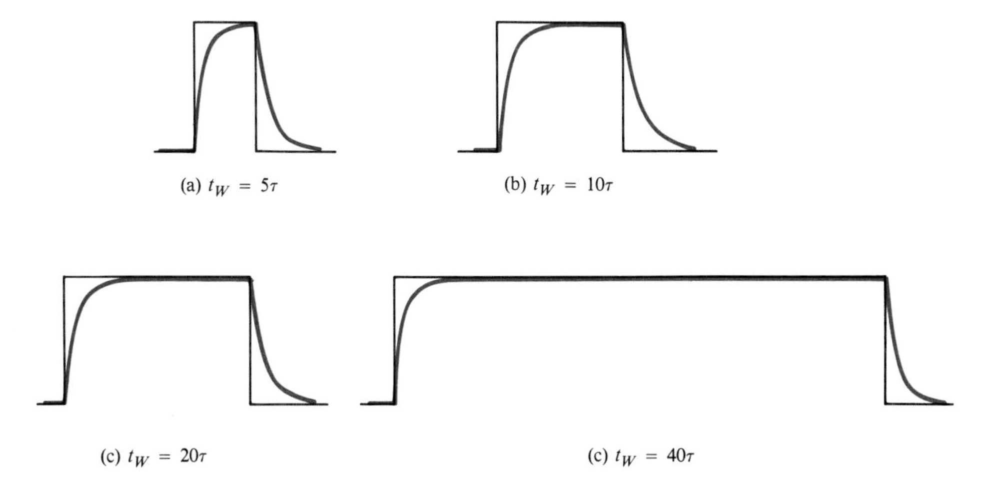

FIGURE 16–5
Variation of an integrator's output pulse shape with input pulse width (the time constant is fixed). Black is input and color is output.

As before, the capacitor charges for the duration of the pulse. However, because the pulse width is less than the time it takes the capacitor to fully charge (5τ), the output voltage will *not* reach the full input voltage before the end of the pulse. *The capacitor only partially charges,* as illustrated in Figure 16–6 for several values of *RC* time constants. Notice that for longer time constants, the output reaches a lower voltage because the capacitor cannot charge as much. Of course, in our examples with a single pulse input, the capacitor fully discharges after the pulse ends.

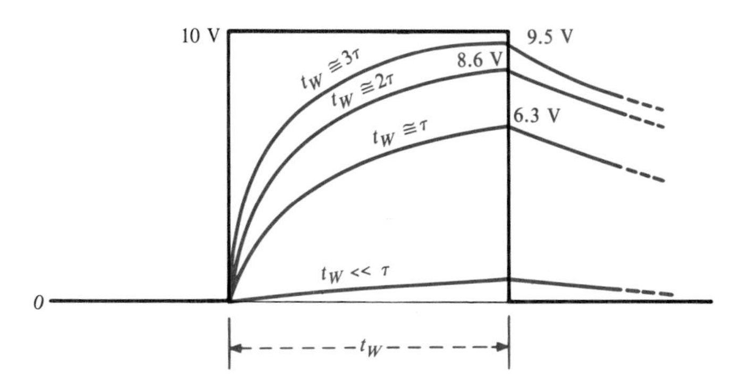

FIGURE 16–6
Capacitor voltage for various time constants which are longer than the input pulse width. Black is input and color is output.

When the time constant is much greater than the input pulse width, the capacitor charges very little, and, as a result, the output voltage becomes almost negligible, as indicated in Figure 16–6.

Figure 16–7 illustrates the effect of reducing the input pulse width for a *fixed* time constant value. As the width is reduced, the output voltage becomes smaller because the capacitor has less time to charge. However, it takes the capacitor the same length of time (5τ) to discharge back to zero for each condition after the pulse is removed.

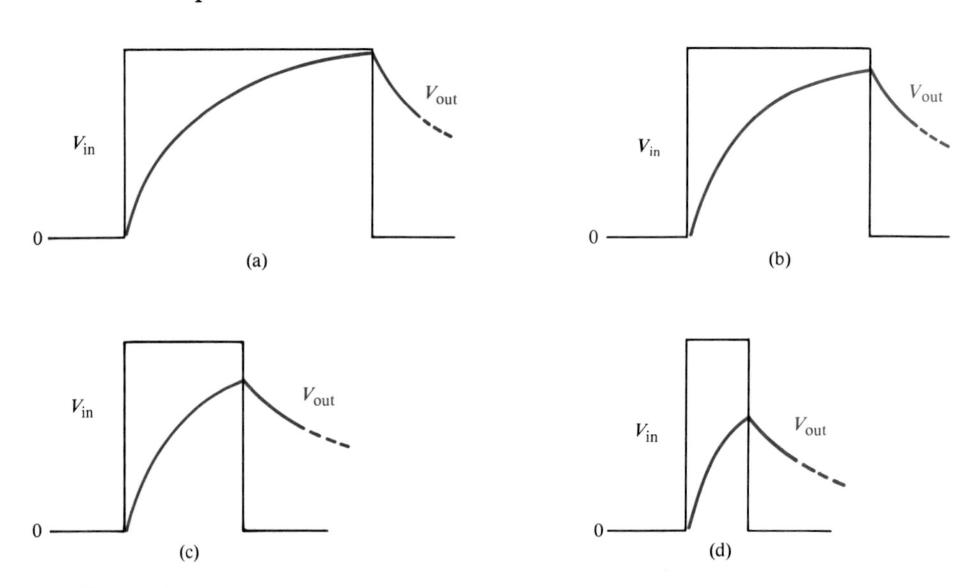

FIGURE 16–7
The capacitor charges less and less as the input pulse width is reduced. The time constant is fixed.

EXAMPLE 16–1

A single 10-V pulse with a width of 100 μs is applied to the integrator in Figure 16–8.

(a) To what voltage will the capacitor charge?
(b) How long will it take the capacitor to discharge if the internal resistance of the pulse source is 50 Ω?
(c) Sketch the output voltage.

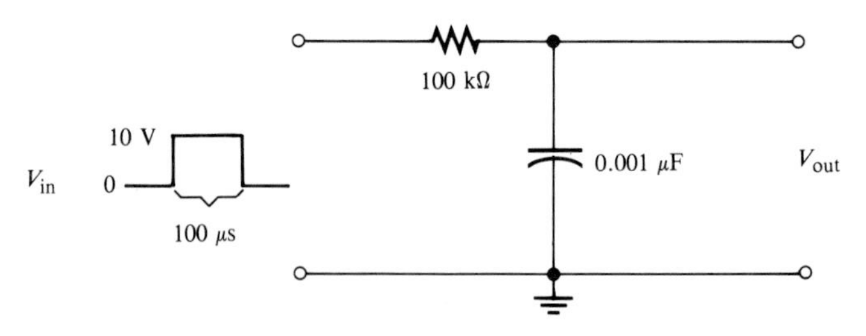

FIGURE 16–8

Solution

(a) The circuit time constant is

$$\tau = RC = (100\text{ k}\Omega)(0.001\ \mu\text{F}) = 100\ \mu\text{s}$$

Notice that the pulse width is exactly equal to the time constant. Thus, the capacitor will charge 63% of the full input amplitude in one time constant, so the output will reach a maximum voltage of 6.3 V.

(b) The capacitor discharges back through the source when the pulse ends. We can neglect the 50-Ω source resistance in series with 100 kΩ. The total discharge time therefore is

$$5\tau = 5(100\ \mu s) = 500\ \mu s$$

(c) The output charging and discharging curve is shown in Figure 16–9.

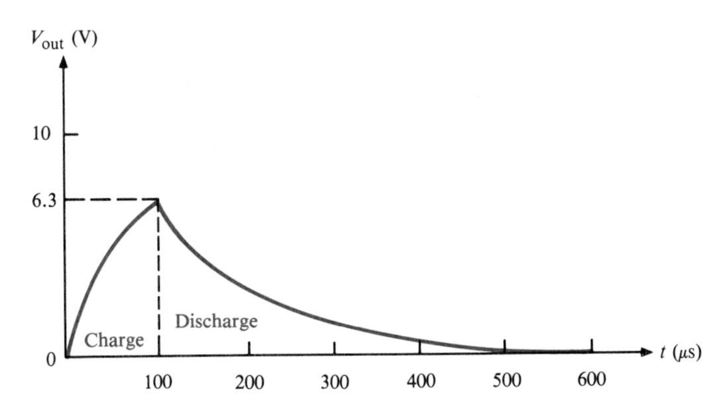

FIGURE 16–9

EXAMPLE 16–2

Determine how much the capacitor in Figure 16–10 will charge when the single pulse is applied to the input.

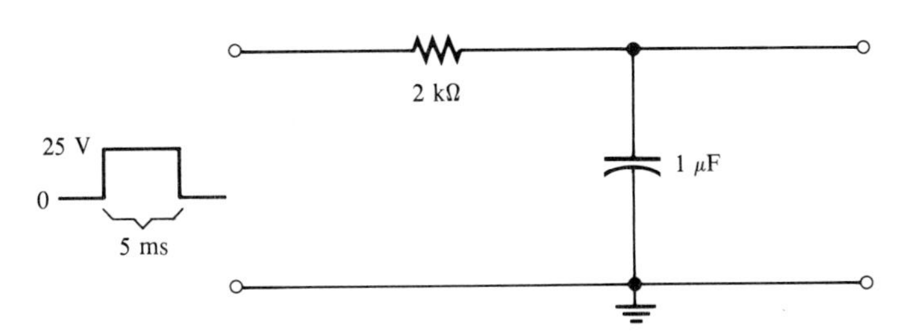

FIGURE 16–10

Solution
Calculate the time constant:

$$\tau = RC = (2\ k\Omega)(1\ \mu F) = 2\ ms$$

Because the pulse width is 5 ms, the capacitor charges for 2.5 time constants (5 ms = 2.5 × 2 ms). We use the exponential formula from Chapter 10 to find the voltage to which the capacitor will charge. The calculation is done as follows:

$$\begin{aligned} v &= V_F(1 - e^{-t/RC}) \quad \text{where } V_F = 25\text{ V} \text{ and } t = 5\text{ ms} \\ &= (25\text{ V})(1 - e^{-5/2}) = (25\text{ V})(1 - e^{-2.5}) \\ &= (25\text{ V})(1 - 0.082) = (25\text{ V})(0.918) \\ &= 22.9\text{ V} \end{aligned}$$

These calculations show that the capacitor charges to 22.9 V during the 5-ms duration of the input pulse. It will discharge back to zero when the pulse goes back to zero.

The calculator sequence for v is

1 − ((5 ÷ 2 = +/−) INV lnx) × 2 5 =

SECTION REVIEW 16–2

1. When an input pulse is applied to an RC integrator, what condition must exist in order for the output voltage to reach full amplitude?
2. For the circuit in Figure 16–11, which has a single input pulse, find the maximum output voltage and determine how long the capacitor will discharge.

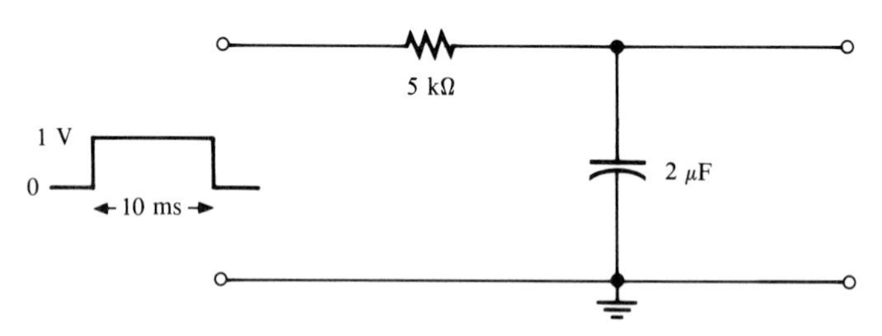

FIGURE 16–11

3. For Figure 16–11, sketch the approximate shape of the output voltage with respect to the input pulse.
4. If the integrator time constant equals the input pulse width, will the capacitor fully charge?
5. Describe the condition under which the output voltage has the approximate shape of a rectangular input pulse.

RESPONSE OF AN *RC* INTEGRATOR TO REPETITIVE PULSES

16–3

In the last section you learned how an RC integrator responds to a *single* pulse input. These basic ideas will be extended in this section to include the integrator response to *repetitive* pulses. In electronic systems, you will encounter repetitive pulse waveforms much more often than single pulses. However, an understanding of the RC integrator response to single pulses is essential to learning how these circuits act with repeated pulses.

If a periodic pulse waveform is applied to an RC integrator, as shown in Figure 16–12, *the output waveshape depends on the relationship of the circuit time constant and the frequency (period) of the input pulses.* The capacitor, of course, charges and discharges in response to a pulse input. *The amount of*

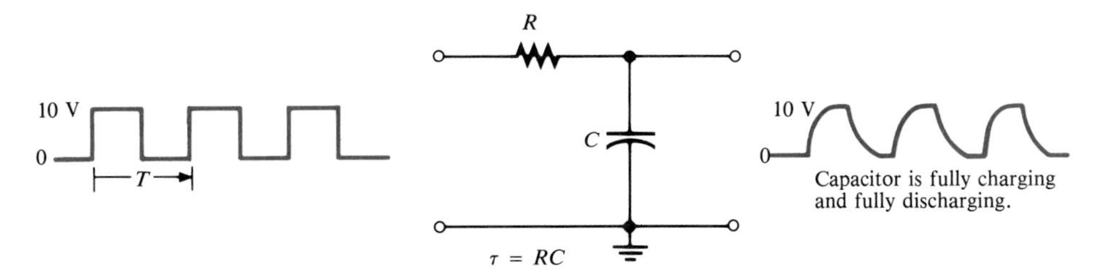

FIGURE 16–12
RC integrator with a repetitive pulse waveform input.

charge and discharge of the capacitor depends both on the circuit time constant and on the input frequency, as mentioned.

If the pulse width and the time between pulses are each equal to or greater than five time constants, the capacitor will fully charge and fully discharge during each period of the input waveform. This case is shown in Figure 16–12.

When the Capacitor Does Not Fully Charge and Discharge

When the pulse width and the time between pulses are *shorter than 5 time constants,* as illustrated in Figure 16–13 for a square wave, the capacitor will *not* completely charge or discharge. We will now examine the effects of this situation on the output voltage of the RC integrator.

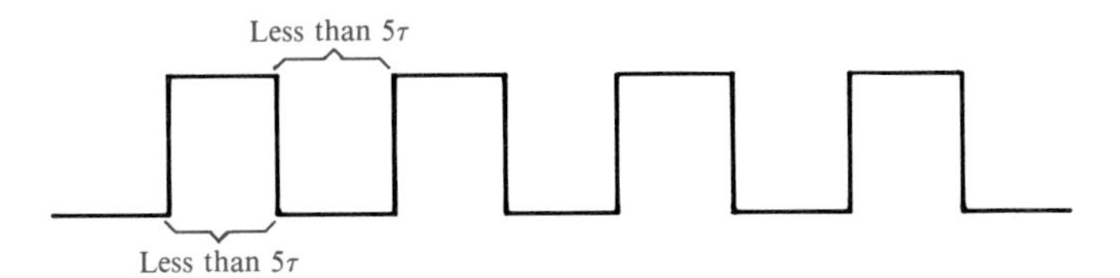

FIGURE 16–13
Waveform that does not allow full charge or discharge of integrator capacitor.

For illustration, let us take an example of an RC integrator with a charging and discharging *time constant equal to the pulse width of a 10-V square wave input,* as in Figure 16–14. This choice will simplify the analysis and will demonstrate the basic action of the integrator under these conditions. At this point, we really do not care what the exact time constant value is because we know from Chapter 10 that an RC circuit charges 63% during one time constant interval.

We will assume that the capacitor in Figure 16–14 begins initially uncharged and examine the output voltage on a pulse-by-pulse basis. The results of this analysis are shown in Figure 16–15.

First Pulse

During the first pulse, the capacitor charges. The output voltage reaches 6.3 V (63% of 10 V), as shown in Figure 16–15.

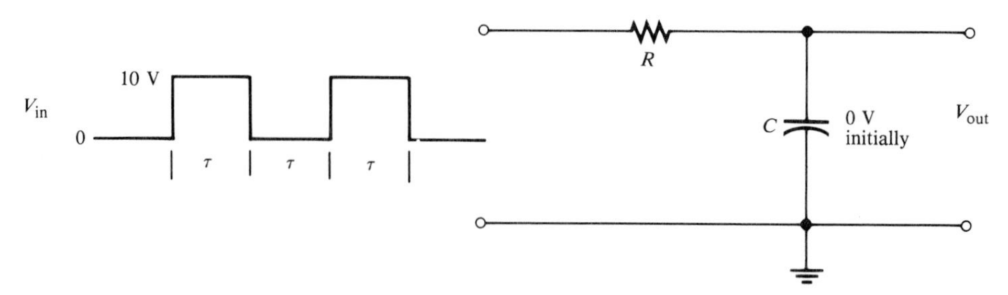

FIGURE 16–14
Integrator with a square wave input having a period equal to two time constants ($T = 2\tau$).

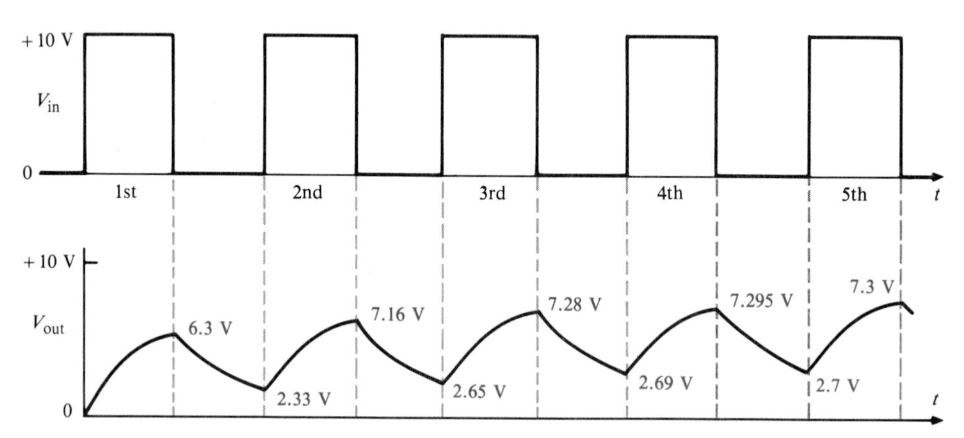

FIGURE 16–15
Input and output for the initially uncharged integrator in Figure 16–14.

Between First and Second Pulses

The capacitor discharges, and the voltage decreases to 37% of the voltage at the beginning of this interval: 0.37(6.3 V) = 2.33 V.

Second Pulse

The capacitor voltage begins at 2.33 V and increases 63% of the way to 10 V. This calculation is as follows: The total charging range is 10 V − 2.33 V = 7.67 V. The capacitor voltage will increase an additional 63% of 7.67 V, which is 4.83 V. Thus, at the end of the second pulse, the output voltage is 2.33 V + 4.83 V = 7.16 V, as indicated in Figure 16–15. Notice that the *average* is building up.

Between Second and Third Pulses

The capacitor discharges during this time, and therefore the voltage decreases to 37% of the voltage by the end of the second pulse: 0.37(7.16 V) = 2.65 V.

Third Pulse

At the start of the third pulse, the capacitor voltage begins at 2.65 V. The capacitor charges 63% of the way from 2.65 V to 10 V: 0.63(10 V − 2.65 V) = 4.63 V. Therefore, the voltage at the end of the third pulse is 2.65 V + 4.63 V = 7.28 V.

Between Third and Fourth Pulses

The voltage during this interval decreases due to capacitor discharge. It will decrease to 37% of its value by the end of the third pulse. The final voltage in this interval is 0.37(7.28 V) = 2.69 V.

Fourth Pulse

At the start of the fourth pulse, the capacitor voltage is 2.69 V. The voltage increases by 0.63(10 V − 2.69 V) = 4.605 V. Therefore, at the end of the fourth pulse, the capacitor voltage is 2.69 V + 4.61 V = 7.295 V. Notice that the values are leveling off as the pulses continue.

Between Fourth and Fifth Pulses

Between these pulses, the capacitor voltage drops to 0.37(7.295 V) = 2.7 V.

Fifth Pulse

During the fifth pulse, the capacitor charges 0.63(10 V − 2.7 V) = 4.6 V. Since it started at 2.7 V, the voltage at the end of the pulse is 2.7 V + 4.6 V = 7.3 V.

Steady State Response

In the preceding discussion, the output voltage gradually built up and then began leveling off. It takes approximately 5τ for the output voltage to build up to a constant *average* value. This interval is the *transient time* of the circuit. Once the output voltage reaches *the average value of the input voltage, a steady state condition* is reached which continues as long as the periodic input continues. This state is illustrated in Figure 16–16 based on the values obtained in the preceding discussion.

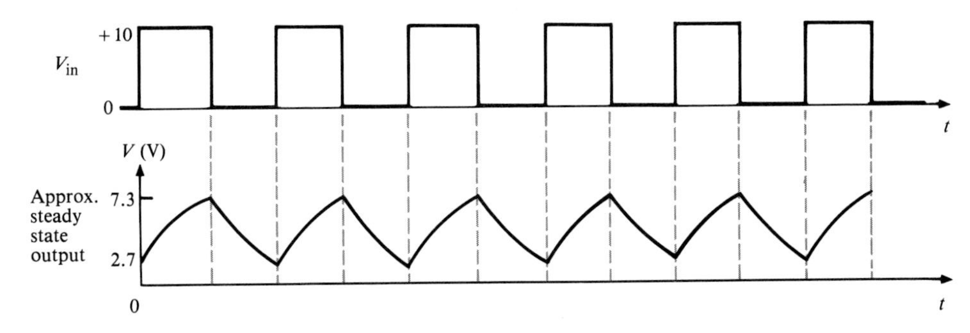

FIGURE 16–16
Output reaches steady state after 5τ.

The transient time for our example circuit is the time from the beginning of the first pulse to the end of the third pulse. The reason for this interval is that the capacitor voltage at the end of the third pulse is 7.28 V, which is about 99% of the final voltage.

The Effect of an Increase in Time Constant

What happens to the output voltage if the *RC* time constant of the integrator is increased with a variable resistor, as indicated in Figure 16–17? As the time constant is increased, the capacitor charges *less* during a pulse and discharges *less* between pulses. The result is a *smaller* fluctuation in the output voltage for increasing values of time constant, as shown in Figure 16–18.

FIGURE 16–17
Integrator with a variable time constant.

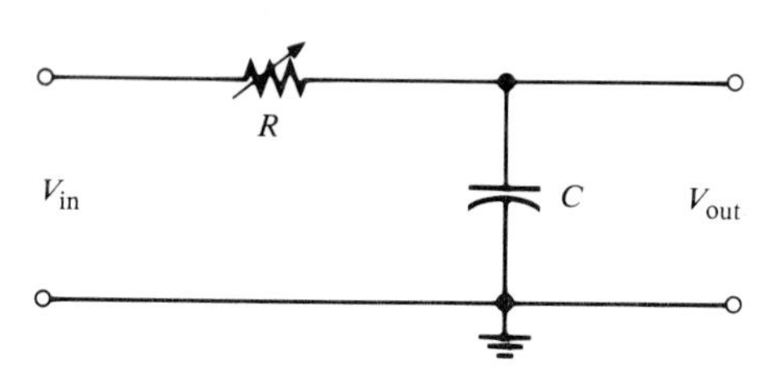

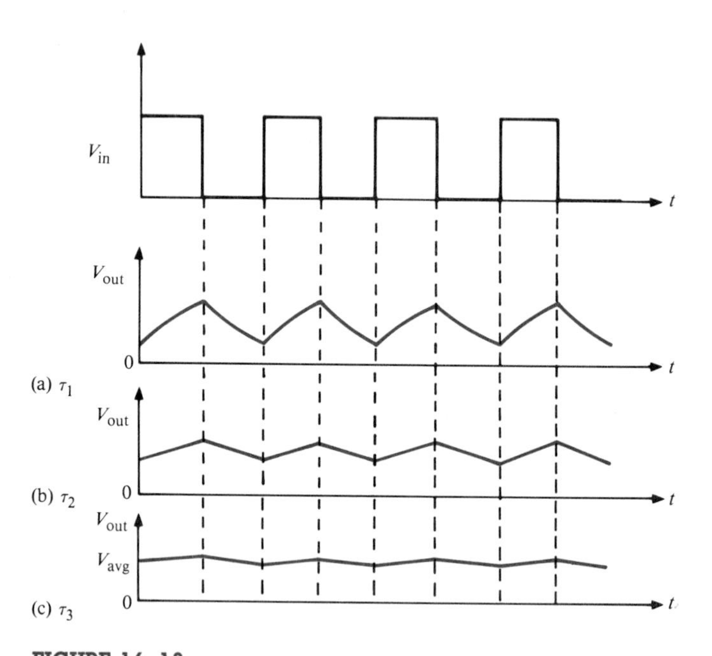

FIGURE 16–18
Effect of longer time constants on the output of an integrator ($\tau_3 > \tau_2 > \tau_1$).

As the time constant becomes extremely long compared to the pulse width, the output voltage approaches a constant dc voltage, as shown in Figure

16–18(c). This value is the *average value* of the input. For a square wave, it is one-half the amplitude.

EXAMPLE 16–3 Determine the output voltage waveform for the first *two* pulses applied to the integrator circuit in Figure 16–19. Assume that the capacitor is initially uncharged.

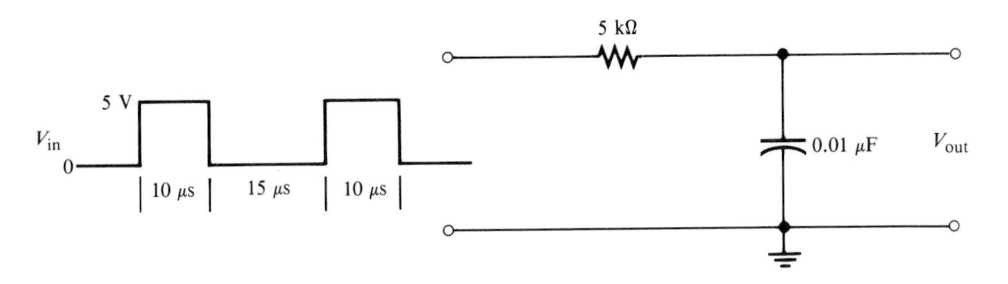

FIGURE 16–19

Solution

First calculate the circuit time constant:

$$\tau = RC = (5\ \text{k}\Omega)(0.01\ \mu\text{F}) = 50\ \mu\text{s}$$

Obviously, the time constant is much longer than the input pulse width or the interval between pulses (notice that the input is not a square wave). Thus, in this case, the exponential formulas must be applied, and the analysis is relatively difficult. Follow the solution carefully.

Step 1: *calculation for first pulse:* Use Equation (10–15) because C is charging. Note that V_F is 5 V, and t equals the pulse width of 10 μs. Therefore,

$$\begin{aligned} v_C &= V_F(1 - e^{-t/RC}) = (5\ \text{V})(1 - e^{-10\ \mu\text{s}/50\ \mu\text{s}}) \\ &= (5\ \text{V})(1 - 0.819) = 0.906\ \text{V} \end{aligned}$$

This result is plotted in Figure 16–20(a).

Step 2: *calculation for interval between first and second pulse:* Use Equation (10–16) because C is discharging. Note that V_i is 0.906 V because C begins to discharge from this value at the end of the first pulse. The discharge time is 15 μs. Therefore,

$$\begin{aligned} v_C &= V_i e^{-t/RC} = 0.906e^{-15\ \mu\text{s}/50\ \mu\text{s}} \\ &= 0.906(0.741) = 0.671\ \text{V} \end{aligned}$$

This result is shown in Figure 16–20(b).

Step 3: *calculation for second pulse:* At the beginning of the second pulse, the output voltage is 0.671 V. During the second pulse, the capacitor will again charge. In this case it does not begin at zero volts. It already has 0.671 V from

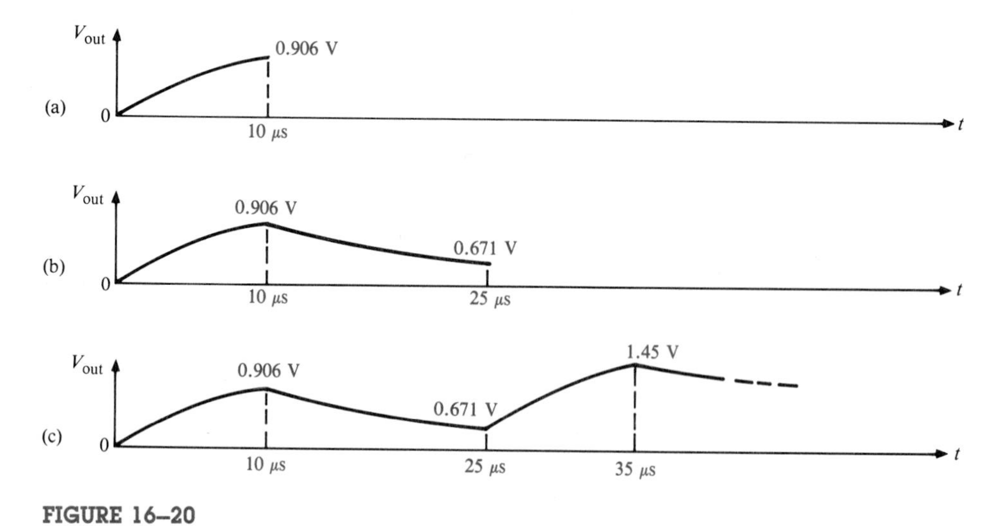

FIGURE 16–20

the previous charge and discharge. To handle this situation, we must use the general formula of Equation (10–13):

$$v = V_F + (V_i - V_F)e^{-t/\tau}$$

Using this equation, we can calculate the voltage across the capacitor at the end of the second pulse as follows:

$$\begin{aligned} v_C &= V_F + (V_i - V_F)e^{-t/RC} \\ &= 5\text{ V} + (0.671\text{ V} - 5\text{ V})e^{-10\ \mu\text{s}/50\ \mu\text{s}} \\ &= 5\text{ V} + (-4.33\text{ V})(0.819) = 5\text{ V} - 3.55\text{ V} \\ &= 1.45\text{ V} \end{aligned}$$

This result is shown in Figure 16–20(c).

Notice that the output waveform builds up on successive input pulses. After approximately 5τ, it will reach its steady state and will fluctuate between a constant maximum and a constant minimum, with an average equal to the average value of the input. We could demonstrate this pattern by carrying the analysis in this example further.

SECTION REVIEW 16–3

1. What conditions allow an *RC* integrator capacitor to fully charge and discharge when a periodic pulse waveform is applied to the input?
2. What will the output waveform look like if the circuit time constant is extremely small compared to the pulse width of a square wave input?
3. When 5τ is greater than the pulse width of an input square wave, the time required for the output voltage to build up to a constant average value is called ________.
4. Define *steady state response.*
5. What does the average value of the output voltage of an integrator equal during steady state?

RESPONSE OF AN *RC* DIFFERENTIATOR TO A SINGLE PULSE

16–4

Figure 16–21 shows a series *RC* circuit. Notice that the output is taken across the *resistor,* thereby distinguishing the *RC* differentiator from the *RC* integrator. This circuit is called a *differentiator* in terms of its pulse response. Recall that in terms of its frequency response, it is a basic *high-pass* filter.

FIGURE 16–21
An *RC* differentiating circuit.

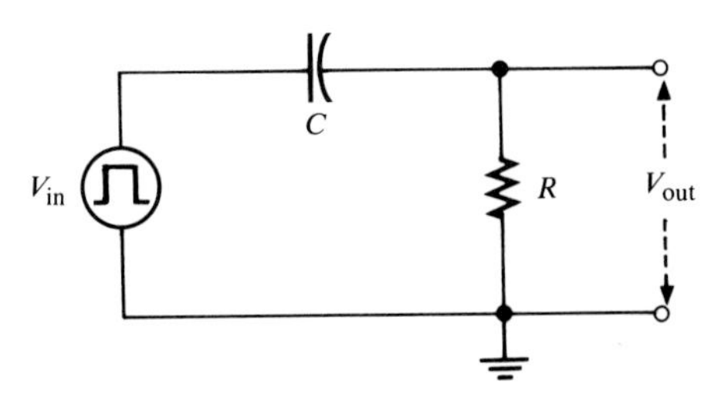

The same action occurs in the differentiator as in the integrator, except the output voltage is taken across the resistor rather than the capacitor. The capacitor charges exponentially at a rate depending on the *RC* time constant. The shape of the differentiator's resistor voltage is determined by the charging and discharging action of the capacitor.

To understand how the output voltage is shaped by a differentiator, we must consider the following:

1. the response to the rising pulse edge,
2. the response between the rising and falling edges, and
3. the response to the falling pulse edge.

Response to the Rising Edge of the Input Pulse

Assume that the capacitor is initially uncharged prior to the rising pulse edge. Prior to the pulse, the input is zero volts. Thus, there are zero volts across the capacitor and also zero volts across the resistor, as indicated in Figure 16–22(a).

Now assume that a 10-V pulse is applied to the input. When the rising edge occurs, point *A* goes to +10 V. Recall that *the voltage across a capacitor cannot change instantaneously,* and thus the capacitor appears instantaneously as a short. Therefore, if point *A* goes instantly to +10 V, then point *B must* also go instantly to +10 V, keeping the capacitor voltage zero for the instant of the rising edge. The capacitor voltage is the voltage from point *A* to point *B*.

The voltage at point *B* with respect to ground is the voltage across the resistor (and the output voltage). Thus, the output voltage suddenly goes to +10 V in response to the rising pulse edge, as indicated in Figure 16–22(b).

Response during Pulse When $t_W \geq 5\tau$

While the pulse is at its high level between the rising edge and the falling edge, the capacitor is charging. When the pulse width is equal to or greater than five time constants ($t_W \geq 5\tau$), the capacitor has time to *fully charge.*

As the voltage across the capacitor builds up exponentially, the voltage across the resistor *decreases* exponentially until it reaches zero volts at the time

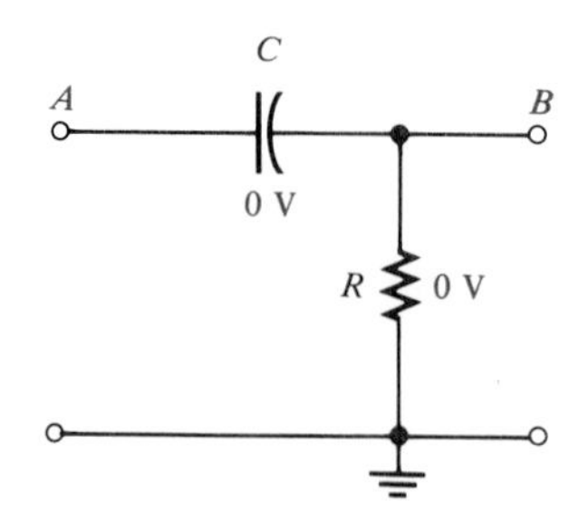

(a) Before pulse is applied

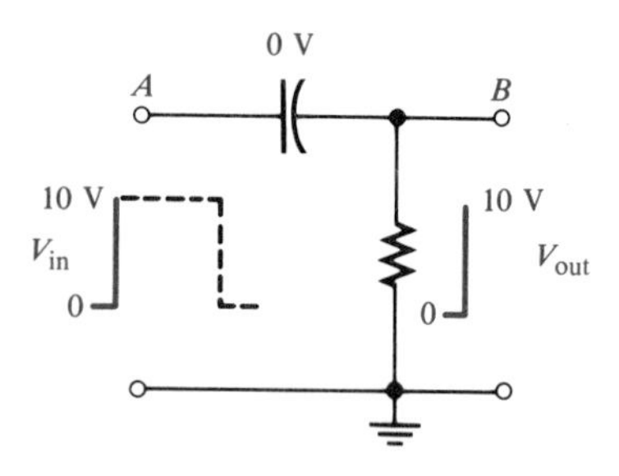

(b) At rising edge of input pulse

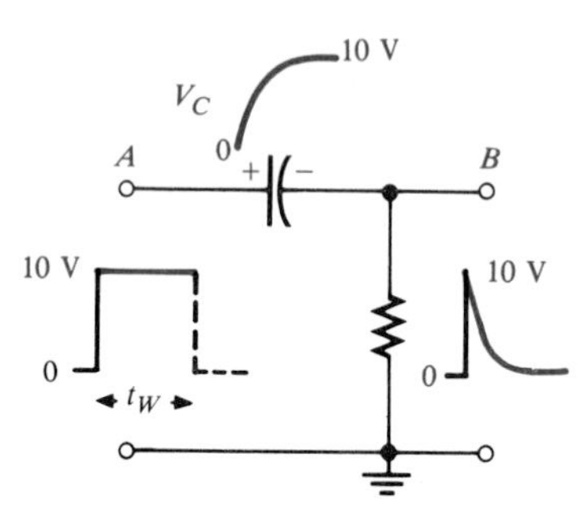

(c) During level part of pulse when $t_W \geqslant 5\tau$

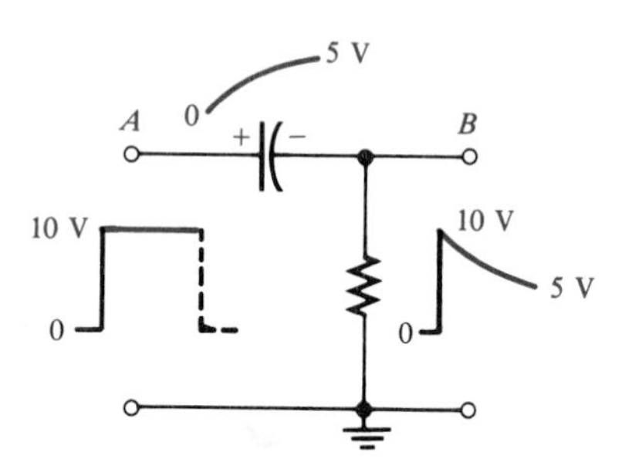

(d) During level part of pulse when $t_W < 5\tau$

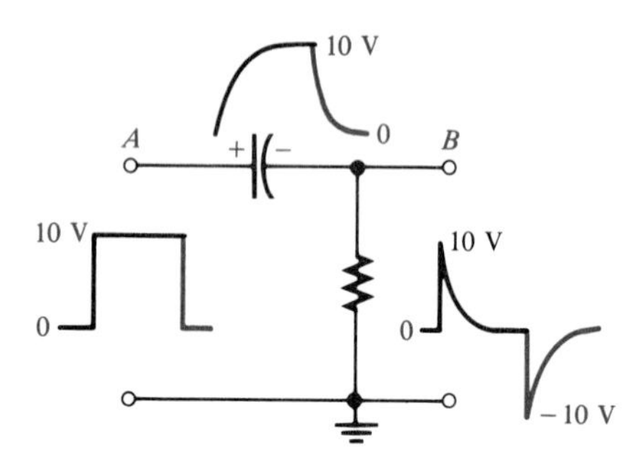

(e) At falling edge of pulse when $t_W \geqslant 5\tau$

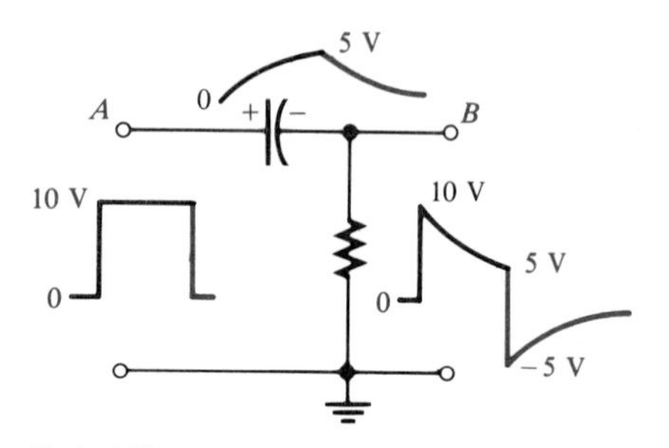

(f) At falling edge of pulse when $t_W < 5\tau$

FIGURE 16–22
Examples of the response of a differentiator to a single input pulse under two conditions: $t_W \geq 5\tau$ and $t_W < 5\tau$.

the capacitor reaches full charge (+10 V in this case). This decrease in the resistor voltage occurs because the sum of the capacitor voltage and the resistor voltage at any instant must be equal to the applied voltage, in compliance with Kirchhoff's voltage law ($v_C + v_R = v_{in}$). This part of the response is illustrated in Figure 16–22(c).

Response during Pulse When $t_W < 5\tau$

When the pulse width is less than five time constants ($t_W < 5\tau$), *the capacitor does not have time to fully charge.* Its partial charge depends on the relation of the time constant and the pulse width.

Because the capacitor does not reach the full +10 V, *the resistor voltage will not reach zero volts* by the end of the pulse. For example, if the capacitor charges to +5 V during the pulse interval, the resistor voltage will decrease to +5 V, as illustrated in Figure 16–22(d).

Response to Falling Edge When $t_W \geq 5\tau$

Let us first examine the case in which the capacitor is *fully charged* at the end of the pulse ($t_W \geq 5\tau$). Refer to Figure 16–22(e). On the falling edge, the input pulse suddenly goes from +10 V back to zero. An instant before the falling edge, the capacitor is charged to 10 V, so point *A* is +10 V and point *B* is 0 V. Since the voltage across a capacitor cannot change instantaneously, when point *A* makes a transition from +10 V to zero on the falling edge, point *B must* also make a 10-V transition from zero to −10 V. This keeps the voltage across the capacitor at 10 V for the instant of the falling edge.

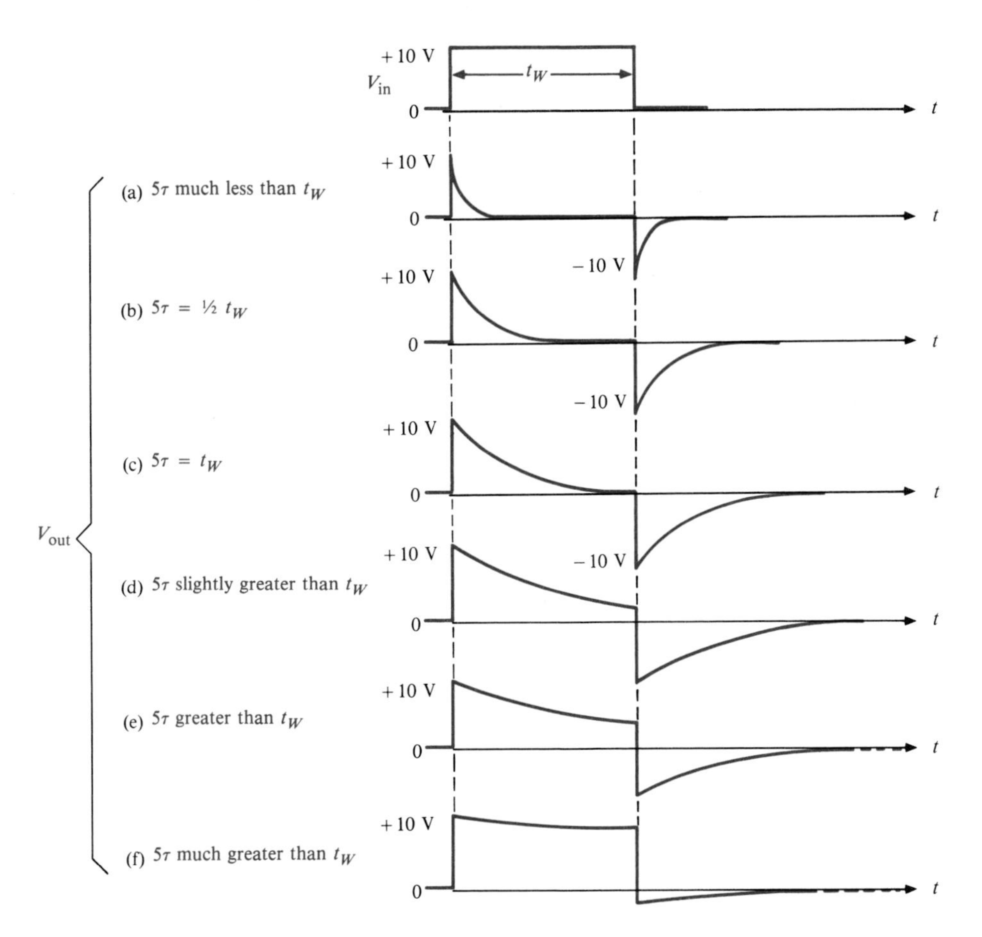

FIGURE 16–23
Effects of a change in time constant on the shape of the output voltage of a differentiator.

The capacitor now begins to *discharge* exponentially. As a result, the resistor voltage goes from -10 V to zero in an exponential curve, as indicated in Figure 16–22(e).

Response to Falling Edge When $t_W < 5\tau$

Next, let us examine the case in which the capacitor is only partially charged at the end of the pulse ($t_W < 5\tau$). For example, if the capacitor charges to $+5$ V, the resistor voltage at the instant before the falling edge is also $+5$ V, because the capacitor voltage plus the resistor voltage must add up to $+10$ V, as illustrated in Figure 16–22(d).

When the falling edge occurs, point A goes from $+10$ V to zero. As a result, point B goes from $+5$ V to -5 V, as illustrated in Figure 16–22(f). This decrease occurs, of course, because the capacitor voltage cannot change at the instant of the falling edge. Immediately after the falling edge, the capacitor begins to discharge to zero. As a result, the resistor voltage goes from -5 V to zero, as shown.

Summary of Differentiator Response to a Single Pulse

Perhaps a good way to summarize this section is to look at the general output waveforms of a differentiator as the time constant is varied from one extreme, when 5τ is much less than the pulse width, to the other extreme, when 5τ is much greater than the pulse width. These situations are illustrated in Figure 16–23. In Part (a) of the figure, the output consists of narrow positive and negative "spikes." In Part (f), the output approaches the shape of the input. Various conditions between these extremes are illustrated in Parts (b) through (e).

EXAMPLE 16–4 Sketch the output voltage for the circuit in Figure 16–24.

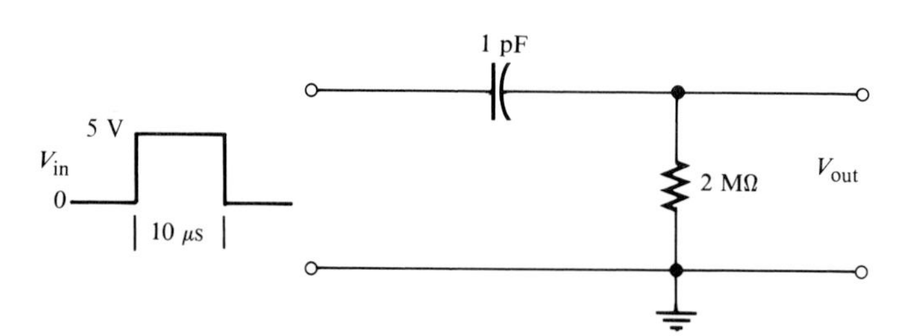

FIGURE 16–24

Solution

First calculate the time constant:

$$\tau = RC = (2\ \text{M}\Omega)(1\ \text{pF}) = 2\ \mu\text{s}$$

In this case, $t_W = 5\tau$, so the capacitor reaches full charge at the end of the pulse.

On the rising edge, the resistor voltage jumps to $+5$ V and then decreases exponentially to zero by the end of the pulse. On the falling edge, the resistor voltage jumps to −5 V and then goes back to zero exponentially. The resistor voltage is, of course, the output, and its shape is shown in Figure 16–25.

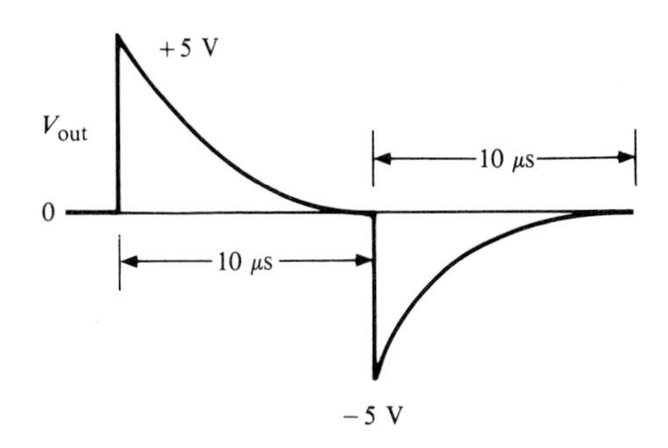

FIGURE 16–25

EXAMPLE 16–5

Determine the output voltage waveform for the differentiator in Figure 16–26.

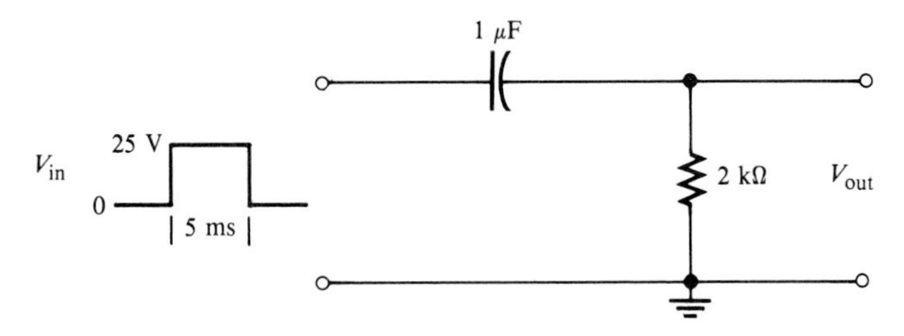

FIGURE 16–26

Solution

First calculate the time constant:

$$\tau = (2\ \text{k}\Omega)(1\ \mu\text{F}) = 2\ \text{ms}$$

On the rising edge, the resistor voltage immediately jumps to +25 V. Because the pulse width is 5 ms, the capacitor charges for 2.5 time constants and therefore does not reach full charge. Thus, we must use the formula for a decreasing exponential, Equation (10–16), in order to calculate to what voltage the output decreases by the end of the pulse:

$$\begin{aligned} v_{out} &= V_i e^{-t/RC} \quad \text{where } V_i = 25\text{ V} \quad \text{and} \quad t = 5\text{ ms} \\ &= 25e^{-5\text{ ms}/2\text{ ms}} = 25(0.082) \\ &= 2.05\text{ V} \end{aligned}$$

This calculation gives us the resistor voltage at the end of the 5-ms pulse width interval.

On the falling edge, the resistor voltage immediately jumps from +2.05 V down to −22.95 V (a 25-V transition). The resulting waveform of the output voltage is shown in Figure 16–27.

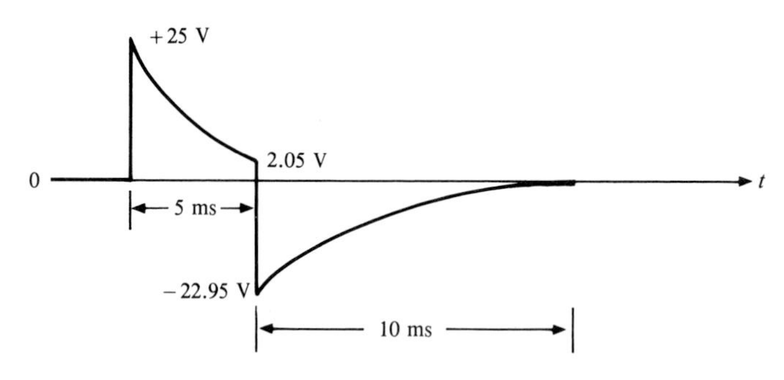

FIGURE 16–27

SECTION REVIEW 16–4

1. Sketch the output of a differentiator for a 10-V input pulse when $5\tau = \frac{1}{2}t_W$.
2. Under what condition does the output pulse shape most closely resemble the input pulse for a differentiator?
3. What does the differentiator output look like when 5τ is much less than the pulse width of the input?
4. If the resistor voltage in a differentiating circuit is down to +5 V at the end of a 15-V input pulse, to what negative value will the resistor voltage go in response to the *falling* edge of the input?

RESPONSE OF AN *RC* DIFFERENTIATOR TO REPETITIVE PULSES

16–5

The differentiator response to a single pulse, covered in the preceding section, is extended in this section to repetitive pulses.

If a periodic pulse waveform is applied to an *RC* differentiating circuit, two conditions again are possible: $t_W \geq 5\tau$ or $t_W < 5\tau$. Figure 16–28 shows the output when $t_W = 5\tau$. As the time constant is reduced, both the positive and the negative portions of the output become narrower. Notice that the *average* value of the output is *zero*.

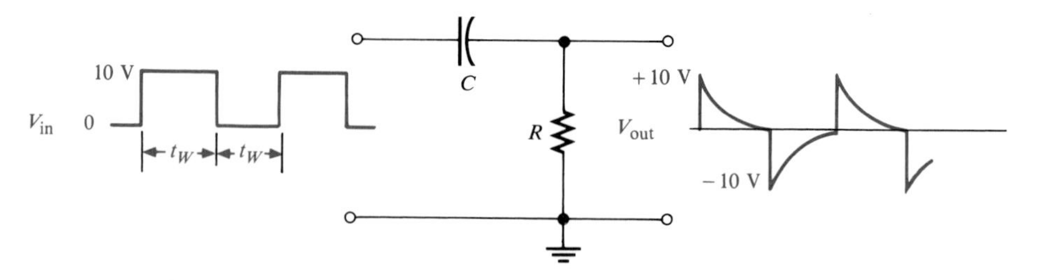

FIGURE 16–28
Example of differentiator response when $t_W = 5\tau$.

Figure 16–29 shows the *steady state* output when $t_W < 5\tau$. As the time constant is increased, the positively and negatively sloping portions become flat-

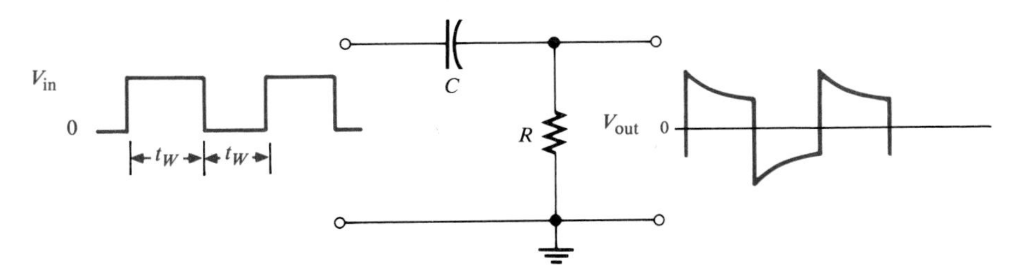

FIGURE 16–29
Example of differentiator response when $t_W < 5\tau$.

ter. For a very long time constant, the output approaches the shape of the input, but with an average value of zero. An average value of zero means that the waveform has equal positive and negative portions. The average value of a waveform is its *dc* component. Because a capacitor blocks dc, the dc component of the input is prevented from passing through to the output.

Like the integrator, the differentiator output takes time (5τ) to reach steady state. To illustrate the response, let us take an example in which the time constant *equals* the input pulse width.

At this point, we do not care what the circuit time constant is, because we know that the resistor voltage will decrease to 37% of its maximum value during one pulse (1τ). We will assume that the capacitor in Figure 16–30 begins initially uncharged, and then we will examine the output voltage on a pulse-by-pulse basis. The results of the analysis to follow are shown in Figure 16–31.

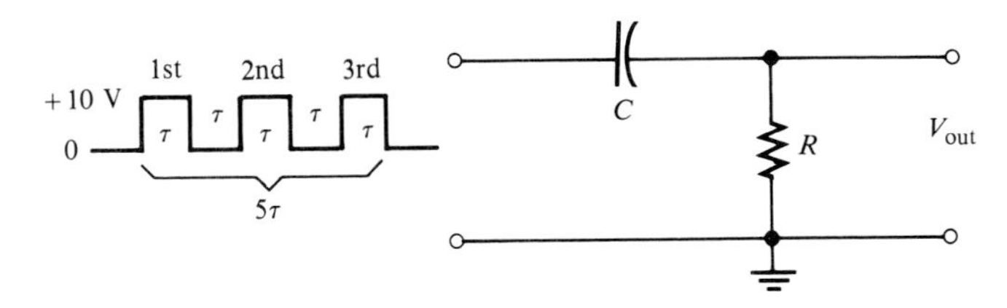

FIGURE 16–30
RC differentiator with $\tau = t_W$.

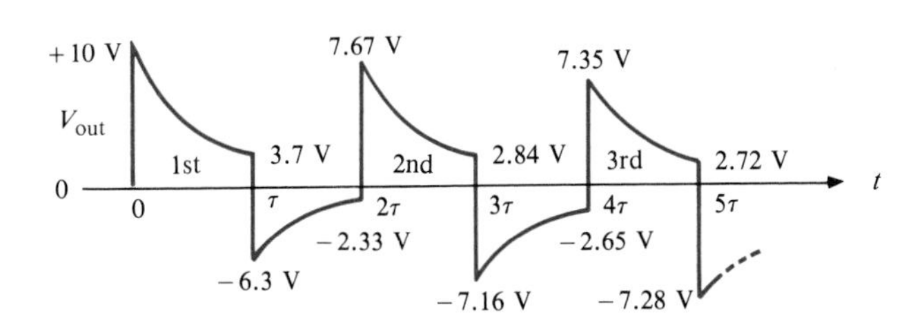

FIGURE 16–31
Differentiator output waveform during transient time for the circuit in Figure 16–30.

First Pulse

On the rising edge, the output instantaneously jumps to +10 V. Then the capacitor partially charges to 63% of 10 V, which is 6.3 V. Thus, the output voltage must decrease to 3.7 V, as shown in Figure 16–31.

On the falling edge, the output instantaneously makes a negative-going 10-V transition to −6.3 V (−10 V + 3.7 V = −6.3 V).

Between First and Second Pulses

The capacitor discharges to 37% of 6.3 V, which is 2.33 V. Thus, the resistor voltage, which starts at −6.3 V, must increase to −2.33 V. Why? Because at *the instant prior to* the next pulse, the input voltage is zero. Therefore, the sum of v_C and v_R must be zero (+2.33 V − 2.33 V = 0). Remember that $v_C + v_R = v_{in}$ at all times.

Second Pulse

On the rising edge, the output makes an instantaneous, positive-going, 10-V transition from −2.33 V to 7.67 V. Then the capacitor charges 0.63 × (10 V − 2.33 V) = 4.83 V by the end of the pulse. Thus, the capacitor voltage increases from 2.33 V to 2.33 V + 4.83 V = 7.16 V. The output voltage drops to 0.37 × (7.67 V) = 2.84 V.

On the falling edge, the output instantaneously makes a negative-going transition from 2.84 V to −7.16 V, as shown in Figure 16–31.

Between Second and Third Pulses

The capacitor discharges to 37% of 7.16 V, which is 2.65 V. Thus, the output voltage starts at −7.16 V and increases to −2.65 V, because the capacitor voltage and the resistor voltage must add up to zero at the instant prior to the third pulse (the input is zero).

Third Pulse

On the rising edge, the output makes an instantaneous 10-V transition from −2.65 V to +7.35 V. Then the capacitor charges 0.63 × (10 V − 2.65 V) = 4.63 V to 2.65 V + 4.63 V = +7.28 V. As a result, the output voltage drops to 0.37 × 7.35 V = 2.72 V.

On the falling edge, the output instantly goes from +2.72 V down to −7.28 V.

After the third pulse, five time constants have elapsed, and the output voltage is close to its steady state. Thus, it will continue to vary from a positive maximum of about +7.3 V to a negative maximum of about −7.3 V, with an average value of zero.

SECTION REVIEW 16–5

1. What conditions allow an *RC* differentiator to fully charge and discharge when a periodic pulse waveform is applied to the input?
2. What will the output waveform look like if the circuit time constant is extremely

small compared to the pulse width of a square wave input?

3. What does the average value of the differentiator output voltage equal during steady state?

THE *RL* INTEGRATOR

16–6

Figure 16–32 shows a series *RL* circuit with *the output taken across the resistor.* This circuit is also known as an *integrator.* Its output waveform under equivalent conditions is the same as that for the *RC* integrator. Recall that in the *RC* case, the output was across the capacitor.

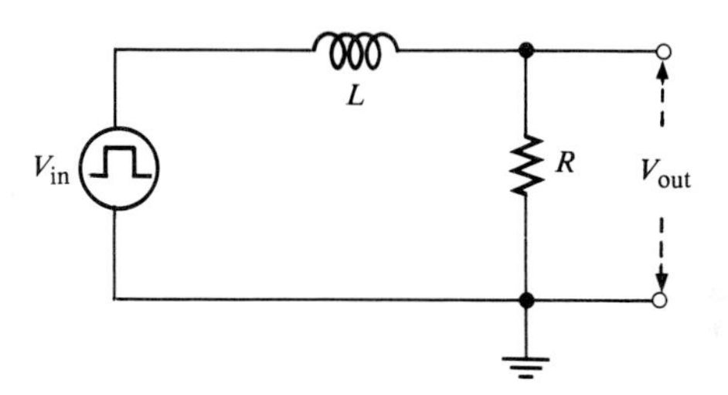

FIGURE 16–32
An *RL* integrating circuit.

As you know, both edges of an ideal pulse are considered to occur instantaneously. Two basic rules for inductor behavior will aid in analyzing *RL* circuit responses to pulse inputs:

1. The inductor appears as an open to an instantaneous change in current and as a short (ideally) to dc.
2. The current in an inductor cannot change instantaneously—it can change only exponentially.

Response of the Integrator to a Single Pulse

When a pulse generator is connected to the input of the integrator and the voltage pulse goes from its low level to its high level, the inductor prevents a sudden change in current. As a result, the inductor acts as an open, and all of the input voltage is across it at the instant of the rising pulse edge. This situation is indicated in Figure 16–33(a).

After the rising edge, the current builds up, and the output voltage follows the current as it increases exponentially, as shown in Figure 16–33(b). The current can reach a maximum of V_p/R if the transient time is shorter than the pulse width ($V_p = 10$ V in this example).

When the pulse goes from its high level to its low level, an induced voltage with reversed polarity is created across the coil in an effort to keep the current equal to V_p/R. The output voltage begins to decrease exponentially, as shown in Figure 16–33(c).

The exact shape of the output depends on the L/R time constant as summarized in Figure 16–34 for various relationships between the time constant and the pulse width. You should note that the response of this *RL* circuit *in*

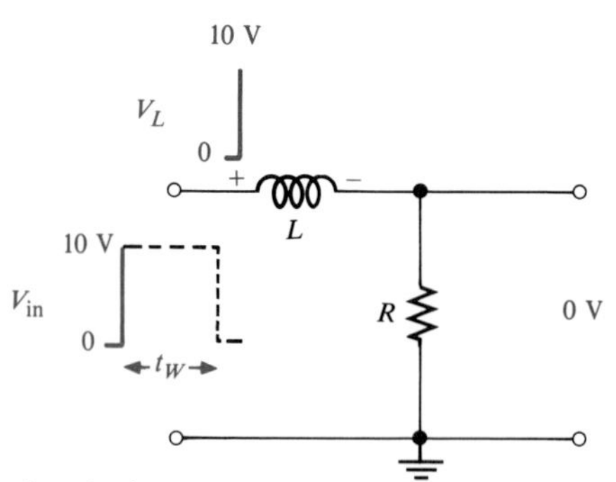

(a) At rising edge of pulse ($i = 0$)

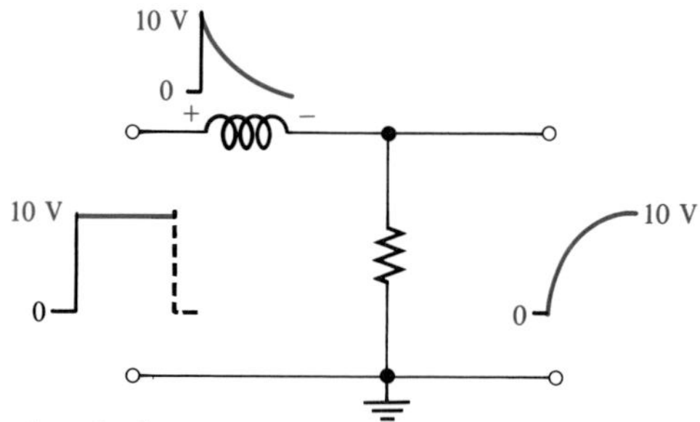

(b) During flat portion of pulse

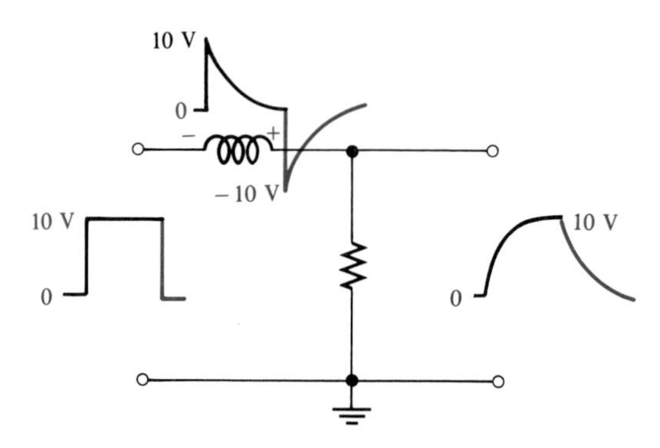

(c) At falling edge of pulse and after

FIGURE 16–33
Illustration of the pulse response of an *RL* integrator ($t_W > 5\tau$).

terms of the shape of the output is identical to that of the *RC* integrator. The relationship of the *L/R* time constant to the input pulse width has the same effect as the *RC* time constant that we discussed earlier in this chapter. For example, when $t_W < 5\tau$, the output voltage will not reach its maximum possible value.

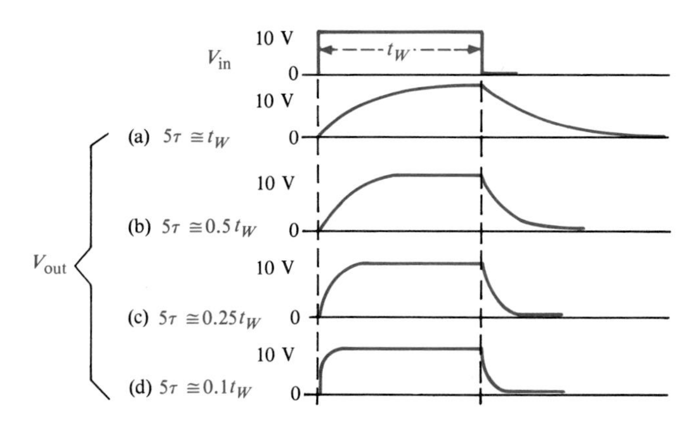

FIGURE 16–34
Illustration of the variation in integrator output pulse shape with time constant.

EXAMPLE 16–6 Determine the maximum output voltage for the integrator in Figure 16–35 when a single pulse is applied as shown.

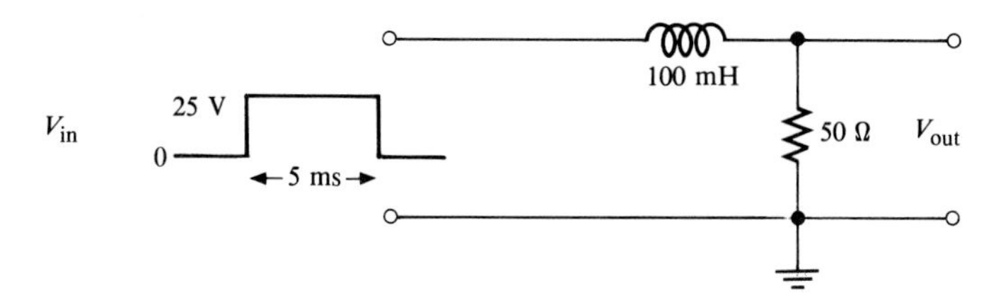

FIGURE 16–35

Solution
Calculate the time constant:

$$\tau = \frac{L}{R} = \frac{100 \text{ mH}}{50 \ \Omega} = 2 \text{ ms}$$

Because the pulse width is 5 ms, the inductor charges for 2.5τ. We must use the exponential formula to calculate the voltage as follows:

$$\begin{aligned} v_{out} &= V_F(1 - e^{-t/\tau}) = 25(1 - e^{-5/2}) \\ &= 25(1 - e^{-2.5}) = 25(1 - 0.082) \\ &= 25(0.918) = 22.95 \text{ V} \end{aligned}$$

EXAMPLE 16–7 A pulse is applied to the *RL* integrator circuit in Figure 16–36. Determine the complete waveshapes and the values for I, V_R, and V_L.

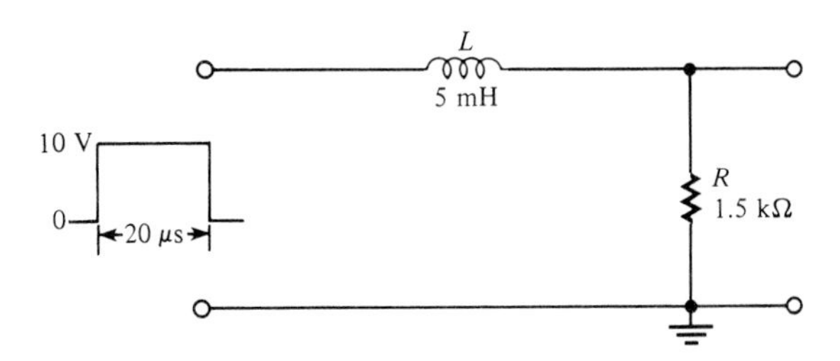

FIGURE 16–36

Solution
The circuit time constant is

$$\tau = \frac{L}{R} = \frac{5 \text{ mH}}{1.5 \text{ k}\Omega} = 3.33 \ \mu\text{s}$$

Since 5τ (16.65 μs) is less than t_W, the current will reach its maximum value and remain there until the end of the pulse.

At the rising edge of the pulse:

$$i = 0 \text{ A}$$
$$v_R = 0 \text{ V}$$
$$v_L = 10 \text{ V}$$

Since the inductor is initially an open, all of the input voltage appears across L.

During the pulse:

i increases exponentially to $V_p/R = 10 \text{ V}/1.5 \text{ k}\Omega = 6.67 \text{ mA}$ in 16.65 μs
v_R increases exponentially to 10 V in 16.65 μs.
v_L decreases exponentially to zero in 16.65 μs.

At the falling edge of the pulse:

$$i = 6.67 \text{ mA}$$
$$v_R = 10 \text{ V}$$
$$v_L = -10 \text{ V}$$

After the pulse:

i decreases exponentially to zero in 16.65 μs.
v_R decreases exponentially to zero in 16.65 μs.
v_L decreases exponentially to zero in 16.65 μs.

The waveforms are shown in Figure 16–37.

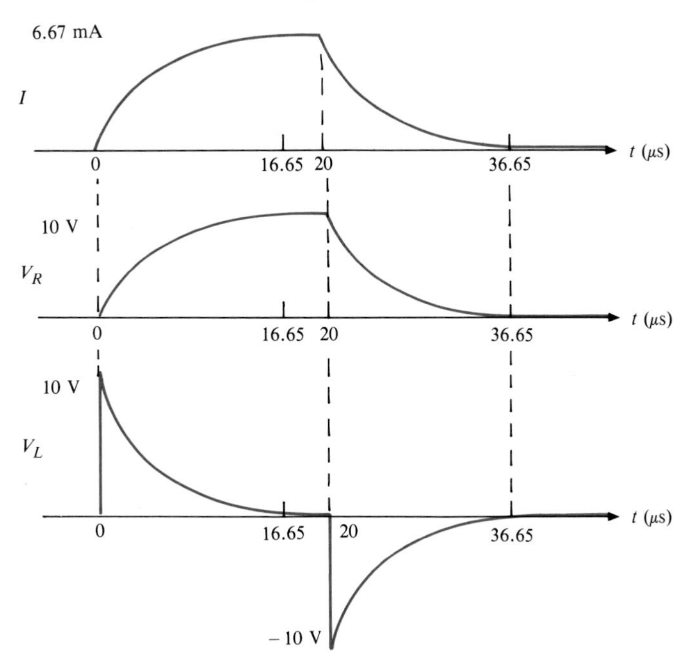

FIGURE 16–37

EXAMPLE 16–8

A 10-V pulse with a width of 1 ms is applied to the integrator in Figure 16–38. Determine the voltage level that the output will reach during the pulse. If the source has an internal resistance of 50 Ω, how long will it take the output to decay to zero? Sketch the output voltage waveform.

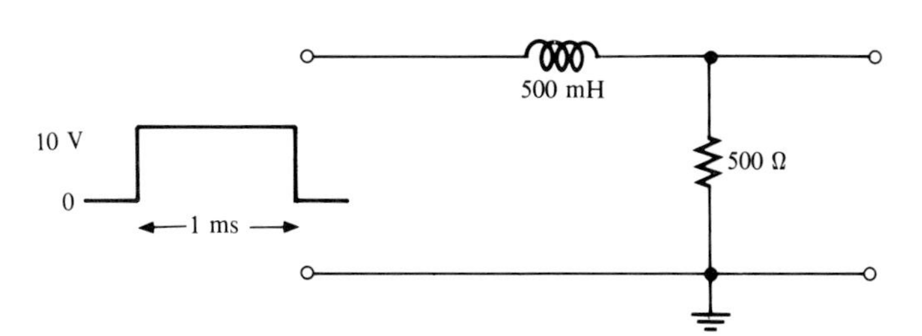

FIGURE 16–38

Solution

During the pulse while L is charging,

$$\tau = \frac{L}{R} = \frac{500 \text{ mH}}{0.5 \text{ k}\Omega} = 1 \text{ ms}$$

Notice that the pulse width is exactly equal to τ. Thus, the output V_R will reach 63% of the full input amplitude in 1τ. Therefore, the output voltage gets to 6.3 V at the end of the pulse.

After the pulse is gone, the inductor discharges back through the 50-Ω source. The total R during discharge is 500 Ω + 50 Ω = 550 Ω; and τ = 500 mH/550 Ω = 909 μs. The source takes 5τ to completely discharge; 5τ = 5(909 μs) = 4545 μs.

The output voltage is shown in Figure 16–39.

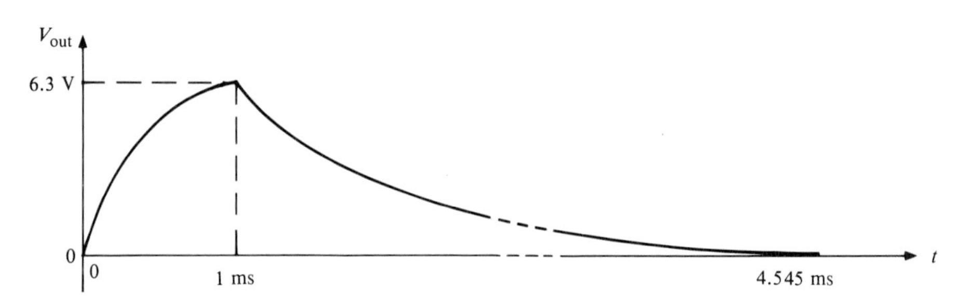

FIGURE 16–39

SECTION REVIEW 16–6

1. In an RL integrator, across which component is the output voltage taken?
2. When a pulse is applied to an RL integrator, what condition must exist in order for the output voltage to reach the amplitude of the input?
3. Under what condition will the output voltage have the approximate shape of the input pulse?

16–7 THE *RL* DIFFERENTIATOR

Figure 16–40 shows a series RL circuit with *the output across the inductor.* Compare this circuit to the RC differentiator in which the output is the resistor voltage.

FIGURE 16–40
An RL differentiating circuit.

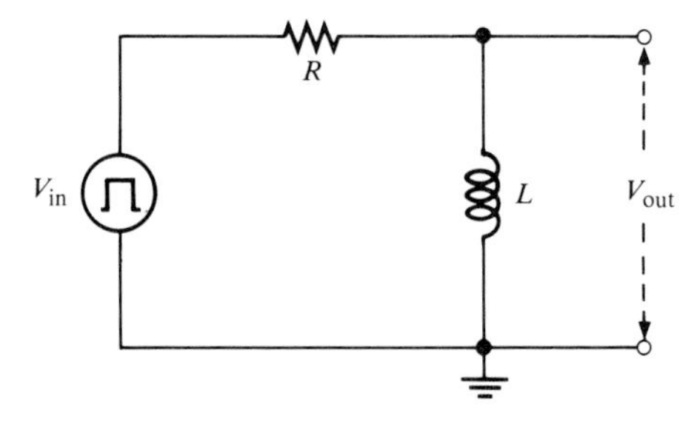

Response of the Differentiator to a Single Pulse

A pulse generator is connected to the input of the differentiator. Initially, before the pulse, there is no current in the circuit. When the input pulse goes from its low level to its high level, the inductor prevents a sudden change in current. It does so, as you know, with an induced voltage equal and opposite to the input.

As a result, L looks like an open, and *all* of the input voltage appears across it at the instant of the rising edge, as shown in Figure 16–41(a) with a 10-V pulse.

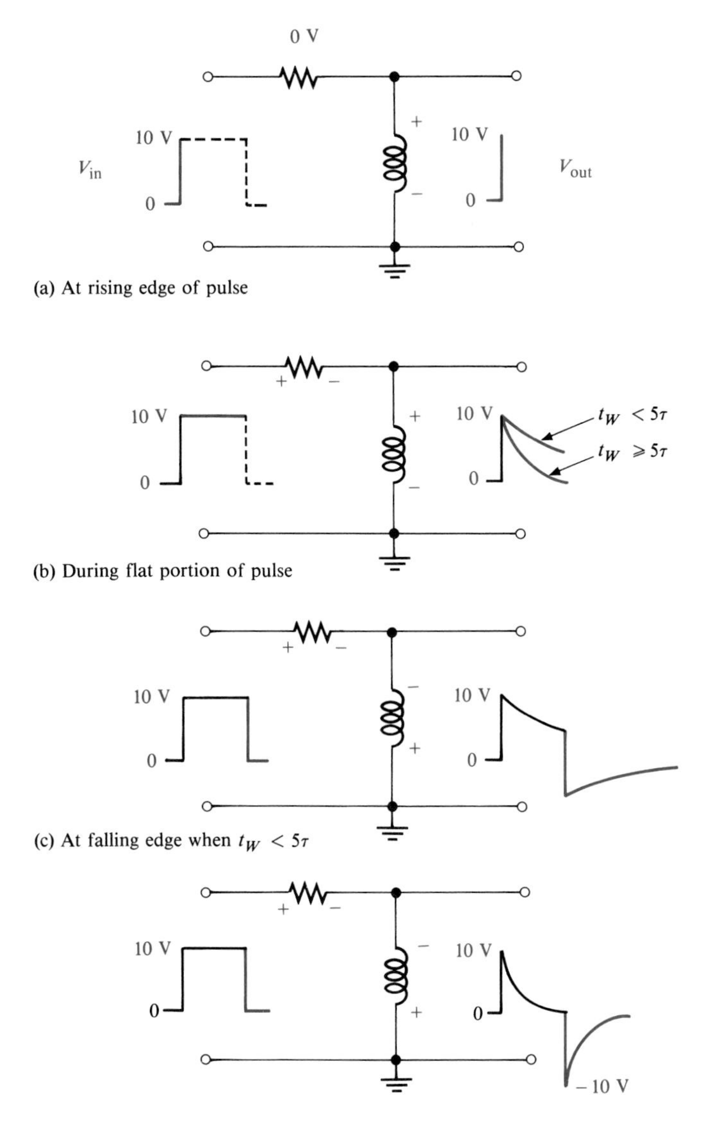

FIGURE 16–41
Illustration of the response of an *RL* differentiator for both time constant conditions.

During the pulse, the current exponentially builds up. As a result, the inductor voltage decreases, as shown in Figure 16–41(b). The rate of decrease, as you know, depends on the L/R time constant. When the falling edge of the input appears, the inductor reacts to keep the current as is, by creating an induced voltage in a direction as indicated in Part (c). This reaction is seen as a sudden negative-going transition of the inductor voltage, as indicated in Parts (c) and (d).

Two conditions are possible, as indicated in Parts (c) and (d). In Part (c), 5τ is greater than the input pulse width, and the output voltage does not have time to decay to zero. In Part (d), 5τ is less than the pulse width, and so the output decays to zero before the end of the pulse. In this case a full, negative, 10-V transition occurs at the trailing edge.

Keep in mind that as far as the input and output waveforms are concerned, the *RL* integrator and differentiator perform the same as their *RC* counterparts.

A summary of the *RL* differentiator response for relationships of various time constants and pulse widths is shown in Figure 16–42.

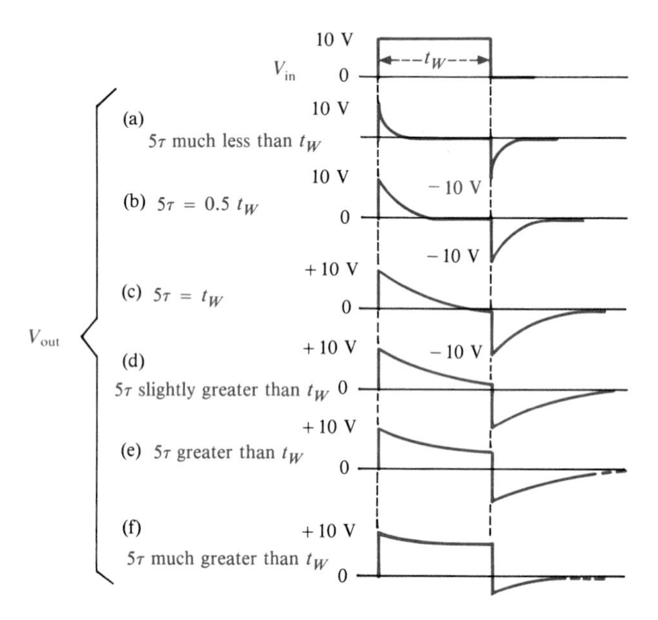

FIGURE 16–42
Illustration of the variation in output pulse shape with the time constant.

EXAMPLE 16–9

Sketch the output voltage for the circuit in Figure 16–43.

FIGURE 16–43

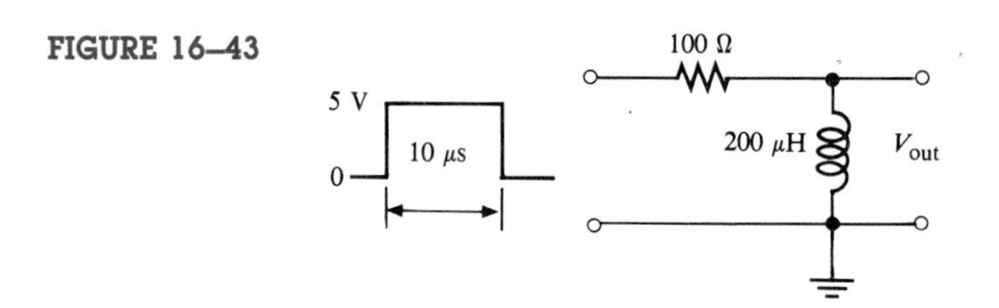

Solution
First calculate the time constant:

$$\tau = \frac{L}{R} = \frac{200\ \mu\text{H}}{100\ \Omega} = 2\ \mu\text{s}$$

In this case, $t_W = 5\tau$, so the output will decay to zero at the end of the pulse.

On the rising edge, the inductor voltage jumps to +5 V and then decays exponentially to zero. It reaches approximately zero at the instant of the falling edge. On the falling edge of the input, the inductor voltage jumps to −5 V and then goes back to zero. The output waveform is shown in Figure 16–44.

FIGURE 16–44

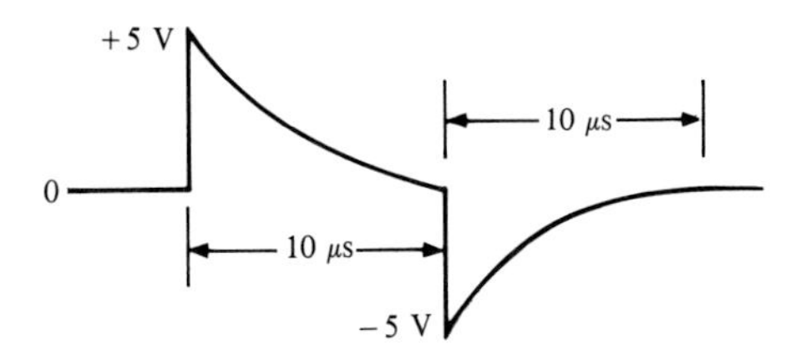

EXAMPLE 16–10

Determine the output voltage waveform for the differentiator in Figure 16–45.

FIGURE 16–45

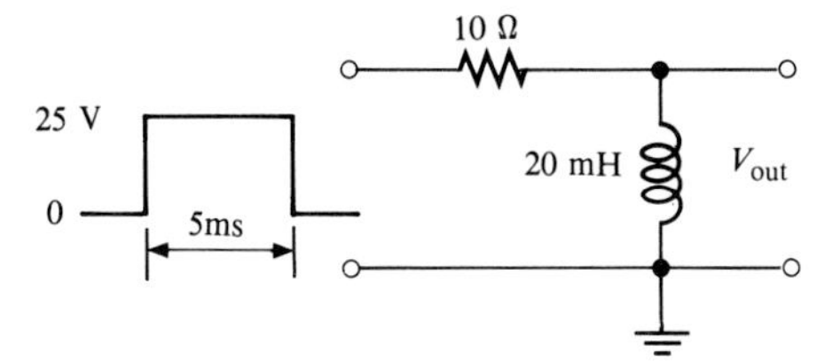

Solution

First calculate the time constant:

$$\tau = \frac{L}{R} = \frac{20\ \text{mH}}{10\ \Omega} = 2\ \text{ms}$$

On the rising edge, the inductor voltage immediately jumps to +25 V. Because the pulse width is 5 ms, the inductor charges for only 2.5τ, so we must use the formula for a decreasing exponential:

$$\begin{aligned} v_L &= V_i e^{-t/\tau} = 25e^{-5/2} \\ &= 25e^{-2.5} = 25(0.082) \\ &= 2.05\ \text{V} \end{aligned}$$

This result is the inductor voltage at the end of the 5-ms input pulse.

On the falling edge, the output immediately jumps from +2.05 V down to −22.95 V (a 25-V negative-going transition). The complete output waveform is sketched in Figure 16–46.

FIGURE 16–46

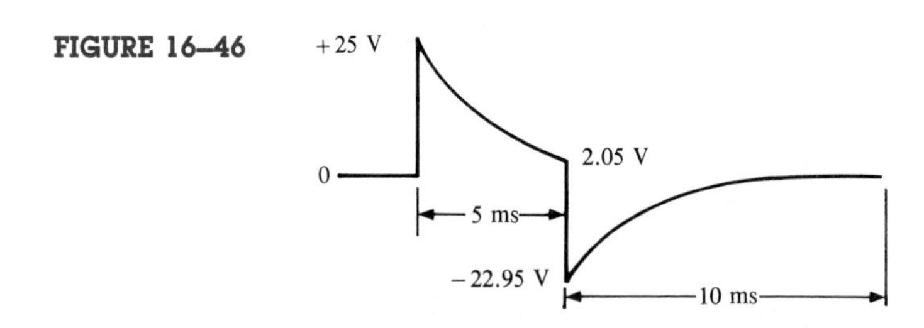

SECTION REVIEW 16–7

1. In an RL differentiator, across which component is the output taken?
2. Under what condition does the output pulse shape most closely resemble the input pulse?
3. If the inductor voltage in an RL differentiator is down to +2 V at the end of a +10-V input pulse, to what negative voltage will the output voltage go in response to the falling edge of the input?

APPLICATION NOTE

For the Application Assignment at the beginning of this chapter, you should use a differentiating circuit with a time constant no greater than 0.1 μs. In one time constant interval, the capacitor will charge 63%, causing the current, and thus the resistor voltage, to decrease to 37% of its peak amplitude.

SUMMARY

Facts

- In an RC integrating circuit, the output voltage is taken across the capacitor.
- In an RC differentiating circuit, the output voltage is taken across the resistor.
- In an RL integrating circuit, the output voltage is taken across the resistor.
- In an RL differentiating circuit, the output voltage is taken across the inductor.
- In an integrator, when the pulse width (t_W) of the input is much less than the transient time, the output voltage approaches a constant level equal to the average value of the input.
- In an integrator, when the pulse width of the input is much greater than the transient time, the output voltage approaches the shape of the input.
- In a differentiator, when the pulse width of the input is much less than the transient time, the output voltage approaches the shape of the input but with an average value of zero.
- In a differentiator, when the pulse width of the input is much greater than the transient time, the output voltage consists of narrow, positive- and negative-going spikes occurring on the leading and trailing edges of the input pulses.

Definitions

- *Transient time*—five time constants; $5RC$ or $5L/R$.
- *Steady state*—the condition of a circuit after an initial transient time.
- *dc component*—the average value of a pulse waveform.

SELF-TEST

1. How do you distinguish an RC differentiating circuit from an integrating circuit?
2. In an RC integrator, to what voltage does the capacitor charge if the 10-V input pulse has a width equal to one time constant?
3. Repeat Question 2 for an RC differentiator.
4. In an RC integrator, under what condition does the output voltage waveform closely resemble that of the input waveform?
5. Repeat Question 4 for an RC differentiator.
6. Under what condition are the positive and negative portions of the differentiator output voltage of equal amplitude?
7. How do you distinguish an RL differentiating circuit from an integrating circuit?
8. Express the maximum possible current in an RL circuit.
9. Under what condition in an RL differentiating circuit does the current reach its maximum possible value?
10. You have two differentiating circuits sitting side-by-side. One is an RC and the other an RL. Both have equal time constants. If you apply the same pulse waveform to each circuit, how can you tell the RC from the RL by only observing their output voltages?

PROBLEM SET A

Section 16–1

16–1 An integrating circuit has $R = 2\ \text{k}\Omega$ in series with $C = 0.05\ \mu\text{F}$. What is the time constant?

16–2 Determine how long it takes the capacitor in an integrating circuit to reach full charge for each of the following series RC combinations:
(a) $R = 50\ \Omega$, $C = 50\ \mu\text{F}$ (b) $R = 3300\ \Omega$, $C = 0.015\ \mu\text{F}$
(c) $R = 22\ \text{k}\Omega$, $C = 100\ \text{pF}$ (d) $R = 5\ \text{M}\Omega$, $C = 10\ \text{pF}$

Section 16–2

16–3 A 20-V pulse is applied to an RC integrator. The pulse width equals *one* time constant. To what voltage does the capacitor charge during the pulse? Assume that it is initially uncharged.

16–4 Repeat Problem 16–3 for the following values of t_W:
(a) 2τ (b) 3τ (c) 4τ (d) 5τ

16–5 Sketch the approximate shape of an integrator output voltage where 5τ is much less than the pulse width of a 10-V square wave input. Repeat for the case in which 5τ is much larger than the pulse width.

16–6 Determine the output voltage for an integrator with a single input pulse, as shown in Figure 16–47. For repetitive pulses, how long will it take this circuit to reach steady state?

FIGURE 16–47

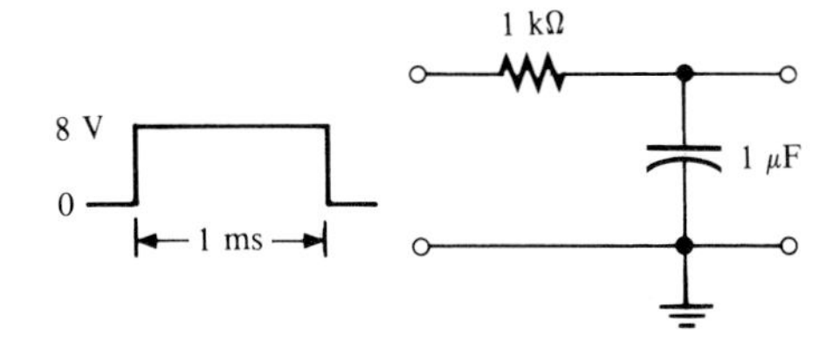

Section 16–3

16–7 Sketch the integrator output in Figure 16–48, showing maximum voltages.

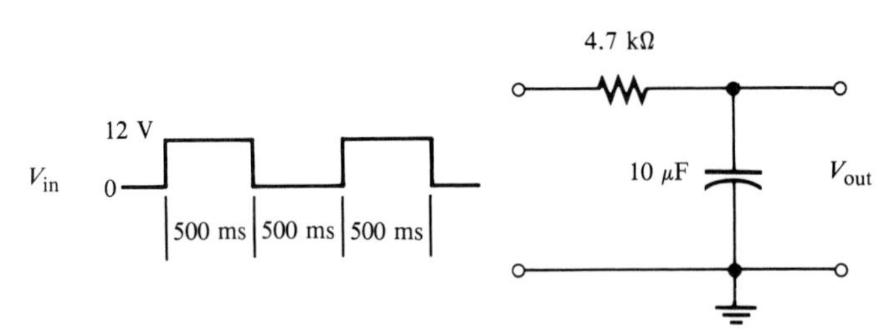

FIGURE 16–48

16–8 A 1-V, 10-kHz pulse waveform with a duty cycle of 25% is applied to an integrator with $\tau = 25$ μs. Graph the output voltage for three initial pulses. C is initially uncharged.

16–9 What is the steady state output voltage of the integrator with a square wave input shown in Figure 16–49?

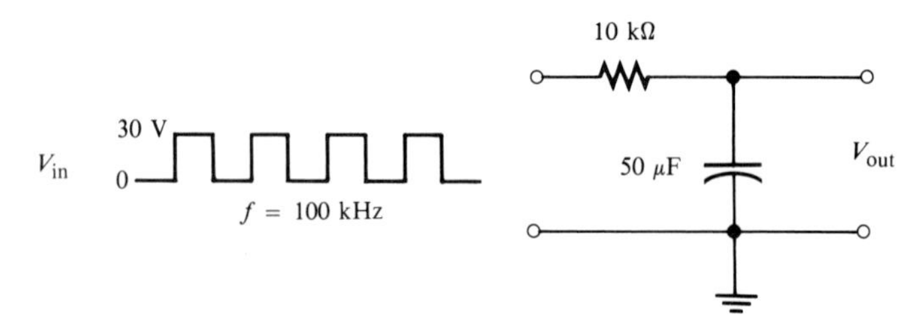

FIGURE 16–49

Section 16–4

16–10 Repeat Problem 16–5 for an *RC* differentiator.

16–11 Redraw the circuit in Figure 16–47 to make it a differentiator, and repeat Problem 16–6.

Section 16–5

16–12 Sketch the differentiator output in Figure 16–50, showing maximum voltages.

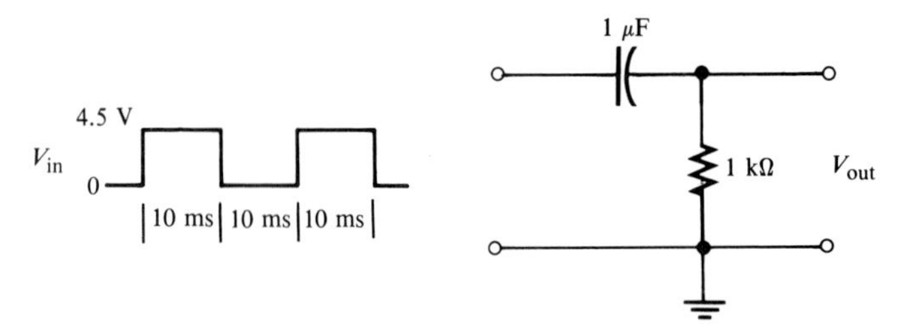

FIGURE 16–50

16–13 What is the steady state output voltage of the differentiator with the square wave input shown in Figure 16–51?

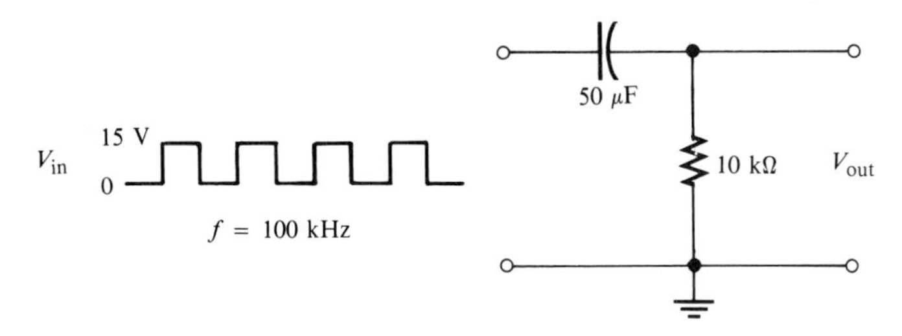

FIGURE 16–51

Section 16–6

16–14 Determine the output voltage for the circuit in Figure 16–52. A single input pulse is applied as shown.

FIGURE 16–52

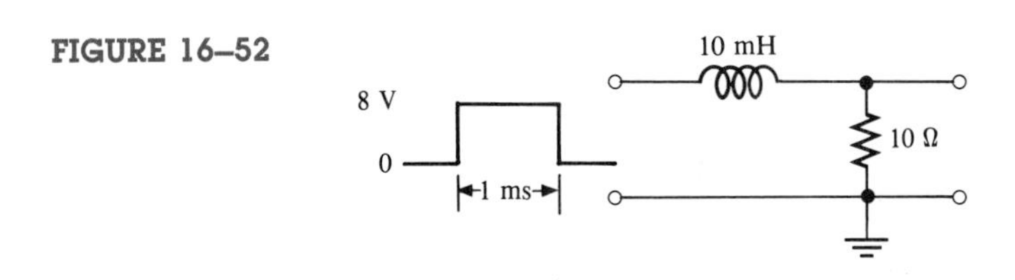

16–15 Sketch the integrator output in Figure 16–53, showing maximum voltages.

FIGURE 16–53

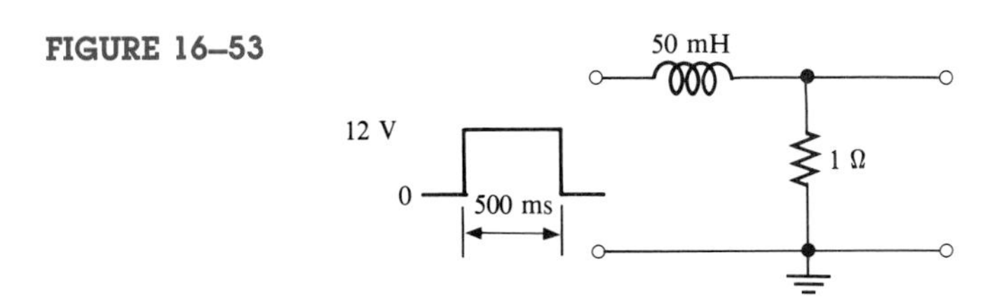

Section 16–7

16–16 (a) What is τ in Figure 16–54?
(b) Sketch the output voltage.

FIGURE 16–54

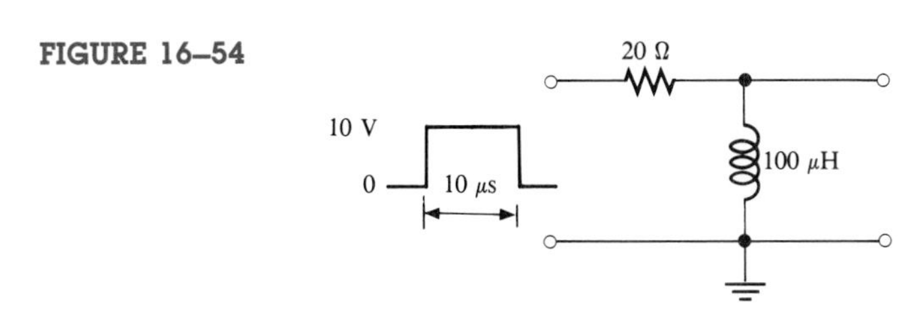

16–17 Draw the output waveform if a periodic pulse waveform with $t_W = 25\ \mu s$ and $T = 60\ \mu s$ is applied to the circuit in Figure 16–54.

PROBLEM SET B

16–18 (a) What is τ in Figure 16–55?
(b) Sketch the output voltage.

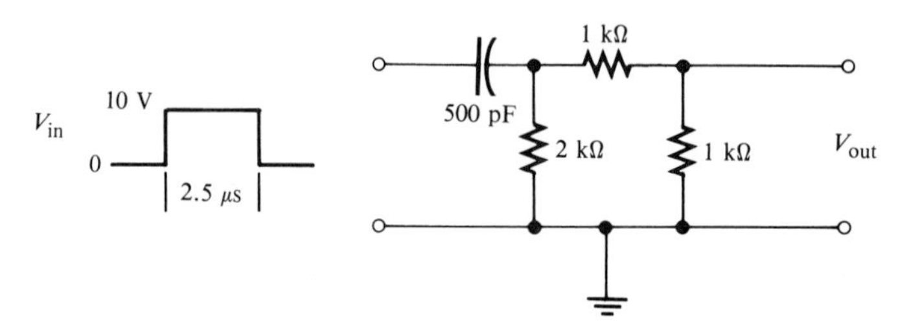

FIGURE 16–55

16–19 (a) What is τ in Figure 16–56?
(b) Sketch the output voltage.

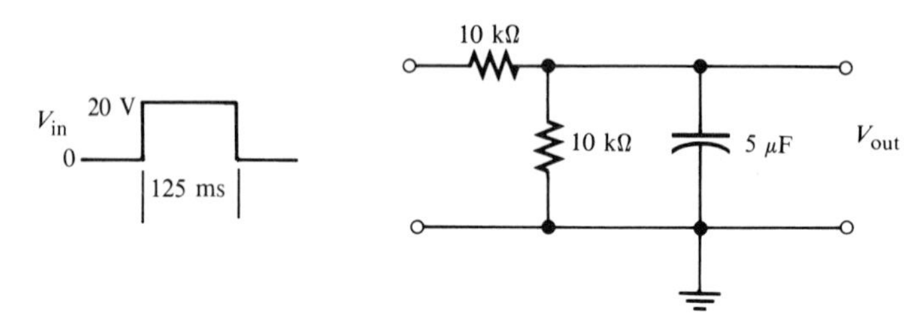

FIGURE 16–56

16–20 Determine the time constant in Figure 16–57. Is this circuit an integrator or a differentiator?

FIGURE 16–57

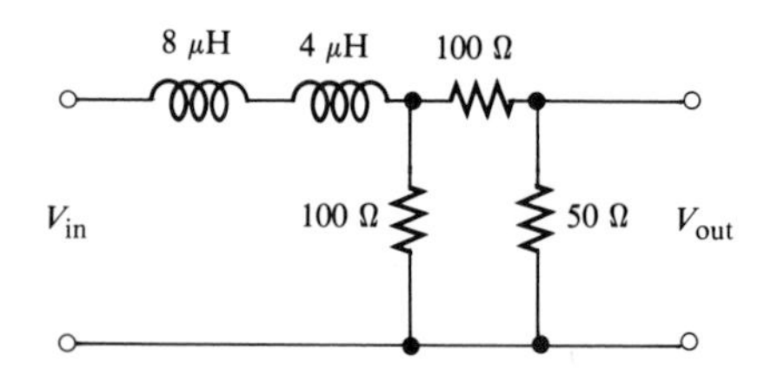

ANSWERS TO SECTION REVIEWS

Section 16–1

1. A series *RC* circuit in which the output is across the capacitor.
2. A voltage applied to the input causes the capacitor to charge. A short across the input causes the capacitor to discharge.

Section 16–2

1. $5\tau \leq t_W$. **2.** 0.63 V, 50 ms. **3.** See Figure 16–58. **4.** No.
5. $5\tau << t_W$ (5τ much less than t_W).

FIGURE 16–58

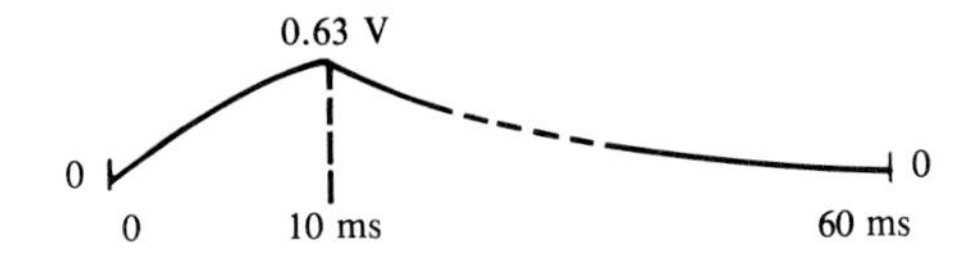

Section 16–3

1. $5\tau \leq t_W$, and $5\tau \leq$ time between pulses. **2.** Like the input.
3. *transient time.* **4.** The response after the transient time has passed.
5. The average value of the input voltage.

Section 16–4

1. See Figure 16–59. **2.** $5\tau >> t_W$. **3.** Positive and negative spikes.
4. −10 V.

FIGURE 16–59

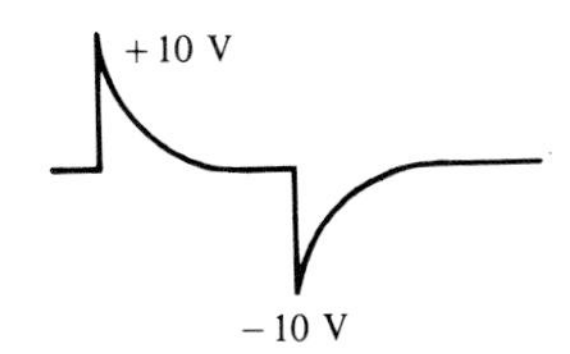

Section 16–5

1. $5\tau \leq t_W$ and $5\tau \leq$ time between pulses. **2.** Positive and negative spikes.
3. 0 V.

Section 16–6

1. Resistor. **2.** $5\tau \leq t_W$. **3.** $5\tau << t_W$.

Section 16–7

1. Inductor. **2.** $5\tau >> t_W$. **3.** −8 V.

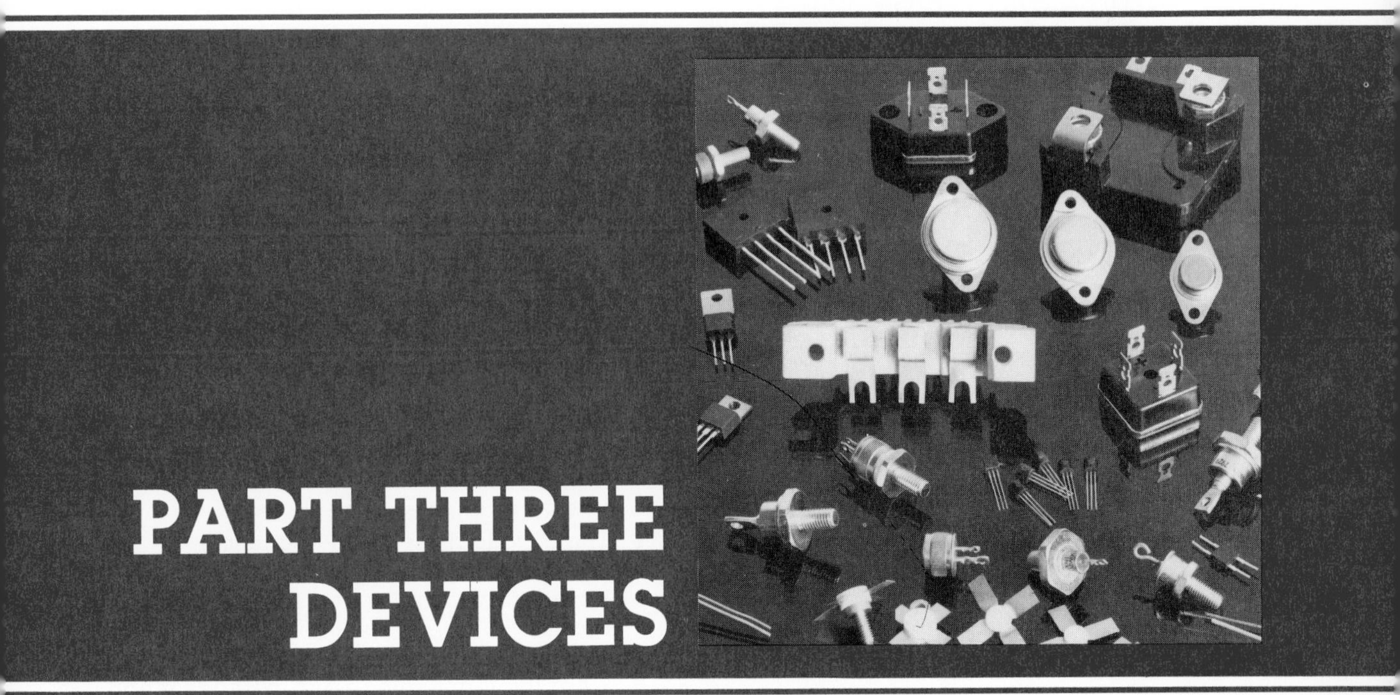

PART THREE DEVICES

Introduction to Semiconductor Devices

To acquire a basic understanding of semiconductors, you must have some knowledge of atomic theory and the structure of semiconductor materials. In this chapter, you will learn about this theory. We will discuss the basic materials used in manufacturing diodes, transistors, and other semiconductor devices and will present PN junctions, an important concept essential for the understanding of diode and transistor operation. Also, other diode characteristics are introduced, and you will learn how to properly use a diode in a circuit.

Specifically, you will learn:

- What an electron shell is.
- What ionization is.
- The basic structure of silicon and germanium.
- How atoms bond together to form crystals.
- How the energy levels within an atom's structure relate to current flow in a material.
- How current flow occurs in a semiconductor material.
- The definition of *P-type* and *N-type* semiconductors.
- How a PN junction is formed.
- The practical use of a PN junction.
- What a diode is.
- How to forward bias and reverse bias a diode.
- What a diode characteristic curve shows.
- What the terms *anode* and *cathode* mean.
- How to identify the terminals of a diode.

17

APPLICATION ASSIGNMENT

Semiconductor diodes are widely used in many types of electronic applications. They are of great importance in power supplies, where they are used as *rectifiers* to convert ac to dc. Your supervisor has removed the diodes from one particular power supply that has malfunctioned and has asked you to check the diodes to see which ones, if any, have failed.

After completing this chapter, you will know how to carry out this assignment. The solution is given in the Application Note at the end of the chapter.

THE ATOMIC STRUCTURE OF SEMICONDUCTORS

17–1

The basic concept of the structure of an atom was studied in Chapter 1. In this chapter, atomic theory is extended to semiconductor materials that are used in electronic devices such as diodes and transistors. This coverage lays the foundation for a good understanding of how semiconductor devices function.

Electron Shells and Orbits

Electrons orbit the nucleus at certain distances from the nucleus. Electrons near the nucleus have less energy than those in more distant orbits. Only *discrete* values of electron energies exist within atomic structures. Therefore, electrons must orbit only at discrete distances from the nucleus.

Each discrete distance (orbit) from the nucleus corresponds to a certain energy level. In an atom, orbits are grouped into energy bands known as *shells*. A given atom has a fixed number of shells. Each shell has a fixed maximum number of electrons at permissible energy levels (orbits). The differences in energy levels within a shell are much smaller than the difference in energy between shells. The shells are designated *K, L, M, N,* and so on, with *K* being closest to the nucleus. This concept is illustrated in Figure 17–1.

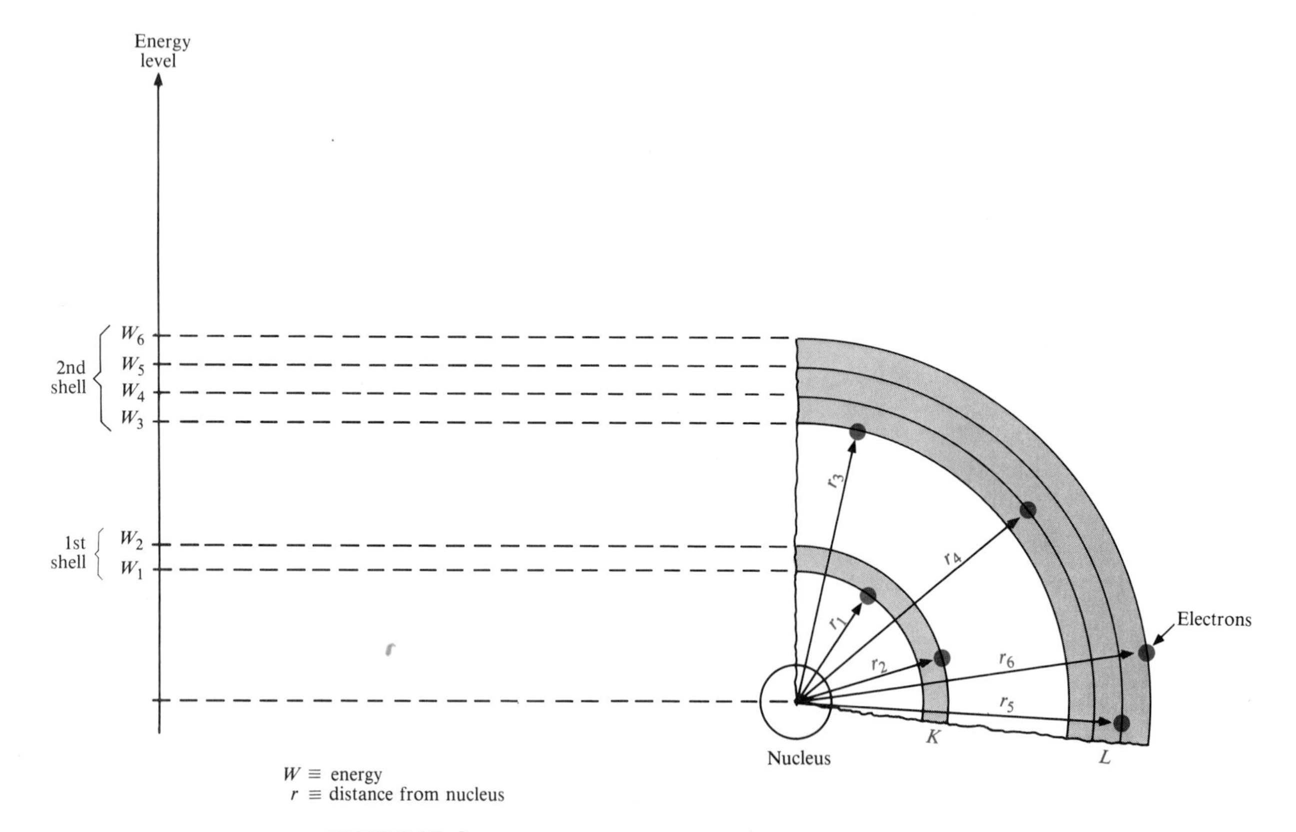

FIGURE 17–1
Energy levels increase as distance from the nucleus of the atom increases.

Valence Electrons

Electrons in orbits farther from the nucleus are less tightly bound to the atom than those closer to the nucleus, because the force of attraction between the positively charged nucleus and the negatively charged electron increases with decreasing distance.

Electrons with the highest energy levels exist in the outermost shell of an atom and are relatively loosely bound to the atom. These *valence electrons* contribute to chemical reactions and bonding within the structure of a material. The *valence* of an atom is the number of electrons in its outermost shell.

Ionization

When an atom absorbs energy from a heat source or from light, for example, the energy levels of the electrons are raised. When an electron gains energy, it moves to an orbit farther from the nucleus. Since the valence electrons possess more energy and are more loosely bound to the atom than inner electrons, they can jump to higher orbits more easily when external energy is absorbed.

If a valence electron acquires a sufficient amount of energy, it can be completely removed from the outer shell and the atom's influence. The departure of a valence electron leaves a previously neutral atom with an excess of positive charge (more protons than electrons). The process of losing a valence electron is known as *ionization,* and the resulting positively charged atom is called a *positive ion.* For example, the chemical symbol for hydrogen is H. When it loses its valence electron and becomes a positive ion, it is designated H^+. The escaped valence electron is called a *free electron.*

When a free electron falls into the outer shell of a neutral hydrogen atom, the atom becomes negatively charged (more electrons than protons) and is called a *negative ion,* designated H^-.

Silicon and Germanium Atoms

Two types of widely used semiconductor materials are *silicon* and *germanium.* Both the silicon and the germanium atoms have four valence electrons. They differ in that silicon has 14 protons in its nucleus and germanium has 32. Figure 17–2 shows the atomic structure for both materials.

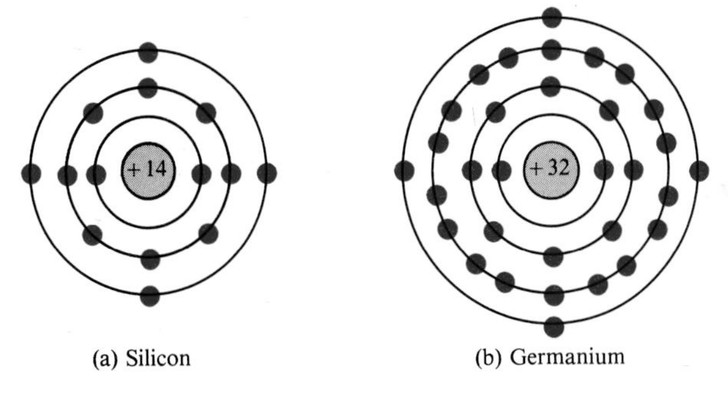

FIGURE 17–2
Silicon and germanium atoms.

SECTION REVIEW 17–1

1. Define *atom.*
2. Name the components of an atom.
3. What is a valence electron?
4. What is a free electron?

ATOMIC BONDING

17–2

When silicon atoms combine into molecules to form a solid material, they arrange themselves in a fixed pattern called a *crystal.* The atoms within the crystal structure are held together by *convalent bonds,* which are created by interaction of the valence electrons of each atom.

Figure 17–3 shows how each silicon atom positions itself with four adjacent atoms. Since an atom can have up to eight electrons in its outer shell, a silicon atom with its four valence electrons shares an electron with each of its four neighbors. This sharing of valence electrons produces the covalent bonds that hold the atoms together, because each shared electron is attracted equally by the two adjacent atoms which share it. Covalent bonding of a pure (intrinsic) silicon crystal is shown in Figure 17–4. Bonding for germanium is similar because it also has four valence electrons.

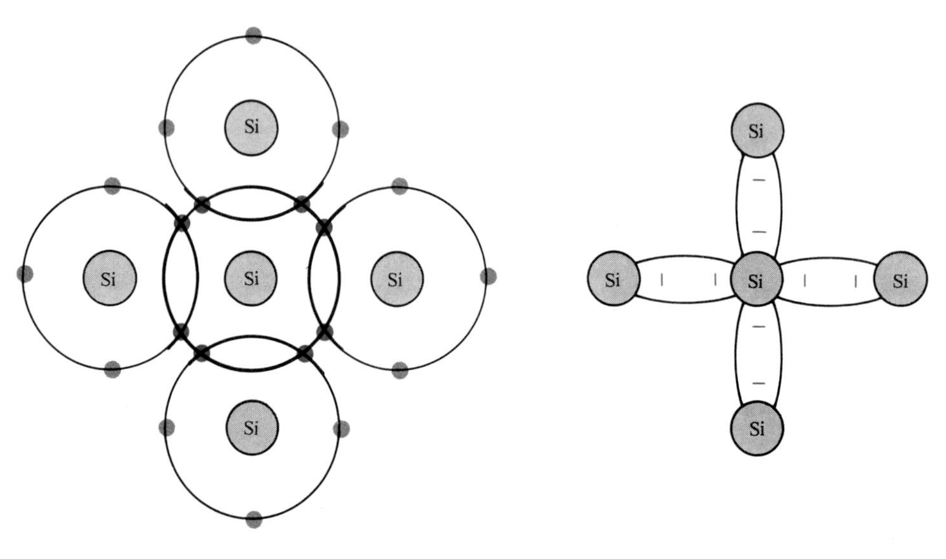

(a) Shared valence electrons form covalent bonds.

(b) Bonding diagram

FIGURE 17–3
Covalent bonds in silicon.

SECTION REVIEW 17–2

1. Name two semiconductor materials.
2. What is a covalent bond?

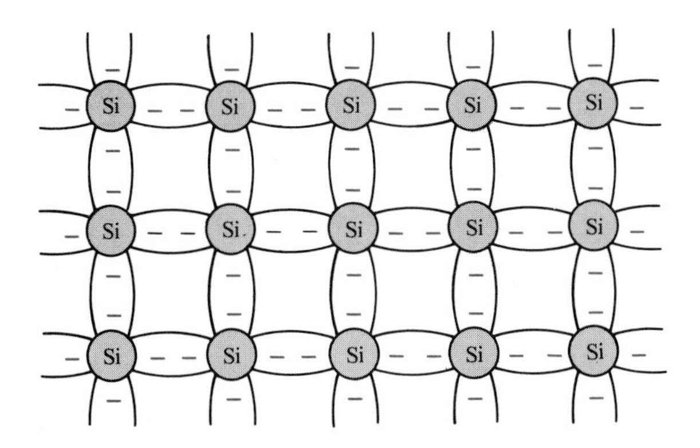

FIGURE 17–4
Covalent bonds in a pure silicon crystal.

CONDUCTION IN SEMICONDUCTOR MATERIALS

17–3

As you have seen, the electrons of an atom can exist only within prescribed energy bands. Each shell around the nucleus corresponds to a certain energy band and is separated from adjacent shells by energy gaps, in which no electrons can exist. This condition, shown in Figure 17–5 for an unexcited silicon atom (no external energy), occurs only at absolute zero temperature.

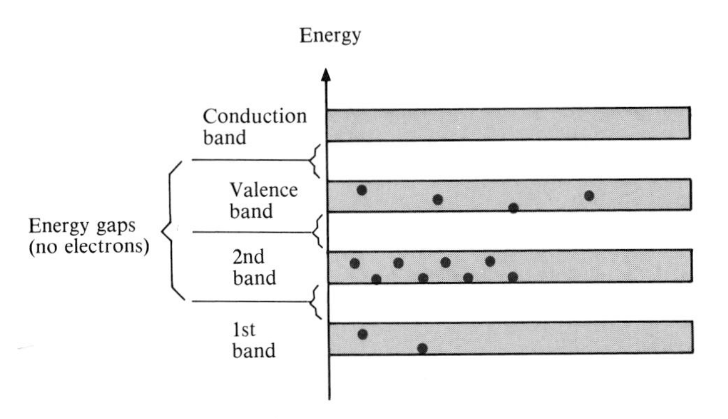

FIGURE 17–5
Energy band diagram for **unexcited** silicon atom.

Conduction Electrons and Holes

A pure silicon crystal at room temperature derives heat (thermal) energy from the surrounding air, causing some valence electrons to gain sufficient energy to jump the gap from the *valence band* into the *conduction band*, becoming free electrons. This situation is illustrated in the energy diagram of Figure 17–6(a) and in the bonding diagram of Figure 17–6(b).

When an electron jumps to the conduction band, a vacancy is left in the valence band. This vacancy is called a *hole*. For every electron raised to the conduction band by thermal or light energy, there is one hole left in the valence

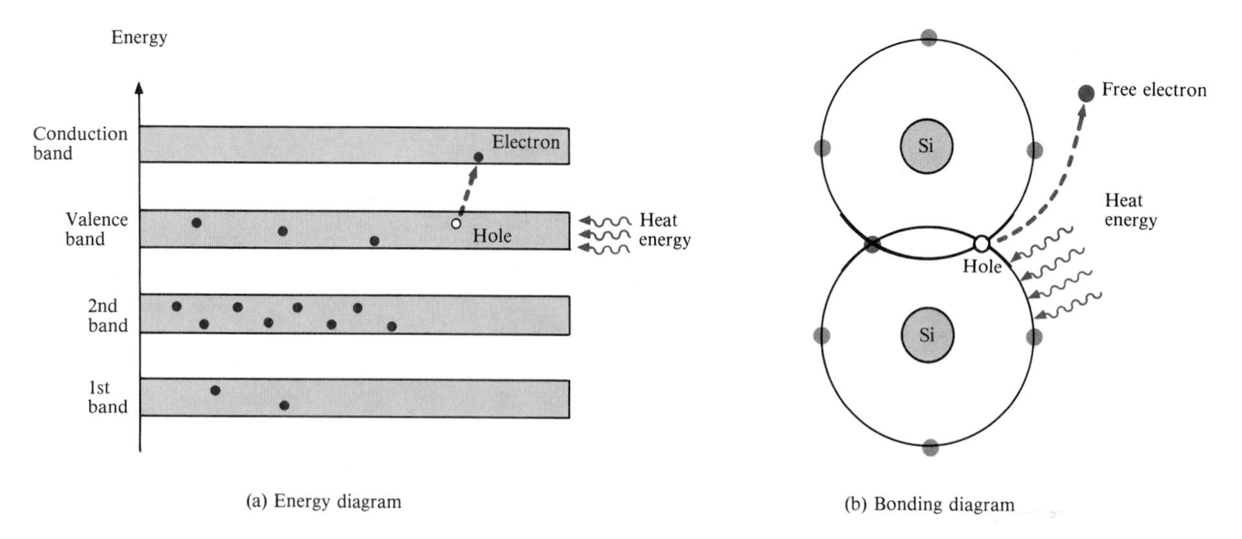

FIGURE 17–6
Creation of an electron-hole pair in an excited silicon atom.

band, creating what is called an *electron-hole pair*. *Recombination* occurs when a conduction band electron loses energy and falls back into a hole in the valence band.

In summary, a piece of pure silicon at room temperature has, at any instant, a number of conduction band (free) electrons that are unattached to any atom and are essentially drifting randomly throughout the material. Also, an equal number of holes are created in the valence band when these electrons jump into the conduction band, as illustrated in Figure 17–7.

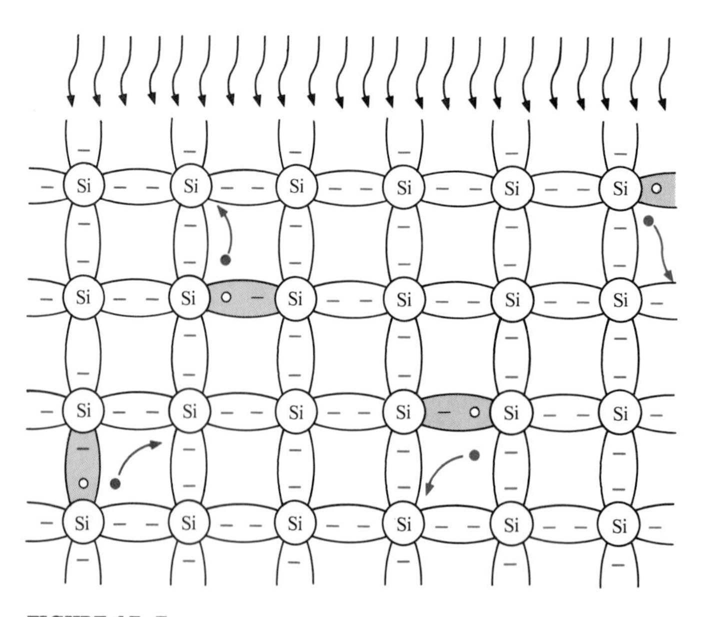

FIGURE 17–7
Electron-hole pairs in a silicon crystal.

Germanium versus Silicon

The situation in a germanium crystal is similar to that in silicon except that, because of its atomic structure, pure germanium has more free electrons than silicon and therefore a higher conductivity. Silicon, however, is the favored semiconductor material and is used far more widely than germanium. One reason for its wide usage is that silicon can be used at a much higher temperature than germanium.

Electron and Hole Current

When a voltage is applied across a piece of silicon, as shown in Figure 17–8, the free electrons in the conduction band are easily attracted toward the positive end. This movement of free electrons is one type of current in a semiconductor material, called *electron current.*

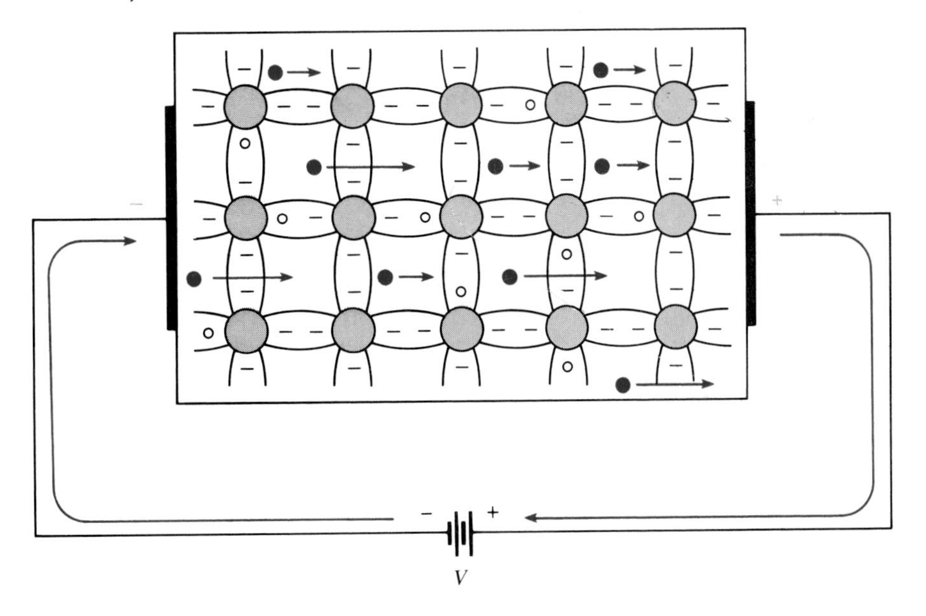

FIGURE 17–8
Free electron current in silicon.

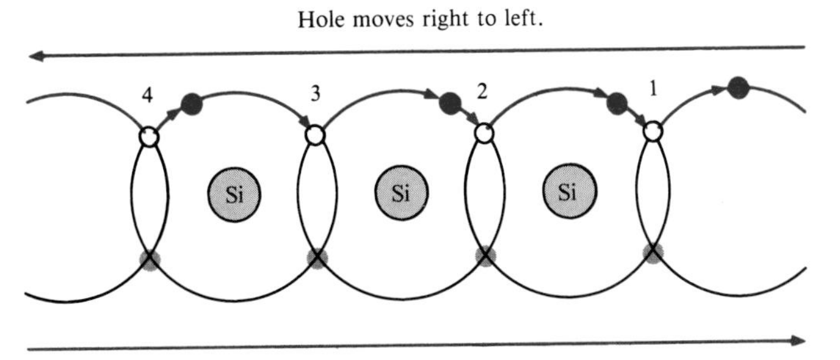

FIGURE 17–9
Hole current in silicon.

Another current mechanism occurs at the valence level, where the holes created by the free electrons exist. Electrons remaining in the valence band are still attached to their atoms and are not free to move randomly in the crystal structure. However, a valence electron can "fall" into a nearby hole, with little change in its energy level, thus leaving another hole where it came from. Effectively the hole has moved from one place to another in the crystal structure, as illustrated in Figure 17–9 (on page 623). This current is called *hole current.*

Semiconductors, Conductors, and Insulators

In a pure (intrinsic) semiconductor, there are relatively few free electrons, so neither silicon nor germanium is very useful in its intrinsic state. They are neither insulators nor good conductors because current in a material depends directly on the number of free electrons.

A comparison of the energy bands in Figure 17–10 for the three types of materials shows the essential differences among them regarding conduction. The energy gap for an insulator is so wide that hardly any electrons acquire enough energy to jump into the conduction band.

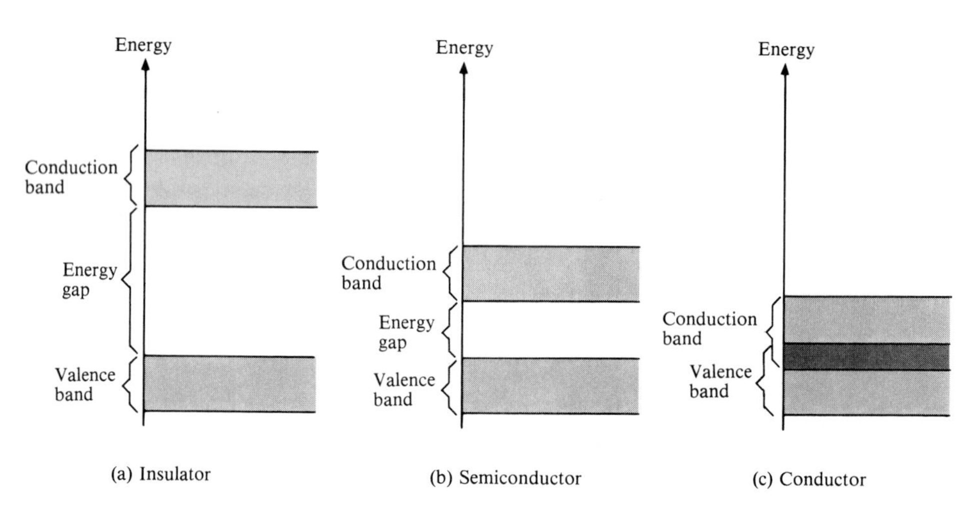

FIGURE 17–10
Energy diagrams for three categories of materials.

The valence band and the conduction band in a conductor (such as copper) overlap so that there are always many conduction electrons, even without the application of external energy. A semiconductor, as Figure 17–10 shows, has an energy gap that is much narrower than that in an insulator.

SECTION REVIEW 17–3

1. In the atomic structure of a semiconductor, within which energy band do free electrons exist? Valence electrons?
2. How are holes created in an intrinsic semiconductor?
3. Why is current established more easily in a semiconductor than in an insulator?

N-TYPE AND P-TYPE SEMICONDUCTORS

17–4

Intrinsic semiconductor materials do not conduct current very well because of the limited number of free electrons in the conduction band. Thus, the resistivity of a semiconductor is much greater than that of a conductor. For example, a one-cubic-centimeter sample of silver has a resistivity of $10^{-6}\Omega\cdot$cm, whereas the resistivity is about 45 $\Omega\cdot$cm for pure germanium and several thousand $\Omega\cdot$cm for pure silicon.

Doping

The resistivities of silicon and germanium can be drastically reduced and controlled by the addition of *impurities* to the pure semiconductor material. This process, called *doping,* increases the number of current carriers (electrons or holes), thus increasing the conductivity and decreasing the resistivity. The two categories of impurities are *N-type* and *P-type.*

N-Type Semiconductor

To increase the number of conduction band electrons in pure silicon, *pentavalent* impurity atoms are added. These are atoms with five valence electrons, such as *arsenic, phosphorus,* and *antimony.*

As illustrated in Figure 17–11, each pentavalent atom (antimony, in this case) forms covalent bonds with four adjacent silicon atoms. Four of the antimony atom's valence electrons are used to form the covalent bonds, leaving *one extra electron.* This extra electron becomes a conduction electron because it is not attached to any atom. The number of conduction electrons can be controlled by the amount of impurity added to the silicon.

Since most of the current carriers are *electrons,* silicon (or germanium) doped in this way is an *N-type* semiconductor material where the *N* stands for the *negative* charge on an electron. The *electrons* are called the *majority carriers* in N-type material. Although the great majority of current carriers in N-type material are electrons, there are some holes. *Holes* in an N-type material are called *minority carriers.*

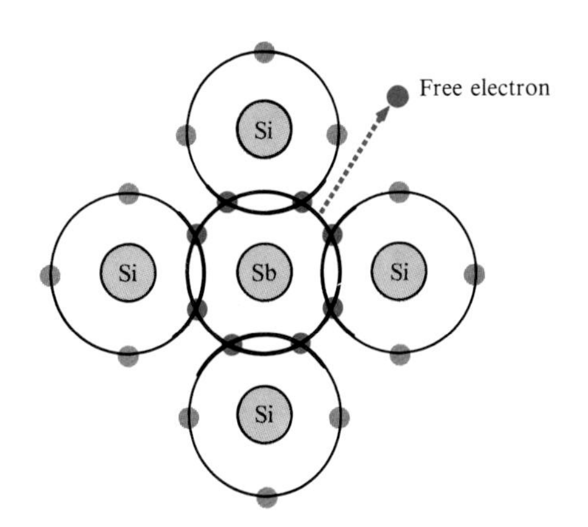

FIGURE 17–11
Pentavalent impurity atom in a silicon crystal. An antimony (Sb) impurity atom is shown in the center.

P-Type Semiconductor

To increase the number of holes in pure silicon, *trivalent* impurity atoms are added. These are atoms with three valence electrons, such as *aluminum, boron,* and *gallium.*

As illustrated in Figure 17–12, each trivalent atom (boron, in this case) forms covalent bonds with four adjacent silicon atoms. All three of the boron atom's valence electrons are used in the covalent bonds; and, since four electrons are required, a *hole* is formed with each trivalent atom. The number of holes can be controlled by the amount of trivalent impurity added to the silicon.

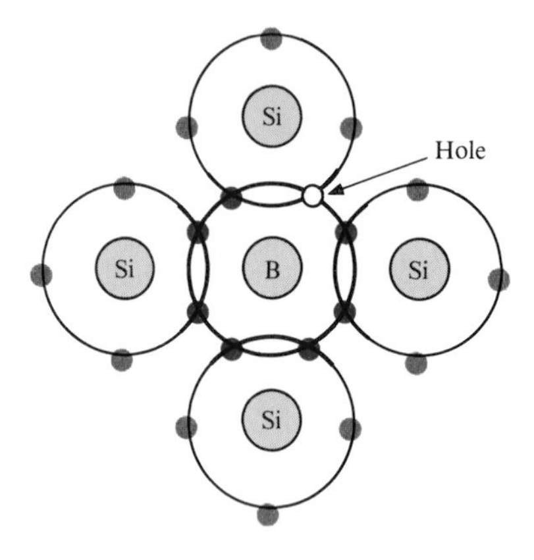

FIGURE 17–12
Trivalent impurity atom in a silicon crystal. A boron (B) impurity atom is shown in the center.

Since most of the current carriers are *holes,* silicon (or germanium) doped in this way is called a *P-type* semiconductor material because holes can be thought of as positive charges. The *holes* are the *majority carriers* in P-type material. Although the great majority of current carriers in P-type material are holes, there are some electrons. *Electrons* in P-type material are called *minority carriers.*

SECTION REVIEW 17–4

1. How is an N-type semiconductor formed?
2. How is a P-type semiconductor formed?
3. What is a majority carrier?

PN JUNCTIONS

17–5

When a piece of silicon is doped so that half is N-type and the other half is P-type, a *PN junction* is formed between the two regions, as shown in Figure 17–13(a). This device is known as a *semiconductor diode.* The N region has many conduction electrons, and the P region has many holes, as shown in Figure 17–13(b). The PN junction is fundamental to the operation not only of diodes but also of transistors and other solid state devices.

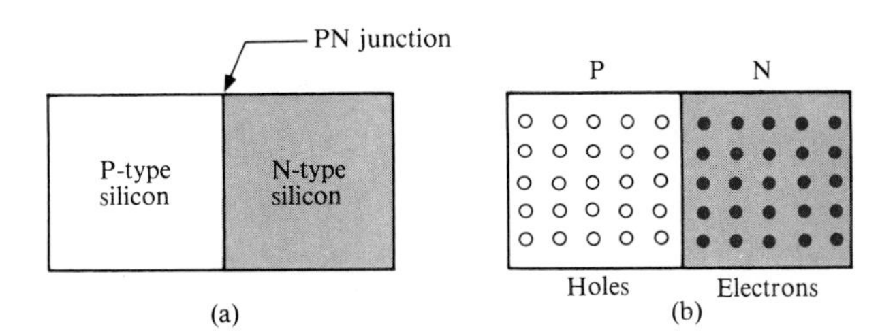

FIGURE 17–13
Basic PN structure at the instant of junction formation.

The Depletion Layer

With no external voltage, the conduction electrons in the N region are aimlessly drifting in all directions. At the instant of junction formation, some of the electrons near the junction *diffuse* across into the P region and recombine with holes near the junction.

For each electron that crosses the junction and recombines with a hole, a pentavalent atom is left with a net positive charge in the N region near the junction, making it a *positive ion*. Also, when the electron recombines with a hole in the P region, a trivalent atom acquires net negative charge, making it a *negative ion*.

As a result of this recombination process, a large number of positive and negative ions builds up near the PN junction. As this buildup occurs, the electrons in the N region must overcome both the attraction of the positive ions and the repulsion of the negative ions in order to migrate into the P region. Thus, as the ion layers build up, the area on both sides of the junction becomes essentially depleted of any conduction electrons or holes and is known as the *depletion layer*. This condition is illustrated in Figure 17–14. When an equilibrium condition is reached, the depletion layer has widened to a point where no more electrons can cross the PN junction.

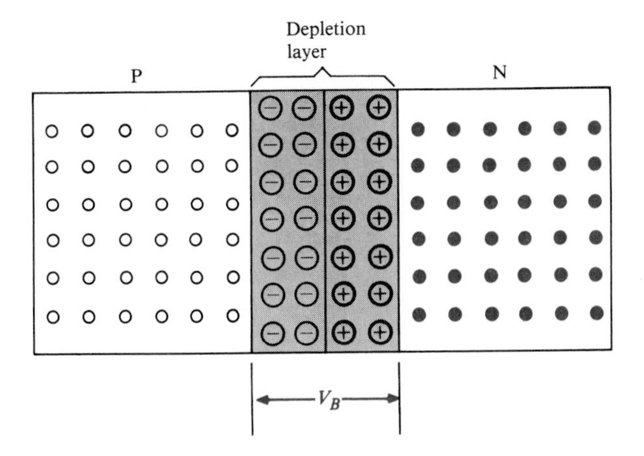

FIGURE 17–14
PN junction equilibrium condition.

The existence of the positive and negative ions on opposite sides of the junction creates a *barrier potential* (V_B) across the depletion layer, as indicated in Figure 17–14. At 25°C, the barrier potential is approximately 0.7 V for silicon and 0.3 V for germanium. As the junction temperature *increases,* the barrier potential *decreases,* and vice versa.

Energy Diagram of the PN Junction

Now, we will look at the operation of the PN junction in terms of its energy level. First consider the PN junction at the instant of its formation. The energy bands of the trivalent impurity atoms in the P-type material are at a slightly higher level than those of the pentavalent impurity atom in the N-type material, as shown in Figure 17–15. They are higher because the core attraction for the valence electrons (+3) in the trivalent atom is less than the core attraction for the valence electrons (+5) in the pentavalent atom. Thus, the trivalent valence electrons are in a slightly higher orbit and, thus, at a higher energy level.

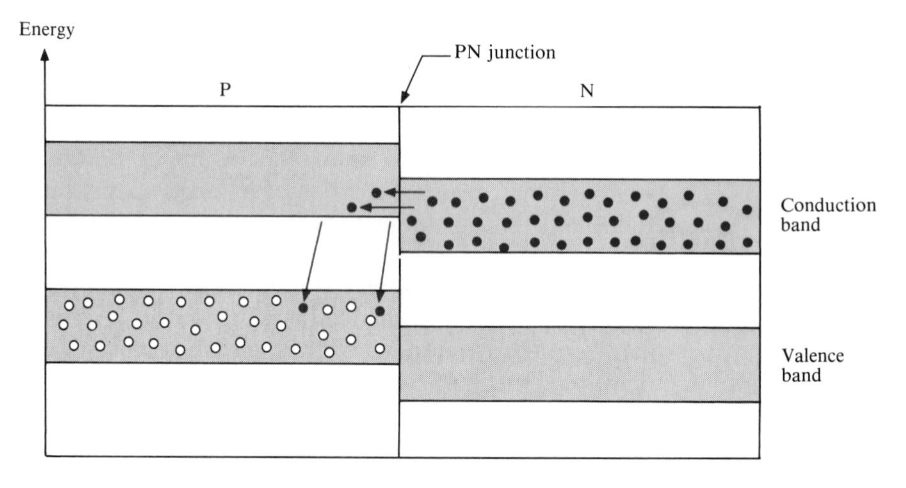

FIGURE 17–15
PN junction energy diagram as diffusion begins.

Notice in Figure 17–15 that there is some overlap of the conduction bands in the P and N regions and also some overlap of the valence bands in the P and N regions. This overlap permits the electrons of higher energy near the top of the N-region conduction band to begin diffusing across the junction into the lower part of the P-region conduction band. As soon as an electron diffuses across the junction, it recombines with a hole in the valence band. As diffusion continues, the depletion layer begins to form. Also, the energy bands in the N region "shift" down as the electrons of higher energy are lost to diffusion. When the top of the N-region conduction band reaches the same level as the bottom of the P-region conduction band, diffusion ceases and the equilibrium condition is reached. This condition is shown in terms of energy levels in Figure 17–16. There is an *energy gradient* across the depletion layer rather than an abrupt change in energy level.

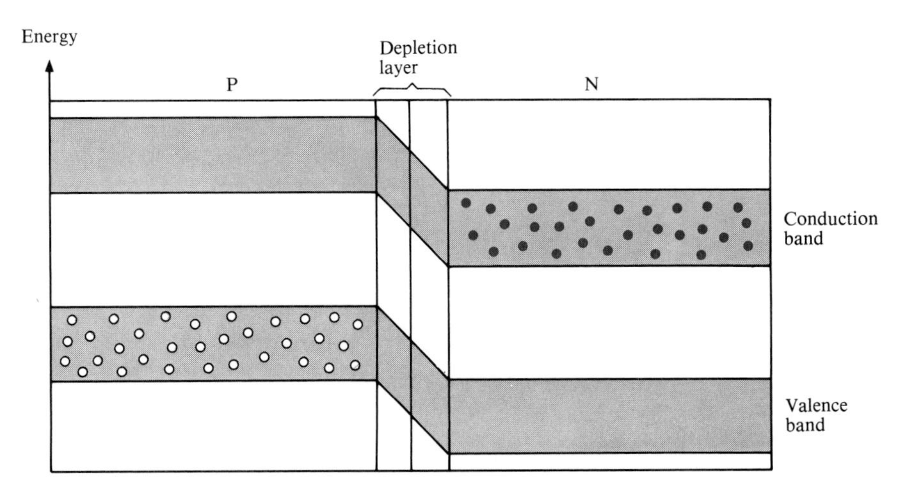

FIGURE 17–16
Energy diagram at equilibrium. The N-region bands are shifted down from their original positions when diffusion begins.

SECTION REVIEW 17–5

1. What is a PN junction?
2. When P and N regions are joined, a depletion layer forms. Describe the characteristics of the depletion layer.
3. The barrier potential for silicon is greater than for germanium (T or F).
4. What is the barrier potential for silicon at 25°C?

BIASING THE DIODE

17–6

As you have seen, there is no current across a PN junction at equilibrium. The primary usefulness of the PN junction diode is its ability to allow current in only *one* direction and to prevent current in the other direction as determined by the *bias*. There are two bias conditions for a PN junction: *forward* and *reverse*. Either of these conditions is created by application of an external voltage of the proper polarity.

Forward Bias

The term *bias* in electronics normally refers to a fixed voltage that sets the operating conditions for a semiconductor device. *Forward bias* is the condition that permits current across a PN junction. Figure 17–17 shows a dc voltage

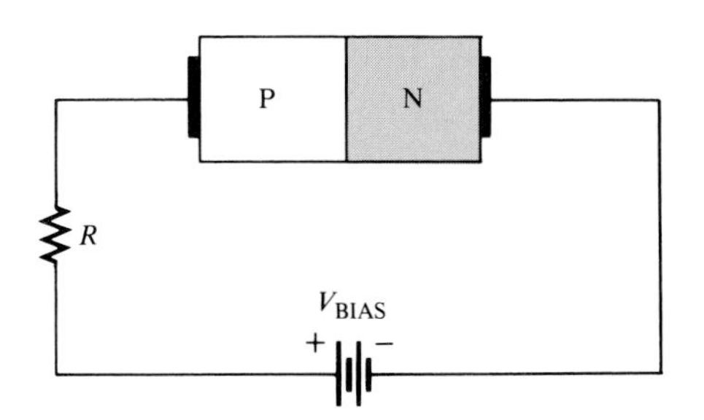

FIGURE 17–17
Forward-bias connection. The resistor limits the forward current in order to prevent damage to the diode.

connected in a direction to forward-bias the diode. Notice that the negative terminal of the battery is connected to the N region (called the *cathode*), and the positive terminal is connected to the P region (called the *anode*).

A discussion of the basic operation of forward bias follows: The negative terminal of the battery pushes the conduction band electrons in the N region toward the junction, while the positive terminal pushes the holes in the P region also toward the junction. Recall from Chapter 2 that like charges repel each other.

When it overcomes the barrier potential, the external voltage source provides the N-region electrons with enough energy to penetrate the depletion layer and cross the junction, where they combine with the P-region holes. As electrons leave the N region, more flow in from the negative terminal of the battery. Thus, current through the N region is the movement of conduction electrons (majority carriers) toward the junction.

Once the conduction electrons enter the P region and combine with holes, they become valence electrons. Then they move as valence electrons from hole to hole toward the positive connection of the battery. The movement of these valence electrons is the same as the movement of holes in the opposite direction. Thus, current in the P region is the movement of holes (majority carriers) toward the junction. Figure 17–18 illustrates current in a forward-biased diode.

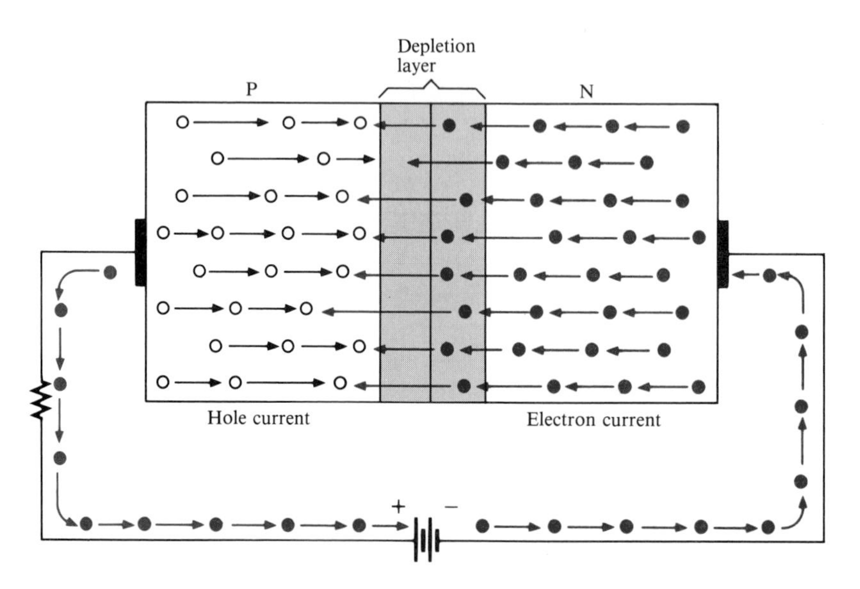

FIGURE 17–18
How current flows in a forward-biased diode.

Effect of Barrier Potential on Forward Bias

The barrier potential of the depletion layer can be envisioned as acting as a small battery that *opposes* bias, as illustrated in Figure 17–19. The resistances R_P and R_N represent the *bulk resistances* of the P and N materials.

The external bias voltage must overcome the barrier potential before the diode conducts, as illustrated in Figure 17–20. Conduction occurs at approximately 0.7 V for silicon and 0.3 V for germanium. Once the diode is conducting

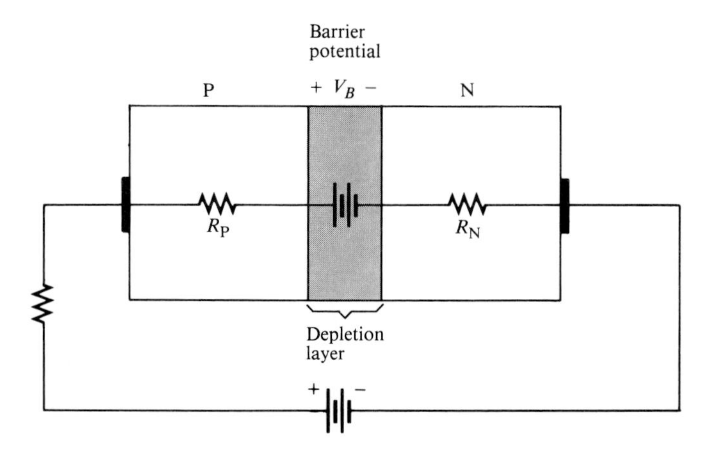

FIGURE 17–19
Barrier potential and bulk resistance equivalent for a PN junction diode.

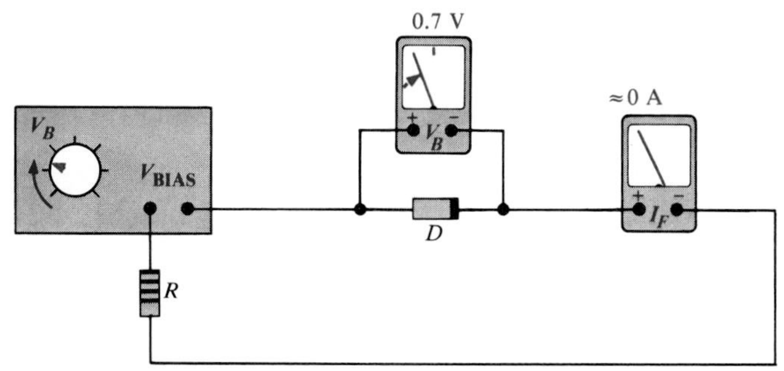

(a) When $V_{BIAS} < V_B$, the diode is essentially nonconducting.

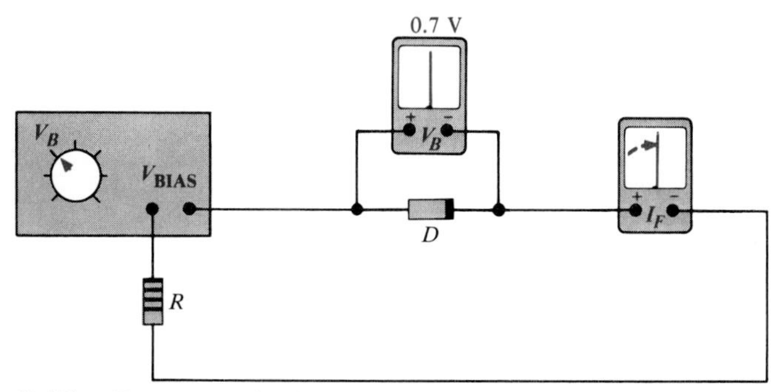

(b) When $V_{BIAS} = V_B$, the diode begins to conduct.

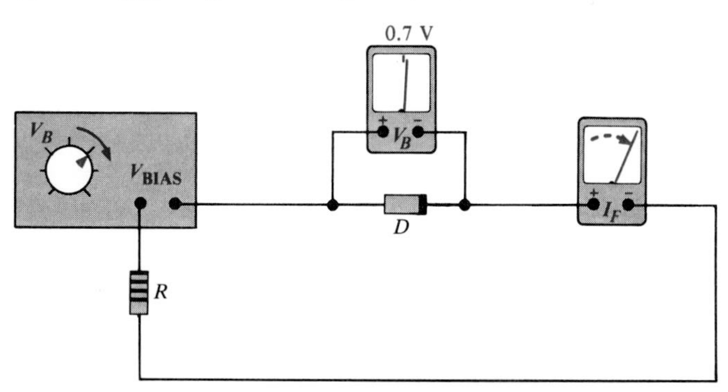

(c) When $V_{BIAS} > V_B$, the diode drop remains at approximately 0.7 V (silicon), but the forward current I_F continues to increase.

FIGURE 17–20
Illustration of diode operation under forward-bias conditions.

in the forward direction, the voltage drop across it remains at approximately the barrier potential and changes very little with changes in forward current (I_F) except for bulk resistance effects, as illustrated in Figure 17–20. The bulk resistances are usually only a few ohms and result in only a small voltage drop when the diode conducts. Often this drop can be neglected.

Energy Diagram for Forward Bias

Forward bias raises the energy levels of the conduction electrons in the N region, allowing them to move into the P region and combine with holes in the valence band. This condition is shown in Figure 17–21.

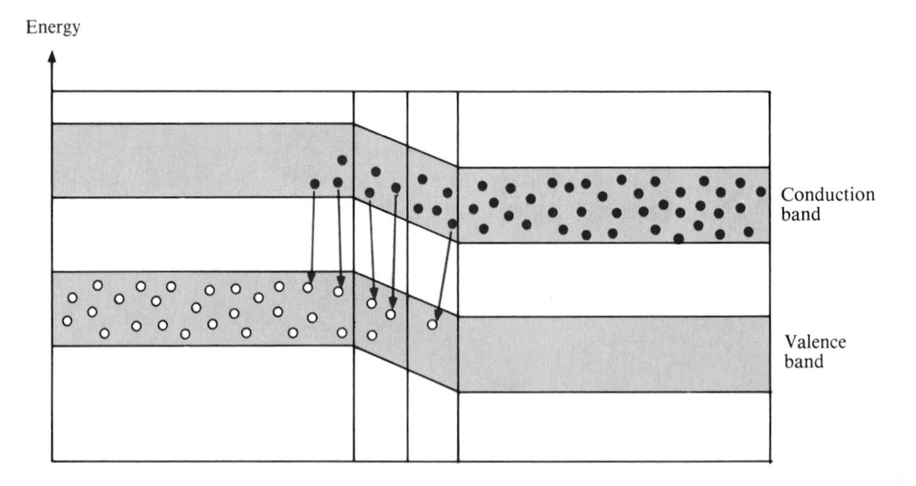

FIGURE 17–21
Energy diagram for forward bias, showing recombination in the depletion layer and in the P region as conduction electrons move across the junction.

Reverse Bias

Reverse bias is the condition that prevents current across the PN junction. Figure 17–22 shows a dc voltage source connected to reverse-bias the diode. Notice that the negative terminal of the battery is connected to the P region, and the positive terminal to the N region.

FIGURE 17–22
Reverse-bias connection.

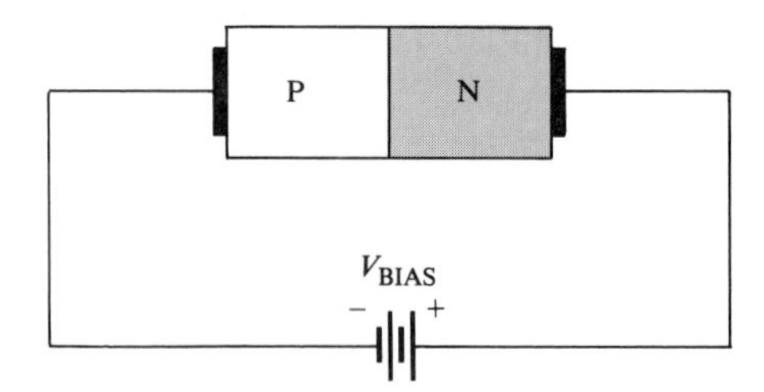

A discussion of the basic operation for reverse bias follows: The negative terminal of the battery attracts holes in the P region away from the PN junction, while the positive terminal also attracts electrons away from the junction. As electrons and holes move away from the junction, the depletion layer widens; more positive ions are created in the N region, and more negative ions are created in the P region, as shown in Figure 17–23(a).

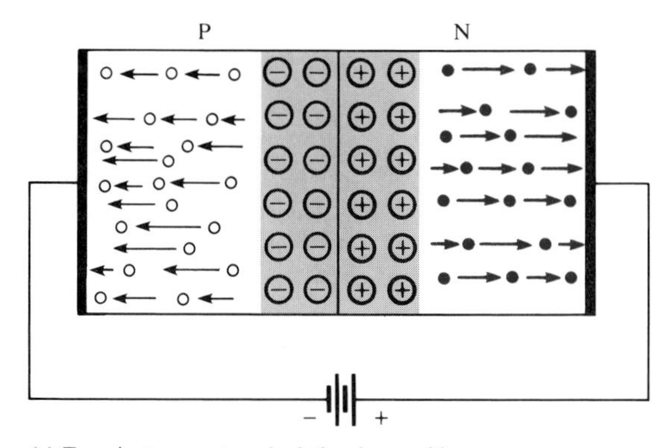

(a) Transient current as depletion layer widens

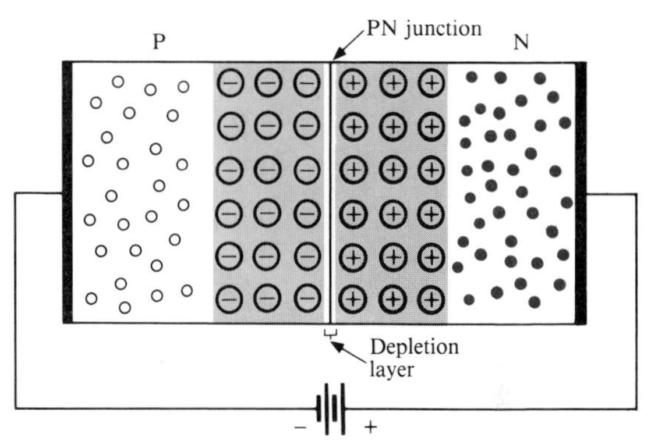

(b) Current ceases when barrier potential equals bias voltage

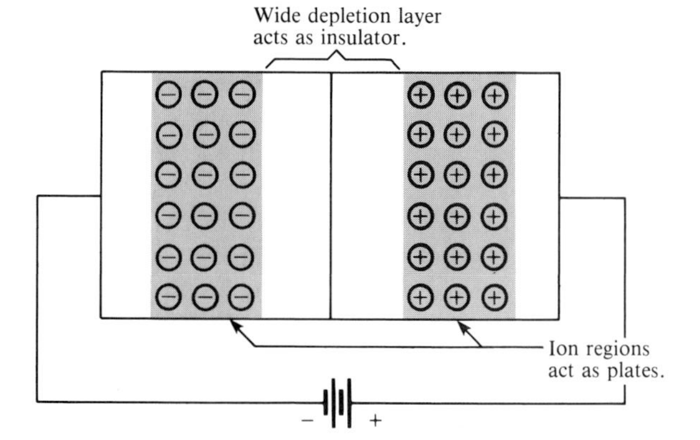

(c) Depletion layer widens as reverse bias increases

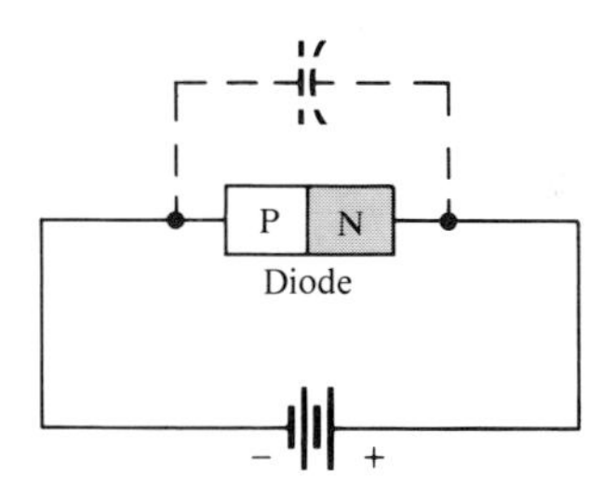

(d) Equivalent circuit with junction capacitance

FIGURE 17–23
Reverse bias.

The depletion layer widens until the potential difference across it equals the external bias voltage. At this point, the holes and electrons stop moving away from the junction, and majority current ceases, as indicated in Figure 17–23(b).

The initial movement of majority carriers away from the junction is called *transient current* and lasts only for a very short time upon application of reverse bias.

When the diode is reverse-biased, the depletion layer effectively acts as an insulator between the layers of oppositely charged ions, forming an effective capacitance, as illustrated in Figure 17–23(c). Since the depletion layer widens with increased reverse-biased voltage, the capacitance decreases, and vice versa. This internal capacitance is called the *depletion-layer capacitance* and can be represented by an equivalent circuit as shown in Figure 17–23(d).

Reverse Leakage Current

As you have learned, *majority current* very quickly becomes zero when reverse bias is applied. There is, however, a very small *leakage current* produced by minority carriers during reverse bias. Germanium, as a rule, has a greater leakage current than silicon. This current is typically in the μA or nA range. A relatively small number of thermally produced electron-hole pairs exist in the depletion layer. Under the influence of the external voltage, some electrons manage to diffuse across the PN junction before recombination. This process establishes a small minority carrier current throughout the material.

The reverse leakage current is dependent primarily on the junction temperature and not on the amount of reverse-biased voltage. A temperature increase causes an increase in leakage current.

Reverse Breakdown

If the external reverse-biased voltage is increased to a large enough value, *avalanche breakdown* occurs. The following describes what happens: Assume that one minority conduction band electron acquires enough energy from the external source to accelerate it toward the positive end of the diode. During its travel, it collides with an atom and imparts enough energy to knock a valence electron into the conduction band. There are now two conduction band electrons. Each will collide with an atom, knocking two more valence electrons into the conduction band. There are now four conduction band electrons which, in turn, knock four more into the conduction band. This rapid multiplication of conduction band electrons, known as an *avalanche effect,* results in a rapid build-up of reverse current.

Most diodes normally are not operated in reverse breakdown and can be damaged if they are. However, a particular type of diode (to be studied later), known as a *zener diode,* is optimized for reverse-breakdown operation.

SECTION REVIEW 17–6

1. Name the two bias conditions.
2. Which bias condition produces majority carrier current?
3. Which bias condition produces a widening of the depletion region?
4. Minority carriers produce the current during avalanche breakdown (T or F).

DIODE CHARACTERISTICS

17–7 Diode Characteristic Curve

As you learned in the last section, a diode conducts current when it is forward-biased if the bias voltage exceeds the barrier potential, and the diode prevents current when it is reverse-biased at less than the breakdown voltage.

Figure 17–24 is a graph of diode current versus voltage. The upper right quadrant of the graph represents the *forward-biased condition.* As you can see, there is very little forward current (I_F) for forward voltages (V_F) below the barrier potential. As the forward voltage approaches the value of the barrier potential (0.7 V for silicon and 0.3 V for germanium), the current begins to increase. Once the forward voltage reaches the barrier potential, the current increases drastically and must be limited by a series resistor. *The voltage across the forward-biased diode remains approximately equal to the barrier potential.*

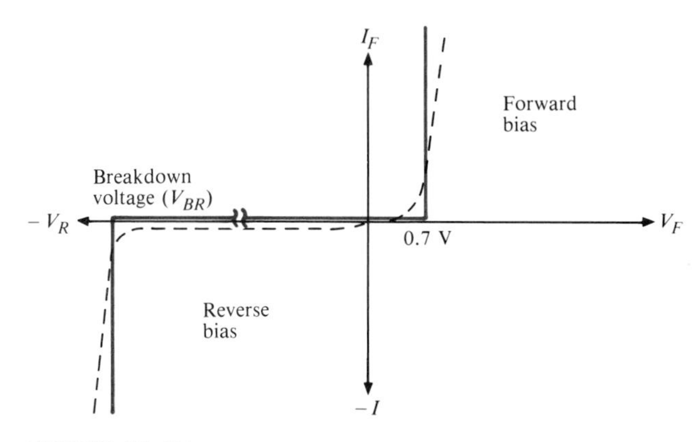

FIGURE 17–24
Diode characteristic curve (silicon).

The lower left quadrant of the graph represents the *reverse-biased condition.* As the reverse voltage increases to the left, the current remains near zero until the breakdown voltage is reached. When breakdown occurs, there is a large reverse current which, if not limited, can destroy the diode. Typically, the breakdown voltage is greater than 50 V for most rectifier diodes.

It should be noted here that most diodes should not be operated in reverse breakdown.

The solid curve in Figure 17–24 represents an *ideal diode.* In a practical diode, some forward conduction occurs before 0.7 V is reached. Also, there is a very small amount of reverse current before breakdown, as indicated by the dashed curve in the figure. In many applications, these nonideal characteristics can be neglected.

Diode Symbol

Figure 17–25(a) is the standard schematic symbol for a general-purpose diode. The arrow points in the direction opposite the electron flow. The two terminals of the diode are the *anode* and *cathode.*

When the anode is positive with respect to the cathode, the diode is *forward-biased* and current is from anode to cathode, as shown in Figure 17–25(b). Remember that when the diode is forward-biased, the barrier potential always appears between anode and cathode, as indicated in the figure. When the anode is negative with respect to the cathode, the diode is *reverse-biased,* as shown in Figure 17–25(c).

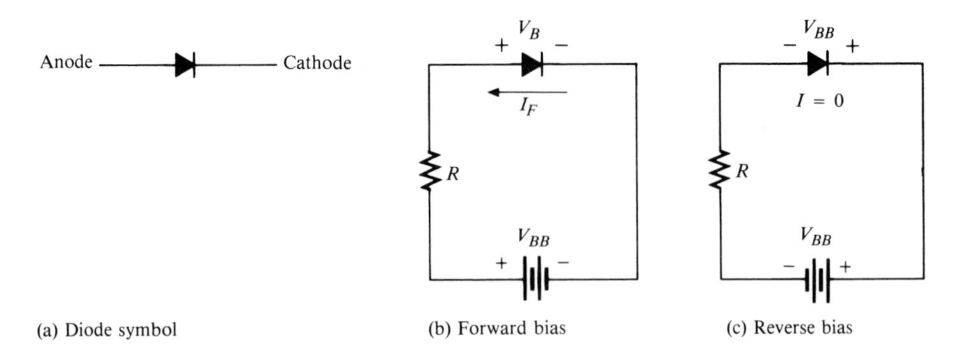

(a) Diode symbol (b) Forward bias (c) Reverse bias

FIGURE 17–25
Diode schematic symbol and bias circuits. V_{BB} is the bias battery voltage, and V_B is the barrier potential.

Some typical diodes are shown in Figure 17–26(a) to illustrate the variety of physical structures. Part (b) illustrates terminal identification.

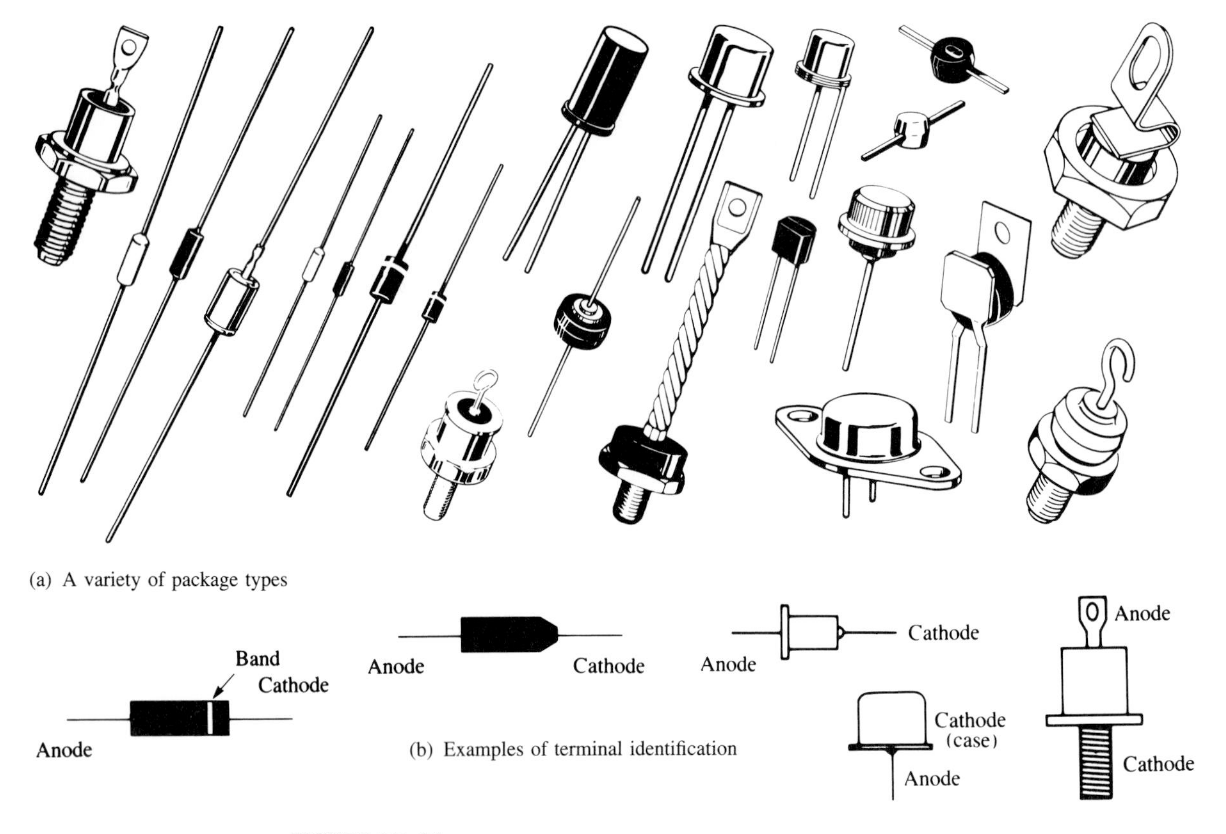

(a) A variety of package types

(b) Examples of terminal identification

FIGURE 17–26
Typical diodes.

Testing the Diode with an Ohmmeter

The internal battery in certain ohmmeters will forward-bias or reverse-bias a diode, permitting a quick and simple check for proper functioning. This check works for 1.5-V ohmmeters such as VOMs, but not for some of the digital multimeters. Many digital multimeters, however, have a diode test position.

To check the diode in the forward direction, connect the positive meter lead to the anode and the negative lead to the cathode, as shown in Figure 17–27(a). When a diode is forward-biased, its internal resistance is low (typically less than 100 Ω).

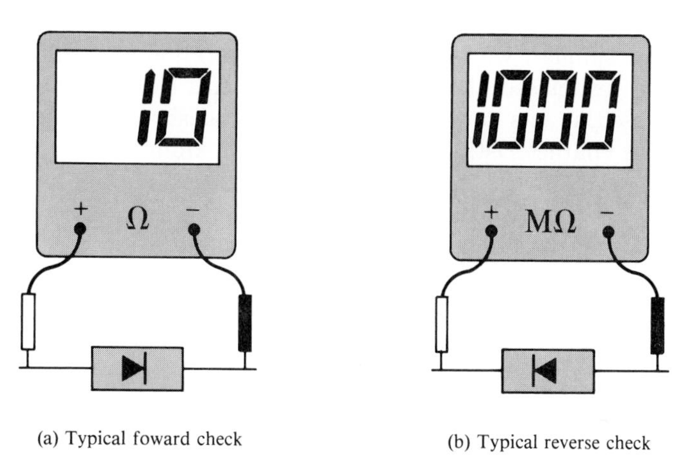

FIGURE 17–27
Checking a semiconductor diode with an ohmmeter.

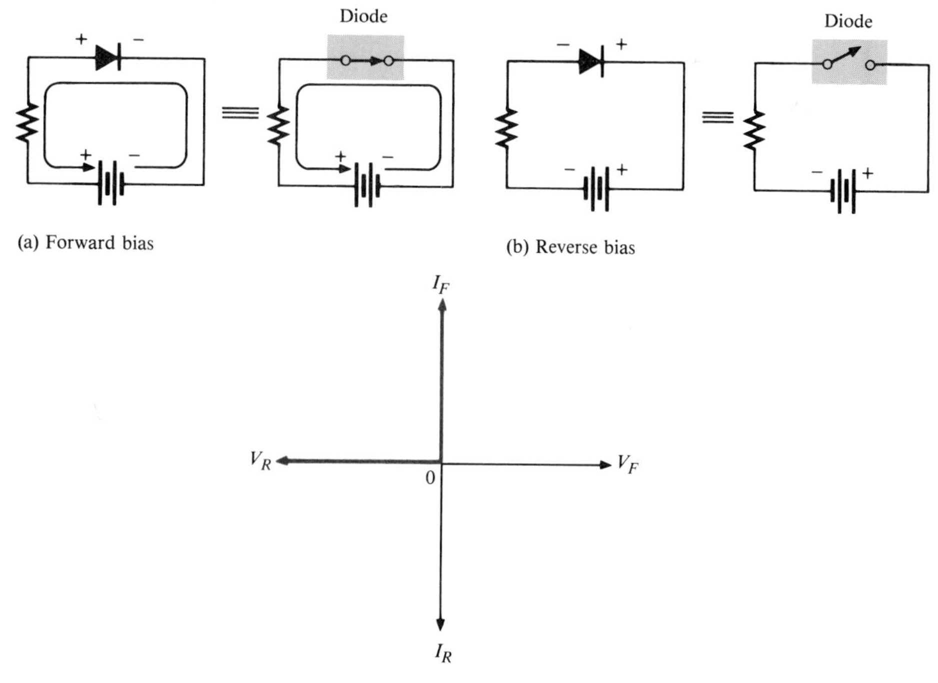

FIGURE 17–28
Ideal approximation of the diode as a switch.

When the meter leads are reversed, as shown in Figure 17–27(b), the internal ohmmeter battery reverse-biases the diode, and a very large resistance value (ideally, infinite) is indicated. The PN junction is *shorted* if a low resistance is indicated in both bias conditions; it is *open* if a very high resistance is read for both checks.

Diode Approximations

The simplest way to visualize diode operation is to think of it as a *switch*. When forward-biased, the diode ideally acts as a closed (on) switch, and when reverse-biased, it acts as an open (off) switch, as shown in Figure 17–28 (on page 637). The characteristic curve for this approximation is also shown. Note that the forward voltage and the reverse current are always zero in the ideal case.

This ideal model, of course, neglects the effect of the barrier potential, the internal resistances, and other parameters. However, in many cases it is accurate enough.

The next higher level of accuracy is the barrier potential model. In this approximation, the forward-biased diode is represented as a closed switch in

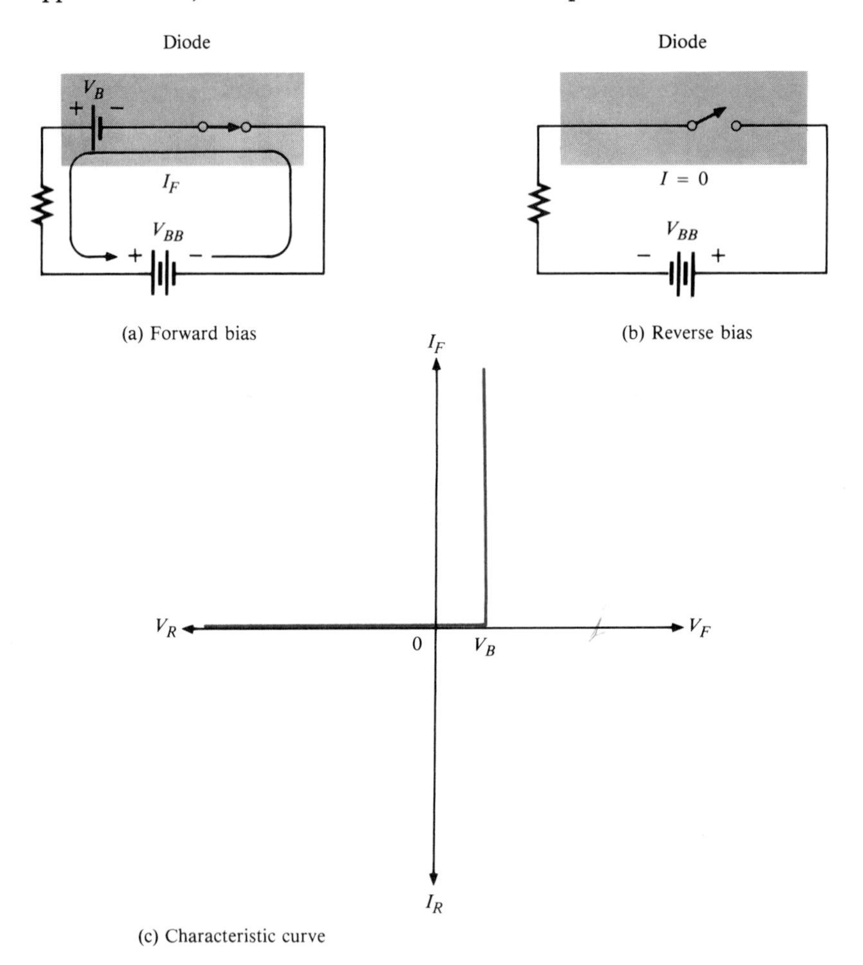

FIGURE 17–29
Diode approximation including barrier potential.

series with a small "battery" equal to the barrier potential V_B (0.7 V for Si and 0.3 V for Ge), as shown in Figure 17–29(a). The positive end of the battery is toward the anode. Keep in mind that the barrier potential cannot actually be measured with a voltmeter; rather it only has the *effect* of a battery when forward bias is applied. The reverse-biased diode is represented by an open switch, as in the ideal case, because the barrier potential does not affect reverse bias, as shown in Figure 17–29(b). The characteristic curve for this ideal model is shown in Part (c). In this book, the barrier potential is included in analysis unless otherwise stated.

One more level of accuracy will be considered at this point. Figure 17–30(a) shows the forward-biased diode model with both the *barrier potential* and the low *forward (bulk) resistance.* Figure 17–30(b) shows how the high *reverse resistance* (leakage resistance) affects the reverse-biased model. The characteristic curve is shown in Part (c).

Other parameters such as junction capacitance and breakdown voltage become important only under certain operating conditions and will be considered where appropriate.

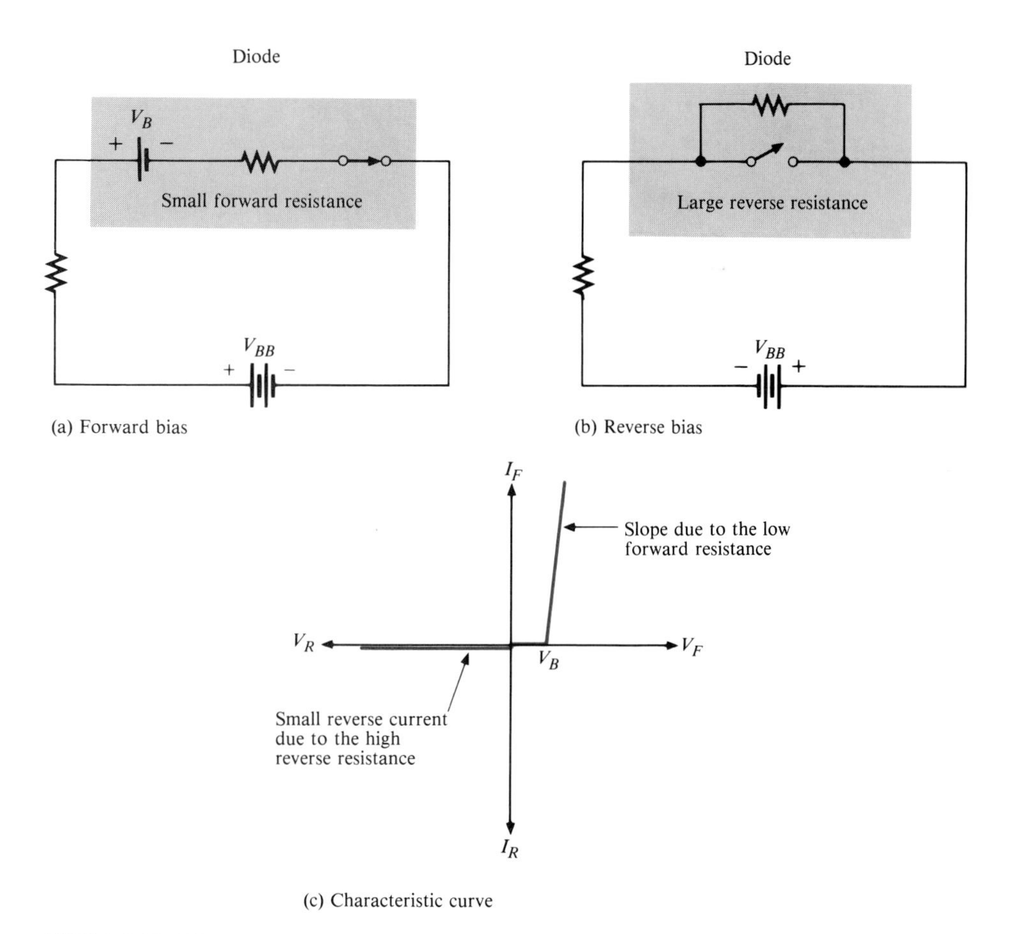

FIGURE 17–30
Diode approximation including barrier potential, forward resistance, and reverse leakage resistance.

SECTION REVIEW 17–7

1. Sketch a rectifier diode symbol and label the terminals.
2. For a normal diode, the forward resistance is quite low and the reverse resistance is very high (T or F).
3. An open switch ideally represents a ________ -biased diode. A closed switch ideally represents a ________ -biased diode.

SUMMARY

Facts

- An atom is described as a nucleus containing protons and neutrons orbited by electrons.
- Protons are positive, neutrons are neutral, and electrons are negative.
- Atomic shells are energy bands.
- The outermost shell containing electrons is the valence shell.
- Silicon and germanium are the predominate semiconductor materials.
- Atoms within a crystal structure are held together with covalent bonds.
- Electron-hole pairs are thermally produced.
- The process of adding impurities to an intrinsic semiconductor to increase and control conductivity is called *doping.*
- A P-type semiconductor is doped with trivalent impurity atoms.
- An N-type semiconductor is doped with pentavalent impurity atoms.
- The depletion layer is a region adjacent to the PN junction containing no majority carriers.
- Forward bias permits majority carrier current through the PN junction.
- Reverse bias prevents majority carrier current.
- A PN structure is called a *diode.*
- Reverse leakage current is due to thermally produced electron-hole pairs.
- Reverse breakdown occurs when the reverse-biased voltage exceeds a specified value.

Definitions

- *Atom*—the smallest particle of an element that retains the characteristics of that element.
- *Positive ion*—an atom that has lost a valence electron.
- *Negative ion*—an atom that has gained an extra valence electron.
- *Free electron*—a conduction band electron.
- *Intrinsic semiconductor*—a pure material with relatively few free electrons.
- *PN junction*—the boundary between N-type and P-type materials.
- *Barrier potential*—the inherent voltage across the depletion layer.

SELF-TEST

1. Which particles in an atom have a negative charge? Positive charge? No charge?
2. Both silicon and germanium have a valence of ________ .
3. What is the valence shell designation in a silicon atom?
4. Distinguish between a neutral atom, a positive ion, and a negative ion.
5. Define *covalent bond.*

6. In what energy band do free electrons exist?
7. Describe what happens when an electron-hole pair is produced.
8. Define *recombination.*
9. What are the two types of current in a semiconductor?
10. What is the essential difference between a semiconductor and an insulator?
11. How does a trivalent impurity modify an intrinsic semiconductor?
12. How does a pentavalent impurity modify an intrinsic semiconductor?
13. What are the majority carriers in an N-type semiconductor?
14. Approximately what value does a bias voltage have to be in order to forward-bias a silicon PN junction?
15. What is the current in the circuit of Figure 17–31 if the total bulk resistance of the silicon diode is 10 Ω?
16. In each circuit of Figure 17–32, indicate whether the diode is forward-biased or reverse-biased.
17. What is the voltage across each silicon diode in Figure 17–32?
18. Indicate which ohmmeter reads low resistance and which reads high resistance in Figure 17–33.

FIGURE 17–31

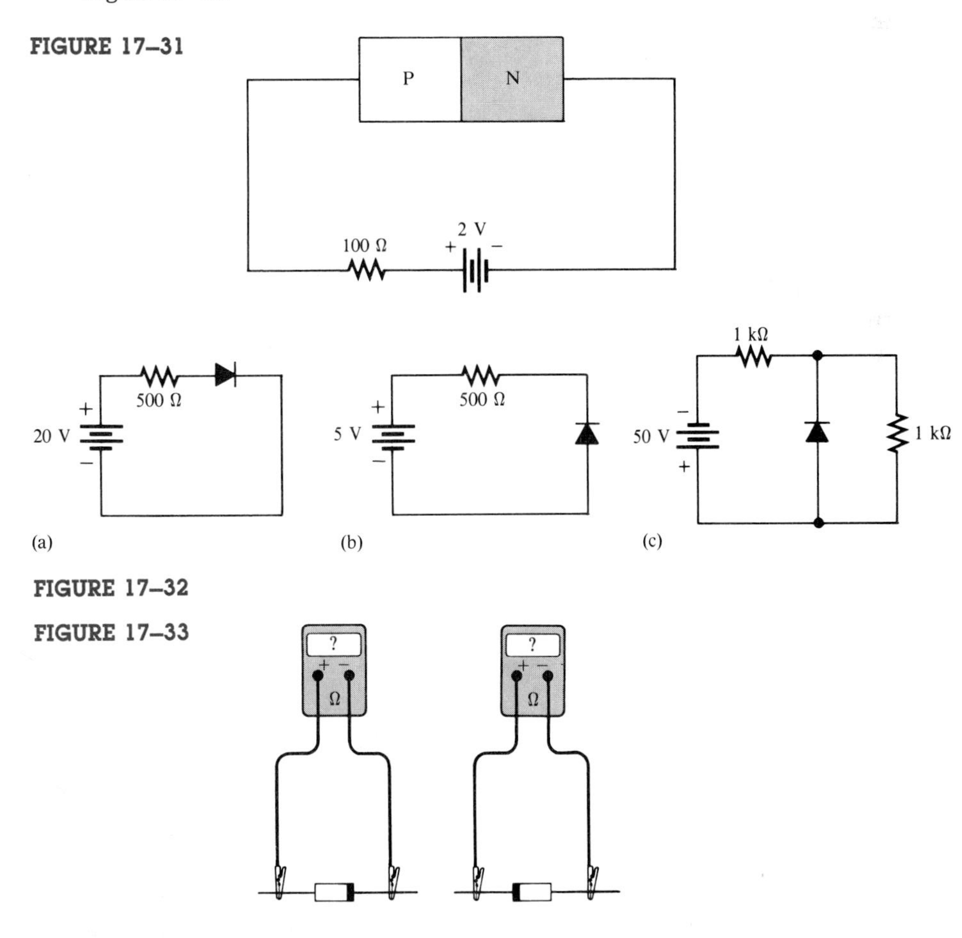

FIGURE 17–32

FIGURE 17–33

ANSWERS TO SECTION REVIEWS

Section 17–1

1. The smallest particle that retains the characteristics of its element.
2. Electrons, protons, neutrons.
3. An electron in the outermost shell (valence band).
4. An electron in the conduction band.

Section 17–2

1. Silicon, germanium. 2. The sharing of electrons with neighboring atoms.

Section 17–3

1. Conduction band, valence band.
2. An electron is thermally raised to the conduction band, leaving a hole in the valence band.
3. The energy gap between the valence band and the conduction band is narrower for a semiconductor.

Section 17–4

1. By the addition of pentavalent atoms to the intrinsic semiconductor material.
2. By the addition of trivalent atoms to the intrinsic semiconductor material.
3. The particle in greatest abundance: electrons in N-type material and holes in P-type material.

Section 17–5

1. The boundary between N-type and P-type materials.
2. Devoid of majority carriers, contains only positive and negative ions. 3. T.
4. 0.7 V.

Section 17–6

1. Forward, reverse. 2. Forward. 3. Reverse. 4. T.

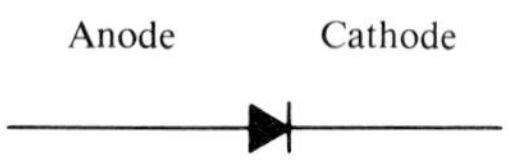

Section 17–7

1. See Figure 17–34. 2. T. 3. Reverse, forward.

Anode Cathode

FIGURE 17–34

Diodes and Applications

18–1 Half-Wave Rectifiers

18–2 Full-Wave Rectifiers

18–3 Rectifier Filters

18–4 Troubleshooting Rectifier Circuits

18–5 Diode Clipping and Clamping Circuits

18–6 Zener Diodes

18–7 Varactor Diodes

18–8 LEDs and Photodiodes

This chapter discusses the applications of diodes in converting ac to dc by the process known as *rectification*. Half-wave and full-wave rectification are introduced, and the basic circuits are studied. The limitations of diodes used in rectifier applications are examined, and diode clipping circuits and dc restoring (clamping) circuits are studied.

In addition to rectifier diodes, zener diodes and their applications in voltage regulation are introduced. Varactor diodes, light-emitting diodes, and photodiodes and their applications also are discussed.

Specifically, you will learn:

- What is meant by *half-wave rectification* and how a basic half-wave rectifier circuit works.
- How to determine the average value of a half-wave signal.
- What is meant by *full-wave rectification*.
- How to determine the average value of a full-wave signal.
- How a center-tapped full-wave rectifier works.
- How a bridge rectifier works.
- How to determine the PIV (peak inverse voltage) across the diodes in the full-wave rectifiers.
- How a capacitor-input filter smooths out a rectified voltage.
- What ripple voltage is.
- How the value of the filter capacitor affects the amount of ripple.
- How an inductor-input filter improves the dc output voltage.
- The characteristics of the zener diode and how it is used as a voltage regulator.
- The meaning of *line regulation* and *load regulation*.
- What a varactor diode is and how it is used as a variable capacitor.
- The basic principles of light-emitting diodes (LEDs) and photodiodes.

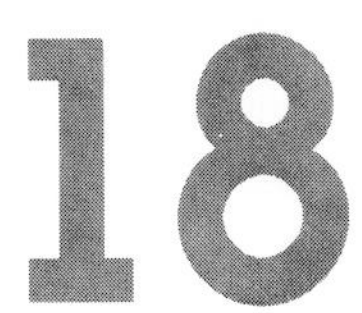

APPLICATION ASSIGNMENT

As a lab technician with a manufacturer of electronic test equipment, you are assigned to a group that is responsible for the development of a new signal generator. Today, your boss asks you to build a circuit that will regulate a dc voltage that can vary from 8 V to 12 V to maintain a nominal 6.8-V output. He tells you no more.

After studying this chapter, you will know what kind of device to use and how to build the circuit. You will also know what additional information you will need. A solution is given in the Application Note at the end of the chapter.

HALF-WAVE RECTIFIERS

18–1

Because of their unique ability to conduct current in only one direction, diodes are used in rectifier circuits. *Rectification* is the process of converting ac to pulsating dc.

In Figure 18–1(a), an ac source is connected to a load resistor through a diode. Let's examine what happens during one cycle of the input voltage.

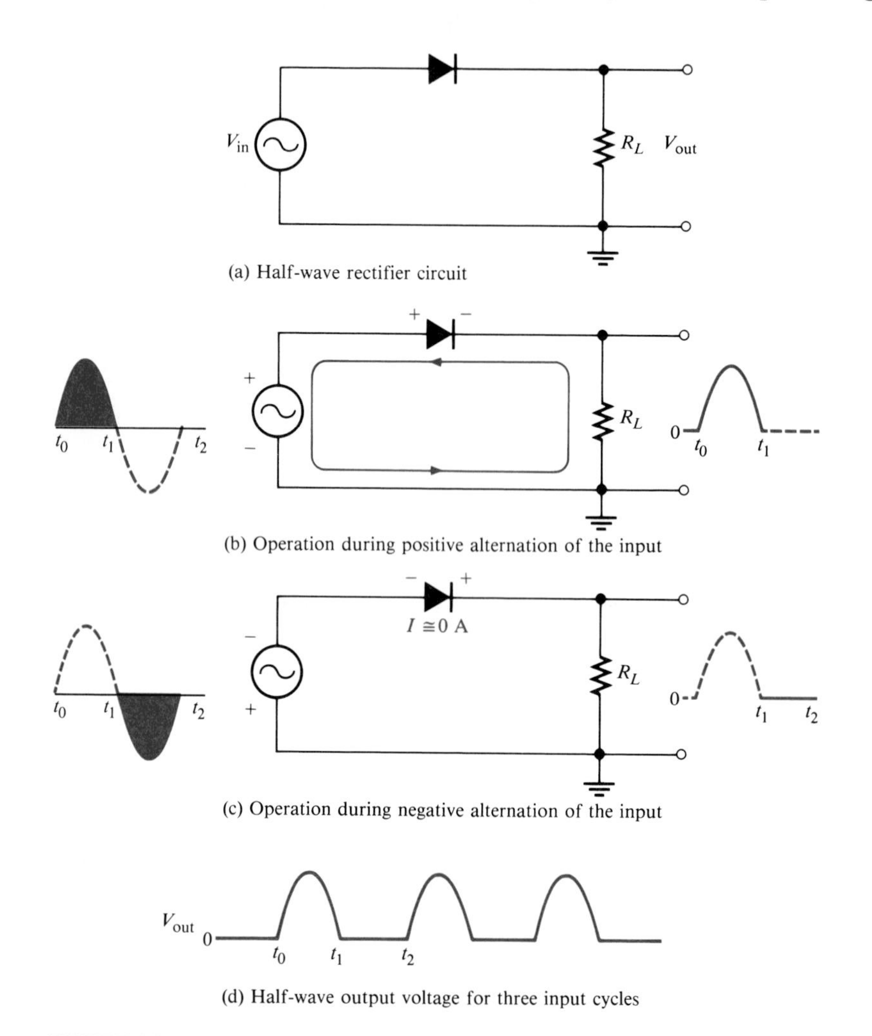

FIGURE 18–1
Operation of half-wave rectifier.

When the sine wave input voltage goes *positive,* the diode is *forward-biased* and conducts current to the load resistor, as shown in Part (b). The current produces a voltage across the load which has the same shape as the positive half-cycle of the input voltage. When the input voltage goes *negative* during the second half of its cycle, the diode is *reversed-biased.* There is no current, so the voltage across the load resistor is 0, as shown in Part (c). The net result is that only the positive half-cycles of the ac input voltage appear across the load, making the output a *pulsating dc voltage,* as shown in Part (d). This process is called *half-wave rectification.*

Average Value of Half-Wave Output

The average value is the value that would be indicated by a dc voltmeter. It can be calculated with the following equation where V_p is the peak value:

$$V_{AVG} = \frac{V_p}{\pi} \tag{18–1}$$

Figure 18–2 shows the half-wave voltage with its average value indicated by the colored dashed line.

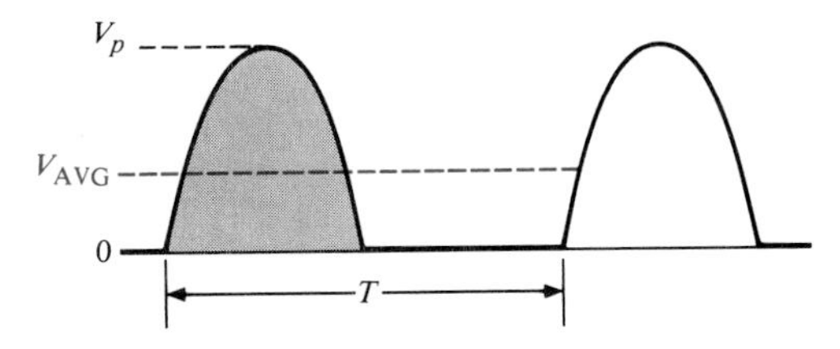

FIGURE 18–2
Average value of half-wave rectified signal.

EXAMPLE 18–1

What is the average (dc) value of the half-wave rectified voltage waveform in Figure 18–3?

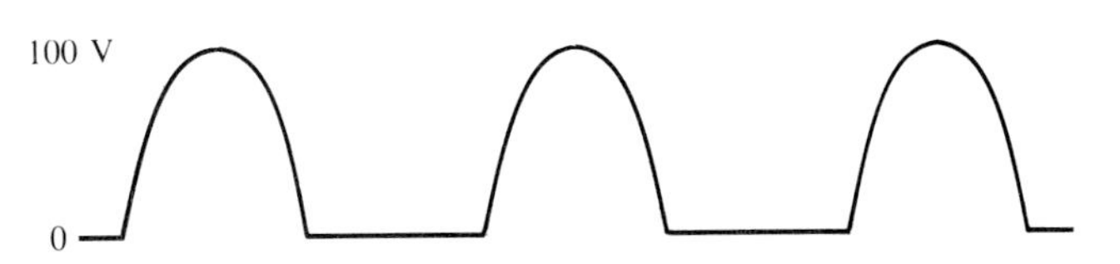

FIGURE 18–3

Solution

$$V_{AVG} = \frac{V_p}{\pi} = \frac{100 \text{ V}}{\pi} = 31.83 \text{ V}$$

Effect of Barrier Potential on Half-Wave Rectifier Output

In the previous discussion, the diode was considered ideal. When the diode barrier potential is taken into account, here is what happens: During the positive half-cycle, the input voltage must overcome the barrier potential before the diode becomes forward-biased. For a silicon diode this results in a half-wave output with a peak value that is 0.7 V less than the peak value of the input

(0.3 V less for a germanium diode), as shown in Figure 18–4. The expression for *peak* output voltage is

$$V_{p(\text{out})} = V_{p(\text{in})} - 0.7\text{ V} \quad (18\text{–}2)$$

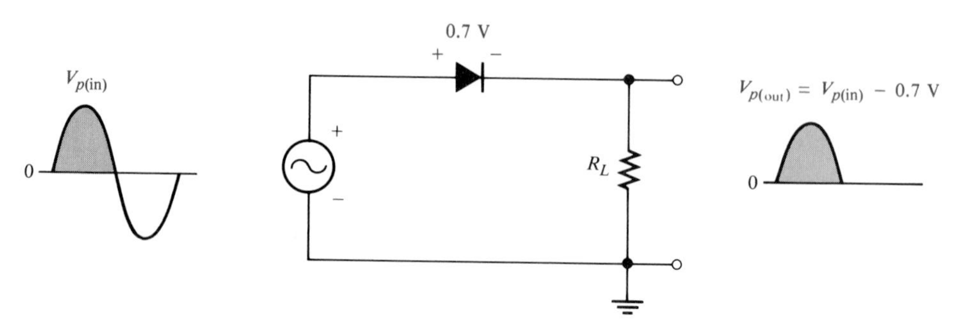

FIGURE 18–4
Effect of barrier potential on half-wave rectified output voltage (silicon diode shown).

In working with diode circuits, you will find that it is often practical to neglect the effect of barrier potential when the peak value of the applied voltage is much greater (at least ten times) than the barrier potential. As mentioned before, we will always use silicon diodes and consider the barrier potential unless stated otherwise.

EXAMPLE 18–2 Determine the peak output voltage of the silicon rectifier circuit in Figure 18–5 for the indicated input voltage.

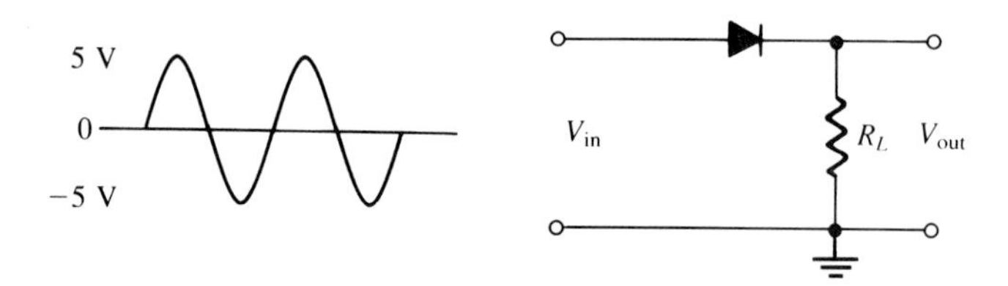

FIGURE 18–5

Solution
The peak half-wave output voltage is

$$V_p = 5\text{ V} - 0.7\text{ V} = 4.3\text{ V}$$

Peak Inverse Voltage (PIV)

The maximum value of reverse voltage, sometimes designated as *peak inverse voltage* (PIV), occurs at the peak of the *negative alternation* of the input cycle when the diode is *reverse-biased.* This condition is illustrated in Figure 18–6.

The PIV equals the peak value of the input voltage, and the diode must be capable of withstanding this amount of repetitive reverse voltage.

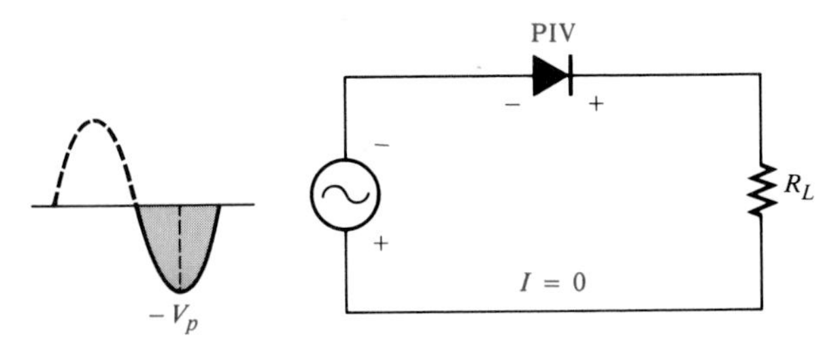

FIGURE 18–6
The PIV occurs at the peak of the half-cycle when the diode is reverse-biased. In this circuit, the PIV occurs at the peak of the negative half-cycle.

SECTION REVIEW 18–1

1. At what point on the input cycle does the PIV occur?
2. For a half-wave rectifier, there is current through the load for approximately what percentage of the input cycle?
3. What is the average value of the voltage shown in Figure 18–7?

FIGURE 18–7

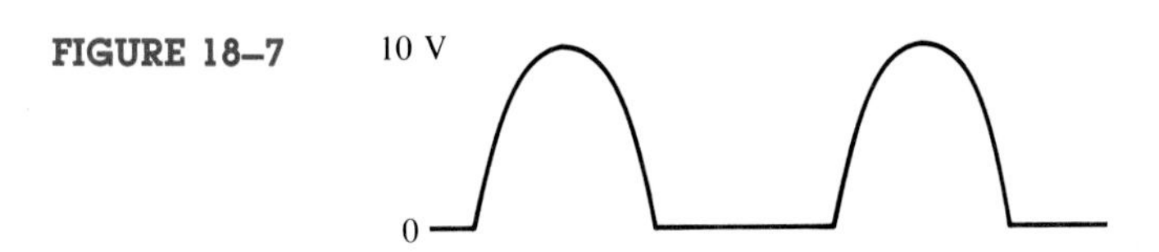

FULL-WAVE RECTIFIERS

18–2

The difference between full-wave and half-wave rectification is that a full-wave rectifier allows unidirectional current to the load during the entire input cycle, and the half-wave rectifier allows this only during one half of the cycle. The result of full-wave rectification is a dc output voltage that pulsates every half-cycle of the input, as shown in Figure 18–8.

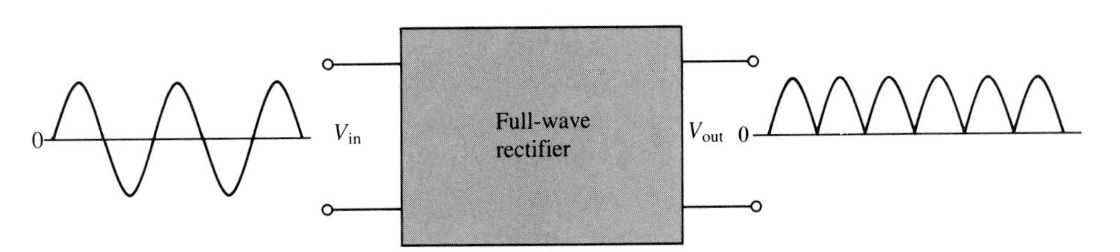

FIGURE 18–8
Full-wave rectification.

The average value for a full-wave rectified voltage is twice that of the half-wave, expressed as follows:

$$V_{AVG} = \frac{2V_p}{\pi} \qquad (18\text{–}3)$$

EXAMPLE 18–3 Find the average value of the full-wave rectified voltage in Figure 18–9.

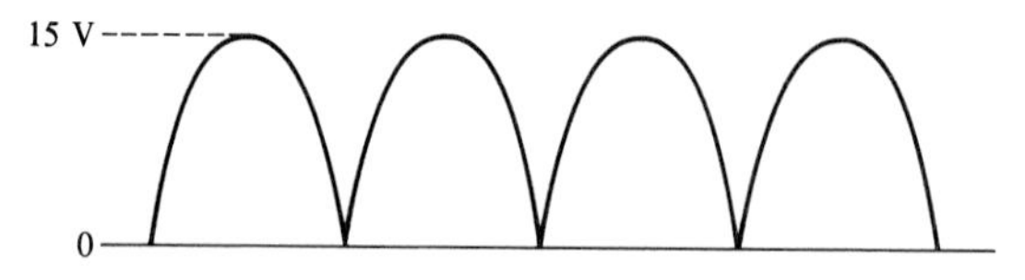

FIGURE 18–9

Solution

$$V_{AVG} = \frac{2V_p}{\pi} = \frac{2(15\text{ V})}{\pi} = 9.55\text{ V}$$

Center-Tapped Full-Wave Rectifier

This type of full-wave rectifier circuit uses two diodes connected to the secondary of a center-tapped transformer, as shown in Figure 18–10. The input signal is coupled through the transformer to the secondary. Half of the secondary voltage appears between the center tap and each end of the secondary winding as shown.

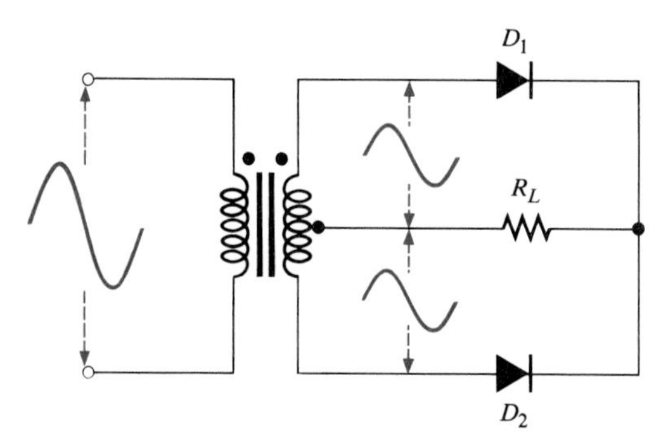

FIGURE 18–10
A center-tapped full-wave rectifier.

For a positive half-cycle of the input voltage, the polarities of the secondary voltages are as shown in Figure 18–11(a). *This condition forward-biases the upper diode D_1 and reverse-biases the lower diode D_2.* The current path is through D_1 and the load resistor, as indicated.

For a negative half-cycle of the input voltage, the voltage polarities on the secondary are as shown in Figure 18–11(b). This condition *reverse-biases D_1 and forward-biases D_2.* The current path is through D_2 and the load resistor, as indicated.

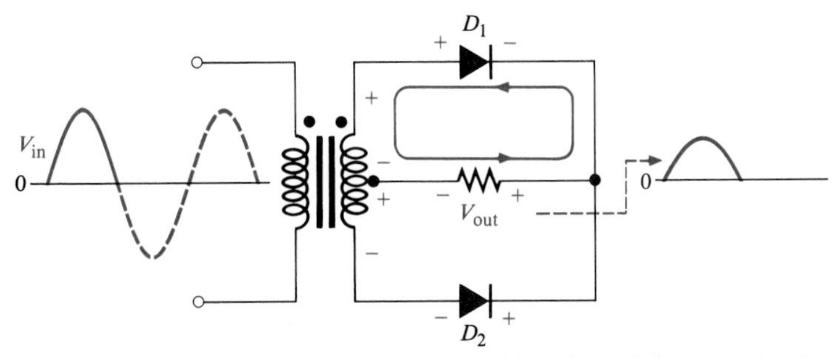

(a) During positive half-cycles, D_1 is forward-biased and D_2 is reverse-biased.

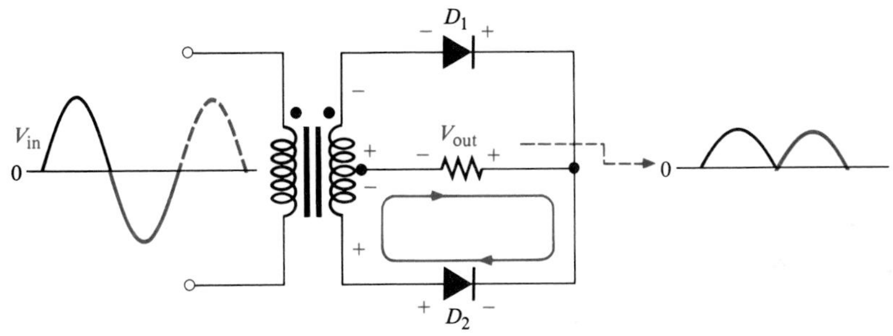

(b) During negative half-cycles, D_2 is forward-biased and D_1 is reverse biased.

FIGURE 18–11
Basic operation of a center-tapped full-wave rectifier. Note that the current through the load resistor is in the same direction during the entire input cycle.

Because the current during both the positive and the negative portions of the input cycle is in the *same direction* through the load, the output voltage developed across the load is a full-wave rectified dc voltage.

Effect of the Turns Ratio on Full-Wave Output Voltage

If the turns ratio of a transformer is 1, the peak value of the rectified output voltage equals half the peak value of the primary input voltage less the barrier potential (diode drop). This value occurs because half of the input voltage appears across each half of the secondary winding.

In order to obtain an output voltage equal to the input (less the barrier potential), you must use a step-up transformer with a turns ratio of 1-to-2 (1:2). In this case, the total secondary voltage is twice the primary voltage, so the voltage across each half of the secondary is equal to the input.

Peak Inverse Voltage (PIV)

Each diode in the full-wave rectifier is alternately forward-biased and then reverse-biased. The maximum reverse voltage that each diode must withstand is the peak value of the total secondary voltage (V_s), as illustrated in Figure 18–12. When the total secondary voltage V_s has the polarity shown, the anode of D_1 is $+V_s/2$ and the anode of D_2 is $-V_s/2$. Since D_1 is forward-biased, its cathode is at the same voltage as its anode ($+V_s/2$, neglecting the barrier potential); this is also the voltage on the cathode of D_2. The total reverse voltage across D_2 therefore is $+V_s/2 - (-V_s/2) = V_s/2 + V_s/2 = V_s$. Since $V_s = 2V_{out}$, the *peak inverse voltage* across either diode in the center-tapped full-wave rectifier is

$$PIV = 2V_{p(out)} \tag{18–4}$$

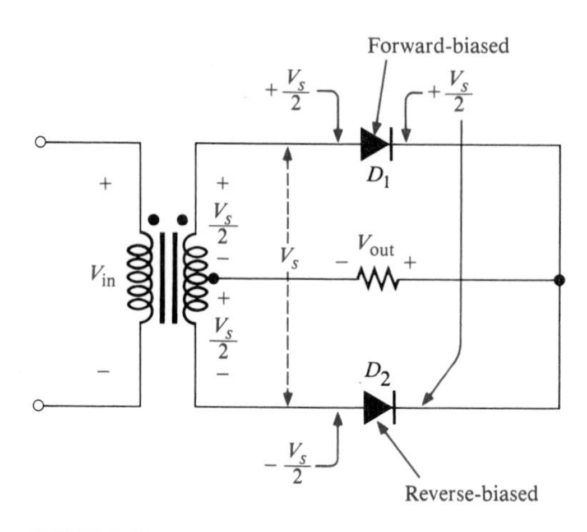

FIGURE 18–12
PIV across a diode is twice the peak value of the output voltage.

EXAMPLE 18–4

(a) Show the voltage waveforms across the secondary winding and across R_L when a 25-V peak sine wave is applied to the primary winding in Figure 18–13.

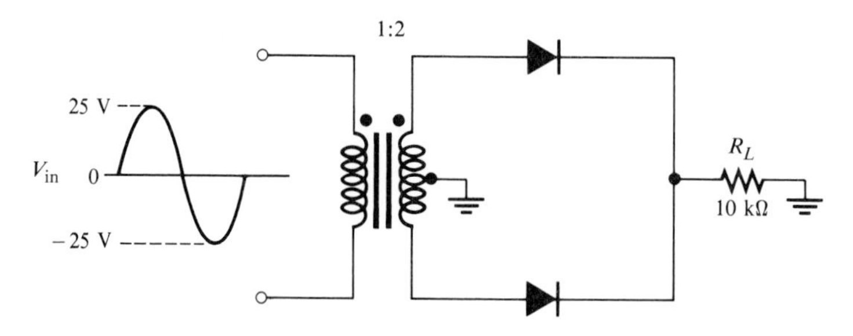

FIGURE 18–13

(b) What minimum PIV rating must the diodes have?

Solution

(a) The waveforms are shown in Figure 18–14.
(b) The total peak secondary voltage is

$$V_s = \left(\frac{N_2}{N_1}\right)V_{in} = (2)25\text{ V} = 50\text{ V}$$

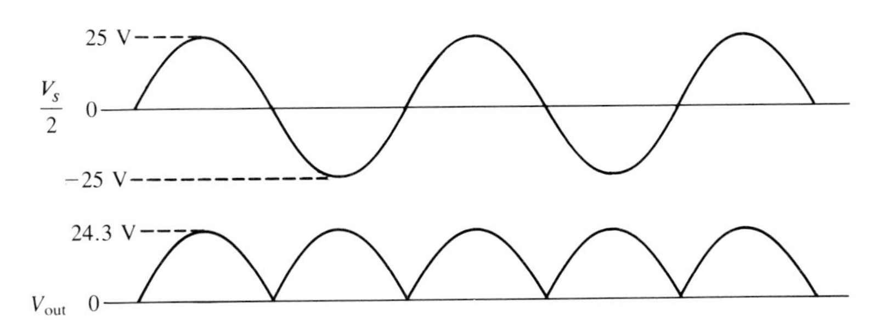

FIGURE 18–14

There is a 25-V peak across each half of the secondary. The output load voltage has a peak value of 25 V, less the 0.7-V drop across the diode. Each diode must have a minimum PIV rating of 50 V − 0.7 V = 49.3 V.

Full-Wave Bridge Rectifier

The full-wave bridge rectifier uses four diodes, as shown in Figure 18–15. When the input cycle is positive as in Part (a), diodes D_1 and D_2 are forward-biased and conduct current in the direction shown. A voltage is developed across R_L which looks like the positive half of the input cycle. During this time, diodes D_3 and D_4 are reverse-biased.

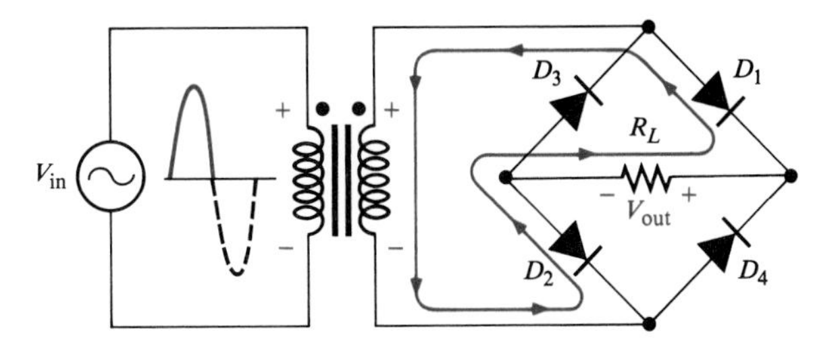

(a) During positive half-cycle of the input, D_1 and D_2 are forward-biased and conduct current. D_3 and D_4 are reverse-biased.

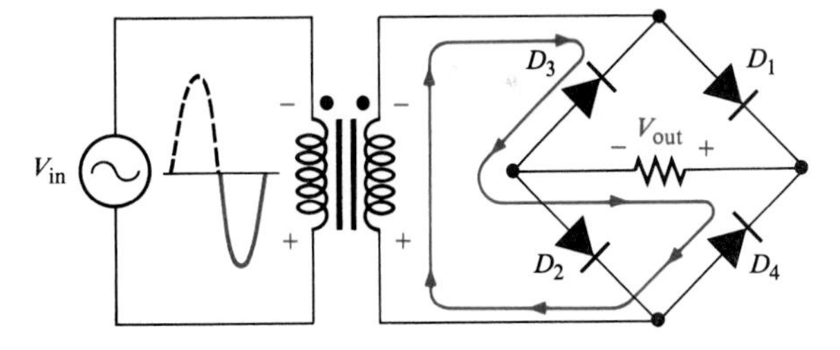

(b) During negative half-cycle, D_3 and D_4 are forward-biased and conduct current. D_1 and D_2 are reverse-biased.

FIGURE 18–15
Operation of full-wave bridge rectifier.

When the input cycle is negative as in Part (b), diodes D_3 and D_4 are forward-biased and conduct current in the same direction through R_L as during the positive half-cycle. During the negative half-cycle, D_1 and D_2 are reverse-biased. A full-wave rectified output voltage appears across R_L as a result of this action.

Bridge Output Voltage

During the positive half-cycle of the secondary voltage, diodes D_1 and D_2 are forward-biased. Neglecting the diode drops, the total secondary voltage V_s appears across the load resistor. The same is true when D_3 and D_4 are forward-biased during the negative half-cycle. Thus,

$$V_{out} = V_s \tag{18–5}$$

As you can see in Figure 18–15, *two* diodes are always in series with the load resistor during both the positive and the negative half-cycles. If these diode drops are taken into account, the output voltage (with silicon diodes) is

$$V_{out} = V_s - 1.4\text{ V} \tag{18–6}$$

Peak Inverse Voltage (PIV)

Assuming that D_1 and D_2 are forward-biased, let us examine the reverse voltage across D_3 and D_4. Visualizing D_1 and D_2 as shorts (ideally), as in Figure 18–16, you can see that D_3 and D_4 have a peak inverse voltage equal to the peak secondary voltage. Since the output voltage is *ideally* equal to the secondary voltage, we have

$$\text{PIV} \cong V_{p(out)} \tag{18–7}$$

The PIV rating of the bridge diodes is half that required for the center-tapped configuration.

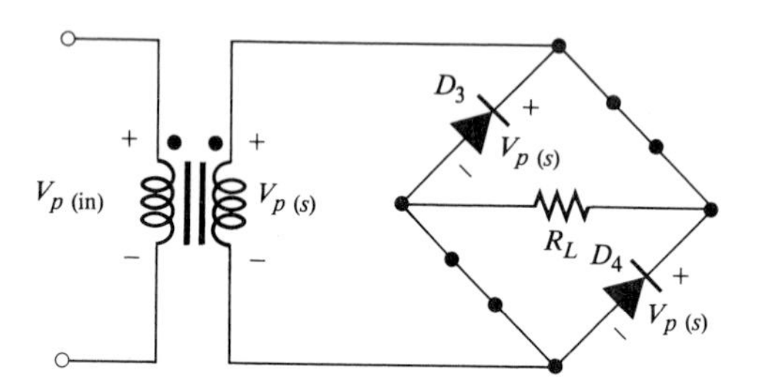

FIGURE 18–16
PIV in a bridge rectifier during the positive half-cycle of the input voltage.

EXAMPLE 18–5

(a) Determine the output voltage for the bridge rectifier in Figure 18–17.
(b) What minimum PIV rating is required for the silicon diodes?

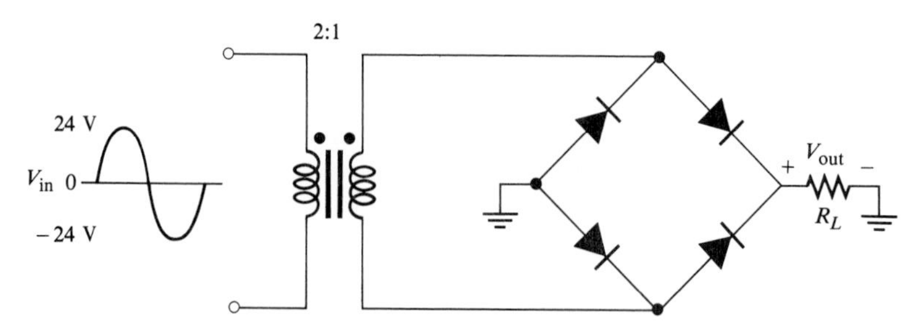

FIGURE 18–17

Solution

(a) The peak output voltage is (taking into account the two diode drops)

$$V_{p(\text{out})} = V_s - 1.4\text{ V} = \left(\frac{N_2}{N_1}\right)V_{p(\text{in})} - 1.4\text{ V} = \left(\frac{1}{2}\right)24\text{ V} - 1.4\text{ V}$$
$$= 12\text{ V} - 1.4\text{ V} = 10.6\text{ V}$$

(b) The PIV for each diode is

$$\text{PIV} \cong V_{p(\text{out})} = 10.6\text{ V}$$

SECTION REVIEW 18–2

1. What is the average value of a full-wave rectified voltage with a peak value of 60 V?
2. Which type of full-wave rectifier has the greater output voltage for the same input voltage and transformer turns ratio?
3. For a given output voltage, the PIV for bridge rectifier diodes is less than for center-tapped rectifier diodes (T or F).

RECTIFIER FILTERS

18–3

In most power supply applications, the standard 60-Hz ac power line voltage must be converted to a sufficiently constant dc voltage. The 60-Hz pulsating dc output of a half-wave rectifier or the 120-Hz pulsating output of a full-wave rectifier must be *filtered* to virtually eliminate the large voltage variations. Figure 18–18 illustrates the filtering concept showing a smooth dc output voltage. The full-wave rectifier voltage is applied to the filter's input, and, ideally, a constant dc level appears on the output.

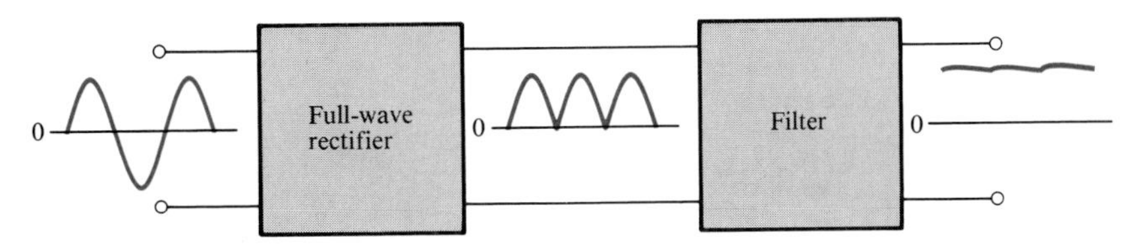

FIGURE 18–18
Basic block diagram of a power supply that converts 60-Hz ac to dc.

Capacitor-Input Filter

A half-wave rectifier with a capacitor-input filter is shown in Figure 18–19. We will use the half-wave rectifier to illustrate the filtering principle; then we will expand the concept to the full-wave rectifier.

During the positive first quarter-cycle of the input, the diode is forward-biased, allowing the capacitor to charge to within a diode drop of the input peak, as illustrated in Figure 18–19(a). When the input begins to decrease below its peak, as shown in Part (b), the capacitor retains its charge and the diode becomes reverse-biased. During the remaining part of the cycle, the capacitor can discharge only through the load resistance at a rate determined by the R_LC

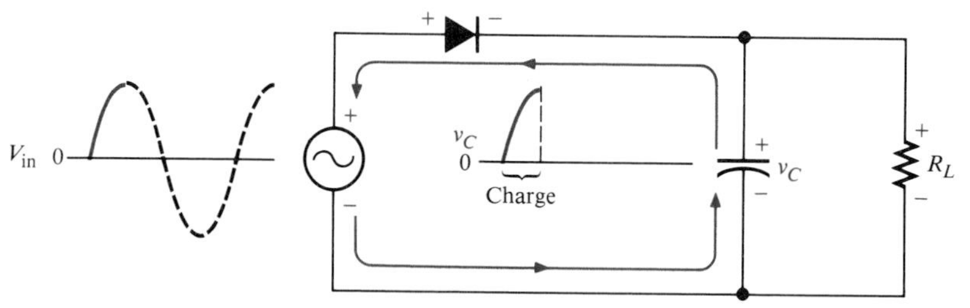

(a) Initial charging of capacitor (diode is forward-biased).

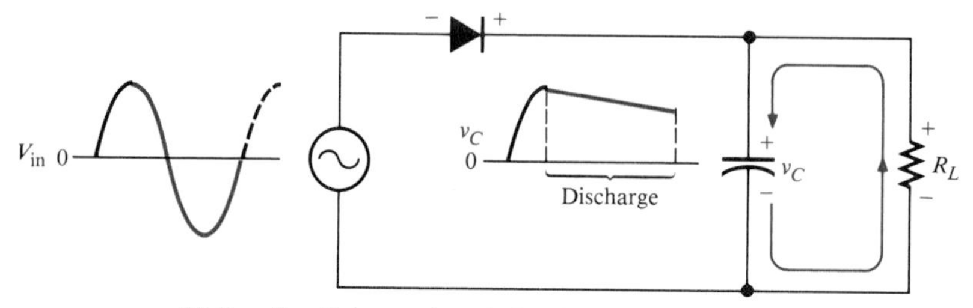

(b) Capacitor discharges through R_L after peak of the positive alternation (diode is reverse-biased).

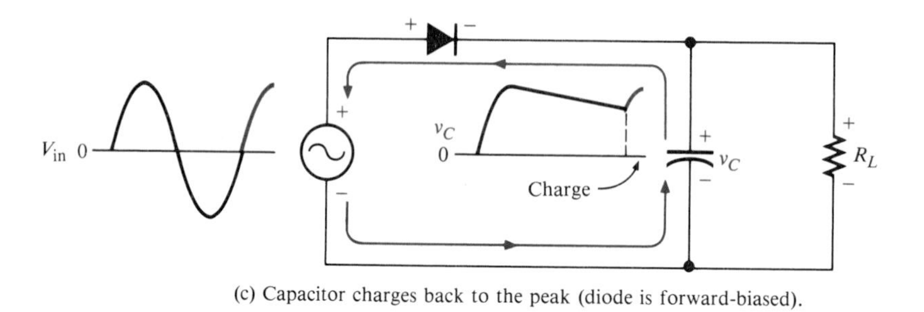

(c) Capacitor charges back to the peak (diode is forward-biased).

FIGURE 18–19
Operation of a half-wave rectifier with a capacitor-input filter.

time constant. The larger the time constant, the less the capacitor will discharge.

During the first quarter of the next cycle, the diode again will become forward-biased when the input voltage exceeds the capacitor voltage by approximately a diode drop, as illustrated in Part (c).

Ripple

As you have seen, the capacitor quickly charges at the beginning of a cycle and slowly discharges after the positive peak (when the diode is reverse-biased). The variation in the output voltage due to the charging and discharging is called the *ripple*. The smaller the ripple, the better the filtering action, as illustrated in Figure 18–20.

For a given input frequency, the output frequency of a full-wave rectifier is twice that of a half-wave rectifier. As a result, a full-wave rectifier is easier

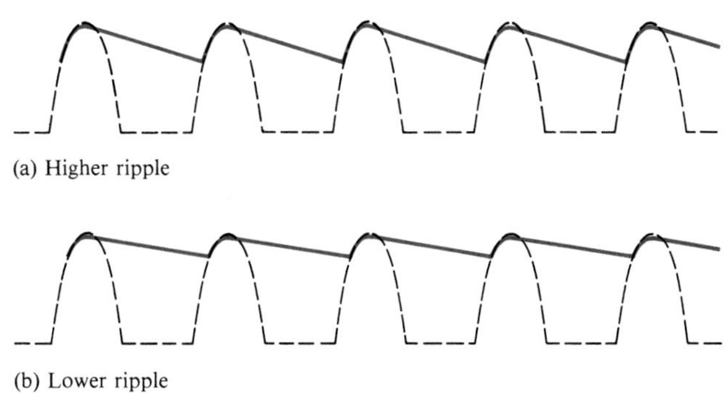

FIGURE 18–20
Half-wave ripple voltage (colored line).

to filter. When filtered, the full-wave rectified voltage has less ripple than does a half-wave signal for the same load resistance and capacitor values. Less ripple occurs because the capacitor discharges less during the shorter interval between full-wave pulses, as shown in Figure 18–21.

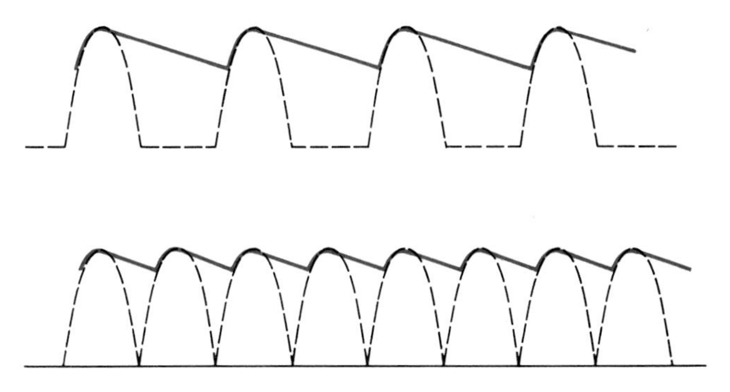

FIGURE 18–21
Comparison of ripple voltages for half-wave and full-wave signals with same filter and same input frequency.

The *ripple factor* is an indication of the effectiveness of the filter and is defined as the ratio of the ripple voltage to the dc (average) value of the filter output voltage. These factors are illustrated in Figure 18–22. The lower the ripple factor, the better the filter. The ripple factor can be decreased by increasing the value of the filter capacitor.

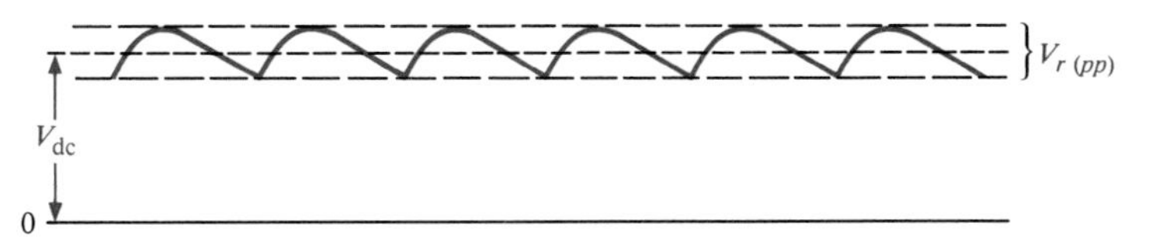

FIGURE 18–22
V_r and V_{dc} determine the ripple factor.

Surge Current in the Capacitor Filter

Before the switch in Figure 18–23(a) is closed, the filter capacitor is uncharged. At the instant the switch is closed, voltage is connected to the bridge and the capacitor appears as a short, as shown. An initial surge of current is produced through the two forward-biased diodes. The worst-case situation occurs when the switch is closed at a peak of the secondary voltage and a maximum surge current $I_{S(max)}$ is produced, as illustrated in the figure.

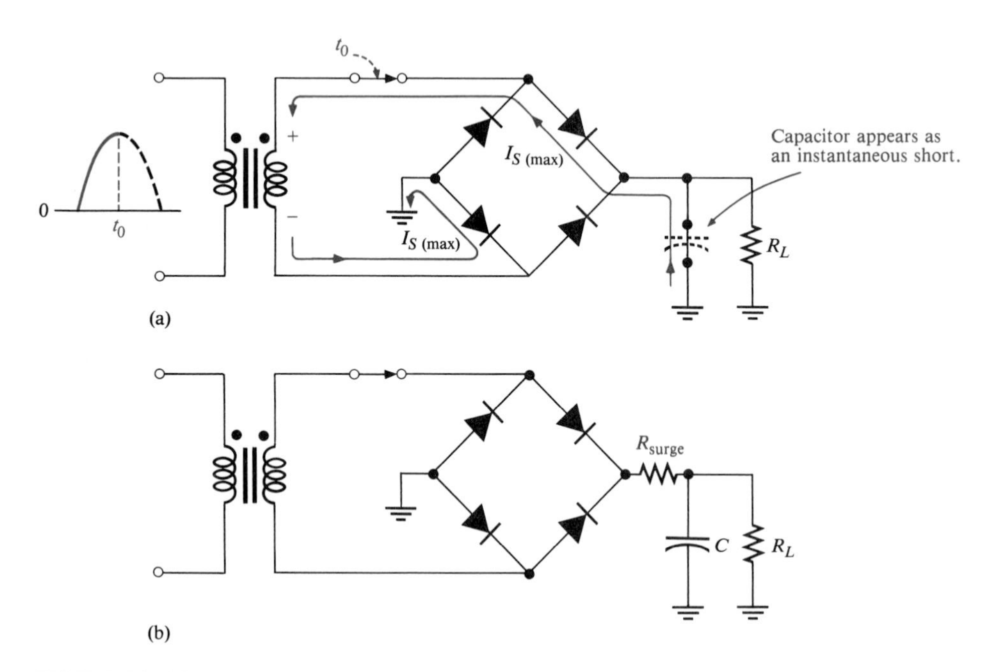

FIGURE 18–23
Surge current in a capacitor-input filter.

It is possible that the surge current could destroy the diodes, and for this reason a surge-limiting resistor is sometimes connected, as shown in Figure 18–23(b). The value of this resistor must be small compared to R_L. Also, the diodes must have a forward current rating such that they can withstand the momentary surge of current.

Inductor-Input Filter

When a choke (inductor) is added to the filter input, as in Figure 18–24, a reduction in the ripple voltage is achieved. The choke has a high reactance at the ripple frequency, and the capacitive reactance is low compared to both X_L and R_L. The two reactances form an ac voltage divider that tends to significantly reduce the ripple from that of a straight capacitor-input filter.

It should be noted at this time that an inductor-input filter produces an output with a dc value approximately equal to the *average* value of the rectified input. The capacitor-input filter, however, produces an output with a dc value approximately equal to the *peak* value of the input.

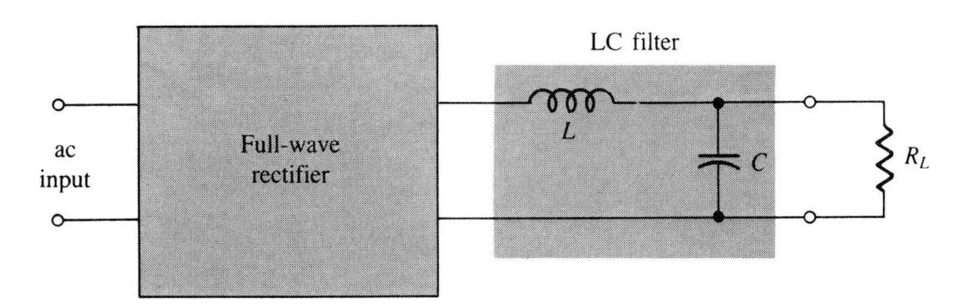

FIGURE 18–24
Rectifier with an *LC* filter.

Another point of comparison is that the amount of ripple voltage in the capacitor-input filter varies inversely with the load resistance. Ripple voltage in the *LC* filter is essentially independent of the load resistance and depends only on X_L and X_C, as long as X_C is sufficiently less than R_L.

π-Type Filter

A one-section π-type filter is shown in Figure 18–25. It can be thought of as a capacitor filter followed by an *LC* filter. It combines the peak filtering action of the single-capacitor filter with the reduced ripple and load independence of the *LC* filter.

FIGURE 18–25
π-type *LC* filter.

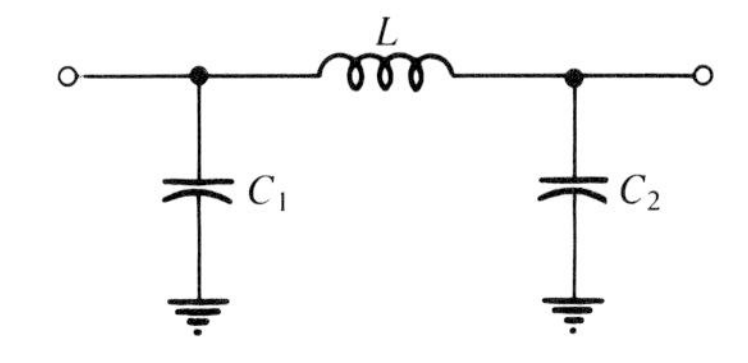

SECTION REVIEW 18–3

1. A 60-Hz sine wave is applied to a half-wave rectifier. What is the output frequency? What is the output frequency for a full-wave rectifier?
2. What causes the ripple voltage on the output of a capacitor-input filter?
3. The load resistance of a capacitor-filtered full-wave rectifier is reduced. What effect does this reduction have on the ripple voltage?
4. Name one advantage of an *LC* filter over a capacitor filter. Name one disadvantage.

18–4 TROUBLESHOOTING RECTIFIER CIRCUITS

Several types of failures can occur in power supply rectifiers. In this section we will examine some possible failures and the effects they would have on a circuit's operation.

Open Diode

A half-wave rectifier with a diode that has opened (a common failure mode) is shown in Figure 18–26. In this case, you would measure 0 V dc across the load resistor, as depicted.

Now consider the full wave, center-tapped rectifier in Figure 18–27. Assume that diode D_1 has failed open. With an oscilloscope connected to the output, as shown in Part (a), you would observe the following: You would see a larger-than-normal ripple voltage at a frequency of 60 Hz rather than 120 Hz. Disconnecting the filter capacitor, you would observe a *half-wave* rectified voltage, as in Part (b). Now let's examine the reason for these observations. If diode

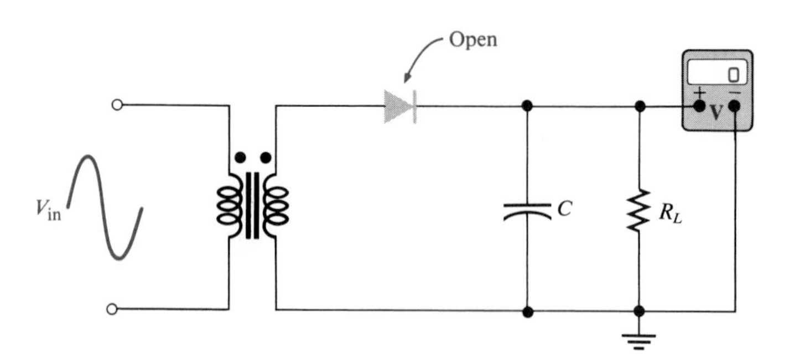

FIGURE 18–26
Test for an open diode in a half-wave rectifier with capacitor-input filter.

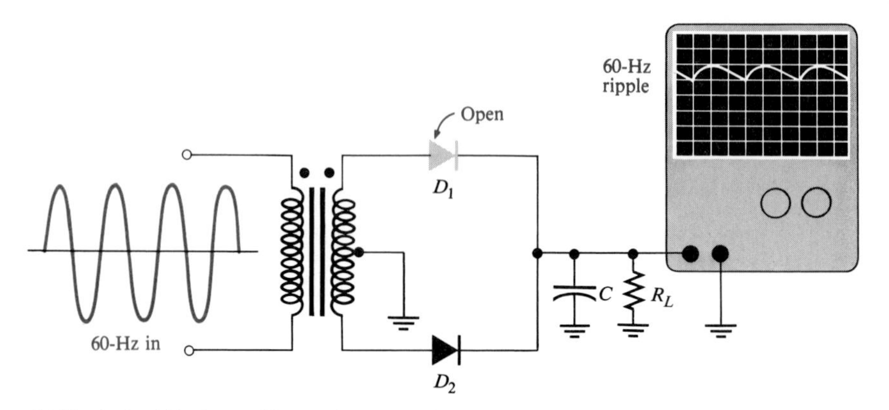

(a) Ripple should be less and have a frequency of 120 Hz.

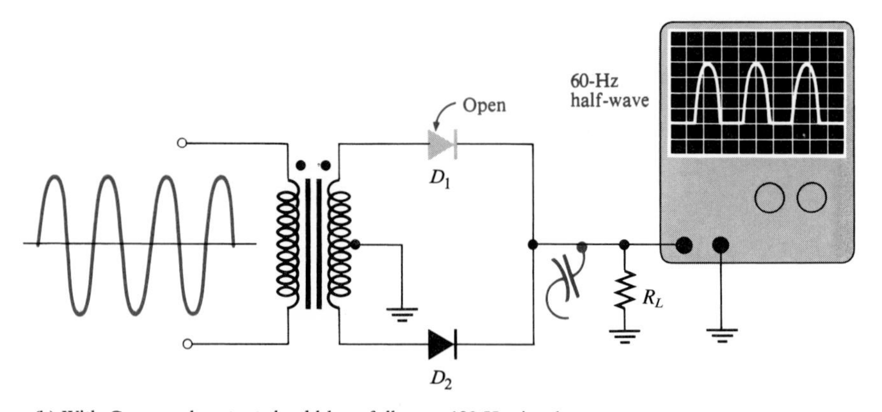

(b) With C removed, output should be a full-wave 120-Hz signal.

FIGURE 18–27
Symptoms of an open diode in a full-wave, center-tapped rectifier.

D_1 is open, there will be current through R_L only during the negative half-cycle of the input signal. During the positive half-cycle, an open path prevents current through R_L. The result is a half-wave voltage, as illustrated.

With the filter capacitor in the circuit, the half-wave signal will allow it to discharge more than it would with a normal full-wave signal, resulting in a larger ripple voltage. Basically, the same observations would be made for an open failure of diode D_2.

An open diode in a bridge rectifier would create symptoms identical to those just discussed for the center-tapped rectifier. As illustrated in Figure 18–28, the open diode would prevent current through R_L during half of the input cycle (in this case, the negative half). As a result, there would be a half-wave output and an increased ripple voltage at 60 Hz, as discussed before.

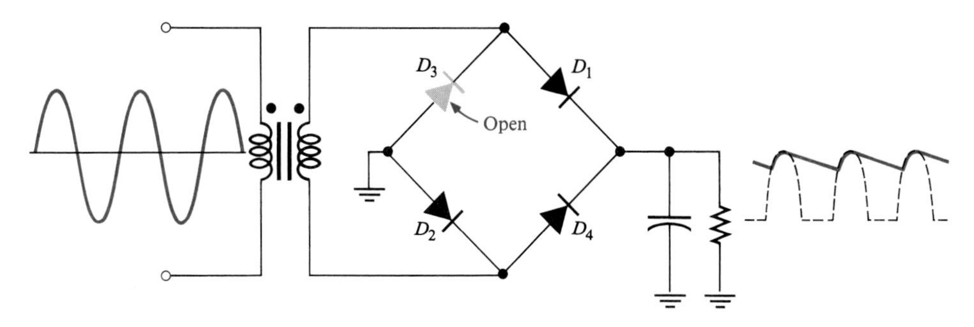

FIGURE 18–28
Effects of an open diode in a bridge rectifier.

Shorted Diode

A shorted diode is one that has failed such that it has a very low resistance in both directions. If a diode suddenly became shorted in a bridge rectifier, it is likely that a sufficiently high current would exist during one half of the input cycle such that the shorted diode itself would burn open or the other diode in series with it would open. The transformer could also be damaged, as illustrated in Figure 18–29 with D_1 shorted.

In Part (a) of the figure, current is supplied to the load through the shorted diode during the first positive half-cycle, just as though it were forward-biased. During the negative half-cycle, the current is shorted through D_1 and D_4, as shown in Part (b). Again, damage to the transformer is possible. It is likely that this excessive current would burn either or both of the diodes open. If only one of the diodes opened, you would still observe a half-wave voltage on the output. If both diodes opened, there would be no voltage developed across the load. These conditions are illustrated in Figure 18–30.

Shorted or Leaky Filter Capacitor

A *shorted capacitor* would most likely cause some or all of the diodes in a full-wave rectifier to open due to excessive current. In any event, there would be no dc voltage on the output.

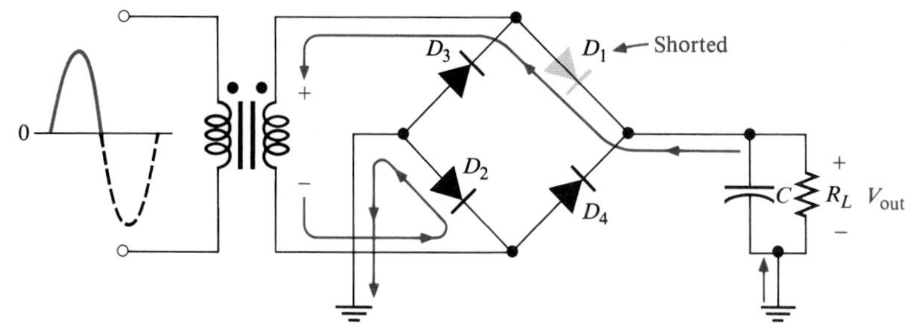

(a) Positive half-cycle: The shorted diode acts as a forward-biased diode, so the load current is normal.

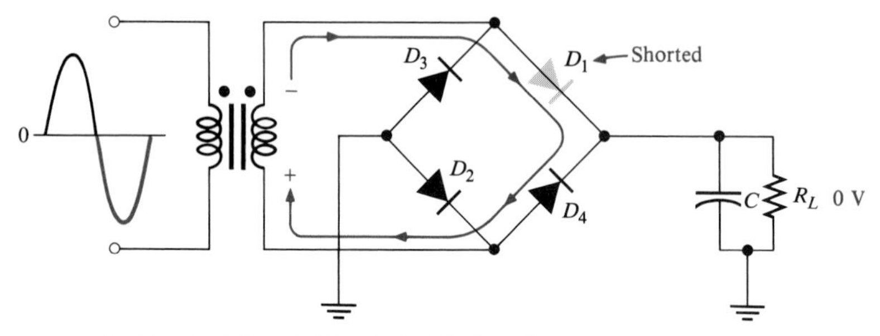

(b) Negative half-cycle: The shorted diode produces a short circuit across the source. As a result D_1, D_4, or the transformer secondary will probably burn open.

FIGURE 18–29
Effects of a shorted diode in a bridge rectifier.

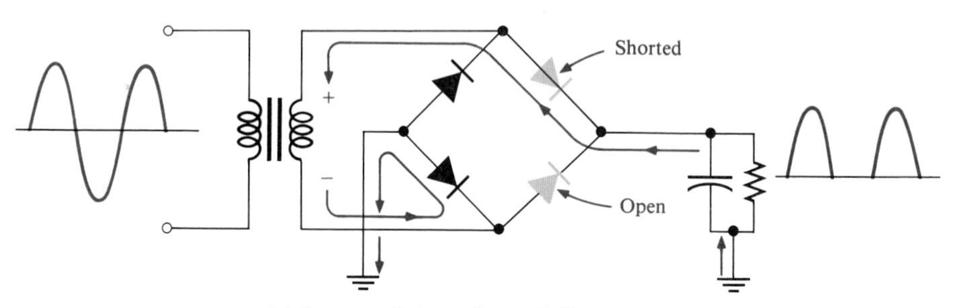

(a) One open diode produces a half-wave output.

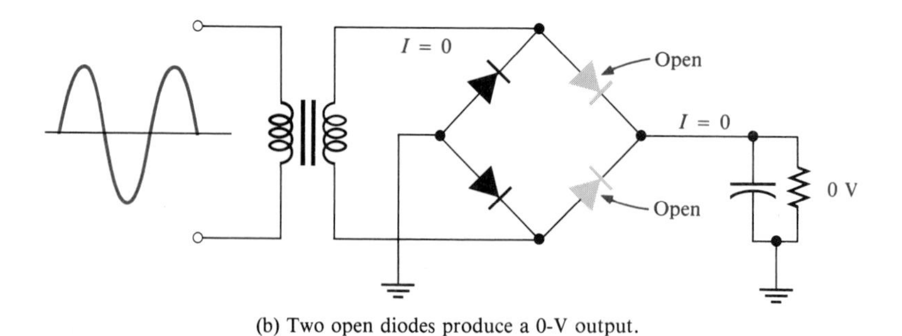

(b) Two open diodes produce a 0-V output.

FIGURE 18–30
Effects of open diodes.

A *leaky capacitor* can be represented by a leakage resistance in parallel with the capacitor, as shown in Figure 18–31(a). The effect of the leakage resistance is to reduce the discharging time constant, causing an increase in ripple voltage on the output, as shown in Figure 18–31(b).

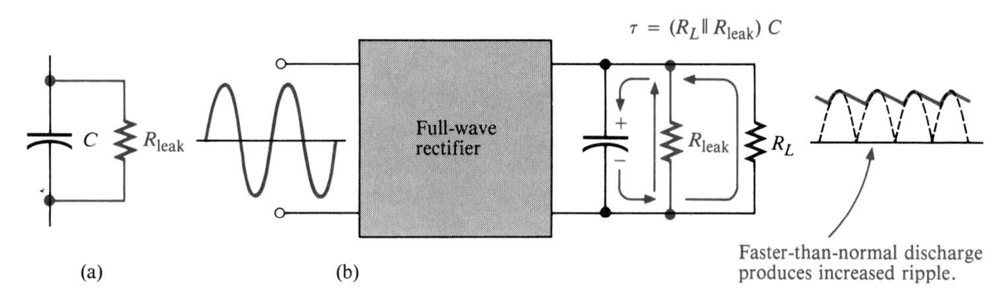

FIGURE 18–31
Effects of a leaky filter capacitor on the output of a full-wave rectifier.

SECTION REVIEW 18–4

1. What effect would an open D_2 produce in the rectifier of Figure 18–27?
2. You are checking a 60-Hz full-wave bridge rectifier and observe that the output has a 60-Hz ripple. What failure(s) do you suspect?
3. You observe that the output ripple of a full-wave rectifier is much greater than normal but its frequency is still 120 Hz. What component do you suspect?

DIODE CLIPPING AND CLAMPING CIRCUITS

18–5

Diode circuits are sometimes used to clip off portions of signal voltages above or below certain levels; these circuits are called *clippers* or *limiters*. Another type of diode circuit is used to restore a dc level to an electrical signal; these are called *clampers*. Both diode circuits will be examined in this section.

Diode Clippers

Figure 18–32(a) shows a diode circuit that clips off the positive part of the input signal. As the input signal goes positive, the diode becomes forward-biased. Since the cathode is at ground potential (0 V), the anode cannot exceed 0.7 V (assuming silicon). Thus, point A is clipped at $+0.7$ V when the input exceeds this value.

When the input goes back below 0.7 V, the diode reverse-biases and appears as an open. The output voltage looks like the negative part of the input, but with a magnitude determined by the R_s and R_L voltage divider as follows:

$$V_{out} = \left(\frac{R_L}{R_s + R_L}\right)V_{in}$$

If R_s is small compared to R_L, then $V_{out} \cong V_{in}$.

Turn the diode around, as in Figure 18–32(b), and the *negative* part of the input is clipped off. When the diode is forward-biased during the negative part of the input, point A is held at -0.7 V by the diode drop. When the input goes

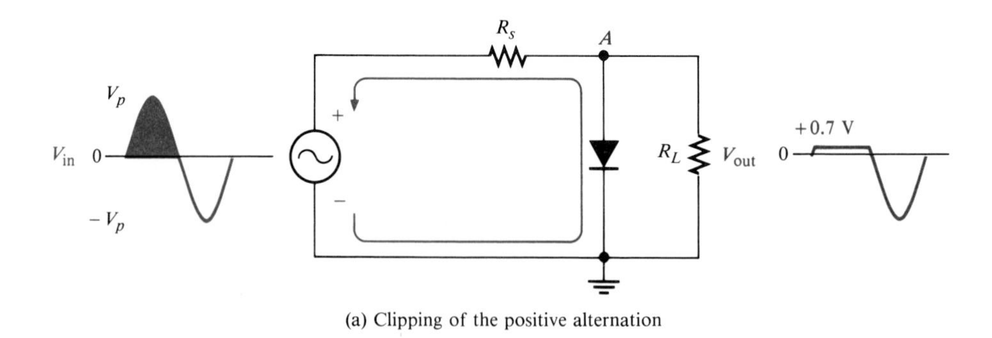

(a) Clipping of the positive alternation

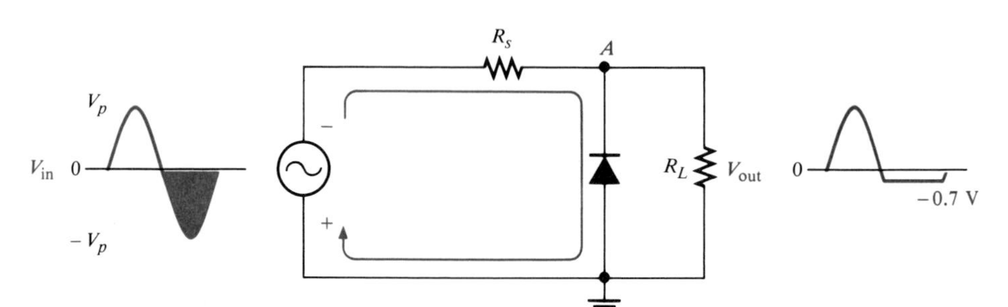

(b) Clipping of the negative alternation

FIGURE 18–32
Diode clipping circuits.

above −0.7 V, the diode is no longer forward-biased and a voltage appears across R_L proportional to the input.

EXAMPLE 18–6

What would you expect to see displayed on an oscilloscope connected as shown in Figure 18–33? The time base on the scope is set to show one and one-half cycles.

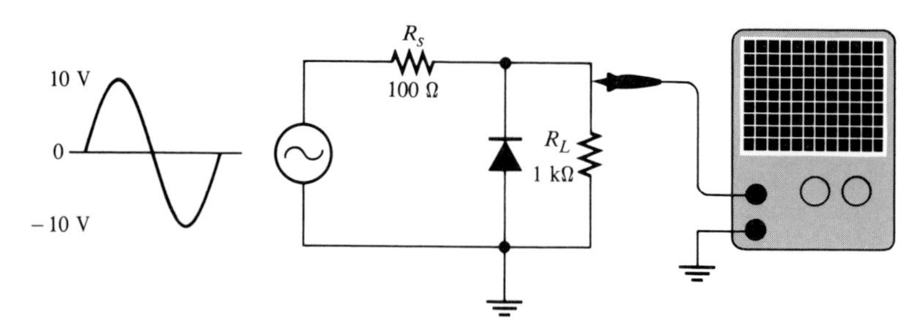

FIGURE 18–33

Solution
The diode conducts when the input voltage goes below − 0.7 V. Thus, we have a negative clipper with a peak output voltage determined by the following equation:

$$V_{p(\text{out})} = \left(\frac{R_L}{R_s + R_L}\right)V_{p(\text{in})} = \left(\frac{1\text{ k}\Omega}{1.1\text{ k}\Omega}\right)10\text{ V}$$
$$= 9.09\text{ V}$$

The scope will display an output waveform as shown in Figure 18–34.

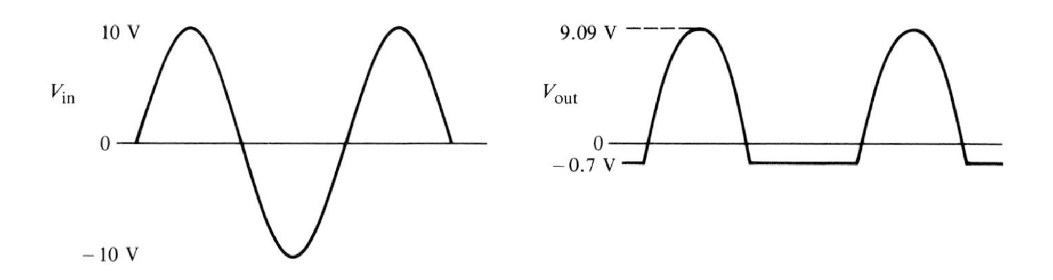

FIGURE 18–34
Waveforms for Figure 18–33.

Adjustment of the Clipping Level

To adjust the level at which a signal voltage is clipped, add a *bias* voltage in series with the diode, as shown in Figure 18–35. The voltage at point A must equal V_{BB} + 0.7 V before the diode will conduct. Once the diode begins to conduct, the voltage at point A is limited to V_{BB} + 0.7 V so that all input voltage above this level is clipped off, as shown in the figure.

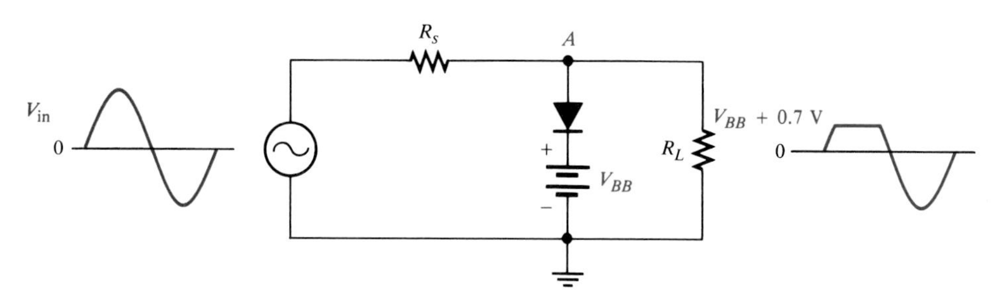

FIGURE 18–35
Positively biased clipper.

If the bias voltage is varied up or down, the clipping level changes correspondingly, as shown in Figure 18–36. If the polarity of the bias voltage is reversed, as in Figure 18–37, voltages above $-V_{BB}$ + 0.7 V are clipped, resulting in an output waveform as shown. The diode is reverse-biased only when the voltage at point A goes below $-V_{BB}$ + 0.7 V.

If it is necessary to clip off voltage below a specified negative level, then the diode and bias battery must be connected as in Figure 18–38. In this case, the voltage at point A must go below $-V_{BB} - 0.7$ V to forward-bias the diode and initiate clipping action, as shown.

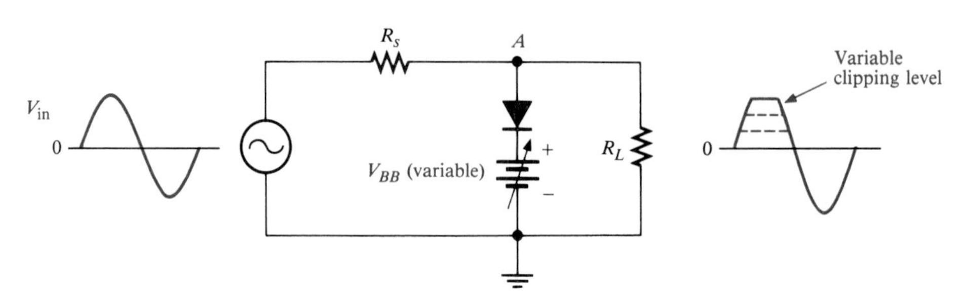

FIGURE 18–36
Positive clipper with variable bias.

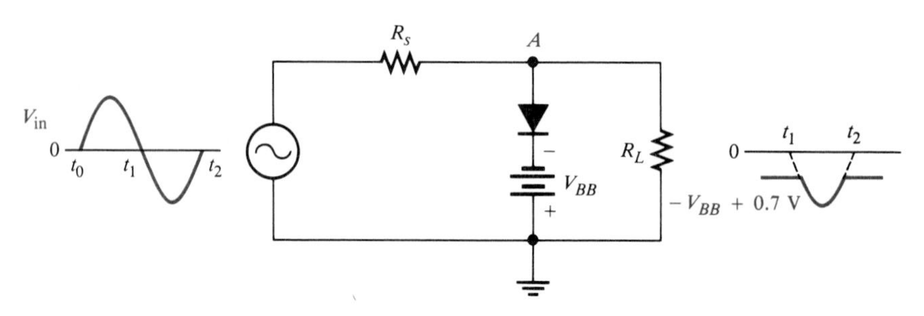

FIGURE 18–37

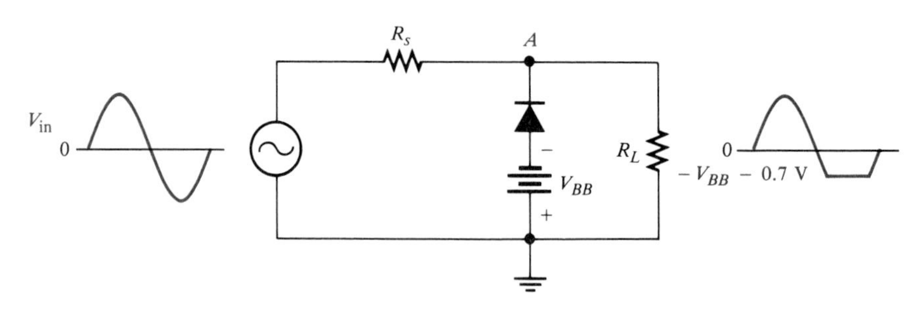

FIGURE 18–38

EXAMPLE 18–7 Figure 18–39 shows a circuit combining a positive-biased clipper with a negative-biased clipper. Determine the output waveform.

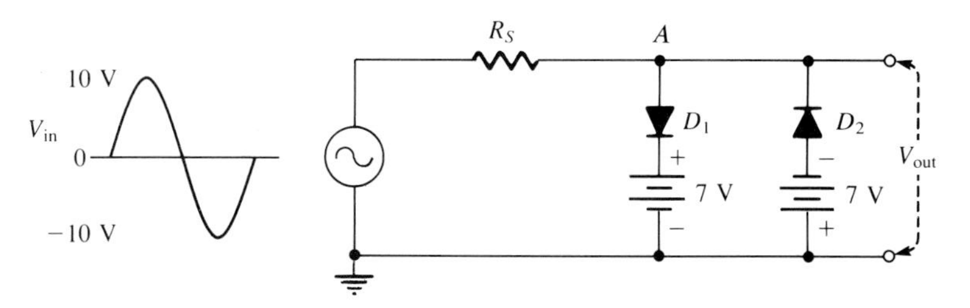

FIGURE 18–39

Solution

When the voltage at point A reaches $+7.7$ V, diode D_1 conducts and clips the waveform at $+7.7$ V. Diode D_2 does not conduct until the voltage reaches -7.7 V. Therefore, positive voltages above $+7.7$ V and negative voltages below -7.7 V are clipped off. The resulting output waveform is shown in Figure 18–40.

FIGURE 18–40
Output waveform for Figure 18–39.

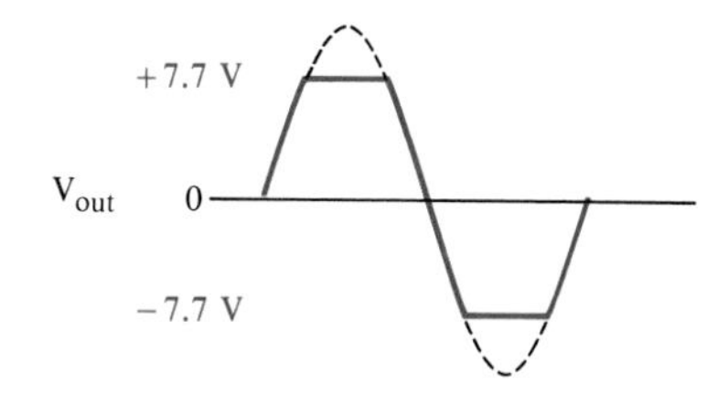

Diode Clampers

The purpose of a clamper is to add a dc level to an ac signal. Clampers are sometimes known as *dc restorers*.

Figure 18–41 shows a diode clamper that inserts a *positive* dc level. To understand the operation of this circuit, consider the first negative half-cycle of the input voltage. When the input initially goes negative, the diode is forward-biased, allowing the capacitor to charge to near the peak of the input ($V_{p(in)} - 0.7$ V), as shown in Figure 18–41(a). Just past the negative peak, the diode becomes reverse-biased because the anode is held near $V_{p(in)}$ by the charge on the capacitor.

The capacitor can discharge only through the high resistance of R_L. Thus, from the peak of one negative half-cycle to the next, the capacitor discharges very little. The amount that is discharged, of course, depends on the value of R_L. For good clamping action, the RC time constant should be at least ten times the period of the input frequency.

The net effect of the clamping action is that the capacitor retains a charge approximately equal to the peak value of the input less the diode drop. The capacitor voltage acts essentially as a battery in series with the input signal, as shown in Figure 18–41(b). The dc voltage of the capacitor adds to the input voltage by superposition, as shown in Figure 18–41(c).

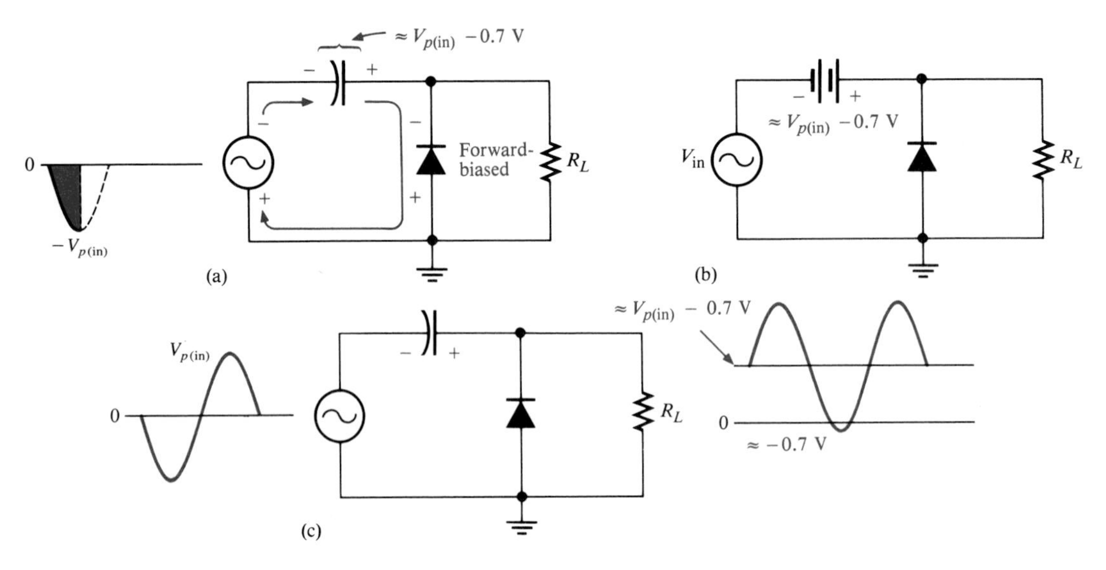

FIGURE 18–41
Positive clamping operation.

If the diode is turned around, a *negative* dc voltage is added to the input signal, as shown in Figure 18–42. If necessary, the diode can be biased to adjust the clamping level.

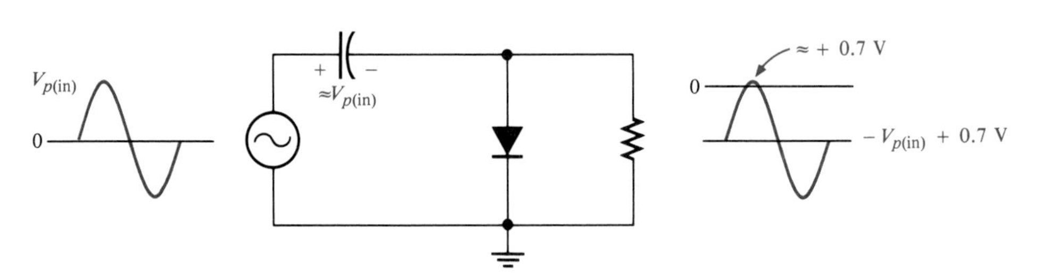

FIGURE 18–42
Negative clamping.

A Clamper Application

A clamping circuit is often used in television receivers as a dc restorer. The incoming composite video signal is normally processed through capacitively coupled amplifiers which eliminate the dc component, thus losing the black and white reference levels and the blanking level. Before being applied to the picture tube, these reference levels must be restored. Figure 18–43 illustrates this process in a general way.

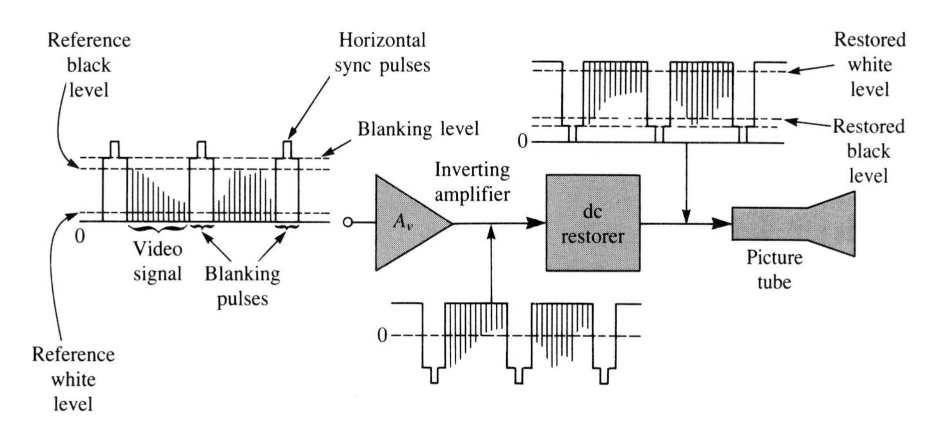

FIGURE 18–43
Clamping circuit (dc restorer) in a TV receiver.

EXAMPLE 18–8

What is the output voltage that you would expect to observe across R_L in the clamping circuit of Figure 18–44? Assume that RC is large enough to prevent significant capacitor discharge.

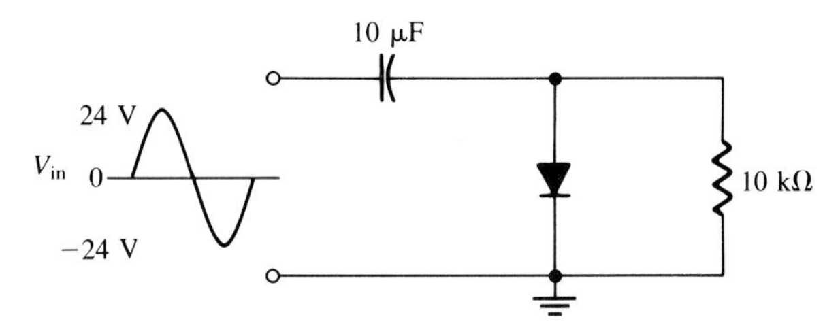

FIGURE 18–44

Solution

Ideally, a negative dc value equal to the input peak less the diode drop is inserted by the clamping circuit:

$$V_{dc} \cong V_{p(in)} - 0.7\text{ V} = 24\text{ V} - 0.7\text{ V} = 23.3\text{ V}$$

Actually, the capacitor will discharge slightly between peaks, and, as a result, the output voltage will have an average value of slightly less than that calculated above.

FIGURE 18–45
Output waveform for Figure 18–44.

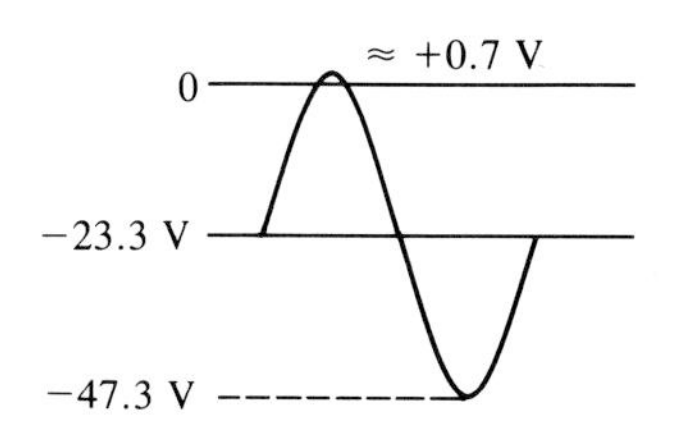

The output waveform goes to approximately 0.7 V above ground, as shown in Figure 18–45 (on page 669).

SECTION REVIEW 18–5

1. Determine the output waveform for the circuit of Figure 18–46.
2. What is the output voltage in Figure 18–47?
3. Sketch the approximate output waveform for a positive clamping circuit having a sine wave input with a peak of 50 V.

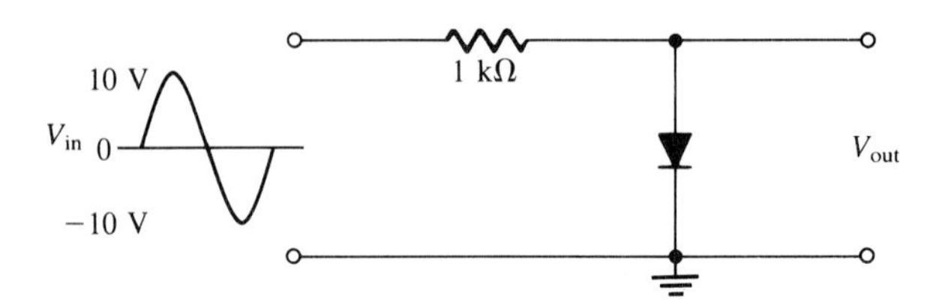

FIGURE 18–46

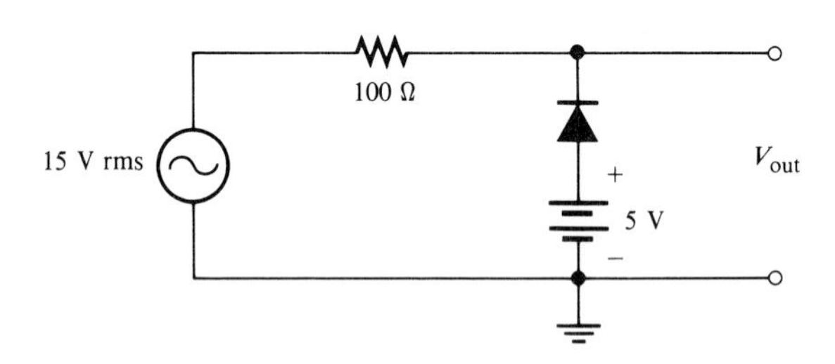

FIGURE 18–47

ZENER DIODES

18–6

The *zener diode* is used for voltage regulation and, like the general-purpose rectifier diode, is important in many power supply applications. The schematic symbol is shown in Figure 18–48.

The zener is a silicon PN junction device that differs from the rectifier diode in that it is optimized for operation in the *reverse breakdown region*. The breakdown voltage of a zener diode is set by carefully controlling the doping level during manufacture. From the discussion of the diode characteristic curve in the last chapter, recall that when a diode reaches reverse breakdown, its voltage remains almost constant even though the current may change drastically. This volt-ampere characteristic is shown again in Figure 18–49.

Zener Breakdown

FIGURE 18–48
Zener diode symbol.

There are two types of reverse breakdown in a zener diode. One is the *avalanche* breakdown that was discussed in Chapter 17; this also occurs in rectifier diodes at a sufficiently high reverse voltage. The other type is *zener* breakdown which occurs in a zener diode at low reverse voltages. A zener diode is heavily doped to reduce the breakdown voltage, causing a very narrow depletion layer. As a result, an intense electric field exists within the depletion layer. Near the break-

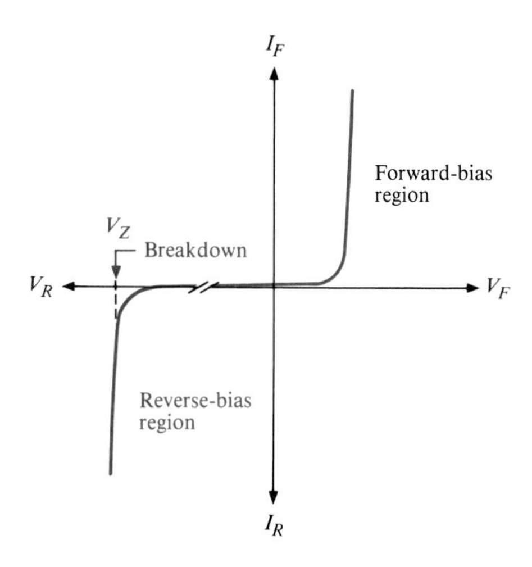

FIGURE 18–49
General diode characteristic.

down voltage (V_Z), the field is intense enough to pull electrons from their valence bands and create current.

Zener diodes with breakdown voltages of less than approximately 5 V operate predominantly in zener breakdown. Those with breakdown voltages greater than approximately 5 V operate predominantly in avalanche breakdown. Both types, however, are called zener diodes. Zeners with breakdown voltages of 1.8 V to 200 V are commercially available.

Breakdown Characteristics

Figure 18–50 shows the reverse portion of the characteristic curve of a zener diode. Notice that as the reverse voltage (V_R) is increased, the reverse current (I_R) remains extremely small up to the "knee" of the curve. At this point, the breakdown effect begins; the zener resistance (R_Z) begins to decrease as the current (I_Z) increases rapidly. From the bottom of the knee, the breakdown voltage (V_Z) remains essentially constant. This *regulating* ability is the key feature of the zener diode: *It maintains an essentially constant voltage across its terminals over a specified range of reverse current values.*

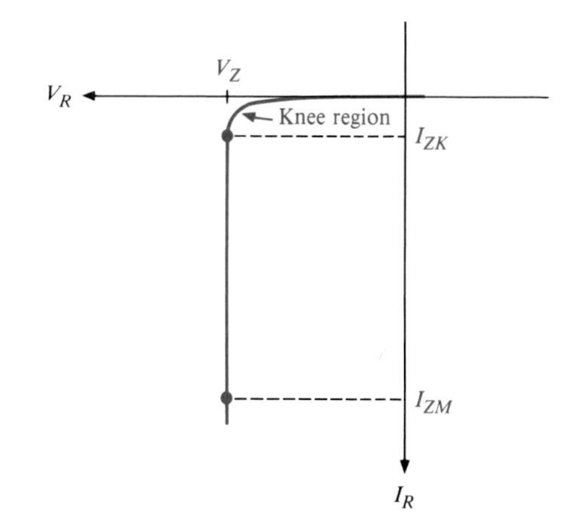

FIGURE 18–50
Reverse characteristic of zener diode.

A *minimum* value of reverse current, I_{ZK}, must be maintained in order to keep the diode in regulation. You can see on the curve that when the reverse current is reduced below the knee of the curve, the voltage changes drastically and regulation is lost. Also, there is a *maximum* current, I_{ZM}, above which the diode may be damaged.

Thus, basically, the zener diode maintains a constant voltage across its terminals for values of reverse current ranging from I_{ZK} to I_{ZM}.

Zener Equivalent Circuit

Figure 18–51(a) shows the ideal approximation of a zener diode in reverse breakdown. It acts simply as a battery having a value equal to the zener voltage. Part (b) represents the practical equivalent of a zener, where the zener resistance R_Z is included. Since the voltage curve is not ideally vertical, a change in reverse current (ΔI_Z) produces a small change in zener voltage (ΔV_Z), as illustrated in Part (c).

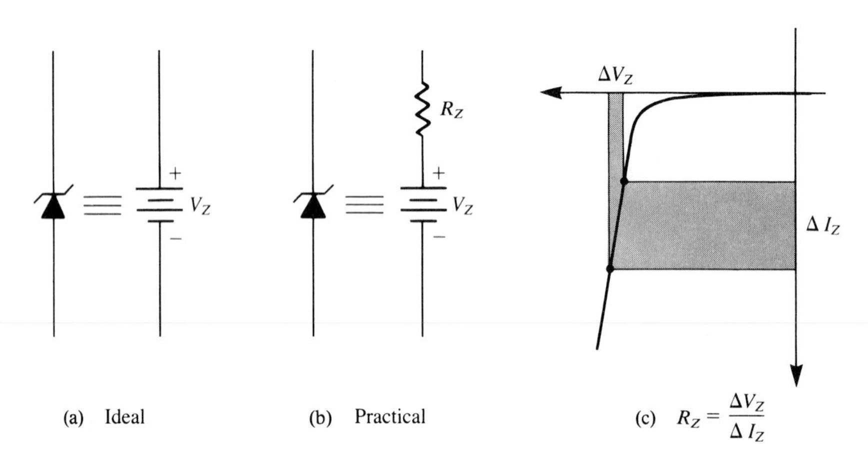

FIGURE 18–51
Zener equivalent circuits.

The ratio of ΔV_Z to ΔI_Z is the resistance, as expressed in the following equation:

$$R_Z = \frac{\Delta V_Z}{\Delta I_Z} \tag{18–8}$$

Normally, R_Z is specified at a particular value of reverse current, I_{ZT}, called the *zener test current.* In most cases, this value of R_Z is approximately constant over the full range of reverse-current values.

EXAMPLE 18–9 A certain zener diode exhibits a 50-mV change in V_Z for a 2-mA change in I_Z. What is the zener resistance?

Solution

$$R_Z = \frac{\Delta V_Z}{\Delta I_Z} = \frac{50 \text{ mV}}{2 \text{ mA}} = 25 \ \Omega$$

EXAMPLE 18–10 A 6.8-V zener diode has a resistance of 5 Ω. What is the actual voltage across its terminals when the current is 20 mA?

Solution
Figure 18–52 represents the diode. The 20-mA current causes a voltage across R_Z as follows:

$$I_Z R_Z = (20 \text{ mA})(5 \ \Omega) = 100 \text{ mV}$$

The polarity is in the same direction as the zener voltage, as shown. Therefore, it adds to V_Z, giving a total terminal voltage of

$$V_Z = V_Z + I_Z R_Z = 6.8 \text{ V} + 100 \text{ mV} = 6.9 \text{ V}$$

Thus, for this particular current value, the terminal voltage is 0.1 V greater than the specified normal zener voltage of 6.8 V. This value, of course, changes with current.

FIGURE 18–52

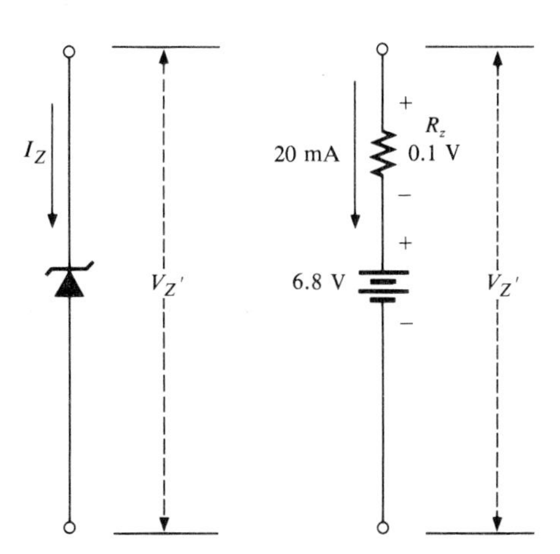

Zener Voltage Regulation

Zener diodes are widely used for voltage regulation. Figure 18–53 illustrates how a zener diode can be used to regulate a varying dc voltage. This process is called *input* or *line regulation.*

As the input voltage varies (within limits), the zener diode maintains an essentially constant voltage across the output terminals. However, as V_{IN} changes, I_Z will change proportionally, and therefore the limitations on the input variation are set by the minimum and maximum current values with which

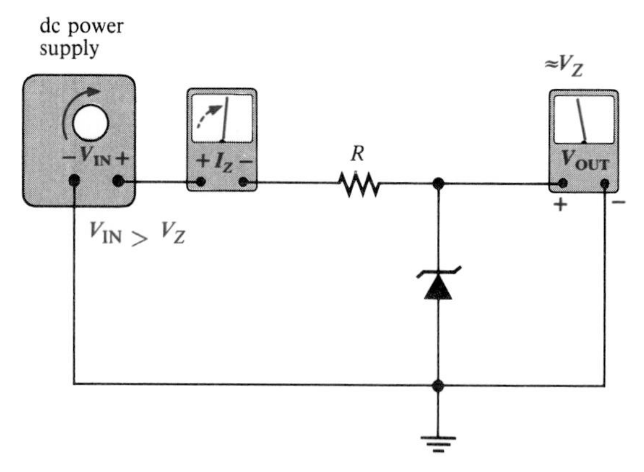

(a) As input voltage increases, V_{OUT} remains constant.

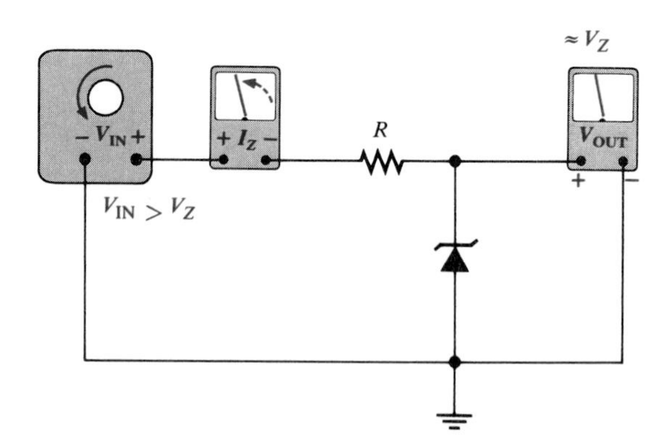

(b) As input voltage decreases, V_{OUT} remains constant.

FIGURE 18–53
Zener regulation of a varying input voltage.

the zener can operate and on the condition that $V_{IN} > V_Z$. R is the series current-limiting resistor.

For example, suppose that the zener diode in Figure 18–54 can maintain regulation over a range of current values from 4 mA to 40 mA. For the *minimum* current, the voltage across the 1-kΩ resistor is $V_R = (4 \text{ mA})(1 \text{ k}\Omega) = 4 \text{ V}$. Since $V_R = V_{IN} - V_Z$, then $V_{IN} = V_R + V_Z = 4 \text{ V} + 10 \text{ V} = 14 \text{ V}$. For the *maximum* current, the voltage across the 1-kΩ resistor is $V_R = (40 \text{ mA})(1 \text{ k}\Omega) = 40 \text{ V}$. Therefore, $V_{IN} = 40 \text{ V} + 10 \text{ V} = 50 \text{ V}$.

FIGURE 18–54

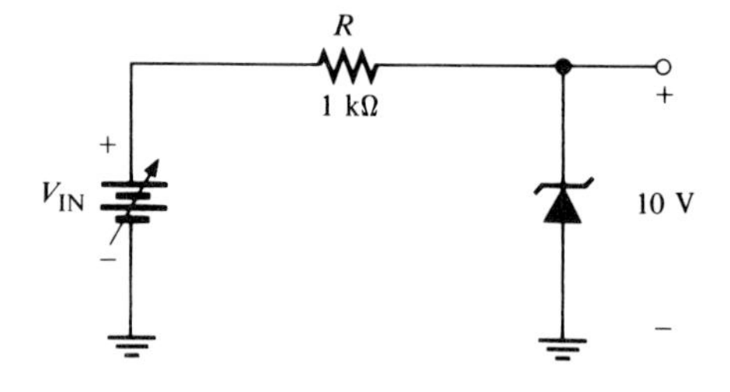

As you can see, this zener diode can regulate an input voltage from 14 V to 50 V and maintain approximately a 10-V output. (The output will vary slightly because of the zener resistance.)

EXAMPLE 18–11 Determine the minimum and the maximum input voltages that can be regulated by the zener diode in Figure 18–55. Assume that $I_{ZK} = 1$ mA, $I_{ZM} = 15$ mA, $V_Z = 5.1$ V, and $R_Z = 10\ \Omega$.

Solution

The equivalent circuit is shown in Figure 18–56. At $I_{ZK} = 1$ mA, the output voltage is

FIGURE 18–55

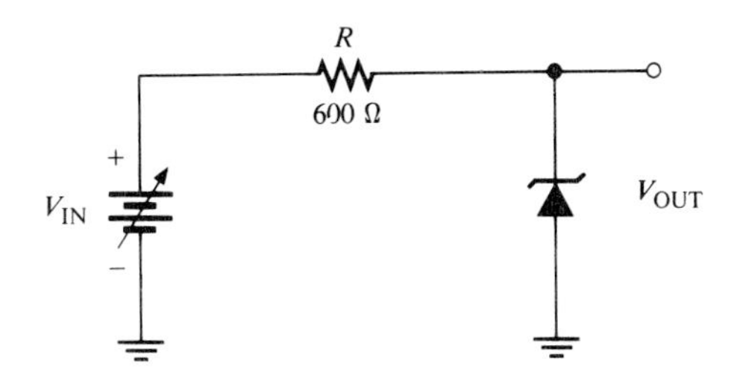

FIGURE 18–56
Equivalent of circuit in Figure 18–55.

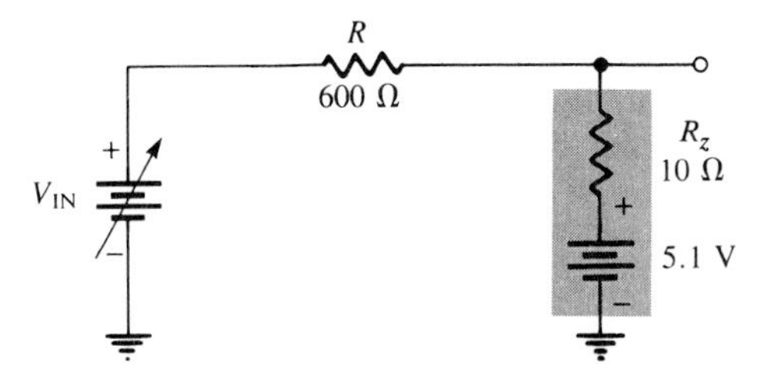

$$V_{OUT} = V_Z + I_{ZK}R_Z = 5.1\text{ V} + (1\text{ mA})(10\ \Omega)$$
$$= 5.1\text{ V} + 0.010\text{ V} = 5.11\text{ V}$$

Therefore,

$$V_{IN(min)} = I_{ZK}R + V_{OUT} = (1\text{ mA})(600\ \Omega) + 5.11\text{ V}$$
$$= 5.71\text{ V}$$

At $I_{ZM} = 15$ mA, the output voltage is

$$V_{OUT} = V_Z + I_{ZM}R_Z = 5.1\text{ V} + (15\text{ mA})(10\ \Omega)$$
$$= 5.1\text{ V} + 0.15\text{ V} = 5.25\text{ V}$$

Therefore,

$$V_{IN(max)} = I_{ZM}R + V_{OUT} = (15\text{ mA})(600\ \Omega) + 5.25\text{ V}$$
$$= 14.25\text{ V}$$

Regulation with a Varying Load

Figure 18–57 shows a zener regulator with a variable load resistor across the terminals. The zener diode maintains a constant voltage across R_L as long as the zener current is greater than I_{ZK} and less than I_{ZM}. This process is called *load regulation.*

No Load to Full Load

When the output terminals are open ($R_L = \infty$), the load current is zero and all of the current is through the zener. When a load resistor is connected, part of the total current is through the zener and part through R_L.

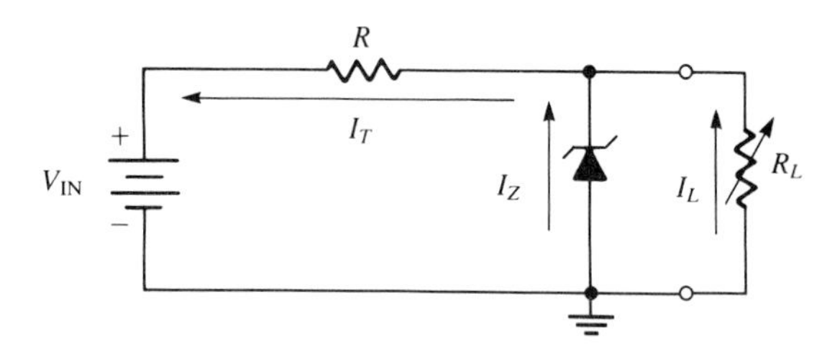

FIGURE 18–57
Zener regulation with a variable load.

As R_L is decreased, I_L goes up and I_Z goes down. The zener diode continues to regulate until I_Z reaches its minimum value, I_{ZK}. At this point the load current is maximum. The following example illustrates.

EXAMPLE 18–12 Determine the minimum and the maximum load currents for which the zener diode in Figure 18–58 will maintain regulation. What is the minimum R_L that can be used? $V_Z = 12$ V, $I_{ZK} = 3$ mA, and $I_{ZM} = 90$ mA. Assume that $R_Z = 0\ \Omega$.

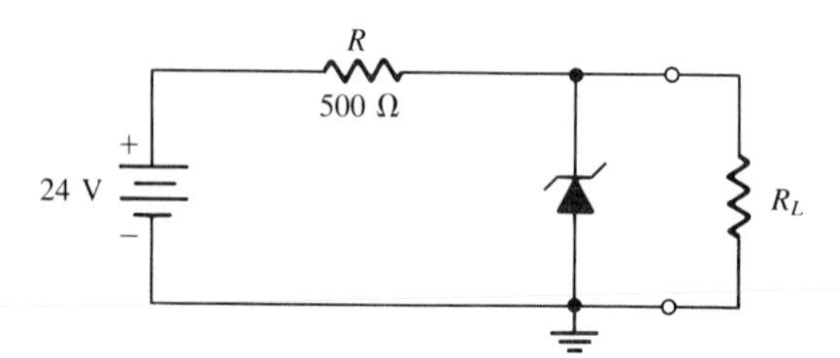

FIGURE 18–58

Solution

When $I_L = 0$ A, I_Z is maximum and equal to the total circuit current I_T:

$$I_Z = \frac{V_{IN} - V_Z}{R} = \frac{24\text{ V} - 12\text{ V}}{500\ \Omega} = 24\text{ mA}$$

Since this is much less than I_{ZM}, 0 A is an acceptable minimum for I_L. That is,

$$I_{L(\min)} = 0\text{ A}$$

The maximum value of I_L occurs when I_Z is minimum, so we can solve for $I_{L(\max)}$ as follows:

$$I_{L(\max)} = I_T - I_{Z(\min)} = 24\text{ mA} - 3\text{ mA} = 21\text{ mA}$$

The minimum value of R_L is

$$R_{L(\min)} = \frac{V_Z}{I_{L(\max)}} = \frac{12\text{ V}}{21\text{ mA}} = 571\ \Omega$$

Percent Regulation

The percent regulation is a figure of merit used to specify the performance of a voltage regulator. It can be in terms of input (line) regulation or load regulation.

The *percent input regulation* specifies how much change occurs in the output voltage for a given change in input voltage. It is usually expressed as a percent change in V_{OUT} for a 1-V change in V_{IN} (%/V).

The *percent load regulation* specifies how much change occurs in the output voltage over a certain range of load current values, usually from minimum current (no load) to maximum current (full load). It is normally expressed as a percentage and can be calculated with the following formula:

$$\text{Percent load regulation} = \frac{V_{NL} - V_{FL}}{V_{FL}} \times 100\% \qquad \textbf{(18–9)}$$

where V_{NL} is the output voltage with no load, and V_{FL} is the output voltage with full load.

EXAMPLE 18–13 A certain regulator has a no-load output voltage of 6 V and a full-load output of 5.82 V. What is the percent load regulation?

Solution

$$\text{Percent load regulation} = \frac{V_{NL} - V_{FL}}{V_{FL}} \times 100\%$$

$$= \frac{6\text{ V} - 5.82\text{ V}}{5.82\text{ V}} \times 100\% = 3.09\%$$

SECTION REVIEW 18–6

1. Zener diodes are normally operated in the breakdown region (T or F).
2. A certain 10-V zener diode has a resistance of 8 Ω at 30 mA. What is the terminal voltage?
3. Explain the difference between input (line) regulation and load regulation.
4. In a zener diode regulator, for what value of load resistance is the zener current a maximum?
5. A zener regulator has an output voltage of 12 V with no load and 11.9 V with full load. What is the percent load regulation?

VARACTOR DIODES

18–7

Varactor diodes are used as *voltage-variable capacitors*. A varactor is basically a *reverse-biased* PN junction that utilizes the inherent capacitance of the depletion layer. The depletion layer, created by the reverse bias, acts as a capacitor *dielectric* because of its nonconductive characteristic. The P and N regions are conductive and act as the capacitor *plates,* as illustrated in Figure 18–59.

When the reverse-bias voltage increases, the depletion layer widens, effectively increasing the dielectric thickness *(d)* and thus *decreasing* the capacitance. When the reverse-bias voltage decreases, the depletion layer narrows, thus *increasing* the capacitance. This action is shown in Figures 18–60(a) and

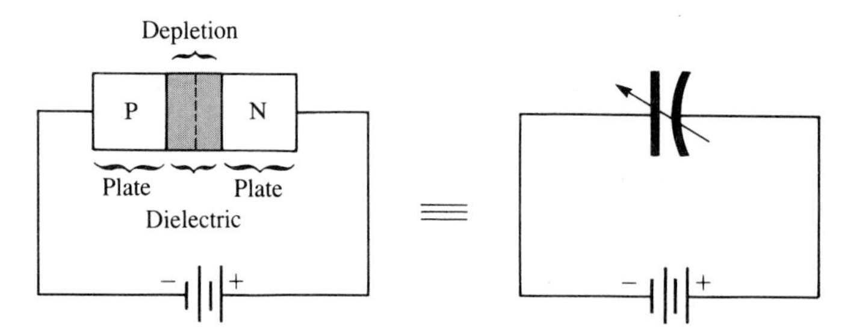

FIGURE 18–59
The reverse-biased varactor diode acts as a variable capacitor.

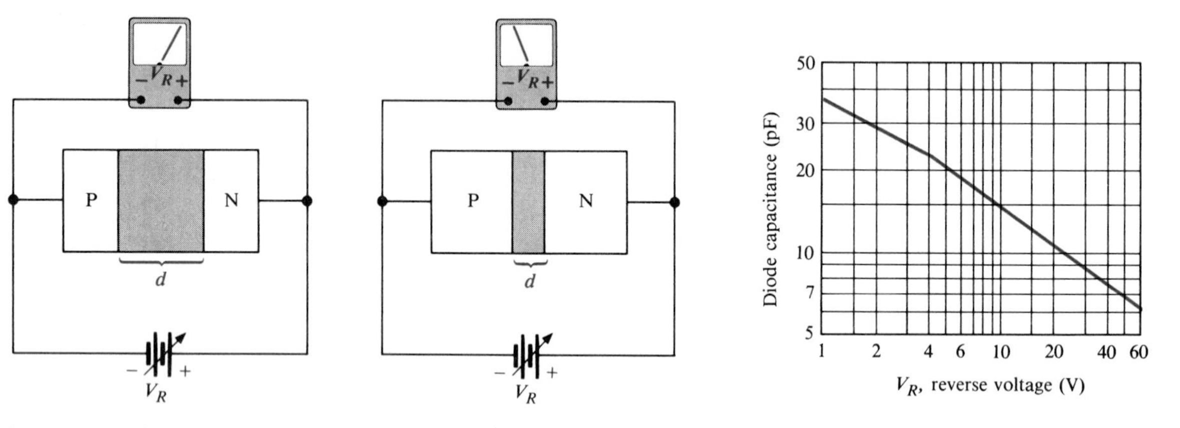

(a) Greater V_R, less capacitance (b) Less V_R, greater capacitance (c) Typical graph of capacitance versus reverse voltage

FIGURE 18–60
Varactor diode capacitance varies with reverse voltage.

18–60(b). A general curve of capacitance versus voltage is shown in Figure 18–60(c).

Recall that capacitance is determined by the *plate area, dielectric constant,* and *dielectric thickness,* as expressed in the following formula:

$$C = \frac{A\epsilon}{d}$$

In a varactor diode, the capacitance parameters are controlled by the method of doping in the depletion layer and the size and geometry of the diode's construction. Varactor capacitances typically range from a few picofarads to a few hundred picofarads.

FIGURE 18–61
Varactor diode.

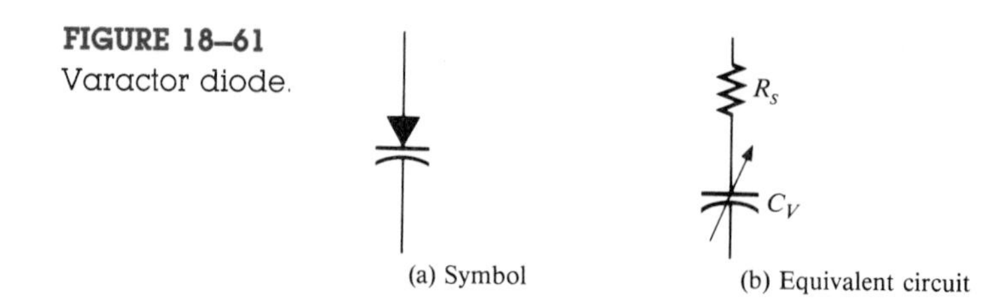

(a) Symbol (b) Equivalent circuit

Figure 18–61(a) shows a common symbol for a varactor, and Part (b) shows a simplified equivalent circuit. R_S is the reverse series resistance, and C_V is the variable capacitance.

Applications

A major application of varactors is in tuning circuits. For example, electronic tuners in TV and other commercial receivers utilize varactors as one of their elements.

When used in a resonant circuit, the varactor acts as a variable capacitor, thus allowing the resonant frequency to be adjusted by a variable voltage level, as illustrated in Figure 18–62 where two varactor diodes provide the total variable capacitance in a parallel resonant (tank) circuit. V_C is a variable dc voltage that controls the reverse bias and therefore the capacitance of the diodes.

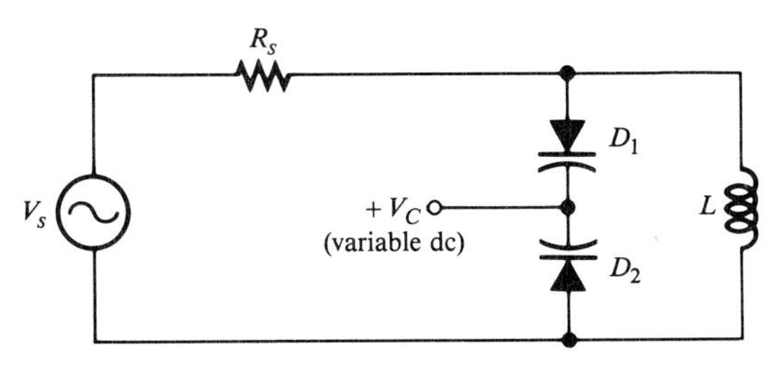

FIGURE 18–62
Varactors in a resonant circuit.

Recall that the resonant frequency of the tank circuit is

$$f_r \cong \frac{1}{2\pi\sqrt{LC}}$$

This approximation is valid for $Q > 10$.

EXAMPLE 18–14 The capacitance of a certain varactor can be varied from 5 pF to 50 pF. The diode is used in a tuned circuit similar to that shown in Figure 18–62. Determine the tuning range for the circuit if $L = 10$ mH.

Solution

The equivalent circuit is shown in Figure 18–63. Notice that the varactor capacitances are in series.

The *minimum* total capacitance is

$$C_{T(\text{min})} = \frac{C_{1(\text{min})}C_{2(\text{min})}}{C_{1(\text{min})} + C_{2(\text{min})}} = \frac{(5\text{ pF})(5\text{ pF})}{10\text{ pF}} = 2.5\text{ pF}$$

The *maximum* resonant frequency therefore is

$$f_{r(\text{max})} = \frac{1}{2\pi\sqrt{LC_{T(\text{min})}}} = \frac{1}{2\pi\sqrt{(10\text{ mH})(2.5\text{ pF})}} \cong 1\text{ MHz}$$

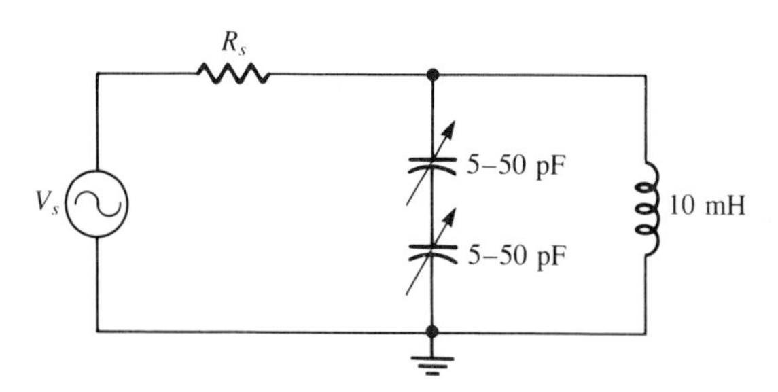

FIGURE 18–63

The *maximum* total capacitance is

$$C_{T(\max)} = \frac{C_{1(\max)}C_{2(\max)}}{C_{1(\max)} + C_{2(\max)}} = \frac{(50 \text{ pF})(50 \text{ pF})}{100 \text{ pF}} = 25 \text{ pF}$$

The *minimum* resonant frequency therefore is

$$f_{r(\min)} = \frac{1}{2\pi\sqrt{LC_{T(\max)}}} = \frac{1}{2\pi\sqrt{(10 \text{ mH})(25 \text{ pF})}} \cong 318 \text{ kHz}$$

SECTION REVIEW 18–7

1. What is the purpose of a varactor diode?
2. Based on the general curve in Figure 18–60(c), what happens to the diode capacitance when the reverse voltage is increased?

LEDs AND PHOTODIODES

18–8

The Light-Emitting Diode (LED)

The basic operation of the LED is as follows: When the device is forward-biased, electrons cross the PN junction from the N-type material and *recombine* with holes in the P-type material. Recall that these free electrons are in the conduction band and at a higher energy level than the holes in the valence band. When recombination takes place, the recombining electrons release energy in the form of *heat* and *light.* A large exposed surface area on one layer of the semiconductor material permits the photons to be emitted as visible light. Figure 18–64 illustrates this process which is called *electroluminescence.*

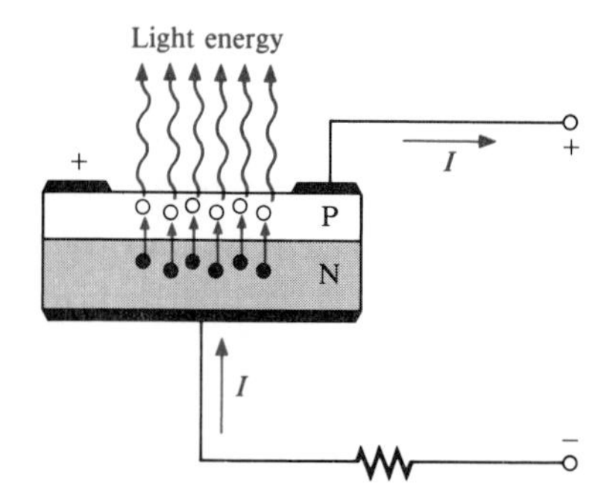

FIGURE 18–64
Electroluminescence in an LED.

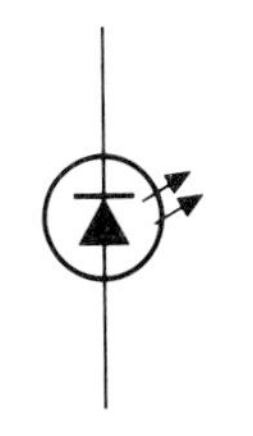

FIGURE 18–65
Symbol for an LED.

The semiconductor materials used in LEDs are *gallium arsenide* (GaAs), *gallium arsenide phosphide* (GaAsP), or *gallium phosphide* (GaP). Silicon and germanium are not used because they are essentially heat-producing materials and are very poor at producing light. GaAs LEDs emit *infrared* (IR) radiation, GaAsP produces either *red* or *yellow* light, and GaP emits *red* or *green* light. The symbol for an LED is shown in Figure 18–65.

The LED *emits* light in response to a sufficient *forward* current, as shown in Figure 18–66(a). The amount of light output is directly proportional to the forward current, as indicated in Part (b). Typical LEDs are shown in Part (c).

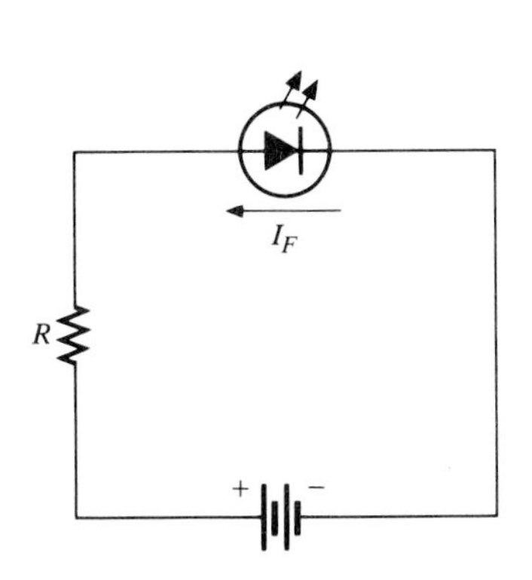

(a) Forward-biased operation

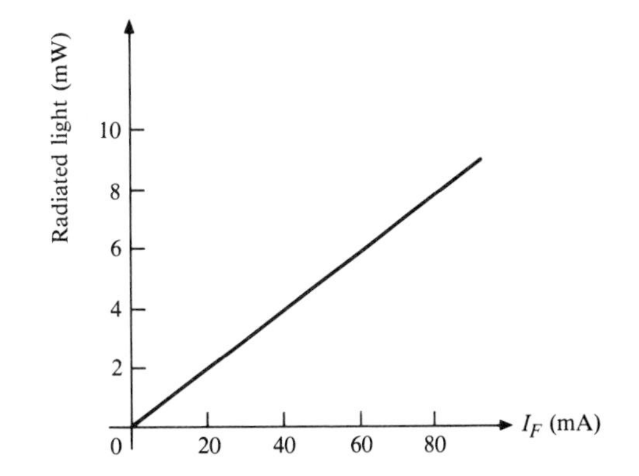

(b) Typical light output versus forward current

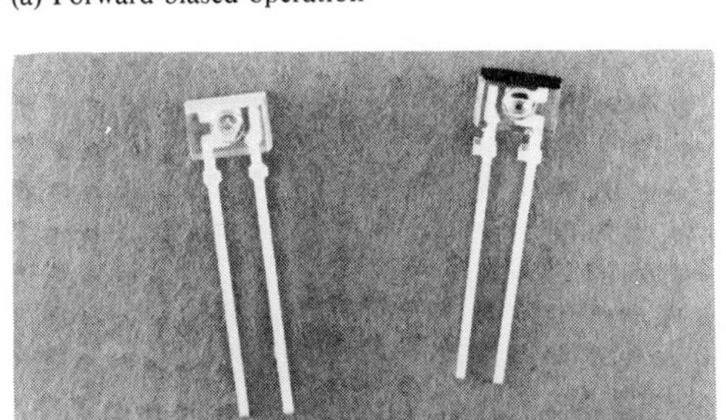

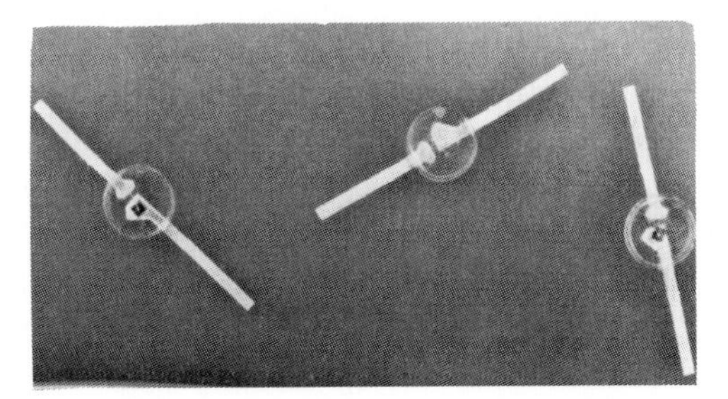

(c)

FIGURE 18–66
LEDs.

Applications

LEDs are commonly used for indicator lamps and readout displays on a wide variety of instruments, ranging from consumer appliances to scientific apparatus. A very common type of display device using LEDs is the seven-segment display. Combinations of the segments form the ten decimal digits. Also, IR-emitting diodes are employed in optical coupling applications, often in conjunction with fiber optics.

The Photodiode

The photodiode is a PN junction device that operates in reverse bias, as shown in Figure 18–67(a). Note the schematic symbol for the photodiode. The photodiode has a small transparent window that allows light to strike the PN junction. Typical photodiodes are shown in Part (b).

Recall that when reverse-biased, a rectifier diode has a very small reverse leakage current. The same is true for the photodiode. The reverse-biased current is produced by thermally generated electron hole pairs in the depletion layer, which are swept across the junction by the electric field created by the reverse voltage. In a rectifier diode, the reverse leakage current increases with temperature due to an increase in the number of electron hole pairs.

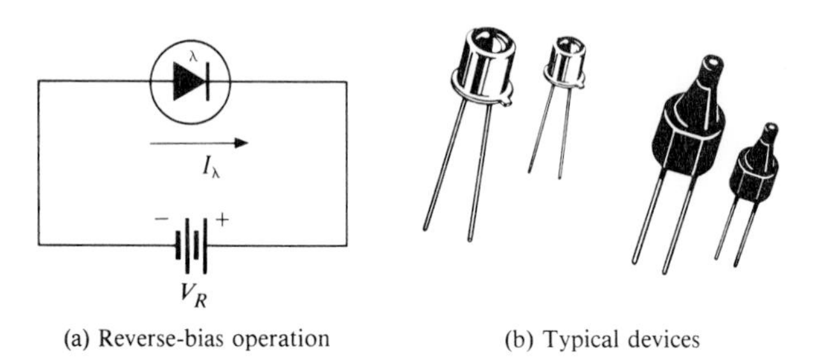

(a) Reverse-bias operation (b) Typical devices

FIGURE 18–67
Photodiode.

A photodiode differs in that the reverse current increases with the light intensity at the PN junction. When there is no incident light, the reverse current I_λ is almost negligible and is called the *dark current.* An increase in the amount of light energy produces an increase in the reverse current, as shown by the graph in Figure 18–68(a). For a given value of reverse-bias voltage, Part (b) shows a set of characteristic curves for a typical photodiode.

From the characteristic curve in Figure 18–68(b), the dark current for this particular device is approximately 25 μA at a reverse bias voltage of 3 V. Therefore, the resistance of the device with no incident light is

$$R_R = \frac{V_R}{I_\lambda} = \frac{3\ \text{V}}{25\ \mu\text{A}} = 120\ \text{k}\Omega$$

At 25,000 lm/m^2, the current is approximately 375 μA at -3 V. The resistance under this condition is

$$R_R = \frac{V_R}{I_\lambda} = \frac{3\ \text{V}}{375\ \mu\text{A}} = 8\ \text{k}\Omega$$

These calculations show that the photodiode can be used as a variable-resistance device controlled by light intensity.

Figure 18–69 illustrates that the photodiode allows essentially no reverse current (except for a very small dark current) when there is no incident light. When a light beam strikes the photodiode, it conducts an amount of reverse current that is proportional to the light intensity.

A simple photodiode application is depicted in Figure 18–70. Here a beam of light continuously passes across a conveyor belt and into a transparent win-

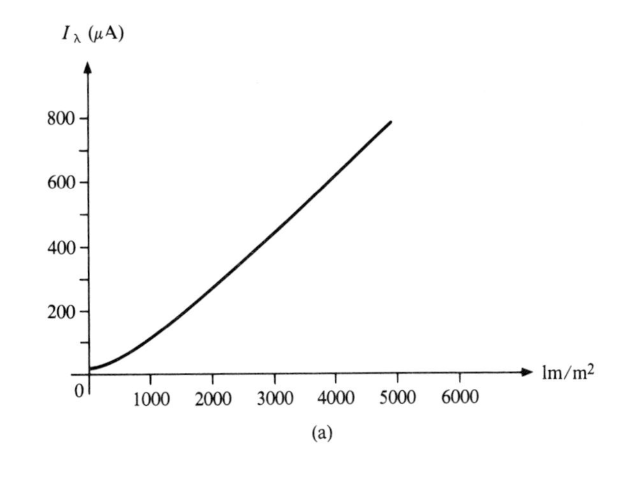

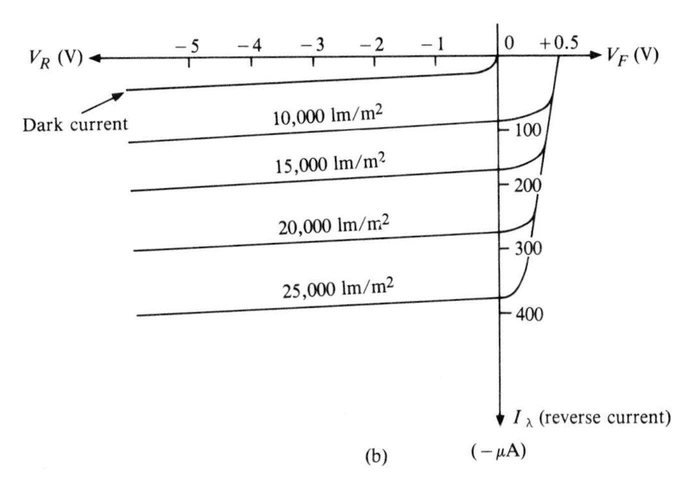

FIGURE 18–68
Typical photodiode characteristics.

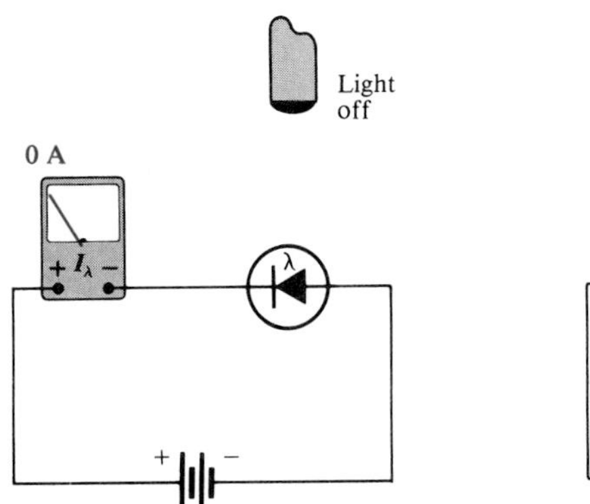

(a) No light, no current

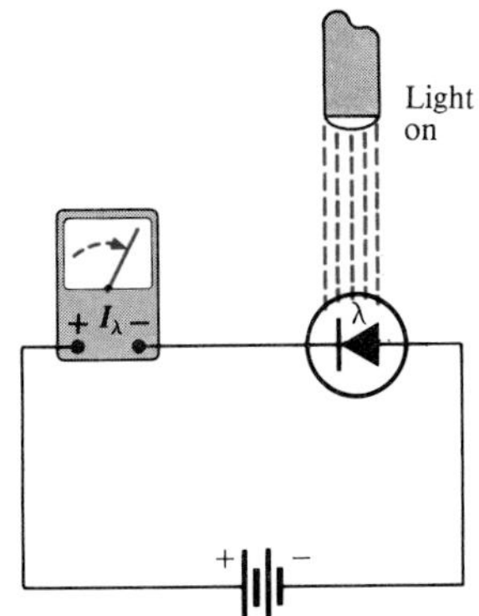

(b) When there is incident light, reverse current flows.

FIGURE 18–69
Operation of a photodiode.

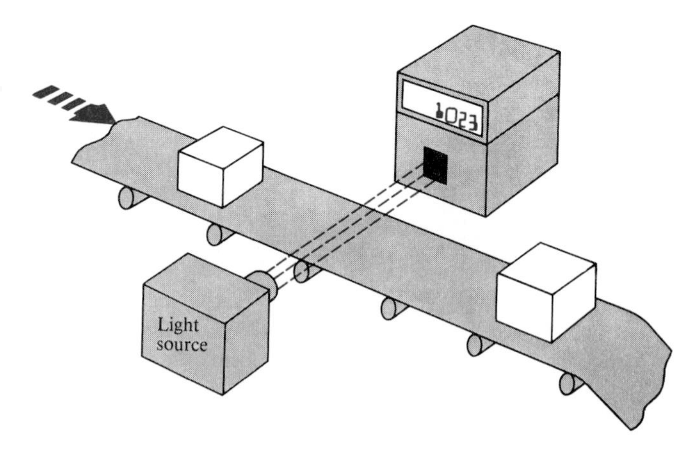

FIGURE 18–70
A photodiode circuit used in a system that counts objects as they pass on a conveyor belt.

dow behind which is a photodiode circuit. When the light beam is interrupted by an object passing by on the conveyor belt, the sudden reduction in diode current activates a control circuit that advances a counter by one. The total count of objects that have passed that point is displayed by the counter. This basic concept can be extended and used for production control, shipping, and monitoring of activity on production lines.

SECTION REVIEW 18–8

1. How does an LED differ from a photodiode?
2. List the semiconductor materials used in LEDs.
3. There is a very small reverse current in a photodiode under no-light conditions. What is this current called?

DIODE DATA SHEETS

18–9

A manufacturer's data sheet specifies *ratings* and *characteristics* for a given type of diode. Table 18–1 shows commonly specified parameters. A data sheet may include some or all of these, depending on how thoroughly a manufacturer wishes to specify the device. Ratings and characteristics fall into three categories: *forward current and voltage, reverse current and voltage,* and *thermal.* Appendix C contains a typical rectifier diode data sheet.

SECTION REVIEW 18–9

1. List the three diode ratings categories.
2. Identify each of the following parameters:
   (a) V_F (b) I_R (c) I_O

APPLICATION NOTE

The Application Assignment requires a 6.8-V zener diode in a circuit of the form shown in Figure 18–53 (excluding the power supply and meters). Additionally, you must know the value of the load so that you can determine the value of the series-limiting resistor *(R)*.

TABLE 18–1
Diode parameters.

Forward	Definition
I_O or $I_{F(avg)}$	Average of maximum forward current
I_{FRM} or $I_{FM(rep)}$	Maximum repetitive peak forward current
I_{FSM} or $I_{FM(surge)}$	Maximum nonrepetitive peak surge current
$V_{F(avg)}$	Average maximum forward voltage drop
$V_{FM(surge)}$	Maximum peak or surge forward voltage drop
V_F	Maximum dc forward voltage drop
Reverse	**Definition**
$I_{R(avg)}$	Maximum average reverse current
I_{RM}	Maximum peak reverse current
I_R	Maximum reverse direct current
V_{RRM} or $V_{RM(rep)}$	Maximum repetitive peak reverse voltage
V_{RWM} or $V_{RM(wkg)}$	Maximum working peak reverse voltage (PIV)
V_{RSM} or $V_{RM(nonrep)}$	Maximum nonrepetitive peak reverse voltage
V_R	Maximum continuous dc reverse blocking voltage
Thermal	**Definition**
θJC	Thermal resistance (W/°C)
T_J	Maximum junction temperature
T_{FA}	Free-air temperature
T_C	Case temperature
i^2t	Maximum current-squared seconds (tells how long a junction can withstand nonrecurring overcurrent)

SUMMARY

Facts

- The single diode in a half-wave rectifier conducts for half of the input cycle.
- The output frequency of a half-wave rectifier equals the input frequency.
- The average (dc) value of a half-wave rectified signal is 0.318 ($1/\pi$) times its peak value.
- The PIV is the maximum voltage appearing across the diode in reverse bias.
- Each diode in a full-wave rectifier conducts for half of the input cycle.
- The output frequency of a full-wave rectifier is twice the input frequency.
- The basic types of full-wave rectifier are center-tapped and bridge.
- The output voltage of a center-tapped, full-wave rectifier is approximately one-half of the total secondary voltage.
- The PIV for each diode in a center-tapped, full-wave rectifier is twice the output voltage.
- The output voltage of a bridge rectifier equals the total secondary voltage.
- The PIV for each diode in a bridge rectifier is half that required for the center-tapped configuration and is approximately equal to the peak output voltage.
- A capacitor-input filter provides a dc output approximately equal to the peak of the input.

- Ripple voltage is caused by the charging and discharging of the filter capacitor.
- The smaller the ripple, the better the filter.
- An *LC* filter provides improved ripple reduction over the capacitor-input filter.
- An inductor-input filter produces a dc output voltage approximately equal to the average value of the rectified input.
- Diode clippers cut off voltage above or below specified levels.
- Diode clampers add a dc level to an ac signal.
- The zener diode operates in reverse breakdown.
- There are two breakdown mechanisms in a zener diode: *avalanche* breakdown and *zener* breakdown.
- When $V_Z < 5$ V, *zener* breakdown is predominant.
- When $V_Z > 5$ V, *avalanche* breakdown is predominant.
- A zener diode maintains an essentially constant voltage across its terminals over a specified range of zener currents.
- Zener diodes are used as shunt voltage regulators.
- Regulation of output voltage over a range of input voltages is called *input* or *line* regulation.
- Regulation of output voltage over a range of load currents is called *load* regulation.
- The *smaller* the percent regulation, the better.
- A *varactor* diode acts as a variable capacitor under reverse-biased conditions.
- The capacitance of a varactor varies inversely with reverse-biased voltage.

Definitions

- *Half-wave rectifier*—a circuit that converts an alternating sine wave into a pulsating dc consisting of one-half of a sine wave for each input cycle.
- *Full-wave rectifier*—a circuit that converts an alternating sine wave into a pulsating dc consisting of both halves of a sine wave for each input cycle.
- *Ripple*—the variation in the dc voltage on the output of a filtered rectifier caused by the slight charging and discharging action of the filter capacitor.
- *Clipping*—cutting off a signal above or below a certain level.
- *Clamping*—inserting a dc level onto an ac signal.
- *Voltage regulation*—the process of maintaining a constant output voltage when there are variations in the input voltage and/or the load.
- *Zener diode*—a diode that operates in reverse breakdown and is used for voltage regulation.
- *Varactor*—a diode that is used as a voltage-variable capacitor.
- *LED*—light-emitting diode that gives off light when forward-biased.
- *Photodiode*—a diode whose reverse resistance changes with incident light.

Formulas

$$V_{AVG} = \frac{V_p}{\pi}$$ Half-wave average value **(18–1)**

$$V_{p(out)} = V_{p(in)} - 0.7\text{ V}$$ Peak half-wave rectifier output **(18–2)**

$$V_{AVG} = \frac{2V_p}{\pi}$$ Full-wave average value **(18–3)**

$\text{PIV} = 2V_{p(\text{out})}$ Diode peak inverse voltage, center-tapped rectifier (neglecting V_B) (18–4)

$V_{\text{out}} = V_s$ Bridge full-wave output (neglecting diode drops) (18–5)

$V_{\text{out}} = V_s - 1.4\text{ V}$ Bridge full-wave output (including diode drops) (18–6)

$\text{PIV} \cong V_{p(\text{out})}$ Diode peak inverse voltage, bridge rectifier (18–7)

$R_Z = \dfrac{\Delta V_Z}{\Delta I_Z}$ Zener resistance (18–8)

$\text{Percent load regulation} = \dfrac{V_{NL} - V_{FL}}{V_{FL}} \times 100\%$ Load regulation (18–9)

Symbols

Diode symbols are shown in Figure 18–71.

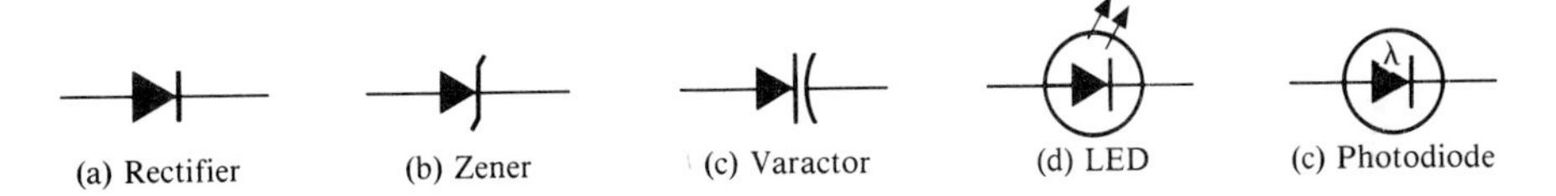

FIGURE 18–71
Diodes

SELF-TEST

1. Define *rectification.*
2. Describe the output voltage from a half-wave rectifier with a sine wave input.
3. If the input frequency of a half-wave rectifier is 60 Hz, what is the output frequency?
4. How many diodes are required for a half-wave rectifier?
5. Describe the output voltage from a full-wave rectifier with a sine wave input.
6. If the input frequency of a full-wave rectifier is 60 Hz, what is the output frequency?
7. Name two major types of full-wave rectifier circuits.
8. Which type of full-wave rectifier requires two diodes, and which type requires four?
9. If one of the diodes in a bridge rectifier opens, what happens to the output?
10. What is the purpose of a capacitor-input filter?
11. For a given full-wave rectifier, what happens to the dc output voltage if the filter capacitance is reduced? Why?
12. How does a zener diode differ from a general-purpose rectifier diode?
13. Name the two types of voltage regulation, and explain each type.
14. What is the principal characteristic of a varactor diode?
15. Under what bias condition is an LED normally operated? Compare the LED to the photodiode.

PROBLEMS

Section 18–1

18–1 Calculate the average value of a half-wave rectified voltage with a peak value of 200 V.

18–2 Sketch the waveforms for the load current and voltage for Figure 18–72. Show the peak values. (The diode is silicon.)

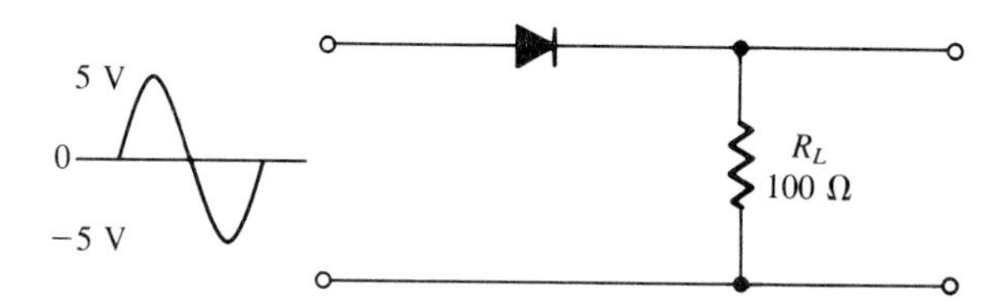

FIGURE 18–72

18–3 Can a diode with a PIV rating of 5V be used in the circuit of Figure 18–72?

18–4 Determine the peak power delivered to R_L in Figure 18–73.

FIGURE 18–73

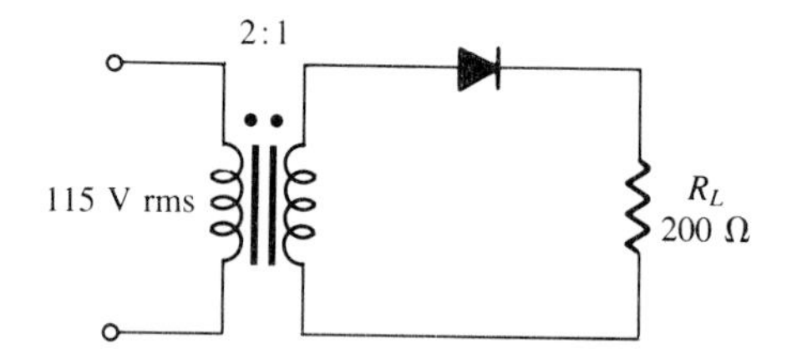

Section 18–2

18–5 Calculate the average value of a full-wave rectified voltage with a peak value of 75 V.

18–6 Consider the circuit in Figure 18–74.
- **(a)** What type of circuit is this?
- **(b)** What is the total peak secondary voltage?
- **(c)** Find the peak voltage across each half of the secondary.
- **(d)** Sketch the voltage waveform across R_L.
- **(e)** What is the peak current through each diode?
- **(f)** What is the PIV for each diode?

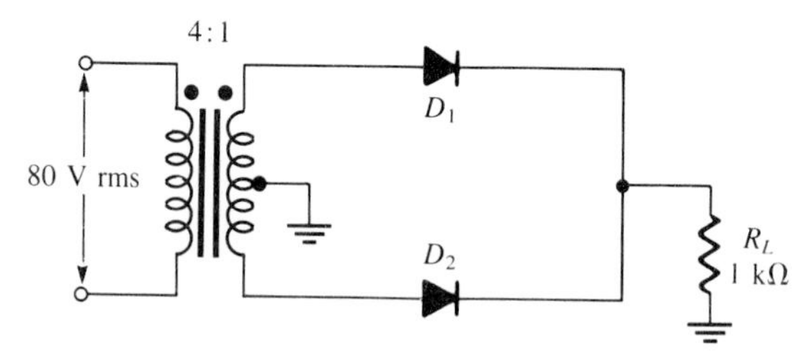

FIGURE 18–74

18–7 Calculate the peak voltage rating of each half of a center-tapped transformer used in a full-wave rectifier that has an average output voltage of 110 V.

18–8 Show how to connect the diodes in a center-tapped rectifier in order to produce a *negative-going* full-wave voltage across the load resistor.

18–9 What PIV rating is required for the diodes in a bridge rectifier that produces an average output voltage of 50 V?

Section 18–3

18–10 The *ideal* dc output voltage of a capacitor-input filter is the (peak, average) value of the rectified input.

18–11 Refer to Figure 18–75 and sketch the following voltage waveforms in relationship to the input waveform: V_{AB}, V_{AD}, V_{BD}, and V_{CD}.

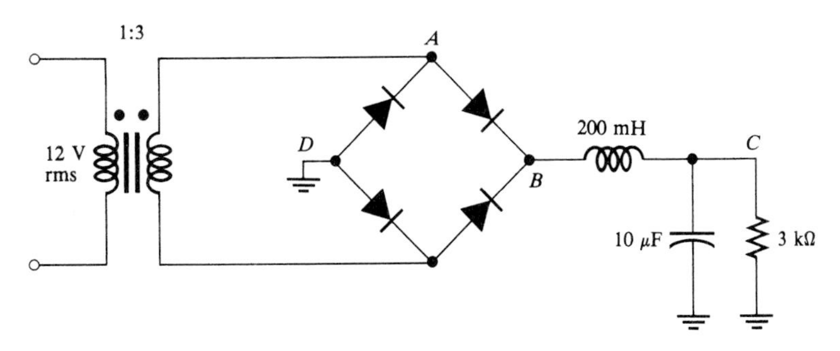

FIGURE 18–75

Section 18–4

18–12 From the meter readings in Figure 18–76, determine if the rectifier circuit is functioning properly. If it is not, determine the most likely failure(s).

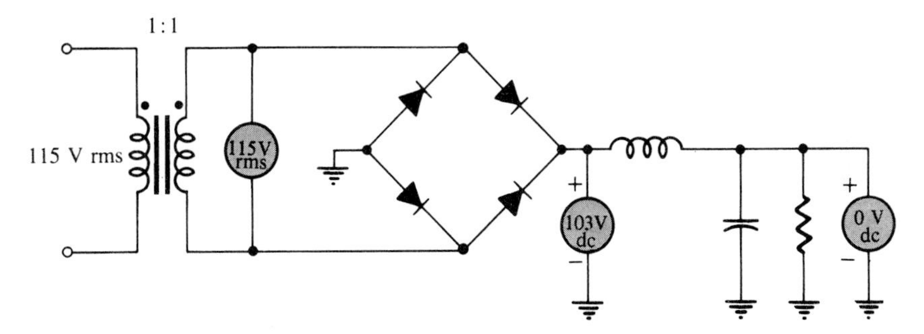

FIGURE 18–76

18–13 Each part of Figure 18–77 shows oscilloscope displays of rectifier output voltages. In each case, determine whether or not the rectifier is functioning properly and, if it is not, what the most likely failure(s) is(are).

Section 18–5

18–14 Sketch the output waveforms for each circuit in Figure 18–78.

18–15 Describe the output waveform of each circuit in Figure 18–79. Assume that the *RC* time constant is much greater than the period of the input.

Section 18–6

18–16 A certain zener diode has a $V_Z = 7.5$ V and an $R_Z = 0.5\ \Omega$. Sketch the equivalent circuit.

18–17 Determine the *minimum* input voltage required for regulation to be established in Figure 18–80. Assume an ideal zener diode with $I_{ZK} = 1.5$ mA and $V_Z = 14$ V.

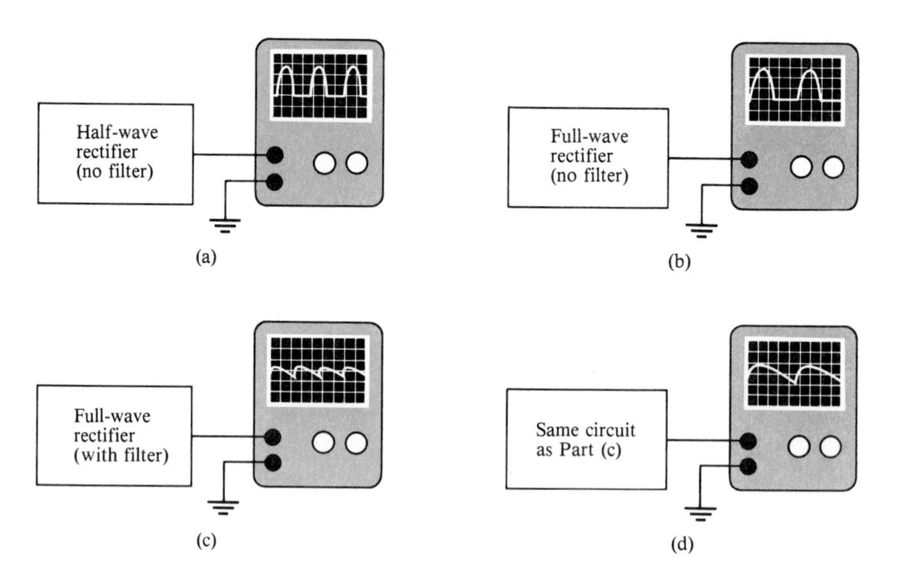

FIGURE 18–77

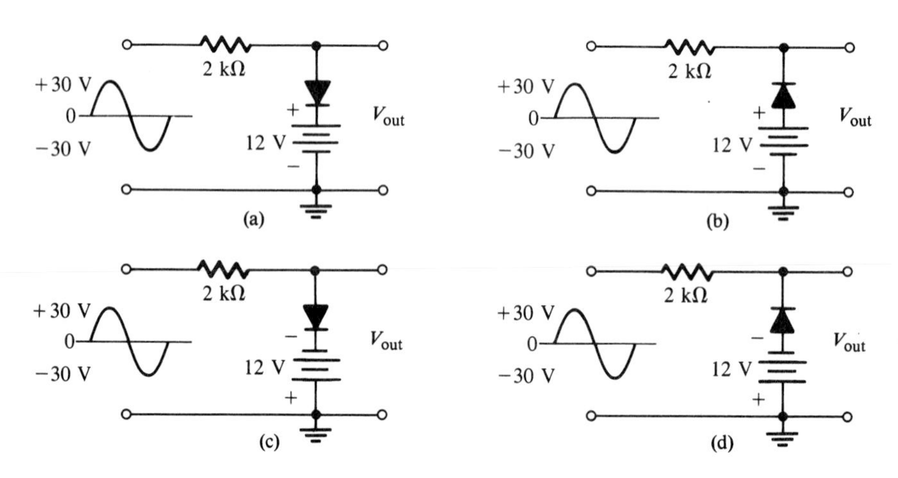

FIGURE 18–78

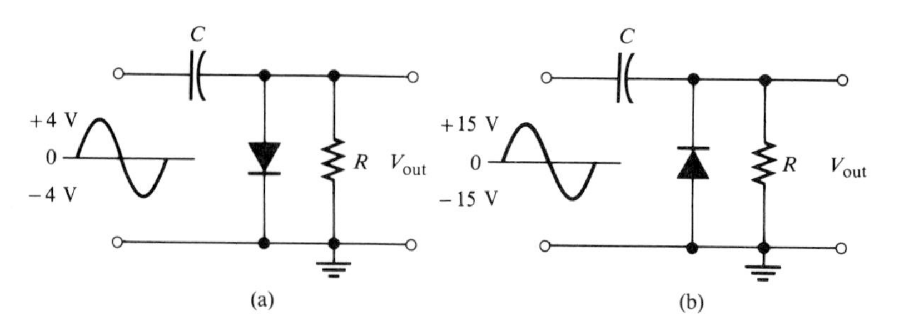

FIGURE 18–79

FIGURE 18–80

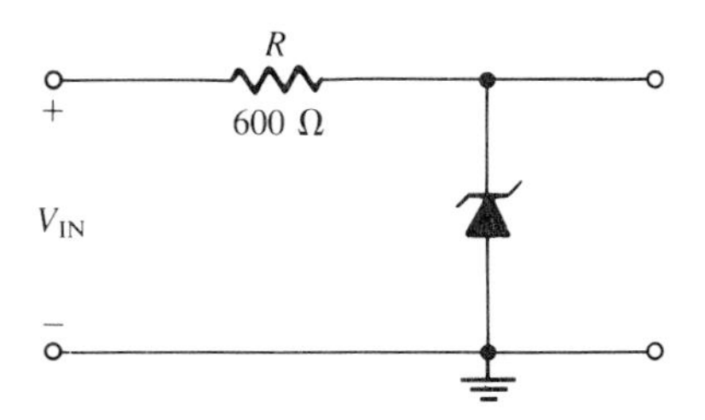

18–18 To what value must R be adjusted in Figure 18–81 to make $I_Z = 40$ mA? Assume that $V_Z = 12$ V and $R_Z = 30\ \Omega$.

FIGURE 18–81

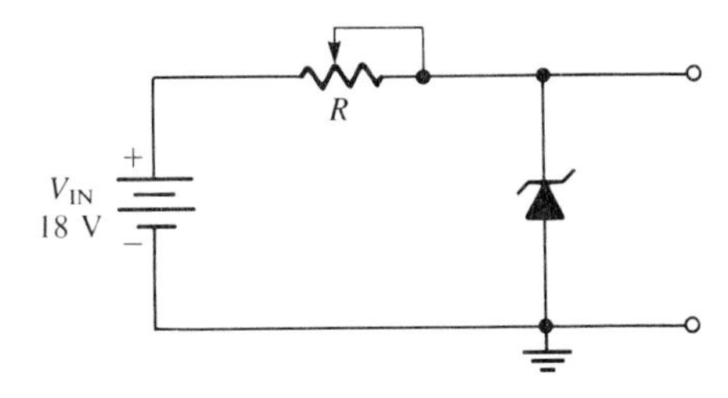

18–19 A loaded zener regulator is shown in Figure 18–82. For $V_Z = 5.1$ V, $I_{ZK} = 5$ mA, $I_{ZM} = 70$ mA, and $R_Z = 12\ \Omega$, determine the minimum and maximum load currents.

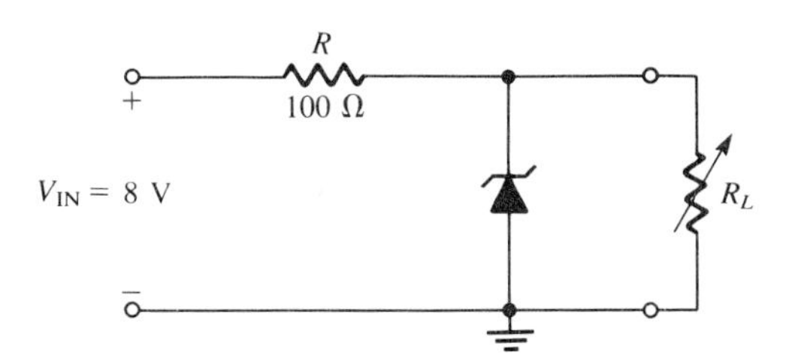

FIGURE 18–82

18–20 Find the percent load regulation in Problem 18–19.

18–21 For the circuit of Problem 18–19, assume that the input voltage is varied from 6 V to 12 V. Determine the percent input regulation with no load and with maximum load.

18–22 The no-load output voltage of a certain zener regulator is 8.23 V, and the full-load output is 7.98 V. Calculate the percent load regulation.

Section 18–7

18–23 Figure 18–83 is a curve of reverse voltage versus capacitance for a certain varactor. Determine the change in capacitance if V_R varies from 5 V to 20 V.

18–24 Refer to Figure 18–83 and determine the value of V_R that produces 25 pF.

18–25 What capacitance value is required for each of the varactors in Figure 18–84 to produce a resonant frequency of 1 MHz?

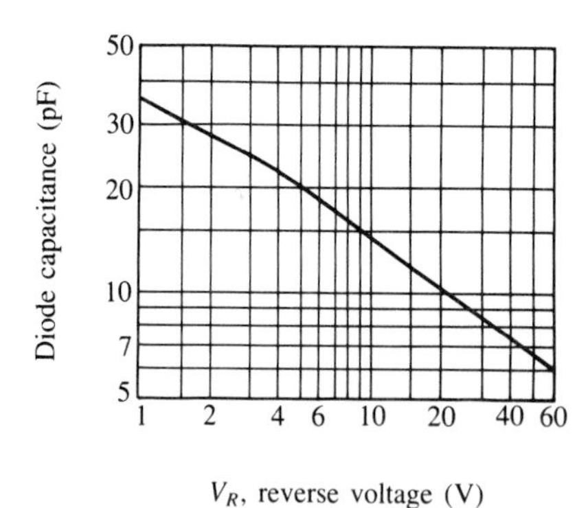

FIGURE 18–83

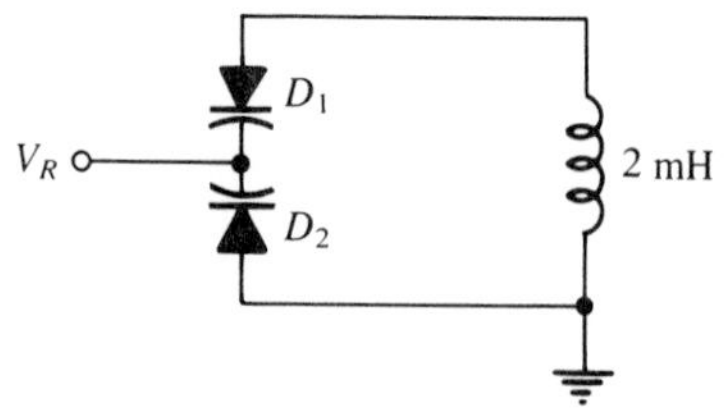

FIGURE 18–84

18–26 At what value must the control voltage be set in Problem 18–25 if the varactors have the characteristic curve in Figure 18–83?

Section 18–8

18–27 When the switch in Figure 18–85 is closed, will the microammeter reading increase or decrease? Assume that D_1 and D_2 are optically coupled.

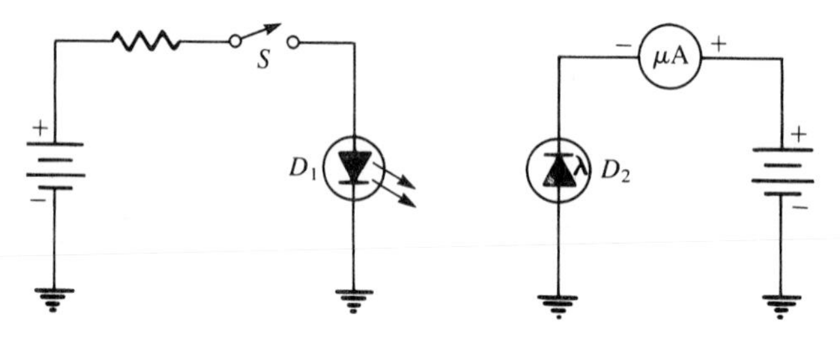

FIGURE 18–85

18–28 With no incident light, a certain amount of reverse current flows in a photodiode. What is this current called?

ANSWERS TO SECTION REVIEWS

Section 18–1

1. 270°. **2.** 50%. **3.** 3.18 V.

Section 18–2

1. 38.2 V. **2.** Bridge. **3.** True.

Section 18–3

1. 60 Hz, 120 Hz. **2.** Capacitor charging and discharging slightly.
3. It increases ripple. **4.** Reduced ripple, less dc voltage output.

Section 18–4

1. Half-wave output. **2.** Open diode. **3.** Leaky filter capacitor.

Section 18–5

1. See Figure 18–86. **2.** See Figure 18–87. **3.** See Figure 18–88.

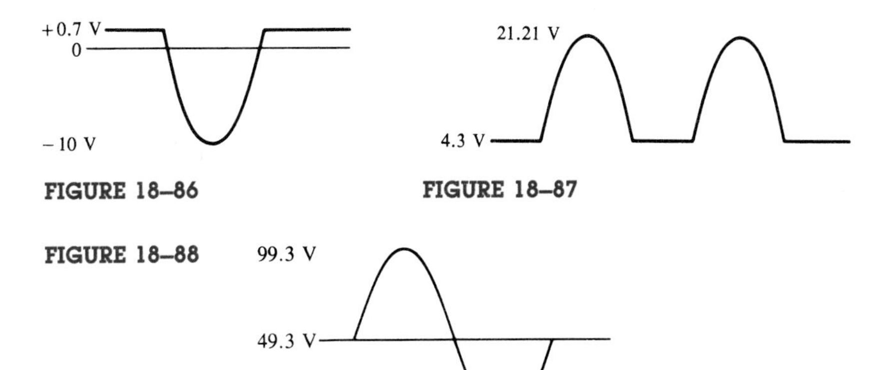

FIGURE 18–86

FIGURE 18–87

FIGURE 18–88

Section 18–6

1. True. **2.** 10.24 V.
3. *Input regulation:* Constant output voltage for varying input voltage. *Load regulation:* Constant output voltage for varying load current.
4. Maximum R_L. **5.** 0.833%.

Section 18–7

1. It is a variable capacitor. **2.** It decreases.

Section 18–8

1. LEDs give off light when forward-biased; photodiodes respond to light when reverse-biased.
2. Gallium arsenide, gallium arsenide phosphide, gallium phosphide.
3. Dark current.

Section 18–9

1. Forward current and voltage; reverse current and voltage; thermal.
2. **(a)** Maximum dc forward voltage drop; **(b)** maximum reverse direct current; **(c)** average maximum forward current.

Transistors and Thyristors

As mentioned earlier, the transistor was invented in 1947 by three scientists at Bell Labs. Since that time, this device has revolutionized electronics by making possible smaller and more reliable circuits and systems. Also, more recent revolutionary developments such as the integrated circuit are an outgrowth of transistor technology.

Two basic types of transistors will be discussed in this chapter: the *bipolar junction transistor (BJT)* and the *field-effect transistor (FET)*. The two major application areas of amplification and switching are introduced. In addition, the unijunction transistor (UJT) will be introduced and common types of thyristors presented.

Specifically, you will learn:

- The basic construction of bipolar transistors.
- The difference between PNP and NPN transistors.
- How to bias a transistor with dc voltages so that it can be operated as an amplifier.
- The transistor currents and how they are related.
- How voltage divider bias works in a transistor circuit.
- What the characteristic curves of a transistor mean.
- The meanings of *cutoff* and *saturation*.
- How a transistor produces voltage gain.
- How a transistor can be used as a switch.
- The significance of transistor parameters.
- How to test a transistor.
- The construction and operation of junction field-effect transistors (JFETs).
- The construction and operation of metal oxide semiconductor field-effect transistors (MOSFETs).
- How a MOSFET functions in the depletion mode and in the enhancement mode.
- How to bias an FET.
- How a unijunction transistor (UJT) works and how it is used in a basic oscillator.
- The basic operation of SCRs, triacs, and diacs, and how they are used as control devices.
- How to recognize the various types of transistor packages and to identify the transistor leads.

19

APPLICATION ASSIGNMENT

A stereo amplifier comes into your shop with one of the channels dead. There is no output signal to the speaker. You suspect that one of the bipolar power transistors has completely failed and thus must be removed and checked. How do you proceed, and what instrument(s) do you need?

After completing this chapter, you will be able to carry out this assignment. A solution is given in the Application Note at the end of the chapter.

BIPOLAR JUNCTION TRANSISTORS (BJTs)

19–1

The bipolar junction transistor is constructed with *three* doped semiconductor regions separated by *two* PN junctions. The three regions are called *emitter, base,* and *collector*. The two types of bipolar transistors are shown in Figure 19–1. One type consists of two N regions separated by a P region (NPN), and the other consists of two P regions separated by an N region (PNP).

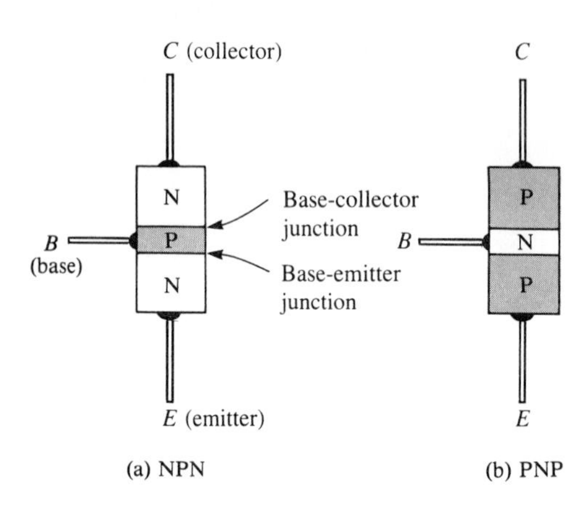

FIGURE 19–1
Construction of bipolar transistors.

The PN junction joining the base region and the emitter region is called the *base-emitter* junction. The junction joining the base region and the collector region is called the *base-collector* junction, as indicated. A wire lead connects to each of the three regions, as shown. These leads are labeled *E, B,* and *C* for emitter, base, and collector, respectively. The base material is lightly doped and very narrow compared to the heavily doped emitter and collector materials. The reason for this is discussed in the next section.

Figure 19–2 shows the schematic symbols for the NPN and PNP bipolar transistors. The term *bipolar* refers to the use of both *holes* and *electrons* as carriers in the transistor structure.

FIGURE 19–2
Transistor symbols.

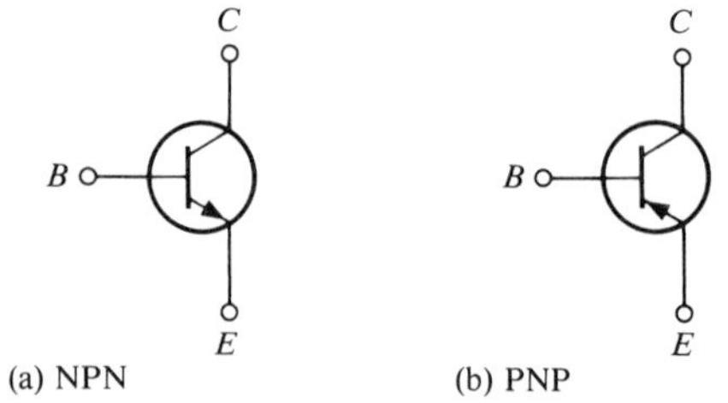

(a) NPN (b) PNP

Transistor Biasing

In order for the transistor to operate properly as an amplifier, the two PN junctions must be correctly *biased* with external voltages. We will use the NPN transistor to illustrate transistor biasing. The operation of the PNP is the same as for the NPN except that the roles of the electrons and holes, the bias voltage polarities, and the current directions are all reversed. Figure 19–3 shows the proper bias arrangement for both NPN and PNP transistors. Notice that in both cases the *base-emitter (BE) junction is forward-biased and the base-collector (BC) junction is reverse-biased.*

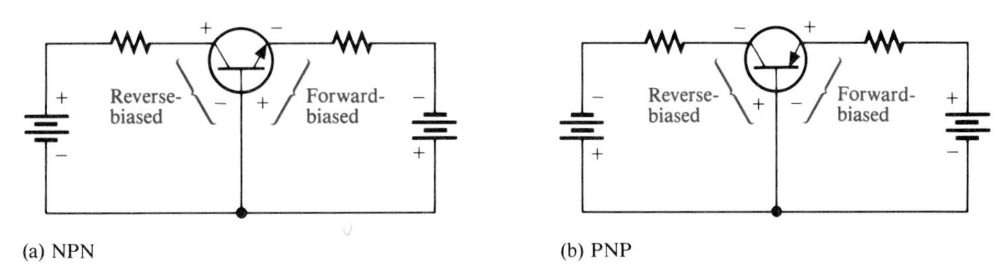

FIGURE 19–3
Forward-reverse bias of a bipolar transistor.

Now, let's examine what happens inside the transistor when it is forward-reverse-biased. The forward bias from base to emitter *narrows* the *BE* depletion layer, and the reverse bias from base to collector *widens* the *BC* depletion layer, as depicted in Figure 19–4(a). The N-type emitter region is teeming with conduction band (free) electrons which easily diffuse across the *BE* junction into the P-type base region, just as in a forward-biased diode.

The base region is lightly doped and very thin so that it has a very limited number of holes. Thus, only a small percentage of all the electrons flowing across the *BE* junction combine with the available holes. These relatively few recombined electrons flow out of the base lead as valence electrons, forming the small base current, I_B, as shown in Figure 19–4(b).

Most of the electrons flowing from the emitter into the base region diffuse into the *BC* depletion layer. Once in this layer, they are pulled across the *BC* junction by the depletion layer field set up by the force of attraction between the positive and negative ions. Actually, you can think of the electrons as being pulled across the reverse-biased *BC* junction by the attraction of the positive ions on the other side, as illustrated in Part (c). The electrons now move through the collector region, out through the collector lead, and into the positive terminal of the external dc source, thereby forming the collector current, I_C, as shown. The amount of collector current depends directly on the amount of base current and is essentially independent of the dc collector voltage.

Transistor Currents

The directions of current in an NPN and a PNP transistor are as shown in Figures 19–5(a) and 19–5(b), respectively. An examination of these diagrams shows that the emitter current is the *sum* of the collector and base currents, expressed as follows:

$$I_E = I_C + I_B \tag{19–1}$$

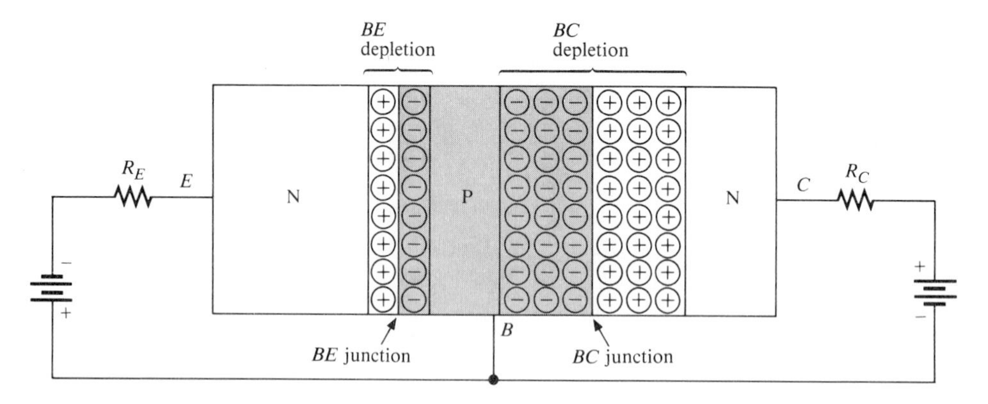

(a) Internal effects of forward-reverse bias

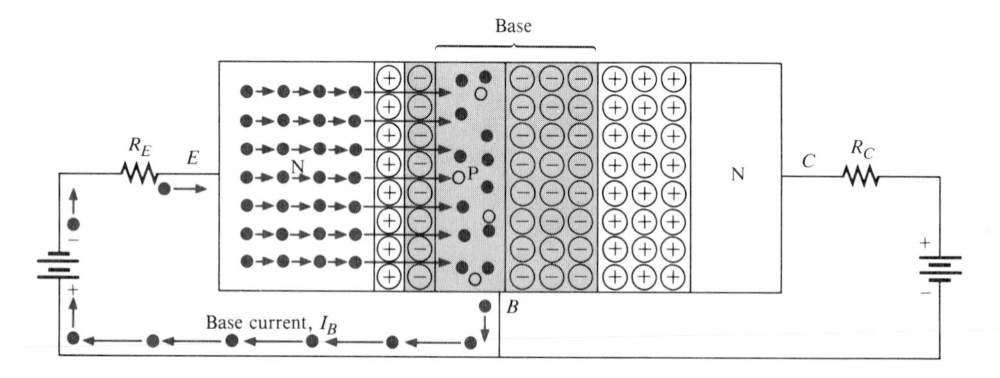

(b) Electron flow across the emitter-base junction

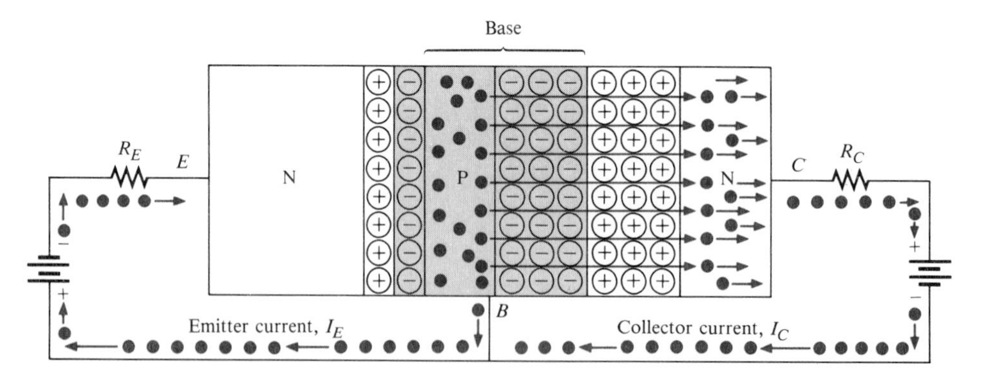

(c) Electron flow across the base-collector junction

FIGURE 19–4
Transistor action.

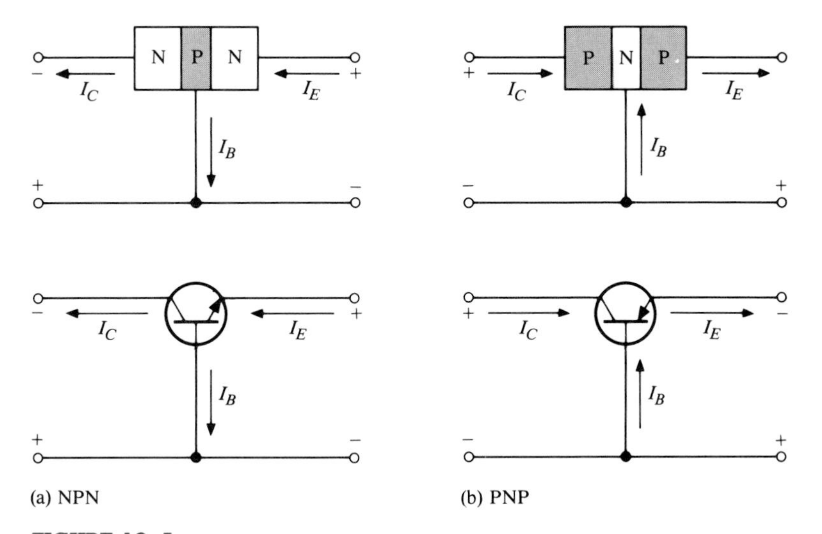

FIGURE 19–5
Current directions in a transistor.

As mentioned before, I_B is very small compared to I_E or I_C. The capital-letter subscripts indicate dc values.

These direct currents are also related by two parameters: the dc alpha (α_{dc}) and the dc beta (β_{dc}). β_{dc} *is the direct current gain* and is usually designated as h_{FE} on the transistor data sheets.

The collector current is equal to α_{dc} times the emitter current:

$$I_C = \alpha_{dc} I_E \tag{19–2}$$

where α_{dc} typically has a value between 0.95 and 0.99.

The collector current is related to the base current by β_{dc}:

$$I_C = \beta_{dc} I_B \tag{19–3}$$

where β_{dc} typically has a value between 20 and 200.

Transistor Voltages

The three dc voltages for the biased transistor in Figure 19–6 are the emitter voltage V_E, the collector voltage V_C, and the base voltage V_B. These voltages are with respect to ground. The collector voltage is equal to the dc supply voltage, V_{CC}, less the drop across R_C:

$$V_C = V_{CC} - I_C R_C \tag{19–4}$$

The base voltage is equal to the emitter voltage plus the base-emitter junction barrier potential (V_{BE}), which is about 0.7 V for a silicon transistor:

$$V_B = V_E + V_{BE} \tag{19–5}$$

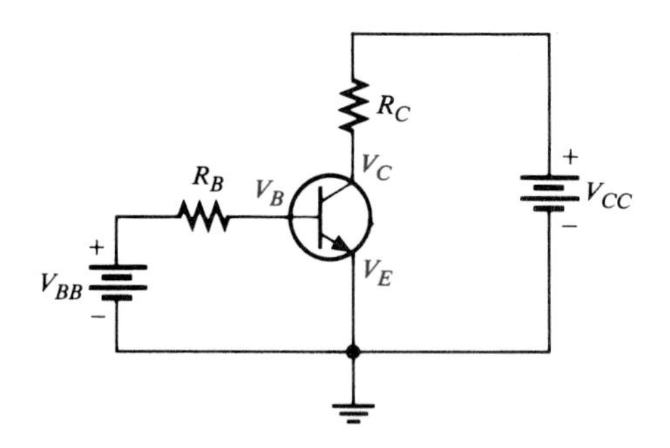

FIGURE 19–6
Bias voltages.

In the configuration of Figure 19–6, *the emitter is the common terminal,* so $V_E = 0$ V.

EXAMPLE 19–1 Find I_B, I_C, I_E, V_B, and V_C in Figure 19–7, where β_{dc} is 50.

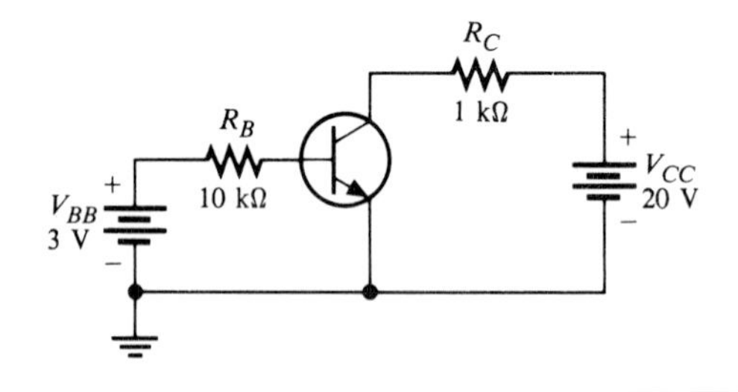

FIGURE 19–7

Solution
Since V_E is ground, $V_B = 0.7$ V. The drop across R_B is $V_{BB} - V_B$, so I_B is calculated as follows:

$$I_B = \frac{V_{BB} - V_B}{R_B} = \frac{3\text{ V} - 0.7\text{ V}}{10\text{ k}\Omega} = 0.23\text{ mA}$$

Now we can find I_C, I_E, and V_C.

$$I_C = \beta_{dc} I_B = 50(0.23\text{ mA}) = 11.5\text{ mA}$$
$$I_E = I_C + I_B = 11.5\text{ mA} + 0.23\text{ mA} = 11.73\text{ mA}$$
$$V_C = V_{CC} - I_C R_C = 20\text{ V} - (11.5\text{ mA})(1\text{ k}\Omega) = 8.5\text{ V}$$

SECTION REVIEW 19–1

1. Name the three transistor terminals.
2. Define *forward-reverse bias.*
3. What is β_{dc}?
4. If I_B is 10 μA and β_{dc} is 100, what is the collector current?

VOLTAGE DIVIDER BIAS

19–2

One of the most common bias arrangements in transistor circuits is the voltage divider bias. This configuration uses only a single dc source to provide forward-reverse bias to the transistor, as shown in Figure 19–8. Resistors R_1 and R_2 form a voltage divider that provides the base bias voltage. Resistor R_E allows the emitter to rise above ground potential.

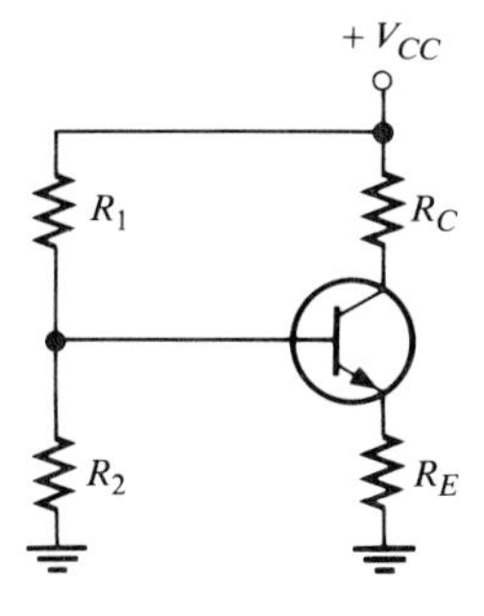

FIGURE 19–8
Voltage divider bias.

The voltage divider is loaded by the resistance viewed from the base of the transistor. In some cases, this loading effect is significant in determination of the base bias voltage. We will now examine this arrangement in more detail.

Input Resistance at the Base

The approximate input resistance of the transistor, viewed from the base of the transistor in Figure 19–9, is derived as follows:

$$R_{IN} = \frac{V_B}{I_B}$$

Neglecting the V_{BE} of 0.7 V, we have

$$V_B \cong V_E = I_E R_E$$

Since $I_E \cong I_C$ when $\alpha_{dc} \cong 1$, then

$$I_E \cong \beta_{dc} I_B$$

Substituting yields

$$V_B \cong \beta_{dc} I_B R_E$$

$$R_{IN} \cong \frac{\beta_{dc} I_B R_E}{I_B}$$

The I_B terms cancel, leaving

$$R_{IN} \cong \beta_{dc} R_E \qquad (19–6)$$

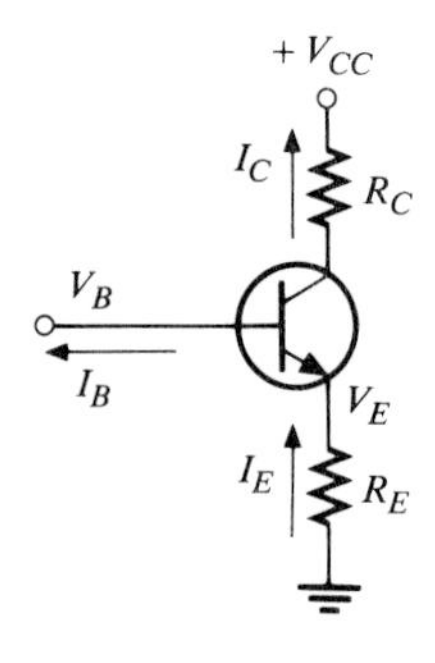

FIGURE 19–9
Circuit for deriving input resistance.

Base Voltage

Now, using the voltage divider formula, the following equation gives the base voltage for the circuit in Figure 19–8:

$$V_B = \left(\frac{R_2 \parallel R_{IN}}{R_1 + R_2 \parallel R_{IN}}\right)V_{CC} \quad \textbf{(19–7)}$$

Generally, if R_{IN} is at least ten times greater than R_2, then Equation (19–7) can be simplified as follows:

$$V_B \cong \left(\frac{R_2}{R_1 + R_2}\right)V_{CC} \quad \textbf{(19–8)}$$

Once you have determined the base voltage, you can determine the emitter voltage V_E (for an NPN transistor) as follows:

$$V_E = V_B - 0.7 \text{ V}$$

EXAMPLE 19–2 Determine V_B, V_E, V_C, V_{CE}, I_B, I_E, and I_C in Figure 19–10.

FIGURE 19–10

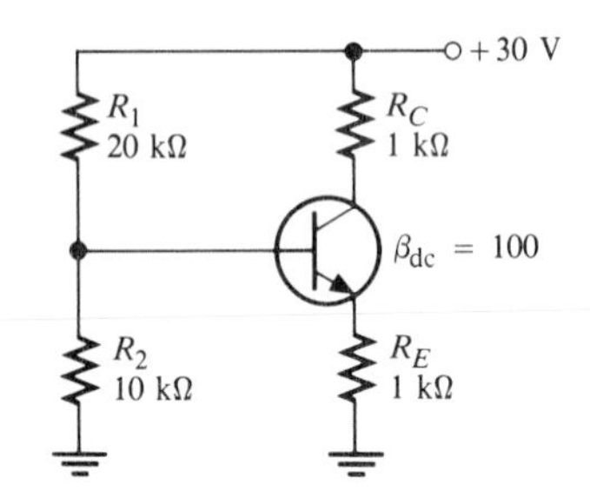

Solution
The input resistance at the base is

$$R_{IN} \cong \beta_{dc}R_E = 100(1 \text{ k}\Omega) = 100 \text{ k}\Omega$$

Since R_{IN} is ten times greater than R_2, the base voltage is approximately

$$V_B \cong \left(\frac{R_2}{R_1 + R_2}\right)V_{CC} = \left(\frac{10 \text{ k}\Omega}{30 \text{ k}\Omega}\right)30 \text{ V} = 10 \text{ V}$$

Therefore, $V_E = V_B - 0.7 \text{ V} \cong 9.3 \text{ V}$.

Now that we know V_E, we can find I_E by Ohm's law:

$$I_E = \frac{V_E}{R_E} = \frac{9.3 \text{ V}}{1 \text{ k}\Omega} = 0.0093 \text{ A} = 9.3 \text{ mA}$$

Since α_{dc} is so close to 1 for most transistors, it is a good approximation to assume that $I_C \cong I_E$. Thus,

$$I_C \cong 9.3 \text{ mA}$$

Using $I_C = \beta_{dc} I_B$ and solving for I_B, we get

$$I_B = \frac{I_C}{\beta_{dc}} \cong \frac{9.3 \text{ mA}}{100} = 0.093 \text{ mA}$$

Now that we know I_C, we can find V_C:

$$V_C = V_{CC} - I_C R_C = 30 \text{ V} - (9.3 \text{ mA})(1 \text{ k}\Omega)$$
$$= 30 \text{ V} - 9.3 \text{ V} = 20.7 \text{ V}$$

Since V_{CE} is the collector-to-emitter voltage, it is the difference of V_C and V_E:

$$V_{CE} = V_C - V_E = 20.7 \text{ V} - 9.3 \text{ V} = 11.5 \text{ V}$$

SECTION REVIEW 19–2

1. How many dc voltage sources are required for voltage divider bias?
2. In Figure 19–10, if R_2 is 5 kΩ, what is the value of V_B?

THE BIPOLAR TRANSISTOR AS AN AMPLIFIER

19–3

The purpose of dc bias is to allow a transistor to operate as an amplifier. Thus, a transistor can be used to increase a small input signal to a much larger value. In this section we will discuss how a transistor acts as an amplifier. The subject of amplifiers is covered in more detail in the next chapter.

Collector Curves

With a circuit such as that in Figure 19–11(a), a set of curves can be generated showing how I_C varies with V_{CE} for various values of I_B. These curves are the *collector characteristic* curves.

Notice that both V_{BB} and V_{CC} are adjustable. If V_{BB} is set to produce a specific value of I_B and V_{CC} is zero, then $I_C = 0$ and $V_{CE} = 0$. Now, as V_{CC} is gradually increased, V_{CE} will increase and so will I_C, as indicated on the portion of the curve between points A and B in Figure 19–11(b).

When V_{CE} reaches approximately 0.7 V, the base-collector junction becomes reverse-biased and I_C reaches its full value determined by the relationship $I_C = \beta_{dc} I_B$. At this point, I_C levels off to an almost constant value as V_{CE} continues to increase. This action appears to the right of point B on the curve. Actually, I_C increases slightly as V_{CE} increases due to widening of the base-collector depletion layer which results in fewer holes for recombination in the base region.

By using other values of I_B, we can produce additional I_C versus V_{CE} curves, as shown in Figure 19–11(c). These curves constitute a *family* of collector curves for a given transistor.

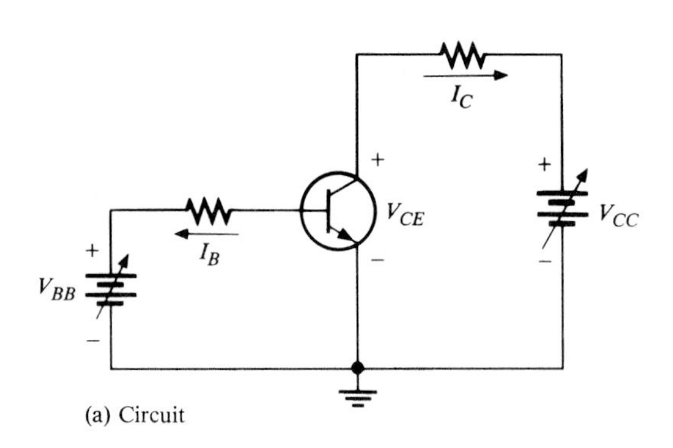

(a) Circuit

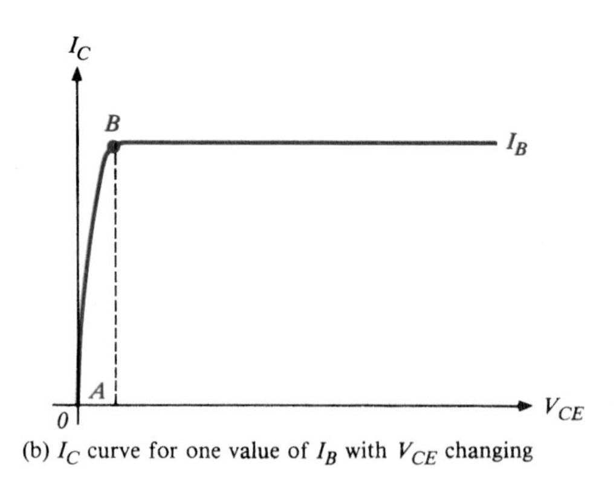

(b) I_C curve for one value of I_B with V_{CE} changing

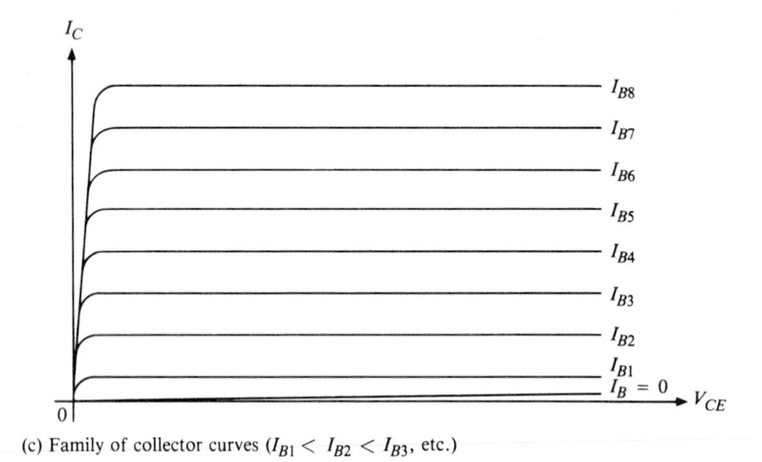

(c) Family of collector curves ($I_{B1} < I_{B2} < I_{B3}$, etc.)

FIGURE 19–11
Collector characteristic curves.

EXAMPLE 19–3 Sketch the family of collector curves for the circuit in Figure 19–12 for I_B = 5 μA to 25 μA in 5-μA increments. Assume that β_{dc} = 100.

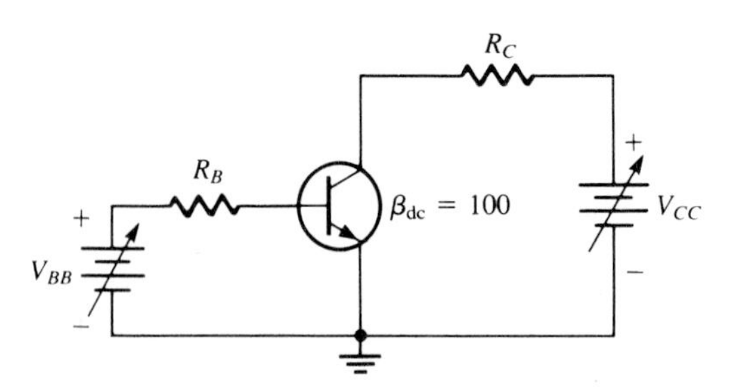

FIGURE 19–12

Solution
Using the relationship $I_C = \beta_{dc} I_B$, values of I_C are calculated and tabulated as follows:

When I_B =	Then I_C =
5 μA	0.5 mA
10 μA	1.0 mA
15 μA	1.5 mA
20 μA	2.0 mA
25 μA	2.5 mA

The resulting curves are plotted in Figure 19–13.

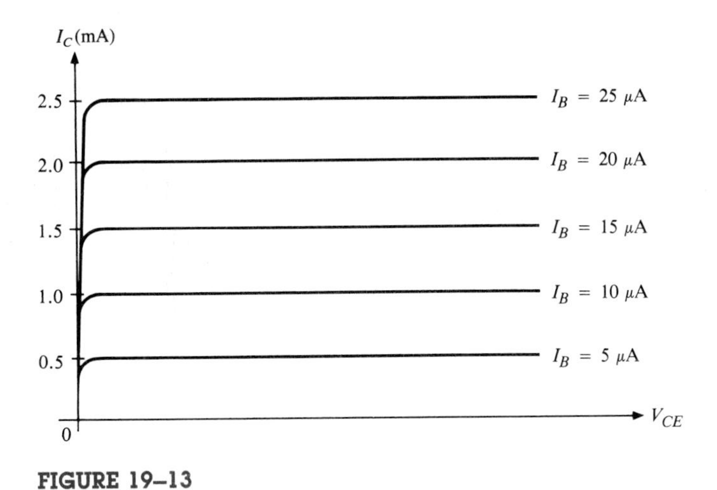

FIGURE 19–13

Cutoff and Saturation

When $I_B = 0$, the transistor is in *cutoff*. Under this condition, there is a very small amount of collector leakage current, I_{CEO}, due mainly to thermally produced carriers. In cutoff, both the base-emitter and the base-collector junctions are *reverse-biased*.

Now let's consider the condition known as *saturation*. When the base current is increased, the collector current also increases and V_{CE} decreases as a result of more drop across R_C.

When V_{CE} reaches a value called $V_{CE(sat)}$, the base-collector junction becomes *forward-biased* and I_C can increase no further even with a continued increase in I_B. At the point of saturation, the relation $I_C = \beta_{dc} I_B$ is no longer valid.

For a transistor, $V_{CE(sat)}$ occurs somewhere below the knee of the collector curves and is usually only a few tenths of a volt for silicon transistors.

Load Line Operation

A straight line drawn on the collector curves between the cutoff and saturation points of the transistor is called the *load line*. Once set up, the transistor always operates along this line. Thus, any value of I_C and the corresponding V_{CE} will fall on this line.

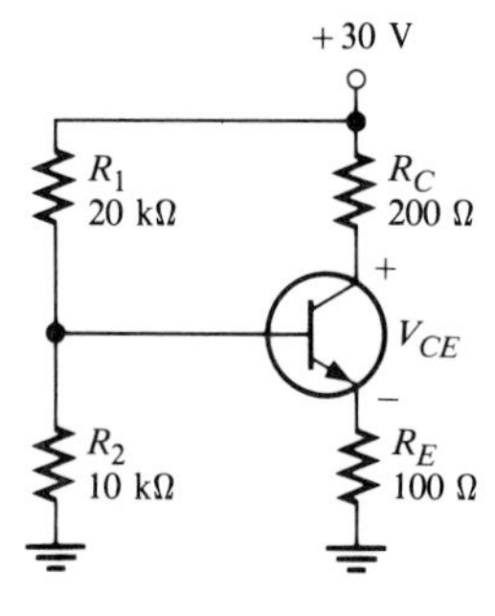

FIGURE 19–14

Now we will set up a load line for the circuit in Figure 19–14, so that we can learn what it tells us about the transistor operation. First the *cutoff point* on the load line is determined as follows: When the transistor is cut off, there is essentially no collector current. Thus, the collector-emitter voltage V_{CE} is equal to V_{CC}. In this case, $V_{CE} = 30$ V.

Next the *saturation point* on the load line is determined. When the transistor is saturated, V_{CE} is approximately zero. (Actually, it is usually a few tenths of a volt, but zero is a good approximation.) Therefore, all of the V_{CC} voltage is dropped across $R_C + R_E$. From this we can determine the saturation value of collector current, $I_{C(sat)}$. This value is the *maximum value* for I_C. We cannot possibly increase it further without changing V_{CC}, R_C, or R_E. In Figure 19–14, the value of $I_{C(sat)}$ is $V_{CC}/(R_C + R_E)$, which is 100 mA.

Next the cutoff and saturation points are plotted on the assumed curves in Figure 19–15, and a straight line, which is the *load line*, is drawn between them.

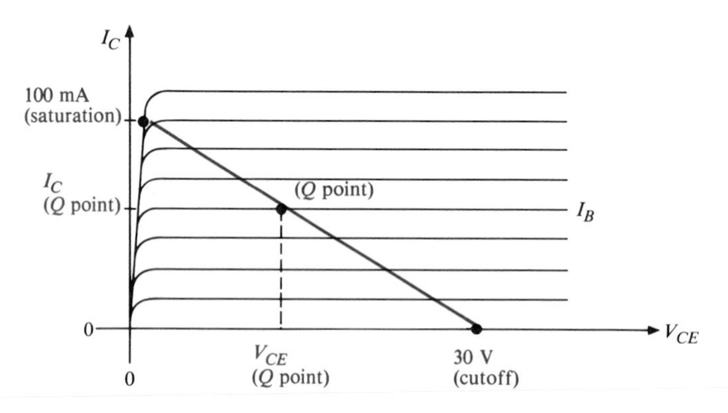

FIGURE 19–15
Load line for the circuit in Figure 19–14.

Q Point

The base current I_B is established by the base bias. The point at which the base current curve intersects the load line is the *quiescent* or *Q point* for the circuit. The coordinates of the Q point are the values for I_C and V_{CE} at that point, as illustrated in Figure 19–15.

Now we have completely described the *dc* operating conditions for the circuit. In the following paragraphs we discuss *ac* conditions.

Signal Operation

The purpose of the circuit in Figure 19–16 is to increase the magnitude of an *input signal.* This increase is called *amplification.* The figure shows an input signal, V_{in}, capacitively coupled to the base. The collector voltage is the *output signal,* as indicated. The input signal voltage causes the base current to vary at the same frequency above and below its dc value. This variation in base current produces a corresponding variation in collector current. However, the variation in collector current is much larger than the variation in base current because of the *current gain* through the transistor. The ratio of the collector current (I_c) to the base current (I_b) is designated β_{ac} (the ac beta) or h_{fe}:

$$\beta_{ac} = \frac{I_c}{I_b} \qquad (19\text{–}9)$$

The value of β_{ac} normally differs slightly from that of β_{dc} for a given transistor. Notice that lower-case subscripts distinguish ac currents and voltages from dc currents and voltages.

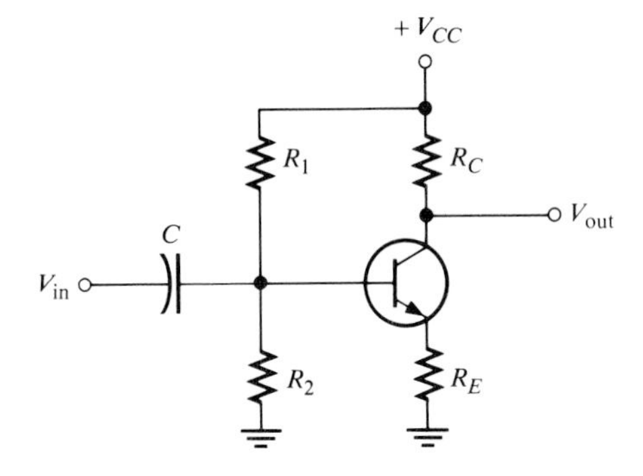

FIGURE 19–16
Voltage divider biased amplifier with capacitively coupled input signal. V_{in} and V_{out} are with respect to ground.

Signal Voltage Gain

Now let us take the amplifier in Figure 19–16 and examine its *voltage gain* with a signal input. The output voltage is the collector voltage. The variation in collector current produces a variation in the voltage across R_C and a resulting variation in the collector voltage, as shown in Figure 19–17.

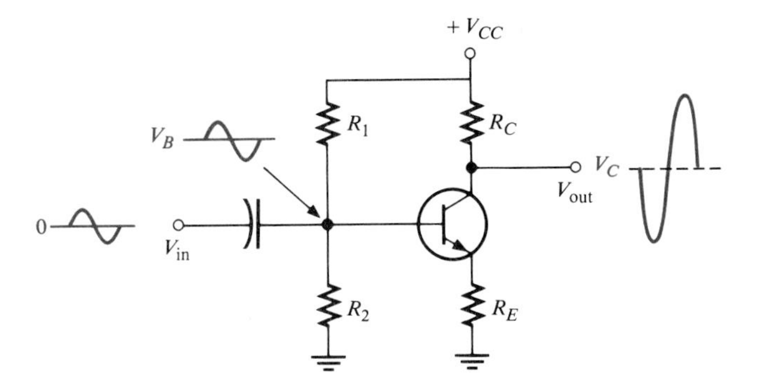

FIGURE 19–17
Voltage amplification.

As the collector current increases, the I_cR_C drop increases. This increase produces a *decrease* in collector voltage because $V_c = V_{CC} - I_cR_C$. Likewise, as the collector current decreases, the I_cR_C drop decreases and produces an *increase* in collector voltage. Therefore, there is a 180° phase difference between the collector current and the collector voltage. The base voltage and collector voltage are also 180° out of phase, as indicated in the figure. This 180° phase difference between input and output is called an *inversion.*

The voltage gain A_v of the amplifier is V_{out}/V_{in}, where V_{out} is the signal voltage at the collector and V_{in} is the signal voltage at the base. Because the base-emitter junction is forward-biased, the signal voltage at the emitter is approximately equal to the signal voltage at the base. Thus, since $V_b \cong V_e$, the gain is approximately V_c/V_e or I_cR_C/I_eR_E. I_c and I_e are very close to the same value because the α_{ac} is close to 1. Therefore, they cancel, giving the voltage gain formula:

$$A_v \cong \frac{R_C}{R_E} \qquad (19\text{–}10)$$

A negative sign on A_v is often used to indicate inversion.

EXAMPLE 19–4

In Figure 19–18, a signal voltage of 50 mV rms is applied to the base.

(a) Determine the output voltage for the amplifier.
(b) Find the dc collector voltage on which the output signal voltage is riding.
(c) Sketch the output waveform.

FIGURE 19–18

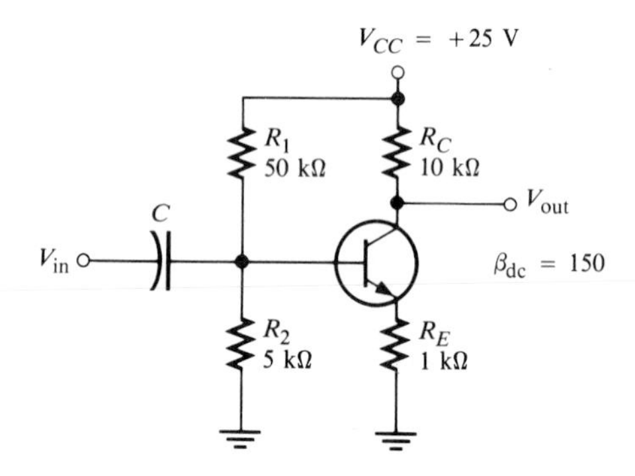

Solution

(a) The signal voltage gain is

$$A_V \cong \frac{R_C}{R_E} = \frac{10\text{ k}\Omega}{1\text{ k}\Omega} = 10$$

The output signal voltage is the input signal voltage times the voltage gain:

$$V_{out} = A_vV_{in} = (10)(50\text{ mV}) = 0.5\text{ V rms}$$

(b) Next we find the dc collector voltage:

$$R_{IN} \cong \beta_{dc}R_E = (150)(1\text{ k}\Omega) = 150\text{ k}\Omega$$

R_{IN} can be neglected since it is more than ten times R_2. Thus,

$$V_B \cong \left(\frac{R_2}{R_1 + R_2}\right)V_{CC} = \left(\frac{5\ \text{k}\Omega}{55\ \text{k}\Omega}\right)25\ \text{V} = 2.27\ \text{V}$$

$$I_C \cong I_E = \frac{V_E}{R_E} = \frac{V_B - 0.7\ \text{V}}{1\ \text{k}\Omega} = 1.57\ \text{mA}$$

$$V_C = V_{CC} - I_C R_C = 25\ \text{V} - (1.57\ \text{mA})(10\ \text{k}\Omega) = 9.3\ \text{V}$$

This value is the dc level of the output. The peak value of the output signal is

$$V_p = 1.414(0.5\ \text{V}) = 0.707\ \text{V}$$

(c) Figure 19–19 shows the waveform.

FIGURE 19–19

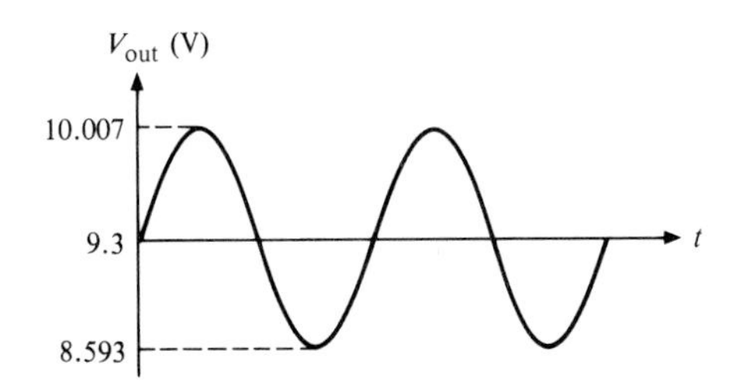

Signal Operation on the Load Line

We can obtain a graphical picture of an amplifier's operation by showing an example of signal variations on a set of collector curves with a load line, as shown in Figure 19–20. Let us assume that the dc Q-point values are as follows: $I_B = 40\ \mu\text{A}$, $I_C = 4\ \text{mA}$, and $V_{CE} = 8\ \text{V}$. The input signal varies the base

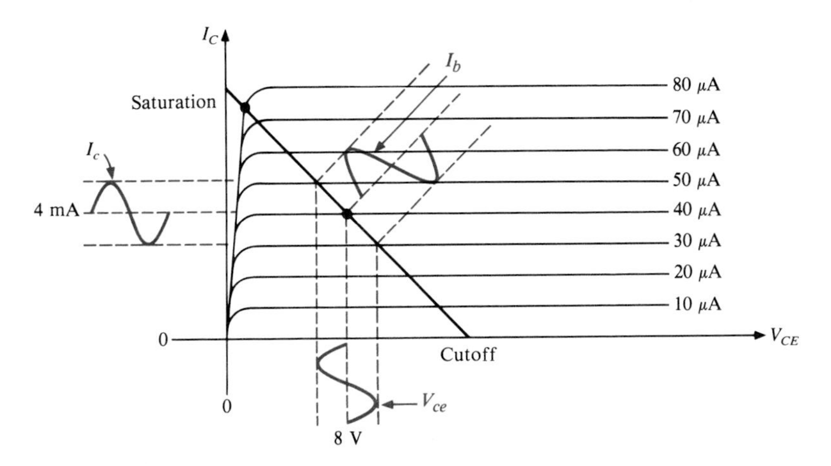

FIGURE 19–20
Signal operation on the load line.

current from a maximum of 50 μA to a minimum of 30 μA. The resulting variations in collector current and V_{CE} are shown on the graph. The operation is linear as long as the variations do not reach cutoff at the lower end of the load line or saturation at the upper end.

SECTION REVIEW 19–3

1. What does the term *amplification* mean?
2. What is the Q point?
3. A certain transistor circuit has $R_C = 47\ \text{k}\Omega$ and $R_E = 2.2\ \text{k}\Omega$. What is the approximate voltage gain?

THE BIPOLAR TRANSISTOR AS A SWITCH

19–4

In the previous section, we discussed the transistor as a linear amplifier. The second major application area is switching applications. When used as an electronic switch, a transistor normally is operated alternately in cutoff and saturation.

Figure 19–21 illustrates the basic operation of a transistor as a switching device. In Part (a), the transistor is cut off because the base-emitter junction is *not* forward-biased. In this condition, there is, ideally, an *open* between collector and emitter. In Part (b), the transistor is saturated because the base-emitter junction is forward-biased and the base current is large enough to cause the collector current to reach its saturated value. In this condition there is, ideally, a *short* between collector and emitter. Actually, a voltage drop of a few tenths of a volt normally occurs.

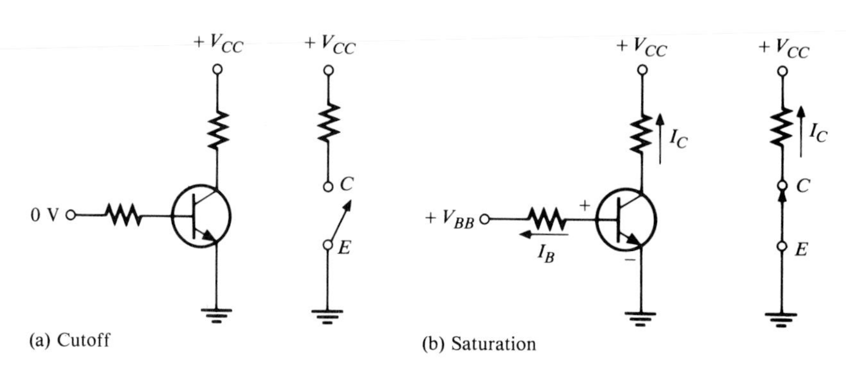

FIGURE 19–21
Ideal switching action of a transistor.

Conditions in Cutoff

As mentioned before, a transistor is in cutoff when the base-emitter junction is *not* forward-biased. All of the currents are approximately zero, and V_{CE} is approximately equal to V_{CC}:

$$V_{CE(\text{cutoff})} \cong V_{CC} \qquad (19\text{–}11)$$

Conditions in Saturation

When the emitter junction is forward-biased and there is enough base current to produce a maximum collector current, the transistor is saturated. Since V_{CE} is approximately zero at saturation, the collector current is

$$I_{C(\text{sat})} \cong \frac{V_{CC}}{R_C} \tag{19–12}$$

The value of base current needed to produce saturation is

$$I_{B(\text{min})} = \frac{I_{C(\text{sat})}}{\beta_{\text{dc}}} \tag{19–13}$$

EXAMPLE 19–5

(a) For the transistor switching circuit in Figure 19–22, what is V_{CE} when $V_{\text{IN}} = 0$ V?
(b) What minimum value of I_B is required to saturate this transistor if the β_{dc} is 200?
(c) Calculate the maximum value of R_B when $V_{\text{IN}} = 5$ V.

FIGURE 19–22

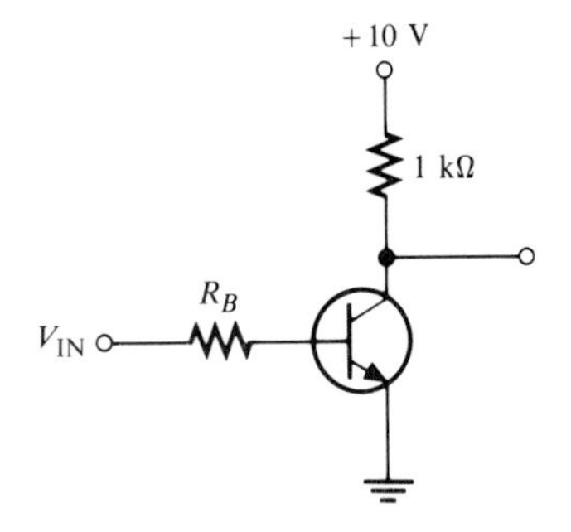

Solution

(a) When $V_{\text{IN}} = 0$ V, the transistor is *off* and $V_{CE} = V_{CC} = 10$ V.
(b) When the transistor is saturated, $V_{CE} \cong 0$ V, so

$$I_{C(\text{sat})} = \frac{V_{CC}}{R_C} = \frac{10\text{ V}}{1\text{ k}\Omega} = 10\text{ mA}$$

$$I_B = \frac{I_{C(\text{sat})}}{\beta_{\text{dc}}} = \frac{10\text{ mA}}{200} = 0.05\text{ mA}$$

This is the value of I_B necessary to drive the transistor to the point of saturation. *Any further increase in I_B will drive the transistor deeper into saturation but will not increase I_C.*

(c) When the transistor is saturated, $V_{BE} = 0.7$ V. The voltage across R_B is $V_{\text{IN}} - 0.7\text{ V} = 4.3$ V. The maximum value of R_B needed to allow a minimum I_B of 0.05 mA is calculated by Ohm's law as follows:

$$R_B = \frac{V_{IN} - 0.7\text{ V}}{I_B} = \frac{4.3\text{ V}}{0.05\text{ mA}} = 86\text{ k}\Omega$$

SECTION REVIEW 19–4

1. When a transistor is used as a switching device, it is operated in either ______ or ______.
2. When does the collector current reach its maximum value?
3. When is the collector current approximately zero?
4. Name the two conditions that produce saturation.
5. When is V_{CE} equal to V_{CC}?

BIPOLAR TRANSISTOR PARAMETERS AND RATINGS

19–5 More about β_{dc}

Because the β_{dc} is a very important bipolar transistor parameter, we need to examine it further. β_{dc} varies with both collector current and temperature. Keeping the junction temperature constant and increasing I_C causes β_{dc} to increase to a maximum. A further increase in I_C beyond this maximum point causes β_{dc} to decrease.

If I_C is held constant and the temperature is varied, β_{dc} changes directly with the temperature. If the temperature goes up, β_{dc} goes up, and vice versa. Figure 19–23 shows the variation of β_{dc} with I_C and junction temperature (T_J) for a typical transistor.

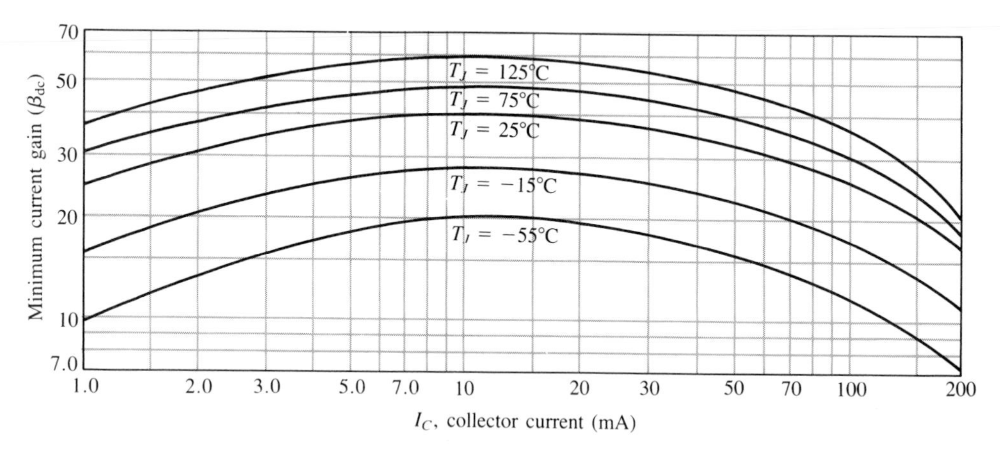

FIGURE 19–23
Variation of β_{dc} with I_C for several temperatures.

A transistor data sheet usually specifies β_{dc} (h_{FE}) at specific I_C values. Even at fixed values of I_C and temperature, β_{dc} varies from device to device for a given type of transistor.

The β_{dc} specified at a certain value of I_C is usually the minimum value, $\beta_{dc(min)}$, although the maximum and typical values are also sometimes specified. A typical transistor data sheet is shown in Appendix C.

Maximum Ratings

Like any other electronic device, the transistor has limitations on its operation. These limitations are stated in the form of *maximum ratings* and are normally specified on the manufacturer's data sheet. Typically, maximum ratings are given for *collector-to-base voltage* V_{CB}, *collector-to-emitter voltage* V_{CE}, *emitter-to-base voltage* V_{EB}, *collector current* I_C, and *power dissipation* P_D.

The product of V_{CE} and I_C must not exceed the maximum power dissipation. Both V_{CE} and I_C cannot be maximum at the same time. If V_{CE} is maximum, I_C can be calculated as

$$I_C = \frac{P_{D(max)}}{V_{CE}} \qquad (19\text{–}14)$$

If I_C is maximum, we can calculate V_{CE} by rearranging Equation (19–14) as follows:

$$V_{CE} = \frac{P_{D(max)}}{I_C} \qquad (19\text{–}15)$$

For a given transistor, a maximum power dissipation curve can be plotted on the collector curves, as shown in Figure 19–24(a). These values are tabulated in Part(b). For this transistor, $P_{D(max)}$ is 0.5 W, $V_{CE(max)}$ is 20 V, and $I_{C(max)}$ is 50 mA. The curve shows that this particular transistor cannot be operated in the shaded portion of the graph. $I_{C(max)}$ is the limiting rating between points *A* and *B*, $P_{D(max)}$ is the limiting rating between points *B* and *C*, and $V_{CE(max)}$ is the limiting rating between points *C* and *D*.

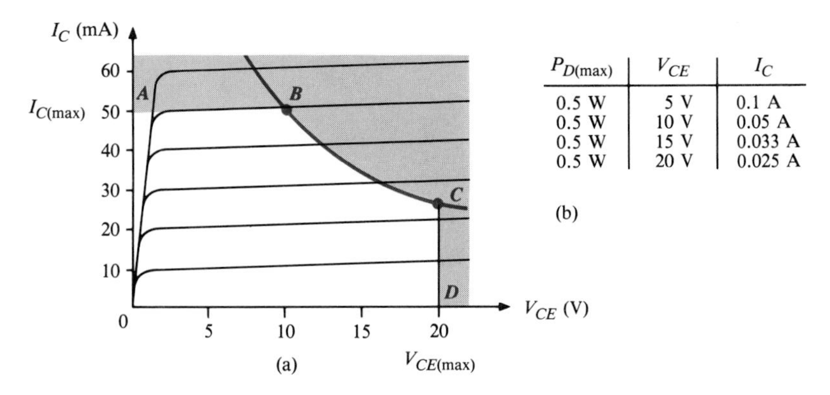

$P_{D(max)}$	V_{CE}	I_C
0.5 W	5 V	0.1 A
0.5 W	10 V	0.05 A
0.5 W	15 V	0.033 A
0.5 W	20 V	0.025 A

FIGURE 19–24
Maximum power dissipation curve.

EXAMPLE 19–6 The silicon transistor in Figure 19–25 has the following maximum ratings: $P_{D(max)} = 0.8$ W, $V_{CE(max)} = 15$ V, $I_{C(max)} = 100$ mA, $V_{CB(max)} = 20$ V, and $V_{EB(max)} = 10$ V. Determine the maximum value to which V_{CC} can be adjusted without exceeding a rating. Which rating would be exceeded first?

FIGURE 19–25

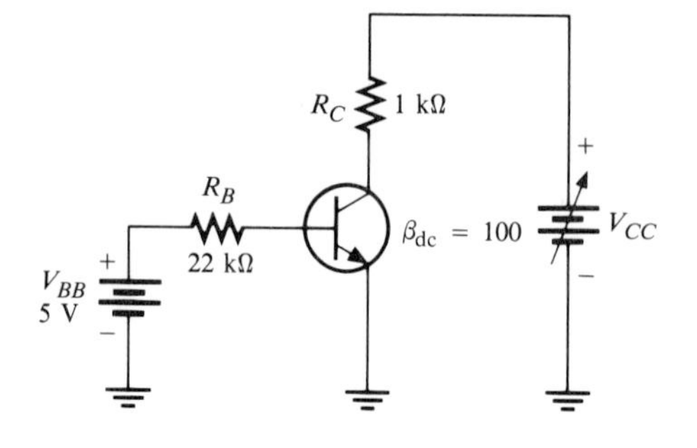

Solution
First find I_B so that I_C can be determined.

$$I_B = \frac{V_{BB} - V_{BE}}{R_B} = \frac{5\text{ V} - 0.7\text{ V}}{22\text{ k}\Omega} = 195.5\ \mu\text{A}$$

$$I_C = \beta_{dc} I_B = (100)(195.5\ \mu\text{A}) = 19.55\text{ mA}$$

I_C is much less than $I_{C(max)}$ and will not change with V_{CC}. It is determined only by I_B and β_{dc}.

The voltage drop across R_C is

$$V_{RC} = I_C R_C = (19.55\text{ mA})(1\text{ k}\Omega) = 19.55\text{ V}$$

Now we can determine the maximum value of V_{CC} when $V_{CE} = V_{CE(max)} =$ 15 V:

$$V_{RC} = V_{CC} - V_{CE}$$

Thus,

$$V_{CC(max)} = V_{CE(max)} + V_{RC} = 15\text{ V} + 19.55\text{ V} = 34.55\text{ V}$$

V_{CC} can be increased to 34.55 V, under the existing conditions, before $V_{CE(max)}$ is exceeded. However, we do not know whether or not $P_{D(max)}$ has been exceeded at this point. Let's find out.

$$P_D = V_{CE(max)} I_C = (15\text{ V})(19.55\text{ mA}) = 0.293\text{ W}$$

Since $P_{D(max)}$ is 0.8 W, it is *not* exceeded when $V_{CC} = 34.55$ V. Thus, $V_{CE(max)}$ is the limiting rating in this case.

It should be noted that if the base current is removed, causing the transistor to turn off, $V_{CE(max)}$ will be exceeded because the entire supply voltage, V_{CC}, will be dropped across the transistor.

SECTION REVIEW 19–5

1. The β_{dc} of a transistor increases with temperature (T or F).
2. Generally, what effect does an increase in I_C have on the β_{dc}?
3. What is the allowable collector current in a transistor with $P_{D(max)} = 0.32$ W when $V_{CE} = 8$ V?

TRANSISTOR TESTING

19–6 Ohmmeter Check of Transistor Junctions

The ohmmeter provides a simple test for open or shorted junctions. *The base-emitter and base-collector junctions are each treated as a diode for this test.* The junction should show a *low* resistance when forward-biased and a very *high* resistance when reverse-biased. The internal battery of the ohmmeter must provide the bias voltage.

Figure 19–26(a) shows a *forward-biased* check of an NPN transistor. Notice the polarities of the ohmmeter leads. Part (b) of the figure demonstrates a *reverse-biased* check. For a PNP transistor, the meter polarities are reversed from those shown in Figure 19–26.

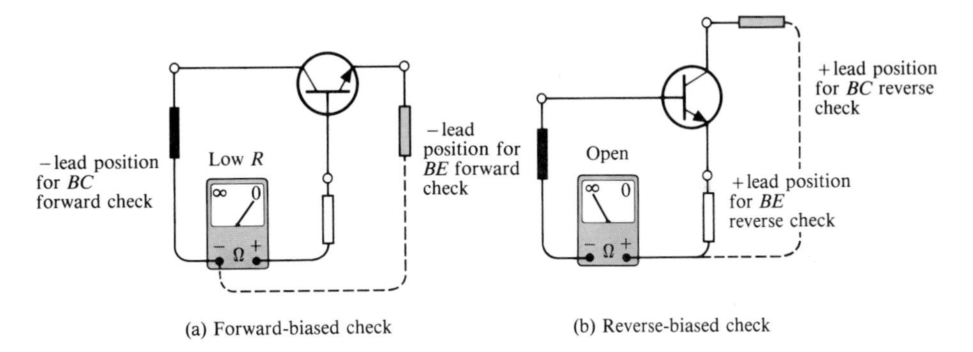

FIGURE 19–26
Ohmmeter check of NPN transistor junctions.

Leakage Measurement

Very small *leakage* currents exist in all transistors and in most cases are small enough to neglect.

When a transistor is connected as shown in Figure 19–27(a) with the base open ($I_B = 0$), it is in cutoff. Ideally $I_C = 0$, but actually there is a small current from collector to emitter, as mentioned earlier, called I_{CEO} (collector-to-emitter current with base open). This leakage current is usually in the nanoampere (nA) range for silicon. A faulty transistor will often have excessive leakage current and can be checked in a transistor tester which connects an ammeter to the transistor as shown.

Another leakage current in transistors is the reverse collector-to-base current, I_{CBO}. This current is measured with the emitter open, as shown in Part (b). If it is excessive, a shorted collector-base junction is likely.

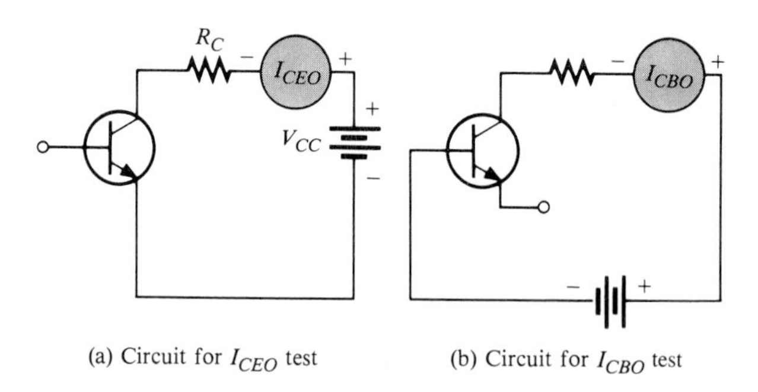

(a) Circuit for I_{CEO} test (b) Circuit for I_{CBO} test

FIGURE 19–27
Leakage-current test circuits.

Gain Measurement

In addition to leakage tests, the typical transistor tester also checks the β_{dc}. A known I_B is applied and the resulting I_C is measured. The reading will indicate the value of the I_C/I_B ratio, although in some units only a relative indication is given.

Most testers provide for an in-circuit β_{dc} check, so that a suspected device does not have to be removed from the circuit for testing. Figure 19–28 shows two typical transistor testers.

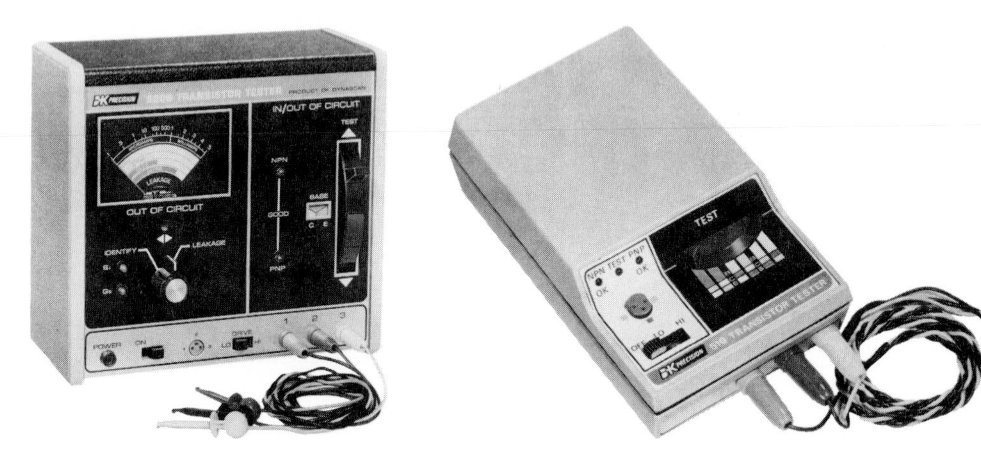

FIGURE 19–28
Transistor tester (courtesy of B&K Precision Dynascan Corp.).

Curve Tracers

The *curve tracer* is an oscilloscope type of instrument that can display transistor characteristics such as a family of collector curves. In addition to measuring and displaying various transistor characteristics, the curve tracer can also display diode curves as well as the β_{dc}. A typical curve tracer is shown in Figure 19–29.

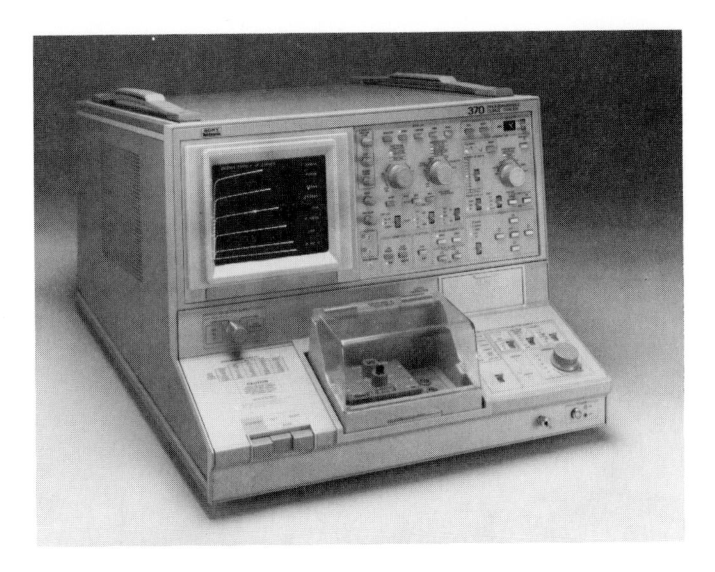

FIGURE 19–29
Curve tracer (courtesy Tektronix, Inc.).

SECTION REVIEW 19–6

1. The positive lead of an ohmmeter is on a transistor's base, the negative lead is on the emitter, and a low resistance reading is obtained. Assuming the transistor is good, is it a PNP or an NPN type?
2. Name two types of transistor leakage currents.

THE JUNCTION FIELD-EFFECT TRANSISTOR (JFET)

19–7

Depending on their structure, JFETs fall into either of two categories: *N-channel* or *P-channel*. Figure 19–30(a) shows the basic structure of an N-channel JFET. Wire leads are connected to each end of the N-channel; the *drain* is at the upper end and the *source* is at the lower end. Two P-type regions are dif-

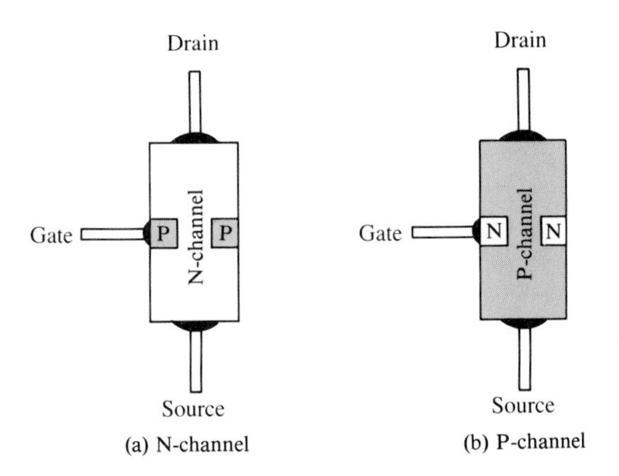

FIGURE 19–30
Structure of the two types of JFET.

fused in the N-type material to form a channel, and *both* P-type regions are connected to the gate lead. In the remaining structure diagrams, the interconnection of *both* P-type regions is omitted for simplicity, with a connection to only one shown. A P-channel JFET is shown in Figure 19–30(b).

Basic Operation

To illustrate the operation of a JFET, bias voltages are shown applied to an N-channel device in Figure 19–31(a). V_{DD} provides a drain-to-source voltage and supplies current from drain to source. V_{GG} sets the reverse-biased voltage between the gate and the source, as shown.

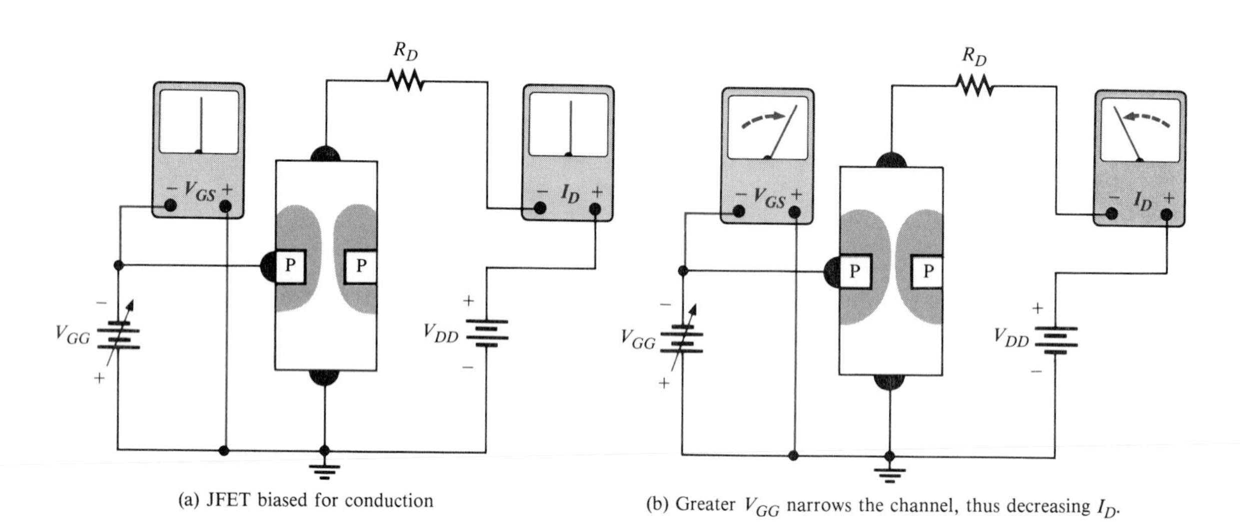

(a) JFET biased for conduction

(b) Greater V_{GG} narrows the channel, thus decreasing I_D.

(c) Less V_{GG} widens channel and increases I_D.

FIGURE 19–31
Effects of V_{GG} on channel width and drain current ($V_{GG} = V_{GS}$).

The JFET is always operated with the gate-to-source PN junction reverse-biased. Reverse biasing of the gate-source junction with a negative gate voltage produces a *depletion* region in the N channel and thus increases its resistance. The channel width can be controlled by varying the gate voltage, and thereby the amount of drain current, I_D, can also be controlled. This concept is illustrated in Figures 19–31(b) and 19–31(c).

The color shaded areas represent the depletion region created by the reverse bias. It is wider toward the drain end of the channel because the reverse-biased voltage between the gate and the drain is greater than that between the gate and the source.

JFET Symbols

The schematic symbols for both N-channel and P-channel JFETs are shown in Figure 19–32. Notice that the arrow on the gate points "in" for N-channel and "out" for P-channel.

FIGURE 19–32
JFET schematic symbols.

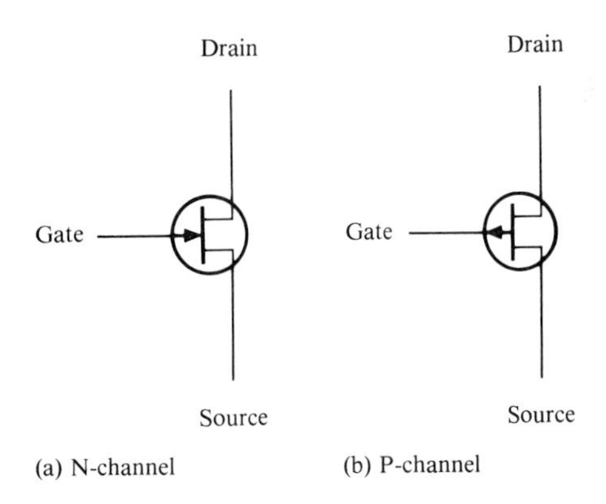

SECTION REVIEW 19–7

1. Name the three terminals of a JFET.
2. An N-channel JFET requires a (positive, negative, or 0) V_{GS}.
3. How is the drain current controlled in a JFET?

JFET CHARACTERISTICS

19–8

Drain Curves and Pinch-Off Voltage

First consider the case where the gate-to-source voltage is 0 (V_{GS} = 0 V). This voltage is produced by shorting the gate to the source, as in Figure 19–33(a). As V_{DD} (and thus V_{DS}) is increased from 0, I_D will increase proportionally, as shown in the graph of Figure 19–33(b) between points *A* and *B*. In this region, the channel resistance is essentially constant because the depletion region is not large enough to have significant effect. This region is called the *ohmic region* because V_{DS} and I_D are related by Ohm's law.

At point *B*, the curve levels off and I_D becomes a relatively constant value called I_{DSS}. It is at this point that the reverse-bias voltage across the *gate-to-drain* junction (V_{GD}) produces a depletion region sufficient to narrow the chan-

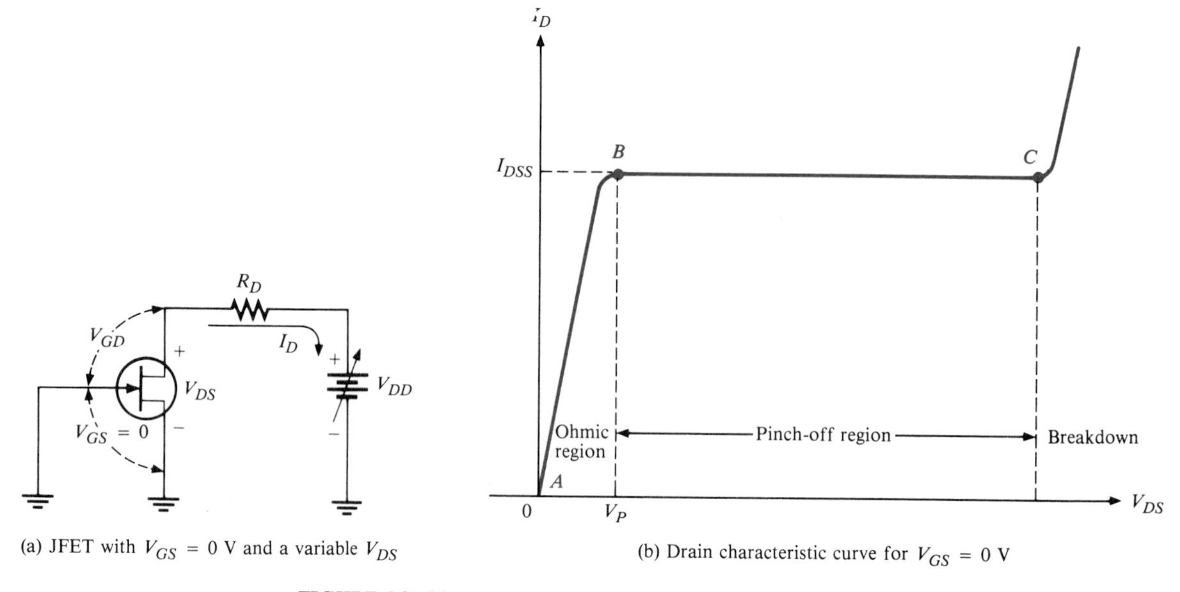

(a) JFET with $V_{GS} = 0$ V and a variable V_{DS}

(b) Drain characteristic curve for $V_{GS} = 0$ V

FIGURE 19–33
Generation of drain characteristic curve for a JFET.

nel so that its resistance begins to *increase* significantly. The value of V_{GD} at this point is called the *pinch-off voltage* V_P. *In the case where the gate bias voltage is zero,* $-V_P = V_{DS}$ *at the point of pinch-off, because* V_{DS} *and* V_{DG} *are equal.* In general, however,

$$V_P = V_{GS} - V_{DS(P)} \tag{19–16}$$

where $V_{DS(P)}$ is the pinch-off value of V_{DS} for a given value of V_{GS}.

V_P is a constant value for a given JFET and represents a fixed parameter. The pinch-off value of V_{DS} is a variable that depends on the gate-to-source bias voltage, V_{GS}.

As you can see from the curve for $V_{GS} = 0$ V in Figure 19–33(b), a continued increase in V_{DS} above point B produces an essentially constant I_D equal to a specific value called I_{DSS}. *I_{DSS} is the maximum value of I_D when $V_{GS} = 0$.* At point C, breakdown occurs and I_D increases rapidly with irreversible damage to the device very likely. JFETs are always operated below the breakdown point and within the pinch-off region (between points B and C).

Now consider a negative gate bias voltage, for example, $V_{GS} = -1$ V, as shown in Figure 19–34(a). As V_{DD} is increased from 0, pinch-off occurs at a lower value of V_{DS}, as shown in Figure 19–34(b). The reason for this is that the pinch-off voltage V_P is constant, and therefore, for a certain negative gate voltage, the drain voltage must only reach a value sufficient to make the gate-to-drain voltage equal to V_P in order to produce pinch-off.

As V_{GS} is set to increasingly negative values, a family of characteristic curves is produced as shown in Figure 19–34(b).

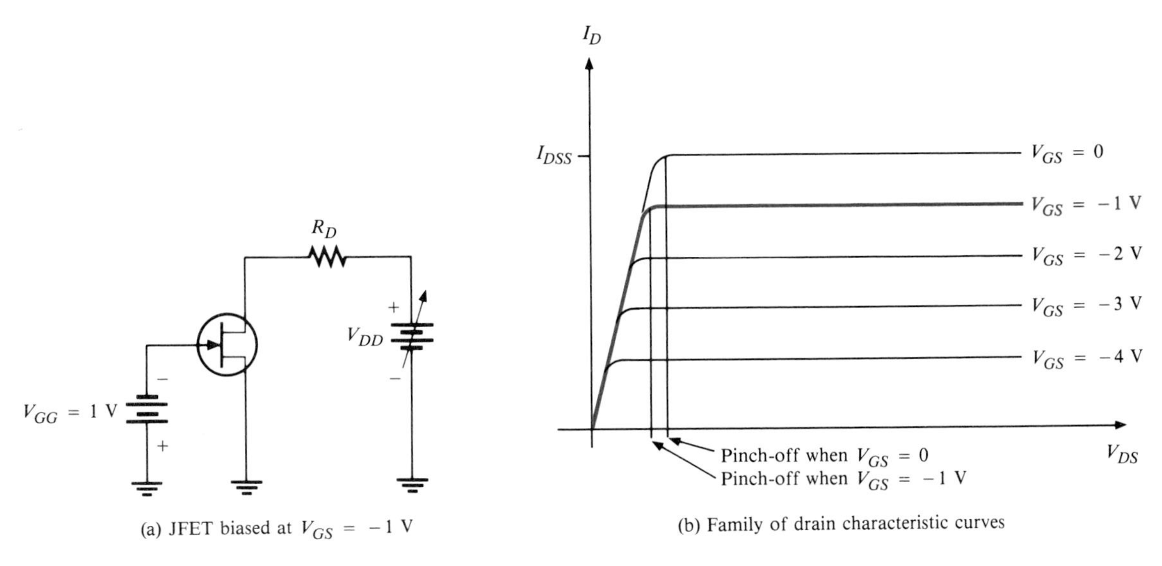

FIGURE 19–34
Pinch-off occurs at a lower V_{DS} when V_{GS} goes from 0 V to -1 V, and to successively more negative values.

Cutoff

As you know, for an N-channel JFET, the more negative V_{GS} is, the smaller I_D in the pinch-off region becomes. When V_{GS} is made sufficiently negative, I_D is reduced to 0 because the depletion region widens to a point where it completely closes the channel. The value of V_{GS} at the cutoff point is designated $V_{GS(off)}$.

For any given N-channel JFET, cutoff occurs when $V_{GS} = -V_P$. Since the pinch-off voltage V_P is a constant for a given JFET, when $V_{GS} = -V_P$ the drain-to-source voltage $V_{DS(P)}$ must be 0. Since there is *no voltage drop* between

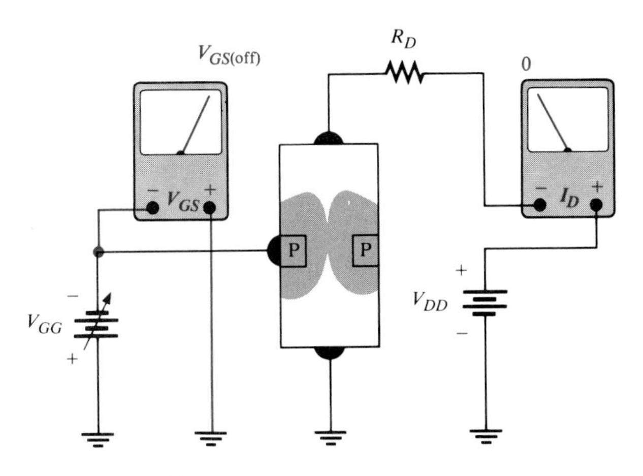

FIGURE 19–35
JFET at cutoff.

the drain and source, I_D *must* be 0. Even though V_{DS} may increase above 0 V, I_D remains essentially constant at near 0 A. Cutoff is illustrated in Figure 19–35 (on page 721).

Do not confuse *cutoff* with *pinch-off. The pinch-off voltage V_P is the value of the gate-to-drain voltage V_{GD} at which the drain current reaches a constant value for a given value of V_{GS}.*

The cutoff voltage $V_{GS(\text{off})}$ is the value of V_{GS} at which the drain current is 0. I_D is 0 only when the magnitude of V_{GS} is equal to or greater than the magnitude of V_P. I_D is nonzero for less negative values of V_{GS}.

EXAMPLE 19–7 For the JFET in Figure 19–36, $V_P = -8$ V and $I_{DSS} = 12$ mA.

(a) Determine the value of V_{DS} at which pinch-off begins.

(b) If the gate is grounded, what is the value of I_D for $V_{DD} = 12$ V when V_{DS} is above pinch-off?

FIGURE 19–36

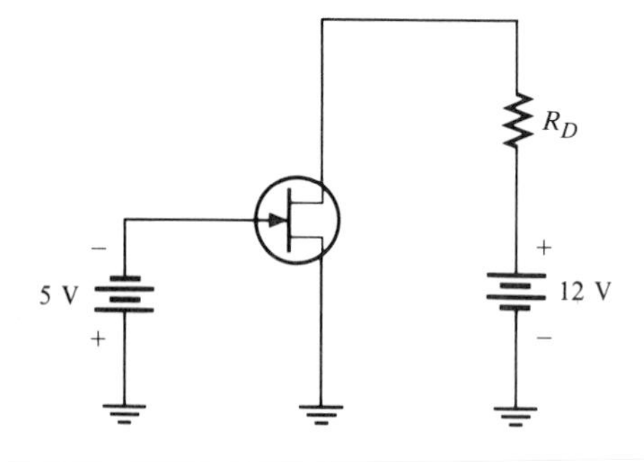

Solution

(a) By rearranging Equation (19–16), we get

$$V_{DS(P)} = V_{GS} - V_P$$

Substituting $V_P = -8$ V and $V_{GS} = -5$ V yields

$$V_{DS(P)} = -5\text{ V} - (-8\text{ V}) = 3\text{ V}$$

This is the value of V_{DS} at the beginning of pinch-off.

(b) When $V_{GS} = 0$ V, $I_D = I_{DSS} = 12$ mA for any value of V_{DS} between pinch-off (8 V) and breakdown.

Why V_P and $V_{GS(\text{off})}$ Are Equivalent

I_D is 0 when $V_{GS} = V_P$. Because a zero drain current corresponds to an *off* condition, the magnitude of V_P is equivalent to the magnitude of $V_{GS(\text{off})}$.

Most FET data sheets give a value only for $V_{GS(\text{off})}$ and not for V_P. But once we know $V_{GS(\text{off})}$, we also know V_P. For example, if $V_{GS(\text{off})} = -5$ V, then $V_P = -5$ V.

JFET Input Resistance and Capacitance

A JFET operates with its gate-source junction reverse-biased. Therefore, the input resistance at the gate is very high. This high input resistance is one advantage of the JFET over the bipolar transistor. (Recall that a bipolar transistor operates with a forward-biased base-emitter junction.)

JFET data sheets often specify the input resistance by giving a value for the *gate reverse current* I_{GSS} at a certain gate-to-source voltage. The input resistance can then be determined using the following equation:

$$R_{IN} = \left| \frac{V_{GS}}{I_{GSS}} \right| \qquad (19\text{–}17)$$

For example, the 2N3970 data sheet lists a maximum I_{GSS} of 250 pA for V_{GS} = 20 V at 25°C. I_{GSS} increases with temperature, so the input resistance decreases.

The input capacitance C_{iss} of a JFET is considerably greater than that of a bipolar transistor because the JFET operates with a reverse-biased PN junction. Recall that a reverse-biased PN junction acts as a capacitor whose capacitance depends on the amount of reverse voltage.

EXAMPLE 19–8

A certain JFET has an I_{GSS} of 1 nA for $V_{GS} = -20$ V. Determine the input resistance.

Solution

$$R_{IN} = \left| \frac{V_{GS}}{I_{GSS}} \right| = \frac{20 \text{ V}}{1 \text{ nA}} = 20{,}000 \text{ M}\Omega$$

SECTION REVIEW 19–8

1. The drain-to-source voltage at the pinch-off point of a particular JFET is 7 V. If the gate-to-source voltage is 0, what is V_P?
2. The V_{GS} of a certain N-channel JFET is increased negatively. Does the drain current increase or decrease?
3. What value must V_{GS} have to produce cutoff in a P-channel JFET with a V_P = 3 V?

THE METAL OXIDE SEMICONDUCTOR FET (MOSFET)

19–9

The MOSFET is the second category of field-effect transistor. It differs from the JFET in that it has no PN junction structure; instead, the gate of the MOSFET is *insulated* from the channel by a silicon dioxide (SiO_2) layer.

There are two basic types of MOSFETs: *depletion-enhancement* (DE) and *enhancement-only* (E).

Depletion-Enhancement (DE) MOSFET

Figure 19–37 illustrates the basic structure of DE MOSFETs. The drain and source are diffused into the substrate material and then connected by a narrow channel adjacent to the insulated gate. Both N-channel and P-channel devices are shown in the figure.

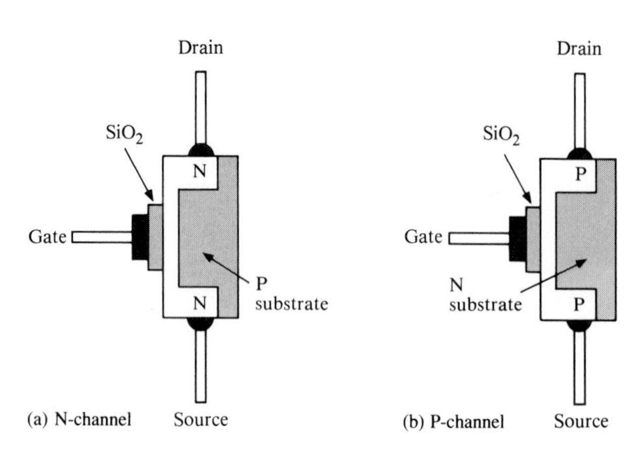

FIGURE 19–37
Basic structure of DE MOSFETs.

We will use the N-channel device to describe the basic operation. The P-channel operation is the same, except the voltage polarities are opposite those of the N-channel.

The DE MOSFET can be operated in either of two modes: the *depletion mode* or the *enhancement mode*. Since the gate is insulated from the channel, either a positive or a negative gate voltage can be applied. The MOSFET operates in the depletion mode when a *negative* gate-to-source voltage is applied and in the enhancement mode when a *positive* gate-to-source voltage is applied.

Depletion Mode

Visualize the gate as one plate of a parallel plate capacitor and the channel as the other plate. The silicon dioxide insulating layer is the dielectric. With a negative gate voltage, the negative charges on the gate repel conduction electrons from the channel, leaving positive ions in their place. Thereby, the N-channel is *depleted* of some of its electrons, thus decreasing the channel conductivity. The greater the negative voltage on the gate, the greater the depletion of N-channel electrons. At a sufficiently negative gate-to-source voltage, $V_{GS(\text{off})}$, the channel is totally depleted and the drain current is 0. This depletion mode is illustrated in Figure 19–38(a).

Like the N-channel JFET, the N-channel DE MOSFET conducts drain current for gate-to-source voltages between $V_{GS(\text{off})}$ and 0. In addition, the DE MOSFET conducts for values of V_{GS} above 0.

Enhancement Mode

With a positive gate voltage, more conduction electrons are attracted into the channel, thus increasing (enhancing) the channel conductivity, as illustrated in Figure 19–38(b).

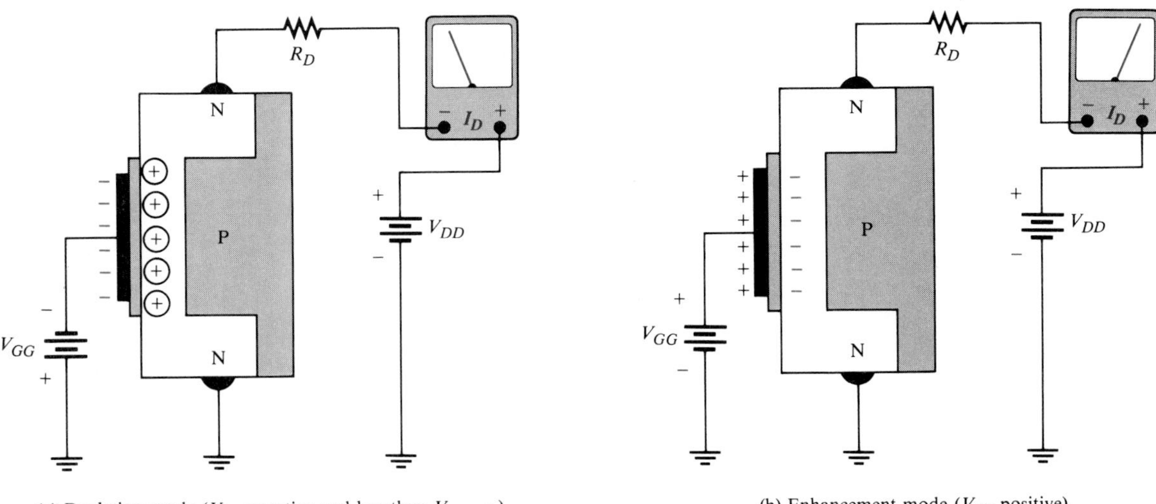

FIGURE 19–38
Operation of N-channel DE MOSFET.

DE MOSFET Symbols

The schematic symbols for both the N-channel and the P-channel depletion-enhancement MOSFETs are shown in Figure 19–39. The substrate, indicated by the arrow, is normally (but not always) connected internally to the source. An inward substrate arrow is for N-channel, and an outward arrow is for P-channel.

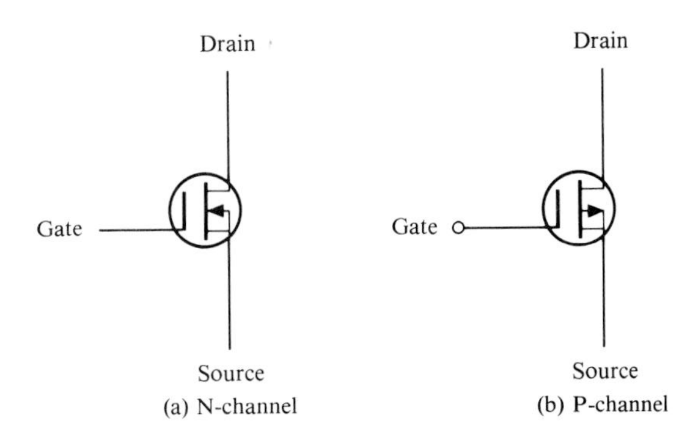

FIGURE 19–39
DE MOSFET schematic symbols.

Enhancement (E) MOSFET

This type of MOSFET operates *only* in the enhancement mode and has no depletion mode. It differs in construction from the DE MOSFET in that it has *no physical channel.* Notice in Figure 19–40(a) that the substrate extends completely to the SiO_2 layer.

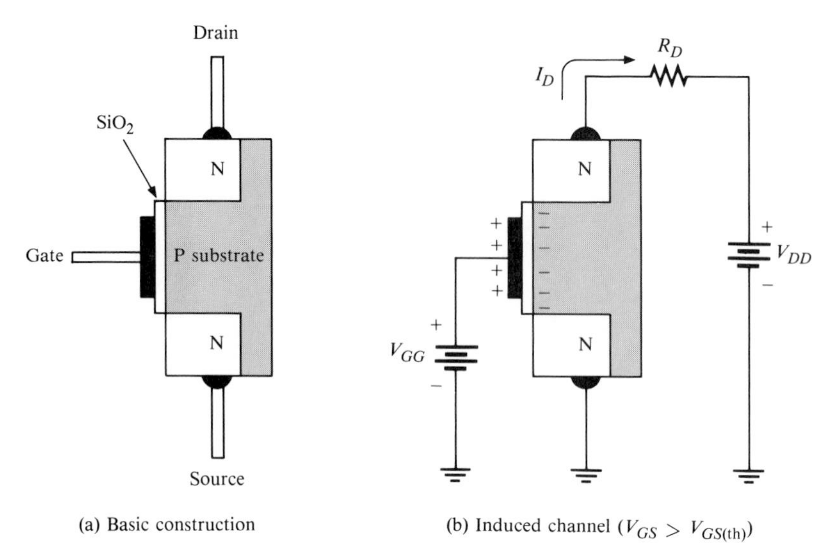

(a) Basic construction

(b) Induced channel ($V_{GS} > V_{GS(th)}$)

FIGURE 19–40
E MOSFET construction and operation.

For an N-channel device, a positive gate voltage above a threshold value, $V_{GS(th)}$, *induces* a channel by creating a thin layer of negative charges in the substrate region adjacent to the SiO_2 layer, as shown in Figure 19–40(b). The conductivity of the channel is enhanced by increasing the gate-to-source voltage, thus pulling more electrons into the channel. For any gate voltage below the threshold value, there is no channel.

The schematic symbols for the N-channel and P-channel E MOSFETs are shown in Figure 19–41. The broken lines symbolize the absence of a physical channel.

FIGURE 19–41
E MOSFET schematic symbols.

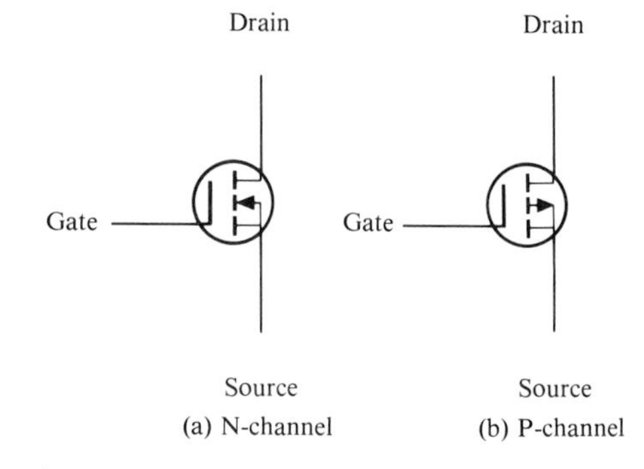

(a) N-channel

(b) P-channel

Handling Precautions

Because the gate of a MOSFET is insulated from the channel, the input resistance is extremely high (ideally infinite). The gate leakage current I_{GSS} for a typical MOSFET is in the pA range, whereas the gate reverse current for a typical JFET is in the nA range.

The input capacitance, of course, results from the insulated gate structure. Excess static charge can be accumulated because the input capacitance combines with the very high input resistance and can result in damage to the device. To avoid excess charge build-up, the following precautions should be taken:

1. MOS devices should be shipped and stored in conductive foam.
2. All instruments and metal benches used in assembly or test should be connected to earth ground (round prong of wall outlets).
3. The assembler's or handler's wrist should be connected to earth ground with a length of wire and a high-value series resistor.
4. Never remove a MOS device (or any other device, for that matter) from the circuit while the power is on.
5. Do not apply signals to a MOS device while the dc power supply is off.

SECTION REVIEW 19–9

1. Name two types of MOSFETs, and describe the major difference in construction.
2. If the gate-to-source voltage in a depletion-enhancement MOSFET is 0, is there current from drain to source?
3. If the gate-to-source voltage in an E MOSFET is 0, is there current from drain to source?

FET BIASING

19–10

Using some of the FET parameters discussed in the previous sections, we will now see how to dc-bias FETs. The purpose of biasing is to select a proper dc gate-to-source voltage to establish a desired value of drain current.

Self-Biasing a JFET

Recall that a JFET must be operated such that the gate-source junction is always reverse-biased. This condition requires a negative V_{GS} for an N-channel JFET and a positive V_{GS} for a P-channel JFET. This can be achieved using the *self-bias* arrangements shown in Figure 19–42.

Notice that the gate is biased at approximately 0 V by resistor R_G connected to ground. The reverse leakage current I_{GSS} does produce a very small voltage across R_G, but this can be neglected in most cases, and it can be assumed that R_G has no voltage drop across it.

For the N-channel JFET in Figure 19–42(a), I_D produces a voltage drop across R_S and makes the source positive with respect to ground. Since $V_G = 0$ and $V_S = I_D R_S$, the gate-to-source voltage is

$$V_{GS} = V_G - V_S = 0 - I_D R_S$$

Thus,

$$V_{GS} = -I_D R_S \qquad (19\text{–}18)$$

For the P-channel JFET shown in Figure 19–42(b), the current through R_S produces a negative voltage at the source, and therefore

$$V_{GS} = +I_D R_S \qquad (19\text{–}19)$$

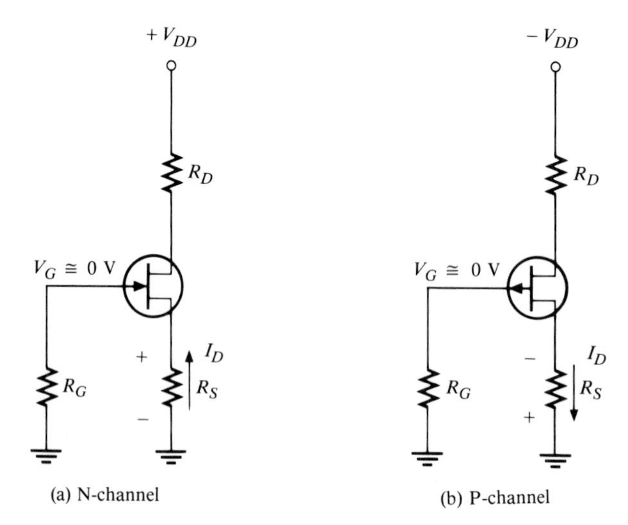

FIGURE 19–42
Self-biased JFETS ($I_S = I_D$ in all FETs).

In the following analysis, the N-channel JFET is used for illustration. Keep in mind that analysis of the P-channel JFET is the same except for opposite polarity voltages.

The drain voltage with respect to ground is determined as follows:

$$V_D = V_{DD} - I_D R_D \tag{19–20}$$

Since $V_S = I_D R_S$, the drain-to-source voltage is

$$V_{DS} = V_D - V_S$$

$$V_{DS} = V_{DD} - I_D(R_D + R_S) \tag{19–21}$$

EXAMPLE 19–9

Find V_{DS} and V_{GS} in Figure 19–43, given that $I_D = 5$ mA.

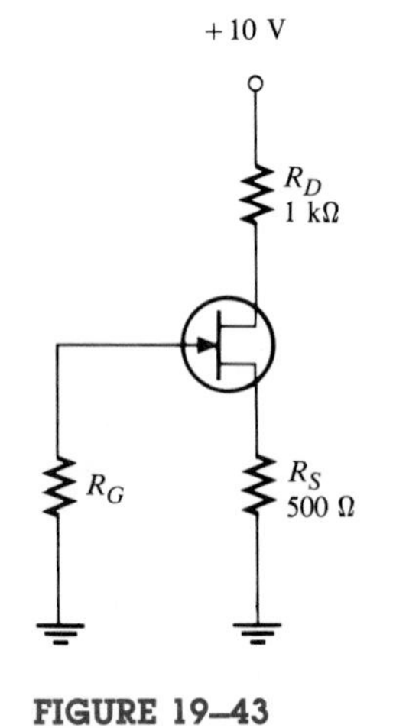

FIGURE 19–43

Solution

$$V_S = I_D R_S = (5 \text{ mA})(500 \ \Omega) = 2.5 \text{ V}$$

$$V_D = V_{DD} - I_D R_D = 10 \text{ V} - (5 \text{ mA})(1 \text{ k}\Omega)$$

$$= 10 \text{ V} - 5 \text{ V} = 5 \text{ V}$$

Therefore,

$$V_{DS} = V_D - V_S = 5 \text{ V} - 2.5 \text{ V} = 2.5 \text{ V}$$

Since $V_G = 0$ V,

$$V_{GS} = V_G - V_S = -2.5 \text{ V}$$

DE MOSFET Bias

Recall that depletion-enhancement MOSFETs can be operated with either positive or negative values of V_{GS}. A simple bias method is to set $V_{GS} = 0$ so that an ac signal at the gate varies the gate-to-source voltage above and below this bias point. A MOSFET with *0 bias* is shown in Figure 19–44.

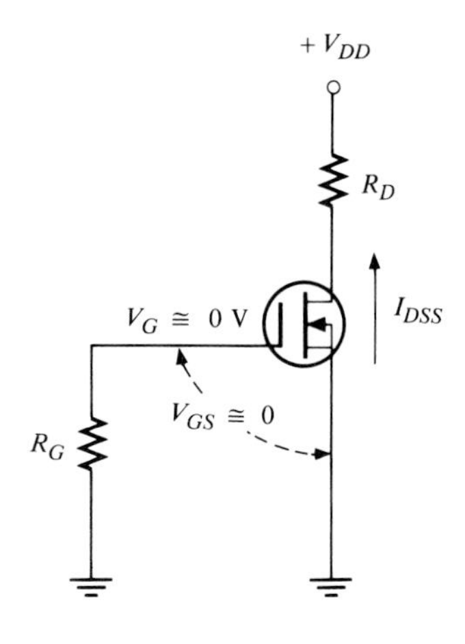

FIGURE 19–44
A zero-biased DE MOSFET.

Since $V_{GS} = 0$, $I_D = I_{DSS}$ as indicated. The drain-to-source voltage is expressed as follows:

$$V_{DS} = V_{DD} - I_{DSS}R_D \tag{19–22}$$

EXAMPLE 19–10 Determine the drain current in the circuit of Figure 19–45. The MOSFET data sheet gives $V_{GS(\text{off})} = -8$ V and $I_{DSS} = 12$ mA.

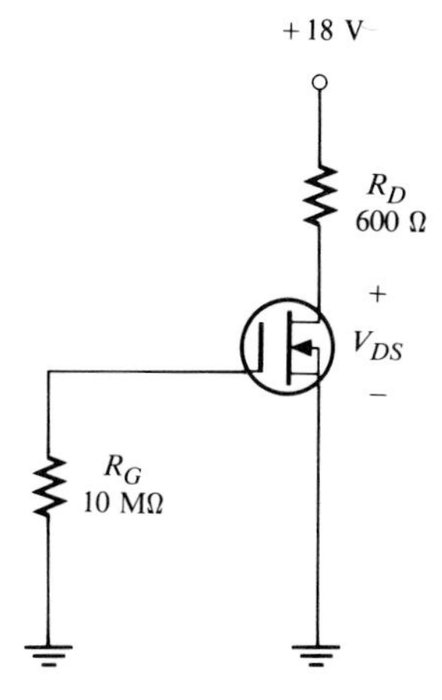

FIGURE 19–45

Solution

Since $I_D = I_{DSS} = 12$ mA, the drain-to-source voltage is calculated as follows:

$$V_{DS} = V_{DD} - I_{DSS}R_D = 18\text{ V} - (12\text{ mA})(600\ \Omega) = 10.8\text{ V}$$

E MOSFET Bias

Recall that enhancement-only MOSFETs must have a V_{GS} greater than the threshold value, $V_{GS(th)}$. Figure 19–46 shows two ways to bias an E MOSFET. An N-channel device is used for illustration. In either bias arrangement, the purpose is to make the gate voltage more positive than the source by an amount exceeding $V_{GS(th)}$.

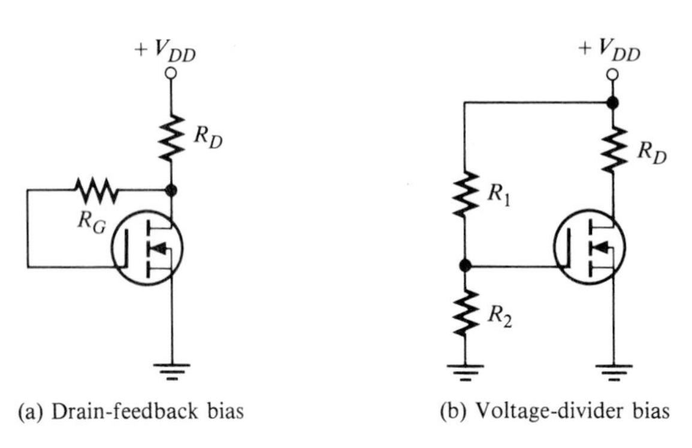

(a) Drain-feedback bias (b) Voltage-divider bias

FIGURE 19–46
E MOSFET biasing arrangements.

In the drain feedback bias circuit in Figure 19–46(a), there is negligible gate current and, therefore, no voltage drop across R_G. As a result, $V_{GS} = V_{DS}$.

EXAMPLE 19–11 Determine the amount of drain current in Figure 19–47. The MOSFET has a $V_{GS(th)}$ of 3 V.

FIGURE 19–47

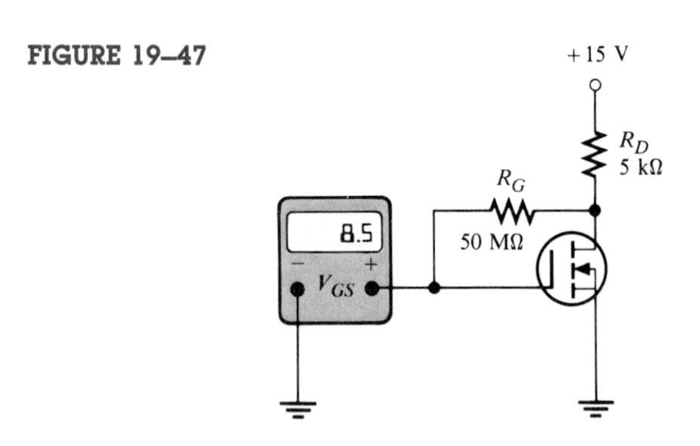

Solution
The meter indicates that $V_{GS} = 8.5$ V. Since this is a drain feedback configuration, $V_{DS} = V_{GS} = 8.5$ V.

$$I_D = \frac{V_{DD} - V_{DS}}{R_D} = \frac{15\text{ V} - 8.5\text{ V}}{5\text{ k}\Omega} = 1.3\text{ mA}$$

Equations for the voltage-divider bias in Figure 19–46(b) are as follows:

$$V_{GS} = \left(\frac{R_2}{R_1 + R_2}\right)V_{DD}$$

$$V_{DS} = V_{DD} - I_D R_D$$

SECTION REVIEW 19–10

1. A P-channel JFET must have a (positive, negative) V_{GS}.
2. In a certain self-biased N-channel JFET circuit, $I_D = 8$ mA and $R_S = 1$ kΩ. Determine V_{GS}.
3. For a DE MOSFET biased at $V_{GS} = 0$, is the drain current equal to 0, I_{GSS}, or I_{DSS}?
4. An N-channel E MOSFET with $V_{GS(th)} = 2$ V must have a V_{GS} in excess of ______ in order to conduct.

UNIJUNCTION TRANSISTORS (UJTs)

19–11

The unijunction transistor (UJT) consists of an emitter and two bases. It has no collector. The UJT is used mainly in switching and timing applications.

Figure 19–48(a) shows the construction of a UJT. The base contacts are made to the N-type bar. The emitter lead is connected to the P-region. The UJT schematic symbol is shown in Figure 19–48(b).

FIGURE 19–48
Unijunction transistor (UJT).

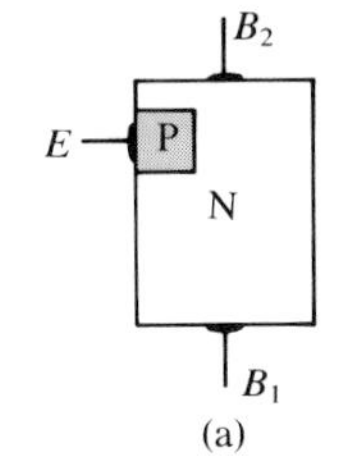

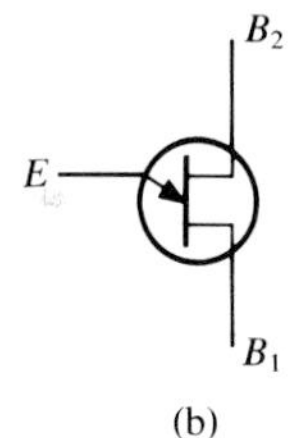

UJT Operation

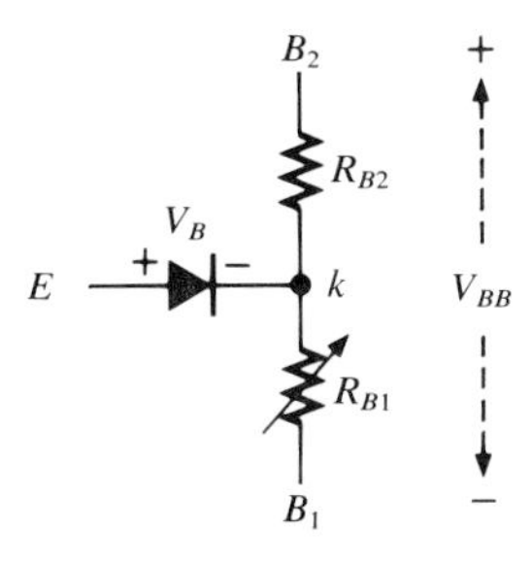

FIGURE 19–49
Equivalent circuit for a UJT.

In normal UJT operation, base 2 (B_2) and the emitter are biased positive with respect to base 1 (B_1). Figure 19–49 will illustrate the operation. This equivalent circuit represents the internal UJT characteristics. The total resistance between the two bases, R_{BB}, is the resistance of the N-type material. R_{B1} is the resistance between point k and base 1. R_{B2} is the resistance between point k and base 2. The sum of these two resistances makes up the total resistance, R_{BB}. The diode represents the PN junction between the emitter and the N-type material.

The ratio R_{B1}/R_{BB} is designated η (the Greek letter eta) and is defined as the *intrinsic standoff ratio.* It takes an emitter voltage of $V_B + \eta V_{BB}$ to turn the UJT on. This voltage is called the *peak voltage.* Once the device is on, resistance R_{B1} drops in value. Thus, as emitter current *increases,* emitter voltage *decreases* because of the decrease in R_{B1}. This characteristic is the *negative resistance* characteristic of the UJT.

As the emitter voltage decreases, it reaches a value called the *valley voltage*. At this point, the PN junction is no longer forward-biased, and the UJT turns off.

An Application

UJTs are commonly used in oscillator circuits. Figure 19–50 shows a typical circuit. Its operation is as follows: Initially the capacitor is uncharged and the UJT is off. When power is applied, the capacitor charges up exponentially. When it reaches the peak voltage, the UJT turns *on* and the capacitor begins to *discharge* as indicated. When the emitter reaches the valley voltage, the UJT turns off and the capacitor begins to charge again. The cycle repeats. Waveforms of the capacitor voltage and the R_1 voltage are shown in the diagram.

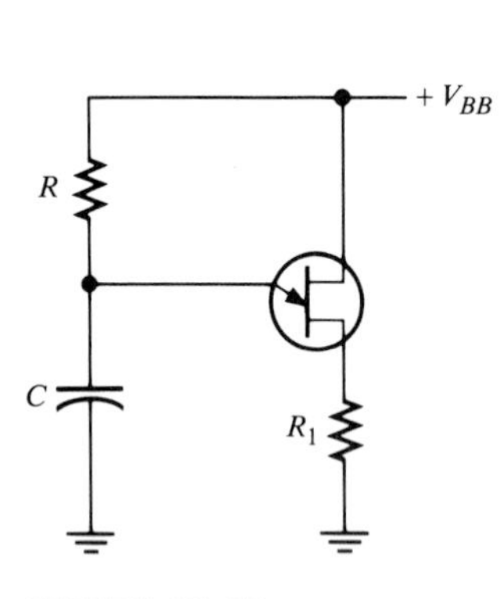

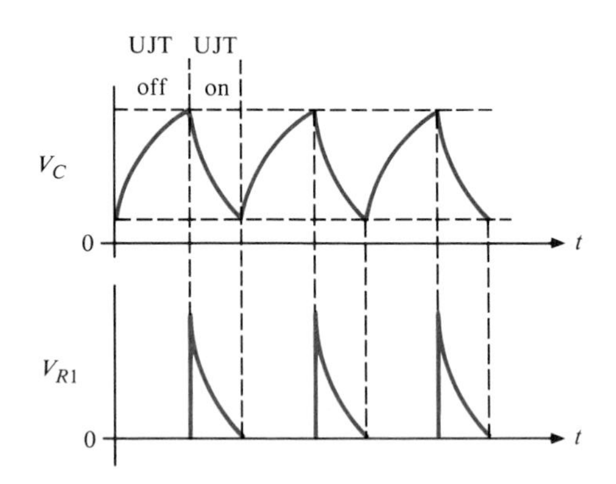

FIGURE 19–50
UJT oscillator circuit.

SECTION REVIEW 19–11

1. What does "UJT" stand for?
2. What are the terminals of a UJT?

THYRISTORS

19–12

Thyristors are a family of devices constructed of four layers of semiconductor material. The types of thyristors covered in this section are the *silicon-controlled rectifier (SCR)*, the *diac*, and the *triac*.

These thyristor devices share certain characteristics. They act as open circuits capable of withstanding a certain rated voltage until triggered. When triggered, they become low-resistance current paths and remain so, even after the trigger is removed, until the current is reduced to a certain level or until they are triggered off, depending on the type of device.

Thyristors can be used to control the amount of ac power to a load and are used in lamp-dimming circuits, motor speed control, ignition systems, and charging circuits, to name a few. UJTs are often used as the trigger devices for thyristors.

Silicon-controlled Rectifiers (SCRs)

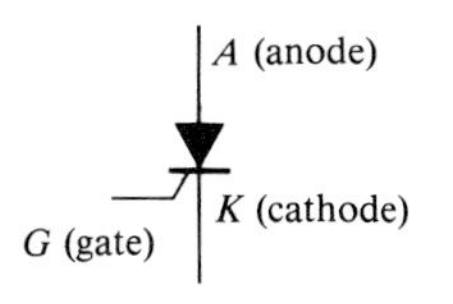

FIGURE 19–51
SCR symbol.

The silicon-controlled rectifier (SCR) has three terminals, as shown in the symbol in Figure 19–51. Like a rectifier diode, it is a unidirectional device, but the conduction of current is controlled by the gate, G. Current is from the cathode K to the anode A when the SCR is on. A positive voltage on the gate will turn the SCR on. The SCR will remain on as long as the current from the cathode to the anode is equal to or greater than a specified value called the *holding current.* When the current drops below the holding value, the SCR will turn off. It can be turned on again by a positive voltage on the gate.

SCR Construction

The SCR is a *four-layer* device. It has two P- and two N-regions, as shown in Figure 19–52(a). This construction can be thought of as an NPN and a PNP transistor interconnected as in Part (b) of the figure. The upper N region is commonly shared by the base and the collector of the transistors.

FIGURE 19–52
Construction of SCR and its transistor analogy.

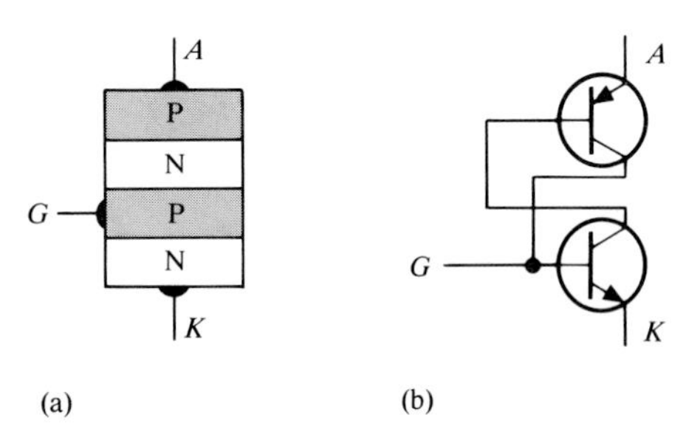

An Application

Figure 19–53 shows an SCR used to rectify and control the average power delivered to a load. The SCR rectifies the 60-Hz ac just as a conventional rectifier diode. It can conduct only during the positive half-cycle. The purpose of the phase controller is to produce a trigger pulse at the SCR gate in order to turn it on during any portion of the positive half-cycle of the input. If the SCR is turned on earlier in the half-cycle, more average power is delivered to the load. If the SCR is fired later in the half-cycle, less average power is delivered.

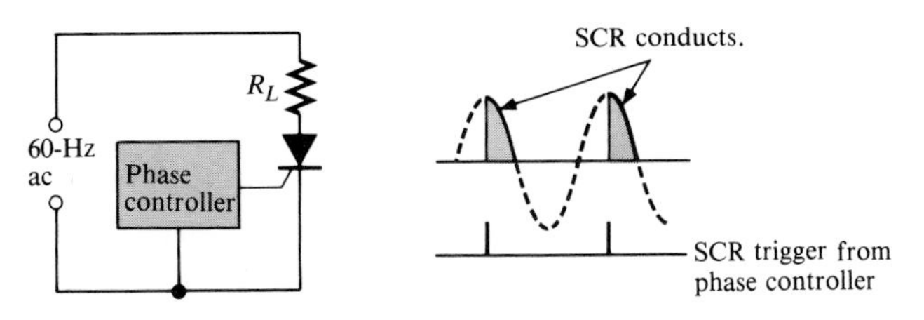

FIGURE 19–53
SCR controls average power to load.

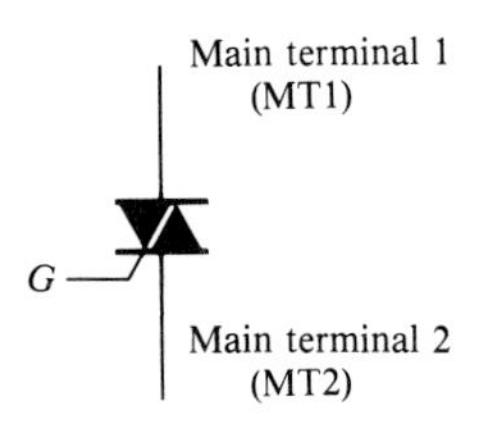

FIGURE 19–54
Triac symbol.

Triacs

The triac is also known as a *bidirectional triode thyristor*. It is equivalent to two SCRs connected to allow current flow in either direction. The SCRs share a common gate. The symbol for a triac is shown in Figure 19–54.

Applications

Like the SCR, triacs are also used to control average power to a load by the method of phase control. The triac can be triggered such that the ac power is supplied to the load for a controlled portion of each half-cycle. During each positive half-cycle of the ac, the triac is off for a certain interval, called the *delay angle* (measured in degrees). Then it is triggered on and conducts current through the load for the remaining portion of the positive half-cycle, called the *conduction angle*. Similar action occurs on the negative half-cycle except that, of course, current is conducted in the opposite direction through the load. Figure 19–55 illustrates this action.

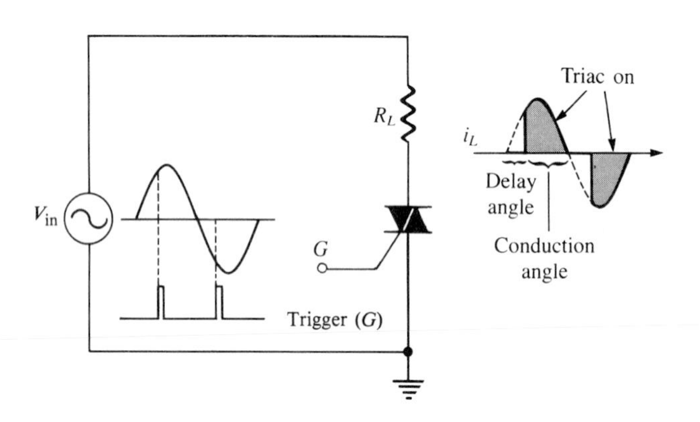

FIGURE 19–55
Basic triac phase control.

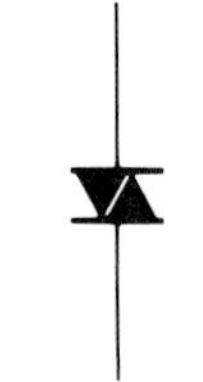

FIGURE 19–56
Diac symbol.

Diacs

The diac is a bidirectional device that does not have a gate. It conducts current in either direction when a sufficient voltage, called the *breakover potential,* is reached across the two terminals. The symbol for a diac is shown in Figure 19–56.

SECTION REVIEW 19–12

1. What does "SCR" stand for?
2. Explain how an SCR basically works.
3. Name a typical application of a triac.
4. A diac is gate-controlled (T or F).

TRANSISTOR PACKAGES AND TERMINAL IDENTIFICATION

19–13

Transistors are available in a wide range of package types for various applications. Those with mounting studs or heat sinks are power transistors. Low- and medium-power transistors are usually found in smaller metal or plastic cases. Still another package classification is for high-frequency devices.

As a technician, you often must check a transistor, either in the circuit or out of the circuit, for possible failure. First, of course, you must be able to recognize the transistor by its physical appearance. Also, you must be able to identify the base, collector, and emitter leads (or gate, drain, and source).

Transistors are found in a range of shapes and sizes. Figure 19–57 shows a few of the most typical case styles. Pin identification and construction diagrams are shown in Figure 19–58 (see page 736) for several types of transistor packages. Generally, the heat sink mounting or stud on power transistors is the collector terminal. On the metal "top hat" cases, the tab is closest to the emitter (or source) lead, and the collector (or drain) is often connected to the case.

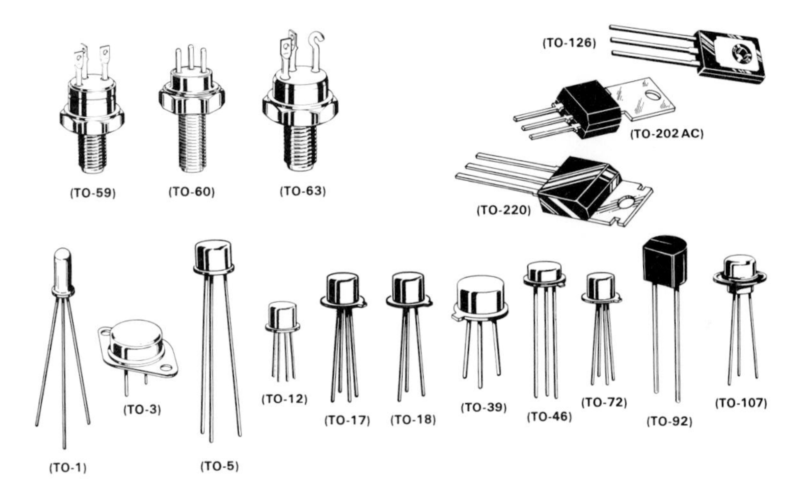

FIGURE 19–57
Some common transistor package configurations (courtesy of Motorola Semiconductor Group).

SECTION REVIEW 19–13

1. Identify the leads on the bipolar transistors in Figure 19–59 (on page 736).

APPLICATION NOTE

For the Application Assignment at the beginning of the chapter, you must remove the suspected transistor completely from the circuit. Then you must use an ohmmeter to check its base-collector and base-emitter junctions in both forward bias and reverse bias, as indicated in Figure 19–26.

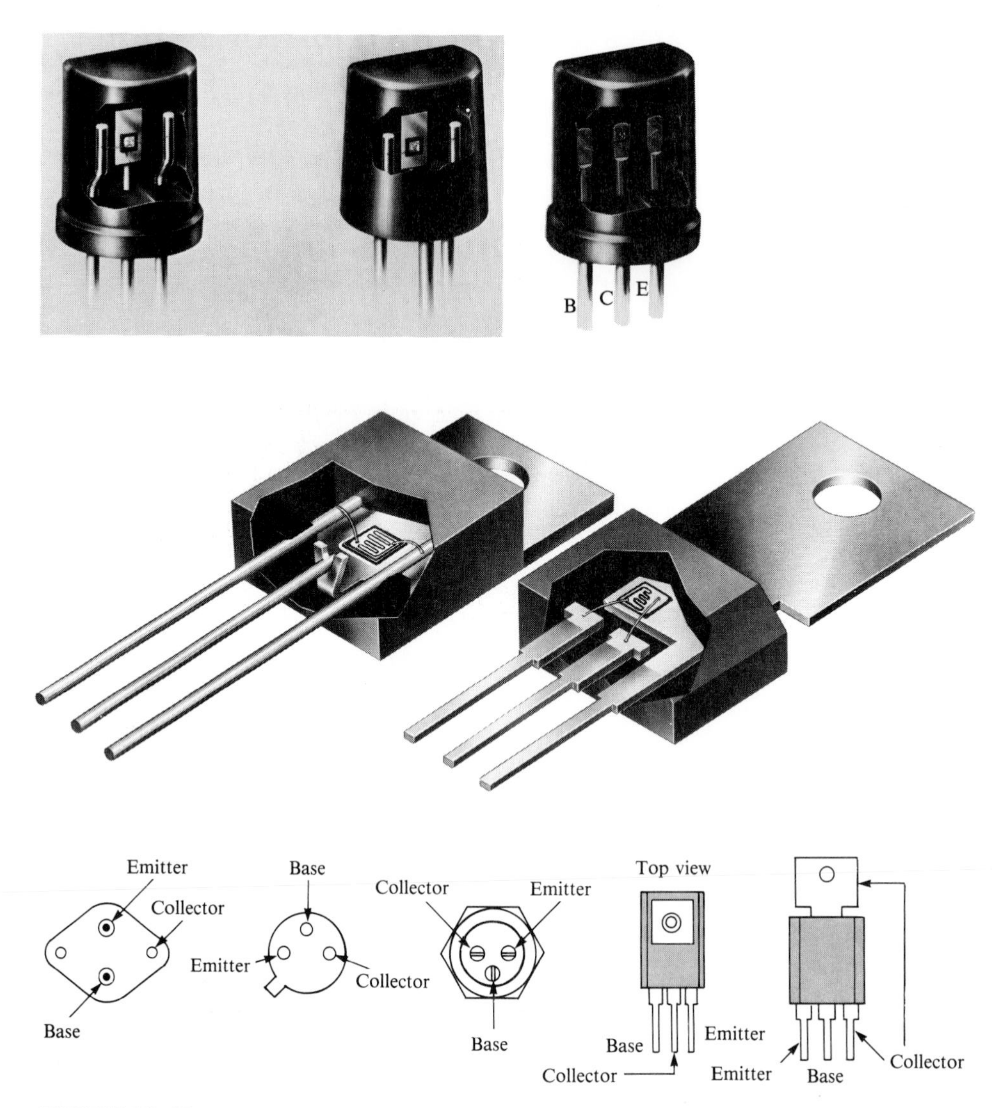

FIGURE 19–58
Typical transistor packages, construction views, and examples of pin arrangements. For FETs, replace base with gate, collector with drain, and emitter with source (courtesy of Motorola Semiconductor Group).

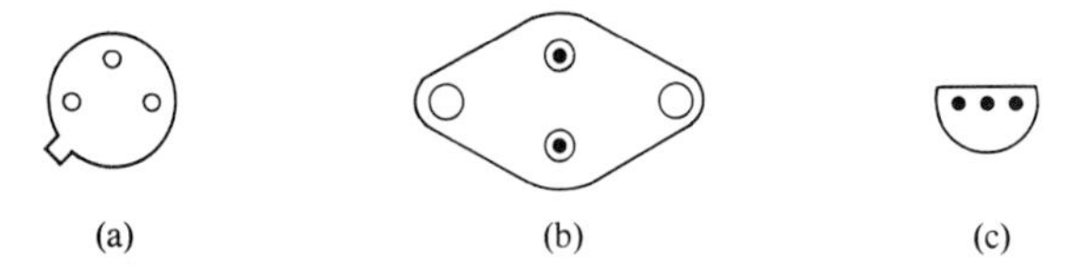

FIGURE 19–59

SUMMARY

Facts

- A bipolar junction transistor (BJT) consists of three regions: emitter, base, and collector.
- The three regions of a BJT are separated by two PN junctions.
- The two types of bipolar transistor are the NPN and the PNP.
- The term *bipolar* refers to two types of current: electron current and hole current.
- A field-effect transistor (FET) has three terminals: source, drain, and gate.
- A junction field-effect transistor (JFET) operates with a reverse-biased gate-to-source PN junction.
- JFETs have very high input resistance due to the reverse-biased gate-source junction.
- JFET current flows between the drain and the source through a *channel* whose width is controlled by the amount of reverse bias on the gate-source junction.
- The two types of JFETs are N-channel and P-channel.
- Metal oxide semiconductor field-effect transistors (MOSFETs) differ from JFETs in that the gate of a MOSFET is insulated from the channel.
- A depletion-enhancement MOSFET (DE MOSFET) can operate with a positive, negative, or zero gate-to-source voltage.
- The DE MOSFET has a physical channel between the drain and the source.
- An enhancement-only MOSFET (E MOSFET) can operate only when the gate-to-source voltage exceeds a threshold value.
- The enhancement-only MOSFET has no physical channel.
- Transistors are used as either amplifying devices or switching devices.

Definitions

- *Alpha* (α)—the ratio of collector current to emitter current.
- *Beta* (β)—the ratio of collector current to base current.
- *Q-point*—the dc operating point of a transistor.
- *Amplification*—an increase in voltage, current, or power.
- *Gain*—the amount of amplification.
- *Cutoff*—the condition in which a transistor acts as an open switch.
- *Saturation*—the condition in which a bipolar transistor acts as a closed switch.

Formulas

$I_E = I_C + I_B$ Bipolar transistor currents (19–1)

$I_C = \alpha_{dc} I_E$ Relationship of collector and emitter currents (19–2)

$I_C = \beta_{dc} I_B$ Relationship of collector and base currents (19–3)

$V_C = V_{CC} - I_C R_C$ Collector voltage (19–4)

$V_B = V_E + V_{BE}$ Base voltage (19–5)

$R_{IN} = \beta_{dc} R_E$ Input resistance at base (19–6)

$V_B = \left(\frac{R_2 \parallel R_{IN}}{R_1 + R_2 \parallel R_{IN}}\right) V_{CC}$ Base voltage with voltage divider bias (19–7)

$V_B \cong \left(\frac{R_2}{R_1 + R_2}\right) V_{CC}$ Approximate base voltage ($R_{IN} >> R_2$) (19–8)

$$\beta_{ac} = \frac{I_c}{I_b} \quad \text{ac beta} \tag{19–9}$$

$$A_V \cong \frac{R_C}{R_E} \quad \text{Voltage gain} \tag{19–10}$$

$$V_{CE(\text{cutoff})} \cong V_{CC} \quad V_{CE} \text{ at cutoff} \tag{19–11}$$

$$I_{C(\text{sat})} \cong \frac{V_{CC}}{R_C} \quad \text{Collector saturation current} \tag{19–12}$$

$$I_{B(\text{min})} = \frac{I_{C(\text{sat})}}{\beta_{dc}} \quad \text{Minimum base current for saturation} \tag{19–13}$$

$$I_C = \frac{P_{D(\text{max})}}{V_{CE}} \quad I_C \text{ for maximum } V_{CE} \tag{19–14}$$

$$V_{CE} = \frac{P_{D(\text{max})}}{I_C} \quad V_{CE} \text{ for maximum } I_C \tag{19–15}$$

$$V_P = V_{GS} - V_{DS(P)} \quad \text{Pinch-off voltage} \tag{19–16}$$

$$R_{IN} = \left|\frac{V_{GS}}{I_{GSS}}\right| \quad \text{JFET input resistance} \tag{19–17}$$

$$V_{GS} = -I_D R_S \quad \text{Self-bias voltage for N-channel JFET} \tag{19–18}$$

$$V_{GS} = +I_D R_S \quad \text{Self-bias voltage for P-channel JFET} \tag{19–19}$$

$$V_D = V_{DD} - I_D R_D \quad \text{Drain voltage} \tag{19–20}$$

$$V_{DS} = V_{DD} - I_D(R_D + R_S) \quad \text{Drain-to-source voltage} \tag{19–21}$$

$$V_{DS} = V_{DD} - I_{DSS} R_D \quad \text{Drain-to-source voltage DE MOSFET} \tag{19–22}$$

Symbols

Transistor and thyristor symbols are given in Figure 19–60.

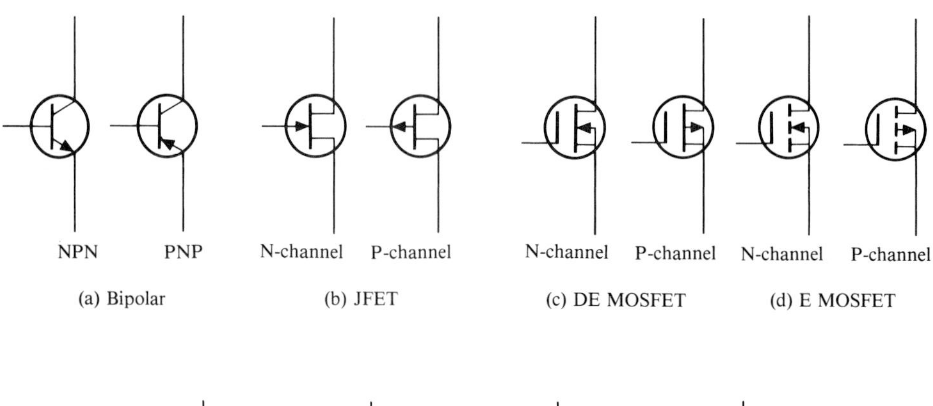

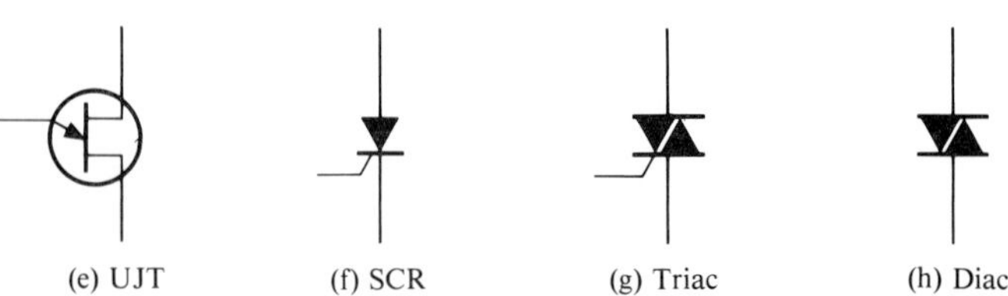

FIGURE 19–60
Transistor and thyristor symbols.

SELF-TEST

1. Name the N-type regions in an NPN transistor.
2. In a PNP transistor, is the base region N or P?
3. For normal operation of a PNP transistor, the base must be (+ or −) with respect to the emitter and (+ or −) with respect to the collector.
4. Name the three currents in a bipolar transistor.
5. Describe the general relationships of the currents named in Question 4.
6. If the beta of a transistor is 30, how much larger is the collector current than the base current?
7. An ohmmeter is connected to a PNP transistor with the positive lead on the base and the negative lead on the emitter. The meter shows a very high value of resistance. Is this a good indication?
8. Name two checks that most transistor testers can perform.
9. How does an increase in the reverse bias between the gate and the source of a JFET affect the current between drain and source?
10. Explain the difference between a DE MOSFET and an E MOSFET.
11. How does a triac differ from an SCR?

PROBLEMS

Section 19–1

19–1 What is the exact value of I_C for $I_E = 5.34$ mA and $I_B = 475\ \mu$A?

19–2 What is the α_{dc} when $I_C = 8.23$ mA and $I_E = 8.69$ mA?

19–3 A certain transistor has an $I_C = 25$ mA and an $I_B = 200\ \mu$A. Determine the β_{dc}.

19–4 In a certain transistor circuit, the base current is 2% of the 30-mA emitter current. Determine the approximate collector current.

19–5 Find I_B, I_E, and I_C in Figure 19–61 given that $\alpha_{dc} = 0.98$ and $\beta_{dc} = 49$.

FIGURE 19–61

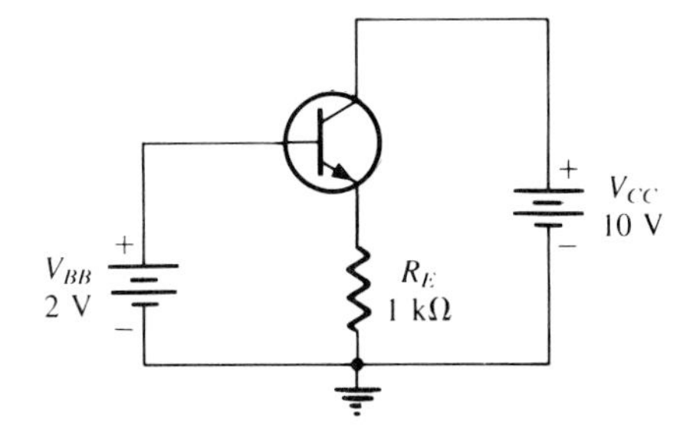

19–6 Determine the terminal voltages of each transistor with respect to ground for each circuit in Figure 19–62. Also determine V_{CE}, V_{BE}, and V_{BC}. $\beta_{dc} = 25$.

Section 19–2

19–7 Determine I_B, I_C, and V_C in Figure 19–63.

19–8 For the circuit in Figure 19–64, find V_B, V_E, I_E, I_C, and V_C.

19–9 In Figure 19–64, what is V_{CE}? What are the Q-point coordinates?

Section 19–3

19–10 A transistor amplifier has a voltage gain of 50. What is the output voltage when the input voltage is 100 mV?

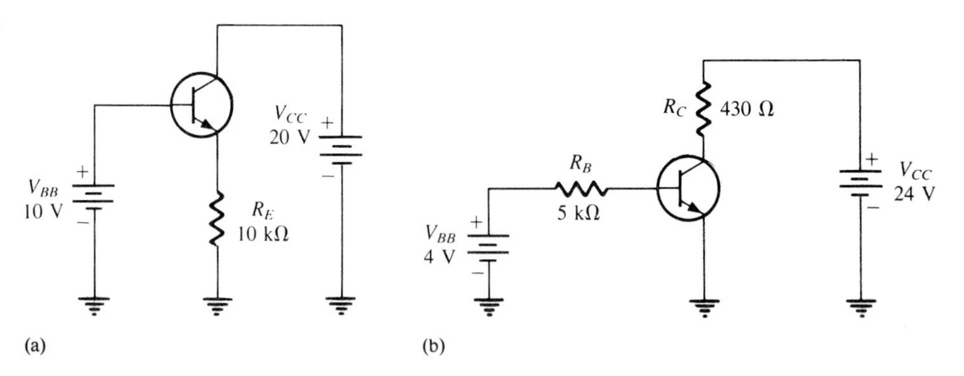

FIGURE 19–62

FIGURE 19–63

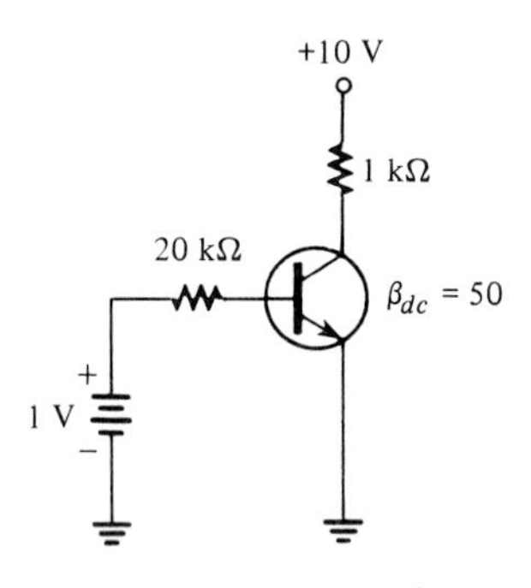

FIGURE 19–64

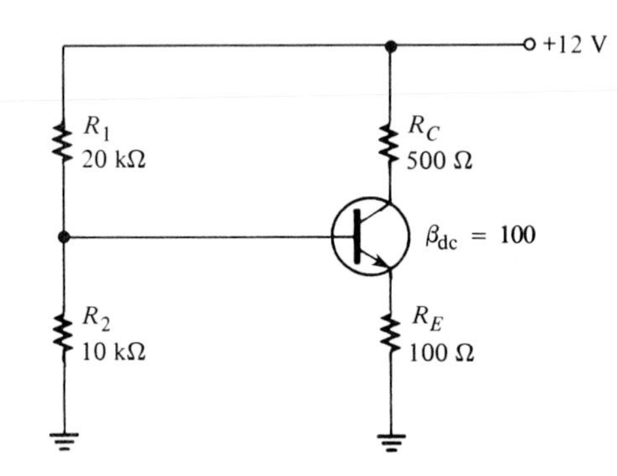

19–11 To achieve an output of 10 V with an input of 300 mV, what voltage gain is required?

19–12 A 50-mV signal is applied to the base of a properly biased transistor with R_E = 100 Ω and R_C = 500 Ω. Determine the signal voltage at the collector.

Section 19–4

19–13 Determine $I_{C(\text{sat})}$ for the transistor in Figure 19–65. What is the value of I_B necessary to produce saturation? What minimum value of V_{IN} is necessary for saturation?

19–14 The transistor in Figure 19–66 has a β_{dc} of 50. Determine the value of R_B required to insure saturation when V_{IN} is 5 V. What must V_{IN} be to cut off the transistor?

FIGURE 19–65

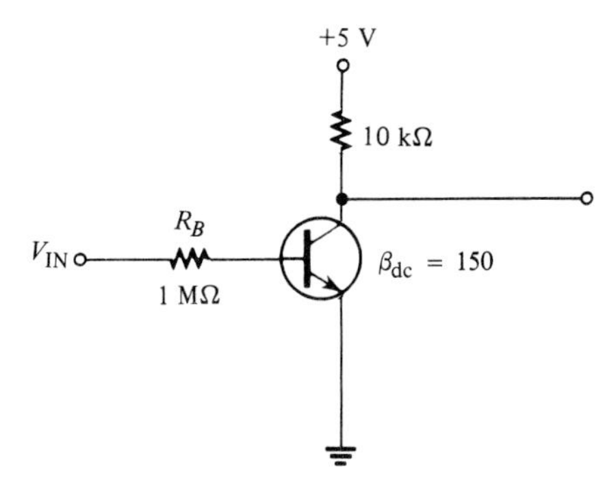

FIGURE 19–66

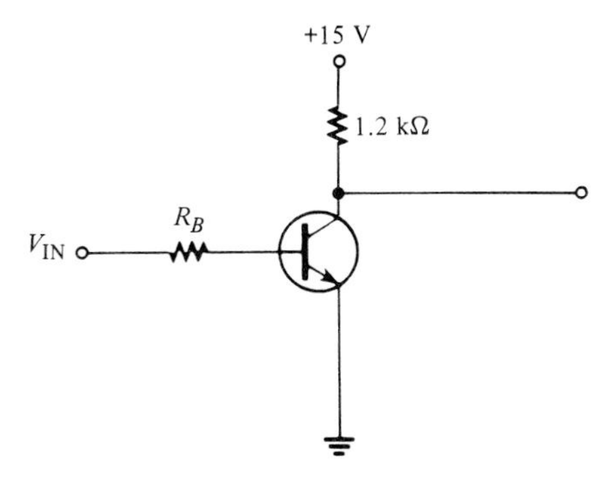

Section 19–5

19–15 If the β_{dc} in Figure 19–62(a) changes from 100 to 150 due to a temperature increase, what is the change in collector current?

19–16 A certain transistor is to be operated at a collector current of 50 mA. How high can V_{CE} go without exceeding a $P_{D(max)}$ of 1.2 W?

Section 19–6

19–17 Which, if any, of the transistors in Figure 19–67 are bad?

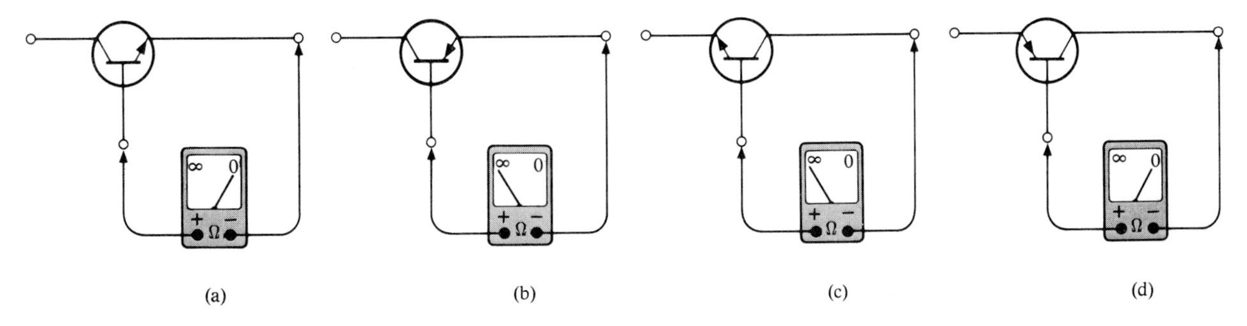

FIGURE 19–67

19–18 Is the transistor in Figure 19–68 an NPN or a PNP device? Assume that it is good. (Refer to Section 19–13.)

FIGURE 19–68

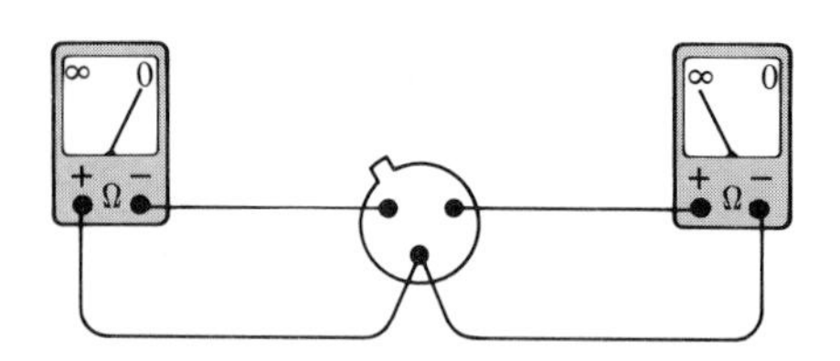

Section 19–7

19–19 The V_{GS} of a P-channel JFET is increased from 1 V to 3 V.
(a) Does the depletion region narrow or widen?
(b) Does the resistance of the channel increase or decrease?

19–20 Why must the gate-to-source voltage of an N-channel JFET always be either 0 or negative?

Section 19–8

19–21 A JFET has a specified pinch-off voltage of -5 V. When $V_{GS} = 0$, what is V_{DS}?

19–22 A certain JFET is biased such that $V_{GS} = -2$ V. What is the value of V_{DS} if V_P is specified to be -6 V?

19–23 A certain JFET data sheet gives $V_{GS(off)} = -8$ V and $I_{DSS} = 10$ mA. Determine the value of V_{DS} at which pinch-off begins when $V_{GS} = -4$ V.

19–24 A certain P-channel JFET has a $V_{GS(off)} = 6$ V. What is I_D when $V_{GS} = 8$ V?

19–25 The JFET in Figure 19–69 has a $V_{GS(off)} = -4$ V. Assume that you increase the supply voltage, V_{DD}, beginning at 0 until the ammeter reaches a steady value. What does the voltmeter read at this point?

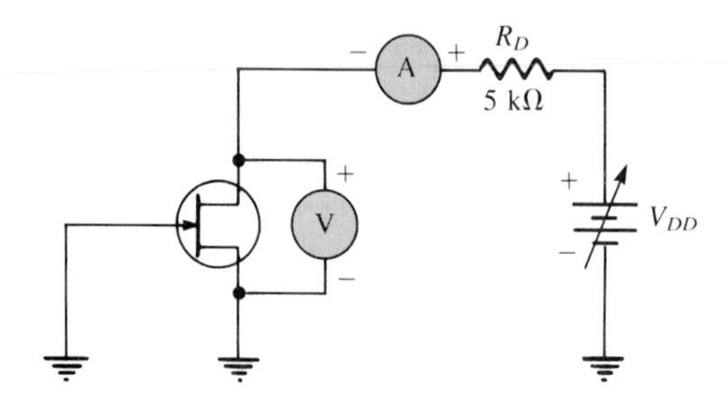

FIGURE 19–69

Section 19–9

19–26 Sketch the schematic symbols for N-channel and P-channel DE MOSFETs and E MOSFETs. Label the terminals.

19–27 Explain why both types of MOSFETs have an extremely high input resistance at the gate.

19–28 A N-channel DE MOSFET with a positive V_{GS} is operating in the ________ mode.

19–29 A certain E MOSFET has a $V_{GS(th)} = 3$ V. What is the minimum V_{GS} for the device to turn on?

Section 19–10

19–30 For each circuit in Figure 19–70, determine V_{DS} and V_{GS}.

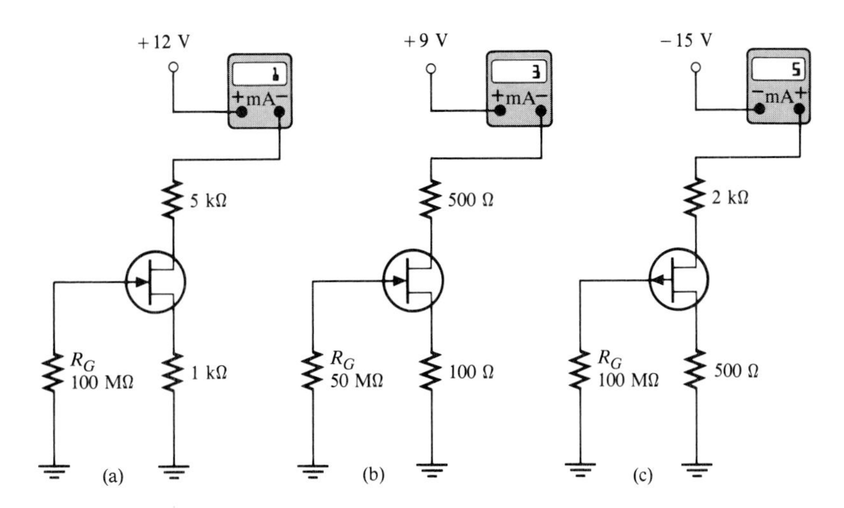

FIGURE 19–70

19–31 Determine in which mode (depletion or enhancement) each DE MOSFET in Figure 19–71 is biased.

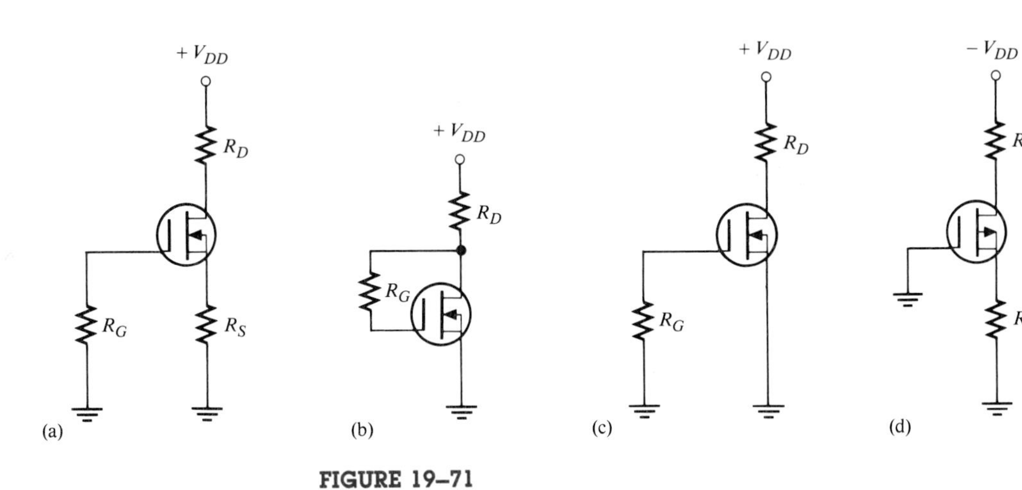

FIGURE 19–71

19–32 Each E MOSFET in Figure 19–72 (on page 744) has a $V_{GS(\text{th})}$ of +5 V or −5 V, depending on whether it is an N-channel or a P-channel device. Determine whether each MOSFET is *on* or *off*.

Section 19–11

19–33 In a UJT, if R_{BB} is 5 kΩ and R_{B1} is 3 kΩ, what is η?

19–34 What is the charging time constant in the oscillator circuit of Figure 19–73?

19–35 How long will the capacitor in Figure 19–73 initially charge if the peak voltage of the UJT is 6.3 V?

19–36 If the valley voltage of the UJT in Figure 19–73 is 2.23 V, what is the peak-to-peak value of the voltage across the capacitor? Sketch its general shape.

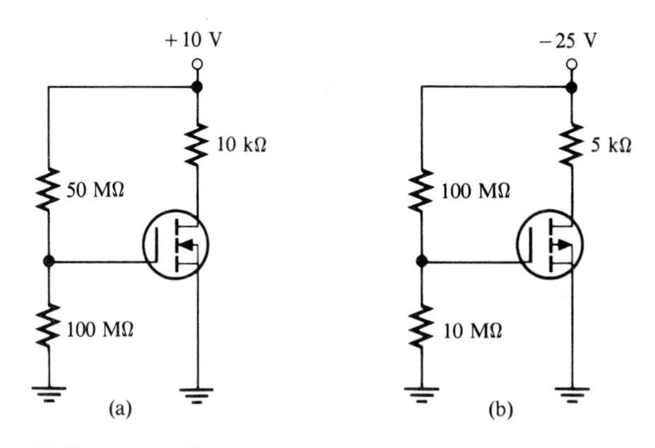

FIGURE 19–72

FIGURE 19–73

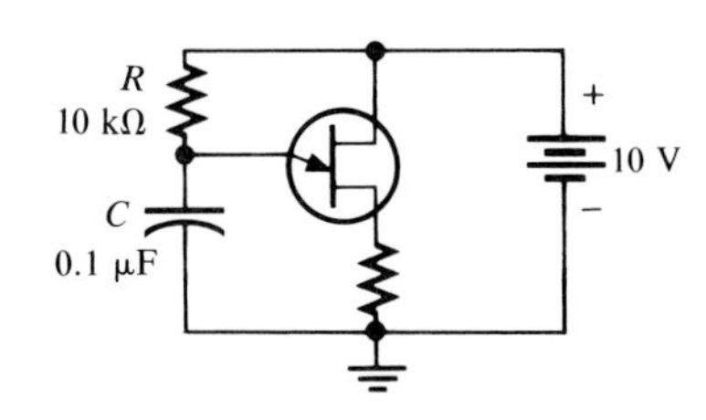

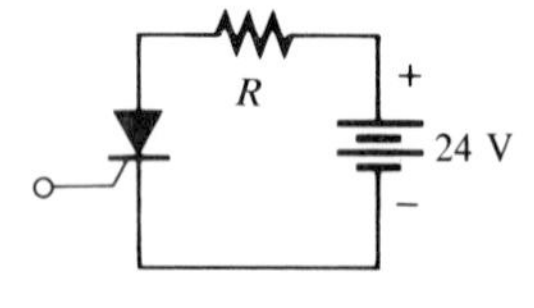

FIGURE 19–74

Section 19–12

19–37 Assume that the holding current for a particular SCR is 10 mA. What is the maximum value of R in Figure 19–74 necessary to keep the SCR in conduction once it is turned on? Neglect the drop across the SCR.

19–38 Repeat Problem 19–37 for a holding current of 500 μA.

ANSWERS TO SECTION REVIEWS

Section 19–1

1. Emitter, base, collector. **2.** The base-emitter junction is forward-biased and the base-collector junction is reverse-biased.

3. Direct current gain. **4.** 1 mA.

Section 19–2

1. One. **2.** 6 V.

Section 19–3

1. The process of increasing the amplitude of an electrical signal.

2. The dc operating point. **3.** 21.36.

Section 19–4

1. Saturation, cutoff. **2.** At saturation. **3.** At cutoff.

4. The base-emitter is forward-biased, and there is sufficient base current.

5. At cutoff.

Section 19–5

1. True. **2.** β_{dc} increases with I_C to a certain value, and then decreases.

3. 40 mA.

Section 19–6

1. NPN. **2.** I_{CBO}, I_{CEO}.

Section 19–7

1. Drain, source, and gate. **2.** Negative. **3.** By V_{GS}.

Section 19–8

1. −7 V. **2.** Decreases. **3.** +3 V.

Section 19–9

1. Depletion-enhancement MOSFET; enhancement only MOSFET. The DE MOSFET has a physical channel; the E MOSFET does not.

2. Yes. **3.** No.

Section 19–10

1. Positive. **2.** −8 V. **3.** I_{DSS} **4.** 2 V.

Section 19–11

1. Unijunction transistor. **2.** Anode, cathode, gate.

Section 19–12

1. Silicon-controlled rectifier. **2.** A positive voltage on the gate turns the SCR on, and the SCR conducts current from anode to cathode. If the current drops below the holding value, the SCR turns off. **3.** Lamp dimmer. **4.** False.

Section 19–13

1. See Figure 19–75.

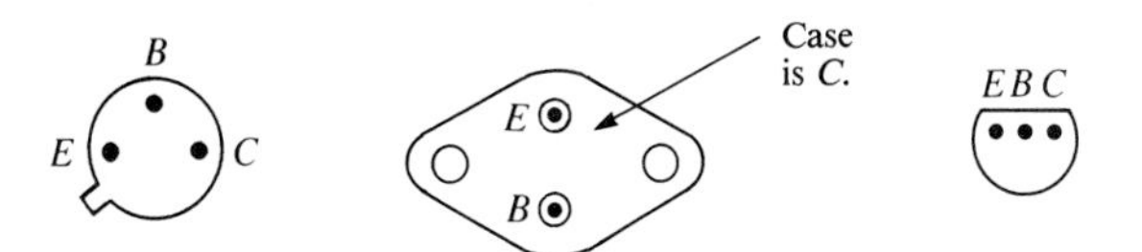

FIGURE 19–75

Amplifiers and Oscillators

As you learned in the previous chapter, the *biasing* of a transistor is purely a dc operation. The purpose of biasing, however, is to establish an operating point (Q point) about which variations in current and voltage can occur in response to an ac input signal.

When very small signal voltages must be amplified, such as from an antenna in a receiver, variations about the Q point of an amplifier are relatively small. Amplifiers designed to handle these small ac signals are called *small-signal* amplifiers. When large swings or variations in voltage and current about the Q point are required for power amplification, *large-signal* amplifiers are used. An example is the power amplifier that drives the speakers in a stereo system.

Regardless of whether an amplifier is in the small- or large-signal category, it will be in one of three configurations of amplifier circuits: common-emitter, common-collector, and common-base. For FETs the configurations are common-source, common-drain, and common-gate. Also, for any of the configurations there are three basic modes of operation possible—class A, class B, and class C—which are covered in this chapter.

Oscillators are also introduced in this chapter. An oscillator is a circuit that produces a sustained sine wave output without an input signal. It operates on the principle of positive feedback.

In this chapter you will learn:

- How circuit parameters affect the voltage gain of common-emitter, common-collector, and common-base amplifiers.
- Other important characteristics of bipolar amplifiers such as current gain, power gain, and input impedance.
- Why common-collector amplifiers are called *emitter-followers.*
- The advantage of Darlington pairs.
- The important characteristics of FET amplifiers.
- How multistage amplifiers are used to increase gain.
- The differences of class A, class B, and class C modes of amplifier operation.
- How to approach amplifier troubleshooting.
- Basics of oscillator operation.
- The basic differences of Colpitts, Hartley, Clapp, crystal, and *RC* oscillators.

20

APPLICATION ASSIGNMENT

The sound is severely distorted in one of the channels of a stereo system. The power amplifiers that drive the speakers are class B push-pull circuits. The power amplifiers are located on a separate plug-in circuit board. You need to pull this board and perform a test to determine if one of the amplifiers has failed.

After completing this chapter, you will be able to complete this assignment. A solution is given in the Application Note at the end of the chapter.

COMMON-EMITTER (CE) AMPLIFIERS

20–1

Figure 20–1 shows a *typical* common-emitter (CE) amplifier. The one shown has voltage divider bias, although other types of bias methods are possible. C_1 and C_2 are *coupling capacitors* used to *pass the signal* into and out of the amplifier such that the source or load will not affect the *dc* bias voltages. C_3 is a *bypass capacitor* that shorts the emitter signal voltage (ac) to ground without disturbing the dc emitter voltage. Because of the bypass capacitor, the emitter is at *signal ground* (but not dc ground), thus making the circuit a *common-emitter* amplifier. The purpose of the bypass capacitor is to increase the signal voltage gain. (The reason why this increase occurs will be discussed shortly.) Notice that the input signal is applied to the base, and the output signal is taken off the collector. All capacitors are assumed to have a reactance of approximately zero at the signal frequency.

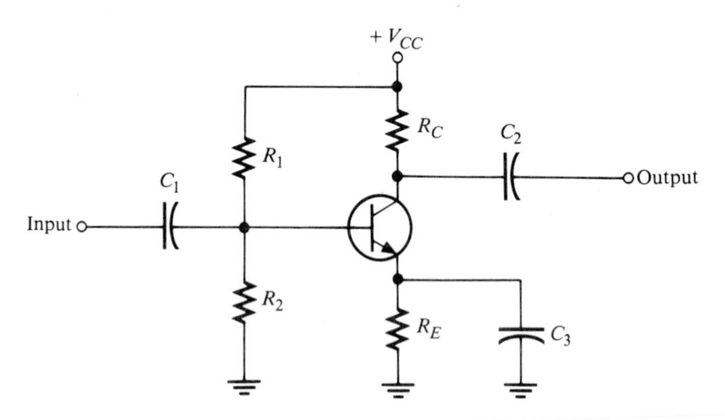

FIGURE 20–1
Typical common-emitter (CE) amplifier.

How a Bypass Capacitor Increases Voltage Gain

The bypass capacitor shorts the signal around the emitter resistor, R_E, in order to increase the voltage gain. To understand why, let us consider the amplifier *without* the bypass capacitor and see what the voltage gain is. The CE amplifier with the bypass capacitor removed is shown in Figure 20–2.

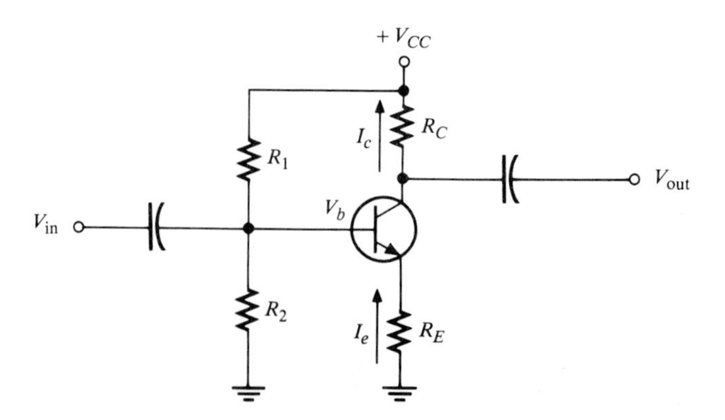

FIGURE 20–2
CE amplifier with bypass capacitor removed.

As before, lower-case subscripts indicate signal (ac) voltages and signal (alternating) currents. The voltage gain of the amplifier is V_{out}/V_{in}. The output signal voltage is

$$V_{out} = I_c R_C$$

The signal voltage at the base is approximately equal to

$$V_b \cong V_{in} \cong I_e(r_e + R_E)$$

where r_e is the *internal emitter resistance* of the transistor. The voltage gain A_v can now be expressed as

$$A_v = \frac{V_{out}}{V_{in}} = \frac{I_c R_C}{I_e(r_e + R_E)}$$

Since $I_c \cong I_e$, the currents cancel and the gain is the ratio of the resistances:

$$A_v = \frac{R_C}{r_e + R_E} \tag{20–1}$$

Keep in mind that this formula is for the CE *without* the bypass capacitor. If R_E is much greater than r_e, then $A_v \cong R_C/R_E$. This gain formula is similar to that presented in the last chapter, where r_e was neglected [Equation (19–10)].

If the bypass capacitor is connected across R_E, it effectively shorts the signal to ground, leaving only r_e in the emitter. Thus, the voltage gain of the CE amplifier with the bypass capacitor shorting R_E is

$$A_v = \frac{R_C}{r_e} \tag{20–2}$$

Now, r_e is a very important transistor parameter because it determines the voltage gain of a CE amplifier in conjunction with R_C. A formula for estimating r_e is given without derivation in the following equation:

$$r_e \cong \frac{25\text{ mV}}{I_E} \tag{20–3}$$

EXAMPLE 20–1

Determine the voltage gain of the amplifier in Figure 20–3 (on page 750) both with and without a bypass capacitor. $\beta_{dc} = \beta_{ac} = 150$.

Solution

First we determine r_e. To do so, we need to find I_E. Thus,

$$V_B \cong \left(\frac{10\text{ k}\Omega}{50\text{ k}\Omega + 10\text{ k}\Omega}\right)10\text{ V} = 1.67\text{ V}$$

$$V_E = V_B - 0.7\text{ V} = 0.97\text{ V}$$

$$I_E = \frac{V_E}{R_E} = \frac{0.97\text{ V}}{1\text{ k}\Omega} = 0.97\text{ mA}$$

$$r_e \cong \frac{25 \text{ mV}}{I_E} \cong \frac{25 \text{ mV}}{0.97 \text{ mA}} = 25.77\ \Omega$$

The voltage gain without a bypass capacitor is

$$A_v = \frac{R_C}{r_e + R_E} = \frac{5 \text{ k}\Omega}{1025.77\ \Omega} = 4.87$$

The voltage gain with the bypass capacitor installed is

$$A_v = \frac{R_C}{r_e} = \frac{5 \text{ k}\Omega}{25.77\ \Omega} = 194$$

As you can see, the voltage gain is greatly increased by the addition of the bypass capacitor. In terms of decibels (dB), the voltage gain is

$$A_v = 20 \log(194) = 45.76 \text{ dB}$$

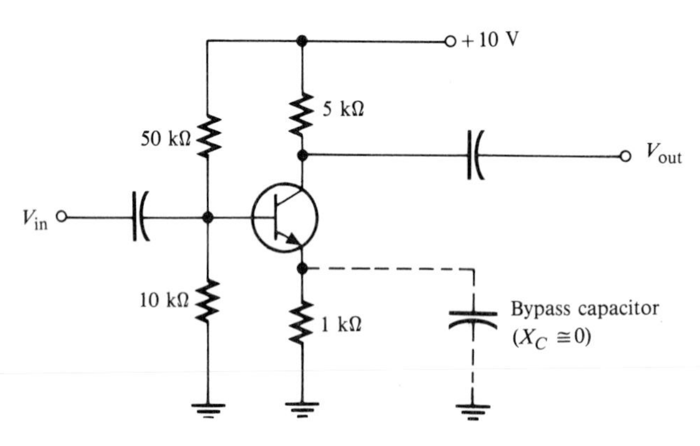

FIGURE 20–3

Phase Inversion

As we discussed in the last chapter, the output voltage at the collector is 180° out of phase with the input voltage at the base. Therefore, the CE amplifier is characterized by a phase inversion between the input and the output. As mentioned before, this inversion is sometimes indicated by a negative voltage gain.

ac Input Resistance

In the last chapter the dc input resistance R_{IN}, viewed from the base of the transistor, was developed. The input resistance "seen" by the *signal* at the base is derived in a similar manner when the emitter resistor is bypassed to ground:

$$R_{in} = \frac{V_b}{I_b}$$

$$V_b = I_e r_e$$

$$I_e \cong \beta_{ac} I_b$$

$$R_{in} \cong \frac{\beta_{ac} I_b r_e}{I_b}$$

The I_b terms cancel, leaving

$$R_{in} \cong \beta_{ac} r_e \qquad (20\text{–}4)$$

Total Input Resistance to CE Amplifier

Viewed from the base, R_{in} is the ac resistance. The actual resistance seen by the source includes that of bias resistors. We will now develop an expression for the total input resistance.

The concept of ac ground was mentioned earlier. At this point it needs some additional explanation because it is important in development of the formula for total input resistance, $R_{in(T)}$.

You have already seen that the bypass capacitor effectively makes the emitter appear as ground to the ac signal, because the X_C of the capacitor is nearly zero at the signal frequency. Of course, to a dc signal the capacitor looks like an open and thus does not affect the dc emitter voltage.

In addition to seeing ground through the bypass capacitor, the signal also sees ground through the dc supply voltage source V_{CC}. It does so because there is *zero* signal voltage at the V_{CC} terminal. Thus, the $+V_{CC}$ terminal effectively acts as *ac ground.* As a result, the two bias resistors R_1 and R_2 appear in parallel to the ac input, because one end of R_2 goes to actual ground and one end of R_1 goes to ac ground (V_{CC} terminal). Also, R_{in} at the base appears in parallel with $R_1 \parallel R_2$. This situation is illustrated in Figure 20–4.

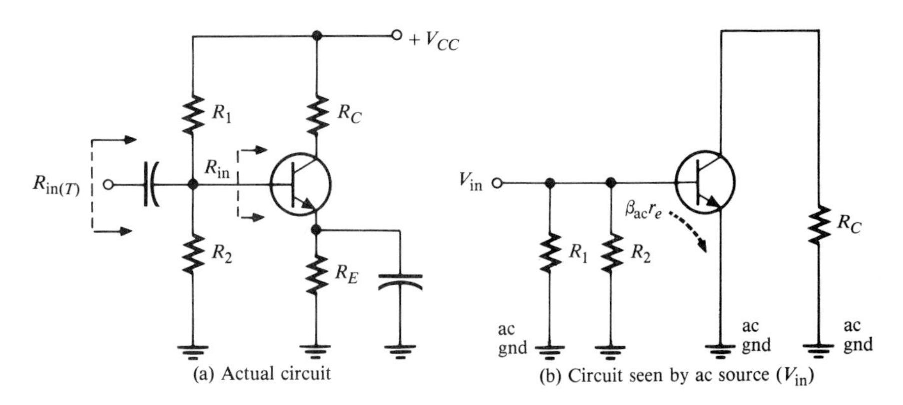

FIGURE 20–4
Total input resistance.

The expression for the total input resistance to the CE amplifier as seen by the ac source is as follows:

$$R_{in(T)} = R_1 \parallel R_2 \parallel R_{in} \qquad (20\text{–}5)$$

R_C has no effect because of the reverse-biased, base-collector junction.

EXAMPLE 20–2 Determine the total input resistance seen by the signal source in the CE amplifier in Figure 20–5. $\beta_{ac} = 150$.

FIGURE 20–5

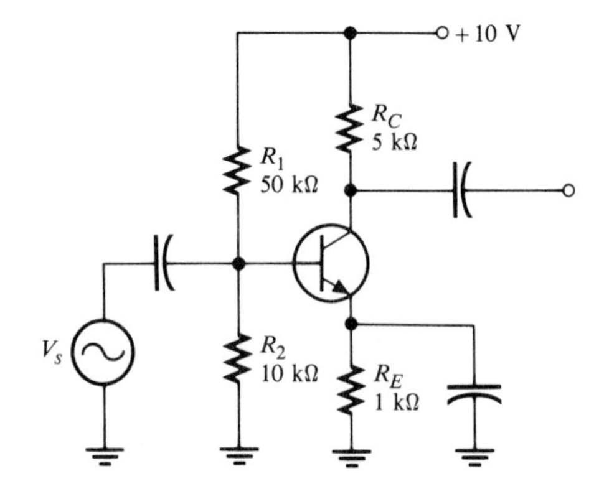

Solution
In Example 20–1, we found r_e for the same circuit. Thus,

$$R_{in} = \beta_{ac} r_e = 150(25.77\ \Omega) = 3865.5\ \Omega$$

$$R_{in(T)} = R_1 \parallel R_2 \parallel R_{in} = 50\ \text{k}\Omega \parallel 10\ \text{k}\Omega \parallel 3865.5\ \Omega = 2.64\ \text{k}\Omega$$

Current Gain

The signal current gain of a CE amplifier is

$$A_i = \frac{I_c}{I_s} \tag{20–6}$$

where I_s is the source current into the amplifier and is calculated by $V_s / R_{in(T)}$.

Power Gain

The power gain of a CE amplifier is the product of the voltage gain and the current gain:

$$A_p = A_v A_i \tag{20–7}$$

EXAMPLE 20–3 Determine the voltage gain, current gain, and power gain for the CE amplifier in Figure 20–6. $\beta_{ac} = 100$.

Solution
First we must find r_e. To do so, we must find I_E. We begin by calculating V_B. Since $R_{IN} = 100\ \text{k}\Omega$, it can be neglected. Thus,

$$V_B \cong \left(\frac{10\ \text{k}\Omega}{110\ \text{k}\Omega}\right) 30\ \text{V} = 2.73\ \text{V}$$

$$I_E = \frac{V_E}{R_E} = \frac{V_B - 0.7\ \text{V}}{R_E} = \frac{2.03\ \text{V}}{1\ \text{k}\Omega} = 2.03\ \text{mA}$$

$$r_e = \frac{25\text{ mV}}{I_E} = \frac{25\text{ mV}}{2.03\text{ mA}} = 12.32\ \Omega$$

The ac voltage gain is

$$A_v = \frac{R_C}{r_e} = \frac{5\text{ k}\Omega}{12.32\ \Omega} = 405.8$$

We determine the signal current gain by first finding $R_{\text{in}(T)}$ to get I_s:

$$R_{\text{in}(T)} = R_1 \parallel R_2 \parallel \beta_{ac} r_e = 100\text{ k}\Omega \parallel 10\text{ k}\Omega \parallel 1.232\text{ k}\Omega$$
$$\cong 1.1\text{ k}\Omega$$
$$I_s = \frac{V_s}{R_{\text{in}(T)}} = \frac{10\text{ mV}}{1.1\text{ k}\Omega} \cong 9.1\ \mu\text{A}$$

Next we must determine I_c:

$$I_c = \frac{V_{\text{out}}}{R_C} = \frac{A_v V_s}{R_C} = \frac{(405.8)(10\text{ mV})}{5\text{ k}\Omega} = 0.812\text{ mA}$$
$$A_i = \frac{I_c}{I_s} = \frac{812\ \mu\text{A}}{9.1\ \mu\text{A}} \cong 89.2$$

The power gain is

$$A_p = A_v A_i = (405.8)(89.2) = 36{,}197$$

The voltage gain and the power gain in decibels are as follows:

$$A_v = 20\log(405.8) = 52.17\text{ dB}$$
$$A_p = 10\log(36{,}197) = 45.59\text{ dB}$$

FIGURE 20–6

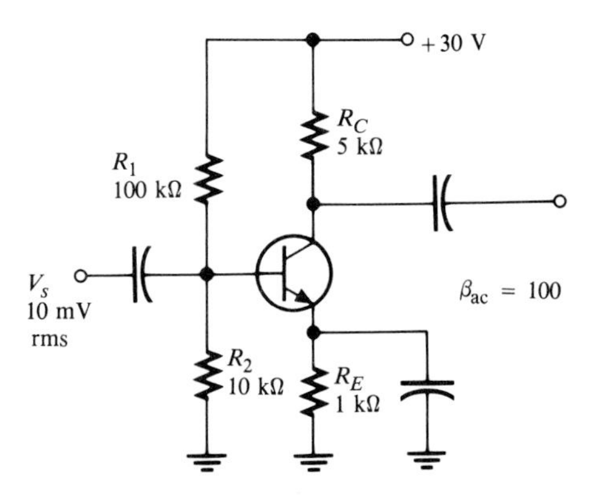

SECTION REVIEW 20–1

1. What is the purpose of the bypass capacitor in a CE amplifier?
2. How is the CE voltage gain determined?
3. If A_v is 50 and A_i is 200, what is the power gain?

COMMON-COLLECTOR (CC) AMPLIFIERS

20–2

The common-collector (CC) amplifier is the second most important of the three basic amplifier configurations. It is commonly referred to as an *emitter-follower.* Figure 20–7 shows an emitter-follower with a voltage divider bias. Notice that the input is applied to the base and the output is taken off the emitter.

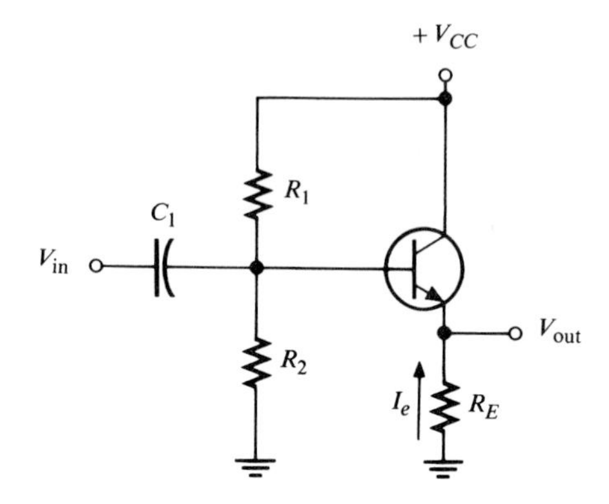

FIGURE 20–7
Typical emitter-follower (common-collector, CC) amplifier.

Voltage Gain

As in all amplifiers, the voltage gain in a CC amplifier is $A_v = V_{out}/V_{in}$. For the emitter-follower, V_{out} is $I_e R_E$, and V_{in} is $I_e(r_e + R_E)$. Therefore, the gain is $I_e R_E/I_e(r_e + R_E)$. The currents cancel, and the gain expression simplifies to

$$A_v = \frac{R_E}{r_e + R_E} \tag{20–8}$$

It is important to notice here that *the gain is always less than 1.* If r_e is much less than R_E, then a good approximation is $A_v \cong 1$.

Since the output voltage is the emitter voltage, it is in phase with the base or the input voltage. As a result, and because the voltage gain is close to 1, the output voltage *follows* the input voltage—thus the term *emitter-follower.*

Input Resistance

The emitter-follower is characterized by a high input resistance, which makes it a very useful circuit. Because of the high input resistance, the emitter follower can be used as a *buffer* to minimize loading effects when one circuit is driving another.

The derivation of the input resistance viewed from the base is similar to that for the CE amplifier. In this case, however, *the emitter resistor is not bypassed:*

$$R_{in} = \frac{V_b}{I_b} = \frac{I_e(r_e + R_E)}{I_b} \cong \frac{\beta_{ac} I_b(r_e + R_E)}{I_b} = \beta_{ac}(r_e + R_E)$$

If R_E is at least ten times larger than r_e, then the input resistance at the base is

$$R_{in} \cong \beta_{ac} R_E \tag{20–9}$$

In Figure 20–7, the bias resistors appear in parallel with R_{in} to the input signal, just as in the voltage divider biased CE amplifier. The total ac input resistance is

$$R_{in(T)} = R_1 \parallel R_2 \parallel R_{in} \tag{20–10}$$

Since R_{in} can be made large with the proper selection of R_E, a much higher input resistance results for this configuration than for the CE circuit.

Current Gain

The signal current gain for the emitter-follower is I_e/I_s; I_s is the signal current and can be calculated as $V_s/R_{in(T)}$. If the bias resistors are large enough to be neglected so that $I_s = I_b$, then the current gain of the amplifier is equal to the current gain of the transistor, β_{ac}. Of course, the same was also true for the CE amplifier. β_{ac} is the maximum achievable current gain in both amplifiers.

$$A_i = \frac{I_e}{I_s} \tag{20–11}$$

Power Gain

The power gain is the product of the voltage gain and the current gain. For the emitter-follower, the power gain is approximately equal to the current gain, because the voltage gain is approximately 1:

$$A_p \cong A_i \tag{20–12}$$

EXAMPLE 20–4 Determine the input resistance of the emitter-follower in Figure 20–8. Also find the voltage gain, current gain, and power gain.

FIGURE 20–8

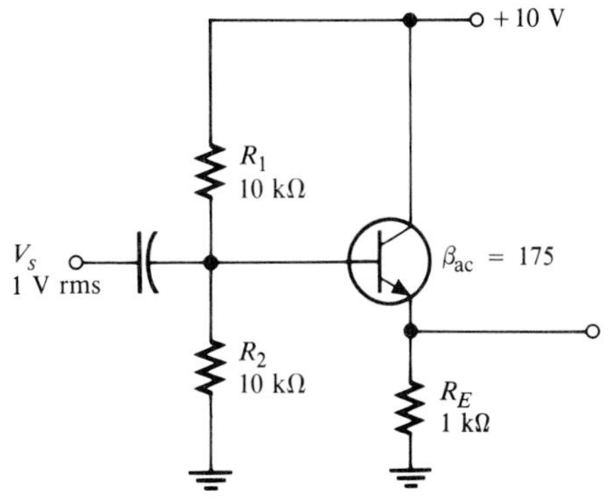

+10 V
R_1 10 kΩ
V_s 1 V rms
$\beta_{ac} = 175$
R_2 10 kΩ
R_E 1 kΩ

Solution

The approximate input resistance viewed from the base is

$$R_{in} \cong \beta_{ac} R_E = (175)(1\ \text{k}\Omega) = 175\ \text{k}\Omega$$

The total input resistance is

$$R_{in(T)} = R_1 \parallel R_2 \parallel R_{in} = 10\ \text{k}\Omega \parallel 10\ \text{k}\Omega \parallel 175\ \text{k}\Omega$$
$$= 4.86\ \text{k}\Omega$$

The voltage gain is, neglecting r_e,

$$A_v \cong 1$$

The current gain is

$$A_i = \frac{I_e}{I_s}$$

$$I_e = \frac{V_e}{R_E} = \frac{A_v V_b}{R_E} \cong \frac{1\ \text{V}}{1\ \text{k}\Omega} = 1\ \text{mA}$$

$$I_s = \frac{V_s}{R_{in(T)}} = \frac{1\ \text{V}}{4.86\ \text{k}\Omega} = 0.21\ \text{mA}$$

$$A_i = \frac{1\ \text{mA}}{0.21\ \text{mA}} = 4.76$$

The power gain is

$$A_p \cong A_i = 4.76$$

The Darlington Pair

As you have seen, β is a major factor in determining the input impedance. The β of the transistor limits the maximum achievable input impedance you can get from a given emitter-follower circuit.

One way to boost input impedance is to use a *Darlington pair,* as shown in Figure 20–9. The collectors of two transistors are connected, and the emitter of the first drives the base of the second. This configuration achieves β multiplication as shown in the following steps:

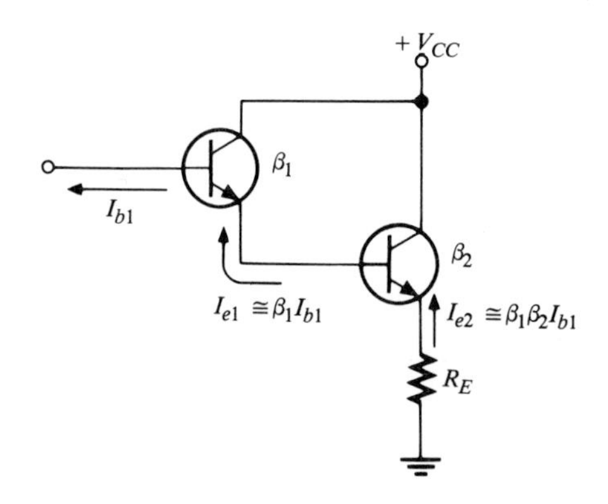

FIGURE 20–9
Darlington pair.

The emitter current of the first transistor is

$$I_{e_1} \cong \beta_1 I_{b_1}$$

This emitter current becomes the base current for the second transistor, producing a second emitter current of

$$I_{e_2} \cong \beta_2 I_{e_1}$$
$$I_{e_2} \cong \beta_1 \beta_2 I_{b1}$$

Therefore, the effective current gain of the Darlington pair is

$$\beta = \beta_1 \beta_2 \tag{20–13}$$

The input impedance is $\beta_1 \beta_2 R_E$.

SECTION REVIEW 20–2

1. What is a common-collector amplifier called?
2. What is its ideal maximum voltage gain?
3. What is the most important characteristic of the CC amplifier?

COMMON-BASE (CB) AMPLIFIERS

20–3

The third basic amplifier configuration is the common-base (CB). A typical common-base circuit is pictured in Figure 20–10. The base is at signal (ac) ground, and the input is applied to the emitter. The output is taken off the collector.

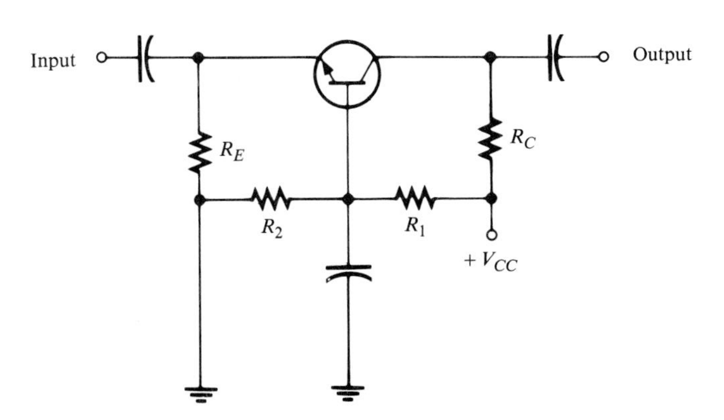

FIGURE 20–10
Typical common-base (CB) amplifier.

Voltage Gain

The input voltage is the emitter voltage V_e. The output voltage is the collector voltage V_c. With this in mind, we develop the voltage gain formula as follows:

$$A_v = \frac{V_c}{V_e} = \frac{I_c R_C}{I_e r_e} \cong \frac{I_e R_C}{I_e r_e}$$

$$A_v = \frac{R_C}{r_e} \tag{20–14}$$

Notice that the gain expression is the same as that for the CE amplifier (when R_E is bypassed).

Input Resistance

The resistance viewed from the emitter appears to the input signal as follows:

$$R_{in} = \frac{V_{in}}{I_{in}} = \frac{V_e}{I_e} = \frac{I_e r_e}{I_e}$$

$$R_{in} = r_e \tag{20–15}$$

Viewed from the source, R_E appears in parallel with R_{in}. However, r_e is normally so small compared to R_E that Equation (20–15) is also valid for the total input resistance, $R_{in(T)}$.

Current Gain

The current gain is the output current I_c over the input current I_e. Since $I_c \cong I_e$, the signal current gain is approximately 1:

$$A_i \cong 1 \tag{20–16}$$

Power Gain

Since the current gain is approximately 1 for the CB amplifier, the power gain is approximately equal to the voltage gain:

$$A_p \cong A_v \tag{20–17}$$

EXAMPLE 20–5 Find the input resistance, voltage gain, current gain, and power gain for the CB amplifier in Figure 20–11.

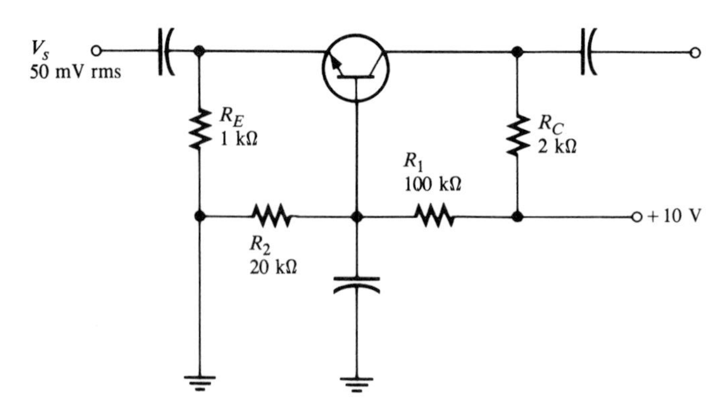

FIGURE 20–11

Solution
First let us find I_E so that we can determine r_e. Then $R_{in} = r_e$. Thus,

$$V_B \cong \left(\frac{R_2}{R_1 + R_2}\right)V_{CC} = \left(\frac{20\text{ k}\Omega}{120\text{ k}\Omega}\right)10\text{V} = 1.67\text{ V}$$

$$V_E = V_B - 0.7\text{ V} = 1.67\text{ V} - 0.7\text{ V} = 0.97\text{ V}$$

$$I_E = \frac{0.97\text{ V}}{1\text{ k}\Omega} = 0.97\text{ mA}$$

$$R_{in} = r_e \cong \frac{25\text{ mV}}{0.97\text{ mA}} = 25.77\ \Omega$$

The signal voltage gain is

$$A_v = \frac{R_C}{r_e} = \frac{2\text{ k}\Omega}{25.77\ \Omega} = 77.6$$

Thus,

$$A_i \cong 1$$

$$A_p \cong 77.6$$

Summary

Table 20–1 summarizes the important characteristics of each of the three amplifier configurations. Also, relative values are indicated for general comparison of the amplifiers.

TABLE 20–1
Comparison of amplifier configurations. The current gains and the input resistance are the maximum achievable values, with the bias resistors neglected.

	CE	CC	CB
Voltage gain, A_v	$\frac{R_C}{r_e}$ High	$\cong 1$ Low	$\frac{R_C}{r_e}$ High
Current gain, $A_{i(max)}$	β_{ac} High	β_{ac} High	$\cong 1$ Low
Power gain, A_p	A_iA_v Very high	$\cong A_i$ High	$\cong A_v$ High
Input resistance, $R_{in(max)}$	$\beta_{ac}r_e$ Low	$\beta_{ac}R_E$ High	r_e Very low

SECTION REVIEW 20–3

1. The same voltage gain can be achieved with a CB as with a CE amplifier (T or F).
2. The CB amplifier has a very low input resistance (T or F).

FET AMPLIFIERS

20–4

Field-effect transistors, both JFETs and MOSFETs, can be used as amplifiers in any of three circuit configurations similar to those for the bipolar transistor. These configurations are *common-source, common-drain,* and *common-gate.*

Transconductance of an FET

Recall that in a bipolar transistor, the base current controls the collector current, and the relationship between these two currents is expressed by the parameter β ($I_C = \beta_{ac} I_b$). In an FET, the gate voltage controls the drain current. An important FET parameter is the *transconductance,* g_m, which is defined as

$$g_m = \frac{I_d}{V_{gs}} \tag{20–18}$$

The transconductance is one factor that determines the voltage gain of an FET amplifier. On data sheets, the transconductance is sometimes called the *forward transadmittance* and is designated y_{fs}.

Common-Source (CS) Amplifiers

A self-biased N-channel JFET with an ac source capacitively coupled to the gate is shown in Figure 20–12. The resistor, R_G, serves two purposes: (1) It keeps the gate at approximately 0 V dc (because I_{GSS} is extremely small), and (2) its large value (usually several megohms) prevents loading of the ac signal source. The bias voltage is created by the drop across R_S. The bypass capacitor, C_2, keeps the source of the FET effectively at ac ground.

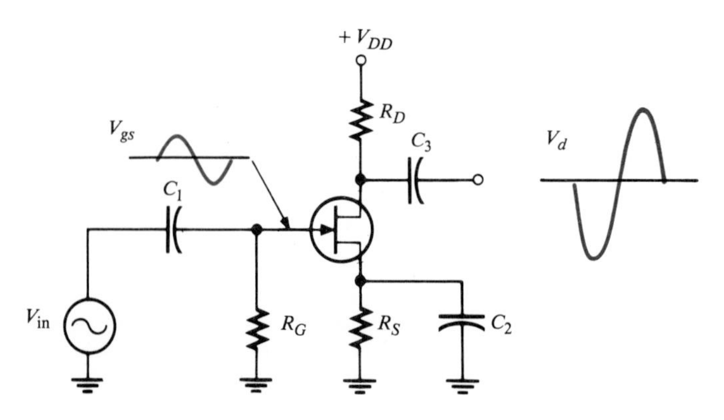

FIGURE 20–12
JFET common-source amplifier.

The signal voltage causes the gate-to-source voltage to swing above and below its Q-point value, causing a swing in drain current. As the drain current increases, the voltage drop across R_D also increases, causing the drain voltage (with respect to ground) to decrease.

The drain current swings above and below its Q-point value *in phase* with the gate-to-source voltage. The drain-to-source voltage swings above and below its Q-point value 180° *out of phase* with the gate-to-source voltage, as illustrated in the figure.

DE MOSFET

A zero-biased N-channel DE MOSFET with an ac source capacitively coupled to the gate is shown in Figure 20–13. The gate is at approximately 0 V dc and the source terminal is at ground, thus making $V_{GS} = 0$ V.

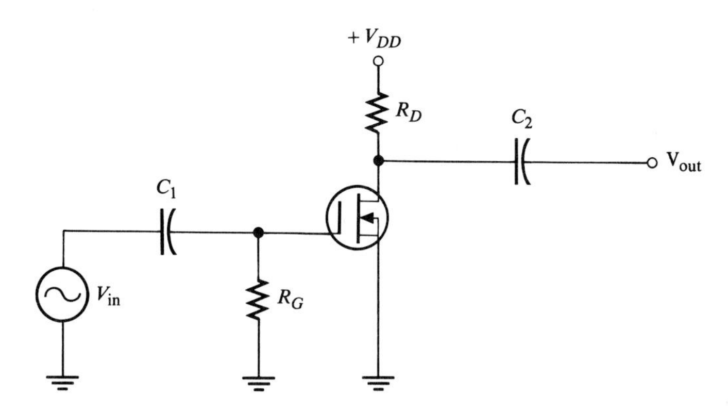

FIGURE 20–13
Zero-biased DE MOSFET common-source amplifier.

The signal voltage causes V_{gs} to swing above and below its 0 value, producing a swing in I_d. The negative swing in V_{gs} produces the depletion mode, and I_d decreases. The positive swing in V_{gs} produces the enhancement mode, and I_d increases.

E MOSFET

A voltage divider-biased, N-channel E MOSFET with an ac signal source capacitively coupled to the gate is shown in Figure 20–14. The gate is biased with a positive voltage such that $V_{GS} > V_{GS(th)}$.

As with the JFET and DE MOSFET, the signal voltage produces a swing in V_{gs} above and below its Q-point value. This swing, in turn, causes a swing in I_d. Operation is entirely in the enhancement mode.

Voltage Gain of the CS Amplifier

Voltage gain, A_v, of an amplifier always equals V_{out}/V_{in}. In the case of the CS amplifier, V_{in} is equal to V_{gs}, and V_{out} is equal to the signal voltage developed across R_D which is $I_d R_D$. Thus,

$$A_v = \frac{I_d R_D}{V_{gs}}$$

Since $g_m = I_d/V_{gs}$, the common-source voltage gain is

$$A_v = g_m R_D \tag{20–19}$$

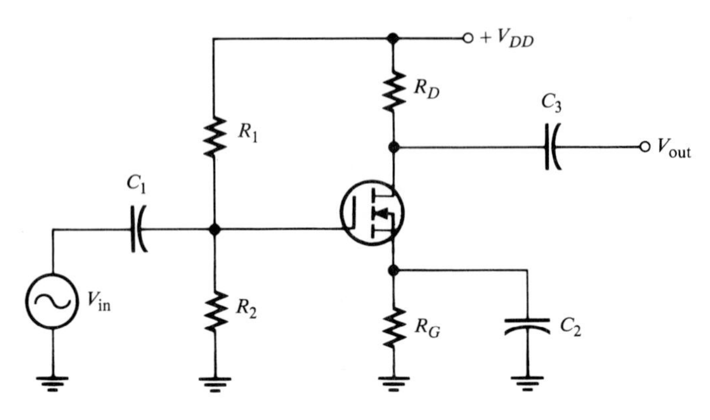

FIGURE 20–14
Common-source E MOSFET amplifier with voltage divider bias.

Input Impedance of the CS Amplifier

Because the input to a CS amplifier is at the gate, the input impedance is extremely high. Ideally, it approaches infinity and can be neglected. As you know, the high input impedance is produced by the reverse-biased PN junction in a JFET and by the insulated gate structure in a MOSFET.

The actual input impedance seen by the signal source is the gate-to-ground resistor R_G in parallel with the FET's input impedance, V_{GS}/I_{GSS}. The reverse leakage current I_{GSS} is typically given on the data sheet for a specific value of V_{GS} so that the input impedance of the device can be calculated.

EXAMPLE 20–6

(a) What is the total output voltage (dc + ac) of the amplifier in Figure 20–15? The g_m is 4500 μS, I_D is 2 mA, $V_{GS(off)}$ is -10 V, and I_{GSS} is 15 nA.

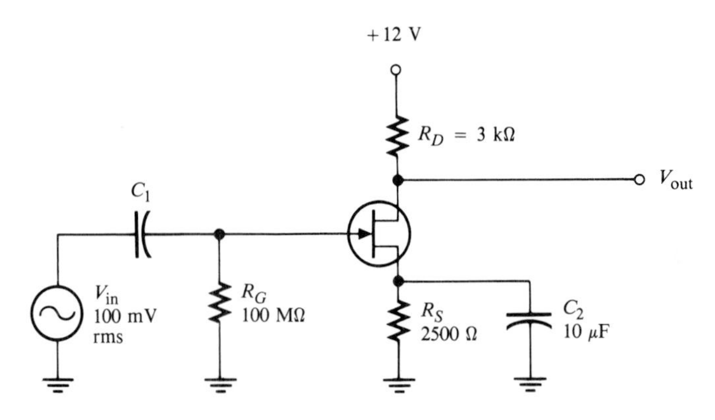

FIGURE 20–15

(b) What is the input impedance seen by the signal source?

Solution

(a) First, find the dc output voltage:

$$V_D = V_{DD} - I_D R_D = 12\text{ V} - (2\text{ mA})(3\text{ k}\Omega) = 6\text{ V}$$

Next, find the ac output voltage by using the gain formula:

$$\frac{V_{out}}{V_{in}} = g_m R_D$$

$$V_{out} = g_m R_D V_{in} = (4500\ \mu\text{S})(3\text{ k}\Omega)(100\text{ mV})$$

$$= 1.35\text{ V rms}$$

The total output voltage is an ac signal with a peak-to-peak value of 1.35 V × 2.828 = 3.82 V, riding on a dc level of 6 V.

(b) The input resistance is determined as follows (since $V_G = 0$ V):

$$V_{GS} = I_D R_S = (2\text{ mA})(2500\ \Omega) = 5\text{ V}$$

The input resistance at the gate of the JFET is

$$R_{IN(gate)} = \frac{V_{GS}}{I_{GSS}} = \frac{5\text{ V}}{15\text{ nA}} = 333\text{ M}\Omega$$

The input impedance seen by the signal source is

$$R_{in} = R_G \parallel R_{IN(gate)} = 100\text{ M}\Omega \parallel 333\text{ M}\Omega = 76.9\text{ M}\Omega$$

Common-Drain (CD) Amplifier

A common-drain JFET amplifier is shown in Figure 20–16 with voltages indicated. Self-biasing is used in this circuit. The input signal is applied to the gate

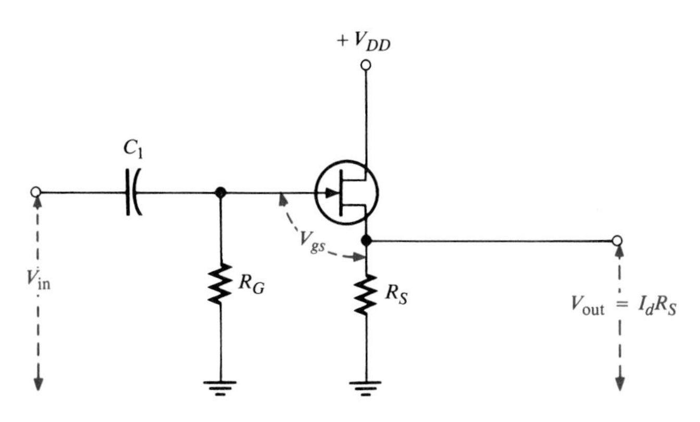

FIGURE 20–16
JFET common-drain amplifier (source-follower).

through a coupling capacitor, and the output is at the source terminal. There is no drain resistor. This circuit, of course, is analogous to the bipolar emitter-follower and is sometimes called a *source-follower.*

Voltage Gain of the CD Amplifier

As in all amplifiers, the voltage gain is $A_v = V_{out}/V_{in}$. For the source-follower, V_{out} is I_dR_S and V_{in} is $V_{gs} + I_dR_S$, as shown in Figure 20–16. Therefore, the gate-to-source voltage gain is $I_dR_S/(V_{gs} + I_dR_S)$. Substituting $I_d = g_mV_{gs}$ into the expression gives the following result:

$$A_v = \frac{g_mV_{gs}R_S}{V_{gs} + g_mV_{gs}R_S}$$

Canceling V_{gs}, we get

$$A_v \cong \frac{g_mR_S}{1 + g_mR_S} \qquad (20\text{–}20)$$

Notice here that the gain is always slightly *less than 1.* If $g_mR_S >> 1$, then a good approximation is $A_v \cong 1$.

Since the output voltage is at the source, it is in phase with the gate (input) voltage.

Input Impedance of the CD Amplifier

Because the input signal is applied to the gate, the input impedance seen by the input signal source is extremely high, just as in the CS amplifier configuration. The gate resistor R_G, in parallel with the input impedance looking in at the gate, is the total input impedance.

EXAMPLE 20–7

(a) Determine the voltage gain of the amplifier in Figure 20–17(a) using the data sheet information in Figure 20–17(b).
(b) Also determine the input impedance.
Assume minimum data sheet values where available.

Solution

(a) From the data sheet, $g_m = y_{fs} = 8000\ \mu S$. The gain is

$$A_v \cong \frac{g_mR_S}{1 + g_mR_S} = \frac{(8000\ \mu S)(10\ k\Omega)}{1 + (8000\ \mu S)(10\ k\Omega)} = 0.988$$

(b) From the data sheet, $I_{GSS} = 10$ nA at $V_{GS} = 15$ V. Therefore,

$$R_{IN(gate)} = \frac{15\ V}{10\ nA} = 1500\ M\Omega$$

$$R_{IN} = R_G \parallel R_{IN(gate)} = 100\ M\Omega \parallel 1500\ M\Omega = 93.75\ M\Omega$$

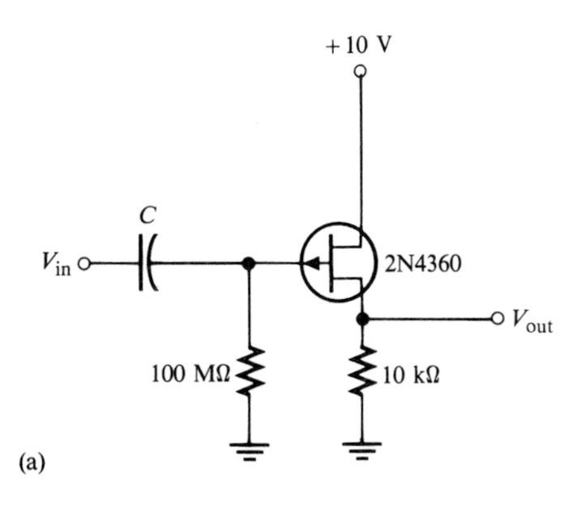

(a)

*ELECTRICAL CHARACTERISTICS (T_A = 25°C unless otherwise noted)

Characteristic	Symbol	Min	Max	Unit
OFF CHARACTERISTICS				
Gate-Source Breakdown Voltage (I_G = 10 μAdc, V_{DS} = 0)	$V_{(BR)GSS}$	20	–	Vdc
Gate-Source Cutoff Voltage (V_{DS} = -10 Vdc, I_D = 1.0 μAdc)	$V_{GS(off)}$	0.7	10	Vdc
Gate Reverse Current	I_{GSS}			
(V_{GS} = 15 Vdc, V_{DS} = 0)		–	10	nAdc
(V_{GS} = 15 Vdc, V_{DS} = 0, T_A = 65°C)		–	0.5	μAdc
ON CHARACTERISTICS				
Zero-Gate Voltage Drain Current (Note 1) (V_{DS} = -10 Vdc, V_{GS} = 0)	I_{DSS}	3.0	30	mAdc
Gate-Source Voltage (V_{DS} = -10 Vdc, I_D = 0.3 mAdc)	V_{GS}	0.4	9.0	Vdc
SMALL-SIGNAL CHARACTERISTICS				
Drain-Source "ON" Resistance (V_{GS} = 0, I_D = 0, f = 1.0 kHz)	$r_{ds(on)}$	–	700	Ohms
Forward Transadmittance (Note 1) (V_{DS} = -10 Vdc, V_{GS} = 0, f = 1.0 kHz)	$\lvert y_{fs} \rvert$	2000	8000	μmhos
Forward Transconductance (V_{DS} = -10 Vdc, V_{GS} = 0, f = 1.0 MHz)	$Re(y_{fs})$	1500	–	μmhos
Output Admittance (V_{DS} = -10 Vdc, V_{GS} = 0, f = 1.0 kHz)	$\lvert y_{os} \rvert$	–	100	μmhos
Input Capacitance (V_{DS} = -10 Vdc, V_{GS} = 0, f = 1.0 MHz)	C_{iss}	–	20	pF
Reverse Transfer Capacitance (V_{DS} = -10 Vdc, V_{GS} = 0, f = 1.0 MHz)	C_{rss}	–	5.0	pF
Common-Source Noise Figure (V_{DS} = -10 Vdc, I_D = 1.0 mAdc, R_G = 1.0 Megohm, f = 100 Hz)	NF	–	5.0	dB
Equivalent Short-Circuit Input Noise Voltage (V_{DS} = -10 Vdc, I_D = 1.0 mAdc, f = 100 Hz, BW = 15 Hz)	E_n	–	0.19	$\mu V/\sqrt{Hz}$

*Indicates JEDEC Registered Data.

Note 1: Pulse Test: Pulse Width ≤ 630 ms, Duty Cycle ≤ 10%.

(b)

FIGURE 20–17

Common-Gate (CG) Amplifier

A typical common-gate amplifier is shown in Figure 20–18. The gate is effectively at ac ground because of capacitor C_2. The input signal is applied at the source terminal through C_1. The output is coupled through C_3 from the drain terminal.

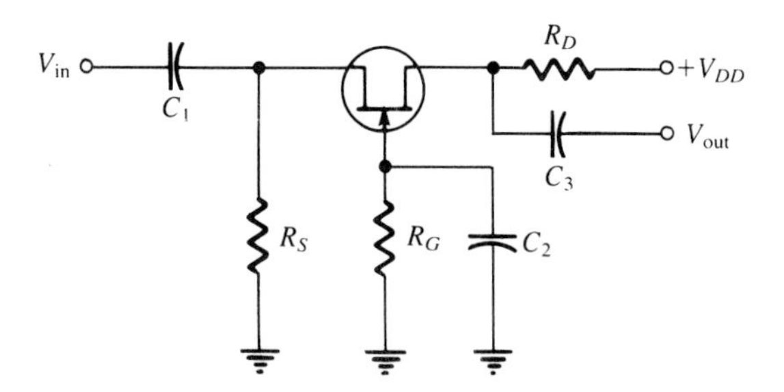

FIGURE 20–18
JFET Common-gate amplifier.

Voltage Gain of the CG Amplifier

The voltage gain from source to drain is developed as follows:

$$A_v = \frac{V_{out}}{V_{in}} = \frac{V_d}{V_{gs}} = \frac{I_d R_D}{V_{gs}} = \frac{g_m V_{gs} R_D}{V_{gs}}$$

$$A_v = g_m R_D \tag{20–21}$$

Notice that the gain expression is the same as for the CS JFET amplifier.

Input Impedance of the CG Amplifier

As you have seen, both the CS and the CD configuration have extremely high input impedances because the gate is the input terminal. In contrast, the common-gate configuration has a low input impedance, as shown in the following steps.

First, the input current (source current) is equal to the drain current:

$$I_{in} = I_d = g_m V_{gs}$$

The input voltage equals V_{gs}:

$$V_{in} = V_{gs}$$

The input impedance at the source terminal therefore is

$$R_{in(source)} = \frac{V_{in}}{I_{in}} = \frac{V_{gs}}{g_m V_{gs}}$$

$$R_{in(source)} = \frac{1}{g_m} \tag{20–22}$$

EXAMPLE 20–8

(a) Determine the voltage gain of the amplifier in Figure 20–19.
(b) Also determine the input impedance.

Solution

(a) This CG amplifier has a load resistor effectively in parallel with R_D, so the effective drain resistance is $R_D \parallel R_L$ and the gain is

$$A_v = g_m(R_D \parallel R_L) = (2500\ \mu S)(10\ k\Omega \parallel 10\ k\Omega) = 12.5$$

(b) The input impedance at the source terminal is

$$R_{in(source)} = \frac{1}{g_m} = \frac{1}{2500\ \mu S} = 400\ \Omega$$

The signal source actually sees R_S in parallel with $R_{in(source)}$, so the total input impedance is

$$R_{in} = 400\ \Omega \parallel 5\ k\Omega = 370\ \Omega$$

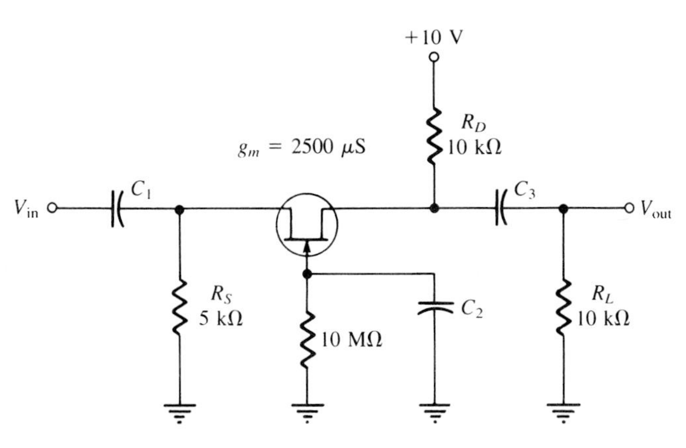

FIGURE 20–19

Summary

A summary of the gain and input impedance characteristics for the three FET amplifier configurations is given in Table 20–2.

TABLE 20–2
Comparison of FET amplifier configurations.

	CS	CD	CG
Voltage gain, A_v	$g_m R_d$	$\dfrac{g_m R_S}{1 + g_m R_S}$	$g_m R_d$
Input impedance, R_{in}	$\left(\dfrac{V_{GS}}{I_{GSS}}\right) \parallel R_G$	$\left(\dfrac{V_{GS}}{I_{GSS}}\right) \parallel R_G$	$\left(\dfrac{1}{g_m}\right) \parallel R_G$

SECTION REVIEW 20–4

1. What factors determine the voltage gain of a CS FET amplifier?
2. A certain CS amplifier has an $R_D = 1\ k\Omega$. When a load resistance of 1 kΩ is capacitively coupled to the drain, how much does the gain change?
3. What is a major difference between a CG amplifier and the other two configurations?

MULTISTAGE AMPLIFIERS

20–5

Several amplifiers can be connected in a *cascaded* arrangement with the output of one amplifier driving the input of the next, as shown in Figure 20–20. Each amplifier in the cascaded arrangement is known as a *stage*. The purpose of a multistage arrangement is to increase the overall gain.

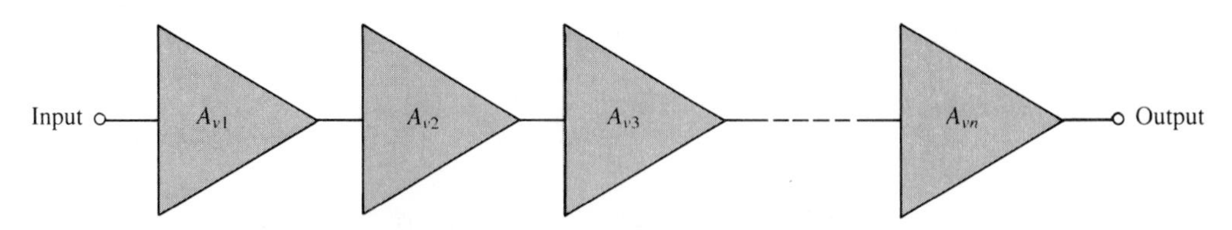

FIGURE 20–20
Cascaded amplifiers.

Multistage Gain

The overall gain A'_v of cascaded amplifiers is the product of the individual gains:

$$A'_v = A_{v1}A_{v2}A_{v3} \cdot \cdot \cdot A_{vn} \tag{20–23}$$

where n is the number of stages.

Decibel Voltage Gain

Amplifier voltage gain is often expressed in decibels (dB) as follows:

$$A_v \text{ (dB)} = 20 \log A_v \tag{20–24}$$

This formula is particularly useful in multistage systems because the overall dB voltage gain is the *sum* of the individual dB gains:

$$A'_v \text{ (dB)} = A_{v1} \text{ (dB)} + A_{v2} \text{ (dB)} + \cdot \cdot \cdot + A_{vn} \text{ (dB)} \tag{20–25}$$

EXAMPLE 20–9 A given cascaded amplifier arrangement has the following voltage gains: $A_{v1} = 10$, $A_{v2} = 15$, and $A_{v3} = 20$. What is the overall gain? Also express each gain in dB and determine the total dB voltage gain.

Solution

$$A'_v = A_{v1}A_{v2}A_{v3} = (10)(15)(20) = 3000$$
$$A_{v1} \text{ (dB)} = 20 \log 10 = 20 \text{ dB}$$
$$A_{v2} \text{ (dB)} = 20 \log 15 = 23.52 \text{ dB}$$
$$A_{v3} \text{ (dB)} = 20 \log 20 = 26.02 \text{ dB}$$
$$A'_v \text{ (dB)} = 20 \text{ dB} + 23.52 \text{ dB} + 26.02 \text{ dB} = 69.54 \text{ dB}$$

Multistage Analysis

The two-stage amplifier in Figure 20–21 illustrates multistage analysis. Notice that both stages are identical CE amplifiers with the output of the first stage *capacitively coupled* to the input of the second stage. Capacitive coupling prevents the dc bias of one stage from affecting that of the other. Also notice that the transistors are designated Q_1 and Q_2.

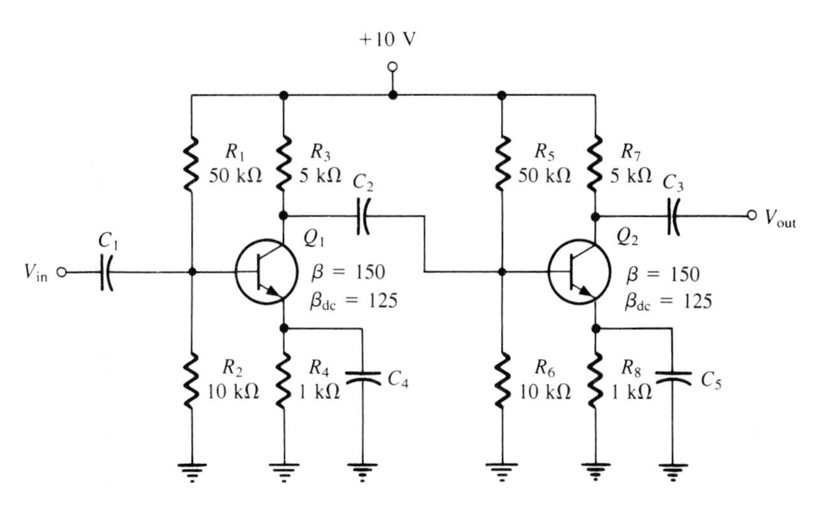

FIGURE 20–21
Two-stage amplifier.

Loading Effects

In determining the gain of the first stage, we must consider the loading effect of the second stage. Because the coupling capacitor C_2 appears as a short to the signal frequency, the total input impedance of the second stage presents an ac load to the first stage.

Looking from the collector of Q_1, the two biasing resistors, R_5 and R_6, appear in parallel with the input impedance at the base of Q_2. In other words, the signal at the collector of Q_1 "sees" R_3 and R_5, R_6, and $R_{in(base)}$ of the second stage all in parallel to ac ground. Thus, the *effective ac collector resistance* of Q_1 is the total of all these in parallel, as Figure 20–22 illustrates.

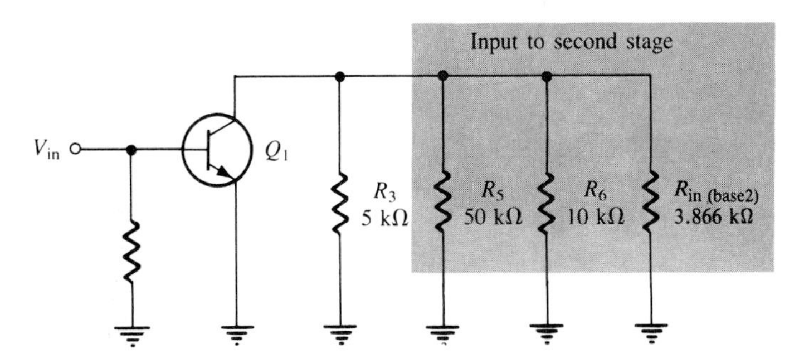

FIGURE 20–22
ac equivalent of first stage in Figure 20–21, showing loading from second stage.

The voltage gain of the first stage is reduced by the loading of the second stage, because the effective ac collector resistance of the first stage is less than the actual value of its collector resistor, R_3. Remember that $A_v = R_C/r_e$ for an unloaded amplifier.

Voltage Gain of the First Stage

The ac collector resistance of the first stage is

$$R_c = R_3 \parallel R_5 \parallel R_6 \parallel R_{in(base2)}$$

Keep in mind that lower-case subscripts denote ac quantities such as for R_c.

You can verify that $I_E = 0.97$ mA, $r_e = 25.77\ \Omega$, and $R_{in(base2)} = 3.866$ kΩ. The effective ac collector resistance of the first stage is as follows:

$$R_c = 5\ \text{k}\Omega \parallel 50\ \text{k}\Omega \parallel 10\ \text{k}\Omega \parallel 3.866\ \text{k}\Omega = 1.73\ \text{k}\Omega$$

Therefore, the base-to-collector voltage gain of the first stage is

$$A_v = \frac{R_c}{r_e} = \frac{1.73\ \text{k}\Omega}{25.77\ \Omega} = 67$$

Voltage Gain of the Second Stage

The second stage has no load resistor, so the ac collector resistance is R_7, and the gain is

$$A_v = \frac{R_7}{r_e} = \frac{5\ \text{k}\Omega}{25.77\ \Omega} = 194$$

Compare this to the gain of the first stage, and notice how much the loading reduced the gain of the first stage.

Overall Voltage Gain

The overall amplifier gain is

$$A_v' = A_{v1}A_{v2} = (67)(194) \cong 13{,}000$$

If an input signal of, say, 100 μV, is applied to the first stage and if the attenuation of the input base circuit is neglected, an output from the second stage of (100 μV)(13,000) = 1.3 V will result. The overall gain can be expressed in dB as follows:

$$A_v' \text{ (dB)} = 20 \log(13{,}000) = 82.28 \text{ dB}$$

SECTION REVIEW 20–5

1. What does the term *stage* mean?
2. How is the overall gain of a multistage amplifier determined?
3. Express a voltage gain of 500 in dB.

CLASS A OPERATION

20–6

When an amplifier, whether it is one of the three configurations of bipolar or an FET type, is biased such that it always operates in the *linear* region where the output signal is an amplified replica of the input signal, it is a *class A amplifier,* as illustrated in Figure 20–23. The discussion and formulas in the previous sections apply to class A operation.

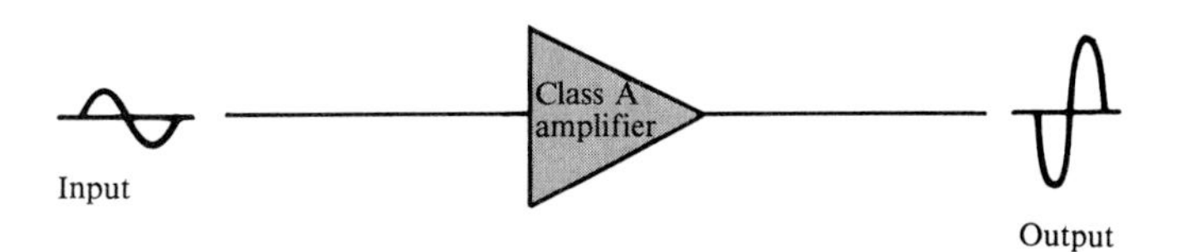

FIGURE 20–23
Class A operation

When the output signal takes up only a small percentage of the total load line excursion, the amplifier is a *small-signal* amplifier. When the output signal is larger and approaches the limits of the load line, the amplifier is a *large-signal* type. Amplifiers are typically operated as large-signal devices when power amplification is the major objective.

Why the Q Point Is Centered for Maximum Output Signal

When the dc operating point (Q point) is at the center of the load line, a maximum class A signal can be obtained. You can see this concept by examining the graph of the load line for a given amplifier in Figure 20–24(a). This graph shows the load line with the Q point at its center. The collector current can vary from its Q-point value, I_{CQ}, up to its saturation value, $I_{C(\text{sat})}$, and down to its cutoff value of zero. Likewise, the collector-to-emitter voltage can swing from its Q-point value, V_{CEQ}, up to its cutoff value, $V_{CE(\text{cutoff})}$, and down to its saturation value of near zero. This operation is indicated in Figure 20–24(b).

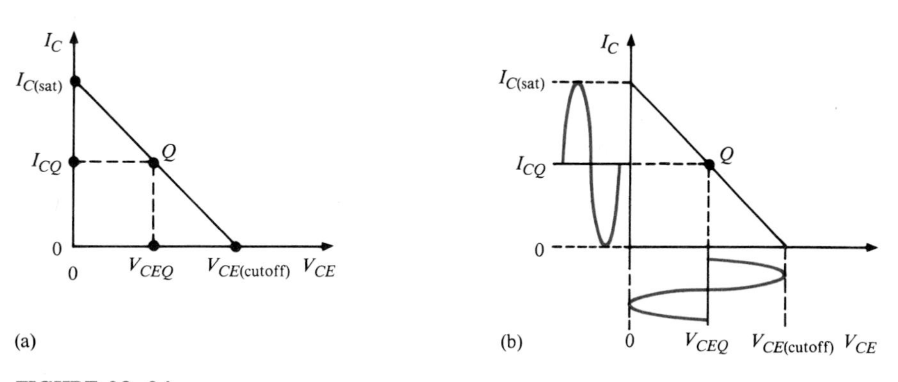

FIGURE 20–24
Maximum class A output occurs when the Q point is centered on the load line.

The *peak* value of the collector current is I_{CQ}, and the *peak* value of the collector-to-emitter voltage is V_{CEQ} in this case. This signal is the maximum that we can obtain from the class A amplifier. Actually, we cannot quite reach saturation or cutoff, so the practical maximum is slightly less.

How a Noncentered Q Point Limits Output Swing

If the Q point is not centered, the output signal is limited. Figure 20–25 shows a load line with the Q point moved away from center toward cutoff. The output variation is limited by cutoff in this case. The collector current can only swing down to near zero and an *equal amount* above I_{CQ}. The collector-to-emitter voltage can only swing up to its cutoff value and an *equal amount* below V_{CEQ}. This situation is illustrated in Figure 20–25(a). If the amplifier is driven any further than this, it will "clip" at cutoff, as shown in Figure 20–25(b).

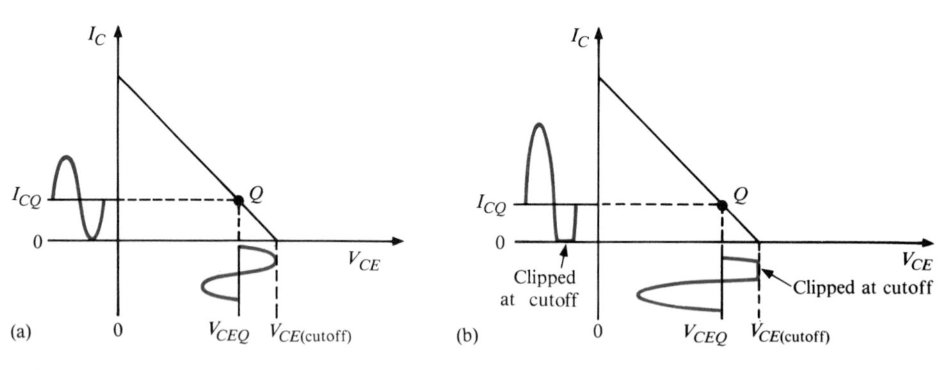

FIGURE 20–25
Q point closer to cutoff.

Figure 20–26 shows a load line with the Q point moved away from center toward saturation. In this case, the output variation is limited by saturation. The collector current can only swing up to near saturation and an *equal amount* below I_{CQ}. The collector-to-emitter voltage can only swing down to its saturation value and an *equal amount* above V_{CEQ}. This situation is illustrated in Figure 20–26(a). If the amplifier is driven any further, it will "clip" at saturation, as shown in Figure 20–26(b).

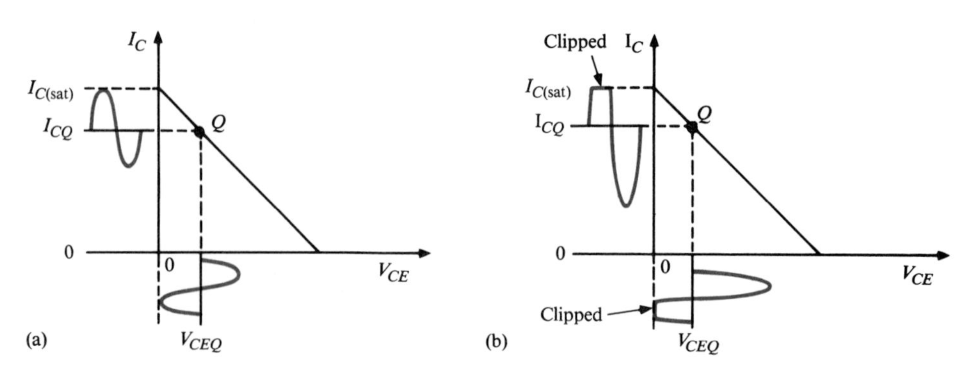

FIGURE 20–26
Q point closer to saturation.

Power Gain

The main purpose of a large-signal amplifier is to achieve power gain. If we assume that the large-signal current gain A_i is approximately equal to β_{dc}, then the power gain for a common-emitter amplifier is

$$A_p = A_i A_v = \beta_{dc} A_v$$

$$A_p = \beta_{dc} \frac{R_c}{r_e} \quad (20\text{–}26)$$

dc Quiescent Power

The power dissipation of a transistor with no signal input is the product of its Q-point current and voltage:

$$P_{DQ} = I_{CQ} V_{CEQ} \quad (20\text{–}27)$$

The quiescent power is the *maximum* power that the class A transistor must handle; therefore, its *power rating* should exceed this value.

Output Power

In general, for any Q-point location, the output power of a CE amplifier is the product of the *rms* collector current and the *rms* collector-to-emitter voltage:

$$P_{out} = V_{ce} I_c \quad (20\text{–}28)$$

Q Point Centered

When the Q point is centered, the maximum collector current swing is I_{CQ}, and the maximum collector-to-emitter voltage swing is V_{CEQ}, as was shown in Figure 20–24(b). The output power therefore is

$$P_{out} = (0.707 V_{CEQ})(0.707 I_{CQ})$$

$$P_{out} = 0.5 V_{CEQ} I_{CQ} \quad (20\text{–}29)$$

This is the *maximum* ac output power from a class A amplifier under signal conditions. Notice that it is one-half the quiescent power dissipation.

Efficiency

Efficiency of an amplifier is the ratio of ac output power to dc input power. The dc input power is the dc supply voltage times the current drawn from the supply.

$$P_{dc} = V_{CC} I_{CC}$$

The average supply current I_{CC} equals I_{CQ}, and the supply voltage V_{CC} is twice V_{CEQ} when the Q point is centered. Therefore, the maximum efficiency is

$$\text{eff}_{max} = \frac{P_{out}}{P_{dc}} = \frac{0.5 V_{CEQ} I_{CQ}}{V_{CC} I_{CC}} = \frac{0.5 V_{CEQ} I_{CQ}}{2 V_{CEQ} I_{CQ}} = \frac{0.5}{2}$$

$$\text{eff}_{max} = 0.25 \quad (20\text{–}30)$$

Thus 25% is the highest possible efficiency available from a class A amplifier and is approached only when the Q point is at the center of the load line.

EXAMPLE 20–10 Determine the following values for a class A amplifier operated with a centered Q point with $I_{CQ} = 50$ mA and $V_{CEQ} = 7.5$ V:

(a) minimum transistor power rating.
(b) ac output power.
(c) efficiency.

Solution
The quiescent dc values are

$$I_{CQ} = 50 \text{ mA}$$
$$V_{CEQ} = 7.5 \text{ V}$$

(a) The maximum power that the transistor must be able to handle is the minimum rating that you would use. Thus,

$$P_{DQ} = I_{CQ}V_{CEQ} = (50 \text{ mA})(7.5 \text{ V}) = 0.375 \text{ W}$$

(b) $P_{out} = 0.5V_{CEQ}I_{CQ} = 0.5(50 \text{ mA})(7.5 \text{ V}) = 0.1875$ W
(c) Since the Q point is centered, the efficiency is at its maximum possible value of 25%.

SECTION REVIEW 20–6

1. What is the optimum Q-point location for class A amplifiers?
2. What is the maximum efficiency of a class A amplifier?
3. A certain amplifier has a centered Q point of $I_{CQ} = 10$ mA and $V_{CEQ} = 7$ V. What is the *maximum* ac output power?

CLASS B PUSH-PULL OPERATION

20–7

When an amplifier is biased such that it operates in the linear region for 180° of the input cycle and is in cutoff for 180°, it is a class B amplifier. This amplifier is illustrated in Figure 20–27, where the output waveform is shown relative to the input.

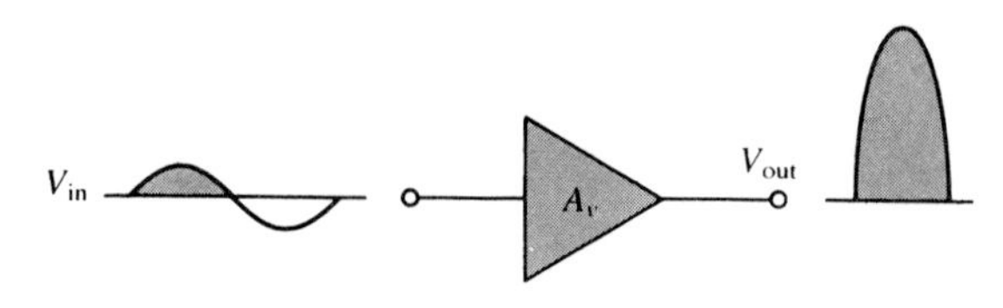

FIGURE 20–27
Class B amplifier (noninverting).

Why the Q Point Is at Cutoff

The class B amplifier is biased at cutoff so that $I_{CQ} = 0$ and $V_{CEQ} = V_{CE(\text{cutoff})}$. It is brought out of cutoff and operates in its linear region when the input signal drives it into conduction. This is illustrated in Figure 20–28 with an emitter-follower circuit.

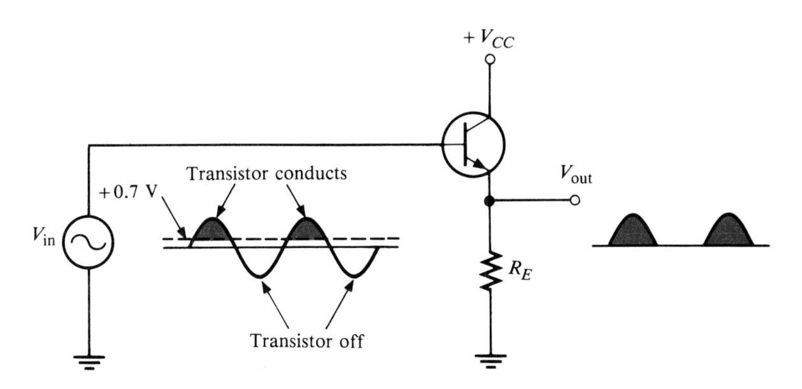

FIGURE 20–28
Common-collector class B amplifier.

Push-Pull Operation

Figure 20–29 shows one type of push-pull class B amplifier using two emitter-followers. This is a *complementary* amplifier, because one emitter-follower uses an NPN transistor and the other a PNP, which conduct on *opposite* alternations of the input cycle.

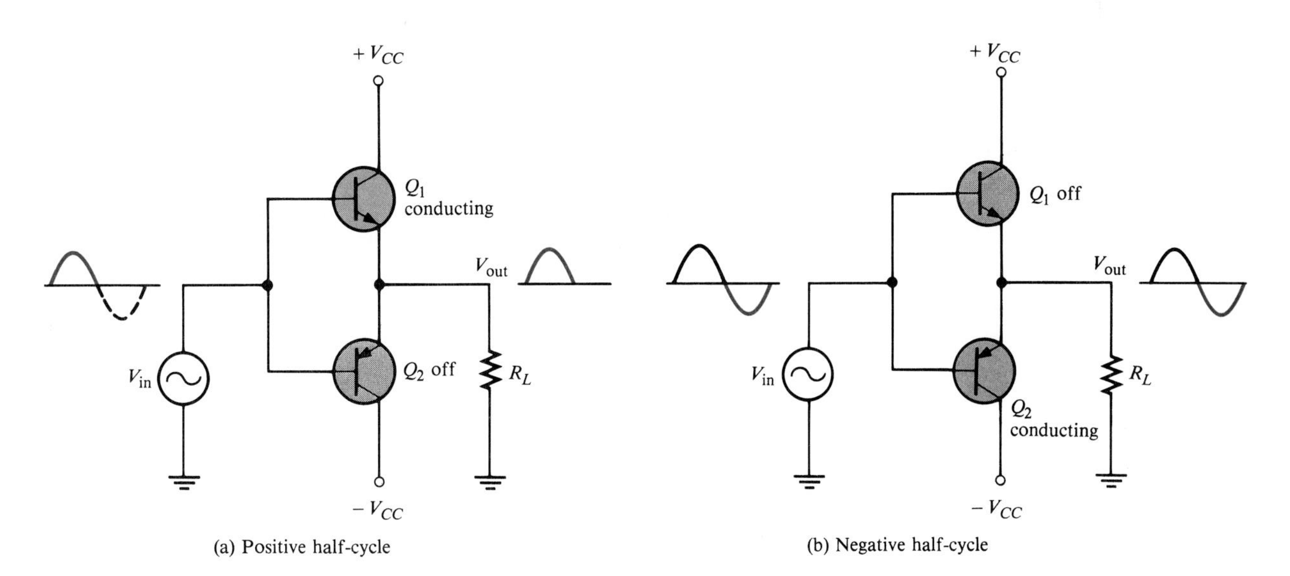

FIGURE 20–29
Class B push-pull operation.

Notice that there is no dc base bias voltage ($V_B = 0$). Thus only the signal voltage drives the transistors into conduction. Q_1 conducts during the positive half of the input cycle, and Q_2 conducts during the negative half.

Crossover Distortion

When the dc base voltage is zero, the input signal voltage must exceed V_{BE} before a transistor conducts. As a result, there is a time interval between the positive and negative alternations of the input when neither transistor is conducting, as shown in Figure 20–30. The resulting distortion in the output waveform is quite common and is called *crossover* distortion.

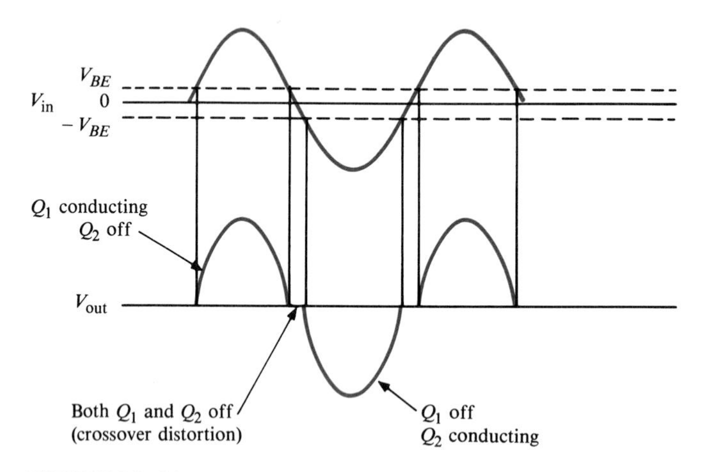

FIGURE 20–30
Illustration of crossover distortion in a class B push-pull amplifier.

Biasing the Push-Pull Amplifier

To eliminate crossover distortion, both transistors in the push-pull arrangement must be biased slightly above cutoff when there is no signal. This can be done with a voltage divider and diode arrangement, as shown in Figure 20–31. When the diode characteristics of D_1 and D_2 are closely matched to the characteristics of the transistor base-emitter junctions, a stable bias is maintained.

Since R_1 and R_2 are of equal value, the voltage with respect to ground at point A between the two diodes is $V_{CC}/2$. Assuming that both diodes and both transistors are identical, the drop across D_1 equals the V_{BE} of Q_1, and the drop across D_2 equals the V_{BE} of Q_2. As a result, the voltage at the emitters is also $V_{CC}/2$, and therefore, $V_{CEQ_1} = V_{CEQ_2} = V_{CC}/2$, as indicated. Because both transistors are biased near cutoff, $I_{CQ} \cong 0$.

ac Operation

Under maximum conditions, transistors Q_1 and Q_2 are alternately driven from near cutoff to near saturation. During the positive alternation of the input signal, the Q_1 emitter is driven from its Q-point value of $V_{CC}/2$ to near V_{CC}, producing a positive peak voltage approximately equal to V_{CEQ}. At the same time, the Q_1 current swings from its Q-point value near zero to near-saturation value, as shown in Figure 20–32(a).

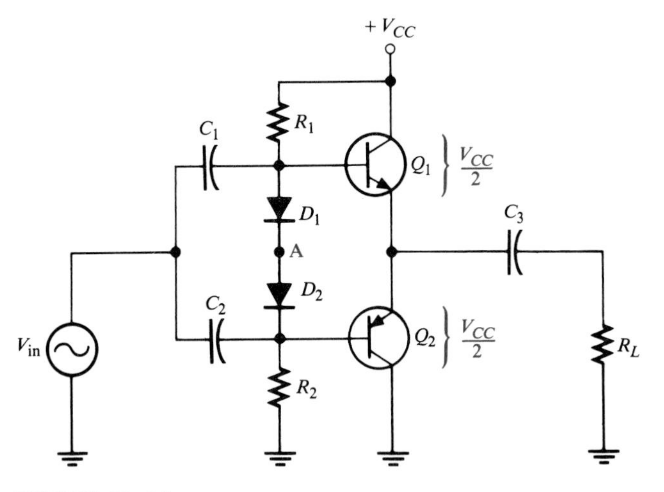

FIGURE 20–31
Biasing the push-pull amplifier to eliminate crossover distortion.

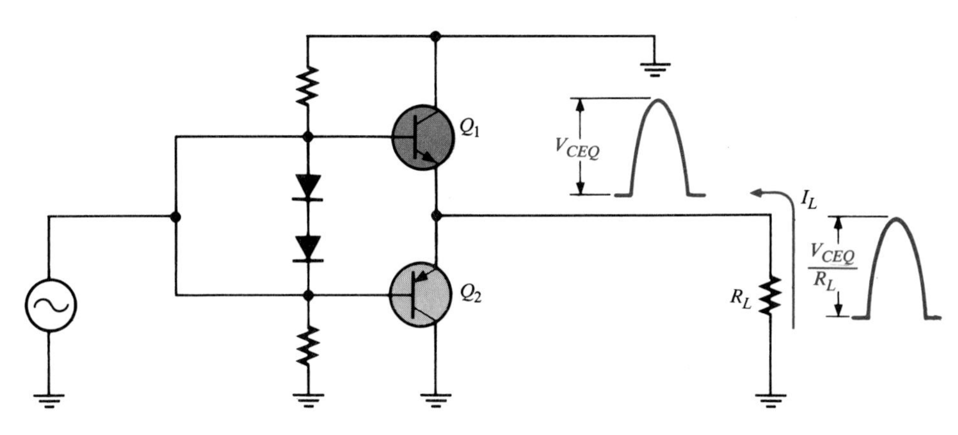

(a) Q_1 conducting with maximum signal output

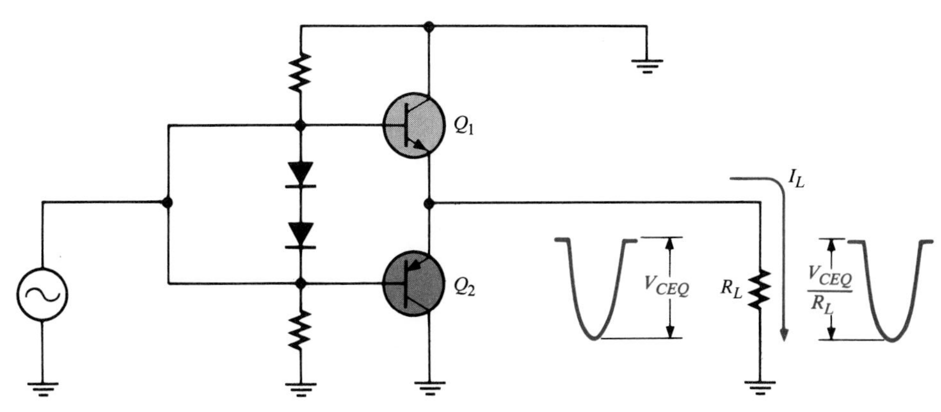

(b) Q_2 conducting with maximum signal output

FIGURE 20–32
ac push-pull operation. Capacitors are assumed to be shorts at the signal frequency, and the dc source is at ac ground.

During the negative alternation of the input signal, the Q_2 emitter is driven from its Q-point value of $V_{CC}/2$ to near zero, producing a negative peak voltage approximately equal to V_{CEQ}. Also, the Q_2 current swings from near zero to near-saturation value, as shown in Figure 20–32(b).

Because the peak voltage across each transistor is V_{CEQ}, the ac saturation current is

$$I_{c(\text{sat})} = \frac{V_{CEQ}}{R_L} \tag{20–31}$$

Since $I_e \cong I_c$ and the output current is the emitter current, the peak output current is also V_{CEQ}/R_L.

EXAMPLE 20–11 Determine the maximum peak values for the output voltage and current in Figure 20–33.

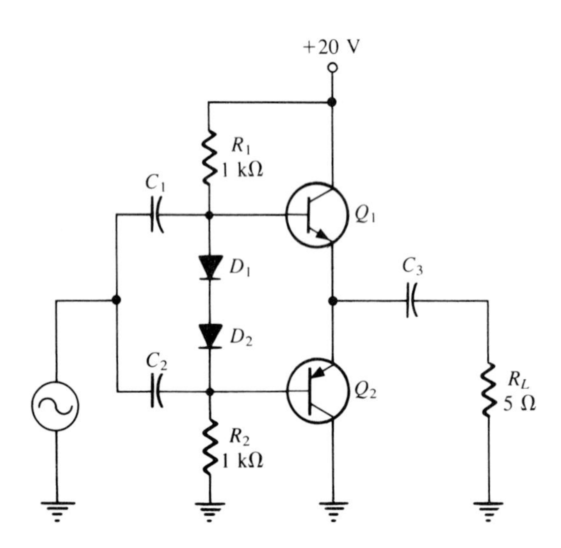

FIGURE 20–33

Solution

The maximum peak output voltage is

$$V_{p(\text{out})} \cong V_{CEQ} = \frac{V_{CC}}{2} = \frac{20\text{ V}}{2} = 10\text{ V}$$

The maximum peak output current is

$$I_{p(\text{out})} \cong I_{c(\text{sat})} = \frac{V_{CEQ}}{R} = \frac{10\text{ V}}{5\ \Omega} = 2\text{ A}$$

Maximum Output Power

It has been shown that the maximum peak output current is approximately $I_{c(sat)}$, and the maximum peak output voltage is approximately V_{CEQ}. The maximum average output power therefore is

$$P_{out} = V_{rms(out)} I_{rms(out)}$$

Since

$$V_{rms(out)} = 0.707 V_{p(out)} = 0.707 V_{CEQ}$$

and

$$I_{rms(out)} = 0.707 I_{p(out)} = 0.707 I_{c(sat)}$$

then

$$P_{out} = 0.5 V_{CEQ} I_{c(sat)}$$

Substituting $V_{CC}/2$ for V_{CEQ}, we get

$$P_{out} = 0.25 V_{CC} I_{c(sat)} \tag{20–32}$$

Input Power

The input power comes from the V_{CC} supply and is

$$P_{dc} = V_{CC} I_{CC}$$

Since each transistor draws current for a half-cycle, the current is a *half-wave* signal with an average value of

$$I_{CC} = \frac{I_{c(sat)}}{\pi}$$

Thus,

$$P_{dc} = \frac{V_{CC} I_{c(sat)}}{\pi} \tag{20–33}$$

Efficiency

The great advantage of push-pull class B amplifiers over class A is a much higher efficiency. This advantage usually overrides the difficulty of biasing the class B push-pull amplifier to eliminate crossover distortion.

The efficiency is again defined as the ratio of ac output power to dc input power:

$$\text{eff} = \frac{P_{out}}{P_{dc}}$$

The maximum efficiency for a class B amplifier is designated eff_{max} and is developed as follows, starting with Equation (20–32):

$$P_{out} = 0.25V_{CC}I_{c(sat)}$$

$$\text{eff}_{max} = \frac{P_{out}}{P_{dc}} = \frac{0.25V_{CC}I_{c(sat)}}{V_{CC}I_{c(sat)}/\pi} = 0.25\pi$$

$$\text{eff}_{max} = 0.785 \quad \textbf{(20–34)}$$

Therefore, the maximum efficiency is 78.5%. Recall that the maximum efficiency for class A is 0.25 (25%).

EXAMPLE 20–12 Find the maximum ac output power and the dc input power of the amplifier in Figure 20–33 of Example 20–11.

Solution

In Example 20–11, $I_{c(sat)}$ was found to be 2 A. Thus,

$$P_{out} = 0.25V_{CC}I_{c(sat)} = 0.25(20 \text{ V})(2 \text{ A}) = 10 \text{ W}$$

$$P_{dc} = \frac{V_{CC}I_{c(sat)}}{\pi} = \frac{(20 \text{ V})(2 \text{ A})}{\pi} = 12.73 \text{ W}$$

SECTION REVIEW 20–7

1. Where is the Q point for a class B amplifier?
2. What causes crossover distortion?
3. What is the maximum efficiency of a push-pull class B amplifier?
4. Explain the purpose of the push-pull configuration for class B.

CLASS C OPERATION

20–8

Class C amplifiers are biased so that conduction occurs for much less than 180°, as Figure 20–34 illustrates. Class C amplifiers are more efficient than either class A or push-pull class B. Thus, more output power can be obtained from class C operation. Because the output waveform is severely distorted, class C amplifiers are normally limited to applications as tuned amplifiers at radio frequencies (rf).

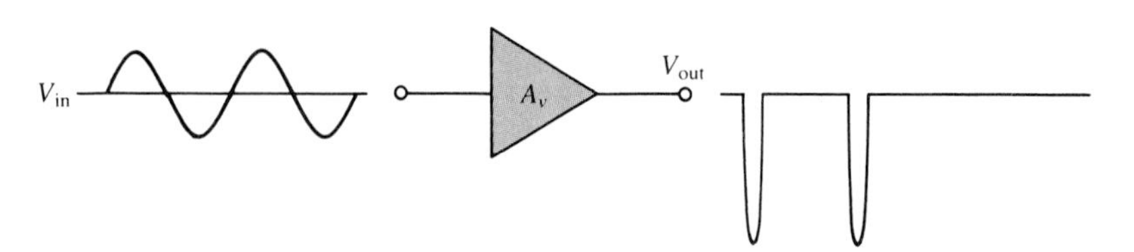

FIGURE 20–34
Class C amplifier (inverting).

Basic Operation

A basic common-emitter class C amplifier with a resistive load is shown in Figure 20–35(a). It is biased *below cutoff* with the $-V_{BB}$ supply. The ac source voltage has a peak value that is slightly greater than $V_{BB} + V_{BE}$ so that the base voltage exceeds the barrier potential of the base-emitter junction for a short time near the positive peak of each cycle, as illustrated in Figure 20–35(b). During this short interval, the transistor is turned on.

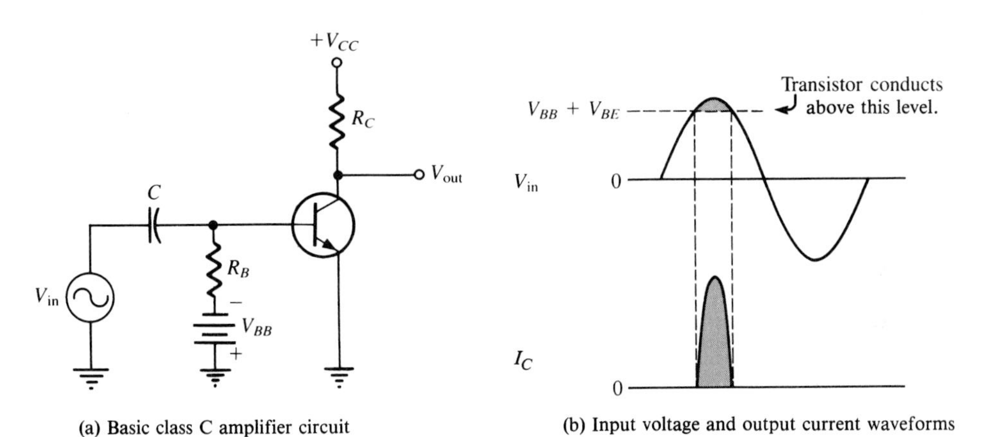

FIGURE 20–35
Class C operation.

When the entire ac load line is used, the maximum collector current is approximately $I_{C(\text{sat})}$, and the minimum collector voltage is approximately $V_{CE(\text{sat})}$.

Power Dissipation

The power dissipation of the transistor in a class C amplifier is low because it is on for only a small percentage of the input cycle. Figure 20–36 shows the collector current pulses. The time between the pulses is the *period (T)* of the ac input voltage.

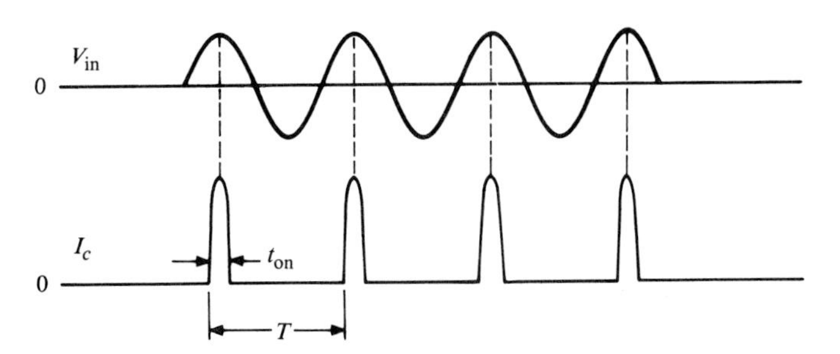

FIGURE 20–36
Collector current pulses in a class C amplifier.

The transistor is on for a short time, t_{on}, and *off* for the rest of the input cycle. Since the power dissipation averaged over the entire cycle depends on the ratio of t_{on} to T and on the power dissipation during t_{on}, it is typically very low.

Tuned Operation

Because the collector voltage (output) is not a replica of the input, the resistively loaded class C amplifier is of no value in linear applications. Therefore, it is necessary to use a class C amplifier with a parallel *resonant circuit* (tank), as shown in Figure 20–37(a). The resonant frequency of the tank circuit is determined by the formula $f_r = 1/(2\pi\sqrt{LC})$.

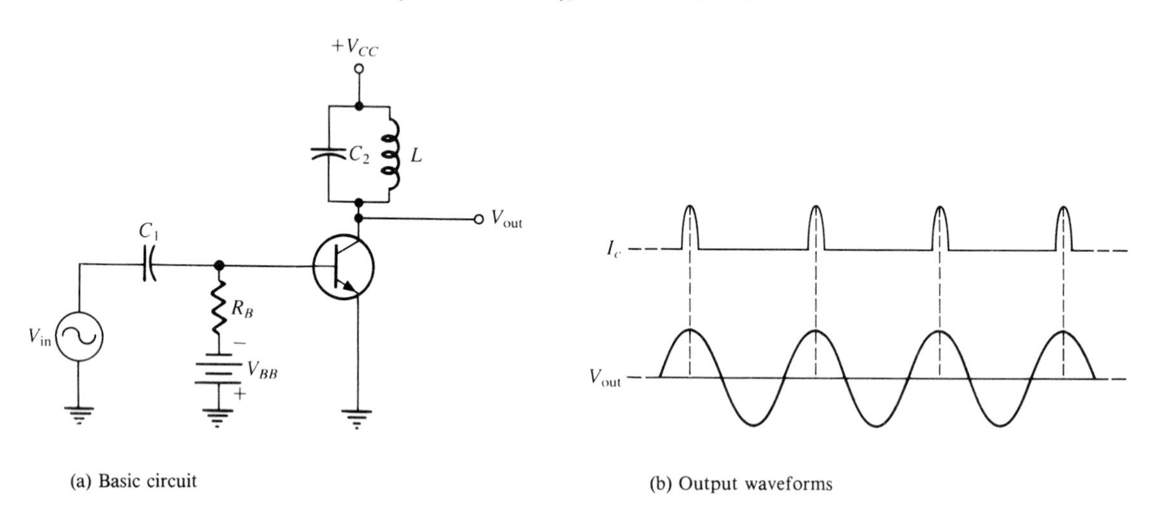

(a) Basic circuit

(b) Output waveforms

FIGURE 20–37
Tuned class C amplifier.

The short pulse of collector current on each cycle of the input initiates and sustains the oscillation of the tank circuit so that an output sine wave voltage is produced, as illustrated in Figure 20–37(b).

The amplitude of each successive cycle of the oscillation would be less than that of the previous cycle because of energy loss in the resistance of the tank circuit, as shown in Figure 20–38(a), and the oscillation would eventually die out. However, the regular recurrences of the collector current pulse re-energizes the resonant circuit and sustains the oscillations at a constant amplitude. When the tank circuit is tuned to the frequency of the input signal, re-energizing occurs on each cycle of the tank voltage, as shown in Figure 20–38(b).

Maximum Output Power

Since the voltage developed across the tank circuit has a peak-to-peak value of approximately $2V_{CC}$, the maximum output power can be expressed as

$$P_{out} = \frac{V_{rms}^2}{R_c} = \frac{(0.707V_{CC})^2}{R_c}$$

$$P_{out} = \frac{0.5V_{CC}^2}{R_c} \qquad (20\text{–}35)$$

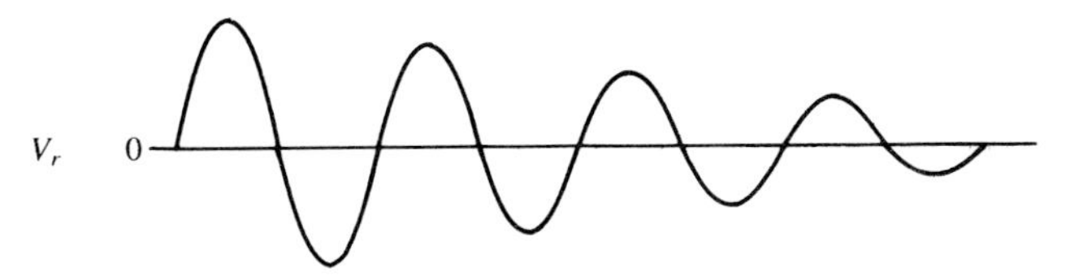

(a) Oscillation dies out due to energy loss.

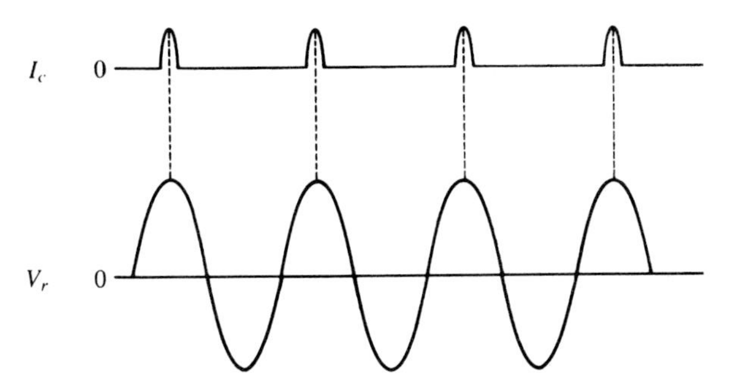

(b) Oscillation is sustained by short pulses of collector current.

FIGURE 20–38
Tank circuit oscillations.

where R_c is the equivalent parallel resistance of the collector tank circuit and represents the parallel combination of the coil resistance and the load resistance. It usually has a low value.

The total power that must be supplied to the amplifier is

$$P_T = P_{out} + P_{D(avg)}$$

Therefore, the efficiency is

$$\text{eff} = \frac{P_{out}}{P_{out} + P_{D(avg)}} \tag{20–36}$$

When $P_{out} >> P_{D(avg)}$, the class C efficiency closely approaches 100%.

EXAMPLE 20–13 A certain class C amplifier has a $P_{D(avg)}$ of 2 mW, a V_{CC} equal to 24 V, and an R_c of 100 Ω. Determine the efficiency.

Solution

$$P_{out} = \frac{0.5V_{CC}^2}{R_c} = \frac{0.5(24\text{ V})^2}{100\ \Omega} = 2.88\text{ W}$$

Therefore,

$$\text{eff} = \frac{P_{out}}{P_{out} + P_{D(avg)}} = \frac{2.88\text{ W}}{2.88\text{ W} + 2\text{ mW}} = 0.9993$$

or

$$\%\text{eff} = 99.93\%$$

SECTION REVIEW 20–8

1. A class C amplifier is normally biased in ______.
2. What is the purpose of the tuned circuit in a class C amplifier?
3. A certain class C amplifier has a power dissipation of 100 mW and an output power of 1 W. What is its efficiency?

TROUBLESHOOTING AN AMPLIFIER

20–9

In working with any circuit, you must first know how it is supposed to work before you can troubleshoot it for a failure. For the multistage amplifier just discussed, since we know what to expect, we can identify and trace a failure.

Let's assume that transistor Q_2 in Figure 20–39 develops an open between the collector and emitter. You, as a technician, do not know what the trouble is, but the absence of a signal at the output indicates that the amplifier has malfunctioned.

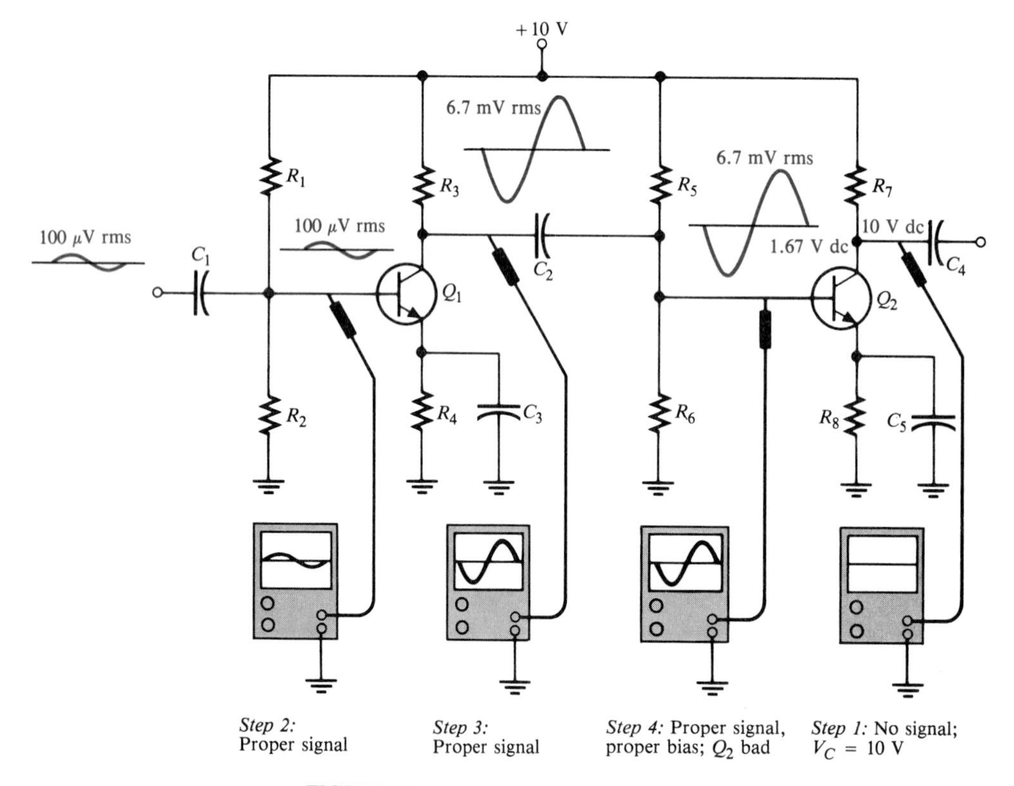

FIGURE 20–39
Using an oscilloscope to troubleshoot an amplifier.

To locate and identify the failure, *trace* the signal from the input to the point where it disappears. This procedure is illustrated in Figure 20–39. With a 100-μV rms test signal applied to the input, an inverted 6.7-mV signal should be at the collector of Q_1. The presence of this signal indicates that the first stage is operating as expected. Next, check the signal at the base of Q_2. If the coupling capacitor C_2 is good, this signal should be the same as the Q_1 collector signal.

Also, you should check the dc bias level at the base of Q_2 to verify correct bias. For this particular amplifier, it should be about 1.67 V.

So far you have found that there is a proper signal and bias level at the base of Q_2, but there is no collector signal. A check of the dc level at the Q_2 collector shows 10 V. Since the signal is not getting through Q_2 and since $V_{C2} = V_{CC}$, a collector-to-emitter open circuit should be suspected. The last step is to replace transistor Q_2 and check the amplifier's performance again. This general signal-tracing method of troubleshooting can be used to isolate most problems in an electronic circuit.

As another example of isolating a component failure in a circuit, we will use a class A amplifier with the output monitored by an oscilloscope, as shown in Figure 20–40. As shown, the amplifier has a normal sine wave output when a sinusoidal input signal is applied.

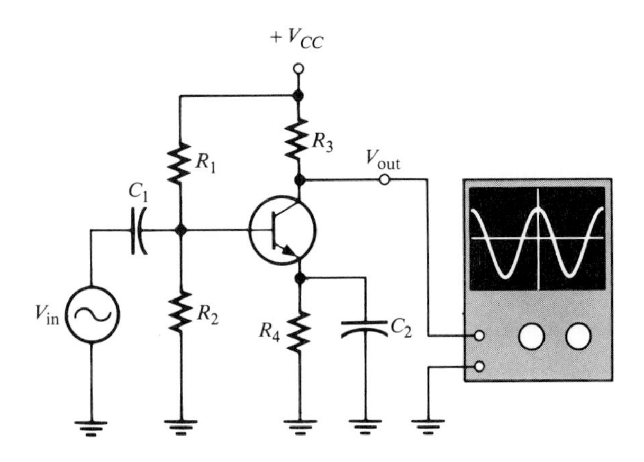

FIGURE 20–40
Class A amplifier with proper output display.

Now, several incorrect output waveforms will be considered and the most likely causes discussed. In Figure 20–41(a), the scope displays a dc level equal to the dc supply voltage, indicating that the transistor is in cutoff. The two possible causes of this condition are (1) the transistor is open from collector to emitter, or (2) R_4 is open, preventing collector and emitter current.

In Figure 20–41(b), the scope displays a dc level at the collector approximately equal to the emitter voltage. The two possible causes of this indication are (1) the transistor is shorted from collector to emitter, or (2) possibly R_2 is open, causing the transistor to be biased in saturation. In the second case, a sufficiently large input signal can bring the transistor out of saturation on its negative peaks, resulting in short pulses on the output.

In Figure 20–41(c), the scope displays an output waveform that is clipped at cutoff. Possible causes of this indication are (1) the Q point has shifted down due to a drastic out-of-tolerance change in a resistor value, or (2) R_1 is open, biasing the transistor in cutoff. In the second case, the input signal is sufficient to bring it out of cutoff for a small portion of the cycle.

In Figure 20–41(d), the scope displays an output waveform that is clipped at saturation. Again, it is possible that a resistance change has caused a drastic

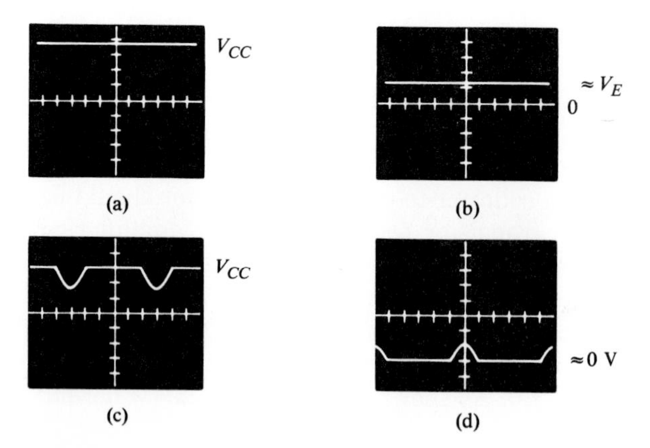

FIGURE 20–41
Oscilloscope displays of output voltage for the amplifier in Figure 20–40, illustrating several types of failures.

shift in the Q point up toward saturation, or R_2 is open, causing the transistor to be biased in saturation, and the input signal is bringing it out of saturation for a small portion of the cycle.

SECTION REVIEW 20–9

1. Assume that the base-emitter junction of Q_2 in Figure 20–39 shorts.
   (a) Will the ac signal at the base of Q_2 change? In what way?
   (b) Will the dc level at the base of Q_2 change? In what way?
2. What would you check for if you noticed clipping at both peaks of the output waveform?
3. A significant loss of gain in the amplifier of Figure 20–40 would most likely be caused by what type of failure?

20–10 OSCILLATORS

Oscillators are circuits that generate an output signal without having an externally applied input signal. They are useful as *signal sources* in many applications. The oscillator is essentially an amplifier in which a portion of the output is fed back to the input. Its operation is based on the principle of *positive feedback*.

The block diagram in Figure 20–42 shows an amplifier with gain A. The output drives a feedback circuit with gain B. The output of the feedback circuit

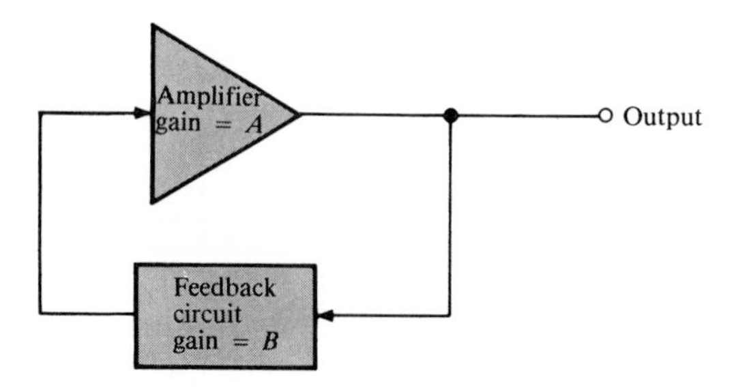

FIGURE 20–42
Block diagram of basic oscillator.

provides the input to the amplifier. This is the basic form of an oscillator. If conditions are proper, the oscillator will continuously produce an output signal by amplifying the feedback signal.

Conditions for Oscillation

There are two conditions that must be met in order for a circuit to oscillate. First, the phase shift through the amplifier and the feedback circuit must be 0°. There must be no phase shift so that the feedback signal will tend to *reinforce* itself rather than cancel. Second, the gain through the amplifier and the feedback circuit must be equal to or greater than 1 ($AB \geq 1$). This gain is called the *loop gain*. If the gain were less than 1, the output signal would decrease and die out.

The *RC* Oscillator

The basic *RC* oscillator shown in Figure 20–43 uses an *RC* network as its feedback circuit. In this case, three *RC* lag circuits each have a phase shift of 60°. The common-emitter transistor contributes a 180° phase shift. The total phase shift through the amplifier and feedback circuit therefore is 360°, which is effectively 0° (no phase shift). The attenuation of the *RC* network and the gain of the amplifier must be such that the overall gain around the feedback loop is equal to 1 at the frequency of oscillation. This circuit will produce a *continuous sine wave output.*

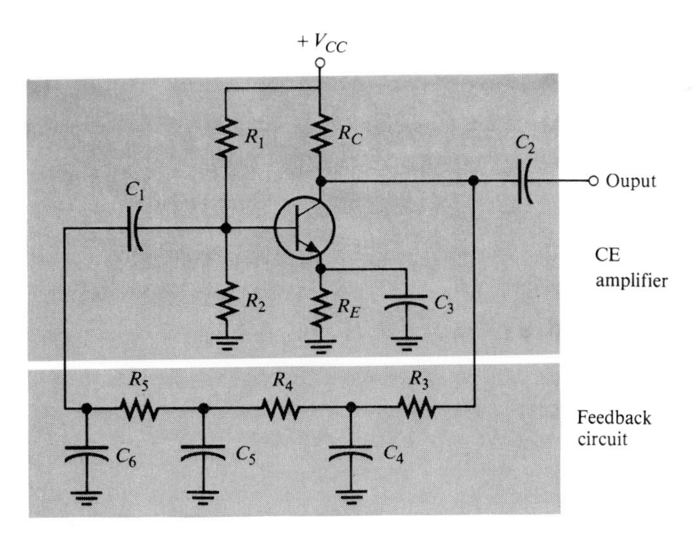

FIGURE 20–43
Basic *RC* oscillator.

The Colpitts Oscillator

One basic type of *tuned* oscillator is the Colpitts, named after its inventor. As shown in Figure 20–44, this type of oscillator uses an *LC* circuit in the feedback loop to provide the necessary phase shift and to act as a filter that passes only

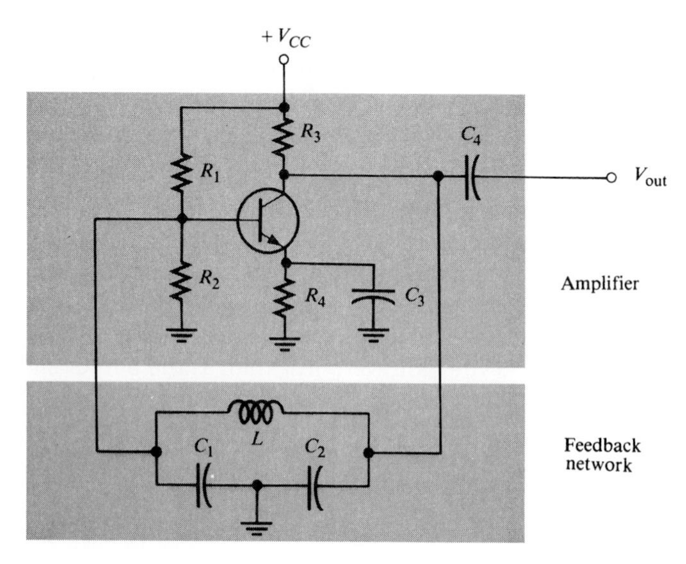

FIGURE 20–44
A basic Colpitts oscillator.

the specified frequency of oscillation. The approximate frequency of oscillation is established by the values of C_1, C_2, and L according to the following familiar formula:

$$f_r \cong \frac{1}{2\pi\sqrt{LC_T}}$$

Because the capacitors effectively appear in series around the tank circuit, the total capacitance is

$$C_T = \frac{C_1C_2}{C_1 + C_2}$$

The Hartley Oscillator

Another basic type of oscillator circuit is the Hartley, which is similar to the Colpitts except that the feedback network consists of two inductors and one capacitor, as shown in Figure 20–45.

The frequency of oscillation is

$$f_r \cong \frac{1}{2\pi\sqrt{L_TC}}$$

The total inductance is the series combination of L_1 and L_2.

The Clapp Oscillator

The Clapp oscillator is similar to the Colpitts except that there is an additional capacitor in series with the inductor, as shown in Figure 20–46. C_1 and C_2 can be selected for optimum feedback, and C_3 can be adjusted to obtain the desired

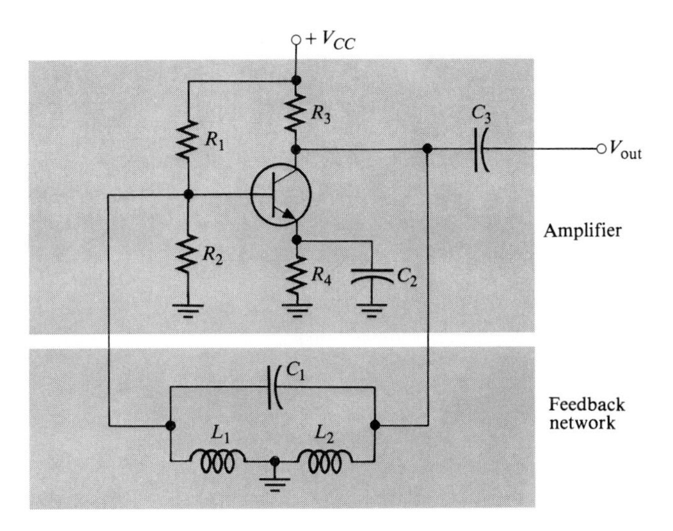

FIGURE 20–45
A basic Hartley oscillator.

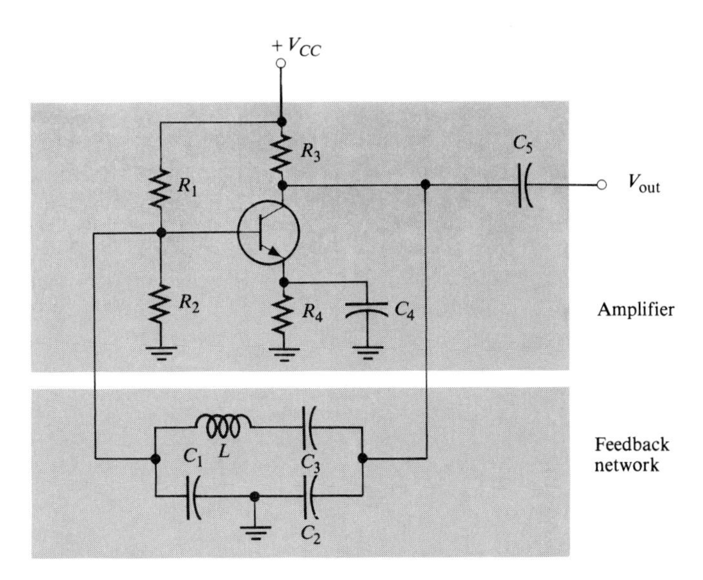

FIGURE 20–46
A basic Clapp oscillator.

frequency of oscillation. Also, a capacitor having a negative temperature coefficient can be used for C_3 to stabilize the frequency of oscillation when there are temperature changes.

The Crystal Oscillator

A crystal oscillator is essentially a tuned-circuit oscillator that uses a *quartz crystal* as the resonant tank circuit. Other types of crystals can be used, but quartz is the most prevalent. Crystal oscillators offer greater frequency stability than other types.

Quartz is a substance found in nature that exhibits a property called the *piezoelectric effect.* When a changing mechanical stress is applied across the crystal to cause it to vibrate, a voltage is developed at the frequency of the mechanical vibration. Conversely, when an ac voltage is applied across the crystal, it vibrates at the frequency of the applied voltage.

The symbol for a crystal is shown in Figure 20–47(a), the electrical equivalent is shown in Part (b), and a typical mounted crystal is shown in Part (c). In construction, a slab of quartz is mounted as shown in Part (d).

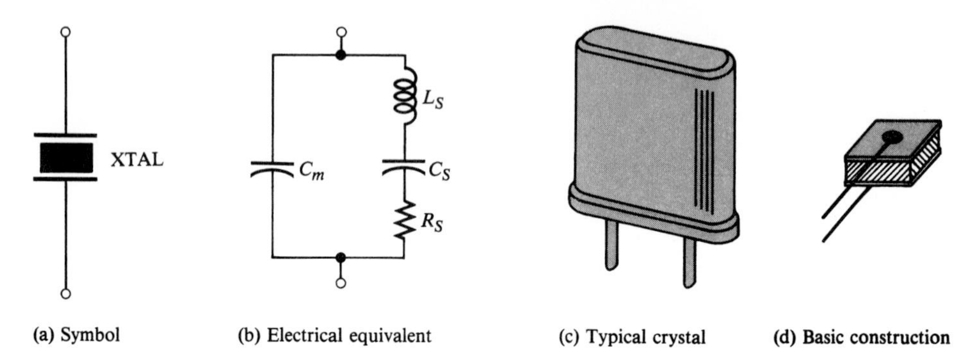

FIGURE 20–47
Quartz crystal.

Series resonance occurs in the crystal when the reactances in the series branch are equal. Parallel resonance occurs, at a higher frequency, when the reactance of L_S equals the reactance of C_m.

A crystal oscillator using the crystal as a series resonant tank circuit is shown in Figure 20–48(a). The impedance of the crystal is *minimum* at the series resonance, thus providing maximum feedback.

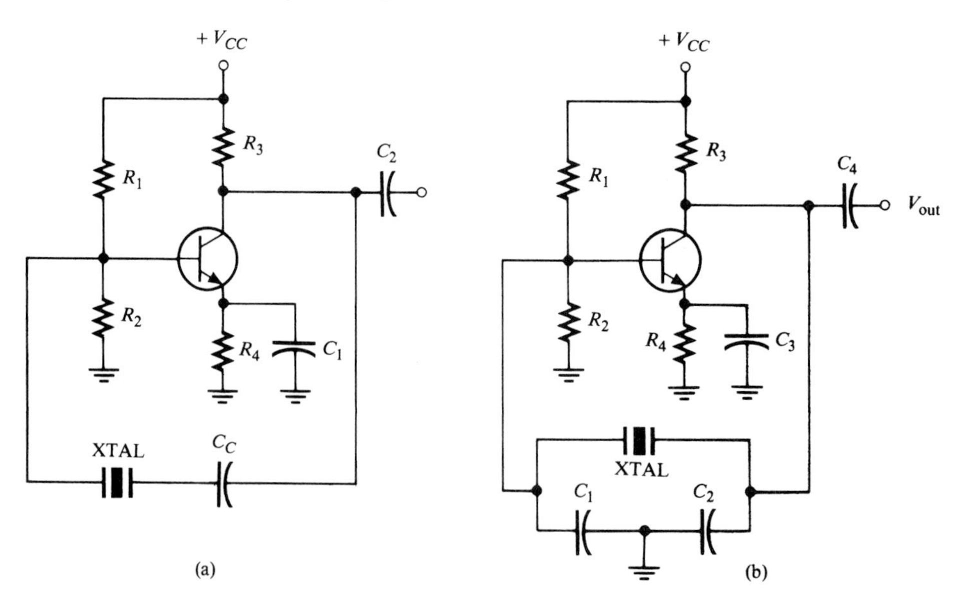

FIGURE 20–48
Basic crystal oscillators.

The capacitor C_c is a tuning capacitor used to fine-tune the frequency. A modified Colpitts configuration, shown in Part (b), uses the crystal in its parallel resonant mode. The impedance of the crystal is maximum at parallel resonance, thus developing the maximum voltage across both C_1 and C_2. The voltage across C_1 is fed back to the input.

SECTION REVIEW 20–10

1. Name four types of oscillators.
2. Describe the basic difference between the Colpitts and Hartley oscillators.

APPLICATION NOTE

One approach to testing the push-pull power amplifiers in the Application Assignment is to inject an input sine wave (be careful not to overdrive the amplifier) and check the output for a full sine wave. If one of the transistors has failed, it is probable that only approximately one-half of the sine wave will be observed. Other possibilities include a failure in the bias network.

SUMMARY

Facts

- The three bipolar transistor amplifier configurations are common-emitter (CE), common-collector (CC), and common-base (CB).
- The general characteristics of a common-emitter amplifier are high voltage gain, high current gain, very high power gain, and low input impedance.
- The general characteristics of a common-collector amplifier are low voltage gain (≤ 1), high current gain, high power gain, and high input impedance.
- The general characteristics of a common-base amplifier are high voltage gain, low current gain (≤ 1), high power gain, and very low input impedance.
- The three FET amplifier configurations are common-source (CS), common-drain (CD), and common-gate (CG).
- The overall voltage gain of a multistage amplifier is the product of the gains of all the individual stages.
- Any of the bipolar or FET configurations can be operated as class A, class B, or class C amplifiers.
- A class A amplifier conducts for the entire 360° of the input cycle.
- A class B amplifier conducts for 180° of the input cycle.
- A class C amplifier conducts for a small portion of the input cycle.

Definitions

- *Voltage gain, A_v*—the ratio of output voltage to input voltage.
- *Power gain, A_p*—the product of voltage gain and current gain.
- *Transconductance, g_m*—the ratio of drain current to gate-to-source voltage in an FET.
- *Oscillator*—a circuit consisting of an amplifier and a phase-shift network connected in a feedback loop; it produces a sustained sinusoidal output without an external input signal. Its operation is based on the principle of *positive feedback.*

Formulas

$A_v = \dfrac{R_C}{r_e + R_E}$ CE voltage gain (unbypassed) (20–1)

$A_v = \dfrac{R_C}{r_e}$ CE voltage gain (bypassed) (20–2)

$r_e \cong \dfrac{25 \text{ mV}}{I_E}$ Internal emitter impedance (20–3)

$R_{in} \cong \beta_{ac} r_e$ CE input impedance (20–4)

$R_{in(T)} = R_1 \parallel R_2 \parallel R_{in}$ CE total input impedance (20–5)

$A_i = \dfrac{I_c}{I_s}$ CE current gain (20–6)

$A_p = A_v A_i$ CE power gain (20–7)

$A_v = \dfrac{R_E}{r_e + R_E}$ CC voltage gain (20–8)

$R_{in} = \beta_{ac} R_E$ CC input impedance (20–9)

$R_{in(T)} = R_1 \parallel R_2 \parallel R_{in}$ CC total input impedance (20–10)

$A_i = \dfrac{I_e}{I_s}$ CC current gain (20–11)

$A_p \cong A_i$ CC power gain (20–12)

$\beta = \beta_1 \beta_2$ Beta for a Darlington pair (20–13)

$A_v = \dfrac{R_C}{r_e}$ CB voltage gain (20–14)

$R_{in} = r_e$ CB input impedance (20–15)

$A_i \cong 1$ CB current gain (20–16)

$A_p \cong A_v$ CB power gain (20–17)

$g_m = \dfrac{I_d}{V_{gs}}$ FET transconductance (20–18)

$A_v = g_m R_D$ CS voltage gain (20–19)

$A_v \cong \dfrac{g_m R_S}{1 + g_m R_S}$ CD voltage gain (20–20)

$A_v = g_m R_D$ CG voltage gain (20–21)

$R_{in(source)} = \dfrac{1}{g_m}$ CG input impedance (20–22)

$A_v' = A_{v1} A_{v2} A_{v3} \cdots A_{vn}$ Multistage gain (20–23)

$A_v \text{ (dB)} = 20 \log A_v$ Voltage gain in dB (20–24)

$A_v' \text{ (dB)} = A_{v1} \text{ (dB)} + A_{v2} \text{ (dB)} + \cdots + A_{vn} \text{ (dB)}$ Multistage dB gain (20–25)

$A_p = \beta_{dc} \dfrac{R_C}{r_e}$ CE large-signal power gain (20–26)

$P_{DQ} = I_{CQ} V_{CEQ}$ Transistor power dissipation (20–27)

$P_{out} = V_{ce} I_c$ CE output power (20–28)

$P_{out} = 0.5 V_{CEQ} I_{CQ}$ CE output power with centered Q point (20–29)

$$\text{eff}_{max} = 0.25 \quad \text{Class A maximum efficiency} \tag{20–30}$$

$$I_{c(sat)} = \frac{V_{CEQ}}{R_L} \quad \text{Class B ac saturation current} \tag{20–31}$$

$$P_{out} = 0.25 V_{CC} I_{c(sat)} \quad \text{Class B output power (maximum)} \tag{20–32}$$

$$P_{dc} = \frac{V_{CC} I_{c(sat)}}{\pi} \quad \text{Class B input power} \tag{20–33}$$

$$\text{eff}_{max} = 0.785 \quad \text{Class B maximum efficiency} \tag{20–34}$$

$$P_{out} = \frac{0.5 V_{CC}^2}{R_c} \quad \text{Class C output power (maximum)} \tag{20–35}$$

$$\text{eff} = \frac{P_{out}}{P_{out} + P_{D(avg)}} \quad \text{Class C efficiency} \tag{20–36}$$

SELF-TEST

1. In a CE amplifier, what effect does a capacitor from emitter to ground have? What is this capacitor called?
2. The collector resistor of a CE amplifier is increased in value. What does this increase do to the voltage gain?
3. Which transistor parameters affect the input impedance of a CE amplifier?
4. How do the output signals of a CE and a CC amplifier differ?
5. Theoretically, what is the largest voltage gain obtainable with a CC amplifier?
6. What is the advantage of a Darlington arrangement?
7. Does the CB amplifier have a lower or higher input resistance than the other two configurations?
8. If power gain were the only consideration, which amplifier configuration would you choose?
9. Define *amplifier efficiency*. Which amplifier configuration is the most efficient?
10. Name the three FET amplifier configurations.
11. What happens in an FET amplifier when an FET with a lower transconductance is substituted into the circuit?
12. What is the primary reason for cascading amplifiers?
13. If three amplifiers, each with a voltage gain of 30, are connected in a multistage arrangement, what is the overall voltage gain?
14. Explain the difference between class A, class B, and class C operation.
15. Describe generally what an oscillator is and how it operates.

PROBLEMS

Section 20–1

20–1 Determine the voltage gain for Figure 20–49.

20–2 Determine each of the dc voltages, V_B, V_C, and V_E, with respect to ground in Figure 20–49.

20–3 Determine the following dc values for the amplifier in Figure 20–50:
(a) V_B (b) V_E (c) I_E
(d) I_C (e) V_C (f) V_{CE}

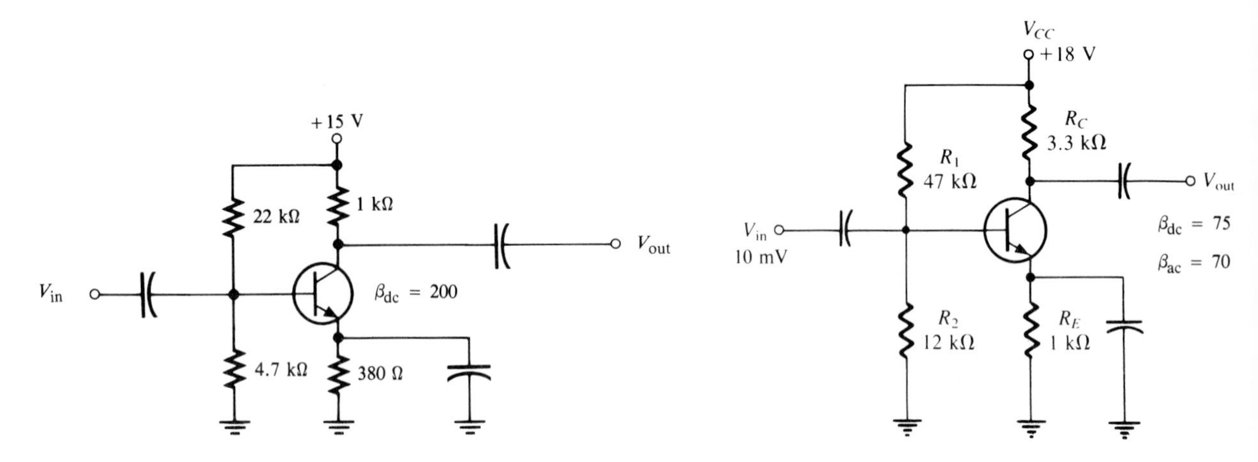

FIGURE 20–49

FIGURE 20–50

20–4 Determine the following ac values for the amplifier in Figure 20–50:
(a) R_{in} (b) $R_{in(T)}$ (c) A_v
(d) A_i (e) A_p

20–5 The amplifier in Figure 20–51 has a variable gain control, using a 100-Ω potentiometer for R_E with the wiper ac grounded. As the potentiometer is adjusted, more or less of R_E is bypassed to ground, thus varying the gain. The total R_E remains constant to dc, keeping the bias fixed. Determine the maximum and minimum gains for this amplifier.

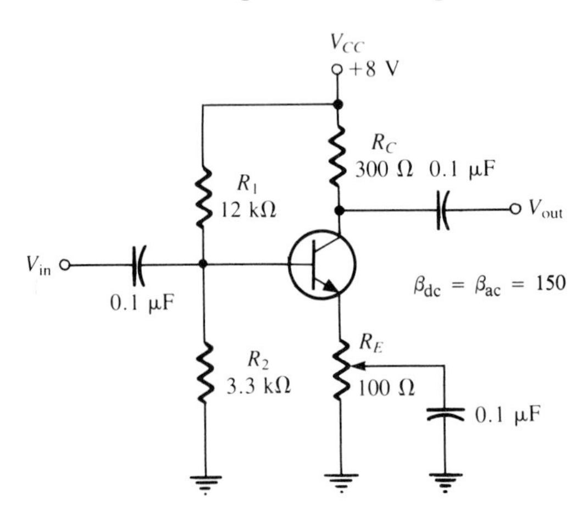

FIGURE 20–51

20–6 If a load resistance of 600 Ω is placed on the output of the amplifier in Figure 20–51, what is the maximum gain?

Section 20–2

20–7 Determine the *exact* voltage gain for the emitter-follower in Figure 20–52.

20–8 What is the total input impedance in Figure 20–52? What is the dc output voltage?

20–9 A load resistance is capacitively coupled to the emitter in Figure 20–52. In terms of signal operation, the load appears in parallel with R_E and reduces the effective emitter resistance. How does this affect the voltage gain?

FIGURE 20–52

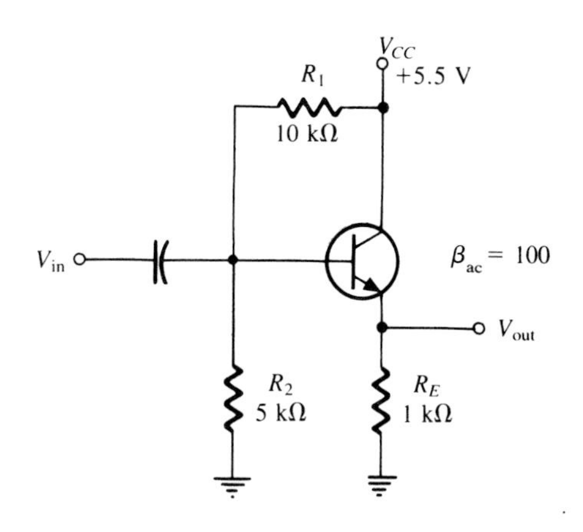

Section 20–3

20–10 What is the main disadvantage of the CB amplifier compared to the CE and the emitter-follower?

20–11 Find R_{in}, A_v, A_i, and A_p for the amplifier in Figure 20–53.

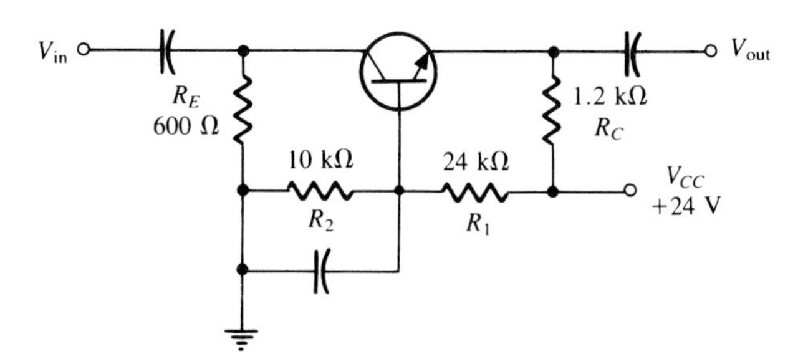

FIGURE 20–53

Section 20–4

20–12 Determine the voltage gain of each CS amplifier in Figure 20–54.

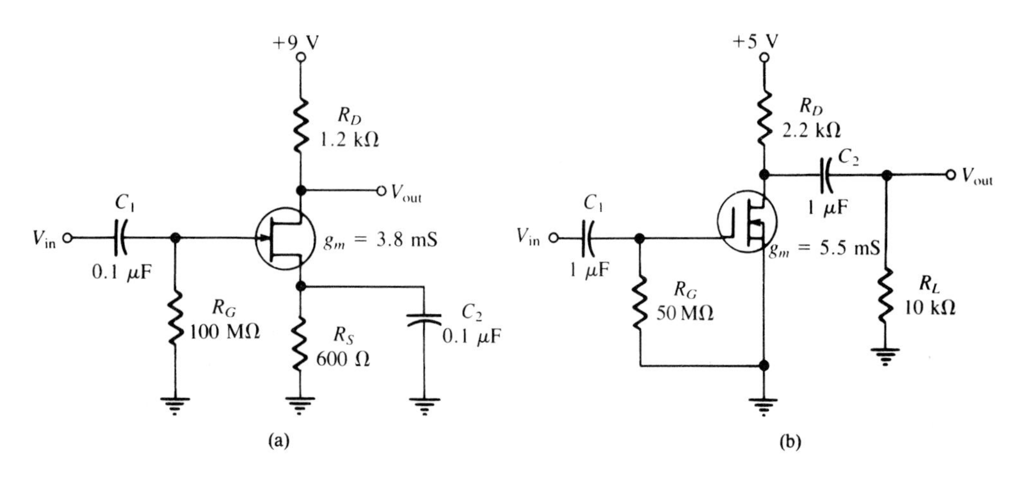

FIGURE 20–54

20–13 Find the gain of each amplifier in Figure 20–55.

20–14 Determine the gain of each amplifier in Figure 20–55 when a 10-kΩ load is capacitively coupled from source to ground.

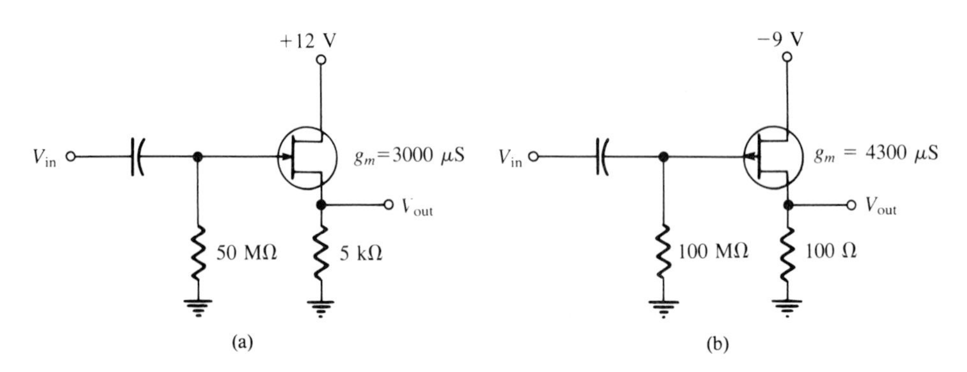

FIGURE 20–55

Section 20–5

20–15 Each of three cascaded amplifier stages has a dB voltage gain of 10. What is the overall dB voltage gain? What is the actual overall voltage gain?

20–16 For the two-stage, capacitively coupled amplifier in Figure 20–56, find the following values:
(a) voltage gain of each stage.
(b) overall voltage gain.
(c) Express the gains found above in dB.

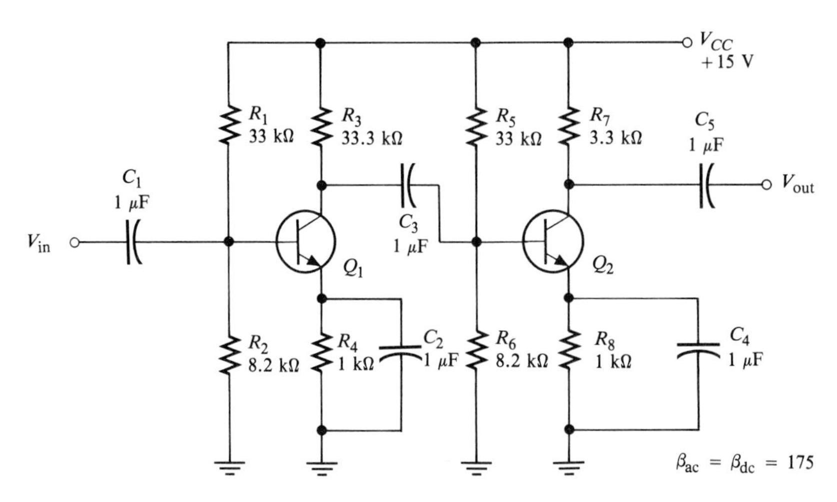

FIGURE 20–56

Section 20–6

20–17 Determine the minimum power rating for each of the transistors in Figure 20–57.

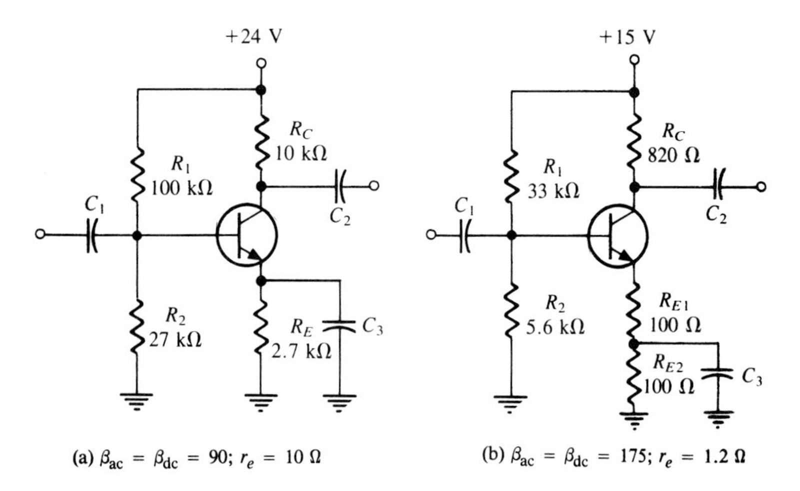

FIGURE 20–57

Section 20–7

20–18 Determine the dc voltages at the bases and emitters of Q_1 and Q_2 in Figure 20–58. Also determine V_{CEQ} for each transistor.

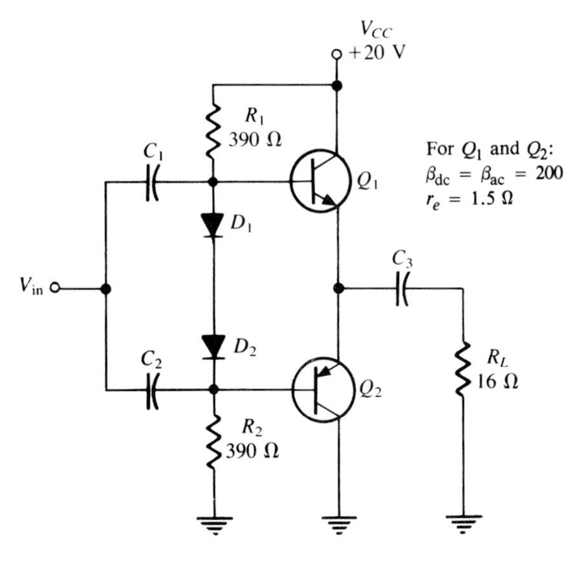

FIGURE 20–58

20–19 Determine the maximum peak output voltage and peak load current for the circuit in Figure 20–58.

20–20 The efficiency of a certain class B push-pull amplifier is 0.71, and the dc input power is 16.25 W. What is the ac output power?

Section 20–8

20–21 What is the resonant frequency of the tank circuit in a class C amplifier with $L = 10$ mH and $C = 0.001\ \mu$F?

20–22 Determine the efficiency of the class C amplifier when $P_{D(avg)} = 10$ mW, $V_{CC} =$ 15 V, and the equivalent parallel resistance in the collector tank circuit is 50 Ω.

Section 20–9

20–23 Sketch the waveforms you would expect to see with a scope across R_L in Figure 20–58 if (a) Q_1 were open from collector to emitter and (b) Q_2 were open from collector to emitter.

20–24 What symptom(s) would indicate each of the following failures under signal conditions in Figure 20–59?
(a) Q_1 open from drain to source. **(b)** R_3 open. **(c)** C_2 shorted.
(d) C_3 shorted. **(e)** Q_2 open from drain to source.

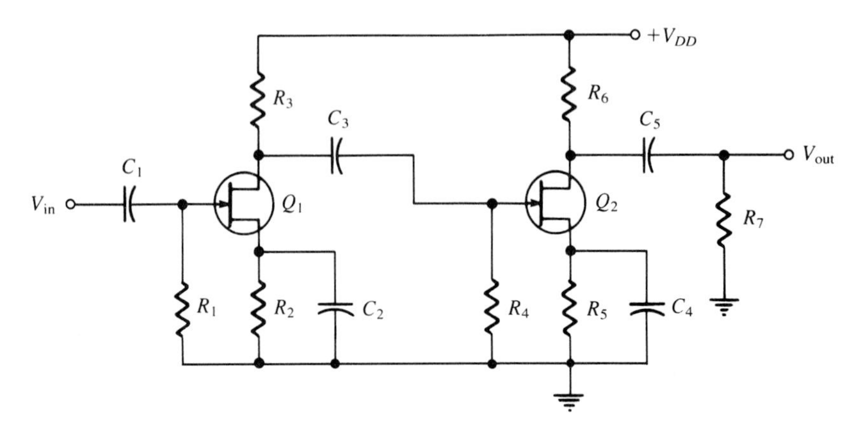

FIGURE 20–59

Section 20–10

20–25 Calculate the frequency of oscillation for each circuit in Figure 20–60, and identify each type of oscillator.

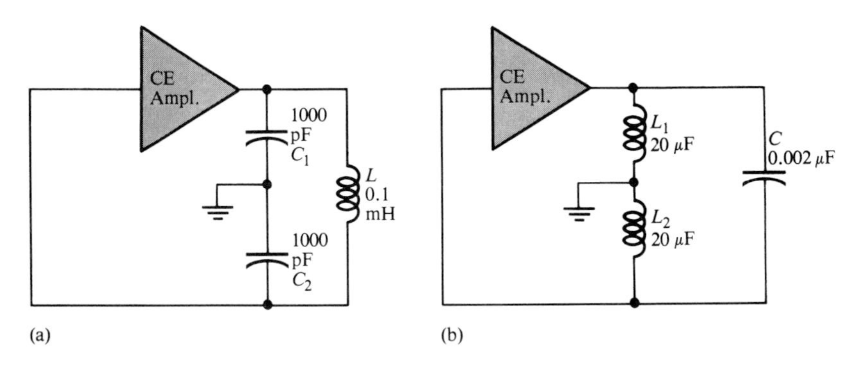

FIGURE 20–60

ANSWERS TO SECTION REVIEWS

Section 20–1

1. To increase voltage gain. **2.** Ratio of collector resistance to total emitter resistance. **3.** 10,000.

Section 20–2

1. Emitter-follower. **2.** 1. **3.** High input impedance.

Section 20–3

1. True. **2.** True.

Section 20–4

1. g_m and R_d. **2.** It is halved. **3.** Common-gate has low input impedance.

Section 20–5

1. One amplifier in a cascaded arrangement. **2.** Product of individual gains.
3. 53.98 dB.

Section 20–6

1. Centered on the load line. **2.** 25%. **3.** 35 mW.

Section 20–7

1. At cutoff. **2.** The barrier potential of the base-emitter junction. **3.** 78.5%.
4. To reproduce both positive and negative alternations of the input signal with greater efficiency.

Section 20–8

1. Cutoff. **2.** To produce a sine wave output. **3.** 90.9%.

Section 20–9

1. **(a)** It will disappear because it is shorted to ground through the base-emitter junction and C_5. **(b)** Yes. It will decrease.
2. Excess input signal voltage. **3.** Open bypass capacitor, C_2.

Section 20–10

1. RC phase-shift, Colpitts, Hartley, crystal.
2. Colpitts uses two capacitors, center-tapped to ground in parallel with an inductor. Hartley uses two coils, center-tapped to ground in parallel with a capacitor.

Operational Amplifiers (Op-Amps)

In the previous chapters of this book, a number of electronic devices were introduced. These devices, such as the diode and the transistor, are individually packaged and are connected in a circuit with other individual devices to form a functional unit. Individually packaged devices are referred to as *discrete components.*

We will now move into the area of *integrated circuits* (ICs), in which many transistors, diodes, resistors, and capacitors are fabricated on a single silicon chip and packaged in a single case to form a functional circuit. The manufacturing process for ICs is complex and beyond the scope of this coverage.

In our study of ICs, we will treat the entire circuit as a device. That is, we will be concerned with what the circuit does from an external point of view and will be less concerned about the internal, component-level operation.

In this chapter you will learn:

- What an operational amplifier (op-amp) is from both an ideal and a practical point of view.
- The basic operation of the differential amplifier.
- The various modes of differential amplifier operation.
- The meaning of *common-mode gain.*
- How differential amplifiers are used as part of an op-amp.
- The important op-amp data sheet parameters.
- The meaning of *common-mode rejection ratio* (CMRR).
- How negative feedback is used with op-amps.
- The basic operation of the noninverting, inverting, and voltage-follower op-amp configurations.

21

APPLICATION ASSIGNMENT

As a recent graduate of a technical school, you have accepted a job with a major electronics company involved in the design, development, and manufacturing of state-of-the-art radar systems for both commercial and military applications. Your first project assignment is with a design and development group working on an advanced terrain-following radar system for a military attack aircraft.

The design engineer asks you to build a high-input impedance, low-output impedance, noninverting amplifier that has a variable voltage gain from 1 to 50. This amplifier is needed to test a certain portion of the radar signal-processing circuitry.

After completing this chapter, you will be able to carry out this assignment. A solution is given in the Application Note at the end of the chapter.

INTRODUCTION TO OPERATIONAL AMPLIFIERS

21–1

Early operational amplifiers (op-amps) were used primarily to perform *mathematical operations* such as addition, subtraction, integration, and differentiation—hence the term *operational.* These early devices were constructed with vacuum tubes and worked with high voltages.

Today's op-amps are *linear integrated circuits* (ICs) that use relatively low supply voltages and are very reliable and inexpensive.

Symbol and Terminals

The standard op-amp symbol is shown in Figure 21–1(a). It has *two input terminals,* called the *inverting input* (−) and the *noninverting input* (+), and one output terminal.

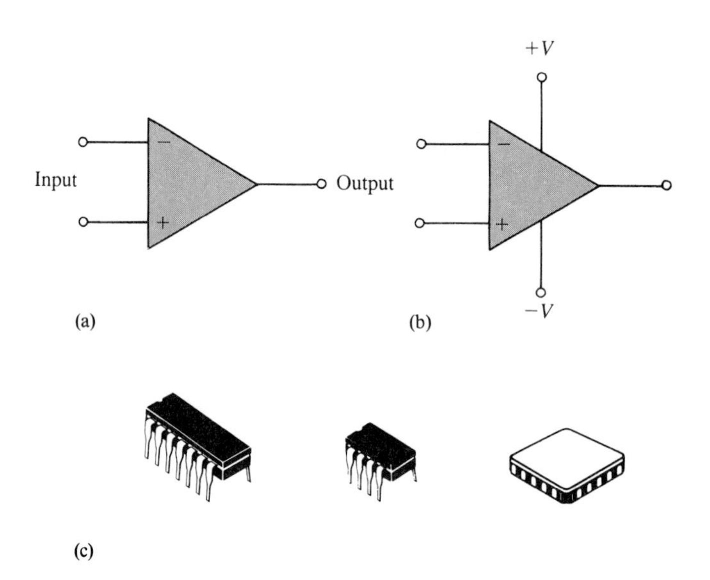

FIGURE 21–1
Op-amp symbols and packages. (courtesy of Motorola Semiconductor)

The typical op-amp operates with two dc supply voltages, one positive and the other negative, as shown in Part (b). Usually these dc voltage terminals are left off the schematic symbol for simplicity but are always understood to be there.

The Ideal Op-Amp

To illustrate what an op-amp is, we will consider its *ideal* characteristics. A practical op-amp, of course, falls short of these ideal standards, but it is much easier to understand and analyze the device from an ideal point of view.

First, the ideal op-amp has *infinite voltage gain* and *infinite bandwidth.* Also, it has an *infinite input impedance* (open), so that it does not draw any power from the driving source. Finally, it has a *0 (zero) output impedance.* These characteristics are illustrated in Figure 21–2. The input voltage V_{in} appears between the two input terminals, and the output voltage is $A_v V_{in}$, as indicated by the symbol for internal voltage source.

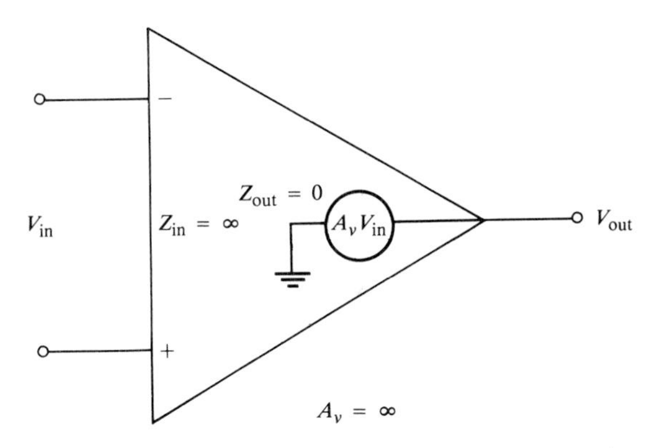

FIGURE 21–2
Ideal op-amp representation.

The concept of infinite input impedance is a particularly valuable analysis tool for the various op-amp configurations, which will be covered later.

The Practical Op-Amp

Although modern IC op-amps approach parameter values that can be treated as ideal in many cases, the ideal device has not been and probably will not be developed even though improvements continue to be made.

Any device has limitations, and the IC op-amp is no exception. Op-amps have both voltage and current limitations. Peak-to-peak output voltage, for example, is usually limited to slightly less than the two supply voltages. Output current is also limited by internal restrictions such as power dissipation and component ratings.

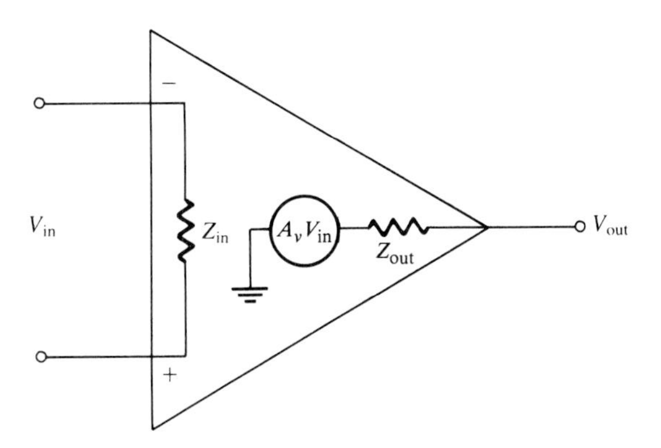

FIGURE 21–3
Practical op-amp representation.

Characteristics of a practical op-amp are high voltage gain, high input impedance, low output impedance, and wide bandwidth, as illustrated in Figure 21–3.

SECTION REVIEW 21–1

1. Sketch the symbol for an op-amp.
2. Describe some of the characteristics of a *practical* op-amp.

THE DIFFERENTIAL AMPLIFIER

21–2

The op-amp, in its basic form, typically consists of two or more *differential amplifier* stages. Because the differential amplifier (diff-amp) is fundamental to the op-amp's internal operation, it is useful to spend some time in acquiring a basic understanding of this type of circuit. A basic diff-amp circuit is shown in Figure 21–4(a), and its block symbol in Part (b).

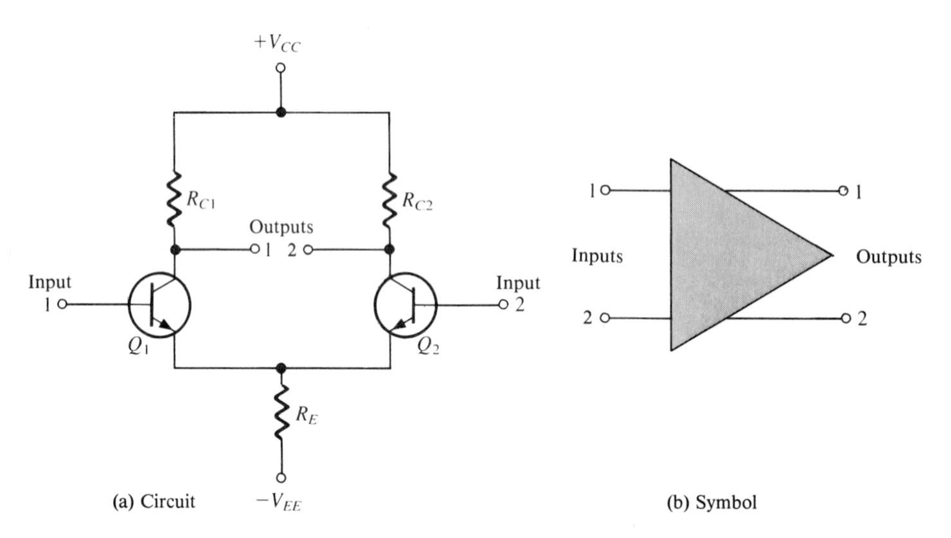

FIGURE 21–4
Basic differential amplifier.

The diff-amp stages that make up part of the op-amp provide *high voltage gain* and *common-mode rejection* (to be defined).

Basic Operation

Although an op-amp typically has more than one diff-amp stage, we will use a single diff-amp to illustrate the basic operation. The following discussion is in relation to Figure 21–5 and consists of a basic dc analysis of the diff-amp's operation.

First, when both inputs are grounded (0 V), the emitters are at −0.7 V, as indicated in Figure 21–5(a). It is assumed that the transistors are identically matched by careful process control during manufacturing so that their dc emitter currents are the same with no input signal. Thus,

$$I_{E1} = I_{E2}$$

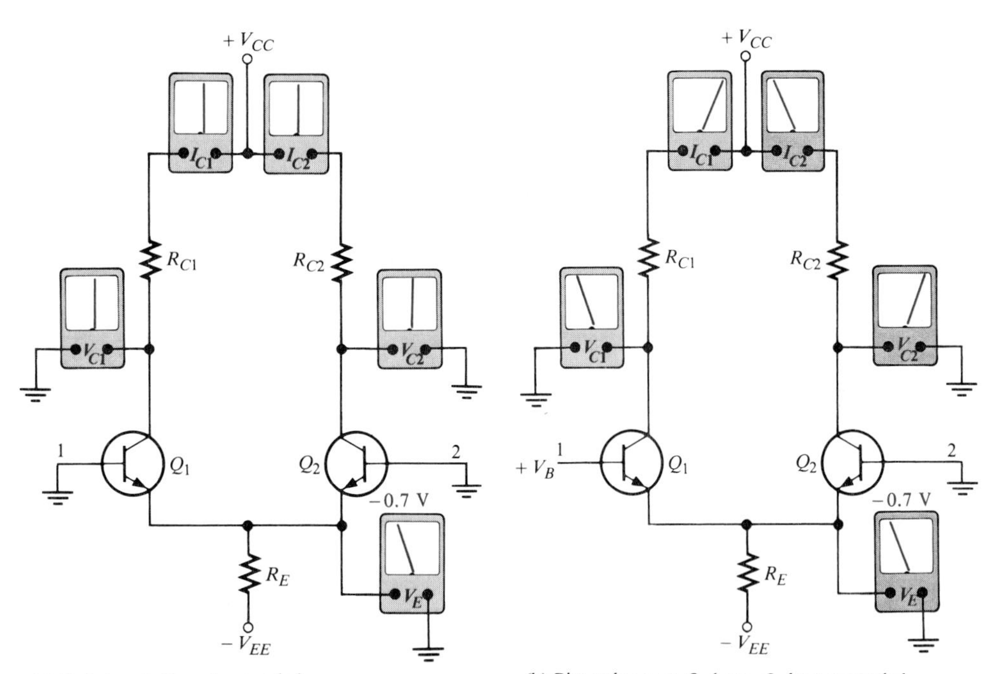

(a) Both inputs (bases) grounded

(b) Bias voltage on Q_1 base, Q_2 base grounded

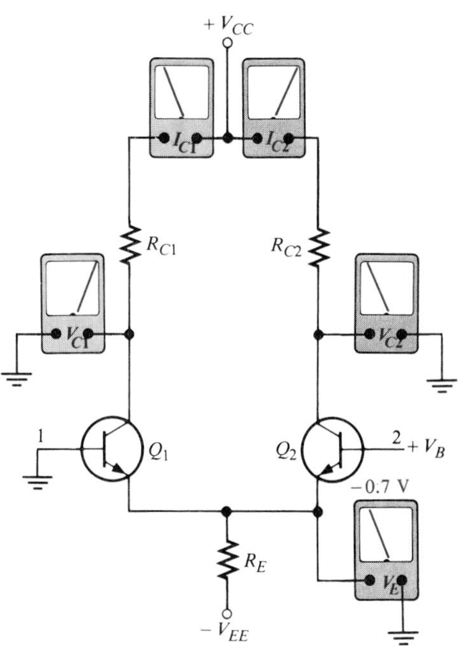

(c) Bias voltage on Q_2 base, Q_1 base grounded

FIGURE 21–5
Basic operation of a differential amplifier. Ground is 0 V.

Since both emitter currents combine through R_E,

$$I_{E1} = I_{E2} = \frac{I_{RE}}{2}$$

where

$$I_{RE} = \frac{V_E - V_{EE}}{R_E}$$

Based on the approximation that $I_C \cong I_E$, it can be stated that

$$I_{C1} = I_{C2} \cong \frac{I_{RE}}{2}$$

Since both collector currents and both collector resistors are equal (when the input voltage is zero),

$$V_{C1} = V_{C2} = V_{CC} - I_{C1}R_{C1}$$

This condition is illustrated in Figure 21–5(a).

Next, input 2 is left grounded, and a positive bias voltage is applied to input 1, as shown in Figure 21–5(b). The positive voltage on the base of Q_1 increases I_{C1} and raises the emitter voltage to

$$V_E = V_B - 0.7 \text{ V}$$

This action *reduces* the forward bias (V_{BE}) of Q_2 because its base is held at 0 V (ground), thus causing I_{C2} to *decrease,* as indicated in Part (b) of the diagram. The net result is that the increase in I_{C1} causes a decrease in V_{C1}, and the decrease in I_{C2} causes an increase in V_{C2}, as shown.

Finally, input 1 is grounded and a positive bias voltage is applied to input 2, as shown in Figure 21–5(c). The positive bias voltage causes Q_2 to conduct more, thus increasing I_{C2}. Also, the emitter voltage is raised. This reduces the forward bias of Q_1, since its base is held at ground, and causes I_{C1} to decrease. The result is that the increase in I_{C2} produces a decrease in V_{C2}, and the decrease in I_{C1} causes V_{C1} to increase as shown.

Modes of Signal Operation

Single-Ended Input

When a diff-amp is operated in this mode, one input is grounded and the signal voltage is applied only to the other input, as shown in Figures 21–6(a) and 21–6(b).

When the signal voltage is applied to input 1, as in Part (a), an inverted, amplified signal voltage appears at output 1 as shown. Also, a signal voltage appears in phase at the emitter of Q_1. Since the emitters of Q_1 and Q_2 are common, this emitter signal becomes an input to Q_2 which functions as a common-base amplifier. The signal is amplified by Q_2 and appears, noninverted, at output 2.

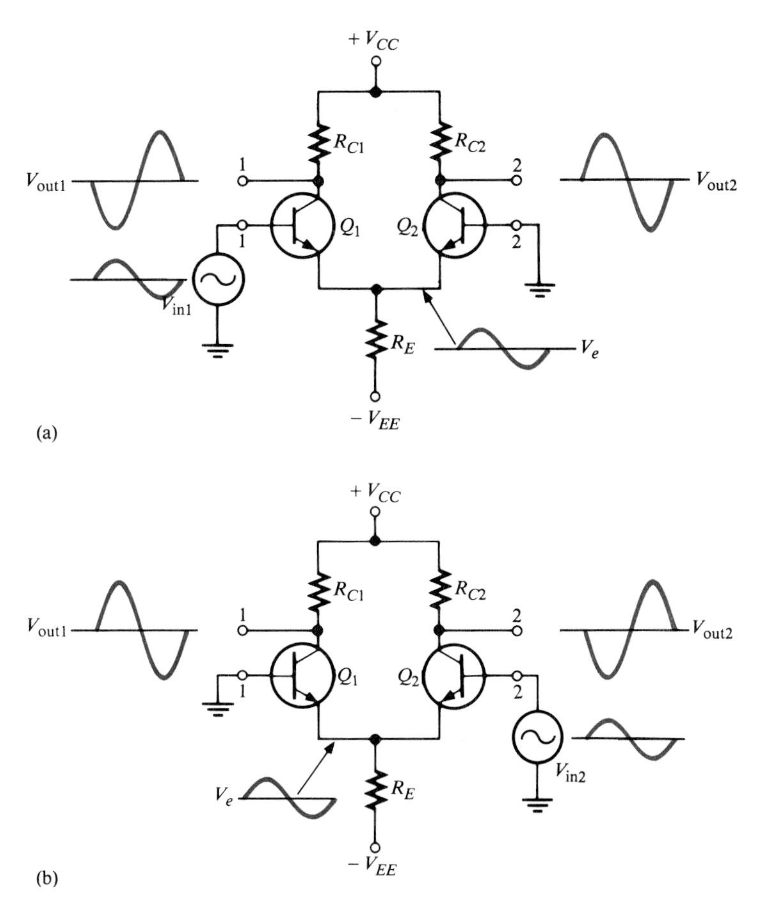

FIGURE 21–6
Single-ended input operation of a differential amplifier.

When the signal is applied to input 2 with input 1 grounded, as in Part (b), an inverted, amplified signal voltage appears at output 2. In this situation, Q_1 acts as a common-base amplifier, and a noninverted, amplified signal appears at output 1.

Differential Input

In this mode, two signals of *opposite polarity* (out of phase) are applied to the inputs, as shown in Figure 21–7(a). This type of operation is also referred to as *double-ended.* Each input affects the outputs, as you will see in the following discussion.

Figure 21–7(b) shows the output signals due to the signal on input 1 acting alone as a single-ended input. Figure 21–7(c) shows the output signals due to the signal on input 2 acting alone as a single-ended input. In Parts (b) and (c), notice that the signals on output 1 are of the same polarity. The same is also true for output 2. By superimposing both output 1 signals and both output 2 signals, we get the total differential operation, as pictured in Figure 21–7(d).

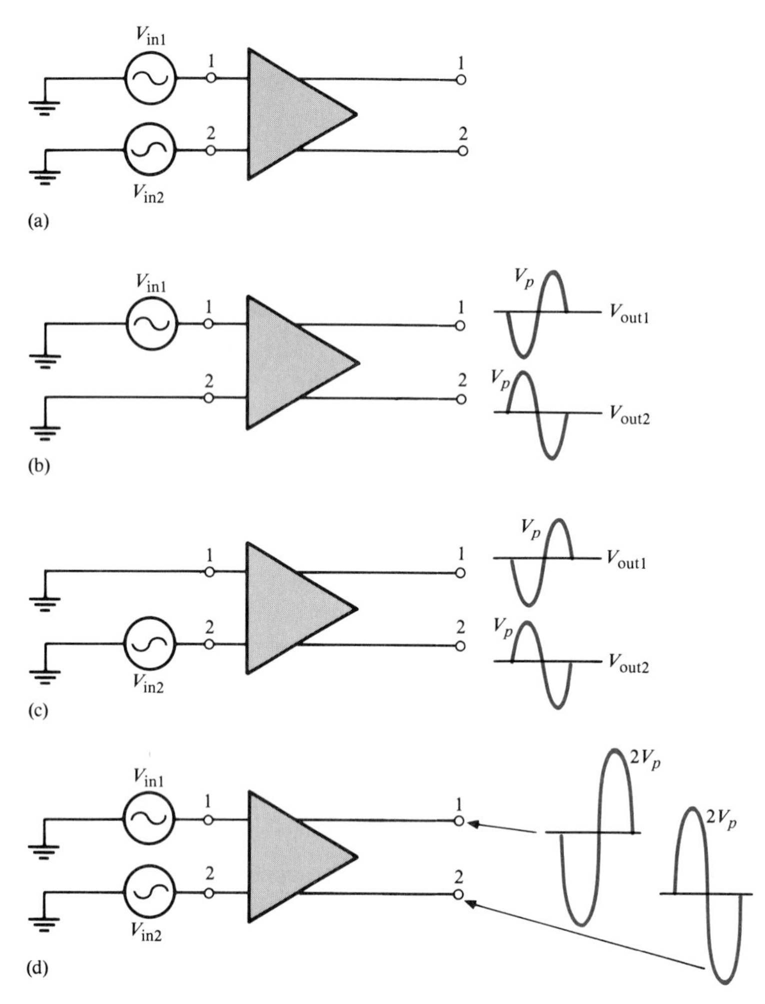

FIGURE 21–7
Differential operation of a differential amplifier.

Common-Mode Input

We will illustrate one of the most important aspects of the operation of a diff-amp by considering the case where two signal voltages of the *same phase, frequency, and amplitude* are applied to the two inputs, as shown in Figure 21–8(a). Again, by considering each input signal as acting alone, you can understand the basic operation.

Figure 21–8(b) shows the output signals due to the signal on only input 1, and Part (c) shows the output signals due to the signal on only input 2. Notice that the component signals on output 1 are of the opposite polarity, and so are those on output 2. When these are superimposed, they cancel, resulting in a zero output voltage, as shown in Figure 21–8(d).

This action is called *common-mode rejection*. Its importance lies in the situation where an *unwanted* signal appears commonly on both diff-amp inputs. Common-mode rejection means that this unwanted signal will not appear on

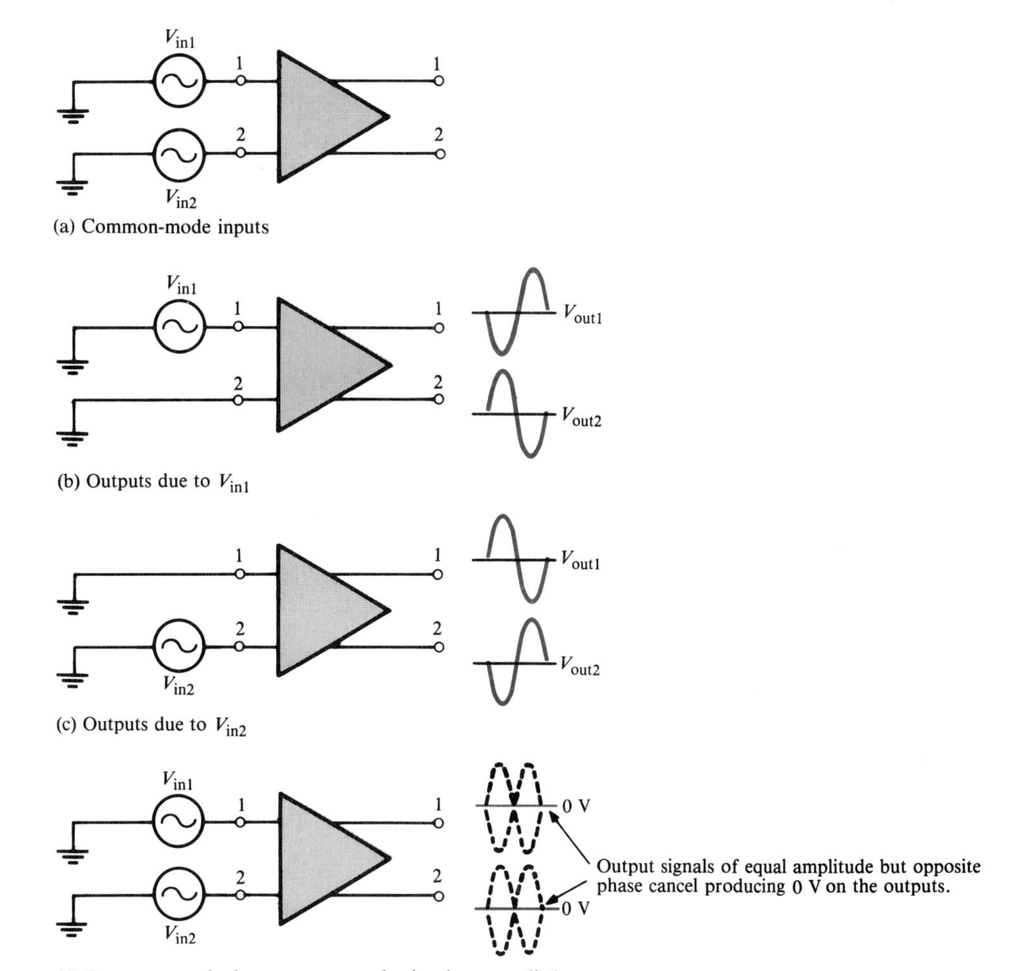

FIGURE 21–8
Common-mode operation of a differential amplifier.

the outputs to distort the *desired* signal. Common-mode signals (noise) generally are the result of the pick-up of radiated energy on the input lines, or adjacent lines, or the 60-Hz power line, or other sources.

In summary, *desired* signals appear on only one input or with opposite polarities on both input lines. These desired signals are amplified and appear on the outputs as previously discussed. Unwanted signals (noise) appearing with the same polarity on both input lines are essentially canceled by the diff-amp and do not appear on the outputs. The *measure* of an amplifier's ability to reject common-mode signals is a parameter called the *common-mode rejection ratio* (CMRR) and is discussed in the following paragraphs.

Common-Mode Gain

Ideally, a diff-amp provides a very high gain for desired signals (single-ended or differential), and 0 (zero) gain for common-mode signals. Practical diff-amps, however, do exhibit a very small common-mode gain (usually much less than 1), while providing a high differential voltage gain (usually several thousand). The higher the differential gain with respect to the common-mode gain, the better the performance of the diff-amp in terms of rejection of common-mode signals. This suggests that a good measure of the diff-amp's performance in rejecting unwanted common-mode signals is the ratio of the differential gain $A_{v(d)}$ to the common-mode gain, A_{cm}. This ratio is called the *common-mode rejection ratio,* CMRR:

$$\text{CMRR} = \frac{A_{v(d)}}{A_{cm}} \quad (21\text{–}1)$$

The higher the CMRR, the better, as you can see. A very high value of CMRR means that the differential gain $A_{v(d)}$ is high and the common-mode gain A_{cm} is low.

The CMRR is often expressed in decibels (dB) as

$$\text{CMRR} = 20 \log \frac{A_{v(d)}}{A_{cm}} \quad (21\text{–}2)$$

EXAMPLE 21–1 A certain diff-amp has a differential voltage gain of 2000 and a common-mode gain of 0.2. Determine the CMRR and express it in dB.

Solution

$$A_{v(d)} = 2000 \quad \text{and} \quad A_{cm} = 0.2$$

$$\text{CMRR} = \frac{A_{v(d)}}{A_{cm}} = \frac{2000}{0.2} = 10{,}000$$

In dB,

$$\text{CMRR} = 20 \log(10{,}000) = 80 \text{ dB}$$

A CMRR of 10,000, for example, means that the desired input signal (differential) is amplified 10,000 times more than the unwanted noise (common-mode). So, as an example, if the amplitudes of the differential input signal and the common-mode noise are equal, the desired signal will appear on the output 10,000 times greater in amplitude than the noise. Thus, the noise or interference has been essentially eliminated.

The following example illustrates further the idea of *common-mode rejection* and the general signal operation of the diff-amp.

EXAMPLE 21–2 The diff-amp shown in Figure 21–9 has a differential voltage gain of 2500 and a CMRR of 30,000. In part (a) of the figure, a single-ended input signal of 500 μV rms is applied. At the same time a 1-V, 60-Hz interference signal appears on both inputs as a result of radiated pick-up from the ac power system.

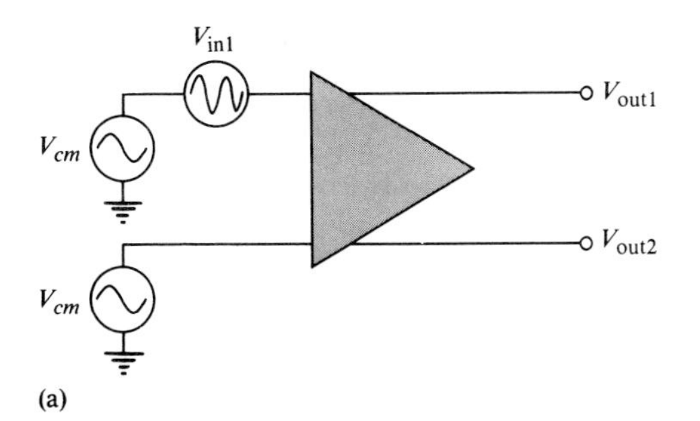

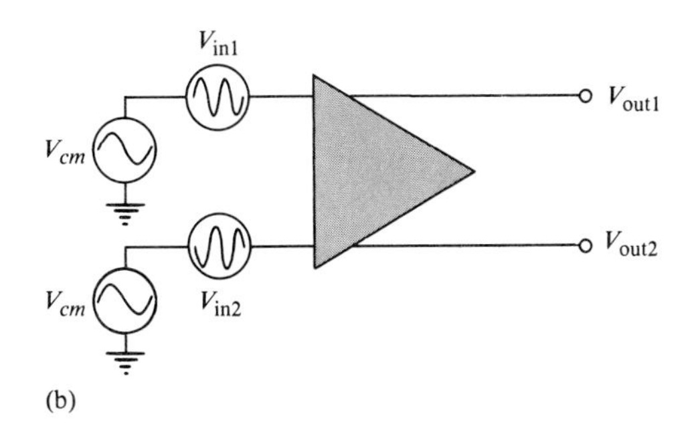

FIGURE 21–9

In Part (b) of the figure, differential input signals of 500 μV rms each are applied to the inputs. The common-mode interference is the same as in Part (a).

(a) Determine the common-mode gain.
(b) Express the CMRR in dB.
(c) Determine the rms output signal for Parts (a) and (b).
(d) Determine the rms interference voltage on the output.

Solution

(a) $\text{CMRR} = \dfrac{A_{v(d)}}{A_{cm}}$

$$A_{cm} = \frac{A_{v(d)}}{\text{CMRR}} = \frac{2500}{30{,}000} = 0.083$$

(b) $\text{CMRR} = 20\log(30{,}000) = 89.5\text{ dB}$

(c) In Figure 21–9(a), the differential input voltage is the *difference* between the voltage on input 1 and that on input 2. Since input 2 is grounded, its voltage is zero. Therefore,

$$V_{\text{in(diff)}} = V_{\text{in1}} - V_{\text{in2}} = 500\ \mu\text{V} - 0\text{ V} = 500\ \mu\text{V}$$

The output signal voltage in this case is taken at output 1:

$$V_{\text{out1}} = A_{v(d)}V_{\text{in(diff)}} = (2500)(500\ \mu\text{V}) = 1.25\text{ V rms}$$

In Figure 21–9(b), the differential input voltage is the difference between the two opposite-polarity, 500-μV signals:

$$V_{\text{in(diff)}} = V_{\text{in1}} - V_{\text{in2}} = 500\ \mu\text{V} - (-500\ \mu\text{V}) = 1000\ \mu\text{V} = 1\text{ mV}$$

The output signal voltage is

$$V_{\text{out1}} = A_{v(d)}V_{\text{in(diff)}} = (2500)(1\text{ mV}) = 2.5\text{ V rms}$$

This shows that a differential input (two opposite-polarity signals) results in a gain that is double that for a single-ended input.

(d) The common-mode input is 1 V rms. The common-mode gain A_{cm} is 0.083. The interference voltage on the output therefore is

$$A_{cm} = \frac{V_{out(cm)}}{V_{in(cm)}}$$

$$V_{out(cm)} = A_{cm}V_{in(cm)} = 0.083(1\text{ V}) = 0.083\text{ V}$$

Simple Op-Amp Arrangement

Figure 21–10 shows two diff-amp stages and an emitter-follower connected to form a very simple op-amp. The first stage can be used with a single-ended or a differential input. The differential outputs of the first stage feed into the differential inputs of the second stage. The output of the second stage is single-ended to drive an emitter-follower to achieve a relatively low output impedance. Both differential stages together provide a high voltage gain and a high CMRR.

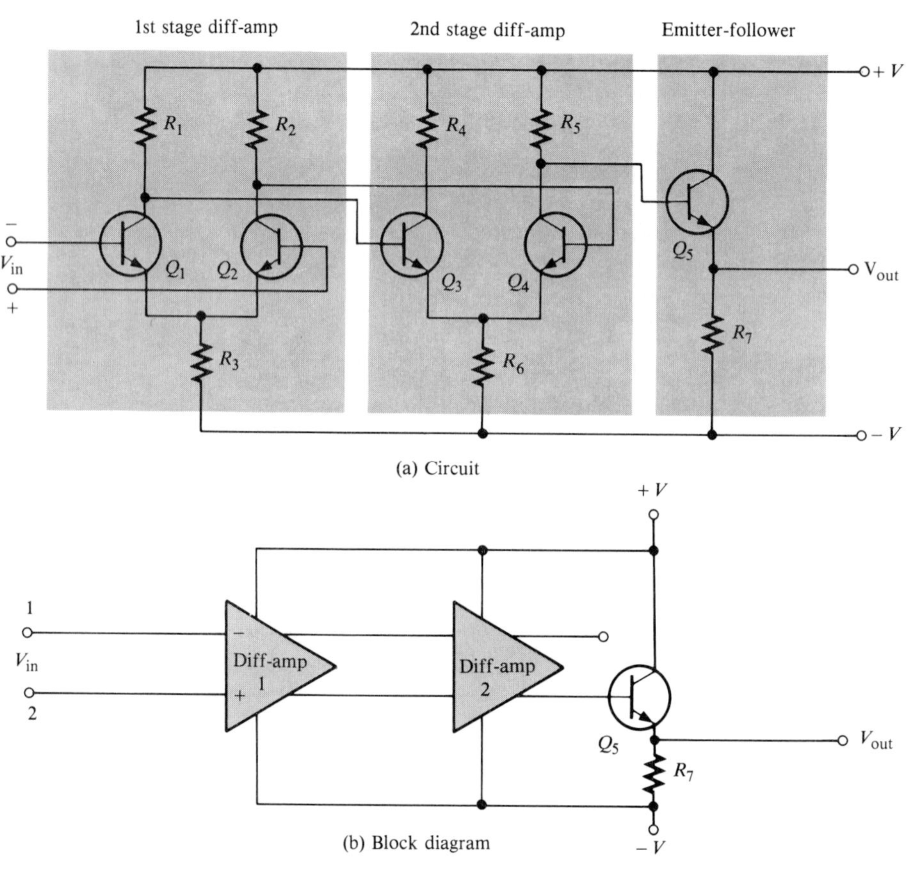

FIGURE 21–10
Internal circuitry of a simple op-amp.

SECTION REVIEW 21–2

1. Distinguish between differential and single-ended inputs.
2. Define *common-mode rejection.*

OP-AMP DATA SHEET PARAMETERS

21–3

In this section, several important op-amp parameters are defined, and four popular IC op-amps are compared in terms of their parameter values.

Input Offset Voltage

The ideal op-amp produces zero volts out for zero volts in. In a practical op-amp, however, a small dc voltage appears at the output when no differential input voltage is applied. Its primary cause is a slight mismatch of the base-emitter voltages of the differential input stage, as illustrated in Figure 21–11(a).

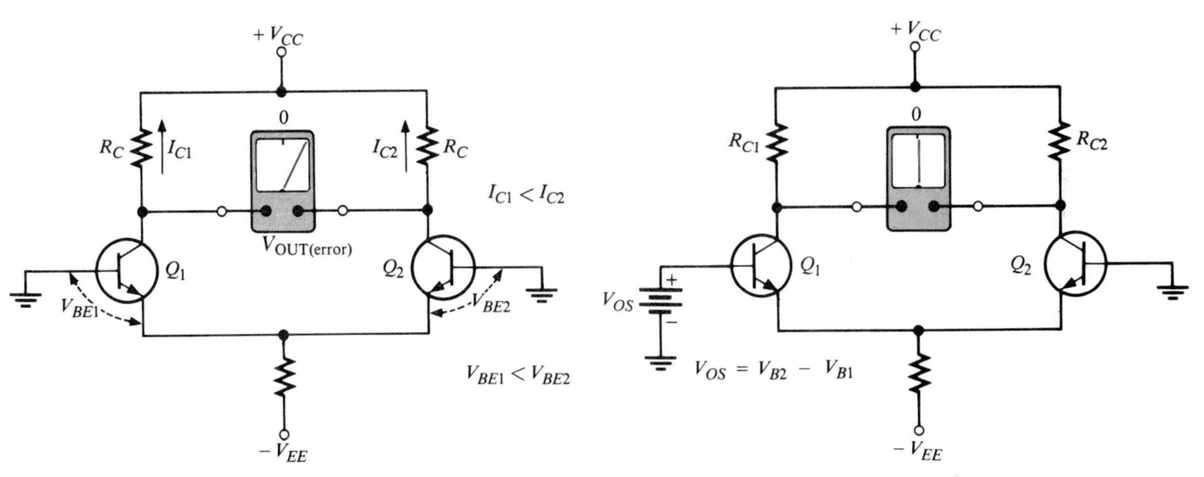

(a) A V_{BE} mismatch causes a small output error voltage.

(b) The input offset voltage is the difference in the voltage between the inputs that is necessary to eliminate the output error voltage (makes $V_{OUT} = 0$)

FIGURE 21–11
Input offset voltage, V_{OS}.

The output voltage of the differential input stage is expressed as

$$V_{OUT} = I_{C2}R_C - I_{C1}R_C \tag{21–3}$$

A small difference in the base-emitter voltages of Q_1 and Q_2 causes a small difference in the collector currents. This results in a nonzero value of V_{OUT}. (The collector resistors are equal.)

As specified on an op-amp data sheet, *the input offset voltage V_{OS} is the differential dc voltage required between the inputs to force the differential output to zero volts,* as demonstrated in Figure 21–11(b).

Typical values of input offset voltage are in the range of 2 mV or less. In the ideal case, it is 0 V.

Input Offset Voltage Drift

The input offset voltage drift is a parameter related to V_{OS} that specifies how much change occurs in the input offset voltage for each degree change in temperature. Typical values range anywhere from about 5 μV per degree Celsius to about 50 μV per degree Celsius.

Usually, an op-amp with a higher nominal value of input offset voltage exhibits a higher drift.

Input Bias Current

You have seen that the input terminals of a diff-amp are the transistor bases and, therefore, the input currents are the base currents.

The input bias current is the direct current required by the inputs of the amplifier to properly operate the first stage. By definition, the input bias current is the *average* of both input currents and is calculated as follows:

$$I_{BIAS} = \frac{I_1 + I_2}{2} \tag{21–4}$$

The concept of input bias current is illustrated in Figure 21–12.

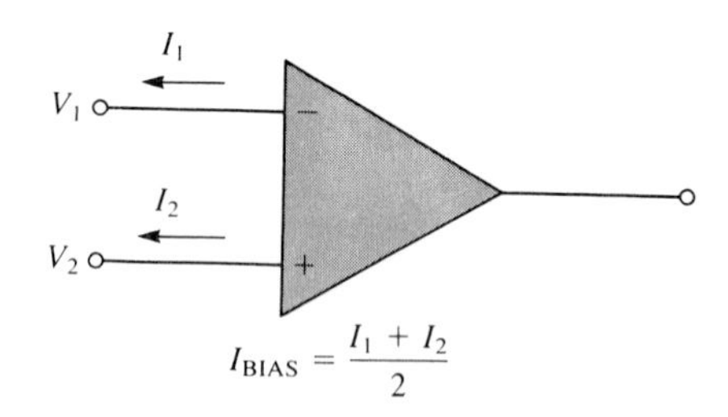

FIGURE 21–12
Input bias current is the average of the two op-amp input currents.

Input Impedance

There are two basic ways of specifying the impedance of an op-amp. The *differential input impedance* is the total resistance between the inverting and the noninverting inputs, as illustrated in Figure 21–13(a). Differential impedance is measured by determining the change in bias current for a given change in differential input voltage. The *common-mode input impedance* is measured from the inputs to ground and is depicted in Figure 21–13(b).

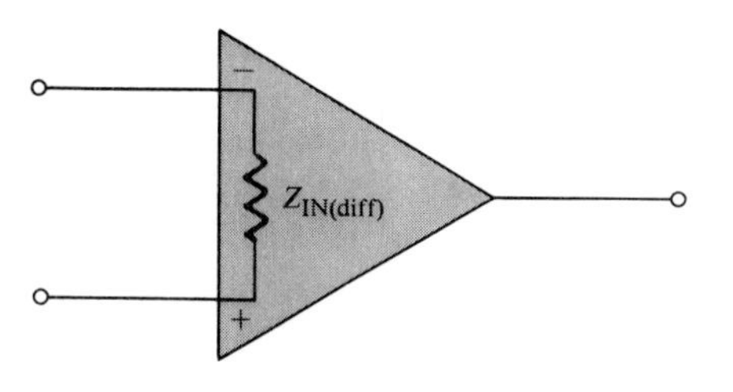

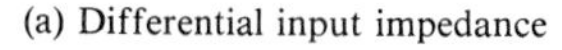

(a) Differential input impedance

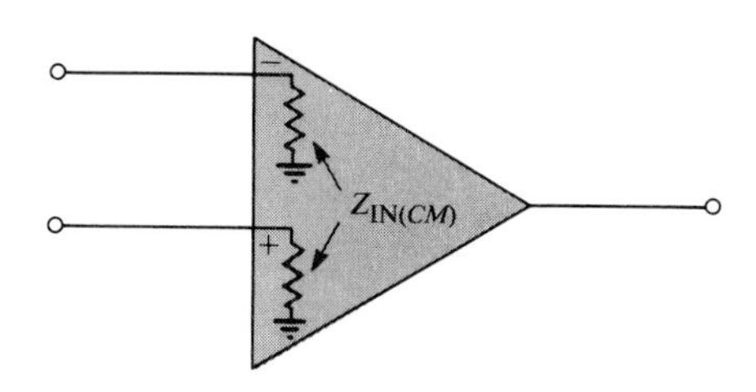

(b) Common-mode input impedance

FIGURE 21–13
Op-amp input impedance.

Input Offset Current

Ideally, the two input bias currents are equal, and thus their difference is zero. In a practical op-amp, however, the bias currents are not exactly equal.

The *input offset current is the difference of the input bias currents,* expressed as

$$I_{OS} = |I_1 - I_2| \tag{21–5}$$

Actual magnitudes of offset current are usually at least an order of magnitude (ten times) less than the bias current.

In many applications the offset current can be neglected. However, high-gain, high-input impedance amplifiers should have as little I_{OS} as possible, because the difference in currents through large input resistances develops a substantial offset voltage, as shown in Figure 21–14.

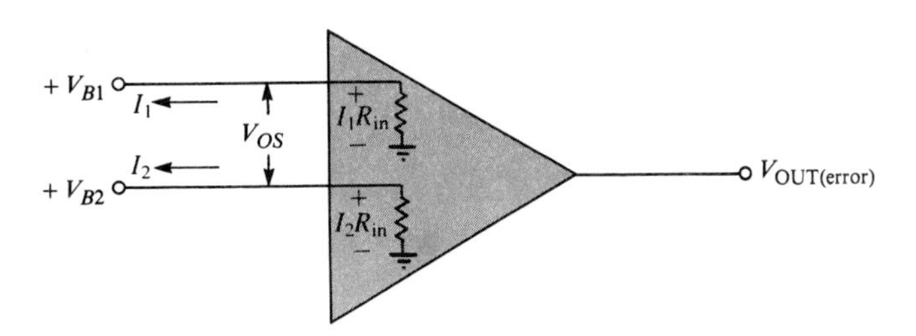

FIGURE 21–14
Effect of input offset current.

The offset voltage developed by the input offset current is

$$V_{OS} = I_1R_{in} - I_2R_{in} = (I_1 - I_2)R_{in}$$

$$V_{OS} = I_{OS}R_{in} \tag{21–6}$$

The error created by I_{OS} is amplified by the gain A_v of the op-amp and appears in the output as

$$V_{OUT(error)} = A_vI_{OS}R_{in} \tag{21–7}$$

The *change* in offset current with temperature is often an important consideration. Values of temperature coefficient in the range of 0.5 nA per degree Celsius are common.

Output Impedance

This parameter is the resistance viewed from the output terminal of the op-amp, as indicated in Figure 21–15.

FIGURE 21–15
Op-amp output impedance.

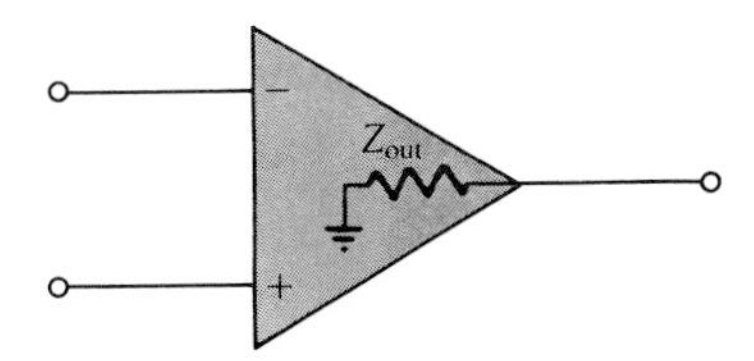

Common-Mode Range

All op-amps have limitations on the range of voltages over which they will operate. The *common-mode range* is the range of input voltages which, when applied to both inputs, will not cause clipping or other output distortion. Many op-amps have common-mode ranges of ±10 V with dc supply voltages of ±15 V.

Open-Loop Voltage Gain, A_{ol}

This is the gain of the op-amp without any external feedback from output to input. A good op-amp has a very high open-loop gain; 50,000 to 200,000 is typical.

Common-Mode Rejection Ratio

The *common-mode rejection ratio* (CMRR), as discussed in conjunction with the diff-amp, is a measure of an op-amp's ability to reject common-mode signals. An infinite value of CMRR means that the output is zero when the same signal is applied to both inputs (common-mode).

An infinite CMRR is never achieved in practice, but a good op-amp does have a very high value of CMRR. As previously mentioned, common-mode signals are undesired interference voltages such as 60-Hz power-supply ripple and noise voltages due to pickup of radiated energy. A high CMRR enables the op-amp to virtually eliminate these interference signals from the output.

The accepted definition of CMRR for an op-amp is the *open-loop gain* (A_{ol}) divided by the common-mode gain:

$$\text{CMRR} = \frac{A_{ol}}{A_{cm}} \tag{21–8}$$

It is commonly expressed in decibels as follows:

$$\text{CMRR} = 20 \log \frac{A_{ol}}{A_{cm}} \tag{21–9}$$

EXAMPLE 21–3 A certain op-amp has an open-loop gain of 100,000 and a common-mode gain of 0.25. Determine the CMRR and express it in dB.

Solution

$$\text{CMRR} = \frac{A_{ol}}{A_{cm}} = \frac{100{,}000}{0.25} = 400{,}000$$

$$\text{CMRR} = 20 \log(400{,}000) = 112 \text{ dB}$$

Slew Rate

The maximum rate of change of the output voltage in response to a step input voltage is the *slew rate* of an op-amp. The slew rate is dependent upon the frequency response of the amplifier stages within the op-amp.

Slew rate is measured with an op-amp connected as shown in Figure 21–16(a). This particular op-amp connection is a unity-gain, noninverting configuration which will be discussed later. It gives a *worst-case* (slowest) slew rate.

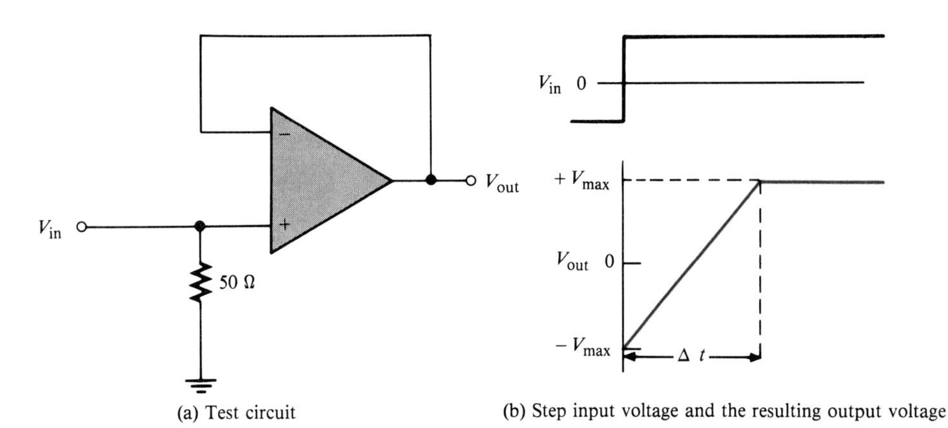

FIGURE 21–16
Measurement of slew rate.

As shown in Part (b), a pulse is applied to the input, and the ideal output voltage is measured. The width of the input pulse must be sufficient to allow the output to "slew" from its lower limit to its upper limit, as shown.

As you can see, a certain time interval, Δt, is required for the output voltage to go from its lower limit $-V_{max}$ to its upper limit $+V_{max}$, once the input step is applied. The slew rate is expressed as

$$\text{Slew rate} = \frac{\Delta V_{out}}{\Delta t} \tag{21–10}$$

where $\Delta V_{out} = +V_{max} - (-V_{max})$. The unit of slew rate is volts per microsecond (V/μs).

EXAMPLE 21–4 The output voltage of a certain op-amp appears as shown in Figure 21–17 in response to a step input. Determine the slew rate.

FIGURE 21–17

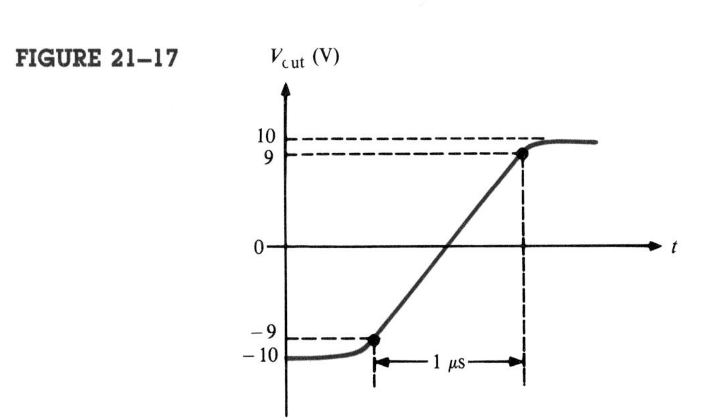

Solution

The output goes from the lower to the upper limit in 1 μs. Since this response is not ideal, the limits are taken at the 90% points, as indicated. Thus, the upper limit is +9 V and the lower limit is −9 V.

The slew rate is

$$\frac{\Delta V}{\Delta t} = \frac{+9\text{ V} - (-9\text{ V})}{1\ \mu\text{s}} = 18\text{ V}/\mu\text{s}$$

Frequency Response

The internal amplifier stages that make up an op-amp have voltage gains limited by junction capacitances. Although the diff-amps used in op-amps are somewhat different from the basic amplifiers discussed, the same principles apply. An op-amp has no internal coupling capacitors, however; therefore, the low frequency response extends down to dc. Frequency-related characteristics will be discussed in the next chapter.

Comparison of Op-Amp Parameters

Table 21–1 provides a comparison of values of some of the parameters just described for four popular IC op-amps. Any values not listed were not given on the manufacturer's data sheet. All values are typical at 25°C.

TABLE 21–1

Parameter	Op-Amp Type			
	MC1741	**LM101**	**LM108**	**LM118**
Input offset voltage	1 mV	1 mV	0.7 mV	2 mV
Input bias current	80 nA	120 nA	0.8 nA	120 nA
Input offset current	20 nA	40 nA	0.05 nA	6 nA
Input impedance	2 MΩ	800 kΩ	70 MΩ	3 MΩ
Output impedance	75 Ω	—	—	—
Open-loop gain	200,000	160,000	300,000	200,000
Slew rate	0.5 V/μs	—	—	70 V/μs
CMRR	90 dB	90 dB	100 dB	100 dB

Other Features

Most available op-amps have two very important features: short-circuit protection and no latch-up. Short-circuit protection keeps the circuit from being damaged if the output becomes shorted, and the no-latch-up feature prevents the op-amp from hanging up in one output state (high or low voltage level) under certain input conditions.

SECTION REVIEW 21–3

1. List ten or more op-amp parameters.
2. List two parameters, not including frequency response, that are frequency dependent.

OP-AMPS WITH NEGATIVE FEEDBACK

21–4

In this section we will discuss several basic ways in which an op-amp can be connected using *negative feedback* to stabilize the gain and increase frequency response.

The extremely high open-loop gain of an op-amp creates an unstable situation because a small noise voltage on the input can be amplified to a point where the amplifier is driven out of its linear region. Also, unwanted oscillations can occur. In addition, the open-loop gain parameter of an op-amp can vary greatly from one device to the next.

Negative feedback takes a portion of the output and applies it back *out of phase* with the input, creating an effective reduction in gain. This closed-loop gain is usually much less than the open-loop gain and independent of it.

Noninverting Amplifier

An op-amp connected as a *noninverting* amplifier with a controlled amount of voltage gain is shown in Figure 21–18. The input signal is applied to the noninverting input. The output is applied back to the inverting input through the feedback network formed by R_i and R_f, creating *negative feedback* as follows:

R_i and R_f form a voltage-divider network which reduces the output V_{out} and connects the reduced voltage V_f to the inverting input. The feedback voltage is expressed as

$$V_f = \left(\frac{R_i}{R_i + R_f}\right)V_{out} \qquad (21\text{–}11)$$

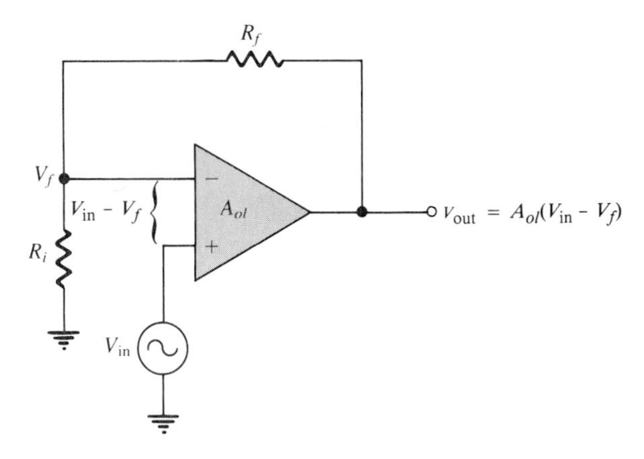

FIGURE 21–18
Noninverting amplifier.

The difference of the input voltage V_{in} and the feedback voltage V_f is the *differential input* to the op-amp, as shown in the figure. This differential voltage is amplified by the open-loop gain of the op-amp (A_{ol}) and produces an output voltage expressed as

$$V_{out} = A_{ol}(V_{in} - V_f) \qquad (21\text{–}12)$$

Letting $R_i/(R_i + R_f) = B$ and then substituting BV_{out} for V_f in Equation (21–12), we get the following algebraic steps:

$$V_{out} = A_{ol}(V_{in} - BV_{out})$$
$$V_{out} = A_{ol}V_{in} - A_{ol}BV_{out}$$
$$V_{out} + A_{ol}BV_{out} = A_{ol}V_{in}$$
$$V_{out}(1 + A_{ol}B) = A_{ol}V_{in}$$

Since the total voltage gain of the amplifier in Figure 21–18 is V_{out}/V_{in}, it can be expressed as

$$\frac{V_{out}}{V_{in}} = \frac{A_{ol}}{1 + A_{ol}B}$$

The product $A_{ol}B$ is usually much greater than 1, so the equation simplifies to

$$\frac{V_{out}}{V_{in}} \cong \frac{A_{ol}}{A_{ol}B}$$

Since

$$A_{cl(NI)} = \frac{V_{out}}{V_{in}}$$

then

$$A_{cl(NI)} \cong \frac{1}{B} \qquad (21\text{–}13)$$

Equation (21–13) shows that the *closed-loop gain* $A_{cl(NI)}$ of the noninverting (NI) amplifier is the reciprocal of the attenuation (B) of the feedback network (voltage divider). It is very interesting to note that the closed-loop gain is not at all dependent on the op-amp's open-loop gain under the condition $A_{ol}B >> 1$. The closed-loop gain can be set by selecting values of R_i and R_f.

EXAMPLE 21–5 Determine the gain of the amplifier in Figure 21–19. The open-loop voltage gain is 100,000.

FIGURE 21–19

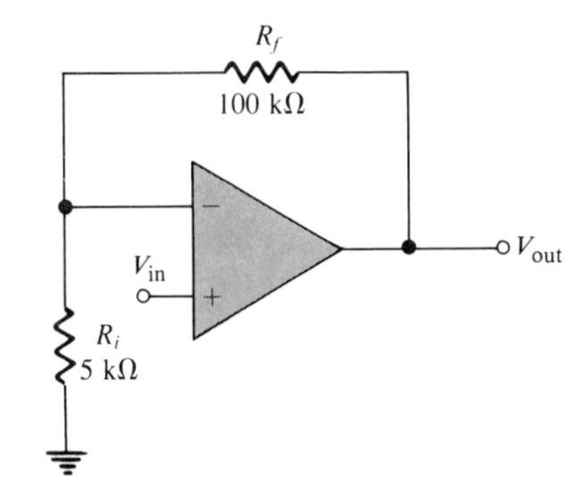

Solution
This is a noninverting op-amp configuration. The feedback gain is

$$B = \frac{R_i}{R_i + R_f} = \frac{5\ \text{k}\Omega}{105\ \text{k}\Omega} = 0.0476$$

Therefore,

$$A_{cl(\text{NI})} \cong \frac{1}{B} = \frac{1}{0.0476} = 21$$

Voltage-Follower

The voltage-follower configuration is a special case of the noninverting amplifier in which all of the output voltage is fed back to the inverting input, as shown in Figure 21–20. As you can see, the straight feedback connection has a voltage gain of approximately 1. The closed-loop voltage gain of a noninverting amplifier is $1/B$ as previously derived. Since $B = 1$, the closed-loop gain of the voltage-follower (VF) is

$$A_{cl(\text{VF})} \cong \frac{1}{B} = 1 \quad (21\text{–}14)$$

FIGURE 21–20
Op-amp voltage-follower.

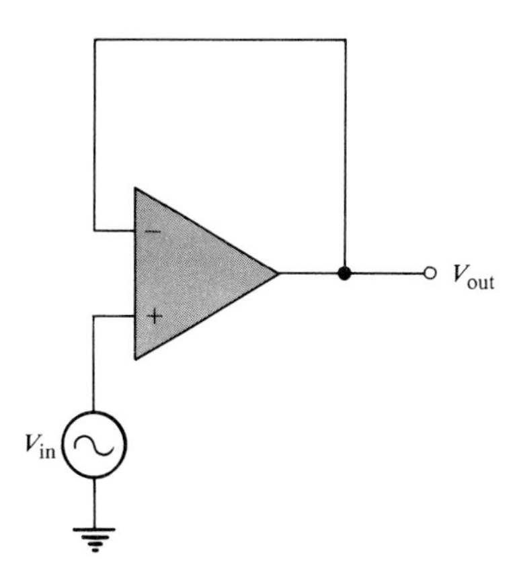

The most important features of the voltage-follower configuration are its *very high input impedance* and its *very low output impedance.* These features make it a nearly ideal buffer amplifier for interfacing high-impedance sources and low-impedance loads.

Inverting Amplifier

An op-amp connected as an *inverting* amplifier with a controlled amount of voltage gain is shown in Figure 21–21. The input signal is applied through a series input resistor R_i to the inverting input. Also, the output is fed back through R_f to the same input. The noninverting input is grounded.

At this point, the ideal op-amp parameters mentioned earlier are very useful in simplifying the analysis of this circuit. In particular, the concept of infi-

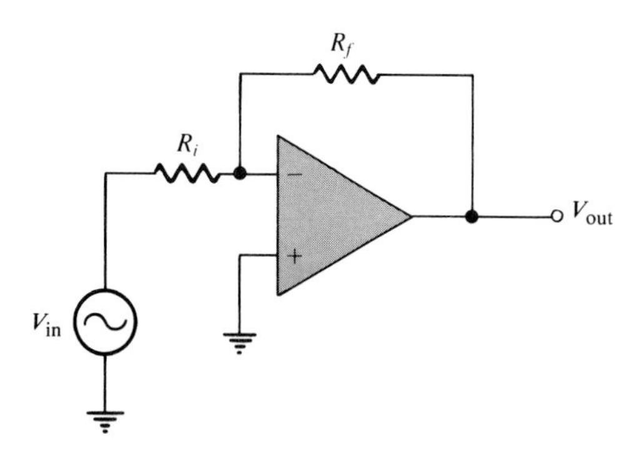

FIGURE 21–21
Inverting amplifier.

nite input impedance is of great value. An infinite input impedance implies *zero current* to the inverting input. If there is zero current through the input impedance, then there must be *no* voltage drop between the inverting and noninverting inputs. That is, the voltage at the inverting $(-)$ input is *zero* because the other input $(+)$ is grounded. This zero voltage at the inverting input terminal is referred to as *virtual ground* and is illustrated in Figure 21–22(a).

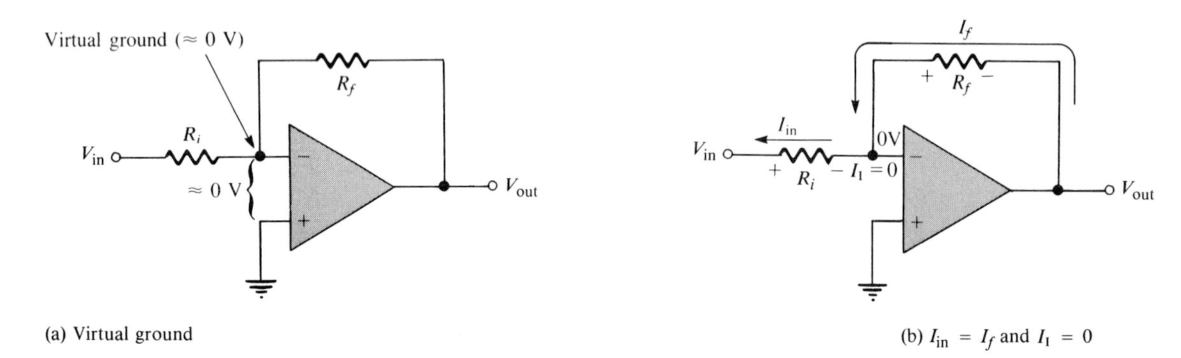

FIGURE 21–22
Virtual ground concept and development of closed-loop voltage gain for the inverting amplifier.

Since there is no current into the inverting input, the current through R_i and the current through R_f are equal, as shown in Figure 21–22(b):

$$I_{in} = I_f$$

The voltage across R_i equals V_{in} because of virtual ground on the other side of the resistor. Therefore,

$$I_{in} = \frac{V_{in}}{R_i}$$

Also, the voltage across R_f equals $-V_{out}$ because of virtual ground, and therefore,

$$I_f = \frac{-V_{out}}{R_f}$$

Since $I_f = I_{in}$,

$$\frac{-V_{out}}{R_f} = \frac{V_{in}}{R_i}$$

Rearranging the terms, we get

$$\frac{V_{out}}{V_{in}} = -\frac{R_f}{R_i}$$

Of course, you recognize V_{out}/V_{in} as the overall gain of the amplifier:

$$A_{cl(I)} = -\frac{R_f}{R_i} \quad (21\text{–}15)$$

Equation (21–15) shows that the closed-loop voltage gain $A_{cl(I)}$ of the inverting (I) amplifier is the ratio of the feedback resistance R_f to the resistance R_i.

The *closed-loop gain* is independent of the op-amp's internal *open-loop gain*. Thus, the negative feedback stabilizes the voltage gain. The negative sign indicates inversion.

EXAMPLE 21–6 Given the op-amp configuration in Figure 21–23, determine the value of R_f required to produce a closed-loop voltage gain of 100.

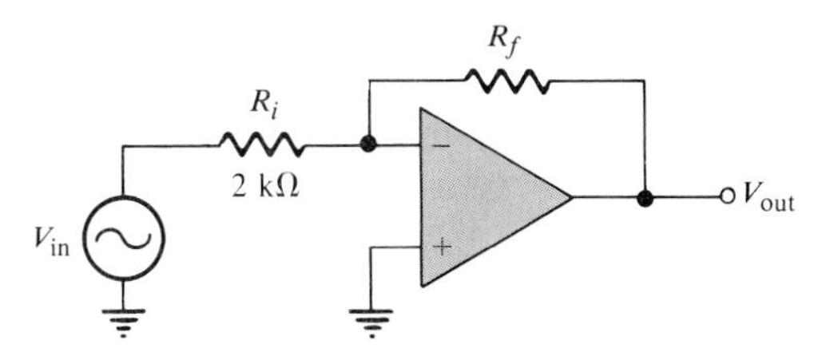

FIGURE 21–23

Solution

$$R_i = 2\ \text{k}\Omega \quad \text{and} \quad A_{cl(I)} = 100$$

$$A_{cl(I)} = \left|\frac{R_f}{R_i}\right|$$

$$R_f = A_{cl(I)}R_i = (100)(2\ \text{k}\Omega) = 200\ \text{k}\Omega$$

SECTION REVIEW 21–4

1. What is the main purpose of negative feedback?
2. The closed-loop voltage gain of each of the op-amp configurations discussed is dependent on the internal open-loop voltage gain of the op-amp (T or F).

3. The attenuation (B) of the negative feedback network of a noninverting op-amp configuration is 0.02. What is the closed-loop gain of the amplifier?

EFFECTS OF NEGATIVE FEEDBACK ON OP-AMP IMPEDANCES

21–5

Data sheets for IC op-amps usually give values for the input impedance (Z_{in}) and the output impedance (Z_{out}). These values are for the op-amp *without any external feedback connections.* Each of the negative feedback configurations discussed in the preceding section produces an overall input and output impedance that is different from the op-amp's internal Z_{in} and Z_{out} (with one exception).

Input and Output Impedances of the Noninverting Amplifier

The input impedance of the noninverting amplifier configuration, $Z_{in(NI)}$, shown in Figure 21–24, is greater than the internal input impedance of the op-amp itself (without feedback) by a factor of $1 + A_{ol}B$, as expressed by the following equation:

$$Z_{in(NI)} = (1 + A_{ol}B)Z_{in} \tag{21–16}$$

FIGURE 21–24
Noninverting amplifier.

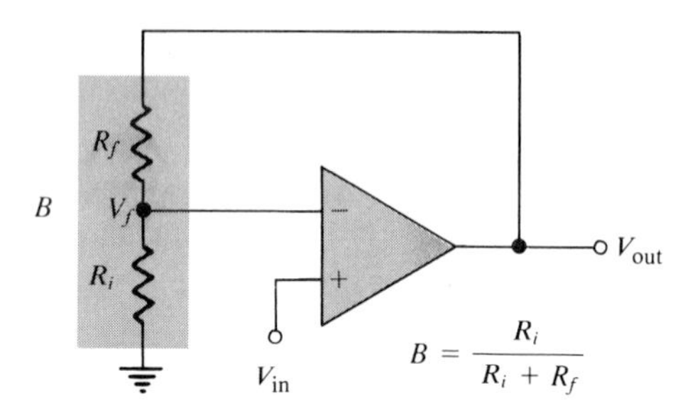

The output impedance with the negative feedback, $Z_{out(NI)}$, is less than the op-amp output impedance by a factor of $1/(1 + A_{ol}B)$, expressed as follows:

$$Z_{out(NI)} = \frac{Z_{out}}{1 + A_{ol}B} \tag{21–17}$$

In summary, the negative feedback in a noninverting configuration increases the input impedance and decreases the output impedance.

EXAMPLE 21–7

(a) Determine the input and output impedances of the amplifier in Figure 21–25. The op-amp data sheet gives $Z_{in} = 2\ \text{M}\Omega$, $Z_{out} = 75\ \Omega$, and $A_{ol} = 200{,}000$.

(b) Find the closed-loop voltage gain.

Solution

(a) The attenuation of the feedback network is

$$B = \frac{R_i}{R_i + R_f} = \frac{10\text{ k}\Omega}{210\text{ k}\Omega} = 0.048$$

$$Z_{\text{in(NI)}} = (1 + A_{ol}B)Z_{\text{in}} = [1 + (200{,}000)(0.048)]2\text{ M}\Omega$$
$$= (1 + 9600)2\text{ M}\Omega = 19{,}202\text{ M}\Omega$$

$$Z_{\text{out(NI)}} = \frac{Z_{\text{out}}}{1 + A_{ol}B} = \frac{75\ \Omega}{1 + 9600} = 0.0078\ \Omega$$

(b) $A_{cl\text{(NI)}} = \dfrac{1}{B} = \dfrac{1}{0.048} = 20.83$

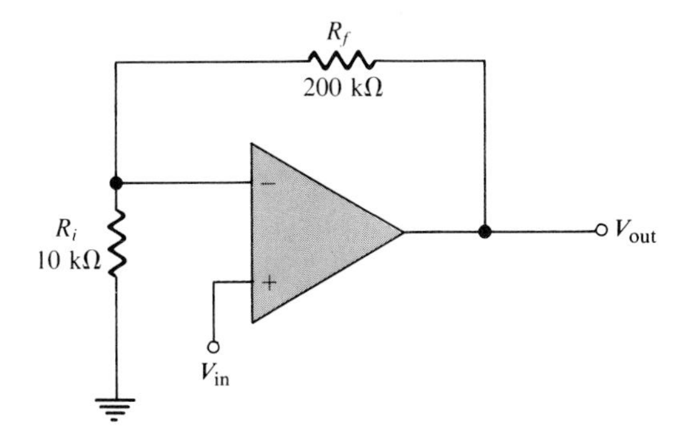

FIGURE 21–25

Voltage-Follower Impedances

Since the voltage-follower is a special case of the noninverting configuration, the same impedance formulas are used with $B = 1$:

$$Z_{\text{in(VF)}} = (1 + A_{ol})Z_{\text{in}} \tag{21–18}$$

$$Z_{\text{out(VF)}} = \frac{Z_{\text{out}}}{1 + A_{ol}} \tag{21–19}$$

As you can see, the voltage-follower input impedance is greater for a given A_{ol} and Z_{in} than for the noninverting configuration with the voltage-divider feedback network. Also, its output impedance is much smaller because B is normally much less than 1 for a noninverting configuration.

EXAMPLE 21–8 The same op-amp in Example 21–7 is used in a voltage-follower configuration. Determine the input and output impedance.

Solution
Since $B = 1$,

$$Z_{\text{in(VF)}} = (1 + A_{ol})Z_{\text{in}} = (1 + 200{,}000)2\text{ M}\Omega$$
$$\cong 400{,}000\text{ M}\Omega$$

$$Z_{out(VF)} = \frac{Z_{out}}{1 + A_{ol}} = \frac{75\ \Omega}{1 + 200{,}000}$$
$$= 0.00038\ \Omega$$

Notice that $Z_{in(VF)}$ is much greater than $Z_{in(NI)}$, and $Z_{out(VF)}$ is much less than $Z_{out(NI)}$ from Example 21–7.

Input and Output Impedances of the Inverting Amplifier

For the inverting amplifier configuration shown in Figure 21–26, the input impedance, $Z_{in(I)}$, approximately equals the external input resistance, R_i:

$$Z_{in(I)} \cong R_i \quad (21\text{–}20)$$

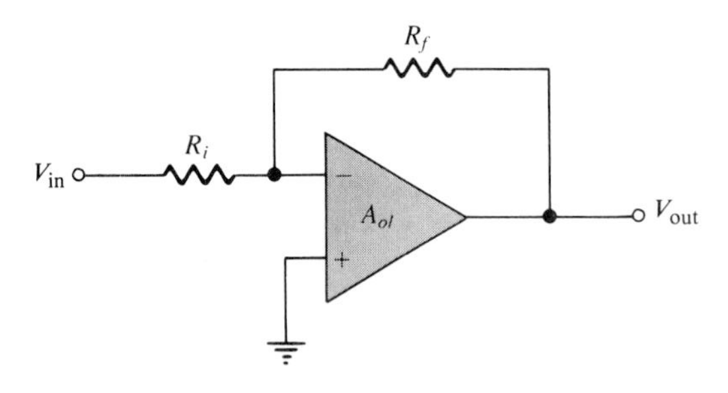

FIGURE 21–26
Inverting amplifier.

The output impedance, $Z_{out(I)}$, approximately equals the internal output impedance of the op-amp, Z_{out}:

$$Z_{out(I)} \cong Z_{out} \quad (21\text{–}21)$$

EXAMPLE 21–9 The op-amp in Figure 21–27 has the following parameters: $A_{ol} = 50{,}000$, $Z_{in} = 4\ \text{M}\Omega$, and $Z_{out} = 50\ \Omega$.

(a) Find the values of the input and output impedances.
(b) Determine the closed-loop voltage gain.

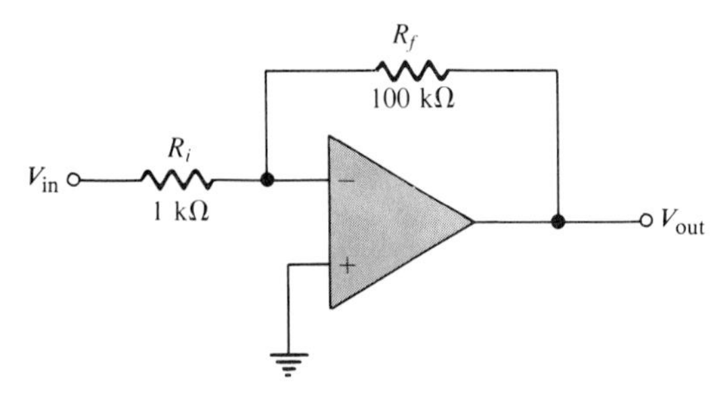

FIGURE 21–27

Solution

(a) $Z_{in(I)} \cong R_i = 1\ \text{k}\Omega$

$Z_{out(I)} \cong Z_{out} = 50\ \Omega$

(b) $$A_{cl(I)} = -\frac{R_f}{R_i} = -\frac{100\ \text{k}\Omega}{1\ \text{k}\Omega} = -100$$

SECTION REVIEW 21–5

1. How does the input impedance of a noninverting amplifier configuration compare to the input impedance of the op-amp itself?
2. Connecting an op-amp in a voltage-follower configuration (increases, decreases) the input impedance.
3. Given that $R_f = 100\ \text{k}\Omega$, $R_i = 2\ \text{k}\Omega$, $A_{ol} = 120{,}000$, $Z_{in} = 2\ \text{M}\Omega$, and $Z_{out} = 60\ \Omega$, determine $Z_{in(I)}$ and $Z_{out(I)}$ for an inverting amplifier configuration.

APPLICATION NOTE

For the Application Assignment at the beginning of the chapter, the type of IC op-amp to be used must be determined. Ask the engineer to specify one. Use a potentiometer as the feedback resistor, R_f, and a fixed resistor for R_i. Choose the value of the pot and the value of R_i so that feedback attenuation, B, gives you a maximum gain of 50 using the following formula:

$$A_{cl(NI)} = \frac{1}{B}$$

When the pot is set at its minimum value of zero, $B = 1$, and thus, $A_{cl(NI)} = 1$. For a gain of 50, B must be 0.02. If R_i is selected to be 1 kΩ, then the maximum value of the pot must be 49 kΩ to produce $B = 0.02$. Select a 50-kΩ pot, which is standard.

SUMMARY

Facts

- The op-amp has three terminals not including power and ground: inverting input (−), noninverting input (+), and output.
- Most op-amps require both a positive and a negative dc supply voltage.
- The ideal (perfect) op-amp has infinite input impedance, zero output impedance, infinite open-loop voltage gain, infinite bandwidth, and infinite CMRR.
- A good practical op-amp has high input impedance, low output impedance, high open-loop voltage gain, and wide bandwidth.
- A diff-amp is normally used for the input stage of an op-amp.
- A differential input voltage appears between the inverting and noninverting inputs of a diff-amp.
- A single-ended input voltage appears between one input and ground (with the other inputs grounded).
- A differential output voltage appears between two output terminals of a diff-amp.
- A single-ended output voltage appears between the output and ground of a diff-amp.

- Common mode occurs when equal, in-phase voltages are applied to both input terminals.
- Input offset voltage produces an output error voltage (with no input voltage).
- Input bias current also produces an output error voltage (with no input voltage).
- Input offset current is the difference between the two bias currents.
- Open-loop voltage gain is the gain of the op-amp with no external feedback connections.
- Slew rate is the rate (in volts per microsecond) that the output voltage of an op-amp can change in response to a step input.
- There are three basic op-amp configurations: inverting, noninverting, and voltage-follower.
- All op-amp configurations (except comparators, to be covered in Chapter 22) employ negative feedback. Negative feedback occurs when a portion of the output voltage is connected back to the inverting input such that it subtracts from the input voltage, thus reducing the voltage gain but increasing the stability and bandwidth.
- A noninverting amplifier configuration has a higher input impedance and a lower output impedance than the op-amp itself (without feedback).
- An inverting amplifier configuration has an input impedance approximately equal to the input resistor R_i and an output impedance approximately equal to the internal output impedance of the op-amp itself.
- The voltage-follower has the highest input impedance and the lowest output impedance of the three configurations.

Definitions

- *Open-loop gain, A_{ol}*—the internal voltage gain of an op-amp without feedback.
- *Closed-loop gain, A_{cl}*—the overall voltage gain of an op-amp with feedback.
- *Common-mode rejection ratio (CMRR)*—a measure of an op-amp's ability to reject signals that appear the same on both inputs.

Formulas

$\text{CMRR} = \dfrac{A_{v(d)}}{A_{cm}}$ Common-mode rejection ratio (diff-amp) **(21–1)**

$\text{CMRR} = 20 \log \dfrac{A_{v(d)}}{A_{cm}}$ Common-mode rejection ratio (dB) **(21–2)**

$V_{OUT} = I_{C2}R_C - I_{C1}R_C$ Differential output **(21–3)**

$I_{BIAS} = \dfrac{I_1 + I_2}{2}$ Input bias current **(21–4)**

$I_{OS} = |I_1 - I_2|$ Input offset current **(21–5)**

$V_{OS} = I_{OS}R_{in}$ Offset voltage **(21–6)**

$V_{OUT(error)} = A_v I_{OS} R_{in}$ Output error voltage **(21–7)**

$\text{CMRR} = \dfrac{A_{ol}}{A_{cm}}$ Common-mode rejection ratio (op-amp) **(21–8)**

$\text{CMRR} = 20 \log \dfrac{A_{ol}}{A_{cm}}$ Common-mode rejection ratio (dB) **(21–9)**

$\text{Slew rate} = \dfrac{\Delta V_{out}}{\Delta t}$ Slew rate **(21–10)**

$$V_f = \left(\frac{R_i}{R_i + R_f}\right)V_{out} \quad \text{Feedback voltage (noninverting)} \quad (21\text{–}11)$$

$$V_{out} = A_{ol}(V_{in} - V_f) \quad \text{Output voltage (noninverting)} \quad (21\text{–}12)$$

$$A_{cl(NI)} \cong \frac{1}{B} \quad \text{Voltage gain (noninverting)} \quad (21\text{–}13)$$

$$A_{cl(VF)} \cong \frac{1}{B} = 1 \quad \text{Voltage gain (voltage-follower)} \quad (21\text{–}14)$$

$$A_{cl(I)} = -\frac{R_f}{R_i} \quad \text{Voltage gain (inverting)} \quad (21\text{–}15)$$

$$Z_{in(NI)} = (1 + A_{ol}B)Z_{in} \quad \text{Input impedance (noninverting)} \quad (21\text{–}16)$$

$$Z_{out(NI)} = \frac{Z_{out}}{1 + A_{ol}B} \quad \text{Output impedance (noninverting)} \quad (21\text{–}17)$$

$$Z_{in(VF)} = (1 + A_{ol})Z_{in} \quad \text{Input impedance (voltage-follower)} \quad (21\text{–}18)$$

$$Z_{out(VF)} = \frac{Z_{out}}{1 + A_{ol}} \quad \text{Output impedance (voltage-follower)} \quad (21\text{–}19)$$

$$Z_{in(I)} \cong R_i \quad \text{Input impedance (inverting)} \quad (21\text{–}20)$$

$$Z_{out(I)} \cong Z_{out} \quad \text{Output impedance (inverting)} \quad (21\text{–}21)$$

SELF-TEST

1. Which of the following characteristics do not *necessarily* apply to an op-amp?
   (a) high gain **(b)** low power
   (c) high input impedance **(d)** low output impedance
   (e) high CMRR **(f)** dc isolation
2. Describe how a diff-amp basically differs from an op-amp.
3. Explain the difference between common-mode input impedance and the differential input impedance.
4. **(a)** Ideally, what value of CMRR is desirable?
   (b) One particular op-amp has a CMRR of 80 dB and another has a CMRR of 100 dB. Which would you use, all other factors being equal?
5. The output voltage of a particular op-amp increases 8 V in 12 μs in response to a step voltage on the input. Determine the slew rate.
6. Identify each of the op-amp configurations in Figure 21–28.

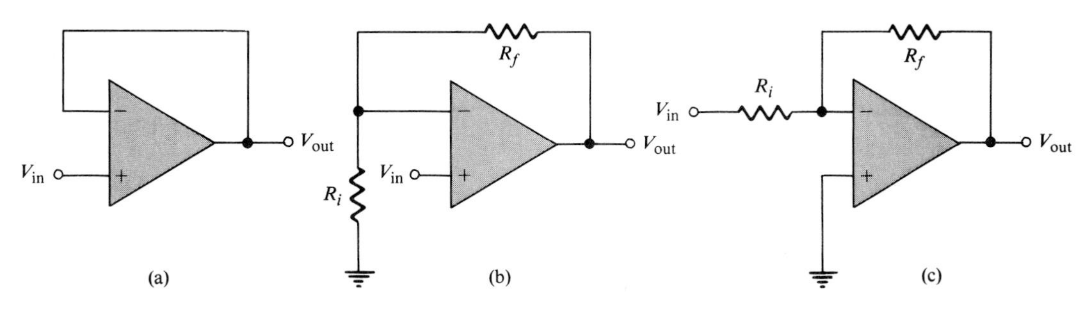

FIGURE 21–28

7. A noninverting amplifier has an R_i of 1 kΩ and an R_f of 100 kΩ. Determine V_f and B, if $V_{out} = 5$ V.
8. What is the closed-loop gain of the amplifier in Question 7?
9. Which of the following characteristics apply to a voltage-follower?
   (a) $A_{cl} > 1$ (b) inverting
   (c) noninverting (d) $A_{cl} \cong 1$
   (e) high Z_{in} (f) high Z_{out}
10. An inverting amplifier has the following circuit values: $R_f = 220$ kΩ, $R_i = 2.2$ kΩ, and $A_{ol} = 25{,}000$. What is the closed-loop gain?
11. Given an op-amp's open-loop gain, can you find its closed-loop gain in a given configuration without any further information?
12. How does the feedback attenuation of a noninverting amplifier differ from that of a voltage-follower?
13. In which configuration would you use a certain given IC op-amp if the only requirement were to achieve the highest possible input impedance?
14. The value of B in a certain noninverting amplifier is 0.025. Determine the closed-loop gain.

PROBLEMS

Section 21–1

21–1 Compare a practical op-amp to the ideal.

21–2 Two IC op-amps are available to you. Their characteristics are as follows:
Op-amp 1: $Z_{in} = 5$ MΩ, $Z_{out} = 100$ Ω, $A_{ol} = 50{,}000$
Op-amp 2: $Z_{in} = 10$ MΩ, $Z_{out} = 75$ Ω, $A_{ol} = 150{,}000$
Choose the one you think is generally more desirable.

Section 21–2

21–3 Identify the type of input and output configuration for each diff-amp in Figure 21–29.

21–4 The dc base voltages in Figure 21–30 are zero. Using your knowledge of transistor analysis, determine the dc differential output voltage. Assume that Q_1 has an $\alpha = 0.98$ and Q_2 has an $\alpha = 0.975$.

Section 21–3

21–5 Determine the bias current, I_{BIAS}, given that the input currents to an op-amp are 8.3 μA and 7.9 μA.

21–6 Distinguish between *input bias current* and *input offset current,* and then calculate the input offset current in Problem 21–5.

21–7 A certain op-amp has a CMRR of 250,000. Convert this to dB.

21–8 The open-loop gain of a certain op-amp is 175,000. Its common-mode gain is 0.18. Determine the CMRR in dB.

21–9 An op-amp data sheet specifies a CMRR of 300,000 and an A_{ol} of 90,000. What is the common-mode gain?

21–10 Figure 21–31 shows the output voltage of an op-amp in response to a step input. What is the slew rate?

21–11 How long does it take the output voltage of an op-amp to go from −10 V to +10 V if the slew rate is 0.5 V/μs?

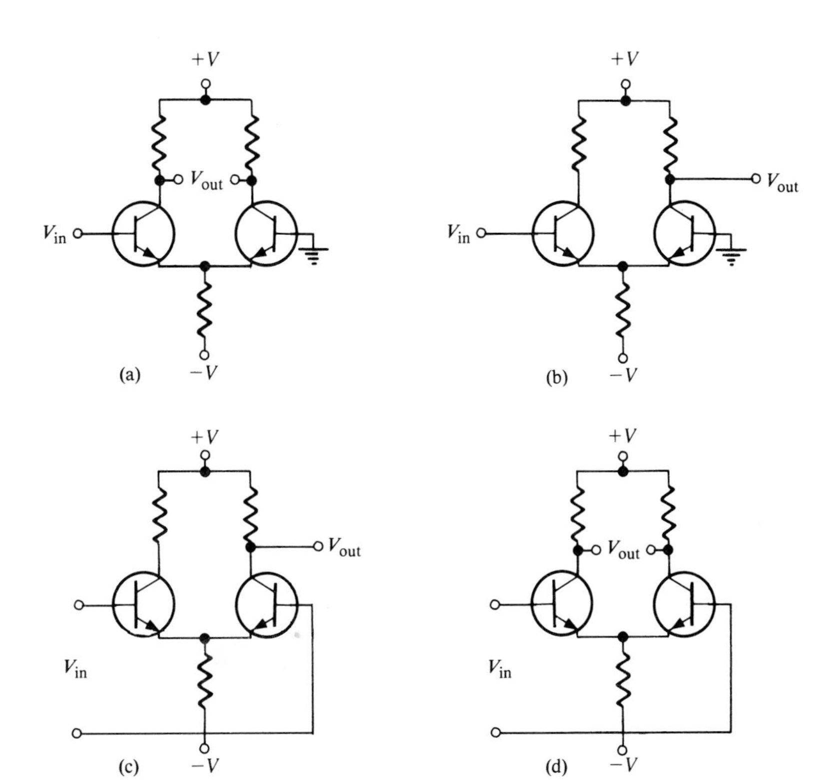

FIGURE 21–29

FIGURE 21–30

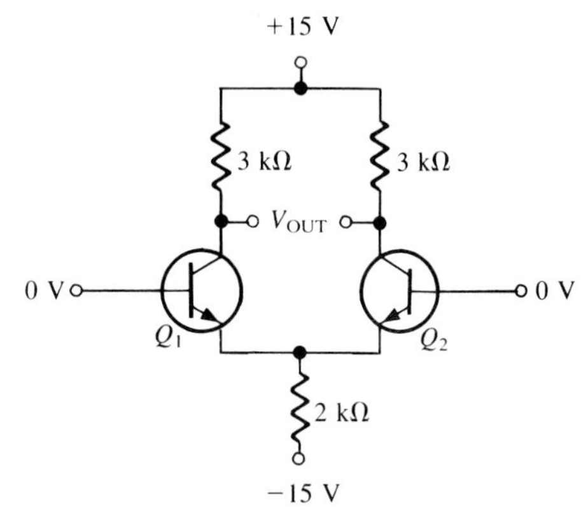

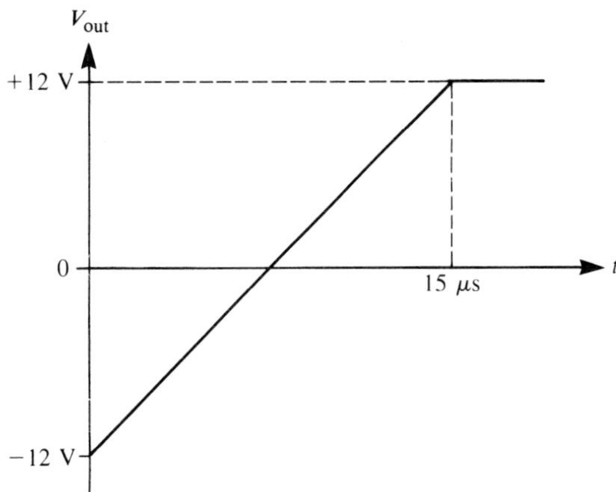

FIGURE 21–31

Section 21–4

21–12 For the amplifier in Figure 21–32, determine the following:
(a) $A_{cl(NI)}$ **(b)** V_{out} **(c)** V_f

FIGURE 21–32

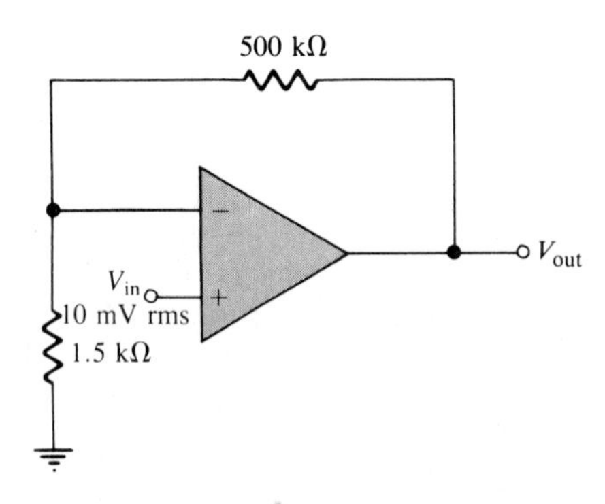

21–13 Determine the closed-loop gain of each amplifier in Figure 21–33.

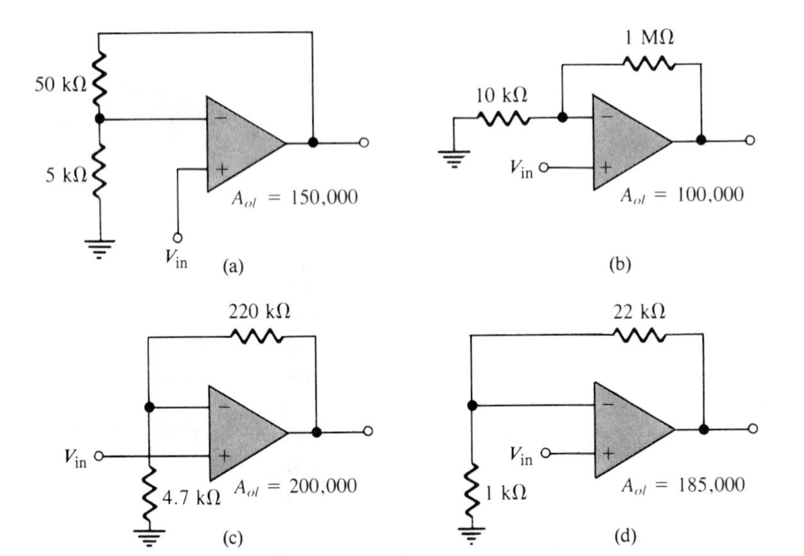

FIGURE 21–33

21–14 Find the value of R_f that will produce the indicated closed-loop gain in each amplifier in Figure 21–34.

21–15 Find the gain of each amplifier in Figure 21–35.

21–16 If a signal voltage of 10 mV rms is applied to each amplifier in Figure 21–35, what are the output voltages and what is their phase relationship with inputs?

21–17 Determine the approximate values for each of the following quantities in Figure 21–36:
(a) I_{in} **(b)** I_f
(c) V_{out} **(d)** closed-loop gain

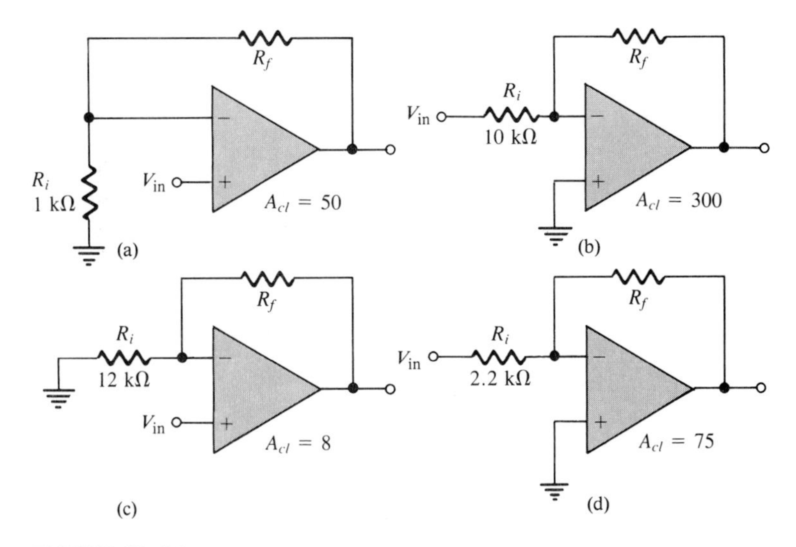

FIGURE 21–34

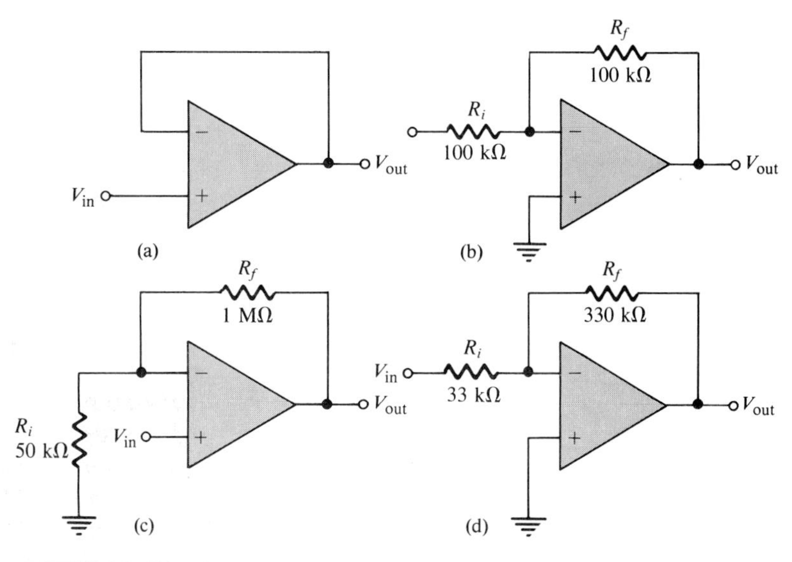

FIGURE 21–35

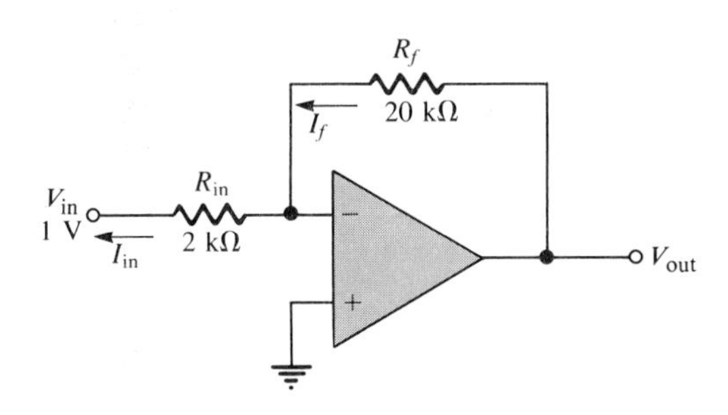

FIGURE 21–36

Section 21–5

21–18 Determine the input and output impedances for each amplifier configuration in Figure 21–37.

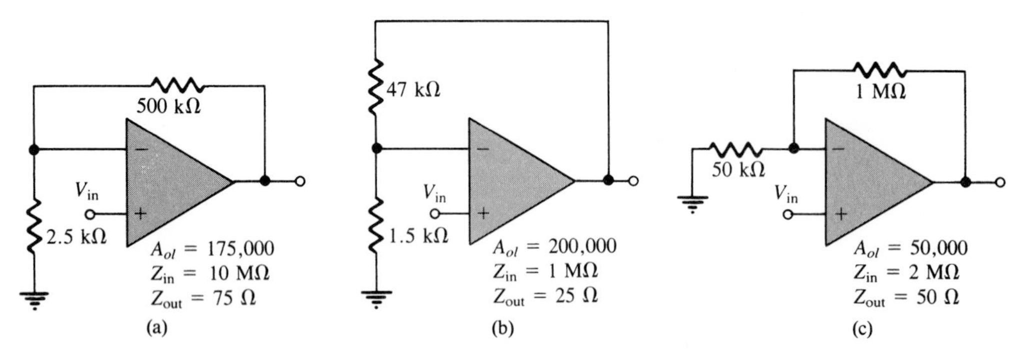

FIGURE 21–37

21–19 Repeat Problem 21–18 for each circuit in Figure 21–38.

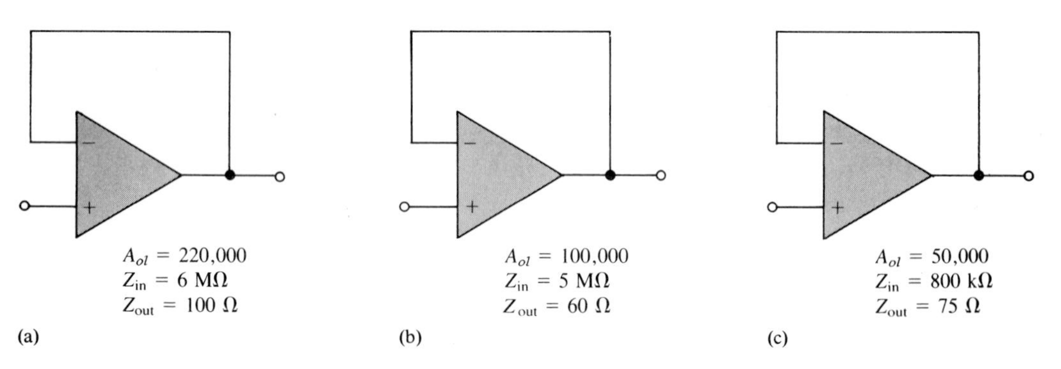

FIGURE 21–38

21–20 Repeat Problem 21–18 for each circuit in Figure 21–39.

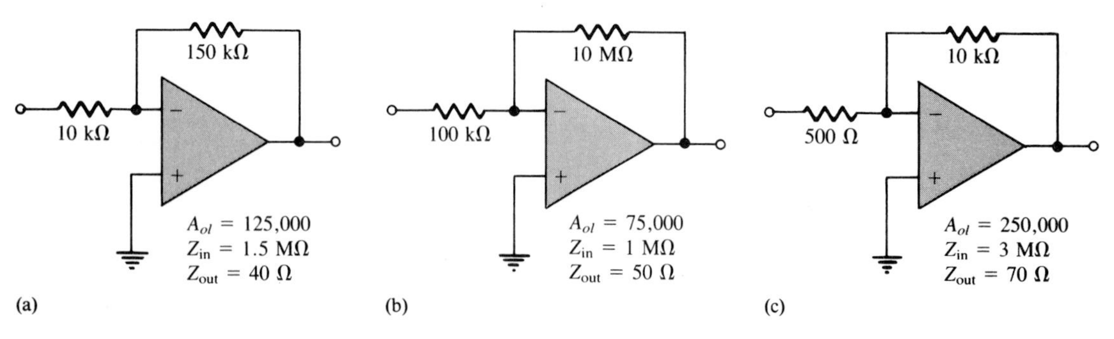

FIGURE 21–39

ANSWERS TO SECTION REVIEWS

Section 21–1

1. See Figure 21–40. **2.** High Z_{in}, low Z_{out}, high voltage gain, wide bandwidth.

FIGURE 21–40

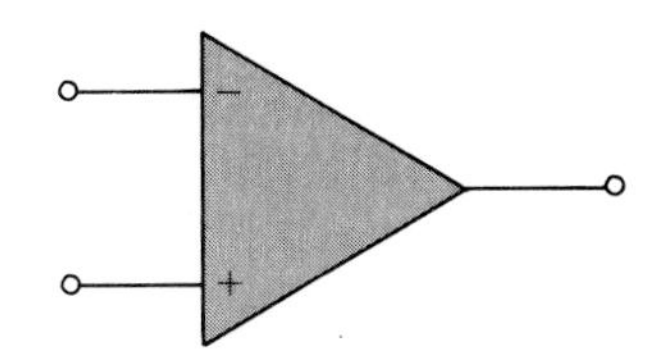

Section 21–2

1. Differential: between two input terminals. Single-ended: from one input terminal to ground (with the other inputs grounded). 2. The ability of an op-amp to reject common-mode signals.

Section 21–3

1. Input bias current, input offset voltage, drift, input offset current, input impedance, output impedance, common-mode range, CMRR, open-loop voltage gain, slew rate, frequency response. **2.** Slew rate and gain.

Section 21–4

1. To stabilize gain. **2.** False. **3.** 50.

Section 21–5

1. Z_{in} of the noninverting amplifier is higher. **2.** Increases. **3.** 2 kΩ, 60 Ω.

Basic Applications of Op-Amps

Op-amps are used in such a wide variety of applications that it is impossible to cover all of them in one chapter or even in one book. Therefore, in this chapter we examine some of the more fundamental applications to illustrate the versatility of the op-amp.

Specifically, you will learn:

- How an op-amp is used as a comparator and how it operates.
- The difference between zero-level detection and nonzero-level detection.
- How an op-amp can be configured as a summing amplifier.
- How to adjust the gain of the summing amplifier.
- How to make a summing amplifier produce an output that is the average of the inputs.
- How to assign various weights to the inputs of a summing amplifier.
- How to use an op-amp as an integrator or differentiator.
- How to connect an op-amp in a configuration that will produce a sawtooth output waveform.
- What a Wien-bridge oscillator is and how it can be implemented with an op-amp.
- The conditions for oscillator start-up and sustained oscillation.
- How an op-amp is used as the active element in low-pass, high-pass, and band-pass filters.
- The significance of a filter pole.
- The meaning of *roll-off* in filter terminology.
- What a three-terminal regulator is.
- How series and shunt regulators work.

22

APPLICATION ASSIGNMENT

As a manufacturing technician in a chemical plant, you are requested by your supervisor to devise an electronic method of determining when the *average* volume of a chemical solution contained in four different storage tanks falls below a specified level. A sensor in each of the storage tanks produces a dc voltage proportional to the remaining volume of solution.

After completing this chapter, you will be able to complete this assignment. A solution is given in the Application Note at the end of the chapter.

COMPARATORS

22–1

Operational amplifiers are often used to *compare* the amplitude of one voltage with another. In this application the op-amp is used in the open-loop configuration, with the input voltage on one input and a *reference voltage* on the other.

Zero-Level Detection

A basic application of the op-amp as a comparator is in determining when an input voltage exceeds a certain level. Figure 22–1(a) shows a *zero-level detector.* Notice that the inverting (−) input is grounded and the input signal voltage is applied to the noninverting (+) input. Because of the high open-loop voltage gain, a very small difference voltage between the two inputs drives the amplifier into saturation, causing the output voltage to go to its limit.

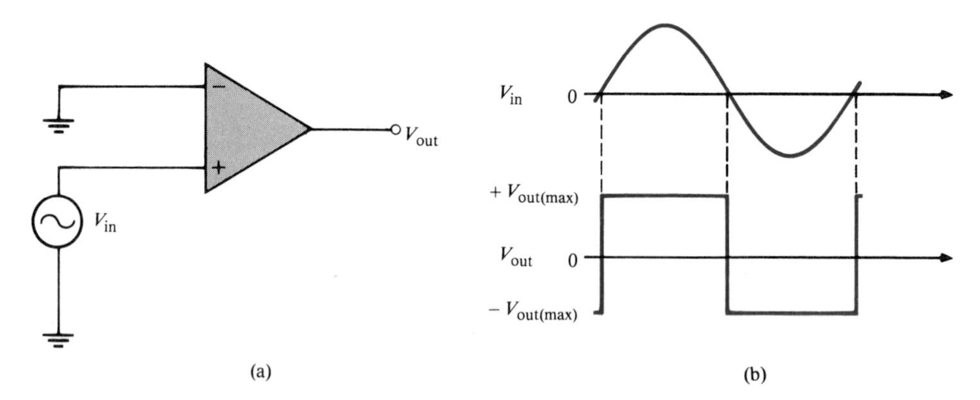

FIGURE 22–1
The op-amp as a zero-level detector.

For example, consider an op-amp having $A_{ol} = 100{,}000$. A voltage difference of only 0.25 mV between the inputs could produce an output voltage of (0.25 mV)(100,000) = 25 V *if the op-amp were capable.* However, since most op-amps have output voltage limitations of less than ±15 V, the device would be driven into saturation.

Figure 22–1(b) shows the result of a sine wave input voltage applied to the noninverting input of the zero-level detector. When the sine wave is *negative,* the output is at its maximum *negative* level. When the sine wave crosses 0, the amplifier is driven to its opposite state and the output goes to its maximum *positive* level, as shown.

As you can see, the zero-level detector can be used as a *squaring circuit* to produce a square wave from a sine wave.

Nonzero-Level Detection

The zero-level detector in Figure 22–1 can be modified to detect voltages other than zero by connecting a *fixed reference voltage,* as shown in Figure 22–2(a). A more practical arrangement is shown in Part (b) using a voltage divider to set the reference voltage as follows:

$$V_{REF} = \frac{R_2}{R_1 + R_2}(+V) \tag{22–1}$$

where $+V$ is the positive op-amp supply voltage.

As long as the input voltage V_{in} is less than V_{REF}, the output remains at the maximum negative level. When the input voltage exceeds the reference voltage, the output goes to its maximum positive state, as shown in Figure 22–2(c) with a sine wave input voltage.

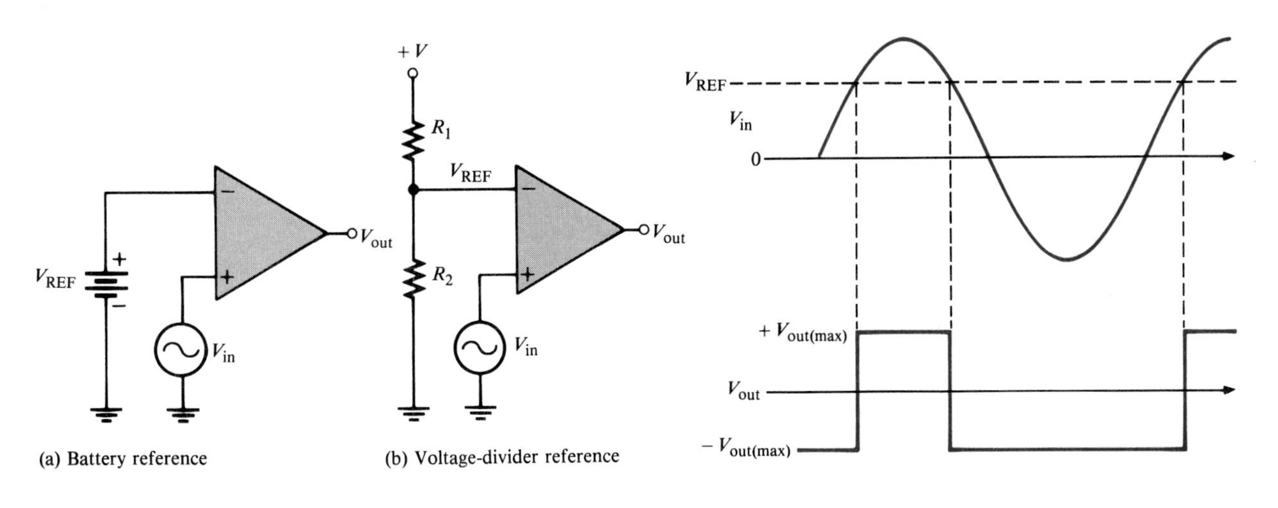

(a) Battery reference (b) Voltage-divider reference (c) Waveforms

FIGURE 22–2
Nonzero-level detectors.

EXAMPLE 22–1 The input signal in Figure 22–3(a) is applied to the comparator circuit in Part (b). Make a sketch of the output showing its proper relationship to the input signal. Assume that the maximum output levels are ± 12 V.

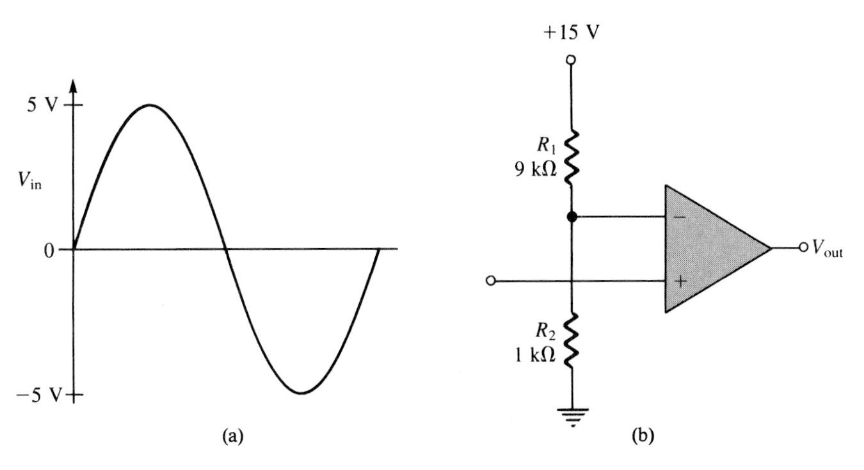

FIGURE 22–3

Solution

The reference voltage is set by R_1 and R_2 as follows:

$$V_{REF} = \frac{R_2}{R_1 + R_2}(+V) = \frac{1\ \text{k}\Omega}{9\ \text{k}\Omega + 1\ \text{k}\Omega}(+15\ \text{V}) = 1.5\ \text{V}$$

As shown in Figure 22–4, each time the input exceeds + 1.5 V, the output voltage switches to its +12 V level, and each time the input goes below +1.5 V, the output switches back to its −12 V level.

FIGURE 22–4

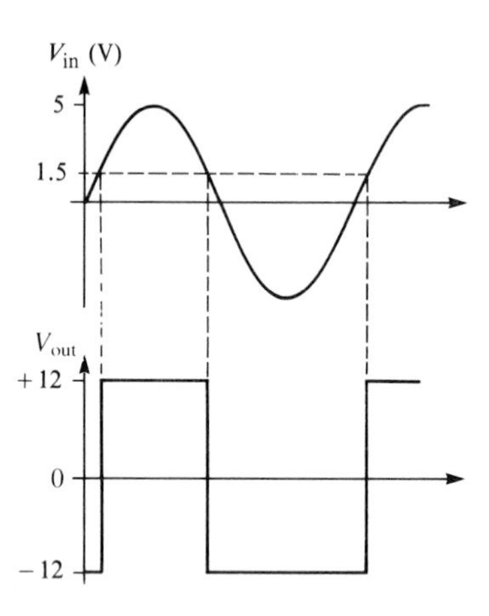

SECTION REVIEW 22–1

1. What is the reference voltage for the comparator in Figure 22–5?
2. Sketch the output waveform for Figure 22–5 when a sine wave with a 5-V peak is applied to the input.

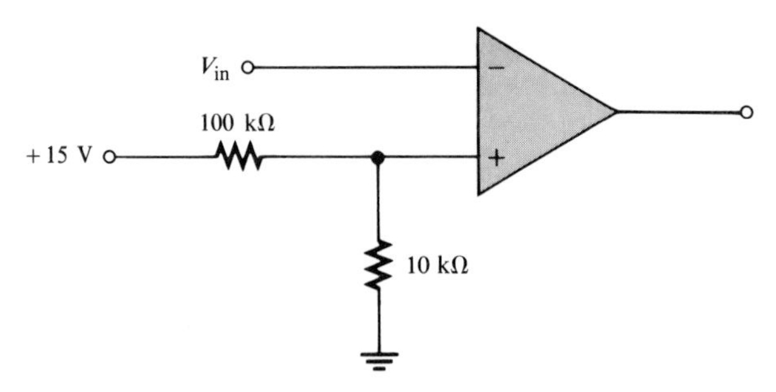

FIGURE 22–5

SUMMING AMPLIFIERS

22–2

The summing amplifier (adder) is a variation of the *inverting* op-amp configuration covered in the last chapter. A two-input summing amplifier is shown in

Figure 22–6, but any number of inputs can be used. The output voltage is proportional to the negative of the algebraic sum of the input voltages.

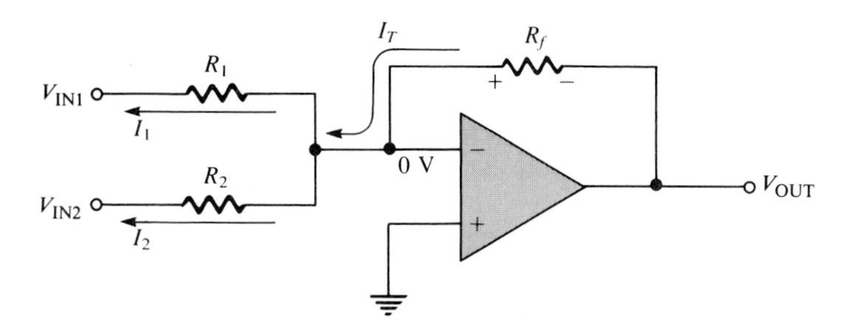

FIGURE 22–6
Two-input, inverting summing amplifier.

The operation of the circuit and derivation of the output expression are as follows: Two voltages, V_{IN1} and V_{IN2}, are applied to the inputs and produce currents I_1 and I_2, as shown in Figure 22–6.

Using the concepts of infinite input impedance and virtual ground, you can see that the inverting input of the op-amp is approximately zero volts, and there is no current into the input. That is, both input currents I_1 and I_2 come together at this summing point, and the total current is through R_f, as indicated:

$$I_T = I_1 + I_2$$

Since $V_{OUT} = -I_T R_f$, the following steps apply:

$$V_{OUT} = -(I_1 + I_2)R_f$$
$$= -\left(\frac{V_{IN1}}{R_1} + \frac{V_{IN2}}{R_2}\right)R_f$$

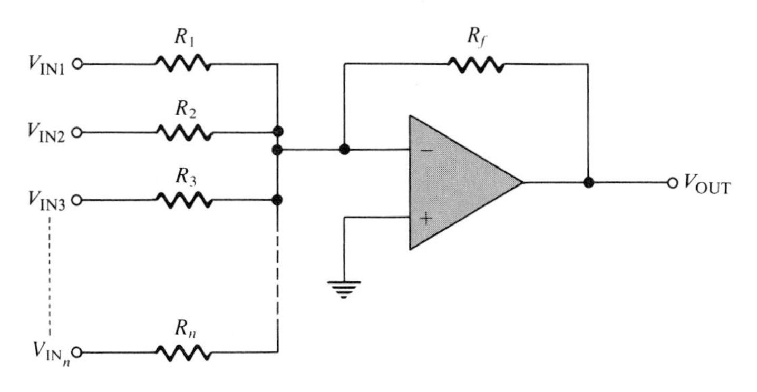

FIGURE 22–7
Summing amplifier with n inputs.

If all three of the resistors are *equal* ($R_1 = R_2 = R_f = R$), then

$$V_{OUT} = -\left(\frac{V_{IN1}}{R} + \frac{V_{IN2}}{R}\right)R$$

$$V_{OUT} = -(V_{IN1} + V_{IN2}) \tag{22–2}$$

Equation (22–2) shows that the output voltage is the *sum* of the two input voltages. A general expression is given in Equation (22–3) for a summing amplifier with n inputs, as shown in Figure 22–7:

$$V_{OUT} = -(V_{IN1} + V_{IN2} + V_{IN3} + \cdots + V_{INn}) \tag{22–3}$$

EXAMPLE 22–2 Determine the output voltage in Figure 22–8.

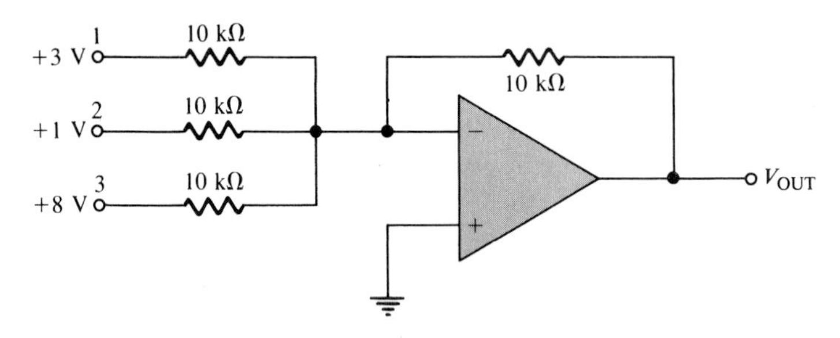

FIGURE 22–8

Solution

$$V_{OUT} = -(V_{IN1} + V_{IN2} + V_{IN3}) = -(3\text{ V} + 1\text{ V} + 8\text{V}) = -12\text{ V}$$

Summing Amplifier with Gain Greater than Unity

When R_f is larger than the input resistors, the amplifier has a gain of R_f/R where R is the value of each input resistor. The general expression for the output is

$$V_{OUT} = -\frac{R_f}{R}(V_{IN1} + V_{IN2} + \cdots + V_{INn}) \tag{22–4}$$

As you can see, the output is the sum of all the input voltages *multiplied* by a constant determined by the ratio R_f/R.

EXAMPLE 22–3 Determine the output voltage for the summing amplifier in Figure 22–9.

Solution

$$R_f = 10\text{ k}\Omega \quad \text{and} \quad R = 1\text{ k}\Omega$$

$$V_{OUT} = -\frac{R_f}{R}(V_{IN1} + V_{IN2}) = -\frac{10\text{ k}\Omega}{1\text{ k}\Omega}(0.2\text{ V} + 0.5\text{ V})$$

$$= -10(0.7\text{ V}) = -7\text{ V}$$

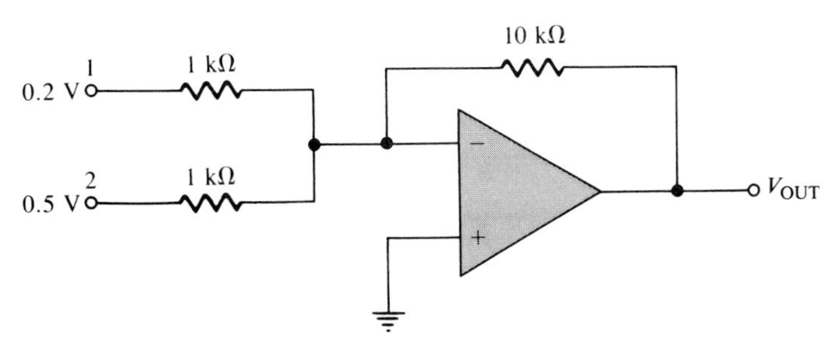

FIGURE 22–9

Averaging Amplifier

A summing amplifier can produce the mathematical average of the input voltages. To do so, set the ratio R_f/R equal to *the reciprocal of the number of inputs.* You know that you obtain the average of several numbers by first adding the numbers and then dividing by the quantity of numbers you have. Examination of Equation (22–4) and a little thought will convince you that a summing amplifier will do the same. The next example will illustrate.

EXAMPLE 22–4 Show that the amplifier in Figure 22–10 produces an output whose magnitude is the average of the input voltages.

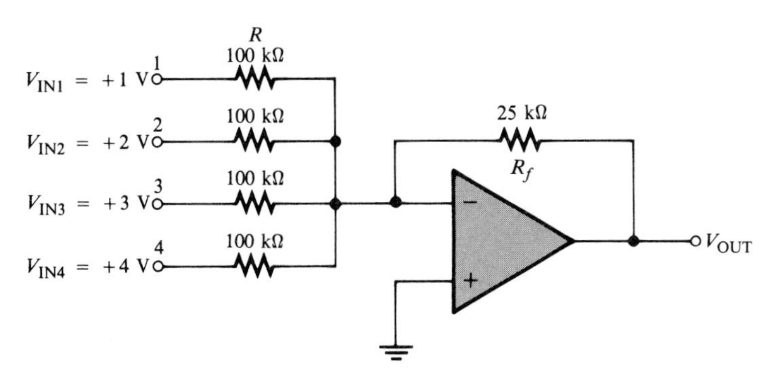

FIGURE 22–10

Solution
The output voltage is

$$
\begin{aligned}
V_{OUT} &= -\frac{R_f}{R}(V_{IN1} + V_{IN2} + V_{IN3} + V_{IN4}) \\
&= -\frac{25\ \text{k}\Omega}{100\ \text{k}\Omega}(1\ \text{V} + 2\ \text{V} + 3\ \text{V} + 4\ \text{V}) = -\frac{1}{4}(10\ \text{V}) \\
&= -2.5\ \text{V}
\end{aligned}
$$

It can easily be shown that the average of the input values is the same as V_{OUT}:

$$\frac{1\text{ V} + 2\text{ V} + 3\text{ V} + 4\text{ V}}{4} = \frac{10\text{ V}}{4} = 2.5\text{ V}$$

Scaling Adder

A different *weight* can be assigned to each input of a summing amplifier by simply adjusting the values of the input resistors. As you have seen, the output voltage can be expressed as

$$V_{OUT} = -\left(\frac{R_f}{R_1}V_{IN1} + \frac{R_f}{R_2}V_{IN2} + \cdots + \frac{R_f}{R_n}V_{INn}\right) \qquad \textbf{(22–5)}$$

The weight of a particular input is set by the ratio of R_f to the input resistance R. For example, if an input voltage is to have a weight of 1, then $R_f = R$. Or, if a weight of 0.5 is required, $R_f = \frac{1}{2}R$. The smaller the value of R, the greater the weight.

EXAMPLE 22–5 For the scaling adder in Figure 22–11 determine the weight of each input voltage and find the output voltage.

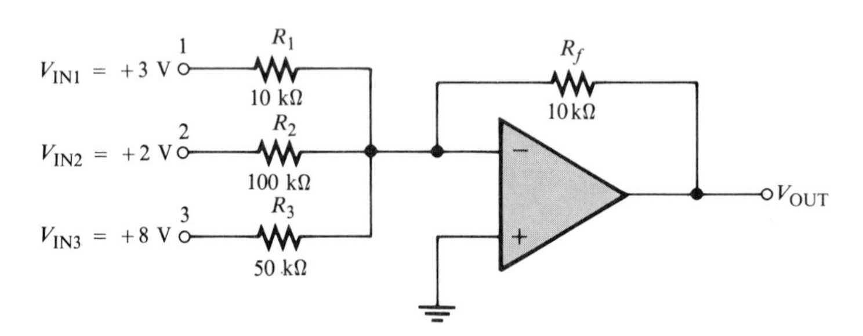

FIGURE 22–11

Solution

Weight of input 1: $\dfrac{R_f}{R_1} = \dfrac{10\text{ k}\Omega}{10\text{ k}\Omega} = 1$

Weight of input 2: $\dfrac{R_f}{R_2} = \dfrac{10\text{ k}\Omega}{100\text{ k}\Omega} = 0.1$

Weight of input 3: $\dfrac{R_f}{R_3} = \dfrac{10\text{ k}\Omega}{50\text{ k}\Omega} = 0.2$

The output voltage is

$$V_{OUT} = -\left(\frac{R_f}{R_1}V_{IN1} + \frac{R_f}{R_2}V_{IN2} + \frac{R_f}{R_3}V_{IN3}\right)$$

$$= -[1(3\text{ V}) + 0.1(2\text{ V}) + 0.2(8\text{ V})]$$
$$= -(3\text{ V} + 0.2\text{ V} + 1.6\text{ V})$$
$$= -4.8\text{ V}$$

SECTION REVIEW 22–2

1. Define *summing point.*
2. What is the value of R_f/R for a five-input averaging amplifier?
3. A certain scaling adder has two inputs, one having twice the weight of the other. If the value of the *lowest* input resistor is 10 kΩ, what is the value of the other input resistor?

INTEGRATORS AND DIFFERENTIATORS

22–3

Integration and differentiation are presented in a way that will give you a very fundamental idea of what they are so you can understand the basic operation of the op-amp integrator and differentiator circuits.

Integration

Basically, integration is the mathematical process of finding the *area under a curve.* Therefore, the op-amp integrator shown in Figure 22–12 produces an output that is proportional to the area under the curve of the input voltage.

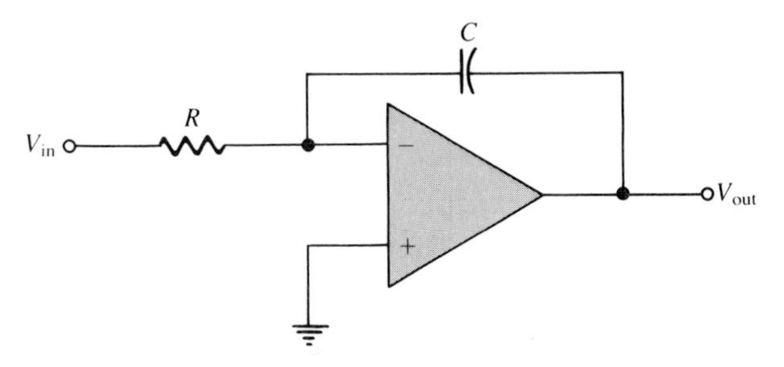

FIGURE 22–12
Op-amp integrator.

To aid in understanding the concept of integration, let's use a rectangular pulse of voltage as the input, as shown in Figure 22–13(a) with amplitude V and width t_W. The total area A under this curve is the height times the width Vt_W. Now, visualize the total area divided into, say, four equal smaller areas, as shown in Part (b). Beginning at $t = 0$ and going to $t = t_1$, the area encompassed up to this point is A_1. At $t = t_2$ the accumulated area is $A_1 + A_2$. At $t = t_3$, the accumulated area is $A_1 + A_2 + A_3$. At $t = t_W$, the total accumulated area is $A_1 + A_2 + A_3 + A_4$. So, as you can see, you can find the total area by summing all of the smaller areas. Integration is the process of summing infinitely small areas under a curve to obtain the total area. Mathematically, the *integral* of the expression for a curve represents the *area* under that curve.

The integral can be plotted graphically, as we will show using the rectangular pulse for simplicity. Starting at $t = 0$ and going to $t = t_1$, the area up to

FIGURE 22–13
Total area under rectangle.

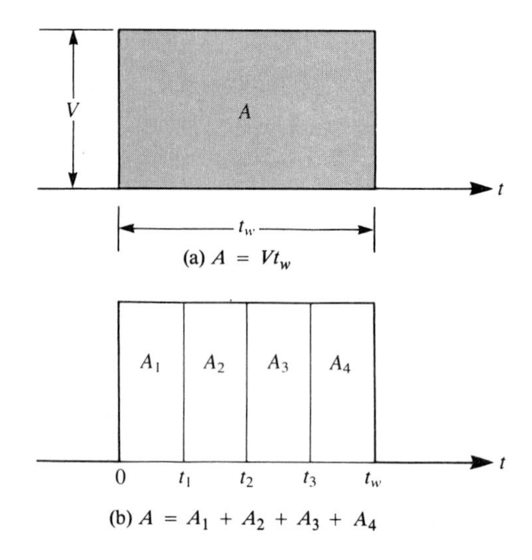

this point can be represented by a point on a graph having a vertical height KA_1, which is proportional to the area, as shown in Figure 22–14. (K is a constant of proportionality.) At each interval of time, another point is added to the graph (we are showing only four points for simplicity). Connecting the points, we get a sloping straight line called a *ramp*.

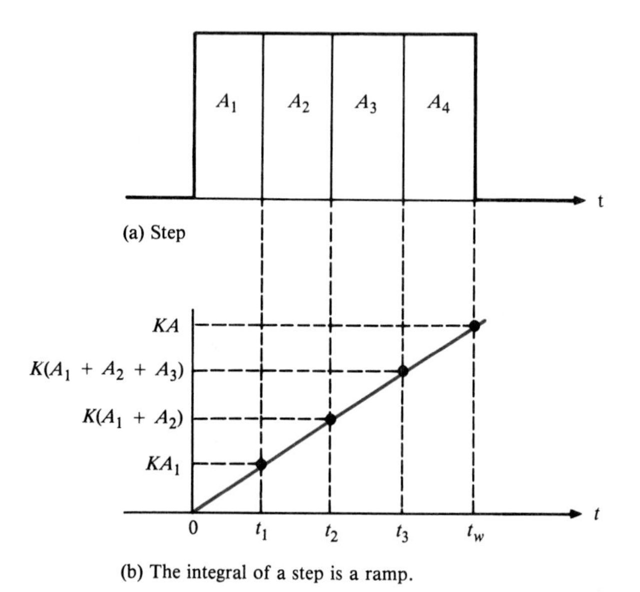

FIGURE 22–14
Integration concept.

Of course, integration is not limited to a rectangular input voltage. Any type of input voltage curve can be integrated, but the rectangular input is perhaps the easiest way to illustrate the principle.

Now let's see how the op-amp integrator in Figure 22–12 performs the integration function. First, remember that the inverting input is at virtual ground. When a positive step input voltage (beginning of a rectangular pulse) is applied to the input, the current through R is

$$I = \frac{V_{IN}}{R}$$

Since the input amplitude has a constant value V_{IN}, the voltage across R is constant, and therefore I is constant.

All of the input current is through the capacitor, charging it in the direction shown in Figure 22–15. Recall that the rate of change of voltage across a capacitor is proportional to the current. Since the current is constant, the capacitor charges at a fixed rate, producing a negative-going ramp because of the inversion.

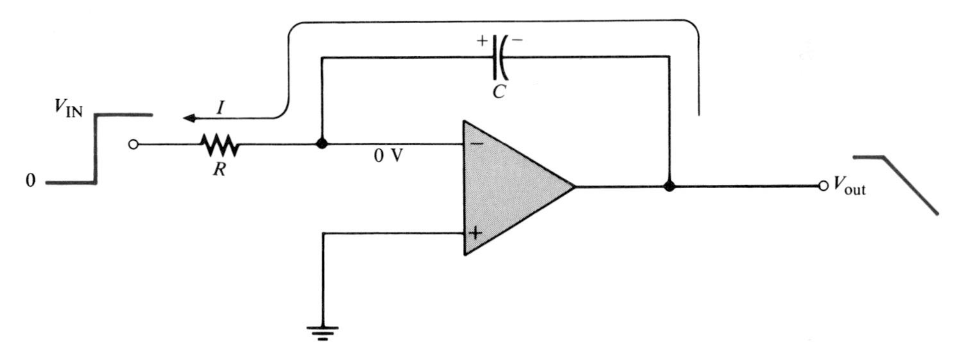

FIGURE 22–15
Inverting integrator with a step input voltage.

Rate of Change

The rate at which the capacitor charges, and therefore the slope of the output ramp, is set by the ratio I/C. Since $I = V_{IN}/R$, the rate of change of the output voltage is

$$\frac{\Delta V_{out}}{\Delta t} = -\frac{V_{IN}}{RC} \qquad (22\text{–}6)$$

where the symbol Δ means "a change in."

EXAMPLE 22–6

(a) Determine the rate of change of the output voltage in response to a single pulse input, as shown for the integrator in Figure 22–16(a). The output voltage is initially zero.

(b) Draw the output waveform.

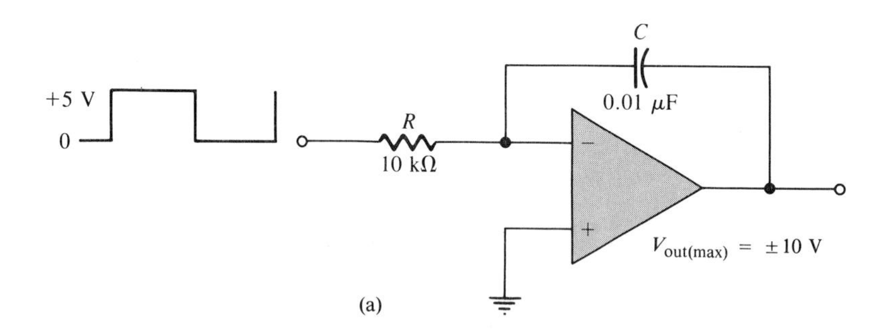

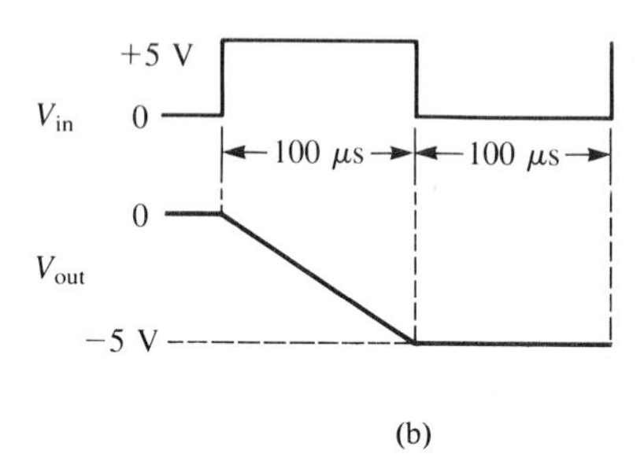

FIGURE 22–16

Solution

(a) The rate of change of the output voltage is

$$\frac{\Delta V_{out}}{\Delta t} = \frac{-V_{IN}}{RC} = -\frac{5 \text{ V}}{(10 \text{ k}\Omega)(0.01\ \mu\text{F})}$$
$$= -50 \text{ kV/s} = -50 \text{ mV}/\mu\text{s}$$

(b) The rate of change was found to be 50 mV/μs in Part (a). When the input is at +5 V, the output is a negative-going ramp. When the input is at 0 V, the output is a constant level. In 100 μs the voltage decreases:

$$\Delta V_{out} = (50 \text{ mV}/\mu\text{s})(100\ \mu\text{s}) = 5 \text{ V}$$

Therefore, the negative-going ramp reaches −5 V at the end of the pulse. The output voltage then remains constant at −5 V for the time that the input is zero. The waveforms are shown in Figure 22–16(b).

Differentiator

An op-amp *differentiator* is shown in Figure 22–17. Notice the arrangement of the resistor and capacitor compared to the integrator.

Differentiation is a process from calculus by which the *rate of change,* $\Delta V/\Delta t$, of a curve at any given point can be determined. It is essentially the inverse of integration. The *derivative* of a curve at a point is its instantaneous rate of change. Therefore, the output of an op-amp differentiator is proportional to the derivative of the input voltage.

We will use two examples to illustrate differentiation.

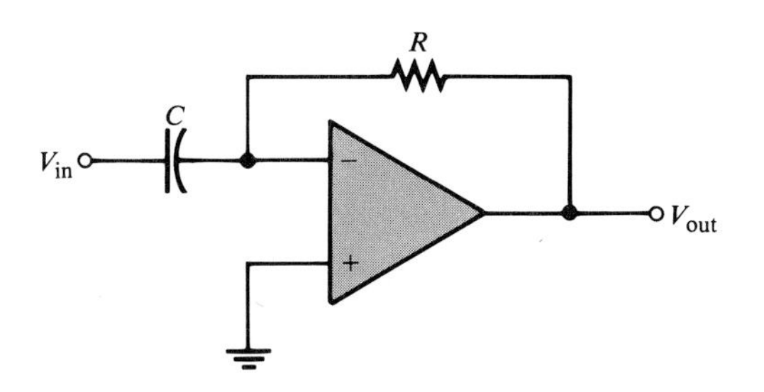

FIGURE 22–17
Op-amp differentiator.

EXAMPLE 22–7 What is the derivative of the triangular waveform in Figure 22–18?

FIGURE 22–18

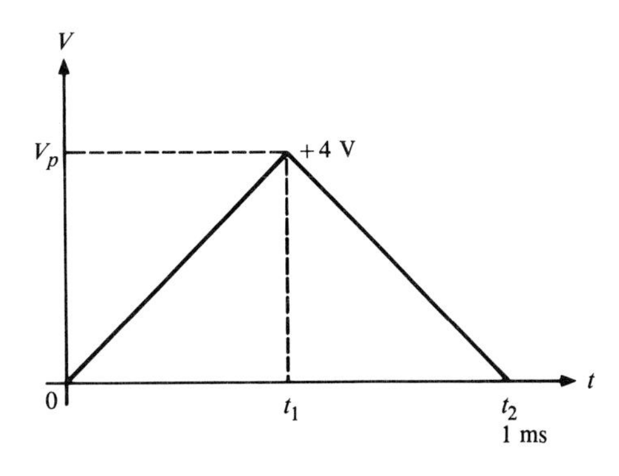

Solution
The derivative is the rate of change. The triangle changes linearly from 0 V to 4 V during the time interval 0 to t_1. Thus,

$$\text{Rate of change} = \frac{\Delta V}{\Delta t} = \frac{4\ \text{V}}{0.5\ \text{ms}} = +8\ \text{V/ms}$$

Therefore, during the interval of 0 to t_1, the derivative is a positive constant.

The triangle changes linearly from 4 V to 0 V during the time interval t_1 to t_2:

$$\text{Rate of change} = \frac{\Delta V}{\Delta t} = \frac{-4\ \text{V}}{0.5\ \text{ms}} = -8\ \text{V/ms}$$

The minus sign indicates a negative slope. Therefore, during the interval t_1 to t_2, the derivative is a negative constant. A graph of the derivative of the triangular wave is shown in Figure 22–19.

FIGURE 22–19
Derivative of a triangular waveform.

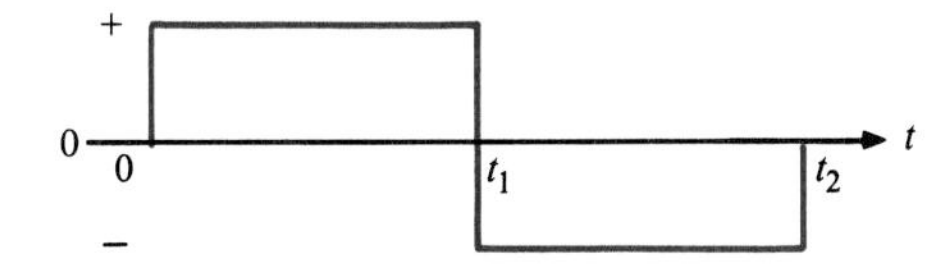

From this example, you can see that a differentiator circuit has a constant output proportional to the slope of a ramp input.

Operation of the Op-Amp Differentiator

First consider the performance of the op-amp differentiator with the application of a triangular input as shown in Figure 22–20. Again, remember that the inverting input is at virtual ground (0 V). When the triangular voltage is applied to the input, the voltage across the capacitor increases at a constant rate. Recall from circuit theory that the current in a capacitor is proportional to the rate of change of the voltage across the capacitor.

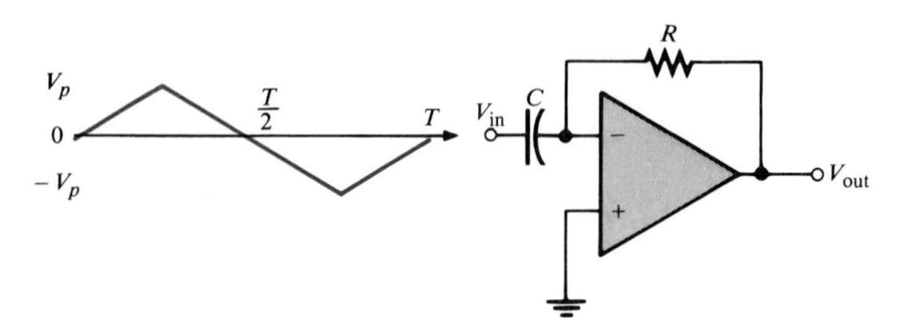

FIGURE 22–20
Differentiator with triangular input voltage.

Since the rate of change of the voltage ramp is constant, the current is also constant and equal to

$$I = \frac{CV_{pp}}{T/2} \tag{22–7}$$

Also, since the output voltage equals the voltage across R, and since I is constant, the output voltage is constant and equal to

$$V_{out} = \pm RC\left(\frac{V_{pp}}{T/2}\right) \tag{22–8}$$

A negative output with an amplitude given by Equation (22–8) occurs during the positive slope of the triangular input due to the inversion of this op-amp. Likewise, a positive output occurs during the negative slope of the input. V_{pp} is the peak-to-peak amplitude of the input, and T is its period.

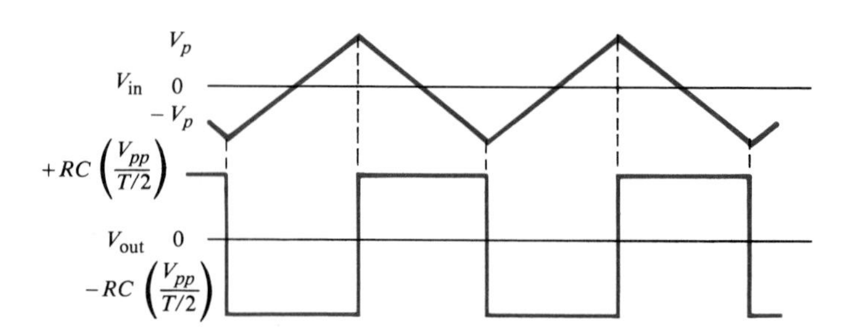

FIGURE 22–21
Input and output waveforms for the differentiator in Figure 22–20.

The waveform diagrams of the input and output for the circuit in Figure 22–20 are shown in Figure 22–21. Of course, other input waveforms are also possible, but the triangular input has been used to illustrate the basic operation.

SECTION REVIEW 22–3

1. Sketch the circuit diagram for an op-amp integrator.
2. An integrator produces a ______ output in response to a step input.
3. A differentiator produces a ______ output in response to a ramp input.

SAWTOOTH GENERATOR

22–4

One way to build a sawtooth generator is with an op-amp integrator that uses a switching device (PUT) in parallel with the feedback capacitor to terminate each ramp at a prescribed level and effectively "reset" the circuit. Figure 22–22(a) shows the implementation.

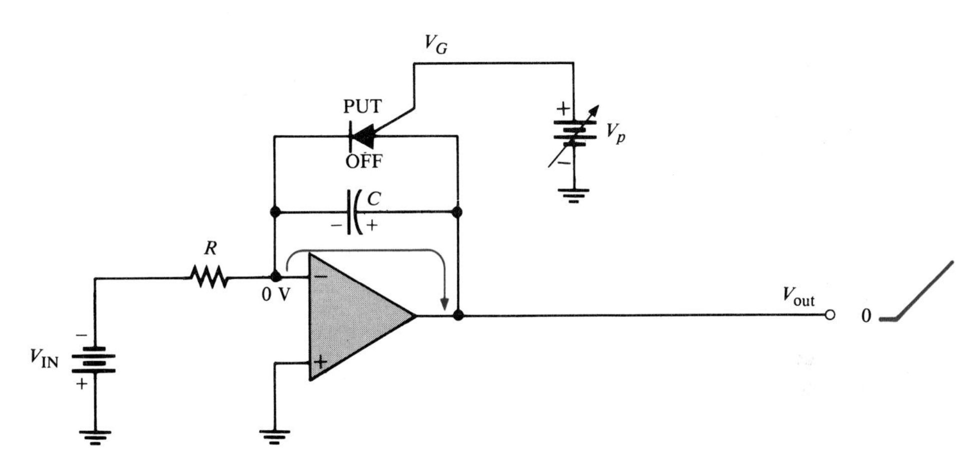

(a) Initially, the capacitor charges, the output ramp begins, and the PUT is off.

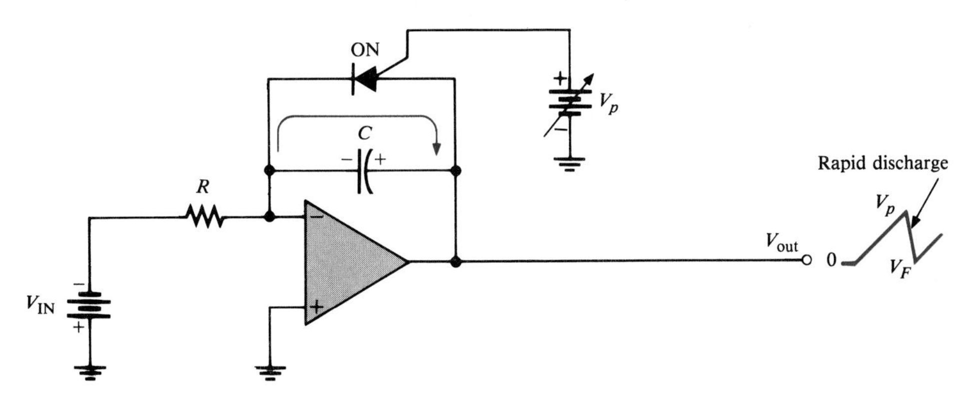

(b) The capacitor rapidly discharges when the PUT momentarily turns on.

FIGURE 22–22
Sawtooth generator operation.

The PUT is a programmable unijunction transistor with a structure similar to that of an SCR except for the placement of the gate. The gate is always biased positively with respect to the cathode. When the anode voltage exceeds the gate voltage by approximately 0.7 V, the PUT turns on. When the anode voltage falls below this level, the PUT turns off. Also, the current must be above the *holding* value to maintain conduction.

The operation of the sawtooth generator begins when the negative dc input voltage, $-V_{IN}$, produces a positive-going ramp on the output. During the time that the ramp is increasing, the circuit acts as a regular integrator.

The PUT triggers on when the output ramp (at the anode) exceeds the gate voltage by 0.7 V. The gate is set to the approximate desired sawtooth peak voltage. When the PUT turns on, the capacitor rapidly discharges, as shown in Figure 22–22(b). The capacitor does not discharge completely to zero because of the PUT's forward voltage, V_F. Discharge continues until the PUT current falls below the holding value. At this point, the PUT turns off and the capacitor begins to charge again, thus generating a new output ramp. The cycle continually repeats, and the resulting output is a repetitive sawtooth waveform, as shown. The sawtooth amplitude and period can be adjusted by varying the PUT gate voltage.

The frequency is determined by the RC time constant of the integrator and the peak voltage set by the PUT. Recall that the charging rate of the capacitor is V_{IN}/RC. The time it takes the capacitor to charge from V_F to V_p is the period of the sawtooth (neglecting the rapid discharge time):

$$T = \frac{V_p - V_F}{|V_{IN}|/RC} \qquad (22\text{–}9)$$

From $f = 1/T$, we get

$$f = \frac{|V_{IN}|}{RC}\left(\frac{1}{V_p - V_F}\right) \qquad (22\text{–}10)$$

EXAMPLE 22–8

(a) Find the amplitude and frequency of the sawtooth output in Figure 22–23. Assume that the forward PUT voltage V_F is approximately 1 V.

(b) Sketch the output waveform.

Solution

(a) First, find the gate voltage in order to establish the approximate voltage at which the PUT turns on:

$$V_G = \frac{R_4}{R_3 + R_4}(+V) = \frac{10\ \text{k}\Omega}{20\ \text{k}\Omega}(15\ \text{V}) = 7.5\ \text{V}$$

This voltage sets the approximate maximum peak value of the sawtooth output (neglecting the 0.7 V):

$$V_p \cong 7.5\ \text{V}$$

The minimum peak value (low point) is

$$V_F \cong 1\text{ V}$$

The period is determined as follows:

$$V_{\text{IN}} = \frac{R_2}{R_1 + R_2}(-V) = \frac{10\text{ k}\Omega}{78\text{ k}\Omega}(-15\text{ V}) = -1.92\text{ V}$$

$$T = \frac{V_p - V_F}{|V_{\text{IN}}|/RC}$$

$$= \frac{7.5\text{ V} - 1\text{ V}}{1.92\text{ V}/(100\text{ k}\Omega)(0.005\ \mu\text{F})} = 1.69\text{ ms}$$

$$f = \frac{1}{1.69\text{ ms}} \cong 592\text{ Hz}$$

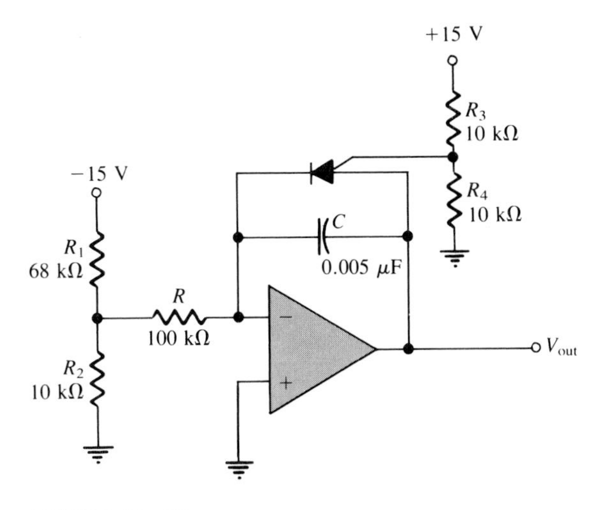

FIGURE 22–23

(b) The output waveform is shown in Figure 22–24.

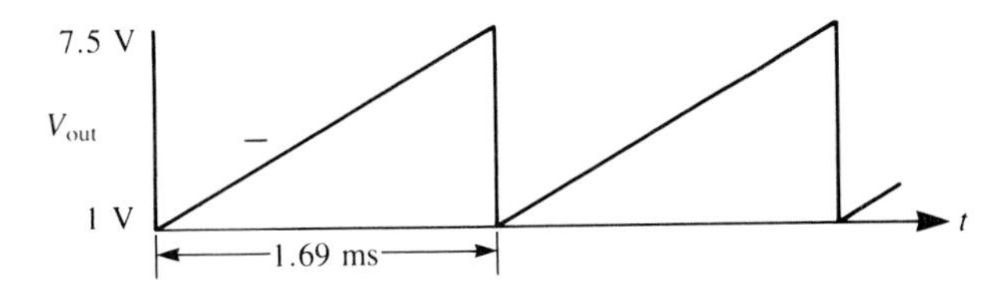

FIGURE 22–24
Output of circuit in Figure 22–23.

SECTION REVIEW 22–4

1. The rate of change of the ramp portion of a sawtooth is 50 V/ms, and the peak-to-peak value is 10 V. How long does it take to reach 10 V?
2. What is the frequency of the sawtooth described in Question 1 if the discharge time is neglected?

WIEN-BRIDGE OSCILLATOR

22–5

One type of sine wave oscillator is called the *Wien-bridge* oscillator. A fundamental part of this oscillator is a *lead-lag* network such as that shown in Figure 22–25(a). R_1 and C_1 together form the *lag* portion of the network; R_2 and C_2 form the *lead* portion. The operation of this circuit is as follows: At lower frequencies, the lead network dominates due to the high reactance of C_2. As the frequency increases, X_{C2} decreases, thus allowing the output voltage to increase. At some specified frequency, the response of the lag network takes over, and the decreasing value of X_{C1} causes the output voltage to decrease.

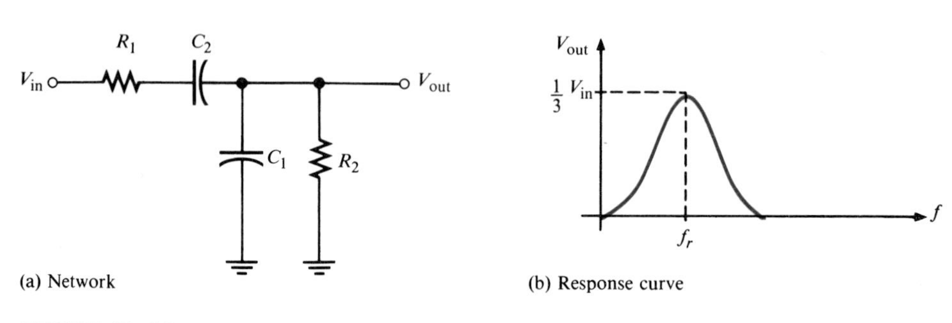

FIGURE 22–25
Lead-lag network and frequency response curve.

Thus, we have a response curve as shown in Figure 22–25(b) where the output voltage peaks at a frequency f_r. At this point, the attenuation of the network is ⅓, expressed as follows:

$$\frac{V_{out}}{V_{in}} = \frac{1}{3} \qquad (22\text{–}11)$$

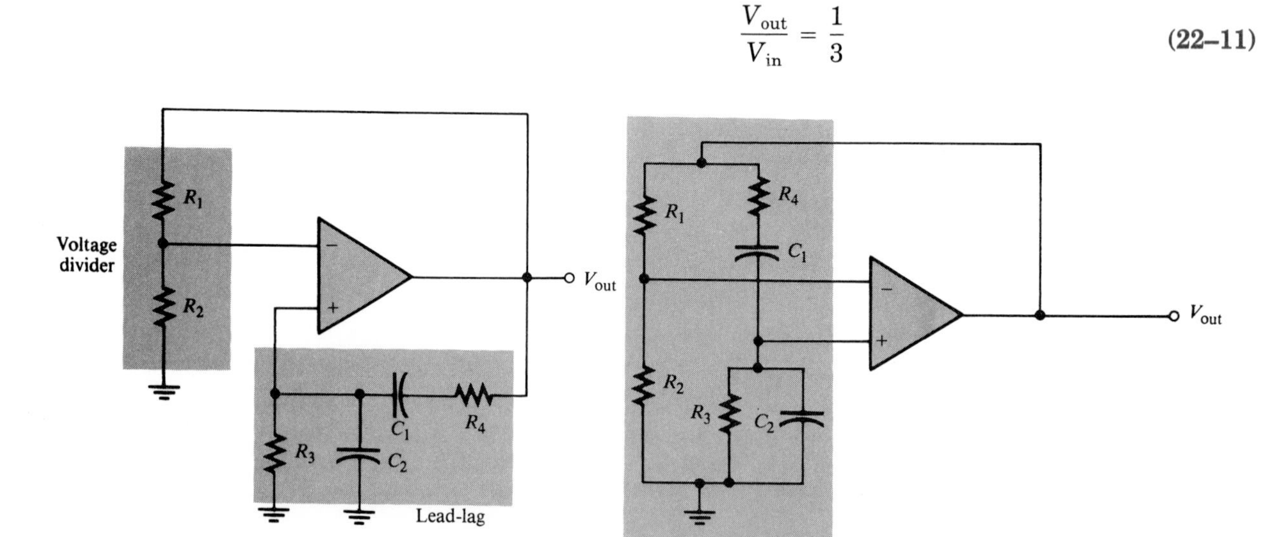

FIGURE 22–26
Wien-bridge oscillator showing two ways to draw the schematic.

The proof of this equation is beyond the scope of this book.

The resonant frequency of the lead-lag network is

$$f_r = \frac{1}{2\pi RC} \tag{22–12}$$

where $R_1 = R_2 = R$ and $C_1 = C_2 = C$.

To summarize, the lead-lag network has a *resonant frequency* f_r at which the phase shift through the network is 0° and the attenuation is ⅓. Below f_r, the lead network dominates and the output leads the input. Above f_r, the lag network dominates and the output lags the input.

The lead-lag network is used in the *positive* feedback loop of an op-amp, as shown in Figure 22–26(a). A voltage divider is used in the *negative* feedback loop.

Basic Circuit

This oscillator circuit can be viewed as a *noninverting* amplifier configuration with the input signal fed back from the output through the lead-lag network. Recall that the *closed-loop* gain is determined by the voltage divider:

$$A_{cl} = \frac{1}{B} = \frac{1}{R_2/(R_1 + R_2)} = \frac{R_1 + R_2}{R_2}$$

The circuit has been redrawn in Figure 22–26(b) to show that the op-amp is connected *across* the Wien bridge. One leg of the bridge is the lead-lag network, and the other is the voltage divider.

The Positive Feedback Conditions for Oscillation

In order for the circuit to produce a sustained sine wave output (oscillate), the phase shift around the positive feedback loop must be 0°, and the gain around the loop must be at least unity (1).

The 0° phase-shift condition is met when the frequency is f_r, because the phase shift through the lead-lag network is 0° and there is no inversion from the noninverting input (+) of the op-amp to the output. This condition is shown in Figure 22–27(a).

The unity-gain condition is met when

$$A_{cl} = 3$$

This offsets the ⅓ attenuation of the lead-lag network, thus making the gain around the positive feedback equal to 1, as depicted in Figure 22–27(b).

To achieve a closed-loop gain of 3,

$$R_1 = 2R_2$$

Then

$$A_{cl} = \frac{1}{B} = \frac{R_1 + R_2}{R_2} = \frac{R_2 + 2R_2}{R_2} = \frac{3R_2}{R_2} = 3$$

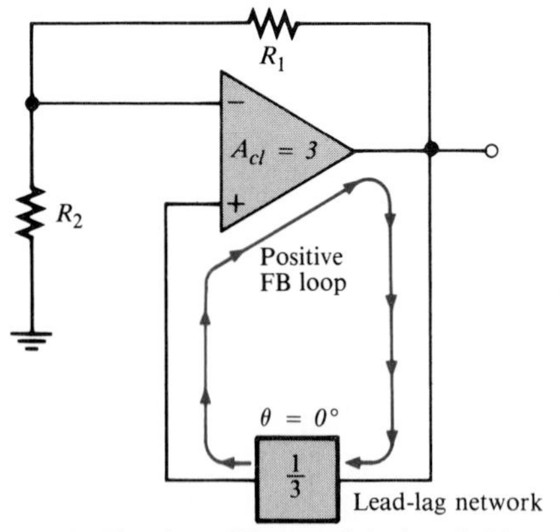

(a) The phase shift around the loop is 0°.

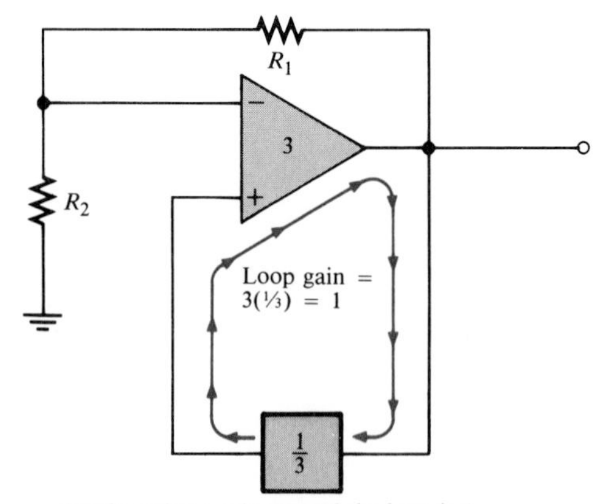

(b) The voltage gain around the loop is 1.

FIGURE 22–27
Conditions for oscillation.

Start-Up Conditions

The unity-gain condition must be met for oscillation to be *sustained.* For oscillation to *begin,* the voltage gain around the positive feedback loop must be *greater than unity* so that the amplitude of the output can build up. Thus, to start with, the loop gain must be more than 1 ($A_{cl} > 3$) until the output signal builds up to a desired level. The gain must then decrease to 1 so that the output signal stays at the desired level, as illustrated in Figure 22–28.

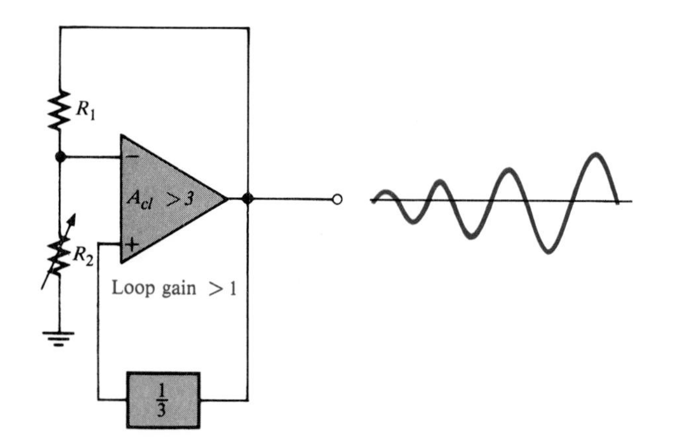

(a) Loop gain greater than 1 causes output to build up.

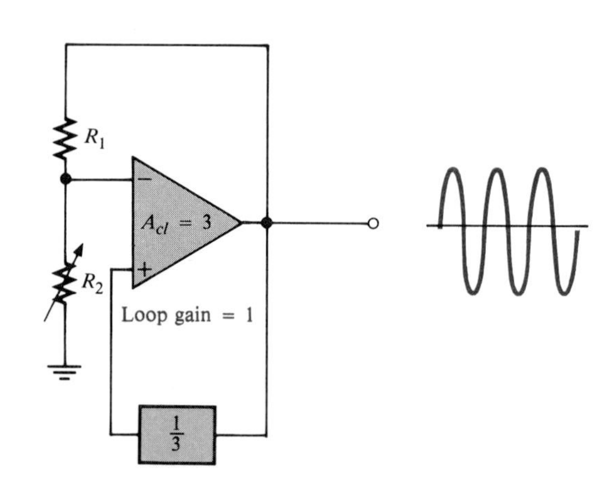

(b) Loop gain of 1 causes sustained constant output.

FIGURE 22–28
Oscillator start-up conditions.

The circuit in Figure 22–29 illustrates a basic method for achieving the conditions described above. Notice that the voltage-divider network has been modified to include an additional resistor R_3 in parallel with a back-to-back zener diode arrangement.

FIGURE 22–29
Self-starting Wien-bridge oscillator

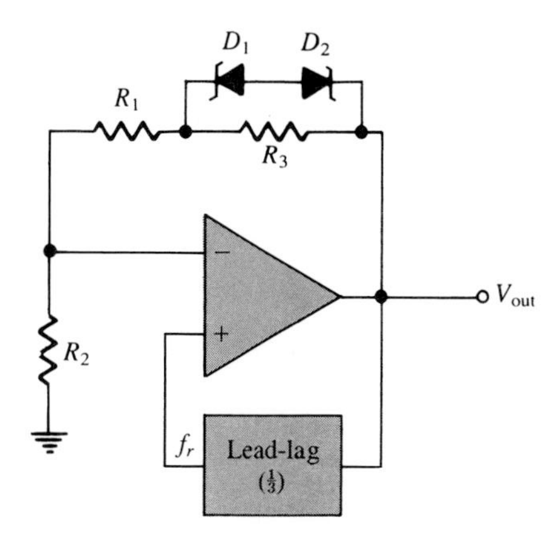

When dc power is first applied, both zener diodes appear as opens. This places R_3 in series with R_1, thus increasing the closed-loop gain as follows ($R_1 = 2R_2$):

$$A_{cl} = \frac{R_1 + R_2 + R_3}{R_2} = \frac{3R_2 + R_3}{R_2} = 3 + \frac{R_3}{R_2}$$

Initially, a small positive feedback signal develops from noise or turn-on transients. The lead-lag network permits only a signal with a frequency equal to f_r to appear *in phase* on the noninverting input. This feedback signal is amplified and continually reinforced, resulting in a buildup of the output voltage.

When the output signal reaches the zener breakdown voltage, the zeners conduct and effectively short out R_3. This lowers the closed-loop gain to 3. At this point the output signal levels off and the oscillation is sustained.

Incidentally, the frequency of oscillation can be adjusted by using gang-tuned capacitors in the lead-lag network.

EXAMPLE 22–9 Determine the frequency of the Wien-bridge oscillator in Figure 22–30. Also verify that oscillations will start and then continue when the output signal reaches 5.4 V.

FIGURE 22–30

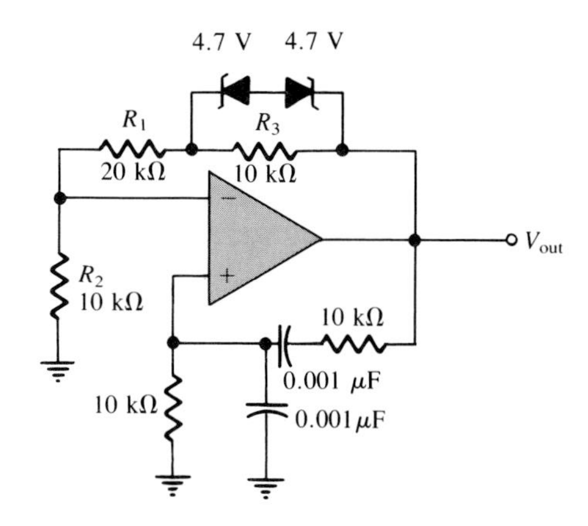

Solution
The frequency is

$$f_r = \frac{1}{2\pi RC} = \frac{1}{2\pi(10\ \text{k}\Omega)(0.001\ \mu\text{F})} = 15.92\ \text{kHz}$$

Initially the closed-loop gain is

$$A_{cl} = \frac{R_1 + R_2 + R_3}{R_2} = \frac{40\ \text{k}\Omega}{10\ \text{k}\Omega} = 4$$

Since $A_{cl} > 3$, the start-up condition is met.

When the output reaches 5.4 V (4.7 V + 0.7 V), the zeners conduct (their forward resistance is assumed small, compared to 20 kΩ), and the closed-loop gain is reached. Thus oscillation is sustained.

$$A_{cl} = \frac{R_1 + R_2}{R_2} = \frac{30\ \text{k}\Omega}{10\ \text{k}\Omega} = 3$$

The Hartley, Colpitts, and Clapp oscillators, discussed in Chapter 20, can also be implemented with op-amps.

SECTION REVIEW 22–5

1. There are two feedback loops in the Wien-bridge oscillator. State the purpose of each.
2. A certain lead-lag network has $R_1 = R_2$ and $C_1 = C_2$. An input voltage of 5 V rms is applied. Its frequency equals the resonant frequency of the network. What is the rms output voltage?

ACTIVE FILTERS

22–6

In this section, we will examine low-pass, high-pass, and band-pass filters with op-amps as the active (gain) elements.

A Low-Pass Active Filter

Figure 22–31 shows a basic active filter and its response curve. Notice that the input circuit is a single low-pass *RC* network, and unity gain is provided by the op-amp with a negative feedback loop. Simply stated, this is a voltage-follower with an *RC* filter between the input signal and the noninverting input.

The voltage at the noninverting input is as follows:

$$V+ = \left(\frac{X_C}{\sqrt{R^2 + X_C^2}}\right)V_{in}$$

Since the gain of the op-amp is 1, the output voltage is equal to $V+$:

$$V_{out} = \left(\frac{X_C}{\sqrt{R^2 + X_C^2}}\right)V_{in} \tag{22–13}$$

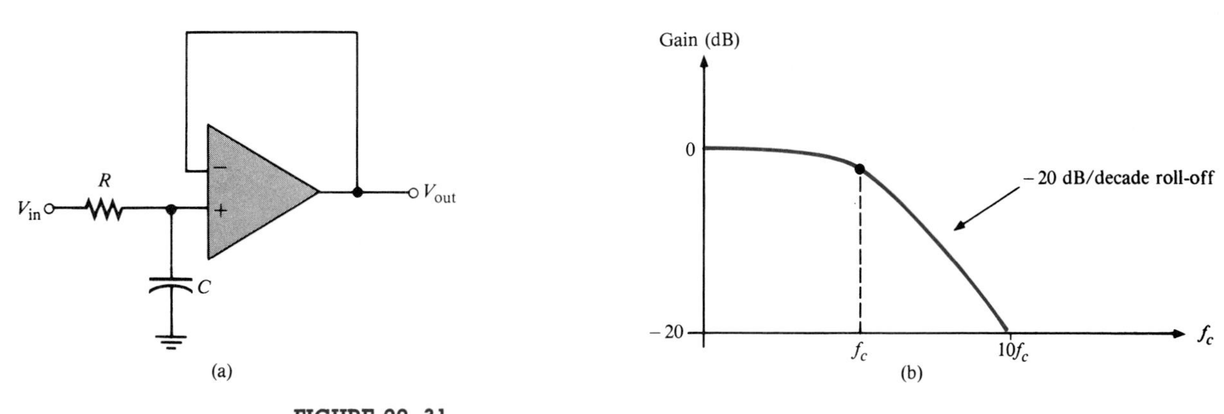

FIGURE 22–31
Single-pole, active low-pass filter and response curve.

A filter with *one RC* network that produces a -20 dB/decade roll-off beginning at f_c is said to be a *single-pole* or *first-order* filter. The term "-20 dB/decade" means that the voltage gain decreases by ten times (-20 dB) when the frequency increases by ten times (decade).

Low-Pass Butterworth Filters

There are several types of active filters, but we will use the *Butterworth* filter to illustrate. A Butterworth filter exhibits a very *flat amplitude* in its pass band. For this reason, this type of filter is sometimes referred to as a *maximally flat filter*.

Figure 22–32(a) shows a *two-pole (second-order) Butterworth* low-pass filter. Since each *RC* network in a filter is considered to be *one-pole*, the *two-pole* filter uses *two RC* networks to produce a roll-off rate of -40 dB/decade, as indicated in Figure 22–32(b). The active filter in Figure 22–32 has unity gain up to near f_c because the op-amp is connected as a voltage-follower.

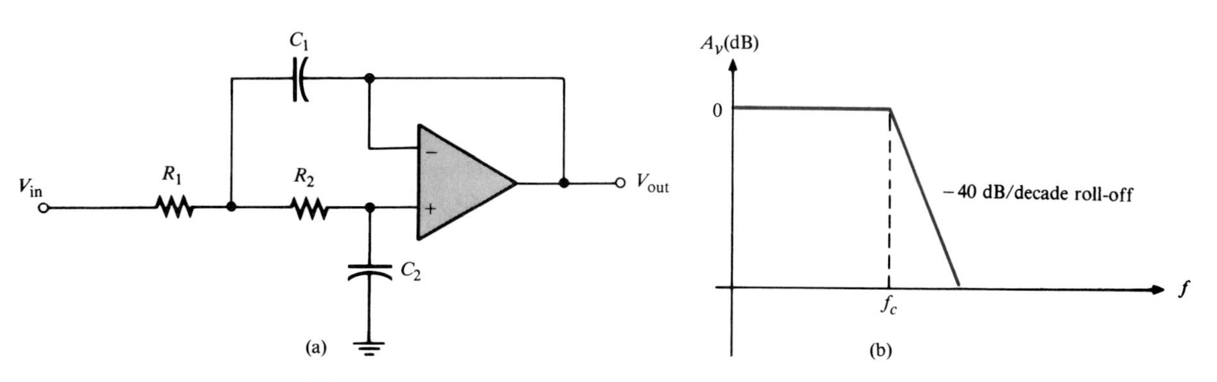

FIGURE 22–32
Two-pole, active low-pass Butterworth filter and its ideal response curve.

One of the *RC* networks is formed by R_1 and C_1, and the other by R_2 and C_2. The critical frequency of this filter can be calculated using the following formula:

$$f_c = \frac{1}{2\pi\sqrt{R_1R_2C_1C_2}} \qquad (22\text{–}14)$$

Figure 22–33(a) shows an example of a two-pole low-pass filter with values chosen to produce the Butterworth response with a critical frequency of 1 kHz. Note that $C_1 = 2C_2$ and $R_1 = R_2$, because these relationships assume a maximally flat Butterworth response with a gain of 0.707 (−3 dB) at f_c. For critical frequencies other than 1 kHz, the capacitance values can be *scaled* inversely with the frequency. For example, as shown in Figures 22–33(b) and 22–33(c), to get a 2-kHz filter, halve the values of C_1 and C_2; for a 500-Hz filter, double the values.

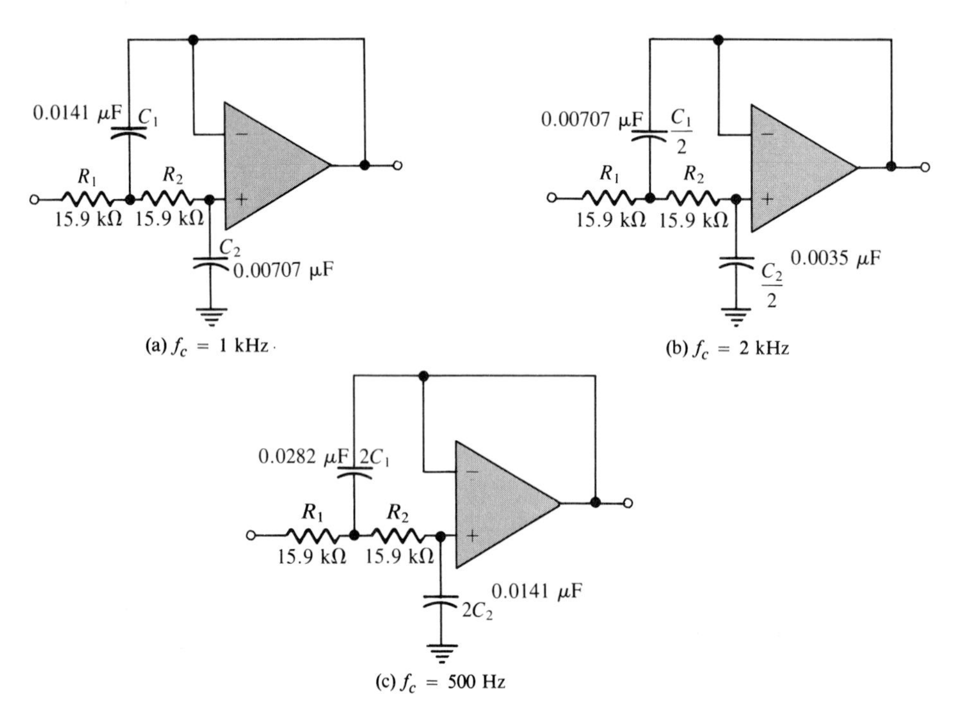

FIGURE 22–33
Butterworth low-pass filters (two-pole).

EXAMPLE 22–10 Calculate the capacitance values required to produce a 3-kHz critical frequency in the filter of Figure 22–34.

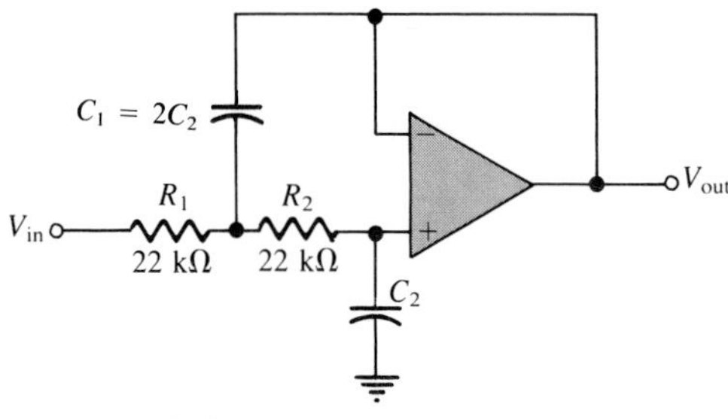

FIGURE 22–34

Solution

The resistor values have already been set at 22 kΩ each. Since these differ from the 1-kHz reference filter, we cannot use the scaling method to get the capacitance values. We will use the formula

$$f_c = \frac{1}{2\pi\sqrt{R_1R_2C_1C_2}}$$

$$f_c^2 = \frac{1}{4\pi^2R_1R_2C_1C_2}$$

Since $C_1 = 2C_2$ and $R_1 = R_2 = R$,

$$f_c^2 = \frac{1}{4\pi^2R^2(2C_2^2)}$$

Solving for C_2, we obtain

$$C_2^2 = \frac{1}{8\pi^2R^2f_c^2}$$

$$C_2 = \frac{1}{\sqrt{2}\,2\pi Rf_c} = \frac{0.707}{2\pi Rf_c} = \frac{0.707}{2\pi(22\text{ k}\Omega)(3\text{ kHz})}$$

$$= 0.0017\ \mu\text{F}$$

$$C_1 = 2C_2 = 2(0.0017\ \mu\text{F}) = 0.0034\ \mu\text{F}$$

A High-Pass Active Filter

In Figure 22–35(a), a high-pass active filter with a 20 dB/decade roll-off is shown. Notice that the input circuit is a single high-pass *RC* network and that unity gain is provided by the op-amp with negative feedback. The response curve is shown in Figure 22–35(b).

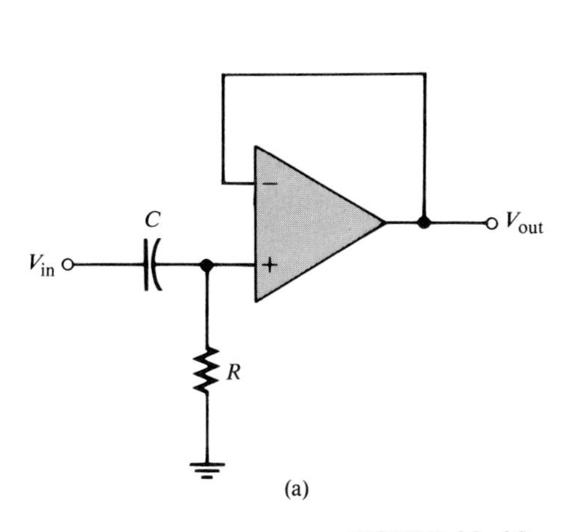

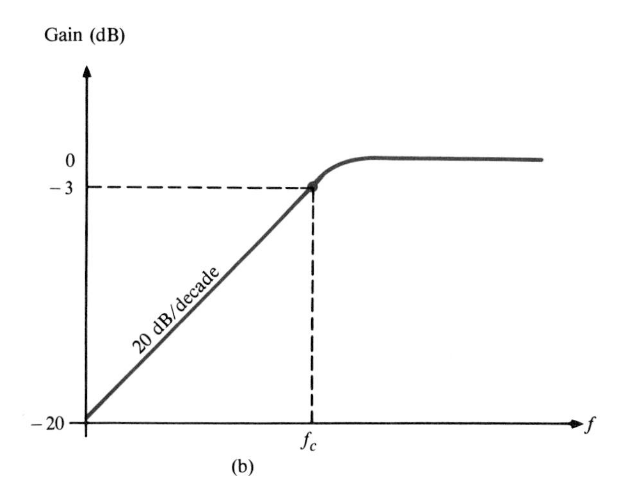

FIGURE 22–35

Single-pole, active high-pass filter and response curve.

Ideally, a high-pass filter passes all frequencies above f_c without limit, as indicated in Figure 22–36(a). In practice, of course, such is not the case. All op-amps inherently have internal RC networks that limit the amplifier's response at high frequencies. Such is the case with the active high-pass filter. There is an upper frequency limit to its response, which essentially makes this type of filter a wide band-pass filter rather than a true high-pass filter, as indicated in Figure 22–35(b). In many applications, the internal high-frequency cutoff is so much greater than the filter's f_c that the internal high-frequency cutoff can be neglected.

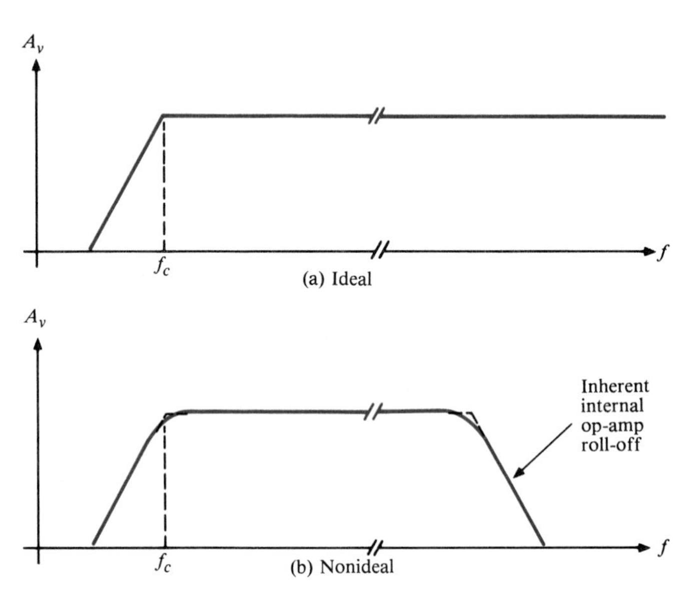

FIGURE 22–36
High-pass filter response.

The voltage at the noninverting input is as follows:

$$V+ = \left(\frac{R}{\sqrt{R^2 + X_C^2}}\right)V_{in}$$

Since the op-amp is connected as a voltage-follower with unity gain, the output voltage is the same as $V+$:

$$V_{out} = \left(\frac{R}{\sqrt{R^2 + X_C^2}}\right)V_{in} \tag{22–15}$$

If the internal critical frequencies of the op-amp are assumed to be much greater than the desired f_c of the filter, the gain will roll off at 20 dB/decade as shown in Figure 22–36(b). This is a single-pole filter because it has one RC network.

High-Pass Butterworth Filters

Like its low-pass counterpart, the high-pass Butterworth is characterized by a very flat amplitude response over its specified range of frequencies.

Figure 22–37 shows a *two-pole Butterworth* active high-pass filter. Notice that it is identical to the corresponding low-pass type, except for the positions of the resistors and capacitors. This filter has a roll-off rate of 40 dB/decade below f_c, and the critical frequency is the same as for the low-pass filter given in Equation (22–14).

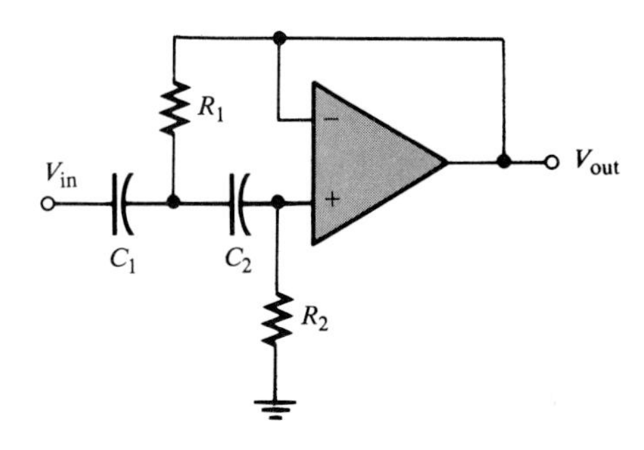

FIGURE 22–37
Two-pole, active high-pass filter.

Figure 22–38 shows a two-pole high-pass filter with values chosen to produce the Butterworth response with a critical frequency of 1 kHz. Note that $R_2 = 2R_1$ and $C_1 = C_2$ because these relationships produce the Butterworth response and assure a gain of 0.707 (−3 dB) at f_c. For frequencies other than 1 kHz, the resistance values can be *scaled* inversely, as was done with the capacitors in the low-pass case.

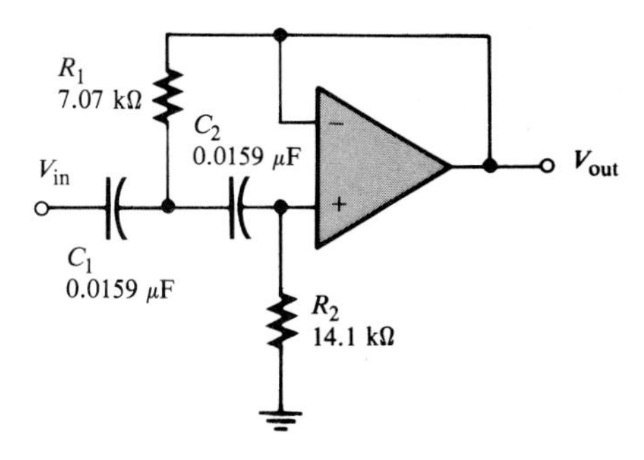

FIGURE 22–38
Two-pole, Butterworth high-pass filter (f_c = 1 kHz).

EXAMPLE 22–11 For the filter of Figure 22–39, calculate the resistance values required to produce a critical frequency of 5.5 kHz.

FIGURE 22–39

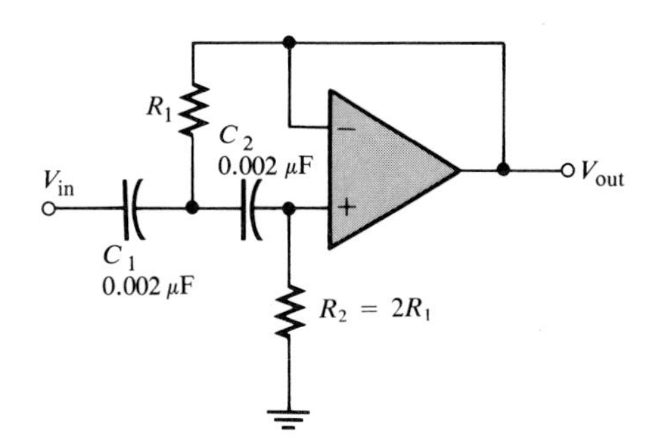

Solution

The capacitor values have been preselected to be 0.0022 μF each. Since these differ from the 1-kHz reference filter, we cannot use the scaling method to get the resistor values.

$$f_c = \frac{1}{2\pi\sqrt{R_1R_2C_1C_2}}$$

$$f_c^2 = \frac{1}{4\pi^2R_1R_2C_1C_2}$$

Since $R_2 = 2R_1$ and $C_1 = C_2 = C$,

$$f_c^2 = \frac{1}{4\pi^2(2R_1^2)C^2}$$

Solving for R_1, we have

$$R_1^2 = \frac{1}{8\pi^2C^2f_c^2}$$

$$R_1 = \frac{1}{\sqrt{2}\,2\pi Cf_c} = \frac{0.707}{2\pi Cf_c}$$

$$= \frac{0.707}{2\pi(0.0022\ \mu\text{F})(5.5\ \text{kHz})} \cong 9.3\ \text{k}\Omega$$

$$R_2 = 2R_1 = 2(9.3\ \text{k}\Omega) = 18.6\ \text{k}\Omega$$

Band-Pass Filter Using High-Pass/Low-Pass Combination

One way to implement a band-pass filter is to use a cascaded arrangement of a high-pass filter followed by a low-pass filter, as shown in Figure 22–40(a). Each of the filters shown is a two-pole configuration so that the roll-off rates of the response curve are ± 40 dB/decade, as indicated in the composite response curve of Part (b). The critical frequency of each filter is chosen so that the response curves overlap, as indicated. The critical frequency of the high-pass filter is lower than that of the low-pass filter.

The lower frequency f_{c1} of the pass band is set by the critical frequency of the high-pass filter. The upper frequency f_{c2} of the pass band is the critical frequency of the low-pass filter. Ideally, the center frequency f_r of the pass band is the geometric average of f_{c1} and f_{c2}. The following formulas express the three frequencies of the band-pass filter in Figure 22–40:

$$f_{c1} = \frac{1}{2\pi\sqrt{R_1R_2C_1C_2}} \tag{22–16}$$

$$f_{c2} = \frac{1}{2\pi\sqrt{R_3R_4C_3C_4}} \tag{22–17}$$

$$f_r = \sqrt{f_{c1}f_{c2}} \tag{22–18}$$

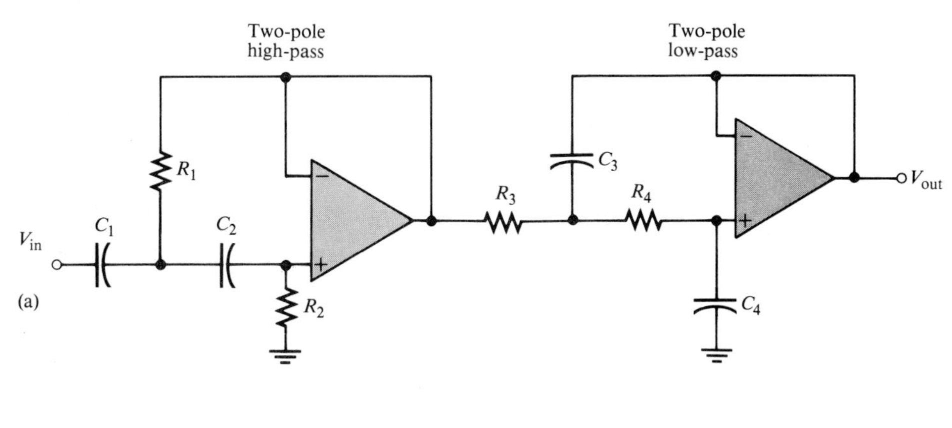

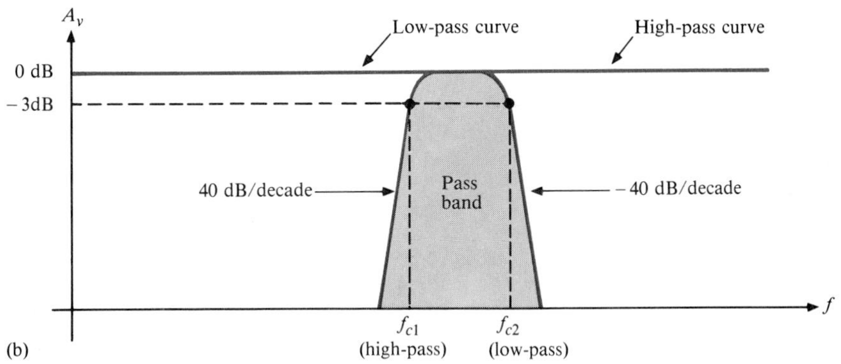

FIGURE 22–40
Band-pass filter formed by combining two-pole, high-pass filter with two-pole, low-pass filter. (The order in which the filters are cascaded does not matter.)

EXAMPLE 22–12

(a) Determine the bandwidth and center frequency for the filter in Figure 22–41.

(b) Sketch the response curve.

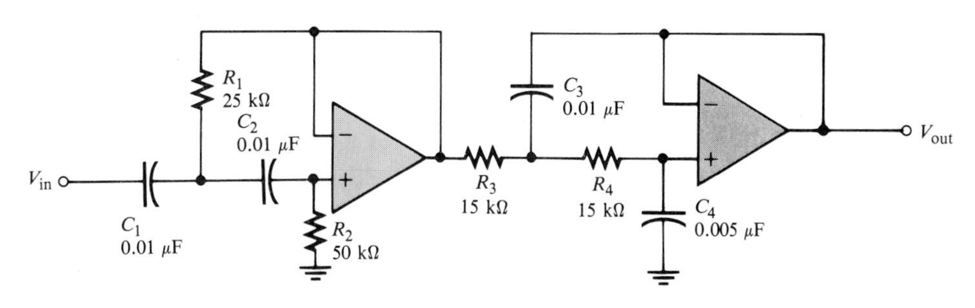

FIGURE 22–41

Solution

(a) The critical frequency of the high-pass filter is

$$f_{c1} = \frac{1}{2\pi\sqrt{R_1R_2C_1C_2}}$$
$$= \frac{1}{2\pi\sqrt{(25\text{ k}\Omega)(50\text{ k}\Omega)(0.01\ \mu\text{F})(0.01\ \mu\text{F})}}$$
$$= 450\text{ Hz}$$

The critical frequency of the low-pass filter is

$$f_{c2} = \frac{1}{2\pi\sqrt{R_3R_4C_3C_4}}$$
$$= \frac{1}{2\pi\sqrt{(15\text{ k}\Omega)(15\text{ k}\Omega)(0.01\ \mu\text{F})(0.005\ \mu\text{F})}}$$
$$= 1.5\text{ kHz}$$
$$BW = f_{c2} - f_{c1} = 1.5\text{ kHz} - 450\text{ Hz} = 1.05\text{ kHz}$$
$$f_r = \sqrt{f_{c1}f_{c2}} = \sqrt{(1.5\text{ kHz})(450\text{ Hz})} = 822\text{ Hz}$$

(b) The response curve is shown in Figure 22–42.

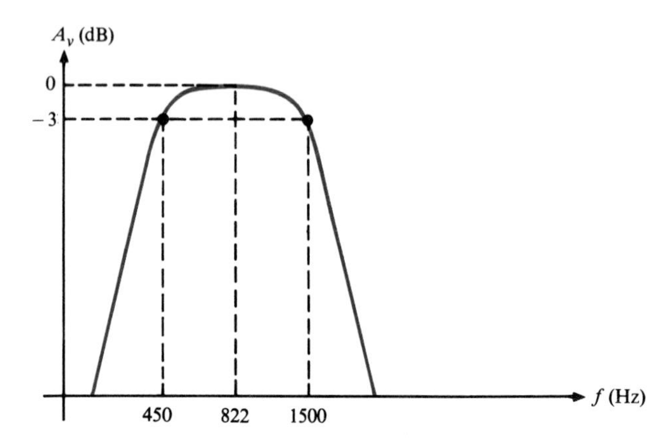

FIGURE 22–42

SECTION REVIEW 22–6

1. In terms of circuit components, what does the term *pole* refer to?
2. What type of response characterizes the Butterworth filter?
3. How does a high-pass Butterworth filter differ in its implementation from a corresponding low-pass version?
4. If the resistance values of a high-pass filter are doubled, what happens to the critical frequency?

THREE-TERMINAL REGULATORS

22–7

Three-terminal linear voltage regulators consist of an input, an output, and a ground terminal. There are two basic types of linear regulator: the *series regulator* and the *shunt regulator.*

Basic Series Regulator

A simple representation of a series linear regulator is shown in Figure 22–43(a), and the basic components are shown in the block diagram in Part (b). Notice that the control element is in *series* with the load between input and output. The output sample circuit senses a change in the output voltage. The error detector compares the sample voltage with a reference voltage and causes the control element to compensate in order to maintain a constant output voltage.

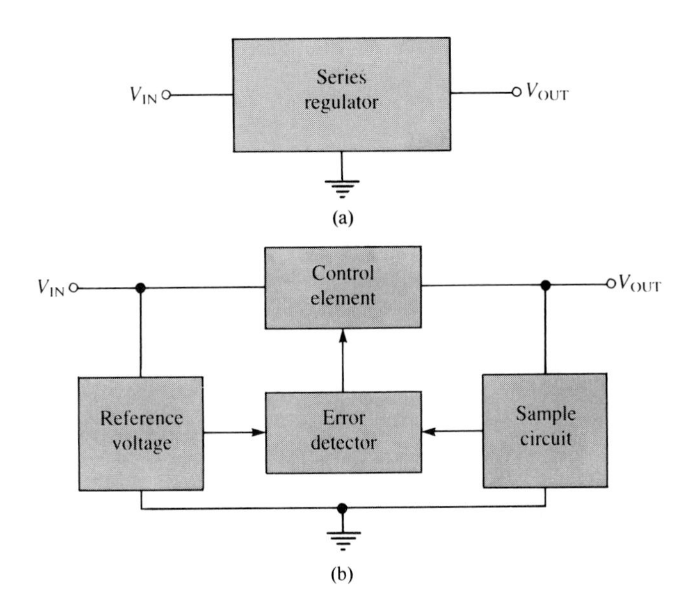

FIGURE 22–43
Block diagram of simple, three-terminal, series voltage regulator.

Regulating Action

A basic op-amp series regulator circuit is shown in Figure 22–44. Its operation is as follows: The resistive voltage divider formed by R_2 and R_3 senses any change in the output voltage. When the output tries to decrease because of a decrease in V_{IN} or because of an increase in I_L, a *proportional* voltage decrease is applied to the op-amp's inverting input by the voltage divider. Since the zener diode holds the other op-amp input at a nearly *fixed* reference voltage V_{REF}, a small difference voltage (error voltage) is developed across the op-amp's inputs. This difference voltage is amplified, and the op-amp's output voltage increases.

This increase is applied to the base of Q_1, causing the emitter voltage V_{OUT} to increase until the voltage to the inverting input again equals the reference (zener) voltage. This action offsets the attempted decrease in output voltage, thus keeping it nearly constant. Q_1 is a power transistor and is often used with a heat sink because it must handle all of the load current.

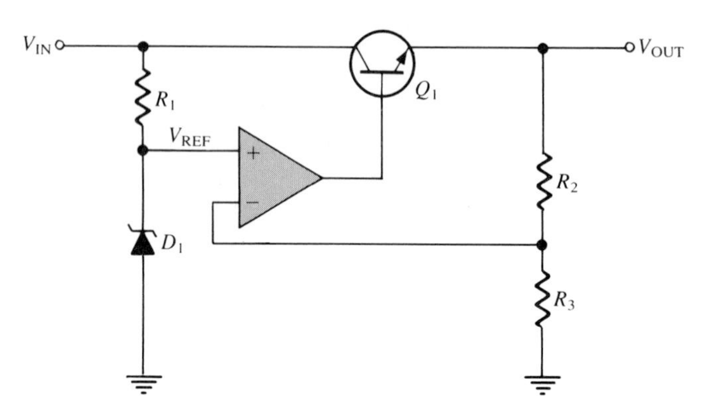

FIGURE 22–44
Basic op-amp series regulator.

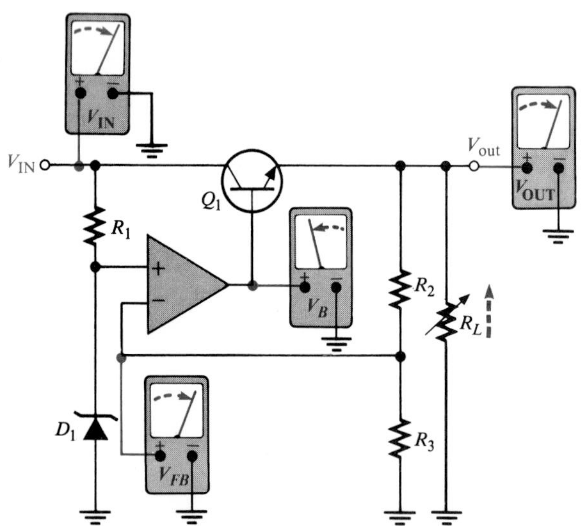

(a) When V_{IN} or R_L increases, V_{OUT} attempts to increase. The feedback voltage, V_{FB}, also attempts to increase, and, as a result, the op-amp's output voltage, V_B, applied to the base of the control transistor, attempts to decrease, thus compensating for the attempted increase in V_{OUT}.

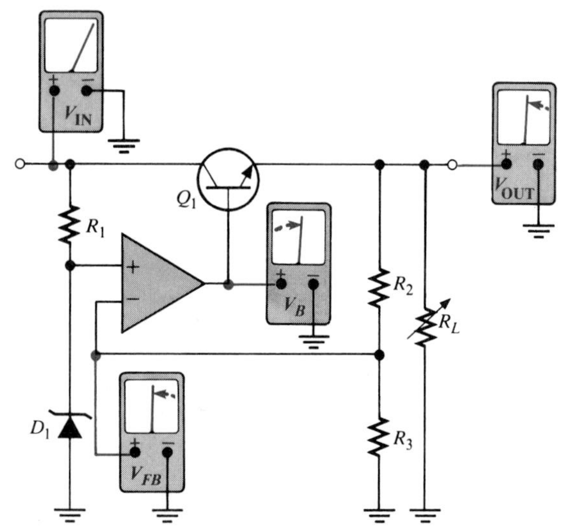

(b) When V_{IN} (or R_L) stabilizes at its new higher value, the voltages are at their original values, thus keeping V_{OUT} constant as result of the negative feedback.

FIGURE 22–45
Illustration of series regulator action that keeps V_{OUT} constant when V_{IN} or R_L changes.

The opposite action occurs when the output tries to increase. Figure 22–45 illustrates the regulating action of the circuit. Percent regulation was discussed in Chapter 18.

The op-amp in Figure 22–44 is actually connected as a noninverting amplifier in which the reference voltage V_{REF} is the input at the noninverting terminal, and the R_2/R_3 voltage divider forms the negative feedback network. The closed-loop voltage gain is

$$A_{cl} = 1 + \frac{R_2}{R_3} \quad (22\text{–}19)$$

Therefore, the regulated output voltage (neglecting the base-emitter voltage of Q_1) is

$$V_{OUT} \cong \left(1 + \frac{R_2}{R_3}\right)V_{REF} \quad (22\text{–}20)$$

From this analysis you can see that the output voltage is determined by the zener voltage and the resistors R_2 and R_3. It is relatively independent of the input voltage, and therefore, regulation is achieved (as long as the input voltage and load current are within specified limits).

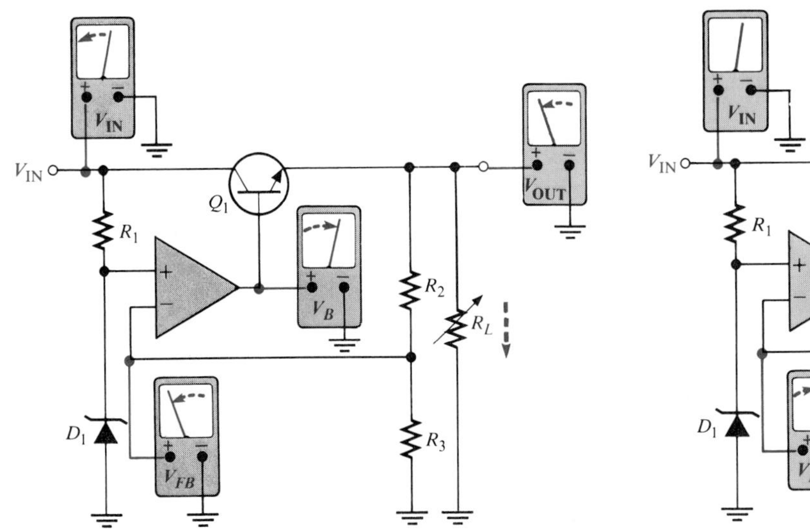

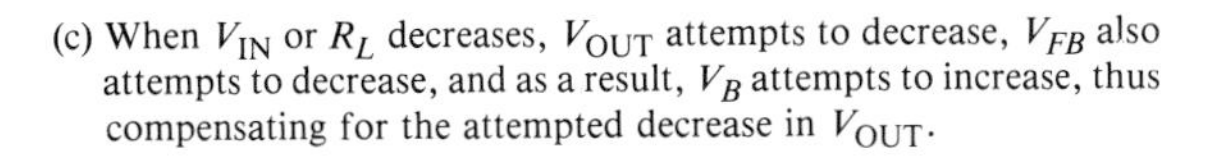
(c) When V_{IN} or R_L decreases, V_{OUT} attempts to decrease, V_{FB} also attempts to decrease, and as a result, V_B attempts to increase, thus compensating for the attempted decrease in V_{OUT}.

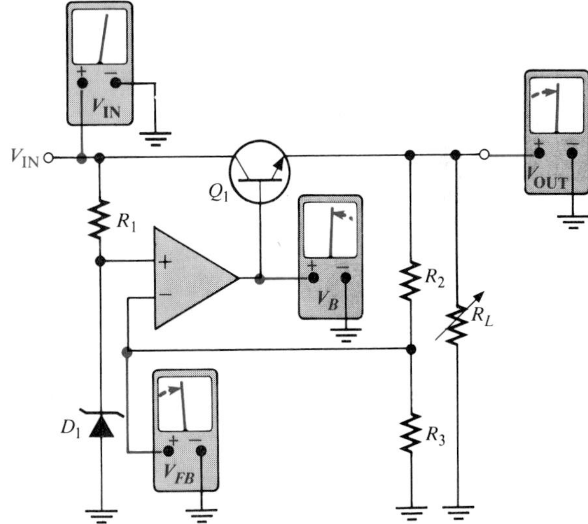

(d) When V_{IN} (or R_L) stabilizes at its new lower value, the voltages are at their original values, thus keeping V_{OUT} constant as a result of the negative feedback.

FIGURE 22–45
(Continued)

EXAMPLE 22–13 Determine the output voltage for the regulator in Figure 22–46.

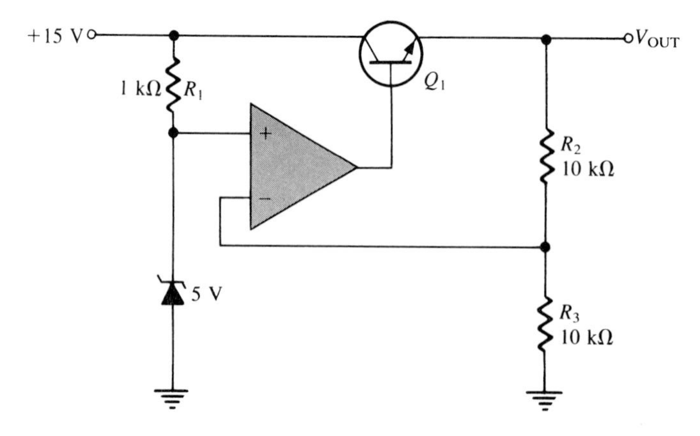

FIGURE 22–46

Solution

$$V_{REF} = 5\text{ V}$$

$$V_{OUT} = \left(1 + \frac{R_2}{R_3}\right)V_{REF} = \left(1 + \frac{10\text{ k}\Omega}{10\text{ k}\Omega}\right)5\text{ V}$$

$$= (2)5\text{ V} = 10\text{ V}$$

Short-Circuit or Overload Protection

If an excessive amount of load current is drawn, the series-pass transistor can be quickly damaged or destroyed. Most regulators employ some type of protection from excess current in the form of a current-limiting mechanism.

Figure 22–47 shows one method of current limiting to prevent overloads called *constant current limiting*. The current-limiting circuit consists of transistor Q_2 and resistor R_4.

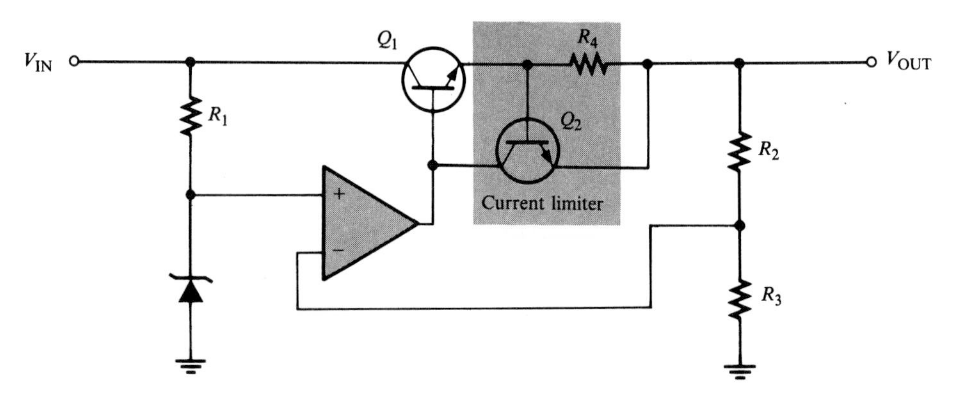

FIGURE 22–47
Series regulator with constant current limiting.

The load current through R_4 creates a voltage from base to emitter of Q_2. When I_L reaches a predetermined maximum value, the voltage drop across R_4 is sufficient to forward-bias the base-emitter junction of Q_2, thus causing it to conduct. Enough Q_1 base current is diverted into the collector of Q_2 so that I_L is limited to its maximum value $I_{L(\max)}$. Since the base-to-emitter voltage of Q_2 cannot exceed about 0.7 V for a silicon transistor, the voltage across R_4 is held to this value, and the load current is limited to

$$I_{L(\max)} = \frac{0.7\ \text{V}}{R_4} \tag{22–21}$$

Basic Shunt Regulator

As you have seen, the control element in the series regulator is the series-pass transistor. A simple representation of a shunt type of linear regulator is shown in Figure 22–48(a), and the basic components are shown in the block diagram in Part (b).

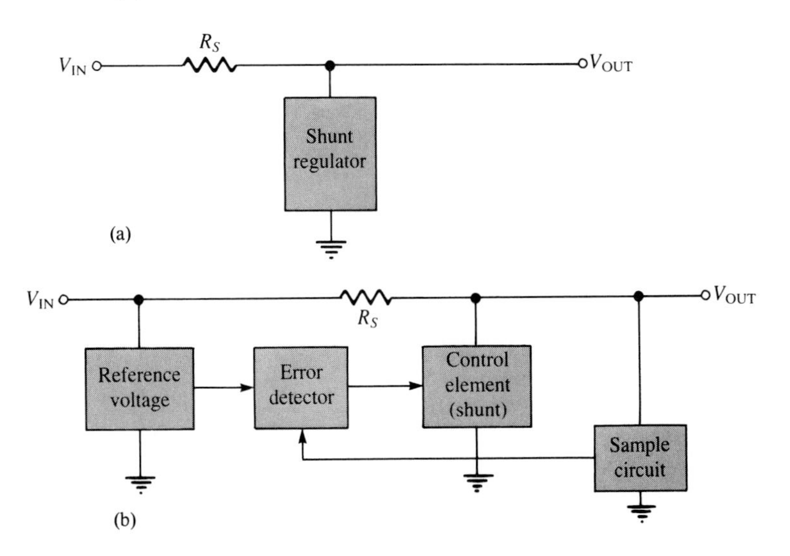

FIGURE 22–48
Block diagrams of simple, three-terminal shunt regulator.

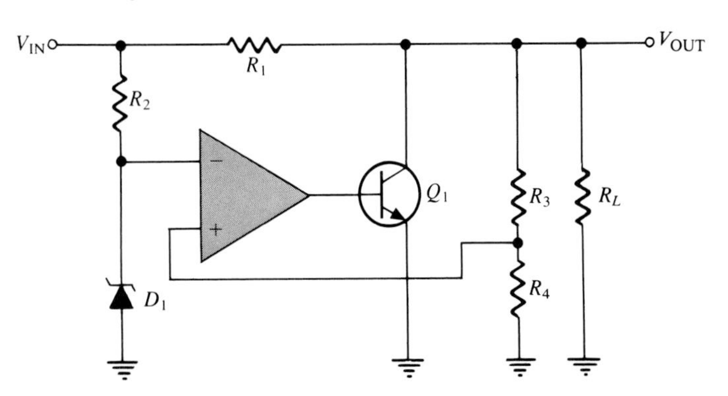

FIGURE 22–49
Basic op-amp shunt regulator.

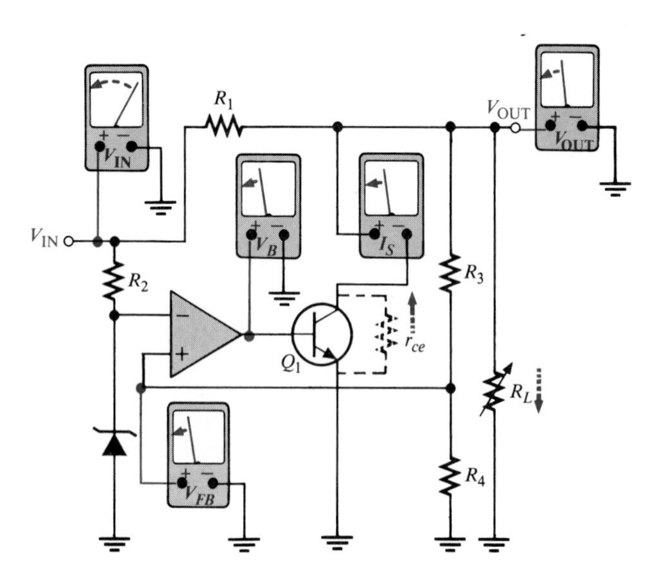

(a) Initial response to a decrease in V_{IN} or R_L

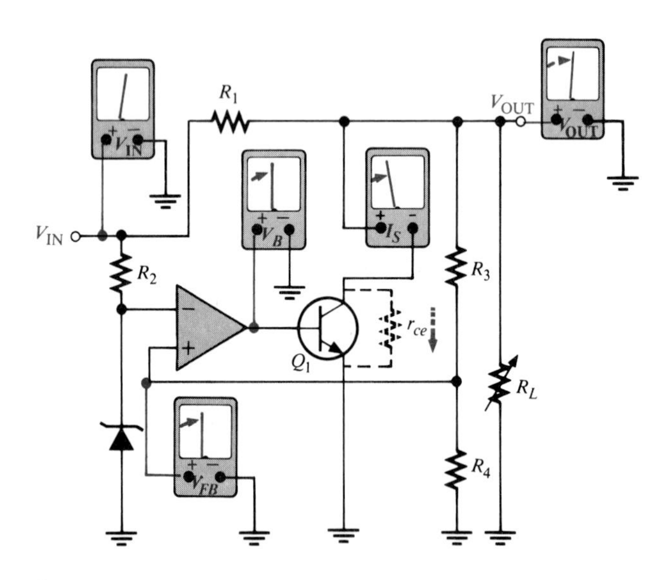

(b) V_{OUT} held constant by the feedback action.

FIGURE 22–50
Sequence of responses when V_{IN} or R_L decreases.

In the basic shunt regulator, the control element is a series resistor (R_1) and a transistor Q_1 in parallel with the load, as shown in Figure 22–49 (on page 871). The operation of the circuit is similar to that of the series regulator, except that regulation is achieved by controlling the current through the parallel transistor Q_1.

When the output voltage tries to decrease due to a change in input voltage, load current, or temperature, the attempted decrease is sensed by R_3 and R_4 and applied to the op-amp's noninverting input. The resulting difference in voltage reduces the op-amp's output, driving Q_1 less, thus reducing its collector current (shunt current) and increasing its effective collector-to-emitter resistance r_{ce}. Since r_{ce} acts as a voltage divider with R_1, this action offsets the attempted decrease in V_{OUT} and maintains it at an almost constant level. The opposite action occurs when the output tries to increase. This regulating action of the shunt element is illustrated in Figure 22–50.

With I_L and V_{OUT} constant, a change in the input voltage produces a change in shunt current (I_S) as follows:

$$\Delta I_S = \frac{\Delta V_{IN}}{R_1}$$

With a constant V_{IN} and V_{OUT}, a change in load current causes an opposite change in shunt current:

$$\Delta I_S = -\Delta I_L$$

This formula says that if I_L increases, I_S decreases, and vice versa.

The shunt regulator is less efficient than the series type but offers inherent short-circuit protection. If the output is shorted ($V_{OUT} = 0$), the load current is limited by the series resistor R_1 to a maximum value as follows ($I_S = 0$):

$$I_{L(\max)} = \frac{V_{IN}}{R_1}$$

EXAMPLE 22–14 In Figure 22–51, what power rating must R_1 have if the maximum input voltage is 12.5 V?

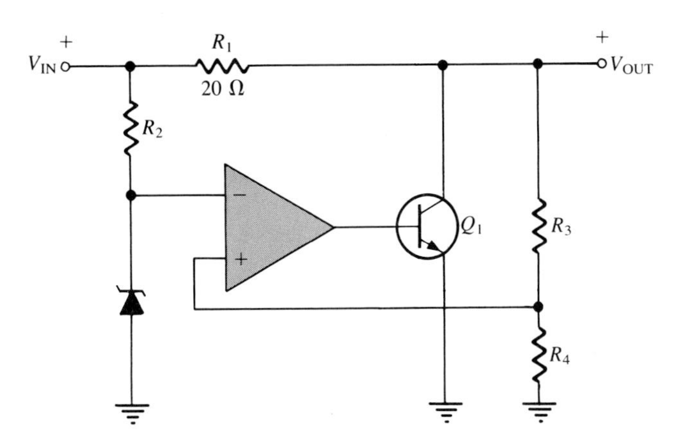

FIGURE 22–51

Solution

The worst-case power dissipation in R_1 occurs when the output is short-circuited. $V_{OUT} = 0$, and when $V_{IN} = 12.5$ V, the voltage dropped across R_1 is

$V_{IN} - V_{OUT} = 12.5$ V.

The power dissipation in R_1 is

$$P_{R1} = \frac{V_{R1}^2}{R_1} = \frac{(12.5\text{ V})^2}{20\ \Omega} = 7.8\text{ W}$$

Therefore, a resistor of at least 10 W should be used.

SECTION REVIEW 22–7

1. How does the control element in a shunt regulator differ from that in a series regulator?
2. Name one advantage and one disadvantage of a shunt regulator over a series type.

APPLICATION NOTE

A four-input, averaging amplifier can be used to sense the average value of the volumes remaining in the tanks. All the input resistors should be of equal value. The output voltage can be adjusted by the ratio of R_f to R.

Connect the output of the averaging amplifier to a comparator with a dc reference voltage selected so that it is equal to the averaging amplifier's output when the total volume in the four tanks is at the minimum specified level. When the average volume drops below the reference level, the output voltage of the averaging amplifier falls below the comparator's reference voltage, thus causing the comparator to change to its opposite state. The comparator drives a display indicator. The suggested arrangement is shown in Figure 22–52.

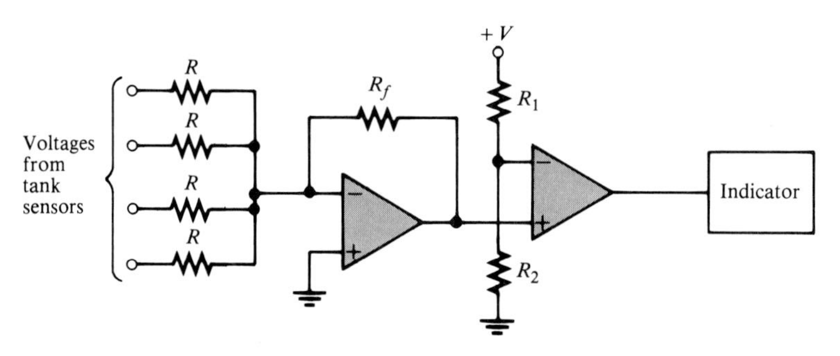

FIGURE 22–52

SUMMARY

Facts

- In an op-amp comparator, when the input voltage exceeds a specified reference voltage, the output changes state.
- The output voltage of a summing amplifier is proportional to the *sum* of the input voltages.
- An *averaging amplifier* is a summing amplifier with a closed-loop gain equal to the reciprocal of the number of inputs.
- In a *scaling adder,* a different weight can be assigned to each input, thus making the input contribute more or contribute less to the output.

- The integral of a step is a ramp.
- The derivative of a ramp is a step.
- In a Wien-bridge oscillator, the closed-loop gain must be equal to 3 in order to have unity gain around the positive feedback loop.
- In filter terminology, a single *RC* network is called a *pole.*
- Each pole in a filter causes the output to roll off (decrease) at a rate of 20 dB/decade.
- Butterworth filters are characterized by a very flat response.
- In a *series* voltage regulator, the control element is a transistor in series with the load.
- In a *shunt* voltage regulator, the control element is a transistor (or zener diode) in parallel with the load.
- The terminals on a three-terminal regulator are *input voltage, output voltage,* and *ground.*

Definitions

- *Integration*—a mathematical process for determining the area under a curve.
- *Differentiation*—a mathematical process for determining the rate of change of a function.
- *Decade*—a tenfold change. When a quantity becomes ten times less or ten times greater, it has changed by a decade.
- *Roll-off*—the decrease in the response of a filter below or above a critical frequency.

Formulas

$$V_{REF} = \frac{R_2}{R_1 + R_2}(+V)$$ Comparator reference (22–1)

$$V_{OUT} = -(V_{IN1} + V_{IN2})$$ Two-input adder (22–2)

$$V_{OUT} = -(V_{IN1} + V_{IN2} + V_{IN3} + \cdots + V_{INn})$$ *n*-input adder (22–3)

$$V_{OUT} = -\frac{R_f}{R}(V_{IN1} + V_{IN2} + \cdots + V_{INn})$$ Adder with gain (22–4)

$$V_{OUT} = -\left(\frac{R_f}{R_1}V_{IN1} + \frac{R_f}{R_2}V_{IN2} + \cdots + \frac{R_f}{R_n}V_{INn}\right)$$ Adder with gain (22–5)

$$\frac{\Delta V_{out}}{\Delta t} = -\frac{V_{IN}}{RC}$$ Rate of change in integrator (22–6)

$$I = \frac{CV_{pp}}{T/2}$$ Differentiator current with triangular input (22–7)

$$V_{out} = \pm RC\left(\frac{V_{pp}}{T/2}\right)$$ Differentiator output with triangular input (22–8)

$$T = \frac{V_p - V_F}{|V_{IN}|/RC}$$ Sawtooth period (22–9)

$$f = \frac{|V_{IN}|}{RC}\left(\frac{1}{V_p - V_F}\right)$$ Sawtooth frequency (22–10)

$$\frac{V_{out}}{V_{in}} = \frac{1}{3}$$ Lead-lag attenuation at f_r (22–11)

$$f_r = \frac{1}{2\pi RC}$$ Lead-lag resonant frequency (22–12)

$$V_{out} = \left(\frac{X_C}{\sqrt{R^2 + X_C^2}}\right)V_{in} \quad \text{Output of one-pole filter (low-pass only)} \quad (22\text{–}13)$$

$$f_c = \frac{1}{2\pi\sqrt{R_1R_2C_1C_2}} \quad \text{Critical frequency of two-pole Butterworth (low-pass)} \quad (22\text{–}14)$$

$$V_{out} = \left(\frac{R}{\sqrt{R^2 + X_C^2}}\right)V_{in} \quad \text{Output of one-pole filter (high-pass only)} \quad (22\text{–}15)$$

$$f_{c1} = \frac{1}{2\pi\sqrt{R_1R_2C_1C_2}} \quad \text{Lower critical frequency} \quad (22\text{–}16)$$

$$f_{c2} = \frac{1}{2\pi\sqrt{R_3R_4C_3C_4}} \quad \text{Upper critical frequency} \quad (22\text{–}17)$$

$$f_r = \sqrt{f_{c1}f_{c2}} \quad \text{Center frequency} \quad (22\text{–}18)$$

$$A_{cl} = 1 + \frac{R_2}{R_3} \quad \text{Closed-loop voltage gain} \quad (22\text{–}19)$$

$$V_{OUT} \cong \left(1 + \frac{R_2}{R_3}\right)V_{REF} \quad \text{Regulator output} \quad (22\text{–}20)$$

$$I_{L(max)} = \frac{0.7\text{ V}}{R_4} \quad \text{For constant current limiting} \quad (22\text{–}21)$$

SELF-TEST

1. Describe briefly what a comparator does.
2. To use a comparator for zero-level detection, to what do you connect the inverting input?
3. How do you change a zero-level detector to a 5-V level detector?
4. Sketch a summing amplifier that will produce an output proportional to the sum of four voltages.
5. How would you increase the gain of the summing amplifier sketched in Question 4?
6. What distinguishes an averaging amplifier from a summing amplifier?
7. What does a scaling adder do?
8. How does the circuitry for an op-amp integrator differ from that for an op-amp differentiator?
9. Which of the op-amp configurations discussed in this chapter do not have feedback?
10. What is the purpose of the PUT in the sawtooth generator circuit of Figure 22–22?
11. Does an oscillator operate with positive or negative feedback?
12. Describe the conditions necessary for oscillation.
13. When the input frequency of a single-pole, low-pass filter increases from 1.5 kHz to 150 kHz, by how many decibels does the output decrease if the critical frequency is 1.5 kHz?
14. What is the purpose of a voltage regulator?

PROBLEMS

Section 22–1

22–1 Determine the output level (maximum positive or maximum negative) for each comparator in Figure 22–53.

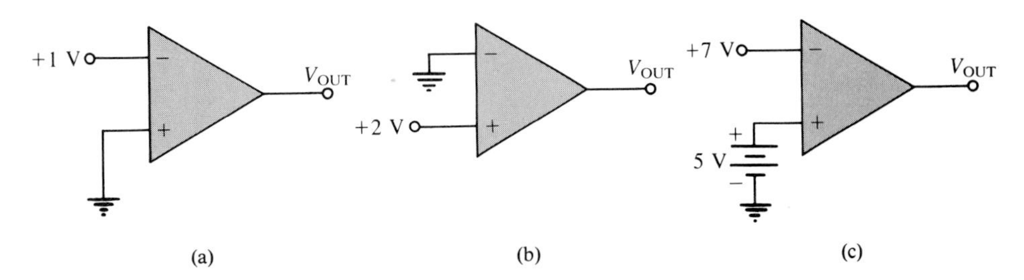

FIGURE 22–53

22–2 A certain op-amp has an open-loop gain of 80,000. The maximum saturated output levels of this particular device are ±12 V when the dc supply voltages are ±15 V. If a differential voltage of 0.15 mV rms is applied between the inputs, what is the peak-to-peak value of the output?

22–3 Sketch the output voltage waveform for each circuit in Figure 22–54 with respect to the input. Show voltage levels.

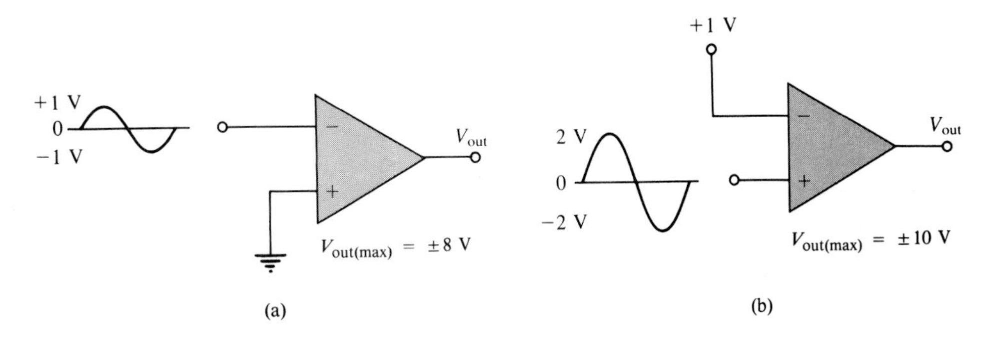

FIGURE 22–54

Section 22–2

22–4 Determine the output voltage for each circuit in Figure 22–55.

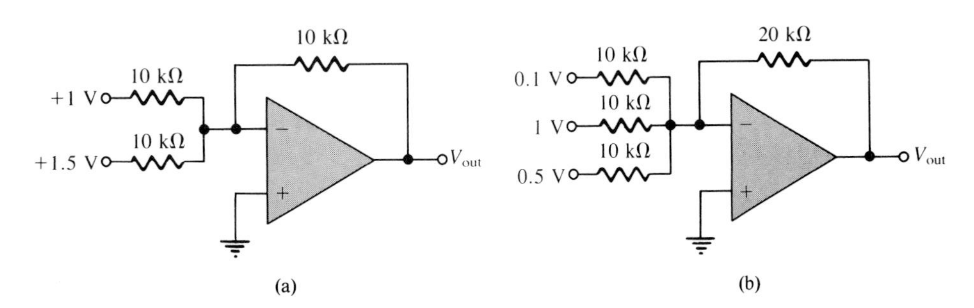

FIGURE 22–55

22–5 Determine the following in Figure 22–56:
(a) V_{R1} and V_{R2} **(b)** current through R_f
(c) V_{OUT}

22–6 Find the value of R_f necessary to produce an output that is 5 times the sum of the inputs in Figure 22–56.

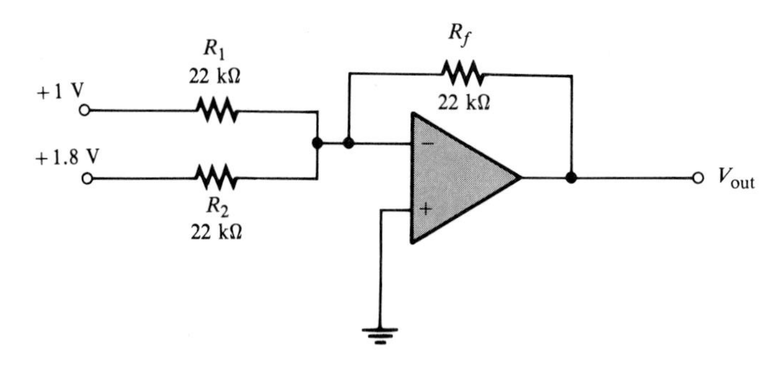

FIGURE 22–56

22–7 Find the output voltage when the input voltages shown in Figure 22–57 are applied to the scaling adder. What is the current through R_f?

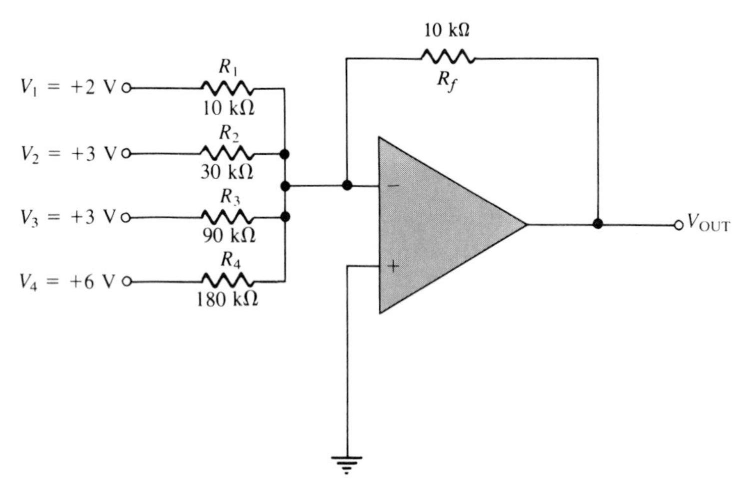

FIGURE 22–57

22–8 Determine the values of the input resistors required in a six-input scaling adder so that the lowest weighted input is 1 and each successive input has a weight *twice* the previous one. Use R_f = 100 kΩ.

Section 22–3

22–9 Determine the rate of change of the output voltage in response to the step input to the integrator in Figure 22–58.

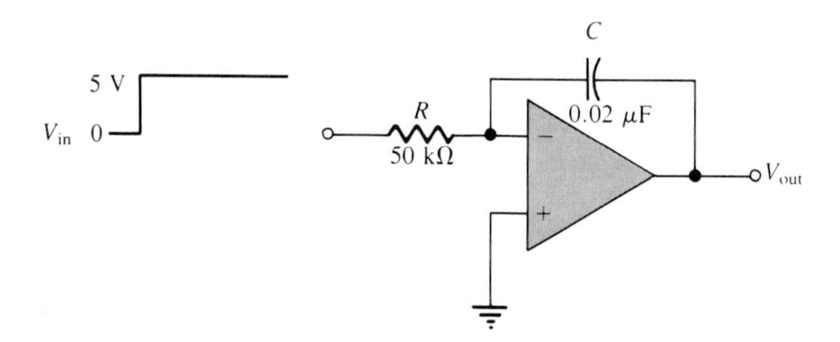

FIGURE 22–58

22–10 A triangular waveform is applied to the input of the circuit in Figure 22–59 as shown. Determine what the output should be, and sketch its waveform in relation to the input.

Section 22–4

22–11 Determine the amplitude and frequency of the output voltage in Figure 22–60. Use 1 V as the forward PUT voltage.

22–12 Modify the sawtooth generator in Figure 22–60 so that its peak-to-peak output is 4 V.

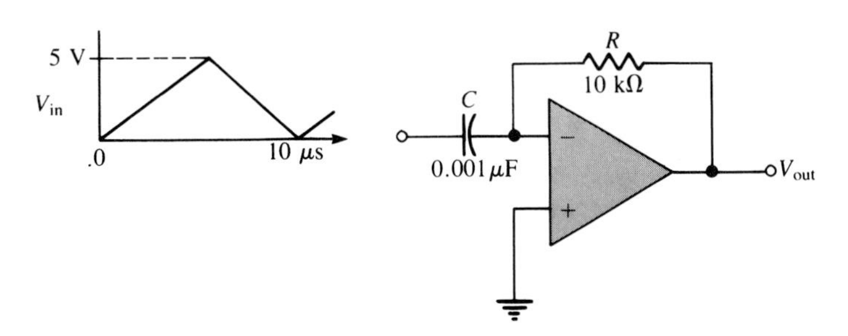

FIGURE 22–59

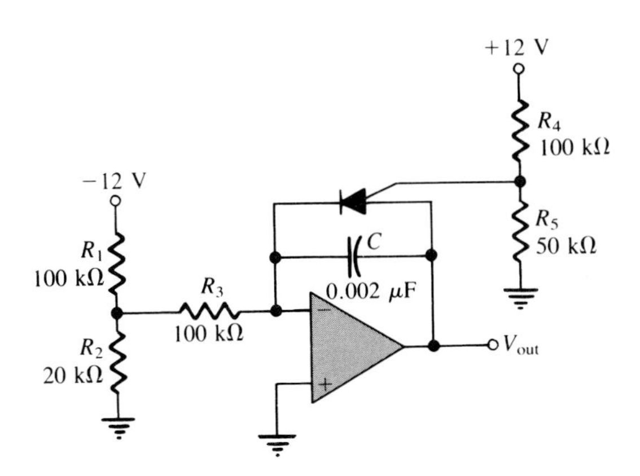

FIGURE 22–60

22–13 A certain sawtooth generator has the following parameter values: $V_{IN} = 3$ V, $R = 4.7$ kΩ, $C = 0.001$ μF, and V_F for the PUT is 1.2 V. Determine its peak-to-peak output voltage if the period is 10 μs.

Section 22–5

22–14 Determine the necessary value of R_2 in Figure 22–61 so that the circuit will oscillate. Neglect the forward resistance of the zener diodes.

22–15 Explain the purpose of R_3 in Figure 22–61.

22–16 What is the initial closed-loop gain in Figure 22–61? At what value of output voltage does A_{cl} change, and to what value does it change?

22–17 Find the frequency of oscillation for the Wien-bridge oscillator Figure 22–61.

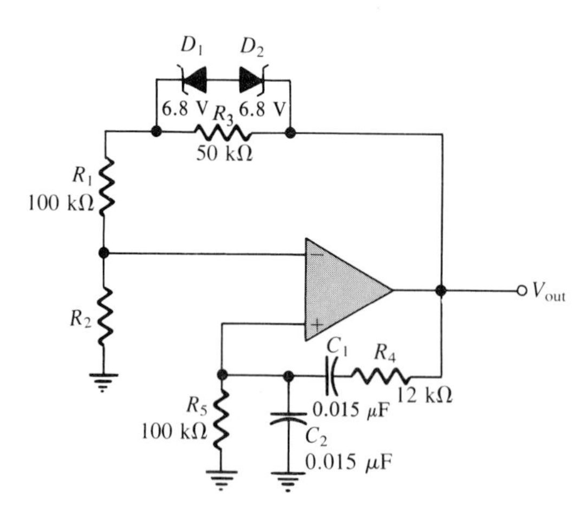

FIGURE 22–61

Section 22–6

22–18 Determine the number of poles in each active filter in Figure 22–62, and identify its type.

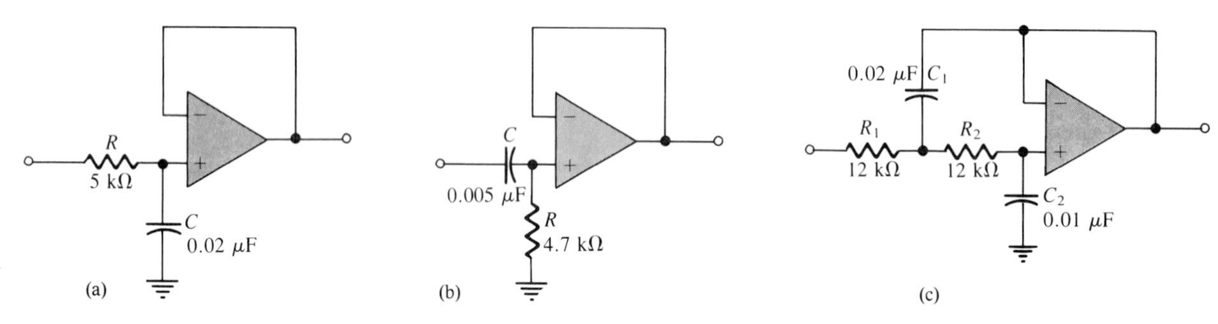

FIGURE 22–62

22–19 Calculate the critical frequencies for the filters in Figure 22–62.

22–20 Determine the bandwidth and center frequency of each filter in Figure 22–63.

Section 22–7

22–21 Determine the output voltage for the series regulator in Figure 22–64.

22–22 If R_3 in Figure 22–64 is doubled, what happens to the output voltage?

22–23 If the zener voltage is 2.7 V instead of 2 V in Figure 22–64, what is the output voltage?

22–24 A series voltage regulator with constant current limiting is shown in Figure 22–65. Determine the value of R_4 if the load current is to be limited to a maximum value of 250 mA. What power rating must R_4 have?

22–25 If R_4 (determined in Problem 22–24) is halved, what is the maximum load current?

22–26 In the shunt regulator of Figure 22–66, when the load current increases, does Q_1 conduct more or less? Why?

22–27 Assume that I_L remains constant and V_{IN} increases by 1 V in Figure 22–66. What is the change in the collector current of Q_1?

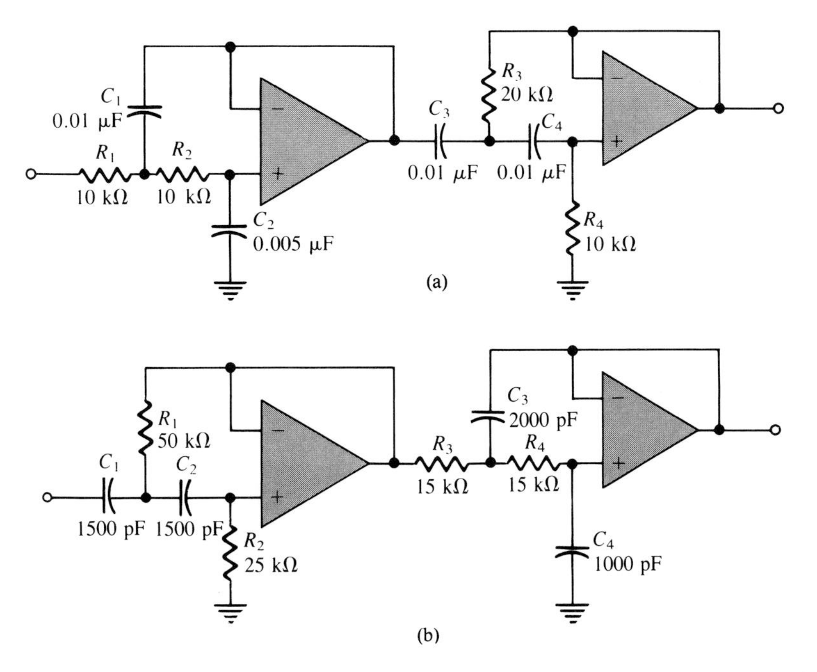

FIGURE 22–63

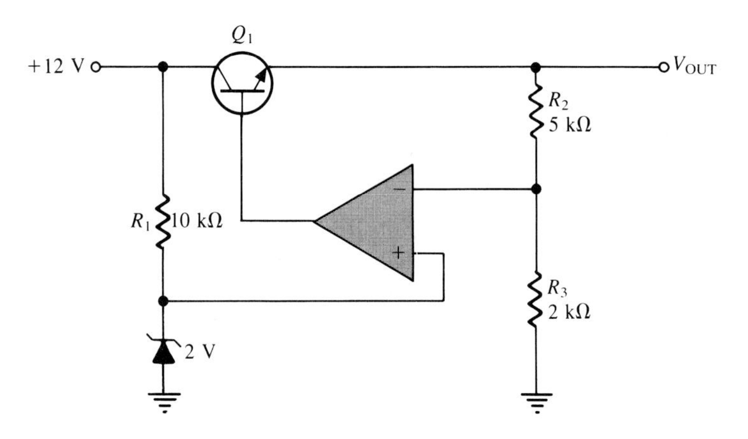

FIGURE 22–64

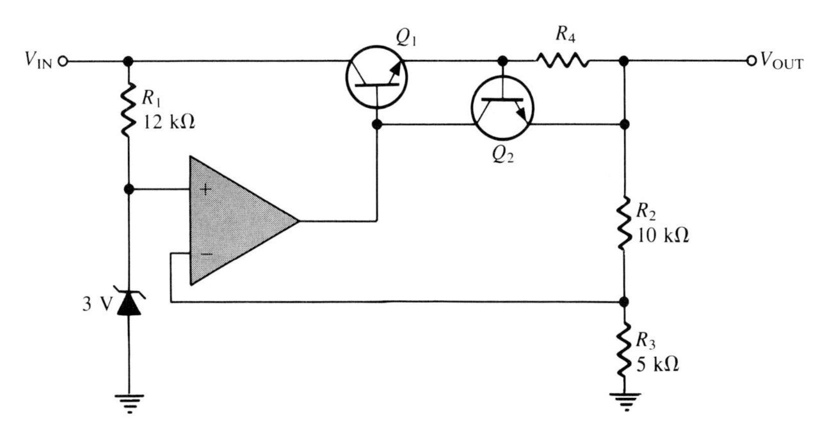

FIGURE 22–65

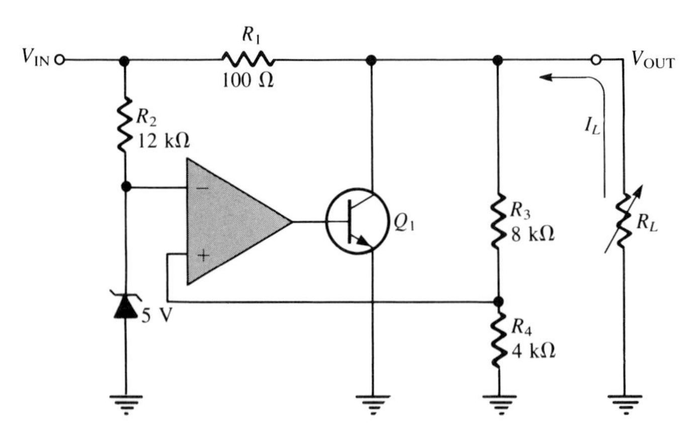

FIGURE 22–66

22–28 With a constant input voltage of 18 V, the load resistance in Figure 22–66 is varied from 1 kΩ to 1.2 kΩ. Neglecting any change in output voltage, how much does the shunt current through Q_1 change?

ANSWERS TO SECTION REVIEWS

Section 22–1

1. 1.36 V. **2.** See Figure 22–67.

FIGURE 22–67

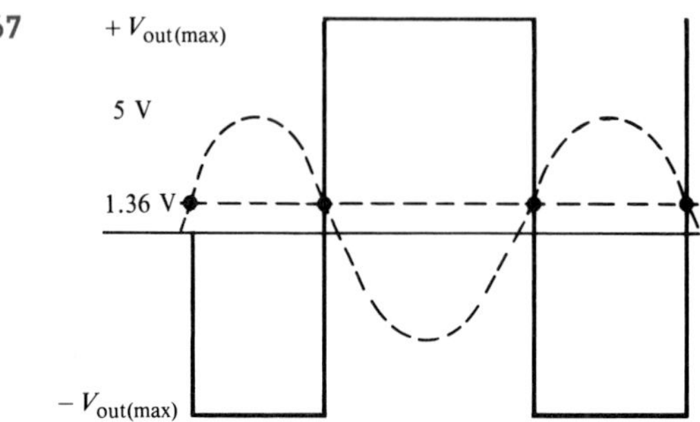

Section 22–2

1. The terminal of the op-amp where the input resistors are commonly connected.

2. ⅕. **3.** 20 kΩ.

Section 22–3

1. See Figure 22–68. **2.** Ramp. **3.** Step.

FIGURE 22–68

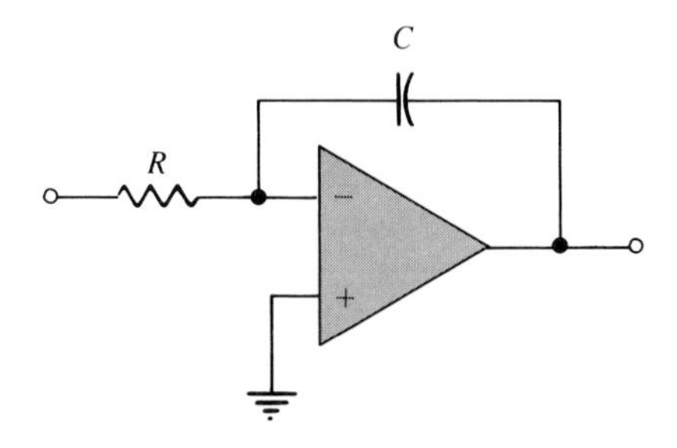

Section 22–4

1. 0.2 ms. **2.** 5 kHz.

Section 22–5

1. Negative feedback loop sets closed-loop gain; positive feedback loop sets frequency.

2. 1.67 V.

Section 22–6

1. A single *RC* network. **2.** Flat in pass band.

3. The *R* and *C* positions are interchanged. **4.** It is halved.

Section 22–7

1. In a shunt regulator, the control element is in parallel with the load rather than in series with the load.

2. A shunt regulator has inherent current limiting, but it is less efficient than a series regulator because part of the load current must be bypassed through the control element.

APPENDIX A
Wire Sizes

Wires are the most common form of conductive material used in electrical applications. They vary in diameter size and are arranged according to standard *gage numbers,* called *American Wire Gage* (AWG) sizes. The larger the gage number is, the smaller the wire diameter is. The AWG sizes are listed in Table A–1.

TABLE A–1
American Wire Gage (AWG) sizes for solid round copper.

AWG #	Area (CM)	Ω/1000 ft at 20C°	AWG #	Area (CM)	Ω/1000 ft at 20C°
0000	211,600	0.0490	19	1,288.1	8.051
000	167,810	0.0618	20	1,021.5	10.15
00	133,080	0.0780	21	810.10	12.80
0	105,530	0.0983	22	642.40	16.14
1	83,694	0.1240	23	509.45	20.36
2	66,373	0.1563	24	404.01	25.67
3	52,634	0.1970	25	320.40	32.37
4	41,742	0.2485	26	254.10	40.81
5	33,102	0.3133	27	201.50	51.47
6	26,250	0.3951	28	159.79	64.90
7	20,816	0.4982	29	126.72	81.83
8	16,509	0.6282	30	100.50	103.2
9	13,094	0.7921	31	79.70	130.1
10	10,381	0.9989	32	63.21	164.1
11	8,234.0	1.260	33	50.13	206.9
12	6,529.0	1.588	34	39.75	260.9
13	5,178.4	2.003	35	31.52	329.0
14	4,106.8	2.525	36	25.00	414.8
15	3,256.7	3.184	37	19.83	523.1
16	2,582.9	4.016	38	15.72	659.6
17	2,048.2	5.064	39	12.47	831.8
18	1,624.3	6.385	40	9.89	1049.0

As the table shows, the size of a wire is also specified in terms of its *cross-sectional area,* as illustrated also in Figure A–1. The unit of cross-sectional area is the *circular mil,* abbreviated CM. One circular mil is the area of a wire with a diameter of 0.001 inch (1 mil). We find the cross-sectional area by expressing the diameter in thousandths of an inch (mils) and squaring it, as follows:

$$A = d^2$$

where A is the cross-sectional area in circular mils and d is the diameter in mils.

FIGURE A–1
Cross-sectional area of a wire.

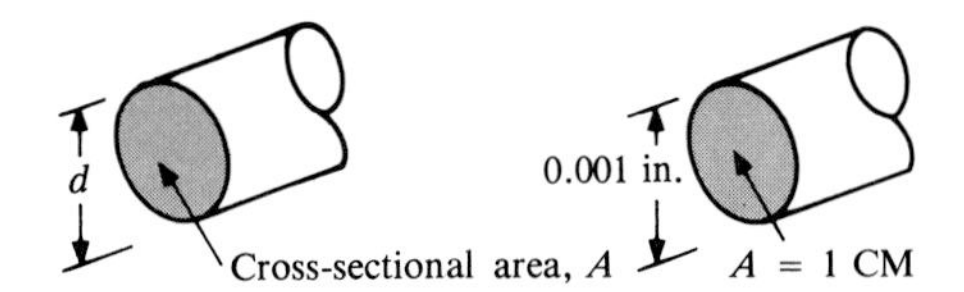

EXAMPLE A–1 What is the cross-sectional area of a wire with a diameter of 0.005 inch?

Solution

$$d = 0.005 \text{ in.} = 5 \text{ mils}$$
$$A = d^2 = 5^2 = 25 \text{ CM}$$

Wire Resistance

Although copper wire conducts electricity extremely well, it still has some resistance, as do all conductors. The resistance of a wire depends on four factors: (1) type of material, (2) length of wire, (3) cross-sectional area, and (4) temperature.

Each type of conductive material has a characteristic called its *resistivity,* ρ. For each material, ρ is a constant value at a given temperature. The formula for the resistance of a wire of length l and cross-sectional area A is

$$R = \frac{\rho l}{A}$$

This formula tells us that resistance increases with resistivity and length, and decreases with cross-sectional area. For resistance to be calculated in ohms, the length must be in feet, the cross-sectional area in circular mils, and the resistivity in CM-Ω/ft.

EXAMPLE A–2 Find the resistance of a 100-ft length of copper wire with a cross-sectional area of 810.1 CM. The resistivity of copper is 10.4 CM-Ω/ft.

Solution

$$R = \frac{\rho l}{A} = \frac{(10.4 \text{ CM-}\Omega\text{/ft})(100 \text{ ft})}{810.1 \text{ CM}} = 1.284\ \Omega$$

Table A–1 lists the resistance of the various standard wire sizes in ohms per 1000 feet at 20°C. For example, a 1000-ft length of 14-gage copper wire has a resistance of 2.525 Ω. A 1000-ft length of 22-gage wire has a resistance of 16.14 Ω. For a given length, the smaller wire has more resistance. Thus, for a given voltage, larger wires can carry more current than smaller ones.

APPENDIX B
Standard Resistance Values

These values are generally available in multiples of 0.1, 1, 10, 100, 1 k, and 1 M.

Resistance Tolerance (±%)

0.1% 0.25% 0.5%	1%	2% 5%	10%	0.1% 0.25% 0.5%	1%	2% 5%	10%	0.1% 0.25% 0.5%	1%	2% 5%	10%	0.1% 0.25% 0.5%	1%	2% 5%	10%	0.1% 0.25% 0.5%	1%	2% 5%	10%	0.1% 0.25% 0.5%	1%	2% 5%	10%
10.0	10.0	10	10	14.7	14.7	—	—	21.5	21.5	—	—	31.6	31.6	—	—	46.4	46.4	—	—	68.1	68.1	68	68
10.1	—	—	—	14.9	—	—	—	21.8	—	—	—	32.0	—	—	—	47.0	—	47	47	69.0	—	—	—
10.2	10.2	—	—	15.0	15.0	15	15	22.1	22.1	22	22	32.4	32.4	—	—	47.5	47.5	—	—	69.8	69.8	—	—
10.4	—	—	—	15.2	—	—	—	22.3	—	—	—	32.8	—	—	—	48.1	—	—	—	70.6	—	—	—
10.5	10.5	—	—	15.4	15.4	—	—	22.6	22.6	—	—	33.2	33.2	33	33	48.7	48.7	—	—	71.5	71.5	—	—
10.6	—	—	—	15.6	—	—	—	22.9	—	—	—	33.6	—	—	—	49.3	—	—	—	72.3	—	—	—
10.7	10.7	—	—	15.8	15.8	—	—	23.2	23.2	—	—	34.0	34.0	—	—	49.9	49.9	—	—	73.2	73.2	—	—
10.9	—	—	—	16.0	—	16	—	23.4	—	—	—	34.4	—	—	—	50.5	—	—	—	74.1	—	—	—
11.0	11.0	11	—	16.2	16.2	—	—	23.7	23.7	—	—	34.8	34.8	—	—	51.1	51.1	51	—	75.0	75.0	75	—
11.1	—	—	—	16.4	—	—	—	24.0	—	24	—	35.2	—	—	—	51.7	—	—	—	75.9	—	—	—
11.3	11.3	—	—	16.5	16.5	—	—	24.3	24.3	—	—	35.7	35.7	—	—	52.3	52.3	—	—	76.8	76.8	—	—
11.4	—	—	—	16.7	—	—	—	24.6	—	—	—	36.1	—	36	—	53.0	—	—	—	77.7	—	—	—
11.5	11.5	—	—	16.9	16.9	—	—	24.9	24.9	—	—	36.5	36.5	—	—	53.6	53.6	—	—	78.7	78.7	—	—
11.7	—	—	—	17.2	—	—	—	25.2	—	—	—	37.0	—	—	—	54.2	—	—	—	79.6	—	—	—
11.8	11.8	—	—	17.4	17.4	—	—	25.5	25.5	—	—	37.4	37.4	—	—	54.9	54.9	—	—	80.6	80.6	—	—
12	—	12	12	17.6	—	—	—	25.8	—	—	—	37.9	—	—	—	56.6	—	—	—	81.6	—	—	—
12.1	12.1	—	—	17.8	17.8	—	—	26.1	26.1	—	—	38.3	38.3	—	—	56.2	56.2	56	56	82.5	82.5	82	82
12.3	—	—	—	18.0	—	18	18	26.4	—	—	—	38.8	—	—	—	56.9	—	—	—	83.5	—	—	—
12.4	12.4	—	—	18.2	18.2	—	—	26.7	26.7	—	—	39.2	39.2	39	39	57.6	57.6	—	—	84.5	84.5	—	—
12.6	—	—	—	18.4	—	—	—	27.1	—	27	27	39.7	—	—	—	58.3	—	—	—	85.6	—	—	—
12.7	12.7	—	—	18.7	18.7	—	—	27.4	27.4	—	—	40.2	40.2	—	—	59.0	59.0	—	—	86.6	86.6	—	—
12.9	—	—	—	18.9	—	—	—	27.7	—	—	—	40.7	—	—	—	59.7	—	—	—	87.6	—	—	—
13.0	13.0	13	—	19.1	19.1	—	—	28.0	28.0	—	—	41.2	41.2	—	—	60.4	60.4	—	—	88.7	88.7	—	—
13.2	—	—	—	19.3	—	—	—	28.4	—	—	—	41.7	—	—	—	61.2	—	—	—	89.8	—	—	—
13.3	13.3	—	—	19.6	19.6	—	—	28.7	28.7	—	—	42.2	42.2	—	—	61.9	61.9	62	—	90.9	90.9	91	—
13.5	—	—	—	19.8	—	—	—	29.1	—	—	—	42.7	—	—	—	62.6	—	—	—	92.0	—	—	—
13.7	13.7	—	—	20.0	20.0	20	—	29.4	29.4	—	—	43.2	43.2	43	—	63.4	63.4	—	—	93.1	93.1	—	—
13.8	—	—	—	20.3	—	—	—	29.8	—	—	—	43.7	—	—	—	64.2	—	—	—	94.2	—	—	—
14.0	14.0	—	—	20.5	20.5	—	—	30.1	30.1	30	—	44.2	44.2	—	—	64.9	64.9	—	—	95.3	95.3	—	—
14.2	—	—	—	20.8	—	—	—	30.5	—	—	—	44.8	—	—	—	65.7	—	—	—	96.5	—	—	—
14.3	14.3	—	—	21.0	21.0	—	—	30.9	30.9	—	—	45.3	45.3	—	—	66.5	66.5	—	—	97.6	97.6	—	—
14.5	—	—	—	21.3	—	—	—	31.2	—	—	—	45.9	—	—	—	67.3	—	—	—	98.8	—	—	—

APPENDIX C
Devices Data Sheets

1N4001 thru 1N4007

Designers▲Data Sheet

"SURMETIC"▲ RECTIFIERS

. . . subminiature size, axial lead mounted rectifiers for general-purpose low-power applications.

Designers Data for "Worst Case" Conditions

The Designers▲ Data Sheets permit the design of most circuits entirely from the information presented. Limit curves — representing boundaries on device characteristics — are given to facilitate "worst case" design.

LEAD MOUNTED SILICON RECTIFIERS

50-1000 VOLTS DIFFUSED JUNCTION

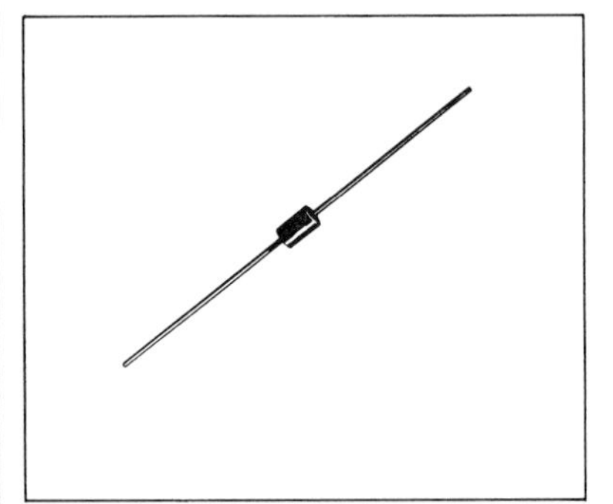

*MAXIMUM RATINGS

Rating	Symbol	1N4001	1N4002	1N4003	1N4004	1N4005	1N4006	1N4007	Unit
Peak Repetitive Reverse Voltage Working Peak Reverse Voltage DC Blocking Voltage	V_{RRM} V_{RWM} V_R	50	100	200	400	600	800	1000	Volts
Non-Repetitive Peak Reverse Voltage (halfwave, single phase, 60 Hz)	V_{RSM}	60	120	240	480	720	1000	1200	Volts
RMS Reverse Voltage	$V_{R(RMS)}$	35	70	140	280	420	560	700	Volts
Average Rectified Forward Current (single phase, resistive load, 60 Hz, see Figure 8, $T_A = 75^oC$)	I_O	1.0							Amp
Non-Repetitive Peak Surge Current (surge applied at rated load conditions, see Figure 2)	I_{FSM}	30 (for 1 cycle)							Amp
Operating and Storage Junction Temperature Range	T_J, T_{stg}	-65 to +175							°C

*ELECTRICAL CHARACTERISTICS

Characteristic and Conditions	Symbol	Typ	Max	Unit
Maximum Instantaneous Forward Voltage Drop (i_F = 1.0 Amp, $T_J = 25^oC$) Figure 1	v_F	0.93	1.1	Volts
Maximum Full-Cycle Average Forward Voltage Drop (I_O = 1.0 Amp, $T_L = 75^oC$, 1 inch leads)	$V_{F(AV)}$	–	0.8	Volts
Maximum Reverse Current (rated dc voltage) $T_J = 25^oC$ $T_J = 100^oC$	I_R	0.05 1.0	10 50	μA
Maximum Full-Cycle Average Reverse Current (I_O = 1.0 Amp, $T_L = 75^oC$, 1 inch leads	$I_{R(AV)}$	–	30	μA

*Indicates JEDEC Registered Data.

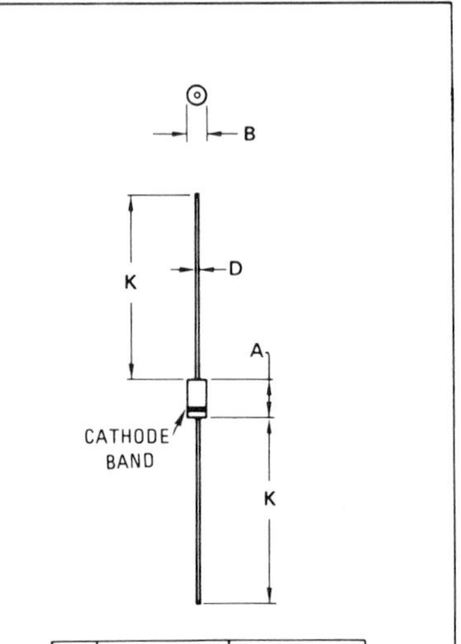

DIM	MILLIMETERS MIN	MILLIMETERS MAX	INCHES MIN	INCHES MAX
A	5.97	6.60	0.235	0.260
B	2.79	3.05	0.110	0.120
D	0.76	0.86	0.030	0.034
K	27.94	–	1.100	–

CASE 59-04

Does Not Conform to DO-41 Outline.

MECHANICAL CHARACTERISTICS

CASE: Transfer Molded Plastic
MAXIMUM LEAD TEMPERATURE FOR SOLDERING PURPOSES: 350°C, 3/8" from case for 10 seconds at 5 lbs. tension
FINISH: All external surfaces are corrosion-resistant, leads are readily solderable
POLARITY: Cathode indicated by color band
WEIGHT: 0.40 Grams (approximately)

▲Trademark of Motorola Inc.

DS 6015 R3

1N4728, A thru 1N4764, A

Designers▲ Data Sheet

ONE WATT HERMETICALLY SEALED GLASS SILICON ZENER DIODES

- Complete Voltage Range – 2.4 to 100 Volts
- DO-41 Package – Smaller than Conventional DO-7 Package
- Double Slug Type Construction
- Metallurgically Bonded Construction
- Nitride Passivated Die

Designer's Data for "Worst Case" Conditions

The Designers▲ Data sheets permit the design of most circuits entirely from the information presented. Limit curves – representing boundaries on device characteristics – are given to facilitate "worst case" design.

1.0 WATT

ZENER REGULATOR DIODES

3.3–100 VOLTS

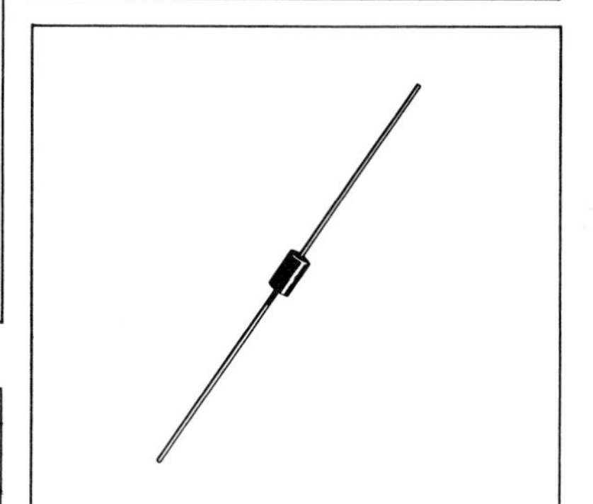

*MAXIMUM RATINGS

Rating	Symbol	Value	Unit
DC Power Dissipation @ $T_A = 50^oC$ Derate above 50^oC	P_D	1.0 6.67	Watt $mW/^oC$
Operating and Storage Junction Temperature Range	T_J, T_{stg}	-65 to +200	^oC

MECHANICAL CHARACTERISTICS

CASE: Double slug type, hermetically sealed glass

MAXIMUM LEAD TEMPERATURE FOR SOLDERING PURPOSES: 230°C, 1/16" from case for 10 seconds

FINISH: All external surfaces are corrosion resistant with readily solderable leads.

POLARITY: Cathode indicated by color band. When operated in zener mode, cathode will be positive with respect to anode.

MOUNTING POSITION: Any

FIGURE 1 – POWER TEMPERATURE DERATING CURVE

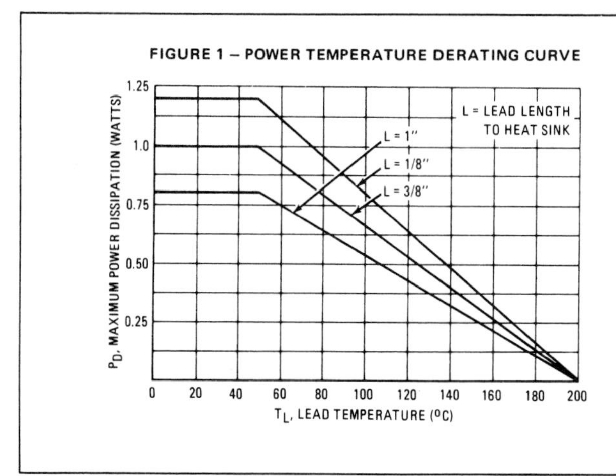

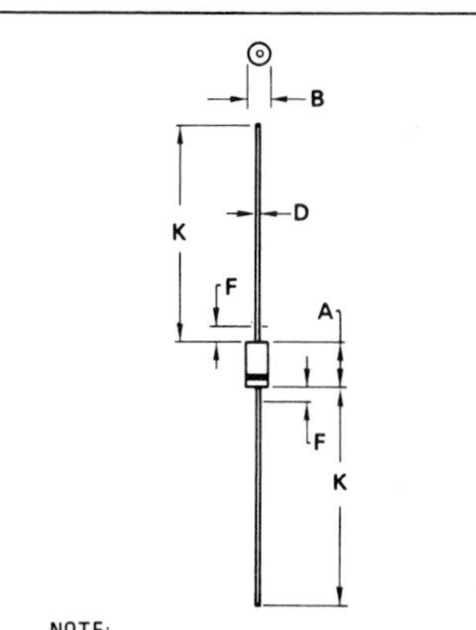

NOTE:
1. POLARITY DENOTED BY CATHODE BAND
2. LEAD DIAMETER NOT CONTROLLED WITHIN "F" DIMENSION.

DIM	MILLIMETERS		INCHES	
	MIN	MAX	MIN	MAX
A	4.07	5.20	0.160	0.205
B	2.04	2.71	0.080	0.107
D	0.71	0.86	0.028	0.034
F	–	1.27	–	0.050
K	27.94	–	1.100	–

All JEDEC dimensions and notes apply.

CASE 59-03 (DO-41)

*Indicates JEDEC Registered Data
▲Trademark of Motorola Inc.

DS 7039 R1

***ELECTRICAL CHARACTERISTICS** (T_A = 25°C unless otherwise noted) V_F = 1.2 V max, I_F = 200 mA for all types.

JEDEC Type No. (Note 1)	Nominal Zener Voltage V_Z @ I_{ZT} Volts (Notes 2 and 3)	Test Current I_{ZT} mA	Maximum Zener Impedance (Note 4) Z_{ZT} @ I_{ZT} Ohms	Z_{ZK} @ I_{ZK} Ohms	I_{ZK} mA	Leakage Current I_R μA Max	V_R Volts	Surge Current @ T_A = 25°C i_r – mA (Note 5)
1N4728	3.3	76	10	400	1.0	100	1.0	1380
1N4729	3.6	69	10	400	1.0	100	1.0	1260
1N4730	3.9	64	9.0	400	1.0	50	1.0	1190
1N4731	4.3	58	9.0	400	1.0	10	1.0	1070
1N4732	4.7	53	8.0	500	1.0	10	1.0	970
1N4733	5.1	49	7.0	550	1.0	10	1.0	890
1N4734	5.6	45	5.0	600	1.0	10	2.0	810
1N4735	6.2	41	2.0	700	1.0	10	3.0	730
1N4736	6.8	37	3.5	700	1.0	10	4.0	660
1N4737	7.5	34	4.0	700	0.5	10	5.0	605
1N4738	8.2	31	4.5	700	0.5	10	6.0	550
1N4739	9.1	28	5.0	700	0.5	10	7.0	500
1N4740	10	25	7.0	700	0.25	10	7.6	454
1N4741	11	23	8.0	700	0.25	5.0	8.4	414
1N4742	12	21	9.0	700	0.25	5.0	9.1	380
1N4743	13	19	10	700	0.25	5.0	9.9	344
1N4744	15	17	14	700	0.25	5.0	11.4	304
1N4745	16	15.5	16	700	0.25	5.0	12.2	285
1N4746	18	14	20	750	0.25	5.0	13.7	250
1N4747	20	12.5	22	750	0.25	5.0	15.2	225
1N4748	22	11.5	23	750	0.25	5.0	16.7	205
1N4749	24	10.5	25	750	0.25	5.0	18.2	190
1N4750	27	9.5	35	750	0.25	5.0	20.6	170
1N4751	30	8.5	40	1000	0.25	5.0	22.8	150
1N4752	33	7.5	45	1000	0.25	5.0	25.1	135
1N4753	36	7.0	50	1000	0.25	5.0	27.4	125
1N4754	39	6.5	60	1000	0.25	5.0	29.7	115
1N4755	43	6.0	70	1500	0.25	5.0	32.7	110
1N4756	47	5.5	80	1500	0.25	5.0	35.8	95
1N4757	51	5.0	95	1500	0.25	5.0	38.8	90
1N4758	56	4.5	110	2000	0.25	5.0	42.6	80
1N4759	62	4.0	125	2000	0.25	5.0	47.1	70
1N4760	68	3.7	150	2000	0.25	5.0	51.7	65
1N4761	75	3.3	175	2000	0.25	5.0	56.0	60
1N4762	82	3.0	200	3000	0.25	5.0	62.2	55
1N4763	91	2.8	250	3000	0.25	5.0	69.2	50
1N4764	100	2.5	350	3000	0.25	5.0	76.0	45

*Indicates JEDEC Registered Data.

NOTE 1 — Tolerance and Type Number Designation. The JEDEC type numbers listed have a standard tolerance on the nominal zener voltage of ±10%. A standard tolerance of ±5% on individual units is also available and is indicated by suffixing "A" to the standard type number.

NOTE 2 — Specials Available Include:

A. Nominal zener voltages between the voltages shown and tighter voltage tolerances,

B. Matched sets.

For detailed information on price, availability, and delivery, contact your nearest Motorola representative.

NOTE 3 — Zener Voltage (V_Z) Measurement. Motorola guarantees the zener voltage when measured at 90 seconds while maintaining the lead temperature (T_L) at 30°C ± 1°C, 3/8" from the diode body.

NOTE 4 — Zener Impedance (Z_Z) Derivation. The zener impedance is derived from the 60 cycle ac voltage, which results when an ac current having an rms value equal to 10% of the dc zener current (I_{ZT} or I_{ZK}) is superimposed on I_{ZT} or I_{ZK}.

NOTE 5 — Surge Current (i_r) Non-Repetitive. The rating listed in the electrical characteristics table is maximum peak, non-repetitive, reverse surge current of 1/2 square wave or equivalent sine wave pulse of 1/120 second duration superimposed on the test current, I_{ZT}, per JEDEC registration; however, actual device capability is as described in Figures 4 and 5.

APPLICATION NOTE

Since the actual voltage available from a given zener diode is temperature dependent, it is necessary to determine junction temperature under any set of operating conditions in order to calculate its value. The following procedure is recommended:

Lead Temperature, T_L, should be determined from

$$T_L = \theta_{LA} P_D + T_A$$

θ_{LA} is the lead-to-ambient thermal resistance (°C/W) and P_D is the power dissipation. The value for θ_{LA} will vary and depends on the device mounting method. θ_{LA} is generally 30 to 40°C/W for the various clips and tie points in common use and for printed circuit board wiring.

The temperature of the lead can also be measured using a thermocouple placed on the lead as close as possible to the tie point. The thermal mass connected to the tie point is normally large enough so that it will not significantly respond to heat surges generated in the diode as a result of pulsed operation once steady-state conditions are achieved. Using the measured value of T_L, the junction temperature may be determined by:

$$T_J = T_L + \Delta T_{JL}.$$

ΔT_{JL} is the increase in junction temperature above the lead temperature and may be found as follows:

$$\Delta T_{JL} = \theta_{JL} P_D$$

θ_{JL} may be determined from Figure 3 for dc power conditions. For worst-case design, using expected limits of I_Z, limits of P_D and the extremes of $T_J(\Delta T_J)$ may be estimated. Changes in voltage, V_Z, can then be found from:

$$\Delta V = \theta_{VZ} \Delta T_J$$

θ_{VZ}, the zener voltage temperature coefficient, is found from Figure 2.

Under high power-pulse operation, the zener voltage will vary with time and may also be affected significantly by the zener resistance. For best regulation, keep current excursions as low as possible.

Surge limitations are given in Figure 5. They are lower than would be expected by considering only junction temperature, as current crowding effects cause temperatures to be extremely high in small spots resulting in device degradation should the limits of Figure 5 be exceeded.

2N4877

MEDIUM-POWER NPN SILICON TRANSISTOR

. . . designed for switching and wide band amplifier applications.

- Low Collector-Emitter Saturation Voltage — $V_{CE(sat)}$ = 1.0 Vdc (Max) @ I_C = 4.0 Amp
- DC Current Gain Specified to 4 Amperes
- Excellent Safe Operating Area
- Packaged in the Compact TO-39 Case for Critical Space-Limited Applications.

4 AMPERE POWER TRANSISTOR

NPN SILICON
60 VOLTS
10 WATTS

* MAXIMUM RATINGS

Rating	Symbol	Value	Unit
Collector-Emitter Voltage	V_{CEO}	60	Vdc
Collector-Base Voltage	V_{CB}	70	Vdc
Emitter-Base Voltage	V_{EB}	5.0	Vdc
Collector Current — Continuous	I_C	4.0	Adc
Base Current	I_B	1.0	Adc
Total Device Dissipation @ T_C = 25°C Derate above 25°C	P_D	10 57.2	Watts mW/°C
Operating and Storage Junction Temperature Range	T_J, T_{stg}	-65 to +200	°C

*Indicates JEDEC Registered Data

THERMAL CHARACTERISTICS

Characteristic	Symbol	Max	Unit
Thermal Resistance, Junction to Case	θ_{JC}	17.5	°C/W

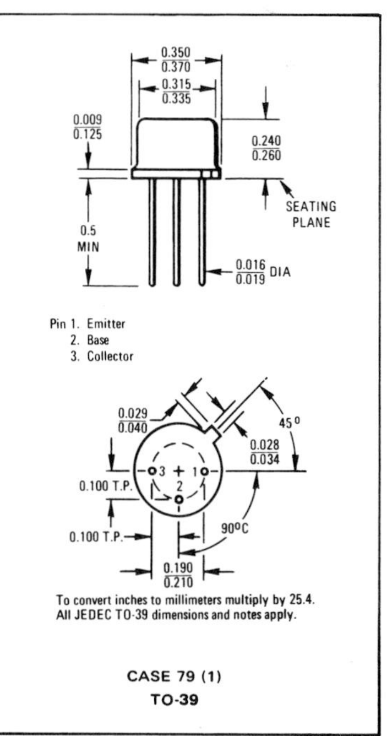

To convert inches to millimeters multiply by 25.4.
All JEDEC TO-39 dimensions and notes apply.

CASE 79 (1)
TO-39

FIGURE 1 — POWER-TEMPERATURE DERATING CURVE

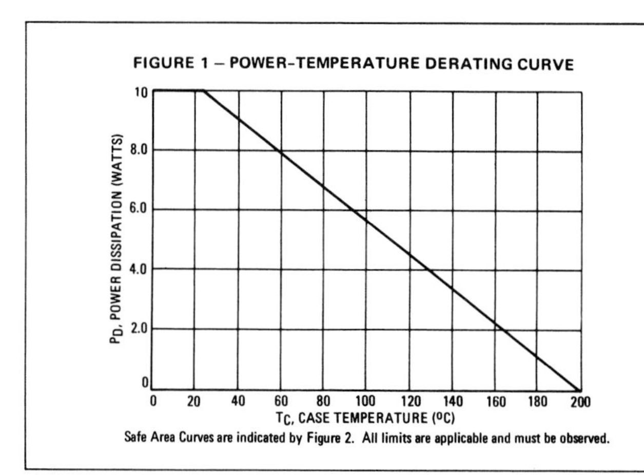

Safe Area Curves are indicated by Figure 2. All limits are applicable and must be observed.

*ELECTRICAL CHARACTERISTICS (T_C = 25°C unless otherwise noted)

Characteristic	Symbol	Min	Max	Unit
OFF CHARACTERISTICS				
Collector-Emitter Sustaining Voltage (1) (I_C = 200 mAdc, I_B = 0)	$V_{CEO(sus)}$	60	–	Vdc
Collector Cutoff Current	I_{CEX}			
(V_{CE} = 70 Vdc, $V_{EB(off)}$ = 1.5 Vdc)		–	100	μAdc
(V_{CE} = 70 Vdc, $V_{EB(off)}$ = 1.5 Vdc, T_C = 100°C)		–	1.0	mAdc
Collector Cutoff Current (V_{CB} = 70 Vdc, I_E = 0)	I_{CBO}	–	100	μAdc
Emitter Cutoff Current (V_{BE} = 5.0 Vdc, I_C = 0)	I_{EBO}	–	100	μAdc
ON CHARACTERISTICS(1)				
DC Current Gain	h_{FE}			–
(I_C = 1.0 Adc, V_{CE} = 2.0 Vdc)		30	–	
(I_C = 4.0 Adc, V_{CE} = 2.0 Vdc)		20	100	
Collector-Emitter Saturation Voltage (I_C = 4.0 Adc, I_B = 0.4 Adc)	$V_{CE(sat)}$	–	1.0	Vdc
Base-Emitter Saturation Voltage (I_C = 4.0 Adc, I_B = 0.4 Adc)	$V_{BE(sat)}$	–	1.8	Vdc
DYNAMIC CHARACTERISTICS				
Current-Gain-Bandwidth Product	f_T			MHz
(I_C = 0.25 Adc, V_{CE} = 10 Vdc, f = 1.0 MHz)		4.0	–	
(I_C = 0.25 Adc, V_{CE} = 10 Vdc, f = 10 MHz)**		30	–	
SWITCHING CHARACTERISTICS				
Rise Time (V_{CC} = 25 Vdc, I_C = 4.0 Adc, I_{B1} = 0.4 Adc)	t_r	–	100	ns
Storage Time (V_{CC} = 25 Vdc, I_C = 4.0 Adc, I_{B1} = I_{B2} = 0.4 Adc)	t_s	–	1.5	μs
Fall Time (V_{CC} = 25 Vdc, I_C = 4.0 Adc, I_{B1} = I_{B2} = 0.4 Adc)	t_f	–	500	ns

*Indicates JEDEC Registered Data.
**Motorola guarantees this value in addition to JEDEC Registered Data.
Note 1: Pulse Test: Pulse Width ≤ 300 μs, Duty Cycle ≤ 2.0%.

FIGURE 2 – ACTIVE-REGION SAFE OPERATING AREA

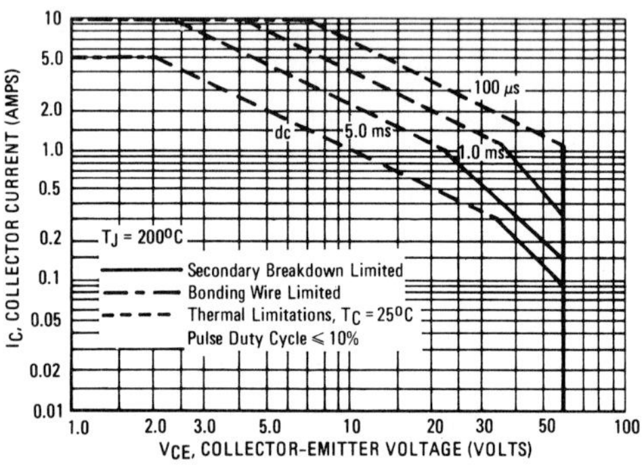

I_C, COLLECTOR CURRENT (AMPS) vs. V_{CE}, COLLECTOR-EMITTER VOLTAGE (VOLTS); curves: dc, 5.0 ms, 1.0 ms, 100 μs; T_J = 200°C; Secondary Breakdown Limited; Bonding Wire Limited; Thermal Limitations, T_C = 25°C; Pulse Duty Cycle ≤ 10%

FIGURE 3 – SWITCHING TIME TEST CIRCUIT

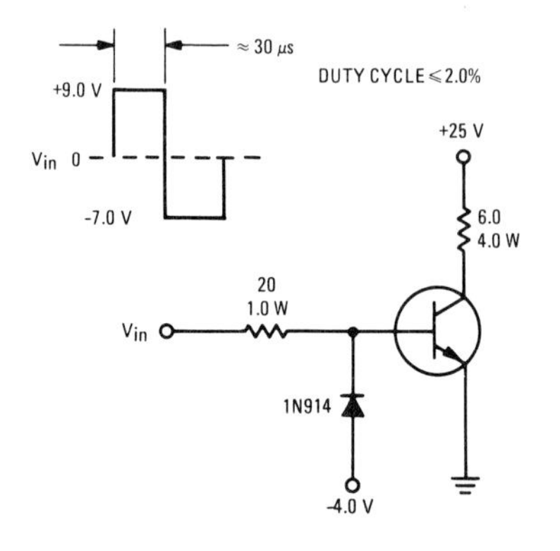

There are two limitations on the power handling ability of a transistor: average junction temperature and second breakdown. Safe operating area curves indicate I_C–V_{CE} limits of the transistor that must be observed for reliable operation; i.e., the transistor must not be subjected to greater dissipation than the curves indicate.

The data of Figure 2 is based on $T_{J(pk)}$ = 200°C; T_C is variable depending on conditions. Second breakdown pulse limits are valid for duty cycles to 10% provided $T_{J(pk)} \leq$ 200°C. At high case temperatures, thermal limitations will reduce the power that can be handled to values less than the limitations imposed by second breakdown. (See AN-415)

MOTOROLA Semiconductor Products Inc.

BOX 20912 • PHOENIX, ARIZONA 85036 • A SUBSIDIARY OF MOTOROLA INC.

6025-2 PRINTED IN USA 3-71 IMPERIAL LITHO 821668 10M

DS 3189

2N1595 thru 2N1599

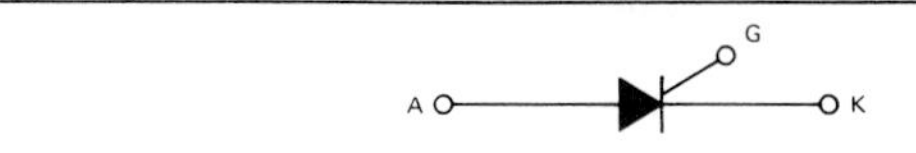

SILICON CONTROLLED RECTIFIERS

1.6 AMPERE RMS
50 thru 400 VOLTS

REVERSE BLOCKING TRIODE THYRISTORS

These devices are glassivated planar construction designed for gating operation in mA/μA signal or detection circuits.

- Low-Level Gate Characteristics — I_{GT} = 10 mA (Max) @ 25°C
- Low Holding Current — I_H = 5.0 mA (Typ) @ 25°C
- Glass-to-Metal Bond for Maximum Hermetic Seal

***MAXIMUM RATINGS** (T_J = 125°C unless otherwise noted).

Rating	Symbol	Value	Unit
Repetitive Peak Reverse Blocking Voltage	V_{RRM}		Volts
2N1595		50	
2N1596		100	
2N1597		200	
2N1598		300	
2N1599		400	
Repetitive Peak Forward Blocking Voltage	V_{DRM}		Volts
2N1595		50	
2N1596		100	
2N1597		200	
2N1598		300	
2N1599		400	
RMS On-State Current (All Conduction Angles)	$I_{T(RMS)}$	1.6	Amps
Peak Non-Repetitive Surge Current (One Cycle, 60 Hz, T_J = -65 to +125°C)	I_{TSM}	15	Amps
Peak Gate Power	P_{GM}	0.1	Watt
Average Gate Power	$P_{G(AV)}$	0.01	Watt
Peak Gate Current	I_{GM}	0.1	Amp
Peak Gate Voltage — Forward	V_{GFM}	10	Volts
Reverse	V_{GRM}	10	
Operating Junction Temperature Range	T_J	-65 to +125	°C
Storage Temperature Range	T_{stg}	-65 to +150	°C

*Indicates JEDEC Registered Data.

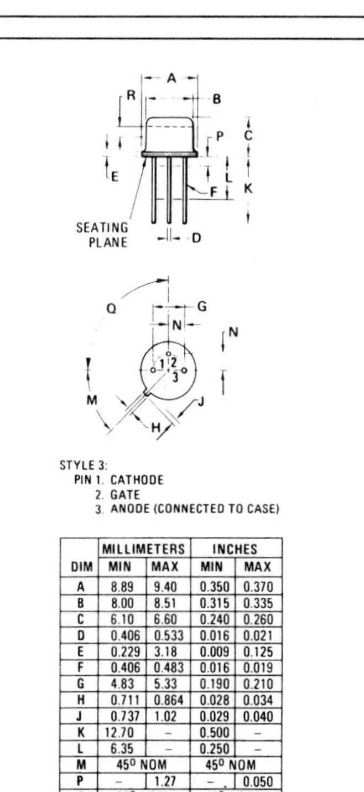

STYLE 3:
PIN 1. CATHODE
2. GATE
3. ANODE (CONNECTED TO CASE)

DIM	MILLIMETERS MIN	MILLIMETERS MAX	INCHES MIN	INCHES MAX
A	8.89	9.40	0.350	0.370
B	8.00	8.51	0.315	0.335
C	6.10	6.60	0.240	0.260
D	0.406	0.533	0.016	0.021
E	0.229	3.18	0.009	0.125
F	0.406	0.483	0.016	0.019
G	4.83	5.33	0.190	0.210
H	0.711	0.864	0.028	0.034
J	0.737	1.02	0.029	0.040
K	12.70	–	0.500	–
L	6.35	–	0.250	–
M	45° NOM		45° NOM	
P	–	1.27	–	0.050
Q	90° NOM		90° NOM	
R	2.54	–	0.100	–

All JEDEC dimensions and notes apply.

CASE 79-02
TO-39

DS 6503 R1

ELECTRICAL CHARACTERISTICS (T_C = 25°C unless otherwise noted).

Characteristic	Symbol	Min	Typ	Max	Unit
*Peak Reverse Blocking Current (Rated V_{RRM}, T_J = 125°C)	I_{RRM}	–	–	1000	μA
*Peak Forward Blocking Current (Rated V_{DRM}, T_J = 125°C)	I_{DRM}	–	–	1000	μA
*Peak On-State Voltage (I_F = 1.0 Adc, Pulsed, 1.0 ms (Max), Duty Cycle ≈ 1%)	V_{TM}	–	1.1	2.0	Volts
*Gate Trigger Current (V_{AK} = 6.0 V, R_L = 12 Ohms)	I_{GT}	–	2.0	10	mA
*Gate Trigger Voltage (V_{AK} = 6.0 V, R_L = 12 Ohms) (V_{AK} = 6.0 V, R_L = 12 Ohms, T_J = 125°C)	V_{GT}	 – 0.2	 0.7 –	 3.0 –	Volts
Reverse Gate Current (V_{GK} = 10 V)	I_{GR}	–	17	–	mA
Holding Current (V_{AK} = 12 V)	I_H	–	5.0	–	mA
Turn-On Time (I_{GT} = 10 mA, I_F = 1.0 A) (I_{GT} = 20 mA, I_F = 1.0 A)	t_{gt}	 – –	 0.8 0.6	 – –	μs
Turn-Off Time (I_F = 1.0 A, I_R = 1.0 A, dv/dt = 20 V/μs, T_J = 125°C)	t_q	–	10	–	μs

*Indicates JEDEC Registered Data.

CURRENT DERATING

FIGURE 1 – CASE TEMPERATURE REFERENCE

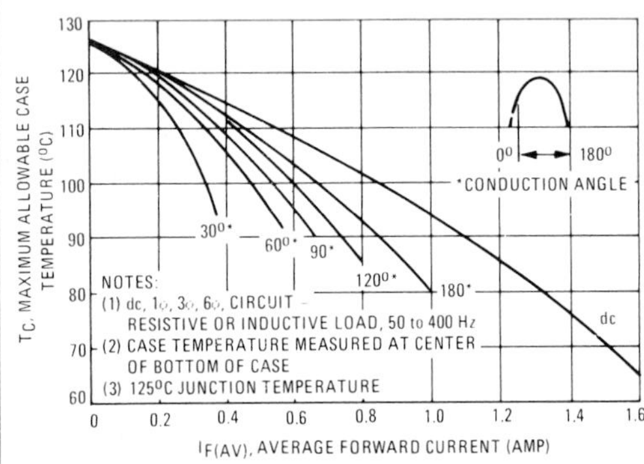

FIGURE 2 – AMBIENT TEMPERATURE REFERENCE

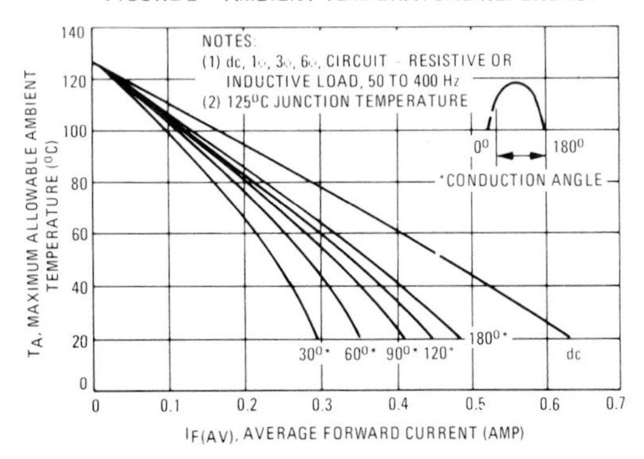

MOTOROLA Semiconductor Products Inc.

BOX 20912 • PHOENIX, ARIZONA 85036 • A SUBSIDIARY OF MOTOROLA INC

10137 PRINTED IN USA (8/77) MPS 10M

TL060, TL060A, TL060B, TL061, TL061A, TL061B, TL062, TL062A, TL062B, TL064, TL064A, TL064B LOW-POWER JFET-INPUT OPERATIONAL AMPLIFIERS

D2392, NOVEMBER 1978 REVISED AUGUST 1985

20 DEVICES COVER MILITARY, INDUSTRIAL, AND COMMERCIAL TEMPERATURE RANGES

- Very Low Power Consumption
- Typical Supply Current . . . 200 μA (per Amplifier)
- Wide Common-Mode and Differential Voltage Ranges
- Low Input Bias and Offset Currents
- Common-Mode Input Voltage Range Includes V_{CC+}
- Output Short-Circuit Protection
- High Input Impedance . . . JFET-Input Stage
- Internal Frequency Compensation (Except TL060)
- Latch-Up Free Operation
- High Slew Rate . . . . 3.5 V/μs Typ

description

The JFET-input operational amplifiers of the TL061 series are designed as low-power versions of the TL081 series amplifiers. They feature high input impedance, wide bandwidth, high slew rate, and low input offset and bias currents. The TL061 series features the same terminal assignments as the TL071 and TL081 series. Each of these JFET-input operational amplifiers incorporates well-matched, high-voltage JFET and bipolar transistors in a monolithic integrated circuit.

Device types with an "M" suffix are characterized for operation over the full military temperature range of −55°C to 125°C, those with an "I" suffix are characterized for operation from −25°C to 85°C, and those with a "C" suffix are characterized for operation from 0°C to 70°C.

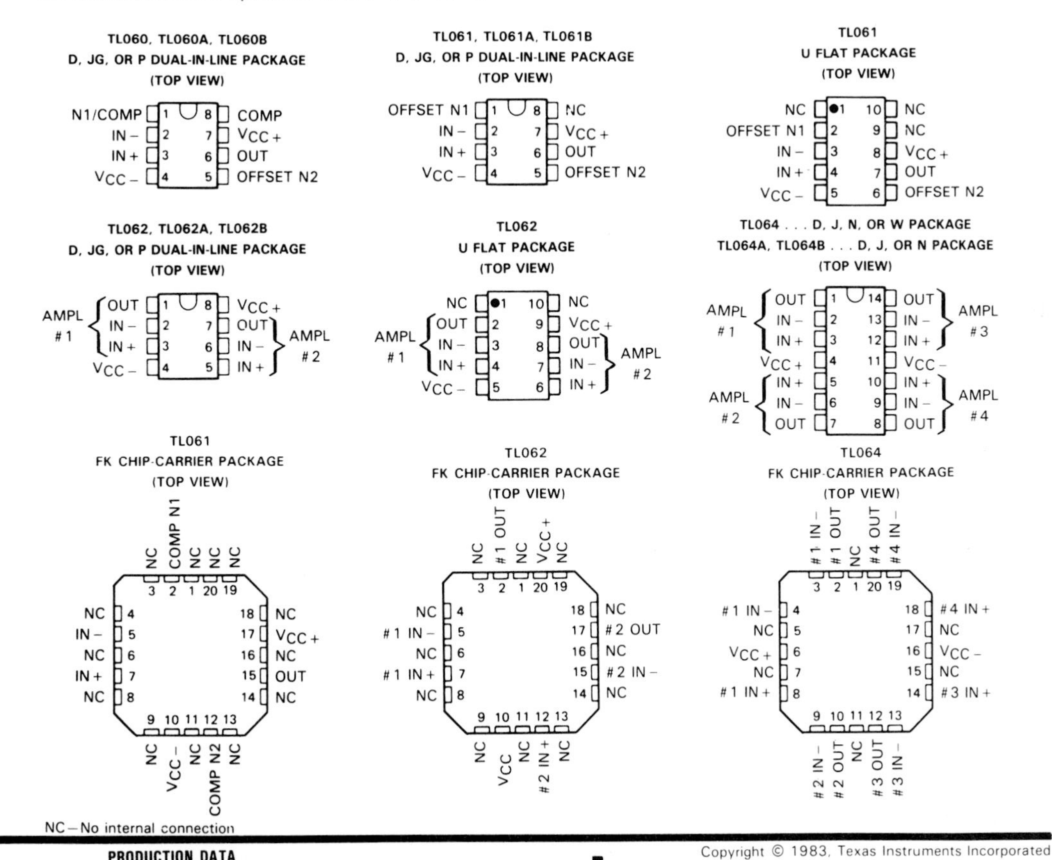

NC—No internal connection

PRODUCTION DATA
This document contains information current as of publication date. Products conform to specifications per the terms of Texas Instruments standard warranty. Production processing does not necessarily include testing of all parameters.

POST OFFICE BOX 225012 • DALLAS, TEXAS 75265

symbol (each amplifier)

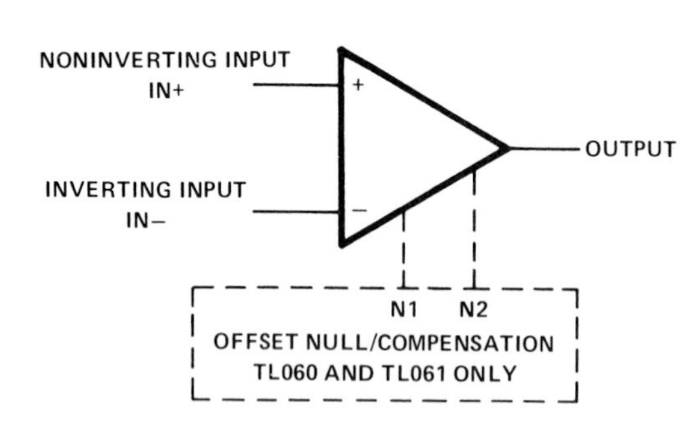

schematic (each amplifier)

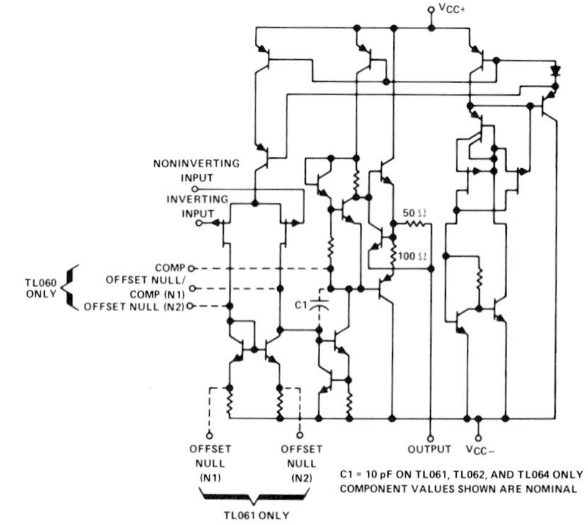

absolute maximum ratings over operating free-air temperature range (unless otherwise noted)

		TL06_M	TL06_I	TL06_C TL06_AC TL06_BC	UNIT
Supply voltage, V_{CC+} (see Note 1)		18	18	18	V
Supply voltage, V_{CC-} (see Note 1)		−18	−18	−18	V
Differential input voltage (see Note 2)		±30	±30	±30	V
Input voltage (see Notes 1 and 3)		±15	±15	±15	V
Duration of output short circuit (see Note 4)		unlimited	unlimited	unlimited	
Continuous total dissipation at (or below) 25°C free-air temperature (see Note 5)	D package		680	680	mW
	FK package	680			mW
	J, JG, N, P, or W package	680	680	680	mW
	U package	675			mW
Operating free-air temperature range		−55 to 125	−25 to 85	0 to 70	°C
Storage temperature range		−65 to 150	−65 to 150	−65 to 150	°C
Lead temperature 1,6 mm (1/16 inch) from case for 60 seconds	J, JG, U, FK, or W package	300	300	300	°C
Lead temperature 1,6 mm (1/16 inch) from case for 10 seconds	D, N, or P package		260	260	°C

NOTES: 1. All voltage values, except differential voltages, are with respect to the midpoint between V_{CC+} and V_{CC-}.
2. Differential voltages are at the noninverting input terminal with respect to the inverting input terminal.
3. The magnitude of the input voltage must never exceed the magnitude of the supply voltage or 15 volts, whichever is less.
4. The output may be shorted to ground or to either supply. Temperature and/or supply voltages must be limited to ensure that the dissipation rating is not exceeded.
5. For operation above 25°C free-air temperature, refer to Dissipation Derating Curves, Section 2. In the J and JG packages, TL06_M chips are alloy mounted; TL06_I, TL06_C, TL06_AC, and TL06_BC chips are glass mounted.

DEVICE TYPES, SUFFIX VERSIONS, AND PACKAGES

	TL060	TL061	TL062	TL064
TL06_M	JG	FK, JG, U	FK, JG, U	FK, J, W
TL06_I	D, JG, P	D, JG, P	D, JG, P	D, J, N
TL06_C	D, JG, P	D, JG, P	D, JG, P	D, J, N
TL06_AC	D, JG, P	D, JG, P	D, JG, P	D, J, N
TL06_BC	D, JG, P	D, JG, P	D, JG, P	D, J, N

operating characteristics, $V_{CC\pm} = \pm 15$ V, $T_A = 25°C$

PARAMETER		TEST CONDITIONS		MIN	TYP	MAX	UNIT
SR	Slew rate at unity gain	$V_I = 10$ V, $C_L = 100$ pF,	$R_L = 10$ kΩ, See Figure 1	1.5	3.5		V/μs
t_r	Rise time	$V_I = 20$ mV,	$R_L = 10$ kΩ,		0.2		μs
	Overshoot factor	$C_L = 100$ pF,	See Figure 1		10%		
V_n	Equivalent input noise voltage	$R_S = 100$ Ω,	f = 1 kHz		42		nV/√Hz

PARAMETER MEASUREMENT INFORMATION

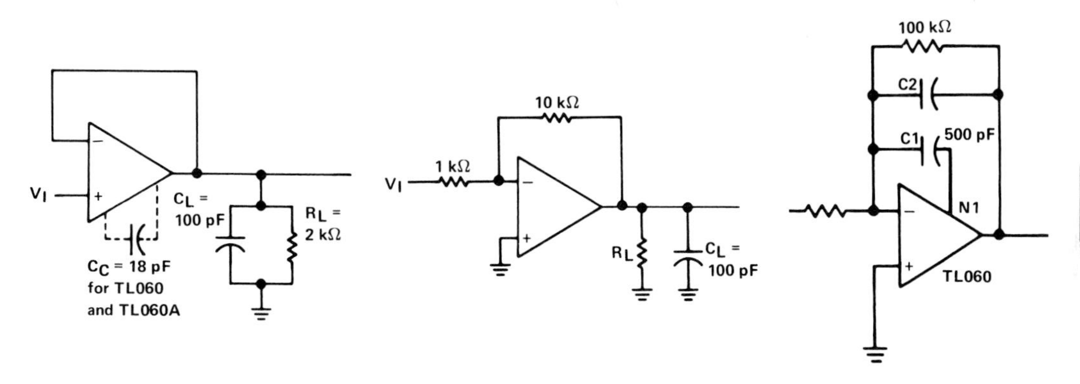

FIGURE 1. UNITY-GAIN AMPLIFIER

FIGURE 2. GAIN-OF-10 INVERTING AMPLIFIER

FIGURE 3. FEED-FORWARD COMPENSATION

INPUT OFFSET VOLTAGE NULL CIRCUITS

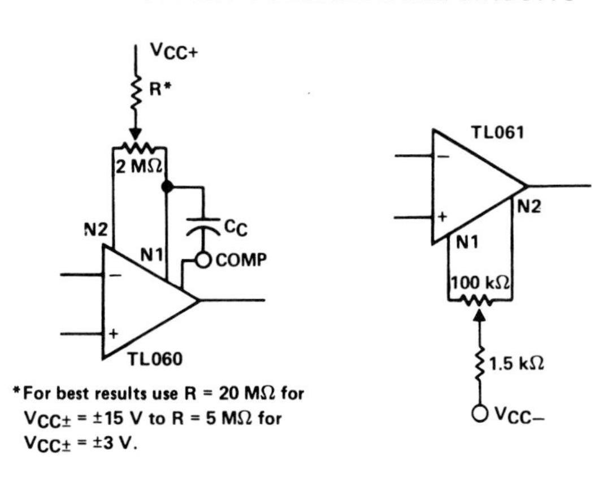

*For best results use R = 20 MΩ for $V_{CC\pm} = \pm 15$ V to R = 5 MΩ for $V_{CC\pm} = \pm 3$ V.

FIGURE 4

FIGURE 5

3

Operational Amplifiers

electrical characteristics, $V_{CC\pm} = \pm 15$ V (unless otherwise noted)

PARAMETER		TEST CONDITIONS†		TL060I TL061I TL062I TL064I			TL060C TL061C TL062C TL064C			TL060AC TL061AC TL062AC TL064AC			TL060BC TL061BC TL062BC TL064BC			UNIT
				MIN	TYP	MAX	MIN	TYP	MAX	MIN	TYP	MAX	MIN	TYP	MAX	
V_{IO}	Input offset voltage	$V_O = 0$,	$T_A = 25°C$		3	6		3	15		3	6		2	3	mV
		$R_S = 50\ \Omega$	T_A = full range			9			20			7.5			5	
α_{VIO}	Temperature coefficient of input offset voltage	$V_O = 0$, $R_S = 50\ \Omega$, T_A = full range			10			10			10			10		μV/°C
I_{IO}	Input offset current‡	$V_O = 0$	$T_A = 25°C$		5	100		5	200		5	100		5	100	pA
			T_A = full range			10			2			2			2	nA
I_{IB}	Input bias current‡	$V_O = 0$	$T_A = 25°C$		30	200		30	200		30	200		30	200	pA
			T_A = full range			20			400			7			7	nA
V_{ICR}	Common-mode input voltage range	$T_A = 25°C$		±11.5	12 to +15		±11	12 to +15		±11.5	12 to +15		±11.5	12 to +15		V
V_{OM}	Maximum peak output voltage swing	$R_L = 10\ k\Omega$,	$T_A = 25°C$	±10	±13.5		±10	±13.5		±10	±13.5		±10	±13.5		V
		$R_L \geq 10\ k\Omega$,	T_A = full range	±10			±10			±10			±10			
A_{VD}	Large-signal differential voltage amplification	$V_O = \pm 10$ V,	$T_A = 25°C$	4	6		3	6		4	6		4	6		V/mV
		$R_L \geq 10\ k\Omega$	T_A = full range	4			3			4			4			
B_1	Unity-gain bandwidth	$R_L = 10\ k\Omega$,	$T_A = 25°C$		1			1			1			1		MHz
r_i	Input resistance	$T_A = 25°C$			10^{12}			10^{12}			10^{12}			10^{12}		Ω
CMRR	Common-mode rejection ratio	$V_{IC} = V_{ICR}$ min, $V_O = 0$, $R_S = 50\ \Omega$, $T_A = 25°C$		80	86		70	86		80	86		80	86		dB
k_{SVR}	Supply voltage rejection ratio $(\Delta V_{CC\pm}/\Delta V_{IO})$	$V_{CC} = \pm 15$ V to ± 9 V, $V_O = 0$, $R_S = 50\ \Omega$, $T_A = 25°C$		80	95		70	95		80	95		80	95		dB
P_D	Total power dissipation (each amplifier)	No load, $V_O = 0$, $T_A = 25°C$			6	7.5		6	7.5		6	7.5		6	7.5	mW
I_{CC}	Supply current (each amplifier)	No load, $V_O = 0$, $T_A = 25°C$			200	250		200	250		200	250		200	250	μA
V_{o1}/V_{o2}	Crosstalk attenuation	$A_{VD} = 100$,	$T_A = 25°C$		120			120			120			120		dB

†All characteristics are measured under optn-loop conditions with zero common-mode voltage unless otherwise specified. Full range for T_A is −25°C to 85°C for TL06_I and 0°C to 70°C for TL06_C, TL06_AC, and TL06_BC.

‡Input bias currents of a FET-input operational amplifier are normal junction reverse currents, which are temperature sensitive as shown in Figure 17. Pulse techniques must be used that will maintain the junction temperature as close to the ambient temperature as possible.

Solutions to Self-Tests

Chapter 1

1. Current: amperes. Voltage: volts. Resistance: ohms. Power: watts. Energy: joules.
2. Amperes: A. Volts: V. Ohms: Ω. Watts: W. Joules: J.
3. Current: I. Voltage: V. Resistance: R. Power: P. Energy: W.
4. **(a)** 10 mA **(b)** 5 kV **(c)** 15 μW **(d)** 20 MΩ.
5. **(a)** 0.000008 A **(b)** 25,000,000 W **(c)** 0.100 V.

Chapter 2

1. A neutral atom with an atomic number of 3 has 3 electrons.
2. Shells are the orbits in which electrons revolve.
3. Semiconductors have fewer free electrons than conductors do, but more than insulators do.
4. The presence of a net positive charge or a net negative charge in a material.
5. They will be pulled toward each other because opposite charges attract.
6. A single electron has a charge of 1.6×10^{-19} coulomb.
7. A positive ion is created when a valence electron escapes from its orbit.
8. Potential difference is voltage, and its unit is the volt.
9. The unit of energy is the joule.
10. Voltage sources.
11. Voltage can exist between two points when there in no current.
12. The repulsive force between the negative voltage source terminal and the free electrons produces movement of the electrons through the material toward the positive terminal, which attracts them.
13. One ampere is the amount of electrical current equal to one coulomb of charge passing a point in one second (1 C/s).
14. Resistors limit current and produce heat.
15. Potentiometers and rheostats are variable resistors.
16. In a linear potentiometer, the resistance varies in direct proportion to the wiper movement. In a tapered device, the resistance is not linearly proportional to the wiper movement.
17. The circuit is open.
18. A current measurement is more difficult than a voltage measurement because the meter must be connected in series, a connection that requires a line to be broken.

Chapter 3

1. Ohm's law states that voltage, current, and resistance are linearly related: $V = IR$.
2. **(a)** The current triples when the voltage is tripled.
   (b) The current decreases by 75% when the voltage is reduced by 75%.
   (c) The current decreases by one-half when the resistance is doubled.
   (d) The current increases by 35% when the resistance is reduced by 35%.

(e) The current increases by four times when the voltage is doubled and the resistance is cut in half.
(f) The current does not change when the voltage is doubled and the resistance is doubled.

3. $I = V/R$.
4. $V = IR$.
5. $R = V/I$.
6. (a) Volts divided by kilohms gives milliamperes: V/kΩ = mA.
   (b) Volts divided by megohms gives microamperes: V/MΩ = μA.
   (c) Milliamperes times kilohms gives volts: mA × kΩ = V.
   (d) Milliamperes times megohms gives kilovolts: mA × MΩ = kV.
   (e) Volts divided by milliamperes gives kilohms: V/mA = kΩ.
   (f) Volts divided by microamperes gives megohms: V/μA = MΩ.
7. Power is the rate at which energy is used.
8. (a) 1000 watts in a kilowatt.
   (b) 1,000,000 watts in a megawatt.
9. Connect the voltmeter across the resistor, and connect the ammeter in series with the resistor. Multiply the voltage across the resistor and the current through it to find the power.
10. (a) $P = VI$
    (b) $P = I^2R$
    (c) $P = V^2/R$.
11. Use a 2-W resistor to handle 1.1 W.
12. An open resistor shows an infinite (∞) reading on the ohmmeter.

Chapter 4

1. See Figure ST–1.
2. The same current (2 A) flows out of all of the resistors.
3. See Figure ST–2.
4. The total resistance is increased by the addition of more resistors in series.
5. The current will increase because the total resistance is reduced.
6. See Figure ST–3.
7. The light is dimmer because the total voltage is reduced and therefore there is less current through the bulb.
8. By Kirchhoff's voltage law, the sum of the voltages is zero.
9. $V_S = 6(5\text{ V}) = 30\text{ V}$.
10. The 10-kΩ resistor has the greatest voltage drop because it has the largest value.
11. One 100-Ω resistor dissipates the most power ($P = V^2/R$).
12. $P_{each} = 10\text{ W}/5 = 2\text{ W}$.
13. Check for an open. The voltage across the open equals the source voltage.
14. Check for a short.

Chapter 5

1. See Figure ST–4.
2. See Figure ST–5.
3. A parallel circuit has multiple current paths. A series circuit has only one current path.
4. 100 Ω.
5. 68 Ω.
6. $R_T = 1\text{ k}\Omega/2 = 500\ \Omega$.

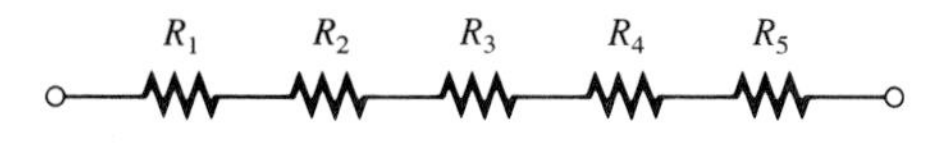

FIGURE ST–1

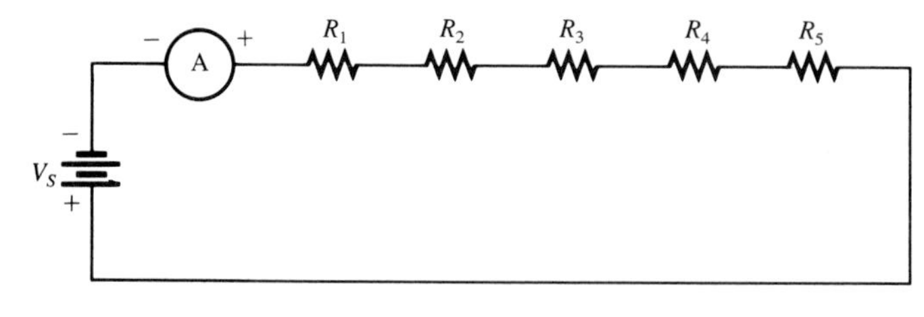

FIGURE ST–2

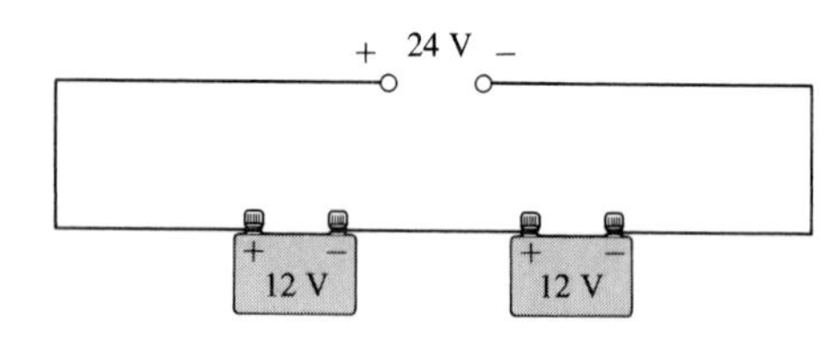

FIGURE ST–3

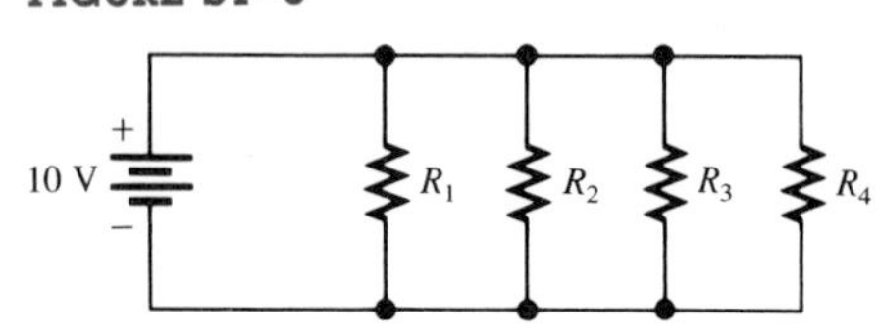

FIGURE ST–4

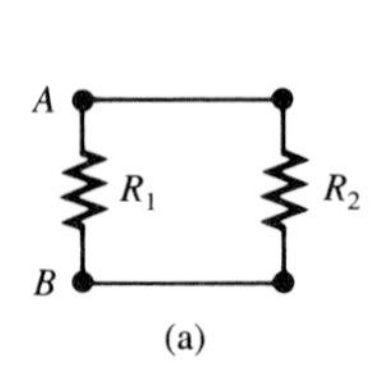

(a)

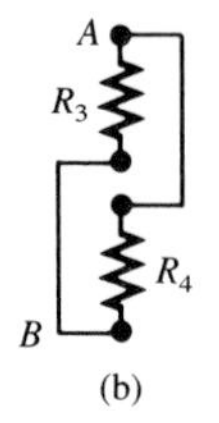

(b)

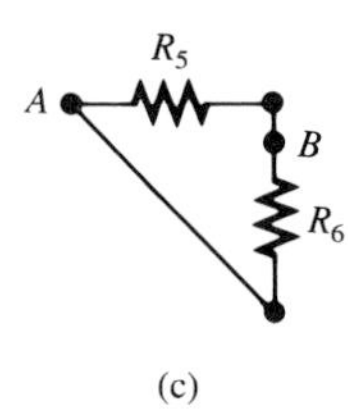

(c)

FIGURE ST–5

7. 9 V.
8. The total resistance decreases as more resistors are added in parallel.
9. The total resistance increases when one of the parallel resistors is removed.
10. $I_T = 5\text{ A} + 3\text{ A} = 8\text{ A}$.
11. The 390-Ω resistor has the most current; the 820-Ω, the least.
12. An increase in total current indicates less total resistance.
13. $P_T = 3(1\text{ W}) = 3\text{ W}$.
14. 10 mA; the other branch currents are unaffected when one branch opens.

Chapter 6

1. See Figure ST–6.
2. $R_T = \dfrac{(R_1 + R_2)(R_3 + R_4 + R_5)}{R_1 + R_2 + R_3 + R_4 + R_5}$.
3. The most current flows through R_1 and R_2.
4. Since there are 6 V across one 1-kΩ resistor, there are also 6 V across the other because they are in series. The voltage across the two 1-kΩ resistors is 12 V, which is also the voltage across the parallel 2.2-kΩ resistor.
5. The 330-Ω resistor has the total current flowing through it. Since 330 Ω is greater than the resistance of the parallel combination of four 1-kΩ resistors (250 Ω), the 330-Ω resistor has the largest voltage drop.
6. Each of the four 1-kΩ resistors in parallel carries 25% of the total current.
7. $V_{BA} = V_B - V_A = 8\text{ V} - 5\text{ V} = 3\text{ V}$.
   $V_{CA} = V_C - V_A = 12\text{ V} - 5\text{ V} = 7\text{ V}$.
8. The voltage will decrease.
9. The 100-kΩ load resistance has the greatest effect.
10. More current will be drawn from the source when a load is connected.
11. There is 0 V across the output of a balanced bridge.
12. A device that indicates current in either direction.
13. Replace all but one source with their internal resistances; then find the currents or voltages due to the remaining source. Repeat for each source, and combine the results.
14. $I = 10\text{ mA} - 8\text{ mA} = 2\text{ mA}$ (in the direction of the larger).
15. See Figure ST–7.
16. The voltmeter probably is loading the circuit.

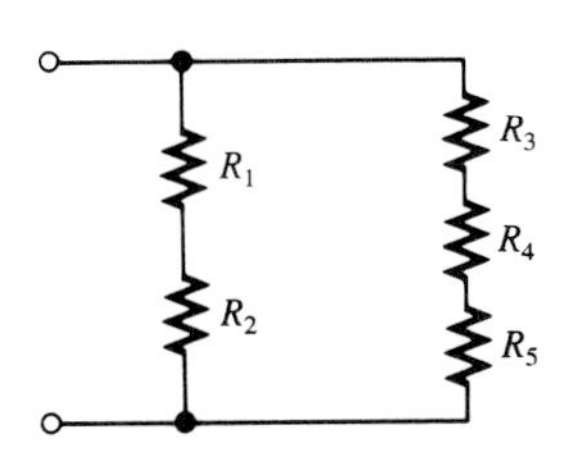

FIGURE ST–6

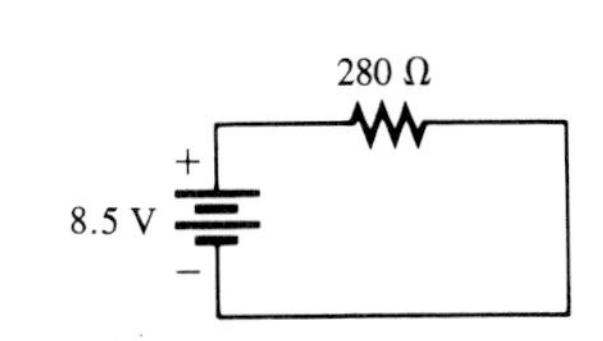

FIGURE ST–7

Chapter 7

1. Like poles repel.
2. A magnetic field consists of the lines of force from the north pole to the south pole of a magnet.
3. Flux density: teslas. Magnetic flux: webers. Magnetomotive force: ampere-turns.
4. Reluctance is the opposition to the establishment of a magnetic field. It is analogous to resistance.
5. Current in the coil of a recording head sets a magnetic field that varies with the direction and magnitude of the current. The magnetic field magnetizes the recording surface proportional to its direction and strength.
6. In a solenoid, the electromagnetic action produces a mechanical movement of a plunger or core. In a relay, the electromagnetic action produces the opening or closing of an electrical contact.
7. Movement of a conductor relative to a magnetic field.
8. True.
9. The induced voltage increases when the number of turns is increased.
10. When the tab cuts through and alters the magnetic field.

Chapter 8

1. Electrodynamometer.
2. To bypass excess current around the meter coil.
3. To drop all voltage above that required for the meter coil.
4. To produce current proportional to the resistance being measured.
5. Four digit positions can indicate from 0 through 9. One digit position can indicate only the digit 1.
6. LED displays are best for low-light conditions.
7. An electron beam sweeps across the screen to produce a pattern.

Chapter 9

1. Alternating current periodically reverses direction of flow. Direct current always flows in one direction.
2. A sine wave has one positive peak and one negative peak during each cycle.

3. A cycle is one repetition of a sine wave, consisting of a positive alternation and a negative alternation.
4. A periodic waveform repeats itself at regular, fixed intervals. A nonperiodic wave does not repeat at regular intervals.
5. The sine wave with the highest frequency, 20 kHz, is changing at the fastest rate.
6. The sine wave with the shortest period, 2 ms, is changing at the fastest rate.
7. 60 Hz = 60 cycles per second.
   Total cycles in 10 s = (60 cycles/s)(10 s) = 600 cycles.
8. $V_{pp} = 2V_p = 2(10\text{ V}) = 20\text{ V}$.
   $V_{rms} = 0.707V_p = 0.707(10\text{ V}) = 7.07\text{ V}$.
9. The rms value is the measure of the heating effect of a sine wave.
10. The average value of a sine wave over a full cycle is zero because the positive portion and the negative portion are equal.
11. The average value is taken over a half-cycle and is 63.7% of the peak value.
12. Check your lab.
13. Rate of rotation; number of poles.
14. AM—amplitude modulation; FM—frequency modulation.
15. 180°
16. (57.3°/rad)(2 rad) = 114.6°
17. $\theta = 45° - 10° = 35°$
18. $v = 15 \sin 32° = 7.95\text{ V}$
19. $V_R = IR = (5\text{ mA})(10\text{ k}\Omega) = 50\text{ V}$
20. $V_s = 6.5\text{ V} + 3.2\text{ V} = 9.7\text{ V}$
21. The one with 50 μs pulses.
22. 50%
23. A sawtooth has two ramps.

Chapter 10

1. (a) True (b) True (c) False (d) False.
2. (a) False (b) True (c) False (d) True.
3. 0.00001 F = 10 μF and is larger than 0.01 μF.
4. 0.0001 μF = 100 pF and is smaller than 1000 pF.
5. Since $Q = CV$, an increase in V produces an increase in Q.
6. Since $W = (1/2)CV^2$, doubling V increases the stored energy by four times.
7. The voltage rating of a capacitor can be increased by increasing the plate separation or by using a dielectric with a greater dielectric constant.
8. All of the following increase the capacitance value:
   (c) Move plates closer;
   (d) Increase plate area;
   (e) Decrease the dielectric thickness.
9. $C_T < 0.05\ \mu$F.
10. $C_T = 4(0.02\ \mu\text{F}) = 0.08\ \mu$F.
11. (a) $v_C = 0$ V at instant of switch closure.
    (b) The capacitor is fully charged at five time constants from switch closure.
    (c) $v_C = V_S$ when the capacitor is fully charged.
    (d) $I = 0$ A when the capacitor is fully charged.
12. The current increases because the capacitive reactance decreases with an increase in frequency.
13. When the frequency is decreased, the capacitor has the greatest voltage because X_C becomes larger than R.
14. The capacitor is very leaky (the leakage resistance is low).

Chapter 11

1. 0.000005 H = 5 μH and is larger than 0.05 μH.
2. Since 0.33 mH = 330 μH, the 33-μH inductance is the smaller.
3. Since $W = (1/2)LI^2$, the stored energy increases when the current increases.
4. Since $W = (1/2)LI^2$, a doubling of the current increases the stored energy by four times.
5. The winding resistance can be decreased by using larger-sized wire.
6. The following result in an increase in inductance:
   (b) Change from an air core to an iron core;
   (c) Increase the number of turns;
   (e) Decrease the length of the core.
7. $L_T < 0.1$ mH.
8. $L_T = 100\ \mu\text{H}/5 = 20\ \mu$H.
9. (a) $v_L = V_S$ at the instant of switch closure.
   (b) $i = 0$ A at the instant of switch closure.
   (c) The current reaches maximum at five time constants from switch closure.
   (d) $v_L = 0$ V at five time constants after switch closure.
   (e) $v_R = 0$ V at instant of switch closure and $v_R = V_S$ at five time constants.
10. The current decreases because X_L increases with an increase in frequency.
11. The resistor voltage is greatest because X_L decreases with a decrease in frequency.
12. The inductor is open.

Chapter 12

1. A changing magnetic field in the primary coil is

coupled to the secondary coil and induces a voltage across the secondary.

2. The turns ratio determines the ratio of secondary voltage to primary voltage.
3. Out of phase.
4. A core provides physical support for the windings and a magnetic path for coupling of the windings.
5. $V_s > V_p$.
6. Step-down: $V_s = 0.5V_p = 0.5(100\text{ V}) = 50\text{ V}$.
7. $P_s = P_p = 10\text{ W}$.
8. $I_s = 3I_p$.
9. $R_{ref} = (1/n)^2R_L = (1)^2R_L = R_L = 1000\ \Omega$.
10. Less.
11. A transformer is used for matching the load resistance to the source resistance by proper selection of the turns ratio using the formula $n = \sqrt{R_L/R_{ref}}$ where $R_{ref} = R_S$.
12. Maximum power is transferred when $R_L = R_S$.
13. $V_s = 0\text{ V}$ because dc voltage cannot be coupled in a transformer.
14. Transformer losses reduce the ideal output voltage.
15. 95% of the primary magnetic flux lines pass through the secondary.

Chapter 13

1. Both V_C and V_R are less than V_s in magnitude and are different in phase.
2. (a) V_R is in phase with I.
   (b) V_C lags I by 90°.
3. Both Z and θ decrease as frequency increases.
4. Z remains the same because X_C is halved and R is doubled.
5. Decrease the frequency to decrease the current.
6. $V_s = \sqrt{V_R^2 + V_C^2} = \sqrt{(10\text{ V})^2 + (10\text{ V})^2} = 14.14\text{ V}$.
7. Frequency must be increased to reduce X_C so that $R > X_C$.
8. The phase angle decreases.
9. The impedance decreases due to a decrease in X_C.
10. $I_T = \sqrt{I_R^2 + I_C^2} = \sqrt{(1\text{ A})^2 + (1\text{ A})^2} = 1.414\text{ A}$.
11. 3 div × 18°/div = 54°.
12. $PF = 1$ means a purely resistive circuit.
13. $P_a = 5\text{ VA}$.
14. $P_a = \sqrt{(100\text{ W})^2 + (100\text{ VAR})^2} = 141.4\text{ VA}$.
15. To avoid exceeding the maximum allowable current.
16. $f_c = 1\text{ kHz}$.

Chapter 14

1. Both V_R and V_L are less than V_s in magnitude and differ in phase angle.
2. (a) V_R is in phase with I. (b) V_L leads I by 90°.
3. Both Z and θ increase as frequency increases.
4. Z does not change.
5. Increase the frequency to reduce the current.
6. $V_s = \sqrt{(10\text{ V})^2 + (10\text{ V})^2} = 14.14\text{ V rms}$.
   $V_s = (1.414)(14.14\text{ V}) = 19.99\text{ V peak}$.
7. The frequency must be decreased to reduce X_L so that $R > X_L$.
8. The phase angle decreases.
9. Z increases due to an increase in X_L.
10. $I_T = \sqrt{(2\text{ A})^2 + (2\text{ A})^2} = 2.83\text{ A}$.
11. 2 div × 18°/div = 36°.
12. 0.9.
13. $P_a = 10\text{ VA}$.
14. $P_a = \sqrt{(10\text{ W})^2 + (10\text{ VAR})^2} = 14.14\text{ VA}$.
15. $BW = 20\text{ kHz}$.

Chapter 15

1. $X_T = 0$.
2. $\theta = 0°$.
3. $Z = R_W = 80\ \Omega$.
4. I leads V_s because the circuit is capacitive ($X_C > X_L$).
5. Since $f_r = 1/2\pi\sqrt{LC}$, an increase in C causes a decrease in f_r.
6. $V_s = V_R = 50\text{ V}$
   $V_L + V_C = 150\text{ V} - 150\text{ V} = 0\text{ V}$.
7. BW increases when the Q is lowered.
8. Because the circuit is predominantly inductive.
9. $I_T = 0\text{ A}$ at parallel resonance.
10. Increase C to decrease f_r.
11. $Q \geq 10$, as a rule of thumb.
12. A reduction in R_{Tp} reduces the Q and thereby increases BW.

Chapter 16

1. The output of an RC differentiator is taken across R. The output of an RC integrator is taken across C.
2. $v_C = 0.63(10\text{ V}) = 6.3\text{ V}$.
3. Same as Problem 2.
4. When the input pulse width is much greater than 5τ.
5. When the input pulse width is much less than 5τ.
6. For $t_W \geq 5\tau$.
7. The output of an RL differentiator is taken across L. The output of an RL integrator is taken across R.
8. $I_{max} = V_s/R$.
9. When the input pulse width is equal to or greater than 5τ.
10. You can't tell the difference by observing the output voltages.

Chapter 17

1. Electrons have negative charge, protons have positive charge, and neutrons have no charge.
2. Silicon and germanium have a valence of 4.
3. The valence shell in silicon is M.
4. A neutral atom has the same number of electrons and protons; a positive ion has more protons than electrons; a negative ion has more electrons than protons.
5. Bonding created by the sharing of valence electrons by two or more atoms.
6. Free electrons exist in the conduction band.
7. When a valence electron acquires sufficient energy to jump into the conduction band, a vacancy (hole) is left in the valence band.
8. Recombination occurs when a free electron falls into a hole in the valence band.
9. Electron current and hole current.
10. The energy gap between the valence band and conduction band of an insulator is much greater than that of a semiconductor.
11. A trivalent impurity adds holes.
12. A pentavalent impurity adds electrons.
13. Electrons are majority carriers in an N-type semiconductor.
14. Approximately 0.7 V.
15. $I = 2\text{ V}/110\ \Omega = 18.18\text{ mA}$.
16. **(a)** Forward **(b)** Reverse **(c)** Forward
17. **(a)** 0.7 V **(b)** 5 V **(c)** 0.7 V
18. The banded end is the cathode of the diode: **(a)** Low R **(b)** High R

Chapter 18

1. Rectification is the process of converting ac voltage to a pulsating dc voltage.
2. A half-wave voltage consists of only the positive or only the negative alternations of the sine wave input.
3. $f_{out} = 60$ Hz.
4. One diode is required for half-wave rectification.
5. A full-wave voltage consists of a single-polarity voltage made up of both alternations of the sine wave input with one alternation inverted.
6. $f_{out} = 120$ Hz.
7. Center-tapped and bridge are two types of full-wave rectifiers.
8. Center-tapped requires two diodes; bridge requires four.
9. The output becomes a half-wave voltage.
10. The capacitor input filter smooths out the voltage variations and produces a dc voltage with ripple.
11. The ripple voltage increases because the capacitor charges and discharges more quickly.
12. The zener diode is operated in reverse breakdown.
13. Line regulation keeps the output constant for a change in input voltage. Load regulation keeps the output constant for a change in the load.
14. The junction capacitance of a varactor diode makes it useful as a variable capacitor.
15. LEDs are operated in forward bias and emit light. Photodiodes are operated in reverse bias and have a reverse current that depends on the amount of incident light.

Chapter 19

1. Collector and emitter are N-type in an NPN.
2. The PNP has an N-type base region.
3. The base must be negative with respect to the emitter and positive with respect to the collector.
4. Collector current, emitter current, and base current.
5. $I_C = \beta I_B$; $I_C = \alpha I_E$; $I_E = I_C + I_B$.
6. I_C is 30 times greater than I_B.
7. Yes, the reading should be high because the base-emitter junction is reverse-biased.
8. Gain and leakage.
9. The drain-to-source current decreases as the reverse bias increases.
10. The DE MOSFET operates in depletion and enhancement modes. The E MOSFET operates in enhancement only.
11. A triac is bidirectional, but an SCR is not.

Chapter 20

1. A capacitor from emitter to ground increases the voltage gain. It is called a *bypass capacitor* because it provides an ac bypass around the emitter resistor.
2. An increase in R_C increases the voltage gain.
3. Beta (β) and r_e.
4. The CE inverts the input; the CC does not.
5. Unity (1).
6. The Darlington arrangement increases the beta (β).
7. Lower.
8. The CE provides the highest power gain, all else being equal.
9. Efficiency is the ratio of output power to input power. Class C is the most efficient.
10. Common-source (CS), common-drain (CD), and common-gate (CG).
11. The voltage gain is reduced.
12. Cascading increases the overall gain.

13. $A_v' = 30 \times 30 \times 30 = 27{,}000$.
14. In class A, the transistor conducts for the entire input cycle (360°). In class B, the transistor conducts for half the input cycle (180°). In class C, the transistor conducts for a very small part of the input cycle.
15. An oscillator is a circuit operating on the principle of positive feedback and producing a continuous sinusoidal or other output with no ac input signal.

Chapter 21

1. (b) Low power (f) dc isolation.
2. An op-amp is typically made up of more than one diff-amp.
3. Differential input impedance is the total impedance between the two input terminals. Common-mode input impedance is the impedance from each input to ground.
4. (a) Infinite
   (b) The one with a CMRR of 100 dB.
5. Slew rate $= \dfrac{\Delta V_{out}}{\Delta t} = \dfrac{8\text{ V}}{12\ \mu\text{s}} = 0.667\text{ V}/\mu\text{s}$.
6. (a) Voltage-follower (b) Noninverting
   (c) Inverting.
7. $B = \dfrac{R_i}{R_i + R_f} = \dfrac{1\text{ k}\Omega}{101\text{ k}\Omega} = 0.0099$.
   $V_f = BV_{out} = (0.0099)5\text{ V} = 0.0495\text{ V} = 49.5\text{ mV}$.
8. $A_{cl(NI)} = \dfrac{1}{B} = \dfrac{1}{0.0099} = 101$.
9. (c) Noninverting (d) $A_{cl} \cong 1$
   (e) High Z_{in}.
10. $A_{cl(I)} = -\dfrac{R_f}{R_i} = -\dfrac{220\text{ k}\Omega}{2.2\text{ k}\Omega} = -100$.
11. No, except for the voltage-follower configuration, which is always approximately 1.
12. In a voltage-follower, the feedback attenuation is unity. In a noninverting amplifier, the feedback attenuation is determined by a voltage divider and is usually much less than unity.
13. Voltage-follower.
14. $A_{cl(NI)} = \dfrac{1}{B} = \dfrac{1}{0.025} = 40$.

Chapter 22

1. A comparator compares an input voltage to a known reference and produces an output transition when they are equal.
2. Ground.
3. Connect the inverting input to a 5-V source.
4. See Figure ST–8.
5. Increase the value of R_f.
6. In an averaging amplifier, the ratio R_f/R is equal to the reciprocal of the number of inputs.
7. In a scaling adder, a specific weight is assigned to each input.
8. In an integrator, C is the feedback element and R is the input element. In a differentiator, these elements are reversed.
9. The comparator does not have feedback.
10. The PUT acts as a switching device to terminate each ramp.
11. An oscillator operates with positive feedback.
12. Unity gain and zero phase shift around the feedback loop.
13. The gain drops 3 dB per one-decade increase in frequency. The two-decade increase in frequency causes the output to drop 6 dB.
14. To keep the dc output voltage constant when either the input voltage or the load changes.

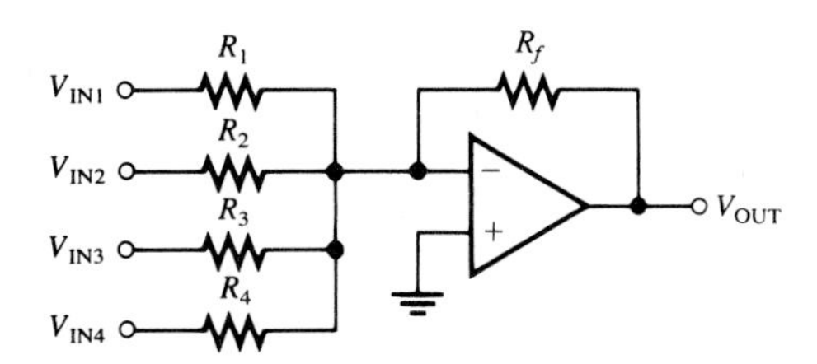

FIGURE ST–8

Answers to Odd-Numbered Problems

Chapter 1

1–1 **(a)** 3×10^3 **(b)** 75×10^3 **(c)** 2×10^6.

1–3 **(a)** $8.4 \times 10^3 = 0.84 \times 10^4 = 0.084 \times 10^5$
(b) $99 \times 10^3 = 9.9 \times 10^4 = 0.99 \times 10^5$
(c) $200 \times 10^3 = 20 \times 10^4 = 2 \times 10^5$.

1–5 **(a)** 0.0000025 **(b)** 5000 **(c)** 0.39.

1–7 **(a)** 126×10^6 **(b)** 5.00085×10^3 **(c)** 6060×10^{-9}.

1–9 **(a)** 20×10^8 **(b)** 36×10^{14} **(c)** 15.4×10^{-15}.

1–11 **(a)** 31 mA **(b)** 5.5 kV **(c)** 200 pF.

Chapter 2

2–1 80×10^{12} C.

2–3 **(a)** 10 V **(b)** 2.5 V **(c)** 4 V.

2–5 20 V.

2–7 **(a)** 75 A **(b)** 20 A **(c)** 2.5 A.

2–9 2 s.

2–11 **(a)** 6500 Ω ± 10% **(b)** 33 Ω ± 10% **(c)** 47,000 Ω ± 20%.

2–13 **(a)** Red, violet, brown
(b) 330 Ω: orange, orange, brown
2.2 kΩ: red, red, red
56 kΩ: green, blue, orange
100 kΩ: brown, black, yellow
39 kΩ: orange, white, orange.

2–15 Current flows through lamp 2.

2–17 Ammeter in series with resistors, with its negative terminal to the negative terminal of source and its positive terminal to one side of R_1. Voltmeter placed across (in parallel with) the source (negative to negative, positive to positive).

2–19 Position 1: V1 = 0, V2 = V_S
Position 2: V1 = V_S, V2 = 0.

2–21 33.33 V.

2–23 AWG #27.

2–25 (b).

2–27 One ammeter in series with battery. One ammeter in series with each resistor (six total).

2–29 See Figure P–1.

Chapter 3

3–1 **(a)** 3 A **(b)** 0.2 A **(c)** 1.5 A.

3–3 15 mA.

3–5 **(a)** 3.33 mA **(b)** 0.55 mA **(c)** 0.588 mA **(d)** 0.5 A **(e)** 6.6 mA.

3–7 **(a)** 2.5 mA **(b)** 2.27 μA **(c)** 8.33 mA.

3–9 **(a)** 10 mV **(b)** 1.65 V **(c)** 15 kV **(d)** 3.52 V **(e)** 0.25 V **(f)** 750 kV **(g)** 8.5 kV **(h)** 3.75 mV.

3–11 **(a)** 81 V **(b)** 500 V **(c)** 125 V.

3–13 **(a)** 2 kΩ **(b)** 3.5 kΩ **(c)** 2 kΩ **(d)** 100 kΩ **(e)** 1 MΩ.

3–15 **(a)** 4 Ω **(b)** 3 kΩ **(c)** 200 kΩ.

3–17 0.417 W.

3–19 **(a)** 1 MW **(b)** 3 MW **(c)** 150 MW **(d)** 8.7 MW.

FIGURE P–1

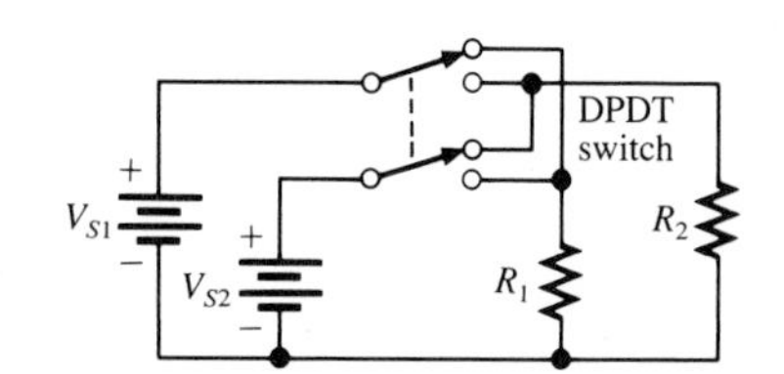

3–21 (a) 2,000,000 μW (b) 500 μW (c) 250 μW (d) 6.67 μW.
3–23 16.5 mW.
3–25 1.175 kW.
3–27 6 W.
3–29 25 Ω.
3–31 0.00186 kWh.
3–33 1 W.
3–35 150 Ω.
3–37 $V = 0$ V, $I = 0$ A; $V = 10$ V, $I = 0.1$ A; $V = 20$ V, $I = 0.2$ A; $V = 30$ V, $I = 0.3$ A; $V = 40$ V, $I = 0.4$ A; $V = 50$ V, $I = 0.5$ A; $V = 60$ V, $I = 0.6$ A; $V = 70$ V, $I = 0.7$ A; $V = 80$ V, $I = 0.8$ A; $V = 90$ V, $I = 0.9$ A; $V = 100$ V, $I = 1$ A.
The graph is a straight line. There is a linear relationship between V and I.
3–39 $R_1 = 0.5\ \Omega$, $R_2 = 1\ \Omega$, $R_3 = 2\ \Omega$.
3–41 10 V, 30 V.
3–43 $I_{MAX} = 3.83$ mA, $I_{MIN} = 3.46$ mA.
3–45 5 V, 20 V.
3–47 2.02 kW/day.
3–49 12 W.

Chapter 4

4–1 See Figure P–2
4–3 0.1 A.
4–5 138 Ω.
4–7 (a) 7.9 kΩ (b) 33 Ω (c) 13.24 MΩ
The series circuit is disconnected from the source, and the ohmmeter is connected across the circuit terminals.
4–9 1126 Ω.
4–11 (a) 0.625 mA (b) 4.26 μA
The ammeter is connected in series.
4–13 0.355 A.
4–15 14 V.
4–17 26 V.
4–19 (a) $V_2 = 6.8$ V
(b) $V_R = 8$ V, $V_{2R} = 16$ V, $V_{3R} = 24$ V, $V_{4R} = 32$ V
The voltmeter is connected across (in parallel with) each resistor for which the voltage is unknown.
4–21 (a) 3.84 V (b) 6.77 V.
4–23 $V_{5.6k\Omega} = 10$ V, $V_{1k\Omega} = 1.79$ V, $V_{560\Omega} = 1$ V, $V_{10k\Omega} = 17.9$ V.
4–25 55 mW.
4–27 The 10-kΩ resistor is shorted.
4–29 5.6 kΩ (3); 1 kΩ (1); 100 Ω (2).
4–31 $V_1 = 4.4$ V, $V_2 = 2$ V, $V_3 = 5.7$ V, $V_4 = 5.7$ V, $V_6 = 5.6$ V
$R_1 = 220\ \Omega$, $R_3 = 285\ \Omega$, $R_4 = 285\ \Omega$, $R_5 = 330\ \Omega$, $R_6 = 280\ \Omega$.
4–33 64.5 mA; 120-Ω resistor burns out first.
4–35 See Figure P–3
4–37 $R_1 = 4700\ \Omega + 680\ \Omega + 68\ \Omega + 1\ \Omega + 1\ \Omega = 5.45\ k\Omega$
$R_2 = 8200\ \Omega + 1500\ \Omega + 120\ \Omega = 9.82\ k\Omega$
$R_3 = 5600\ \Omega + 820\ \Omega + 120\ \Omega + 10\ \Omega = 6.55\ k\Omega$
$R_4 = 6800\ \Omega + 1000\ \Omega + 180\ \Omega + 100\ \Omega + 100\ \Omega = 8.18\ k\Omega$
All resistors are ⅛ W.

Chapter 5

5–1 See Figure P–4
5–3 12 V, 5 mA.
5–5 1350 mA.
5–7 $I_2 = I_3 = 7.5$ mA. An ammeter in series with each resistor in each branch.

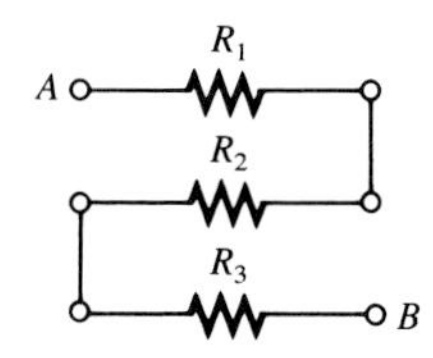

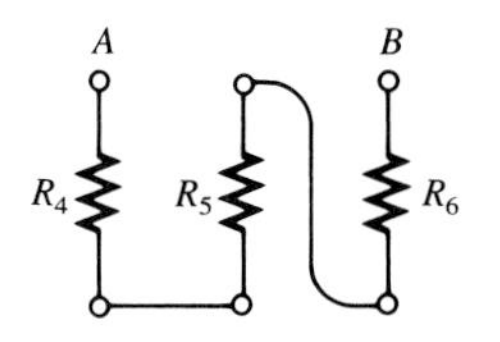

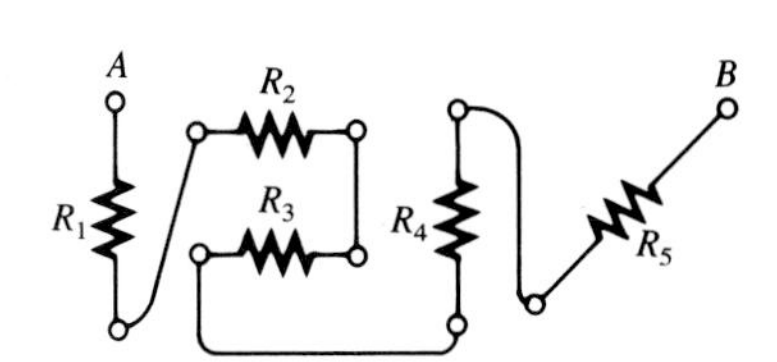

FIGURE P–2

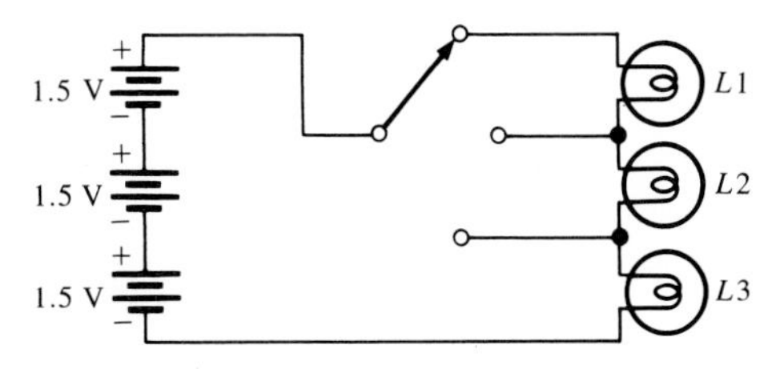

FIGURE P–3

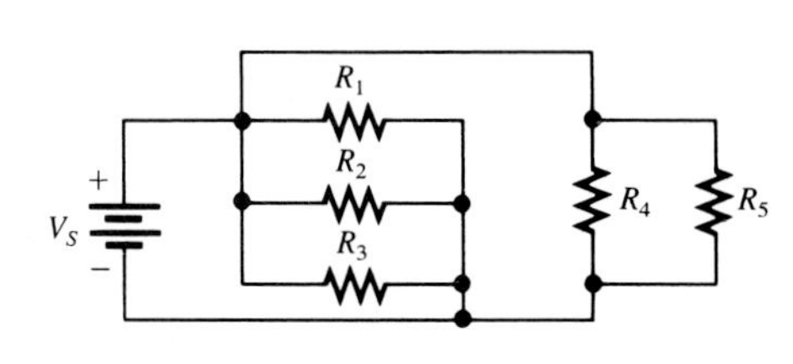

FIGURE P–4

5–9 **(a)** 359 Ω **(b)** 25.6 Ω **(c)** 819 Ω **(d)** 996 Ω.

5–11 2 kΩ.

5–13 **(a)** 0.909 A **(b)** 76 mA.

5–15 Circuit (a).

5–17 $I_1 = 2.19$ A, $I_2 = 0.81$ A.

5–19 200 mW.

5–21 0.682 A, 4.092 A.

5–23 The 500-Ω resistor is open.

5–25 11.36 mA.

5–27 **(a)** 10 **(b)** 10 kΩ **(c)** 5 mA **(d)** 50 V.

5–29 The 4.7-kΩ resistor is shorted.

5–31 $R_2 = 750$ Ω, $R_4 = 423$ Ω.

5–33 Checks for a single open resistor:
Ohmmeter connected to pins 1 and 2: 767 Ω is correct. A 1-kΩ or a 3.3-kΩ reading indicates an open resistor.
Ohmmeter connected to pins 3 and 4: 159.5 Ω is correct. A 270-Ω or a 390-Ω reading indicates an open resistor.
Ohmmeter connected to pins 5 and 6: 198 kΩ is correct. A 247-kΩ reading indicates that R_5 is open; a 227-kΩ reading indicates that R_6 is open. A 279-kΩ reading indicates that R_7 is open; a 323-kΩ reading indicates that R_8 is open.

Chapter 6

6–1 R_2, R_3, and R_4 are in parallel, and this parallel combination is in series with both R_1 and R_5.

6–3 See Figure P–5.

6–5 2003 Ω.

6–7 **(a)** 128 Ω **(b)** 790.5 Ω.

6–9 **(a)** $I_1 = I_4 = 11.7$ mA
$I_2 = I_3 = 5.85$ mA
$V_1 = 0.655$ V
$V_2 = V_3 = 0.585$ V
$V_4 = 0.257$ V
(b) $I_1 = 3.8$ mA
$I_2 = 0.618$ mA
$I_3 = 1.27$ mA
$I_4 = 1.91$ mA
$V_1 = 2.58$ V
$V_2 = V_3 = V_4 = 0.42$ V.

6–11 $V_A = 25$ V, $V_B = 15.13$ V, $V_C = 5.92$ V.

6–13 7.5 V unloaded, 7.29 V loaded.

6–15 56-kΩ load.

6–17 360 Ω.

6–19 20.99 mA.

6–21 $R_{TH} = 18$ kΩ, $V_{TH} = 2.7$ V.

6–23 No, the meter should read 4.39 V.

6–25 The 7.5-V and 5-V readings are incorrect, indicating that the 3-kΩ resistor is open.

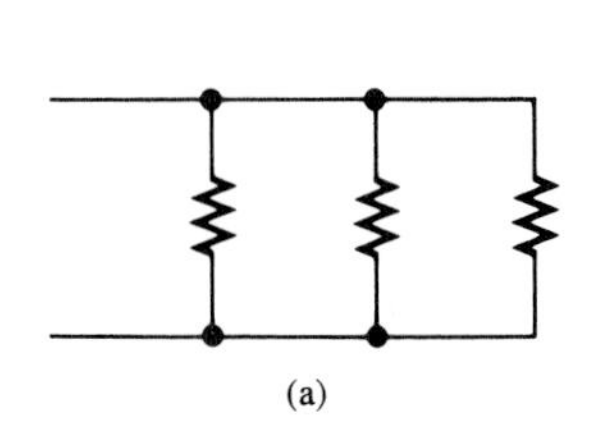

(a)

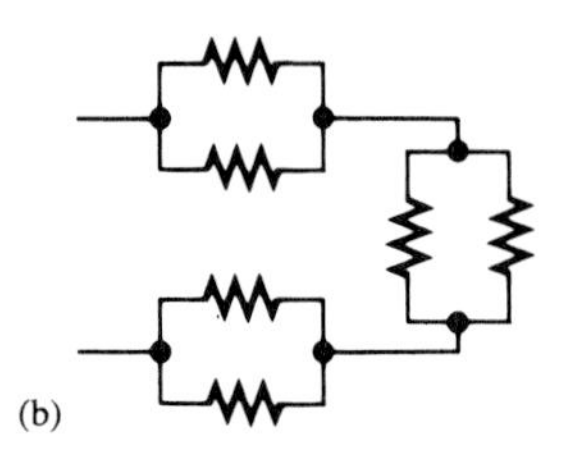

(b)

FIGURE P–5

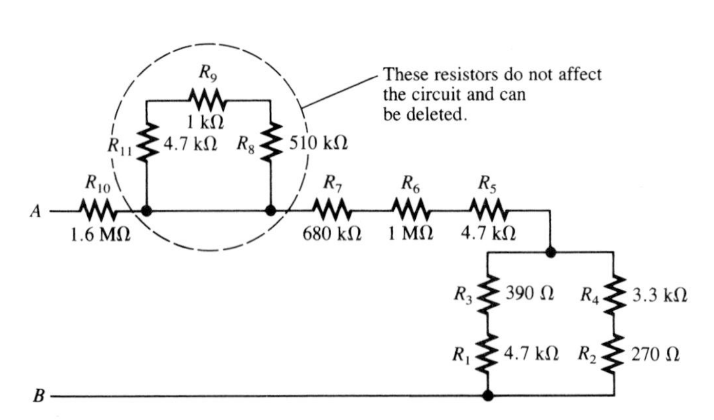

FIGURE P–6

6–27 See Figure P–6.
6–29 $R_T = 5.76\text{ k}\Omega$, $V_A = 3.3\text{ V}$, $V_B = 1.7\text{ V}$, $V_C = 0.85\text{ V}$.
6–31 $V_1 = 1.61\text{ V}$, $V_2 = 6.7\text{ V}$, $V_3 = 1.77\text{ V}$, $V_4 = 3.4\text{ V}$, $V_5 = 0.387\text{ V}$, $V_6 = 2.63\text{ V}$, $V_7 = 0.387\text{ V}$, $V_8 = 1.77\text{ V}$, $V_9 = 1.61\text{ V}$.
6–33 109.9 Ω.
6–35 $R_1 = 180\ \Omega$, $R_2 = 60\ \Omega$.
6–37 0.844 mA.
6–39 11.73 V.

Chapter 7

7–1 Decreases.
7–3 Reverses.
7–5 7500 At/m.
7–7 1 mA.
7–9 To electrically connect the loop to the external circuit.
7–11 See Figure P–7.

Chapter 8

8–1 Half-scale.
8–3 100 μA.
8–5 **(a)** 0 A **(b)** 0 A **(c)** 9 mA **(d)** 99 mA **(e)** 0.999 A.
8–7 2 MΩ.
8–9 60 kΩ.
8–11 10 kΩ.
8–13 See Figure P–8
8–15 Actual voltage is 20.41 V, which is more than the scale setting can handle. Thus, a 1 appears in the overflow (left-most) digit of the display, giving a reading of 10.41 V. The technician must increase the scale range.
8–17 1.8 V.
8–19 V_A and V_B readings are incorrect; R_5 or R_6 is open.
8–21 Measure the voltage across each resistor individually; then calculate I and P for each.

Chapter 9

9–1 **(a)** 1 Hz **(b)** 5 Hz **(c)** 20 Hz **(d)** 1000 Hz **(e)** 2 kHz **(f)** 100 kHz.
9–3 2 μs.
9–5 **(a)** 8.48 V **(b)** 24 V **(c)** 7.64 V.
9–7 $V_p = 25\text{ V}$, $V_{pp} = 50\text{ V}$, $V_{\text{rms}} = 17.68\text{ V}$, $V_{\text{avg}} = 15.93\text{ V}$.
9–9 120 Hz.
9–11 Sine wave with peak at 75° leads by 15°. Sine wave with peak at 100° lags by 10°. $\theta = 25°$.
9–13 **(a)** 7.32 V **(b)** 15.40 V **(c)** 21.66 V **(d)** 26.57 V **(e)** 26.57 V **(f)** 16.22 V **(g)** −26.57 V **(h)** −16.22 V.
9–15 30°: 12.99 V
45°: 14.49 V
90°: 12.99 V
180°: −7.5 V
200°: −11.49 V
300°: −7.5 V.
9–17 See Figure P–9
9–19 **(a)** 7.07 mA **(b)** 6.37 mA **(c)** 10 mA **(d)** 20 mA **(e)** 10 mA.
9–21 $t_r = 3.5\text{ ms}$, $t_f = 3.5\text{ ms}$, $t_W = 12.5\text{ ms}$, $V = 5\text{ V}$.
9–23 **(a)** −0.375 V **(b)** 3.01 V.
9–25 **(a)** 50 kHz **(b)** 10 Hz.
9–27 25 kHz.

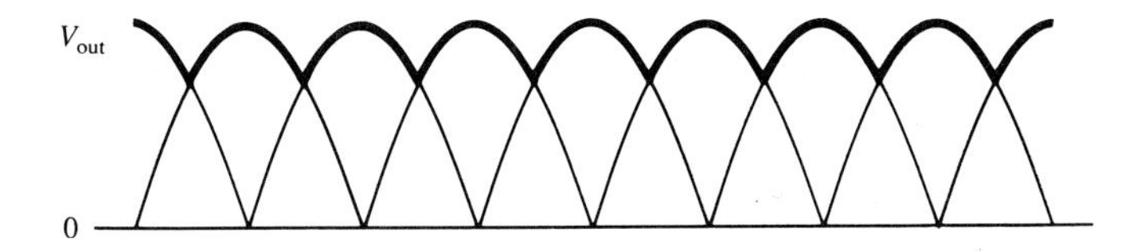

FIGURE P–7

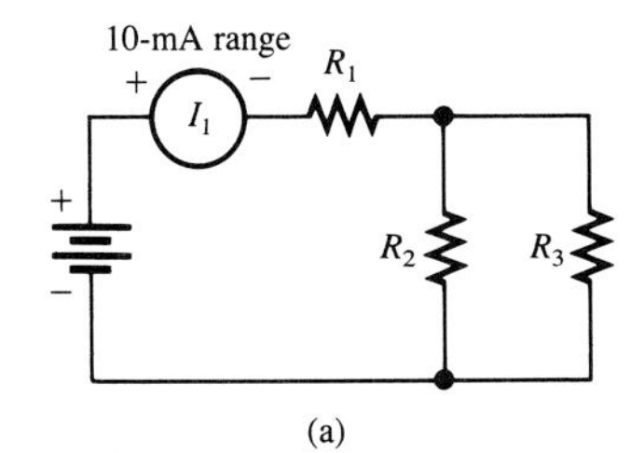

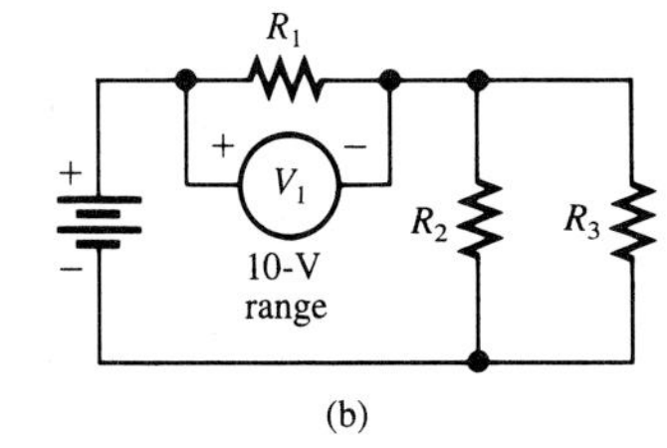

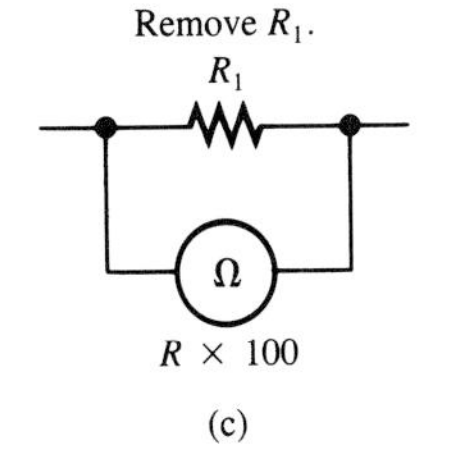

FIGURE P–8

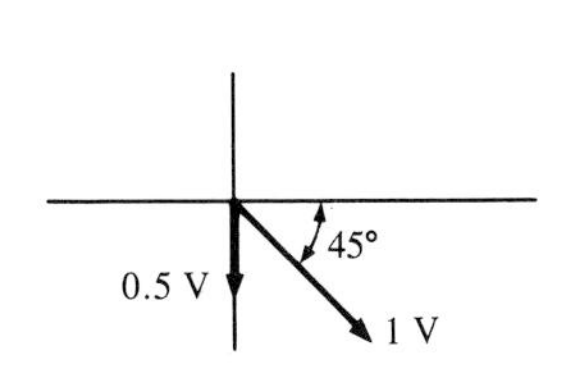

FIGURE P–9

9–29 V_{RL} is a 300-V peak-to-peak sine wave riding on a 200-V dc level. $I_{max} = 3.5$ A, $V_{avg} =$ 200 V.

Chapter 10

10–1 (a) 5 μF (b) 1 μC (c) 10 V.
10–3 (a) 0.001 μF (b) 0.0035 μF (c) 0.00025 μF.
10–5 2 μF.
10–7 8.9 pF.
10–9 A 12.5-pF increase.
10–11 Ceramic.
10–13 (a) 0.02 μF (b) 0.047 μF (c) 0.001 μF (d) 220 pF.
10–15 (a) 0.67 μF (b) 68.97 pF (c) 2.7 μF.
10–17 (a) 1057 pF (b) 0.121 μF.
10–19 (a) 2.5 V (b) 3.33 V (c) 4 V.
10–21 (a) 12.5 ms (b) 247.5 μs (c) 11 μs (d) 250 μs.
10–23 (a) 9.2 V (b) 1.24 V (c) 0.46 V (d) 0.168 V.
10–25 (a) 17.97 V (b) 12.79 V (c) 6.61 V.
10–27 (a) 31.83 Ω (b) 111 kΩ (c) 49.7 Ω.
10–29 200 Ω.
10–31 $P_{true} = 0$ W, $P_r = 3.39$ mVAR.
10–33 0 Ω.
10–35 Shorted.
10–37 1.71 ms.
10–39 1.59 ms.
10–41 (a) Charges to 3.2 V in 10 ms, then discharges to 0 V in 215 ms. (b) Charges to 3.2 V in 10 ms, then discharges to 2.85 V in 5 ms, then charges toward 20 V.
10–43 0.0052 μF

Chapter 11

11–1 (a) 1000 mH (b) 0.25 mH (c) 0.01 mH (d) 0.5 mH.
11–3 3536 turns.
11–5 0.05 J.
11–7 155 μH.
11–9 7.14 μH.
11–11 (a) 4.33 H (b) 50 mH (c) 0.57 μH.
11–13 (a) 1 μs (b) 2.13 μs (c) 2 μs.
11–15 (a) 5.52 V (b) 2.03 V (c) 0.75 V (d) 0.27 V (e) 0.1 V.
11–17 (a) 136 kΩ (b) 1.57 kΩ (c) 0.0179 Ω.
11–19 $I_T = 10.05$ A, $I_{L2} = 6.7$ A, $I_{L3} = 3.35$ A.
11–21 100.5 VAR.
11–23 (a) Infinite resistance (b) Zero resistance (c) Lower R_W.
11–25 2.86 mA.
11–27 26.1 mA.

Chapter 12

12–1 3
12–3 (a) Positive at top (b) Positive at bottom (c) Positive at top (d) Positive at bottom.
12–5 50 turns.
12–7 (a) Same polarity, 100 V rms (b) Opposite polarity, 100 V rms.
12–9 240 V.
12–11 33.33 mA.
12–13 27.3 Ω.
12–15 0.5
12–17 (a) 6 V (b) 0 V (c) 40 V.
12–19 94.5 W.
12–21 0.98
12–23 25 kVA.
12–25 Section 1: 2; Section 2: 0.5; Section 3: 0.25
12–27 (a) 48 V (b) 25 V.
12–29 (a) 0.05 (b) 41.67 A (c) 2.08 A.
12–31. (a) The meter ground shorts out the lower 100-Ω resistor, so the full secondary voltage is measured by the meter; 20 V. (b) The meter is measuring half of the secondary voltage; 10 V.

Chapter 13

13–1 8 kHz, 8 kHz.
13–3 (a) 269.3 Ω (b) 1166 Ω.
13–5 (a) 37.1 mA (b) 4.29 mA.
13–7 $I_T = 12$ mA, $V_{C(0.1)} = 1.27$ V, $V_{C(0.2)} =$ 0.637 V, $V_R = 0.6$ V, $\theta = 72.48°$ (V_s lagging I_T).
13–9 1.03 kΩ.
13–11 (a) 433.1 Ω, 54.7° (b) 249.8 Ω, 70.5° (c) 155.7 Ω, 78° (d) 79.2 Ω, 83.9°.
13–13 $I_{C1} = 118.2$ mA, $I_{C2} = 50.3$ mA, $I_{R1} = 36.4$ mA, $I_{R2} = 44.4$ mA, $I_T = 186.9$ mA, $\theta =$ 64.4° (V_s lagging I_T).
13–15 (a) 3.9 kΩ (b) 20 μA (c) 15.7 μA (d) 25.6 μA (e) 38.1° (V_s lagging I_T).
13–17 $V_{C1} = 8.77$ V, $V_{C2} = 3.24$ V, $V_{C3} = 3.24$ V $V_{R1} = 2.1$ V, $V_{R2} = 1.14$ V.
13–19 $I_T = 82.7$ mA, $I_{C2R1} = 15.3$ mA, $I_{C3} = 67.5$ mA, $I_{R2R3} = 6.35$ mA.

13–21 4.03 VA.
13–23 0.91
13–25 **(a)** 0.06° **(b)** 5.74° **(c)** 45.1° **(d)** 84.32°.
13–27 **(a)** 89.96° **(b)** 86.4° **(c)** 57.86° **(d)** 9.04°.
13–29 See Figure P–10
13–31 Figure 13–67: 995 Hz
Figure 13–68: 1.59 kHz.
13–33 **(a)** I_{LA} = 4.4 A, I_{LB} = 3.06 A
(b) P_{rA} = 509.2 VAR, P_{rB} = 210.5 VAR
(c) $P_{true(A)}$ = 822.8 W, $P_{true(B)}$ = 640.5 W
(d) P_{aA} = 967.6 VA, P_{aB} = 674.2 VA.
13–35 12 Ω, 13.3 μF.
13–37 0.0796 μF.

Chapter 14

14–1 15 kHz.
14–3 **(a)** 111.8 Ω **(b)** 1.8 kΩ.
14–5 **(a)** 17.38 Ω **(b)** 63.97 Ω **(c)** 126.23 Ω **(d)** 251.61 Ω.
14–7 **(a)** 89.4 mA **(b)** 2.78 mA.
14–9 37°.
14–11 See Figure P–11
14–13 7.69 Ω.
14–15 2387 Hz.
14–17 **(a)** 274 Ω **(b)** 89.3 mA **(c)** 159 mA **(d)** 182.5 mA **(e)** 60.7° (I_T lagging V_s).
14–19 V_{R1} = 7.92 V, V_{R2} = 20.84 V.
14–21 I_T = 36 mA, I_L = 33.2 mA, I_{R2} = 13.9 mA.
14–23 1.28 W, 0.96 VAR.
14–25 PF = 0.386, P_{true} = 0.347 W, P_r = 0.692 VAR, P_a = 0.9 VA.
14–27 See Figure P–12
14–29 See Figure P–13
14–31 5.83 V
14–33 92.9 mA
14–35 **(a)** 394 mA **(b)** 221.6 mA **(c)** 333.33 mA **(d)** 325.8 mA

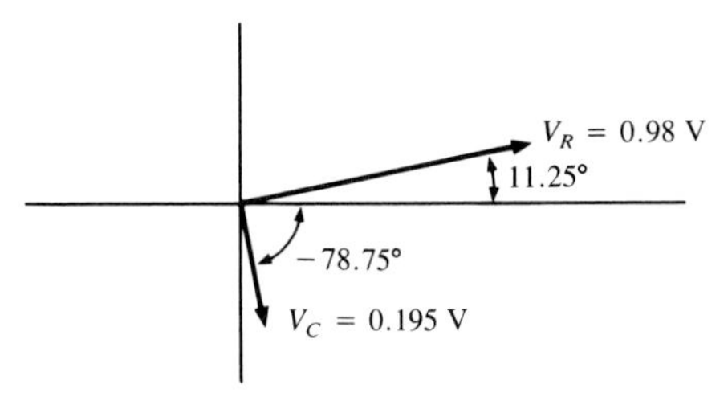

(a) For Figure 13–67

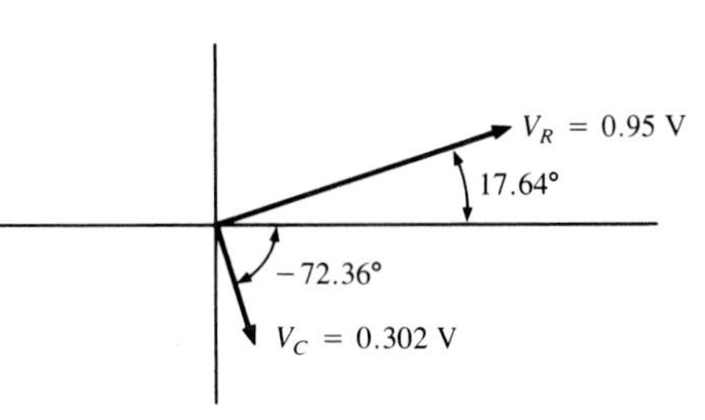

(b) For Figure 13–68

FIGURE P–10

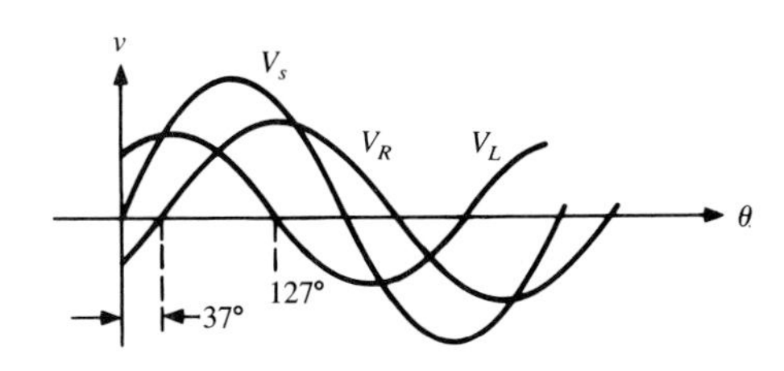

FIGURE P–11

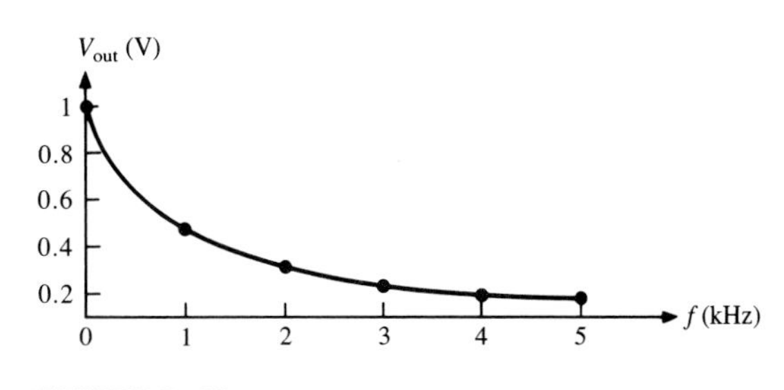

FIGURE P–12

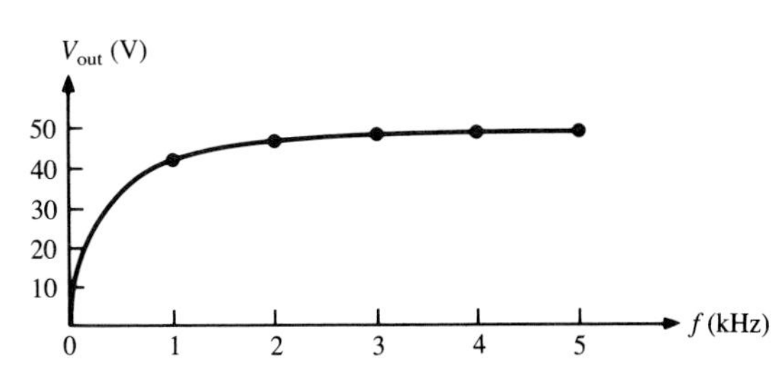

FIGURE P–13

Chapter 15

15–1 479.6 Ω, 88.8° (V_s lagging I), 479.5 Ω capacitive.
15–3 Impedance increases.
15–5 See Figure P–14
15–7 $Z = 200\ \Omega$, $X_C = X_L = 2\ \text{k}\Omega$.
15–9 0.5 A.
15–11 **(a)** 1.12 kHz **(b)** 2.4 kHz.
15–13 99.697 Ω.
15–15 $I_R = 50$ mA, $I_L = 4.4$ mA, $I_C = 8.3$ mA, $I_T = 50.15$ mA.
15–17 Infinity.
15–19 $I_T = 0.012\ \mu$A, $I_C = I_L = 193\ \mu$A.

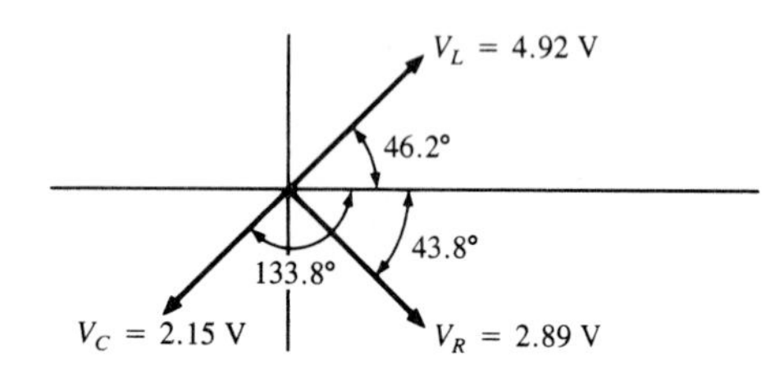

FIGURE P–14

15–21 400 Hz.
15–23 1.99 mH, 0.2 μF.
15–25 **(a)** 0 dB **(b)** −4.4 dB **(c)** −3 dB **(d)** −14 dB.
15–27 None.
15–29 $f_{r(\text{series})} = 4.11$ kHz, $V_{\text{out}} = 9.97$ V, $f_{r(\text{parallel})} =$ 2.6 kHz, $V_{\text{out}} \cong 10$ V.

Chapter 16

16–1 100 μs.
16–3 12.6 V.
16–5 See Figure P–15
16–7 See Figure P–16
16–9 15-V dc level with a very small charge/discharge fluctuation.
16–11 Exchange positions of R and C. See Figure P–17. 5 ms.
16–13 Approximately the same shape as input, but with an average value of 0 V.
16–15 See Figure P–18
16–17 See Figure P–19
16–19 **(a)** 25 ms **(b)** See Figure P–20.

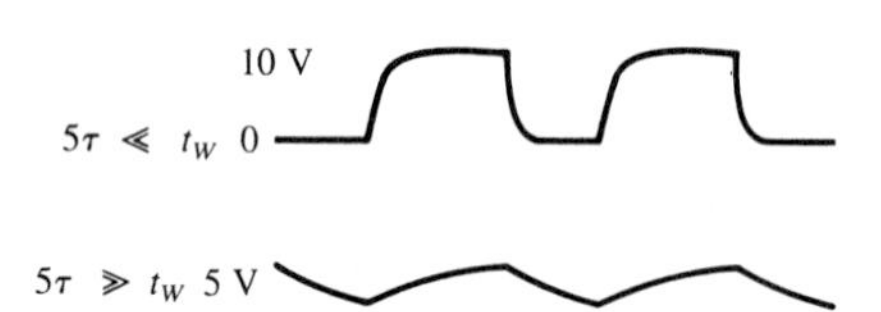

FIGURE P–15

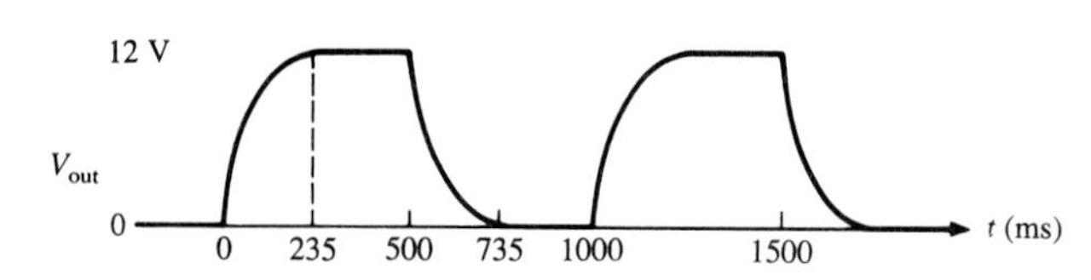

FIGURE P–16

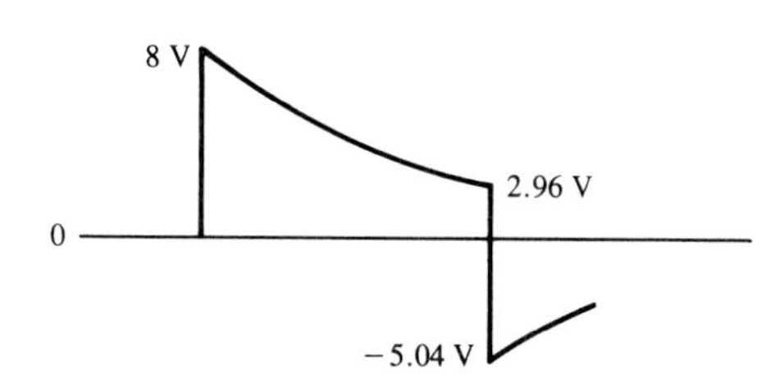

FIGURE P–17

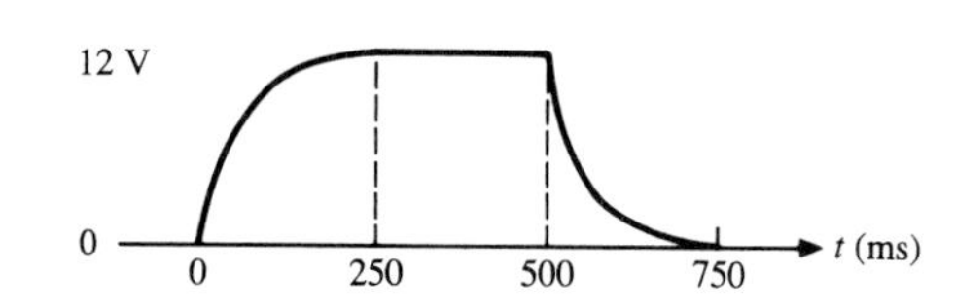

FIGURE P–18

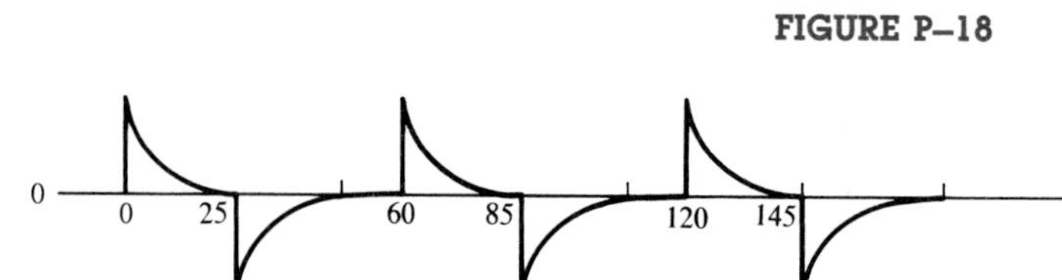

FIGURE P–19

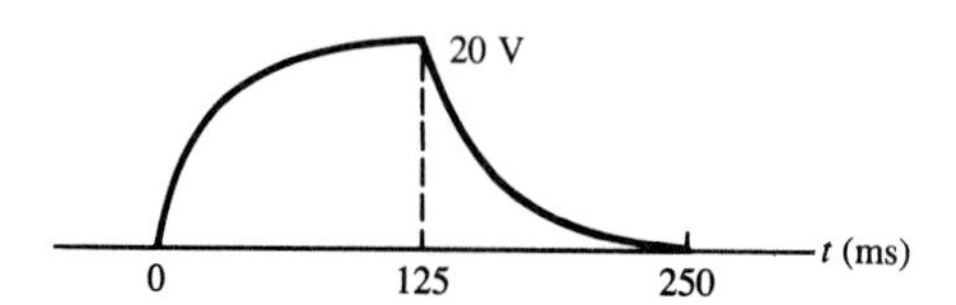

FIGURE P–20

Chapter 18

18–1 63.66 V.
18–3 Yes.
18–5 47.75 V.
18–7 172.79 V.
18–9 78.54 V.
18–11 See Figure P–21
18–13 **(a)** Good **(b)** Bad, open diode
(c) Good **(d)** Bad, open diode.
18–15 **(a)** A sine wave with a positive peak at +0.7 V, a negative peak at −7.3 V, and a dc value of −3.3 V.
(b) A sine wave with a positive peak at +29.3 V, a negative peak at −0.7 V, and a dc value of +14.3 V.
18–17 14.9 V.
18–19 $I_{L(\text{min})} = 0$ A, $I_{L(\text{max})} = 20.9$ mA.
18–21 10.7%, 8.7%.
18–23 A decrease of 10 pF.
18–25 25.34 pF each.
18–27 Increase.

Chapter 19

19–1 4.865 mA.
19–3 125.
19–5 $I_B = 26\ \mu$A, $I_E = 1.3$ mA, $I_C = 1.274$ mA.
19–7 $I_B = 15\ \mu$A, $I_C = 750\ \mu$A, $V_C = 9.25$ V.
19–9 $V_{CE} = 1.8$ V. Q point: $I_C = 17$ mA, $V_{CE} = 1.8$ V.
19–11 33.33
19–13 0.5 mA, 3.33 μA, 4.03 V.
19–15 0.003 mA.
19–17 **(a)** Good **(b)** Good
(c) Bad **(d)** Bad.
19–19 **(a)** Narrows **(b)** Increases.
19–21 5 V.
19–23 **(a)** 4 V **(b)** 10 mA.
19–25 4 V.
19–27 The gate is insulated from the channel by an SiO_2 layer.
19–29 3 V.
19–31 **(a)** Depletion **(b)** Enhancement
(c) Zero bias **(d)** Depletion.
19–33 0.6
19–35 1 ms.
19–37 2.4 kΩ.

Chapter 20

20–1 204.1
20–3 **(a)** 3.25 V **(b)** 2.55 V

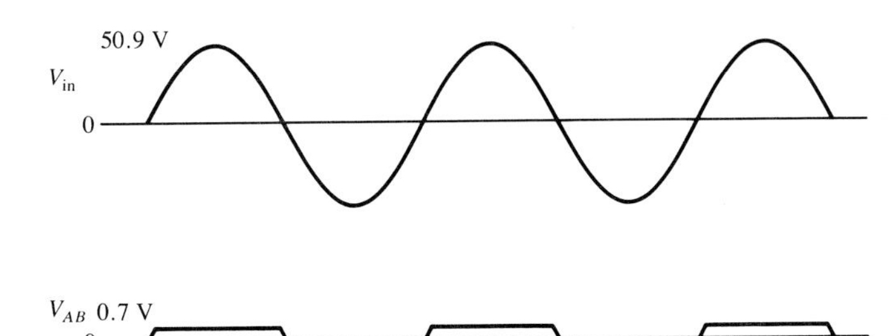

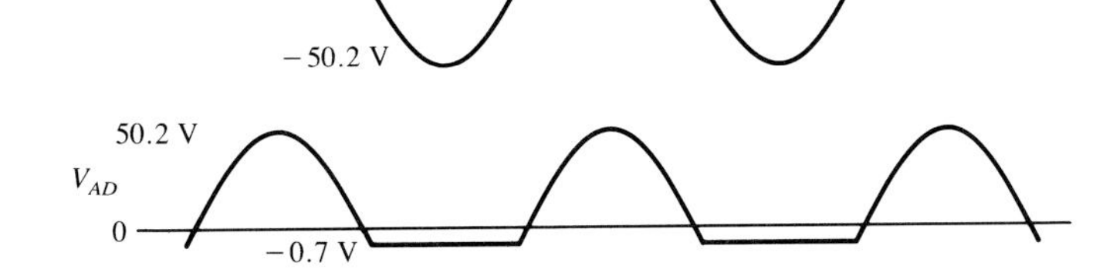

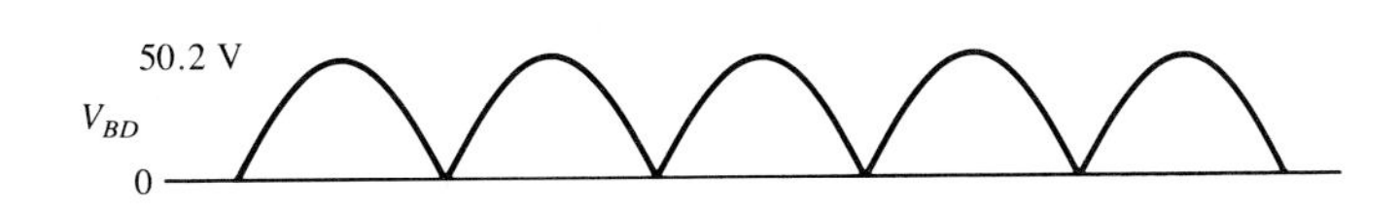

FIGURE P–21

(c) 2.55 mA (d) ≈ 2.55 mA
(e) 9.59 V (f) 7.04 V.

20–5 $A_{v(\max)} = 92.3$, $A_{v(\min)} = 2.91$

20–7 0.978

20–9 A_v is reduced slightly.

20–11 $R_{in} = 2.36\ \Omega$, $A_v = 508.5$, $A_i = 1$, $A_p = 508.5$

20–13 (a) 0.938 (b) 0.301

20–15 30 dB, 31.62

20–17 (a) 7.71 mW (b) 53.56 mW.

20–19 10 V, 625 mA.

20–21 50.33 kHz.

20–23 See Figure P–22

20–25 (a) 711.8 kHz (b) 562.7 kHz.

Chapter 21

21–1 *Practical op-amp:* High open-loop gain, high input impedance, low output impedance, large bandwidth, high CMRR.
Ideal op-amp: Infinite open-loop gain, infinite input impedance, zero output impedance, infinite bandwidth, infinite CMRR.

21–3 (a) Single-ended input, differential output
(b) Single-ended input, single-ended output
(c) Differential input, single-ended output
(d) Differential input, differential output.

21–5 8.1 μA.

21–7 107.96 dB.

21–9 0.3

21–11 40 μs.

21–13 (a) 11 (b) 101 (c) 47.81 (d) 23.

21–15 (a) 1 (b) −1 (c) 21 (d) −10.

21–17 (a) 0.5 mA (b) 0.5 mA
(c) −10 V (d) −10.

21–19 (a) $Z_{in(VF)} = 1.32 \times 10^{12}\ \Omega$
$Z_{out(VF)} = 0.455\ m\Omega$
(b) $Z_{in(VF)} = 5 \times 10^{11}\ \Omega$
$Z_{out(VF)} = 0.6\ m\Omega$
(c) $Z_{in(VF)} = 40{,}000\ M\Omega$
$Z_{out(VF)} = 1.5\ m\Omega$.

Chapter 22

22–1 (a) Maximum negative
(b) Maximum positive
(c) Maximum negative.

22–3 See Figure P–23

22–5 (a) $V_{R1} = 1$ V, $V_{R2} = 1.8$ V
(b) 127.27 μA (c) −2.8 V.

22–7 −3.66 V, 366 μA.

22–9 −5 mV/μs.

22–11 3 V peak-to-peak, 3.33 kHz.

22–13 6.38 V.

22–15 When dc power is first applied, both zener diodes appear as opens. As a result, R_3 is placed in series with R_1, thus increasing the closed-loop gain to a value greater than unity to assure that oscillation will begin.

22–17 884.2 Hz.

22–19 (a) 1.59 kHz (b) 6772.6 Hz
(c) 937.83 Hz.

22–21 7 V.

22–23 9.45 V.

22–25 500 mA.

22–27 10 mA.

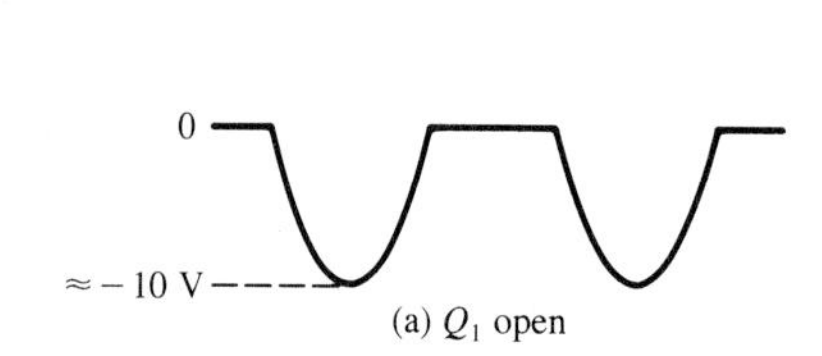

(a) Q_1 open

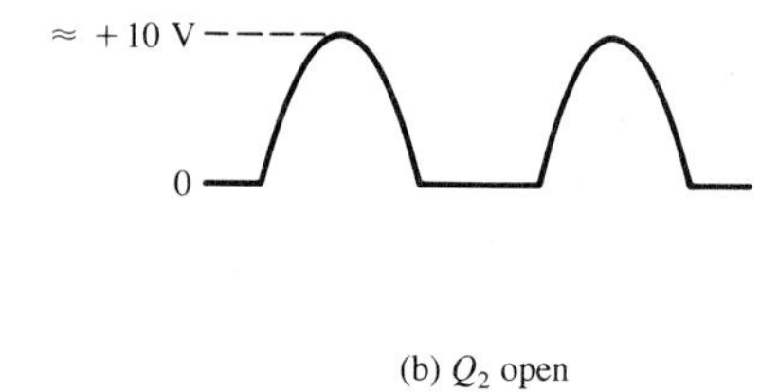

(b) Q_2 open

FIGURE P–22

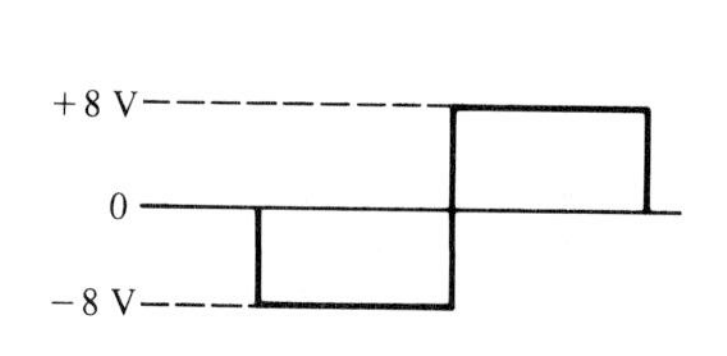

(a)

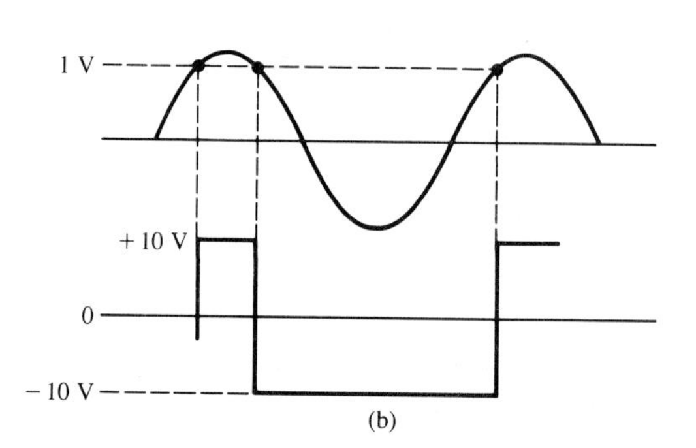

(b)

FIGURE P–23

Glossary

Active filter A frequency-selective circuit consisting of active devices such as transistors or op-amps.

Admittance A measure of the ability of a reactive circuit to permit current. The reciprocal of impedance.

Alpha (α) The ratio of collector current to emitter current in a bipolar junction transistor.

Alternating current (ac) Current that reverses direction in response to a change in voltage polarity.

Ammeter An electrical instrument used to measure current.

Ampere (A or amp) The unit of electrical current.

Ampere-hour (Ah) The measure of the capacity of a battery to supply electrical current.

Amplification The process of increasing the power, voltage, or current of an electrical signal.

Amplifier An electronic circuit having the capability of amplification and designed specifically for that purpose.

Amplitude The voltage or current value of an electrical signal that usually implies the maximum value.

Analog Characterizing a linear process in which a variable takes on a *continuous* set of values within a given range.

Anode The more positive terminal of a diode or other electronic device.

Apparent power The power that *appears* to be being delivered to a reactive circuit. The product of volts and amperes with units of VA (volt-amperes).

Arc tangent An inverse trigonometric function meaning "the angle whose tangent is." Also called *inverse tangent* and designated $\tan^{-1}$.

Asynchronous Having no fixed time relationship, such as two waveforms that are not related to each other in terms of their time variations.

Atom The smallest particle of an element possessing the unique characteristics of that element.

Atomic number The number of electrons in a neutral atom.

Atomic weight The number of protons and neutrons in the nucleus of an atom.

Attenuation The process of reducing the power, voltage, or current value of an electrical signal. It can be thought of as negative amplification.

Audio Related to ability of the human ear to detect sound. Audio frequencies (af) are those that can be heard by the human ear, typically from 20 Hz to 20 kHz.

Autotransformer A transformer having only one coil for both its primary and its secondary.

AWG American Wire Gage, a standardization of wire sizes according to the diameter of the wire.

Band-pass filter A filter that passes frequencies within a certain range and blocks all others.

Band-stop filter A filter that blocks frequencies within a certain range.

Bandwidth (*BW*) The characteristic of certain electronic circuits that specifies the usable range of frequencies that pass from input to output.

Base One of the semiconductor regions in a bipolar transistor.

Baseline The normal level of a pulse waveform. The level in the absence of a pulse.
Battery An energy source that uses a chemical reaction to convert chemical energy into electrical energy.
Beta (β) The ratio of collector current to base current in a bipolar junction transistor.
Bias The application of dc voltage to a diode, transistor, or other electronic device to produce a desired mode of operation.
Binary Characterized by two values or states.
Bipolar Characterized by two PN junctions.
BJT Bipolar junction transistor.
Bode plot An idealized graph of the gain, in dB, versus frequency. Used to illustrate graphically the response of an amplifier or filter circuit.
Branch One of the current paths in a parallel circuit.
Buffer A circuit used between a driving circuit and a load in order to reduce the amount of loading on the driving circuit.

Capacitance The ability of a capacitor to store electrical charge.
Capacitor An electrical device possessing the property of capacitance.
Cascade An arrangement of circuits in which the output of one circuit becomes the input to the next.
Cathode The more negative terminal of a diode or other electronic device.
Center tap (CT) A connection at the midpoint of the secondary of a transformer.
Charge An electrical property of matter in which an attractive force or a repulsive force is created between two particles. Charge can be positive or negative.
Chassis The metal framework or case upon which an electrical circuit or system is constructed.
Chip A tiny piece of semiconductor material upon which an integrated circuit is constructed.
Choke An inductor. The term is used more commonly in connection with inductors used to block or *choke off* high frequencies.
Circuit An interconnection of electrical components designed to produce a desired result.
Circuit breaker A resettable protective electrical device used for interrupting current in a circuit when the current has reached an excessive level.
Circular mil (CM) The unit of the cross-sectional area of a wire. A wire with a diameter of 0.001 inch has a cross-sectional area of one circular mil.
Clamper A circuit that adds a dc level to an ac signal.
Class A A category of amplifier circuit that conducts for the entire input cycle and produces an output signal that is a replica of the input signal.
Class B A category of amplifier circuit that conducts for half of the input cycle.
Class C A category of amplifier circuit that conducts for a very small portion of the input cycle.
Clipper A circuit that removes part of a waveform above and/or below a specified level.
Closed circuit A circuit with a complete current path.
Closed-loop An op-amp configuration in which the output is connected back to the input.
CMRR Common-mode rejection ratio. A measure of an op-amp's ability to reject common-mode signals. The ratio of open-loop gain to common-mode gain.
Coefficient of coupling A constant associated with transformers that specifies the magnetic field in the secondary as a result of that in the primary.
Coil A common term for an inductor or for the primary or secondary winding of a transformer.
Collector One of the semiconductor regions of a bipolar junction transistor.
Common-base A BJT amplifier configuration in which the base is the common (grounded) terminal.
Common-collector A BJT amplifier configuration in which the collector is the common (grounded) terminal.
Common-emitter A BJT amplifier configuration in which the emitter is the common (grounded) terminal.
Comparator A circuit that compares an input voltage to a reference voltage and produces an output indicating the relationship (less or greater than).
Computer An electronic digital system that can be programmed to perform various tasks, such as mathematical computations, at extremely high speed and that can store large amounts of data.
Conductance The ability of a circuit to allow current. It is the reciprocal of resistance. The units are siemens (S).
Conductor A material that allows electrical current to flow with relative ease. An example is copper.
Core The material within the windings of an inductor that influences the electromagnetic characteristics of the inductor.

Coulomb (C) The unit of electrical charge.
Covalent Related to the bonding of two or more atoms by the interaction of their valence electrons.
Critical frequency The frequency at which the response of an amplifier or filter is 3 dB less than at midrange.
CRT Cathode ray tube.
Crystal The pattern or arrangement of atoms forming a solid material.
Current The rate of flow of electrons.
Cutoff The nonconducting state of a transistor.
Cycle The repetition of a pattern.

Darlington A two-transistor arrangement that produces a multiplication of current gain.
Decade A tenfold change in the value of a quantity.
Decibel (dB) The unit of the logarithmic expression of a ratio, such as power or voltage.
Degree The unit of angular measure corresponding to 1⁄360 of a complete revolution.
Derivative The instantaneous rate of change of a function, determined mathematically.
Diac A semiconductor device that can conduct current in either of two directions when properly activated.
Dielectric The insulating material used between the plates of a capacitor.
Differential amplifier (diff-amp) An amplifier that produces an output proportional to the difference of two inputs.
Differentiator A circuit (usually *RC*) that produces an output which approaches the mathematical derivative of the input.
Digital Characterizing a nonlinear process in which a variable takes on discrete values within a given range.
Digital multimeter (DMM) A meter that measures voltage, current, and resistance, and produces a digital readout of the measured value.
Diode An electronic device that permits current flow in only one direction.
Direct current (dc) Current that flows in only one direction.
Discrete device An individual electrical or electronic component that must be used in combination with other components to form a complete functional circuit.
Doping The process of imparting impurities to an intrinsic semiconductor material in order to control its conduction characteristics.
Duty cycle A characteristic of a pulse waveform that indicates the percentage of time that a pulse is present during a cycle.

Effective value A measure of the heating effect of a sine wave. Also known as the rms (root mean square) value.
Efficiency A measure of energy loss in a circuit in the process of converting electrical energy from one form to another.
Electrical Related to the use of electrical voltage and current to achieve desired results.
Electromagnetic Related to the production of a magnetic field by an electrical current in a conductor.
Electron The basic particle of electrical charge in matter. The electron possesses negative charge.
Electronic Related to the movement and control of free electrons in semiconductors or vacuum devices.
Element One of the unique substances that make up the known universe. Each element is characterized by a unique atomic structure.
Emitter One of the three regions in a bipolar transistor.
Emitter-follower A popular term for a common-collector amplifier.
Energy The ability to do work.
Exponent The number of times a given number is multiplied by itself. The exponent is called the *power*, and the given number is called the *base*.

Falling edge The negative-going transition of a pulse.
Fall time The time interval required for a pulse to change from 90% of its amplitude to 10% of its amplitude.
Farad (F) The unit of capacitance.
Feedback The process of returning a portion of a circuit's output signal to the input in such a way as to create certain specified operating conditions.
Field The invisible forces that exist between oppositely charged particles (electric field) or between the north and south poles of a magnet or electromagnet (magnetic field).
Field-effect transistor (FET) A type of transistor that uses an induced electric field to control current.
Filter A type of electrical circuit that passes certain frequencies and rejects all others.
Flux The lines of force in a magnetic field.

Flux density The amount of flux per unit area in a magnetic field.
Free electron A valence electron that has broken away from its parent atom and is free to move from atom to atom within the atomic structure of a material.
Frequency A measure of the rate of change of a periodic function. The electrical unit of frequency is the hertz (Hz).
Full wave The entire ac cycle, consisting of both positive and negative alternations.
Function generator An electronic test instrument that is capable of producing several types of electrical waveforms, such as sine, triangular, and square waves.
Fuse A protective electrical device that burns open when excessive current flows in a circuit.

Gain The amount by which an electrical signal is increased or amplified.
Gate One of the three terminals of an FET.
Generator An energy source that produces electrical signals.
Germanium A semiconductor material.
Giga A prefix used to designate 10^9 (one thousand million).
Ground In electrical circuits, the common or reference point. It can be chassis ground or earth ground.

Half wave One half-cycle of an ac signal.
Harmonics The frequencies contained in a composite waveform which are integer multiples of the repetition frequency (fundamental).
Henry (H) The unit of inductance.
Hertz (Hz) The unit of frequency. One hertz equals one cycle per second.
High-pass filter A filter that passes higher frequencies and rejects lower frequencies.
Hole The absence of an electron.
Hypotenuse The longest side of a right triangle.

Impedance The total opposition to current in a reactive circuit. The unit is ohms (Ω).
Induced voltage Voltage produced as a result of a changing magnetic field.
Inductance The property of an inductor whereby a change in current causes the inductor to produce an opposing voltage.
Inductor An electrical device having the property of inductance. Also known as a *coil* or a *choke*.
Infinite Having no bounds or limits.
Input The voltage, current, or power applied to an electrical circuit to produce a desired result.
Instantaneous value The value of a variable at a given instant in time.
Insulator A material that does not allow current under normal conditions.
Integrated circuit (IC) A type of circuit in which all of the components are constructed on a single, tiny piece of semiconductor material.
Integrator A type of *RC* or *RL* circuit that produces an output which approaches the mathematical integral of the input.
Ionization The removal or addition of an electron from or to an atom so that the resulting atom (ion) has a net positive or negative charge.

Joule (J) The unit of energy.
Junction The point at which two different types of semiconductor materials are joined.
Junction field-effect transistor (JFET) A type of FET that operates with a reverse-biased junction to control current in a channel.

Kilo A prefix used to designate 10^3 (one thousand).
Kilowatt-hour (kWh) A common unit of energy used mainly by utility companies.
Kirchhoff's laws A set of circuit laws that describe certain voltage and current relationships in a circuit.

Lag A condition of the phase or time relationship of waveforms in which one waveform is behind the other in phase or time.
Lead A wire or cable connection to an electrical or electronic device or instrument. Also, a condition of the phase or time relationship of waveforms in which one waveform is ahead of the other in phase or time.
Leading edge The first step or transition of a pulse.
Light-emitting diode (LED) A type of diode that emits light when forward current flows.
Linear Characterized by a straight-line relationship.
Line regulation The percent change in output voltage for a given change in line (input) voltage.
Load The device upon which work is performed.
Loading The amount of current drawn from the output of a circuit through a load.
Load regulation The percent change in output voltage for a given change in load current.
Low-pass filter A filter that passes lower frequencies and rejects higher ones.

Magnetic Related to or possessing characteristics of magnetism. Having a north and a south pole with lines of force extending between the two.
Magnetomotive force (mmf) The force produced by a current in a coiled wire in establishing a magnetic field. The unit is ampere-turns (At).
Magnitude The value of a quantity, such as the number of amperes of current or the number of volts of voltage.
Mega A prefix designating 10^6 (one million).
Micro A prefix designating 10^{-6} (one-millionth).
Milli A prefix designating 10^{-3} (one-thousandth).
Modulation The process whereby a signal containing information (such as voice) is used to modify the amplitude (AM) or the frequency (FM) of a much higher frequency sine wave (carrier).
MOSFET Metal-oxide semiconductor field-effect transistor.
Multimeter An instrument used to measure current, voltage, and resistance.
Mutual inductance The inductance between two separate coils, such as in a transformer.

Nano A prefix designating 10^{-9} (one-thousand-millionth).
Negative feedback The return of a portion of the output signal to the input such that it is out of phase with the input signal.
Network A circuit.
Neutron An atomic particle having no electrical charge.
Node A point or junction in a circuit where two or more components connect.

Ohm (Ω) The unit of resistance.
Ohmmeter An instrument for measuring resistance.
Open circuit A circuit in which there is not a complete current path.
Operational amplifier (op-amp) A special type of amplifier exhibiting very high gain, very high input impedance, very low output impedance, and good rejection of common-mode signals.
Oscillator An electronic circuit that produces a time-varying output signal without an external input signal using positive feedback.
Oscilloscope A measurement instrument that displays signal waveforms on a screen.
Output The voltage, current, or power produced by a circuit in response to an input or to a particular set of conditions.

Parallel The relationship in electric circuits in which two or more current paths are connected between the same two points.
Peak value The maximum value of an electrical waveform, particularly in relation to sine waves.
Period The time interval of one cycle of a periodic waveform.
Periodic Characterized by a repetition at fixed intervals.
Permeability A measure of the ease with which a magnetic field can be established within a material.
Phase The relative displacement of a time-varying waveform in terms of its occurrence.
Pico A prefix designating 10^{-12} (one-billionth).
Positive feedback The return of a portion of the output signal to the input such that it is in phase with the input.
Potentiometer A three-terminal variable resistor.
Power The rate of energy consumption.
Power factor The relationship between volt-amperes and true power or watts. Volt-amperes multiplied by the power factor equals true power.
Power supply An electronic instrument that produces voltage, current, and power from the ac power line or batteries in a form suitable for use in various applications to power electronic equipment.
Primary The input winding of a transformer.
Proton A positively charged atomic particle.
Pulse A type of waveform that consists of two equal and opposite steps in voltage or current, separated by a time interval.
Pulse width The time interval between the opposite steps of an ideal pulse. Also, the time between the 50% points on the leading and trailing edges of a nonideal pulse.
Push-pull A type of class B amplifier in which one output transistor conducts for one half-cycle and the other conducts for the other half-cycle.
Programmable unijunction transistor (PUT) A device similar to the SCR that can be set up to turn on at a predetermined anode voltage.

***Q* point** The dc operating (bias) point of an amplifier.
Quality factor (*Q*) The ratio of reactive power to true power in a coil or a resonant circuit.

Radian A unit of angular measurement. There are 2π radians in a complete revolution. One radian equals 57.3°.

Ramp A type of waveform characterized by a linear increase or decrease in voltage or current.
Reactance The opposition of a capacitor or an inductor to sinusoidal current. The unit is ohms (Ω).
Reactive power The rate at which energy is stored and alternately returned to the source by a reactive component.
Recombination The process of a free electron falling into a hole in the valence band of an atom.
Rectifier An electronic circuit that converts ac into pulsating dc.
Regulator An electronic circuit that maintains an essentially constant output voltage with a changing input voltage or load.
Reluctance The opposition to the establishment of a magnetic field in an electromagnetic circuit.
Resistance Opposition to current. The unit is ohms (Ω).
Resistivity The resistance that is characteristic of a given material.
Resistor An electrical component possessing resistance.
Resonance In an *LC* circuit, the condition when the impedance is minimum (series) or maximum (parallel).
Response In electronic circuits, the reaction of a circuit to a given input.
Rheostat A two-terminal, variable resistor.
Right angle A 90° angle.
Ringing An unwanted oscillation on a waveform.
Ripple The fluctuation in the dc output voltage of a rectifier circuit.
Rise time The time interval required for a pulse to change from 10% of its amplitude to 90% of its amplitude.
Rising edge The positive-going transition of a pulse.
Roll-off The decrease in the gain of an amplifier or filter above or below the critical frequencies.
Root mean square (rms) The value of a sine wave that indicates its heating effect. Also known as *effective value*.

Saturation The state of a bipolar junction transistor in which the collector has reached a maximum and is independent of the base current.
Sawtooth A type of electrical waveform composed of ramps.
Schematic A symbolized diagram of an electrical or electronic circuit.
Silicon-controlled rectifier (SCR) A device that can be triggered on to conduct current in one direction.
Secondary The output winding of a transformer.
Semiconductor A material that has a conductance value between that of a conductor and that of an insulator. Silicon and germanium are examples.
Series In an electrical circuit, a relationship of components in which the components are connected such that they provide a single current path between two points.
Short A zero resistance connection between two points.
Siemen (S) The unit of conductance.
Signal A time-varying electrical waveform.
Silicon A semiconductor material used in transistors.
Sine wave A type of alternating electrical waveform.
Slope The vertical change in a line for a given horizontal change.
Source Any device that produces energy.
Steady state An equilibrium condition in a circuit.
Step A voltage or current transition from one level to another.
Susceptance The ability of a reactive component to permit current flow. The reciprocal of reactance.
Switch An electrical or electronic device for opening and closing a current path.
Synchronous Having a fixed time relationship.

Tangent A trigonometric function that is the ratio of the opposite side of a right triangle to the adjacent side.
Tank A parallel resonant circuit.
Tapered Nonlinear, such as a tapered potentiometer.
Temperature coefficient A constant specifying the amount of change in the value of a quantity for a given change in temperature.
Terminal An external contact point on an electronic device.
Tesla (T) The unit of flux density. Also, webers per square meter.
Thyristor A class of four-layer semiconductor devices.
Time constant A fixed time interval, set by R, C, and L values, that determines the time response of a circuit.

Tolerance The limits of variation in the value of an electrical component.

Trailing edge The last edge to occur in a pulse.

Transformer An electrical device that operates on the principle of electromagnetic induction. It is used for increasing or decreasing an ac voltage and for various other applications.

Transient A temporary or passing condition in a circuit. A sudden and temporary change in circuit conditions.

Transistor A semiconductor device used for amplification and switching applications in electronic circuits.

Triac A bidirectional thyristor that can be triggered into conduction.

Triangular wave A type of electrical waveform that consists of ramps.

Trigger The activating input of some electronic devices.

Troubleshooting The process and technique of identifying and locating faults in an electronic circuit.

True power The average rate of energy consumption. In an electrical circuit, true power occurs only in the resistance, and it represents a net energy loss.

Turns ratio The ratio of the number of secondary turns to the number of primary turns in the transformer windings.

Unijunction transistor (UJT) A type of transistor consisting of an emitter and two bases.

Unity gain A gain of one.

Valence Related to the outer shell or orbit of an atom.

Varactor A diode whose junction capacitance varies with reverse voltage. The device is optimized to make use of this characteristic.

Volt The unit of voltage or electromotive force (emf).

Voltage The amount of energy available to move a certain number of electrons from one point to another in an electrical circuit.

Voltage regulator An electronic circuit that maintains an essentially constant output voltage with changes in input voltage and load.

Watt (W) The unit of true power.

Waveform The pattern of variations of a voltage or a current.

Weber The unit of magnetic flux.

Winding The loops of wire or coil in an inductor or transformer.

Wiper The variable contact in a potentiometer or other device.

Zener diode A type of diode that operates in reverse breakdown to provide voltage regulation.

Index

WE VALUE YOUR OPINION—PLEASE SHARE IT WITH US

Merrill Publishing and our authors are most interested in your reactions to this textbook. Did it serve you well in the course? If it did, what aspects of the text were most helpful? If not, what didn't you like about it? Your comments will help us to write and develop better textbooks. We value your opinions and thank you for your help.

Text Title ______________________________ Edition __________

Author(s) ______________________________

Your Name (optional) ______________________________

Address ______________________________

City ____________________ State __________ Zip __________

School ______________________________

Course Title ______________________________

Instructor's Name ______________________________

Your Major ______________________________

Your Class Rank _______ Freshman _______ Sophomore _______Junior _______ Senior

_______ Graduate Student

Were you required to take this course? _______ Required _______Elective

Length of Course? _______ Quarter _______ Semester

1. Overall, how does this text compare to other texts you've used?

   _______ Superior _______Better Than Most _______ Average _______Poor

2. Please rate the text in the following areas:

	Superior	*Better Than Most*	*Average*	*Poor*
Author's Writing Style	_______	_______	_______	_______
Readability	_______	_______	_______	_______
Organization	_______	_______	_______	_______
Accuracy	_______	_______	_______	_______
Layout and Design	_______	_______	_______	_______
Illustrations/Photos/Tables	_______	_______	_______	_______
Examples	_______	_______	_______	_______
Problems/Exercises	_______	_______	_______	_______
Topic Selection	_______	_______	_______	_______
Currentness of Coverage	_______	_______	_______	_______
Explanation of Difficult Concepts	_______	_______	_______	_______
Match-up with Course Coverage	_______	_______	_______	_______
Applications to Real Life	_______	_______	_______	_______

3. Circle those chapters you especially liked:
   1 2 3 4 5 6 7 8 9 10 11 12 13 14 15 16 17 18 19 20
   What was your favorite chapter? ______
   Comments:

4. Circle those chapters you liked least:
   1 2 3 4 5 6 7 8 9 10 11 12 13 14 15 16 17 18 19 20
   What was your least favorite chapter? ______
   Comments:

5. List any chapters your instructor did not assign. ______________________________

6. What topics did your instructor discuss that were not covered in the text? ______________

   ____

7. Were you required to buy this book? ______ Yes ______ No

   Did you buy this book new or used? ______ New ______ Used

   If used, how much did you pay? __________

   Do you plan to keep or sell this book? ______ Keep ______ Sell

   If you plan to sell the book, how much do you expect to receive? ______

   Should the instructor continue to assign this book? ______ Yes ______ No

8. Please list any other learning materials you purchased to help you in this course (e.g., study guide, lab manual).

   ____

9. What did you like most about this text? ______________________________

   ____

10. What did you like least about this text? ______________________________

    ____

11. General comments:

May we quote you in our advertising? ______ Yes ______ No

Please mail to: Boyd Lane
College Division, Research Department
Box 508
1300 Alum Creek Drive
Columbus, Ohio 43216

Thank you!